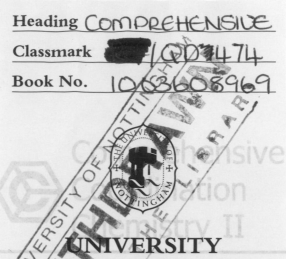

Comprehensive Coordination Chemistry II

FROM BIOLOGY TO NANOTECHNOLOGY

EDITORS-IN-CHIEF

Jon A McCleverty
University of Bristol, UK

Thomas J Meyer
Los Alamos National Laboratory, New Mexico, USA

Volume 3
COORDINATION CHEMISTRY OF THE S, P, AND F METALS

VOLUME EDITOR

G F R Parkin
Columbia University, New York, USA

ELSEVIER
PERGAMON

AMSTERDAM   BOSTON   HEIDELBERG   LONDON   NEW YORK   OXFORD
PARIS   SAN DIEGO   SAN FRANCISCO   SINGAPORE   SYDNEY   TOKYO

UK   Elsevier Ltd., The Boulevard, Langford Lane, Kidlington, Oxford, OX5 IGB, UK

USA   Elsevier Inc., 525 B Street, Suite 1900, San Diego, CA 92101-4495, USA

First edition 2004

Library of Congress Catalogue Card Number: 2003113706

A catalogue record for this book is available from the British Library

ISBN: 0-08-0437486 (Set)

ISBN: 0-08-0443257 (Volume 3)

This book is printed on acid-free paper

Printed in the United Kingdom

Contents

Contents of all Volumes

Preface

Comprehensive Coordination Chemistry (CCC), published in 1987, was intended to give a contemporary overview of the field. The goal was to provide both a convenient first source of information and a stimulus for further advances in the field.

In *Comprehensive Coordination Chemistry II* (CCC2) we have adopted the same general approach. Developments in coordination chemistry since 1982 are surveyed in an authoritative and critical manner taking into account important new trends in biology, materials science, and other areas.

As in many areas of science, it is impossible to provide a totally comprehensive review – the field has grown enormously in the past 20 years. Consequently, our intention is to provide the readers of the series with the most reliable and informative background information on particular areas of coordination chemistry based on key primary and secondary references. In doing so we recognize that those readers will be researchers at all levels including students, non-experts from other areas of science, and industrial chemists. Our hope is that CCC2 will provide a clear overview, at a state-of-the-art level, of those areas that the Editors-in-Chief and the Volume Editors believe to be especially important and/or of high relevance to future developments.

Before proceeding further, it is necessary to define what we include as "coordination" chemistry to set the terms of reference for what follows. For CCC (1987) this was taken to include the synthesis and properties of the products of association of Brønsted bases with a Lewis acid. This definition excluded most organometallic compounds. This definition is still useful, the arbitrary limitation being retained that any coordination compound in which the number of metal–carbon bonds is at least half the coordination number of the metal is deemed to be "organometallic" and nominally outside the scope of coverage. This includes η^n-hydrocarbon ligands but exceptions have been made for complexes containing CO, CNR, NO, and related π-acid ligands.

The emergence of supramolecular chemistry in the early 1980s led to *Comprehensive Supramolecular Chemistry*, published in 1997, which contains much of interest to coordination chemists. Coverage of this area in CCC2 is restricted to developments since 1990. The growth in both bioinorganic and materials chemistries since 1980 has been remarkable. Coordination chemistry has played a key role in their development. They appear prominently in CCC2 where we have attempted to highlight important developments and document fundamental advances.

CCCII comprises ten volumes, of which the last contains only subject indexes. The first two volumes describe the development of new ligands since the 1980s, which complements Volume 2 in CCC. They also include new techniques of synthesis and characterization, with a special emphasis on the burgeoning physical techniques which are increasingly applied to the study of coordination compounds. Developments in theory, computation methods, simulation, and useful software are reported. The volumes conclude with a series of case studies, which illustrate how synthesis, spectroscopy, and other physical techniques have been successfully applied in unravelling some significant problems in coordination chemistry.

Volumes 3–6 describe developments in the coordination chemistry of the metallic elements since 1982 (*s*, *p*, and *f*-block metals, transition metals of Groups 3–6; 7–8; 9–12). These volumes correspond to Volumes 3, 4, and 5 in CCC. A review of technetium coordination chemistry was unavailable when CCC was published, and a complete account of its development from the earliest discoveries to present-day applications is incorporated in the new work. In these volumes space limitations restrict the material that can be presented. The information that appears has been selected to give a near comprehensive coverage of new discoveries, new interpretations of experiment and theory, and applications, where relevant.

The style of reporting follows that used in CCC (1987). In the element chapters, discussion of element properties of bioinorganic and industrial relevance is deliberately limited in scope. These issues are addressed separately in subsequent volumes and are extensively cross-referenced.

In the nanoscale regime (1–100 nm) materials exhibit size-sensitive properties and offer significant prospects for achieving molecular-level control of catalysis, sensors, molecular circuitry, and other applications. This prospect has led to a surge of interest and increasing research in nanoscience and nanotechnology. These are areas in which coordination chemistry plays a substantial role. With this in mind, the synthesis, structure, and physical properties of coordination-complex-based super- and supramolecules, clusters, and nano-particles are presented in depth in Volume 7. This volume describes species ranging from "traditional" monomeric complexes to ligand-stabilized multimetallic assemblies, metal or semiconductor nanoparticles, dendrimers, other polymer-based assemblies, and mesogenic materials. It also reports on the electron transfer, photochemical and photophysical, optical, and magnetic characteristics of these sometimes-remarkable materials. The emphasis in this volume is experimental with some supporting theoretical discussion.

Volume 8 is devoted to the coordination chemistry of metal ions that are involved in biological processes. Throughout this volume, relevant biochemical issues are discussed, but the focus is primarily on structure, function, and properties of the metal centers in biomolecules. Relevant synthetic models and/or functional mimics are included, but the majority of complexes prepared as potential models are discussed in Volumes 2–6.

Volume 9 is concerned with actual and potential applications of metal coordination complexes. Major developments since the 1980s in the uses of coordination compounds have occurred in catalysis and medicine. There have been important developments of coordination chemistry in the technology of dyes and optical materials, for solar energy harvesting, for hydrometallurgical extraction, and in providing MOCVD precursors for new electronic materials. As mentioned above, the last volume in the series contains the indexes.

There are organizational differences between CCC2 and CCC (1987). One is having editors responsible for individual volumes. Another is a greatly increased emphasis on coordination chemistry in medicine and industry, which is an expected result of a maturing area of science. Only Volume 6 was devoted to this subject in CCC (1987).

We are extremely grateful to our editorial colleagues for their invaluable aid in selecting authors and for their participation in helping to define both the contents and organization of CCC2. The authors for this series were deliberately chosen to reflect the geographically diverse nature of the field and include contributors from academe, national laboratories, and industry. We are extremely grateful to them for their hard work and for the insights they have provided in presenting an extraordinary range of topics.

We would also like to acknowledge the editorial and publishing staff at Elsevier for their professional and comprehensive support of the editors and authors in the production of this substantial series. We particularly mention Angela Greenwell, Sandra Migchielsen, Jerome Michalczyk, and Wendy Tomlinson. Their patience, good humor, and professionalism have been constant in the sometimes difficult moments required to bring this complex project involving many authors from around the world to fruition.

Finally, we, the Editors-in-Chief, hope that the readers of this second work in the *Comprehensive Coordination Chemistry* series will find it as useful and informative as the first. It is our hope that the field of coordination chemistry and those who use and advance it will be major beneficiaries of our efforts and those of our authors.

Jon A. McCleverty
Bristol, UK
March 2003

Thomas J. Meyer
Los Alamos, USA
March 2003

Introduction to Volume 3

Volume 3 describes the Coordination Chemistry of the s-, p-, and f-block metals.

Chapter 1 is concerned with the 1s and 2s metals and describes trends in the development of their chemistry since the mid-1980s, such as the increased use of sterically bulky ligands, recognition of importance of non-ionic interactions, reappraisal of the "spectator" role of s-block ions, and the application of computational methods. Biological roles of these elements are discussed in Volume 8.

Chapter 2 is concerned with the chemistry of scandium, yttrium, and the lanthanides and is discussed according to the nature of the ligand in which the donor is from Groups 14–17. Divalent and tetravalent lanthanide chemistry is also described.

Chapter 3 describes the chemistry of the actinides, including the historical development. The chemistry described is subdivided according to whether the actinide is early (thorium to plutonium) or late (transplutonium elements). Within this subdivision, the chemistry is further classified according to the oxidation state of the metal (ranging from +3 to +7), and the type of donor (ranging from elements of Groups 15–17). The chapter also contains information pertaining to element separation and aspects of nuclear technology (which is not discussed in Volume 9 and therefore represents a departure from the format of *Comprehensive Coordination Chemistry*).

Chapter 4 describes the chemistry of aluminum and gallium. In addition to aluminum(III) and gallium(III) coordination complexes, this chapter also focuses on complexes with aluminum–aluminum and gallium–gallium bonds, and also describes cyclogallenes and metalloaromaticity.

Chapter 5 describes the chemistry of indium and thallium, including subvalent compounds of indium(II), thallium(II), and thallium(I). Applications of indium and thallium complexes are also described.

Chapter 6 describes the chemistry of arsenic, antimony, and bismuth, including a discussion of the role that these elements play in the environment and biology and medicine. Applications of these complexes are also discussed.

Chapter 7 describes the chemistry of germanium, tin, and lead according to M^{IV} and M^{II} oxidation states. Within this classification, the chemistry is further subdivided according to ligand type, which ranges from elements of Groups 13–17.

G F R Parkin
New York, USA
March 2003

Coordination Chemistry: The Past, Present, and Possible Future

Some thoughts gleaned by the Editors-in-Chief from conversations with the International Advisory Board

In the past 20 years, inorganic chemistry has been greatly enriched by the continuing development of organometallic chemistry and the entry of new thinking from an organic perspective. However, the field continues to evolve and there is a growing emphasis on coordination chemistry and the fundamental principles that guide it. The driving forces for the evolution have come from bioinorganic and biomimetic chemistry, and the growing interest in materials. The interest in new materials has been further fueled by an enhanced understanding of the underlying principles and their extension to smaller and smaller domains, some remarkable advances in synthesis, and a marriage between device and materials physics on the one hand, and basic science on the other. The role of the chemist is growing with the advent of nanoscience and nanotechnology and their promise of materials with properties tailored at the molecular level. There are many new directions in coordination chemistry, in molecular magnetism, supramolecular chemistry, non-silicon-based devices, precursors for vapor phase deposition, and single molecule-based photonic devices and sensors.

Theory is playing a significant role. Density functional theory is enabling a deeper understanding of the electronic structure of simple and complex molecules. Theory is being used to calculate spectroscopic parameters, accurately predict structures, and to understand chemical reactivity.

Among some of the successes of coordination chemistry up to now have been:

- tuning of variable valency via ligand control of reductions potentials;
- tuning of spin states;
- isomer preference of oxidation states and valence/geometry recognition;
- water oxidation by a ruthenium shuttle;
- oxygen atom transfer from water, per-acids, and oxo-metal reagents;
- correlating excited state properties of metal complexes with electronic and molecular structure;
- photo-induced electron and energy transfer in metal complex-based molecular assemblies;
- thioether coordination and activation of homolog-specific transformations;
- recognition of the importance of metal nitrosyl complexes in the "biology" and "physiology" of NO;
- spontaneous polynucleation via oximato and phenolato bridging ligands;
- characterization of vanadate esters of carbohydrates;
- unraveling of modes of actions of some metaloenzymes;
- development of metalacycles and the insertion of unsaturates into metalacycles;
- recognition of "non-innocence" as a significant factor in systems where ligands and metals are both redox active;
- isolation and characterization of radical anion ligand complexes, and recognition of their role in biology;
- custom design of cluster oxo-anions and rationalization of their structural parameters, and creation of super-large cluster ions modeling pieces of oxide surfaces;
- increased understanding of structure–function relationships through structural solutions of metallo-enzymes and other bio-molecules;

- application of density functional theory to the elucidation of electronic and molecular structure;
- providing an understanding of the localized-to-delocalized transition in mixed-valence chemistry.

The above represents a long and impressive list of achievements. But the subject continues to develop and grow rapidly. So what are the current "hot topics"? These may include:

- bio-transformations, particularly hydrogen evolution, conversion of nitrogen into ammonia, multi-electron transfer processes, and methane oxidation – all under ambient conditions; and water oxidation;
- all aspects of materials chemistry where the unique properties of transition metals can be exploited;
- metal complexes in supramolecular assemblies for use in catalysis and in optical and magnetic devices;
- use of metal complexes in aqueous solutions (avoidance of organic solvents in synthesis and catalysis, particularly with respect to industrial processes); it is no exaggeration to say that a large part of life processes are basically pH-controlled in aqueous solution;
- metal complexes in biology – either for (i) medical purposes such as chemotherapy or (ii) the identification of metal complex cores in biological functions such as their role as "acids" in aqueous media;
- development of ligand design to facilitate supramolecular systems and designed self-assembly;
- use of coordination compounds as optical triggers and probes, particularly with respect to long-distance electron tunneling in proteins;
- metal binding by carbohydrates.

Having moved from the past to the present, what then is the likely future of coordination chemistry? Very clearly current fashions and realities indicate a multidisciplinary development, particularly in the life sciences. Nature continually presents chemists with surprises, and coordination chemistry will continue to respond to the challenges of bio-mimicry, and associated developments in catalysis and materials science. It is a fact that only 5% of the bacteria in soil are identified and the roles of metals in bacteria are, in general, not evaluated. This could be of considerable industrial and agricultural significance in the future. Enzymology has been high profile for 30 years already, and is likely to be a major preoccupation for another 300! The roles of metals and their coordination environments will continue to be elucidated and modeled in ways that could contribute to the "green chemistry" revolution.

Coordination chemistry will continue to strengthen its role as a central expertise and discipline for materials science. It is critical to the development of new materials for nanoscience and nanotechnology. In materials science, light-driven processes are of enormous importance and processes based on molecular-level phenomena may provide the basis for photonics and information storage in the future. In catalysis, the use of metals will grow, particularly when control of asymmetric processes is mastered.

International Advisory Board

Contributors to Volume 3

H. V. Rasika Dias
The University of Texas at Arlington, USA

G. Reid
University of Southampton, Southampton, UK

G. H. Robinson
The University of Georgia, Athens, GA, USA

R. D. Rogers
The University of Alabama, Tuscaloosa, AL, USA

3.1
Group 1s and 2s Metals

T. P. HANUSA
Vanderbilt University, Nashville, TN, USA

3.1.1 INTRODUCTION AND REVIEW OF COORDINATION PROPERTIES

Even though they occupy adjacent columns of the periodic table and possess marked electronic similarities, the 12 members of the *s*-block elements nevertheless form coordination compounds of surprising diversity. The alkali (Group 1, Li to Fr) and alkaline-earth (Group 2, Be to Ra) metals share n^{sx} valence electron configurations in their elemental state ($x = 1$, alkali metals; $x = 2$, alkaline-earth metals), and have low ionization potentials. Consequently, they all display—with some important exceptions—only +1 (for Group 1) and +2 (for Group 2) oxidation states. The highly electropositive nature of the metals also means that their bonds to other elements are strongly polar, and compounds of the *s*-block elements are often taken as exemplars of ionic bonding.

The uniform chemistry that these electronic similarities might imply is strongly modulated by large variations in radii and coordination numbers. The change from four-coordinate Li^+ (0.59 Å) to 12-coordinate Cs^+ (1.88 Å)[1] represents more than a three-fold difference in size; the change from four-coordinate Be^{2+} (0.27 Å) to 12-coordinate Ba^{2+} (1.61 Å) is nearly six-fold. With noble gas electron configurations for the ions, bonding in *s*-block compounds is largely nondirectional, and strongly influenced by ligand packing around the metals. Although to a first approximation the geometries of many mononuclear s-block coordination complexes are roughly spherical, the presence of multidentate and sterically bulky ligands can produce highly irregular structures.

One of the consequences of the large increase in the number of structurally characterized compounds reported since the publication of *Comprehensive Coordination Chemistry* (*CCC*, 1987) is that some of the long-standing expectations for Group 1 and 2 chemistry need to be qualified. A conventional generalization holds that the coordination number (c.n.) of a complex should rise steadily with the size of the metal ion, and there is in fact abundant data to support this assumption for small monodentate ligands. For example, analysis of water-coordinated ions indicates that the most common c.n. for Be^{2+},[2] Mg^{2+},[3] and Ca^{2+} are four, six, and six to eight, respectively.[4] When more complex aggregates or those containing sterically bulky or macrocyclic ligands are considered, however, the relationship between ion size and c.n. is weakened; e.g., lithium is found with a c.n. of eight in the now-common $[(12\text{-crown-4})_2Li]^+$ ion (first structurally authenticated in 1984),[5] whereas barium is only three-coordinate in $\{[Ba[N(SiMe_3)_2]_2\}_2$.[6] Similarly, the standard classification of s-block ions as hard (type a) Lewis acids leads to the prediction that ligands with hard donor atoms (e.g., O, N, halogens) will routinely be preferred over softer (type b) donors. This is often true, but studies of the "cation-π" interaction (see Section 3.1.2.2) have demonstrated that the binding of s-block ions to "soft" donors can be quite robust; the gas-phase interaction energy of the K^+ ion with benzene, for example, is greater than that to water.[7] Furthermore, the toxicity of certain barium compounds may be related to the ability of the Ba^{2+} to coordinate to "soft" disulfide linkages, even in the presence of harder oxygen-based residues.[8]

The alkali- and alkaline-earth metals are widespread on earth (four of the eight most common elements in the earth's crust are s-block elements) and their compounds are ubiquitous in daily life. Considering that an estimated one-third of all proteins require a metal ion for their structure or function,[4] and that the most common metals in biological systems are from these two families (Na^+, K^+, Mg^{2+}, Ca^{2+}), the importance of the Group 1 and 2 elements to biology cannot be overestimated.

In the last 20 years, interest in current and potential applications of these elements in oxide- or sulfide-containing materials such as the superconducting cuprates,[9] ferroelectric ceramics,[10,11] and phosphor systems has also sharply increased. There has been a correspondingly intensive search for molecular precursors to these species that could be used in chemical vapor deposition (CVD), sol-gel, or spray pyrolysis methods of fabrication.[12-14] All of these factors mean that the coordination compounds of the s-block metals are becoming increasingly important to many branches of chemistry and biology, and the reported chemistry for these elements is vast. Although the number of compounds known for each metal varies substantially, only francium (Fr), all of whose isotopes are radioactive and short-lived (the longest is ^{223}Fr with $t_{1/2} = 22$ min, thereby making it the most unstable of the first 103 elements), has no reported coordination complexes.

The number of reports of new compounds has increased to the point that it is no longer possible to provide exhaustive coverage of them within the confines of a reasonably sized work. As one example, there were as of the end of the year 2000 over 1,100 crystallographically characterized coordination compounds containing an s-block element and one or more coordinated water molecules; fewer than 150 of these structures were reported before 1985.

3.1.2 TRENDS SINCE THE MID-1980s

During the last third of the twentieth century, the coordination chemistry of the s-block elements gained new-found recognition as being essential to the development of materials science and biology, and eminently worthy of study on its own merits. Prior to the 1967 discovery by Petersen of the ability of crown ethers to form robust complexes with even the largest alkali- and alkaline-earth metals,[15] the prospects for an extensive coordination chemistry of the s-block elements appeared dim. The "macrocyclic revolution" generated new interest in Group 1 and 2 complexes, however, and the early developments with ligands such as the crown ethers, cryptands, and calixarenes were documented in *CCC* (1987). More recent advances in the chemistry of macrocyclic s-block complexes have been described in *Comprehensive Supramolecular Chemistry*.

The development of s-block metal chemistry in the last 15 years has been accelerated by several other trends, including the expanded use of sterically bulky ligands, the growing recognition that a strictly electrostatic view of the interaction of the Group 1 and Group 2 metals with their ligands is too limiting, and that "cation-π" interactions have an important role to play in their chemistry. Associated with the last item is the acknowledgment that *s*-block ions are not necessarily passive counterions in complexes of the main group and transition metals, but may critically alter the structure of these species. Finally, the increasing power of computers and the emergence of density functional theory methods of computation have made calculations on *s*-block species more common, more accurate, and more important than ever before as a probe of bonding and structure and as a guide to reactivity. Each of these trends in examined in turn below.

3.1.2.1 Increased Use of Sterically Bulky Ligands

Although Li^+, Be^{2+} and Mg^{2+} are about the size of first row transition metals (e.g., Fe^{2+}) or the lighter *p*-block ions (Ge^{2+}, P^{2+}), Na^+ and Ca^{2+}, with radii of approximately 1.0 Å, are roughly the size of the largest trivalent lanthanides. The radii of Cs^+ and Ba^{2+} are comparable to those of polyatomic cations such as NH_4^+ and PH_4^+.[16] Not only does the large radii of the *s*-block metals accommodate high coordination numbers, but in the presence of sterically compact ligands (e.g., -NH_2, -OMe, halides), extensive oligomerization or polymerization will also occur, leading to the formation of nonmolecular compounds of limited solubility or volatility.

The demand for sources of the *s*-block metal ions that would be useful for materials synthesis[12] or in biological applications has led to a large increase in the use of ligands that are sterically bulky and/or contain internally chelating groups. The resulting compounds are often monomers or low oligomers (dimers, trimers), and their well-defined stoichiometries and reproducible behavior have aided attempts to develop a consistent picture of *s*-block metal reactivity, down to the level of individual metal–ligand bonds. The many clathrate and calixarene complexes described in *CCC* (1987) and *Comprehensive Supramolecular Chemistry* are well-known examples of the influence of steric effects on Group 1 and 2 metal compounds. Numerous cases are known in nonmacrocyclic systems as well; e.g., the oligomeric $[KOCH_3]_x$ is soluble only in water and alcohols, but $[K(\mu_3\text{-}OBu^t)]_4$ is a cubane-like tetramer[17,18] that is soluble in ether and aromatic hydrocarbons. Similarly, the amides $M(NR_2)_2$ (M = Mg, Ca, Sr, Ba) are nonmolecular solids with ionic lattices when R = H, but are discrete dimers $[M(NR_2)_2]_2$ when R = $SiMe_3$, and are soluble in hydrocarbons.[19]

Metal centers that are coordinated with sterically bulky groups usually have lower formal coordination numbers than their counterparts with smaller ligands, sometimes as small as three for Cs^+ and Ba^{2+}. In such cases, secondary intramolecular contacts between the ligand and metal can occur. These can be subtle, as in the agostic interactions between the $SiMe_3$ groups on amido ligands and metal centers (e.g., in $[(Me_3Si)_2N]_3LiMg$)[20] or more obvious, as in the cation-π interactions discussed in the next section. In any case, further progress with the *s*-block metals can be expected to make even greater use of sterically demanding substituents, including those with internally chelating groups.

3.1.2.2 Recognition of the Importance of Non-ionic Interactions

The conventional approach to understanding bonding in *s*-block coordination complexes views the metal–ligand interactions as essentially electrostatic; i.e., that the metals can be considered as nonpolarizable mono- or dipositive ions, with the ligands arranged around them to maximize cation/anion contacts and minimize intramolecular steric interactions. Even this "simple" analysis can lead to structures that are quite complex, but it has been clear since the 1960s that a more sophisticated analysis of bonding must be used in some cases. The gaseous Group 2 dihalides (MF_2 (M = Ca, Sr, Ba); MCl_2 (M = Sr, Ba); BaI_2),[21–23] for example, are nonlinear, contrary to the predictions of electrostatic bonding. An argument based on the "reverse polarization" of the metal core electrons by the ligands has been used to explain their geometry, an analysis that makes correct predictions about the ordering of the bending for the dihalides (i.e., Ca < Sr < Ba; F < Cl < Br < I).[22,23] Other *ab initio* calculations on Group 1 complexes M^+L_2 (M = K, Rb, Cs; L = NH_3, H_2O, HF) that have employed quasirelativistic pseudopotentials and flexible, polarized basis sets indicate that bent L—M—L arrangements are favored energetically over linear structures for M = Rb, Cs.[24] The source of the bending has been ascribed to polarization of the cation by the ligand field,[24] although whether the noble-gas cores of the metal cations are polarizable enough to account for the observed bending has been questioned.[25] The "reverse polarization" analysis can be recast in molecular orbital terms; i.e., bending leads to a reduction in the antibonding character in the HOMO. This interpretation has been examined in detail with calculations on RaF_2.[26]

An alternative explanation for the bending in ML_2 species has focused on the possibility that metal d orbitals might be involved. Support for this is provided by calculations that indicate a wide range of small molecules, including MH_2, MLi_2, $M(BeH)_2$, $M(BH_2)_2$, $M(CH_3)_2$, $M(NH_2)_2$, $M(OH)_2$, and MX_2 ($M = Ca$, Sr, Ba) should be bent, at least partially as an effect of metal d-orbital occupancy.[24,27–31] The energies involved in bending are sometimes substantial (e.g., the linearization energy of $Ba(NH_2)_2$ is placed at ca. 28 kJ mol^{-1}).[29] Complexes of Ba^{2+} with three NH_3, H_2O, or HF ligands have been computed to prefer pyramidal over trigonal-planar arrangements, although the pyramidalization energy is less than 1 kcal mol^{-1}. Spectroscopic confirmation of the bending angles in most of these small molecules is not yet available, however.

However fascinating these effects from incipient covalency might be, they are of low energy, and may be masked by steric effects or crystal packing forces in solid-state structures. A different sort of noncovalent influence that has gained recognition in the past two decades is the so-called "cation–π interaction," which describes the involvement of cations with a ligand's π-electrons (usually, but not necessarily, those in an aromatic ring).[7] Table 1 lists some observed and calculated binding energies for monocations and various π-donors. Note particularly that the interaction energy of benzene with the "hard" K^+ ion (19.2 kcal mol^{-1}), for example, is even slightly greater than to water in the gas phase. The interaction energy falls in the order $Li^+ > Na^+ > K^+ > Rb^+$, which is expected for an ionic interaction, but the binding order is more a marker of the strength of the interaction, rather than evidence of an ionic origin for the effect. Several factors are thought to contribute to the cation-π phenomenon, including induced dipoles in aromatic rings, donor-acceptor and charge transfer effects, and the fact that sp^2-hybridized carbon is more electronegative than is hydrogen.

The cation-π interaction is believed to be operative in many biological systems, such as K^+-selective channel pores,[32] and Na^+-dependent allosteric regulation in serine proteases.[33] There are also coordination complexes of the s-block elements that display pronounced M^{n+}-arene interactions to coordinated ligands. Many examples could be cited; representative ones are provided by the reaction of $Ga(mesityl)_3$ or $In(mesityl)_3$ (mesityl = 2,4,6-$Me_3C_6H_2$) with CsF in acetonitrile, which yields $[\{Cs(MeCN)_2\}\{mes_3GaF\}]_2 \cdot 2MeCN$ and $[\{Cs(MeCN)_2\}\{mes_3InF\}]_2 \cdot 2MeCN$, respectively. A similar reaction with $Ga(CH_2Ph)_3$ gives $[Cs\{(PhCH_2)_3GaF\}]_2 \cdot 2MeCN$. The structures are constructed around $(CsF)_2$ rings and display Cs—phenyl interactions (see Figure 1).[34] In the structure of $Na[Nd(OC_5H_3Ph_2-2,6)_4]$, formed from $NdCl_3$ and $Na(OC_5H_3Ph_2-2,6)$ in 1,3,5-tri-t-butylbenzene at 300 °C, the sodium is coordinated to three bridging oxygen atoms and exhibits cation-π interactions with three phenyl groups.[35]

Table 1 Monovalent ion–molecule binding energies (gas-phase).

Ion	*Molecule*	*Binding energy* (ΔH, kcal mol^{-1})
Li^+	C_6H_6	38.3 (exp.)
Li^+	C_6H_6	43.8 (calc.)
Na^+	C_6H_6	28.0 (exp.)
Na^+	C_6H_6	24.4 (calc.)
K^+	C_6H_6	19.2 (exp.)
K^+	C_6H_6	19.2 (calc.)
$K^+ \cdot C_6H_6$	C_6H_6	18.8 (exp.)
$K^+ \cdot (C_6H_6)_2$	C_6H_6	14.5 (exp.)
$K^+ \cdot (C_6H_6)_3$	C_6H_6	12.6 (exp.)
K^+	H_2O	17.9 (exp.)
Rb^+	C_6H_6	15.8 (calc.)
NH_4^+	C_6H_6	19.3 (exp.)
NMe_4^+	C_6H_6	9.4 (exp.)

Source: Ma (1997)[7]

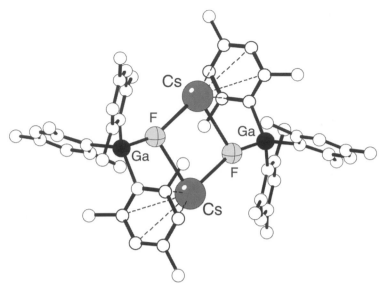

Figure 1   The structure of [Cs{(PhCH₂)₃GaF}]₂, illustrating the cation-π interactions.

3.1.2.3   Reappraisal of the "Spectator" Role of *s*-Block Ions

Considering the prevalence of cation-π interactions, it is not surprising that in some cases *s*-block ions may play an important role in modifying the structure and bonding of metal complexes. This represents a more direct kind of interaction than is usually credited to the ions when they are viewed as "spectator" species, i.e., simply as countercharges to complex anions. In many cases, verification of the "nonspectator" role of *s*-block species requires structural authentication through X-ray crystallography, so it is natural that a growing awareness of the importance of such interactions has coincided with the increase in crystallographically characterized compounds during the last two decades.

The consequences of the interaction vary significantly, and only a few examples are detailed here; others can be found throughout this chapter. At one level, cation-π interactions can be responsible for the existence of coordination polymers by serving as interanionic bridges, e.g., reaction of $La_2[OC_6H_3(Pr^i)_2\text{-}2,6]_6$ with two equivalents of $Cs[OC_6H_3(Pr^i)_2\text{-}2,6]$ in THF yields the base-free caesium salt $Cs^+[La(OC_6H_3(Pr^i)_2\text{-}2,6)_4]^-$.[36] The latter is an oligomer, in which the caesium ions, supported only by π-interactions (Cs⁺–ring plane = 3.6 Å), bind the lanthanum aryloxide anions together (see Figure 2). Similar interactions are observed in $(Cs_2)^{2+}[La(OC_6H_3\text{-}(Pr^i)_2\text{-}2,6)_5]^{2-}$.[37]

In other cases, intramolecular interactions with *s*-block metal ions may materially change the nature of the associated complexes. Although it involves organometallic complexes, examination

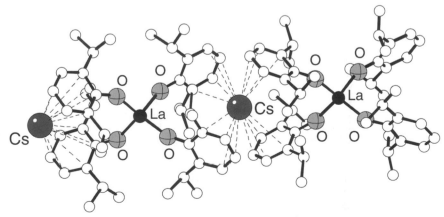

Figure 2   The structure of base-free oligomer $Cs^+[La(OC_6H_3(Pr^i)_2\text{-}2,6)_4]^-$, supported only by cation-π interactions.

of several such cases is instructive. The sodium metal reduction of $[(2,4,6-(Pr^i)_3C_6H_2)_2C_6H_3]GaCl_2$ in Et_2O gives red–black crystals of a compound with the molecular formula $Na_2[Ga(2,4,6-(Pr^i)_3C_6H_2)_2C_6H_3]_2$.[38] X-ray crystallographic analysis indicates that the compound has a dimeric structure with a 2.319(3) Å Ga–Ga separation. Based on several criteria, including the presence of two-coordinate gallium and the relatively short bond, an argument has been made that the compound contains a Ga≡Ga triple bond, i.e., that the compound could be viewed as containing the $[RGa≡GaR]^{2-}$ ion. Discussion over the appropriateness of this description has been extensive; arguments in favor of a high Ga—Ga bond order (≥ 2.5)[39,40] and those preferring a lower value (≤ 2)[41–44] have used a variety of computational tests to substantiate their viewpoints. Early in the debate it was observed, however, that the sodium "counterions" are in a strategic position in the molecule; i.e., where they can engage in a π-interaction between phenyl rings (Na–ring plane (2.75–2.81 Å) (see Figure 3).[45] It has since been recognized that the Na^+-arene interaction is responsible for at least some of the short Ga—Ga distance; calculations cannot reproduce the metal separation if the anion is modeled simply as isolated $[HGaGaH]^{2-}$ or $[MeGaGaMe]^{2-}$ units.[39,46]

It is clear that the presence of Na^+ is critical to the existence of the molecule; if potassium is substituted for sodium in the reduction of $[(2,4,6-(Pr^i)_3C_6H_2)_2C_6H_3]GaCl_2$, the very different $K_2[Ga_4(C_6H_3-2,6-(2,4,6-(Pr^i)_3C_6H_2)_2)_2]$ moiety is isolated (see Figure 4).[47] The almost square Ga_4 ring is capped on both sides by K^+ ions that are at somewhat different distances from the plane (3.53, 3.82 Å). The potassium ions are clearly involved with phenyl groups on the ligands at distances of 3.1 Å. It is apparent that the identity of the alkali metal cation is critical to the formation of the compounds, and that it is incorrect to view the *s*-block ions as freely interchangeable.

There are other examples of Group 1 ions involved in other main-group systems, many of which are organometallic species and outside the scope of this chapter. There are also compounds in which an *s*-block ion serves as both a linker in a coordination polymer and as an integral part of a metal aggregate, such as the $[K(18-crown-6)]_3KSn_9$ cluster (see Figure 5).[48]

3.1.2.4   Application of Computational Methods to Complexes

The enormous increase in readily available computing power since the 1980s has greatly affected the study of *s*-block metal complexes. A long-standing assumption that the Group 1 and 2 metal ions (especially the former) could be successfully modeled as point charges in molecular orbital

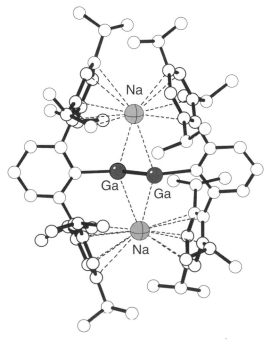

Figure 3   Na–phenyl contacts in $Na_2[Ga(2,4,6-(Pr^i)_3C_6H_2)_2C_6H_3]_2$.

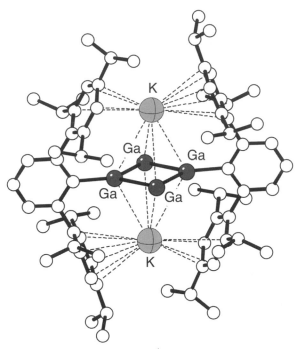

Figure 4   The structure of K$_2$[Ga$_4$(C$_6$H$_3$-2,6-(2,4,6-(Pr^i)$_3$C$_6$H$_2$)$_2$)$_2$], illustrating the K$^+$–phenyl interactions.

calculations has been shown to be increasingly inadequate. Schleyer first demonstrated with calculations on organolithium complexes that attempts to understand the bonding and reactivity of *s*-block complexes severely test the performance of *ab initio* computational methods.[49,50] Owing to their lack of valence electrons, alkali and alkaline-earth complexes are formally electron deficient and conformationally "floppy," and only small energies (often 1–2 kcal mol^{-1}) are required to alter their geometries by large amounts (e.g., bond angles by 20° or more). In such cases, the inclusion of electron correlation effects becomes critical to an accurate description of the structure of the molecules. Traditional Hartree–Fock approaches, especially when combined with small or minimal basis sets, are generally inadequate for these complexes. Some of the

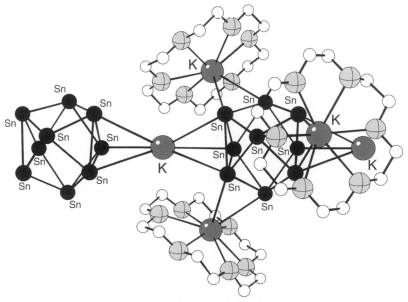

Figure 5   The structure of the tin aggregate, [K(18-crown-6)]$_3$KSn$_9$.

quantitative or semiquantitative agreement claimed in the past between observed and calculated energies and structures must now be ascribed to fortuitous cancellation of errors.

Density functional theory (DFT) methods, which implicitly incorporate electron correlation in a computationally efficient form, have found wide use in main-group chemistry.[51–53] In general, they have been more successful than Hartree–Fock techniques in dealing with organoalkali and organoalkaline-earth molecules, and there is growing evidence of their successful use with coordination complexes. Nevertheless, a wide range of computational techniques continues to be used in *s*-block element chemistry, from molecular modeling and semiempirical methods, to high-level coupled cluster and DFT approaches. Representative samples of the application of computational investigations to *s*-block coordination compounds are found in the sections below.

3.1.3 MACROCYCLIC COMPOUNDS

As noted in Section 3.1.2, the introduction of the crown ethers in the late 1960s gave legitimacy to the concept of stable coordination complexes of the alkali metals. Their presence, and that of many other macrocyclic counterparts (e.g., porphyrins) and three-dimensional chelators (e.g., cryptands, calixarenes) is now pervasive in both alkali and alkaline-earth coordination chemistry, and the literature on these complexes is vast. Early work in this area was summarized in *CCC* (1987), and examined in a more focused manner in *Comprehensive Supramolecular Chemistry*. It is not the intent of this section to repeat such material, but rather to highlight new developments since the mid-1990s. In some cases, specialist reviews are available on these subjects; they will be noted where relevant.

3.1.3.1 Porphyrins and Phthalocyanines

The *s*-block metal most commonly complexed to a porphyrin is magnesium, and many such compounds have been prepared in the course of studies on models for bacteriochlorophyll.[54] These include the metallotetraphenylporphyrin cation radical ($MgTPP^{+*}$), obtained as its perchlorate salt,[55] and the neutral MgTPP, isolated as an adduct with (1-methylimidazole),[56] 4-picoline,[56] piperidine,[56] water,[57,58] and methanol.[58] Related magnesium porphyrin derivatives have been prepared in the study of photosynthetic reaction centers; e.g., the tetrakis(4-methoxyphenyl) H_2O adduct,[59] and octaethylporphyrinato dimers, whose strength of coupling (reflected also in UV/vis spectra) is strongly dependent on the polarity of the solvent.[60] The tetraphenylporphyrin framework does not undergo significant structural change on oxidation, thus making neutral molecules realistic models for radical cationic species.

MgTPP has also been examined as a substrate for constructing "porphyrin sponges," i.e., lattice clathrates that can reversibly absorb and release guest molecules.[61–65] Such guests as methyl benzoate,[62] propanol and (*R*)-phenethylamine) have been structurally authenticated; other examples are known.[64]

Porphyrin complexes of *s*-block metals other than magnesium have received less attention. Reaction of free-base porphyrins (H_2Por = octaethylporphyrin (H_2OEP), *meso*-tetra-phenylporphyrin, *meso*-tetra-*p*-tolylporphyrin, *meso*-tetrakis(4-*t*-butylphenyl)porphyrin, and *meso*-tetrakis (3,4,5-trimethoxyphenyl)porphyrin (H_2TMPP)) with two equivalents of $MN(SiMe_3)_2$ (M = Li, Na, K) in THF or dimethoxyethane (DME) yields $M_2(THF)_4Por$ and $M_2(DME)_2Por$, respectively. The lithium derivatives crystallize from THF, DME, and diacetone alcohol as 1:1 $[LiQ_n][Li(Por)]$ salts (Q = THF, $n = 4$; Q = DME, diacetone alcohol (DAA), $n = 2$).[66] The lithium TMPP derivative crystallizes from acetone, and consists of $[Li(TMPP)]^-$ and a $[Li(DAA)_2]^+$ counterion; the octaethylporphyrin derivative is isolated as the $[Li(THF)_4]^+$ $[LiOEP]^-$ salt.[67] ^7Li NMR spectroscopy and conductivity measurements indicate that these ionic structures are retained in polar solvents; in relatively nonpolar solvents, symmetrical ion-paired structures are observed.

The solid state structure of the centrosymmetric dilithium tetraphenylporphyrin bis(diethyletherate) differs from the salt-like compounds, in that the $[Li(Et_2O)]^+$ moiety is coordinated to both faces of the porphyrin in a square pyramidal fashion (Li–N = 2.23–2.32 Å).[68] A related motif is found in the case of sodium octaethylporphyrinate; X-ray crystallography reveals two $Na(THF)_2$ moieties symmetrically bound to all four nitrogen atoms, one on each face of the porphyrin ring (Na—N (av) = 2.48 Å). The structure of the potassium derivative $K_2(py)_4(OEP)$ is similar (K—N (av) = 2.84 Å).[66]

Although attempts to prepare the neutral lithium octaethylporphyrin radical ([Li(OEP)·]) have been unsuccessful, neutral π-radicals of three Li porphyrins, tetraphenylporphyrin [Li(TPP)·], tetra(pentafluorophenyl)porphyrin [Li(PFP)·], and tetra(3,5-bis-*tert*-butylphenyl)porphyrin [Li(TBP)·] are available from the dilithium porphyrins by oxidation with ferrocenium hexafluorophosphate in THF or dichloromethane.[69] The resulting lithium porphyrin radicals have been isolated by crystallization; [Li(TPP)·] is insoluble in acetone and in nonpolar solvents, whereas [Li(PFP)·] and [Li(TBP)·] are soluble in acetone, with the latter slightly soluble even in toluene and benzene. The UV/vis spectra of the radicals have been studied in acetonitrile solutions, which display negligible Λ_M values; this indicates that the compounds exist as tight ion pairs. The absence of hyperfine splitting for [Li(TPP)·] and [Li(PFP)·] at room temperature in solution and in the solid state suggests that they exist in the $^2A_{1u}$ ground state, which has low spin density on the *meso*-carbons and the nitrogen atoms.

Crystallization of [Li(TPP)·] from dichloromethane and diethyl ether yields purple crystals; the solid state structure indicates that the lithium atom is bound in the plane of the porphyrin. The porphyrin macrocycle is slightly ruffled, with opposite pyrrolic carbons up to 0.3 Å above or below the mean porphyrin plane.[69]

Several examples of porphyrin complexes of calcium are now known. Activated calcium in THF reacts with H_2OEP at room temperature, producing the bimetallic complex $Et_8N_4Ca_2(THF)_4$ in 73% yield. Subsequent reaction of the calcium complex with $Et_8N_4Li_4(THF)_4$ in THF generates the calcium–lithium complex $Et_8N_4CaLi_2(THF)_3$. Both have been structurally characterized.[70] 5,10,15,20-Tetrakis(4-*t*-butylphenyl)porphyrin (H_2L) reacts with activated calcium to give CaL, which in turn reacts with pyridine with or without added NaI or $CaI_2(THF)_4$ to give $CaL(Py)_3$, $[CaNaL(Py)_6]I$ and $Ca_3L_2(MeCN)_4I_2$, respectively. In $CaL(Py)_3$, the calcium is seven-coordinate, and is displaced from the N_4 plane of the porphyrin. $Ca_3L_2(MeCN)_4I_2$ is a double-decker sandwich compound with the outer two calcium atoms coordinated by four porphyrin N atoms, two acetonitriles and an iodide (see Figure 6). The results indicate that in polar aprotic solvents, calcium porphyrin derivatives can be stable.[71]

Phthalocyanine ligands, structurally related to porphyrins, confer distinctive optoelectronic properties on their complexes. Lithium phthalocyanine (LiPc) forms stacks in the solid state with a Li—Li′ distance of 3.245 Å,[72] this is longer than in the metal (3.04 Å), but less than the sum of the van der Waals thicknesses of the rings (see Figure 7). The extra electron left from removing two hydrogen atoms and replacing them with Li^+ is delocalized in the central ring of the

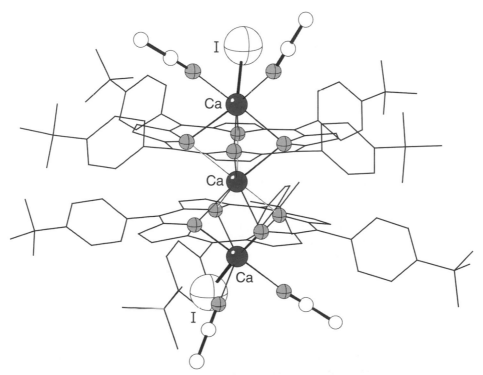

Figure 6 The double decker sandwich porphyrin complex $Ca_3L_2(MeCN)_4I_2$.

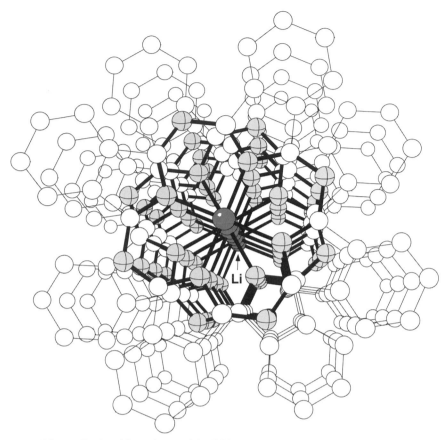

Figure 7 Stacking observed in lithium phthalocyanine (LiPc).

macrocycle.[73] Despite the stacking of the molecules, and the anticipated overlap of the π orbitals of the Pc ligand,[74] LiPc is in fact a semiconductor[75] with an optical gap of 0.5 eV, and not a one-dimensional conductor. Magnetic susceptibility, heat capacity, and optical conductivity measurements indicate that LiPc should be considered a Mott–Hubbard insulator.[76] The localized electrons behave as an $S = 1/2$ antiferromagnetic spin chain. The related iodinated compound LiPcI is EPR silent, reflecting the loss of unpaired electrons. It is an intrinsic semiconductor, with diamagnetic susceptibility.[76]

Magnesium phthalocyanine (MgPc) is a blue semiconductor with a thin film optical band gap of 2.6 eV;[77] its X-phase exhibits an intense near-IR-absorption.[78] It has attracted attention as a material for laser printer photoreceptors,[79] optical disks based on GaAsAl laser diodes,[80] and photovoltaic devices.[81] Crystalline $MgPc/(H_2O)_2 \cdot (N\text{-methyl-2-pyrrolidone})_2$ exhibits a near-IR absorption whose spectral shape is similar to that of the X-phase.[78,82] The near-IR absorption has been interpreted from the standpoint of exciton coupling effects. Structures have been calculated for both MgPc and its radical anion doublet ($MgPc^-$), using *ab initio* (6–31G(d,p)) and semiempirical (INDO/1) SCF approaches. The anion displays first-order Jahn-Teller distortion, and the effect that varying the degree of distortion has on the computed anion spectrum has been examined.[83]

3.1.3.2   Group 16 Ligands

3.1.3.2.1   Crown ethers

Crown ether complexes of the *s*-block metals number in the many hundreds,[84] and reviews focused on them, including their use in separation chemistry[85–87] and selective ion extractions,[88,89] are extensive.[90–96] Growing interest has been expressed in the use of macrocyclic ethers in the design of electroactive polymers.[97]

The 12-crown-4 ring is often complexed with lithium,[98] and the sandwich $[(12\text{-crown-}4)_2\text{Li}]^+$ ion is common, although examples with Na^+,[99–106] K^+,[106] Rb^+,[106] and Mg^{2+}[107] ions are known. The centrosymmetric dimer $[\text{Li}(12\text{-crown-}4)]_2^{2+}$, in which each lithium ion forms an intermolecular Li—O bond with a neighboring crown ether molecule (Li—O = 2.01 Å) in a rectangular four-membered Li_2O_2 ring has been described.[108]

Cation-coordinating macrocycles have been used to form amorphous electrolytes; if the cavity of the macrocycle is larger than that of the cation, the resulting complex is a glass that has a subambient glass transition temperature and high ionic conductivity.[109,110] Coordination of the lithium ion in $\text{Li}[\text{CF}_3\text{SO}_2\text{N}(\text{CH}_2)_3\text{OCH}_3]$ by 12-crown-4, for example, lengthens the Li—N distance to 2.01 Å, which indicates a weakening of the interaction between the lithium cation and the $[\text{CF}_3\text{SO}_2\text{N}(\text{CH}_2)_3\text{OCH}_3]^-$ anion.[111] Such an environment may facilitate ionic conductivity.

Molecular conductors have been constructed by using supramolecular cations as counterions to complex anions. For example, the charge-transfer salt $\text{Li}_{0.6}(15\text{-crown-}5)[\text{Ni}(\text{dmit})_2]_2\cdot\text{H}_2\text{O}$ (dmit = 2-thioxo-1,3-dithiol-4,5-dithiolate) exhibits both electron and ion conductivity: the stacks of the Ni complex provide a pathway for electron conduction, and stacks of the crown ethers provide channels for Li-ion motion.[112] The μ-crown cation $\{[\text{Li}(12\text{-crown-}4)](\mu\text{-}12\text{-crown-}4)[\text{Li}(12\text{-crown-}4)]\}^{2+}$ has been generated as the counterion to $[\text{Ni}(\text{dmit})_2]^{2-}$.[106] The salt displays a room temperature conductivity of $30\,\text{S cm}^{-1}$ and exhibits a semiconductor–semiconductor phase transition on the application of pressure or on lowering the temperature.

The 15-crown-5 ring binds a larger range of *s*-block ions than does 12-crown-4, and simple $[\text{M}(15\text{-crown-}5)]^+$ or $[\text{L}_n\text{M}(15\text{-crown-}5)]^+$ ($\text{L} = \text{H}_2\text{O}$, halide, ether, acetonitrile, etc.) complexes are common. Sandwich species of the form $[(15\text{-crown-}5)_2\text{M}]^+$ ($\text{M} = \text{K}^+$,[113,114] Cs^+,[115] Ba^{2+},[116]) are known, including the chloride-bridged species $\{[\text{Li}(15\text{-crown-}5)](\mu\text{-Cl})[\text{Li}(15\text{-crown-}5)]\}^+$.[117]

The reaction of lithium chloride with 15-crown-5 in THF produces an extended chain structure consisting of alkali metals and bridging halogens. The repeating units, $\text{Li}(\mu\text{-Cl})\text{Li}(15\text{-crown-}5)$, are connected by additional bridging Cl atoms. One lithium has close contacts with one Cl (2.34 Å) and all five oxygen atoms of 15-crown-5, and the other Li is close to three Cl (2.35–2.38 Å) and one oxygen of 15-crown-5 (see Figure 8). With the use of hydrated lithium chloride, the lithium is coordinated to all five oxygen atoms of the crown as well as to an additional oxygen atom from H_2O in a distorted pentagonal pyramidal geometry. The Cl^- counteranion is isolated from the Li^+ cation, and is hydrogen-bonded to the coordinated water molecule.[118]

The reaction of NaBr or KBr with 15-crown-5 and TlBr_3 in ethanol produces the unusual self-assembled cations $[\{\text{M}(15\text{-crown-}5)\}_4\text{Br}]^{3+}$, whose formation has been templated by the bromide anion. The crystal structure of $[\{\text{Na}(15\text{-crown-}5)\}_4\text{Br}][\text{TlBr}_4]$ reveals that the bromide is surrounded by four Na(15-crown-5) units with crystallographically imposed D_{2d}-symmetry (Na–Br = 2.89 Å; cf. 2.98 in NaBr) (see Figure 9). A folded network of TlBr_4^- anions surrounds the cations.[119]

The 18-crown-6 ether is widely represented among the *s*-block elements, and is found in a large range of compounds, either as the simple $[(18\text{-crown-}6)\text{M}]^+$ ion, coordinated with various anions $((18\text{-crown-}6)\text{ML}; \text{L} = \text{H}_2\text{O}$, ethers, alcohols, HMPA, NH_3, etc.) or as the sandwich species $[(18\text{-crown-}6)_2\text{M}]^+$. It is often thought to fit best with K^+ or Sr^{2+}, but Rb^+ can sit in the center

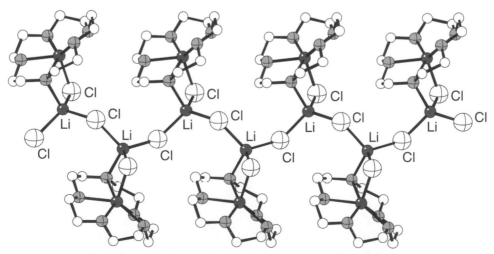

Figure 8 The structure of the LiCl/15-crown-5 polymer.

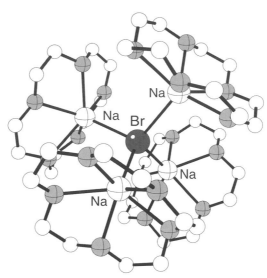

Figure 9   The solid state structure of the $[\{Na(15\text{-}crown\text{-}5)\}_4Br]^+$ cation.

of the crown, occupying a crystallographic inversion site (Rb—O bond length of 2.82–2.87 Å).[120] "Club sandwiches" of the form $[(18\text{-}crown\text{-}6)Cs(18\text{-}crown\text{-}6)Cs(18\text{-}crown\text{-}6)]^{2+}$ have been described; the central 18-crown-6 ring displays longer coordination interactions (Cs—O = 3.51 Å (av)) than the end crowns (Cs–O = 3.27 Å (av)) (see Figure 10).[121,122]

The study of luminescence has often involved alkali metal crown complexes. Luminescent copper(I) halide complexes have been isolated from the reaction of elemental copper with NH_4X (X = I, Br or SCN), RbI and 18-crown-6 in MeCN. Halo- or pseudohalo-cuprate(I) anions crystallize with the geometrically rigid crown ether cation $[Rb(18\text{-}crown\text{-}6)]^+$. The complexes $[\{Rb(18\text{-}crown\text{-}6)\}_2MeCN][Cu_4I_6]$, $[Rb(18\text{-}crown\text{-}6)][Rb(18\text{-}crown\text{-}6)(MeCN)_3]_2[\{Rb(18\text{-}crown\text{-}6)\}_6 Cu_4I_7][Cu_7I_{10}]_2$, $\{[Rb(18\text{-}crown\text{-}6)][Cu_3I_3Br]\}_\infty$ and $\{[Rb(18\text{-}crown\text{-}6)][Cu_2(SCN)_3]\}_\infty$ have been characterized. The first three complexes display temperature-sensitive emission spectra in the solid state.[123] The structure of the second is unusually complex: one $[Rb(18\text{-}crown\text{-}6)]^+$ cation and two $[Rb(18\text{-}crown\text{-}6)(MeCN)_3]^+$ cations, the bulky supramolecular cation $[\{Rb(18\text{-}crown\text{-}6)\}_6 Cu_4I_7]^{3+}$ (see Figure 11) and the crown-like $[Cu_7I_{10}]^{3-}$ cluster are present.[123]

Luminescence and electronic energy transport characteristics of supramolecular $[M(18\text{-}crown\text{-}6)_4 MnBr_4][TlBr_4]_2$ (M = Rb, K) complexes (see Figure 12) were studied in the expectation that $[MnBr_4]^{2-}$ ions would be effective luminescent probes for solid state (18-crown-6) rotation-conformational

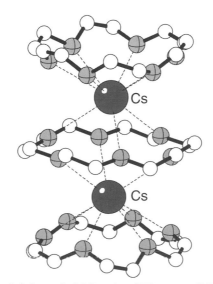

Figure 10   The structure of the "club sandwich" cation $[(18\text{-}crown\text{-}6)Cs(18\text{-}crown\text{-}6)Cs(18\text{-}crown\text{-}6)]^{2+}$.

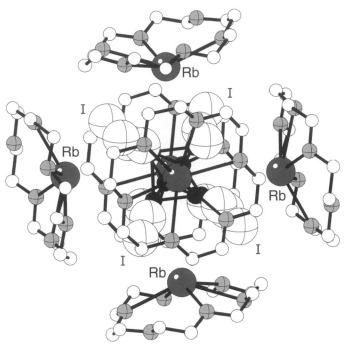

Figure 11 The structure of the supramolecular cation $[\{Rb(18\text{-}crown\text{-}6)\}_6Cu_4I_7]^{3+}$.

motion. Luminescence and excitation spectra are normal when M = Rb (a strong emission at 77 K with λ_{max} of 535 nm is observed, with weak room temperature luminescence), but when M = K, an unusual orange emission with $\lambda_{max} \approx 570$ nm is observed; it has been attributed to crystal defects.[124]

When reduced, fullerene can be supported by $[K(18\text{-}crown\text{-}6)]^+$. Paramagnetic red-black $[K(18\text{-}crown\text{-}6)]_3[C_{60}]$ is prepared by dissolving potassium in molten 18-crown-6, followed by addition of C_{60}, or by reducing C_{60} with potassium in DMF followed by reaction with 18-crown-6. In the solid state, the potassium ions bind to the six oxygen atoms of the crown ethers; two potassium ions are η^6-bonded to opposite 6-membered rings on C_{60}^{3-}, whereas the third is bound to a crown ether as well as to two toluene molecules (see Figure 13).[125]

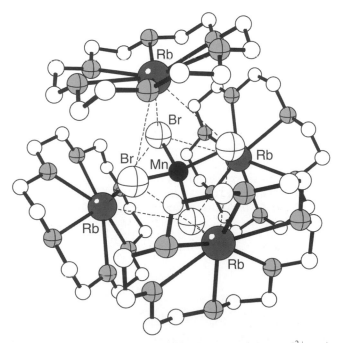

Figure 12 The structure of the $[Rb(18\text{-}crown\text{-}6)_4MnBr_4]^{2+}$ cation.

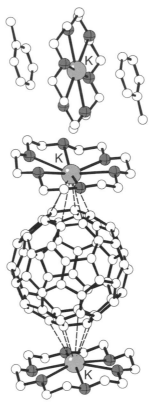

Figure 13   The structure of [K(18-crown-6)]$_3$[C$_{60}$].

In the solid state, the macrocyclic complex Rb$_3$(18-crown-6)$_3$Cu$_2$[N(CN)$_2$]$_5$ includes polymeric dicyanoamidocuprate(I) anions, and the Cu atoms are coordinated at the nitrile nitrogens (Cu—N = 1.89–2.07 Å). There are two types of Cu atoms with different environments, planar-trigonal and tetrahedral. The [Rb(18-crown-6)]$^+$ units form puckered planes about 11 Å apart (see Figure 14).[126]

Large crown ethers have been investigated for their sometimes unexpected ion selectivities. The structural origins of the selectivity of Rb$^+$ ion over other alkali metal ions by tribenzo-21-crown-7 has been elucidated from single-crystal X-ray structures of Cs[tribenzo-21-crown-7]NO$_3$, {[Rb(4,4-bis-*t*-butylbenzo,benzo-21-crown-7)(dioxane)]$_2$(μ-dioxane)}Cl, and Na[4,4-bis-*t*-butylbenzo,benzo-21-crown-7]ReO$_4$. Different crown conformations are observed for each structure. Molecular mechanics calculations on the conformers suggest that the selectivity found for the crown for Rb$^+$ and Cs$^+$ over the smaller Na$^+$ can be largely attributed to the energetically unfavorable conformation that must be adopted to achieve heptadentate coordination with optimum Na—O distances. The selectivity for Rb$^+$ over Cs$^+$ may be a consequence of stronger Rb—O bonds, which outweigh the small (0.7–0.9 kcal mol^{-1}) steric preference for Cs$^+$ over Rb$^+$.[127]

Alkali metal picrates have been used to measure formation constants for crown ethers in solution, but the selectivity of benzo crown ethers for metal picrates, relative to the analogous chlorides, nitrates, perchlorates, and thiocyanates, may vary significantly. Apparently, π–π interactions between the picrate ions and the aromatic ring(s) on the crown are responsible for the difference. The importance of the "picrate effect" rises as the number of benzo groups in the crown ether is increased, and it varies with their location in the macrocycle. The dependence of the picrate ^{1}H NMR chemical shift on the metal cation and/or macrocycle identity has been used to study picrate-crown ether π-stacking in large crown ether (18, 21, and 24-membered) complexes.[128]

3.1.3.2.2   *Cryptands and related species*

The *s*-block metals are commonly complexed with the macrocyclic cryptands, sepulchrates, and related species[129] to form large, non-interacting cations that are used to stabilize a variety of anions, such as metal clusters (e.g., Ge$_5$$^{2-}$,[130] Ge$_9$$^{3-}$,[131] Ge$_{18}$$^{6-}$,[132] Sn$_5$$^{2-}$,[133] Sn$_9$$^{3-}$,[134,135]

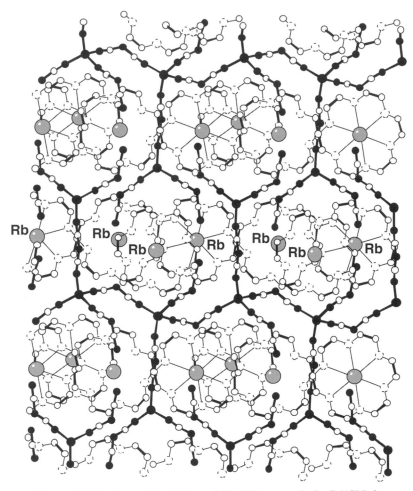

Figure 14 Section of the lattice of $Rb_3(18\text{-crown-}6)_3Cu_2[N(CN)_2]_5$.

$Sn_2Se_6{}^{4-}$,[136] $K_2Sn_2Te_6{}^{2-}$,[136] $Pb_9{}^{3-}$,[134,137] $Pb_9{}^{4-}$,[137] $Pb_2S_3{}^{2-}$,[138] $Pb_2Se_3{}^{2-}$,[138] $PbTe_3Tl^{3-}$,[138] $Pb_2Te_3{}^{2-}$,[139] $As_2S_4{}^{2-}$,[140] $As_4Se_6{}^{2-}$,[140] $As_{10}S_3{}^{2-}$,[140] $Sb_2Se_4{}^{2-}$,[140] Bi_3Ga^{2-},[141] Bi_3In^{2-},[141] $Bi_5In_4{}^{3-}$,[141] $Se_{10}Sn_4{}^{4-}$,[142] $Se_2Tl_2{}^{2-}$,[143] $Te_2Tl_2{}^{2-}$,[143] and $MoAs_8{}^{2-}$).[144]

The relative inertness of cryptands has made them especially useful for the isolation of otherwise highly reactive or unstable anions. For example, the reaction between RbO_3 and 18-crown-6 in liquid ammonia permits the isolation of the crystalline ozonide complex [Rb(18-crown-6)]$O_3\cdot NH_3$.[145] The use of cryptands is required to isolate complexes derived from the less stable LiO_3 and NaO_3 in liquid ammonia; crystalline ozonide complexes {Li[2.1.1]}O_3 ([2.2.1] = 4,7,13,18-tetraoxa-1,10-diazabicyclo[8.5.5]eicosane) and {Na[2.2.2]}O_3 ([2.2.2] = 4,7,13, 16,21,24-hexaoxa-1,10-diazabicyclo[8.8.8]hexacosane) can be obtained that contain the bent $O_3{}^-$ anion.[146] The diamagnetic $Bi_2{}^{2-}$ anion has been isolated as its [K([2.2.2]crypt)] salt.[147] Each "naked" anion (Bi—Bi = 2.8377(7) Å) is surrounded by eight [K-crypt]$^+$ cations, and it is notable that the dianion has been stabilized without the bulky substituents usually required for isolation of multiply bonded main-group species (see Figure 15).[148]

The fulleride dianion has been isolated in the solid state as [K([2.2.2]crypt)]$_2$[C$_{60}$]; its structure consists of alternating layers of ordered $C_{60}{}^{2-}$ anions and [K([2.2.2]crypt)]$^+$ cations.[149] The complete separation of the anions (>13.77 Å) by the cations allows EPR and magnetic susceptibility measurements on the isolated fulleride.

3.1.3.2.3 *Calixarenes*

Calixarenes, the cyclic oligomers formed from condensation reactions between *para*-substituted phenols and formaldehyde, are inexpensive compounds that are stable to both basic and acidic media.[150,151]

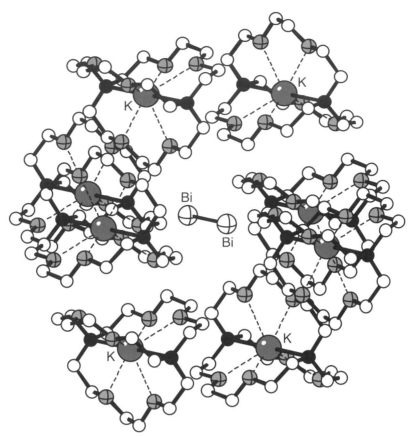

Figure 15   The [K([2.2.2]crypt)] salt of the Bi_2^{2-} anion.

Their ability to complex both neutral and ionic species has driven their employment as complexing agents, extractants,[152–156] in chemical sensing (detection) devices,[157–159] and as catalysts.[160,161]

Calixarenes excel in the complexation of large ions, and this has been exploited in the development of ligands for radium.[162] ^{223}Ra ($t_{1/2} = 11.4$ d) is an α-particle emitter that has been evaluated for use in cell-directed therapy of cancer. Such use requires that it be attached to a monoclonal antibody or related targeting protein with high specificity, and that the complex exhibit kinetic stability at physiological pH in the presence of much greater concentrations of other potentially binding ions such as Mg^{2+} and Ca^{2+}. The lipophilic acrylic polyether carboxylic acid, bis-1,8-(2'-carboxy-3-naphthoxy)-3,6-dioxaoctane, exhibits selectivity for Ra^{2+} over Ba^{2+}, but does not have adequate binding stability to serve in radiotherapy.[163]

Bifunctional radium-selective ligands together with effective linkers to the protein antibody have been developed from the 1,3-alkoxycalix[4]arene-crown-6 cavity, which has a high selectivity for Cs^+ over K^+.[164] Modified with proton-ionizable crowns with carboxylate sidearms to enhance the binding of alkaline-earth ions, the two ionizable calixarene-crowns, *p-t*-butylcalix[4]arene-crown-6-dicarboxylic acid (see Figure 16(a)) and *p-t*-butylcalix[4]arene-crown-6-dihydroxamic acid (see Figure 16(b)), are able to extract greater than 99.9% of radium in the presence of Mg^{2+}, Ca^{2+}, Sr^{2+}, and Ba^{2+}. The lariat arms prevent radium from escaping from the cavity, and the complexes display kinetic stability in the presence of serum-abundant metal ions including Na^+, K^+, Mg^{2+}, Ca^{2+}, and Zn^{2+} at relatively high concentrations (10^{-2} M) and pH 7.4.

The ability of calixarenes to bind large metal ions with high kinetic stability is important in the search for complexants for radionuclides such as ^{137}Cs ($t_{1/2} = 30.2$ yr) and ^{85}Sr ($t_{1/2} = 65$ d) from the reprocessing of exhausted nuclear fuel.[165] There has been considerable interest in caesium-complexed calix[4]-bis-crowns as selective Cs-carriers.[166] Transport isotherms of trace level ^{137}Cs through supported liquid membranes containing calix[4]-bis-crowns have been determined as a function of the ionic concentration of the aqueous feeder solutions, and 1,3-calix[4]-bis-*o*-benzo-crown-6 appears to be much more efficient in decontamination than mixtures of crown ethers and acidic exchangers, especially in highly acidic media.[167]

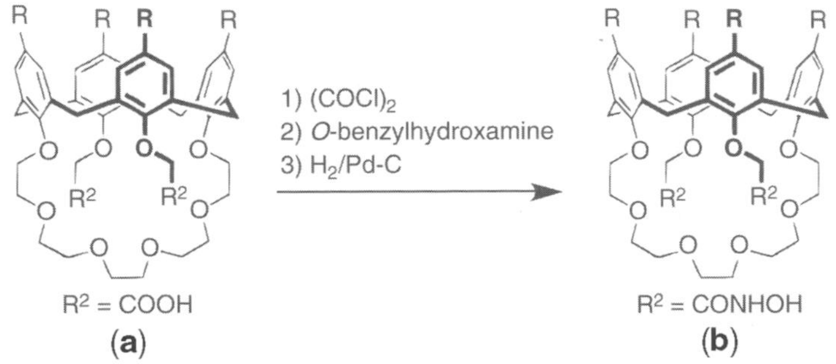

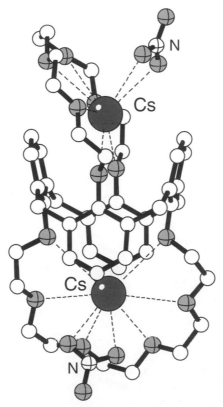

Figure 16 Two ionizable calixarene-crowns used to complex Ra²⁺.

The complexing properties of 1,3-calix[4]-bis-crown-6 towards Cs⁺ ions have been studied by [133]Cs and ¹H-NMR spectroscopy. Crystal structures of caesium complexes indicate that the cations are bound in the polyether loops (e.g., the dinitrato complex, see Figure 17), and suggest that the ligand is preorganized for Cs⁺ ion complexation. This may explain the high selectivity displayed toward the cation.[168] Caesium ions are also observed to bind to the polyether loops in the substituted calixarenes prepared from the base-catalyzed reactions of calix[4]crown-6 with TsO(CH₂CH₂O)₂X(OCH₂CH₂)₂OTs [X = o-C₆H₄, 2,3-naphthalenediyl].[169] Similar caesium binding is observed in the binuclear complex formed from 1,3-calix[4]-bis-crown-6 and caesium iodide. The two Cs⁺ ions are located at the center of a coordination site defined by the six oxygen atoms of the crown-ether chains, and are bonded to six oxygen atoms and iodide counterions; they also interact with the two closest benzene rings.[170]

Cone diallyloxybis-crown-4 calix[6]arene and its 1,2,3-alternate stereoisomer have been isolated in 11% and 15% yields, respectively, by bridging a 1,4-diallyloxy calix[6]arene with triethylene glycol di-p-tosylate, 4-MeOC₆H₄SO₂OCH₂(CH₂OCH₂)₂CH₂OSO₂C₆H₄-4-Me. Both conformers form 1:1 complexes with all alkali metal ions, but are structurally preorganized such that each exhibits a strong preference for the caesium ion. The structure of the complex between the cone

Figure 17 Dinitrato derivative of Cs⁺ and 1,3-calix[4]-bis-crown-6.

calixarene and caesium tetraphenylborate reveals cooperative complexation of caesium by both crown-4-ethers (see Figure 18). The association constants of caesium and rubidium ions with the cone stereoisomer are 20–50 times greater than that for the 1,2,3-alternate stereoisomer; cooperative binding of cations by the two crown ether moieties is not possible for the latter. The Cs^+/Na^+ selectivity factor for the cone isomer is found to be 1,500, while that of the 1,2,3-alternate stereoisomer is 140.[171]

1,3-Dialkoxycalix[4]arene-crown-6 ligands are obtained in the fixed 1,3-alternate conformation in 63–85% yield by the reaction of the corresponding 1,3-dialkoxycalix[4]arenes with pentaethylene glycol ditosylate in acetonitrile in the presence of Cs_2CO_3. The corresponding cone conformer of the diisopropyl derivative has been synthesized via selective demethylation of the 1,3-dimethoxycalix-crown and subsequent dialkylation. Extraction with alkali metal picrates reveals a strong preference of the ligands for Cs^+; greater than 99.8% of Cs^+ can be removed at pH = 0 from solutions that are 4 M in Na^+. Thermodynamic measurements obtained for the complexation of the diisopropyl derivative indicate a high stability constant in methanol (log $\beta = 6.4 \pm 0.4$). The entropy of complexation ($T\Delta S = -15\,kJ\,mol^{-1}$) is less negative than for other crown ethers, and probably derives from the preorganization of the ligand. Both X-ray crystallographic and solution NMR studies confirm that the cation is positioned between the two aromatic rings.[172]

In an interesting variation on the use of calixarenes to complex caesium ions, when $[HNC_5H_5]_2[UO_2Cl_4]$ is treated with *t*-Bu-calix[6]arene (H_6L) in pyridine, no reaction is observed, even after refluxing for 12 hours. When one equivalent of caesium triflate is added to the mixture, however, the pale yellow color of the solution immediately turns deep red, and a heterotrimetallic complex of the *t*-Bu-calix[6]arene can be isolated. The crystal structure of the compound reveals that two uranyl cations and a caesium atom are coordinated to the macrocycle (see Figure 19).[173] The two uranyl cations are bound in an external fashion to the macrocycle through the deprotonated oxygens of the phenolate groups. The caesium cation is bound to the two protonated oxygens of the calixarene that do not form bonds with uranium, and is also bound in an approximately η^6-fashion to the faces of the two phenolic rings (mean Cs–centroid distance = 3.35 Å). NMR experiments (^1H and ^{133}Cs) indicate that the caesium cation interacts with H_6L in pyridine and changes its conformation, which is critical for subsequent binding of the uranyl cation.

Calix[6]- and calix[8]-arene amides have been found to be efficient ionophores for the selective extraction of strontium from highly acidic radioactive solutions.[174] Often low concentrations of strontium ion (ca. 10^{-3} M) must be removed in the presence of much higher alkali metal ions (e.g., $[Na^+] = 4$ M), and therefore ligands with high Sr^{2+}/Na^+ selectivity are desirable.[175] Strontium complexes of calixarene amides, in particular, have been studied as part of the search for high alkaline-earth selectivity. A *p-t*-butylcalix[6]arene hexaamide forms a 1:1 complex with strontium picrate, whereas related *p-t*-butylcalix[8]arene and *p*-methoxycalix[8]arene octaamides encapsulate two strontium cations each. The binding geometries of the metal cations depend on the ligand size and whether a chloride or picrate counteranion is present.[176] The higher Sr^{2+}/Na^+ selectivity shown by calix[8]arene derivatives compared to those of calix[6]- and calix[4]-arene

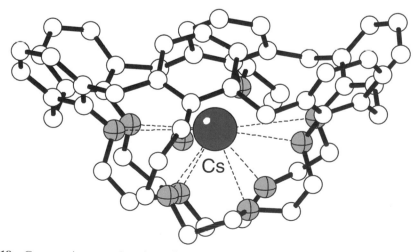

Figure 18   Cooperative complexation of caesium by both crown-4-ethers in a cone calixarene.

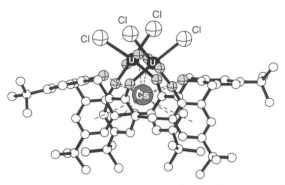

Figure 19 Cooperative binding of two uranyl cations and a caesium atom within a Bu^t-calix[6]arene.

amides appears to be mainly a consequence of the low binding ability of the larger calixarene ligands towards the sodium cation, which in turn stems from its small size relative to the calixarene cavity.

Various homo- and heterometallic aggregates can be constructed within calixarene frameworks. Tetralithiation of *p-t*-butylcalix[4]arene (H$_4$L) in the presence of wet HMPA affords the monomeric complex (Li$_4$LLiOH$_4$·HMPA), in which LiOH is incorporated into an Li$_5$O$_5$ core based on a square pyramid of Li atoms. When the same reaction is conducted with dry HMPA, a dimeric LiOH-free species containing an Li$_8$O$_8$ core formed by the edge-sharing of two square pyramids of Li atoms is generated (see Figure 20).[177] The deprotonation of substituted (Pr^i and Bu^j) calix[8]arenes (H$_8$L) with Bu^{n}Li in DMF followed by reaction with anhydrous SrBr$_2$ yields the discrete, structurally authenticated molecular complexes Li$_4$Sr$_2$(H$_2$L)(O$_2$CC$_4$H$_9$)$_2$(DMF)$_8$ (the Pr^i derivative is depicted in Figure 21). The heterometallic Li$_4$Sr$_2$ cores fit within the flexible cavities of the calix[8]arene.[178]

Cation-π interactions, which are frequently encountered in calixarenes complexes, are observed in three related potassium complexes of calix[6]arenes, [K$_2$(MeOH)$_5$]{*p*-H-calix[6]arene-2H}, [K$_2$(MeOH)$_4$]{*p-t*-butylcalix[6]arene-2H} and [K$_2$(H$_2$O)$_5$]{*p*-H-calix[6]arene-2H}. The crystal

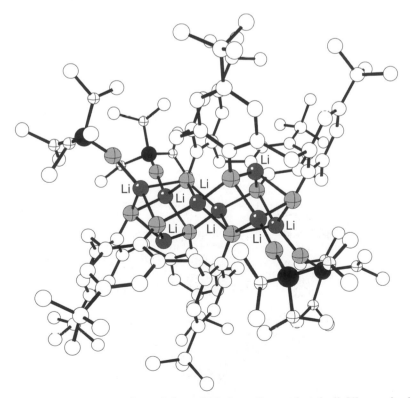

Figure 20 Octalithium aggregate formed from lithiation of *p*-tert-butylcalix[4]arene in dry HMPA.

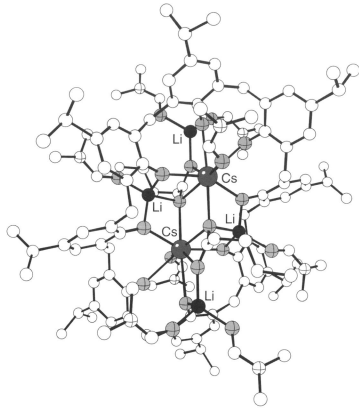

Figure 21 Structure of the strontium derivative $Li_4Sr_2(H_2L)(O_2CC_4H_9)_2(DMF)_8$ formed from $Pr^icalix[8]$ arene (H_8L).

structure of each complex indicates that the doubly deprotonated macrocyclic ligand incorporates two K^+ ions and adopts the double partial cone conformation. The structures of the first two are similar in that one K^+ ion is positioned near the center of the cavity of the macrocycle and binds to four phenolic oxygens and two methanol ligands, while the other K^+ ion binds to either phenolic oxygen and four methanols, or to three methanols. In the structure of the aqua complex, each of the two K^+ ions are mirror-related and linked to each other through three bridging waters. Close contact between K^+ ion and aryl rings is observed in all three structures.[179]

3.1.3.2.4 *Alkalides and electrides*

Alkalides and electrides are salts in which alkali metals (Na, K, Rb, Cs; Li derivatives are unknown) and electrons, respectively, are the anionic species. The formal M(–I) oxidation state of the alkalide ions gives them closed subshell ns^2 electron configurations, and the extra electron gives them large effective sizes, ~2.7 Å for Na^- to ~3.5 Å for Cs^-.[180] The crystal structures of known electrides are similar to the corresponding alkalides except that the anionic sites are empty.[181] The field of alkalides and electrides expanded tremendously in the 1980s and 1990s through the work of Dye and co-workers, and the first structurally characterized alkalide $(Cs^+(18\text{-}crown\text{-}6)Na^-)$[182] and electride $(Cs^+(18\text{-}crown\text{-}6)2e^-)$[183] came from his group. The area has been reviewed in *Comprehensive Supramolecular Chemistry*[184] and other summaries are available.[185,186]

Recent work has helped to refine the understanding of the physical and magnetic properties of these systems. The synthesis, structure, polymorphism, and electronic and magnetic properties of the electride Rb(cryptand[2.2.2])e^- have been described. Depending on the manner of preparation and the temperature, the antiferromagnetic electride can display a range of electrical conductivity, from poor $(<10^{-4}\,S\,cm^{-1})$—consistent with localized electrons—to near-metallic electrical conductivity.[187] Studies of the phase transitions in $Cs^+(18\text{-}crown\text{-}6)_2e^-$ with NMR, EPR, and variable-temperature powder X-ray diffraction indicates that it undergoes a slow irreversible transition above 230 K from a crystalline low temperature phase to a disordered Curie–Weiss paramagnetic high temperature phase.[188]

Ligands other than crowns and cryptands, which are the most common complexants of the cations in alkalides and electrides, have begun to receive more investigation. The properties of the lithium–sodium–methylamine system ($LiNa(CH_3NH_3)_n$) have been examined as a function of n. The phase diagram (established with DTA) shows the presence of a compound with $n \approx 6$, which melts congruently at 168.5 ± 0.5 K. A combination of EPR and alkali-metal NMR spectroscopies and static magnetic susceptibilities data indicate that the sodide $LiNa(CH_3NH_2)_4$ could be considered a type of near-metal in the liquid state, with a conductivity similar to that of $Li(CH_3NH_2)_4$ (conductivity of around $400\,S\,cm^{-1}$).[189]

By using compounds that have only C—N linkages and no amine hydrogens, alkalides with improved thermal stability can be generated. Thus when 4,7,13,16,21,24-hexamethyl-1,4-7,10,13,16,21,24-octaazabicyclo[8.8.8]hexacosane (i.e., the fully methylated aza analog of cryptand[2.2.2]) is allowed to react with NaK or K in MeNH$_2$, the corresponding sodide or potasside salt is formed. Characterized with thin film reflectance spectral data, SQUID (Superconducting QUantum Interference Device) measurements, and crystallography, these represent the first alkalides that are stable at, and even slightly above, room temperature ($\sim$50–60 °C).[190] An interesting extension of this concept led to the examination of 3^6 adamanzane as a ligand. The reaction of protonated 3^6 adamanzane glycolate with Na in liquid NH_3 converts the glycolate into the disodium salt $NaOCH_2COONa$, releasing H_2 and 3^6AdzH^+ cations. These subsequently recombine with Na^- anions to form the complex $3^6AdzH^+Na^-$ (Equation(1)):

$$AdzH^+HOCH_2COO^- + 3Na \rightarrow AdzH^+Na^- + NaOCH_2CONa \downarrow + 0.5H_2 \uparrow \qquad (1)$$

The sodide, stable to −25 °C, has been dubbed an "inverse sodium hydride" (see Figure 22).[191] The strategy of using kinetically trapped cations in polyaza cages may lead to new classes of stabilized alkalides and electrides.

Parallels have been proposed between the dissolution of the alkali metals in nonaqueous solvents and the interactions of alkali metals with zeolites.[192,193] The sorption of sodium or potassium vapor into dehydrated zeolites produces intensely colored compounds, ranging from burgundy red to deep blue, depending upon the metal concentration. A combination of EPR,

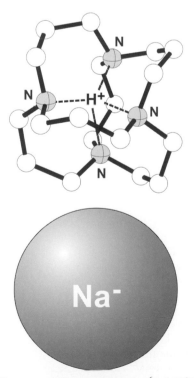

Figure 22 The structure of the "inverse sodium hydride" $3^6AdzH^+Na^{\cdot-}$. The sodide anion is drawn approximately to scale.

magnetic susceptibility, and powder neutron diffraction measurements has been applied to the characterization of the sodium- and potassium-based "cluster crystals" in zeolites Y and A, respectively.[194] It has been suggested that molecular wires might be constructed in the one-dimensional channels of alkali metal-loaded zeolite L.[195,196]

3.1.4 NONMACROCYCLIC COMPLEXES

3.1.4.1 Hydroborates

In this section, only monoboron ligands (e.g., BH_3, BH_4^-) are considered, and not those containing higher boranes or B—B bonded units. Until the mid-1980s, the tetrahydroborate anion BH_4^- ("borohydride") was considered to be an essentially noncoordinating anion in alkali metal chemistry. This was a reasonable conclusion, as its MBH_4 salts crystallize with ionic lattices (e.g., $NaBH_4$ has the NaCl structure)[197] and are high-melting solids. The beryllium compound $Be(BH_4)_2$ was known to be polymeric in the solid state,[198] although it is monomeric in the gas phase with six-coordinate Be (i.e., each BH_4 group is η^3-).[199–201] Disagreements over its conformation led to many, sometimes conflicting, computational studies.[202–212] The chemistry of the alkaline-earth borohydrides has been reviewed.[213]

In conjunction with auxiliary ligands, a wide variety of coordination geometries have now been structurally authenticated for the *s*-block borohydride ions, including η^1-, η^2-, and η^3- modes (see Table 2). In some cases, the compounds are prepared by recrystallizing the $M(BH_4)_n$ salts or their THF adducts in the presence of the supporting ligand (e.g., pyridine, trimethyltriazacyclononane[214] bipy (2,2′-bipyridyl), pyrazolylborates,[215] diglyme, and 18-crown-6[216]). In the case of pyridine and substituted pyridine solvates, IR and ^{11}B-NMR data do not provide definitive information about the coordination modes of the BH_4^- ligands.[214]

Table 2 Structurally characterized borohydride complexes of the *s*-block metals.

Complex	$M-\mu H$(Å)	References
$Py_3Li(\eta^2\text{-}BH_4)$	1.97, 2.05	214
$(p\text{-benzyl-py})_3Li(\eta_3\text{-}BH_4)$	2.08–2.31	214
$[(2,4,6\text{-}Me_3)py]_3Li(\eta^2\text{-}BH_4)$	1.81	214
$(DME)_2Li(\eta^2\text{-}BH_4)$	2.02	740
$(PMEDTA)Li(\eta^2\text{-}BH_4)$	1.92–2.04	214
$[(c\text{-}1,3,5\text{-}(MeNCH_2)_3)Li(\mu:\eta^3\text{-}BH_4)]_2$	1.92–2.08	214
$[(TMEDA)Li(\mu:\eta^2\text{-}BH_4)]_2$	2.02–3.12	741
$(18\text{-crown-}6)[Li(\eta^1\text{-}BH_4)]_2$	1.72	742
$[HC(3,5\text{-}Me_2pz)_3]Li(\eta^3\text{-}BH_4)$	2.06–3.12	215
$(THF)_3Li(\eta^3\text{-}BH_4)$	3.11–3.12	740
$\{[H_2C(3,5\text{-}Me_2pz)_2]Li(\mu:\eta^3\text{-}BH_4)\}_2$	2.04–3.18	215
$\{[4,4'\text{-}Me_2bipy]Li(\mu:\eta^3\text{-}BH_4)\}_2$	1.91–2.23	215
$[(Et_2O)_2Li(\mu:\eta^4\text{-}BH_4)]_n$	1.97–2.34	740,743
$[(Bu^tMeO)_2Li(\mu:\eta^4\text{-}BH_4)]_n$	2.08–3.16	740
$[(triglyme)Li(\eta^2\text{-}BH_4)]_n$	2.05	740
$[(1,3\text{-dioxolane})Li(\mu:\eta^2\text{-}BH_4)]_n$	1.98–2.07	740
$[(PMEDTA)Na(\mu:\eta^3\text{-}BH_4)]_n$	2.41–2.74	214
$[(PMEDTA)Na(\mu:\eta^3\text{-}BH_4)]_n$	2.64	214
$[(\mu\text{-}Bu^tO)_2Be(\eta^2\text{-}BH_4)]_2Be$	1.46–1.61	744
$(THF)_3Mg(\eta^2\text{-}BH_4)$	2.45	745
$(diglyme)Mg(\eta^2\text{-}BH_4)_2$	1.94–2.00	746
$(diglyme)2Ca(\eta^2\text{-}BH_4)_2$	2.45–2.48	747
$(\eta^3\text{-diglyme})(\eta^2\text{-diglyme})Ca(\eta^2\text{-}BH_4)(\eta^3\text{-}BH_4)$	2.33–2.60	748
$[(THF)_2Sr(\mu:\eta^4\text{-}BH_4)_2]_n$	2.65–2.92	740
$(diglyme)_2Sr(\eta^3\text{-}BH_4)_2$		216
$(18\text{-crown-}6)Sr(\eta^3\text{-}BH_4)$		216
$(diglyme)_2Ba(\eta^3\text{-}BH_4)_2$		216
$[(THF)_2Ba(\mu:\eta^2\text{-}BH_4)_2]_n$	2.80–2.90	740
$(18\text{-crown-}6)Ba(\eta^3\text{-}BH_4)_3$		216

The coordination chemistry of other $BH_n^{(3-n)}$ ligands has been developed with the *s*-block elements. The reaction of Na metal with $Me_2NH \cdot BH_3$ in THF yields $Na[(H_3B)_2NMe_2]$, which can be isolated as a THF solvate $\{Na[(H_3B)_2NMe_2]\}_5 \cdot THF$, and which reduces aldehydes, ketones, acyl chlorides, and esters to the corresponding alcohols.[217] Addition of 15-crown-5 or benzo-15-crown-5 to a THF solution of $Na[(H_3B)_2NMe_2]$ yields $Na[(H_3B)_2NMe_2] \cdot 15$-crown-5 and $Na[(H_3B)_2N-Me_2] \cdot$benzo-15-crown-5, respectively. $\{Na[(H_3B)_2NMe_2]\}_5 \cdot THF$ crystallizes in an extended three-dimensional lattice, in which the Na atoms are coordinated by six to nine hydridic hydrogens. $Na[(H_3B)_2NMe_2] \cdot$benzo-15-crown-5 is a molecular compound in the solid state, with a $[(\mu\text{-}H\text{-}BH_2)_2\text{-}N(CH_3)_2]^-$ ligand; only one hydrogen atom of each BH_3 group coordinates to the Na center.

The reaction between NaSH and $THF \cdot BH_3$ under dehydrogenation conditions or between anhydrous Na_2S and $THF \cdot BH_3$ produces $Na[H_3B\text{-}\mu_2\text{-}S(B_2H_5)]$, whose structure has been examined with SCF calculations.[218] Addition of $NaBH_4$ to $Na[H_3B\text{-}\mu_2\text{-}S(B_2H_5)]$ in diglyme or triglyme generates the $[S(BH_3)_4]^{2-}$ ion, which has been crystallized as $[Na(triglyme)]_2[S(BH_3)_4]$. The latter contains a μ_4-sulfur atom, i.e., $[(triglyme)Na(\mu\text{-}H)BH_2(\mu\text{-}H)_2BH]_2(\mu_4\text{-}S)$ $(Na\text{—}S = 2.75, 2.92 \text{ Å})$ (see Figure 23).

The "superhydride" anion (BEt_3H^-) has been isolated and crystallographically characterized as its sodium derivative. Two moles of sodium superhydride $(NaHBEt_3)$ and one mole of mesitylene form a crystalline compound that has been characterized with differential scanning calorimetry and single-crystal X-ray diffraction. The molecule consists of a central dimeric $Na_2(\mu\text{-}HBEt_3)_2$ core; each sodium is also coordinated to another bridging $HBEt_3$ anion that is in turn bound to another sodium atom. The two terminal sodium ions display η^6- cation-π interactions to two mesitylene rings (see Figure 24).[219]

Deprotonation of $Me_2NH \cdot BH_3$ with Bu^nLi generates lithium(dimethylamino)trihydroborate, $Li(Me_2NBH_3)$, which is found to be unstable in most ethers, reversibly decomposing into lithium hydride and $Li(Me_2N\text{-}BH_2\text{-}NMe_2\text{-}BH_3)$. Five solvates of (TMEDA, dioxane, 1,3-dioxolane, 1,3,5-trioxane, 12-crown-4) were characterized by X-ray diffraction. Only the TMEDA adduct displays $Li\text{—}H\text{—}B$ interactions; i.e., two (TMEDA)Li units are connected to one another by $Li(\mu\text{-}H)BH_2\text{—}NMe_2$ bridges.[220]

The reaction of $2,6\text{-}Mes_2C_6H_3Li$ and BBr_3 produces the $2,6\text{-}Mes_2C_6H_3BBr_2$ derivative in good yield. Reduction of the latter with KC_8 affords potassium 9-borafluorenyl salts. The product isolated from reduction with four equivalents of KC_8 in diethyl ether, crystallizes from THF/hexane as a centrosymmetric dimer in which two THF molecules solvate each potassium ion and also interact with the 9-H and 9-Me groups of the 9-borafluorenyl rings. The product from the reduction of the arylboron dibromide with excess KC_8 in benzene also forms a centrosymmetric dimer; each potassium ion is η^6-coordinated by benzene in addition to the solvation by 9-H and 9-Me groups from the 9-borafluorenyl rings.[221]

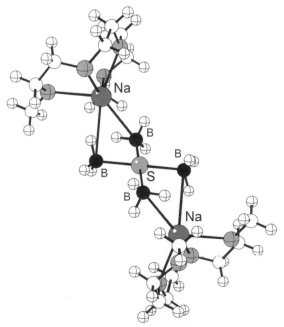

Figure 23 The structure of $[(triglyme)Na(\mu\text{-}H)BH_2(\mu\text{-}H)_2BH]_2(\mu_4\text{-}S)$.

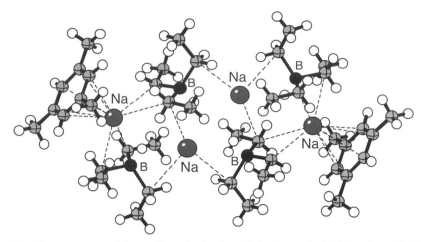

Figure 24   The structure of the sodium derivative of the "superhydride" anion, Na[HBEt$_3$].

The reaction of 3,5-bis(trifluoromethyl)pyrazole and KBH$_4$ in toluene at 115–120 °C produces white K[H$_2$B(3,5-(CF$_3$)$_2$pz)$_2$], characterized with ^{1}H, ^{13}C, and ^{19}F spectroscopy. It has a polymeric structure resulting from close contacts between the potassium ion and the nitrogens, some of the fluorines of the CF$_3$ groups on the 3- and 5-positions, and a hydrogen of the BH$_2$ moieties. Each potassium center is coordinated to three nitrogens (2.82–3.04 Å), five fluorines (2.81–3.33 Å) and to a hydrogen (K—H = 2.77 Å) (see Figure 25). The potassium salt has been used in the synthesis of Cu(II) and Zn(II) derivatives.[222]

3.1.4.2   Group 14 Ligands

In this section are covered complexes that contain *s*-block–(Si, Ge, Sn, Pb) bonds as their primary point of interest. Cluster compounds that contain Group 14–Group 14 bonds, or in which an

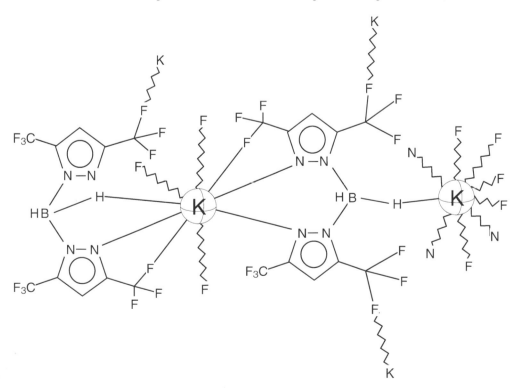

Figure 25   The structure of K[H$_2$B(3,5-(CF$_3$)$_2$pz)$_2$].

alkali or alkaline-earth metal is involved peripherally (e.g., the $Na(THF)_2$ units in the tin cube $[(Si(Bu^t)_3)_6Sn_8](Na(THF)_2)_2)$[223] are not considered here.

A variety of monomeric species with terminal *s*-block–Group 14 bonds are known and have been structurally characterized (see Table 3). $(THF)_3Li–E(SiMe_3)_3$ complexes are commonly prepared from the reaction of $E(SiMe_3)_4$ with MeLi, which eliminates $SiMe_4$ (E = Si,[224] Ge,[225] Sn[226]). A one-pot synthesis is available for the preparation of $(THF)_3Li–Sn(SiMe_3)_3$ by the reaction of chlorotrimethylsilane and lithium wire; the resulting $Li(THF)_n \cdot SiMe_3$ is treated with

Table 3 Structurally authenticated *s*-block complexes with group 14 ligands.

Complex	$M—E_{term}$ (Å)	References
$[(THF)_3Li–Si(SiMe_3)_3] \cdot 0.5Si(SiMe_3)_4$	2.68	224
$(THF)_3Li–Si(SiMe_3)_3$	2.61, 2.67 (at 153 K); 2.67 (at 213 K)	224,749
$(THF)_3Li–SiPh_3$	2.67	749
$(THF)_3Li–SiPh(NEt_2)_2$	2.63	749
$(THF)_3Li–SiPh_2(NEt_2)$	2.68	749
$HC\{SiMe_2N(4-CH_3C_6H_4)\}_3SnLi(THF)_3$	2.89, 2.97	750
$MeSi\{SiMe_2NBu^t\}_3PbLi(THF)$	2.83	750
$(THF)_2Li[(Si[(NCH_2Bu^t)_2C_6H_4-1,2])Si(SiMe_3)_3]$	2.61	751
$[(THF)_2Li]_2–Si[Si(Pr^i)_3]_2$	2.55 (av)	229
$(C_7H_8)_3K–Sn(CH_2CMe_3)_3$	3.55	752
$Na_8(OCH_2CH_2OCH_2CH_2OMe)_6(SiH_3)_2$	3.05	232
$[Li\{Me_2Si(H)NBu^t\}]_3$	1.95 (av)	233
$Mg_2\{Me_2Si(H)NBu^t\}_4$	3.14, 2.27	233
$(THF)_3Li–SiPh_2(NPh_2)$	2.73	753
$(TMEDA)Mg–(SiMe_3)_2$	2.65, 2.67	754
$(CO)_4Fe–Sn(\mu-OBu^t)_3Sr(\mu-OBu^t)_3Sn–Fe(CO)_4$	3.29	755
$(CO)_5Cr–Sn(\mu-OBu^t)_3Ba(\mu-OBu^t)_3Sn–Cr(CO)_5$	3.49	755
$(TMEDA)BrMg–SiMe_3$	2.63	756
$(PMDTA)BrMg–SiMe_3$	2.65	756
$[Me_3SiSiMe_2Li]_4$	2.68 (av)	757
$Cl_2Sn\{O(SiPh_2O)_2\}_2-\mu(Li(THF)_2)_2$	3.01	758
$(THF)_3Li–Ge[(Bu^tSi(OSiMe_2NPh)_3]_3$	2.90	759
$(p\text{-dioxane})LiSi(2-(Me_2NCH_2)C_6H_4)_2$ $Si(2-(Me_2NCH_2)C_6H_4)_2Si(2-(Me_2NCH_2)$ $C_6H_4)_2Li(p\text{-dioxane}) \cdot (p\text{-dioxane}) \cdot cyclohexane$	2.54, 2.55	760
$(THF)_2Li(2-(Me_2NCH_2)C_6H_4)_2Si–Si$ $(2-(Me_2NCH_2)C_6H_4)_2Li(THF)_2$	2.57 (av)	760
$(PMDTA)Li–SnPh_3$	2.86, 2.88	228
$(THF)_3Li–Si(SiMe_3)_2SiMe_2(Bu^t)$	2.68	761
$[Li–Si(PhMe_2Si)_2Me]_2$	2.66, 2.78	762
$[Li–Si(PhMe_2Si)_2Ph]_2$	2.63, 2.77	762
$(THF)_3Li–Ge(SiMe_3)_3$	2.67	763
$(THF)_2Li–Ge(2-Me_2NC_6H_4)_3$	2.60	764
$(PMDTA)Li–Ge(SiMe_3)_3$	2.65	763
$[Na–Si(SiMe_3)_3]_2$	2.99 (av)	231
$[Na–Si(SiMe_3)_3]_2 \cdot C_6H_6$	3.02 (av)	231
$[K–Si(SiMe_3)_3]_2$	3.39 (av)	231
$(C_6D_6)_3K–Si(SiMe_3)_3$	3.34 (av)	231
$[Rb–Si(SiMe_3)_3]_2 \cdot toluene$	3.52–3.62	231
$[Cs–Si(SiMe_3)_3]_2 \cdot toluene$	3.77–3.85	231
$[Cs–Si(SiMe_3)_3]_2 \cdot biphenyl \cdot n\text{-pentane}$	3.68–3.81	231
$[Cs–Si(SiMe_3)_3]_2 (\mu\text{-THF})$	3.67–3.73	231
$(\eta^6\text{-}C_7H_8)Na–Sn(SiMe_3)_3 \cdot n\text{-pentane}$	3.07	231
$(THF)_3Li–Sn(SiMe_3)_3$	2.87	227
$(THF)_2Na–Si(Bu^t)_3$	2.92	230
$(PMDTA)Na–Si(Bu^t)_3$	2.97	230
$(PMDTA)Li–PbPh_3$	2.86	234
$(C_6D_6)_3K–Si(Bu^t)_3$	3.38	230
$[Li–Si(Bu^t)_3]_2$	2.63, 2.67	230
$[Na–Si(Bu^t)_3]_2$	3.06, 3.07	230
$(THF)_4Ca–(SnMe_3)_2$	3.27	765

a 1:4 molar equivalent of tin(IV) chloride.[227] Multinuclear NMR spectroscopy has been used to verify the correspondence between solution and solid state structures; direct observation of $^1J(^7\text{Li}-^{117,119}\text{Sn}) = 412\,\text{Hz}$ in a solution of pentamethyldiethylenetriamine (PMDTA)Li–SnPh$_3$, for example, confirmed that the Sn–Li bond observed in the solid state (2.87 Å (av)) persists in solution.[228]

Treating the persilyl-substituted 1,1-bis(triisopropylsilyl)-2,3-bis(trimethylsilyl)silirene with Li in THF results in cleavage of two Si–C bonds in the three-membered ring, yielding Li$_2$Si[Si(Pr^i)$_3$]$_2$ and Me$_3$SiC≡CSiMe$_3$ (see Scheme 1). A THF adduct was structurally characterized by X-ray crystallography and NMR spectroscopy ($^1J(^{29}\text{Si}-^6\text{Li}) = 15.0\,\text{Hz}$)). Reaction of the dilithiosilane with Br$_2$Si[Si(Pr^i)$_3$]$_2$ in THF gave the disilene (Pr$^i_3$Si)$_2$Si=Si(Si(Pr$^i_3$)$_2$ in almost quantitative yield.[229]

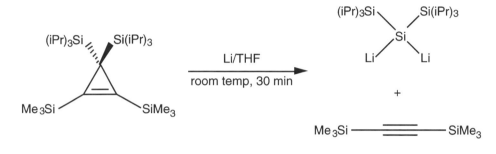

Scheme 1

Derivatives of the bulky "supersilyl" ligand ($-$SiBu$^t_3$) can be isolated from the reaction of lithium, sodium, or potassium with Bu$^t_3$SiX halides in alkanes or THF, and optionally in the presence of donors including ethers, amines, and aromatic hydrocarbons. Ethers can be exchanged for stronger donors like PMDTA, 18-crown-6, and cryptand-222.[230] Alkali metal supersilyl complexes are extremely water- and air-sensitive; and because the $-$SiBu$^t_3$ ligand is strongly basic, complexes such as (18-crown-6)NaSiBu$^t_3$ and (cryptand-222)(Na,K)SiBu$^t_3$ are unstable owing to their tendency to deprotonate the crown and cryptand, respectively. Even weak bases such as THF or benzene can be deprotonated, with concomitant formation of HSiBu$^t_3$. The complexes couple in the presence of AgNO$_3$ to generate the disilane Bu$^t_3$Si—SiBu$^t_3$, and react with Me$_3$SiX to form Me$_3$Si—SiBu$^t_3$. At 100 °C, Bu$^t_3$SiX oxidizes M–SiBu$^t_3$ (Equation(2)):

$$\text{MSiBu}^t_3 + \text{Bu}^t_3\text{SiX} \rightarrow 2\text{Bu}^t_3\text{Si}\cdot + \text{MX} \tag{2}$$

The supersilyl radical will abstract a hydrogen atom from alkanes to form HSiBu$^t_3$ and R·; the latter can then generate secondary products.[230]

The reaction between Zn[Si(SiMe$_3$)$_3$]$_2$ and potassium, rubidium, and caesium in heptane affords the donor-free derivatives.[231] The use of boiling *n*-heptane with sodium can be used to produce NaSi(SiMe$_3$)$_3$. NMR and Raman spectra, structural analysis, and calculations on model systems indicate that the bonding is largely ionic in the compounds. The unsolvated species [MSi-(SiMe$_3$)$_3$]$_2$ are cyclic dimers with almost planar M$_2$Si$_2$ rings; various benzene and toluene solvates are known also.

Large aggregates containing M—Si bonds have been formed by a variety of methods. Treatment of SiH$_4$ with dispersed Na in diglyme at 100 °C yields [Na$_8$(OCH$_2$CH$_2$OCH$_2$CH$_2$OMe)$_6$-(SiH$_3$)$_2$].[232] The eight sodium atoms form a cube, the faces of which are capped by the alkoxo oxygen atoms of the three ligands, which in turn are each bound to four sodium atoms (Na—O = 2.30–2.42 Å). The sodium and oxygen atoms generate an approximate rhombododeca-hedron. Six of the eight sodium atoms are five-fold coordinated to oxygen atoms, and the coordination of the other two Na atoms is completed by SiH$_3^-$ ions, which have inverted C_{3v} symmetry (i.e., Na—Si—H angle 58–62°) (see Figure 26). Despite the strange appearance of the latter, *ab initio* calculations (MP4/6-31G(d)) on the simplified model compound (NaOH)$_3$NaSiH$_3$ indicate that the form with inverted hydrogens is 6 kJ mol^{-1} lower in energy than the uninverted form; electrostatic (rather than agostic) interactions (H$^{\delta-}$—Na$^{\delta+}$) between the SiH$_3$ group and the three adjacent sodium atoms stabilize the inverted form.

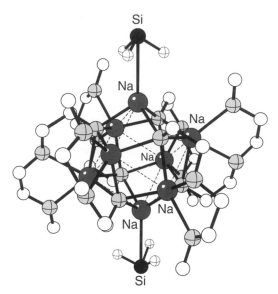

Figure 26 The structure of [Na$_8$(OCH$_2$CH$_2$OCH$_2$CH$_2$OMe)$_6$(SiH$_3$)$_2$], with its inverted SiH$_3$ moieties.

Agostic Si—H–Li contacts (i.e. less than 3.0 Å) are not present in the solvent-free X-ray crystal structure of [Li{Me$_2$Si(H)NBu^t}]$_3$; shorter Bu^{t}CH$_3$–Li distances (2.8 Å) are observed instead. In toluene solution at −80 °C, solution, the ^{1}H–^{6}Li HOESY (Heteronuclear Overhauser Enhancement Spectroscopy) spectrum of Li[Me$_2$Si(H)NBu^t] confirms that the trimeric species lacks Si—H–Li interactions, but two other species with strong Si—H–Li interactions are present also. Computations on the model system Li-HN-SiH$_3$ suggest that Si—H–Li interactions are energetically favored and result in increased Si—H distances and decreased Si—H frequencies.[233] In contrast, the related magnesium-based compound Mg$_2$[Me$_2$ Si(H)NBu^t]$_4$ does display evidence of agostic Si—H–Mg contacts in the solid state. The two short agostic Si—H–Mg interactions (Mg—H = 2.2, 2.5 Å) are represented by two low ν(Si—H) frequencies in the IR spectrum (2,040, 1,880 cm^{-1}).[233]

The bonding between Li and Sn and Pb has been investigated computationally with SCF/HF (Self-Consistent Field/Hartree-Fock) methods in the model compounds LiSnPh$_3$ and LiPbPh$_3$.[234] The $6pz$ orbital of the lead in LiPbPh$_3$ is oriented toward the lithium atom, and hence the Pb—Li bond involves the Pb($6pz + 6s$)–Li($2s$) orbitals, and not the Pb($6s$)–Li($2s$) orbitals only. The other lobe of the $6pz$ orbital is directed between the phenyl rings on the lead, and is able to interact with them. The compression of the phenyl ligands toward the z-axis is evident in the C(phenyl)– Pb–C(phenyl) angle of 102° calculated for the model compound, which compares favorably with the corresponding angle of 94.3° observed crystallographically in the complex (PMDTA)Li–PbPh$_3$.

3.1.4.3   Group 15 Ligands

3.1.4.3.1   *Nitrogen donor ligands*

(i) Amides, especially bis(trimethylsilyl)amides

Whereas the parent amides of the *s*-block metals (M(NH$_2$)$_{1,2}$) possess either ionic lattices or framework structures,[197] replacement of the hydrogen atoms with groups of increasing size leads to molecular complexes. Group 1 and 2 amido complexes are remarkably versatile reagents, and are used both in organic synthesis[235] and to prepare a wide variety of derivatives of other main-group and transition metal complexes. Issues in the coordination chemistry of amidolithium reagents that are concerned chiefly with applications in synthetic organic chemistry are summarized elsewhere.[236–242]

A widely used class of *s*-block amido complexes is that containing the bis(trimethylsilyl) group, −N(SiMe)$_2$. Examples of these compounds were synthesized by Wannagat in the 1960s,[243] but all the *s*-block metals (excepting Fr and Ra) are now represented. The chemistry of the heavy alkaline-earth derivatives has been extensively developed by Westerhausen;[244] this chemistry has been reviewed.[19]

Synthesis of the *s*-block metal bis(trimethylsilyl)amides is varied, and representative routes are summarized in Table 4. The lithium compound can be prepared by reaction of LiBu^n with hexamethyldisilazane (pK_a = 25.8) (Table 4, Equation (1)),[245] and the sodium derivative is available by transmetallation of the metal with Hg[N(SiMe$_3$)$_2$]$_2$ at room temperature (Table 4, Equation (2)).[246] The monomeric Be[N(SiMe$_3$)$_2$]$_2$ can be prepared from the reaction of (Et$_2$O)-BeCl$_2$ with Na[N(SiMe$_3$)$_2$] Table 4, Equation (4)),[247] whereas reaction of MgBu$^n_2$ with HN(SiMe$_3$)$_2$ produces Mg[N(SiMe$_3$)$_2$]$_2$ in quantitative yield (Table 4, Equation (5)).[248] The heavy alkaline-earth derivatives are available from reaction of Ca, Sr, and Ba with M[N(SiMe$_3$)$_2$]$_2$ (M = Sn, Hg) (Table 4, Equations (2) and (3))[249] or with HN(SiMe$_3$)$_2$ in THF (Table 4, Equation 6) or THF/NH$_3$ (Table 4, Equations (7a)–(7c)). The approach relies on the solubility of ammonia in ethereal solvents (especially tetrahydrofuran) (Table 4, Equation (7a)), the dissolution of the metal in the saturated ammonia/THF solution (Table 4, Equation (7b)), and the reactivity of the dissolved metal with the parent amine (Table 4, Equation (7c)). The ammonia is a catalyst in the process.[250] Halide metathesis with MN(SiMe$_3$)$_2$ (M = Na, K) has been used for the calcium, strontium, and barium derivatives (Table 4, Equation (8)),[249] Other reactions involving metal alkoxides[251] and paratoluenesulfonates[252] have been reported.

Synthesis of the compounds is sensitive to the reaction conditions and identity of the metals. For example, the dimeric [MN(SiMe$_3$)$_2$]$_2$ (M = Rb, Cs) species form from the reaction of the metal in refluxing neat hexamethyldisilazane (Table 4, Equation (9)).[253] If caesium is first dissolved in liquid ammonia, however, addition of HN(SiMe$_3$)$_2$ results in the formation of a mono(trimethylsilyl)amido complex, [CsHN(SiMe$_3$)$_2$]$_4$.[254] The latter is a tetrameric species in which one N—Si bond of the original amine has been broken, and the remaining N—Si bond is short enough (1.59(1) Å) to suggest the presence of some degree of multiple bonding (see Figure 27). The formula of heterometallic amido complexes can also be difficult to predict; e.g., addition of LiN(SiMe$_3$)$_2$ to Ca[N(SiMe$_3$)$_2$]$_2$ in THF produces the heterometallic species [Ca{N(SiMe$_3$)$_2$}{μ-N(SiMe$_3$)$_2$}$_2$Li(THF)]; if the barium amido complex is used instead, [Ba{N(SiMe$_3$)$_2$}$_2$(THF)$_3$][Li$_2${μ-N(SiMe$_3$)$_2$}$_2$(THF)$_2$] can be isolated.[255]

A large number of structurally characterized bis(trimethylsilyl)amido complexes now exist; Table 5 gives a representative selection of monometallic homoleptic compounds, both base-free and with coordinated ethers. Other examples are known with coordinated fluorobenzenes,[256] isonitriles,[257] methylated pyridines,[258] various amines (TMEDA, PMDTA, TMPDA (Tetramethylpropylenediamine), BzNMe$_2$ (benzyldimethylamine)),[259] Ph$_3$PO,[260] (Bu^{n}O)(Pr^i)CO,[261] and 1,3-(Pr^i)$_2$-3,4,5,6-tetrahydropyrimid-2-ylidene.[262] Their structures illustrate the complex interactions between metal size, ligand bulk, and molecular structure that exist with these metals. For example, among the alkali metal base-free species, the unsolvated Li derivative crystallizes as a cyclic trimer,[263] whereas the Na salt is found both as a trimer[264] and as infinite chains of [Na-N(SiMe$_3$)$_2$–]$_n$ units.[265] The potassium,[266] rubidium, and caesium[253] derivatives exist as discrete dimers in the solid state, constructed around planar [M—N–]$_2$ frameworks.

In the Group 2 derivatives, less diversity is found: the beryllium complex is monomeric,[267,268] probably for steric reasons, but the Mg–Ba compounds are dimers.[6] Although the M—Si

Table 4 Synthetic methods for the preparation of *s*-block bis(trimethylsilyl)amido complexes.

Reaction	No.
LiR + HN(SiMe$_3$)$_2$ → LiN(SiMe$_3$)$_2$ + RH	(1)
m/2 Hg[N(SiMe$_3$)$_2$]$_2$ + M → M[N(SiMe$_3$)$_2$]$_m$(S)$_x$ + m/2 Hg	(2)
m/2 Sn[N(SiMe$_3$)$_2$]$_2$ + M → M[N(SiMe$_3$)$_2$]$_m$(S)$_x$ + m/2 Sn	(3)
2 Na[N(SiMe$_3$)$_2$] + (Et$_2$O)BeCl$_2$ → Be[N(SiMe$_3$)$_2$]$_2$ + 2 NaCl	(4)
Mg(Bu^n)$_2$ + 2 HN(SiMe$_3$)$_2$ → Mg[N(SiMe$_3$)$_2$]$_2$ + 2 (Bu^n)H	(5)
M(NH$_3$)$_m$ + n HNR$_2$ → M(NR$_2$)$_n$(NH$_3$)$_{m-x}$ + n/2 H$_2$ + x NH$_3$	(6)
NH$_{3(g)}$ + THF → NH$_{3(sat)}$	(7)
M + m NH$_{3(sat)}$ → M(NH$_3$)$_m$(S)$_x$	(7)
M(NH$_3$)$_m$(S)$_x$	(7)
$\quad$ + n HNR$_2$ → M(NR$_2$)$_n$(S)$_x$ + n/2 H$_2$ + m NH$_3$	
2 M^I[N(SiMe$_3$)$_2$] + M^{II}X$_2$ → M^{II}[N(SiMe$_3$)$_2$]$_2$ + 2 M^{I}X	(8)
M^I + xsHNR$_2$ → M^{I}N(SiMe$_3$)$_2$ + 1/2 H$_2$	(9)

M^I = alkali metal; M^{II} = alkaline-earth metal; X = halide; S = coordinated solvent.

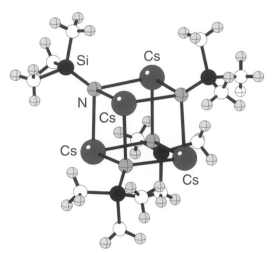

Figure 27 The structure of the tetrameric mono(trimethylsilyl)amido complex, [CsHN(SiMe$_3$)$_2$]$_4$.

distances change as expected with the increasing size of the cation, the difference between terminal and bridging N—Si distances decreases with the larger metals.

Solution and gas-phase structures do not always match those in the solid state: unsolvated bis(trimethylsilyl)amido]lithium and -sodium are dimeric in hydrocarbon solutions, for example, and LiN(SiMe$_3$)$_2$ remains a dimer in the gas phase,[269] although the sodium[270] and magnesium[271] derivatives are primarily or exclusively monomeric. Mg[N(SiMe$_3$)$_2$]$_2$ will dissociate into monomers in aromatic solvents,[249] whereas the M[N(SiMe$_3$)$_2$]$_2$ (M = Ca, Sr, Ba) derivatives are dimeric in solution, as evidenced by cryoscopic molecular weights.[249]

The correlation between the Si—N distance and the Si—N—Si' angle in metal silylamides has been the subject of considerable discussion.[265,272–274] Partial multiple bonding between Si and N, in which the lone pair electrons on nitrogen are delocalized onto silicon, has been invoked as an explanation for the Si—N distance/Si—N—Si' angle relationships, but steric interactions may also play a critical role in determining the geometries. In fact, the Si—Si' separation is relatively constant at ~3.0 Å for a variety of transition metal, main group, and *f*-element bis(trimethylsilyl) amide complexes,[266] and is probably of steric origin; i.e., bis(trimethylsilyl)amide complexes containing short Si—N bonds must necessarily possess relatively large Si—N—Si' angles to avoid violating a minimum SiMe$_3$–SiMe$_3'$ separation.

The trimethylsilyl groups are associated with hydrocarbon solubility of the base-free and ether adducts. This is a particularly remarkable property, in that the metal atoms in many of the structures are exposed to external molecules, and are not appreciably shielded by the trimethylsilyl group of the amido ligands (see Figure 28). Strong ion pairing interaction between the M^{n+} and the N(SiMe$_3$)$_2^-$ ions probably contributes to their solubility.[266]

Extensive NMR data (^{1}H, ^{13}C, ^{15}N, ^{29}Si) exist for these species;[19] the δ(N) and δ(SiMe$_3$) shifts are sensitive to the identity of the metal center. In the case of the alkaline-earth compounds, differences in the chemical shifts and coupling constants for the bridging and terminal –SiN(Me)$_2$ groups disappear for the barium derivatives, an indication that they are in fast exchange.

Owing to the widespread use of Lewis bases to activate alkyllithium and amidolithium reagents in organic synthesis, the relationship between solvation and aggregation in the reactivity of lithium complexes is of considerable importance. The conventional interpretation of the activation phenomenon holds that bases disrupt the oligomers in which alkyllithium and amidolithium complexes are typically found (i.e., tetramers for CH$_3$Li, hexamers for Bu^{n}Li, etc.), and it is the monomers that are the reactive species. Experimental evidence to the contrary, however, has been provided by Collum and co-workers,[275–306] who have examined the molecularity of solvated amidolithiums with ^{6}Li, ^{15}N, and ^{13}C-NMR spectroscopy, and have argued that strongly coordinated ligands do not necessarily promote higher reactivity, nor do similar reaction rates automatically imply similar mechanisms.[307,308]

The structural chemistry and solution behavior of lithium diisopropylamide (LiN(CHMe$_2$)$_2$, LDA) typifies the complexities of these systems. Although the base-free compound crystallizes as infinite helical chains with two-coordinate lithium and four-coordinate nitrogen,[309] it is isolated from *N,N,N',N'*-tetramethylethylenediamine (TMEDA)/hexane mixtures as an infinite array of

Table 5 Bond lengths (Å) in solid state homoleptic *s*-block bis(trimethylsilyl)amide complexes and base adducts.

Complex	M–N_{term} (Å)	M–N_{term} (Å)	N_{term}–Si (Å)	N_{brid}–Si (Å)	References
$[LiN(SiMe_3)_2]_3$		1.98–2.03	1.73		263,766
$[(Et_2O)LiN(SiMe_3)_2]_2$		2.04	1.70		767,768
$[(THF)LiN(SiMe_3)_2]_2$		2.03	1.69		245,769,770
$[(BzNMe_2)LiN(SiMe_3)_2]_2$		2.03–2.09	1.71		259
$[(\eta^1\text{-}DME)LiN(SiMe_3)_2]_2$		2.01–2.05	1.70		259
$(PMDTA)LiN(SiMe_3)_2$	1.99		1.67		259
$(12\text{-}crown\text{-}4)LiN(SiMe_3)_2$	1.97		1.68		771
$(TMEDA)LiN(SiMe_3)_2$	1.89		1.68		259
$[(p\text{-}dioxane)LiN(SiMe_3)_2]_2$		2.03/2.04	1.70/1.70		259
$\{p\text{-}dioxane\cdot[LiN(SiMe_3)_2]_2\}_\infty$					
$[NaN(SiMe_3)_2]_3$		2.36–2.40	1.70		264,772
$[NaN(SiMe_3)_2]_\infty$		2.36 (av)	1.69		265
$[(THF)NaN(SiMe_3)_2]_2$		2.40	1.68		773
$[(p\text{-}dioxane)_2NaN(SiMe_3)_2]_n$	2.38		1.67		774
$\{[(\eta^1\text{-}TMPDA)NaN(SiMe_3)_2]_2\}_\infty$		2.43	1.68		259
$[KN(SiMe_3)_2]_2$		2.77/2.80	1.68		266
$[KN(SiMe_3)_2]_2\cdot toluene$		2.75/2.80	1.67		775
$[KN(SiMe_3)_2]_2\cdot1,3\text{-}diisopropyl\text{-}$		2.76/2.84	1.67		262
3,4,5,6-tetrahydropyrimid-2-ylidene					
$(p\text{-}dioxane)_2KN(SiMe_3)_2$	2.70		1.64		776
$[RbN(SiMe_3)_2]_2$		2.88/2.96	1.67		253
$[(p\text{-}dioxane)_3RbN(SiMe_3)_2]_n\cdot$		2.95/3.14	1.67		774
$(p\text{-}dioxane)_n$					
$[CsN(SiMe_3)_2]_2$		3/07/3.15	1.67		253
$[CsN(SiMe_3)_2]_2\cdot toluene$		3.02/3.14	1.68		253
$[(p\text{-}dioxane)_3CsN(SiMe_3)_2]_n\cdot$		3.07/3.39	1.67		774
$(p\text{-}dioxane)_n$					
$[Be[N(SiMe_3)_2]_2$	1.56		1.73		267,268
$Mg[N(SiMe_3)_2]_2$ (g)	1.91		1.70		271
$\{[Mg[N(SiMe_3)_2]_2\}_2$ (s)	1.98	3.15	1.71	1.77	777
$(THF)_2Mg[N(SiMe_3)_2]_2$	2.02		1.71		246
$[(p\text{-}dioxane)Mg[N(SiMe_3)_2]_2$	1.95		1.70		778
$\{[Ca[N(SiMe_3)_2]_2\}_2$	2.27	2.47	1.70	1.73	779
$(THF)_2Ca[N(SiMe_3)_2]_2$	2.30		1.69		447
$(DME)Ca[N(SiMe_3)_2]_2$	2.27		1.68		779
$\{[Sr[N(SiMe_3)_2]_2\}_2$	2.24	2.64	1.69	1.71	780
$(THF)_2Sr[N(SiMe_3)_2]_2$	2.46		1.67		781
$\{(p\text{-}dioxane)Sr[N(SiMe_3)_2]_2\}_n$	2.45		1.69		782
$(DME)_2Sr[N(SiMe_3)_2]_2$	2.54		1.69		780
$\{[Ba[N(SiMe_3)_2]_2\}_2$	2.58	2.82	1.69	1.70	6
$(THF)_2Ba[N(SiMe_3)_2]_2$	2.59		1.68		6
$\{(THF)Ba[N(SiMe_3)_2]_2\}_2$	2.60	2.83/2.90	1.69	1.71	6
$(THF)_3Ba[N(SiMe_3)_2]_2$	2.64		1.67		255

dimers linked by bridging (nonchelating) TMEDA ligands (see Figure 29).[295] Lithium-6 and ^{15}N-NMR spectroscopic studies indicate that LDA in neat TMEDA exists as a cyclic dimer bearing a single η^1-coordinated TMEDA ligand on each lithium. However, the equilibrium between solvent-free LDA and the TMEDA-solvated dimer shows a strong temperature dependence; TMEDA coordinates readily only at low temperature, and high TMEDA concentrations are required to saturate the lithium coordination spheres at ambient temperature.[295]

NMR spectroscopic studies of ^{6}Li-^{15}N labeled lithium hexamethyldisilazide in solvents including THF, 2-methyltetrahydrofuran (2-MeTHF), 2,2-dimethyltetrahydrofuran (2,2-Me$_2$THF), diethyl ether (Et$_2$O), *t*-butyl methyl ether (Bu^{t}OMe), *n*-butyl methyl ether (Bu^{n}OMe), tetra-hydropyran (THP), methyl *i*-propyl ether (Pr^{i}OMe), and trimethylene oxide (oxetane) have been used to characterize the nature of the solvated species. Mono-, di-, and mixed-solvated dimers can be identified in the limit of slow solvent exchange, but ligand exchange is too fast to observe

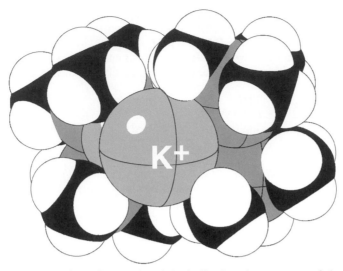

Figure 28 Space-filling drawing of [KN(SiMe₃)₂]₂, indicating the exposure of the potassium atom.

bound and free diisopropyl ether (Pr^i_2O), *t*-amyl methyl ether (Me₂(Et)COMe), and 2,2,5,5-tetramethyltetrahydrofuran (2,2,5,5-Me₄THF). Relative free energies and enthalpies of LiN(SiMe₃)₂ dimer solvation display an approximately inverse correlation of binding energy and ligand steric demand, but there is no simple correlation between reduced aggregation state and increasing strength of the lithium–solvent interaction. Contributions from solvation enthalpy and entropy, with the added complication of variable solvation numbers (higher with more sterically compact solvents) affect the measured free energies of aggregation.[285]

The structures of lithiated phenylacetonitrile and 1-naphthylacetonitrile have been studied in THF and HMPA–THF solution. In pure THF, ^{7}Li-NMR line width studies suggest that these species exist as contact ion pairs. In the presence of 0.25–2 equivalents of HMPA, HMPA-solvated monomeric and dimeric contact ion pairs can be identified with ^{31}P and ^{7}Li-NMR spectroscopy, but with four to six equivalents of added HMPA, NMR spectra provide direct evidence for the formation of HMPA-solvated separated ion pairs.[310]

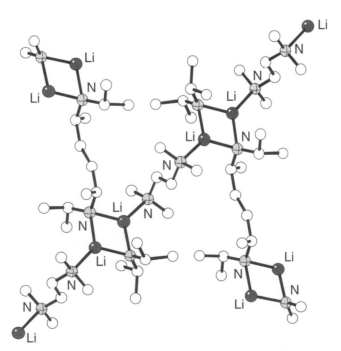

Figure 29 Portion of the solid state structure of TMEDA-adducted LiN(CHMe₂)₂.

The complexity of the solvation of lithium bases has also been demonstrated by studies of LiNPr$^i_2$ mediated ester enolization of Bu^t-cyclohexanecarboxylate in four different solvents (THF, Bu^{t}OMe, HMPA/THF and DMPU/THF (DMPU = 1,3-dimethyl-3,4,5,6-tetrahydro-2(1H)-pyrimidone)).[311] Even when experiments are designed to exclude mixed aggregate effects, four different mechanisms with nearly identical rates are observed, involving:

    (i)   disolvated monomers in THF,
    (ii) monosolvated dimers in Bu^{t}OMe,
    (iii) both monosolvated monomer and tetrasolvated dimers in HMPA/THF, and
    (iv) mono- and disolvated monomers in DMPU/THF.

Both monomeric and aggregated species (e.g., "open" dimers in Bu^{t}OMe and triple ions in HMPA/THF) are reactive. In related work in which aggregate formation was maximized, it was shown that the rates of enolization in the presence of the mixed aggregates are much lower and solvent dependent.[312] The autoinhibition correlates with the relative stabilities of the mixed aggregates; the stabilities do not, however, correlate in a straightforward manner with the ligating properties of the solvent.

A particularly interesting development in heterometallic amido complexes are the so-called "inverse crown ether" complexes. These 8-membered (M—N—Mg—N)$_2$ rings (M = Li, Na, K)[20,313,314] act as polymetallic hosts to anionic species. Most common are oxo or peroxo species, in which the O^{2-} or O$_2^{2-}$ ions are derived from molecular oxygen, although hydride encapsulation has also been described.[315] Larger 12-membered (NaNMgNNaN)$_2^{2+}$ or 24-membered (KNMgN)$_6^{2+}$ variants, which function as single or multiple traps for arene-based anions, are also known.[316]

In a typical preparation of an "inverse crown", the reaction of n-butyllithium with dibutylmagnesium and oxygenated 2,2,6,6-tetramethylpiperidine (LH) affords the complex [L$_4$Li$_2$Mg$_2$O]. The same reaction with n-butylsodium in place of n-butyllithium and HN(SiMe$_3$)$_2$ in place of tetramethylpiperidine yields [(Me$_3$Si)$_2$N]$_4$Na$_2$Mg$_2$(O$_2$)$_x$(O)$_{1-x}$.[313] The structure of the related [(Me$_3$Si)$_2$N]$_4$Li$_2$Mg$_2$(O$_2$)$_x$(O)$_{1-x}$ contains a side-on bonded peroxide molecule occupying a square-planar site. The four disordered metal atoms achieve a three-coordinate geometry by bridging to amido N atoms, producing an eight-membered ring (see Figure 30; an oxide atom replaces the peroxide molecule in approximately 30% of the molecules within the bulk lattice; it is positioned at the center of the O(1)—O(1′) bond).[20]

Compounds that are both heterometallic and contain mixed-amido ligand sets are rare. A novel example is found from the reaction of the magnesium amide Mg(TMP)$_2$ (TMP = 2,2,6,6-tetramethylpiperidide (TMP)) with the lithium amide LiN(SiMe$_3$)$_2$ in hydrocarbon solution. A sterically promoted hydrogen transfer/amine elimination process occurs, yielding the mixed lithium–magnesium, mixed amide (R,R/S,S)-{LiMg(TMP)[CH$_2$Si(Me)$_2$N(SiMe$_3$)]}$_2$.[317] Its molecular structure is dimeric, composed of dinuclear (LiNMgN) monomeric fragments with pendant Me$_2$SiCH$_2$ arms that bind intramolecularly through the methylene CH$_2$ unit to the Mg center

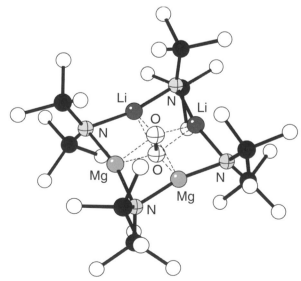

Figure 30   The structure of the "inverse crown ether" complex [(Me$_3$Si)$_2$N]$_4$Li$_2$Mg$_2$(O$_2$)$_x$(O)$_{1-x}$.

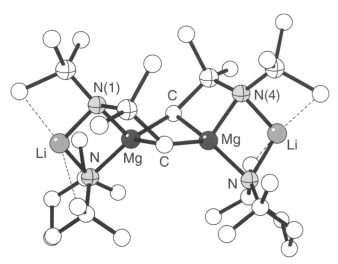

Figure 31 The structure of the lithium–magnesium, mixed amide $(R,R/S,S)$-{LiMg(TMP)[CH$_2$Si(Me)$_2$N-(SiMe$_3$)]}$_2$.

(see Figure 31). Dimerization occurs through intermolecular bonds from the methylene CH$_2$ unit to the Mg center of the other monomeric fragment. The formally two-coordinate lithium atoms are engaged in agostic interactions with one TMP and the methyl of a bis(trimethylsilylamido) group. Interestingly, the nitrogen atoms N(1) and N(4) of two N(SiMe$_3$)$_2$ groups are chiral; each binds to one Li and one Mg center, and to one SiMe$_3$ and one SiMe$_2$CH$_2$ group. Only the enantiomeric R,R and S,S pair has been observed.

(ii) Imidoalkalis

The imidoalkalis, RR′C=NM, constitute a large family of compounds that encompass M—N, M—O and M—C rings and aggregates of various sizes. They were the first group of compounds in which the concepts of "ring-stacking" and "ring-laddering" were systematically developed.[318–323] They are useful precursors to imido complexes of *p*-block and transition metal elements, for example (Equation (3)):

$$R''_3SiCl + RR'C{=}NLi \rightarrow R'RC{=}NSiR''_3 + LiCl \qquad (3)$$

With the bulky base HMPA, a dimeric imidolithium can be isolated (i.e., (Bu$_2^t$C=NLi·HMPA)$_2$ (Li—N = 1.92, 1.95 Å)[324]). With less sterically demanding bases, more complex aggregates can form. For example, (Ph$_2$C=NLi·Py)$_4$ has a cubane structure that can be viewed as two stacked (LiN)$_2$ four-membered ring systems (see Figure 32).[324] The cube is considerably distorted, with three distinct ranges of Li—N distances (2.03, 2.08, and 3.16 Å (av)).Somewhat greater uniformity is observed in the cubic imidolithium complex {Li[N=C (Bu^t)CHCHC(SiMe$_3$)(CH$_2$)$_2$CH$_2$]}$_4$, formed from the reaction of 2-trimethylsilylcyclohexenyllithium with (Bu^t)CN (Li—N = 1.99–2.04 Å).[325]

Larger clusters are also known; among the first to be structurally authenticated was {Li[N=C-(Bu^t)$_2$]}$_6$ (Figure 33);[326] the structure of {Li[N=C(NMe$_2$)$_2$]}$_6$ is similar. Both are based on folded chair-shaped Li$_6$ rings held together by triply bridging N=CR$_2$ groups; the μ_3-imino ligands function as three-electron donors, forming one two-center Li—N bond and one three-center Li$_2$N bond to isosceles triangles of bridged metal atoms. The mean Li—Li distance in the metal rings is 2.35 Å in {Li[N=C(Bu^t)$_2$]}$_6$, and the mean dihedral angle between Li$_6$ chair seats and backs is 85°. The N atoms of the bridging N=CR$_2$ groups are roughly equidistant from the three bridged Li atoms (Li—N = 2.06 Å (av)).[326] The stacked ring motif is found in the related hexameric compounds [LiN=C(C$_6$H$_5$)C(CH$_3$)$_3$]$_6$ (generated from (Bu^t)C≡N and PhLi or C$_6$H$_5$C≡N and Bu^{t}Li) and [LiN=C(C$_6$H$_5$)NMe$_2$]$_6$ (generated from C$_6$H$_5$C≡N and LiNMe$_2$ or Me$_2$NC≡N and LiC$_6$H$_5$).[318,327]

A rare example of a heterometallic imido complex is the triple-stacked Li$_4$Na$_2$[N=C(Ph)-(Bu^t)]$_6$, prepared from the reaction of PhLi and PhNa with (Bu^t)NC (see Figure 34). The molecule has six metal sites in a triple-layered stack of four-membered M—N rings, with the outer rings containing lithium and the central ring containing sodium; the latter is four-coordinate.[328]

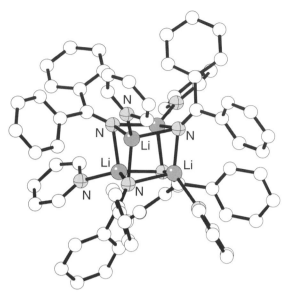

Figure 32   The structure of (Ph$_2$C=NLi·py)$_4$, constructed from stacked (LiN)$_2$ rings.

(iii)   Thiocyanates

Thiocyanate (SCN$^-$), and to a lesser extent, selenocyanate (SeCN$^-$) ions are among the classic ambidentate ligands. The conventional expectation is that they should display *N*–coordination to the "class a" *s*-block metals, and the majority of crystallographically characterized examples support this. Thiocyanate ions are often found in complexes with macrocyclic ligands, such as crown ethers,[329–358] azacrowns,[359–374] cryptands,[375] glymes,[376–379] hemispherands,[380] paracyclophanes,[381] and other polydentate oxygen ligands.[382–386] Nonmacrocyclic thiocyanate compounds are found with water,[335,387] pyridines,[388–391] THF,[392] acetates,[393] HMPA,[394–397] TMPDA,[398] phenolate[399] and other more complex ligands.[400,401] Homoleptic examples also exist (e.g., [Ca(NCS)$_6$]$^{4-}$,[334] [Rb(NCS)$_4$]$^{3-}$, [Cs(NCS)$_4$]$^{3-}$[402]). In the case of the complex of (*E*)-9,10-diphenyl-2,5,8,11,14,17-hexaoxaoctadec-9-ene with NaSCN, a unique Na$_3$(μ_3-NCS) arrangement is found (see Figure 35).[403]

Notwithstanding these examples, the preference for *N*–coordination is not absolute. *S*–bound NCS$^-$ ligands are found bound to potassium in crown ether complexes.[369,404] For example, the 3′,5′-difluoro-4′-hydroxybenzyl-armed monoaza-15-crown-5 ether forms *F*-bridged polymer-like

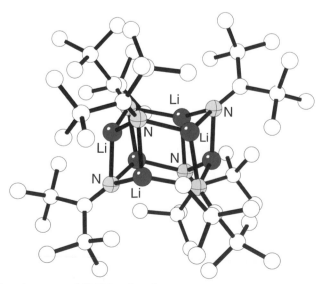

Figure 33   The structure of {Li[N=C(Bu^t)$_2$]}$_6$, based on folded chair-shaped Li$_6$ rings.

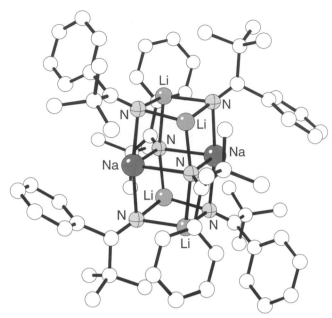

Figure 34 The structure of the bimetallic triple-stacked complex Li$_4$Na$_2$[N=C(Ph)(Bu^t)]$_6$.

complexes with MSCN (M = K, Rb) in the solid state and in solution. The thiocyanate ligands are *S*-bound and terminal in the potassium complex, *S*-bound and bridging in the rubidium derivative (see Figure 36).[405]

Bridging SCN ligands will naturally have both *N*– and *S*–bound ends. Such ligands are found in the 1:1 TMEDA complex of lithium thiocyanate, (LiSCN·TMEDA)$_n$ prepared from solid NH$_4$SCN and Bu^{n}Li in TMEDA/hexane. In hydrocarbon solvents, it assumes a dimeric, purely *N*—Li bridged structure, but in the solid state it exists as a polymeric solid with lithium ions linked by SCN ligands (Li—N = 1.96 Å; Li—S = 2.57 Å).[406] Similar bridged structures are adopted by SCN ions spanning potassium complexed azacrown ligands,[407] and barium-complexed bipyridine.[408]

A more complex arrangement is generated from the reaction of solid NH$_4$SCN with solid NaH and HMPA (1:1:2 molar ratios) in toluene, which produces both the discrete dimer [NaNCS-(μ-HMPA)(HMPA)]$_2$ with bridging and terminal HMPA ligands and *N*-bound NCS$^-$ ligands, and the polymeric (NaNCS·HMPA)$_n$, which has both *N*– and *S*–bound NCS$^-$ ligands (Na—N = 2.33 Å; Na—S = 2.89 Å).[395] The detection of radicals with EPR spectroscopy suggests that the reaction proceeds via an electron-transfer mechanism.

Heterobimetallic species containing μ-SCN ligands are found in the polymeric chain structures [K(18-crown-6)(η^5-C$_5$Me$_5$)$_2$Yb(NCS)$_2$]$_\infty$,[409] [K(18-crown-6)(μ-SCN)]$_2$UO$_2$(NCS)$_2$(H$_2$O),[410] and

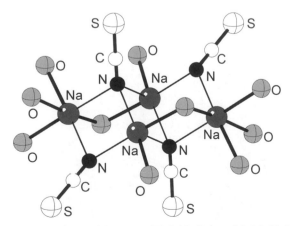

Figure 35 The structure of complex formed between (*E*)-9,10-diphenyl-2,5,8,11,14,17-hexaoxaoctadec-9-ene and NaSCN.

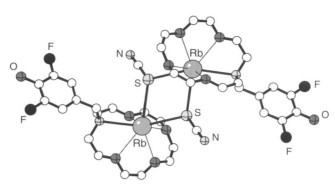

Figure 36 Portion of the polymeric lattice of an *S*-bound thiocyanate rubidium crown ether complex.

$[K(18\text{-crown-6})]_2K[Bi(SCN)_6]$, which exists in two isomeric forms.[411] In the yellow form, octahedral $[Bi(SCN)_6]^{3-}$ anions are surrounded by four $K[K(18\text{-crown-6})]_2$ units in a layered arrangement; all the thiocyanates are *N*–bound to potassium. In the yellow form, the thiocyanatobismuth anions have four *S*-bonded and two *trans N*-bonded thiocyanate ligands, and are arranged in parallel columns separated by K^+ cations; each potassium has two *trans S*-bound and four *N*-bound thiocyanate ligands. Two $[K(18\text{-crown-6})]^+$ units are located between the columns.[411]

Structurally characterized selenocyanate complexes of the *s*-block metals are much rarer, but as befits the softer nature of selenium relative to sulfur, all known examples are *N*–bound.[411–414]

(iv) N-donor stabilized metal-centered radicals

The 1,4-di-*t*-butyl-1,4-diazabutadiene ligand $((Bu^t)_2DAB)$ is able to stabilize a ligand-centered radical with lithium,[415] or triplet biradicals with magnesium, calcium, and strontium.[416] The dark green lithium species can be generated from $(t\text{-}Bu)_2DAB$ and sonicated lithium in hexane; activated magnesium will react with $(Bu^t)_2DAB$ in THF to form the deep red $Mg[(Bu^t)_2DAB]_2$. EPR measurements ($g_{av} = 2.0034$ for Li[415]; 2.0036 for Mg[417]), DFT calculations,[418] and X-ray single-crystal structural studies have been used to characterize the products. The lithium compound, for which calculations suggest the metal–ligand interaction is primarily ionic, displays a tetrahedral coordination geometry (see Figure 37). One $(Bu^t)_2DAB$ ligand (involving N1 and N4) has a geometry consistent with a radical anion (Li—N(mean) = 1.995 Å; N—Li—N = 88.3(3)°); the other is bound as a simple neutral ligand (Li—N(mean) = 3.141 Å; N—Li N = 79.5(2)°). In the magnesium compound, the metal center is also tetrahedrally coordinated, but the molecule

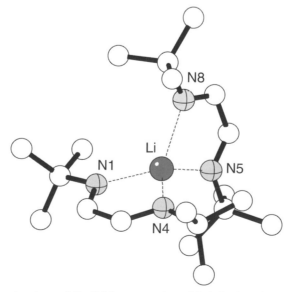

Figure 37 The structure of the lithium complex of 1,4-di-*t*-butyl-1,4-diazabutadiene.

possesses nearly *2mm* symmetry, and both ligands display features consistent with radical anions (Mg—N = 2.070(7), 2.067(7) Å; Mg—N = 83.1(3), 82.4(3)°). It should be noted that it is possible to prepare beryllium, calcium and barium complexes of the diazabutadiene ligand that contain doubly-reduced ligands; the complexes [Be(1,4-bis(4-methylphenyl)-2,3-diphenyl-1,3-diazabuta-diene)(THF)]$_2$ (black-green),[419] [Ca(NRCPh=CPh=NR-*N,N'*)(DME)$_2$] (R = C$_6$H$_4$-4-Me) (orange) and [Ba$_2$(DME)$_3$(NPhCPh=CPhNPh)$_2$DME] (red) have been structurally characterized.[420]

(v) Other nitrogen donor ligands

The pyrazolate anion, [C$_3$H$_3$N$_2$]$^-$, is isoelectronic with the cyclopentadienyl anion, and has been incorporated into *s*-block metal complexes. Reaction of the disubstituted pyrazoles 3,5-R$_2$pzH (R = Bu^t or Ph) with KH in THF yields [(3,5-R$_2$pz)K(THF)$_n$]$_6$ [R = Bu^t, *n* = 0; R = Ph, *n* = 1]. The crystal structure of the hexameric drum-shaped diphenylpyrazolato derivative (see Figure 38) indicates that each potassium is bonded in an η^2-manner to one pyrazolato ligand, in an η^1-fashion to two adjacent ligands, and to a single THF ligand. The complex serves as a pyrazolato transfer reagent to titanium and tantalum.[421]

The hexadentate ligand bis{3-[6-(2,2'-bipyridyl)]pyrazol-1-yl}hydroborate contains two terden-tate chelating arms joined by a [BH$_2$]$^-$ spacer. Its potassium derivative crystallizes in the form of a double helix, in which each metal ion is six-coordinated by a terdentate arm from each of the two (pyrazol-1-yl)hydroborate ligands; the two ligands are bridging, and folded at the –BH$_2$– linkage (see Figure 39).[422]

The reaction of magnesium bromide with potassium 3,5-di-tert-butylpyrazolate, K[Bu$^t_2$pz], in toluene affords Mg$_2$(Bu$^t_2$pz)$_4$; in THF the product is Mg$_2$(Bu$^t_2$pz)$_4$(THF)$_2$.[423] Both of these are dinuclear complexes with two bridging pyrazolato and two chelating η^2-pyrazolato ligands. TMEDA binds to the unsolvated compound or displaces THF from the solvate to give Mg(Bu$^t_2$pz)$_2$ (TMEDA), which is a mononuclear species with two η^2-pyrazolates and chelating TMEDA.

Reaction of CaBr$_2$ with two equivalents of K[Bu$^t_2$pz] in THF yields Ca(Bu$^t_2$pz)$_2$(THF)$_2$, which on treatment with pyridine, TMEDA, PMDETA, triglyme, and tetraglyme generates the adducts Ca(Bu$^t_2$pz)$_2$(Py)$_3$, Ca(Bu$^t_2$pz)$_2$(TMEDA), Ca(Bu$^t_2$pz)$_2$(PMDETA), Ca(Bu$^t_2$pz)$_2$(triglyme), and Ca(Bu$^t_2$pz)$_2$(tetraglyme), respectively. A related series of THF, PMDETA, triglyme, and tetra-glyme adducts of the 3,5-dimethylpyrazolate anion can also be prepared. The Bu^t pyrazolato complexes are mononuclear, with η^2-bound ligands. The compounds have been investigated for

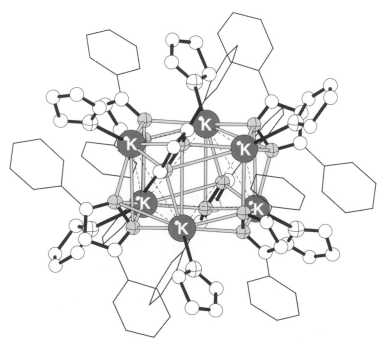

Figure 38 The structure of the hexameric complex [(3,5-Ph$_2$pz)K(THF)$_n$]$_6$.

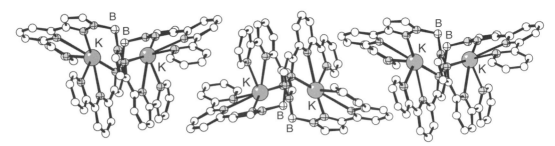

Figure 39 The structure of the potassium derivative of bis{3-[6-(2,2′-bipyridyl)]pyrazol-1-yl}hydroborate.

their potential use in CVD, but only $Ca(Bu^t_2pz)_2$(triglyme) displays appreciable volatility, subliming at 160 °C (0.1 mmHg), and even then, decomposition by triglyme ligand loss is competitive with sublimation. In other complexes, unfavorable steric interactions between the pyrazolato ligands and the neutral donors leads to the loss of the latter on heating before sublimation occurs.[424]

In the synthesis of barium 3,5-dimethylpyrazolate from barium metal and the pyrazole, silicone joint grease and *cyclo*-$(Me_2SiO)_4$ are cleaved, forming the complex $[(THF)_6Ba_6(3,5\text{-dimethylpyr-}$ $azolato)_8\{(OSiMe_2)_2O\}_2]$. Two $[O(SiMe_2O)_2]^{2-}$ bidentate chelating siloxane anions are coordinated above and below a nearly hexagonal Ba_6^{2+} layer that has been compared to the (110) layer in cubic body-centered metallic barium (see Figure 40). Eight σ/π coordinated pyrazolate anions flank the barium layer, and six coordinated THF molecules are located at the periphery of the Ba_6 plane.[425]

In its *s*-block complexes, the amidinate ligand, $[RC(NR')(NR'')]^{-}$,[426] displays bonding arrangements that are strongly metal- and substituent-dependent. It is commonly found as a delocalized, bidentate moiety, forming four-membered M—N—C—N rings. Among the structurally authenticated monometallic species are $[PhC(NSiMe_3)(N\text{-myrtanyl})]Li\cdot TMEDA$ (which is chiral by virtue of the myrtanyl group),[427] $[PhC(NSiMe_3)_2]_2Be$,[428] $[PhC(NSiMe_3)_2]_2Mg(THF)_2$,[429] $[PhC(NSi\text{-}$ $Me_3)_2]_2Mg(NCPh)_2$,[430] $[2\text{-Py}(CH_2)_2NC(p\text{-MePh})NPh]_2Mg$ (with a intramolecularly coordinated pendant pyridine),[431] $[PhC(NSiMe_3)_2]_2Ca(THF)_2$,[432] and $[PhC(NSiMe_3)_2]_2Ba(THF)(DME)$.[433]

Extreme steric bulk on the amidinate ligand can force it to become monodentate, as in the lithium complex derived from *N,N′*-diisopropyl(2,6-dimesityl)benzamidine. The TMEDA adduct (see Figure 41) displays a short N—Li bond (1.94 Å), and the N—C—N bond lengths suggests that the amidinate is not fully delocalized (C1—N1 = 1.32 Å and C1—N2 = 1.34 Å).[434]

Bimetallic complexes are represented by $\{[N(SiMe_3)C(Ph)NC(Ph)=C(SiMe_3)_2Li](CNPh)\}_2$[435] $\{[\mu_2\text{-}N,N'\text{-di}(p\text{-tolyl})formamidinato\text{-}N,N,N']Li(Et_2O)\}_2$,[436] $\{[N(SiMe_3)C(Ph)N(CH_2)_3NMe_2]Li\}_2$ (with a γ-pendant amine functionality),[437] $[MeC_6H_4C(NSiMe_3)_2Li(THF)]_2$ and $[PhC(NSiMe_3)_2\text{-}$ $Na(Et_2O)]_2\cdot Et_2O$, whose M–M′ distances (2.42 and 2.74 Å, respectively) are 80% and 73% of those in the metals,[438] and $[4\text{-MeC}_6H_4C(NSiMe_3)_2\cdot Li(NCC_6H_4Me\text{-}4)]_2$. The latter has two four-coordinated Li cations bound in an *N,N′*-bidentate π-fashion to one amidinato anion, thereby forming a double diazaallyl Li bridge, and in a monodentate σ-fashion to a nitrogen lone pair of a

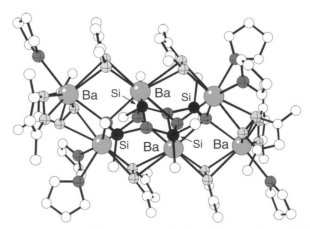

Figure 40 The structure of $[(THF)_6Ba_6(3,5\text{-dimethylpyrazolato})_8\{(OSiMe_2)_2O\}_2]$.

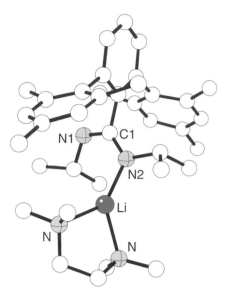

Figure 41 The structure of the TMEDA adduct of the lithium derivative of *N*,*N*-diisopropyl(2,6-dimesityl)-benzamidine.

second amidinate (see Figure 42). In solution, the different Li-amidinate bonding arrangements ($\sigma \rightleftharpoons \pi$) rapidly interconvert.[439]

More elaborate polymetallic species are also known. The reaction of benzonitrile with LiN(SiMe$_3$)$_2$ in hexane affords the trimeric complex [Li$_3$\{PhC(NSiMe$_3$)$_2$\}$_3$(NCPh)].[440] Two of the lithium cations are coordinated by three nitrogen atoms of two phenylamidinate anions, and the other cation is ligated by four nitrogen atoms of two chelating phenylamidinate anions and an adducted benzonitrile molecule (see Figure 43). The tetra- and hexametallic boraamidinate complexes \{[Bu^{n}B(NBu^t)$_2$]Li$_2$\}$_2$ and \{[MeB(NBu^t)$_2$]Li$_2$\}$_3$ have been synthesized from the reaction of B[N(H)(Bu^t)]$_3$ with three equivalents of BuLi or LiMe, respectively. The tetrametallic derivative possesses a 10-atom Li$_4$B$_2$N$_4$ framework isostructural with the Li$_4$Si$_2$N$_4$ core in \{Li$_2$[Me$_2$Si(N-Bu^t)$_2$]\}$_2$.[441] The larger aggregate is constructed around a distorted Li$_6$[N(Bu^t)]$_6$ hexagonal prism,

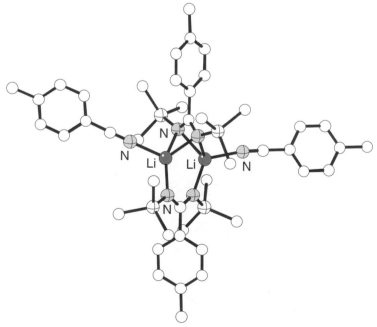

Figure 42 The structure of [4-MeC$_6$H$_4$C(NSiMe$_3$)$_2$·Li(NCC$_6$H$_4$Me-4)]$_2$.

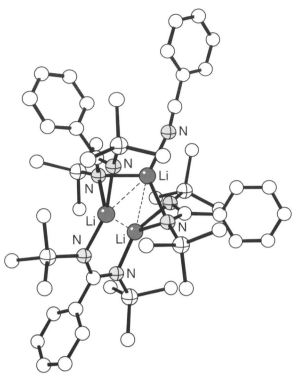

Figure 43   The structure of the trimeric complex [Li$_3${PhC(NSiMe$_3$)$_2$}$_3$(NCPh)].

in which alternate Li$_2$N$_2$ rings are *N,N'*-capped by a BMe unit, generating a tricapped hexagonal prismatic cluster (see Figure 44).[442]

The reaction of lithium cyclopentylamide in the presence of traces of H$_2$O generates the large aggregate [{(*c*-C$_5$H$_9$)N(H)}$_{12}$(O)Li$_{14}$].[443] In contrast to the ladders and rings that dominate the analysis of many amido and imido complexes, the molecule contains a salt-like, distorted, face-centered cube of lithium cations surrounding its central oxo anion (Li—O = 1.89 Å).

The elegantly simple cations [Ba(NH$_3$)$_n$]$^{2+}$ are generated in the course of reducing the fullerenes C$_{60}$ and C$_{70}$ with barium in liquid ammonia. The X-ray crystal structure of [Ba(NH$_3$)$_7$]C$_{60}$·NH$_3$ reveals a monocapped trigonal antiprism around the metal, with an ordered C$_{60}$ dianion.[444] An even more highly coordinated barium cation, [Ba(NH$_3$)$_9$]$^{2+}$ (Ba—N = 2.89–2.97 Å), was identified from the reduction of C$_{70}$. The coordination geometry around barium is a distorted tricapped trigonal prism; the fullerene units are linked in slightly zigzagging linear chains by single C—C bonds (1.53 Å) (see Figure 45).[445]

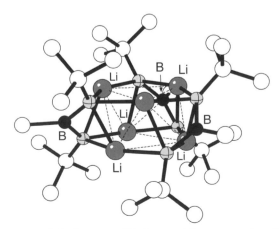

Figure 44   The structure of the hexametallic boraamidinate complex {[MeB(NBu$^t_2$)$_2$]Li$_2$}$_3$.

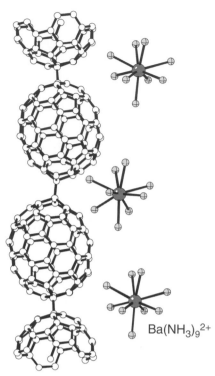

$Ba(NH_3)_9^{2+}$

Figure 45 Portion of the polymeric structure of $[Ba(NH_3)_9]C_{70}\cdot 7NH_3$.

3.1.4.3.2 *Phosphorus donor ligands*

The interaction of *s*-block metals and phosphorus-based ligands takes many forms, from neutral phosphines and anionic diphosphides $[PR]^{2-}$, to mixed (P,N) and (P,S) donors. The area has been one of intensive investigation, and there is practical interest in the use of such compounds in organophosphorus synthesis. The area has witnessed considerable growth: prior to 1985, fewer than five compounds containing an *s*-block element bonded to phosphorus had been structurally authenticated; by the end of 2000, the total was over 200. Extensive reviews of the area have appeared.[19,322,446–448]

(i) Neutral phosphines

As expected from the mismatch between the type a *s*-block metals and the type b character of neutral phosphines, adducts of the *s*-block metals with coordinated PR_3 groups are rare; examples are found among organometallic species,[449] but not usually with coordination complexes. Pendant phosphines that form part of a chelating ligand can be isolated, however, such as in the tetrameric sodium enolate $\{Na[(Pr^i)_2PC(H){=}C(O)Ph]\}_4$, with a Na–P interaction at 2.85 Å.[450] and the chelating diphenylphophino arms found in $\{Li[\mu\text{-}OC(CMe_3)_2CH_2PR_2]\}_2$ (R = Me, Ph) (Li–P distance = 2.50, 2.65 Å).[451]

(ii) Monosubstituted phosphido complexes

Group 1 and Group 2 compounds containing the dinegative PR^{2-} unit are not common, but complex structures can be formed from them. Magnesium forms several types with $Mg_{2n}P_{2m}$ cores. The magnesiation of tri(*t*-butyl)silylphosphine with $Mg(Bu^n)_2$ in THF yields tetrameric $[(THF)MgPSi(Bu^t)_3]_4$.[452] The central Mg_4P_4 cube is only slightly distorted, with Mg—P ranging from 2.55 to 2.59 Å. When the reaction is conducted in toluene, the larger aggregate $Mg_6[P(H)Si(Bu^t)_3]_4P[Si(Bu^t)_3]_2$ is generated. A Mg_4P_2 octahedron is at the center, with the phosphorus atoms in a *trans* position. The Mg–Mg edges are bridged by the $P(H)Si(Bu^t)_3$ substituents. A refluxing toluene solution of Mg_6-$[P(H)Si(Bu^t)_3]_4P[Si(Bu^t)_3]_2$ will eliminate $H_2PSi(Bu^t)_3$ and precipitate the hexameric $[MgPSi(Bu^t)_3]_6$.

Its structure is based on a Mg_6P_6 hexagonal drum, with Mg—P distances varying between 2.47 and 2.51 Å in the six-membered Mg_3P_3 ring, and 2.50–2.60 Å between the two rings (see Figure 46).[453] Large $M_n(PR)_m$ aggregates have been synthesized with lithium reagents; e.g., dilithiation of H_2PSi-$(Pr^i)_3$ produces the hexadecametallic cluster $Li_{16}(PSi(Pr^i)_3)_{10}$.[454] It is a doubly capped Archimedean antiprism with ten phosphorus centers and a lithium atom located on each deltahedral face.

The parent phosphide anion (PH_2^-) has been structurally authenticated only in lithium complexes (i.e., $Li(PH_2)(DME)_2$ (Li—P = 2.60 Å),[455] $[Li(PH_2)(DME)]_n$ (Li—P = 2.60 Å)[456,457]) but examples of $M(P(H)R)_n$ compounds are more common. Among these are aryl-substituted species $MP(H)Ar$, where Ar = phenyl,[458] mesityl,[459–461] 2,4,6-tri-*t*-butylphenylphosphide,[462,463] 2,6-dimesitylphenylphosphide,[464,465] and 2,6-dimesitylphenylphosphide.[462] The latter ligand forms a bimetallic sandwich complex with 18-crown-6 that displays cation-π interactions (Cs—P = 3.42 Å; Cs—C distance = 3.48–3.77 Å) (see Figure 47).[462] Other primary phosphido complexes include those with –P(H)CH₃,[466] –P(H)(Bu^t),[467] –P(H)(*c*-C₆H₁₁).[468,469] and –P(H)Si (Pr^i)₃.[244,453,470–473]

The triisopropylsilylphosphido ligand has been used to generate a variety of polymetallic species. For example, metalation of $H_2PSi(Pr^i)_3$ with $Ca(THF)_2[N(SiMe_3)_2]_2$ in tetrahydropyran (THP) in a molar ratio of 3:2 yields $(Me_3Si)_2NCa[\mu-P(H)Si(Pr^i)_3]_3Ca(THP)_3$; the complex contains a trigonal-bipyramidal Ca_2P_3 core, with the metal atoms on the apices;[473] a presumably similar complex (based on NMR evidence) can be made with THF as the supporting ether.[474] A heteroleptic complex, $(THF)_2Ba[N(SiMe_3)_2][P(H)Si(Bu^t)_3]$, can be derived from the equimolar reaction of (tri-*tert*-butyl-silyl)phosphine with $(THF)_2Ba[N(SiMe_3)_2]_2$ in toluene. Addition of a second equivalent of H_2PSi-$(Bu^t)_3$ to the latter affords $(THF)Ba_3(PSi(Bu^t)_3)_2[P(H)Si(Bu^t)_3]_2$ (see Figure 48); it can be viewed as a Ba_4P_4 heterocubane, with two opposite faces capped by $(THF)Ba[P(H)Si(Bu^t)_3]_2$ units.[473] An Sr_4P_4 cube serves as the core of $[Sr(THF)_2(\mu-PHSi(Pr^i)_3)_2\{Sr_2(\mu_4-PSi(Pr^i)_3)_2\}_2Sr(\mu-PHSi(Pr^i)_3)_2(THF)_2]$, which is formed by elimination of $PH_2Si(Pr^i)_3$ from $Sr(THF)_4(PH(Pr^i)_3)_2$ in toluene.[244]

Complexes are also known that contain both PR^{2-} and PHR^- functionalities.[475] The silaphosphane $R_2Si(PH_2)[P(H)SiPh_3]$ (R_2Si = (2,4,6-$(Pr^i)_3C_6H_2)(Bu^t)Si$) reacts with Bu^nLi with loss of butane to form the tetrametallic dimer $\{Li_2(PSiPh_3)[P(H)(\mu-SiR_2)]\}_2$. On further treatment with LiCl the latter generates an $Li_{10}P_8$ aggregate, a "dimer of dimers," in which two of the tetrametallic dimers are joined at a central $(LiCl)_2$ ring (see Figure 49).[476] Treatment of ethyl tris(triisoproplysilylphosphino)silane), $EtSi\{P[Si(Pr^i)_3]H\}_3$, with either Bu^nLi or Bu^nNa leads to distinctly different products. The lithium derivative, $\{EtSi[P(Si(Pr^i)_3)Li]_3\}_2$, has a distorted

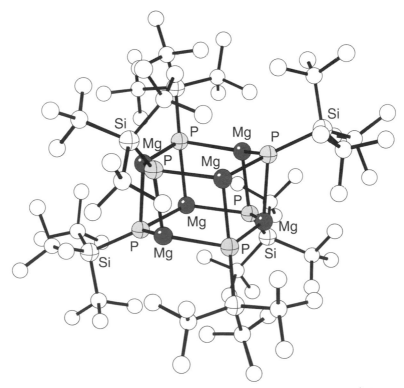

Figure 46 The structure of the hexameric complex $[MgPSi(Bu^t)_3]_6$.

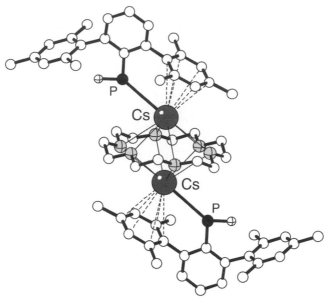

Figure 47 The structure of the 18-crown-6 complex of caesium 2,6-dimesitylphenylphosphide.

rhombic dodecahedral $Si_2P_6Li_6$ framework, whereas the product from the sodium reaction has an open polyhedral structure, with two bridging sodium atoms, each coordinated η^2 to a molecule of the toluene solvent (see Figure 50). It is thought that the mismatch between Si—P (2.2 Å) and Na—P (2.8 Å) distances prohibits the formation of a closed structure.[476]

(iii) Disubstituted phosphido complexes

Disubstituted phosphido ligands $[PRR']^-$ have been incorporated into numerous *s*-block complexes. Many examples are known, and are covered in detail in the previously cited reviews; only

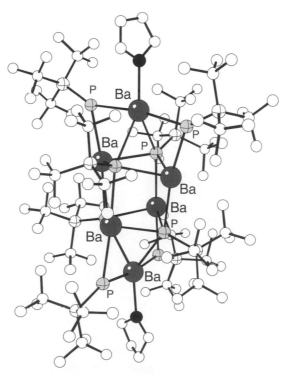

Figure 48 The structure of $(THF)Ba_3(PSi(Bu^t)_3)_2[P(H)Si(Bu^t)_3]_2$.

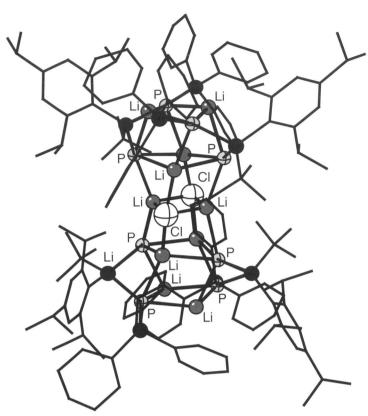

Figure 49   The structure of the Li_8P_8 aggregate derived from LiCl and $\{Li_2(PSiPh_3)[P(H)(\mu\text{-}SiR_2)]\}_2$.

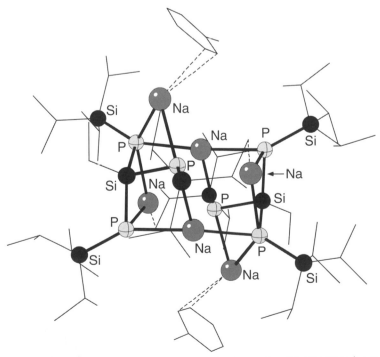

Figure 50   The structure of the open polyhedral complex $\{EtSi[P(Si(Pr^i)_3)Na]_3\}_2$.

representative systems need be presented here. A particularly well-studied class of these complexes are those with trialkylsilyl groups as substituents, especially the bis(trimethyl)silyl moiety.[19] Synthetic methods for these compounds vary, from direct metalation (e.g., the reaction of $HPN(SiMe_3)_2$ with KH or $KO(Bu^t)$ produces $KP(SiMe_3)_2$; $(Bu^n)_2Mg$ and $HPN(SiMe_3)_2$ generate $Mg[P(SiMe_3)]_2)$ to metathesis (e.g., $M[N(SiMe_3)_2]_2$ (M = Mg, Ca, Sr, Ba) react with $HP(SiR_3)$ to produce the corresponding phosphides). Representative complexes are listed in Table 6. In donor solvents (ether, THF), the complexes are monomeric; in hydrocarbons, monomer/dimer equilibria can be observed in NMR spectra.

Other series of phosphido complexes are well-established, including those containing the ligands $P(Bu^t)_2^-$[477,478] and PPh_2^-. The compact size of the latter permits extensive bridging to occur, generating polymeric structures. $\{Li(Et_2O)PPh_2\}_\infty$, $\{Li(THF)_2PPh_2\}_\infty$, and the related $\{Li(THF)P(cyclohexyl)_2\}_\infty$, for example, all have infinite –Li–P—Li—P chains in the solid state, with distorted tetrahedral coordination around phosphorus.[479] A similar arrangement is found in $[Li(DME)PPh_2]_\infty$.[477] Sodium diphenylphosphide, $NaPPh_2$, prepared from sodium and $PClPh_2$ in refluxing dioxane, crystallizes from dioxane as $[Na_4(\mu\text{-dioxane})_8/2(\mu\text{-diox-ane})(PPh_2)_4]_\infty$, which includes a dioxane molecule suspended within eight-membered Na_4P_4 rings (Na—P = 2.88–3.00 Å); the rings, in turn, are linked by additional dioxane molecules to form a network structure (Figure 51).[480] The solid-state structure of $[K(dioxane)_2PPh_2]_\infty$ displays π-interactions between the potassium ions and aryl rings of a neighboring PPh_2 anion, thereby generating a three-dimensional framework.[481]

The effect of steric bulk in limiting oligomerization is evident in the structures of $[(TMEDA)\text{-}LiPPh_2]_2$, which is a dimer, and $(PMDTA)LiPPh_2$, which is a monomer (Li—P = 2.57 Å). Lithium-7 and ^{31}P-NMR spectroscopic studies indicate that both solid-state structures are retained in arene solvents, although the TMEDA adduct dissociates to some extent.[482] Alternatively, extra bulk on the phenyl rings will limit oligomerization; thus unlike the oligomeric $\{Li(Et_2O)_2PPh_2\}_\infty$, $[Li(OEt_2)P(mesityl)_2]_2$ is a discrete dimer.[459]

The ladder structures found in s-block amides are common structural motifs in phosphides as well. Solvent free $[LiP(SiMe_3)_2]_6$ displays such an arrangement in the solid state, with four five-coordinate and two four-coordinate P atoms and four three-coordinate and two two-coordinate Li atoms; the Li—P distances range from 2.38 Å to 2.63 Å (see Figure 52).[483] $Li_4(\mu_2\text{-}PR_2)_2(\mu_3\text{-}PR_2)_2(THF)_2$, formed from the reaction of $P(SiMe_3)_3$ with Bu^nLi in THF, has a fused tricyclic $(LiP)_4$ ladder skeleton. The Li atoms are three-coordinate, with each of the two terminal lithiums bound to two P atoms and one THF, while the two internal lithiums have three phosphorus atoms as neighbors (see Figure 53).[484] In solution, there is no NMR evidence for ^7Li–^{31}P

Table 6 Selected bond lengths (Å) and angles (°) of the complexes $(L)M[P(SiR3)_2]_n$ as well as chemical $^{31}P[^1H]$ shifts of the bis(phosphanides).

Compound	$\delta ppm(^{31}P[^1H])$	M—P(Å)	P—M—P(°)	References
$[LiP(SiMe_3)_2]_2$		2.45–2.50	107.0	475
$[LiP(SiMe_3)_2]_6$		2.38–2.62	104.1–114.0	483
$[(THF)_2LiP(SiMe_3)_2]_2$	−298	2.62	100.0	484
$Li_4(\mu_2\text{-}PR_2)_2(\mu_3\text{-}PR_2)_2(THF)_2$	−298	2.44–2.64	105.4–149	484
$[(DME)LiP(SiMe_3)_2]_2$		2.56	104.3	457
$\{[(Me_3Si)_2PK(THF)]_2\}_n$	−293	3.32–3.43	99.0, 140.2	783
$\{[(Me_3Si)_2PRb(THF)]_2\}_n$	−287	3.42–3.49	98.4, 139.3	783
$\{[(Me_3Si)_2PCs(THF)]_2\}_n$	−270	3.58, 3.64	95.3, 165.1	783
$((Me_3Si)_2PMg[\mu\text{-}P(SiMe_3)_2]_2)_2Mg$	−243, −275	2.45, 2.55, 2.60–2.68	116.9–122.2, 144.7 (av)	784
$(THF)_2Mg[P(SiMe_3)_2]_2$	−295	2.50	143.6	504
$(DME)Mg[P(SiMe_3)_2]_2$	−296	2.49	122.5	785
$(THF)_4Ca[P(SiMe_3)_2]_2$	−282	2.91, 2.92	175.2	786
$(TMTA)_2Ca[P(SiMe_3)_2]_2$	−277	2.99	110.2	787
$(THF)_4Sr[P(SiMe_3)_2]_2$	−274	3.04, 3.01	174.2	788
$(THF)_4Sr[P(SiMe_2Pr^i)_2]_2$	−290	3.089	168.5	789
$(THF)_4Ba[P(SiMe_3)_2]_2$	251	3.158,3.190	174.9	787
$(THF)_4Ba[P(SiMe_2Pr^i)_2]_2$	−274	3.200,3.184	139.9	790
$(DME)_3Ba[P(SiMe_2CH_2)_2]_2$	−289	3.333	178.9	791

Source: Alkaline-earth compounds: Westerhausen.19

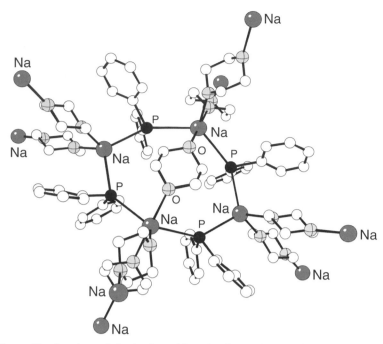

Figure 51   Portion of the lattice of [Na$_4$(μ-dioxane)$_8$/2(μ-dioxane)(PPh$_2$)$_4$]$_\infty$.

coupling. Other ladder structure are known, including [(THF)LiCl·(THF)$_2$LiP(Bu^t)$_2$]$_2$[485] and [R-PLi$_2$(F-R)]$_2$ (R = diisopropyl-(2,4,6-triisopropylphenyl)silane.[486]

Compounds containing a variety of substituted heteroallyl-like ligands, with PCP (bis(phosphino)methanides),[487–491] PNP (bis(phosphino)amides),[492] NCP,[493] PSiP (diphosphasilaallyl),[494,495] and PPP (triphosphides)[496,497] frameworks are known. Some of these complexes can adopt a variety of coordination modes; in Rb(18-crown-6)(N(PPh$_2$)$_2$), the phosphinoamide is P,P-ligated,[492] whereas in (THF)$_3$LiN(PPh$_2$)$_2$, the ligand is bound η^2-P,N. Phosphorous-31 and ^{6}Li-NMR spectroscopy indicate that the structure of the latter in THF solution is similar to that in the solid state; dynamic ^{31}P-NMR spectroscopic measurements indicate that an 8.1 kcal mol^{-1} rotation barrier exists around the P—N bonds.[498] The utility of some of these complexes as ligand transfer reagents has been investigated. Thus from the reactions of ZrCl$_4$ with {(TMEDA)-Li[CH(PMe$_2$)(SiMe$_3$)]}$_2$, a mixture of the compounds Cl$_{(4-n)}$Zr[CH(PMe$_2$)(SiMe$_3$)]$_n$ ($n = 1$–4) has been characterized with NMR spectroscopy. With (TMEDA){Li[C(PMe$_2$)(SiMe$_3$)$_2$]}$_2$, only the disubstituted product Cl$_2$Zr[C(PMe$_2$)(SiMe$_3$)$_2$]$_2$ is obtained.[499,500]

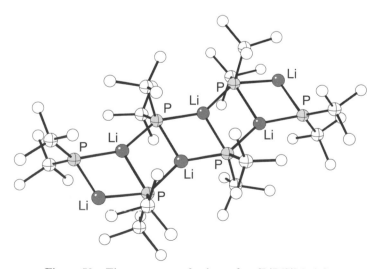

Figure 52   The structure of solvent-free [LiP(SiMe$_3$)$_2$]$_6$.

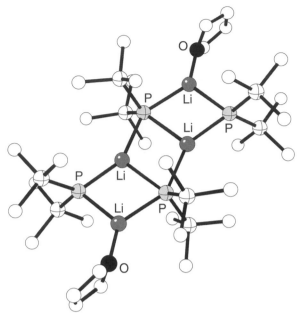

Figure 53 The structure of $Li_4(\mu_2\text{-PR}_2)_2(\mu_3\text{-PR}_2)_2(THF)_2$.

3.1.4.3.3 Arsenic donor ligands

Early reports describing the preparation of *s*-block arsenides such as $Ca(AsH_2)_2$ and $Ca(AsHMe)_2$ provided few details on their properties.[501] That situation has changed with the use of silyl-substituted arsenic ligands, and compounds containing them display considerable structural diversity. This area has been reviewed.[502]

Sometimes arsines themselves are the starting materials for *s*-block derivatives. The preparation of $[Li(AsH_2)(DME)_2]$ from AsH_3 has been reported,[455] and in the solid state it possesses a trigonal bipyramidal metal center with the AsH_2 residue in an equatorial site (Li—As = 2.699 Å). The reaction of Bu^nLi with $PhAsH_2$ in toluene–THF produces the primary arsenide $[PhAsHLi\cdot 2THF]_\infty$. It crystallizes in the form of helical polymers; the right-hand enantiomorph was identified in the lattice (see Figure 54).[503] The reaction of $(n,s\text{-butyl})_2Mg$ with bis(trimethylsilyl)arsine in THF yields $[(Me_3Si)_2As]_2Mg(THF)_2$. The magnesium atom is in a distorted tetrahedral environment (Mg— As = 2.59, 2.60 Å; Mg—O = 2.05, 2.06 Å), whereas the environment around arsenic is pyramidal.[504]

Many arsenic derivatives have been synthesized from bis(trimethylsilyl)amido complexes. The 2:1 reaction of $AsH_2SiPr^i_3$ with $Ba[N(SiMe_3)_2]_2\cdot 2(THF)$ in THF affords the bis(arsenide) complex $[Ba(THF)_3\{\mu\text{-As(H)SiPr}^i_3\}_3BaAs(H)SiPr^i_3(THF)_2]$.[473] $M(NR_2)_2$ (M = Ca, Sr; R = Me_3Si) react with $HAsR_2$ in THF to give $(R_2As)_2M(THF)_4$.[505] Both $(R_2As)_2Ca(THF)_4$ and $(R_2As)_2Sr(THF)_4$ exist as colorless *trans*-isomers with a nearly linear As—M—As moiety; however, the light-sensitive Sr analog contains two different configurations for the As atoms. One As atom is surrounded in a nearly trigonal planar manner with an Sr—As bond length of 3.10 Å, whereas the other arsenic atom has an angle sum of 338° and an Sr—As distance of 3.15 Å.

Metallation of $As[SiMe_2(Bu^t)]_2H$ with $Ba[N(SiMe_3)_2]_2\cdot 4(THF)$ generates $Ba[As(SiMe_2-(Bu^t))_2]_2\cdot 4(THF)$ (see Figure 55), which in the solid state exists as a distorted pentagonal bipyr-amid with apical arsenic atoms and a vacant equatorial site shielded by the trialkylsilyl groups. The As—Ba—As′ angle is 140.8°. When the compound is recrystallized from toluene, THF ligands are lost to give the dimeric $\{Ba[As(SiMe_2(Bu^t))_2]_2(THF)\}_2$ (see Figure 56), which contains four-coordinate Ba centers.[506] The magnesiation of $AsH_2(SiPr^i_3)$ in THF yields $[Mg(THF)AsSi-Pr^i_3]_4$(see Figure 57), constructed around a Mg_4As_4 cube.[507]

Reaction of $AsH_2(SiPr^i_3)$ with $M[N(SiMe_3)_2]_2\cdot 2(THF)$ (M = Ca or Sr) produces $M[As(H)Si-Pr^i_3]_2(THF)_4$, which is in equilibrium with the dimers $M[As(H)SiPr^i_3][\mu\text{-As(H)SiPr}^i_3]_3M(THF)_3$ by elimination of THF. Reaction of the equilibrium mixtures with diphenylbutadiyne gives the metal bis(THF)bis(2,5-diphenylarsolide) species. A mechanism based on intermolecular $H/SiPri_3$ exchange has been proposed that explains the formation of both the $[As(SiPr^i_3)_2]^-$ and 2,5-diphenyl-3,4-bis(phenylethynyl)arsolide anions; the latter was isolated as a solvent-separated ion pair with the binuclear $[(THF)_3Ca\{\mu\text{-As(H)SiPr}^i_3\}_3Ca(THF)_3]^+$ cation (see Figure 58).[508]

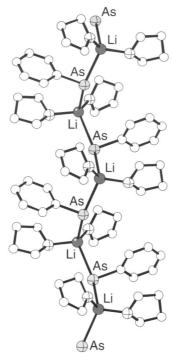

Figure 54   Portion of the helical structure of [PhAsHLi·2THF]$_\infty$.

The solid state structure of the lithium arsenide {Li$_2$[μ_2-As(SiMe$_3$)$_2$][μ_3-As(SiMe$_3$)$_2$](THF)}$_2$ reveals a [LiAs]$_4$ ladder-like framework with four antiparallel adjacent As–Li rungs (see Figure 59). The two Li atoms of the central ring each bridge three As centers, while the two Li atoms on the outer rings each span two As atoms and are coordinated to one THF molecule. The Li arsenide [((Me$_3$Si)$_2$As)Li(THF)$_2$]$_2$ crystallizes as a centrosymmetric dimer constructed around a four-membered As—Li—As—Li ring; each Li atom is coordinated to two molecules of THF.[509]

Reaction of two equivalents of LiAsH$_2$(DME) with (2,4,6-triisopropylphenyl)$_2$SiF$_2$ at 20 °C in THF gives (2,4,6-triisopropylphenyl)$_2$SiFAsHLi in quantitative yield.[510] The product reacts with [Pr$^i_3$Si]SO$_2$CF$_3$ to give (2,4,6-triisopropylphenyl)$_2$SiFAsHSiPr$^i_3$, which upon lithiation with Bu^{n}Li

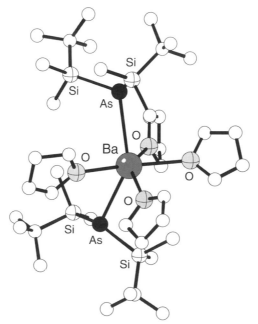

Figure 55   The structure of Ba[As(SiMe$_2$(Bu^t))$_2$]$_2$·4(THF).

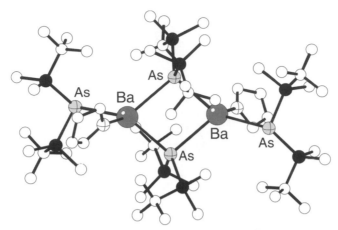

Figure 56 The structure of {Ba[As(SiMe$_2$(Bu^t))$_2$]$_2$(THF)}$_2$.

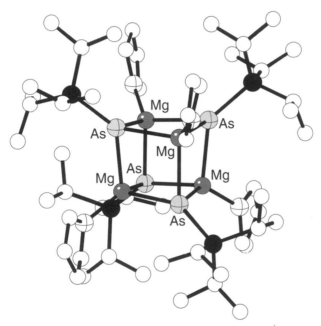

Figure 57 The structure of [Mg(THF)AsSiPr$^i_3$]$_4$.

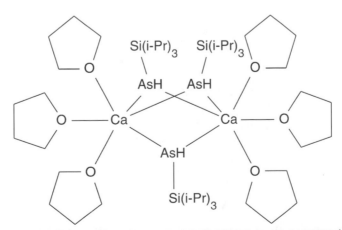

Figure 58 Schematic structure of the binuclear cation [(THF)$_3$Ca{μ-As(H)SiPr$^i_3$}$_3$Ca(THF)$_3$]$^+$.

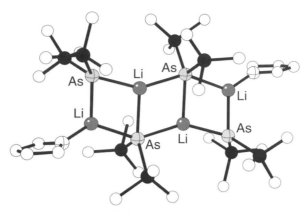

Figure 59   The structure of $\{Li_2[\mu_2\text{-}As(SiMe_3)_2][\mu_3\text{-}As(SiMe_3)_2](THF)\}_2$.

in THF-hexane produces (2,4,6-triisopropylphenyl)$_2$SiFAs$\{Li(THF)_2\}\{SiPr^i_3\}$ (see Figure 60). The latter undergoes thermal elimination of LiF and THF in toluene at 80 °C to produce the arsanylidenesilane (2,4,6-triisopropylphenyl)$_2$Si = AsSiPr$^i_3$.

A variety of large clusters have been formed by lithiation of arsine derivatives. Treatment of AsRH$_2$ (R = SiMe$_2$C(Pr^i)Me$_2$) with Li$_2$O-containing Bu^{n}Li generates the orange-yellow (RAs)$_{12}$Li$_{26}$O aggregate, which is based on a slightly distorted As$_{12}$ icosahedron with all faces capped by lithium. Four Li$^+$ cations are located in the center of the roughly spherical framework; the ions are encapsulate a Li$_2$O molecule, thereby generating an octahedral [Li$_6$O]$^{4+}$ core (see Figure 61).[511] Partial lithiation of AsRH$_2$ in the presence of LiOH leads to the isolation of the intermediate species [Li$_{20}$O(RAsH)$_6$-(RAs)$_6$], which contains wheel-like [Li$_{18}$As$_{12}$] ladder structures with [Li$_2$O] units acting as the "stabilizing axis" of the wheel; it is isotypic with a phosphorus analogue.[454] Dilithiation of AsH$_2$SiPr$^i_3$ produces the decameric cluster Li$_{16}$(AsSiPr$^i_3$)$_{10}$. It forms a doubly capped Archimedean antiprism with ten arsenic centers and a lithium atom located on each deltahedral face.[454]

3.1.4.4   Group 16 Ligands

3.1.4.4.1   Oxygen donor ligands

The generally high oxophilicity of the alkali and alkaline-earth elements gives oxygen donor ligands a prominent place in *s*-block coordination chemistry. Oxygen donor ligands, of which

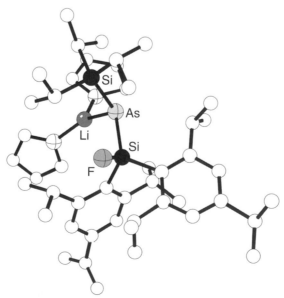

Figure 60   The structure of (2,4,6-triisopropylphenyl)$_2$SiFAs$\{Li(THF)_2\}\{SiPr^i_3\}$.

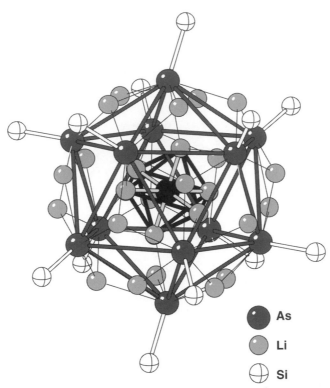

As

Li

Si

Figure 61 The structure of the $(RAs)_{12}Li_{26}O$ aggregate $(R = SiMe_2CPr^iMe_2)$. The alkyl groups have been removed for clarity.

carboxylates, diketonates, enolates, and alkoxides and aryloxides are considered here, are found in a vast array of *s*-block complexes. The interest in their properties encompasses inorganic synthesis (many *s*-block complexes are used as transfer reagents for *p*-, *d*-, and *f*-block complexes), materials chemistry (e.g., precursors to metal oxides), and the biological realm (water and oxygen-containing organic functional groups play a critical role in defining the structure and reactions of many proteins and enzymes).

(i) Carboxylates

Compared to other *s*-block metal complexes with anionic oxygen donors, the metal carboxylates are particularly robust; although often hygroscopic, they will not decompose upon absorption of water, and are as a class quite thermally stable. Their handling is thus simpler than alkoxides and aryloxides. The bonding arrangements of calcium carboxylates were reviewed in the early 1980s,[512] and the chemistry of Group 2 carboxylates and thiocarboxylates, and their potential applications as reagents in CVD have been extensively reviewed.[13]

Carboxylates can form structurally complex units, often assisted by the presence of bridging water molecules, which can generate three-dimensional networks. For example, in the crystal structure of $[Mg\{C_2H_4(CO_2Et)_2\}_3]^{2+}[MgCl_4]$, the cations are linked by other diethyl succinate ligands to form a linear polymer. In the cation, each magnesium atom is octahedrally coordinated by six carbonyl oxygen atoms of ethyl succinate molecules.[513] A more complex infinite two-dimensional structure exists in barium diethyl 1,3-dithiepane-2-ylidenemalonate via intermetallic coordination of the dicarboxylic groups. The metal center adopts a nine-coordinate geometry with three different carboxylate bonding arrangements (chelated monodentate, bidentate η^2, monodentate) and two water molecules. Each ligand is associated with five barium atoms, in the form of a layer structure (see Figure 62).[514]

Even ostensibly monomeric complexes can display close contacts in the solid state that will affect their reactivity. Bis(*trans*-but-2-enoato)calcium forms discrete molecules with intermolecular Ca—O contacts (2.36 Å) that are only slightly longer than the intramolecular bonds

---

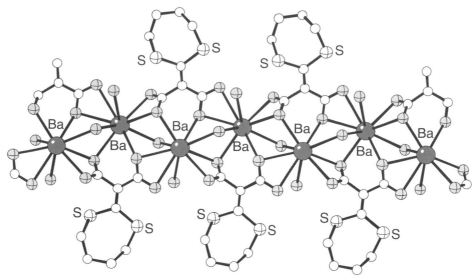

Figure 62 Portion of the lattice of barium diethyl 1,3-dithiepane-2-ylidenemalonate, illustrating the various carboxylate binding modes present.

(Ca—O = 2.30 Å). The closeness of the molecules help explains why under irradiation with γ-rays from ^{60}Co, a solid-state cyclodimerization reaction in induced, producing *cis,trans*-nepetic acid, one of four possible diastereomers.[515] Related radiation-induced chemistry is displayed by aquated (3-butenoato)calcium, $Ca(CH_2=CHCH_2CO_2)_2(H_2O)$, synthesized from 3-butenoic acid and calcium carbonate. The carboxylate is a two-dimensional coordination polymer with nearly parallel vinyl groups and short –C=C–C=C– contacts of 3.73 Å and 3.90 Å (see Figure 63).[516]

Alkali metal ions are commonly used to complex carboxylic acids of biological importance for structural analysis. Examples of such derivatives include 2-epimutalomycin-potassium dihydrate and 28-epimutalomycin-potassium (metabolites from mutalomycin fermentation),[517] the rubidium salt of CP-80,219 (an antibiotic related to dianemycin),[518] kijimicinate-rubidium hexane solvate (a polyether antibiotic),[519] griseocheline-calcium (an antifungal antibiotic),[520] the sodium salt of the antibiotic A204A,[521] and the potassium salt of the polyether antibiotic monensin A dihydrate.[522] Comparison of the conformation of the latter with the sodium derivative reveals structural

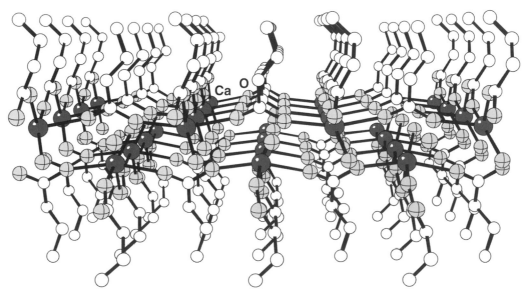

Figure 63 Portion of the lattice of $Ca(CH_2=CHCH_2CO_2)_2(H_2O)$, illustrating the close packing of the vinyl groups.

features that help explain the selectivity of monensin for Na^+ over K^+. In the K^+ derivative, the conformation of the dioxaspiro-fused ring is altered and, according to the results of molecular mechanics calculations, raised in energy compared to the highly conserved Na^+-complexed form.[521]

Liquid carboxylate salts of the Group 2 elements are readily prepared via the reaction of polyether carboxylic acids such as 2-[2-(2-methoxy)ethoxy]ethoxyacetic acid (MEEA) and (2-[2-(2-methoxyethoxy)ethoxy]acetic acid (MEEAA) with metal hydroxides, carbonates, and alkoxides.[523–525] These salts can be used directly in organometallic deposition processes to prepare ceramic films. They have also been used as solvents for transition metal or lanthanide nitrates and acetates in liquid precursors for polymetallic ceramics. Such monocarboxylates are generally highly soluble in a range of solvents including H_2O, methanol, acetone, THF, $CHCl_3$, and CH_2Cl_2.

Unlike the carboxylates prepared from MEAA and MEEAA, the calcium, strontium, and barium derivatives of the dicarboxylic 3,6,9–trioxaundecanedioic acid (TODD) are solids at room temperature, and display little solubility in solvents less polar than H_2O and MeOH.[525] In the solid state, the calcium compound $Ca(TODD)(H_2O)_2$ forms a discrete dinuclear unit in which bridging carboxylates span the two calcium centers and all the oxygen atoms except one belonging to a carboxylate group are coordinated to the metal centers (see Figure 64). The ligand is twisted into a fan-like shape in order to allow the oxygen atoms to approach the metal center closely enough to bond (Ca—O(ether) = 2.46–2.63 Å).

Potassium, rubidium and caesium thiocarboxylates (MS_2CR; M = K, Rb, Cs; R = Me, Et, Pr^i, cyclohexyl, Ph, 2- and $4-MeC_6H_4$, 4-MeO and $4-ClC_6H_4$, $2,4,6-Me3C_6H_2$) have been synthesized by reaction of thiocarboxylic acid or its O-trimethylsilyl esters with KF, RbF, and CsF.[526,527] The structures of several of the derivatives have been determined, including potassium benzene-, 2-methoxybenzene-, and 4-methoxybenzenecarbothioates, rubidium and caesium 2-methoxybenzene-carbothioate, potassium 2-methoxybenzenecarboselenoate, and rubidium 2-methoxybenzenecar-botelluroate. The metal derivatives have a dimeric structure in which the O and/or S atom is associated with the metal of the opposite molecule. In the 2-methoxybenzenecarbothioates, dimeric metal thiolate units held together by both the chelating thiocarboxylate groups and the *o*-methoxy functionalities form the motif for polymeric structures (see Figure 65).[527] The $C(sp^2)$—S distances of the thiocarboxylate groups range from 1.70–1.72 Å; this value is close enough to that of a C—S single bond to indicate that the negative charges may be partially localized on the sulfur atoms.

(ii) Diketonates

Metal diketonates, the salts of β-diketones and β-ketoimines, are of interest for their usefulness as reagents (principally the alkali metal derivatives) and as potential precursors (the alkaline-earth derivatives) for CVD.[528] The latter are discussed in additional detail in Section 3.1.4.4.1(vi), and Group 2 β-diketonates used for this purpose have been extensively reviewed.[13] Considerable interest has been expressed in fluorinated derivatives, which can display substantially increased volatility relative to the hydrocarbon compounds. Commonly used β-diketones and β-ketoimines

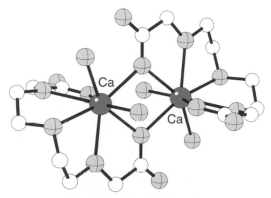

Figure 64 The structure of the calcium derivative of 3,6,9-trioxaundecanedioic acid.

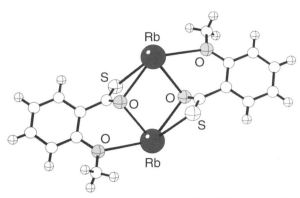

Figure 65   The dimeric motif in the solid state structure of rubidium 2-methoxybenzenecarbothioate.

and their abbreviations are listed in Table 7. Note that the 2,2,6,6,-tetramethylheptane-3,5-dionate anion is variously abbreviated in the literature as THD, TMHD, and DPM.

The deprotonation of β-diketones and β-ketoimines can be accomplished by a variety of methods including reaction with aqueous solutions of metal chlorides,[529] hydroxides, carbonates,[530] or ethoxides.[531,532] For example, reaction of Mg(OEt)$_2$ and [Ca(OEt)(EtOH)$_4$]$_n$ with HTMHD (1:2) yields the homoleptic β-diketonate compounds M(TMHD)$_2$; the calcium complex

Table 7   Commonly used β-diketones and β-diketonimines in the preparation of metal diketonates.

β-diketone and β-diketonimine	Formula	Anion abbreviation
pentane-2,4-dione	$CH_3C(O)CH_2C(O)CH_3$	acac
1,1,1-trifluoropentane-2,4-dione	$CF_3C(O)CH_2C(O)CH_3$	tfa, tfac
1,1,1,5,5,5-hexafluoropentane-2,4-dione	$CF_3C(O)CH_2C(O)CF_3$	Hfa, hfac
1,1,1-trifluoro-5,5-dimethylhexane-2,4-dione	$CF_3C(O)CH_2C(O)C(CH_3)_3$	tpm
1,1,1,5,5,6,6,6-octafluorohexane-2,4-dione	$CF_3C(O)CH_2C(O)CF_2CF_3$	ofhd
2,2,6,6-tetramethylheptane-3,5-dione	$(CH_3)_3CC(O)CH_2C(O)C(CH_3)_3$	dpm, tmhd, thd
1,1,1,2,2-pentafluoro-6,6-dimethylheptane-3,5-dione	$(CH_3)_3CC(O)CH_2C(O)CF_2CF_3$	ppm
1,1,1,5,5,6,6,7,7,7-decafluoroheptane-2,4-dione	$CF_3C(O)CH_2C(O)CF_2CF_2CF_3$	dfhd
1,1,1,2,2,3,3-heptafluoro-7,7-dimethyloctane-4,6-dione	$(CH_3)_3CC(O)CH_2C(O)CF_2CF_2CF_3$	hpm, fod
1,1,1,2,2,3,3,7,7,8,8,9,9,9-tetradecafluorononane-4,6-dione	$CF_3CF_2CF_2C(O)CH_2C(O)CF_2CF_2CF_3$	tdfn, tdfnd
1,1-dimethyl-8-methoxyoctane-3,5-dione	$(CH_3)_3CC(O)CH_2C(O)CH_2CH_2CH_2OCH_3$	dmmod
1,1,1-trichloropentane-2,4-dione	$CCl_3C(O)CH_2C(O)CH_3$	tclac
1,3-diphenylpropane-1,3-dione	$PhC(O)CH_2C(O)Ph$	dpp, Ph$_2$acac
2,2-dimethyl-5-N-(2-methoxyethoxyethylimino)-3-hexanone	$(CH_3)_3CC(O)CH_2C(N(CH_2)_2OCH_3)CH_3$	miki
2,2-dimethyl-5-N-(2-(2-methoxy)ethoxy-ethylimino)-3-hexanone	$(CH_3)_3CC(O)CH_2C(N(CH_2(N(CH_2CH_2O)_2CH_2)CH_3)$	diki
2,2-dimethyl-5-N-(2-(2-ethoxy)ethoxy)-ethoxyethylimino)-3-hexanone	$(CH_3)_3CC(O)CH_2C(N(CH_2CH_2O)_3CH_3)CH_3$	triki
5-N-(2-methoxyethylimino)-2,2,6,6-tetramethyl-3-heptanone	$(CH_3)_3CC(O)CH_2C(N(CH_2)_2OCH_3)C(CH_3)_3$	dpmiki
5-N-(2-(2-methoxy)ethoxyethylimino)-2,2,6,6-tetramethyl-3-heptanone	$(CH_3)_3CC(O)CH_2C(N(CH_2CH_2O)_2CH_3)C(CH_3)_3$	dpdiki
5-N-(2-(2-(-2-ethoxy)ethoxy)ethoxyethylimino)-2,2,6,6-tetramethyl-3-heptanone	$(CH_3)_3CC(O)CH_2C(N(CH_2CH_2O)_3CH_3)C(CH_3)_3$	Dptriki

Source: Wojtczak (1996).[13]

is a rare example of a homoleptic β-diketonate complex, and has a triangular core with 6-coordinate calcium atoms (see Figure 66).[533,534] Metathetical reaction of the sodium salts of THD with hydrated barium chloride has been used to prepare hydrated diketonates.[535]

Metalla-β-diketonates of the s-block compounds are sensitive compounds, and despite the interest in their application to CVD processes, differences in handling procedures have led to conflicting reports of reactivity and volatility. The use of auxiliary Lewis bases, including NH_3, THF, pyridine, and HTMHD has been claimed to increase the volatility of β-diketonates, ostensibly by causing deoligomerization of aggregates.[536-538] However, in the course of studies of the often ill-characterized "Ba(THD)$_2$" with the potential base NEt$_3$, the peroxo complex $[Ba_6(THD)_{10}(H_2O)_4(OH)_2(O_2)][HNEt_3]_2$ was isolated and structurally authenticated (see Figure 67).[539] In this complex, which is based on an octahedron of Ba atoms and contains a μ_4-peroxo group (O—O = 1.48 Å), the oxygen source is evidently ambient air and/or water.

At times, only small changes in steric bulk are enough to affect the aggregation of complexes. For example, crystallographic examination of M(THD)$_2$(THF)$_4$ (M = Sr or Ba) has shown them to have mononuclear structures; interestingly, the analogous diethyl ether and tetrahydropyran complexes ([M(THD)$_2$(THP)$_2$]$_2$ and [M(THD)$_2$(Et$_2$O)]$_2$, respectively) are dinuclear.[540] Adduct formation with polyethers or polyamines[541-543] has been used in attempts to inhibit aggregation and consequently raise volatility, and the improvement in properties can be substantial. Complexation of tetraglyme to Ba(TDFND)$_2$ for example, lowers the melting point by 96 °C and the sublimation temperature by 60 °C, so that the adduct sublimes at 90 °C and 10^{-2} torr.[544,545] In the structure of the adduct, the tetraglyme is found wrapped around the metal, enforcing its mononuclearity and lowering intermolecular forces (see Figure 68).[546]

(iii) Enolates

Metal enolates of the s-block metals encompass a variety of forms, depending on the nature of the R group and the presence of auxiliary bases coordinated to the metal. Some are of a simple nature, e.g., [ArC(CH$_2$O)O][K(18-crown-6)] (Ar = Ph, 2-MeOC$_6$H$_4$, 1,3,5-Me$_3$C$_6$H$_2$), [PhC(=CHMe)O—K(18-crown-6)], and [PhC(CH$_2$)O][K(crypt-2,2,2)],[547] and not discussed further, and some derivatives are mentioned in this Chapter in the context of other chemistry; e.g., the ketoiminato "lariat crowns" investigated for use in CVD chemistry.[548] (See also Section 3.1.4.4.1(vi)) A tetrameric sodium complex with a chelating phosphine ligand[450] has also been discussed elsewhere (see Section 3.1.4.3.2). In this section, focus is placed on aggregates of enolates, from dimers to larger clusters.

Dimeric enolates have been isolated in the course of studies on the effects of H-bonding on deuteration; the trimethylethylenediamine adducts of [(Z)-MeCH=C(OLi)NMe$_2$]$_2$ and [CH$_2$=C(OLi)CMe$_3$]$_2$ have μ_2-enolate moieties.[549] The N,N'-dimethyl-N,N'-(1,3-propanediyl)urea

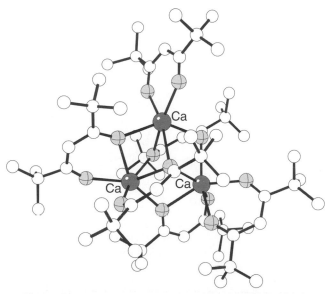

Figure 66 The structure of the trimeric Ca(TMHD)$_2$.

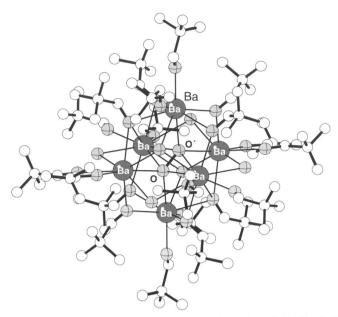

Figure 67   The structure of the peroxo complex [Ba$_6$(THD)$_{10}$(H$_2$O)$_4$(OH)$_2$(O$_2$)][HNEt$_3$]$_2$.

solvate of lithium 1,3-(Bu^t)$_2$-1,3-butadienolate also displays a central four-membered (Li—O)$_2$ ring, and curiously, the Li—C$=$O– angle is bent (153, 156°), and not linear as theoretical considerations would predict.[550] The magnesium complex ((Z)-BrMgOC(CMe$_3$)$=$CH-Me)$_2$·(Et$_2$O)$_2$ is likewise dimeric, and its bridging enolate O atoms form slightly puckered (Mg—O)$_2$ rings.[551]

The Hauser base reagents Pr$^i_2$NMgCl and Pr$^i_2$NMgBr react with a variety of enolizable ketones to yield magnesium enolates. They cannot be isolated in the presence of THF, but when diethyl ether is used as a solvent and in the presence of HMPA, the halomagnesium enolate compounds {(Bu^t)C($=$CH$_2$)OMgBr·HMPA}$_2$ and Me$_2$CHC($=$CMe$_2$)OMgBr·HMPA can be identified. The former is a dimer, with enolate bridges, whereas the latter is a simple monomer. A computational study (HF/6-31G(d)) indicated that the enolate anion [H(CH$_2$=)CO]$^-$ will bridge in preference to the halides F$^-$, Cl$^-$, and Br$^-$, and that the amido anions Me$_2$N$^-$, (H$_3$Si)$_2$N$^-$, and (Me$_3$Si)$_2$N$^-$ are favored over the chloride anion in three-coordinate dimer systems. The presence of solvents may switch the bridging preferences, however.[552]

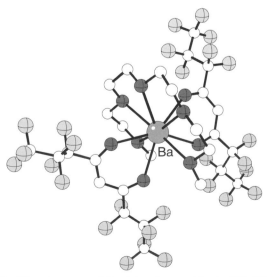

Figure 68   The structure of the tetraglyme adduct of Ba(TDFND)$_2$.

Mixed enolate/amides have been found with the same core structure as pure enolates; e.g., MeC(O)CMe$_3$ reacts with LiN(SiMe$_3$)$_2$ and DME to afford Li$_2$(CH$_2$C(O)CMe$_3$)[N(Si-Me$_3$)$_2$]·2DME, and Me$_2$CO$_2$CMe$_3$ reacts with NaN(SiMe$_3$)$_2$ and TMEDA to yield Na$_2$-(MeCH$_2$CO$_2$CMe$_3$)[N(SiMe$_3$)$_2$]·2TMEDA. The two compounds, prepared as models in stereoselectively enhanced enolate reactions, have slightly puckered rings and predictably longer Li—N than Li—O bonds (2.07 Å and 1.87 Å, respectively, in the lithium complex).[553]

Aggregates consisting of lithium halides (LiBr, LiI) with lithium amides or enolates (i.e., LiOC(Pr^i)=CMe$_2$) have been identified as either heterodimers or -trimers (for amido species) or heterodimers only (for the enolate complexes).[554] An *ab initio* and semiempirical PM3 theoretical study of model systems shows that solvated heterodimers between LiBr and either LiNH$_2$ or LiOC(H)=CH$_2$ are favored over the respective homodimeric species, and that a stable eight-membered ring transition state exists for the enolization step between LiCl·LiNH$_2$ and acetaldehyde. The dissociation of donor solvents is computed to require more energy for heterodimers than for homodimers.

A triple anion complex containing enolate, amide, and halide functionalities can be isolated from the mixture of *n*-butyl bromide, hexamethyldisilazane, TMEDA, Bu^{n}Li and pinacolone (Bu^{t}COMe). The resulting solution of LiBr, LiN(SiMe$_3$)$_2$, LiOC(Bu^t)=CH$_2$, and TMEDA produces crystals of Li$_4$(μ_4-Br)(μ-OC(Bu^t)=CH$_2$)$_2$(μ-N(SiMe$_3$)$_2$)(TMEDA)$_2$, which, instead of forming a ladder-type structure, consists of a planar butterfly of four lithium atoms bonded to a μ_4-Br; the stability of this arrangement has been studied with semi-empirical (PM3) and *ab initio* HF/LANL2DZ computations.[555]

The most common aggregate above the dimers are the cubes, which are known for lithium, sodium and magnesium. Condensation of lithium pinacolate and pivalaldehyde produces an aldolate that in the presence of pyridine leads to the isolation of tetrameric 4:3 and 4:4 enolate-pyridine complexes; these are constructed around Li$_4$O$_4$ cubes (see Figure 69).[556]

The reaction of tetramethyl-1,3-cyclobutanedione with R$_3$SiLi (R = Me$_3$Si or Et) and Et$_3$GeLi results in the opening of the cyclobutanedione ring to give the corresponding β-ketoacylsilane lithium enolates, which after aqueous workup afford the β-ketoacylsilanes Me$_2$CHCOCMe$_2$COR. The lithium enolate itself (R = SiMe$_3$) is constructed around a Li$_4$O$_4$ cube, with chelating enolate anions (see Figure 70). *Ab initio* calculations (HF/6-31G(d)) were used to demonstrate that Li$^+$ complexation in the enolate weakens the hyperconjugative interactions between the O lone pair (n$_o$) and the σ–C–Si orbital, and is responsible for the two new transitions observed in the UV/vis spectra; one is red-shifted, and the other blue-shifted relative to the absorptions of the corresponding β-ketoacylsilanes.[557]

Cocrystallization of either LiOCMe$_3$ or KOCMe$_3$ with preformed potassium or lithium enolate derived from MeCOCMe$_3$ in the presence of THF yields a novel aggregate composed of four

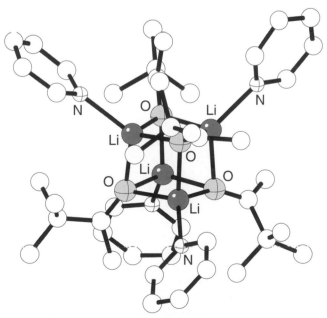

Figure 69 The structure of tetrakis((μ_3-*tert*-butylethenolato-*O,O,O*)-pyridine-lithium).

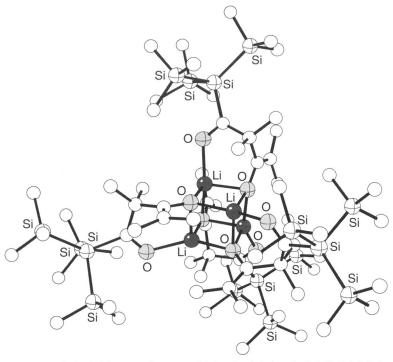

Figure 70   The structure of the lithium enolate tetrakis(μ_3-1-tris(trimethylsilyl)silyl-2,2,4-trimethylpentane-1,3-dionato-O,O')-tetra-lithium.

enolate residues, four *t*-butoxides, four Li$^+$, five K$^+$, a KOH residue and five THF molecules (see Figure 71). The polymetallic compound is based on a square-based pyramid of potassium ions, with each edge bridged by an O atom. Triply bridging lithium ions span the oxygens on the sides of the pyramid.[558]

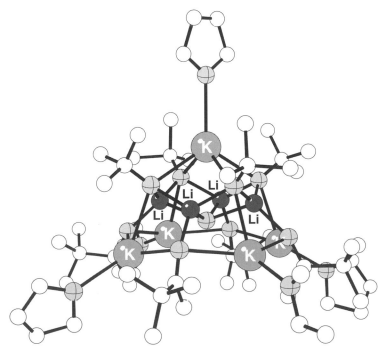

Figure 71   The structure of tetrakis((μ_4-*tert*-butyloxo-O,O,O,O)-(μ_3-1-methyleneneopentyloxo-O,O,O))-(μ_4-hydroxo)-pentakis(tetrahydrofuran)-tetra-lithium-penta-potassium.

The heterometallic aggregate [Li$_2$Na$_4${OC(=CH$_2$)(Bu^t)}$_6$((Pr^i)$_2$NH)$_2$] has been prepared from sodium diisopropylamide, sodium diisopropylamide, and pinacolone, and has been structurally authenticated by X-ray crystallography. It possesses a distorted triple-layered stack consisting of a (Na—O)$_2$ ring sandwiched between two Li—O—Na—O rings. Interestingly, two edges of the face-sharing cubes are absent; the Na–O separations are 2.88 Å, whereas other Na–O contacts in the molecule range from 2.29–2.42 Å (see Figure 72). The difference in size between Li and Na, combined with contacts between the H$_2$C=C moieties of the enolates and metals, are probably responsible for the open stack structure. The fact that coordinated diisopropylamine is found at each end of the cluster demonstrates that the free base itself (as distinct from the metallated amide) may influence the regioselectivities of incoming reactants.[559]

A complete thermochemical analysis has been described for the aldol reaction of lithiopinacolonate with pivalaldehyde in hexane at 25 °C and in cyclohexane at 25 °C and 6 °C.[560] Reactions were performed in the presence and absence of THF, TMEDA, and DME. The heats of reaction of pivalaldehyde with the hexameric lithiopinacolonate, the tetrameric and dimeric enolate-ligand complexes as well as heats of interaction of the hexameric enolate with the ligands were determined, and it was found that the tetrameric lithium aldolate product does not complex with any of the three ligands in hydrocarbon solution. Caution was raised about proposed mechanisms based on data not gathered under modern synthetic reaction conditions. The unsolvated hexameric enolate, (Me$_3$CCOCH$_2$Li)$_6$, has a classic drum shape and nearly S_6 symmetry (see Figure 73).[561,562]

The reaction products between Mg[N(SiMe$_3$)$_2$]$_2$ and 2,4,6-trimethylacetophenone in hexane solution have been determined with ^{1}H-NMR spectroscopic analysis of the solids precipitated from solution.[563] Only enolate and unenolized ketone are present in the solids, and the absence of any amide suggests formation of a magnesium bis(enolate). Formation of the latter (in preference to an amido(aldolate)) probably reflects a combination of steric crowding and electronic characteristics of the ketone that retard the addition reaction. The structure of the isolated bis(enolate), Mg$_4${OC(2,4,6-trimethylphenyl)=CH$_2$}$_8${O=C(2,4,6-trimethylphenyl)Me}$_2$(C$_6$H$_5$Me)$_2$, reveals four metals in a linear arrangement with six bridging and two terminal enolates and two terminal ketones; three orthogonal (Mg—O)$_2$ rings are thereby formed (see Figure 74).

(iv) Alkoxides and aryloxides

Alkali and alkaline-earth derivatives of phenol and the lower alcohols (methoxides, ethoxides) have been known for more than a century. Traditional applications for them include use as lubricants (e.g., lithium greases), polymerization catalysts, and surfactant stabilizers. Based on their solubility and volatility (both generally low), alkaline-earth alkoxides were presumed to be

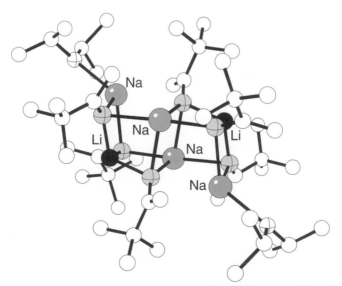

Figure 72 The structure of [Li$_2$Na$_4${OC(=CH$_2$)(Bu^t)}$_6$((Pr^i)$_2$NH)$_2$].

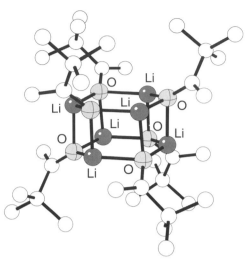

Figure 73   The structure of the unsolvated hexameric enolate $(Me_3CCOCH_2Li)_6$.

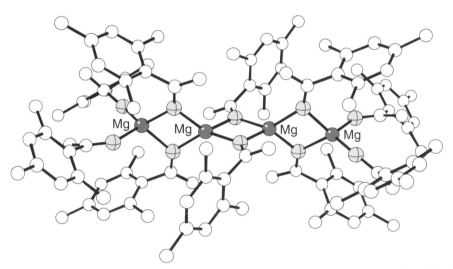

Figure 74   The structure of $Mg_4\{OC(2,4,6\text{-trimethylphenyl})=CH_2\}_8\{O=C(2,4,6\text{-trimethylphenyl})Me\}_2$ $(C_6H_5Me)_2$.

oligomeric or polymeric substances, although until the last quarter of the twentieth century comparatively little was known of their structural chemistry.[564]

The interest expressed in ion transport across biological membranes in the 1970s[565,566] and early 1980s, and the discovery of superconducting oxides in the mid-1980s[567] initiated intensive study of *s*-block alkoxides, especially in their role as precursors to metal oxides and halides.[12] The area has expanded considerably since the 1980s, and extensive reviews of various subjects in alkoxide chemistry, especially the alkaline-earth derivatives, are now available.[13,564,568–570]

Standard approaches to metal alkoxide synthesis (see Table 8) involve the direct reaction of the bulk metal in the neat alcohol, sometimes under reflux conditions (Table 8, Equation (1)). In practice, this generally works well for Group 1 species, but side reactions stemming from the heterogeneous nature of the system often complicate the reactions with the Group 2 metals; oxo-alkoxide complexes are often formed from the latter.[571–574] Degradation of the alkoxide ligand may be involved, but in the reaction of *t*-butanol with barium metal, for example, a diolato ligand is found coordinated to the metal (Equation(4)):

$$Ba + Bu^tOH \rightarrow H_3Ba_6(O)(OCMe_3)_{11}(OCEt_2CH_2O)(THF)) \tag{4}$$

Table 8 Synthetic methods for the preparation of *s*-block alkoxide and aryloxide complexes.

Reaction	No.
$M + n \text{ HOR} \rightarrow M(OR)_n(HOR)_x + n/2 \text{ H}_2$	(1)
$M + n \text{ HOR} \rightarrow M(OR)_n(S)_x + n/2 \text{ H}_2$	(2)
$M + n \text{ HOR} + NH_3 \rightarrow M(OR)_n(NH_3)_x + n/2 \text{ H}_2$	(3)
$NH_{3(g)} \rightarrow NH_{3(sat)}$	(4a)
$M + m \text{ NH}_{3(sat)} \rightarrow M(NH_3)_m(S)_x$	(4b)
$M(NH_3)_m(S)_x + n \text{ HOR} \rightarrow M(OR)_n(S)_x + n/2 \text{ H}_2 + m \text{ NH}_3$	(4c)
$n \text{ M}^{I}OR + M^{II}X_2 \rightarrow M^{II}(OR)_nX_{2-n} + n \text{ M}^{I}X$	(5)
$MH_n + n \text{ HOR} \rightarrow M(OR)_n(S)_x + n \text{ H}_2$	(6)
$M(NR_2)_n + n \text{ HOR}' \rightarrow M(OR')_n(S)_x + n \text{ NHR}_2$	(7)
$(OBu^t)_2(Sr,Ba) + 2 \text{ Sn}(OBu^t)2 \rightarrow Sn(\mu\text{-}OBu^t)_3(Sr,Ba)(\mu\text{-}OBu^t)_3Sn$	(8a)
$Ba(OCH_2CH_2OMe)_2 + 4 \text{ Cu}(THD)(OCH_2CH_2OMe) \rightarrow BaCu_4(OCH_2CH_2OMe)_6(THD)_4$	(8b)
$M^{II} + 2Pr^iOH + 2M(OPr^i)_3 \rightarrow M^{II}\{M(OPr^i)_4\}_2 + H_2$	(9a)
$Ba + 4 \text{ Zr}(OPr^i)_4 \cdot Pr^iOH \rightarrow Ba\{Zr_2(OPr^i)_9\}_2 + 2HOP_r^i + H_2$	(9b)

M^I = alkali metal; M^{II} = alkaline-earth metal; X = halide; S = coordinated solvent.

Since the diolate is not formed in toluene, the THF solvent is its most likely source.[573] Metal-coordinated alcohol molecules frequently accompany the isolated alkoxides, and the rate and possible the yield of reactions are dependent on the cleanliness of the surface of the metal (the use of activated metals can be helpful),[575,576] the acidity of the alcohol, and the solubility of the resulting complex. A variation on this method allows the metal to react with the alcohol in polar coordinating solvents, such as ethers (Table 8, Equation (2)); oligomerization of the resultant complex, depending on how tightly the solvent is held, can complicate the isolation of binary alkoxides. Solvent-free metal vapor synthesis has been used to form calcium and barium aryloxides.[251]

Syntheses that exploit the solubility of the alkaline-earth metals in liquid ammonia have proven practical for alkoxide work, as they generate high yields, reaction rates, and purity (Table 8, Equation (3)). In a refinement of this approach, Caulton and co-workers have used dissolved ammonia in an ethereal solvent, usually THF, to effect the production of a number of alkoxides of barium,[573,577] and this method has also been examined with calcium and strontium (Table 8, Equations (4a) to (4c)).[578] Displacement reactions using alkali metal alkoxides and alkaline-earth dihalides (Table 8, Equation (5)),[579–581] and between alkaline-earth hydrides or amides and alcohols (Table 8, Equations (6) and (7)),[251] have been examined, but alkali-metal halide impurities, incomplete reactions, and unexpected equilibria and byproducts can affect the usefulness of these approaches.

Heterometallic alkoxides of calcium, strontium, and barium with transition or posttransition metals have been formed by reactions with the preformed alkoxides (e.g., Table 8, Equations (8a) and (8b)).[582,583] A variation on this approach generates the Group 2 alkoxides *in situ* by reaction of the metal with an alcohol and/or another alkoxide (e.g., Table 8, Equations (9a) and (9b)).[564,584]

Alkoxide and aryloxide complexes are moisture-sensitive, and are usually colorless or white solids with melting points substantially above room temperature. An interesting exception to this generalization is provided by the brown barium alkoxides $Ba[O(CH_2CH_2O)_nCH_3]_2$ ($n = 2$, 3), formed from the reaction of elemental barium with a stoichiometric quantity of the polyether alcohol in THF.[585] They are soluble in diethylether, THF and aromatic hydrocarbons, and are liquid at room temperature.

The number of structurally characterized homo- and heterometallic alkoxides and aryloxides of the *s*-block elements is now in the hundreds, and the previously cited reviews should be consulted for extensive listings.[13,564,568–570] A large amount of structural diversity exists in the compounds, and monomers up to nonametallic clusters and polymeric species are represented. In this regard, the large molecular aggregate $Ca_9(OCH_2CH_2OMe)_{18}(HOCH_2CH_2OMe)_2$ is particularly interesting. Isolated from the reaction of calcium metal with methoxyethanol,[586] it represents a transition between the polymeric Group 2 alkoxides of the lower alcohols (e.g., $Ca(OMe)_2]_x$, $Ba(OEt)_2]_x$) and the mono- or dinuclear complexes formed with bulkier alcohols. The central $Ca_9(\mu_3\text{-}O)_8(\mu_2\text{-}O)_8O_{20}$ section displays three six-coordinate metals and six seven-coordinate metals that can be viewed as filling octahedral holes in two close-packed oxygen layers (average Ca—(μ_3-O) = 2.39 Å; Ca—(μ_2-O) = 2.29 Å; Ca—(O_{ether}) = 2.60 Å) (see Figure 75). As such, the Ca—O substructure mimics part of the CdI_2 lattice. The particular size of the aggregate has been suggested as representing thermodynamic compromise between maximizing the coordination number of the calcium atoms and the number of independent particles.

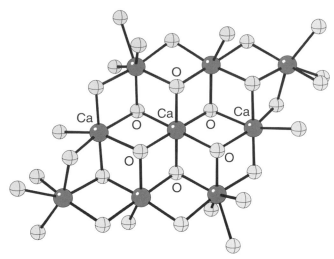

Figure 75   The structure of the calcium–oxygen core of $Ca_9(OCH_2CH_2OMe)_{18}(HOCH_2CH_2OMe)_2$.

The variety of M—O bonding modes available (e.g., μ_1, μ_2, μ_3) and the potential contribution from M—O π-bonding in s-block alkoxides and aryloxides have made quantitative and even qualitative predictions of structures problematic. The difficulties encountered when analyzing the bonding with terminal oxygen-based ligands are compounded with bridging ligands, and there are few generalizations available for accurately predicting their structural details.[587] Nevertheless, some trends have become clearer as the number of crystallographically characterized alkoxides and mono(alkoxides) of low nuclearity has increased. The ability of the larger metals to accommodate more ligands in their coordination spheres and hence to form more extensive aggregates is counterbalanced by their lower Lewis acidity. It is not axiomatic, for example, that barium compounds will have higher coordination numbers than their calcium analogs.[581] This inherent electronic effect can be reinforced by the presence of sterically demanding ligands.

An illustration of these principles is provided by the Group 2 aryloxide complexes. The calcium and barium derivatives of 2,6-di-t-butyl-4-methylphenol (BHT) were the first monomeric alkoxides of the heavier alkaline-earth metals to be structurally characterized;[251,581] the strontium derivative has also been crystallographically examined.[576] The dominating effect of the extremely bulky BHT ligands is evident from the fact that the three aryloxides are isostructural monomers, even though there is a substantial change in metal radii from 1.00 Å (Ca^{2+}) to 1.35 Å (Ba^{2+}). The complexes display distorted trigonal bipyramidal geometries, with two of the THF ligands lying on the axes in a nearly linear arrangement (O—M—O' = 177–179°). The remaining THF and the two aryloxide ligands lie in the equatorial plane (see Figure 76). Although metal radii have usually been thought to play a key role in determining metal–ligand geometries, it appears that the ligand charge (and the operation of either "primary" and "dative" metal–ligand bonding)[588] may be just as critical. The aryloxide complexes display a nonadditive relationship between metal radii and metal–alkoxide distances; the increase in M—OR distance from calcium to barium (2.20 Å to 2.40 Å; $\Delta = 0.20$ Å) is considerably smaller than the increase in metal radii (0.93 Å to 1.30 Å; $\Delta = 0.37$ Å). The M—THF distances, however, do vary approximately as the metal radii (2.40 Å to 2.73 Å; $\Delta = 0.33$ Å). The packing of ligands around the metal may also serve to control coordination geometries.[580]

Ba(BHT)$_2$(THF)$_3$ reacts readily with BaI$_2$ in THF to produce the dimeric mono(alkoxide) complex [IBa(BHT)(THF)$_3$]$_2$ (see Figure 77).[580] The coordination geometry around the barium atoms is distorted octahedral, with the two bridging iodides, the BHT, and a THF ligand in one plane, and two additional THF molecules lying above and below the plane (Ba—I (I') = 3.44 (3.59) Å and Ba—OAr = 2.41 Å). The compound could be viewed as having been formed by the fusion of two coordinately unsaturated "IBa(BHT)(THF)$_3$" fragments, but the fact that a stable tetrasolvate (i.e., IBa(BHT)(THF)$_4$) has not been isolated is telling, in that the monomeric mono(aryloxide) ICaBHT(THF)$_4$ can be synthesized, despite being constructed around a smaller metal center.[580]

The flexibility of the s-block coordination sphere is particularly evident with heterometallic clusters. This is strikingly evident in the mixed Li-heavier alkali metal t-butoxides [(Bu^{t}O)$_8$Li$_4$M$_4$] (M = Na, K, Rb, Cs), which form a structurally authenticated homologous series.[589,590] They are

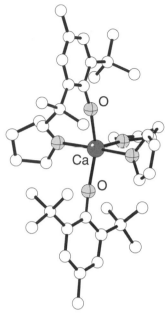

Figure 76 The structure of Ca(BHT)$_2$(THF)$_3$.

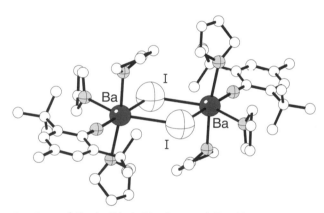

Figure 77 The structure of the iodide-bridged mono(alkoxide) dimer, [IBa(BHT)(THF)$_3$]$_2$.

constructed around (M$^+$)$_4$ planes, capped with chelating (O$_4$Li$_2$)$^{2-}$ dianions, and contain a "breastplate structure" that involves μ_3-Li, μ_4-M, μ_3-O, and μ_4-O atoms (see Figure 78). *Ab initio* calculations indicate that the assembly of the triple ion sandwich is exothermic. The flexibility of this "breastplate" framework has been suggested to be an important, but underappreciated, structural motif in heterometallic cluster chemistry. In other heterometallic alkoxides, where closed structure forms are not present, it can be difficult to describe the coordination geometry in terms of regular polyhedra.[568] For example, all the metal atoms in the alkane-soluble [BaZr$_2$(OPr^i)$_{10}$]$_2$ are six-coordinate; the Zr atoms can be viewed as occupying the centers of two face-sharing octahedra, with the Ba atoms connected by one [μ_2-OPr^i]$^-$ and two [μ_3- OPr^i]$^-$ ligands from each Zr (see Figure 79). The barium atoms have geometries only loosely related to octahedra, as the Ba—O distances range from 2.55 Å to 2.90 Å, with "trans" O—Ba—O′ angles varying from 114.6° to 167.6°.

(v) Other oxygen donor ligands

Calixarene-based M$_5$ [calix[4]arene sulfonates]·xH$_2$O (M = Na (x = 12); K (x = 8); Rb (x = 5); Cs (x = 4)) have been used to construct supramolecular assemblies. They have been structurally characterized, and exist as bilayers of anionic truncated pyramids in the "cone" configuration;

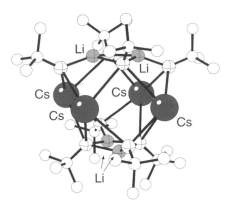

Figure 78   The structure of $(Bu^tO)_8Li_4Cs_4$.

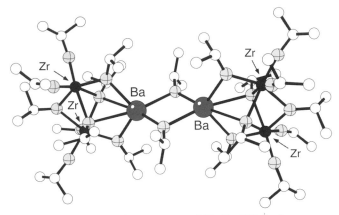

Figure 79   The structure of $[BaZr_2(OPr^i)_{10}]_2$.

their structures have been compared to those for clay minerals.[591] On the addition of pyridine *N*-oxide and lanthanide ions, the calixarenes assemble in a parallel alignment with spherical and helical tubular structures. Alkali ions (e.g., Na^+) assist in stabilizing the tubular assemblies (formed with various M^{3+} lanthanide ions) by coordinating to the sulfonate groups of calixarenes in adjacent turns of the helix.[592] Many examples of the self-assembly in aqueous solutions of bowl-shaped sodium *p*-sulfonatocalix[4,5]arenes are now known, and extensive reviews are available.[593,594] The complexes have found uses in selective isolation of Keggin ions,[595] chiral recognition,[596] and fullerene selectivity.[597]

A water-soluble sulfonated crown ether (see Figure 80) has been prepared and used as an ion size selection reagent. In the synergistic extraction of alkaline earth ions with 4-benzoyl-3-methyl-1-phenyl-5-pyrazolone and trioctylphosphine oxide in cyclohexane, addition of the sulfonated crown ether shifted the extraction for the larger ions to a higher pH level, thereby improving the separation.[598]

The first structurally authenticated molecular peroxide of an *s*-block element was reported in the form of the dodecameric lithium *t*-butyl peroxide, $\{Li[\eta^2\text{-}O_2(Bu^t)]\}_{12}$.[599] Isolated from the reaction of Bu^tLi and molecular oxygen, the lithium ion bridges the two peroxide oxygens, lengthening the O—O bond to 1.48 Å (see Figure 81). Quantum chemical calculations on the reaction between MeLi with LiOOH to give MeOLi and LiOH were used as a model for the formation of the compound, which can be viewed as an intermediate in alkoxide generation. The oxygen-scavenging properties of alkali metal-containing organometallic compounds, which can result in encapsulated oxide or peroxide ions, have been reviewed.[600]

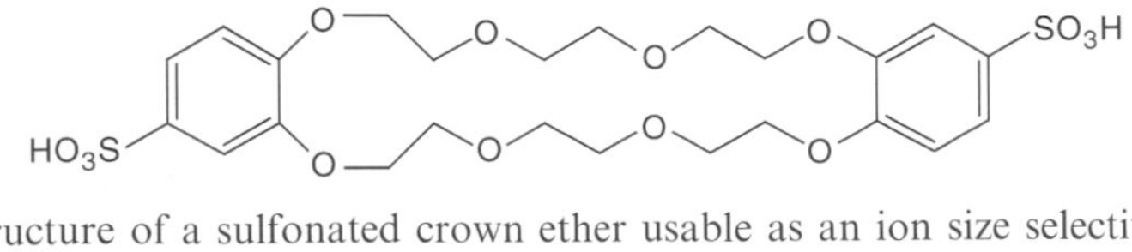

Figure 80   The structure of a sulfonated crown ether usable as an ion size selective masking reagent.

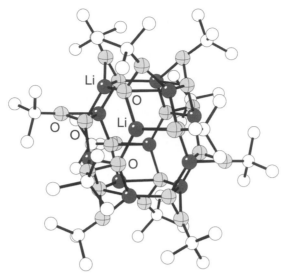

Figure 81 The structure of the lithium tert-butyl peroxide, $\{Li[\eta^2\text{-}O_2(Bu^t)]\}_{12}$.

(vi) Chemical vapor deposition

The technique of CVD; sometimes abbreviated as MOCVD (metalorganic chemical vapor deposition) has been under intensive development for the *s*-block elements, and particularly the alkaline-earth metals, since the late 1980s.[13,601–604] The production of complex oxides of calcium, strontium, and barium, such as the perovskite-based titanates $(Sr,Ba)TiO_3$ and superconducting cuprates (e.g., $YBa_2Cu_3O_{7-x}$),[9,605] has been a focus of much of this research. In addition, cerium-doped alkaline-earth sulfides (e.g., $(Ca,Sr,Ba)S$, $CaGa_2S_4$)[606] are of interest as phosphors for electroluminescent devices, as are sodium and potassium dopants for color modification.[607] Volatile sources of magnesium are required for CVD-doping of the Group 13 nitrides that serve as the basis for blue and green light-emitting and laser diodes.[608–611]

Owing to the low volatility or unfavorable deposition properties of alkaline-earth compounds such as the simple metal alkoxides and acac (acetylacetonate) derivatives, they are usually unsuitable as precursors to electronic materials. The approaches used to obtain reagents more useful as precursors for CVD work have generally focused on the reduction of intermolecular forces by the use of sterically bulky ligands (e.g., $[THD]^-$, $[OCH(CMe_3)_2]^-$),[531,533,612–614] fluorinated derivatives (e.g., $[OC_4F_9]^-$),[615,616] ligands with internally chelating groups (e.g., $O(CH_2CH_2OCH_3)^-$) or "lariats" (e.g., $(RCOCHC(NR')Me)_2$ (R = *t*-Bu; R' = $(CH_2CH_2O)_2Me$)),[548,617] polyethers (e.g., $Ba(H\text{-}FA)_2{\cdot}CH_3O$ $(CH_2CH_2O)_6$-*n*-C_4H_9 (see Figure 82)),[618] and the addition of Lewis bases (e.g., OR_2, NH_3) to alkoxides to form adducts.[619] Coordination compounds such as amides,[258] amidinates,[620] and pyrazolates[423] have been proposed as alternatives to the use of the organometallic reagents $(Cp_2Mg, (MeCp)_2Mg)$ that dominate magnesium CVD. Some of the currently used compounds in *s*-block CVD are described in more detail in appropriate sections elsewhere in this Chapter. Extensive review articles on alkaline-earth CVD are also available.[12,14,621]

The demands for high purity reagents and consistent gas-phase behavior are especially critical in *s*-block chemistry, and consistent pictures of CVD reactivity are not always easy to obtain. Problems with residues in deposited materials are common. For example, fluorinated compounds are favored for their increased volatility, but their use can lead to the deposition of metal fluorides that contaminate the deposited oxides.[622] Films of the high-*T*c superconductor $(TlO)_mBa_2Ca_{n-1}Cu_nO_{2n+2}$ ($m = 1, 2; n = 1, 2, 3$) generated under MOCVD conditions from $Ba(H\text{-}FA)_2$(tetraglyme), Ca(dipivaloylmethanate)$_2$, and $Cu(acac)_2$, for example, are contaminated with both BaF_2 and CaF_2; extra processing is required to remove the fluorides.[9] Handling problems also occur with fluorinated compounds; for example, partial hydration of $Ca(HFA)_2$ samples with attendant lowering of their vapor pressure is difficult to avoid.[623]

Owing to the sensitivity of alkaline-earth alkoxides and their derivatives to air and moisture, their application to systems of practical interest is not always straightforward. Early reports on the thermal behavior of the widely-used barium oxide precursor "$Ba(THD)_2$," for example, indicated that ∼25–40% of the material remains unsublimed under oxide forming conditions.[624,625] Such reports are now thought to reflect the use of partially hydrolyzed and/or

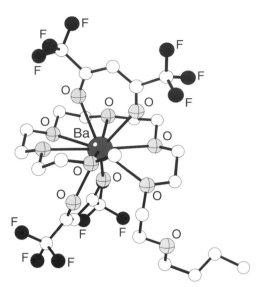

Figure 82   The structure of Ba(HFA)$_2$·CH$_3$O(CH$_2$CH$_2$O)$_6$-n-C$_4$H$_9$.

adducted material, which would reduce volatilities, as later thermal gravimetric analysis indicated that only a 5–6% residue remains after heating to 410 °C.[531]

The solid state structure of "Ba(THD)$_2$" has also been the subject of controversy. An X-ray crystal structure of the commercially available "anhydrous Ba(THD)$_2$" was found to be a partially hydrolyzed pentabarium aggregate, described as Ba$_5$(THD)$_9$(OH)(H$_2$O)$_3$.[626] Re-analysis of the structural parameters has suggested that it be formulated as the even more degraded species Ba$_5$(THD)$_5$(HTHD)$_4$(O)(OH)$_3$.[627] Rigorously anhydrous Ba(THD)$_2$ has been the subject of several structural investigations,[531,628,629] the most accurate of which appears to be the low temperature study of Drake and co-workers.[531] The complex crystallizes as a centrosymmetric tetramer, with the seven-coordinate barium atoms extensively bridged by THD ligands. Two of the barium atoms (Ba(1) and Ba(1)′) are coordinated by one terminal THD ligand; four other coordination modes of the remaining β-diketonates are also observed in the structure (see Figure 83). Some Ba—O bonding interactions are long (up to 3.14 Å) and weak, which may account for the ease with which the tetramer is disrupted by Lewis bases to yield dimeric species such as [Ba$_2$(THD)$_4$L$_2$]$_2$ (L = NH$_3$, Et$_2$O).[619,630]

Considerable care is obviously needed in handling Group 2 alkoxides and related compounds reliably in CVD applications. In some cases, the use of mixtures of precursors with different ligands (e.g., β-diketonates, β-ketoesterates, alkoxides) can lead to ligand exchange reactions that improve stability toward moisture.[621] The development of aerosol-assisted CVD (AACVD) has meant that the volatility of precursors is no longer as critical, and wider varieties of precursors can be used.[631–633]

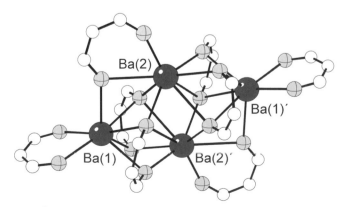

Figure 83   The structure of Ba$_4$(THD)$_8$; the t-butyl groups have been removed from the THD ligands for clarity.

3.1.4.4.2 Sulfur donor ligands

Driven in part by interest in potential materials applications,[634–637] the number of *s*-block complexes containing ligands with sulfur donors has increased tremendously since the 1980s. The structures of hundreds of such compounds are now known, even though the first crystal structure of a lithium thiolate did not appear until 1985.[638]

The *s*-block thiolates $M(SR)_n$ are generally synthesized by one of three routes: hydrogen elimination from a metal hydride (used for Na and K, Equation (5a));[639] alkane elimination from an alkyllithium (Equation (5b));[640] and metallation (used for Na–Cs, Equation (5c)).[641]

$$MH + HSR \rightarrow MSR + H_2 \tag{5a}$$

$$Bu^nLi + HSR \rightarrow LiSR + C_4H_{10} \tag{5b}$$

$$M + HSR \rightarrow MSR + 0.5H_2 \tag{5c}$$

Such compounds take a variety of forms including monomers, dimers, trimers, cubes, fused cubes, and large rings and polymers; predictably, the degree of aggregation is dictated to a large extent by the steric bulk of the thiolate and associated ligands. The area of *s*-block thiolates has received repeated comprehensive review,[642–646] and simple thiolates and most carboxythiolates,[527,647] including intramolecularly stabilized species, are not covered further here.

The reaction of [tris(3-*p*-tolylpyrazolyl)hydroborate)]MgMe with H_2S produces the monomeric hydrosulfido complex [tris(3-*p*-tolylpyrazolyl)hydroborate)]MgSH, which has been structurally authenticated; the Mg—S bond length is 2.35 Å.[648] Other monodentate sulfur ligands include 2-(1-methylethyl)-1,3-dimethyl-1,3,2-diazaphosphorinane 2-sulfide, whose lithium complex has been modeled with molecular orbital calculations,[649] 1,3-dimethyl-2-benzylide-2-thioxo-1,3,2-diazaphosphorinane-S,S),[650] and *N*-diisopropoxythiophosporylthiobenzamine.[651,652]

Polysulfide linkages have been incorporated into several *s*-block complexes, with the S_6^{2-} anion being especially common. Sulfur powder reacts with Bu^nLi, lithium metal or solid lithium hydride in toluene/TMEDA, or with $LiBH_4$ with THF to afford $Li_2S_6{\cdot}(TMEDA)_2$.[653] The same compound can also be prepared directly from the reaction of Li_2S_2 with TMEDA in toluene.[654] The compound contains a central Li_2S_2 ring (Li—S = 2.49 Å) (see Figure 84), and in donor solvents, the S_6^{2-} residue cleaves to give the blue $S_3^{\cdot-}$ radical anion. A tetraethylethylenediamine (TEEDA) analogue to the TMEDA complex can be prepared by a Li_2S_2/TEEDA/toluene combination, but if the triply coordinating PMDETA is substituted instead, a bridging tetrasulfido unit is formed that has a zigzag chain structure.[655] The sodium PMDETA counterpart, formed from sodium hydride and sulfur in toluene/PMDETA, again contains an S_6^{2-} residue that is bound in a manner similar to that found in the Li/TMEDA and Li/TEEDA aggregates (Na—S = 2.82, 2.91 Å).[656]

The hexasulfido moiety can be found in a chain form even with a large metal if crown ethers are present as supporting ligands. Thus reaction of 18-crown-6 and K_2S_5 in acetonitrile leads to $K(18\text{-crown-6})]_2S_6{\cdot}2MeCN$, from which the acetonitrile is easily lost. The S_6^{2-} anion is suspended as a *transoid* chain between two crown ether-coordinated potassium atoms (K—S = 3.08 Å).[657] If an even larger metal is used, however, the hexasulfido anion reverts to its ring binding mode, e.g., a

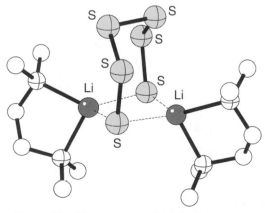

Figure 84 The structure of $Li_2S_6{\cdot}(TMEDA)_2$.

multidecker stack involving crown ethers and polysulfide anions can be isolated from the reaction of dibenzo-18-crown-6, Cs_2CO_3, and sulfur in H_2S-saturated acetonitrile.[658] Molecules of Cs_4 (dibenzo-18-crown-6)$_3$(S_6)$_2$·2MeCN are built of stacks of three crown ether molecules and two hexasulfide chains with the cations located between them (Cs—S = 3.46–3.55 Å) (see Figure 85). The conformation of the hexasulfide chains is all-*cis*.[658]

Dithiocarboxylic acids and their trimethylsilyl esters readily react with potassium, rubidium, and caesium fluorides to give MS_2CR complexes (M = K, Rb, Cs; R = Me, Et, Pr^i, cyclohexyl, Ph, 2- and 4-MeC_6H_4, 4-MeO and 4-ClC_6H_4, 2,4,6-$Me_3C_6H_2$) in moderate to good yields. The ammonium derivatives Me_4NS_2CR can be prepared by the reaction of NaS_2CR with Me_4NCl. The structures of KS_2CEt, $RbS_2SC_6H_4Me$-4, and $CsS_2CC_6H_4Me$-4 have been characterized crystallographically. They have a dimeric structure, (RCSSM)$_2$, in which the two dithiocarboxylate groups are chelated to the metal cations that are located on the upper and lower sides of the plane involving the two opposing dithiocarboxylate groups (e.g., Figure 86; Rb—S = 3.47 Å (av)). The K$^+$ ions display cation-π interactions with the tolyl fragment of a neighboring molecule, whereas the Rb$^+$ and Cs$^+$ cations interact with two neighboring tolyl fragments. The sodium salt was found to be a monomer, with η^1-bound ligands. (Na—S = 1.81 Å; S—Na—S = 116.9°).[526]

A variety of heterocycles will react to form complexes with M–S interactions, including 2-sulfanylbenzothiazole,[659] 1-methyl-1H-tetrazole-5-thiol,[660] benzoxazole-2(1H)-thione,[661] 5-(1-naphthylamino)-1,2,3,4-thiatriazole,[662] 2-mercaptobenzoxazole,[663] and 5-mercapto-1-naphthyltetrazole).[662] Some of these ligands undergo alkali–promoted rearrangement on complexation, such as the transformation of 5-amino-substituted thiatriazoles into 5-thio-substituted tetrazoles. For example, when solid Ba(OH)$_2$ suspended in toluene containing HMPA reacts with 5-(1-naphthylamino)-1,2,3,4-thiatriazole, the monomeric Ba[5-mercapto-1-naphthyltetrazole-N,S]$_2$·3HMPA complex is formed.[662]

Complexes derived from 2-mercaptopyrimidine, 2-mercaptothiazoline, and 2-mercaptobenzimidazole have been described.[664] The latter is a dimer in which each lithium is chelated by an N—C—S unit of the organic dianion; the two end lithium atoms of the dimer are each coordinated to two terminal HMPA molecules, whereas the two central lithium atoms are linked by two μ-HMPA molecules (see Figure 87). The reasons for the difference between this structure and those displayed by the other complexes (which in the case of the 2-mercaptothiazoline complex contains direct S—Li bonding) have been examined with *ab initio* calculations. The option of generating a strong C=N bond in the 2-mercaptothiazoline complex rather than a weaker C=S bond apparently drives the lithium–sulfur interaction.

Thiocarbamates and dithiocarbonates, which have attracted interest as possible sources of metal sulfides, are known with a variety of s-block metals, including lithium,[665] calcium,[392,666] strontium,[666] and barium.[666] Dithiocarbonates have been formed by CS_2 insertion into metal alkoxide bonds.[666] Magnesium-isothiocyanate and -carbodiimide insertion products, e.g., Mg(SCPhN(Bu^t))$_2$(THF)$_2$, Mg(SCPhNPh)$_2$(THF)$_2$, Mg(Pr^{i}NCRN(Pr^i))$_2$(THF)$_2$ (R = Ph, Et or Pr^i) and Mg(Bu^{t}NCEtN(Bu^t))$_2$(THF)$_2$ have been synthesized by the reaction between MgR$_2$ (R = Ph, Et or Pr^i) and various isothiocyanates and carbodiimides in THF solution. The reaction of Mg(SCPhN(Bu^t))$_2$(THF)$_2$ with an excess of PhNCO cyclotrimerizes the latter to afford (PhNCO)$_3$·THF.[667]

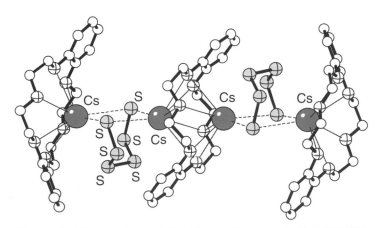

Figure 85   The structure of Cs$_4$(dibenzo-18-crown-6)$_3$(S$_6$)$_2$·2MeCN.

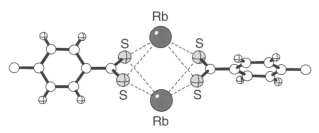

Figure 86 The structure of $(RbS_2C_6H_4Me-4)_2$.

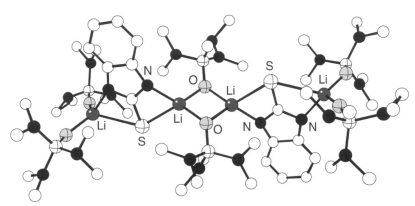

Figure 87 The structure of bis((μ_2-2-mercaptobenzimidazolinato-*N,N,S,S*)-(μ_2-hexamethylphosphoramido-*O,O*)-bis(hexamethylphosphoramide)-dilithium).

Larger rings with S–M contacts have been prepared by a variety of routes. The reaction of the sodium salt of $Ph_2P(S)NHP(S)Ph_2$ with either triglyme or tetraglyme affords the corresponding adducts $Na\{Ph_2P(S)NP(S)Ph_2\}$(glyme), which are monomeric species with all four oxygen atoms of the glyme moieties coordinated to the Na^+ cations. The sulfur atoms of the $[Ph_2P(S)NP(S)Ph_2]^-$ anion are bound in a symmetrical fashion to the sodium cations, forming six-membered S—P—N—P—S—Na rings (Na—S = 2.84–3.08 Å).[668] The reaction of 1,3-dimethyl-2-iminoimidazoline with KMe gives the corresponding potassium imidate, from which (1,3-dimethylimidazolin-2-imino)CS_2K is obtained in almost quantitative yield. Its crystal structure (see Figure 88) contains a framework in which rings of K_2 units bridged by the four sulfur atoms of two thiolate ligands are connected by N and S bridges.[669] The addition of MeLi or BuLi to alkyl isothiocyanates produces Li thioamidates $\{Li[RCS(NR')]\}_n$. When R = Bu and R' = Bu^t, the unsolvated hexamer $\{Li[BuCS(NBu^t)]\}_2$ is obtained. In contrast, the solvated derivatives $\{Li \cdot THF[MeCS(NBu^t)]\}_\infty$ and $\{Li \cdot 2THF[MeCS(NMe)]\}_\infty$ form single-strand polymers.[670]

More complex S, N, O interactions are found in the polymeric $(Cs(5,5$-dimethyl-4-oxoimidazolidine-2-thione)OH)$_\infty$, which was isolated in an attempt to prepare the Cs salt of the monoanion of 5,5-dimethyl-4-oxoimidazolidine-2-thione. The polymeric complex consists of layers of $(Cs(5,5$-dimethyl-4-oxoimidazolidine-2-thione)OH)$_\infty$ along the crystallographic [010] plane. Each Cs atom displays eight-fold coordination with four different thione molecules and three hydroxy molecules and 3 OH$^-$ groups as surrounding ligands (see Figure 89).[671]

Dropwise addition of Bu^{n}Li to a slight excess of the isothiocyanate Bu^{t}NCS in hexane yields hexagonal prisms of Li[CS(NBu^t)(Bu^n)]. X-ray crystallography reveals that the molecule is constructed around hexagonal Li_6S_6 aggregates, in which the Bu^{n}CN(Bu^t) bridges form a paddle-wheel arrangement with D_{3d} symmetry (see Figure 90).[672]

Ab initio calculations on mono- and di-lithiated derivatives of thiourea have been used to predict that Li atom(s) will bridge N and S centers, leading to lengthening of the C—bond and shortening of one or both of the C—bonds in thiourea. Structurally authenticated di-lithiated diphenylthiourea, [PhNLiC(=NPh)SLi·2HMPA]$_2$, contains monomeric units with S—Li and N(μ_2—Li)N bonds, these monomers then being linked by N: → Li coordination (see Figure 91).[673]

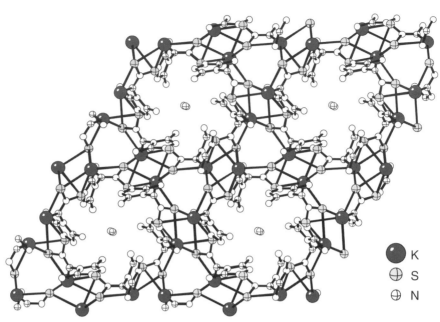

● K
⊕ S
⊕ N

Figure 88 Portion of the structure of dipotassium (μ_4-bis(1,3-dimethylimidazoline-2-dithiocarbiminato)). Acetonitrile of crystallization is in the center of the rings.

An investigation of the lithiation and CS_2-insertion reactions of (Ph)(pyridyl)CH$_2$ precursors (reactions used in the synthesis of ketene dithioacetals) lead to the monolithiated complexes (Ph)(pyridyl)CHCS$_2$Li·TMEDA (R$_1$ = Ph, R$_2$ = pyridyl) and (H)(2-methylpyrazine)CHC-S$_2$Li·TMEDA (R$_1$ = H, R$_2$ = 2-methylpyrazine). Interestingly, attempted second lithiation of the former complex fails to give the anticipated (Ph)(pyridyl)C=CS$_2$Li$_2$, but synthetic and ^{1}H-NMR spectroscopic evidence indicates that the 2-methylpyriazine complex can be lithiated further.[674]

Bis(pentamethylcyclopentadienyl)phosphine (C$_5$Me$_5$)$_2$PH reacts with S under basic conditions to give the corresponding dithiophosphinate salt Li[(C$_5$Me$_5$)$_2$PS$_2$] (see Figure 92), which is formed via the intermediate (C$_5$Me$_5$)$_2$P(S)H. The corresponding transition metal dithiophosphinate is formed on treatment with Co(II) chloride.[675]

Reaction of 2-mercapto-1-methylimidazole and 2-mercapto-1-mesitylimidazole with LiBH$_4$ in toluene produces the corresponding lithiated derivatives. The bidentate coordination of the sulfur atoms to lithium is augmented by donation from one of the B—H groups of the imidazole ligands. In the mesityl derivative, Li—B = 2.80, 3.09 Å (values for two independent molecules);

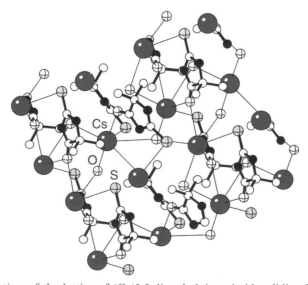

Figure 89 Portion of the lattice of (Cs(5,5-dimethyl-4-oxoimidazolidine-2-thione)OH)$_\infty$.

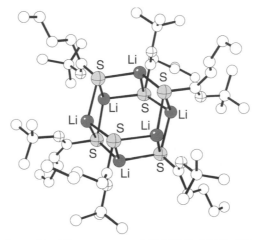

Figure 90 The structure of {Li[CS(Bu^t)N(Bu^n)]}$_6$.

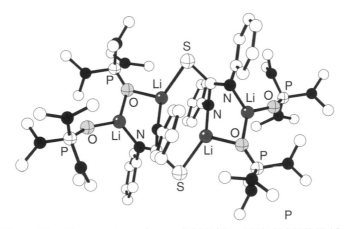

Figure 91 The structure of urea, [PhNLiC(=NPh)SLi·2HMPA]$_2$.

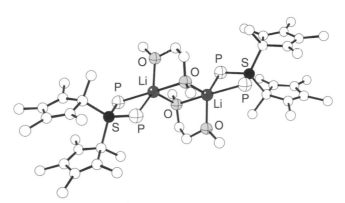

Figure 92 The structure of {(DME)Li[(C$_5$Me$_5$)$_2$PS$_2$]}$_2$.

Li–H distance = 1.86, 2.34 Å (see Figure 93). Similar coordination is observed in thallium and zinc derivatives.[676]

3.1.4.4.3 *Selenium and tellurium donor ligands*

Only after the publication of *CCC* (1987) were well-characterized coordination compounds of the *s*-block elements containing bonds to Se or Te described. As a rule selenium or tellurium donor

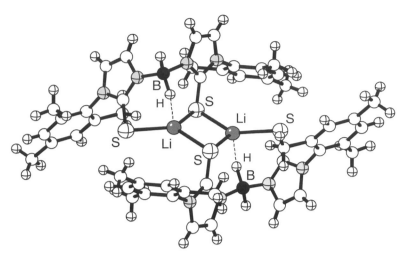

Figure 93 The structure of lithium bis((μ_2-bis(2-mercapto-1-mesitylimidazolyl)dihydrogenborate-H).

ligands possess large, sterically demanding substituents that confer kinetic stability on the complexes.

Direct insertion of elemental Se into the Li—C bond of $(THF)_3LiC(SiMe_3)_3$ in DME produces $(DME)LiSeC(SiMe_3)_3$.[677] The related centrosymmetric dimer $[(DME)LiSeSi(SiMe_3)_3]_2$ has also been structurally characterized (Li—Se = 2.57, 2.62 Å).[678] Interestingly, Te is displaced from $(THF)_2LiTeSi(SiMe_3)_3$ by Se in THF at $-55\,°C$ in a novel chalcogen metathesis reaction (Equation(6)):

$$(THF)_2LiTeSi(SiMe_3)_3 + Se \rightarrow (THF)_2LiSeSi(SiMe_3)_3 + Te \qquad (6)$$

Use of the analogous germanium $(THF)_2LiTeGe(SiMe_3)_3$ reagent gives only intractable materials when treated with Se.[677]

The 1:1 reaction of $[PhC\equiv CLi]_n$ with Se metal in THF/TMEDA gives the monomeric insertion product $PhC\equiv CSeLi·TMEDA·THF$.[679] Selenium also inserts into the Li—N bond of lithium 2,2,6,6-tetramethylpiperidide to form lithium 2,2,6,6-tetramethylpiperidinoselenolate. In the complex, one Li atom is coordinated tetrahedrally by two molecules of THF and two Se atoms, whereas the other Li atom exhibits an approximately rectangular-planar coordination by two (N, Se)-chelating groups. The Li atoms are bridged by two Se atoms, thus forming a planar Li_2Se_2 core (see Figure 94).[680] *Ab initio* Hartree–Fock calculations indicate that a hypothetical nonchelated

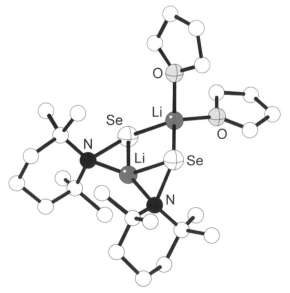

Figure 94 The structure of lithium 2,2,6,6-tetramethylpiperidinoselenolate.

dimer would be of distinctly higher energy than the observed form—apparently charge transfer from Se to N supports the (N, Se)-chelation.

The –Se(aryl) moiety is commonly incorporated into complexes; for example, 2,2'-bipyridine (bipy) forms coordination complexes with lithium benzeneselenolate and lithium pyridine-2-selenolate. The compounds can be recrystallized from THF, in which they are highly soluble. Dimeric {Li(bipy)(SePh)}$_2$ contains two lithium ions bridged by a pair of symmetrical benzene-selenolate ligands, with bidentate bipy ligands bound to each lithium ion (Li—Se = 2.55, 2.59 Å). In {Li(bipy)(NC$_5$H$_4$Se)}$_2$, each lithium ion is coordinated to a bidentate bipy ligand, one bridging selenium atom, and the nitrogen atom from the second bridging pyridine-2-selenolate ligand, thus forming an eight-membered [Li—Se—C—N–]$_2$ ring with a chair conformation (Li—Se = 2.62 Å).[681]

Monomeric (THF)$_3$LiSe(2,4,6-tri-*t*-butylphenyl) can be produced from the reaction of HSe(2,4,6-tri-*t*-butylphenyl) and Bu^{n}Li[682] or by the reduction of bis(2,4,6-tri-*t*-butylphenyl) diselenide (R$_2$Se$_2$) by LiBEt$_3$H;[683] in the complex, the lithium center is bound to the three THF molecules and the selenium in a pseudotetrahedral manner (Li—Se = 2.57 Å). The lithium selenolate reacts with Bu$^t_2$PCl, CH$_2$Cl$_2$, Me$_3$SiCl, Me$_3$SnCl, and Ph$_3$PAuCl to give Bu$^t_2$PSeR, RSeCH$_2$Cl (or (RSe)$_2$CH$_2$), Me$_3$SiSeR, Me$_3$SnSeR and Ph$_3$PAuSeR, respectively.[683]

Lithiation of HSe(2,4,6-(Me$_3$C)$_3$C$_6$H$_2$) with BuLi in the presence of one equivalent of THF produces [Li(THF)Se(2,4,6-(Me$_3$C)$_3$C$_6$H$_2$)]$_3$. The molecule has a Li$_3$Se$_3$ core, with pyramidal coordination at Se and almost planar Li coordination (sum of angles around Li = 355–360°).[684]

Direct reaction of potassium or rubidium metal with the sterically encumbering selenol HSeC$_6$H$_6$-2,6-Trip$_2$ (Trip = 2,4,6-Pr$^i_3$C$_6$H$_2^-$) stabilizes the dimeric selenates MSeC$_6$H$_3$-2,6-Trip$_2$ (M = K, Rb), The compounds, characterized with ^{1}H-, ^{13}C-, and ^{77}Se-NMR and IR spectroscopy, crystallize as toluene solvates with M$_2$Se$_2$ cores. Each potassium or rubidium interacts in a π-fashion with two *ortho*-aryl groups and also σ-bonds to the chalcogens. The π-interaction is retained even in the presence of donor solvents (Et$_2$O).[685]

Reaction of the sterically encumbered silylselanol HSeSi(SiMe$_3$)$_3$ with either Bu$_2$Mg or the bis(trimethylsilyl)amides of Ca, Sr, and Ba in the presence of Lewis bases yields crystalline alkaline-earth selenolates. The Mg selenolate has been crystallized as a tris((dimethylphosphino)-methyl)-*t*-butylsilane complex, whereas the Ca, Sr, and Ba complexes have been isolated as TMEDA adducts. The magnesium complex is constructed around a six-membered P—Mg—P—C—S—C–ring that is puckered in a chair conformation. The strontium complex has Sr—Se = 2.94 Å with a linear Se—Sr—Se' core.[686]

The reaction of MgBr$_2$ with two equivalents of the sterically demanding Li[Se(2,4,6-Bu$^t_3$C$_6$H$_2$)] in THF generates the mononuclear Mg{Se(2,4,6-Bu$^t_3$C$_6$H$_2$)}$_2$(THF)$_2$ (Mg—Se = 2.53 Å; Se—Mg—Se' = 122.2°) in good yield.[687] The treatment of SrI$_2$(THF)$_5$ with two equivalents of K[Se(2,4,6-Bu$^t_3$C$_6$H$_2$)] in THF produces Sr(Se(2,4,6-Bu$^t_3$C$_6$H$_2$))$_2$(THF)$_4$·2THF in good yield. The latter displays a distorted octahedral environment at the metal center (Sr—Se = 3.07 Å; Se—Sr—Se' = 171.9°).[688]

The Mg phenylselenolate complex [tris(3-*p*-tolylpyrazolyl)borate]MgSePh was synthesized by the reactions of [tris(3-*p*-tolylpyrazolyl)borate]MgMe with PhSeH and Ph$_2$Se$_2$. The solid state structure indicates that the magnesium is coordinated to three nitrogen atoms of the pyrazolate ligand and to the SePh ligand (Mg—Se = 2.50 Å).[689] The structurally characterized monomeric hydroselenido complex [tris(3-*p*-tolylpyrazolyl)borate]MgSeH was synthesized by the reaction of [tris(3-*p*-tolylpyrazolyl)borate]MgMe with H$_2$Se (Mg—SeH = 2.465(2) Å). The complex reacts with [tris(3-*p*-tolylpyrazolyl)borate]MgMe to give the dinuclear bridging selenido complex {[tris(3-*p*-tolylpyrazolyl)borate]Mg}$_2$Se; the Mg—Se—Mg' moiety is linear.[648]

The structurally authenticated compounds Ba{Se(2,4,6-tri-*t*-butylphenyl)}$_2$(THF)$_4$ (monomer, Ba—Se = 3.28 Å), [Ba(18-crown-6)(HMPA)$_2$][Se(2,4,6-tri-*t*-butylphenyl)]$_2$ (solvent separated ion triple), [Ba{Se(2,4,6-triisopropylphenyl)}$_2$(Py)$_3$(THF)]$_2$ (dimer, μ-Ba—Se = 3.30, 3.42 Å) and [Ba{Se(2,4,6-triisopropylphenyl)}$_2$(18-crown-6)] (monomer, Ba—Se = 3.23, 3.24 Å) have been prepared by reductive insertion of Ba (dissolved in NH$_3$) into the Se—Se bond of corresponding diorganodiselenides.[690] Various heteroatomic aggregates containing selenium and displaying interactions with alkali metals have been described and structurally characterized, including (TMEDA)$_2$Li$_2$Se$_4$ (distorted trigonal prismatic Li$_2$Se$_4$ core),[691] Li$_2$Se$_5$(PMDETA)$_2$ (Li ions bound to terminal Se atoms of a zigzag chain),[655] (Et$_4$N)$_3$Na[Ru(CO)$_2$(Se$_4$)$_2$]$_2$ (two [Ru(CO)(Se$_4$)$_2$]$_2^-$ ions are bound to an octahedrally coordinated Na$^+$ atom, with Na—Se = 3.00 Å (av)),[692] and [(NMe$_4$)$_3$KSn$_2$Se$_6$]$_\infty$ (K ions link Sn$_2$Se$_6^{4-}$ units, K—Se = 3.34 Å (av)).[136]

The reaction between NiCl$_2$·6H$_2$O and (SeCH$_2$CH$_2$Se)$^{2-}$ in EtOH generates the [Ni(SeCH$_2$CH$_2$Se)$_2$]$^{2-}$ ion, which has been isolated as its potassium and tetramethylammonium salts. The potassium ions in $K_2[Ni(SeCH_2CH_2Se)_2] \cdot 2EtOH$ display close contacts between the Se atoms of the ligands (K—Se distance = 3.39–3.60 Å) in the anions, and the ethanolic oxygen atoms (K–O distance = ca. 2.8 Å).[693] The 1:1 reaction of $Ph_2P(Se)N(SiMe_3)_2$ with potassium *t*-butoxide in THF at room temperature produces $(K[Ph_2P(Se)NSiMe_3] \cdot THF)_2$. The dimeric complex contains two four-membered N—P—Se—K rings fused to a central K_2Se_2 ring (K—Se = 3.37–3.42 Å).[694] The bimetallic complex $[(Py)_2Yb(SeC_6H_5)_2(\mu\text{-}SeC_6H_5)_2Li(Py)_2]$ contains a Li—Se—Yb^{III}—Se ring (Li—Se = 2.57, 2.69 Å); the phenylselenolato ligands are the only anionic ligands on the ytterbium center.[695]

Sodium polyselenide reacts with Ph_2PCl in THF/EtOH to give a mixture of products, including the oligomeric $Na_2[Ph_2PSe_2]_2 \cdot THF \cdot 5H_2O$, which consists of a central polymeric core built up of $Na(H_2O)_6$ and $Na(H_2O)_3(THF)(Se)$ units (Na—Se = 2.98 Å) with additional hydrogen bonds to $[Ph_2PSe_2]^-$.[696] Selenium and $[(Bu^tNH)P(\mu\text{-}N\text{-}Bu^t)_2(NH\text{-}Bu^t)]$ react to form *cis*-$[(Bu^tNH)(Se)P(\mu\text{-}N\text{-}Bu^t)_2(Se)(NH\text{-}Bu^t)]$, which will react with $KN(SiMe_3)_2$ to produce $\{[(THF)K[Bu^tN(Se)P(\mu\text{-}N\text{-}Bu^t)_2P(Se)NBu^t]K(THF)_2]_2\}_\infty$. It forms an infinite network of twenty-membered $K_6Se_6P_4N_4$ rings involving two types of K–Se interactions (K—Se = 3.26–3.42 Å) (see Figure 95).[697]

The bulky (tris(trimethylsilyl)silyl)tellurido ligand, $-TeSi(SiMe_3)_3$, has been incorporated into a variety of *s*-block coordination compounds. Metalation of $HTe\{Si(SiMe_3)_3\}$ with BuLi yields pale yellow $LiTe[Si(SiMe_3)_3]$; crystals obtained from cyclopentane solution indicate that the compound is a hexamer.[698] It forms a centrosymmetric but distorted hexagonal prism that is built up alternately of Li and $Te\{Si(SiMe_3)_3\}$; the $(LiTe)_6$ rings are slightly puckered and adopt a chair conformation (see Figure 96).[698]

Tellurium inserts the Li—Si bond of $(THF)_3LiSi(SiMe_3)_3$ in THF to produce colorless $(THF)_2LiTeSi(SiMe_3)_3$, which crystallizes as a dimer with a planar Li—Te—Li′—Te′ ring and two tris(trimethylsilyl)silyl substituents in a *trans* position. The same compound can be prepared from $(THF)Li[N(SiMe_3)_2]$ and $HTeSi(SiMe_3)_3$; it has been crystallographically characterized as the dimeric mono-THF solvate, $[(THF)LiTeSi(SiMe_3)_3]_2$.[677] 1,2-Dimethoxyethane displaces the THF from $(THF)_2LiTeSi(SiMe_3)_3$ to form the dimeric $(DME)LiTeSi(SiMe_3)_3$, which can also be formed directly from the reaction of $LiSi(SiMe_3)_3 \cdot 1.5(DME)$ and Te in DME.[699] Reduction of the ditelluride $(SiMe_3)TeTeSi(SiMe_3)_3$ with Na/Hg in THF yields colorless crystals of the sodium derivative $(THF)_{0.5}NaTeSi(SiMe_3)_3$. Tellurolysis of $MN(SiMe_3)_2$ (M = Li, Na) or $KOCMe_3$ with the tellurol $HTeSi(SiMe_3)_3$ in hexane gives the toluene-soluble, base free tellurolate derivatives $MTeSi(SiMe_3)_3$ (M = Li, Na, K). A TMEDA derivative of the potassium complex $(TMEDA)K\text{-}TeSi(SiMe_3)_3$ can also be isolated.[700]

Reduction of $(2,4,6\text{-}Me_3C_6H_2)_2Te_2$ and $(2,4,6\text{-}Pr^i_3C_6H_2)_2Te_2$ with two equivalents of $LiEt_3BH$ in THF produces the lithium tellurolates $(2,4,6\text{-}Me_3C_6H_2)_2TeLi(THF)_{1.5}$ and $(2,4,6\text{-}Pr^i_3C_6H_2)_2\text{-}TeLi(THF)_{2.5}$, respectively. Direct Te insertion into the C—Li bond of $(2,4,6\text{-}Bu^t_3C_6H_2)Li(THF)_3$ in THF produces the structurally authenticated $(2,4,6\text{-}Bu^t_3C_6H_2)TeLi(THF)_3$, while a similar reaction between elemental tellurium and $(o\text{-}C_6H_4CH_2NMe_2)Li$ yields the chelating tellurolate $(o\text{-}C_6H_4CH_2NMe_2)TeLi(DME)$. The action of Na/Hg amalgam on THF solutions of

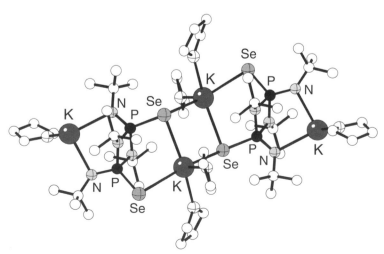

Figure 95 The structural motif of $\{[(THF)K[Bu^tN(Se)P(\mu\text{-}N\text{-}Bu^t)_2P(Se)NBu^t]K(THF)_2]_2\}_\infty$.

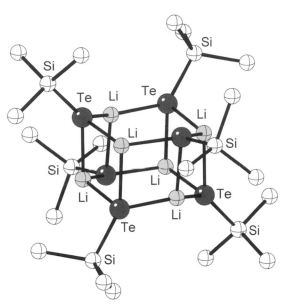

Figure 96 The structure of $\{LiTe[Si(SiMe_3)_3]\}_6$.

$(2,4,6\text{-}Me_3C_6H_2)_2Te_2$ or $(2,4,6\text{-}Pr^i_3C_6H_2)_2Te_2$ followed by work-up with TMEDA or DME leads to the sodium tellurolates. Treatment of $(2,4,6\text{-}Pr^i_3C_6H_2)_2Te_2$ with two equivalents of $K[Bu^s_3BH]$ in THF gives the polymeric tellurolate, $(2,4,6\text{-}Pr^i_3C_6H_2)_2TeK(THF)_{1.33}$. The infinite ladder-like structure is disrupted by reaction with 18-crown-6, which generates a monomeric derivative.[701]

Reaction of $HTeSi(SiMe_3)_3$ with Bu_2Mg in hexane gives the base-free, homoleptic tellurolate $Mg[TeSi(SiMe_3)_3]_2$ as colorless plates; recrystallization in THF gives the bis(THF) adduct.[686,702] The analogous pyridine adduct was prepared by the same reactions in pyridine; both the pyridine and tellurolate ligands are displaced by 12-crown-4 to form $[Mg(12\text{-}C\text{-}4)_2][TeSi(SiMe_3)_3]_2$. Reactions between two equivalents of $HTeSi(SiMe_3)_3$ and $M[N(SiMe_3)_3]_2(THF)_2$ (M = Ca, Sr, Ba) in hexane or toluene (for Ca) gives high yields of the corresponding tellurolate complexes, which can be isolated as THF adducts; pyridine adducts can also be formed.[686,702,703] The calcium compound displays a high field shift of the $^{125}Te\{^1H\}$-NMR resonance ($-2,204$ ppm vs. Me_2Te), and its crystal structure contains calcium in a distorted octahedral environment (Ca—Te = 3.19 Å; Ca—O = 2.36 and 2.41 Å); the two tellurido ligands are in a *trans* position. The $(Py)_5Ba[TeSi(SiMe_3)_3]_2$ complex displays a distorted pentagonal bypyramidal environment, with Ba—Te = 3.38 Å, Te—Ba—Te = 171.9°.[686]

3.1.4.5 Group 17 Ligands

Halides of the *s*-block metals such as NaCl and $CaCl_2$ are among the most widely known of all compounds, and their uses are legion.[704] The molecular structure and spectra of the Group 1 and 2 halides have been reviewed.[705–707] The use of the halides as reagents means that numerous *s*-block compounds with terminal halide ligands are known[708] and in many cases have been structurally characterized (e.g., adducts with nitrogen and mixed nitrogen-oxygen bases such as MeCN, Py, en (ethylene-1,2-diamine), dien (diethylenetriamine), TMEN (tetramethylenediamine), DMF, phen, bipy, terpy, substituted pyridines, water, and ROH (R = Me, Pr, Bu).[709–718] Often their structures and bond distances are adequately rationalized with fundamental electrostatic arguments. The focus of this section is on features of molecular halide complexes that are distinctive or have provided new insights into bonding and reactivity.

The molecular structures of the gas-phase alkaline-earth dihalides have been a source of continuing interest. Gas-phase electron diffraction (GED) and mass spectrometric (MS) measurements confirm that the vapor phase structure of $BeCl_2$ at 274 °C is mostly that of a linear monomer, with a thermal average Be—Cl distance of 1.798(4) Å. A small amount (2.5%) of a halide-bridged dimer form is also present, however, with Be—Cl_{term} = 1.83(1) Å, Be—Cl_{bridg} = 1.97(2), and Cl_{term}—Be—Cl_{bridg} = 134(4)°.[719] Dimers (ca. 1%) have also been detected in MgX_2 vapor.[720]

The structures of the gaseous dihalides of calcium, strontium, and barium have been reviewed.[705,707] As noted in Section 3.1.2.2, CaF_2, SrF_2, $SrCl_2$, and BaF_2, $BaCl_2$, and $BaBr_2$ have permanent molecular dipole moments in the gas phase. The structure of RaF_2, although not experimentally known, is also calculated to be bent (F—Ra—F = 118°; Ra—F = 2.30 Å).[26] Such bending defies interpretation using simple VSEPR (Valence Shell Electron Pair Repulsion) theory,[24,721] and there is not yet agreement on the most satisfactory rationale for it. Some degree of covalency, whether in the guise of polarization arguments, or more explicitly in molecular orbital terms, is evidently required as part of the explanation.

A series of $M_xX_y(THF)_n$ adducts for M = Li, Be, Mg, Ca, Sr, and Ba have been structurally authenticated (see Table 9). Although the distances mostly follow the trends expected from additivity of metal and ligand radii, maximum coordination numbers are not always sterically dictated;[580] Ca^{2+} and Sm^{3+} are almost exactly the same size (1.0 Å),[1] for example, yet calcium coordinates to only four THF ligands in addition to the two iodides, whereas the $[SmI_2(THF)_5]^+$ cation crystallizes from THF with two iodides and five THFs in the plane of the Sm atom.[722] Another THF ligand could bind to the calcium without undue steric crowding.

In conjunction with other molecular halide complexes with various O-donor ligands such water and acetone, correspondences between the metal coordination number and solid-state structures have been identified.[723] For the barium iodides, a progression is observed from the parent nonmolecular BaI_2 ($PbCl_2$ lattice type)[724] to framework (e.g., $[BaI_2(\mu\text{-}H_2O)_2]_n$), layer ($[BaI_2(\mu\text{-}H_2O)(OCMe_2)]_n$), chain ($[Ba(\mu\text{-}I)_2(THF)_3]_n$) and finally monomeric structures ($BaI_2(THF)_5$). Coming almost full circle, the latter can be partially hydrolyzed to form $BaI(OH)(H_2O)_4$, in which a three-dimensional network between molecules is constructed via H bonds. The iodohydroxide is a possible intermediate in the generation of sol-gels, leading ultimately to $[Ba(OH)_2\text{-}(H_2O)_x]$.[725]

The propensity for halides to bridge can lead to the formation of larger aggregates, particularly with the highly polarizing Li^+ cation. The three lithium heterocubanes that have been described, i.e., $[LiCl(HMPTA)]_4$,[726] $[LiBr(Et_2O)]_4$,[727] and $[LiI(NEt_3)]_4$,[728] are formed by special and/or adventitious routes. The iodo complex, for example, is isolated from the reaction of $LiN(SiMe_3)_2$ with the metastable GaI or AlI in the presence of NEt_3; it cannot be obtained directly from a mixture of LiI and NEt_3. The energetics of the formation of $[LiI(NEt_3)]_4$ and its stability with respect to solid LiI have been examined with DFT calculations. These suggest that the activation energy of the solvation of solid LiI to give monomeric $[LiI(NEt_3)]$ as an intermediate is too high, and consequently the presence of energetic donors such as the metastable monovalent Ga or Al is required.

"Opening" the cube leads to ladder-like structures such as Li_4Cl_4(azetidine)$_2$[N-(3-aminopropyl)-azetidine]$_2$,[729] Li_4I_4(2,4,6-trimethylpyridine)$_6$,[730] and Li_4Br_4(2,6-dimethylpyridine)$_6$.[731] The latter, prepared by recrystallizing LiBr in pyridine, has a structure typical for the class; i.e., a stepped tetramer (see Figure 97). A more complex species, {LiLLiClLiL(THF)}$_2$, is formed as a side product of the reaction of $[LiL(THF)_n]$ (L = N,N-dimethyl-N'-trimethylsilylethane-1,2-diamide)

Table 9 Bond lengths (Å) in tetrahydrofuran adducts of metal halides.[a]

Complex	Coord. No	$M–X_{term}$ (Å)	$M–X_{brid}$ (Å)	M—O (Å)	References
$[(THF)_3Li(\mu\text{-}Cl)Li(THF)_3]^+$	4		2.25	1.93–1.96	792
$[(THF)_3Li]_3(\mu_3\text{-}Cl)]^{2+}$	4		2.25	1.90	793
$[(THF)_3Li(\mu\text{-}Br)Li(THF)_3]^+$	4		2.29–2.51	1.96–2.02	794
$(THF)_3LiI$	4	2.74		1.92–1.95	793
$[(THF)_3Mg(\mu\text{-}Cl)_3Mg(THF)_3]^+$	6		2.51	2.08	795
$(THF)_2BeCl_2$	4	1.98		1.65	796
$(THF)_4MgCl_2$	6	2.45		2.09–3.12	797
$(THF)_4MgBr_2$	4 (6)	2.63 (2.80 from adjacent mol.)		3.13	798
$(THF)_4MgBr_2$	6	2.66		3.14	799
$(THF)_4CaI_2$	6	3.11		2.34	580
$(THF)_5SrI_2$	7	3.23		2.54–2.63	688
$[(THF)_3Ba(\mu\text{-}I)_2]_n$	7		3.46–3.52	2.72–2.76	723
$(THF)_5BaI_2$	7	3.38		2.72	723

[a] Distances that vary by less than 0.03 Å have been averaged.

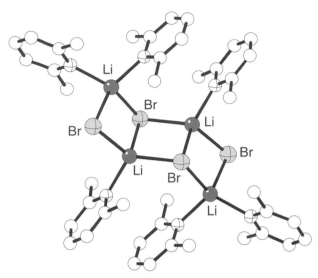

Figure 97 The structure of $Li_4Br_4(2,6\text{-dimethylpyridine})_6$.

with $LuCl_3(THF)_2$ in THF. $\{LiLLiClLiL(THF)\}_2$ has a LiCl bonded between LiL and LiL(THF) units, thereby generating a "three-rung ladder", which is further connected by Li—Cl bonds to a second LiLLiClLiL(THF) moiety (see Figure 98).[732] Finally, from the reaction of $GeBr_4$ with two equivalents of $Li_2[1\text{-naphthylamide}]$ in THF–Et_2O, the infinite corrugated ladder $(LiBr \cdot THF)_\infty$ can be isolated.[733] It can be regarded as the product of the association of cubes of $(LiBr \cdot THF)_4$ (see Figure 99). The more straightforward reaction of LiX (X = Cl, Br, I) with TMEDA gives adducts with the empirical formulas $(LiCl)_3(TMEDA)_2$, $[Li(TMEDA)Br]_2$, and $[Li(TME\text{-}DA)I]_2$;[734] the bromide and iodide are conventional μ-μ'-dihalo bridged dimers with four-coordinate N_2LiX_2 environments around lithium, but the chloride forms polymeric sheets based on a double cubane Li_6Cl_6 unit that is solvated by chelating and bridging TMEDAs (see Figure 100).

The strongly polarizing power of the Li^+ cation can cause otherwise noncoordinating anions to be incorporated into complexes. For example, the reaction of NH_4PF_6 with Bu^nLi in toluene containing PMDETA yields $[(PMDETA)LiPF_6]_2$, in which PF_6^- anions bridge $(PMDETA)Li^+$ units via Li—F interactions (1.91 Å and 3.14 Å) (see Figure 101). *Ab initio* MO calculations indicate that the charge on the N of PMDETA is more negative ($-0.59\,e^-$) than on the F of PF_6^- ($-0.49\,e^-$), but that a second PMDETA ligand could not fit around the small Li^+ cation, thus leaving it open to bind to the hexafluorophosphate anion.[735]

An unusual class of compounds at the border between solid state and organometallic chemistry have been prepared that contain the $(C_5R_5)TiF_2$ fragment as a stabilizing agent. The reaction in

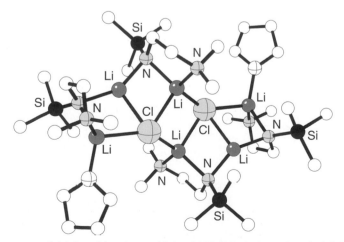

Figure 98 The structure of bis(μ_4-chloro)-tetrakis(μ_2-*N*,*N*-dimethyl-*N*-trimethylsilyl-1,2-diaminoethane)-bis(tetrahydrofuran)-hexalithium.

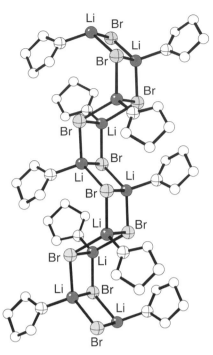

Figure 99 Portion of the infinite ladder of (LiBr·THF)$_\infty$.

THF of $(C_5Me_5)TiF_3$ and either metallic sodium or calcium in the presence of mercury forms the sparingly hydrocarbon-soluble clusters $(C_5Me_5)_{12}Ti_{14}Na_{18}F_{48}(THF)_6$ and $[(C_5Me_5)TiF_2]_6$-$CaF_2(THF)_2$, respectively.[736] The former contains a complex central $[(TiF_3)_2(NaF)_{18}]$ core, in which the sodium atoms are either five- or six-coordinate to THF and/or F atoms. The average Na—F distance in the complex (2.32 Å) is remarkably close to that in solid NaF (2.31 Å), although there is no real structural resemblance between the two systems. The structural correspondence to fluorite is not strong in the calcium complex either, but like CaF_2, the central metal is eight-coordinate, and the average Ca—F distance is close to that in fluorite (see Figure 102).

Compounds similar to these (i.e., $\{(C_5Me_5)TiF_3\}_4CaF_2$ and $\{(C_5Me_4Et)TiF_3\}_4CaF_2$) have been formed from the reaction between $(C_5Me_5)TiF_3$ or $(C_5Me_4Et)TiF_3$ in the presence of CaF_2

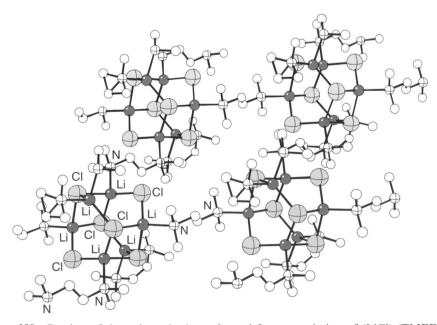

Figure 100 Portion of the polymeric sheets formed from association of (LiCl)$_3$(TMEDA)$_2$.

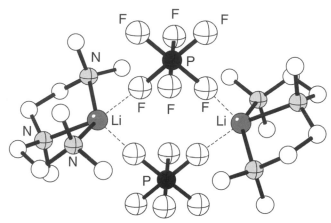

Figure 101 The structure of $[(PMDETA)LiPF_6]_2$.

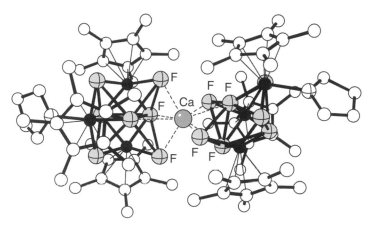

Figure 102 The structure of $[(C_5Me_5)TiF_2]_6CaF_2(THF)_2$.

(prepared *in situ* from Me_3SnF and $CaCl_2$).[737] Recrystallization of the latter two compounds in the presence of HMPA results in the formation of $\{(C_5Me_5)TiF_3\}_4(CaF_2)(HMPA)$ and $\{(C_5Me_4Et)TiF_3\}_4(CaF_2)(HMPA)$, respectively. When the latter is dissolved in $CDCl_3$, the HMPA dissociates completely, a process studied with ^1H-, ^{19}F-, and variable-temperature NMR. A molecule of solvent probably occupies the site on the organometallic complex vacated by the HMPA, and a temperature-dependent equilibrium exists between the solvent-solvated species and $\{(C_5Me_4Et)TiF_3\}_4CaF_2$. With increasing temperature, the equilibrium is entropy-shifted to the non-solvated form.[738] A related system containing lithium has been formed from the reaction of two equivalents of $(C_5Me_5)TiF_3$ with LiF (generated from Me_3SnF and LiCl) in THF. In the solid state, the lithium atom is coordinated by four F atoms ($Li—F = 1.90\,\text{Å}$ (av)); in solution, it is in equilibrium with $Li[(C_5Me_5)_2Ti_2F_7]$ and $(C_5Me_5)_2Ti_2F_6$.[739]

3.1.5 REFERENCES

1. Shannon, R. D. *Acta Crystallogr., Sect. A.* **1976**, *32*, 751.
2. Bock, C. W.; Glusker, J. P. *Inorg. Chem.* **1993**, *32*, 1242.
3. Bock, C. W.; Kaufman, A.; Glusker, J. P. *Inorg. Chem.* **1994**, *33*, 419.
4. Katz, A. K.; Glusjer, J. P.; Beebe, S. A.; Bock, C. W. *J. Am. Chem. Soc.* **1996**, *118*, 5752.
5. Hope, H.; Olmstead, M. M.; Power, P. P.; Xiaojie, X. *J. Am. Chem. Soc.* **1984**, *106*, 819.
6. Vaartstra, B. A.; Huffman, J. C.; Streib, W. E.; Caulton, K. G. *Inorg. Chem.* **1991**, *30*, 121.
7. Ma, J. C.; Dougherty, D. A. *Chem. Rev.* **1997**, *97*, 1303.
8. Murugavel, R.; Baheti, K.; Anantharaman, G. *Inorg. Chem.* **2001**, *40*, 6870.
9. Malandrino, G.; Richeson, D. S.; Marks, T. J.; DeGroot, D. C.; Schindler, J. L.; Kannewurf, C. R. *Appl. Phys. Lett.* **1991**, *58*, 182.
10. Chandler, C. D.; Roger, C.; Hampden-Smith, M. J. *Chem. Rev.* **1993**, *93*, 1205.

11. Frey, M. H.; Payne, D. A. *Chem. Mater.* **1995**, *7*, 123.
12. Bradley, D. C. *Chem. Rev.* **1989**, *89*, 1317.
13. Wojtczak, W. A.; Fleig, P. F.; Hampden-Smith, M. J. *Adv. Organomet. Chem.* **1996**, *40*, 215.
14. Matthews, J. S.; Rees, W. S., Jr. *Adv. Inorg. Chem.* **2000**, *50*, 173.
15. Pedersen, C. J. *J. Am. Chem. Soc.* **1967**, *89*, 7017.
16. Jenkins, H. D. B.; Thakur, K. P. *J. Chem. Educ.* **1979**, *56*, 576.
17. Weiss, E.; Alsdorf, H.; Kuehr, H. *Angew. Chem., Int. Ed. Engl.* **1967**, *6*, 801.
18. Chisholm, M. H.; Drake, S. R.; Naiini, A. A.; Streib, W. E. *Polyhedron* **1991**, *10*, 337.
19. Westerhausen, M. *Coord. Chem. Rev.* **1998**, *176*, 157.
20. Kennedy, A. R.; Mulvey, R. E.; Rowlings, R. B. *J. Am. Chem. Soc.* **1998**, *120*, 7816.
21. Wharton, L.; Berg, R. A.; Klemperer, W. *J. Chem. Phys.* **1963**, *39*, 2023.
22. Kasparov, V. V.; Ezhov, Y. S.; Rambidi, N. G. *J. Struct. Chem.* **1979**, *20*, 260.
23. Guido, M.; Gigli, G. *J. Chem. Phys.* **1976**, *65*, 1397.
24. Kaupp, M.; Schleyer, P. v. R. *J. Phys. Chem.* **1992**, *96*, 7316.
25. von Szentpály, L.; Schwerdtfeger, P. *Chem. Phys. Lett.* **1990**, *170*, 555.
26. Lee, E. P. F.; Soldan, P.; Wright, T. G. *Inorg. Chem.* **2001**, *40*, 5979.
27. Kaupp, M.; Schleyer, P. v. R.; Stoll, H.; Preuss, H. *J. Am. Chem. Soc.* **1991**, *113*, 6012.
28. Kaupp, M.; Schleyer, P. v. R.; Stoll, H.; Preuss, H. *J. Chem. Phys.* **1991**, *94*, 1360.
29. Kaupp, M.; Schleyer, P. v. R. *J. Am. Chem. Soc.* **1992**, *114*, 491.
30. Seijo, L.; Barandiarán, Z.; Huzinaga, S. *J. Chem. Phys.* **1991**, *113*, 3762.
31. Hassett, D. M.; Marsden, C. J. *J. Chem. Soc., Chem. Commun.* **1990**, 667.
32. Nakamura, R. L.; Anderson, J. A.; Gaber, R. F. *J. Biol. Chem.* **1997**, *272*, 1011.
33. Dang, Q. D.; Guinto, E. R.; Di Cera, E. *Nature Biotechnol.* **1997**, *15*, 146.
34. Werner, B.; Kräuter, T.; Neumüller, B. *Organometallics* **1996**, *15*, 3746.
35. Deacon, G. B.; Feng, T. C.; Junk, P. C.; Skelton, B. W.; White, A. H. *J. Chem. Soc., Dalton Trans.* **1997**, 1181.
36. Clark, D. L.; Hollis, R. V.; Scott, B. L.; Watkin, J. G. *Inorg. Chem.* **1996**, *35*, 667.
37. Clark, D. L.; B, D. G.; Feng, T. C.; Hollis, R. V.; Scott, B. L.; Skelton, B. W.; Watkin, J. G.; White, A. H. *Chem. Commun.* **1996**, 1729.
38. Su, J.; Li, X.-W.; Crittendon, R. C.; Robinson, G. H. *J. Am. Chem. Soc.* **1997**, *119*, 5471.
39. Xie, Y.; Schaefer, H. F.; Robinson, G. H. *Chem. Phys. Lett.* **2000**, *317*, 174.
40. Grutzmacher, H.; Fassler, T. F. *Chem. Eur. J.* **2000**, *6*, 2317.
41. Grunenberg, J.; Goldberg, N. *J. Am. Chem. Soc.* **2000**, *122*, 6045.
42. Allen, T. L.; Fink, W. H.; Power, P. P. *Dalton* **2000**, 407.
43. Molina, J. M.; Dobado, J. A.; Heard, G. L.; Bader, R. F. W.; Sundberg, M. R. *Theor. Chem. Acc.* **2001**, *105*, 365.
44. Grunenberg, J. *J. Chem. Phys.* **2001**, *115*, 6360.
45. Cotton, F. A.; Cowley, A. H.; Feng, X. *J. Am. Chem. Soc.* **1998**, *120*, 1795.
46. Takagi, N.; Schmidt, M. W.; Nagase, S. *Organometallics* **2001**, *20*, 1646.
47. Twamley, B.; Power, P. P. *Angew. Chem., Int. Ed. Engl.* **2000**, *39*, 3500.
48. Fassler, T. F.; Hoffmann, R. *Angew. Chem., Int. Ed. Engl.* **1999**, *38*, 543.
49. Sannigrahi, A. B.; Kar, T.; Niyogi, B. G.; Hobza, P.; Schleyer, P. v. R. *Chem. Rev.* **1990**, *90*, 1061.
50. Streitwieser, A.; Bachrach, S. M.; Dorigo, A.; Schleyer, P. v. R. *Lithium Chemistry* **1995**, 1.
51. Cotton, F. A.; Cowley, A. H.; Feng, X. *J. Am. Chem. Soc.* **1998**, *120*, 1795.
52. Dunne, J. P.; Fox, S.; Tacke, M. *J. Mol. Struct.-THEOCHEM* **2001**, *543*, 157.
53. Koch, W.; Holthausen, M. C. *A Chemist's Guide to Density Functional Theory*, 2nd ed. **2001**, Wiley: New York.
54. Ong, C. C.; Rodley, G. A. *Chem. New Zealand* **1985**, *49*, 7.
55. Barkigia, K. M.; Spaulding, L. D.; Fajer, J. *Inorg. Chem.* **1983**, *22*, 349.
56. McKee, V.; Choon, O. C.; Rodley, G. A. *Inorg. Chem.* **1984**, *23*, 4242.
57. Choon, O. C.; McKee, V.; Rodley, G. A. *Inorg. Chim. Acta* **1986**, *123*, L11.
58. McKee, V.; Rodley, G. A. *Inorg. Chim. Acta* **1988**, *151*, 233.
59. Yang, S.; Jacobson, R. A. *Inorg. Chim. Acta* **1991**, *190*, 129.
60. Brancato-Buentello, K. E.; Scheidt, W. R. *Angew. Chem., Int. Ed. Engl.* **1997**, *36*, 1456.
61. Byrn, M. P.; Curtis, C. J.; Khan, S. I.; Sawin, P. A.; Tsurumi, R.; Strouse, C. E. *J. Am. Chem. Soc.* **1990**, *112*, 1865.
62. Byrn, M. P.; Curtis, C. J.; Goldberg, I.; Hsiou, Y.; Khan, S. I.; Sawin, P. A.; Tendick, S. K.; Strouse, C. E. *J. Am. Chem. Soc.* **1991**, *113*, 6549.
63. Byrn, M. P.; Curtis, C. J.; Goldberg, I.; Huang, T.; Hsiou, Y.; Khan, S. I.; Sawin, P. A.; Tendick, S. K.; Terzis, A.; Strouse, C. E. *Mol. Cryst. Liq. Cryst. Sci. Technol., Sect. A* **1992**, *211*, 135.
64. Byrn, M. P.; Curtis, C. J.; Hsiou, Y.; Khan, S. I.; Sawin, P. A.; Tendick, S. K.; Terzis, A.; Strouse, C. E. *J. Am. Chem. Soc.* **1993**, *115*, 9480.
65. Byrn, M. P.; Curtis, C. J.; Hsiou, Y.; Khan, S. I.; Sawin, P. A.; Terzis, A.; Strouse, C. E. *Compr. Supramol. Chem.* **1996**, *6*, 715.
66. Arnold, J.; Dawson, D. Y.; Hoffman, C. G. *J. Am. Chem. Soc.* **1993**, *115*, 2707.
67. Arnold, J. *J. Chem. Soc., Chem. Commun.* **1990**, 976.
68. Dawson, D. Y.; Arnold, J. *J. Porphyrins Phthalocyanines* **1997**, *1*, 121.
69. Gebauer, A.; Dawson, D. Y.; Arnold, J. *Dalton* **2000**, 111.
70. Bonomo, L.; Dandin, O.; Solari, E.; Floriani, C.; Scopelliti, R. *Angew. Chem., Int. Ed. Engl.* **1999**, *38*, 914.
71. Bonomo, L.; Lehaire, M.-L.; Solari, E.; Scopelliti, R.; Floriani, C. *Angew. Chem., Int. Ed. Engl.* **2001**, *40*, 771.
72. Sugimoto, H.; Mori, M.; Masuda, H.; Taga, T. *J. Chem. Soc., Chem. Commun.* **1986**, 1986.
73. Turek, P.; André, J.-J.; Giraudeau, A.; Simon, J. *Chem. Phys. Lett.* **1987**, *134*, 471.
74. Orti, E.; Brédas, J. L.; Clarisse, C. *J. Chem. Phys.* **1990**, *92*, 1228.
75. Latte, B.; Assmann, B.; Homborg, H. *Z. Anorg. Allg. Chem.* **1997**, *623*, 1281.
76. Dumm, M.; Dressel, M.; Nicklas, M.; Lunkenheimer, P.; Loidl, A.; Weiden, M.; Steglich, F.; Assmann, B.; Homborg, H.; Fulde, P. *Eur. Phys. J. B: Condens. Matter Phys.* **1998**, *6*, 317.
77. Krishnakumar, K. P.; Menon, C. S. *J. Solid State Chem.* **1997**, *128*, 27.

78. Endo, A.; Matsumoto, S.; Mizuguchi, J. *J. Phys. Chem. A* **1999**, *103*, 8193.
79. Loutfy, R. O.; Hor, A. M.; DiPaola-Baranyi, G.; Hsiao, C. K. *J. Imag. Sci.* **1985**, *29*, 116.
80. Daidoh, T.; Matsunaga, H.; Iwata, K. *Nippon Kagaku Kaishi* **1988**, 1090.
81. Riad, S. *Thin Solid Films* **2000**, *370*, 253.
82. Matsumoto, S.; Endo, A.; Mizuguchi, J. *Z. Kristallogr.* **2000**, *215*, 182.
83. Cory, M. G.; Hirose, H.; Zerner, M. C. *Inorg. Chem.* **1995**, *34*, 2969.
84. Lehn, J.-M.; Ball, P. *New Chemistry* **2000**, 300.
85. Wai, C. M. In *Recent Progress in Actinides Separation Chemistry*, Proceedings of the Workshop on Actinides Solution Chemistry, WASC '94, Tokai, Japan, Sept. 1–2, 1994; 1997, p 81.
86. Khopkar, S. M.; Gandhi, M. N. *J. Sci. Ind. Res.* **1996**, *55*, 139.
87. Heumann, K. G. *Top. Curr. Chem.* **1985**, *127*, 77.
88. Bartsch, R. A.; Lu, J.; Ohki, A. *J. Inclusion Phenom. Mol. Recognit. Chem.* **1998**, *32*, 133.
89. Bond, A. H.; Dietz, M. L.; Chiarizia, R. *Ind. Eng. Chem. Res.* **2000**, *39*, 3442.
90. Hilgenfeld, R.; Saenger, W. *Host Guest Complex Chem., Macrocycles* **1985**, 43.
91. Mandolini, L. *Pure Appl. Chem.* **1986**, *58*, 1485.
92. Raevskii, O. A. *Koord. Khim.* **1990**, *16*, 723.
93. Sato, M.; Akabori, S. *Trends Org. Chem.* **1990**, *1*, 213.
94. Cacciapaglia, R.; Mandolini, L. *Pure Appl. Chem.* **1993**, *65*, 533.
95. Tahara, R.; Morozumi, T.; Suzuki, Y.; Kakizawa, Y.; Akita, T.; Nakamura, H. *J. Inclusion Phenom. Mol. Recognit. Chem.* **1998**, *32*, 283.
96. Arion, V.; Revenco, M.; Gradinaru, J.; Simonov, Y.; Kravtsov, V.; Gerbeleu, N.; Saint-Aman, E.; Adams, F. *Rev. Inorg. Chem.* **2001**, *21*, 1.
97. Fabre, B.; Simonet, J. *Coord. Chem. Rev.* **1998**, *178–180*, 1211.
98. Sachleben, R. A.; Moyer, B. A. *ACS Symposium Series* **1999**, *716*, 114.
99. Putzer, M. A.; Magull, J.; Goesmann, H.; Neumueller, B.; Dehnicke, K. *Chem. Ber.* **1996**, *129*, 1401.
100. Putzer, M. A.; Neumueller, B.; Behnicke, K.; Magull, J. *Chem. Ber.* **1996**, *129*, 715.
101. Henschel, D.; Wijaya, K.; Moers, O.; Blaschette, A.; Jones, P. G. *Z. Naturforsch. B: Chem. Sci.* **1997**, *52*, 1229.
102. Wijaya, K.; Henschel, D.; Moers, O.; Blaschette, A.; Jones, P. G. *Z. Naturforsch. B: Chem. Sci.* **1997**, *52*, 1219.
103. Putzer, M. A.; Neumuller, B.; Dehnicke, K. *Z. Anorg. Allg. Chem.* **1997**, *623*, 539.
104. Zhang, H.; Wang, X.; Zhang, K.; Teo, B. K. *Inorg. Chem.* **1998**, *37*, 3490.
105. Putzer, M. A.; Neumueller, B.; Dehnicke, K. *Z. Anorg. Allg. Chem.* **1998**, *624*, 57.
106. Akutagawa, T.; Hasegawa, T.; Nakamura, T.; Takeda, S.; Inabe, T.; Sugiura, K.-i.; Sakata, Y.; Underhill, A. E. *Inorg. Chem.* **2000**, *39*, 2645.
107. Wesemann, L.; Trinkaus, M.; Englert, U.; Mueller, J. *Organometallics* **1999**, *18*, 4654.
108. Dietrich, A.; Neumuller, B.; Dehnicke, K. *Z. Anorg. Allg. Chem.* **1999**, *625*, 619.
109. Dillon, R. E. A.; Stern, C. L.; Shriver, D. F. *Chem. Mater.* **2001**, *13*, 2516.
110. Dillon, R. E.; Shriver, D. F. *Mater. Res. Soc. Symp. Proc.* **1998**, *496*, 505.
111. Dillon, R. E. A.; Stern, C. L.; Shriver, D. F. *Chem. Mater.* **2000**, *12*, 1122.
112. Nakamura, T.; Akutagawa, T.; Honda, K.; Underhill, A. E.; Coomber, A. T.; Friend, R. H. *Nature (London)* **1998**, *394*, 159.
113. Blake, A. J.; Gould, R. O.; Li, W.-S.; Lippolis, V.; Parsons, S.; Radek, C.; Schroder, M. *Angew. Chem., Int. Ed.* **1998**, *37*, 293.
114. Bulychev, B. M.; Bel'skii, V. K. *Zh. Neorg. Khim.* **1997**, *42*, 260.
115. Marsh, R. E. *Acta Crystallogr., Sect. B* **1999**, *B55*, 931.
116. Junk, P. C.; Steed, J. W. *J. Chem. Soc., Dalton Trans.* **1999**, 407.
117. Lange, D.; Klein, E.; Bender, H.; Niecke, E.; Nieger, M.; Pietschnig, R.; Schoeller, W. W.; Ranaivonjatovo, H. *Organometallics* **1998**, *17*, 2425.
118. Boulatov, R.; Du, B.; Meyers, E. A.; Shore, S. G. *Inorg. Chem.* **1999**, *38*, 4554.
119. Fender, N. S.; Kahwa, I. A.; White, A. J. P.; Williams, D. J. *J. Chem. Soc., Dalton Trans.* **1998**, 1729.
120. Domasevitch, K. V.; Ponomareva, V. V.; Rusanov, E. B.; Gelbrich, T.; Sieler, J.; Skopenko, V. V. *Inorg. Chim. Acta* **1998**, *268*, 93.
121. Domasevitch, K. V.; Ponomareva, V. V.; Rusanov, E. B. *J. Chem. Soc., Dalton Trans.* **1997**, 1177.
122. Domasevitch, K. V.; Rusanova, J. A.; Vassilyeva, O. Y.; Kokozay, V. N.; Squattrito, P. J.; Sieler, J.; Raithby, P. R. *J. Chem. Soc., Dalton Trans.* **1999**, 3087.
123. Rusanova, J. A.; Domasevitch, K. V.; Vassilyeva, O. Y.; Kokozay, V. N.; Rusanov, E. B.; Nedelko, S. G.; Chukova, O. V.; Ahrens, B.; Raithby, P. R. *Dalton* **2000**, 2175.
124. Fender, N. S.; Fronczek, F. R.; John, V.; Kahwa, I. A.; McPherson, G. L. *Inorg. Chem.* **1997**, *36*, 5539.
125. Fassler, T. F.; Hoffmann, R.; Hoffmann, S.; Worle, M. *Angew. Chem., Int. Ed. Engl.* **2000**, *39*, 2091.
126. Manskaya, Y. A.; Domasevich, K. V.; Polyakov, V. R.; Kokozei, V. N.; Vasil'eva, O. Y. *Russ. J. Gen. Chem. (Engl. Transl.)* **1999**, *69*, 97.
127. Bryan, J. C.; Sachleben, R. A.; Lavis, J. M.; Davis, M. C.; Burns, J. H.; Hay, B. P. *Inorg. Chem.* **1998**, *37*, 2749.
128. Talanova, G. G.; Elkarim, N. S. A.; Talanov, V. S.; Hanes, R. E., Jr.; Hwang, H.-S.; Bartsch, R. A.; Rogers, R. D. *J. Am. Chem. Soc.* **1999**, *121*, 11281.
129. Graf, E.; Hosseini, M. W.; Ruppert, R.; Kyritsakas, N.; De Cian, A.; Fischer, J.; Estournes, C.; Taulelle, F. *Angew. Chem., Int. Ed. Engl.* **1995**, *34*, 1115.
130. Campbell, J.; Schrobilgen, G. J. *Inorg. Chem.* **1997**, *36*, 4078.
131. Fassler, T. F.; Schutz, U. *Inorg. Chem.* **1999**, *38*, 1866.
132. Xu, L.; Sevov, S. C. *J. Am. Chem. Soc.* **1999**, *121*, 9245.
133. Somer, M.; Carrillo-Cabrera, W.; Peters, E.-M.; Peters, K.; Kaupp, M.; von Schnering, H.-G. *Z. Anorg. Allg. Chem.* **1999**, *625*, 37.
134. Fassler, T. F.; Hunziker, M. *Z. Anorg. Allg. Chem.* **1996**, *622*, 837.
135. Fassler, T. F.; Hoffmann, R. *Z. Kristallogr. New Cryst. Struct.* **2000**, *215*, 139.
136. Campbell, J.; Devereux, L. A.; Gerken, M.; Mercier, H. P. A.; Pirani, A. M.; Schrobilgen, G. J. *Inorg. Chem.* **1996**, *35*, 2945.

137. Campbell, J.; Dixon, D. A.; Mercier, H. P. A.; Schrobilgen, G. J. *Inorg. Chem.* **1995**, *34*, 5798.
138. Borrmann, H.; Campbell, J.; Dixon, D. A.; Mercier, H. P. A.; Pirani, A. M.; Schrobilgen, G. J. *Inorg. Chem.* **1998**, *37*, 6656.
139. Park, C.-W.; Salm, R. J.; Ibers, J. A. *Can. J. Chem.* **1995**, *73*, 1148.
140. Smith, D. M.; Park, C.-W.; Ibers, J. A. *Inorg. Chem.* **1996**, *35*, 6682.
141. Xu, L.; Sevov, S. C. *Inorg. Chem.* **2000**, *39*, 5383.
142. Campbell, J.; DiCiommo, D. P.; Mercier, H. P. A.; Pirani, A. M.; Schrobilgen, G. J.; Willuhn, M. *Inorg. Chem.* **1995**, *34*, 6265.
143. Borrmann, H.; Campbell, J.; Dixon, D. A.; Mercier, H. P. A.; Pirani, A. M.; Schrobilgen, G. J. *Inorg. Chem.* **1998**, *37*, 1929.
144. Eichhorn, B. W.; Mattamana, S. P.; Gardner, D. R.; Fettinger, J. C. *J. Am. Chem. Soc.* **1998**, *120*, 9708.
145. Korber, N.; Jansen, M. *J. Chem. Soc., Chem. Commun.* **1990**, 1654.
146. Korber, N.; Jansen, M. *Chem. Ber.* **1996**, *129*, 773.
147. Xu, L.; Bobev, S.; El-Bahraoui, J.; Sevov, S. C. *J. Am. Chem. Soc.* **2000**, *122*, 1838.
148. Tokitoh, N.; Arai, Y.; Okazaki, R.; Nagase, S. *Science (Washington, D.C.)* **1997**, *277*, 78.
149. Faessler, T. F.; Spiekermann, A.; Spahr, M. E.; Nesper, R. *Angew. Chem., Int. Ed. Engl.* **1997**, *36*, 486.
150. Pochini, A.; Ungaro, R. *Compr. Supramol. Chem.* **1996**, *2*, 103.
151. Gutsche, C. D. *Calixarenes* **2001**, *2001*, 1.
152. Gutsche, C. D. *ACS Symposium Series* **2000**, *757*, 2.
153. Milbradt, R.; Bohmer, V. *Calixarenes* **2001**, *2001*, 663.
154. Ludwig, R. *Fresenius J. Anal. Chem.* **2000**, *367*, 103.
155. Kolarik, Z. *Miner. Process. Extract. Metall. Rev.* **2000**, *21*, 89.
156. Wipff, G. *Calixarenes* **2001**, *2001*, 312.
157. Hayashita, T.; Teramae, N.; Kuboyama, T.; Nakamura, S.; Yamamoto, H.; Nakamura, H. *J. Inclusion Phenom. Mol. Recognit. Chem.* **1998**, *32*, 251.
158. Diamond, D.; Nolan, K. *Anal. Chem.* **2001**, *73*, 22A.
159. Dakanale, E.; Levi, S.; Rosler, S.; Auge, J.; Hartmann, J.; Henning, B. *GIT Lab. J.* **1998**, *2*, 118.
160. Cacciapaglia, R.; Mandolini, L. *Calix. Action* **2000**, 241.
161. Steyer, S.; Jeunesse, C.; Armspach, D.; Matt, D.; Harrowfield, J. *Calixarenes* **2001**, *2001*, 513.
162. Chen, X.; Ji, M.; Fisher, D. R.; Wai, C. M. *Inorg. Chem.* **1999**, *38*, 5449.
163. McDowell, W. J.; Case, G. N.; Bartsch, R. A.; Czech, B. P. *Solvent Extr. Ion Exch.* **1986**, *4*, 411.
164. Dijkstra, P. J.; Brunink, J. A. J.; Bugge, K. E.; Reinhoudt, D. N.; Harkema, S.; Ungaro, R.; Ugozzoli, F.; Ghidini, E. *J. Am. Chem. Soc.* **1989**, *111*, 7567.
165. Asfari, Z.; Bressot, C.; Vicens, J.; Hill, C.; Dozol, J. F.; Rouquette, H.; Eymard, S.; Lamare, V.; Tournois, B. *ACS Symp. Ser.* **1996**, *642*, 376.
166. Lamare, V.; Dozol, J.-F.; Fuangswasdi, S.; Arnaud-Neu, F.; Thuery, P.; Nierlich, M.; Asfari, Z.; Vicens, J. *J. Chem. Soc., Perkins Trans. 2* **1999**, 271.
167. Barboso, S; Casnati, A; Dozol, J.-F.; Pochini, A.; Ungaro, R. *Chim. Ind.* **2000**, *82*, 423.
168. Asfari, Z.; Naumann, C.; Vicens, J.; Nierlich, M.; Thuery, P.; Bressot, C.; Lamare, V.; Dozol, J.-F. *New J. Chem.* **1996**, *20*, 1183.
169. Asfari, Z.; Thuery, P.; Nierlich, M.; Vicens, J. *Tetrahedron Lett.* **1999**, *40*, 499.
170. Thuery, P.; Nierlich, M.; Lamare, V.; Dozol, J.-F.; Asfari, Z.; Vicens, J. *Acta Crystallogr., Sect. C: Cryst. Struct. Commun.* **1996**, *C52*, 2729.
171. Blanda, M. T.; Farmer, D. B.; Brodbelt, J. S.; Goolsby, B. J. *J. Am. Chem. Soc.* **2000**, *122*, 1486.
172. Casnati, A.; Pochini, A.; Ungaro, R.; Ugozzoli, F.; Arnaud, F.; Fanni, S.; Schwing, M.-J.; Egberink, R. J. M.; de Jong, F.; Reinhoudt, D. N. *J. Am. Chem. Soc.* **1995**, *117*, 2767.
173. Leverd, P. C.; Berthault, P.; Lance, M.; Nierlich, M. *Eur. J. Inorg. Chem.* **1998**, 1859.
174. Dozol, J.-F.; Ungaro, R.; Casnati, A. (Commissariat a l'Energie Atomique, France). Preparation and Use of Calixarene Acetamido Derivatives for Selective Removal of Strontium from Aqueous Solutions. In *PCT Int. Appl.* WO 2001012586, 2001, p 52.
175. Ungaro, R.; Arduini, A.; Casnati, A.; Pochini, A.; Ugozzoli, F. *Pure Appl. Chem.* **1996**, *68*, 1213.
176. Casnati, A.; Baldini, L.; Pelizzi, N.; Rissanen, K.; Ugozzoli, F.; Ungaro, R. *Dalton* **2000**, 3411.
177. Davidson, M. G.; Howard, J. A. K.; Lamb, S.; Lehmann, C. W. *Chem. Commun.* **1997**, 1607.
178. Clague, N. P.; Crane, J. D.; Moreton, D. J.; Sinn, E.; Teat, S. J.; Young, N. A. *J. Chem. Soc., Dalton Trans.* **1999**, 3535.
179. Murayama, K.; Aoki, K. *Inorg. Chim. Acta* **1998**, *281*, 36.
180. Huang, R. H.; Huang, S. Z.; Dye, J. L. *J. Coord. Chem.* **1998**, *46*, 13.
181. Dye, J. L.; Wagner, M. J.; Overney, G.; Huang, R. H.; Nagy, T. F.; Tomanek, D. *J. Am. Chem. Soc.* **1996**, *118*, 7329.
182. Dye, J. L. *J. Phys. Chem.* **1984**, *88*, 3842.
183. Dawes, S. B.; Ward, D. L.; Huang, R. H.; Dye, J. L. *J. Am. Chem. Soc.* **1986**, *108*, 3534.
184. Wagner, M. J.; Dye, J. L. *Compr. Supramol. Chem.* **1996**, *1*, 477.
185. Dye, J. L. *Macromol. Symp.* **1998**, *134*, 29.
186. Dye, J. L. *Inorg. Chem.* **1997**, *36*, 3816.
187. Xie, Q.; Huang, R. H.; Ichimura, A. S.; Phillips, R. C.; Pratt, W. P., Jr.; Dye, J. L. *J. Am. Chem. Soc.* **2000**, *122*, 6971.
188. Wagner, M. J.; Ichimura, A. S.; Huang, R. H.; Phillips, R. C.; Dye, J. L. *J. Phys. Chem. B* **2000**, *104*, 1078.
189. DeBacker, M. G.; Mkadmi, E. B.; Sauvage, F. X.; Lelieur, J.-P.; Wagner, M. J.; Concepcion, R.; Kim, J.; McMills, L. E. H.; Dye, J. L. *J. Am. Chem. Soc.* **1996**, *118*, 1997.
190. Kim, J.; Ichimura, A. S.; Huang, R. H.; Redko, M.; Phillips, R. C.; Jackson, J. E.; Dye, J. L. *J. Am. Chem. Soc.* **1999**, *121*, 10666.
191. Redko, M. Y.; Vlassa, M.; Jackson, J. E.; Misiolek, A. W.; Huang, R. H.; Dye, J. L. *J. Am. Chem. Soc.* **2002**, *124*, 5928.
192. Edwards, P. P.; Woodall, L. J.; Anderson, P. A.; Armstrong, A. R.; Slaski, M. *Chem. Rev.* **1993**, 305.

193. Barker, P. D.; Anderson, P. A.; Dupree, R.; Kitchin, S.; Edwards, P. P.; Woodall, L. J. *Mater. Res. Soc. Symp. Proc.* **1996**, *431*, 191.
194. Woodall, L. J.; Anderson, P. A.; Armstrong, A. R.; Edwards, P. P. *J. Chem. Soc., Dalton Trans.* **1996**, 719.
195. Anderson, P. A.; Woodall, L. J.; Porch, A.; Armstrong, A. R.; Hussain, I.; Edwards, P. P. *Mater. Res. Soc. Symp. Proc.* **1995**, *384*, 9.
196. Edwards, P. P.; Anderson, P. A.; Woodall, L. J.; Porch, A.; Armstrong, A. R. *Mater. Sci. Eng., A* **1996**, *A217/218*, 198.
197. Wells, A. F. *Structural Inorganic Chemistry*, 5th ed. **1984**, Clarendon: Oxford, U.K.
198. Marynick, D. S.; Lipscomb, W. N. *Inorg. Chem.* **1972**, *11*, 820.
199. Gundersen, G.; Hedberg, L.; Hedberg, K. *J. Chem. Phys.* **1973**, *59*, 3777.
200. Brendhaugen, K.; Haaland, A.; Novak, D. P. *Acta Chem. Scand., Ser. A* **1975**, *A29*, 801.
201. Gaines, D. F.; Walsh, J. L.; Hillenbrand, D. F. *J. Chem. Soc., Chem. Commun.* **1977**, 224.
202. Marynick, D. S.; Lipscomb, W. N. *J. Amer. Chem. Soc.* **1973**, *95*, 7211.
203. Marynick, D. S. *J. Chem. Phys.* **1976**, *64*, 3080.
204. Dewar, M. J. S.; Rzepa, H. S. *J. Am. Chem. Soc.* **1978**, *100*, 777.
205. Marynick, D. S. *J. Am. Chem. Soc.* **1979**, *101*, 6876.
206. Kirillov, Y. B.; Boldyrev, A. I.; Klimenko, N. M. *Koord. Khim.* **1980**, *6*, 1503.
207. Trindle, C.; Datta, S. N. *Proc. Indian Acad. Sci.: Chem. Sci.* **1980**, *89*, 175.
208. Kirillov, Y. B.; Klimenko, N. M.; Zakzhevskii, V. G. *Zh. Strukt. Khim.* **1983**, *24*, 158.
209. Zyubin, A. S.; Kaupp, M.; Charkin, O. P.; Shloer, P. R. *Zh. Neorg. Khim.* **1993**, *38*, 677.
210. Derecskei-Kovacs, A.; Marynick, D. S. *Chem. Phys. Lett.* **1994**, *228*, 252.
211. Saeh, J. C.; Stanton, J. F. *J. Am. Chem. Soc.* **1997**, *119*, 7390.
212. Mire, L. W.; Wheeler, S. D.; Wagenseller, E.; Marynick, D. S. *Inorg. Chem.* **1998**, *37*, 3099.
213. Konoplev, V. N.; Mal'tseva, N. N.; Khain, V. S. *Koord. Khim.* **1992**, *18*, 1143.
214. Giese, H.-H.; Habereder, T.; Nöth, H.; Ponikwar, W.; Thomas, S.; Warchhold, M. *Inorg. Chem.* **1999**, *38*, 4188.
215. Reger, D. L.; Collins, J. E.; Matthews, M. A.; Rheingold, A. L.; Liable-Sands, L. M.; Guzei, L. A. *Inorg. Chem.* **1997**, *36*, 6266.
216. Bremer, M.; Nöth, H.; Thomann, M.; Schmidt, M. *Chem. Ber.* **1995**, *128*, 455.
217. Noth, H.; Thomas, S. *European Journal of Inorganic Chemistry* **1999**, 1373.
218. Binder, H.; Loos, H.; Borrmann, H.; Simon, A.; Flad, H. J.; Savin, A. *Z. Anorg. Allg. Chem.* **1993**, *619*, 1353.
219. Koester, R.; Schuessler, W.; Boese, R.; Blaeser, D. *Chem. Ber.* **1991**, *124*, 2259.
220. Noeth, H.; Thomas, S.; Schmidt, M. *Chem. Ber.* **1996**, *129*, 451.
221. Grigsby, W. J.; Power, P. P. *J. Am. Chem. Soc.* **1996**, *118*, 7981.
222. Dias, H. V. R.; Gorden, J. D. *Inorg. Chem.* **1996**, *35*, 318.
223. Wiberg, N.; Lerner, H.-W.; Wagner, S.; Nöth, H.; Seifert, T. *Z. Naturforsch., B: Chem. Sci.* **1999**, *54*, 877.
224. Heine, A.; Herbst-Irmer, R.; Sheldrick, G. M.; Stalke, D. *Inorg. Chem.* **1993**, *32*, 2694.
225. Brook, A. G.; Abdesaken, F.; Soellradl, H. *J. Organomet. Chem.* **1986**, *299*, 9.
226. Preuss, F.; Wieland, T.; Perner, J.; Heckmann, G. *Z. Naturforsch., B: Chem. Sci.* **1992**, *47*, 1355.
227. Cardin, C. J.; Cardin, D. J.; Clegg, W.; Coles, S. J.; Constantine, S. P.; Rowe, J. R.; Teat, S. J. *J. Organomet. Chem.* **1999**, *573*, 96.
228. Reed, D.; Stalke, D.; Wright, D. S. *Angew. Chem.* **1991**, *103*, 1539.
229. Sekiguchi, A.; Ichinohe, M.; Yamaguchi, S. *J. Am. Chem. Soc.* **1999**, *121*, 10231.
230. Wiberg, N.; Amelunxen, K.; Lerner, H. W.; Schuster, H.; Nöth, H.; Krossing, I.; Schmidt-Amelunxen, M.; Seifert, T. *J. Organomet. Chem.* **1997**, *542*, 1.
231. Klinkhammer, K. W. *Chem. Eur. J.* **1997**, *3*, 1418.
232. Pritzkow, H.; Lobreyer, T.; Sundermeyer, W.; van Eikema Hommes, N. J. R.; Schleyer, P. v. R. *Angew. Chem., Int. Ed. Engl.* **1994**, *33*, 216.
233. Goldfuss, B.; Schleyer, P. v. R.; Handschuh, S.; Hampel, F.; Bauer, W. *Organometallics* **1997**, *16*, 5999.
234. Armstrong, D. R.; Davidson, M. G.; Moncrieff, D.; Stalke, D.; Wright, D. S. *J. Chem. Soc., Chem. Commun.* **1992**, 1413.
235. Evans, D. A. Stereoselective Alkylation Reactions of Chiral Metal Enolates. In *Asymmetric Synthesis*; Morrison, J. D., Ed., Academic Press: New York, **1983**, Vol. 3, Chapter 1, pp 1–110.
236. Gorrell, I. B. *Annu. Rep. Prog. Chem., Sect. A: Inorg. Chem.* **1995**, *91*, 3.
237. Gorrell, I. B. *Annu. Rep. Prog. Chem., Sect. A: Inorg. Chem.* **1996**, *92*, 3.
238. Gorrell, I. B. *Annu. Rep. Prog. Chem., Sect. A: Inorg. Chem.* **1997**, *93*, 3.
239. Gorrell, I. B. *Annu. Rep. Prog. Chem., Sect. A: Inorg. Chem.* **1998**, *94*, 3.
240. Gorrell, I. B. *Annu. Rep. Prog. Chem., Sect. A: Inorg. Chem.* **1999**, *95*, 3.
241. Gorrell, I. B. *Annu. Rep. Prog. Chem., Sect. A: Inorg. Chem.* **2000**, *96*, 5.
242. Gorrell, I. B. *Annu. Rep. Prog. Chem., Sect. A: Inorg. Chem.* **2001**, *97*, 5.
243. Wannagat, U. *Pure Appl. Chem.* **1969**, *19*, 329.
244. Westerhausen, M.; Birg, C.; Krofta, M.; Mayer, P.; Seifert, T.; Nöth, H.; Ptitzner, A.; Nilges, T.; Deiseroth, H.-J. *Z. Anorg. Allg. Chem.* **2000**, *626*, 1073.
245. Engelhardt, L. M.; Jolly, B. S.; Junk, P. C.; Raston, C. L.; Skelton, B. W.; White, A. H. *Aust. J. Chem.* **1986**, *39*, 1337.
246. Bradley, D. C.; Hursthouse, M. B.; Ibrahim, A. A.; Abdul Malik, K. M.; Motevalli, M.; Moseler, R.; Powell, H.; Runnacles, J. D.; Sullivan, A. C. *Polyhedron* **1990**, *9*, 2959.
247. Bürger, H.; Forker, C.; Goubeau, J. *Monatsh. Chem.* **1965**, *96*, 597.
248. Henderson, K. W.; Allan, J. F.; Kennedy, A. R. *Chem. Commun.* **1997**, 1149.
249. Westerhausen, M. *Inorg. Chem.* **1991**, *30*, 96.
250. Kuhlman, R. L.; Vaartstra, B. A.; Caulton, K. G. *Inorg. Synth.* **1997**, *31*, 8.
251. Hitchcock, P. B.; Lappert, M. F.; Lawless, G. A.; Royo, B. *J. Chem. Soc., Chem. Commun.* **1990**, 1141.
252. Frankland, A. D.; Lappert, M. F. *J. Chem. Soc., Dalton Trans.* **1996**, 4151.
253. Neander, S.; Behrens, U. *Z. Anorg. Allg. Chem.* **1999**, *625*, 1429.

254. Tesh, K. F.; Jones, B. D.; Hanusa, T. P.; Huffman, J. C. *J. Am. Chem. Soc.* **1992**, *114*, 6590.
255. Davies, R. P. *Inorg. Chem. Commun.* **2000**, *3*, 13.
256. Williard, P. G.; Liu, Q.-Y. *J. Org. Chem.* **1994**, *59*, 1596.
257. Coles, M. P.; Swenson, D. C.; Jordan, R. F.; Young, V. G., Jr. *Organometallics* **1997**, *16*, 5183.
258. Sebestl, J. L.; Nadasdi, T. T.; Heeg, M. J.; Winter, C. H. *Inorg. Chem.* **1998**, *37*, 1289.
259. Henderson, K. W.; Dorigo, A. E.; Liu, Q.-L.; Williard, P. G. W. *J. Am. Chem. Soc.* **1997**, *119*, 11855.
260. Armstrong, D. R.; Davidson, M. G.; Davies, R. P.; Mitchell, H. J.; Oakley, R. M.; Raithby, P. R.; Snaith, R.; Warren, S. *Angew. Chem., Int. Ed. Engl.* **1996**, *35*, 1942.
261. Williard, P. G.; Liu, Q.-Y.; Lochmann, L. *J. Am. Chem. Soc.* **1992**, *114*, 348.
262. Alder, R. W.; Blake, M. E.; Bortolotti, C.; Bufali, S.; Butts, C. P.; Linehan, E.; Oliva, J. M.; Orpen, A. G.; Quayle, M. J. *Chem. Commun.* **1999**, 241.
263. Rogers, R. D.; Atwood, J. L.; Grüning, R. *J. Organomet. Chem.* **1978**, *157*, 229.
264. Driess, M.; Pritzkow, H.; Skipinski, M.; Winkler, U. *Organometallics* **1997**, *16*, 5108.
265. Grüning, R.; Atwood, J. L. *J. Organomet. Chem.* **1977**, *137*, 101.
266. Tesh, K. F.; Hanusa, T. P.; Huffman, J. C. *Inorg. Chem.* **1990**, *29*, 1584.
267. Clark, A. H.; Haaland, A. *J. Chem. Soc. D* **1969**, 912.
268. Clark, A. H.; Haaland, A. *Acta Chem. Scand.* **1970**, *24*, 3024.
269. Fjeldberg, T.; Hitchcock, P. B.; Lappert, M. F.; Thorne, A. J. *J. Chem. Soc., Chem. Commun.* **1984**, 822.
270. Green, J. C.; Payne, M.; Seddon, E. A.; Andersen, R. A. *J. Chem. Soc., Dalton Trans.* **1982**, 887.
271. Fjeldberg, T.; Andersen, R. A. *J. Mol. Struct.* **1984**, *125*, 287.
272. Lappert, M. F.; Power, P. P.; Sanger, A. R.; Srivastava, R. C. *Metal and Metalloid Amides* **1980**, Halsted Press: New York.
273. Brauer, D. J.; Bürger, H.; Liewald, G. R. *J. Organomet. Chem.* **1983**, *248*, 1.
274. Bartlett, R. A.; Power, P. P. *J. Am. Chem. Soc.* **1987**, *109*, 6509.
275. Rutherford, J. L.; Collum, D. B. *J. Am. Chem. Soc.* **2001**, *123*, 199.
276. Remenar, J. F.; Collum, D. B. *J. Am. Chem. Soc.* **1998**, *120*, 4081.
277. Remenar, J. F.; Lucht, B. L.; Kruglyak, D.; Romesberg, F. E.; Gilchirst, J. H.; Collum, D. B. *J. Org. Chem.* **1997**, *62*, 5748.
278. Remenar, J. F.; Collum, D. B. *J. Am. Chem. Soc.* **1997**, *119*, 5573.
279. Remenar, J. F.; Lucht, B. L.; Collum, D. B. *J. Am. Chem. Soc.* **1997**, *119*, 5567.
280. Sun, X.; Kenkre, S. L.; Remenar, J. F.; Gilchrist, J. H.; Collum, D. B. *J. Am. Chem. Soc.* **1997**, *119*, 4765.
281. Aubrecht, K. B.; Collum, D. B. *J. Org. Chem.* **1996**, *61*, 8674.
282. Lucht, B. L.; Bernstein, M. P.; Remenar, J. F.; Collum, D. B. *J. Am. Chem. Soc.* **1996**, *118*, 10707.
283. Lucht, B. L.; Collum, D. B. *J. Am. Chem. Soc.* **1996**, *118*, 3529.
284. Lucht, B. L.; Collum, D. B. *J. Am. Chem. Soc.* **1996**, *118*, 2217.
285. Lucht, B. L.; Collum, D. B. *J. Am. Chem. Soc.* **1995**, *117*, 9863.
286. Romesberg, F. E.; Collum, D. B. *J. Am. Chem. Soc.* **1995**, *117*, 2166.
287. Carlier, P. R.; Lucht, B. L.; Collum, D. B. *J. Am. Chem. Soc.* **1994**, *116*, 11602.
288. Romesberg, F. E.; Collum, D. B. *J. Am. Chem. Soc.* **1994**, *116*, 9198.
289. Romesberg, F. E.; Collum, D. B. *J. Am. Chem. Soc.* **1994**, *116*, 9187.
290. Lucht, B. L.; Collum, D. B. *J. Am. Chem. Soc.* **1994**, *116*, 7949.
291. Lucht, B. L.; Collum, D. B. *J. Am. Chem. Soc.* **1994**, *116*, 6009.
292. Bernstein, M. P.; Collum, D. B. *J. Am. Chem. Soc.* **1993**, *115*, 8008.
293. Romesberg, F. E.; Bernstein, M. P.; Gilchrist, J. H.; Harrison, A. T.; Fuller, D. J.; Collum, D. B. *J. Am. Chem. Soc.* **1993**, *115*, 3475.
294. Bernstein, M. P.; Collum, D. B. *J. Am. Chem. Soc.* **1993**, *115*, 789.
295. Bernstein, M. P.; Romesberg, F. E.; Fuller, D. J.; Harrison, A. T.; Collum, D. B.; Liu, Q. Y.; Williard, P. G. *J. Am. Chem. Soc.* **1992**, *114*, 5100.
296. Romesberg, F. E.; Collum, D. B. *J. Am. Chem. Soc.* **1992**, *114*, 2112.
297. Gilchrist, J. H.; Collum, D. B. *J. Am. Chem. Soc.* **1992**, *114*, 794.
298. Hall, P. L.; Gilchrist, J. H.; Collum, D. B. *J. Am. Chem. Soc.* **1991**, *113*, 9571.
299. Hall, P. L.; Gilchrist, J. H.; Harrison, A. T.; Fuller, D. J.; Collum, D. B. *J. Am. Chem. Soc.* **1991**, *113*, 9575.
300. Romesberg, F. E.; Gilchrist, J. H.; Harrison, A. T.; Fuller, D. J.; Collum, D. B. *J. Am. Chem. Soc.* **1991**, *113*, 5751.
301. Galiano-Roth, A. S.; Kim, Y. J.; Gilchrist, J. H.; Harrison, A. T.; Fuller, D. J.; Collum, D. B. *J. Am. Chem. Soc.* **1991**, *113*, 5053.
302. Kim, Y. J.; Bernstein, M. P.; Roth, A. S. G.; Romesberg, F. E.; Williard, P. G.; Fuller, D. J.; Harrison, A. T.; Collum, D. B. *J. Org. Chem.* **1991**, *56*, 4435.
303. Gilchrist, J. H.; Harrison, A. T.; Fuller, D. J.; Collum, D. B. *J. Am. Chem. Soc.* **1990**, *112*, 4069.
304. Galiano-Roth, A. S.; Collum, D. B. *J. Am. Chem. Soc.* **1989**, *111*, 6772.
305. DePue, J. S.; Collum, D. B. *J. Am. Chem. Soc.* **1988**, *110*, 5518.
306. DePue, J. S.; Collum, D. B. *J. Am. Chem. Soc.* **1988**, *110*, 5524.
307. Collum, D. B. *Acc. Chem. Res.* **1992**, *25*, 448.
308. Collum, D. B. *Acc. Chem. Res.* **1993**, *26*, 227.
309. Barnett, N. D. R.; Mulvey, R. E.; Clegg, W.; O'Neil, P. A. *J. Am. Chem. Soc.* **1991**, *113*, 8187.
310. Carlier, P. R.; Lo, C. W. S. *J. Am. Chem. Soc.* **2000**, *122*, 12819.
311. Sun, X.; Collum, D. B. *J. Am. Chem. Soc.* **2000**, *122*, 2452.
312. Sun, X.; Collum, D. B. *J. Am. Chem. Soc.* **2000**, *122*, 2459.
313. Kennedy, A. R.; Mulvey, R. E.; Rowlings, R. B. *Angew. Chem., Int. Ed. Engl.* **1998**, *37*, 3180.
314. Kennedy, A. R.; Mulvey, R. E.; Roberts, B. A.; Rowlings, R. B.; Raston, C. L. *Chem. Commun.* **1999**, 353.
315. Gallagher, D. J.; Henderson, K. W.; Kennedy, A. R.; O'Hara, C. T.; Mulvey, R. E.; Rowlings, R. B. *Chem. Commun.* **2002**, 376.
316. Andrews, P. C.; Kennedy, A. R.; Mulvey, R. E.; Raston, C. L.; Roberts, B. A.; Rowlings, R. B. *Angew. Chem., Int. Ed. Engl.* **2000**, *39*, 1960.

317. Barr, L.; Kennedy, A. R.; MacLellan, J. G.; Moir, J. H.; Mulvey, R. E.; Rodger, P. J. A. *Chem. Commun.* **2000**, 1757.
318. Armstrong, D. R.; Barr, D.; Snaith, R.; Clegg, W.; Mulvey, R. E.; Wade, K.; Reed, D. *J. Chem. Soc., Dalton Trans.* **1987**, 1071.
319. Mulvey, R. E. *Chem. Soc. Rev.* **1998**, *27*, 339.
320. Armstrong, D. R.; Barr, D.; Clegg, W.; Hodgson, S. M.; Mulvey, R. E.; Reed, D.; Snaith, R.; Wright, D. S. *J. Am. Chem. Soc.* **1989**, *111*, 4719.
321. Barr, D.; Clegg, W.; Hodgson, S. M.; Lamming, G. R.; Mulvey, R. E.; Scott, A. J.; Snaith, R.; Wright, D. S. *Angew. Chem.* **1989**, *101*, 1279.
322. Mulvey, R. E. *Chem. Soc. Rev.* **1991**, *20*, 167.
323. Gregory, K.; Schleyer, P. v. R.; Snaith, R. *Adv. Inorg. Chem.* **1991**, *37*, 47.
324. Barr, D.; Snaith, R.; Clegg, W.; Mulvey, R. E.; Wade, K. *J. Chem. Soc., Dalton Trans.* **1987**, 2141.
325. Hitchcock, P. B.; Lappert, M. F.; Leung, W.-P.; Liu, D.-S.; Mak, T. C. W.; Wang, Z.-X. *J. Chem. Soc., Dalton Trans.* **1999**, 1263.
326. Clegg, W.; Snaith, R.; Shearer, H. M. M.; Wade, K.; Whitehead, G. *J. Chem. Soc., Dalton Trans.* **1983**, 1309.
327. Barr, D.; Clegg, W.; Mulvey, R. E.; Snaith, R.; Wade, K. *J. Chem. Soc., Chem. Commun.* **1986**, 295.
328. Barr, D.; Clegg, W.; Mulvey, R. E.; Snaith, R. *J. Chem. Soc., Chem. Commun.* **1989**, 57.
329. Bianchi, A.; Giusti, J.; Paoletti, P.; Mangani, S. *Inorg. Chim. Acta* **1986**, *117*, 157.
330. Huixiang, Z.; Yunxiu, S.; Guangdi, Y.; Cheng, S.; Jin Changchun, J. *Gaodeng Xuexiao Huaxue Xuebao (Chem. J. Chin. Uni.)* **1986**, *7*, 721.
331. Lockhart, J. C.; McDonnell, M. B.; Clegg, W.; Stuart-Hill, M. N. *J. Chem.Soc., Perkin Trans.* **1987**, *2*, 639.
332. Dale, J.; Eggestad, J.; Fredriksen, S. B.; Groth, P. *Chem. Commun.* **1987**, 1391.
333. Groth, P. *Acta Chem. Scand. Ser. A* **1987**, *41*, 355.
334. Wei, Y. Y.; Tinant, B.; Declercq, J. P.; van Meerssche, M.; Dale, J. *Acta Crystallogr., Sect.C* **1987**, *43*, 1274.
335. Wei, Y. Y.; Tinant, B.; Declercq, J. P.; van Meerssche, M.; Dale, J. *Acta Crystallogr., Sect.C* **1988**, *44*, 77.
336. Wei, Y. Y.; Tinant, B.; Declercq, J. P.; van Meerssche, M.; Dale, J. *Acta Crystallogr., Sect.C* **1988**, *44*, 73.
337. Wei, Y. Y.; Tinant, B.; Declercq, J. P.; van Meerssche, M.; Dale, J. *Acta Crystallogr., Sect.C* **1988**, *44*, 68.
338. Raevskii, O. A.; Tkachev, V. V.; Atovmyan, L. O.; Zubareva, V. E.; Bulgak, I. I.; Batyr, D. G. *Koord. Khim.* **1988**, *14*, 1697.
339. Buchanan, G. W.; Kirby, R. A.; Charland, J. P. *J. Am. Chem. Soc.* **1988**, *110*, 2477.
340. Suwinska, K.; Lipkowski, J. *J. Inclusion Phenom. Macrocyclic Chem.* **1988**, *6*, 237.
341. Reiss, C. A.; Goubitz, K.; Heijdenrijk, D. *Acta Crystallogr., Sect. C: Cryst. Str.Commun.* **1990**, *46*, 1084.
342. Buchanan, G. W.; Kirby, R. A.; Charland, J. P. *Can. J. Chem.* **1990**, *68*, 49.
343. Tkachev, V. V.; Atovmyan, L. O.; Zubareva, V. E.; Raevskii, O. A. *Koord. Khim.* **1990**, *16*, 443.
344. Olsher, U.; Krakowiak, K. E.; Dalley, N. K.; Bradshaw, J. S. *Tetrahedron* **1991**, *47*, 2947.
345. Olsher, U.; Frolow, F.; Dalley, N. K.; Weiming, J.; Yu, Z.-Y.; Knobeloch, J. M.; Bartsch, R. A. *J. Am. Chem. Soc.* **1991**, *113*, 6570.
346. Buchanan, G. W.; Kirby, R. A.; Charland, J. P.; Ratcliffe, C. I. *J. Org. Chem.* **1991**, *56*, 203.
347. Buchanan, G. W.; Mathias, S.; Lear, Y.; Bensimon, C. *Can. J. Chem.* **1991**, *69*, 404.
348. Buchanan, G. W.; Mathias, S.; Bensimon, C.; Charland, J. P. *Can. J. Chem.* **1992**, *70*, 981.
349. Sachleben, R. A.; Burns, J. H. *J. Chem. Soc., Perkin Trans* **1992**, *2*, 1971.
350. Czech, B. P.; Zazulak, W.; Kumar, A.; Olsher, U.; Feinberg, H.; Cohen, S.; Shoham, G.; Dalley, N. K.; Bartsch, R. A. *J. Heterocycl. Chem.* **1992**, *29*, 1389.
351. Delgado, M.; Wolf, R. E., Jr.; Hartman, J. R.; McCafferty, G.; Yagbasan, R.; Rawle, S. C.; Watkin, D. J.; Cooper, S. R. *J. Am. Chem. Soc.* **1992**, *114*, 8983.
352. Burns, J. H.; Sachleben, R. A.; Davis, M. C. *Inorg. Chim. Acta* **1994**, *223*, 125.
353. Cheng, S. *Chin. J. Struct. Chem.* **1997**, *16*, 24.
354. Driega, A. B.; Buchanan, G. W.; Bensimon, C. *Can. J. Chem.* **1998**, *76*, 142.
355. Chekhlov, A. N.; Martynov, I. V. *Dokl. Akad. Nauk SSSR* **1999**, *367*, 70.
356. Bryan, J. C.; Sachleben, R. A. *Acta Crystallogr., Sect. C* **2000**, *56*, 1104.
357. Hirayama, F.; Zabel, V.; Saenger, W.; Vogtle, F. *Acta Crystallogr., Sect. C* **1985**, *41*, 61.
358. Koch, K. R.; Niven, M. L.; Sacht, C. *J. Coord. Chem.* **1992**, *26*, 161.
359. Groth, P. *Acta Chem. Scand. Ser. A* **1985**, *39*, 68.
360. Bradshaw, J. S.; McDaniel, C. W.; Skidmore, B. D.; Nielsen, R. B.; Wilson, B. E.; Dalley, N. K.; Izatt, R. M. *J. Heterocycl. Chem.* **1987**, *24*, 1085.
361. He, G.-X.; Kikukawa, K.; Ohe, H.; Machida, M.; Matsuda, T. *J. Am. Chem. Soc.* **1988**, *110*, 603.
362. Akabori, S.; Kumagai, T.; Habata, Y.; Sato, S. *J. Chem. Soc., Chem. Commun.* **1988**, 661.
363. Akabori, S.; Kumagai, T.; Habata, Y.; Sato, S. *J. Chem. Soc., Perkin Trans.* **1989**, *1*, 1497.
364. Abou-Hamdan, A; Lincoln, S. F.; Snow, M. R.; Tiekink, E. R. T. *Aust. J. Chem.* **1988**, *41*, 1363.
365. Goubitz, K.; Reiss, C. A.; Heijdenrijk, D. *Acta Crystallogr., Sect. C : Cryst. Str.Commun.* **1990**, *46*, 1087.
366. Lutze, G.; Tittelbach, F.; Graubaum, H.; Ramm, M. *Phosphorus, Sulfur Silicon Relat. Elem.* **1994**, *91*, 81.
367. Zhang, L.-J.; Lin, H.-K.; Bu, X.-H.; Chen, Y.-T.; Liu, X.-L.; Miao, F.-M. *Inorg. Chim. Acta* **1995**, *240*, 257.
368. Zhang, L.-J.; Liu, X.-L.; Ma, S.-K.; Zhou, W.-H.; Miao, F.-M. *Chin. J. Struct. Chem. (Jiegou Huaxue)* **1996**, *15*, 15.
369. Habata, Y.; Akabori, S. *J. Chem. Soc., Dalton Trans.* **1996**, 3871.
370. Kubo, K.; Yamamoto, E.; Kato, N.; Mori, A. *Acta Crystallogr., Sect. C* **1999**, *55*, 1819.
371. Yamamoto, E.; Kubo, K.; Kato, N.; Mori, A. *Acta Crystallogr., Sect. C: Cryst. Str. Commun.* **2000**, *56*, 329.
372. Chekhlov, A. N. *Koord. Khim.* **2000**, *26*, 163.
373. Chekhlov, A. N. *Koord. Khim.* **2000**, *26*, 151.
374. Buchanan, G. W.; Driega, A. B.; Yap, G. P. A. *Can. J. Chem.* **2000**, *78*, 316.
375. Lincoln, S. F.; Horn, E.; Snow, M. R.; Hambley, T. W.; Brereton, I. M.; Spotswood, T. M. *J. Chem. Soc., Dalton Trans.* **1986**, 1075.
376. Wei, Y. Y.; Tinant, B.; Declercq, J. P.; van Meerssche, M.; Dale, J. *Acta Crystallogr., Sect. C* **1987**, *43*, 1076.
377. Wei, Y. Y.; Tinant, B.; Declercq, J. P.; van Meerssche, M.; Dale, J. *Acta Crystallogr., Sect. C* **1987**, *43*, 1080.
378. Wei, Y. Y.; Tinant, B.; Declercq, J. P.; van Meerssche, M.; Dale, J. *Acta Crystallogr., Sect. C* **1987**, *43*, 1270.

379. Wei, Y. Y.; Tinant, B.; Declercq, J. P.; van Meerssche, M.; Dale, J. *Acta Crystallogr., Sect. C* **1987**, *43*, 1279.
380. Tucker, J. A.; Knobler, C. B.; Goldberg, I.; Cram, D. J. *J. Org. Chem.* **1989**, *54*, 5460.
381. Bauer, H.; Matz, V.; Lang, M.; Krieger, C.; Staab, H. *Chem. Ber.* **1994**, *127*, 1993.
382. Metzger, E.; Aeschimann, R.; Egli, M.; Suter, G.; Dohner, R.; Ammann, D.; Dobler, M.; Simon, W. *Helv. Chim. Acta* **1986**, *69*, 1821.
383. Tinant, B.; Declercq, J.-P.; Weiler, J.; De Man, X. *Acta Crystallogr., Sect. C* **1989**, *45*, 1050.
384. Iimori, T.; Still, W. C.; Rheingold, A. L.; Staley, D. L. *J. Am. Chem. Soc.* **1989**, *111*, 3439.
385. Thomas, L. M.; Ramasubbu, N.; Bhandary, K. K. *Biopolymers* **1994**, *34*, 1007.
386. Tkachev, V. V.; Raevskii, O. A.; Luk'yanov, N. V.; Van'kin, G. I.; Yurchenko, R. I.; Yurchenko, V. G.; Solotnov, A. F.; Pinchuk, A. M.; Galenko, T. G.; Ivanova, T. A.; Atovmyan, L. O. *Izv. Akad. Nauk, Ser. Khim.* **1992**, 2784.
387. Lu, T.-H.; Lin, J.-L. L.; Lan, W.-J.; Chung, C.-S. *Acta Crystallogr., Sect. C* **1997**, *53*, 1598.
388. Lipkowski, J.; Soldatov, D. *Supramol. Chem.* **1993**, *3*, 43.
389. Lipkowski, J.; Soldatov, D. *J. Coord. Chem.* **1993**, *28*, 265.
390. Lipkowski, J.; Soldatov, D. V. *J. Inclusion Phenom. Macrocyclic Chem.* **1994**, *18*, 317.
391. Bohland, H.; Hanay, W.; Noltemeyer, M.; Meller, A.; G, S. H. *Fresenius Z. Anal. Chem.* **1998**, *361*, 725.
392. Purdy, A. P.; George, C. F. *Main Group Chem.* **1996**, *1*, 229.
393. Chow, M.-Y.; Mak, T. C. W. *Inorg. Chim. Acta* **1992**, *202*, 231.
394. Barr, D.; Doyle, M. J.; Drake, S. R.; Raithby, P. R.; Snaith, R. *J. Chem. Soc., Chem. Commun.* **1988**, 1415.
395. Barr, D.; Doyle, M. J.; Drake, S. R.; Raithby, P. R.; Snaith, R. *Polyhedron* **1989**, *8*, 215.
396. Barr, D.; Brooker, A. T.; Doyle, M. J.; Drake, S. R.; Raithby, P. R.; Snaith, R.; Wright, D. S. *J. Chem. Soc., Chem. Commun.* **1989**, 893.
397. Barr, D.; Doyle, M. J.; Drake, S. R.; Raithby, P. R.; Snaith, R.; Wight, D. S. *Inorg. Chem.* **1989**, *28*, 1767.
398. Armstrong, D. R.; Khandelwal, A. H.; Raithby, P. R.; Snaith, R.; Stalke, D.; Wright, D. S. *Inorg. Chem.* **1993**, *32*, 2132.
399. Raithby, P. R.; Reed, D.; Snaith, R.; Wright, D. S. *Angew. Chem., Int. Ed. Engl.* **1991**, *30*, 1011.
400. Seebach, D.; Buerger, H. M.; Plattner, D. A.; Nesper, R.; Faessler, T. *Helv. Chim. Acta* **1993**, *76*, 2581.
401. Tamburini, S.; Vigato, P. A.; Casellato, U.; Graziani, R. *J. Chem. Soc., Dalton Trans.* **1989**, 1993.
402. Mori, H.; Tanaka, S.; Mori, T.; Maruyama, Y. *Bull. Chem. Soc. Jpn.* **1995**, *68*, 1136.
403. Tinant, B.; Declercq, J.-P.; Weiler, J. *J. Chem. Soc., Perkin Trans.* **1994**, *2*, 1539.
404. Kejian, X; Huijie, L; Huixiang, Z; Yunxiu, S. *Chin. Sci. Bull. (Engl. Transl.)* **1986**, *31*, 95.
405. Habata, Y.; Saeki, T.; Akabori, S.; Zhang, X. X.; Bradshaw, J. S. *Chem. Commun.* **2000**, 1469.
406. Barr, D.; Doyle, M. J.; Mulvey, R. E.; Raithby, P. R.; Snaith, R.; Wright, D. S. *J. Chem. Soc., Chem. Commun.* **1988**, 145.
407. Arnold, K. A.; Viscariello, A. M.; Kim, M.; Gandour, R. D.; Fronczek, F. R.; Gokel, G. W. *Tetrahedron Lett.* **1988**, *29*, 3025.
408. Watson, W. H.; Grossie, D. A.; Voegtle, F.; Mueller, W. M. *Acta Crystallogr., Sect. C: Cryst. Struct. Commun.* **1983**, *C39*, 720.
409. Labahn, T.; Mandel, A.; Magull, J. *Z. Anorg. Allg. Chem.* **1999**, *625*, 1273.
410. Wang, M.; Zheng, P.; Zhang, J.; Chen, Z.; Shen, J.; Yang, Y. *Acta Crystallogr., Sect. C: Cryst. Struct. Commun.* **1987**, *C43*, 1544.
411. Crispini, A.; Errington, R. J.; Fisher, G. A.; Funke, F. J.; Norman, N. C.; Orpen, A. G.; Stratford, S. E.; Struve, O. *J. Chem. Soc., Dalton Trans.* **1994**, 1327.
412. Brodersen, K.; Cygan, M.; Hummel, H.-U. H. *Z. Naturforsch.,Teil B* **1984**, *39*, 582.
413. Farrugia, L. J.; Carmalt, C. J.; Norman, N. C. *Inorg. Chim. Acta* **1996**, *248*, 263.
414. Hanay, W.; Bohland, H.; Noltemeyer, M.; Schmidt, H.-G. *Mikrochim. Acta* **2000**, *133*, 197.
415. Gardiner, M. G.; Hanson, G. R.; Henderson, M. J.; Lee, F. C.; Raston, C. L. *Inorg. Chem.* **1994**, *33*, 2456.
416. Corvaja, C.; Pasimeni, L. *Chem. Phys. Lett.* **1976**, *39*, 261.
417. Clopath, P.; Von Zelewsky, A. *Helv. Chim. Acta* **1973**, *56*, 980.
418. Kaltsoyannis, N. *J. Chem. Soc., Dalton Trans.* **1996**, 1583.
419. Thiele, K.-H.; Lorenz, V.; Thiele, G.; Zoennchen, P.; Scholz, J. *Angew. Chem.* **1994**, *106*, 1461.
420. Lorenz, V.; Neumueller, B.; Thiele, K.-H. *Z. Naturforsch., B: Chem. Sci.* **1995**, *50*, 71.
421. Yelamos, C.; Heeg, M. J.; Winter, C. H. *Inorg. Chem.* **1998**, *37*, 3892.
422. Fleming, J. S.; Psillakis, E.; Couchman, S. M.; Jeffery, J. C.; McCleverty, J. A.; Ward, M. D. *J. Chem. Soc., Dalton Trans.* **1998**, 537.
423. Pfeiffer, D.; Heeg, M. J.; Winter, C. H. *Angew. Chem., Int. Ed. Engl.* **1998**, *37*, 2517.
424. Pfeiffer, D.; Heeg, M. J.; Winter, C. H. *Inorg. Chem.* **2000**, *39*, 2377.
425. Steiner, A.; Lawson, G. T.; Walfort, B.; Leusser, D.; Stalke, D. *J. Chem. Soc., Dalton Trans.* **2001**, 219.
426. Barker, J.; Kilner, M. *Coord. Chem. Rev.* **1994**, *133*, 219.
427. Averbuj, C.; Tish, E.; Eisen, M. S. *J. Am. Chem. Soc.* **1998**, *120*, 8640.
428. Niemeyer, M.; Power, P. P. *Inorg. Chem.* **1997**, *36*, 4688.
429. Walther, D.; Gebhardt, P.; Fischer, R.; Kreher, U.; Gorls, H. *Inorg. Chim. Acta* **1998**, *281*, 181.
430. Westerhausen, M.; Hausen, H. D. *Z. Anorg. Allg. Chem.* **1992**, *615*, 27.
431. Kincaid, K.; Gerlach, C. P.; Giesbrecht, G. R.; Hagadorn, J. R.; Whitener, G. D.; Shafir, A.; Arnold, J. *Organometallics* **1999**, *18*, 5360.
432. Westerhausen, M.; Schwarz, W. *Z. Naturforsch., B: Chem. Sci.* **1992**, *47*, 453.
433. Westerhausen, M.; Schwarz, W. *Z. Anorg. Allg. Chem.* **1993**, *619*, 1455.
434. Schmidt, J. A. R.; Arnold, J. *Chem. Commun.* **1999**, 2149.
435. Caro, C. F.; Hitchcock, P. B.; Lappert, M. F.; Layh, M. *Chem. Commun.* **1998**, 1297.
436. Cotton, F. A.; Haefner, S. C.; Matonic, J. H.; Wang, X.; Murillo, C. A. *Polyhedron* **1996**, *16*, 541.
437. Doyle, D.; Gun'ko, Y. K.; Hitchcock, P. B.; Lappert, M. F. *Dalton* **2000**, 4093.
438. Stalke, D.; Wedler, M.; Edelmann, F. T. *J. Organomet. Chem.* **1992**, *431*, C1.
439. Eisen, M. S.; Kapon, M. *J. Chem. Soc., Dalton Trans.* **1994**, 3507.
440. Gebauer, T.; Dehnicke, K.; Goesmann, H.; Fenske, D. *Z. Naturforsch., B: Chem. Sci.* **1994**, *49*, 1444.

441. Brauer, D. J.; Buerger, H.; Liewald, G. R. *J. Organomet. Chem.* **1986**, *308*, 119.
442. Brask, J. K.; Chivers, T.; Schatte, G. *Chem. Commun.* **2000**, 1805.
443. Clegg, W.; Horsburgh, L.; Dennison, P. R.; Mackenzie, F. M.; Mulvey, R. E. *Chem. Commun.* **1996**, 1065.
444. Himmel, K.; Jansen, M. *Inorg. Chem.* **1998**, *37*, 3437.
445. Brumm, H.; Peters, E.; Jansen, M. *Angew. Chem., Int. Ed. Engl.* **2001**, *40*, 2069.
446. Becker, G.; Eschbach, B.; Kaeshammer, D.; Mundt, O. *Z. Anorg. Allg. Chem.* **1994**, *620*, 29.
447. Westerhausen, M. *Trends Organomet. Chem.* **1997**, *2*, 89.
448. Karsch, H. H.; Graf, V.; Reisky, M. *Phosphorus, Sulfur Silicon Relat. Elem.* **1999**, *144–146*, 553.
449. Hanusa, T. P. *Chem. Rev.* **1993**, *93*, 1023.
450. Fryzuk, M. D.; Gao, X.; Rettig, S. J. *Can. J. Chem.* **1995**, *73*, 1175.
451. Engelhardt, L. M.; Harrowfield, J. M.; Lappert, M. F.; MacKinnon, I. A.; Newton, B. H.; Raston, C. L.; Skelton, B. W.; White, A. H. *J. Chem. Soc., Chem. Commun.* **1986**, 846.
452. Westerhausen, M.; Krofta, M.; Mayer, P.; Warchhold, M.; Nöth, H. *Inorg. Chem.* **2000**, *39*, 4721.
453. Westerhausen, M.; Krofta, M.; Pfitzner, A. *Inorg. Chem.* **1999**, *38*, 598.
454. Driess, M.; Hoffmanns, U.; Martin, S.; Merz, K.; Pritzkow, H. *Angew. Chem., Int. Ed. Engl.* **1999**, *38*, 2733.
455. Becker, G.; Eschbach, B.; Mundt, O.; Reti, M.; Niecke, E.; Issberner, K.; Nieger, M.; Thelen, V.; Nöth, H.; Waldhoer, R.; Schmidt, M. *Z. Anorg. Allg. Chem.* **1998**, *624*, 469.
456. Jones, R. A.; Koschmieder, S. U.; Nunn, C. M. *Inorg. Chem.* **1987**, *26*, 3610.
457. Becker, G.; Hartmann, H. M.; Schwarz, W. *Z. Anorg. Allg. Chem.* **1989**, *577*, 9.
458. Hey, E.; Engelhardt, L. M.; Raston, C. L.; White, A. H. *Angew. Chem.* **1987**, *99*, 61.
459. Bartlett, R. A.; Olmstead, M. M.; Power, P. P.; Sigel, G. A. *Inorg. Chem.* **1987**, *26*, 1941.
460. Hey, E.; Weller, F. *J. Chem. Soc., Chem. Commun.* **1988**, 782.
461. Niediek, K.; Neumueller, B. *Z. Anorg. Allg. Chem.* **1993**, *619*, 885.
462. Rabe, G. W.; Liable-Sands, L. M.; Incarvito, C. D.; Lam, K.-C.; Rheingold, A. L. *Inorg. Chem.* **1999**, *38*, 4342.
463. Rabe, G. W.; Heise, H.; Liable-Sands, L. M.; Guzei, I. A.; Rheingold, A. L. *Dalton* **2000**, 1863.
464. Rabe, G. W.; Kheradmandan, S.; Yap, G. P. A. *Inorg. Chem.* **1998**, *37*, 6541.
465. Rabe, G. W.; Kheradmandan, S.; Liable-Sands, L. M.; Guzei, I. A.; Rheingold, A. L. *Angew. Chem., Int. Ed. Engl.* **1998**, *37*, 1404.
466. Becker, G.; Eschbach, B.; Mundt, O.; Seidler, N. *Z. Anorg. Allg. Chem.* **1994**, *620*, 1381.
467. Beswick, M. A.; Hopkins, A. D.; Kerr, L. C.; Mosquera, M. E. G.; Palmer, J. S.; Raithby, P. R.; Rothenberger, A.; Wheatley, A. E. H.; Wright, D. S.; Stalke, D.; Steiner, A. *Chem. Commun.* **1998**, 1527.
468. Hey-Hawkins, E.; Kurz, S. *Phosphorus, Sulfur Silicon Relat. Elem.* **1994**, *90*, 281.
469. Koutsantonis, G. A.; Andrews, P. C.; Raston, C. L. *J. Chem. Soc., Chem. Commun.* **1995**, 47.
470. Westerhausen, M.; Schneiderbauer, S.; Knizek, J.; Nöth, H.; Pfitzner, A. *Eur. J. Inorg. Chem.* **1999**, 2215.
471. Westerhausen, M.; Krofta, M.; Mayer, P. *Z. Anorg. Allg. Chem.* **2000**, *626*, 2307.
472. Wiberg, N.; Worner, A.; Fenske, D.; Nöth, H.; Knizek, J.; Polborn, K. *Angew. Chem., Int. Ed. Engl.* **2000**, *39*, 1838.
473. Westerhausen, M.; Digeser, M. H.; Krofta, M.; Wiberg, N.; Nöth, H.; Knizek, J.; Ponikwar, W.; Seifert, T. *Eur. J. Inorg. Chem.* **1999**, 743.
474. Westerhausen, M.; Loew, R.; Schwarz, W. *J. Organomet. Chem.* **1996**, *513*, 213.
475. Driess, M.; Pritzkow, H. *Z. Anorg. Allg. Chem.* **1996**, *622*, 1524.
476. Driess, M.; Huttner, G.; Knopf, N.; Pritzkow, H.; Zsolnai, L. *Angew. Chem., Int. Ed. Engl.* **1995**, *34*, 316.
477. Stieglitz, G.; Neumueller, B.; Dehnicke, K. *Z. Naturforsch., B: Chem. Sci.* **1993**, *48*, 156.
478. Rabe, G. W.; Riede, J.; Schier, A. *Acta Crystallogr., Sect. C: Cryst. Struct. Commun.* **1996**, *C52*, 1350.
479. Bartlett, R. A.; Olmstead, M. M.; Power, P. P. *Inorg. Chem.* **1986**, *25*, 1243.
480. Kuhl, O.; Sieler, J.; Baum, G.; Hey-Hawkins, E. *Z. Anorg. Allg. Chem.* **2000**, *626*, 605.
481. Kuhl, O.; Sieler, J.; Hey-Hawkins, E. *Z. Kristallogr.* **1999**, *214*, 496.
482. Mulvey, R. E.; Wade, K.; Armstrong, D. R.; Walker, G. T.; Snaith, R.; Clegg, W.; Reed, D. *Polyhedron* **1987**, *6*, 987.
483. Hey-Hawkins, E.; Sattler, E. *J. Chem. Soc., Chem. Commun.* **1992**, 775.
484. Hey, E.; Hitchcock, P. B.; Lappert, M. F.; Rai, A. K. *J. Organomet. Chem.* **1987**, *325*, 1.
485. Westerhausen, M.; Hartmann, M.; Schwarz, W. *Inorg. Chim. Acta* **1998**, *269*, 91.
486. Driess, M.; Rell, S.; Pritzkow, H.; Janoschek, R. *Chem. Commun.* **1996**, 305.
487. Brauer, D. J.; Hietkamp, S.; Stelzer, O. *J. Organomet. Chem.* **1986**, *299*, 137.
488. Karsch, H. H.; Richter, R.; Deubelly, B.; Schier, A.; Paul, M.; Heckel, M.; Angermeier, K.; Hiller, W. *Z. Naturforsch., B: Chem. Sci.* **1994**, *49*, 1798.
489. Karsch, H. H.; Reisky, M. *Eur. J. Inorg. Chem.* **1998**, 905.
490. Karsch, H. H.; Deubelly, B.; Mueller, G. *J. Organomet. Chem.* **1988**, *352*, 47.
491. Clegg, W.; Izod, K.; McFarlane, W.; O'Shaughnessy, P. *Organometallics* **1998**, *17*, 5231.
492. Ellermann, J.; Bauer, W.; Schuetz, M.; Heinemann, F. W.; Moll, M. *Monatsh. Chem.* **1998**, *129*, 547.
493. Westerhausen, M.; Digeser, M. H.; Schwarz, W. *Inorg. Chem.* **1997**, *36*, 521.
494. Bender, H. R. G.; Niecke, E.; Nieger, M. *J. Am. Chem. Soc.* **1993**, *115*, 3314.
495. Niecke, E.; Klein, E.; Nieger, M. *Angew. Chem.* **1989**, *101*, 792.
496. Kovacs, I.; Krautscheid, H.; Matern, E.; Sattler, E.; Fritz, G.; Hoenle, W.; Borrmann, H.; von Schnering, H. G. *Z. Anorg. Allg. Chem.* **1996**, *622*, 1564.
497. Wiberg, N.; Woerner, A.; Lerner, H.-W.; Karaghiosoff, K.; Fenske, D.; Baum, G.; Dransfeld, A.; Von Rague Schleyer, P. *Eur. J. Inorg. Chem.* **1998**, 833.
498. Kremer, T.; Hampel, F.; Knoch, F. A.; Bauer, W.; Schmidt, A.; Gabold, P.; Schuetz, M.; Ellermann, J.; Schleyer, P. v. R. *Organometallics* **1996**, *15*, 4776.
499. Karsch, H. H.; Grauvogl, G.; Kawecki, M.; Bissinger, P.; Kumberger, O.; Schier, A.; Mueller, G. *Organometallics* **1994**, *13*, 610.
500. Karsch, H. H.; Deubelly, B.; Grauvogl, G.; Lachmann, J.; Mueller, G. *Organometallics* **1992**, *11*, 4245.
501. Stone, F. G. A.; Burg, A. J. *J. Am. Chem. Soc.* **1954**, *76*, 386.
502. Izod, K. *Adv. Inorg. Chem.* **2000**, *50*, 33.
503. Beswick, M. A.; Lawson, Y. G.; Raithby, P. R.; Wood, J. A.; Wright, D. S. *J. Chem. Soc., Dalton Trans.* **1999**, 1921.

504. Westerhausen, M.; Pfitzner, A. *J. Organomet. Chem.* **1995**, *487*, 187.
505. Westerhausen, M.; Schwarz, W. *Z. Naturforsch., B: Chem. Sci.* **1995**, *50*, 106.
506. Westerhausen, M.; Digeser, M. H.; Knizek, J.; Schwarz, W. *Inorg. Chem.* **1998**, *37*, 619.
507. Westerhausen, M.; Makropoulos, N.; Piotrowski, H.; Warchhold, M.; Nöth, H. *J. Organomet. Chem.* **2000**, *614–615*, 70.
508. Westerhausen, M.; Birg, C.; Piotrowski, H. *Eur. J. Inorg. Chem.* **2000**, 2173.
509. Jones, L. J. III; McPhail, A. T.; Wells, R. L. *J. Coord. Chem.* **1995**, *34*, 119.
510. Driess, M.; Pritzkow, H. *Angew. Chem.* **1992**, *104*, 350.
511. Driess, M.; Pritzkow, H.; Martin, S.; Rell, S.; Fenske, D.; Baum, G. *Angew. Chem., Int. Ed. Engl.* **1996**, *35*, 986.
512. Einspahr, H.; Bugg, C. E. *Acta Crystallogr.* **1981**, *B37*, 1044.
513. Sobota, P.; Szafert, S. l. *Inorg. Chem.* **1996**, *35*, 1778.
514. Kim, K. M.; Lee, S. S.; Jung, O.-S.; Sohn, Y. S. *Inorg. Chem.* **1996**, *35*, 3077.
515. Cho, T. H.; Chaudhuri, B.; Snider, B. B.; Foxman, B. M. *Chem. Commun.* **1996**, 1337.
516. Vela, M. J.; Snider, B. B.; Foxman, B. M. *Chem. Mater.* **1998**, *10*, 3167.
517. Fehr, T.; Kuhn, M.; Loosli, H. R.; Ponelle, M.; Boelsterli, J. J.; Walkinshaw, M. D. *J. Antibiot.* **1989**, *42*, 897.
518. Dirlam, J. P.; Presseau-Linabury, L.; Koss, D. A. *J. Antibiot.* **1990**, *43*, 727.
519. Takahashi, Y.; Nakamura, H.; Ogata, R.; Matsuda, N.; Hamada, M.; Naganawa, H.; Takita, T.; Iitaka, Y.; Sato, K.; Takeuchi, T. *J. Antibiot.* **1990**, *43*, 441.
520. Scharfenberg-Pfeiffer, D.; Czugler, M. *Pharmazie* **1991**, *46*, 781.
521. Pangborn, W.; Duax, W.; Langs, D. *Acta Crystallogr., Sect. C: Cryst. Struct. Commun.* **1987**, *C43*, 890.
522. Pangborn, W.; Duax, W.; Langs, D. *J. Am. Chem. Soc.* **1987**, *109*, 2163.
523. Walker, E. H., Jr.; Apblett, A. W. *Ceram. Trans.* **1999**, *94*, 205.
524. Apblett, A. W.; Georgieva, G. D.; Mague, J. T. *Can. J. Chem.* **1997**, *75*, 483.
525. Bahl, A. M.; Krishnaswamy, S.; Massand, N. M.; Burkey, D. J.; Hanusa, T. P. *Inorg. Chem.* **1997**, *36*, 5413.
526. Kato, S.; Kitaoka, N.; Niyomura, O.; Kitoh, Y.; Kanda, T.; Ebihara, M. *Inorg. Chem.* **1999**, *38*, 496.
527. Niyomura, O.; Kato, S.; Kanda, T. *Inorg. Chem.* **1999**, *38*, 507.
528. Vaartstra, B. A.; Gardiner, R. A.; Gordon, D. C.; Ostrander, R. L.; Rheingold, A. L. *Mater. Res. Soc. Symp. Proc.* **1994**, *335*, 203.
529. Belcher, R.; Cranley, C. R.; Majer, J. R.; Stephen, W. I.; Uden, P. C. *Anal. Chim. Acta* **1972**, *60*, 109.
530. Arunasalam, V. C.; Baxter, I.; Drake, S. R.; Hursthouse, M. B.; Malik, K. M. A.; Miller, S. A. S.; Mingos, D. M. P.; Otway, D. J. *J. Chem. Soc., Dalton Trans.* **1997**, 1331.
531. Drake, S. R.; Otway, D. J.; Hursthouse, M. B.; Abdul Malik, K. M. *J. Chem. Soc., Dalton Trans.* **1993**, 2883.
532. Darr, J. A.; Drake, S. R.; Hursthouse, M. B.; Malik, K. M. A.; Miller, S. A. S.; Mingos, D. M. P. *J. Chem. Soc., Dalton Trans.* **1997**, 945.
533. Arunasalam, V. C.; Drake, S. R.; Hursthouse, M. B.; Malik, K. M. A.; Miller, S. A. S.; Mingos, D. M. P. *J. Chem. Soc., Dalton Trans.* **1996**, 2435.
534. Haenninen, T.; Mutikainen, I.; Saanila, V.; Ritala, M.; Leskelae, M.; Hanson, J. C. *Chem. Mater.* **1997**, *9*, 1234.
535. Drozdov, A. A.; Troyanov, S. I.; Pisarevsky, A. P.; Struchkov, Y. T. *Polyhedron* **1994**, *13*, 1445.
536. Buriak, J. M.; Cheatham, L. K.; Graham, J. J.; Gordon, R. G.; Barron, A. R. *Mater. Res. Soc. Symp. Proc.* **1991**, *204*, 545.
537. Matsuno, S.; Uchikawa, F.; Yoshizaki, K. *Jpn. J. Appl. Phys., Part 2* **1990**, *29*, L947.
538. Buriak, J. M.; Cheatham, L. K.; Gordon, R. G.; Graham, J. J.; Barron, A. R. *Eur. J. Solid State Inorg. Chem.* **1992**, *29*, 43.
539. Auld, J.; Jones, A. C.; Leese, A. B.; Cockayne, B.; Wright, P. J.; O'Brien, P.; Motevalli, M. *J. Mater. Chem.* **1993**, *3*, 1203.
540. Paw, W.; Baum, T. H.; Lam, K.-C.; Rheingold, A. L. *Inorg. Chem.* **2000**, *39*, 2011.
541. Drake, S. R.; Hursthouse, M. B.; Abdul Malik, K. M.; Miller, S. A. S.; Otway, D. J. *Inorg. Chem.* **1993**, *32*, 4464.
542. Gardiner, R. A.; Gordon, D. C.; Stauf, G. T.; Vaartstra, B. A.; Ostrander, R. L.; Rheingold, A. L. *Chem. Mater.* **1994**, *6*, 1967.
543. Darr, J. A.; Drake, S. R.; Otway, D. J.; Miller, S. A. S.; Mingos, D. M. P.; Baxter, I.; Hursthouse, M. B.; Malik, K. M. A. *Polyhedron* **1997**, *16*, 2581.
544. Shamlian, S. H.; Hitchman, M. L.; Cook, S. L.; Richards, B. C. *J. Mater. Chem.* **1994**, *4*, 81.
545. Malandrino, G.; Fragala, I. L.; Neumayer, D. A.; Stern, C. L.; Hinds, B. J.; Marks, T. J. *J. Mater. Chem.* **1994**, *4*, 1061.
546. Nash, J. A. P.; Barnes, J. c.; Cole-Hamilton, D. J.; Richards, B. C.; Cook, S. L.; Hitchman, M. L. *Adv. Mater. Opt. Electron.* **1995**, *5*, 1.
547. Veya, P.; Floriani, C.; Chiesi-Villa, A.; Rizzoli, C. *Organometallics* **1994**, *13*, 214.
548. Schulz, D. L.; Hinds, B. J.; Stern, C. L.; Marks, T. J. *Inorg. Chem.* **1993**, *32*, 249.
549. Laube, T.; Dunitz, J. D.; Seebach, D. *Helv. Chim. Acta* **1985**, *68*, 1373.
550. Amstutz, R.; Dunitz, J. D.; Laube, T.; Schweizer, W. B.; Seebach, D. *Chem. Ber.* **1986**, *119*, 434.
551. Williard, P. G.; Salvino, J. M. *J. Chem. Soc., Chem. Commun.* **1986**, 153.
552. Allan, J. F.; Clegg, W.; Henderson, K. W.; Horsburgh, L.; Kennedy, A. R. *J. Organomet. Chem.* **1998**, *559*, 173.
553. Williard, P. G.; Hintze, M. J. *J. Am. Chem. Soc.* **1990**, *112*, 8602.
554. Henderson, K. W.; Dorigo, A. E.; Liu, Q.-Y.; Williard, P. G.; Schleyer, P. v. R.; Bernstein, P. R. *J. Am. Chem. Soc.* **1996**, *118*, 1339.
555. Henderson, K. W.; Dorigo, A. E.; Williard, P. G.; Bernstein, P. R. *Angew. Chem., Int. Ed. Engl.* **1996**, *35*, 1322.
556. Pospisil, P. J.; Wilson, S. R.; Jacobsen, E. N. *J. Am. Chem. Soc.* **1992**, *114*, 7585.
557. Apeloig, Y.; Zharov, I.; Bravo-Zhivotovskii, D.; Ovchinnikov, Y.; Struchkov, Y. *J. Organomet. Chem.* **1995**, *499*, 73.
558. Williard, P. G.; MacEwan, G. J. *J. Am. Chem. Soc.* **1989**, *111*, 7671.
559. Henderson, K. W.; Williard, P. G.; Bernstein, P. R. *Angew. Chem., Int. Ed. Engl.* **1995**, *34*, 1117.
560. Arnett, E. M.; Fisher, F. J.; Nichols, M. A.; Ribeiro, A. A. *J. Am. Chem. Soc.* **1990**, *112*, 801.
561. Williard, P. G.; Carpenter, G. B. *J. Am. Chem. Soc.* **1985**, *107*, 3345.
562. Williard, P. G.; Carpenter, G. B. *J. Am. Chem. Soc.* **1986**, *108*, 462.

563. Allan, J. F.; Henderson, K. W.; Kennedy, A. R.; Teat, S. J. *Chem. Commun.* **2000**, 1059.
564. Bradley, D. C.; Mehrotra, R. C.; Gaur, D. P. *Metal Alkoxides* **1978**, Academic Press: New York.
565. Johnson, S. M.; Herrin, J.; Liu, S. J.; Paul, I. C. *J. Am. Chem. Soc.* **1970**, *92*, 4428.
566. Smith, G. D.; Duax, W. L. *J. Am. Chem. Soc.* **1976**, *98*, 1578.
567. Bednorz, J. G.; Müller, K. A. Z. *Z. Phys. B.* **1986**, *64*, 189.
568. Caulton, K. G.; Hubert-Pfalzgraf, L. G. *Chem. Rev.* **1990**, *90*, 969.
569. Hubert-Pfalzgraf, L. G. *Polyhedron* **1994**, *13*, 1181.
570. Mehrotra, R. C.; Singh, A.; Sogani, S. *Chem. Soc. Rev.* **1994**, *23*, 215.
571. Caulton, K. G.; Chisholm, M. H.; Drake, S. R.; Folting, K.; Huffman, J. C. *Inorg. Chem.* **1993**, *32*, 816.
572. Caulton, K. G.; Chisholm, M. H.; Drake, S. R.; Huffman, J. C. *J. Chem. Soc., Chem. Commun.* **1990**, 1498.
573. Caulton, K. G.; Chisholm, M. H.; Drake, S. R.; Folting, K. *J. Chem. Soc., Chem Commun.* **1990**, 1349.
574. Bock, H.; Hauck, T.; Naether, C.; Roesch, N.; Staufer, M.; Haeberlen, O. D. *Angew. Chem., Int. Ed. Engl.* **1995**, *34*, 1353.
575. McCormick, M. J.; Moon, K. B.; Jones, S. R.; Hanusa, T. P. *J. Chem. Soc., Chem. Commun.* **1990**, 778.
576. Drake, S. R.; Otway, D. J.; Hursthouse, M. B.; Abdul Malik, K. M. *Polyhedron* **1992**, *11*, 1995.
577. Caulton, K. G.; Chisholm, M. H.; Drake, S. R.; Streib, W. E. *Angew. Chem.* **1990**, *102*, 1492.
578. Drake, S. R.; Otway, D. J. *J. Chem. Soc., Chem. Commun.* **1991**, 517.
579. Tesh, K. F.; Hanusa, T. P. *J. Chem. Soc., Chem. Commun.* **1991**, 879.
580. Tesh, K. F.; Burkey, D. J.; Hanusa, T. P. *J. Am. Chem. Soc.* **1994**, *116*, 2409.
581. Tesh, K. F.; Hanusa, T. P.; Huffman, J. C.; Huffman, C. J. *Inorg. Chem.* **1992**, *31*, 5572.
582. Veith, M.; Kaefer, D.; Huch, V. *Angew. Chem.* **1986**, *98*, 367.
583. Bidell, W.; Shklover, V.; Berke, H. *Inorg. Chem.* **1992**, *31*, 5561.
584. Vaartstra, B. A.; Huffman, J. C.; Streib, W. E.; Caulton, K. G. *Inorg. Chem.* **1991**, *30*, 3068.
585. Rees, W. S., Jr.; Moreno, D. A. *J. Chem. Soc., Chem. Commun.* **1991**, 1759.
586. Goel, S. C.; Matchett, M. A.; Chiang, M. Y.; Buhro, W. E. *J. Am. Chem. Soc.* **1991**, *113*, 1844.
587. Drake, S. R.; Streib, W. E.; Chisholm, M. H.; Caulton, K. G. *Inorg. Chem.* **1990**, *29*, 2707.
588. Haaland, A. *Angew. Chem., Int. Ed. Engl.* **1989**, *28*, 992.
589. Armstrong, D. R.; Clegg, W.; Drummond, A. M.; Liddle, S. T.; Mulvey, R. E. *J. Am. Chem. Soc.* **2000**, *122*, 11117.
590. Clegg, W.; Liddle, S. T.; Drummond, A. M.; Mulvey, R. E.; Robertson, A. *Chem. Commun.* **1999**, 1569.
591. Atwood, J. L.; Coleman, A. W.; Zhang, H.; Bott, S. G. *J. Inclusion Phenom. Mol. Recognit. Chem.* **1989**, *7*, 203.
592. Steed, J. W.; Johnson, C. P.; Barnes, C. L.; Juneja, R. K.; Atwood, J. L.; Reilly, S.; Hollis, R. L.; Smith, P. H.; Clark, D. L. *J. Am. Chem. Soc.* **1995**, *117*, 11426.
593. Atwood, J. L.; Barbour, L. J.; Hardie, M. J.; Raston, C. L. *Coord. Chem. Rev.* **2001**, *222*, 3.
594. Hardie, M. J.; Raston, C. L. *Dalton* **2000**, 2483.
595. Drljaca, A.; Hardie, M. J.; Raston, C. L. *J. Chem. Soc., Dalton Trans.* **1999**, 3639.
596. De Mendoza, J. *Chem. Eur. J.* **1998**, *4*, 1373.
597. Haino, T.; Yanase, M.; Fukazawa, Y. *Angew. Chem., Int. Ed. Engl.* **1998**, *37*, 997.
598. Umetani, S.; Sasaki, T.; Matsui, M.; Tsurubou, S.; Kimura, T.; Yoshida, Z. *Anal. Sci.* **1997**, *13*, 123.
599. Boche, G.; Moebus, K.; Harms, K.; Lohrenz, J. C. W.; Marsch, M. *Chem. Eur. J.* **1996**, *2*, 604.
600. Wheatley, A. E. H. *Chem. Soc. Rev.* **2001**, *30*, 265.
601. Stringfellow, G. B. *Organometallic Vapor Phase Epitaxy: Theory and Practice* **1989**, Academic press: San Diego, CA.
602. Hitchman, M. L.; Jensen, K. F., Eds. *Chemical Vapor Deposition*; Academic Press: New York, 1993.
603. Rees, W. S., Jr., Ed., *CVD of Nonmetals* VCH: New York, 1996.
604. Pierson, H. O. *Handbook of Chemical Vapor Deposition: Principles, Technology, and Applications;* 2nd ed. **1999**, Noyes: Norwich, NY.
605. Geballe, T. H.; Hulm, J. K. *Science (Washington, D.C.)* **1988**, *239*, 367.
606. Braithwaite, N.; Weaver, G. *Electronic Materials* **1990**, Butterworth: London.
607. Tiitta, M.; Niinisto, L. *Chem. Vap. Deposition* **1997**, *3*, 167.
608. Mohammad, S. N.; Salvador, A. A.; Morkoc, H. *Proc. IEEE* **1995**, *83*, 1306.
609. Morkoc, H.; Mohammad, S. N. *Science (Washington, D.C.)* **1995**, *267*, 51.
610. Gunshor, R. L.; Nurmikko, A. V. *MRS Bull.* **1995**, *20*, 15.
611. Cao, X. A.; Pearton, S. J.; Ren, F. *Crit. Rev. Solid State Mater. Sci.* **2000**, *25*, 279.
612. Shinohara, K.; Munakata, F.; Yamanaka, M. *Jpn. J. Appl. Phys., Part 2* **1988**, *27*, L1683.
613. Purdy, A. P.; George, C. F.; Callahan, J. H. *Inorg. Chem.* **1991**, *30*, 2812.
614. Thompson, S. C.; Cole-Hamilton, D. J.; Gilliland, D. D.; Hitchman, M. L.; Barnes, J. C. *Adv. Mater. Opt. Electron.* **1992**, *1*, 81.
615. Purdy, A. P.; Berry, A. D.; Holm, R. T.; Fatemi, M.; Gaskill, D. K. *Inorg. Chem.* **1989**, *28*, 2799.
616. Richards, B. C.; Cook, S. L.; Pinch, D. L.; Andrews, G. W.; Lengeling, G.; Schulte, B.; Juergensen, H.; Shen, Y. Q.; Vase, P.; Freltoft, T.; Spee, C. I. M. A.; Linden, J. L.; Hitchman, M. L.; Shamlian, S. H.; Brown, A. *Physica C (Amsterdam)* **1995**, *252*, 229.
617. Schulz, D. L.; Hinds, B. J.; Neumayer, D. A.; Stern, C. L.; Marks, T. J. *Chem. Mater.* **1993**, *5*, 1605.
618. Neumayer, D. A.; Studebaker, D. B.; Hinds, B. J.; Stern, C. L.; Marks, T. J. *Chem. Mater.* **1994**, *6*, 878.
619. Rees, W. S. J.; Carris, M. W.; Hesse, W. *Inorg. Chem.* **1991**, *30*, 4479.
620. Sadique, A. R.; Heeg, M. J.; Winter, C. H. *Inorg. Chem.* **2001**, *40*, 6349.
621. Hubert-Pfalzgraf, L. G.; Guillon, H. *Appl. Organomet. Chem.* **1998**, *12*, 221.
622. Gupta, A.; Jagannathan, E. I.; Cooper, E. A.; Giess, E. A.; Landman, J. I.; Hussey, B. W. *Appl. Phys. Lett.* **1988**, *52*, 2077.
623. Bradley, D. C.; Hasan, M.; Hursthouse, M. B.; Motevalli, M.; Khan, O. F. Z.; Pritchard, R. G.; Williams, J. O. *J. Chem. Soc., Chem. Commun.* **1992**, 575.
624. Yuhya, S.; Kikuchi, K.; Yoshida, M.; Sugawara, K.; Shiohara, Y. *Mol. Cryst. Liq. Cryst.* **1990**, *184*, 231.
625. Kim, S. H.; Cho, C. H.; No, K. S.; Chun, J. S. *J. Mater. Res.* **1991**, *6*, 704.
626. Turnipseed, S. B.; Barkley, R. M.; Sievers, R. E. *Inorg. Chem.* **1991**, *30*, 1164.
627. Rees, W. S., Jr., Alkaline Earth Metals: Inorganic Chemistry. *The Encyclopedia of Inorganic Chemistry*; King, R. B., Ed., 1994; Vol. 1, pp 67–87.

628. Drozdov, A. A.; Trojanov, S. I. *Polyhedron* **1992**, *11*, 2877.
629. Gleizes, A.; Sans-Lenain, S.; Medus, D. *C. R. Hebd. Seances Acad. Sci., Ser 2* **1991**, *313*, 761.
630. Rossetto, G.; Polo, A; Benetollo, F; Porchia, M; Zanella, P *Polyhedron* **1992**, *11*, 979.
631. Kim, B.-R.; Hwang, S.-C.; Lee, H.-G.; Shin, H.-S. *Kor. J. Chem. Eng.* **2000**, *17*, 524.
632. Yoon, J.-G.; Kyoo Oh, H.; Jong Lee, S. *Phys. Rev. B: Condens. Matter* **1999**, *60*, 2839.
633. Kunze, K.; Bihry, L.; Atanasova, P.; Hampden-Smith, M. J.; Duesler, E. N. *Chem. Vap. Deposition* **1996**, *2*, 105.
634. Yuta, M. M.; White, W. B. *J. Electrochem. Soc.* **1992**, *139*, 2347.
635. Kumta, P. N.; Risbud, S. H. *J. Mater. Sci.* **1994**, *29*, 1135.
636. Kondo, K.; Okuyama, H.; Ishibashi, A. *Appl. Phys. Lett.* **1994**, *64*, 3434.
637. Kondo, K.; Ukita, M.; Yoshida, H.; Kishita, Y.; Okuyama, H.; Ito, S.; Ohata, T.; Nakano, K.; Ishibashi, A. *J. Appl. Phys.* **1994**, *76*, 2621.
638. Aslam, M.; Bartlett, R. A.; Block, E.; Olmstead, M. M.; Power, P. P.; Sigel, G. E. *J. Chem. Soc., Chem. Commun.* **1985**, 1674.
639. Chadwick, S.; Ruhlandt-Senge, K. *Chem. Eur. J.* **1998**, *4*, 1768.
640. Ruhlandt-Senge, K.; Englich, U.; Senge, M. O.; Chadwick, S. *Inorg. Chem.* **1996**, *35*, 5820.
641. Niemeyer, M.; Power, P. P. *Inorg. Chem.* **1996**, *35*, 7264.
642. Setzer, W. N.; Schleyer, P. v. R. *Adv. Organomet. Chem.* **1985**, *24*, 353.
643. Pauer, F.; Power, P. P. *Lithium Chem.* **1995**, 295.
644. Janssen, M. D.; Grove, D. M.; Van Koten, G. *Prog. Inorg. Chem.* **1997**, *46*, 97.
645. Ruhlandt-Senge, K. *Comments Inorg. Chem.* **1997**, *19*, 351.
646. Englich, U.; Ruhlandt-Senge, K. *Coord. Chem. Rev.* **2000**, *210*, 135.
647. Tatsumi, K.; Matsubara, I.; Inoue, Y.; Nakamura, A.; Cramer, R. E.; Tagoshi, G. J.; Golen, J. A.; Gilje, J. W. *Inorg. Chem.* **1990**, *29*, 4928.
648. Ghosh, P.; Parkin, G. *Chem. Commun.* **1996**, 1239.
649. Kranz, M.; Denmark, S. E.; Swiss, K. A.; Wilson, S. R. *J. Org. Chem.* **1996**, *61*, 8551.
650. Denmark, S. E.; Swiss, K. A.; Wilson, S. R. *J. Am. Chem. Soc.* **1993**, *115*, 3826.
651. Solov'ev, V. N.; Chekhlov, A. N.; Martynov, I. V. *Koord. Khim.* **1991**, *17*, 618.
652. Solov'ev, V. N.; Chekhlov, A. N.; Zabirov, N. G.; Martynov, I. V. *Dokl. Akad. Nauk* **1992**, *323*, 1132.
653. Banister, A. J.; Barr, D.; Brooker, A. T.; Clegg, W.; Cunnington, M. J.; Doyle, M. J.; Drake, S. R.; Gill, W. R.; Manning, K.; Raithby, P. R.; Snaith, R.; Wade, K.; Wright, D. S. *J. Chem. Soc., Chem. Commun.* **1990**, 105.
654. Tatsumi, K.; Inoue, Y.; Nakamura, A.; Cramer, R. E.; VanDoorne, W.; Gilje, J. W. *Angew. Chem.* **1990**, *102*, 455.
655. Tatsumi, K.; Kawaguchi, H.; Inoue, K.; Tani, K.; Cramer, R. E. *Inorg. Chem.* **1993**, *32*, 4317.
656. Besser, S.; Herbst-Irmer, R.; Stalke, D.; Brooker, A. T.; Snaith, R.; Wright, D. S. *Acta Crystallogr., Sect. C: Cryst. Struct. Commun.* **1993**, *C49*, 1482.
657. Bacher, A. D.; Mueller, U.; Ruhlandt-Senge, K. *Z. Naturforsch., B: Chem. Sci.* **1992**, *47*, 1673.
658. Schnock, M.; Boettcher, P. *Z. Naturforsch., B: Chem. Sci.* **1995**, *50*, 721.
659. Andrews, P. C.; Koutsantonis, G. A.; Raston, C. L. *J. Chem. Soc., Dalton Trans.* **1995**, 4059.
660. Cea-Olivares, R.; Jimenez-Sandoval, O.; Hernandez-Ortega, S.; Sanchez, M.; Toscano, R. A.; Haiduc, I. *Heteroat. Chem.* **1995**, *6*, 89.
661. Banbury, F. A.; Davidson, M. G.; Raithby, P. R.; Stalke, D.; Snaith, R. *J. Chem. Soc., Dalton Trans.* **1995**, 3139.
662. Banbury, F. A.; Davidson, M. G.; Martin, A.; Raithby, P. R.; Snaith, R.; Verhorevoort, K. L.; Wright, D. S. *J. Chem. Soc., Chem. Commun.* **1992**, 1152.
663. Mikulcik, P.; Raithby, P. R.; Snaith, R.; Wright, D. S. *Angew. Chem.* **1991**, *103*, 452.
664. Armstrong, D. R.; Mulvey, R. E.; Barr, D.; Porter, R. W.; Raithy, P. R.; Simpson, T. R. E.; Snaith, R.; Wright, D. S.; Gregory, K.; Mikulcik, P. *J. Chem. Soc., Dalton Trans.* **1991**, 765.
665. Ball, S. C.; Cragg-Hine, I.; Davidson, M. G.; Davies, R. P.; Edwards, A. J.; Lopez-Solera, I.; Raithby, P. R.; Snaith, R. *Angew. Chem., Int. Ed. Engl.* **1995**, *34*, 921.
666. Bezougli, I. K.; Bashall, A.; McPartlin, M.; Mingos, D. M. P. *J. Chem. Soc., Dalton Trans.* **1998**, 2671.
667. Srinivas, B.; Chang, C.-C.; Chen, C.-H.; Chiang, M. Y.; Chen, I. T.; Wang, Y.; Lee, G.-H. *J. Chem. Soc., Dalton Trans.* **1997**, 957.
668. Blake, A. J.; Darr, J. A.; Howdle, S. M.; Poliakoff, M.; Li, W.-S.; Webb, P. B. *Journal of Chem. Crystallogr.* **1999**, *29*, 547.
669. Kuhn, N.; Fawzi, R.; Steimann, M.; Wiethoff, J. *Z. Anorg. Allg. Chem.* **1997**, *623*, 1577.
670. Chivers, T.; Downard, A.; Parvez, M. *Inorg. Chem.* **1999**, *38*, 5565.
671. Arca, M.; Demartin, F.; Devillanova, F. A.; Garau, A.; Isaia, F.; Lippolis, V.; Verani, G. *Inorg. Chem.* **1998**, *37*, 4164.
672. Chivers, T.; Downard, A.; Yap, G. P. A. *Inorg. Chem.* **1998**, *37*, 5708.
673. Armstrong, D. R.; Mulvey, R. E.; Barr, D.; Snaith, R.; Wright, D. S.; Clegg, W.; Hodgson, S. M. *J. Organomet. Chem.* **1989**, *362*, C1.
674. Ball, S. C.; Cragg-Hine, I.; Davidson, M. G.; Davies, R. P.; Raithby, P. R.; Snaith, R. *Chem. Commun.* **1996**, 1581.
675. Ebels, J.; Pietschnig, R.; Nieger, M.; Niecke, E.; Kotila, S. *Heteroat. Chem.* **1997**, *8*, 521.
676. Kimblin, C.; Bridgewater, B. M.; Hascall, T.; Parkin, G. *Dalton* **2000**, 891.
677. Bonasia, P. J.; Christou, V.; Arnold, J. *J. Am. Chem. Soc.* **1993**, *115*, 6777.
678. Flick, K. E.; Bonasia, P. J.; Gindelberger, D. E.; Katari, J. E. B.; Schwartz, D. *Acta Crystallogr., Sect. C: Cryst. Struct. Commun.* **1994**, *C50*, 674.
679. Beswick, M. A.; Harmer, C. N.; Raithby, P. R.; Steiner, L.; Tombul, M.; Wright, D. S. *J. Organomet. Chem.* **1999**, *573*, 267.
680. Nothegger, T.; Wurst, K.; Probst, M.; Sladky, F. *Chem. Ber. Recl.* **1997**, *130*, 119.
681. Khasnis, D. V.; Buretea, M.; Emge, T. J.; Brennan, J. G. *J. Chem. Soc., Dalton Trans.* **1995**, 45.
682. Ruhlandt-Senge, K.; Power, P. P. *Inorg. Chem.* **1991**, *30*, 3683.
683. Du Mont, W. W.; Kubiniok, S.; Lange, L.; Pohl, S.; Saak, W.; Wagner, I. *Chem. Ber.* **1991**, *124*, 1315.
684. Ruhlandt-Senge, K.; Power, P. P. *Inorg. Chem.* **1993**, *32*, 4505.
685. Niemeyer, M.; Power, P. P. *Inorg. Chim. Acta* **1997**, *263*, 201.

686. Gindelberger, D. E.; Arnold, J. *Inorg. Chem.* **1994**, *33*, 6293.
687. Ruhlandt-Senge, K. *Inorg. Chem.* **1995**, *34*, 3499.
688. Ruhlandt-Senge, K.; Davis, K.; Dalal, S.; Englich, U.; Senge, M. O. *Inorg. Chem.* **1995**, *34*, 2587.
689. Ghosh, P.; Parkin, G. *Polyhedron* **1997**, *16*, 1255.
690. Ruhlandt-Senge, K.; Englich, U. *Chem. Eur. J.* **2000**, *6*, 4063.
691. Worden, T. A. J.; Wright, D. S.; Steiner, A. *Polyhedron* **1998**, *17*, 4011.
692. Draganjac, M.; Dhingra, S.; Huang, S. P.; Kanatzidis, M. G. *Inorg. Chem.* **1990**, *29*, 590.
693. Marganian, C. A.; Baidya, N.; Olmstead, M. M.; Mascharak, P. K. *Inorg. Chem.* **1992**, *31*, 2992.
694. Chivers, T.; Parvez, M.; Seay, M. A. *Inorg. Chem.* **1994**, *33*, 2147.
695. Berardini, M.; Emge, T. J.; Brennan, J. G. *J. Chem. Soc., Chem. Commun.* **1993**, 1537.
696. Pilkington, M. J.; Slawin, A. M. Z.; Williams, D. J.; Woollins, J. D. *Polyhedron* **1991**, *10*, 2641.
697. Chivers, T.; Krahn, M.; Parvez, M. *Chem. Commun.* **2000**, 463.
698. Becker, G.; Klinkhammer, K. W.; Massa, W. *Z. Anorg. Allg. Chem.* **1993**, *619*, 628.
699. Becker, G.; Klinkhammer, K. W.; Lartiges, S.; Boettcher, P.; Poll, W. *Z. Anorg. Allg. Chem.* **1992**, *613*, 7.
700. Bonasia, P. J.; Gindelberger, D. E.; Dabbousi, B. O.; Arnold, J. *J. Am. Chem. Soc.* **1992**, *114*, 5209.
701. Bonasia, P. J.; Arnold, J. *J. Organomet. Chem.* **1993**, *449*, 147.
702. Gindelberger, D. E.; Arnold, J. *J. Am. Chem. Soc.* **1992**, *114*, 6242.
703. Becker, G.; Klinkhammer, K. W.; Schwarz, W.; Westerhausen, M.; Hildenbrand, T. *Z. Naturforsch., B: Chem. Sci.* **1992**, *47*, 1225.
704. Büchel, K. H.; Moretto, H.-H.; Woditsch, P. Alkali and Alkaline Earth Metals and their Compounds. In *Industrial Inorganic Chemistry*, Wiley-VCH: Weinheim, **2000**; Chapter 3.1, pp 213–246.
705. Hargittai, M. *Coord. Chem. Rev.* **1988**, *91*, 35.
706. Beattie, I. R. *Angew. Chem., Int. Ed. Engl.* **1999**, *38*, 3294.
707. Hargittai, M. *Chem. Rev.* **2000**, *100*, 2233.
708. Snaith, R.; Wright, D. S. *Lithium Chem.* **1995**, 227.
709. Waters, A. F.; White, A. H. *Aust. J. Chem.* **1996**, *49*, 27.
710. Waters, A. F.; White, A. H. *Aust. J. Chem.* **1996**, *49*, 147.
711. Skelton, B. W.; Waters, A. F.; White, A. H. *Aust. J. Chem.* **1996**, *49*, 137.
712. Kepert, D. L.; Waters, A. F.; White, A. H. *Aust. J. Chem.* **1996**, *49*, 117.
713. Skelton, B. W.; Waters, A. F.; White, A. H. *Aust. J. Chem.* **1996**, *49*, 99.
714. Waters, A. F.; White, A. H. *Aust. J. Chem.* **1996**, *49*, 87.
715. Waters, A. F.; White, A. H. *Aust. J. Chem.* **1996**, *49*, 73.
716. Waters, A. F.; White, A. H. *Aust. J. Chem.* **1996**, *49*, 61.
717. Kepert, D. L.; Skelton, B. W.; Waters, A. F.; White, A. H. *Aust. J. Chem.* **1996**, *49*, 47.
718. Waters, A. F.; White, A. H. *Aust. J. Chem.* **1996**, *49*, 35.
719. Girichev, A. G.; Giricheva, N. I.; Vogt, N.; Girichev, G. V.; Vogt, J. *J. Mol. Struct.* **1996**, *384*, 175.
720. Berkowitz, J.; Marquart, J. R. *J. Chem. Phys.* **1962**, *37*, 1853.
721. McGrady, G. S.; Downs, A. J. *Coord. Chem. Rev.* **2000**, *197*, 95.
722. Evans, W. J.; Bloom, I.; Grate, J. W.; Hughes, L. A.; Hunter, W. E.; Atwood, J. L. *Inorg. Chem.* **1985**, *24*, 4620.
723. Fromm, K. M. *Angew. Chem. Int. Ed. Engl.* **1997**, *36*, 2799.
724. Brackett, E. B.; Brackett, T. E.; Sass, R. L. *J. Phys. Chem.* **1963**, *67*, 2132.
725. Fromm, K. M.; Goesmann, H. *Acta Crystallogr., Sect. C: Cryst. Struct. Commun.* **2000**, *C56*, 1179.
726. Barr, D.; Clegg, W.; Mulvey, R. E.; Snaith, R. *J. Chem. Soc., Chem. Commun.* **1984**, 79.
727. Neumann, F.; Hampel, F.; Schleyer, P. v. R. *Inorg. Chem.* **1995**, *34*, 6553.
728. Doriat, C.; Koeppe, R.; Baum, E.; Stoesser, G.; Koehnlein, H.; Schnoeckel, H. *Inorg. Chem.* **2000**, *39*, 1534.
729. Jockisch, A.; Schmidbaur, H. *Inorg. Chem.* **1999**, *38*, 3014.
730. Raston, C. L.; Robinson, W. T.; Skelton, B. W.; Whitaker, C. R.; White, A. H. *Aust. J. Chem.* **1990**, *43*, 1163.
731. Raston, C. L.; Whitaker, C. R.; White, A. H. *Inorg. Chem.* **1989**, *28*, 163.
732. Deacon, G. B.; Forsyth, C. M.; Junk, P. C.; Skelton, B. W.; White, A. H. *J. Chem. Soc., Dalton Trans.* **1998**, 1381.
733. Edwards, A. J.; Paver, M. A.; Raithby, P. R.; Russell, C. A.; Wright, D. S. *J. Chem. Soc., Dalton Trans.* **1993**, 3265.
734. Raston, C. L.; Skelton, B. W.; Whitaker, C. R.; White, A. H. *Aust. J. Chem.* **1988**, *41*, 1925.
735. Armstrong, D. R.; Khandelwal, A. H.; Raithby, P. R.; Kerr, L. C.; Peasey, S.; Shields, G. P.; Snaith, R.; Wright, D. S. *Chem. Commun.* **1998**, 1011.
736. Liu, F.-Q.; Stalke, D.; Roesky, H. W. *Angew. Chem., Int. Ed. Engl.* **1995**, *34*, 1872.
737. Pevec, A.; Demsar, A.; Gramlich, V.; Petricek, S.; Roesky, H. W. *J. Chem. Soc., Dalton Trans.* **1997**, 2215.
738. Demsar, A.; Pevec, A.; Petricek, S.; Golic, L.; Petric, A.; Bjorgvinsson, M.; Roesky, H. W. *J. Chem. Soc., Dalton Trans.* **1998**, 4043.
739. Demsar, A.; Pevec, A.; Golic, L.; Petricek, S.; Petric, A.; Roesky, H. W. *Chem. Commun.* **1998**, 1029.
740. Giese, H.-H.; Nöth, H.; Schwenk, H.; Thomas, S. *Eur. J. Inorg. Chem.* **1998**, 941.
741. Armstrong, D. R.; Clegg, W.; Colquhoun, H. M.; Daniels, J. A.; Mulvey, R. E.; Stephenson, I. R.; Wade, K. *J. Chem. Soc., Chem. Commun.* **1987**, 630.
742. Antsyshkina, A. S.; Sadikov, G. G.; Porai-Koshits, M. A.; Konoplev, V. N.; Silina, T. A.; Sizareva, A. S. *Koord. Khim.* **1994**, *20*, 274.
743. Heine, A.; Stalke, D. *J. Organomet. Chem.* **1997**, *542*, 25.
744. Morosin, B.; Howatson, J. *J. Inorg. Nucl. Chem.* **1979**, *41*, 1667.
745. Lobkovskii, E. B.; Titov, L. V.; Psikha, S. B.; Antipin, M. Y.; Struchkov, Y. T. *Zh. Strukt. Khim.* **1982**, *23*, 172.
746. Lobkovskii, E. V.; Titov, L. V.; Levicheva, M. D.; Chekhlov, A. N. *Zh. Strukt. Khim.* **1990**, *31*, 147.
747. Hanecker, E.; Moll, J.; Nöth, H. *Z. Naturforsch., B: Anorg. Chem., Org. Chem.* **1984**, *39B*, 424.
748. Lobkovskii, E. B.; Chekhlov, A. N.; Levicheva, M. D.; Titov, L. V. *Koord. Khim.* **1988**, *14*, 543.
749. Dias, H. V. R.; Olmstead, M. M.; Ruhlandt-Senge, K.; Power, P. P. *J. Organomet. Chem.* **1993**, *462*, 1.
750. Hellmann, K. W.; Gade, L. H.; Gevert, O.; Steinert, P.; Lauher, J. W. *Inorg. Chem.* **1995**, *34*, 4069.
751. Gehrhus, B.; Hitchcock, P. B.; Lappert, M. F.; Slootweg, J. C. *Chem. Commun.* **2000**, 1427.
752. Hitchcock, P. B.; Lappert, M. F.; Lawless, G. A.; Royo, B. *J. Chem. Soc., Chem. Commun.* **1993**, 554.

753. Kawachi, A.; Tamao, K. *J. Am. Chem. Soc.* **2000**, *122*, 1919.
754. Roesch, L.; Pickardt, J.; Imme, S.; Boerner, U. *Z. Naturforsch., B: Anorg. Chem., Org. Chem.* **1986**, *41B*, 1523.
755. Veith, M.; Weidner, S.; Kunze, K.; Kaefer, D.; Hans, J.; Huch, V. *Coord. Chem. Rev.* **1994**, *137*, 297.
756. Goddard, R.; Krueger, C.; Ramadan, N. A.; Ritter, A. *Angew. Chem., Int. Ed. Engl.* **1995**, *34*, 1030.
757. Sekiguchi, A.; Nanjo, M.; Kabuto, C.; Sakurai, H. *Organometallics* **1995**, *14*, 2630.
758. Abrahams, I.; Motevalli, M.; Shah, S. A. A.; Sullivan, A. C. *J. Organomet. Chem.* **1995**, *492*, 99.
759. Veith, M.; Schutt, O.; Huch, V. *Angew. Chem. Int. Ed. Engl.* **2000**, *39*, 601.
760. Belzner, J.; Dehnert, U.; Stalke, D. *Angew. Chem.* **1994**, *106*, 2580.
761. Apeloig, Y.; Yuzefovich, M.; Bendikov, M.; Bravo-Zhivotovskii, D.; Klinkhammer, K. *Organometallics* **1997**, *16*, 1265.
762. Sekiguchi, A.; Nanjo, M.; Kabuto, C.; Sakurai, H. *Angew. Chem., Int. Ed. Engl.* **1997**, *36*, 113.
763. Freitag, S.; Herbst-Irmer, R.; Lameyer, L.; Stalke, D. *Organometallics* **1996**, *15*, 2839.
764. Kawachi, A.; Tanaka, Y.; Tamao, K. *Eur. J. Inorg. Chem.* **1999**, 461.
765. Westerhausen, M. *Angew. Chem.* **1994**, *106*, 1585.
766. Mootz, D.; Zinnius, A.; Böttcher, B. *Angew. Chem., Int. Ed. Engl.* **1969**, *8*, 378.
767. Engelhardt, L. M.; May, A. S.; Raston, C. L.; White, A. H. *J. Chem. Soc., Dalton Trans.* **1983**, 1671.
768. Lappert, M. F.; Slade, M. J.; Singh, A.; Atwood, J. L.; Rogers, R. D. *J. Am. Chem. Soc.* **1983**, *105*, 302.
769. Mack, H.; Frenzen, G.; Bendikov, M.; Eisen, M. S. *J. Organomet. Chem.* **1997**, *549*, 39.
770. Davies, R. P. *Inorg. Chem. Commun.* **2000**, *3*, 13.
771. Power, P. P.; Xiaojie, X. *J. Chem. Soc., Chem. Commun.* **1984**, 358.
772. Knizek, J.; Krossing, I.; Nöth, H.; Schwenk, H.; Seifert, T. *Chem. Ber.* **1997**, *130*, 1053.
773. Karl, M.; Seybert, G.; Massa, W.; Harms, K.; Agarwal, S.; Maleika, R.; Stelter, W.; Greiner, A.; Heitz, W.; Neumuller, B.; Dehnicke, K. *Z. Anorg. Allg. Chem.* **1999**, *625*, 1301.
774. Edelmann, F. T.; Pauer, F.; Wedler, M.; Stalke, D. *Inorg. Chem.* **1992**, *31*, 4143.
775. Williard, P. G. *Acta Crystallogr.* **1988**, *C44*, 270.
776. Domingos, A. M.; Sheldrick, G. M. *Acta Crystallogr., Sect. B* **1974**, *30*, 517.
777. Westerhausen, M.; Schwarz, W. *Z. Anorg. Allg. Chem.* **1992**, *609*, 39.
778. Her, T. Y.; Chang, C. C.; Lee, G. H.; Peng, S. M.; Wang, Y. *J. Chin. Chem. Soc. (Taipei)* **1993**, *40*, 315.
779. Westerhausen, M.; Schwarz, W. *Z. Anorg. Allg. Chem.* **1991**, *604*, 127.
780. Westerhausen, M.; Schwarz, W. *Z. Anorg. Allg. Chem.* **1991**, *606*, 177.
781. Westerhausen, M.; Hartmann, M.; Makropoulos, N.; Wieneke, B.; Wieneke, M.; Schwarz, W.; Stalke, D. *Z. Naturforsch., B: Chem. Sci.* **1998**, *53*, 117.
782. Cloke, F. G. N.; Hitchcock, P. B.; Lappert, M. F.; Lawless, G. A.; Royo, B. *J. Chem. Soc., Chem. Commun.* **1991**, 724.
783. Englich, U.; Hassler, K.; Ruhlandt-Senge, K.; Uhlig, F. *Inorg. Chem.* **1998**, *37*, 3532.
784. Westerhausen, M.; Digeser, M. H.; Wieneke, B.; Nöth, H.; Knizek, J. *Eur. J. Inorg. Chem.* **1998**, 517.
785. Westerhausen, M.; Schwarz, W. *Z. Anorg. Allg. Chem.* **1994**, *620*, 304.
786. Westerhausen, M.; Schwarz, W. *Z. Anorg. Allg. Chem.* **1996**, *622*, 903.
787. Westerhausen, M.; Schwarz, W. *J. Organomet. Chem.* **1993**, *463*, 51.
788. Westerhausen, M. *J. Organomet. Chem.* **1994**, *479*, 141.
789. Westerhausen, M.; Digeser, M. H.; Nöth, H.; Knizek, J. *Z. Anorg. Allg. Chem.* **1998**, *624*, 215.
790. Westerhausen, M.; Lang, G.; Schwarz, W. *Chem. Ber.* **1996**, *129*, 1035.
791. Westerhausen, M.; Hartmann, M.; Schwarz, W. *Inorg. Chem.* **1996**, *35*, 2421.
792. Bazhenova, T. A.; Ivleva, I. N.; Kulikov, A. V.; Shestakov, A. F.; Shilov, A. E.; Antipin, M. Y.; Lysenko, K. A.; Struchkov, Y. T. *Russ. J. Coord. Chem. (Transl. of Koord. Khim.)* **1995**, *21*, 674.
793. Nöth, H.; Waldhör, R. *Z. Naturforsch., B* **1998**, *53*, 1525.
794. Schnepf, A.; Schnockel, H. *Angew. Chem., Int. Ed. Engl.* **2001**, *40*, 712.
795. Bogdanovic, B.; Janke, N.; Krueger, C.; Mynott, R.; Schlichte, K.; Westeppe, U. *Angew. Chem.* **1985**, *97*, 972.
796. Bel'skii, V. K.; Strel'tsova, N. R.; Bulychev, B. M.; Ivakina, L. V.; Storozhenko, P. A. *Zh. Strukt. Khim.* **1987**, *28*, 166.
797. Huang, Q.; Qian, Y.; Zhuang, J.; Tang, Y. *Chin. J. Struct. Chem. (Jiegou Huaxue)* **1987**, *6*, 43.
798. Sarma, R.; Ramirez, F.; McKeever, B.; Chaw, Y. F.; Marecek, J. F.; Nierman, D.; McCaffrey, T. M. *J. Am. Chem. Soc.* **1977**, *99*, 5289.
799. Metzler, N.; Nöth, H.; Schmidt, M.; Treitl, A. *Z. Naturforsch., B: Chem. Sci.* **1994**, *49*, 1448.

Comprehensive Coordination Chemistry II
ISBN (set): 0-08-0437486

Volume 3, (ISBN 0-08-0443257); pp 1–92

3.2
Scandium, Yttrium, and the Lanthanides

S. COTTON
Uppingham School, UK

3.2.1 SCANDIUM

3.2.1.1 Introduction

Scandium is still a neglected element. It is the most expensive metal in its period (caused by the fact that its even distribution in the earth means that there are no rich ores) and its chemistry is virtually exclusively that of the +3 oxidation state, so that it is not classed as a transition metal and is often "silent" to spectroscopy and not amenable to study by many of the usual spectroscopic tools of the coordination chemist. Chemists have often either tended to assume that complexes of Sc^{3+} are just like those of the tripositive ions of the $3d$ transition metals or that they resemble lanthanide complexes. Neither of these assumptions is correct—how incorrect we are now realizing. Scandium chemistry is starting to exhibit characteristics all of its own, and possibly the burgeoning use of scandium compounds in organic synthesis may drive a real expansion of scandium chemistry.

Several "early" structures of scandium compounds that were reported, such as $[ScCl_3(THF)_3]$[1] and $[Sc(acac)_3]$[2] (acac = acetylacetonate) featured a coordination number of six; this, taken together with the coordination number of six exhibited by its oxide and halides, was probably responsible for the view, often unstated, that scandium compounds were generally six-coordinate. It is now becoming clear that, since Sc^{3+} is a larger ion than any of the succeeding $3d$ transition metal ions (the ionic radius of six-coordinate Sc^{3+} is 0.745Å[3], contrast Ti^{3+} 0.670 Å), it infrequently exhibits a coordination number greater than six in its complexes. On the other hand, it is smaller than lutetium, the last lanthanide (ionic radii of six-coordinate Sc^{3+}, La^{3+}, and Lu^{3+} are 0.745Å, 1.032 Å, and 0.861 Å respectively), and thus tends to exhibit lower coordination numbers than the lanthanides; although there are sometimes similarities with lutetium,[4] this point should not be over emphasized. (This point is demonstrated in Table 1, which shows comparative coordination numbers of Sc, La, and Lu in a number of typical binary compounds and complexes.) Whilst it could fairly be stated[5] in 1987 that "in those crystal structures that are known, Sc^{3+} is predominantly six-coordinated," the position has now changed; all the coordination numbers between three and nine are confirmed by X-ray diffraction work.

The last major review of the coordination chemistry of scandium appeared in 1987; whilst other reviews have appeared concerned with scandium chemistry,[6,7] with its structural chemistry,[8] and with the role of scandium inorganic synthesis,[9,10] this section is concerned with covering the area from the previous review, though for the sake of readability, there will be occasional reference to earlier work.

3.2.1.2 Group 14 Ligands

Some simple alkyl and aryl compounds of metals in normal oxidation states can be considered as honorary coordination compounds.

Triphenylscandium was the first well-characterized organoscandium compound with a σ-bonded ligand, but its structure has been unknown, although believed to be polymeric. The THF adduct, $[ScPh_3(THF)_2]$, stable at $-35\,^{\circ}C$, is monomeric, however, and has a trigonal

Table 1 Coordination numbers (C.N.) of a number of related Sc, La, and Lu complexes.

Compound	Sc compound/complex formula	C.N.	La compound/complex formula	C.N.	Lu compound/complex formula	C.N.
Oxide	Sc_2O_3	6	La_2O_3	7	Lu_2O_3	6
Fluoride	ScF_3	6	LaF_3	$9+2$	LuF_3	9
Chloride	$ScCl_3$	6	$LaCl_3$	9	$LuCl_3$	6
Bromide	$ScBr_3$	6	$LaBr_3$	9	$LuBr_3$	6
Iodide	ScI_3	6	LaI_3	8	LuI_3	6
Acetylacetonate	$Sc(acac)_3$	6	$La(acac)_3(H_2O)_2$	8	$Lu(acac)_3(H_2O)$	7
EDTA complex	$Sc(EDTA)(H_2O)_2^-$	8	$La(EDTA)(H_2O)_3^-$	9	$Lu(EDTA)(H_2O)_2^-$	8
THF adduct of trichloride	$ScCl_3(THF)_3$	6	$[LaCl(\mu\text{-}Cl)_2(THF)_2]_n$	8	$LuCl_3(THF)_3$	6
Terpy complex of nitrate	$Sc(NO_3)_3(terpy)$	8.5	$La(NO_3)_3(terpy)(H_2O)_2$	11	$Lu(NO_3)_3(terpy)$	9
Aqua ion	$[Sc(H_2O)_7]^{3+}$	7	$[La(H_2O)_9]^{3+}$	9	$[Lu(H_2O)_8]^{3+}$	8
Bis(trimethylsilyl) amide	$Sc(N(SiMe_3)_2)_3$	3	$La(N(SiMe_3)_2)_3$	3	$Lu(N(SiMe_3)_2)_3$	3
Ph_3PO complex of nitrate	$Sc(\eta^2\text{-}NO_3)_3(Ph_3PO)_2$	8	$La(\eta^1\text{-}NO_3)(\eta^2\text{-}NO_3)(Ph_3PO)_2$	9	$[Lu(\eta^2\text{-}NO_3)_2(Ph_3PO)_4]NO_3$	8

bipyramidal structure with axial THF molecules.[11] Its instability suggests that, on warming, loss of THF molecules occurs, with concomitant formation of polyhapto-linkages and oligomerization. A stable octahedral permethylate species exists in the form of the $[ScMe_6]^{3-}$ ion, isolated as its tris(TMEDA) salt from the reaction of $ScCl_3$ with MeLi in the presence of excess TMEDA.[12] There is good evidence for $[Sc(CH_2SiMe_3)_3(THF)_2]$;[13] although it has not been isolated or reported, there seems no reason why the alkyl $[Sc\{CH(SiMe_3)_2\}_3]$ should not exist,[14,15] by analogy with $[M\{CH(SiMe_3)_2\}_3]$ (M = Y, La, Pr, Nd, Sm, Lu) and with the corresponding silylamides.

Certain porphyrin derivatives with metal–carbon σ bonds are discussed in Section 3.2.1.3.4.

It should be noted that there are organometallic compounds involving polyhapto ligands that contain linkages such as Sc—Te that are not found in coordination compounds. In addition, a number of recent reviews on the organometallic chemistry of the lanthanides include reference to scandium compounds.[16–20]

3.2.1.3 Group 15 Ligands

3.2.1.3.1 Ammonia and amines

Anhydrous ScX_3 (X = Cl, Br, I) reacts with gaseous ammonia forming ammine complexes[21] such as $ScX_3 \cdot 5NH_3$ (X = Cl, Br) and $ScX_3 \cdot 4NH_3$. (X = Cl, Br, I). Nothing is known about the structures of any of these compounds but recently the first ammine complex to be characterized,[22] $(NH_4)_2[Sc(NH_3)I_5]$ has been obtained as pink crystals from the reaction of NH_4I and metallic scandium in a sealed tube at 500 °C. Scandium has octahedral coordination with Sc—N = 3.29(2) Å and Sc—I = 2.856(5)–2.899(5) Å. There are no further reports concerning simple complexes of bipy and phen (bipy = 2,2′-bipyridyl; phen = 1,10-phenanthroline) with scandium, but a little work has been carried with terdentate ligands. The complex $[Sc(terpy)(NO_3)_3]$, (terpy = 2,2′,:6′, 2″-terpyridyl) formed when scandium nitrate reacts with terpy in MeCN, has effective "8.5"-coordination with one very asymmetrically bidentate nitrate; in contrast to the later lanthanides (see Section 3.2.2.3.4) where reaction of the hydrated lanthanide nitrates in MeCN gives nine-coordinate $[Ln(terpy)(NO_3)_3]$.[23,24] Another terdentate ligand, 4-amino-bis(2,6-(2-pyridyl))-1,3,5-triazine (abptz), forms a complex $[Sc(abptz)(NO_3)_3]$ which contains eight-coordinate scandium with one nitrate monodentate.[25]

3.2.1.3.2 Thiocyanates

Compared with the lanthanides (Section 3.2.2.3.7) little study has been made of these. However, like the corresponding lanthanide complexes with this counter-ion, $[Bu_4N]_3[Sc(NCS)_6]$ has octahedrally coordinated scandium.[26]

3.2.1.3.3　Amides

Like the succeeding $3d$ transition metals scandium forms a three-coordinate silylamide $Sc[N(SiMe_3)_2]_3$ but unlike them its solid state structure is pyramidal, not planar, in which respect it resembles the lanthanides and uranium. Like the silylamides of $3d$ metals, however, it does not form adducts with Lewis bases, presumably on steric grounds. However, the less congested amide $[Sc\{N(SiHMe_2)_2\}_3]$ forms a THF adduct $[Sc\{N(SiHMe_2)_2\}_3(THF)]$, which has distorted tetrahedral coordination of scandium, with short $Sc\cdots Si$ contacts in the solid state; this is in contrast to the five-coordinate $[Ln\{N(SiHMe_2)_2\}_3(THF)_2]$.[27] Another amide, a triamidoamine complex with four-coordination of scandium (see Scheme 1) distils on heating the corresponding "ate" complex.[28]

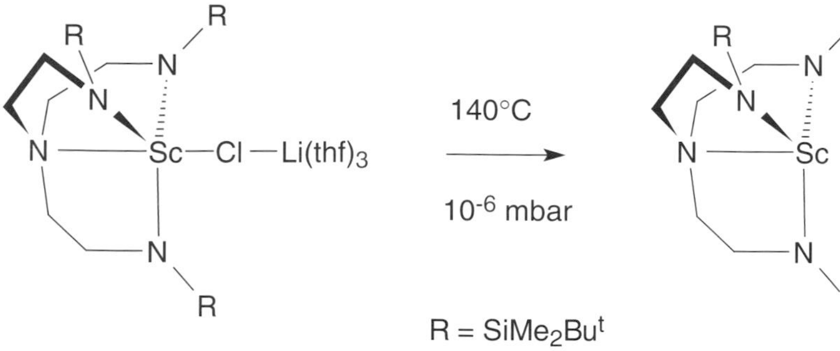

R = SiMe₂Buᵗ

Scheme 1

Scandium (and similar yttrium) benzamidinate complexes $[\{RC_6H_4C(NSiMe_3)_2\}_2ScCl(THF)]$ ($R = H$, MeO) have been reported[29] and are believed to have octahedral structures (**1**).

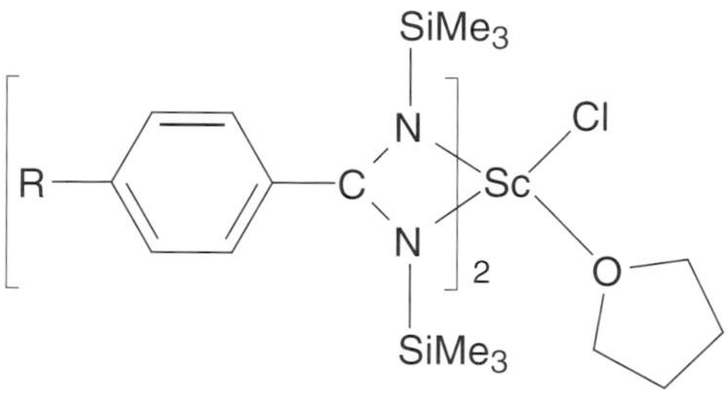

(1)

The dichloro-lithium adduct (**2**) has also been characterized crystallographically,[30] but the lithium-free halides are better synthons for a range of hydrocarbyls and hydride, $[\{PhC(NSiMe_3)_2\}_2ScR]$ ($R = CH_2SiMe_3$, 2,4,6-$Me_3C_6H_2$, CH_2SiMe_2Ph, H). Unlike the corresponding Cp^*_2ScR systems, these show no signs of σ-bond metathesis on heating in hydrocarbon solvents. On the other hand, $[\{PhC(NSiMe_3)_2\}_2Sc(CH_2SiMe_3)]$ undergoes hydrogenolysis on reaction with H_2 (1 atm) in benzene or alkanes:

$$[\{PhC(NSiMe_3)_2\}_2Sc(CH_2SiMe_3)] + H_2 \rightarrow [\{\{PhC(NSiMe_3)_2\}_2ScH\}_2] + SiMe_4 \tag{1}$$

The IR spectrum of the hydride (**3**) shows a band due to $\nu(Sc\text{—}H)$ for bridging hydrogens at $1,283\,cm^{-1}$ (shifted to $907\,cm^{-1}$ on deuteration) whilst the structure of the hydride features Sc—H bonds of $1.87\,\text{Å}$ to $2.00\,\text{Å}$. This inserts $PhC\equiv CPh$ forming the alkenyl $[\{PhC(NSiMe_3)_2\}_2Sc(C(Ph)\text{=}CH(Ph))]$.

3.2.1.3.4　Compounds of porphyrins and other macrocyclic ligands

A range of porphyrins and phthalocyanines exist; syntheses often involve the high-temperature routes typical of the transition metals; thus when $ScCl_3$ is refluxed with H_2TTP (H_2TTP = meso-tetratolylporphyrin) in 1-chloronaphthalene, $Sc(TTP)Cl$ is formed.[31] This has the expected square pyramidal structure with Sc $0.68\,\text{Å}$ above the N_4 basal plane (Sc—Cl $= 2.32\,\text{Å}$; Sc—N $= 2.17$–$2.18\,\text{Å}$).

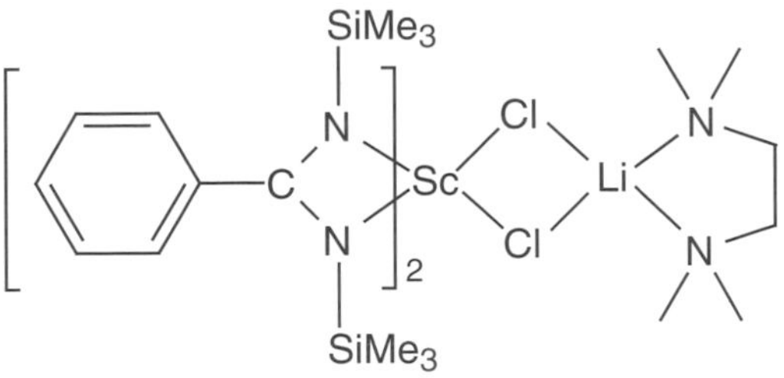

(2)

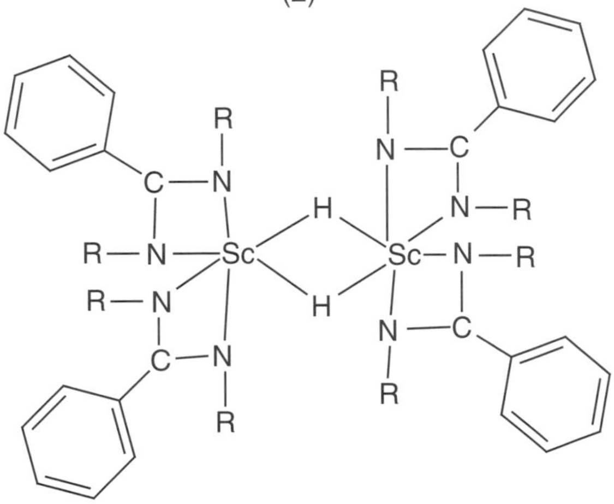

(3)

Inadvertent hydrolysis of $Sc(TTP)(C_5Me_5)$ leads to the oxo-bridged dimer $[Sc(TTP)]_2(\mu\text{-}O)$; this has a bent (110°) Sc—O—Sc bridge. Recently a high-yield low-temperature route[32,33] to $Sc(OEP)Cl$ ($H_2OEP = 2,3,7,8,12,13,17,18$-octaethylporphyrin) has been utilized in toluene solution at 100 °C:

$$Li_2(OEP) + ScCl_3(THF)_3 \rightarrow Sc(OEP)Cl + 2LiCl + 3THF \qquad (2)$$

Though moisture-sensitive in solution, $Sc(OEP)Cl$ forms air-stable red crystals. The chloride can be replaced by alkoxy, alkylamide, alkyl, and cyclopentadienyl groups. The structures have been reported of two σ-alkyls $Sc(OEP)R$ ($R = Me$, $CH(SiMe_3)_2$), both of which have the anticipated square pyramidal structure with Sc out of the basal plane (by 0.66 Å in the methyl structure) and $Sc\text{—}N \approx 2.16$ Å. The bond lengths for $[Sc(OEP)\{Me\}]$—Sc—C 3.246(3) Å—(Sc—N = 2.151, 2.152, 2.157, 2.158 Å; average of 2.1545 Å) and for $[Sc(OEP)\{CH(SiMe_3)_2\}]$—Sc—C 3.243(8) Å—(Sc—N = 2.142, 2.151, 2.162, 2.196 Å; average of 2.163 Å) show very similar Sc—C bond lengths in an uncrowded environment. Hydrolysis of all $Sc(OEP)X$ derivatives produces a dimeric hydroxy derivative $[(OEP)Sc(\mu\text{-}OH)_2Sc(OEP)]$.

In the presence of scandium perchlorate, 2,6-diacetylpyridine condenses with *m*-phenylenediamine to form a macrocyclic complex $ScL(ClO_4)_3 \cdot 4H_2O$ (L is shown as (4)). The structure has not been determined, but the perchlorates are not coordinated.[34] The reaction probably proceeds via the (isolable) complex $Sc(diacetylpyridine)_2(ClO_4)_3 \cdot 7H_2O$.

The template reaction of 2,6-diacetylpyridine with 3,3'-diaminodipropylamine in the presence of $ScCl_3$ or $Sc(ClO_4)_3$ gives complexes of a 14-membered N_4 macrocycle (5).[35]

3.2.1.3.5 *Compounds with P- and N, P-donor ligands*

These are as yet rare, given the limited ability of a hard acid like Sc^{3+} to bind to a soft base like a tertiary phosphine. Recent developments include the synthesis of $ScCl_2(THF)[N(SiMe_2CH_2P\,Pr^i_2)_2]$

(4)   (5)

from $ScCl_3(THF)_3$ and $LiN(SiMe_2CH_2PPr^i_2)_2$. This has to be carried out in toluene since the reaction in THF is a failure, recalling the lack of reactivity of lithium alkyls and aryls with $Cp*_2ScCl(THF)$. The THF complex, which has a *mer*-octahedral structure ((6); XRD data) loses its THF on pumping.[34]

(6)

The chlorines can be replaced by alkyl groups using RLi (but not RMgX or R_2Zn) to afford alkyls $ScR_2[N(SiMe_2CH_2PPr^i_2)_2]$ ((7): $R = Me$, Et, CH_2SiMe_3) which are very hydrocarbon-soluble and have to be recrystallized from $(Me_3Si)_2O$. They have trigonal bipyramidal five-coordinate structures (XRD data; $R = Et$, CH_2SiMe_3); despite the fact that they are formally 12-electron compounds, there is no evidence for agostic interactions between scandium and β-hydrogens.

Another compound featuring Sc—P bonding is the phosphomethanide $Sc[C(PMe_2)_2X]_3$ ((8); $X = SiMe_3$, PMe_2)

$$Sc(CF_3SO_3)_3 + 3Li[C(PMe_2)_2X] \rightarrow Sc[C(PMe_2)_2X]_3 + 3CF_3SO_3Li \qquad (3)$$

where the ligands bind in a fashion intermediate between σ-chelating and π-type (η^3)-coordination.[35]

3.2.1.4 Group 16 Ligands

3.2.1.4.1 Salts and the aqua ion

(i) Anhydrous and hydrated salts

A recent synthesis[38] of anhydrous $Sc(ClO_4)_3$ from hydrated scandium chloride proceeds via an orange adduct $[Sc(ClO_4)_3 \cdot 0.25Cl_2O_6]$ that loses the Cl_2O_6 in vacuo at 95 °C (see Scheme (2)).

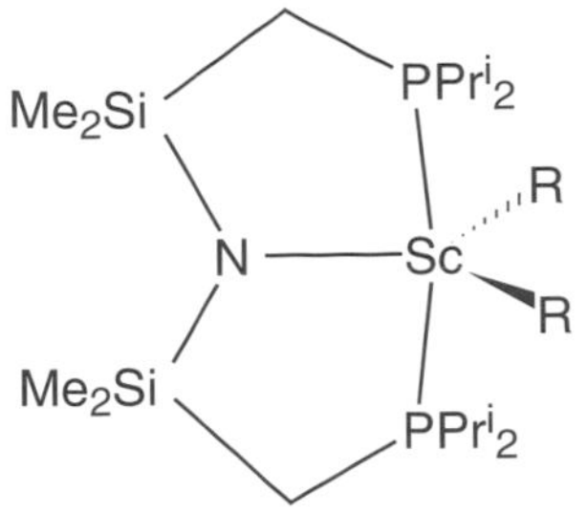

(7)

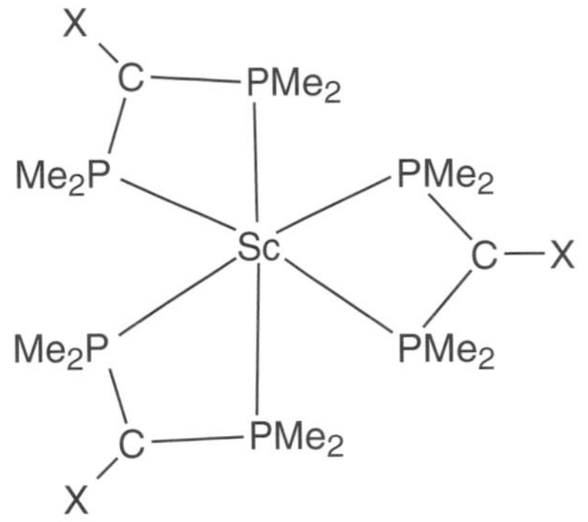

(8)

Infrared and Raman spectra are similar to those of anhydrous gallium perchlorate and are interpreted in terms of the presence of bridging bidentate perchlorates.

$$ScCl_3 \cdot 3.7\,H_2O \xrightarrow[-15^\circ]{Cl_2O_6} Sc(ClO_4)_3 \cdot 0.25\,Cl_2O_6 \xrightarrow[-Cl_2O_6]{95^\circ C} Sc(ClO_4)_3$$

Scheme 2

Grey anhydrous scandium triflate, $[Sc(O_3SCF_3)_3]$ (triflate = trifluoromethane sulfonate), has been obtained[39] by dehydration of the hydrate at 190–200 °C; the hydrated salt was itself obtained from the reaction of hydrated scandium chloride and dilute triflic acid. $[Sc(O_3SCF_3)_3]$, in which triflate is believed to act as a bidentate ligand (similar to perchlorate in $Sc(ClO_4)_3$), is not isomorphous with the lanthanide analogues.

The anhydrous scandium carboxylates $Sc(OCOR)_3$ (R = H, CH_3) have long been known to have polymeric structures with six-coordinate scandium.[40] Scandium formate has a 2-D polymeric structure whilst in the acetate, chains of Sc^{3+} ions are bridged by acetate groups with essentially octahedral coordination of scandium. Similar bridging and six-coordination is found in the chloroacetate. Scandium propynoate (R is $C\equiv CH$) crystallizes anhydrous from aqueous solution[41] and has an infinite three-dimensional structure, again with six-coordinate scandium ($Sc-O = 2.081$–$2.091(2)$ Å). On γ-irradiation, it changes color from colorless to orange, a change accompanied by a gradual disappearance of the $\nu(C\equiv C)$ stretching mode in the IR spectrum, indicating conversion to a poly(propynoate).

Hydrated scandium perchlorate has long been known but it is only recently that $Sc(ClO_4)_3 \cdot 6H_2O$ has been shown[42] to be isomorphous with the lanthanide analogues thus containing $[Sc(OH_2)_6]^{3+}$ ions, although neither details of the structure nor bond lengths have not been reported. On dehydration $Sc(ClO_4)_3 \cdot 6H_2O$ forms $Sc(OH)(ClO_4)_2 \cdot H_2O$, which has a sheet structure with octahedral coordination of scandium. A number of other hydrated scandium salts of uncertain structure have been known for many years[43–45] such as the very hygroscopic $Sc(NO_3)_3 \cdot nH_2O$ ($n = 2, 3, 4$) and $Sc(BrO_3)_3 \cdot 3H_2O$; the formulae of these complexes indicate that anion coordination is likely. However, $[Sc(NO_3)_3(H_2O)_2]$ and $[Sc(NO_3)_3(H_2O)_3]$ molecules, eight and nine-coordinate respectively, have been encapsulated inside crown ethers[46–49] indicating a likely structure for the coordination sphere in the hydrated nitrate.

It was only in 1995 that the first structure of a hydrated salt of scandium containing only water molecules in its coordination sphere was reported.[50] Refluxing scandium oxide with triflic acid leads to the isolation of hydrated scandium triflate $Sc(O_3SCF_3)_3 \cdot 9H_2O$. It is isomorphous with the hydrated lanthanide triflates, containing tricapped trigonally prismatic coordinate scandium in the $[Sc(H_2O)_9]^{3+}$ ions, with $Sc-O$ (vertices) = $2.171(9)$ Å and $Sc-O$ (face capped) $2.47(2)$ Å.

The fact that this structure is observed for all $M(O_3SCF_3)_3 \cdot 9H_2O$ (M = Sc, Y, La–Lu), irrespective of ionic radius, reflects the role of hydrogen bonding between the coordinated water molecules and the triflate groups in stabilizing the structure and has no implications for the coordination number of scandium in aqueous solution. The high coordination number of nine is the maximum yet observed for scandium.

Reaction of refluxing aqueous $HC(SO_2CF_3)_3$ with scandium oxide yields the triflide salt $[Sc(OH_2)_7](C(SO_2CF_3)_3)_3$. The crystal structure, though complicated by disorder, was solved to an R value of 0.095 and revealed[51] two scandium-containing sites, with coordination geometries described as either distorted capped trigonal prismatic (80%) or distorted pentagonal bipyramid (20%). For the main site Sc—O distances fall in the range 2.113(13)–3.222(10) Å, averaging 2.17 Å. In contrast, the ytterbium analogue contains eight-coordinate $[Yb(OH_2)_8]^{3+}$ ions.

Recrystallization of the scandium halides (except the fluoride) from slightly acidified aqueous solution (to prevent the hydrolysis that could otherwise result in the hydroxy-bridged dimers discussed below) gives the hydrated salts $ScX_3 \cdot 7H_2O$ (X = Cl, Br) and $ScI_3 \cdot 8H_2O$. X-ray diffraction studies show them all to contain $[Sc(OH_2)_7]^{3+}$ ions; in the chloride and iodide, the coordination geometry is essentially pentagonal bipyramidal, whilst in the iodide there is a substantial distortion. Axial Sc—O bonds tend to be shorter than those in the pentagonal plane; thus, in the chloride, the axial Sc—O distance is 2.098 Å; the equatorial distances average 2.183 Å.[52] In view of the tendency of many of the lanthanide halides to contain coordinated halide ions, the chlorides in particular, the absence of halide from the coordination sphere of scandium in these compounds is remarkable.

Overall then, the existence of the three scandium-containing ions $[Sc(OH_2)_x]^{3+}$ ($x = 6, 7, 9$) in five different solid salts indicates that in itself X-ray diffraction data on solids cannot be relied upon to indicate the coordination number in aqueous solution. Ultimately, it is the solubility of a particular salt that determines which compound crystallizes out from aqueous solution.

Three other salts where water shares the coordination sphere of scandium with anions have been studied. Yellow needles obtained[53] from the reaction of freshly precipitated $Sc(OH)_3$ with picric acid proved to be a 1:1 adduct with picric acid, *trans*-$[Sc(OH_2)_4(pic)_2](pic)(Hpic) \cdot 8.2H_2O$ (Hpic = $HOC_6H_2(NO_2)_3$-2,4,6); scandium is present as part of a six-coordinate cation, with Sc$\cdots$OH$_2$ = 2.100(9), 2.102(9), 2.113(9) and 2.121(8) Å, and Sc—O 2.019(8), 2.046(8) Å.

In contrast, hydrated scandium tosylate, $Sc(SO_3C_6H_4CH_3$-4$)_3 \cdot 6H_2O$ contains[54] *cis*-coordinated tosylate ligands in a six-coordinate cation having the structure *cis*-$[Sc(OH_2)_4(SO_3C_6H_4CH_3-4)_2]^+$ with Sc$\cdots$OH$_2$ 2.097(4), 2.118(4), 2.119(4) and 2.132(4) Å; Sc—O 2.021(4), 2.067(4) Å). In contrast, the later lanthanides form square antiprismatic $[Ln(OH_2)_6(SO_3C_6H_4CH_3-4)_2]^+$ (Ln = Sm, Gd, Ho, Er, Yb, Y) cations). In the final example, $ScCl_3(H_2O)_3$ molecules have been encapsulated in a cryptand ligand, rather as hydrated scandium nitrate is trapped by crown ethers. In [H$_2$L] *mer*-$ScCl_3(H_2O)_3 \cdot 3H_2O$ (L = cryptand-2,2,2)[55] the Sc$\cdots$OH$_2$ distances are 2.078(10), 2.132(9), and 2.155(9) Å. All these six-coordinate compounds have average Sc$\cdots$OH$_2$ distances of around 2.11 Å; these are in line with a predicted value for Sc$\cdots$OH$_2$ in six coordination of 2.10–2.11 Å, extrapolating from the Ti$\cdots$OH$_2$ distances of 2.018–2.046 Å found[56] in the tosylate salt of the $[Ti(OH_2)_6]^{3+}$ ion, making allowance for the radius of the scandium ion being 0.075 Å greater.[3]

In addition to these compounds, a number of dimeric salts containing the di-μ-hydroxy bridged species $[(H_2O)_5Sc(\mu$-OH$)_2Sc(H_2O)_5]^{4+}$ ions (9) have been characterized.

(9)

Attempted synthesis[57] of scandium benzene sulfonate from $ScCl_3$ and sodium benzene sulfonate led to the isolation of the dimer $[(H_2O)_5Sc(\mu$-OH$)_2Sc(H_2O)_5]$ $(C_6H_5SO_3)_4 \cdot 4H_2O$. This contains seven-coordinate scandium with an approximately pentagonal bipyramidal coordination geometry. The axial Sc$\cdots$OH$_2$ bonds average 2.146 Å and the equatorial Sc$\cdots$OH$_2$ bonds average 3.227 Å, whilst the Sc$\cdots$OH bridge distances of 2.072 Å are shorter than the terminal Sc—O distances of 2.111–2.125 Å in the $[Sc(OH)_6]^{3-}$ ion (see Section 2.1.4.3) and certainly do not give any evidence for congestion in the coordination sphere of scandium. Crystallization of hydrated scandium chloride from *n*- and *iso*- propanol (where presumably some hydrolysis occurred) gives the related substance $[(H_2O)_5Sc(\mu$-OH$)_2Sc(H_2O)_5]Cl_4 \cdot 2H_2O$ whilst the analoguous $[(H_2O)_5Sc(\mu$-OH$)_2Sc(H_2O)_5]Br_4 \cdot 2H_2O$ has been made from scandium bromide.[58] In these two compounds, the coordination geometry has been described as close to a monocapped trigonal prism, but most significantly the average

Sc$\cdots$OH$_2$ (see Table 2) and the Sc—OH bond lengths are similar in all three of these dimers. An independent structure of [(H$_2$O)$_5$Sc(μ-OH)$_2$Sc(H$_2$O)$_5$]Cl$_4$ has been reported.[59]

These [(H$_2$O)$_5$Sc(μ-OH)$_2$Sc(H$_2$O)$_5$]$^{4+}$ ions are significant in that they are believed to be formed in the first stage of the hydrolysis of the scandium aqua ion, and the coordination geometry involving scandium bound to seven water molecules and hydroxide ions can clearly derive from a [Sc(OH$_2$)$_7$]$^{3+}$ ion, in the way that [(H$_2$O)$_4$Fe(μ-OH)$_2$Fe(H$_2$O)$_4$]$^{4+}$ is believed to relate to the [Fe(OH$_2$)$_6$]$^{3+}$ ion.

The complex [(picolinato)$_2$(H$_2$O)Sc(μ-OH)$_2$Sc(OH$_2$)(picolinato)$_2$] shares with the preceding compounds double hydroxy bridges and seven-coordinate scandium.[60] Here the Sc$\cdots$OH$_2$ distance is 2.172(2) Å whilst the Sc—O bridge distances are 2.063(2) Å and 2.080(1) Å, similar to those in the three hydroxy-bridged aqua species. The average Sc—OH$_2$ bond lengths in these four compounds that contain seven-coordinate scandium lie in the range 2.175–2.194 Å, significantly longer than in the six-coordinate complexes.

(ii) The scandium aqua ion

For many years, there was no concrete evidence for the coordination number of scandium in aqueous solution; it has been generally assumed, without any solid evidence, that the scandium aqua ion was [Sc(OH$_2$)$_6$]$^{3+}$. Possibly this was done by analogy with the ions of the succeeding transition metals, all of which form [M(OH$_2$)$_6$]$^{3+}$(M = Ti–Co), yet one cogent argument against this is the fact that Sc^{3+} has an ionic radius substantially greater than that of the succeeding 3d metals (0.745 Å in six-coordination), intermediate between that of Ti^{3+} (0.670 Å) and the lanthanides (La^{3+} = 1.032 Å; Lu^{3+} = 0.861 Å). In solution, the lanthanides form [Ln(OH$_2$)$_9$]$^{3+}$ aqua ions for the early lanthanides and [Ln(OH$_2$)$_8$]$^{3+}$ aqua ions for the later ones. Purely on steric grounds, a coordination number intermediate between 6 and 8 might be predicted for scandium.

Japanese workers used vibrational spectroscopy in an important study.[61] They compared the Raman spectra of acidic (pH < 2) solutions and glasses of Sc^{3+}(aq) and Al^{3+}(aq), the latter being a known example of six coordination, finding significant differences in both the number and polarization of bands. Glasses of [Al(OH$_2$)$_6$]$^{3+}$ show three bands (one polarized) in the region expected for metal–ligand stretching vibrations; three are predicted (ν_1 (a_{1g}), ν_2 (e_g) and ν_5 (t_{2g})) for an octahedral ion. Glasses of Sc^{3+}(aq) show four bands (two polarized) in the corresponding region, at 450 (p), 410, 375 and 310 (p) cm^{-1}. In contrast to the three Raman-active metal–ligand stretching vibrations of the octahedral [Sc(OH$_2$)$_6$]$^{3+}$ ion, a pentagonal bipyramidal [Sc(OH$_2$)$_7$]$^{3+}$ ion with D_{5h} local symmetry would give rise to five bands, two with a_1' symmetry that would be expected to be polarized. At the pH of the measurements, significant amounts of a dimeric ion [(H$_2$O)$_5$Sc(μ-OH)$_2$Sc(H$_2$O)$_5$]$^{4+}$ are unlikely, and in fact any species of lower symmetry would give rise to more bands. An important point is that spectra of Sc(ClO$_4$)$_3$ yield similar results to those obtained from the chloride, indicating that anion coordination is not significant under these conditions.

In a subsequent study from the same workers, X-ray diffraction data obtained from scandium perchlorate solutions[62] indicated a Sc—O distance of 2.180(7) Å with a coordination number of 7.4(4). The sharpness of the peak at 2.180 Å suggested that all Sc—O bond lengths in the solution were comparable, rather than falling into two groups, as found in the crystal structure of the [Sc(OH$_2$)$_9$]$^{3+}$ ion. X-ray absorption fine structure (XAFS) data from scandium triflate solutions also indicate a Sc—O distance of 2.18(2) Å, but no coordination number could be unambiguously obtained from the XAFS measurements, though the data were not consonant with a tricapped trigonal prismatic nine-coordinate structure. This Sc—O distance of 2.18 Å, nearly 0.1 Å longer than that found for six-coordination, fits well with the Sc–water distances found in the seven-coordinate [Sc(OH$_2$)$_7$]$^{3+}$ species.

Considering all the data, it seems most likely that the predominant species present in rather acidic aqueous solution is a pentagonal bipyramidal [Sc(OH$_2$)$_7$]$^{3+}$ ion, with an average Sc—O distance around 2.18 Å.

(iii) Hydroxide complexes

M$_3$[Sc(OH)$_6$] (M = K, Rb) complexes have been isolated from the reaction of scandium nitrate with amide and nitrate ions in supercritical ammonia.[63] X-ray diffraction studies confirm the presence of octahedral [Sc(OH)$_6$]$^{3-}$ ions. The Sc—O bond distance is 2.111(9) Å in Rb$_3$[Sc(OH)$_6$] and 2.120(4)–2.125(4) Å in K$_3$[Sc(OH)$_6$].

3.2.1.4.2 *Complexes of other group 16-donor ligands*

Scandium forms complexes with a wide range of polar O-donor ligands, such as pyridine N-oxide, dimethyl sulfoxide, triphenylphosphine- and arsine oxides, and hexamethylphosphoramide being examples. In general, the complexes reported have tended to fall into two main series with ScL_3X_3 typical where X is a coordinating anion like chloride and ScL_6X_3 (the norm when noncoordinating anions like perchlorate are present). Where nitrate features as a ligand, higher coordination numbers are possible, because of the small "bite" angle of bidentate nitrate group. X-ray data are still rather limited.

The structures[1,54] of *mer*-$Sc(THF)_3Cl_3$ and *mer*-$Sc(H_2O)_3Cl_3$ have recently been complemented by $Sc(THF)_2(H_2O)Cl_3$, synthesized[64] by reaction of the stoichiometric amount of water with $Sc(THF)_3Cl_3$. Compound (**10**) shows that the *mer*- geometry is retained; the Sc—Cl distance *trans*- to water is shorter (Sc—Cl of 2.399 Å) than the mutually *trans*-Sc—Cl distances (2.477(3) Å and 2.478(3) Å).

(**10**)

It did not prove possible to isolate the "missing" member of the series, [$Sc(THF)(H_2O)_2Cl_3$] by the same route.

Scandium triflate reacts with Ph_3PO in ethanol forming $Sc(OTf)_3(Ph_3PO)_4$, (OTf = triflate) which has the structure *trans*-[$Sc(OTf)_2(Ph_3PO)_4$]OTf, featuring octahedrally coordinated scandium, with monodentate triflates. In [$Sc(OTf)_2(Ph_3PO)_4$]OTf, Sc—O (PPh$_3$) distances lie in the range 2.051(2)–2.090(2) Å, with Sc—O(Tf) 2.111(2) Å and 2.138(2) Å. It has some catalytic activity for reactions like the nitration of anisole, but less than that of the hydrated scandium triflate.[65]

Numerous scandium nitrate complexes with phosphine oxides, [$Sc(NO_3)_3(Ph_3PO)_2$], [$Sc(NO_3)_2(Ph_2MePO)_4$](NO_3), [$Sc(NO_3)_3(Ph_2MePO)_2$], [$Sc(NO_3)_3(Me_3PO)_2(EtOH)$], and [$Sc(Me_3$-$PO)_6$]($NO_3$)$_3$ have been synthesized.[66] [$Sc(NO_3)_3(Ph_3PO)_2$] has an eight-coordinate structure with symmetrically bidentate nitrates; Sc—O(PPh$_3$) being 2.047–2.068(7) Å and Sc—O (nitrate) lying in the range 3.205(8)–2.311(7) Å. This complex is obtained from all stoichiometries of reaction mixture, but with Ph_2MePO, two different complexes can be made. With a 1:1 or 2:1 molar ratio, [$Sc(NO_3)_3(Ph_2MePO)_3$] is obtained, which appears to have all nitrates coordinated, although it is not clear if they are all bidentate. With a 4:1 (or higher) ligand:scandium ratio, the product is $Sc(NO_3)_3(Ph_2MePO)_4$. In solution, it gives a broad resonance in the ^{31}P-NMR spectrum at room temperature, which separates on cooling into separate signals characteristic of $Sc(NO_3)_3(Ph_2$-$MePO)_4$ and $Sc(NO_3)_3(Ph_2MePO)_3$. In the solid state, it has the structure [$Sc(NO_3)_2(Ph_2Me$-$PO)_4$](NO_3), again with symmetrically bidentate nitrates; Sc—O(P) being 2.088–2.099(6) Å and Sc—O (nitrate) lying in the range 2.311(6)–2.425(6) Å, the nitrate groups being *trans*- to each other. Reaction of scandium nitrate with a large excess of Me_3PO results in [$Sc(Me_3PO)_6$](NO_3)$_3$, the IR spectra of which indicate only ionic nitrate groups; there is NMR evidence of it dissociating to a species such as [$Sc(Me_3PO)_5(NO_3)$](NO_3)$_2$ in solution in the absence of excess ligand. [$Sc(NO_3)_3(Me_3PO)_2(EtOH)$] is obtained from scandium nitrate and Me_3PO reacting in a 1:12 ratio in ethanol.

In contrast to the reaction with Ph_3PO, where only [$Sc(NO_3)_3(Ph_3PO)_2$] can be isolated, Ph_3AsO forms 2:1 and 3:1 complexes.[67] Reaction between scandium nitrate and Ph_3AsO in acetone gives [$Sc(NO_3)_3(Ph_3AsO)_2$], whilst reaction in ethanol affords $Sc(NO_3)_3(Ph_3AsO)_3$, shown by X-ray diffraction to be seven-coordinate [$Sc(NO_3)_2(Ph_3AsO)_3$]NO_3. The polyhedron can be regarded as derived from a trigonal bipyramid, if the coordinated bidentate nitrates (which lie in the equatorial plane) are thought of as occupying a single site. Sc—O nitrate distances fall in the range 3.250–3.267 Å. The equatorial Sc—O distance is 1.999 Å and the axial ones 1.996–2.030 Å. Addition of further Ph_3AsO does not displace any more nitrate groups. On reaction between scandium nitrate and Me_3AsO in cold ethanol, [$Sc(Me_3AsO)_6$](NO_3)$_3$ is obtained; the cation has octahedrally coordianted scandium with Sc—O in the range 2.064–2.100 Å. There is evidence for nitrates entering the coordination sphere in solution in the absence of excess Me_3AsO.[55] Scandium NMR spectroscopy has been successfully applied to these nitrate complexes.

Several halide complexes have been isolated in the solid state,[68] namely [ScCl(Me$_3$PO)$_5$]Cl$_2$, [Sc(Me$_3$PO)$_6$]X$_3$ (X = Br, I), [ScX$_2$(Ph$_3$PO)$_4$]X, [ScX$_2$(Ph$_3$AsO)$_4$]X (X = Cl, Br, I), [Sc(Me$_3$As-O)$_6$]X$_3$ (X = Cl, Br, I), [ScCl$_3$(Ph$_2$MePO)$_3$], and [ScBr$_2$(Ph$_3$MePO)$_4$]Br, whilst others have been identified in solution by multinuclear NMR. The structure of *trans*-[ScBr$_2$(Ph$_3$PO)$_4$]Br shows a linear Br—Sc—Br arrangement with typical Sc—O distances of approximately 2.07 Å, and Sc—Br of 2.652(1) Å and 2.661(1) Å. *trans*-[ScCl$_2$(Ph$_3$AsO)$_4$]Cl similarly has an octahedral geometry at scandium, with Sc—O in the range 2.059–2.089(7) Å and Sc—Cl distances, at 2.545 Å and 2.562(4) Å, rather long in comparison with Sc—Cl distances in other chloro complexes like [ScCl$_3$(THF)$_3$]. [Sc(Me$_3$AsO)$_6$]Br$_3$ has Sc—O distances in the range 2.08(2)–2.11(2) Å, closely comparable with those in the nitrate salt. The picture that emerges from these studies is one in which chloride has a considerably greater affinity for Sc^{3+} than do the heavier halogens.

Reaction of anhydrous scandium triflate with HMPA gives Sc(HMPA)$_4$(CF$_3$SO$_3$)$_3$, which contains *trans*-[Sc(HMPA)$_4$(CF$_3$SO$_3$)$_2$]$^+$ ions, similar compounds being formed by most lanthanides (Ce–Lu).[69] Though Sc—O bond lengths have not been published, it was reported that the P—O bond length is relatively long compared to those in most of the lanthanide compounds, a possible reflection of the strength of the Sc—O bond.

Unlike the lanthanides, which form tetrahydro-2-pyrimidone (pu) complexes [Lnpu$_8$](CF$_3$SO$_3$)$_3$ with square antiprismatic eight-coordination, scandium yields [Scpu$_6$](CF$_3$SO$_3$)$_3$, which unexpectedly has trigonal antiprismatic coordination of scandium, possibly partly induced by side-on interactions between pairs of pu ligands.[70] Complexes with various N-oxides ScL$_6$(CF$_3$SO$_3$)$_3$ (L = pyridine N-oxide, 2-picoline N-oxide, 3-picoline N-oxide, 4-picoline N-oxide) and ScL$_5$(H$_2$O)(CF$_3$SO$_3$)$_3$ have been synthesized, all of these presumably contain octahedrally coordinated scandium.[71]

Six- and seven-coordination is found in the scandium picrate complex of *trans*-1,4-dithiane-1,4-dioxide (TDHD), [Sc$_6$(pic)$_6$(TDHD)$_3$(OH)$_{10}$(H$_2$O)$_2$](pic)$_2$(H$_2$O)$_{10}$ which has hexameric clusters of scandium ions, joined by TDHD bridges.[72] Four of the scandiums are six-coordinate and two are seven-coordinate. Complexes with naphthyridine N-oxide (napyo) have been synthesized.[73] Those reported are Sc(napyo)$_2$(NO$_3$)$_3$, Sc(napyo)$_4$(ClO$_4$)$_3$, Sc(napyo)$_4$(NCS)$_3$, and Sc(napyo)$_3$Cl$_3$; these are respectively 1:1, 1:3, 1:3, and non-electrolytes in solution. The chloride is thus presumably six-coordinate, but no diffraction data are reported. 1,10-Phenanthroline *N*-oxide complexes Sc(phenNO)$_4$(NCS)$_3$ and Sc(phenNO)$_3$Cl$_3$, with presumably similar structures, have also been made.[74] Octahedral coordination is found in [ScCl$_3$(DME)(MeCN)] and [ScCl$_3$(diglyme)].[75]

The coordination of the nitrate groups in the complexes of phosphine and arsine oxides already discussed is generally symmetrically bidentate. However, in Rb$_2$[Sc(NO$_3$)$_5$] there are three bidentate and two monodentate nitrates,[76] resulting in eight-coordination for scandium, in contrast to (NO$^+$)$_2$[Sc(NO$_3$)$_5$]$^{2-}$, where one monodentate and four bidentate nitrates give nine-coordinate scandium.[77]

3.2.1.4.3 Alkoxides

Scandium alkoxides represent a class of compound as yet with poorly characterized structures. Attempted synthesis of the isopropoxide Sc(OPr^i)$_3$ results in the production[78] of a pentanuclear compound [Sc$_5$O(OPr^i)$_{13}$], similar to those formed by yttrium, ytterbium, and indium. It has the structure [Sc$_5$(μ_5-O)(μ_3-OPr^i)$_4$(μ_2-OPr^i)$_4$(OPr^i)$_5$]. Anodic oxidation of scandium in aliphatic alcohols has been used as a pathway to scandium alkoxides [Sc(OR)$_3$] (R = Me, Et) and [Sc$_5$O-(OPr^i)$_{13}$].[79] The elusive structure of the methoxide [Sc(OMe)$_3$] is not yet known but it does appear not to contain an oxo ligand, and to be a polymer. Alcoholysis of [Sc$_5$O(OPr^i)$_{13}$] leads to [Sc(OR)$_3$] (R = Me, Et, Bu^n), [Sc$_5$O(OBu^s)$_{13}$], and [Sc$_5$O(OPr^i)$_8$(OBu^t)$_5$]. Some scandium alkoxo-aluminates [Sc(Al(OR)$_4$)$_3$] (R = Et, Bu^n) have also been reported.

3.2.1.4.4 Diketonates and other chelating ligands

Since the determination[2] of the structure of tris(acetylacetonato)scandium (Sc—O of 2.061–2.082 Å, average Sc—O 2.070(9) Å), several other diketonates have been examined, though no detailed structures have been reported. The compound *mer*-[Sc(CF$_3$COCHCOCH$_3$)$_3$] is isostructural with the aluminium, gallium, rhodium, and iridium analogues,[80] as well as those of the 3*d* metals V–Co whilst the dipivaloylmethanide [Sc(Me$_3$COCHCOMe$_3$)$_3$] is isostructural with the iron and indium analogues.[81] Scandium α-diketonates such as [Sc(tropolonate)$_3$] form adducts in solution with trioctylphosphine oxide (TOPO); β-diketonates like [Sc(acac)$_3$] do not.[82] In the gas phase, [Sc(acac)$_3$] has C_3 symmetry, in contrast to the D_3 symmetry in the crystal.[83]

Scandium β-diketonates [Sc(tmod)$_3$] and [Sc(mhd)$_3$] (tmod = 2,2,7-trimethyloctane-3,5-dionate; mhd = 6-methylheptane-2,4-dionate) have been studied as liquid-injection MOCVD precursors for Sc$_2$O$_3$ with a view to their use in the synthesis of the pyroelectric material Pb(Sc$_{0.5}$Ta$_{0.5}$)O$_3$ (PST).[84] Acetylpyrazolones also act as chelating ligands. In a comparative study of the coordinating ability of the ligand HPMTFP (1-phenyl-3-methyl-4-trifluoroacetyl-pyrazolone-5),[85] it was found that scandium formed a seven-coordinate adduct [Sc(PMTFP)$_3$(OPPh$_3$)] whilst neodymium forms the eight-coordinate [Nd(PMTFP)$_3$(OPPh$_3$)$_2$]. (The Sc—OP distance quoted in the paper (1.198 Å) appears to be in error; other Sc—O distances are in the range 2.12–3.27 Å.)

3.2.1.4.5 *Crown ether complexes and related systems*

Although some crown ether complexes of scandium were made at the time that the lanthanide complexes were first made, it is only recently that they have been investigated systematically and definitive structural information has become available, mainly due to Willey and co-workers.[55,86,89–92]

Resemblance to the lanthanides extends to the existence of two series; "inner sphere" complexes exist in which scandium is bound directly to the oxygens in the crown ether ring, whilst in a second class, a crown ether is present in the second coordination sphere of the scandium, hydrogen-bonded to a scandium aqua-species. Scandium nitrate imitates the later lanthanide nitrates in only forming "outer sphere" complexes with the larger rings such as 18-crown-6, although it seems possibly that scandium nitrate might complex directly with a small crown like 12-crown-4, but this does not seen to have been explored.

Although the structure of hydrated scandium nitrate itself has not been reported, both 8- and 9-coordinate [Sc(NO$_3$)$_3$(H$_2$O)$_2$] and [Sc(NO$_3$)$_3$(H$_2$O)$_3$] molecules have been encapsulated by crown ethers, and several structures have been reported for such "outer sphere" complexes.[43–49] In the [[Sc(NO$_3$)$_3$(H$_2$O)$_3$]·18-crown-6] complex, Sc—O(water) distances are 2.120(6), 3.221(6), and 2.303(15) Å, whilst the Sc—O(nitrate) distances of 3.227(6), 3.240(17), 3.243(5), 2.342(7), 2.348(16), and 2.366(7) Å show similar trends. In contrast, the Sc···OH$_2$ distances in [Sc(NO$_3$)$_3$(H$_2$O)$_2$]·15-crown-5 are 2.120 and 2.143 Å, whilst the Sc—O (ether) distances span a range of 2.195–3.245 Å. Similarly, in [Sc(NO$_3$)$_3$(H$_2$O)$_2$].benzo-15-crown-5, the Sc···OH$_2$ distances are 2.123(3) and 2.143(3) Å, and the Sc—O (ether) distances range from 2.158(3) to 3.248(3)Å. It can be seen that the Sc—O bond lengths in the nine-coordinate [Sc(NO$_3$)$_3$(H$_2$O)$_3$] complex span a wider range than in the [Sc(NO$_3$)$_3$(H$_2$O)$_2$] complexes, suggesting that there is some steric congestion here, and comparison of the Sc···O (H$_2$O) distances indicates that the third water molecule in [[Sc(NO$_3$)$_3$(H$_2$O)$_3$]·18-crown-6] is loosely held.

In contrast, a cryptand ligand has been used[55] to encapsulate and isolate a mixed aqua/chloro species in [H$_2$L] *mer*-ScCl$_3$(H$_2$O)$_3$.3 H$_2$O (L = cryptand-2,2,2). The scandium-containing species has Sc—O 2.078(10), 2.132(9) and 2.155(9) Å with Sc—Cl 2.413(6), 2.419(4), and 2.419(5) Å, and is closely comparable to the familiar *mer*-ScCl$_3$(THF)$_3$.

Direct scandium–crown ether complexation could give rise to two types of complex, firstly "extra-cavity" complexes, in which the coordinated scandium lies outside the ligand cavity, in a kind of half-sandwich structure, and secondly "intra-cavity" complexes, where scandium fits into the cavity within the crown ether ring. To date, the latter are more common, and generally feature a ScCl$_2$ moiety threaded through the ligand cavity. A ^{45}Sc-NMR study of crown ethers led to the characterization of [ScCl$_2$(crown)]$^+$ (crown = 15-crown-5, dibenzo-24-crown-8, dibenzo-30-crown-10, 1-aza-15-crown-5, and 1-aza-18-crown-6) whilst [ScCl$_2$(12-crown-4)]$^+$ and [Sc(12-crown-4)$_2$]$^+$ have both been established.[86] Reaction of ScCl$_3$(MeCN)$_3$ with 15-crown-5 affords a 1:1 complex.[87] No structural information is available, so it might have a half-sandwich molecular structure [ScCl$_3$(15-crown-5)] or alternatively be [ScCl$_2$(15-crown-5)]$^+$Cl$^-$. However, in the presence of CuCl$_2$, the "threaded" complex {[ScCl$_2$(15-crown-5)]$^+$}$_2$[CuCl$_4$$^{2-}$] is obtained. The cation (**11**) has pentagonal bipyramidal seven-coordination of scandium; five oxygens form an approximately planar equatorial belt with a near linear (176.4–179.8°) Cl—Sc—Cl unit threaded through the crown ether ring roughly at right angles to the O$_5$ plane, with Sc—O distances falling into a narrow range of 2.09–2.15 Å. Here CuCl$_2$ has acted as a halide ion abstractor, making it possible for insertion of the ScCl$_2$ moiety into the crown ether cavity.

Quite independently, a systematic study has explored the reaction of SbCl$_5$ as a chloride ion extractor with acetonitrile solutions of ScCl$_3$(THF)$_3$. This generates a solvated [ScCl$_2$]$^+$ unit,

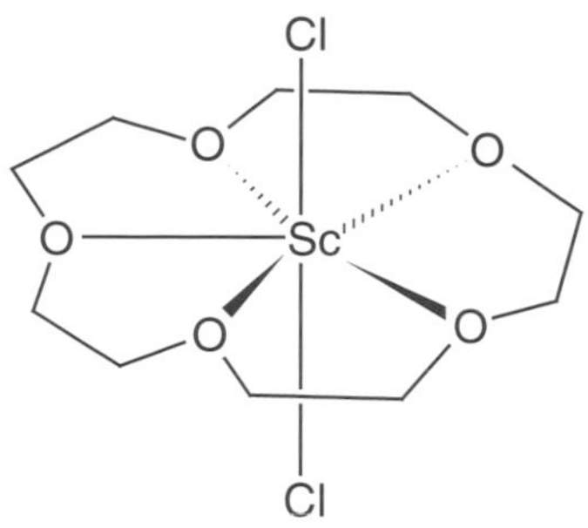

(11)

which reacts with the crown ether, replacing the weakly coordinated nitrile (and THF) ligands (see Scheme (3)).

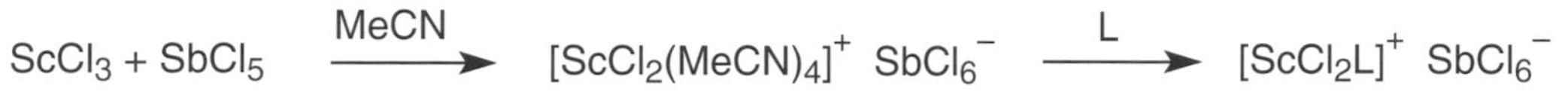

$$ScCl_3 + SbCl_5 \xrightarrow{\text{MeCN}} [ScCl_2(MeCN)_4]^+ \, SbCl_6^- \xrightarrow{\text{L}} [ScCl_2L]^+ \, SbCl_6^-$$

Scheme 3

Reaction of $ScCl_3(THF)_3$ with $SbCl_5$ in the absence of crown ether affords $[ScCl_2(THF)_4]^+[SbCl_5(thf)]^-$; in contrast, Y and La form $[MCl_2(THF)_5]^+[SbCl_5(THF)]^-$ (M = Y, La).[88]

The first crown ether complex made by this route to be reported[89] was $[ScCl_2(18\text{-crown-}6)]^+$-$[SbCl_6]^-$. The crystal structure of this compound shows that one oxygen atom in the crown ether remains uncoordinated to the metal, confirming the preference of scandium for pentagonal bipyramidal coordination; Sc—O distances fall in the range 2.190–3.229 Å. Subsequently, the related $[ScCl_2(\text{crown})]^+[SbCl_6]^-$ (crown = 15-crown-5,[90] benzo-15-crown-5,[90] and dibenzo-18-crown-6[91]) complexes have been isolated; all are believed to have a pentagonal bipyramidal coordination geometry, which has been confirmed crystallographically for the benzo-15-crown-5 complex.

In contrast, the 1.8 Å diameter of the cavity in the tetradentate 12-crown-4 ring is too small for a $[ScCl_2]^+$ ion to fit, so an "extra-cavity" (half-sandwich) structure **(12)** has been suggested for $[ScCl_2(12\text{-crown-}4)(MeCN)]^+SbCl_6^-$, though no diffraction data are available. Unlike the "intra-cavity" crown ether complexes already mentioned, a MeCN is believed to coordinate, again reflecting scandium's preference for seven-coordination.[90]

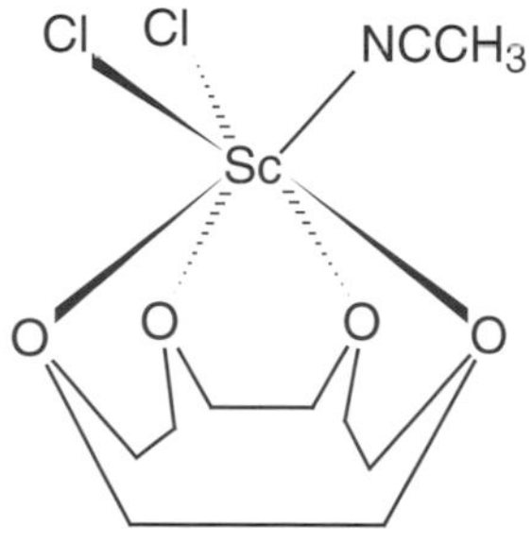

(12)

Larger oxacrown rings have more donor atoms and are also more flexible. Reaction of $ScCl_3(THF)_3$ and 1 mole of $SbCl_5$ with 1 mole of a larger crown ether affords the complexes $[ScCl_2(\text{dibenzo-24-crown-8})(H_2O)]^+$ $SbCl_6^-\cdot2MeCN$ and $[ScCl_2(\text{dibenzo-30-crown-10})(H_2O)_2]^+SbCl_6^-\cdot MeCN\cdot H_2O$. Both cations again feature seven-coordinate scandium in a pentagonal bipyramidal environment.[91] The water molecules (probably arising from either charcoal or solvent used in recrystallization) play an important role in the structure, apart from coordinating to the metal, they also hydrogen-bond to oxygens in the crown ether that are not coordinated to scandium. Whereas a smaller crown ether like 18-crown-6, not to mention 15-crown-5 or benzo-15-crown-5, can occupy five equatorial sites round scandium with little ring distortion and concomitant strain, this is not possible for the larger ethers with eight or ten oxygens in the ring. It should also be noted that whilst the dibenzo-30-crown-10 ligand has a large enough cavity to accommodate two scandium ions, it does not do so. The scandium–water distances in these compounds are intermediate in length between those found in the six and seven-coordinate aqua complexes, probably because the macrocyclic ligands have the flexibility to ensure a small "bite" for each donor atom.

Use of excess $SbCl_5$ permits the extraction of further chlorides from the crown ether complexes.[91] Thus reaction of $[ScCl_3(THF)_3]$ in MeCN with 3 moles of $SbCl_5$ and dibenzo-18-crown-6 leads to the complex $[ScCl(dibenzo-18-crown-6)(MeCN)]^{2+}(SbCl_6^-)_2$; this reacts with a large excess of $SbCl_5$ forming a species identified by NMR as $[Sc(dibenzo-18-crown-6)(MeCN)_2]^{3+}$.

Another route to complete halide extraction involves the use of $SbCl_5$ and $SbCl_3$ together (see Scheme (4)). The cation contains a Sc^{3+} ion sandwiched between two 12-crown-4 molecules in approximately square antiprismatic eight-coordination with Sc—O distances in the range 2.160(8)–3.274(9) Å, averaging 3.212 Å.[92]

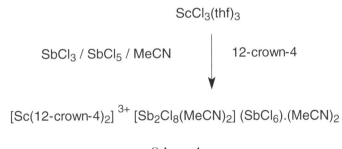

ScCl₃(thf)₃

SbCl₃ / SbCl₅ / MeCN   12-crown-4

$[Sc(12\text{-crown-}4)_2]^{3+}$ $[Sb_2Cl_8(MeCN)_2]$ $(SbCl_6).(MeCN)_2$

Scheme 4

The structure of a di-μ-oxo bridged calix[3]arene complex has been reported[93] whilst other complexes have been isolated with p-t-butylcalix[n]arenes (n = 4, 6, 8). p-t-Butylcalix[4]arene forms a 2:1 complex whilst p-t-butylcalix[6]arene and p-t-butylcalix[8]arene form 1:1 complexes, but these compounds at present lack structural characterization.[94]

3.2.1.5 Mixed Group 15 and 16 Donors

In contrast with the lanthanides and the 3d metals, until recently no structure had been reported of a complex of scandium with either EDTA or DTPA (DTPA = diethylenetriaminepentaacetic acid). $NH_4[Sc(EDTA)(H_2O)_2]\cdot3H_2O$ has eight-coordinate scandium.[95] In this compound, EDTA is, as might be expected, hexadentate, whilst there are two coordinated water molecules, in contrast to the norm of one for 3d metals such as Fe^{3+} (but fewer than for the early lanthanides, again in line with expectations based on ionic radii). Bond lengths are 2.1434, 2.1440, 2.1464, 2.1665, 3.2514, and 3.2848 Å (Sc—O) and 2.4568 and 2.4582 Å (Sc—N). The scandium–oxygen bond lengths involving the coordinated water molecules (3.25–3.28 Å) are similar to those in aqua species mentioned in Section 3.2.1.4.1. Eight-coordinate scandium is also found in the structure of $MnSc(DTPA)\cdot4H_2O$, which has octadentate DTPA occupying the coordination sphere of scandium with no coordinated water molecules,[96] in contradistinction to the nine-coordinate magnetic resonance imaging (MRI) contrast agent $[Gd(DTPA)(H_2O)]^{2-}$; scandium is bound to five oxygens (Sc—O 2.116(3)–2.197(3) Å) and three nitrogens (Sc—N 2.377(3)–2.489(3) Å), the distances being similar to those in the EDTA complex.

These two compounds nicely illustrate the tendency of Sc^{3+} to adopt coordination numbers intermediate between those of the M^{3+} ions in the first transition series and those found for the Ln^{3+} ions.

3.2.1.6 Complexes of Group 17 Ligands

3.2.1.6.1 *Binary halides and simple complexes*

Although the binary halides in the solid state all have giant structures[6] (all but the fluoride (WO_3 structure) having the $FeCl_3$ (BiI_3) structure), they exhibit molecular structures in the gas phase and many complexes are known.

A neutron-diffraction refinement of the $ScCl_3$ structure (using $ScCl_3$ prepared by reductive chlorination of Sc_2O_3 at 900 °C) gives an average Sc—Cl distance of 2.52 Å (Sc—Sc 3.68Å).[97]

In the gas phase, most recent electron-diffraction data at 1,750 K indicate isolated ScF_3 molecules to be planar or nearly so[98] with Sc—F $= 1.847$ Å; studies on ScI_3 indicate[99] the presence of both monomers and dimers in the vapor at 1,050K; the monomers have Sc—I 2.62 Å with a I—Sc—I angle of 117(3)°. Most recently, density functional theory calculations[100] on $ScCl_3$ and Sc_2Cl_6 dimers favor a planar D_{3h} monomeric structure for the former (Sc—Cl 3.285 Å) and a D_{2h} dichloro-bridged structure for the dimer with Sc—Cl (terminal) of 3.260 Å and Sc—Cl of 2.475 Å. Electron-diffraction studies on $ScCl_3$ vapor indicate a slight distortion to a C_{3v} pyramidal structure having Sc—Cl of 3.291 Å and Cl—Sc–Cl angle of 119.8°; in comparison with six-coordinate $ScCl_3$ in the solid state, the shorter Sc—Cl bond lengths in the three- and four-coordinate vapor phase species are to be expected.

$KScF_4$ has a layered structure[101] in which scandium acquires six-coordination by edge-sharing alternate *cis*- and *trans*- corners with Sc—F terminal bond lengths in the range 1.941–1.983 Å and bridging distances of 2.009–2.038 Å. Na_3ScF_6 has the cryolite structure[102] with an average Sc—F distance of 2.007 Å. High-pressure studies of Na_3ScF_6 (cryolite structure) indicate little change in the octahedral coordination of scandium up to 27.9 kbar.[103] Sr_2ScF_7 has seven-coordinate scandium, however.[104] Rb_2KScF_6, synthesized by heating stoichiometric amounts of RbF, KF, and ScF_3 at 700 °C, exists in three different crystalline forms.[105] The lowest temperature (monoclinic) form has a structure related to cryolite; on warming to 223 K, it transforms to a tetragonal structure and on further warming to 252 K changes to a cubic elpasolite structure. K_2NaScF_6 also has the cubic elpasolite structure.[106] Ba_2ScCl_7, synthesized from a 1:1 molar mixture of $BaCl_2$ and $ScCl_3$ at 580 °C, has the structure $Ba_2[ScCl_6]Cl$ in which the Sc—Cl bond lengths in the octahedral anion fall in the range 2.42–2.52 Å (average 2.48Å).[107] Na_3ScCl_6, made by heating NaCl and $ScCl_3$ together in a stoichiometric ratio, has the cryolite structure and is isotopic with Na_3LnCl_6 (Ln = Dy-Lu, Y)[108] whilst $Cs_2LiScCl_6$ has the elpasolite structure with Sc—Cl of 2.476–2.481 Å.[106] Heating a 1:1 mixture of NaCl and $ScCl_3$ together at 630 °C affords $NaScCl_4$, isostructural with $NaLuCl_4$; Na^+ and Sc^{3+} ions occupy 1/4 of the octahedral sites between the layers of chloride ions, the structure being made up of *cis*-edge-sharing $ScCl_6$ octahedra.[109] Reaction of lanthanide oxides with NH_4Cl is a classic route to anhydrous lanthanide chlorides; DTA study of the reaction of Sc_2O_3 with NH_4Cl indicates that $(NH_4)_3ScCl_6$, $(NH_4)_2ScCl_5(H_2O)$, and $(NH_4)_3Sc_2Cl_9$ are formed as intermediates.[110]

Na_3ScBr_6 has the Na_3CrCl_6 structure.[111] Study of the $CsI–ScI_6$ phase diagram identified the compounds Cs_3ScI_6 and $Cs_3Sc_2I_9$, whilst ScI_4^- ions are believed to exist in the molten phase.[112] Raman spectra of the vapor over $CsI–ScI_3$ melts exhibited bands at 127 and 153 cm^{-1}, assigned to stretching vibrations in $CsScI_4$ and ScI_3 molecules respectively, whilst Sc—I stretching frequencies of 119 cm^{-1} and 129 cm^{-1} have been assigned to ScI_6^{3-} and ScI_4^- ions respectively.

3.2.1.6.2   *Other halide complexes*

The reduced chlorides, ScCl, Sc_5Cl_8, Sc_7Cl_{10}, Sc_7Cl_{12}, synthesized by chemical transport reactions from $Sc/ScCl_3$ mixtures, have been well established for years and have been reviewed.[113] They mostly contain chains of edge-sharing octahedra, with $Sc_6Cl_{12}^{3-}$ octahedra in Sc_7Cl_{12}. Sc_2Br_3 is also known. Some attention has now been given to iodides. The metallic diiodide is the only intermediate in the Sc/ScI_3 system,[114] it is readily prepared by the reaction of Sc with ScI_3 in the range 550–870 °C. The actual composition of the compound is $Sc_{0.93}I_2$; it has a cation-deficient CdI_2 structure, with Sc—I of 2.934 Å. In the field of iodide complexes, reaction of Sc, ScI_3, and LiI or NaI at 750–850 °C has been found to give the intensely air-sensitive $LiScI_3$ and $Na_{0.5}ScI_3$.[115] Both contain chains of confacial octahedra. $LiScI_3$ has essentially undistorted ScI_6 trigonal antiprisms (with Sc—I 2.91 Å) centered upon scandium atoms (Sc—Sc 3.384Å) as in the previously established $RbScX_3$ (X = Cl, Br) and $CsScX_3$ (X = Cl, Br, I). $Na_{0.5}ScI_3$, however, has some pairing of the scandiums; alternate Sc–Sc distances are 3.278 and 3.572 Å; the larger average Sc—Sc separation reflects reduction in bonding electron population compared with $LiScI_3$. $LiScI_3$ is weakly paramagnetic and $Na_{0.5}ScI_3$ has a small temperature-independent paramagnetism.[116]

Metallothermic reduction of $ScCl_3$ by caesium in the presence of carbon gives $Cs_4[Sc_6C]Cl_{13}$;[117] it contains an isolated Sc_6C cluster surrounded by 18 bridging chlorides. Heating ScI_3 and MI_2 (M = Co, Ni) with scandium at 750–950 °C gives clusters Sc_7MI_{12} having a $Sc(Sc_6MI_{12})$ structure. Other clusters $Sc(Sc_6XCl_{12})$ (X = B,N) and $Sc(Sc_6YI_{12})$ (Y = B,C) have also been reported.[118]

Table 2 A summary of key properties of the lanthanides, transition metals, and Group I metals.

	4f	3d	Group I
Electron configurations of ions	Variable	Variable	Noble gas
Stable oxidation states	Usually +3	Variable	1
Coordination numbers in complexes	Commonly 8–10	Usually 6	Often 4–6
Coordination polyhedra in complexes	Minimise repulsion	Directional	Minimise repulsion
Trends in coordination numbers	Often constant in block	Often constant in block	Increase down group
Donor atoms in complexes	"Hard" preferred	"Hard" and "soft"	"Hard" preferred
Hydration energy	High	Usually moderate	Low
Ligand exchange reactions	Usually fast	Fast and slow	Fast
Magnetic properties of ions	Independent of environment	Depends on environment and ligand field	None
Electronic spectra of ions	Sharp lines	Broad lines	None
Crystal field effects in complexes	Weak	Strong	None
Organometallics in Low oxidation states	Few	Common	None
Multiply bonded atoms in complexes	None	Common	None

3.2.2   YTTRIUM AND THE LANTHANIDES

3.2.2.1   Introduction

The lanthanides exhibit a number of features in their chemistry that differentiate them from the *d*-block metals:

(i)   A wide range of coordination numbers (generally 6–12, but numbers of 2–4 are known).

(ii)   Coordination geometries are determined by ligand steric factors rather than crystal field effects.

(iii)   They generally form labile "ionic" complexes that undergo facile exchange of ligand (though when multidentate complexing agents like DTPA are used, high stability constants obtain).

(iv)   The 4f orbitals in the Ln^{3+} ion do not participate directly in bonding. Their spectroscopic and magnetic properties are thus largely uninfluenced by the ligand.

(v)   Small crystal-field splittings and very sharp electronic spectra in comparison with the *d*-block metals.

(vi)   They prefer anionic ligands with donor atoms of rather high electronegativity (e.g., O,F).

(vii)   They readily form hydrated complexes (on account of the high hydration energy of the small Ln^{3+} ion) and this can cause uncertainty in assigning coordination numbers.

(viii)   Insoluble hydroxides precipitate at neutral pH unless complexing agents are present.

(ix)   The chemistry is largely that of one (3+) oxidation state.

(x)   They do not form $Ln=O$ or $Ln\equiv N$ multiple bonds of the type known for many transition metals and certain actinides.

A summary of key properties of the lanthanides, transition metals, and Group 1 and 2 metals appears in Table 2.

Since the appearance of the previous review in this series,[5] there have appeared three textbooks with a substantial content of lanthanide chemistry[119–121] as well as a review (to late 1992) summarizing the coordination chemistry of scandium, yttrium, and the lanthanides[6] as well as a text covering many important areas of lanthanide chemistry.[122] Other books have appeared with valuable preparative details of both starting materials and key compounds[123] and reviews of important areas extending into coordination chemistry,[124,125] applications of lanthanides in synthetic organic chemistry,[126,127] the biochemistry of the elements[128] and various historical perspectives.[129]

The burgeoning growth in lanthanide chemistry within the last 20 years has many reasons, but three areas responsible for this may be mentioned. Interest in mixed-oxide "warm superconductors" from the late 1980s has stimulated research in volatile materials such as alkoxides and diketonates; medicinal applications including magnetic resonance imaging agents has driven work with complexes of poly(aminocarboxylate) complexes; the increasing use of reagents such as SmI_2 as a one-electron reductant; and lanthanide triflates as Lewis acids in synthetic organic chemistry.

3.2.2.2 Group 14 Ligands

3.2.2.2.1 *Alkyls*

Reaction of lanthanide trichlorides with $LiCH(SiMe_3)_2$ tends to result in the formation of chlorine-containing complexes (see Scheme (5)). A different synthetic route, eliminating the presence of halide, has been adopted to synthesize $[Ln\{CH(SiMe_3)_2\}_3]$, involving replacement of aryloxide groups in $[Ln(OC_6H_3Bu^t_2\text{-}2,6)_3]$, thus obviating the possibility of chloride retention

$$[Ln\{OC_6H_3Bu^t_2\text{-}2,6\}_3] + LiCH(SiMe_3)_2 \rightarrow [Ln\{CH(SiMe_3)_2\}_3] + 3LiOC_6H_3Bu^t_2\text{-}2,6 \qquad (4)$$

In the initial report of these compounds, $[La\{CH(SiMe_3)_2\}_3]$ and $[Sm\{CH(SiMe_3)_2\}_3]$ were both synthesized by this route.[130] They have pyramidal structures, similar to those found in the silylamides $[Ln\{N(SiMe_3)_2\}_3]$, with La—C of 2.515(9) Å and Sm—C of 2.33(2) Å; such pyramidal structures may be adopted in the solid state to minimize nonbonding interactions involving the ligands. Bond lengths can be compared with those in the analogous[131] $U\{CH(SiMe_3)_2\}_3]$ where U—C = 2.48(2) Å; the U—C distance is 0.03 shorter than La—C, on ionic radius grounds a discrepancy of 0.01 Å is expected.

The structure of the pyramidal three-coordinate $[Y\{CH(SiMe_3)_2\}_3]$ has also been reported;[132] whilst other members of the series to have been synthesized are $[Ln\{CH(SiMe_3)_2\}_3]$ (Ln = Pr, Nd, Sm, Er and Lu)[133–135] Single-crystal-absorption and linear dichroism spectra have been reported and analyzed for Ln = Pr, Nd, Sm, and Er.[134,135]

Chemically $[Ln\{CH(SiMe_3)_2\}_3]$ compounds behave as Lewis acids. They are of course attacked by moisture, but also react with nucleophiles such as amines and phenols to form the corresponding lanthanide silylamides and aryloxides[130]

$$[Sm\{CH(SiMe_3)_2\}_3] + HOC_6H_2Bu^t_2\text{-}2,6\text{-Me4} \rightarrow [Sm\{OC_6H_2Bu^t_2\text{-}2,6\text{-Me4}\}_3] \qquad (5)$$

$$[Sm\{CH(SiMe_3)_2\}_3] + HN(SiMe_3)_2 \rightarrow [Sm\{N(SiMe_3)_2\}_3] \qquad (6)$$

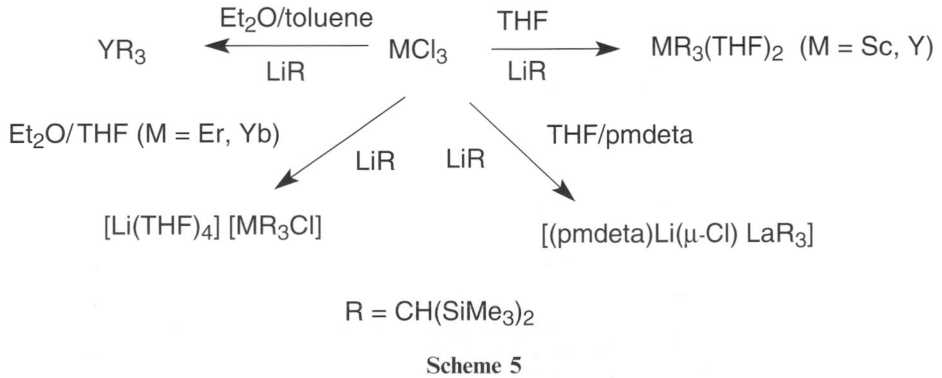

$$YR_3 \xleftarrow[\text{LiR}]{\text{Et}_2\text{O/toluene}} MCl_3 \xrightarrow[\text{LiR}]{\text{THF}} MR_3(THF)_2 \ (M = Sc, Y)$$

Et$_2$O/ THF (M = Er, Yb)   LiR   LiR   THF/pmdeta

$[Li(THF)_4] [MR_3Cl]$ $[(pmdeta)Li(\mu\text{-}Cl) LaR_3]$

$R = CH(SiMe_3)_2$

Scheme 5

They are also attacked by methyllithium forming bridged species[136]

$$[Ln\{CH(SiMe_3)_2\}_3] + MeLi + pmdeta \rightarrow [(pmdeta)Li(\mu - Me)Ln\{CH(SiMe_3)_2\}_3](Ln = La, Sm) \quad (7)$$

The samarium compound has a roughly linear bridge and a long Li—CH_3 bond (2.42 Å) but a short Sm—CH_3 bond (2.33(3) Å), thus the bridge can be termed asymmetric. The Sm—C $(CH(SiMe_3)_2)_3)$ bonds (2.49(3), 2.52(3), and 2.53(3) Å) are similar in length to those in $[La\{CH-(SiMe_3)_2\}_3]$. Even more striking examples of electrophilic behavior have been reported for $[Lu\{CH(SiMe_3)_2\}_3]$. It, but not $[La\{CH(SiMe_3)_2\}_3]$, reacts with KCl in ether, forming $[(Et_2O)K-(\mu-Cl)Lu\{CH(SiMe_3)_2\}_3]$. It has been remarked[137] that it is noteworthy that solvation of the potassium and coordination of the chloride to potassium and lutetium compensate for the loss of lattice energy (The KBr analogue can be prepared similarly). Ether is removed on gentle heating in vacuo, the product being $[K(\mu-Cl)Lu\{CH(SiMe_3)_2\}_3]$, which dissolves in toluene forming $[(\eta^6-C_7H_8)_2K(\mu-Cl)Lu\{CH(SiMe_3)_2\}_3]$. This compound contains a rather bent K—Cl—Lu bridge (145.9°) whilst the Lu—Cl distance is 2.515 Å. The Lu—C distances are 2.324(10), 2.349(10), and 2.357(8) Å; allowing for the difference in radii, these resemble those in $[La\{CH(SiMe_3)_2\}_3]$. Reaction of YCl_3 with $LiCH(SiMe_3)_2$ in diethyl ether leads to $[(Et_2O)_3Li(\mu-Cl)Y\{CH(SiMe_3)_2\}_3]$,[138] where Y—C is 2.423(12) Å, similar to those in the bridged lanthanum and lutetium compounds.

The CH_2SiMe_3 ligand is less sterically demanding so that $[Ln(CH_2SiMe_3)_3]$ are coordinatively unsaturated. Thus reaction of ytterbium chips with Me_3SiCH_2I in THF gives $[Yb(CH_2Si-Me_3)_3(THF)_2]$. This has the expected trigonal bipyramidal structure with axial THF molecules (Yb—O averages 2.330 Å). The average Yb—C distance is 2.374 Å.[139] Similarly, Yb reacts with Me_3CCH_2I in THF to form trigonal bipyramidal $[Yb(CH_2CMe_3)_3(THF)_2]$.[140] Unlike the case with $-CH_2SiMe_3$, four of the bulkier $-CH(SiMe_3)_2$ ligands cannot be so readily accommodated round a lanthanide ion; just as the isolobal $-N(SiMe_3)_2$ ligand cannot replace the fourth chlorine in $ThCl_4$, leading to the isolation of $[ThCl\{N(SiMe_3)_2\}_3]$, the chloride in $[YbCl\{CH(SiMe_3)_2\}_3]^-$ cannot be replaced. The $[YbCl\{CH(SiMe_3)_2\}_3]^-$ ion has a distorted tetrahedral structure with Yb—C distances of 2.372(16), 2.373(24) and 2.391(20) Å and Yb—Cl of 2.486 Å; the environment is similar to that of lanthanum in $[(pmdeta)Li(\mu-Cl)La\{CH(SiMe_3)_2\}_3]$. The La—C distances (2.55(2), 2.58(2), and 2.60(2) Å; mean value 2.57(3) Å) and C—La—C angles (average 108.8°) in the latter are, however, very similar to those in the three-coordinate pyramidal $[La\{CH(SiMe_3)_2\}_3]$ (La—C 2.515 Å; 109.9°) suggesting that the chloride bridge causes minimal distortion. $Li[Er\{CH(SiMe_3)_2\}_4]$ can, however, be obtained by heating a solution of $[Li(THF)_4][ErCl\{CH(SiMe_3)_2\}_3]$ in hexane, possibly by a disproportionation reaction.[141] A complex, $[\{(Me_3SiCH_2)_x(Me_3CO)_{1-x}Y(\mu-OCMe_3)_4-[Li(THF)_4](\mu_4-Cl)\}[Y(CH_2SiMe_3)_4]$, containing the tetrahedral $[Y(CH_2SiMe_3)_4]^-$ ion has been isolated from the reaction of YCl_3 with $LiOCMe_3$ and $LiCH_2SiMe_3$.[142] Y—C distances are in the range 2.403(8) Å to 2.420(9) Å, averaging 2.41(2) Å (which, allowing for the ionic radius differences, are very similar to the Yb—C distances in $[Yb\{CH(SiMe_3)_2\}_3Cl]^-$.

Simple methyls $Ln(CH_3)_3$ would be coordinatively unsaturated. However, fully characterized anionic species have been obtained as $[Ln(CH_3)_6]^{3-}$ anions by the reaction of excess (6.5 mols) MeLi with $LnCl_3$ in ether in the presence of 3 moles of a chelating ligand, either tetramethylethylenediamine or tris(1,2-dimethoxyethane) (L-L), the compounds having the formulae $[Li(L-L)]_3[M(CH_3)_6]$[143,144]

$$MCl_3 + 6MeLi + 3L\text{-}L \rightarrow [Li(L\text{-}L)]_3[M(CH_3)_6] + 3LiCl \quad (8)$$

The structures of three of the compounds have been determined; they show essentially octahedral coordination of the lanthanide with bond angles around 90°. Average lanthanum–carbon distances are 2.563(18) Å (Ho); 2.57(2) Å (Er), and 2.53(2) Å (Lu). More complicated lanthanide methyl species have been synthesized by another route, involving reaction of main-group methyls, Lewis acids, with lanthanide alkoxides[145] and amides,[146] a process of the type implicated in the lanthanide-catalyzed polymerization of conjugated dienes

$$Ln(OBu^t)_3 + 3AlMe_3 \rightarrow [Ln(\mu\text{-}OBu^t)(\mu\text{-}Me)AlMe_2_3](Ln = Pr, Nd, Y) \quad (9)$$

$$NdCl_3 + 3LiNMe_2 \rightarrow [Nd(NMe_2)_3(LiCl)_3] + 3MMe_3 \rightarrow [Nd(NMe_2)_3(MMe_3)_3](M = Al, Ga) \quad (10)$$

The amide products do not have the anticipated symmetrical structure, instead one MMe_3 group does not form a μ-methyl bridge. Reaction with excess MMe_3 gives heterometallic peralkyls $[Ln\{(\mu-Me)_2Me_2\}_3]$; a partially-exchanged product has been fully characterized.[146] The heterometallic peralkyls (M = Al) form inclusion compounds where Al_2Me_6 molecules are trapped in channels between the $[Ln\{(\mu-Me)_2AlMe_2\}_3]$ molecules.[147]

Triaryls of the heavier lanthanides have been synthesized by a reaction that does not involve salt elimination. Extended reaction at room temperature between powdered Ln (Ln = Ho, Er, Tm, Lu) and Ph_2Hg in the presence of catalytic amounts of LnI_3 affords the σ-aryls *fac*-$LnPh_3(THF)_3$. With Eu and Yb the divalent compounds $LnPh_2(THF)_2$ are obtained.[148,149] The structures of the erbium and thulium compounds show them to have molecular structures with octahedral coordination of the lanthanides with bond lengths of Er—C = 2.412, 2.440, and 2.442 Å and Tm—C = 2.416, 2.421, and 2.425 Å. There is some indication of steric crowding indicated by C—Ln—C angles of 99.2–103.5° (Er) and 99.8–104.2° (Tm), whilst the Ln—C bond lengths also seem slightly long in comparison with $PhGdCl_2.4THF$, as discussed below.

Yellow $YbPh_3(THF)_3$ is a minor product of the reaction between $(C_{10}H_8)Yb(THF)_2$ and Ph_2Hg along with the mixed valence $Yb_2Ph_5(THF)_4$ system. $Yb_2Ph_5(THF)_4$, which has the structure $Ph_2(THF)Yb(\mu-Ph)_3Yb(THF)_3$, has been viewed as an association of $Yb^{II}Ph_2(THF)$ and $Yb^{III}Ph_3(THF)_3$ though there is some η^2 character in some of the bridging interactions. The terminal Yb^{III}—C distances average 2.42 Å, in line with the values for $[LnPh_3(THF)_3]$ (Ln = Er, Tm); though the bridging Yb—C distances are, as expected rather longer (averaging 2.60 Å), the Yb^{III}—C distances are in two cases slightly shorter than the Yb^{II}—C distances.

Lanthanide triphenyls are now firmly established for the heavier lanthanides (Ho–Lu), but it remains to be seen if $[LnPh_3(THF)_4]$ is feasible for the lighter metals (and whether $[ScPh_3(THF)_3]$ can be isolated).

Monophenyls have been isolated by using a deficit of reagent. Thus reaction between $LnCl_3$ and PhLi (0.5 mol) in THF gives $PhLnCl_2 \cdot nTHF$ (Sm, Gd $n = 4$; Pr $n = 3$); the seven-coordinate gadolinium compound has a Gd—C distance of 2.416(24) Å. This is relatively short compared to the triphenyls, evidence for possible crowding in them.[150]

Six-coordination is also found in $[(dmp)YbCl_2(N-Meim)_2py].toluene$[151] (dmp = 2,6-dimesitylphenyl; N-Meim = N-methylimidazole) whilst distorted trigonal bipyramidal five-coordination exists in $[Ln(Dnp)Cl_2(THF)_2]$ (Dnp = 2,6-di(1-naphthyl)phenyl); Ln = Y, Yb, Tm).[152] Donor-functionalized terphenyl derivatives have also been made.[153] $[[(Danip)Yb(\mu_2-Cl)_2(\mu_3-Cl)Li(thf)]_2]$ and $[[(Danip)Ln(\mu_2-Cl)_2(\mu_2-Cl)Li(thf)]_2]$ (Ln = Y, Sm) have structures based on LiCl-bridged $(DanipLnCl_2)$ units stabilized through additional coordination of two methoxy groups to lanthanum. (Danip = 2,6-di(o-anisol)phenyl.)

The first structural characterization of cationic lanthanide alkyl complexes has been achieved.[154] $[Ln(CH_2SiMe_3)_3(THF)_2]$ (Ln = Y, Lu) react with $B(C_6X_5)_3$ (X = H, F) in the presence of crown ethers forming $[Ln(CH_2SiMe_3)_2(CE)(THF)_n]^+[B(C_6X_5)_3(CH_2SiMe_3)]^-$ (CE = [12]-crown-4, $n = 1$; CE = [15]-crown-5, [18]-crown-6, $n = 0$). In all these complexes, the crown ethers utilise all their donor atoms.

In THF but in the absence of crown ether, $[Lu(CH_2SiMe_3)_3(THF)_2]$ reacts with $B(C_6F_5)_3$ forming $[Lu(CH_2SiMe_3)_2(THF)_3]^+[B(C_6X_5)_3(CH_2SiMe_3)]^-$. In $[Ln(CH_2SiMe_3)_2([12]-crown-4)-(THF)]^+[B(C_6X_5)_3(CH_2SiMe_3)]^-$, Lu—C distances are 2.340 and 2.354 Å, whilst Lu—O distances are in the range 2.406–2.503 Å. The Lu—O (THF) distance is 2.307 Å. In $[Ln(CH_2SiMe_3)_2([15]-crown-5)]^+[B(C_6X_5)_3(CH_2SiMe_3)]^-$, Lu—C distances are 2.345 Å and 2.364 Å, whilst Lu—O distances are in the range 2.359–2.421 Å, whilst in $[Ln(CH_2SiMe_3)_2([18]-crown-6)]^+[B(C_6X_5)_3-(CH_2SiMe_3)]^-$, Lu—C distances are 2.366 Å and 2.371 Å, whilst Lu—O distances are in the range 2.399–2.524 Å.

A number of lanthanide carbene derivatives, $[ErL_3Cl_3]$, $[Y(L)\{(N(SiMe_3)_2\}_3(THF)]$, and *trans*-$[Y(L)_2\{(N(SiMe_3)_2\}_3]$ (L = 1,3-dimethylimidazolin-2-ylidene) have been synthesized.[155]

3.2.2.3 Group 15 Ligands

3.2.2.3.1 *Ammonia and other monodentate neutral ligands*

Because of the basicity of the ligand, and consequent inability to form isolable complexes in supercritical solution, ammonia complexes have scarcely been studied. However, the first homoleptic lanthanide ammine complexes, $[Yb(NH_3)_8][Cu(S_4)_2].NH_3$, $[Yb(NH_3)_8]$-

$[Ag(S_4)_2].2NH_3$, and $[La(NH_3)_9][Cu(S_4)_2]$ have been synthesized by reactions in aqueous ammonia.[156]

Direct reaction of the lanthanide halides with pyridine gives pyridine complexes of the lanthanides,[157] with the synthesis of $[YCl_3Py_4]$ and $[LnCl_3Py_4].0.5Py$ (Ln = La, Er). These all have pentagonal bipyramidal structures, with two chlorines occupying the axial positions. In the yttrium compound, the axial Y—Cl distances are virtually identical at 2.5994(7) Å and 2.6006 (7) Å, with the equatorial distance being 2.6388 (7) Å; the Y—N distances are in the range 2.487–2.578(2) Å. In the lanthanum and erbium compounds, the pattern in M—Cl distances is similar, with axial distances of 2.652 Å and 2.661(1) Å and an equatorial distance of 2.679 Å in the lanthanum compound; and with axial distances of 2.5578(8) Å and 2.5840 Å and an equatorial distance of 2.6211(8) Å in the erbium compound. It will be interesting to see whether this stoichiometry is maintained to the end of the series, as this would be a rare example of the same structure persisting with decreasing ionic radius (compare $[LnL_2(NO_3)_3]$ (L = phen, bipy)). $[MI_3py_4]$ (M = Ce, Nd) have been used as starting materials in the syntheses of terpy complexes.[158]

Piperazine has been reported to form 8:1 complexes with lanthanide perchlorates but as yet there is no structural information.[159] A number of complexes of N-methylimidazole (N-Meim) have been made, $[SmI_3(THF)_3]$ reacts with N-Meim forming[160] square-antiprismatic $[Sm(Meim)_8]I_3$, whilst $[YX_2(N\text{-}Meim)_5]^+X^-$, (X = Cl, Br); $[YCl_2(N\text{-}Meim)_5]^+[YCl_4(N\text{-}Meim)_2]^-$; and $[Ce(NO_3)_3(N\text{-}Meim)_4]$ have also been characterized.[161]

3.2.2.3.2 *Nitrile complexes*

Among nitriles, CH_3CN in particular has been widely employed as a solvent in nonaqueous lanthanide chemistry, but little is known about its complexes until recently.

A NMR study of lanthanum nitrate solutions in MeCN led to the identification of a number of complexes including $[La(NO_3)_3(MeCN)_4]$ and $[La(NO_3)_3(MeCN)_3(H_2O)]$, but they were not isolated.[162,163] A few complexes such as $[Eu(MeCN)_x(BF_4)_3]$ ($x \sim 3$) have previously been prepared, in this case by oxidation of metallic Eu by $NOBF_4$, but have lacked structural characterization.[164]

However, in the 1990s syntheses were reported[165–167] for MeCN complexes $[Ln(MeCN)_n]X_3$. Routes have included the reaction of $LnCl_3$ with $AlCl_3$ in MeCN, reaction of the labile complexes $La(OSO)_x(AsF_6)_3$ with MeCN, and ultrasonication of mixtures of the lanthanide metal with $AlCl_3$ and MeCN in C_2Cl_6; among others, a series $[Ln(MeCN)_9](AlCl_4)_3 \cdot MeCN$ (Ln = La, Pr, Nd, Sm–Tb, Ho, Yb) has been made. The structures of $[La(MeCN)_9](AsF_6)_3 \cdot MeCN$; $[Sm(MeCN)_9](AsF_6)_3.3MeCN$; $[Ln(MeCN)_9](AlCl_4)_3.MeCN$ (Ln = Pr, Sm) and $[Yb(MeCN)_8](AlCl_4)_3$ have all been reported. The $[Ln(MeCN)_9]^{3+}$ ion has the familiar trigonal prismatic coordination of the lanthanide; La—N bond lengths in $[La(MeCN)_9](AsF_6)_3 \cdot MeCN$ fall in the range 2.575(9)–2.650(5) Å whilst in $[Sm(MeCN)_9](AsF_6)_3.3MeCN$ the Sm—N bonds span 2.510(5)–2.546(5) Å. $[Yb(MeCN)_8](AlCl_4)_3$ has dodecahedral eight-coordination of ytterbium with Yb—N bonds between 2.367(5) Å and 2.422 (4) Å.

A number of adducts of the silylamides $[Ln(N(SiMe_3)_2)_3]$ form nitrile adducts; they are discussed in Section 3.2.2.3.5.

3.2.2.3.3 *2,2'-Bipyridyl, 1,10-phenanthroline and other bidentate neutral ligands*

Many complexes of 2,2'-Bipyridyl (bipy) and 1,10-phenanthroline (phen) have been examined in detail in a series of papers by White and his co-workers.

Both 1:1 and 2:1 complexes of lanthanide chlorides with bipy have lately received detailed and extensive crystallographic study.[168,169] Compounds synthesized by reaction in ethanol and characterized include $[(bipy)Ln(OH_2)_6]Cl_3$ (Ln = Ho-Lu, Y), $[(bipy)Ln(OH_2)_6]Cl_3 \cdot bipy \cdot 2H_2O$ (Ln = Er-Lu, Y), $[(bipy)Ln(OH_2)_4Cl_2]Cl \cdot H_2O$, $[(bipy)(EtOH)_2Cl_2La(\mu\text{-}Cl)_2LaCl_2(EtOH)_2(bipy)]$, $[(bipy)_2La(OH_2)_4Cl]Cl_2.2H_2O$, $[(bipy)_2Pr(OH_2)Cl_3] \cdot 0.5$ EtOH, $[(bipy)_2Ln(OH_2)_2Cl_2]Cl$ (Ln = Pr, Er), Pr, Er), $[(bipy)_2Ln(OH_2)Cl_3] \cdot EtOH$ (Ln = Nd, Eu), $[(bipy)_2Cl_2La(\mu\text{-}Cl)_2LaCl_2(bipy)_2] \cdot EtOH$, and $[(bipy)_2YbCl_3]$. For the 1:1 complexes of early lanthanides like La and Pr, the tendency seems to be for the formation of neutral complexes containing all available chloride ions and frequently binuclear in composition, usually containing at least two solvent molecules per lanthanide. By later in the series, mononuclear species tend to become more normal, and chloride ions are often excluded from the coordination sphere by solvent molecules. In the case of the 2:1 complexes, nine-coordination is possible at the start of the series in $[(bipy)_2Ln(OH_2)_4Cl]Cl_2.2H_2O$, whilst by the end of the series ytterbium is seven-coordinate in $[(bipy)_2YbCl_3]$. M—N bond lengths contract from 2.724–2.787 Å in the lanthanum compound to 2.439–2.479 Å in the ytterbium compound. The situation has been described as "a multidimensional jigsaw puzzle which still requires a great deal of work for its complete description."

Bipyridyl complexes with lanthanide bromides do not seem to have received attention, but the first lanthanide iodide complexes were reported in 1999. In a comparative study with UI_3, complexation of bipy with (Ln = Ce, Nd) was investigated.[170] UI_3 and CeI_3 both form 1:1 and 1:2 complexes in solution, with a 1:3 complex at high bipy concentrations, whereas NdI_3 only forms a 1:2 complex. The structure of $[CeI_3(bipy)_2(py)].5py.bipy$ was reported; it has eight-coordinate Ce, with Ce—N (bipy) = 2.67(3) Å, Ce—I = (average 3.23(3) Å) and Ce—N 2.678(9) Å. Reaction of $[NdI_3(py)_4]$ and bipy in attempts to make the Nd analogue resulted in dimeric $[(bipy)_3Nd(\mu\text{-}OH)_2Nd(bipy)_3]I_4\cdot3Py$, presumably due to inadvertent hydrolysis.[171]

The reaction between bipy and $La(NO_3)_3.6H_2O$ has been studied in MeCN solution by multi-nuclear NMR; species identified in solution include $La(MeCN)_4(NO_3)_3$, $La(MeCN)_2(bipy)(NO_3)_3$ and of course the final product $La(bipy)_2(NO_3)_3$.[163]

Complexes $Ln(bipy)_2(NO_3)_3$ have been studied in more detail than the other bipy complexes; all appear to have 10-coordinate structures with all nitrates present as bidentate ligands. The coordination geometry has been variously described as a bicapped dodecahedron and as a sphenocorona. Unlike the La complex, the Lu complex does not possess disorder about the twofold axis. Lu—N distances are 2.46–2.67(1) Å and Lu—O distances in the range 2.426(9)–2.556(9) Å.[172] $[Y(bipy)_2(NO_3)_3]$ is, like $[Y(phen)_2(NO_3)_3]$, isostructural with its lanthanide analogues.[173] Similarly, the structure of $[Nd(bipy)_2(NO_3)_3]$ has been shown to be isostructural with $[Ln(bipy)_2(NO_3)_3]$ (Ln = Y, La, Lu).[174] A number of compounds $Ln(bipy)_3(NO_3)_3$ (Ln = Ce, Pr, Nd, Yb) have been reported;[175] the neodymium complex was shown to be $[Nd(bipy)_2(NO_3)_3]$. bipy, with the third bipy molecule not associating with the neodymium-containing complex.[174]

In the presence of 15-crown-5, reaction of lanthanum nitrate with bipy in MeOH–MeCN led to $[La(bipy)(NO_3)_3(H_2O)_2(MeOH)]\cdot15$-crown-5, which has 11-coordinate lanthanum.[176] The increase in coordination number in comparison with $[La(bipy)_2(NO_3)_3]$ results in an slight (and possibly not statistically significant) increase in La—N distance from 2.66 Å to 2.70 Å, and in La—O from 2.56–2.63 Å to 2.69 Å. The presence of the coordinated MeOH molecule supplies a fifth hydrogen atom so that hydrogen bonds can be formed to all the crown ether oxygens. In another synthesis in the presence of a crown ether, a 10-coordinate complex, $[La(bipy)(NO_3)_3(H_2O)_2]\cdot$benzo-15-crown-5, is obtained.[177]

Among carboxylate complexes, $[Eu(o\text{-}ABA)_3(bipy)]\cdot$bipy (o-ABA = o-aminobenzoate) is dimeric with four bridging carboxylates, one chelating carboxylate, and a bipyridyl ligand affording eight-coordinate europium.[178] $[\{Pr(O_2CCMe_3)_3(bipy)\}_3]$ has two bidentate bridging and two tridentate cyclic bridging carboxylates; praesodymium is nine-coordinate.[179] The dimethoxybenzoate complex $[La(2,3\text{-}DMOBA)_3(2,2'\text{-}bpy)]$ is a dimer with the lanthanum atoms bridged by four carboxylates. The central La atom is nine-coordinate, having distorted mono-capped square-antiprism geometry.[180] $[La_2(O_2CC{\equiv}CH)_6(bipy)_2(H_2O)_2]\cdot4H_2O.2$ bipy does not undergoes solid-state polymerization when exposed to ^{60}Co γ-rays.[181]

A considerable number of bipy (and phen) adducts of lanthanide dithiocarbamates $[Ln(S_2CNR_2)_3(L)]$ have been synthesized, usually by one-pot syntheses; the interest here is in their potential as precursors to lanthanide sulfides. $[Ln(S_2CNMe_2)_3(bipy)]$ (Ln = La, Pr, Nd, Sm-Yb, Y) and $[Ln(S_2CNEt_2)_3(bipy)]$ (Ln = La, Pr, Nd, Sm-Lu, Y) have been synthesized.[182] The structure of $[Er(S_2CNEt_2)_3(bipy)]$ has been determined[183] as has that of $[Eu(bipy)(S_2CNEt_2)_3]$.[184] The synthesis of $[Sm(S_2CNEt_2)_3(L)]$ (L = phen, bipy) and the structure of $[Sm(S_2CNEt_2)_3(bipy)]$ have been reported.[185] The related $[Eu(L)(S_2PBu^i_2)_3]$ (L = phen, bipy) have monomeric structures with distorted dodecahedral coordination of europium.[186]

Among phenanthroline complexes, one feature present in many solid-state structures is π—π stacking between the planar phenanthroline rings. The first complex with a noncoordinating anion to be characterized fully was $[Ce(phen)_4(MeCN)_2](ClO_4)_3\cdot3MeCN$, which has 10-coordinate cerium in a bicapped square antiprismatic geometry.[187]

As with bipy, a considerable number of chloride complexes have been examined in detail.[169] Compounds characterized by crystallography include $[(phen)_2La(OH_2)_5]Cl_3\cdot phen\cdot4H_2O$, $[(phen)_2La(OH_2)_5]Cl_3\cdot MeOH\cdot H_2O$, $[(bipy)_2Ln(OH_2)_4Cl]Cl_2\cdot2H_2O$, $[(phen)_2Lu(OH_2)_4]Cl_3\cdot2H_2O$, $[(phen)_2Ln(OH_2)_3Cl]Cl_2\cdot H_2O$ (Ln = Dy, Er, Y), and $[(phen)_2Ln(OH_2)Cl_3]\cdot MeOH$ (Ln = La, Pr, Nd, Eu). Nine-coordination is possible for compounds like $[(phen)_2Ln(OH_2)_5]Cl_3\cdot phen\cdot4H_2O$

whilst by the end of the lanthanide series eight-coordination is more normal in [(phen)$_2$-Lu(OH$_2$)$_4$]Cl$_3$·2H$_2$O and [(phen)$_2$Er(OH$_2$)$_3$Cl]Cl$_2$·H$_2$O. An apparent 3:1 complex, La(phen)$_3$Cl$_3$·9H$_2$O has been shown to be [La(phen)$_2$(OH$_2$)$_5$]Cl$_3$·4H$_2$O.phen.[188]

The first structure of a [Ln(phen)$_2$(NO$_3$)$_3$] complex was reported in 1992 for the lanthanum compound.[162] It closely resembled the established bipy analogue in that the three nitrate groups were bidentate and the lanthanum was 10-coordinate. The structural information was complemented by a multinuclear solution (^{1}H-, ^{13}C-, ^{17}O-, and ^{139}La) NMR study. The structure of the other "extreme" member of the series, the lutetium complex, was reported in 1996.[172] Unlike the La complex, but like [Lu(bipy)$_2$(NO$_3$)$_3$], the study was not complicated by disorder. The complexes appear to form an isomorpohous and isostructural series. On moving from the lanthanum to the lutetium compound, the Ln—N distances decrease from 2.646(3)–2.701(3) Å (La) to 2.462(8)–2.479(8) Å (Lu), and the range of Ln—O distances decreases from 2.580(3)–2.611(3) Å for the lanthanum compound to 2.364(8)–2.525(6) Å for the lutetium complex. Several structures have subsequently been reported of other [Ln(phen)$_2$(NO$_3$)$_3$] systems.[173,189–191] [Ln(phen)$_2$(NO$_3$)$_3$] (Ln = Pr,[189] Nd,[189,192] Sm,[189] Eu,[189,190] Dy,[189] Y[173,191]) are isostructural; the individual complex molecules associate by π–π stacking into one dimensional chains which themselves arrange into pseudo-one-dimensional close packed patterns.[189] Luminescence spectra of Eu^{3+}-containing [Ln(phen)$_2$(NO$_3$)$_3$] (Ln = Y, La, Nd, Lu) have been investigated.[190] [Y(phen)$_2$(NO$_3$)$_3$] is, like [Y(bipy)$_2$(NO$_3$)$_3$] isostructural with its lanthanide analogues,[72] as well as with its lanthanide analogues, with Y–O distance in the range 2.477–2.516 Å and Y—N bonds of 2.492 Å and 2.549 Å. π—π stacking between the rings of neighboring phen ligands with an interplanar separation of 3.51 Å leads them to associate into one-dimensional chains.[191] A complex analyzing as La(phen)$_4$-(NO$_3$)$_3$·3H$_2$O is [La(phen)$_2$(H$_2$O)$_2$(NO$_3$)$_2$]· NO$_3$·2phen·H$_2$O with intermolecular π—π stacking.[192]

In another phenanthroline complex, the structure of [phenH][La(NO$_3$)$_4$(H$_2$O)(phen)].H$_2$O features what are becoming the familiar π—π interactions, this time between the [phenH]$^+$ ions.[193] Other phen complexes reported are new dinuclear species [(phen)$_2$(H$_2$O)$_2$Ln(μ-OH)$_2$Ln-(H$_2$O)$_2$(phen)$_2$](NO$_3$)$_4$·2phen (Ln = Er, Lu).[194]

Reaction of lutetium acetate and phen in ethanol afforded Lu(O$_2$CCH$_3$)$_3$·(phen)·H$_2$O which is dinuclear [(phen)(CH$_3$COO)Lu(μ-O$_2$CCH$_3$)$_4$Lu(CH$_3$COO)(phen)]. The terminal acetate groups are symmetrically bidentate; two of the bridging acetates are symmetrical, the other pair bridging in such a way that one of the two oxygens is bound to both lutetiums and the other is bound only to one. Lu—N is 2.486(4)–2.551(5) Å and terminal Lu—O bonds 2.323–2.503(5)(5) Å.[195] Similarly, Ln(OAc)$_3$.phen (Ln = La, Ce) is[196] dimeric [(AcO)(phen)Ln(μ-OAc)$_4$Ln(phen)(OAc)] with two types of bridging acetate. The cerium compound gives a triplet EPR spectrum at 4.2 K, showing a bridging interaction between the two ceriums (D = 0.21 cm^{-1}); magnetic measurements to low temperatures on both compounds confirm weak interactions.

A phenanthroline complex of europium caproate, [Eu(O$_2$CC$_5$H$_{11}$)$_3$(phen)], has a dimeric structure in which each europium is bound to a chelating phenanthroline and one bidentate carboxylate as well as additionally to two bidentate bridging carboxylates and two tridentate bridging carboxylates, giving nine-coordinate europium. Some subtle splitting of bands in the fluorescence spectrum has been ascribed to vibronic interactions.[197] As already discussed in the section on bipy, a number of complexes of the type [Ln(phen)(S$_2$CNR$_2$)$_3$] have been synthesized and studied, with structures determined for the eight-coordinate [Ln(phen)(S$_2$CNEt$_2$)$_3$] (Ln = Eu,[198] Yb[184]). Improved syntheses are reported for [Ln(S$_2$CNR$_2$)$_3$(phen)] (Ln = Eu, Er; NR$_2$ = NEt$_2$, NMeCy; N(CH$_2$)$_5$).[199] The synthesis of [Sm(S$_2$CNEt$_2$)$_3$(phen)] has been reported.[185]

Solvothermal synthesis of [La(en)$_4$Cl]In$_2$Te$_4$ has been reported; the cation has monocapped square antiprismatic coordination.[200] Reaction of Y$_2$S$_3$ with NH$_4$I in en at 568 K yields crystals of eight-coordinate [Y(en)$_4$](SH)$_{2.72}$I$_{0.28}$; a similar reaction in anhydrous ammonia yields an uncharacterized complex, probably an ammine.[201]

Stability constants of 1:1 and 2:1 complexes with diethylenetriamine (dien) have been determined by potentiometry.[202]

3.2.2.3.4 Complexes of terpyridyl and other tridentate ligands

There has been a great awakening of interest in complexes of 2,2′:6′,2″-terpyridyl (terpy), paralleling developments in its coordination chemistry with the *d*-block metals. Tridentate N-donor ligands are efficient in separating actinides from lanthanides selectively by solvent extraction. 2,4,6-tris-2-pyridyl-1,3,5-triazine (tptz) and terpyridyl (terpy) and their derivatives have been popular ligands for study. Phenyl-substituted terpyridyls have been studied as extractants with lower solubility in aqueous phases.

Reaction of aqueous lanthanide chlorides with alcoholic solutions of terpy affords complexes $Ln(terpy)Cl_3.xH_2O$ (Ln = La–Nd, $x = 8$; Ln = Sm, $x = 7.6$; Ln = Eu, $x = 7.45$; Ln = Gd, $x = 7.1$; Ln = Tb–Er, Yb–Lu, $x = 7$; Ln = Tm, Y, $x = 6$). They contain $[Ln(terpy)Cl(H_2O)_n]^{2+}$ ions (Ln = La–Nd, $n = 5$; Ln = Sm–Lu, $n = 4$).

On proceeding from La to Nd in the series of $[Ln(terpy)Cl(H_2O)_5]^{2+}$ ions, there are the expected contractions in Ln—Cl, from 2.903(2) Å to 2.855(1) Å; in average Ln—N distance, from 2.688 Å to 2.616 Å; and in average Ln—O distance, from 2.561 Å to 2.505 Å. Similarly, in the $[Ln(terpy)Cl(H_2O)_4]^{2+}$ ions, on passing from Sm to Lu, there are decreases in Ln—Cl, from 2.794(2) Å to 2.665(2) Å; in the average Ln—N distance, from 2.513 Å to 2.457 Å; and in average Ln—O distance, from 2.475 Å to 2.317 Å. In addition, a compound $Sm(terpy)Cl_3\cdot H_2O$ was found to be dimeric $[(terpy)(H_2O)Cl_2Sm(\mu\text{-}Cl)_2Sm(H_2O)Cl_2(terpy)]$, containing eight-coordinate samarium.[203]

Bromide complexes of terpy have been studied[204] with two families, nine-coordinate $[Ln(terpy)(H_2O)_6]Br_3.H_2O$ (Ln = La–Er) and eight-coordinate $[Ln(terpy)(H_2O)_5]Br_3.3H_2O$ (Ln = Tm–Lu). These to some extent resemble the chlorides, though there is no halide in the coordination sphere in any of these compounds. For the $[Ln(terpy)(H_2O)_6]^{3+}$ ions, the Ln—N distances contract from 2.656–2.684 Å (La) to 2.39–2.44 Å (Lu), with similar contractions being evident in the Ln—O distances. Towards the end of the lanthanide series, there is a tendency for a hydroxy-bridged dimeric species to be formed in preference, but usually acidification converts it into the mononuclear complex. The dimers contain bridging hydroxy groups, not halides as found in $[(terpy)(H_2O)Cl_2Sm(\mu\text{-}Cl)_2Sm(H_2O)Cl_2(terpy)]$, reflecting the decreasing coordinating power of bromide. Indeed, overall the absence of any bromide coordination should be noted, but these complexes were obtained by crystallization from solutions of hydrated $LnBr_3$ and terpy in ethanol, and in view of the recent isolation of 2:1 terpy complexes of lanthanide iodides with all iodides coordinated, syntheses using other stoichiometries, anhydrous bromides, and less coordinating solvents would be expected to afford complexes such as $[Ln(terpy)_2Br_2]Br$.[204]

Terpy reacts with solutions of LnI_3 in anhydrous pyridine forming[205] eight-coordinate $[Ln(terpy)_2I_2]I$ (Ln = Ce, Nd) and a nine-coordinate uranium analogue $[U(terpy)_2I_2(py)]I$ has been made. $[Ce(terpy)_2I_2]I$ crystallizes from slightly damp solvents forming nine-coordinate $[Ce(terpy)_2I_2(H_2O)]I$. Ln—N bond lengths average at 2.63(2) Å for the Ce complex and 2.60 (2) Å for the Nd complex; Ln—I distances average 3.182(3) Å and 3.153(4) Å respectively. Proton-NMR competition experiments indicate that terpy has a stronger affinity for U^{III} than for Ce^{III} or Nd^{III}. A comparison of structural data for $[Ce(terpy)_2I_2(H_2O)]I$ and $[U(terpy)_2I_2(py)]I$ indicates that the average U–N distance is shorter by about 0.05 Å, though on size grounds they would be expected to be very similar. It has been suggested that this shortening reflects a π back-bonding interaction between the 5f orbitals of uranium and the terpyridyl ligand which is absent in the lanthanide complex.[205]

The classic nine-coordinate $[Ln(terpy)_3](ClO_4)_3$ complexes have been reinvestigated.[206] Reaction of terpy with lanthanide perchlorates in MeCN affords $[La(terpy)_3](ClO_4)_3\cdot2MeCN\cdot0.67H_2O$, $[Ln(terpy)_3](ClO_4)_3\cdot MeCN\cdot H_2O$ (Ln = Ce, Pr, Sm, Eu), and $[Ln(terpy)_3](ClO_4)_3$ (Eu, Lu, Y); all of these have nine-coordinate $[Ln(terpy)_3]^{3+}$ cations. The Ln—N bond lengths show the expected contraction; Ce—N distances fall into the range 2.622–2.679 Å whilst Lu—N distances are 2.437–2.553 Å.[206] Complexes of 4-alkylated terpyridyls, $[Ln(4\text{-}Rterpy)_3](ClO_4)_3$ (Ln = La, Eu Tb; R = Et, Bu^t) have been synthesized; introducing these alkyl groups considerably increases the luminescence efficiency, possibly due to the bulk of the alkyl group preventing approach of deactivating molecules, as well as their electronic effects.[207]

Complexes of bridged terpy ligands have been examined:[208] thus europium has tricapped trigonal prismatic nine-coordination in $[EuL_3](ClO_4)_3$ (L = 3,3':5,3'-dimethylenetripyridyl); the ligand contains ethylene bridges between the pyridine rings in the ligand molecules, giving a more stable triple helical $[EuL_3]^{3+}$ complex species, again with D_3 symmetry as found round europium in $[Eu(terpy)_3](ClO_4)_3$.

Previous reports of nitrate complexes were limited to synthetic studies[209] reporting 1:1 complexes with various degrees of hydration. A large number of complexes have been reported in the literature in recent years; interest has been reawakened by the possibility of using related ligands in the separation of lanthanide fission products from actinides in the reprocessing of nuclear fuel rods. It is now clear that the situation is complex and that various stoichiometries are obtainable, depending not least upon the solvent used. For the majority of complexes, a 1:1 terpy: lanthanide ratio obtains. The first of the recent studies was carried out by Bensimon and Frechette.[210] Addition of terpy to a solution of $La(NO_3)_3\cdot6H_2O$ in MeCN was followed by ^1H-, ^{17}O- and ^{139}La-NMR investigations. A number of species were identified to be present in solution, namely

[La(terpy)(NO$_3$)$_3$(MeCN)],   [La(terpy)(NO$_3$)$_3$(H$_2$O)],   [Ln(terpy)(NO$_3$)$_4$(MeCN)]$^-$, [Ln(terpy)(NO$_3$)$_4$(H$_2$O)]$^-$, and [Ln(terpy)$_2$(NO$_3$)$_2$]$^+$. The complex Ln(terpy)$_{1.5}$(NO$_3$)$_3$ was isolated from a MeCN solution with a 8:1 terpy:La ratio and was shown by X-ray diffraction to possess the structure [Ln(terpy)$_2$(NO$_3$)$_2$]$^+$[Ln(terpy)(NO$_3$)$_4$]$^-$. In the 10-coordinate cation, La—O are 2.627(11)–2.646(11) Å whilst La—N distances lie in the range 2.649(12)–2.685(10) Å with one outlier at 2.736(13) Å. In the 11-coordinate anion, La—N are 2.705(13), 2.709(13), and 2.769(12) Å; La—O fall into a bracket between 2.608(13)–2.684 (13) Å. Subsequently it has been found[211] that reactions of hydrated lanthanide nitrates with up to four moles of terpy in CH$_3$CN afford complexes with a similar 1.5:1 stoichiometry, [Ln(terpy)$_2$-(NO$_3$)$_2$]$^+$[Ln(terpy)(NO$_3$)$_4$]$^-$ (Ln = Nd, Sm, Tb, Dy, Ho). In these, however, the lanthanide is 10-coordinate in both the cation and the anion, one nitrate group in the anion being monodentate, in contrast to the anion in the 1.5:1 complex of the larger lanthanum ion, where all nitrates are bidentate; this is evidently a consequence of congestion arising around the smaller lanthanide ions. In the [Nd(terpy)$_2$(NO$_3$)$_2$]$^+$ cation, Nd–N distances are 2.592(7)–2.651(6) Å and Nd—O 2.526(5)–2.598(6) Å, whilst in [Nd(terpy)(NO$_3$)$_4$]$^-$, Nd—N distances are 2.591(7)–2.600(6) Å, the monodentate Nd—O bond is 2.434 (7) Å, and the Nd—O distances in the bidentate groups are rather widely spread at 2.548(6)–2.644(7) Å. In the [Ho(terpy)$_2$(NO$_3$)$_2$]$^+$ cation, Ho—N distances are 2.538(7)–2.605(7) Å and Ho—O 2.405(6)–2.598(6) Å, whilst in [Ho(terpy)(NO$_3$)$_4$]$^-$, Ho—N distances are 2.531(7)–2.544(7) Å, the monodentate Ho—O bond is 2.323(6)Å and the Ho—O distances in the bidentate groups are even more widely spread at 2.435(7)–2.580(7) Å. The increasing asymmetry in the bond to the nitrate groups suggests that congestion is again increasing. A [Sm(terpy)(NO$_3$)$_4$]$^-$ anion has also been isolated from acidified solution as the (terpyH$_2$)$^{2+}$ salt.[212] This is an 11-coordinate anion with four bidentate nitrates (in contrast to the three bidentate and one monodentate group in the the [Sm(terpy)$_2$(NO$_3$)$_2$]$^+$ salt) with eight Sm—O distances ranging from 2.494(5) Å to 2.742(5) Å (averging 2.56 Å), contrasting with the six bidentate Sm—O distances of 2.497(13)–2.613(13) Å and one monodentate distance of 2.370(14) Å in the [Sm(terpy)$_2$(NO$_3$)$_2$]$^+$ salt. The Sm—N distances average 2.637 Å in the 11-coordinate anion ((terpyH$_2$)$^{2+}$ salt) and 2.569 Å in the 10-coordinate anion ([Sm(terpy)$_2$(NO$_3$)$_2$]$^+$ salt). Clearly factors such as the energetics of ionic packing can have a pronounced effect on coordination geometry, in the absence of the strong crystal-field effects that would apply in a transition-metal analogue. The [Ce(terpy)(NO$_3$)$_4$]$^-$ ion is found in (Hpy)[Ce(NO$_3$)$_4$(terpy)]py, formed by addition of pyridine to a day-old mixture of cerium nitrate and terpy in MeCN.[213] The coordination polyhedron of the Ce atom is irregular; the cerium is 11-coordinate, with four bidentate nitrates, the Ce—N distances being 2.682(2), 2.685(2), and 2.705(2) Å. Seven of the Ce—O distances span 2.537(2)–2.712(2) Å, the eighth being 2.942(3) Å.[213]

A study by Semenova and White covered the whole lanthanide series. In this they examined the reaction of the hydrated nitrates with one mole of terpy in MeCN, followed by recrystallization of the initial complex from water. They found[214] that the earlier members of the lanthanide series form 10-coordinate [Ln(terpy)(NO$_3$)$_2$(H$_2$O)$_3$]NO$_3$ (Ln = La–Gd) and the later lanthanides form nine-coordinate [Ln(terpy)(NO$_3$)$_2$(H$_2$O)$_2$]NO$_3$·2H$_2$O (Ln = Tb, Lu, Y). These compounds result from solvolysis by water of an initial [Ln(terpy)(NO$_3$)$_3$(H$_2$O)$_x$] species (see below), displacing one nitrate group. Detailed structures have been reported for the La, Gd, Tb, Lu, and Y complexes. They show a girdle of ligands comprising a virtually planar terdentate terpy ligand and two or three water molecules coordinated round the "waist" of the metal, with bidentate nitrates completing the coordination sphere above and below the metal. In the [Ln(terpy)(NO$_3$)$_2$(H$_2$O)$_3$]$^{3+}$ ions, the range of Ln—N distances contracts from 2.632(4)–2.688(3) Å (Ln = La) to 2.52(1)–2.56(1) Å (Ln = Gd), with a similar pattern in the Ln—O distances. The range of Ln—N distances in the [Tb(terpy)(NO$_3$)$_2$(H$_2$O)$_2$]$^{3+}$ ion is 2.50(1)–2.52(1) Å, contracting to 2.469(6)–2.474(6) Å in [Lu(terpy)(NO$_3$)$_2$(H$_2$O)$_2$]$^{3+}$ ion.[214] The structure of [Gd(terpy)(NO$_3$)$_2$(H$_2$O)$_3$]NO$_3$ has been reported independently.[215]

When the hydrated nitrates react directly with terpy in MeCN, without added water, the initial products are [Ln(terpy)(NO$_3$)$_3$(H$_2$O)$_x$]. Compounds Ln(terpy)(NO$_3$)$_3$·H$_2$O (Ln = La, Pr, Er, Yb) and Ln(terpy)(NO$_3$)$_3$ (Ln = Gd, Yb) have been reported, but these have lacked structural information.[216,217] In a more detailed study,[211] reaction of the hydrated lanthanide nitrates with terpy in CH$_3$CN was found to afford [Nd(terpy)(NO$_3$)$_3$(H$_2$O)], [Ln(terpy)(NO$_3$)$_3$(H$_2$O)]·terpy (Ln = Ho, Er, Tm, Yb), and [Yb(terpy)(NO$_3$)$_3$], all of which feature solely bidentate nitrates. In [Tm(terpy)(NO$_3$)$_3$(H$_2$O)], one nitrate is monodentate. In 10-coordinate [Nd(terpy)(NO$_3$)$_3$(H$_2$O)], there is quite a bit of asymmetry in the Nd—N distances (2.586(10)–2.703(13) Å); the Nd—O (water) distance is 2.488(8) Å and the Nd—O nitrate distances span the range 2.530(9)–2.632(9) Å, averaging 2.567 Å. In [Ln(terpy)(NO$_3$)$_3$(H$_2$O)]·terpy (Ln = Ho, Er, Tm, Yb), the binding of the terpy ligand to the lanthanide is more symmetrical. [Tm(terpy)(NO$_3$)$_3$(H$_2$O)] has a Tm—O bond length for the monodentate group, of 3.251 Å, 0.12 Å, shorter than any of the Tm—O distances in the bidentate groups, whilst there is considerable asymmtery in a bidentate group in [Ln(terpy)-(NO$_3$)$_3$(H$_2$O)]·terpy, where the Ho—O distances are 2.431(8) Å and 2.725(11) Å, such that the nitrate has been considered intermediate between mono- and bidentate. The smallest of the lanthanide ions covered in this study, ytterbium, forms a complex with no water in its coordination sphere, but three bidentate nitrates. In [Yb(terpy)(NO$_3$)$_3$], Yb—O distances range from 2.364(8) Å to 2.406(10) Å, apart from one at 2.456(9) Å, whilst Yb—N distances are 2.417(7)–2.419(8) Å.[211] Reaction of yttrium nitrate with terpy in MeCN yields two yttrium complexes.[218] Reaction with two moles of MeCN followed by crystallization gave crystals of nine-coordinate [Y(terpy)(NO$_3$)$_3$(H$_2$O)]·terpy·3MeCN containing two bidentate and one monodentate nitrate groups. Y—N distances average 2.497 Å, Y—OH$_2$ is 2.311 Å, and Y—O (nitrate) distances are 2.311 Å for the monodentate group and range from 2.390 Å to 2.504 Å (avaerage 2.445 Å) for the bidentate groupings. Layering of more dilute solutions with ether formed [Y(terpy)(NO$_3$)$_3$(H$_2$O)], with 10-coordinate molecules with three bidentate nitrates. Here Y—N distances average 2.519 Å and the Y—O distances within the bidentate nitrates average 2.508Å within a range between 2.414 Å and 2.523 Å, with an outlier at 2.736 Å, over 0.2 Å greater than the others, indicating considerable congestion in the coordination sphere.[218]

In contrast to the early lanthanides, where the same complex is obtained from synthesis in either acetonitrile or ethanol, solvent affects the structure for the 1:1 complexes of later lanthanides. The first indication of this was the discovery that the reaction of hydrated erbium nitrate with terpy in ethanol affords [Er(terpy)(NO$_3$)$_3$.(C$_2$H$_5$OH)] which contains both bidentate and monodentate nitrates as well as a coordinated ethanol.[219] Erbium is nine-coordinate in this complex. Er—N bond lengths (2.456(8)–2.485(8)Å) are very similar to those in the eight-coordinate [Er(terpy)Cl(H$_2$O)$_4$]Cl$_2$·3H$_2$O; two nitrates are bidentate, with Er—O distances averaging 2.416 Å, some 0.14 Å longer than that of the Er—O bond involving the monodentate nitrate Er—O (3.278(7) Å). A second oxygen in this monodentate nitrate is hydrogen bonded to the coordinated ethanol molecule (Er—O = 2.333(8)Å. Another feature of this structure is short C—H···O interactions involving nitrate oxygens and hydrogens of the terpyridyl ligand in other molecules.

Reaction of later lanthanide nitrates with terpy in MeCN solution affords nine-coordinate [Ln(terpy)(NO$_3$)$_3$] (Ln = Yb, Lu). Solvolysis of a nitrate group in [Yb(terpy)(NO$_3$)$_3$] is stereoselective, the nitrate *trans*- to the terpy ligand being replaced by ethanol and by water, with the formation of [Yb(terpy)(NO$_3$)$_3$(EtOH)] (which has one monodentate nitrate) and [Yb(terpy)-(NO$_3$)$_2$(H$_2$O)$_2$]NO$_3$.2H$_2$O respectively. A similar effect is noted in the lutetium analogue with the isolation of an unusual complex [Lu(terpy)(NO$_3$)$_2$(H$_2$O)(EtOH)](NO$_3$), where both water and ethanol are bound to lutetium in preference to nitrate coordination, as well as the ethanol solvate [Lu(terpy)(NO$_3$)$_3$(EtOH)]. In [Lu(terpy)(NO$_3$)$_3$], Lu—N distances are 2.379–2.407 Å and Lu—O bonds fall into the range 2.350–2.440 Å, so that even with the smallest lanthanide, all the nitrates are essentially bidentate. Replacement of a bidentate nitrate *trans*- to the terpy by a coordinated ethanol and a monodentate nitrate causes a certain reorganization in the coordination sphere, with Lu—N bonds increasing by about 0.06 Å; the Lu—O (water) distance is 3.279(3) Å and the Lu—O(ONO$_2$) distance is 3.279(3) Å, showing that the Lu—O bond is about 0.1 Å shorter than for an oxygen atom in a bidentate nitrate group.[24]

The 11-coordinate [La(NO$_3$)$_3$(terpy)(MeOH)$_2$] has been isolated from the reaction of lanthanum nitrate with terpy in methanol, the lanthanum being 11-coordinate, with three bidentate nitrates, the La—N distances being 2.688 (3), 2.700 (3), and 2.715(3) Å. Five of the La—O (nitrate) distances span 2.596(3)–2.727(3) Å, the sixth being 2.926(3) Å; La—O (methanol) bonds are 2.560(2) Å and 2.580 (2) Å.[214] In the isomorphous [Ce(NO$_3$)$_3$(terpy)(MeOH)$_2$], whose polyhedron is described as an icosahedron with two vertices replaced by one, the Ce—N distances are 2.682(2), 2.685(2), and 2.705(2) Å. Five of the Ce—O (nitrate) distances span 2.578(2)–2.712(2) Å, the sixth being 2.942(3) Å; Ce—O (methanol) bonds are 2.537(2) Å and 2.559(2) Å.[213] Thus a wide range of 1:1 complexes are known, though the picture is not yet complete, and some findings have yet to be reported. The role of solvent is critical in these syntheses.

Mononuclear [Yb(O$_2$CCCl$_3$)$_3$.(terpy).MeOH] contains coordinated methanol and two unidentate carboxylates and one bidentate carboxylate affording eight coordinate ytterbium and [Ln(O$_2$CCCl$_3$)$_3$·(terpy)·(H$_2$O)] (Ln = La-Nd) are dimeric with nine-coordinate lanthanides. A compound [Lu(O$_2$CCCl$_3$)$_3$·(terpy)·(H$_2$O)] contains both the foregoing geometries.[195]

Several reports have appeared of the syntheses and structures of mixed-ligand quaternary complexes involving terpy, [Yb(acac)(terpy)(NO$_3$)$_2$],[220] [Nd(hfac)(terpy)(NO$_3$)$_2$(H$_2$O)],[221]

[Ln(hfac)$_2$(terpy)(NO$_3$)] (Ln = Gd, Dy, Er, Tb, Yb),[221] [Nd(terpy)(dbm)(NO$_3$)$_2$],[221] and [Ln(terpy)-(dbm)$_2$(NO$_3$)] (Ln = Pr, Ho)[222] all of which were crystallographically characterized. (Hhfac = hexafluoroacetylacetone; Hdbm = dibenzoylmethane.) A full report has appeared on a series of these compounds; [Ln(terpy)(NO$_3$)$_2$(acac)(H$_2$O)$_n$] (Ln = La, Pr, n = 1; Ln = Nd-Lu, n = 0).[223] They exhibit three different solid-state structures. [La(terpy)(NO$_3$)$_2$(acac)(H$_2$O)] is 10-coordinate, with two bidentate nitrates and a chelating diketonate and a coordinated water; [Pr(terpy)(NO$_3$)$_2$(acac)(H$_2$O)] is nine-coordinate, with one monodentate and one bidentate nitrate and a chelating diketonate and a coordinated water; and the remainder are [Ln(terpy)(NO$_3$)$_2$-(acac)]. (Nd–Lu) with two bidentate nitrates and a chelating diketonate.

Complexes of 2,4,6-tris-2-pyridyl-1,3,5-triazine (tptz) have been reinvestigated. Tptz acts as a terdentate ligand in the structures of [Eu(tptz)Cl$_3$(MeOH)$_2$]·2MeOH (eight-coordinate) and [Pr(tptz)(OAc)$_3$]$_2$·2MeOH (10-coordinate). The latter[224] features the presence of double (η^2,μ-1,1) acetate bridges increasingly familiar from other lanthanide carboxylate complexes. Pr—N distances fall into a range 2.674(6)–2.717(6) Å, rather larger than those in the nine-coordinate [Pr(terpy)Cl(H$_2$O)$_5$]$^{2+}$ ion. In the [Eu(tptz)Cl$_3$(MeOH)$_2$] molecule, Eu—N distances are 2.555(4) Å to the central nitrogen and 2.616(4)–2.645 (4) Å for the nitrogens in the pyridine rings.[224] Although it has proved difficult to obtain good quality crystals from MeCN solution, the structure of [Sm(tptz)(NO$_3$)$_3$(H$_2$O)]·2H$_2$O has been determined and shows it to contain 10-coordinate Sm with all nitrates bidentate.[216] Sm—N distances are 2.571(4) Å to the central nitrogen and 2.631(4)–2.644(5) Å for the nitrogens in the pyridine rings, showing a similar pattern to the europium complex. The bond to the water molecule, at 2.420(4) Å, is as usual in the terpy analogues, considerably shorter than the Sm—O (nitrate) bonds, which lie in a range 2.492(4)–2.615 (4) Å.

A very detailed study has been made[226] of lanthanide nitrate complexes of 4-amino-bis(2,6-(2-pyridyl))-1,3,5-triazine (abptz). Many of these compounds resemble corresponding terpy complexes and gradually a picture is being built up of how the structure and stoichiometry depend on factors such as the radius of the metal ion, composition of the reaction mixture, and the solvent employed. Complexes [La(abptz)(NO$_3$)$_3$(H$_2$O)$_2$] (11-coordinate), [Ln(abptz)-(NO$_3$)$_3$(H$_2$O)] (Ln = La, Pr–Sm; 10-coordinate), [Ln(abptz)(NO$_3$)$_3$(H$_2$O)] (Ln = Yb, Y; nine-coordinate, with one monodentate nitrate), [Ln(abptz)(NO$_3$)$_2$(H$_2$O)$_3$]NO$_3$ (Ln = Nd, Sm; 10-coordinate) and [Ln(abptz)(NO$_3$)$_2$(H$_2$O)$_2$]NO$_3$ (Ln = Eu-Lu; nine-coordinate) have all been crystallographically characterized.[226] These compounds exhibit the trends in bond distances already commented on in discussing terpyridyl complexes, but some general points are relevant here. The fact that 10- and 11-coordinate [La(abptz)(NO$_3$)$_3$(H$_2$O)$_n$] (n = 1, 2) can be obtained, unlike subsequent lanthanides which, at most, adopt just 10-coordination (although no data were reported for cerium) shows that the position of lanthanum as the largest lanthanide ion sometimes causes it to behave atypically. A comparison between the two lanthanum complexes shows the decrease in coordination number from 11 to 10 is accompanied by a decrease in La–O (water) distance from 2.589(4) Å and 2.610(4) Å to 2.483(5) Å; similarly, the La–N distances decrease from 2.739(4), 2.755(5), and 2.805(6) Å in the 11-coordinate compound to 2.576(5), 2.625(5), and 2.641(5) Å in 10-coordinate [La(abptz)(NO$_3$)$_3$(H$_2$O)]. The range of La—O (nitrate) distances decreases from 2.628(4)–2.702(5) Å, with an outlier at 2.805(6) Å in [La(abptz)(NO$_3$)$_3$(H$_2$O)$_2$], to 2.525(5)–2.620(5) Å in [La(abptz) (NO$_3$)$_3$(H$_2$O)], indication of the congestion in the coordination sphere in the 11-coordinate complex. Across the family of 10-coordinate [La(abptz)(NO$_3$)$_3$(H$_2$O)] (La = La, Pr, Nd, Sm) species, there is a general tendency towards contraction of bond lengths with increasing atomic number, but with irregularities, partly because the Nd complex is not isomorphous with the others. Compounds [Ln(abptz)(NO$_3$)$_2$(H$_2$O)$_3$]NO$_3$ (Ln = Nd, Sm; 10-coordinate) and [Ln(abptz)(NO$_3$)$_2$(H$_2$O)$_2$] NO$_3$ (Ln = Eu–Lu) are clearly analogous to the terpy complexes studied by White and co-workers.[214] They can be regarded as having a girdle round the "waist" of the metal comprised of the abptz ligand and two or three water molecules, with two bidentate nitrates coordinated to the metal in axial positions. A comparison of the structures of the two 10-coordinate species [Nd(abptz)(NO$_3$)$_3$(H$_2$O)] and [Nd(abptz)(NO$_3$)$_2$(H$_2$O)$_3$] NO$_3$ is informative. Average Nd—N distances are 2.605 Å and 2.603 Å respectively; average Nd—O(H$_2$O) distances are 2.470 Å and 2.485 Å respectively, closely comparable, but the Nd—O (nitrate) distances for [Nd(abptz)(NO$_3$)$_3$(H$_2$O)] fall within a range of 2.546(4)–2.597(4) Å, averaging 2.568 Å, whilst for [Nd(abptz)(NO$_3$)$_2$(H$_2$O)$_3$]NO$_3$ the respective values are 2.588(6), 2.602(5), 2.608(5), and 2.728(7) Å, averaging 2.632 Å. There is thus congestion brought about by replacing a compact nitrate group by two water molecules, leading to an increased tendency for one nitrate group to become monodentate. This view is supported by the value of 2.849(14) Å for the "long" Sm—O distance in [Sm(abptz)(NO$_3$)$_2$(H$_2$O)$_3$]NO$_3$. At this point, further contraction in the lanthanide radius evidently results in ejection of a water from the coordination sphere (rather than the presence of three waters and a monodentate nitrate) with the formation of $[Ln(abptz)(NO_3)_2(H_2O)_2]NO_3$ (Ln = Eu–Lu). These exhibit the usual trend of decreasing bond lengths with increasing atomic number until in the case of the lutetium complex there is some evidence for steric strain again. The compounds $[Ln(abptz)(NO_3)_3(H_2O)]$ (Ln = Yb, Y) are nine-coordinate, with one monodentate nitrate, rather than the alternative $[Ln(abptz)(NO_3)_3]$, with three monodentate nitrates, adopted for $[Ln(terpy)(NO_3)_3]$ for Yb, Lu and some other later lanthanides, suggesting that subtle factors are at work in the adoption of this structure. Extensive hydrogen-bonding networks are present in the structures of all these compounds. Complexes of 2,6-bis(5,6-dialkyl-1,2,4-triazin-3-yl)pyridines have attracted attention as extremely effective selective complexing agents for the actinides, and a few studies of model lanthanide complexes have been reported.[227–229] The $[Ln((Pr^nbtp)_3]^{3+}$ (Ln = Sm, Tm, Yb) cations are nine-coordinate in a number of salts, displacing weakly coordinating anions such as iodide,[227] whilst the similar $[Ce(Rbtp)_3]I_3$ (R = Me, Pr^n) have been reported,[228] comparative studies showing that Ce has a lesser affinity for Pr^nbtp than uranium(III). These $[LnL_3]^{3+}$ species have not been isolated for the early lanthanides, however; they instead form some novel dinuclear complexes $[Ln_2L_2(NO_3)_6]$ (R = Me; Ln = La, Pr, Nd, Sm).[229] In these compounds, the lanthanides are bound to three bidentate nitrates and one terdentate ligand, with one oxygen atom additionally coordinating to the other lanthanide, affording 10-coordination overall. Species $[NdL(NO_3)_3(EtOH)]$ (R = Et) and $[[NdL_2(NO_3)_2]^+]_2[Nd(NO_3)_5]$ (R = Bu^i) have also been characterized. This plethora of stoichiometric variation is striking, and it would be interesting to know whether it occurs in solution too. Complexes (2:1) are formed by 2,6-bis(benzimidazol-2′-yl)pyridine, $[Ln(bzimpy)_2(NO_3)_2]NO_3$, which exhibit strong luminescence (Ln = Eu, Tb).[230,231] In $[Eu(mbzimpy)_3](ClO_4)_3$ europium has tricapped trigonal prismatic nine-coordination (mbzimpy = 2,6-bis(1-methylbenzimidazol-2-yl)pyridine); fluorescence spectra indicate that this geometry is maintained in solution. The cation has approximately C_3 symmetry in the solid state and fluorescence spectra support this. $[Eu(mbzimpy)_3](ClO_4)_3$ has Eu—N distances in the range 2.576(5)–2.613(7) Å.[232] Similar complexes $[Ln(mbzimpy)_3](ClO_4)_3$ can be made for other lanthanides (Ln = La, Gd, and Tb), but for lutetium only $[Lu(mbzimpy)_3(H_2O)(MeOH)](ClO_4)_3$ can be isolated; the eight-coordinate cation has Lu—N distances of 2.37(1)–2.46(1) Å, Lu—O (H_2O) = 3.29(1) Å and Lu—O (MeOH) = 2.35(1) Å. Spectrophotometric titrations indicate that 1:1, 1:2, and 1:3 complexes are formed in solution throughout the lanthanide series but the $[Ln(mbzimpy)_3]^{3+}$ ion is less stable for heavier (and smaller) lanthanides Ho, Yb, and Lu. A similar effect occurs with the related terdentate ligands 2,6-bis(1-X-benzimidazol-2-yl)pyridine (X = Pr, 3,5-dimethoxybenzyl).[233] The 1:1 and 1:2 complexes formed by this ligand show the normal thermodynamic behavior associated with electrostatic effects, but the tris complexes $[Ln(mbzimpy)_3]^{3+}$ display unusual selectivity for the mid-lanthanide ions.[234] The triple-helical structure found in the crystal structure of the Eu complex appears to be retained in solution for the others, with control of the coordination cavity caused by intrastrand π—π stackings maximized at Gd. Only the 1:2 complex could be isolated for Yb, a hydroxy-bridged dimer $[(mbzimpy)_2Yb(H_2O)(\mu\text{-}OH)_2Yb(mbzimpy)_2](ClO_4)_4$, with eight-coordinate Yb. This has Yb—N distances of 2.420(5) Å to 2.567(5) Å and Yb—O distances 3.231(4)–3.235(5) Å. A spectroscopic study of $[Ln(mbzimpy)(NO_3)_3(MeOH)]$ (Ln = Eu, Tb) indicates that the ligand has similar photophysical properties to terpy.[235] 2,6-Bis(1′-ethyl-5′-methyl-benzimidazol-2′-yl)pyridine (L) reacts with lanthanide perchlorates in a similar way, forming mononuclear triple-helical complexes $[LnL_3](ClO_4)_3$ (Ln = Eu, Gd, Tb). $[EuL_3](ClO_4)_3 \cdot 4MeCN$ has slightly distorted tricapped trigional prismatic coordination of Eu, with Eu—N distances of 2.53(2)–2.66(2) Å. The presence of the ethyl groups causes a slide of the strands which distorts the trigonal symmetry, as shown in the luminescence spectra.[236] When two terdentate bzimpy ligands are linked by a "spacer" to discourage formation of a mononuclear complex, self-assembly leads to a triple helical binuclear complex, $[Ln_2L_3](ClO_4)_6$ (Ln = La, Eu, Gd, Tb, Lu; L = bis[1-methyl-2-(6′-[1″-(3,5-dimethoxybenzyl)-benzimidazol-2″-yl]pyrid-2′-yl)benzimidazol-5-yl]methane). The structure of $[Eu_2L_3](ClO_4)_6 \cdot 9MeCN$ shows europium has tricapped trigonal prismatic nine-coordination with Eu—N distances of 2.54(3)–2.64(3) Å, averaging 2.59(6) Å, the europium site having pseudo-D_3 symmetry. Luminescence studies support this view but confirm that secondary interactions with lattice waters in $[Eu_2L_3](ClO_4)_6 \cdot nH_2O$ (n = 2 or 9) destroy this high symmetry.[237,238]

Another approach to forming these helical species is to use ligands in essence derived from the bis(benzimidazol-2′-yl)pyridine family, with the benzimidazole rings replaced with carboxamide moities, so that they remain tridentate ligands, but with a NO_2 donor group. The purpose of this was to obtain more strongly luminescent species. Thus N,N,N′,N′-tetraethylpyridine-2,6-dicarboxamide (L) forms $[LnL_3]^{3+}$ complex ions across the lanthanide series.[239] Structures for $[LaL_3](ClO_4)_3$ and $[EuL_3](CF_3SO_3)_3$ show that each ligand strand is meridionally tri-coordinated, producing tricapped trigonal prismatic coordination of the lanthanides with near D_3 symmetry. Ranges of Ln—O distances are 2.470(5)–2.527(6) Å for the La complex and 2.392(5)–2.426(5) Å for the europium complex; corresponding ranges for the Ln—N bonds are 2.679(7)–2.731(7) Å and 2.547(6)–2.569(9) Å for the La and Eu complexes respectively. The lanthanum complex is more distorted and fluorescence studies on Eu^{3+} doped into the La, Gd, and Lu complexes suggest that the ligand cavity is a better match for the heavier lanthanide ions.[239] An alternative approach, using a simpler ligand, diethyl pyridine-2,6-dicarboxylate, showed that it formed complexes with 1:1, 1:2, and 1:3 stoichiometries in solution. These have low stabilities and the 1:3 complexes cannot be isolated in the solid state. The 2:1 complex $[EuL_2(CF_3SO_3)_2(OH_2)](CF_3SO_3)$ contains nine-coordinate europium; the amide ligands are indeed tridentate, with Eu—N distances of 2.542(8)–2.573(7) Å; Eu—O (ester) of 2.458(7)–2.561(6) Å, Eu—O (water) 2.392(8) Å and Eu—O (triflate) 2.361(6)–2.412(8) Å. The Eu—O distances for the ester oxygens are especially long, reflecting the weakness of the coordination.[240] Using segmental ligands from the bis(1-alkyl-2-[6′-(N,N-diethylcarbamoyl)pyridin-2′-yl]benzimidazol-5-yl)methane (L) family, which contain two linked N_2O donors, triple-stranded helicate complexes $[Ln_2L_3](ClO_4)_3$ are obtained; their structure is confirmed by X-ray diffraction on the Tb member and the triple helical structure is maintained in solution to judge from NMR and ES-MS measurements.[241] Quantum yield determinations on the Eu complex indicate luminescence 50 times stronger than in corresponding compounds containing benzimidazole groups rather than carboxamide groups. Use of lanthanide triflates as starting materials tends to result in double-stranded complexes containing coordinated triflate. Use of the segmental ligand bis(1-ethyl-2-[6′-(carboxy)pyridin-2′-yl]benzimidazol-5-yl)methane, a molecule of which contains two distinct N_2O donor groups, leads to the formation of complexes $[Ln_2L_3](ClO_4)_3$, again across the whole lanthanide series. The three helical ligands wrap round the metal ions, again achieving pseudo-D_3 symmetries; NMR spectra indicate that time-averaged D_3 symmetries are maintained in solution.[242]

A detailed study has been reported of lanthanide nitrate complexes of the terdentate triazole ligand 2,6-bis(5-methyl-1,2,4-triazol-3-yl)-pyridine (DMTZP) which behaves as a planar terdentate ligand resembling terpy, tptz, and their fellows. Complexes isolated and examined crystallographically include $[La(DMTZP)(NO_3)(H_2O)_5](NO_3)_2$, $[Ln(DMTZP)(NO_3)_3(H_2O)]$ (Ln = Nd, Sm, Tb); $[Ho(DMTZP)(NO_3)_3(H_2O)]$ (with one monodentate nitrate), and $[Ln(DMTZP)(NO_3)_3]$ (Ln = Er, Yb) have all been characterized crystallographically.[243] The tendency towards decreasing coordination number with smaller lanthanide ion noted elsewhere is marked here, but one particularly unusual feature is the presence of only one coordinated nitrate group in the 10-coordinate complex $[La(DMTZP)(NO_3)(H_2O)_5](NO_3)_2$. Normally only one nitrate is replaced, usually lying *trans*- to the terdentate ligand, and it would be interesting to know whether this is caused by the particular ligand type or whether there is some other cause. The La—N bond lengths are 2.713(18), 2.724(17), and 2.774(12) Å (average 2.737 Å) in comparison with values of 2.576(5), 2.625(5), and 2.641(5) Å, average 2.614 Å in 10-coordinate $[La(abptz)(NO_3)_3(H_2O)]$, suggesting there is congestion here. A comparison between the 10-coordinate species $[Nd(DMTZP)(NO_3)_3(H_2O)]$ and $[Nd(abptz)(NO_3)_3(H_2O)]$ indicates longer Nd—N bonds in the former (2.652 Å vs. 2.605 Å) but longer Nd—O (nitrate) distances in the latter (2.549 Å for the DMTZP complex vs. 2.5671 Å for the abptz complex) which may reflect the strength of metal–ligand bonding.[243]

Several complexes of the tripodal tetradentate ligands tpza and tpa (tpa = tris[(2-pyridyl)methyl] amine; tpza = tris[(2-pyrazinyl)methyl]amine) have been made.[244] $[Ln(tpa)Cl_3]$ (Ln = Eu, Tb, Lu) have seven-coordination, with average Ln—N distances decreasing from 2.585 Å (Eu) to 2.520 Å (Lu) and M—Cl distances changing from 2.664 Å (Eu) to 2.602 Å (Lu). The larger neodymium ion can accommodate a coordinated solvent molecule in the eight-coordinate Nd analogue, $[Nd(tpa)Cl_3(MeOH)]$, where the increase in coordination number causes an increase in average Nd—Cl distance to 2.760 Å but more significantly results in one of the Nd—N bonds in the "equivalent" pyridyl arms, at 2.7021(14) Å, being appreciably longer than the other two (2.601(2) Å and 2.620(2) Å). Reaction of the lanthanide perchlorates with tpza in MeCN affords $[Ln(tpza)(H_2O)_3(MeCN)_3](ClO_4)_3$ (Ln = La, Nd, Eu). The structure of the Nd compound shows that it has 10-coordinate neodymium, with Nd—N (tpza) in the range 2.719(7) Å to 2.770(8) Å, Nd—N (MeCN) of 2.620(9)–2.698(8) Å and Nd—O 2.424(6) Å to 2.543(6) Å. $[MI_3(THF)_4]$ (M = La, U) reacts with[245] tpza forming $[M(tpza)I_3(NCMe)]$ and $[M(tpza)I_3(THF)]$. The tripodal tpza is tetradentate, so that these complexes are all eight-coordinate. In the MeCN adducts, the M—N (pyrazine) distances are very similar, whilst the U—N (MeCN) distance is 0.05 Å shorter than the distance in the corresponding La complex, whereas in $[M(tpza)I_3(THF)]$, the M—N distance is 0.05 Å shorter than in the La complex. These findings are interpreted in terms of some covalent contribution to the U—N bonding.

Lanthanide ions react with the tripodal ligand tris(2-benzimidazol-2-ylmethyl)amine (ntb) forming bis complexes (even with a deficit of ligand) where the ligand encapsulates the metal to the exclusion of chloride from the coordination sphere. The $[Ln(ntb)_2]^{3+}$ ions have been studied crystallographically in $[Ln(ntb)_2](ClO_4)_3$ (Ln = La, Nd, Eu) and $[Lu(ntb)_2]Cl_3$ and display strong π—π interactions between the benzimidazole rings.[246] On the other hand, reaction with lanthanide nitrates[247] gives $[Ln(ntb)(NO_3)_3]\cdot H_2O$ (Ln = La, Ce, Nd, Sm–Dy, Er) in which tetradentate ntb and three bidentate nitrates afford 10-coordinate monomers. The structures of the Ce and Er compounds have been determined. Average Ce—O and Er—O distances are 2.597 Å and 2.473 Å respectively; Ce—N and Er—O (imine) distances are 2.639 Å and 2.525 Å; Ce—N and Er—O (tertiary) distances are 2.825 Å and 2.673 Å. Eu^{3+} and Tb^{3+} complexes have been synthesized[248] of tpa (tpa = tris(2-pyridylmethyl)amine) and of a chiral tris(2-pyridylmethyl)amine ligand (L), $[Ln(L)](CF_3SO_3)_3$ (Ln = Eu and Tb). Emission spectra of these complexes show that the spectra are most sensitive in both intensity and in line shape to the presence of a wide range of anions $(X = I^-, Br^-, Cl^-, F^-, ClO_4^-, NO_3^-, SCN^-, CH_3CO_2^-, HSO_4^-,$ and $H_2PO_4^-)$. These effects were more noticeable when using the chiral ligand, whose europium complex works most effectively as a nitrate-specific luminescent sensor.

Reaction of ethanolic lanthanide nitrates with the hexadentate ligand tpen (tpen = tetrakis-(2-pyridylmethyl)-1,2-ethylenediamine) gives the complexes $Ln(tpen)(NO_3)_3\cdot 3H_2O$ (Ln = La, Tb).[249] These are $[Ln(tpen)(NO_3)_2]NO_3$ with the lanthanide coordinated to six nitrogens from tpen and two bidentate nitrates. The mean La—N bond length is 2.720(21) Å and the mean La—O bond is 2.583(13) Å; corresponding values for the terbium complex are Tb—N = 2.624(29) Å and Tb—O = 2.492(33) Å. Investigations on complexation of tpen and Ln^{3+} confirms that these complexes are also stable in aqueous solution.[250] The tetradentate 2,2':6',2'':6',2'''-quaterpyridine (qtpy) and hexadentate 2,2':6',2'':6',2''':6',2'''':6',2'''''-sexipyridine (spy) ligands form 1:1 complexes with yttrium and europium nitrates respectively. Reaction of yttrium chloride with hot methanolic qtpy gives $[Y(qtpy)Cl_3(H_2O)_6]$ of unknown structure, and $Y(qtpy)(NO_3)_3(H_2O)_2$, which contains nine-coordinate $[Y(qtpy)(NO_3)_2(H_2O)]^+$ cations. The quaterpyridine ligand is essentially planar (recalling the behavior of terpy) and the water molecule is also coordinated in the same plane; the two symmetrically bidentate nitrate groups coordinate above and below the plane of the qtpy. The Y—O (water) distance is 2.321(8) Å and Y—N distances fall in the narrow range 2.464(8)–2.469(9) Å whilst Y—O 2.403(7)–2.434(8) Å.[251] Reaction of spy with hot methanolic europium nitrate leads to $[Eu(spy)(NO_3)_2]NO_3$ (the ionic nitrate can be exchanged for PF_6). This contains 10-coordinate $[Eu(spy)(NO_3)_2]^+$ cations where the hexadentate spy twists itself helically round the metal. The two coordinated nitrates are bidentate (Eu—O 2.504(16)–2.558(16) Å, whilst the Eu—N distances fall in the range 2.536(14)–2.589(10) Å.[252] There is obviously considerable scope here for further investigation.

3.2.2.3.5 *Amides and pyrazolides*

Developments here have included a considerable broadening in the range of alkylamide ligands investigated. Several more $[Ln(N(SiMe_3)_2)_3]$ (Ln = Y, Ce, Dy, Er, and Yb) have had their structures determined, in addition to earlier structures of the Nd and Eu complexes.[253–256] Reactions of LnI_3 with $LiN(SiMe_3)_2$ have been investigated.[257] With LaI_3 and SmI_3, a mixture of $[LnI(N((SiMe_3)_2)_2]$ and $[Ln(N((SiMe_3)_2)_3]$ obtains; in contrast, only $[YbI(N((SiMe_3)_2)_2]$ can be isolated for ytterbium. In the case of Ln, the dimeric $[(THF)((Me_3Si)_2N)_2Ln(\mu-I)_2La(N((SiMe_3)_2)_2(THF)]$ has been crystallized. Anhydrous lanthanide triflates are good alternatives to $LnCl_3$ for making $[Ln(N(SiMe_3)_2)_3]$ (Ln = La, Nd, Sm, Er) which in turn react forming acyclic Schiff base complexes.[258] $[Ln(N(SiMe_3)_2)_3]$ (Ln = Y, La-Nd, Sm, Eu, Tb-Ho, Tm–Yb) react with CyNC (Cy = cyclohexyl) forming five-coordinate adducts $[Ln(N(SiMe_3)_2)_3(CyNC)_2]$. The crystal structure of the Nd complex confirms the trigonal bipyramidal geometry.[259] The Nd—N distances are 2.331–2.341 Å and Nd—C 2.737(6) Å. The Nd—N distance can be compared with that of 3.29(2) Å in the parent $[Nd(N(SiMe_3)_2)_3]$, indicating a degree of congestion, especially in the long Nd—C distances. $[Y(N(SiMe_3)_2)_3]$ forms a tbp (trigonal bipyramidal) adduct $[Y(N(SiMe_3)_2)_3(PhCN)_2]$ with axial nitriles;[253] in a preparation under similar conditions, the corresponding bis(trimethylsilyl) methyl is isolated as the four-coordinate adduct $[Y(CH(SiMe_3)_2)_3(\mu-Cl)Li(Et_2O)_3]$. The structure of three-coordinate $[Yb(N(SiMe_3)_2)_3]$ (Yb—N 2.183 (5) Å) has been reported, analogous to other lanthanide homologues.[260] It was one of several products from the reaction of $YbCl_3$ and $NaN(SiMe_3)_2$ in a 1:2 molar ratio, along with $[Yb(N(SiMe_3)_2)_2(\mu-Cl)(THF)]_2$ and

[Yb$_3$Cl$_4$O((N(SiMe$_3$)$_2$)$_3$(THF)$_3$]. The latter is a trinuclear cluster with a triangle of ytterbiums, where Yb is octahedarlly coordinated by an amide nitrogen, a THF, two μ_2-chlorine atoms and a μ_3-chlorine and μ_3-oxygen. Extended heating of [Yb(N(SiMe$_3$)$_2$)$_2$(μ-Cl)(THF)]$_2$ in heptane at 90 °C results in the removal of THF, affording the base-free [Yb(N(SiMe$_3$)$_2$)$_2$(μ-Cl)]$_2$, which contains four-coordinate ytterbium. Removal of the THF molecules from the solvate means that the base-free compound exhibits structural unsaturation with a shortening of Yb—N and Yb—Cl from 2.185 Å and 2.683 Å respectively in [Yb(N(SiMe$_3$)$_2$)$_2$(μ-Cl)(THF)]$_2$ to 2.144 Å and 2.623 Å respectively in [Yb(N(SiMe$_3$)$_2$)$_2$(μ-Cl)]$_2$. The peroxo complex [Yb$_2$(N(SiMe$_3$)$_2$)$_4$(μ-O$_2$)(THF)$_2$] has also been examined.[260] Several others have investigated the reaction of LnCl$_3$ with a deficit of Li N(SiMe$_3$)$_2$ that affords [Ln(N(SiMe$_3$)$_2$)$_2$(μ-Cl)(THF)]$_2$ (Ln = Eu,[261] Gd,[261] Yb[261], Ce,[262] Nd,[262] and Sm[263]).

[((Me$_3$Si)$_2$N)$_2$(THF)Sm(μ-X)$_2$Sm(THF)((N(SiMe$_3$)$_2$)$_2$] (X = Cl, Br) have symmetrical halogen bridges.[264] LnCl$_3$ reacts with two moles of Na[N(SiMe$_3$)$_2$] forming dimeric [((Me$_3$Si)$_2$N)$_2$(THF)Ln(μ-Cl)$_2$Ln(THF)((N(SiMe$_3$)$_2$)$_2$] (Ln = Ce, Nd); they can also be prepared by redistribution reactions between LnCl$_3$ and [Ln(N(SiMe$_3$)$_2$)$_3$]. The structure of the Nd compound was determined.[264] The three-coordinate amides [Ln(N(SiMe$_3$)$_2$)$_3$] do not react with tris(t-butyl)methanol (tritox-H) but the less hindered [Ln(N(SiHMe$_2$)$_2$)$_3$] undergo clean solvolysis affording [Ln(tritox$_3$-(THF)] (Ln = Y, Nd, Dy, Er). Structures have been reported for four-coordinate [Ln(tritox)$_3$(THF)] (Ln = Y, Nd) and five-coordinate [Ln(N(SiHMe$_2$)$_2$)$_3$(THF)$_2$] (Ln = Nd, Y).[265]

[Ln{N(SiHMe$_2$)$_2$)$_3$(THF)$_2$] (Ln = Y, La-Lu) are isostructural, with trigonal bipyramidal coordination; structures have now been determined for the La and Lu compounds, in addition to the Nd and Y analogues previously reported.[266] The structure of [Sm(N(SiHMe$_2$)$_2$)$_3$(THF)$_2$] is tbp with axial THF ligands.[267] [Y(N(SiHMe$_2$)$_2$)$_3$(THF)$_2$] readily reacts with [p-t-butylcalix[4]arene]H$_4$ at room temperature exchanging all the amide ligands forming a dimeric calix[4]arene complex.[268] [Ln(N(SiHMe$_2$)$_2$)$_3$(THF)$_2$] (Ln = Y, La) react with substituted bisoxazolinate salts forming mono and bis(bisoxazolinate) complexes.[269] [Nd{N(SiMe$_3$)$_2$)$_3$] and [Nd{N(SiHMe$_2$)$_2$}$_3$(THF)$_2$] have been anchored to the internal walls of microporous MCM-41 (mesoporous silicate) as potential catalyst precursors.[270] The amides [Ln(N(SiHMe$_2$)$_2$)$_3$(THF)$_2$] (Ln = Y, La) has been grafted onto MCM-41 then used as precursors for immobilized [Ln(fod)$_n$] (fod = 1,1,1,2,2,3,3-heptafluoro-7,7-dimethyl-4,6-octanedionate) species, which are promising catalysts for the Danishefsky heterogeneous Diels–Alder reaction. Untethered [Ln(N(SiHMe$_2$)$_2$)$_3$(THF)$_2$] can be used to make [Ln(fod)$_3$] in good yield.[271] Amides [Ln{N(SiMe$_3$)(2,6-Pr$^i_2$C$_6$H$_3$)}$_2$(THF)$_2$] have been reported to have distorted tetrahedral structures.[272] [Yb{N(SiMe$_3$)(2,6-Pr$^i_2$C$_6$H$_3$)}$_2$(THF)$_2$] partially desolvates in vacuo forming [Yb{N (SiMe$_3$)(2,6-Pr$^i_2$C$_6$H$_3$)}$_2$(THF)]. Use of a highly fluorinated amide has permitted the synthesis of [Sm(N(SiMe$_3$)(C$_6$F$_5$))$_3$] (**13**) in which there are many Sm···F and agostic interactions.

The neodymium compound [(η^6-C$_6$H$_5$Me)Nd(N(C$_6$F$_5$)$_2$)$_3$] has a η^6-bonded toluene molecule with a distorted piano-stool geometry.[273] [Sm(N(C$_6$F$_5$)(SiMe$_3$))$_3$(THF)] (**14**) has the distorted tetrahedral

(**13**)

coordination of samarium but with additional close Sm···F contacts and an "agostic" Sc—C contact.[274] The amide [Er{NBu^t(SiMe$_2$H)}$_3$], which has an unexpectedly high vapor pressure, displays three agostic Er—H—Si interactions in the solid state.[275] [Ln(N(SiMe$_3$)$_2$)$_3$] (Ln = Y, La) react with trisilanols forming dimeric lanthanide silasesquionanes; these can be converted into monomers by suitable donor ligands.[276] [Y(N(SiMe$_3$)$_2$)$_3$] reacts with HN(QPPh$_2$)$_2$ (Q = S, Se) forming [Y(N(QPPh$_2$)$_2$)$_3$]; the ligands are bound η^3- through two sulfur atoms and a nitrogen atom.[277] [Ln{(N(SiMe$_3$)$_2$)$_3$] (Ln = La, Sm, and Y) have been studied as highly active catalysts for the Tischenko reaction dimerizing aldehydes to esters, e.g., converting benzaldehyde to benzyl benzoate.[278] Some [Ln(N(SiMe$_3$)$_2$)$_3$] (Ln = Yb, Y) have been used as synthons for lanthanide silsesquioxanes.[279] [Ln(N(SiMe$_3$)$_2$)$_3$] are good synthons for the preparation of binaphtholate complexes M$_3$[Ln(binol)$_3$] (M = Li, Na).[280] [Ln(N(SiMe$_3$)$_2$)$_3$] (Ln = Sm, Nd) are useful precursors for the synthesis of seven-coordinate Schiff base complexes.[281] [Ln(N(SiMe$_3$)$_2$)$_3$] form ylidic olefin adducts. The amide ligand-N(SiMe$_3$)Ph is less bulky than -N(SiMe$_3$)$_2$, to judge by the isolation of THF adducts [Ln{N(SiMe$_3$)Ph}$_3$(THF)$_x$] (Ln = La, x = 2; Ln = Nd-Lu, Y, x = 1); the neodymium compound having tetrahedral coordination. -N(SiMe$_3$)(C$_6$H$_3$Pr$^i_2$-2,6) is bulkier, so that only of these ligands can be introduced in [NdCl(N(SiMe$_3$)Ph)$_2$(THF)].[282] When the steric demands of the amide group are less than for -N(SiMe$_3$)$_2$, adduct formation or attachment of another amide group can complete the coordination sphere. Thus four-coordinate amides [Ln(NPr$^i_2$)$_3$(THF)] and [Li(THF)Ln(NPr$^i_2$)$_4$]^{283} (Ln = La, Y, Lu) have been made. [Li(THF)$_4$][Yb(NPh$_2$)$_4$] has tetrahedral coordination of ytterbium.[284] whilst [Li(THF)$_4$] [Ln(NPh$_2$)$_4$] (Ln = Er, Yb) are unexpected products of the reaction between LnCl$_3$ and LiNPh$_2$ (two moles) and LiCH$_2$CH$_2$PPh$_2$.[285] [(Pr$^i_2$N)$_2$Nd(μ-NPr$^i_2$)$_2$Li(THF)] has four-coordinate neodymium;[286] [Nd(NPr$^i_2$)$_3$] reacts with AlMe$_3$ affording [Nd(NPr$^i_2$){(μ-NPr$^i_2$)(μ-Me)AlMe$_2$}{(μ-Me)$_2$AlMe$_2$}]. A number of amido metallates have been synthesized.[286] [Na(12-crown-4)$_2$][Ln((N(SiMe$_3$)$_2$)$_3$(OSiMe$_3$)] (Ln = Sm, Eu, Yb, Lu); [Na(THF)$_3$Ln((N(SiMe$_3$)$_2$)$_3$(C≡CPh)] (Ln = Ce, Sm, Eu), and [Na(THF)$_6$][(Me$_3$Si)$_2$N)$_2$Lu(μ-NH$_2$)(μ-NSiMe$_3$)-Lu((N-(SiMe$_3$)$_2$)$_2$] all have tetrahedral coordination of the lanthanide. Even in the presence of an excess of NdCl$_3$, the NdCl$_3$–LiNPr$^i_2$ reaction yields mainly [Nd(NPr$^i_2$)$_3$(THF)] but with some [(Pr$^i_2$N)$_2$Nd(NPr$^i_2$)$_2$Li(THF)].[287] The structure shows terminal Nd—N links at 3.283(17)–3.291(16) Å and Nd—N bridges of 2.393 (15) Å and 2.406(16) Å. Analogues [(Pr$^i_2$N)$_2$Ln(NPr$^i_2$)$_2$Li(THF)] (Ln = La, Y, Yb) have also been described.[283] [Nd(NPr$^i_2$)$_3$(THF)] reacts with AlMe$_3$ forming [(Pr$^i_2$N)Nd[(μ-NPr$^i_2$)(μ-Me)AlMe$_2$][(μ-Me)$_2$AlMe$_2$].[288] A cluster, [Yb$_3$Br$_4$O((N(SiMe$_3$)$_2$)$_3$(THF)$_3$], has been obtained from YbBr$_3$ and NaN(SiMe$_3$)$_2$.[289] Four-coordination is found[290] in (THF)$_3$Li(μ-Cl)NdR$_3$ (R = N(SiMe$_3$)$_2$, OC(Bu^t)$_3$). Yb and HgPh$_2$ react with (2-MeOC$_6$H$_4$)NHSiMe$_3$ and (2-PhOC$_6$H$_4$) NHSiMe$_3$ forming the unexpected Yb^{III} compounds [((2-MeOC$_6$H$_4$)NSiMe$_3$)$_2$Yb(μ-OMe)$_2$Yb((2-MeOC$_6$H$_4$)NSiMe$_3$)$_2$] and [Yb((2-PhOC$_6$H$_4$)NHSiMe$_3$)$_2$(OPh)(THF)], due to a C—O bond cleavage in the aryl ether.[291]

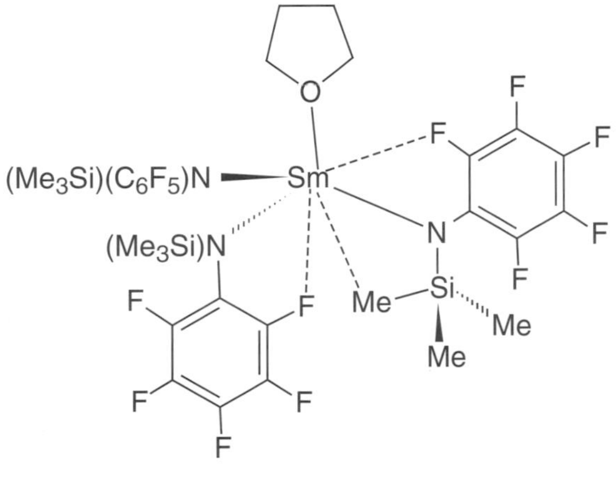

(14)

In contrast to the three-coordinate silylamides and diisopropylamides, the lanthanides do not form simple homoleptic dimethylamides. Reaction of NdCl$_3$ with LiNMe$_2$ gives an adduct [Nd(NMe$_2$)$_3$(LiCl)$_3$], which with MMe$_3$ (M = Al, Ga) gives[292] bridged [Nd(NMe$_2$)$_3$(MMe$_3$)$_3$]. A peralkyl [Nd{(μ-Me)$_2$(GaMe$_2$)}$_3$] has also been reported.

A systematic study has been made of lanthanide arylamides.[293] Reaction of NdCl$_3$ with KNHAr (Ar = Ph) leads to [Nd(NHPh)$_3$·3KCl], reminiscent of the adducts of the dimethylamides. Using the bulkier 2,6-Me$_2$C$_6$H$_3$ group, anionic species [K(THF)$_6$]$_2$[{Ln(μ-NH(2,6-Me$_2$C$_6$H$_3$))(NH-(2,6-Me$_2$C$_6$H$_3$)$_3$)$_2$]   (Ln = Nd,   Sm)   and   [K(DME)(THF)$_3$][{Y$_2$(μ-NH(2,6-Me$_2$C$_6$H$_3$))$_2$(μ-Cl)(NH(2,6-Me$_2$C$_6$H$_3$))$_4$}(THF)$_2$] are formed. The even more demanding 2,6-Pr$^i_2$C$_6$H$_3$ group leads to octahedral [Nd(NH-2,6-Pr$^i_2$C$_6$H$_3$)$_3$(THF)$_3$] and trigonal bipyramidal [Ln(NH-2,6-Pr$^i_2$C$_6$H$_3$)$_3$(THF)$_2$]   (Ln = Y,   Yb). Unsolvated   dimeric   compounds   [{Ln(μ-NH(2,6-2,6-Pr$^i_2$C$_6$H$_3$))(NH(2,6-Pr$^i_2$C$_6$H$_3$)$_2$)$_2$]   (Ln = Y,   Yb) are produced by amine exchange with [Ln{N(SiMe$_3$)$_2$}$_3$].[294] Phenylamides have been used to make polynuclear lanthanide amides including   [Ln$_2$Br$_4$(μ-NHPh)$_2$(THF)$_5$]   (Ln = Sm,   Gd);   [Ln$_4$(μ_4-O)(NHPh)$_3$(OSiMe$_2$NPh)$_6$-Na$_5$(THF)$_7$].THF (Ln = Gd, Yb) have also been reported.

Reaction of [SmCl$_3$(THF)$_3$] with LiNR$_2$ (R = Cy, Pr^i) give products dependent upon R, [(Cy$_2$N)$_2$Sm(μ-Cl)(THF)]$_2$ and [(Pr$^i_2$N)$_2$SmCl$_3$Li(TMEDA)$_2$]. The former can be converted into monomeric [Sm(NCy$_2$)$_3$(THF)].[295] A new family of compounds of anionic lanthanide silylamides has been reported ((**15**); Ln = Sm, Eu, Gd–Tm, Lu; (L)$_3$ = (THF)$_3$, (Et$_2$O)$_3$, and (THF)$_2$(Et$_2$O)) which features four-coordinate lanthanides.[296] They do not afford neutral tris(complexes) on heating.

$$\overline{[\text{Ln}\{\text{N(SiMe}_2\text{CH}_2\text{CH}_2\text{SiMe}_2)\}_3(\mu\text{-Cl})\text{Li(L)}_3}$$

(**15**)

A chelating bis(silylamide) ligand forms yttrium complexes, including a five-coordinate methyl (**16**).[297]

(**16**)

K[Er(η^2-Bu$^t_2$pz)$_4$] and [K(18-crown-6)(DME)(C$_6$H$_5$Me)] [Er(η^2-Bu$^t_2$pz)$_4$] are the first homoleptic lanthanide pyrazolides; in the unsolvated compound, the potassium ions are η^3-bonded to two pyrazolides.[298] Reaction of Bu$^t_2$pzH with lanthanide metals and mercury at 220 °C affords [Ln(η^2-Bu$^t_2$pz)$_3$]   (Ln = Nd,   Sm)   and   [Yb$_2$(η^2-Bu$^t_2$pz)$_5$],   the   latter   having   the   structure [(η^2-Bu$^t_2$pz)Yb^{II}(μ-η^2: η^2-Bu$^t_2$pz)$_2$Yb^{III}(η^2-Bu$^t_2$pz)$_2$].[299] Other 3,5-di-*t*-butylpyrazolides to be reported include [Y(η^2-Bu$^t_2$pz)$_3$L$_2$] (L = Py, THF), [Er(η^2-Bu$^t_2$pz)$_3$L$_2$] (L = Py, 4-*t*-BuPy, 4-*n*-BuimH), [Er(η^2-Bu$^t_2$pz)$_3$(Bu$^t_2$pzH)] and [Lu(η^2-Bu$^t_2$pz)$_3$(4-*t*-Bupy)$_2$]. Several of these have been shown to have monomeric structures though the unsolvated [Y(η^2-Bu$^t_2$pz)$_3$] is evidently oligomeric.[300] Study has been made[301] of compounds using a β-diketiminate ligand, (L-L)$^-$ (L-L = (R)NC(Ph)C(H)C(Ph)N(R) (where R = Me$_3$Si); reaction of LnCl$_3$ or LnI$_3$ with Na(L-L) gives [Ln(L-L)$_2$Cl] (Ln = Ce, Pr, Nd, Sm, Eu, Yb) and [Tm(L-L)$_2$I]. [Ce(L-L)$_2$Cl] reacts with LiCHR$_2$ forming [Ce(L-L)$_2$(CHR$_2$)]; this compound and [Nd(L-L)$_2$Cl] both have monomeric structures. The N-substituted guanidinates [M{CyNC{N(SiMe$_3$)$_2$}NCy}$_2$(μ-Cl)$_2$LiS$_2$] (M = Sm, Yb: S = Et$_2$O, 1/2 TMEDA) have been synthesized and offer a route to solvent-free alkyls and amides, of which [Sm{CyNC{N(SiMe$_3$)$_2$}NCy}$_2$(CH(SiMe$_3$)$_2$)] and [Yb{CyNC{N(SiMe$_3$)$_2$}-NCy}$_2$(N(SiMe$_3$)$_2$)] have been characterized.[302]

Benzamidinates have been studied, often using this ligand in association with cyclopentadienyl type ligands. Yttrium (and scandium) complexes have been investigated[303] and monomeric [(PhC(N(Si-Me$_3$)$_2$)$_3$)Ln] complexes have been obtained.[304] [(PhC(N(SiMe$_3$)$_2$)$_2$)YCl(THF)] can be turned into [(PhC(N(SiMe$_3$)$_2$)$_2$)YR(THF)] (R = BH$_4$, N(SiMe$_3$)$_2$, CH$_2$Ph, CH(SiMe$_3$)$_2$, etc.) and, by hydrogenolysis of [(PhC(N(SiMe$_3$)$_2$)$_2$)Y(CH(SiMe$_3$)$_2$)(THF)], into dimeric hydrides [{(PhC(N(SiMe$_3$)$_2$)$_2$)Y(μ-H)}$_2$].[305]

3.2.2.3.6 *Polypyrazolylborates*

The following abbreviations are used here: Tp = tris(pyrazolyl)borate; Tp^{Bupy} = tris(3-{(4-t-butyl)-pyrid-2-yl}-pyrazol-1-yl)borate; Tp^{cpd-} = Hydrotris[3-(carboxypyrrolidido)pyrazol-1-yl]borate; Tp^{Me2} = tris-(3,5-dimethylpyrazolyl)borate; Tp^{Me2,4Et} = tris(3,5-dimethyl-4-ethylpyrazolyl) borate; Tp^{Ms} = tris(3-mesitylpyrazolyl)borate; Tp^{Ms*} = bis(3-mesitylpyrazolyl)(5-mesitylpyrazolyl) borate; Tp^{Ph2} = tris(3,5-diphenylpyrazolyl)borate; Tp^{2-pyr} = tris(3-(2-pyridyl)pyrazolyl)borate; Tp^{Th2} = tris(3,5-dithienyl-pyrazolyl)borate; Bp$^{(COC)Py}$ = bis[3-(2-pyrid-2-yl)5-(methoxymethyl) pyrazol-1-yl]hydroborate.

F-element poly(pyrazolyl)borates have been reviewed.[306,307] [LnTp$_3$] have been synthesized (Ln = Sc, Y, La–Nd, Sm–Lu); the structure of the Pr and Nd compounds show the lanthanides to be nine-coordinate, though an eight-coordinate structure is adopted by later metals (e.g., Yb).[308] Working with a 2:1 ratio of reactants, complexes [Ln(Tp)$_2$Cl(THF)] are formed. The structure of the corresponding hydrate is known [Y(Tp)$_2$Cl(H$_2$O)].[309] Using totally moisture-free conditions, seven-coordinate [Ln(Tp)$_2$Cl] is formed, but this readily adds water forming [Ln(Tp)$_2$Cl(H$_2$O)].[310] Reaction of LnCl$_3$.6H$_2$O with KTp affords [Ln(Tp)$_2$(L)Cl] (Ln = Lu, Nd, L = HPz; Ln = Lu, L = H$_2$O); with N-methylpyrazole, [LnCl{Tp}$_2$(N-Mepz)] are obtained, these reacting with Na(quin) (quin = 8-quinolinate) to form [Ln{Tp}$_2$(quin)]. [Nd (Tp)$_2$(L)Cl] (L = Hpz, H$_2$O) have square antiprismatic coordination of neodymium.[311] Reaction of [Y(NCS)$_3$(phen)$_3$] with KTp leads to the isolation of [{Y(Tp)(NCS)(phen)(μ-OH)}$_2$].[312] [Ln(Tp)$_2$(L)] (L = salicylaldehydate or 5-methoxysalicylaldehydate; Ln = Y, Pr to Lu) have been synthesized. Several picolinate N-oxide complexes [Ln(Tp)$_2$(ONC$_5$H$_4$CO$_2$-2)] have been synthesized; the structure of the eight-coordinate terbium compound shows the picolinate N-oxide to chelate.[313] [Eu(Tp)$_2$(5-methoxysalicylaldehydate)] has the dodecahedral eight-coordination characteristic of europium.[314] Structures are reported of [Eu(Tp)$_2$(μ-O$_2$CPh)]$_2$, [Eu(Tp)(μ-O$_2$CPh)$_2$]$_2$, [Eu(Tp)$_2$(OC(Me)CHC(Me)O)], and [Eu(Tp)$_2$(OC(Ph)CHC(Ph)O)], as well as [(Gd(Tp)$_2$)$_2$(μ-1,4-(O$_2$C)$_2$C$_6$H$_4$)].[315] A few simple mono Tp complexes are known,[316] such as [Y(Tp)X$_2$(THF)$_2$] (X = Cl, Br) and [Nd(Tp)I$_2$(THF)$_2$] (X-ray). A very wide range of compounds, many of Sm, Eu, and Yb, in both the +2 and +3 states, have been synthesized with Tp^{Me2}. Use of the lanthanide triflate in synthesis leads to Ln(Tp^{Me2})$_2$(OTf). Triflate coordinates for the lighter lanthanides, as in [Nd(Tp^{Me2})$_2$(O$_3$SCF$_3$)] whereas the smaller lanthanides form ionic [Ln(Tp^{Me2})$_2$]O$_3$SCF$_3$, as in the ytterbium compound.[317,318] The triflates form mono(MeCN) adducts, the MeCN being bound in [La(Tp^{Me2})$_2$(MeCN)]O$_3$SCF$_3$ but not in [Nd(Tp^{Me2})$_2$(O$_3$SCF$_3$)].MeCN. This can be regarded as a type of ionization isomerism.[319] Dodeca-hedral eight-coordination is found in these and in [La(Tp^{Me2})$_2$(NO$_3$)]. Reaction of KTp^{Me2} with LnCl$_3$ in THF gives [Ln(Tp^{Me2})$_2$X], from which the chloride can be replaced by other ligands such as diketonates. A range of Sm(Tp^{Me2})$_2$X compounds,[320] where X = F, Cl, have monomeric seven-coordinate molecular structures, but for X = I and BPh$_4$, the compounds are [Sm(Tp^{Me2})$_2$]$^+$X$^-$. A comparison of the ytterbium(II) and (III) compounds [Yb(Tp^{Me2})$_2$] and [Yb(Tp^{Me2})$_2$]O$_3$SCF$_3$ show six-coordination for ytterbium in both cases, with Yb—N bond lengths ca. 0.16 Å longer in the Yb^{II} compound.[318] The samarium(II) poly(pyrazolyl)borate [Sm(Tp^{Me2})$_2$] undergoes a one-electron oxidation with [Hg(C≡CPh)$_2$] forming monomeric seven-coordinate [Sm(Tp^{Me2})$_2$(C≡CPh)].[321] This undergoes a remarkable rearrangement at 105 °C in benzene solution with the exchange between a pyrazole ring and a alkynyl group, forming [Sm(Tp^{Me2})((HB(Me$_2$pz)$_2$(C≡CPh))(Me$_2$pz)]. Reaction of [Sm(Tp^{Me2})$_2$Cl] with NaOR (R = Ph, C$_6$H$_4$-4-Bu^t) affords [Sm(Tp^{Me2})$_2$(OR)].[322] Analogues with the heavier chalcogenides, [Sm(Tp^{Me2})$_2$(SR) (R = Ph, CH$_2$Ph), [Sm(Tp^{Me2})$_2$(SR) (R = Ph, C$_6$H$_4$-4-Bu^t), and [Sm(Tp^{Me2})$_2$(TePh)] (as well as analogues with the Tp^{Me2Et} ligand) have been made by reductive cleavage of dichalcogenides with [Sm(Tp^{Me2})$_2$]. (Tp^{Me2Et} = tris(3,5-dimethyl-4-ethylpyrazolyl)borate). The structures of [Sm(Tp^{Me2})$_2$(SR) (R = Ph, C$_6$H$_4$-4-Bu^t) display a distortion due to twisting of a ligand about a B—N bond and π-stacking of a pyrazolyl ring with a phenyl group. Samarium superoxide complexes with the hydrotris(3,5-dimethylpyrazolyl)borate ligand, as well as a compound with a μ_3-oxo group have been investigated.[323,324] Mono Tp complexes have been made,[325,326] from the reaction of one mole of KTp^{Me2} with LnCl$_3$ in THF (Ln = Y, Nd). [Nd(Tp^{Me2})I$_2$(THF)] only has one THF bound, in contrast to the Tp analogue, showing the greater steric demands of the Tp^{Me2} ligand. The halogens can be replaced to afford alkyls and aryls [Nd(Tp^{Me2})R$_2$(THF)] (R is, for example, CH$_2$SiMe$_3$, Ph) which on hydrogenolysis give catalytically active hydrides.[326] Using a 2:1 reaction stoichiometry, (Tp^{2-pyr}) forms 2:1 complexes such as [Sm(Tp^{2-pyr})$_2$]$^+$ with icosahedral 12-coordination of samarium.[327,328] [Eu(trop)(NO$_3$)(Tp^{2-pyr})] contains a hexadentate pyrazolylborate and bidentate tropolone and nitrate ligands giving eight-coordination. In [Tb(dbm)$_2$(Tp^{2-pyr})], one arm of the pyrazolylborate is pendant, so it is tetradentate, giving eight-coordinate terbium.[329,330] Reaction of KTp$^{2\text{-pyr}}$ with hydrated LnCl$_3$ in methanolic NH$_4$PF$_6$ remarkably leads to fluoride abstraction and the isolation of [Ln(Tp$^{2\text{-pyr}}$)F-(MeOH)$_2$]PF$_6$ (Ln = Sm, Eu, Gd, Tb, Ho, Yb), the bulk of the podand inhibiting the formation of a fluoride bridged oligomer. In the absence of NH$_4$PF$_6$, nitrate complexes such as [Ln(Tp$^{2\text{-pyr}}$)(NO$_3$)$_2$] can be isolated.[331] Hydrotris[3-(carboxypyrrolidido)pyrazol-1-yl]borate (Tp$^{\text{cpd}-}$) is a potentially hexadentate N$_3$O$_3$ donor, and forms complexes [Ln(Tp$^{\text{cpd}-}$)$_2$]PF$_6$ (Ln = La, Nd, Sm) in which one ligand is acting as a N$_3$O$_3$ donor and the other as a N$_2$O$_2$ donor.[332] Tris[3-{(4-t-butyl)pyrid-2-yl}-pyrazol-1-yl]-hydroborate (1-) (Tp^{Bupy}) is potentially hexadentate and bis[3-(2-pyrid-2-yl)5-(methoxymethyl)pyrazol-1-yl]hydroborate (1-) (Bp$^{(\text{COC})\text{Py}}$) is potentially tetradentate. Owing to nitrates being bidentate, [La(Tp^{Bupy})(NO$_3$)$_2$] has 10-coordinate lanthanum, as has [La(Bp$^{(\text{COC})\text{Py}}$)$_2$(NO$_3$)], but in [La(Bp$^{(\text{COC})\text{Py}}$)$_2$(CF$_3$SO$_3$)] the presence of a monodentate triflate causes lanthanum to be nine-coordinate.[333] A number of studies have been made of ligands that are not tris(pyrazolyl)borates. [Yb(B(pz)$_4$)$_3$].EtOH has eight-coordinated ytterbium, one pyrazolylborate being bidentate.[334]

A number of [Ln{H$_2$B(Me$_2$-pz)$_2$}$_3$] (Ln = Y, Ce, Sm, Yb) have been investigated whilst a study of luminescence spectra of [Tb{H$_2$B(Me$_2$-pz)$_2$}$_3$] in the solid state or in nonpolar solvents is consistent with the maintenance of a trigonal prismatic arrangement of nitrogens about Tb with three further weak B-H···Tb interactions; on dissolution in polar solvents the spectrum changes.[335]

Using Bp$^{2\text{-pyr}}$, the bis(pyrazolyl)borate analogue of Tp$^{2\text{-pyr}}$, complexes [Ln(Bp$^{2\text{-pyr}}$)$_2$(NO$_3$)] (Ln = Eu, Gd, Tb) and [Eu(Bp$^{2\text{-pyr}}$)$_2$(DMF)]ClO$_4$ have been isolated.[336,337]

Eight-coordinate [Sm{B(pz)$_4$}$_2$(sal)] shows stereochemical nonrigid behavior in solution.[338]

3.2.2.3.7 *Thiocyanate*

The thiocyanate group has been studied in detail as a ligand for lanthanide(III) ions in recent years. Recent extensive studies by Japanese workers in particular have shown that the stoichiometry and structure of the anionic complex obtained depends greatly upon factors such as the counter ion used, upon the solvent employed, etc. The complexes are generally made by reaction of the lanthanide thiocyanate with the alkylammonium thiocyanate in a suitable solvent (e.g., an alcohol) or mixture of solvents. [NEt$_4$]$_3$[Ln(NCS)$_6$].solvent (M = Er,Yb; solvent = C$_6$H$_6$, C$_6$H$_5$F, C$_6$H$_5$Cl, C$_6$H$_5$CH$_3$) have octahedrally coordinated lanthanides;[339] similarly [NBu$^n_4$]$_3$[Ln(NCS)$_6$] (M = Y, Pr-Yb) are known to be six-coordinate, exemplified in the structure of the neodymium compound.[340] Reaction of lanthanide thiocyanates with tetramethylammonium thiocyanate in methanol-water, followed by slow crystallization, gives [NEt$_4$]$_4$[Ln(NCS)$_7$(H$_2$O)] (Ln = La-Nd, Dy, Er) which have a cubic eight-coordinate geometry round the lanthanides.[341] A synthesis in which crystallization is achieved by evaporation in a desiccator over benzene affords [NEt$_4$][Ln(NCS)$_4$(H$_2$O)$_4$] (Ln = Nd, Eu) with square-antiprismatic coordination.[342] In contrast, if the water is removed by forming an azeotrope with benzene-ethanol and vacuum evaporation, followed by crystallization of a methanolic solution in benzene vapor, crystals of [NEt$_4$]$_4$[Ln(NCS)$_7$].benzene (Ln = La, Pr) are obtained, which have a capped trigonal prismatic geometry.[343] Reaction of the lanthanide thiocyanates with tetramethylammonium thiocyanate in methanol-water, followed by slow crystallization usually affords [NMe$_4$]$_3$[Ln(NCS)$_6$(MeOH)(H$_2$O)] (Ln = La-Nd, Sm–Dy, Er), which have lanthanides in a square antiprismatic environment. However, by working in a water-free environment, [NMe$_4$]$_4$[Ln(NCS)$_7$] (Ln = Dy, Er, Yb) were obtained, in which the coordination geometry most closely approximates to a pentagonal bipyramid.[344] Crystallization of a methanolic solution of the reactants in benzene vapor, gives crystals of [NMe$_4$]$_5$[Ln(NCS)$_8$].2C$_6$H$_6$.[345]

3.2.2.4 Group 15 Ligands Involving Phosphorus

A small but increasing selection of these compounds has been reported.

[Nd(P(SiMe$_3$)$_2$)$_3$(THF)$_2$] has a trigonal bipyramidal structure,[346] whilst unsolvated [Y(P(SiMe$_3$)$_2$)$_3$] is a dimer, [Yb(P(SiMe$_3$)$_2$)$_2$(μ-P(SiMe$_3$)$_2$)]$_2$, with tetrahedral coordination of yttrium. The average Y—P terminal bond length is 2.677 Å and the bridging Y—P distances average 2.848 Å.[347,348] Sm(O$_3$SCF$_3$)$_3$ reacts with LiPBu$^t_2$ forming [{((THF)$_2$Li(μ-PBu$^t_2$)$_2$}$_2$Sm] with tetrahedral coordination of samarium by four phosphorus atoms.[349] Other compounds such as $[La(N(SiMe_3)_2)_2(Ph_3PO)_2(PPh_2]$[350] have involved the presence of other donor groups, sometimes as chelating ligands where the other donor atom in the chelate is a "harder" base and assists coordination, as in compounds such as $[Nd(OCBu^t_2CH_2PMe_2)_3]$[351] or the organometallic $[Y(C_3H_5)\{N(SiMe_2CH_2PMe_2\}_2(\mu\text{-}Cl)_2]$.[352]

3.2.2.5 Group 16 Ligands Involving Oxygen

3.2.2.5.1 *The aqua ion and hydrated salts*

The nature of the lanthanide aqua ion has been the source of much study but now appears to be fairly well understood. Much of the evidence (though not the most recent) is covered in an authoritative review.[353]

To summarize, the coordination number of $[Ln(H_2O)_n]^{3+}$(aq) is believed to be nine for the early lanthanides (La–Eu) and eight for the later metals (Dy–Lu), with the intermediate metals exhibiting a mixture of species. The nine-coordinate species are assigned tricapped trigonal prismatic structures and the eight-coordinate species square antiprismatic coordination. A considerable amount of spectroscopic data has led to this conclusion, using a range of techniques. A early deduction for coordination numbers in solution was that the visible spectrum of Nd^{3+}(aq) and $[Nd(H_2O)_9]^{3+}$ ions in solid neodymium bromate are very similar to each other and quite different to those of Nd^{3+} ions in eight-coordinate environments. Among the recent data, ^{17}O-NMR data for water exchange by the hydrated ion were interpreted in terms of a constant coordination number across the series, as rate constants varied smoothly with atomic number.[354] In contrast, virtually all other measurements point to a decrease. Neutron-scattering measurements on the solutions of the aqua ions indicate a decrease in coordination number.[355] A neutron-diffraction study of $Nd(ClO_4)_3$ and $Sm(ClO_4)_3$ in solution indicates coordination numbers of 9.0 and 8.5 respectively, indicating that there are both eight and nine-coordinate species present for samarium.[356] Values of ~7.9 have been obtained for Dy^{3+} and Lu^{3+}; a molecular dynamics simulation study of lanthanide ions in aqueous solution found that it was necessary to include allowance for polarization of the water molecule by lanthanide ions to get good agreement with this.[357] EXAFS spectra of aqueous solutions of lanthanum perchlorate are in close agreement with solid $[Ln(H_2O)_9](CF_3SO_3)_3$, suggesting a coordination number of nine for La^{3+}(aq).[358] In a study of chloride complexation by the lanthanides, hydration numbers of the aqua ions were deduced by EXAFS, values being 9.2 (La), 9.3 (Ce), 9.5 (Nd), 9.3 (Eu), 8.7 (Yb), and 9.7 (Y).[359] In contrast to the actinides, ability to coordinate chloride decreases across the series, with Yb^{3+} showing no tendency to do so in 14 M LiCl. Information from hydration studies of lanthanide and actinide(III) ions by laser-induced fluorescence spectroscopy has been combined with other techniques to indicate a change in hydration number from nine to eight in the Eu–Tb and Bk–Es regions of the series.[360,361] A luminescence study of lanthanide complexes reveals a linear correlation between the decay constant and the number of coordinated water molecules,[362] used to calculate first coordination sphere hydration numbers of 9.0, 9.1, 8.3, and 8.4 for Sm^{3+}, Eu^{3+}, Tb^{3+}, and Dy^{3+}, respectively. X-ray scattering of yttrium chloride solutions at pH 1.2 indicate about eight water molecules in the first coordination sphere. It has been known for some time that the amount of $[Ln(H_2O)_9]^{3+}$ (Ln is, for example, Gd, Eu) increases as the water content of solutions decreases. This has been rationalized by considering outer-sphere complex formation as well.[363] A molecular dynamics simulation[364] for water exchange between $[Ln(H_2O)_n]^{3+}$ and bulk water reveals very fast exchange between the bulk water and the hydrated samarium ion to maintain the equilibrium between $[Sm(H_2O)_9]^{3+}$ and $[Sm(H_2O)_8]^{3+}$. Monte Carlo simulations of Ln^{3+}(aq) ions have been reported,[365] agreeing with a change in coordination number from nine to eight in mid-series and with a dissociative mechanism for the ennea-aqua ions. An investigation of the transport properties of the trivalent lanthanide (and actinide ions) using radiochemical methods indicated a change in hydration number in each series.[366]

Hydrated salts are readily prepared by reaction of the lanthanide oxide or carbonate with the acid. Salts of non-coordinating anions most often crystallize as salts $[Ln(OH_2)_9]X_3$ (X is, for example. bromate, triflate, ethylsulfate, tosylate). A study of a series of triflates, $[Ln(OH_2)_9](CF_3SO_3)_3$, (Ln = La-Nd, Sm–Dy, Yb, Lu), have been examined in detail.[367] Their structures resemble the corresponding bromates and ethylsulfates in being hexagonal, all containing the tricapped trigonal prismatic $[Ln(OH_2)_9]^{3+}$ ion, even for the later lanthanides. The crystals of the triflate contain columns of $[Ln(OH_2)_9]^{3+}$ cations and $CF_3SO_3^-$ ions, with the columns linked by a three-dimensional network of hydrogen bonds. This is presumably a factor that stabilizes this structure and favours its isolation, even for the later lanthanides (Gd–Lu), where the eight-coordinate $[Ln(OH_2)_8]^{3+}$ ion predominates in solution (as will be seen, this ion does crystallize with certain other counter ions), and similar arguments are likely to apply to the ethylsulfates and bromates. The lanthanide–water distances for the positions capping the prism faces and at the vertices are different. Moreover, on crossing the series from La to Lu, the Ln—O distance decreases from 2.611 Å to 2.519 Å for the three face-capping oxygens but change more steeply from 2.513 Å to 3.287 Å for the six apical oxygens.[367] The series of ethylsulfates, $[Ln(OH_2)_9](C_2H_5SO_4)_3$, have been studied at 298 K and 171 K more recently, with comparable results.[368] However, changing the counter-ion can affect the aqua species isolated. In contrast, the perchlorate salts are $[Ln(OH_2)_6](ClO_4)_3$ and $[terpyH_2]_2[Tb(OH_2)_8]_7Cl_7 \cdot 8/3H_2O$ contains[369] a "pure" aqua ion formed even in the presence of a large excess of chloride ions. Other eight-coordinate lanthanides are found in $[[Eu(OH_2)_8]_2(V_{10}O_{28}) \cdot 8H_2O]$[370,371] and eight-coordinate ytterbium in ytterbium triflide, $[Yb(OH_2)_8][C(O_2SCF_3)_2]_3$ (the scandium analogue being seven-coordinate).[51] Another encapsulated inside a crown ether is $[Lu(OH_2)_8]^{3+}$ in $[Lu(OH_2)_8]Cl_3 \cdot 1.5(12\text{-crown-4}) \cdot 2H_2O$, though this is the normal CN for the lutetium aqua ion.[372] Early lanthanides (La–Nd) form $[Ln(OH_2)_9](p\text{-MeC}_6H_4SO_3)_3$ with the usual trigonal prismatic coordination but for the later lanthanide ions the tosylates Ln $(p\text{-MeC}_6H_4SO_3)_3 \cdot 9H_2O$ contain $[Ln(OH_2)_6(p\text{-MeC}_6H_4SO_3)_2]^+$ ions (Ln = Sm, Gd, Dy, Ho, Er, Yb, Y) in distorted dodecahedral eight-coordination.[373,374]

It has been known for over 20 years that the hydrated lanthanide perchlorates $Ln(ClO_4)_3 \cdot 6H_2O$ contain octahedral $[Ln(H_2O)_6]^{3+}$ ions in the solid state. Their isolation, however, must reflect a balance of factors such as solubility and hydrogen-bonding in the solid state, as the close similarity between the EXAFS spectra of aqueous solutions of lanthanum perchlorate and the spectrum of solid $[Ln(H_2O)_9](CF_3SO_3)_3$, indicates a coordination number of nine for $La^{3+}(aq)$ in the perchlorate solutions.[358] Similarly, the fact that erbium is eight-coordinate in $[[Er(OH_2)_8]\text{-}(ClO_4)_3 \cdot (dioxan) \cdot 2H_2O]$ shows the fine balance here.[375] Lower hydrates than the hexahydrates exist. Partial dehydration of $Ln(ClO_4)_3 \cdot 6H_2O$ (Ln = Er, Lu) gives $Lu(ClO_4)_3 \cdot 3H_2O$ and $Er(ClO_4)_3 \cdot H_2O$, both of which have eightfold coordination of the lanthanide.[376] Similarly, $Yb(ClO_4)_3 \cdot H_2O$ has been shown to have bi- and tridentate perchlorate groups giving eight-coordinate ytterbium.[377] The anhydrous perchlorates exist in high and low-temperature forms, both with nine-coordinate lanthanides, with terdentate perchlorates.[378,379] Anhydrous $Ln(ClO_4)_3$ (Ln = La-Er, Y) have a UCl_3-type structure, confirmed by powder,[380] and single-crystal studies.[381] The anhydrous triflates $[Ln(O_3SCF_3)_3]$ (Ln = Dy, Ho) decompose to LnF_3 on heating; their IR spectra suggest that triflate acts as a bidentate bridging ligand.[382] Stepwise thermal decomposition of several lanthanide triflates has been examined, with LnF_3 as the eventual product.[383] The structures of several methanesulfonates have been reported;[384] in all cases methanesulfonate acts as a bidentate ligand. $[Ln(O_3SMe)_3 \cdot 2H_2O]$ (Ln = Ce, Sm, Tb) are isostructural, with eight-coordinate lanthanides, whilst ytterbium is seven-coordinate in $[Yb(O_3SMe)_3]$.

Among the hydrated nitrates, there is a clear decrease in coordination number as the ionic radius of the lanthanide increases.[385] All nitrate groups are coordinated as bidentate ligands in these compounds, but the number of waters of crystallization is no guide to how many are actually bound to the metal. Thus compounds $Ln(NO_3)_3 \cdot 6H_2O$ are known for La–Dy and Y. Of these, the lanthanum and cerium compounds are $[Ln(NO_3)_3 \cdot (H_2O)_5] \cdot H_2O$ (Ln = La, Ce) with 11-coordinate lanthanides whilst the others are $[Ln(NO_3)_3 \cdot (H_2O)_4] \cdot 2H_2O$ (Ln = Pr–Dy, Y) with 10-coordination for the metal. Under different conditions, a series of pentahydrates $Ln(NO_3)_3 \cdot 5H_2O$ is obtained (Ln = Eu, Dy–Yb), which also contain $[Ln(NO_3)_3 \cdot (H_2O)_4]$ molecules. Lutetium forms $Ln(NO_3)_3 \cdot 4H_2O$ and $Ln(NO_3)_3 \cdot 3H_2O$, isolated under very similar conditions, both of which contain nine-coordinate $[Lu(NO_3)_3 \cdot (H_2O)_3]$ molecules.[385]

The chlorides of La and Ce, $LnCl_3 \cdot 7H_2O$ are dimeric $[(H_2O)_7Ln(\mu\text{-Cl})_2Ln(OH_2)_7]Cl_4$ with what has been described as singly capped square antiprismatic coordination of the metals whilst $LnCl_3 \cdot 6H_2O$ (Ln = Nd–Lu) have antiprismatic $[LnCl_2(H_2O)_6]^+$ ions with the coordinated chlorides on opposite sides of the polyhedron. There are extensive hydrogen bonding networks involving both coordinated and non-coordinated chlorides and water molecules.[386] Thermal dehydration of $CeCl_3 \cdot 7H_2O$ established[387] the existence of $CeCl_3 . xH_2O$ (x = 6, 3, 2, 1); the hexahydrate has monomeric $[CeCl_2(H_2O)_6]^+$ ions whilst the trihydrate has a structure based on a chain of $[CeCl_{4/2}Cl(H_2O)_3]$ units. Three different stoichiometries exist for the hydrated bromides, which tend to be deliquescent. Lanthanum and cerium form $LnBr_3 \cdot 7H_2O$, which are isomorphous with the corresponding chlorides in being dimeric $[(H_2O)_7Ln(\mu\text{-Br})_2Ln(OH_2)_7]Br_4$. Hexahydrates $LnBr_3 \cdot 6H_2O$ (Ln = Pr–Dy) again resemble heavier rare earth chlorides in being

[LnBr$_2$(H$_2$O)$_6$]Br (though this resemblance does not extend to the end of the lanthanide series for the bromides). A comparison between corresponding chlorides and bromides indicates that the Ln—O distances, though shorter, show more sensitivity to the lanthanide contraction than the Ln–halogen distances. The heaviest lanthanides form LnBr$_3$·8H$_2$O (Ln = Ho–Lu) which are [Ln(H$_2$O)$_8$]Br$_3$, with no bromide coordinated, and which resemble a structure found in the hydrated iodides of the heavier lanthanides.[388] Study of the hydrated iodides has been difficult because of their deliquescence and their tendency to oxidation in air. Recent diffraction studies,[389] however, indicated that the iodides of the earlier metals are LnX$_3$·9H$_2$O (Ln = La–Ho), containing the familiar tricapped trigonal prismatic [Ln(OH$_2$)$_9$]$^{3+}$ ions. Compared with the triflates and other complexes containing the tricapped trigonal prismatic nine-coordinate species, the spread of Ln—O distances is much smaller, falling in the range 2.552–2.576 Å for Ln = La and similarly between 2.403 Å and 2.405 Å for the holmium compound. The average Ln—O distance decreases from 2.55 Å (Ln = La) to 2.40 Å in the holmium compound. Coordinated water molecules all interact with iodides ions, not water molecules. For the heavier lanthanides, LnX$_3$·10H$_2$O (Ln = Er-Lu) contain square antiprismatic [Ln(OH$_2$)$_8$]$^{3+}$ ions; similarly here the coordinated water molecules tend to have iodides for nearest neighbors, rather than hydrogen bonding to the lattice waters. No iodide ions are coordinated in either phase, unlike the lanthanide chlorides and bromides. The structures of the anhydrous acetates have been discussed in a series of papers. Reaction of the oxide or carbonate with acetic acid affords hydrated salts [Ln(CH$_3$COO)$_3$.nH$_2$O] (a number of hydrates can be isolated, but usually Ln = La–Ce, n = 1.5; Ln = Pr–Sm, n = 3; Ln = Gd–Lu, Y, n = 4) which can decompose to basic salts on attempted dehydration. Anhydrous lanthanum acetate has been synthesized from reaction of La$_2$O$_3$ and CH$_3$COONH$_4$ in a melt (under different conditions (NH$_4$)$_3$[La(CH$_3$COO)$_6$] is obtained).[390] It contains 10-coordinate lanthanum, involving both chelating and bridging acetates, the latter having one oxygen bound to two different lanthanum ions and the second oxygen just bound to one. La—O distances lie in the range 2.474(3) Å to 2.794(3) Å, with an average of 2.615 Å. Ce(CH$_3$COO)$_3$ is isostructural. Both nine- and 10-coordinate praseodymium are found in Pr(CH$_3$COO)$_3$, synthesized by dehydration of Pr(CH$_3$COO)$_3$·1.5H$_2$O at 180 °C.[391] For the two types of nine-coordinate Pr sites, the average Pr—O distances are 2.535 Å and 2.556 Å, whilst for the 10-coordinate site the average Pr—O distance is 2.611 Å. Like La(CH$_3$COO)$_3$ this has a three-dimensional network structure, whereas the later metals adopt chain structures in Ln(CH$_3$COO)$_3$. For later lanthanides, anhydrous acetates have been synthesized by crystallization from diluted acetic acid solutions of their oxides and caesium acetate at 120 °C. Holmium acetate adopts a structure shared with other Ln(CH$_3$COO)$_3$ (Ln = Sm–Er, Y). Here holmium occupies two slightly different eight-coordinate sites, with average Ho—O distances of 2.370 Å and 2.381 Å. Ln(CH$_3$COO)$_3$ (Ln = Tm–Lu) have the structure exemplified by Lu(CH$_3$COO)$_3$, in which Lu is seven-coordinate (average Lu–O 3.275 Å). On heating, both these structures change to the six-coordinate Sc(CH$_3$COO)$_3$ structure, this drop in coordination number being accompanied by an acetate group switching to a symmetrical bridging mode.[392] A number of different hydrated acetates have been characterized. The hydrated acetates, unlike the acetates of transition metals like Cr^{III}, Fe^{III}, and Ru^{III}, do not adopt oxo-centred structures with M$_3$O cores, presumably owing to the inability of the lanthanides to form π-bonds. A study[393] of the "maximally hydrated" acetates (obtained by crystallization of neutral aqueous solutions at room temperature) has identified three main series, though it certainly seems that for any given lanthanide there may be two or more phases with nearly equal stabilities under ambient conditions. The sesquihydrates [Ln(CH$_3$COO)$_3$·1½ H$_2$O] (Ln = La-Pr) have structures with acetate bridged chains crosslinked by further acetate bridges; the lanthanides being nine and 10-coordinated by bridging and chelating acetates. The monohydrates [Ln(CH$_3$COO)$_3$·H$_2$O] (Ln = Ce–Nd) have one-dimensional polymeric structures with acetate bridges, the lanthanides being nine-coordinate in this case, whilst the tetrahydrates [Ln(CH$_3$COO)$_3$·4H$_2$O] (Ln = Sm-Lu) are acetate-bridged dimers, the lanthanides again being nine-coordinate. Some of the acetate bridges in these compounds are asymmetric, again featuring one oxygen being bound to two lanthanide ions and the other oxygen bound to one. In addition to these compounds, crystallization of europium acetate from acidic solution leads to the isolation of dimeric species [Eu$_2$(CH$_3$COO)$_6$(H$_2$O)$_4$]·2CH$_3$COOH and [Eu$_2$(CH$_3$COO)$_6$(H$_2$O)$_2$(CH$_3$COOH)$_2$], the latter being related to the former by replacement of two coordinated water molecules by two monodentate acetic acids.

[Sm(CH$_3$COO)$_3$·3H$_2$O]·CH$_3$COOH has nine-coordinate samarium.[394] Hydrated praseodymium propionate, Pr(C$_3$H$_7$COO)$_3$·3H$_2$O, contains two distinctly different praseodymium sites. Each praseodymium is coordinated by four bidentate bridging carboxylates; additionally one is bound to three water molecules, the other is additionally bound to two bidentate propionates.[395] Sm(OAc)$_3$·AcOH has nine-coordinate samarium.[396] The structural and thermal behavior of a series of hydrated neodymium alkanoates have been studied.[397] The butanoate $Nd(C_3H_7COO)_3(H_2O)$ has a zigzag chain structure containing nine-coordinate neodymium in capped square antiprismatic coordination, there being two different neodymium sites. One neodymium is bound to four bridging tridentate carboxylates and one bridging bidentate group, as well as to two waters; the other neodymium is coordinated to four bridging tridentate carboxylates, one bridging bidentate carboxylate, and one chelating carboxylate. The higher homologues have a similar structure. They display a thermotropic mesophase, which has been identified as a smectic A phase.

Some series of hydrated trifluoro- and trichloroacetates have been studied too.[398] $[Ln(CF_3COO)_3 \cdot 3H_2O)]$ (Ln = La, Ce) are two-dimensional polymers containing eight-coordinate lanthanides, whilst $[Ln(CF_3COO)_3 \cdot 3H_2O)]$ (Ln = Pr–Lu) have seven-coordinate lanthanides in dimeric units with four bridging carboxylates. $[La(CCl_3COO)_3 \cdot 5H_2O]$ and $[Ce(CCl_3COO)_3 . 3H_2O]$ are linear one-dimensional polymers with nine- and eight-coordinate lanthanides respectively. $[Ln(CCl_3COO)_3 \cdot 2H_2O)]$ (Ln = Pr–Lu) are also linear polymers with a bridging water molecule rather than just carboxylate bridges. In these compounds the carboxylate groups act as bridging bidentate ligands or as terminal monodentate groups. Ethanol adducts of the trichloroacetates have also been characterized. $[Ln(CCl_3COO)_3 \cdot 3EtOH]$ (Ln = La, Yb) are dimers with eight-coordinate lanthanides; four carboxylates bridge whilst the other four coordination positions of the lanthanide are occupied by three ethanol molecules and one carboxylate.[398] A number of lanthanide salts of unsaturated acids, including acrylic, methacrylic, maleic, and fumaric acids, have been synthesized. Anhydrous europium methacrylate, $[Eu(H_2C=C(Me)CO_2)_3]$, has a chain structure in which Eu is eight-coordinate, one carboxylate acting as a bidentate bridging ligand and the other two also bridging, but with one oxygen bound to two europium, the chelate-bridging mode familiar through a number of lanthanide carboxylate structures. Both this compound and the acrylates $[Ln(H_2C=CHCO_2)_3]$ (Ln = Eu, Tb) undergo radical-induced polymerization.[399] $[La_2(O_2CC\equiv CH)_6(H_2O)_4] \cdot 2H_2O$ undergoes solid-state polymerization when exposed to ^{60}Co γ-rays.[400] Fast-ion bombardment of lanthanide acetate and malonate salts[401] leads to clustering as a result of ion–molecule reactions to give in the general case the ion $[(LnO)_x(RCO_2)_yO_z]^+$ with x reaching four, y reaching three, and z reaching four. Binary clusters $[(LnO)_xO_y]^+$ are also found, with x reaching high values. Some combinations of x, y, and z are associated with markedly high levels of stability, e.g., in $[(PrO)_2(CH_3CO_2)O]^+$, $[(PrO)_4(CH_3CO_2)O_2]^+$, $[(TbO)_4(CH_3CO_2)]^+$, $[Ho(CH_3CO_2)]^+$, and $[(LaO)(malonate)O_4]^+$. Other carboxylates studied include $[Ce(O_2CR)_3(H_2O)_2]$ (RCO_2 = laurate, stearate, octate); they can be used to make photosensitive polythene films;[402] $[Gd(CF_3CO_2)_3 \cdot 3H_2O]$ is a dimer with bridging trifluoroacetates.[403] Praesodymium is also nine-coordinate in $[Pr(O_2CR)_3(H_2O)]_\infty$ (R = 2,6-difluorobenzoate),[404] whilst $Pr(O_2CH_2CH_2CH_3)_3 . 3H_2O$ has an infinite chain structure in which pairs of praesodymiums are bridged by four tridentate bridging propionates,[405] and $[Dy(C_5H_4N-2-CO_2)_3(H_2O)_2]$ has eight-coordinate dysprosium.[406] Among a variety of other salts to be studied are the dialkylcarbamates $[Ln_4(O_2CNPr^i_2)_{12}]$ (Ln = Nd, Gd, Ho, Yb) with a steady contraction in Ln—O distances, averaging 14%, across the lanthanide series.[407] A variety of sulfate structures exists, where both water and sulfates are bound to the metal, usually with nine-coordination.[408] Erbium is six-coordinate in $Er_2(SO_4)_3$, seven-coordinate in $Er(SO_4)(HSO_4)$ and eight-coordinate in $Er(HSO_4)_3$. EXAFS studies of aqua complexes and polyaminepolycarboxylate complexes have also been used to obtain coordination numbers.[409] Thiocyanates contain $Ln(NCS)_3(H_2O)_6$ (Ln = La-Dy) and $Ln(NCS)_3(H_2O)_5$ (Ln = Sm–Eu) molecules. Lanthanum thiocyanate reacts with hexamethylene tetramine forming $[La(NCS)_3 \cdot 2[N_4(CH_2)_6] \cdot 9H_2O]$, which contains $[La(NCS)_2 \cdot (H_2O)_7]^+$ ions.[410] Structural and magnetic properties are reported for $[Me_2NH_2][MCl_4(H_2O)_2]$ (M = Nd, Pr), which contain edge-connected $[MCl_{4/2}Cl_2(H_2O)_2]$ trigondodecahedra.[411,412]

$Pr(ClO_3)_3 \cdot 2H_2O$ has praseodymium coordinated by seven different chlorates and two water molecules.[413] Lanthanum is nine-coordinate in $Rb[La(OAc)_4]$ in which each La is coordinated by six different acetate groups in a chain structure.[414] Thulium nitrilotriacetate, $[Tm(NTA)(H_2O)_2] \cdot 2H_2O$, has a ladder-like structure.[415]

Mixing aqueous solutions of $PrCl_3$ and $Pr(NO_3)_3$ gives green crystals of $PrCl_2(NO_3).5H_2O$, which contains $[PrCl_2(OH_2)_6]^+$ and $[PrCl_2(NO_3)_2(OH_2)_4]^-$.[416]

$Ce(NO_3)_5(H_3O)_2(H_2O)$ has[417] 12-coordinate cerium (six bidentate nitrates). whilst in polymeric $[La(O_2P(OMe)_2)_3]$, lanthanum is octahedrally coordinated[418] by six oxygens from six bridging ligands. The role of the counter ion in structure is indicated[419] by $Rb_2[Y(NO_3)_5]$ (eight-coordinate), as the (NO) salt is 10-coordinate.

Carbonate complexes have been increasingly studied because of their importance in mobilizing the actinides in the environment; lanthanide complexes are nonradioactive models for the later actinides. The first mononuclear lanthanide carbonate complex[420] has been structurally characterized in the form of $[N(CH_2)_3]_5[Nd(CO_3)_4]$. In contrast, the anions in $[Co(NH_3)_6][Sm(CO_3)_3-(H_2O)]\cdot 4H_2O$ are linked in a zigzag chain structure with $\mu\text{-}\eta^2\text{-}\eta^1$ carbonate bridges affording nine-coordinate samarium.[421] Similar bridges are found in the one-dimensional chains[422] in $[Co(NH_3)_6]_6[K_2(H_2O)_{10}][Nd_2(CO_3)_8]_2\cdot 20H_2O$, formed on mixing solutions of $[Co(NH_3)_6]Cl_3$, $Nd(NO_3)_3$, and K_2CO_3.

3.2.2.5.2 *Phosphine and arsine oxides, and other neutral monodentate donors*

Much more is now known about the structures of these compounds. Lanthanide triflates reacts with Ph_3PO in ethanol forming $Ln(OTf)_3(Ph_3PO)_4$ (Ln = La, Nd, Er, Lu). All have the structure $[Ln(OTf)_2(Ph_3PO)_4]OTf$; the erbium and lutetium compounds have octahedral coordination, like the Sc analogue, but the complexes of the larger La and Nd have seven-coordination with one triflate being bidentate (from IR evidence, the transition between six and seven-coordination seems to occur at Sm. The Nd—O (PPh_3) distances in $[Nd(OTf)_2(Ph_3PO)_4]OTf$ are in the range 2.304(2)–2.339(2) Å, with Nd—O(Tf) 2.408(2) Å (monodentate) and 2.6553–2.624(2) Å (bidentate); in $[Lu(OTf)_2(Ph_3PO)_4]OTf$, Lu—O (PPh_3) distances lie in the range 2.156(5)–2.199(5) Å, with Lu—O(Tf) 3.202(6) Å and 3.232(5). In a study of complexes of yttrium halides with phosphine oxides,[423] it was found that $YF_3\cdot 1/2\,H_2O$ shows no signs of reaction with these ligands, but a wide range of complexes have been isolated with the other halides. The complexes actually isolated are $[YX_2(Ph_3PO)_4]Z$ (X = Cl, Br, I; Z = X or PF_6); $[YX_3(Ph_2MePO)_3]$; $[YCl_2(Ph_2MePO)_4]PF_6$; $[YCl(Ph_3PO)_5](SbCl_6)_2$; $[Y(Me_3PO)_6]X_3$ (X = Cl, Br, I); $[YX_2(Ph_3AsO)_4]X$ (X = Cl, Br, I_5); $[Y(Me_3AsO)_6]Cl_3$; and $[YCl_2(L\text{-}L)_4]Cl$ (L-L = $o\text{-}C_6H_4(P(=O)Ph_2)_2$ or $Ph_2P(O)CH_2P(O)Ph_2$) have all been synthesized and characterized by multinuclear NMR. Most of these compounds were obtained by reaction of the hydrated halide with the tertiary phosphine or arsine oxide in ethanol, acetone, or CH_2Cl_2, using a range of stoichiometries. Reaction of YCl_3 with Ph_2MePO in ethanol or acetone gives $[YCl_3(Ph_2MePO)_3]$ irrespective of ratio, but in the presence of NH_4PF_6, $[YCl_2(Ph_2McPO)_4]PF_6$ was isolated. In the presence of the chloride ion abstractor $SbCl_5$, $[YCl_2(Ph_3PO)_4]Cl$ forms $[YCl(Ph_3PO)_5](SbCl_6)_2$. Structures have been reported for $[YCl_2(Ph_3PO)_4]Cl\cdot 2\frac{1}{2}EtOH\cdot H_2O$, $[YBr_2(Ph_3PO)_4]PF_6\cdot Et_2O$, and $[Y(Me_3PO)_6]Br_3$. $[YCl_2(Ph_3PO)_4]Cl\cdot 2\frac{1}{2}EtOH\cdot H_2O$ contains a *trans*-octahedral cation with Y—Cl = 2.613(2) Å and 2.625(1) Å and Y—O distances in the range 3.223(3)–3.233(3) Å. A similar geometry is found in the *trans*-$[YBr_2(Ph_3PO)_4]$ cation, with Y—Br = 2.775(1) Å and 2.794(1) Å and Y—O distances in the range 3.220(4)–3.233(5) Å. In $[Y(Mc_3PO)_6]Br_3$, coordination approximates to the octahedron, with Y—O distances between 3.214(5) Å and 3.233(5) Å and O—Y—O angles in the range 89.6–91.4°. Octahedral coordination of yttrium seems to be the rule for all these compounds. $CeCl_3\cdot 6H_2O$ reacts with Me_3PO in MeOH forming $[Ce(Me_3PO)_4Cl_3]\cdot 4H_2O$; on slow crystallization of a $MeNO_2$ solution in air, crystals of $[Ce(Me_3PO)_4(H_2O)_4]Cl_3\cdot 3H_2O$ result. The coordination sphere can be described as a triangulated dodecahedron; this can be described as two interpenetrating tetrahedra with a flattened tetrahedron formed by the four phosphine oxides and an elongated tetrahedron formed by the four coordinated waters.[424] $[Ln(Ph_3PO)_5Cl](FeCl_4)_2$ (Ln = La-Nd, Sm–Er, Y) have been made; it is possibly surprising to have five of these round a lanthanide.[425] A number of yttrium nitrate complexes, $[Y(NO_3)_3(L)_3]$ (L = Ph_3PO, Ph_2MePO, Me_3PO), $[Y(NO_3)_3(L)_2(EtOH)]$ (L = Ph_3PO, Ph_2MePO), $[Y(NO_3)_3(Me_3PO)_2(H_2O)]$, and $[Y(NO_3)_2(Ph_3PO)_4](NO_3)$ have been synthesized from the reactions of yttrium nitrate and the ligand, and their ^{89}Y NMR spectra reported.[426] Whilst only one complex is generally isolated from a particular solution, ^{31}P-NMR studies show that a mixture of species is frequently present, and complexes with different stoichiometries can be isolated by altering reaction conditions. Thus reaction of $Y(NO_3)_3\cdot 6H_2O$ with one or two moles of Ph_3PO in boiling ethanol gives $[Y(NO_3)_3(Ph_3PO)_2(EtOH)]$ whilst four moles of triphenylphosphine oxide gives $Y(NO_3)_3(Ph_3PO)_3$, and reaction with six moles of Ph_3PO in cold ethanol affords $Y(NO_3)_3(Ph_3PO)_4$, the latter being $[Y(NO_3)_2(Ph_3PO)_4]NO_3$. Structures have been determined for $[Y(NO_3)_3(L)_3]$ (L = Ph_3PO, Ph_2MePO, Me_3PO) and $[Y(NO_3)_3(Ph_3PO)_2(EtOH)]$. All of these have nine-coordinate yttrium with bidentate nitrates, but if the nitrate groups are conceived of as occupying one coordination position, the geometry can be described as *mer*-octahedral for the Ph_3PO and Ph_2MePO complexes, and *fac*- for the Me_3PO complex. In $[Y(NO_3)_3(Ph_3PO)_3]$, Y—O (P) distances are 3.269, 3.283, and 3.284(5) Å and Y—O (N) distances lie in the range 2.403(5)–2.506(5) Å whilst in $[Y(NO_3)_3(Me_3PO)_3]$ they are similar, Y—O (P) distances being

3.262(7), 3.279(7), and 3.281(7) Å and Y—OM(N) being 2.441(7)–2.520(7) Å.[426] A detailed study has been made of the reactions of lanthanide nitrates with a large excess of Ph_3PO in acetone.[427] Early lanthanides (La—Nd) form $[Ln(NO_3)_3(Ph_3PO)_4]$, which in the solid state have the structure $[Ln(\eta^2\text{-}NO_3)_2(\eta^1\text{-}NO_3)(Ph_3PO)_4]$; in solution ($Me_2CO$ or CH_2Cl_2) they dissociate into $[Ln(NO_3)_3(Ph_3PO)_3]$ and Ph_3PO and reaction in ethanol gives only $[Ln(NO_3)_3(Ph_3PO)_3]$. The crystal structure of $[Ln(NO_3)_3(Ph_3PO)_4]$ shows that lanthanum is nine-coordinate with two bidentate nitrates (La—O = 2.656(8) Å and 2.695(8) Å) and one monodentate nitrate (La—O = 2.585(17) Å); all four phosphine oxides are coordinated with La—O 2.449(8) Å and 2.488(7) Å. The Ce, Pr, and Nd complexes appear to have the same structure. The corresponding reaction in ethanol affords $[Ln(NO_3)_3(Ph_3PO)_3]$, of which series the structure of the lanthanum complex has been determined. $[La(NO_3)_3(Ph_3PO)_3]\cdot CHCl_3\cdot EtOH$ has three bidentate nitrate groups, adopting the *mer*-pseudooctahedral structure (considering nitrate to occupy one coordination position). La—O(P) distances lie in the range 2.373 Å to 2.427 Å and La—O(N) distances are 3.283–2.681 Å. The slightly shorter La–O(P) distances in the tetrakis complex suggest greater steric demand by the triphenylphosphine oxide ligand. For the nitrates of Sm–Gd, reaction with Ph_3PO in either ethanol or acetone results in only $[Ln(NO_3)_3(Ph_3PO)_3]2Me_2CO$; 4:1 complexes do not seem isolable, possibly because of solubility factors. The corresponding reaction with later lanthanides (Tb–Lu) in cold ethanol gives $[Ln(NO_3)_2(Ph_3PO)_4]NO_3$, whilst $[Ln(NO_3)_3(Ph_3PO)_3]\cdot 2\text{-}Me_2CO$ are readily obtained from propanone. In the case of lutetium, the structure of $[Lu(NO_3)_2(Ph_3PO)_4]NO_3$ has been determined. The symmetrical bidentate nitrates are mutually *trans*- with Lu—O 2.429(10) Å and 2.406(9) Å; Lu—O(P) distances are 2.181(9) Å and 3.220(9) Å. In other studies, structures have been determined for a number of $[Ln(NO_3)_3(Ph_3PO)_2(EtOH)]$.[428] The cerium complex has Ce–O(P) distances of 2.369(2) Å and 2.385(2) Å; Ce–O (nitrate) are 2.549(3), 2.563(3), 2.572(3), 2.575(3), 2.580(3), and 2.596(3) Å; Ce–O (EtOH) is 2.515 Å. Other structures have been reported for the neodymium,[429] samarium,[430] and europium[431] analogues. As noted earlier a number of nine-coordinate lanthanides are also found in $[Ln(NO_3)_3(Ph_3PO)_3]$. An alternative synthesis of $[Ce(NO_3)_3(Ph_3PO)_3]$ is from the reaction of $(NH_4)_2[Ce(NO_3)_6]$ and Ph_3PO in propanone which tends to lead to reduction.[432] The Ce–O (NO_3) distances range from 2.58(1) Å to 2.63(1) Å whilst Ce–O ($OPPh_3$) distances are shorter at 2.39(1)–2.42(1) Å. Similarly $[Sm(NO_3)_3(Ph_3PO)_3]\cdot 2$acetone and $[Sm(NO_3)_3(Ph_3QO)_2(EtOH)]\cdot$acetone (Q = P, As) were all shown to have nine-coordinate samarium.[433]

A similar study of reactions of the lanthanide nitrates with the slightly less bulky diphenylmethylphosphine oxide has been made.[434] Reaction in a 1:2.5 molar ratio in acetone gives $[Ln(\eta^2\text{-}NO_3)_3(Ph_2MePO)_3]$ for all lanthanides, which are monomers. If the bidentate nitrates are thought of as occupying one coordination position, the structure of nine-coordinate $[La(NO_3)_3(Ph_2MePO)_3]$ corresponds to a *fac*-octahedron, in contrast to the pseudomeridional structure of the Ph_3PO analogue (again tending to suggest the lessened steric effects associated with Ph_2MePO). La—O(P) distances are 2.407, 2.418, and 2.436 Å, whilst La—O(N) bend lengths are in the range 2.584 Å to 2.641 Å; if it is assumed that the other lanthanides form complexes with similar structures, this is a rare case of the same structure type being adopted across the whole lanthanide series, despite the reduction in size of the Ln^{3+} ion. On adding small amounts of Ph_2MePO to solutions of $[Ln(NO_3)_3(Ph_2MePO)_3]$ (Ln = La, Ce), an additional NMR resonance is detected but there is no increase in conductance, indicating a 4:1 species is obtained (at very high Ph_2MePO concentrations, the conductivity increases and another new resonance appears, possibly due to the nonisolable $[Ln(NO_3)_2(Ph_2MePO)_5]^+$). A 4:1 complex has been isolated from solution for lanthanum only and the structure of $[La(\eta^2\text{-}NO_3)_3(Ph_2MePO)_4]$ determined. It is ten-coordinate, with all nitrates bidentate, in contrast to $[Ln(NO_3)_3(Ph_3PO)_4]$. The La–O bond lengths are, as expected, longer than those in nine-coordinate $[La(NO_3)_3(Ph_2MePO)_3]$, with La—O(P) = 2.462 Å and 2.513 Å and La—O(N) lengths in the range 2.649–2.708 Å. $[Ln(NO_3)_3(Ph_2MePO)_3]$ (Ln = Pr-Tb) are unaffected in solution by excess ligand but others (Ln = Dy-Lu) tend to dissociate into $[Ln(NO_3)_2(Ph_2MePO)_4]^+$; compounds $[Ln(\eta^2\text{-}NO_3)_2(Ph_2MePO)_4]PF_6$ have been isolated for these metals. The structure of the ytterbium compound shows that the eight-coordinate cation contains a rough YbO_4 square involving the four phosphine oxide ligands (Yb—O(P) = 2.186–3.222 Å) with bidentate nitrates attached to ytterbium above and below the plane of the square (Yb—O (N) = 2.410–2.452 Å).

When Ph_3AsO was reacted with lanthanum nitrate, reaction in acetone solution led to $[La(NO_3)_2(Ph_3AsO)_4]NO_3$ and $[La(NO_3)_3(Ph_3AsO)_3]$, depending upon the stoichiometry of the mixture. From ethanolic solution, $[La(NO_3)_3(Ph_3AsO)_2(EtOH)]$ was obtained. $[La(NO_3)_2(Ph_3AsO)_4]NO_3$ has eight-coordinate lanthanum, with bidentate nitrates *trans*- to each other on opposite sides of the YbO_4 unit formed by the lanthanum and four-coordinated arsine oxides, similar to the yttrium analogue. La—O(As) distances are 2.340–2.361 Å whilst La—O(N) distances are 2.635–2.656 Å.

[La(NO$_3$)$_3$(Ph$_3$AsO)$_2$(EtOH)] has the pseudomeridional coordination described for several [Ln(NO$_3$)$_3$(Ph$_3$PO)$_2$(EtOH)] species and also known for the Sm analogue. La–O(N) distances fall in the range 2.581–2.664 Å and La–O(As) are 2.324–2.347 Å and La–O(EtOH) is 2.552 Å. Reaction of lanthanum nitrate with Me$_3$AsO in acetone yields [La(Me$_3$AsO)$_6$](NO$_3$)$_3$, believed to be octahedral like the Sc and Y analogues, and decomposing to La(Me$_3$AsO)$_4$(NO$_3$)$_3$, which has two bidentate nitrates and one monodentate one, like the Ph$_3$PO analogue. Reaction in ethanol affords [La(NO$_3$)$_3$(Me$_3$AsO)$_2$(H$_2$O)].[435] [Eu(NO$_3$)$_3$(Ph$_3$AsO)$_3$]·4H$_2$O again has the pseudomeridional structure.[436] The structure of the mixed-metal compound [La(NO$_3$)$_2$(Ph$_3$PO)$_4$] [Ni(C$_4$N$_2$S$_2$)$_2$].2MeOH has been reported.[437]

Nd$_2$(S$_2$O$_6$)$_3$.14H$_2$O has each neodymium bound to six water molecules and to three oxygens from different dithionates; in Nd$_2$(S$_2$O$_6$)$_3$(Ph$_3$PO)$_4$.8H$_2$O, each neodymium is eight-coordinate, bound to two phosphine oxides, four water molecules, and two dithionates (one monodentate, one a bridging ligand).[438] Mass spectra are reported[439] of the dithionates [Ln$_2$(S$_2$O$_3$)$_3$] and their Ph$_3$PO complexes, together with the structure of [Pr$_2$(S$_2$O$_3$)$_3$(Ph$_3$PO)$_6$(H$_2$O)$_6$].

An unusual route has been described to synthesize hexamethylphosphoramide complexes of lanthanum. A mixture of lanthanum metal, NH$_4$NCS, and HMPA in toluene reacts when subjected to ultrasonication followed by heating, forming monomeric La(NCS)$_3$(HMPA)$_4$. Using the appropriate ammonium salt, Y(NCS)$_3$(HMPA)$_3$, LaBr$_3$(HMPA)$_4$, and La(NO$_3$)$_3$(HMPA)$_3$ were similarly obtained; La(NO$_3$)$_3$(HMPA)$_3$ has a nine-coordinate structure with bidentate nitrates.[440] The complexes *mer*-[LnCl$_3$(HMPA)$_3$] have been established for many years, a recent example being *mer*-[YbCl$_3$(HMPA)$_3$][441] Now the synthesis of isomorphous *fac*-[LnCl$_3$(HMPA)$_3$] (Ln = La, Pr, Nd, Sm, Eu, Gd; full structure for Sm) has been reported[442] to set alongside the *mer*-isomers; the isomerization was followed in solution by NMR, and is believed to occur by an associative mechanism. [Ln(HMPA)$_6$](BrO$_4$)$_3$ are isomorphous with [Ln(HMPA)$_6$]X$_3$ (Ln = La–Lu; X = ClO$_4$, ReO$_4$).[443,444] Applications of a solution of SmI$_2$ in HMPA as a one-electron reductant in organic syntheses have doubtless prompted studies of samarium complexes of HMPA. [SmI$_3$(HMPA)$_4$], prepared from Sm and CH$_2$I$_2$ in HMPA/THF, scavenges traces of water forming [Sm(HMPA)$_2$(H$_2$O)$_5$]I$_3$·2 HMPA and [Sm(H$_2$O)$_4$(HMPA)$_3$]I$_3$, both with pentagonal bipyramidally coordinated samarium[445]. A similar compound, [Sm(H$_2$O)$_3$(HMPA)$_4$]I$_3$ has been isolated as a by-product from a reaction mixture.[446]

Other samarium complexes of hexamethylphosphoramide to have their structures reported are [Sm(H$_2$O)$_5$(HMPA)$_2$]I$_3$(HMPA)$_2$, [Sm(H$_2$O)$_3$(HMPA)$_4$]I$_3$, [SmCl(H$_2$O)$_4$(HMPA)$_2$]Cl$_2$.THF, [SmCl(HMPA)$_5$](BPh$_4$)$_2$, [Sm(O$_3$SCF$_3$)$_2$(HMPA)$_4$](O$_3$SCF$_3$), [Sm(O$_3$SCF$_3$)$_3$(H$_2$O)(HMPA)$_3$], and [Sm(hmpa)$_3$(η^2-NO$_3$)$_3$].[447]

Syntheses and structures are also reported for the Sm^{III} complexes [SmBr$_3$(HMPA)$_2$(THF)] and [SmBr$_2$(HMPA)$_4$]Br.THF.[448] In contrast (but in keeping with the lower stability of Tm^{II}) [TmI$_2$(DME)$_3$] reacts with HMPA forming [TmI$_3$(HMPA)$_4$]; this recrystallizes from pyridine as (depending on conditions) [TmI$_2$(HMPA)$_4$]I.5Py or [TmI(Py)(HMPA)$_4$]I$_2$.[449]

Isolated studies have been made before of the lanthanide nitrate complexes of dimethylsulfoxide, but now a single study has been made of the whole series.[450] The earlier metals (La—Sm) form 4:1 complexes whilst smaller metal ions (Eu—Lu, Y) form 3:1 complexes. There is no evidence to support earlier suggestions that both 3:1 and 4:1 species can exist for the same lanthanide (e.g., Gd). Ln(DMSO)$_4$(NO$_3$)$_3$ (Ln = La-Sm) are 10-coordinate in the solid state. Ln—O bond lengths are 2.451–2.488 Å (La—O(DMSO)) and (La—O 2.647–2.738 Å (La—O(NO$_3$)) whereas Sm—O distances are 2.360–2.409 Å (La—O(DMSO)) and Sm—O 2.540–2.749 Å (La—O(NO$_3$)). With the heavier metals nine-coordinate Ln(DMSO)$_3$(NO$_3$)$_3$ (Ln = Eu–Lu, Y) species are formed. In Eu(DMSO)$_3$(NO$_3$)$_3$ Eu—O bond lengths are 2.314–2.352 Å (Eu—O(DMSO)) and Eu—O 2.478–2.541 Å (Eu—O(NO$_3$)) whereas Lu—O distances are 3.215–3.235 Å (Lu—O(DMSO)) and (La—O 2.359–2.475 Å (Lu—O(NO$_3$)). All nitrates are bidentate. The structure of [Y(DMSO)$_3$(NO$_3$)$_3$] shows it to be nine-coordinate, like the Eu, Er, and Lu analogues; the degree of asymmetry in the Y—O (nitrate) bond varies; in one group, Y—O distances are 2.439(6) Å and 2.458(6) Å, whilst in the other two nitrates, distances are 2.445(6) Å and 2.502(7) Å and 2.415(7) Å and 2.469(7) Å. Y—O (DMSO) distances are 3.259(6), 3.276(6), and 2.301(5) Å.[451] Eight DMSO molecules can fit round lanthanum as in Ln(DMSO)$_8$[Cr(NCS)$_6$], with La—O distances in the range 2.46–2.51 Å.[452] The EXAFS spectra of [Ln(DMSO)$_8$](CF$_3$SO$_3$)$_3$ in both the solid state and in DMSO solution are very similar, indicating the same coordination geometry in both; La—O distances deduced are 2.486 Å and 2.504 Å respectively.[358]

Tetrahydrofuran complexes of the lanthanide chlorides have attracted considerable attention. The anhydrous trichlorides are themselves very useful starting materials in the synthesis of compounds such as alkoxides and aryloxides, alkylamides, and organometallic compounds in general; however, they are difficult to prepare from the hydrated chlorides and are also difficult to prepare by dehydration of the hydrated chlorides.[453-455] Some routes such as dehydration of the hydrated halides with $SOCl_2$ or triethylorthophosphate have given hydrated complexes $[LnCl_3(H_2O)(THF)]_n$. The THF complexes therefore have considerable utility as synthons. A range of stoichiometries is known. Reaction of the lanthanide metals with $HgCl_2$ in THF has been employed but presents problems in separating excess metal.[456,457] One synthesis reported is from reaction of the metals and Me_3SiCl in MeOH.[458] Perhaps the best route is sonication of lanthanide powders and C_2Cl_6 in THF.[459] The formulae and structures of these complexes present considerable diversity. Five different stoichiometries of $LnCl_3(THF)_x$ (x is, for example, 2, 2.5, 3, 3.5, 4) and six different structure types have been identified in these complexes. The compound obtained not only depends upon the lanthanide and the reaction stoichiometry but upon reaction conditions. The pattern across the series reflects an overall decrease in coordination number from eight (La) to six (Lu). Lanthanum is unique in forming $[LaCl_3(THF)_2]$ which has a single-stranded polymer—La(μ-Cl)$_3$(THF)$_2$La(μ-Cl)$_3$(THF)$_2$La— with *cis*-THF molecules and square antiprismatic eight-coordination of lanthanum.[459] Bridging La—Cl distances are necessarily long, at 2.870(3)–2.968(3) Å, and La—O distances 2.549(7)–2.595(7) Å. $[LnCl_3(THF)_2]$(Ce–Nd) are different, although again polymeric, in this case seven-coordinate $\cdots$ LaCl(thf)$_2$(μ-Cl)$_2$ LnCl(thf)$_2$(μ-Cl)$_2$...... The compound $[PrCl_3(THF)_2]$ has Pr—Cl (terminal) 2.633(1) Å and 2.808(1)–2.850(2) Å for the bridging chlorines; Pr—O are 2.472(3)–2.498(4) Å. A third type, found for Nd–Gd, are monomeric seven-coordinate $[LnCl_3(THF)_4]$, whilst Gd–Tm form a nominal $[LnCl_3(THF)_{3.5}]$, which in fact has an ionic structure $[LnCl_2(THF)_5]^+[LnCl_4(THF)_2]^-$, containing a seven-coordinate cation and octahedrally coordinated six-coordinate anion, both with *trans*- geometries. In the cation of $[ErCl_2(THF)_5]^+[ErCl_4(THF)_2]^-$, Er—Cl is 2.554(3) Å whilst Er—O distances range from 2.353(6) Å to 2.402(9) Å; in the anion, Er—Cl distances are 2.585(3) Å to 2.594(3) Å and Er—O 3.294(7) Å. Ytterbium forms a dimeric $[Cl_2(THF)_2Yb(\mu$-Cl)$_2Yb(THF)_2Cl_2]$, whilst both ytterbium and lutetium form a monomeric octahedral $[LnCl_3(THF)_3]$ (Ln = Yb, Lu) long familiar with scandium. In $[YbCl_3(THF)_3]$, Yb—Cl distances are 2.513(4)–2.533(3) Å and Yb—O are 3.254–2.337(8) Å. Structures have been reported for many individual compounds and far-IR spectra of the complexes have been correlated with structural type.[459]

An independent report of the structure of $ErCl_3(THF)_{3.5}$ has appeared,[460] showing it to be the expected $[ErCl_2(THF)_5]^+[ErCl_4(thf)_2]^-$; the structure of $[EuCl_3(THF)_4]$ has also been determined again.[461] $[YCl_3(THF)_{3.5}]$ is confirmed to be $[trans$-YCl$_2$(THF)$_5][trans$-YCl$_4$(THF)$_2]$ whilst $[YCl_3(THF)_2]$ has a chain structure with double chlorine bridges, having pentagonal bipyramidal coordination.[462] In another important paper reporting the structures of a number of complexes of THF and related ligands; $[PrCl(\mu$-Cl)$_2$(THF)$_2]_n$, $[Nd(\mu$-Cl)$_3$(THF)(H$_2$O)]$_n$ and $[GdCl_3(THF)_4]$ were all obtained from the dehydration of the hydrated chloride with thionyl chloride; their structures were reported and patterns in the structures in the series $[LnCl_3(THF)_n]$ (n = 2, 3, 3.5, and 4) discussed.[463] The chain structure of $[NdCl_3(THF)_2]$ has been examined[464] and $[DyCl_3(THF)_{3.5}]$ has been shown[465] to be $[DyCl_2(THF)_5]^+[DyCl_4(THF)_2]^-$. $LaCl_3(THF)(H_2O)$ is a polymer with eight-coordinate lanthanum, $[La(\mu$-Cl)$_3$(THF)(H$_2$O)]$_n$, isostructural with the Ce and Nd analogues.[466] Sometimes the structures of other ether complexes have been determined. Thus the structures of both $[DyCl_3(DME)_2]$ and $[DyCl_2(THF)_5]^+[DyCl_4(THF)_2]^-$ have been reported.[467] Although most work has been concentrated on the chlorides, reports of other THF complexes have appeared. Reaction of lanthanum metal with CH_2X_2 (X = Br, I) under ultrasound conditions in THF affords $LaX_3(THF)_4$; recrystallization of $[LaBr_3(THF)_4]$ from 1,2-dimethoxyethane (DME) or bis(2-methoxyethyl)ether (diglyme) affords dimeric $[LaBr_2(\mu$-Br)(DME)$_2]_2$ and $[LaBr_2(diglyme)_2]^+[LaBr_4(diglyme)]^-$. Lanthanides react with hexachloroethane in DME forming $[LnCl_3(DME)_2]$ (Ln = La, Nd, Er, Yb); similar reaction in MeCN affords $[YbCl_2(MeCN)_5]_2^+[YbCl_3(MeCN)(\mu$-Cl)$_2YbCl_3(MeCN)]$. Yb reacts with 1,2-dibromoethane in THF or DME forming $[YbBr_3(THF)_3]$ or $[YbBr_3(DME)_2]$.[468] La and ICH_2CH_2I in THF react on

Table 3 Lanthanide chloride complexes with THF.

	La	Ce	Pr	Nd	Sm	Eu	Gd	Tb	Dy	Ho	Er	Tm	Yb	Lu
						Structure types characterized								
Type	1	2	2	2,3	3	3	3,4	4	4	4	4	4	5,6	6
C.N. of metal	8	7	7	7,7	7	7	7,7+6	7+6	7+6	7+6	7+6	7+6	6,6	6

Description of types: 1. La(THF)$_2$(μ-Cl)$_3$Ln(THF)$_2$(μ-Cl)$_3$La 2. LaCl(THF)$_2$(μ-Cl)$_2$LnCl(THF)$_2$(μ-Cl)$_2$La 3. $[LnCl_3(THF)_4]$ 4. $[LnCl_2(THF)_5] + [LnCl_4(THF)_2]^-$ 5. $[Cl_2(THF)_2Ln(\mu$-Cl)$_2Ln(THF)_2Cl_2]$ 6. $[LnCl_3(THF)_3]$.

exposure in sunlight forming [LaI$_2$(THF)$_5$]I$_3$.[469] Structures of [SmCl$_3$(THF)$_4$], [ErCl$_2$(THF)$_5$]$^+$-[ErCl$_4$(THF)$_2$]$^-$, [ErCl$_3$(DME)$_2$], and [Na(18-crown-6)(THF)$_2$]$^+$[YbBr$_4$(thf)$_2$]$^-$ were also reported in this work. [NdBr$_3$(THF)$_4$] has a pentagonal bipyramidal structure and has been studied in the context of butadiene polymerization.[470] La reacts with C$_2$H$_4$I$_2$ in THF forming [LaI$_3$(THF)$_4$].[471] The structures of [LnI$_2$(THF)$_5$]$^+$[LnI$_4$(thf)$_2$]$^-$ has been reported, where Ln = Sm[472], Yb[473]; the former was produced by O$_2$ oxidation of solutions of SmI$_2$ in THF. New types of THF complex, [Pr(THF)$_4$(NO$_3$)$_3$] and [Ln(THF)$_3$(NO$_3$)$_3$] (Ln = Ho, Yb) as well as the dimethoxyethane complexes [Ln(DME)$_2$(NO$_3$)$_3$] (Ln = Pr, Ho) have been reported.[474]. The structure of [Ce(DME)$_2$(NO$_3$)$_3$] has also been reported.[475]

Thiocyanate complexes have been synthesized by metathesis, from LnCl$_3$ and KNCS in THF, followed by filtering off the KCl. They appear to have the same formula, Ln(NCS)$_3$(THF)$_4$, across the series, but with a significant difference. Thus the ytterbium compound is a monomer, having a pentagonal bipyramidal structure with two axial thiocyanates. There is quite a lot of variation on Yb—O distances; if the Yb—O bond is inserted between two THF ligands, then the Yb—O distance is 2.36–2.40 Å; if THF inserted between a NCS and a THF, then Yb—O is in the range 3.22–3.25 Å, indicating the importance of steric effects. Yb—N distances are 3.22–2.31 Å.[476] For earlier lanthanides, the same stoichiometry Ln(NCS)$_3$(THF)$_4$ obtains, but there is association by Ln···SNC—Ln bridges so lanthanides are in eight-coordinate square antiprismatic coordination. Ln—S interactions are in the region of 3.10 Å (Nd) to 3.26 Å (Er), increasing in length as Ln gets smaller suggesting that the interaction weakens as steric crowding increases. These compounds are obtained for all Ln from Pr to Er.[477] Reactions of LnCl$_3$ with SnCl$_4$ in THF gives [*trans*-LnCl$_2$(THF)$_4$]$^+$[SnCl$_5$THF)]$^-$ (Ln = Ce, Gd, Yb), containing a cation having the familiar pentagonal bipyramidal coordination.[478]

Among complexes of urea derivatives, [Ln(pu)$_8$](OTf)$_3$ (Ln = La–Lu except Pm, Y) have been synthesized and the structures of the Nd–Ho, Yb, and Y compounds determined.[479] [Sm(pu)$_8$](O$_3$SCF$_3$)$_3$ (pu = tetrahydr-2-pyriminidone) has samarium in square antiprismatic eight-coordination.[480] A number of lactam complexes have been studied. Two families of lactam complexes [Ln(ε-caprolactam)$_8$](CF$_3$SO$_3$)$_3$ (Ln = La–Eu) and [Ln(ε-caprolactam)$_7$]-(CF$_3$SO$_3$)$_3$ (Ln = Gd, Tb, Dy, Yb, Lu);[481] [Ln(δ-valerolactam)$_8$](ReO$_4$)$_3$ (Ln = Pr, Nd, Sm, and Eu) and [Ln(δ-valerolactam)$_7$](ReO$_4$)$_3$ (Ln = Tb)[482] whose stoichoiometries appear to reflect the lanthanide contraction have been synthesized. The cation in [Pr(ε-caprolactam)$_8$]-(CF$_3$SO$_3$)$_3$ has slightly distorted dodecahedral geometry whilst in [Eu(ε-caprolactam)$_8$](ReO$_4$)$_3$ it is square antiprismatic. [Sm(NO$_3$)$_3$(N-butylcaprolactam)$_3$] contains samarium in a distorted tricapped trigonal prismatic environment.[483] Among the δ-valerolactam complexes [Ln(δ-valer-olactam)$_8$](ClO$_4$)$_3$ (Ln = Pr–Ho) and [Ln(δ-valerolactam)$_7$](ClO$_4$)$_3$ (Ln = Er–Lu, Y), the neodymium complex has been found to have square antiprismatic eight-coordination.[484] Crystallization of LnCl$_3$ from neat ε-caprolactone and ε-caprolactone/THF mixtures[485,486] affords a variety of complexes, including [MCl(μ-Cl)$_2$(THF)$_2$]$_\infty$ (M = Ce, Nd), [TbCl$_2$(THF)$_5$]$^+$[TbCl$_4$(thf)$_2$]$^-$, YCl$_3$(ε-caprolactone)$_3$, and [M(ε-caprolactone)$_8$]$^{3+}$ [Cl$_3$M(μ-Cl)$_3$MCl$_3$]$^{3-}$ (M = Nd, Sm). Lanthanides react with iodine in propan-2-ol affording pentagonal bipyramidal [LnI$_3$(HOPr^i)$_4$] (Ln = La, Ce, Nd).[487]

3.2.2.5.3 Diketonates

Although there has been no similar development in diketonate chemistry remotely resembling the outbust of shift reagent work in the 1970s, research has continued to progress, with potential applications such as precursors for high-temperature superconductors and chemical vapor deposition agents. Synthetic approaches have become more sophisticated, with direct syntheses from convenient starting materials like the oxides, or the avoidance of water (which can be hard to remove from adducts) are two ideas. Volatile adducts with molecules like glyme have been promising new developments. It has long been recognized that conventional synthetic methods for the acetylacetonates yield hydrates, [Ln(acac)$_3$(H$_2$O)$_n$], from which the water cannot be removed without some decomposition. Reaction of [Y{N(SiMe$_3$)$_2$}$_3$] with Hacac gives hydrocarbon-soluble [Y(acac)$_3$]$_n$; from NMR measurements, $n \sim 4$. Attempted slow crystallization and inadvertent hydrolysis led to [Y$_4$(OH)$_2$(acac)$_{10}$], a molecule with a diamond shaped Y$_4$ core, having μ_4-OH groups above and below the plane, and each acac terminal, affording eight-coordinate Y. Controlled vacuum thermolysis (85 °C) of [Y(acac)$_3$(H$_2$O)$_3$] gives a product that can be crystallized from benzene to form [Y$_4$(OH)$_2$(acac)$_{10}$]·C$_6$H$_6$. It was suggested that hydrogen-bonding between water molecules and acac oxygens in [Y(acac)$_3$(H$_2$O)$_3$] leads to the loss of acacH on

thermolysis.[488] [Y(acac)$_3$] reacts with carboxylate alumoxanes in a *chimie douce* route to YAG
that affords the advantage of greater processability of the pre-ceramic.[489] A solid state synthesis
has been reported[490] for [Pr(acac)$_3$] from anhydrous PrCl$_3$ and Macac (M = Li, Na). [Ce(acac)$_3$-
(H$_2$O)$_2$]·H$_2$O is isomorphous with the Eu and Y analogues, with square antiprismatic coordination of cerium.[491] The coordination geometry in [Ln(acac)$_3$(phen)] (Ln = Ce, Pr) is described as
slightly distorted square antiprismatic.[492] Compounds [Ln(acac)$_3$(Ph$_3$PO)$_3$] have been reported for
most of the lanthanides, but there is no structural information as of 2003. If all three phosphine
oxides were coordinated, these would be stereochemically congested molecules.[493]

Y(acac)$_3$ can be brominated with N-bromosuccinimide at carbon-3 forming Y(3-Bracac)$_3$,
isolated as a monohydrate. NMR spectroscopy shows that in solution two rings chelate via O
and Br, the other by two oxygens; the molecule is fluxional.[494]

The luminescence of [Tb(acac)$_3$(phen)] doped into alumina decreases with increasing oxygen
concentration and has potential as an oxygen sensor.[495] [Yb(acac)$_2$(OAc)(OH$_2$)]$_2$ has dodecahedral eight-coordination of ytterbium.[496] Mass spectra of Ln(acac)$_3$ give evidence that compounds of Eu, Sm, and Yb undergo oxidation state change from Ln^{III} to Ln^{II}. Ce and Gd do
not.[497] [Y(PhCOCHCOPh)$_3$] and its MeCN adduct have been synthesized and evaluated as a
precursor for thin oxide films by chemical beam epitaxy.[498] Reaction of [Gd(tmhd)$_3$]$_2$ with various
polyethers affords a range of monomeric and dimeric glyme complexes such as [Gd(tmhd)$_3$(diglyme)] and [{Gd(tmhd)$_3$(triglyme)] which exhibit good volatility and thermal stability.[499]
(Htmhd = 2,2,6,6-tetramethyl-3,5-heptanedione.) In [{Er(tmhd)$_3$}$_2$tetraglyme] the bridging tetraglyme binds to each erbium through two oxygens, completing distorted square antiprismatic
environments for erbium.[500] Diketonates [Y(tmhd)$_3$·H$_2$O]$_2$, [Y(tmhd)$_3$], and [Y(tmod)$_3$]$_2$
(tmod = 2,2,7-trimethylocatane-3,5-dionate) are also possible CVD compounds.[501] Checked syntheses of [Y(tmhd)$_3$(H$_2$O)] and [Y(tmhd)$_3$] have been published.[502] Structures of the triboluminescent
complexes [Ln(tmhd)$_3$(4-Me$_2$Npy)] have been determined.[503]

[Ln(tmhd)$_3$(Me$_2$phen)] (Ln = La, Eu, Tb, Ho) has two square antiprismatic isomers in the unit
cell. Emissions from both isomers can be discerned in the fluorescence spectrum of the Eu
compound and shows unusually high splitting of the $^5D_0 \rightarrow {}^7F_0$ transition.[504] Eu(tmhd)$_3$(terpy)
is nine-coordinate, again with two slightly different molecules present in the crystal, its luminescence spectrum shows a broad but unresolved $^5D_0 \rightarrow {}^7F_0$ transition, even at 77 K. Average Eu—O
and Eu—N distances are 2.380 Å and 2.645 Å respectively, whereas for the second isomer they are
2.385 Å and 2.663 Å.[505]

The versatility of diketonates like Ln(tmhd)$_3$ (Ln = Eu, Y) is well illustrated by their ability to form
carbene adducts.[506] An eight-coordinate diketonate [Eu(dbm)$_3$(bath)] (bath = bathophenanthroline)
(dbm = dibenzoylmethanide) has found application as a high-efficiency emitter in an electroluminescent device.[507] EXAFS measurements on [Ln(hfac)$_3$(OH$_2$)$_2$] (hfac = hexafluoroacetylacetonate;
Ln = Pr, Eu) indicate a coordination number of about 11, suggesting that some Ln···F interactions
are present.[508] A one-pot synthesis of [Eu(hfac)$_3$·L] (L = terpy, diglyme) from Eu$_2$O$_3$ and Hhfac in the
presence of L has been described[509] as has a one-step route[510] to Ln(diketonate)$_3$ (diketonate is, for
example, acac, tfa, dpm, etc.) via lanthanide methyls prepared *in situ* from LaCl$_3$ and MeLi. La$_2$O$_3$
and Hhfac react together with tetraglyme in hexane forming [La(hfac)$_3$(tetraglyme)], an air stable and
volatile (95 °C, 10^{-4} mm Hg) potential MOCVD precursor.[511] Similar compounds [La(hfac)$_3$(monoglyme)·H$_2$O], [La(hfac)$_3$(diglyme)], and [La(hfac)$_3$(triglyme)] have also been prepared.[512]
(Hhfac = 1,1,1,5,5,5-hexafluoropentane-2,4-dione; monoglyme = Me(OCH$_2$CH$_2$)OMe; diglyme = Me(OCH$_2$CH$_2$)$_2$ OMe; triglyme = Me(OCH$_2$CH$_2$)$_3$OMe; tetraglyme = Me(OCH$_2$CH$_2$)$_4$OMe;
terpy = 2,2′:6′,2″-terpyridyl). Glyme complexes [La(hfac)$_3$·diglyme] and [La(hfac)$_3$·triglyme] are
highly volatile potential MOCVD precursors[513] as are [Gd(hfac)$_3$·monoglyme] and [Gd(hfac)$_3$·di-
glyme].[514] Eu$_2$O$_3$ reacts directly with Hhfac (Hhfa = hexafluoroacetylacetone) in the presence of
tridentate ligands L (L = terpy, diglyme, and bis(2-methoxyethyl)ether) to afford [Eu(hfa)$_3$L]. The
volatile and thermally stable [Eu(hfa)$_3$(diglyme)] has a capped square antiprismatic geometry.[515]
[Y(hfac)$_3$] reacts with monoglyme and diglyme forming monomeric adducts [Y(hfac)$_3$(glyme)]
which are eight and nine-coordinate respectively.[516] In contrast, triglyme and tetraglyme form the
ionic substances [Y(hfac)$_2$(glyme)]$^+$[Y(hfac)$_4$]$^-$. Sublimation of [Y(hfac)$_2$(triglyme)]$^+$[Y(hfac)$_4$]$^-$ in
the presence of "adventitious" water yields the outer-sphere glyme complex [[Y(hfac)$_3$(OH$_2$)$_2$](triglyme)] which has an infinite chain structure. Eight-coordinate [Ln(hfac)$_3$(diglyme)] (Ln = La, Nd,
Sm, Eu, Gd; hfac = hexafluoroacetylacetonate; diglyme = CH$_3$OCH$_2$CH$_2$OCH$_2$CH$_2$OCH$_3$) have
been synthesized by the reaction of Ln$_2$O$_3$ with Hhfac and diglymc in toluene.[517,518] Under these
conditions CeO$_2$ does not form an isolable complex but [Ln(hfac)$_3$(diglyme)] (Ln = Ce, Tb) can
be made by a substitution of another diketonate complex, reacting [Ln(acac)$_3$] with a slight excess
of Hhfac and diglyme. The Nd, Eu, Sm, and Gd compounds undergo reaction with metallic potassium

to form $[LnF(hfac)_3K(diglyme)]_2$, in which the lanthanide is still in the 3+ state, fluorine having been abstracted from an hfac ligand. In these compounds, the lanthanide is still eight-coordinate, bound to two bridging fluorides, three oxygens from bridging hfac ligands, and also to a terdentate triglyme.[519] $[\{Ce(fod)_3(tetraglyme\}_2]$ (tetraglyme = tetraethylene glycol dimethyl ether) has been synthesized; because of its volatility and stability, it shows promise as a source of thick ceria films through MOCVD.[520]

A 1:1 complex of $[Eu(fod)_3]$ and Michler's ketone emits red luminescence under daylight illumination.[521] Chemical shifts in the NMR spectra of adducts of $[Ln(fod)_3(bipy)]$ and $[Ln(fod)_3(phen)]$ are dipolar in origin.[522] $[Eu(ttfa)_3(phen)]$ (ttfa = thenoyltrifluoroacetonate) has been doped into organically modified silicate matrices; on reaction with hexamethyldisilazane, a composite phosphor with high emission intensity was formed.[523] The first perfluoroacetylacetonate complexes have been reported.[524]

$[Eu(OAc)_3]$ reacts with acacF-$_7$H to form nine-coordinate $[Ln(acacF-_7)_3(CH_3COOH)_3]$ in which the acetic acid molecules are coordinated by the C=O group; one carbonyl group of the diketonate is hydrogen bonded to an acetic acid molecule (and the corresponding Eu—O bond is slightly lengthened). The acetic acids are readily displaced by phosphine oxides with the formation of complexes such as eight-coordinate $[Ln(acacF-_7)_3(OPPh_3)_2]$. A number of dimeric tetraglyme complexes $[Ce_2(diketonate)_6(tetraglyme)]$ (tetraglyme = $CH_3O(CH_2CH_2O)_4CH_3$; diketonate = etbd; 1,1,1,5,5,5-hexafluoropentane-2,5-dionate, 1,1,1,2,2,3,3-heptafluoro-7,7-dimethyloctane-4,6-dionate) have been made.[525] Syntheses of methylpivaloylacetates of Y, Ba and Cu have been reported[526] including the structure of $[Y(mpa)_3(bipy)]$ (mpaH = methylpivaloylacetate) whilst tris(tropolonato) samarium has been synthesized electrochemically.[527] Fluorinated diketonate complexes $[Ln(RCOCHCOR')_3]$ (R, R' are, for example, CF_3, C_3F_7, Bu^t, etc.) have been shown to be efficient and rapid transport agents for potassium benzyloxocarbonylamino acidates.[528]

Lanthanide β-diketonates of fluorinated ligands form 1:1 complexes with aminoacids, enabling their extraction, transport, and chiral recognition.[529] $[Eu(bta)_3(H_2O)_2]$ and $[Eu(bta)_3(Ph_3PO)_2]$ (bta = benzoyl-1,1,1-trifluoroacetylacetonate) both have dodecahedral eight-coordination of europium.[530] [Tris(4,4,4-trifluoro-1-phenyl-1,3-butanedionato)(1,10-phenanthroline-N-oxide)europium] exhibits very high luminescence quantum yields at room temperature and shows promise as a light-conversion molecular device.[531] Europium and terbium β-diketonates have been used in organic electroluminescent devices.[532] Luminescence spectra of a number of nitrogen base adducts of $[Ln(diketonate)_3]$ (diketonate = acac, fod, PhCOCHCOPh) have been used to obtain information about the symmetry around the lanthanide ion.[533] The adducts $[Eu(btfa)_3 \cdot (bipy)]$ and $[Eu(bzac)_3.(bipy)]$ (btfa = 4,4,4-trifluoro-1-phenyl-2,4-butanedione; bzac = 1-phenyl-2,4-butanedione) have been reported, as well as the structure of $[Eu(btfa)_3 \cdot (bipy)]$; the fluorinated compound shows a higher quantum yield in fluorescence.[534] The volatility and the effect of the conjugated system of the ligand upon the luminescence of adducts of europium β-diketonates has been examined, with a view to the synthesis of electroluminescent devices.[535]

Coordination polyhedra for the diketonates and their adducts remain a fertile area of research. In addition to compounds already discussed, many other structures have been published. Structures have been reported for the neutral diketonates $[Eu(dbm)_3(bipy)]$,[536] $[Eu(C_6H_5COCHCOCF_3)_3(Ph_3PO)_2]$[537] and $[Ho(Me_3CCOCHCOCMe_3)_3(pivalic acid)]$[538] $[Gd(hfpd)_3(Me_2CO)(H_2O)]$ has square antiprismatic coordination (hfpd = 1,1,1,5,5,5,-hexafluoropentane-2,4-dione) whilst $[Gd_3(\mu_3-OH)_3(\mu_2-H_2O)_2(H_2O)_4(hfpd)_8]$ has a structure with a M_4O_6 core.[539] $[M(dpm)_3(H_2O)]$ (M = Y, Gd) reacts with hmteta (hmteta = hexamethyltriethylenetetramine, $Me_2N(CH_2CH_2N-Me)_2CH_2CH_2NMe_2$) forming dimeric $[(dpm)_3M(\mu-hmteta)M(dpm)_3]$ in which only three of the four nitrogen atoms in the amine are bound to yttrium.[540] $[Eu(TAN)_3(bipy)]$ (TAN = 4,4,4-trifluoro-1-(2-naphthyl)-1,3-butanedionate) crystallizes in two forms with slightly different coordination polyhedra, one bicapped trigonal prismatic, the other square antiprismatic.[541] A new β-diketonate ligand, 1,3-di-(2-furyl)-1,3-propanedione(dfp), has been used to make the complex $[Eu(dfp)_3(phen)]$, a red emitter fabricated into a double layer electroluminescent device.[542] Other $[Eu(R_1COCHCOR_2)_3(phen)]$ (R_1, R_2 are, for example, Me, Ph) have been doped into blue-emitting conjugated polymers, producing pure red-emitting LEDs.[543] Enhanced luminescence has been noted for some Nd and Eu β-diketonates in polymer matrices.[544] Achiral $[Ln(diketonate)_3]$ systems form 1:1 complexes with amino alcohols that give CD spectra dependent upon the absolute configuration of the substrate.[545] Heterobimetallic compounds have been studied lately on account of their structures and magnetic properties. They are also potential MOCVD single-source precursors. The series $[Ni(salen)La(hfa)_3]$ (Ln = Y, La-Yb) which have similar structures to several of the known $[Cu(salen)La(hfa)_3]$ compounds sublime without decomposition in vacuo.[546]

Co-crystallization of a mixture of [Y(hfa)$_3$] and [Cu(acac)$_2$] affords [Y(hfa)$_3$(H$_2$O)$_2$Cu(acac)$_2$] (hfa = hexafluoroacetylacetonate) in which the individual metal diketonate complexes are linked by hydrogen bonds.[547] Heating in vacuum induces ligand exchange and the liberation of gaseous [Cu(hfa)$_2$]. Structures of heterodinuclear complexes show lanthanide-copper distances of approximately 3.2Å and some tetrahedral distortion around copper. There is a small ferromagnetic interaction between the lanthanide and copper ions ($J = 0.8\,\text{cm}^{-1}$).[548]

Reactions between copper or lanthanide tmhd complexes and copper or barium aminoalkoxides have been investigated[549] and the structures of [PrCu(η^2-tmhd)$_3$(μ-η^2-O(CH$_2$)$_2$NMe$_2$)$_2$] and of [Y$_2$(η^2-tmhd)$_4$(μ-η^2-OCH(CH$_2$NMe$_2$)$_2$)$_2$] determined.

[CuLGd(hfac)$_2$]$_2$(H$_3$L = 1-(2-hydroxybenzamido)-2-(2-hydroxy-3-methoxybenzilidineamino)ethane) is a cyclic Gd$_2$Cu$_2$ complex with a $S = 8$ ground state due to ferromagnetic coupling between Gd and Cu.[550]

The first lanthanide β-ketoiminate complexes have been made[551,552] including [Yb(Bu^t-CO.CH.C(Bu^t)NPr^n)$_3$]. Fluorine-free ketoiminates [Ln(miki)$_3$] (Ln = Ce, Nd, Er) (**17**) are highly volatile and low-melting fluorine free precursors for MOCVD of lanthanide oxide thin films.

(17)

Anionic complexes have also attracted attention. Na[Er(pta)$_4$] (pta = pivaloyltrifluoroacetylacetonate) contains tetragonally antiprismatic coordination of erbium.[553] [NH$_4$][Ce(etbd)$_4$] (etbd = 1-ethoxy-4,4,4-trifluorobutane-1,3-dionate), has distorted square antiprismatic coordination of Ce.[554] (Et$_4$N)[Eu(dbm)$_4$] is triboluminescent (emits light when fractured)—an effect generally associated with non-centric space groups. It had been claimed that this compound was an exception,[555] but this has shown not to be the case.[556] Interest has been shown[557] in second-order nonlinear optical Langmuir–Blodgett films based on [Eu(dbm)$_4$]$^-$. Salts of the [Ln(ttfa)$_4$]$^-$ ion have attracted attention. The synthesis and structure of (E)-N-ethyl-4-(2-(4′-dimethylaminophenyl)ethenyl) pyridinium [La(ttfa)$_4$][558] are reported. M[Eu(ttfa)$_4$] (ttfa = thenoyltrifluoroacetonato) are soluble in common organic solvents and thus suitable for doping into polymer films to make light-emitting diodes.[559] The structure and fluorescence spectrum of the ethylpyridinium salt of [Eu(ttfa)$_4$]$^-$ has been determined.[560] Study of luminescence from (N,N-distearyldimethylammonium)[Eu(ttfa)$_4$] shows enhancement of the intensity of luminescence from the $^5D_1$ excited state relative to the $^5D_0$ state in monolayers compared with either solutions or the crystalline state.[561] Second-harmonic generation from monolayers of hemicyanine salts of [Eu(ttfa)$_4$]$^-$ has been reported.[562] Another anionic complex has been used to prepare a photoactive bilayer lipid membrane.[563]

3.2.2.5.4 *Alkoxides and aryloxides*

Many alkoxides in particular have been known since the 1960s, but interest in them has been stimulated recently by their potential use as precursors for deposition of metal oxides using the sol-gel or MOCVD process. A review covering the literature to 1990 has appeared.[564] Traditionally, alkoxides are made by salt-elimination reactions of lanthanide chlorides with alkali metal alkoxides (or aryloxides) which sometimes causes chloride retention

$$6\text{NdCl}_3 + 18\text{NaOPr}^i \rightarrow \text{Nd}_6(\text{OPr}^i)_{17}\text{Cl} + 18\text{NaCl} \tag{11}$$

but increasingly other sources of the lanthanide, like amide complexes, are being used[565] in order to avoid possible chloride retention (and –ate ion formation)

$$\text{Ln}\{\text{N(SiMe}_3)_2\}_3 + 3\text{LiOR} \rightarrow \text{Ln(OR)}_3 + 3\text{LiN(SiMe}_3)_2 \tag{12}$$

A route useful in a few cases is the reaction between lanthanide metal chips and the alcohol (usually isopropanol) in the presence of $HgCl_2$ catalyst

$$Ln + 3Pr^iOH \rightarrow Ln(OR)_3 + 3/2H_2 \tag{13}$$

$$10Y + 30HOCH_2CH_2OCH_3 \rightarrow [Y(OCH_2CH_2OCH_3)_3]_{10} + 15H_2 \tag{14}$$

Simple methoxides like $La(OMe)_3$ are ill-defined structurally but are oligomeric. Reaction of erbium chips with isopropanol as described affords principally the oxo-centred cluster $Er_5O(OPr^i)_{13}$ (with a square-pyramidal core) but there is evidence that gentle work-up gives an alkoxide "$Ln(OPr^i)_3$" without an oxogroup.[566]

Reaction of $Nd\{N(SiMe_3)_2\}_3$ with diisopropylmethanol in hexane in the presence of THF gives the binuclear $[Nd(OCHPr^i_2)_3(THF)]_2$

$$2Nd\{N(SiMe_3)_2\}_3 + 6LiOCHPr^i_2 + 2THF \rightarrow [Nd_2(OCHPr^i_2)_6(THF)_2] + 6LiN(SiMe_3)_2 \tag{15}$$

This has the structure $[(Pr^i_2CHO)_2(THF)Nd(\mu\text{-}OCHPr^i_2)_2Nd(THF)(OCHPr^i_2)_2]$.

The THF ligands are readily exchanged for pyridine to form $[Nd_2(OCHPr^i_2)_6(Py)_2]$, whilst reaction with 1,2-dimethoxyethane (DME) gives $[Nd_2(OCHPr^i_2)_6(\mu\text{-}DME)]$, in which the binuclear alkoxy-bridged dimer units survive, linked to each other by bridging dimethoxyethane molecules.[567] Neodymium has trigonal bipyramidal coordination in all these compounds. Reaction of $Ln\{N(SiMe_3)_2\}_3$ with neopentanol gives neopentoxides $[Ln(OCH_2Bu^t)_3]_4$ (Ln = La, Nd) that are tetramers based on a square of lanthanides with each lanthanide bound to one terminal and four bridging alkoxides. Nonbonding La—La distances are 3.85 Å in the lanthanum compound; terminal La—O distances are ~2.16 Å and bridging La—O distances ~2.37–2.44 Å. IR spectra in both the solid state and solution show absorption bands at ~2,680 cm^{-1}, ascribed to Ln···H—C agostic interactions.[568] Using more bulky alkoxide groups, trinuclear complexes which are not oxo-centered have been obtained, using the route of alcoholysis of the amide, in the form of $Ln_3(OR)_9(ROH)_2$ (Ln — Y, R = Bu^t, Am^t; Ln = La, R = Bu^t). Increasing ligand bulk further enables the isolation of dimers $[Ln(OR)_3]_2$ (Ln = Y, R = CMe$_2$Pr^i, CMeEtPr^i, CEt$_3$; Ln = La; R = CMe$_2$Pr^i, CMeEtPr^i). $La_3(OBu^t)_9$ $(Bu^tOH)_2$ has the structure $[La_3(\mu_3\text{-}OBu^t)_2(\mu\text{-}OBu^t)_3(OBu^t)_4(Bu^tOH)_2]$ in which the La$_3$ triangle is capped by two μ_3-OBu^t groups; lanthanum is octahedrally coordinated.[569] Reaction of NdCl$_3$ with NaOBu^t in THF affords a THF-solvated t-butoxide of neodymium, shown to be $[Nd_3(\mu_3\text{-}OR)_2(\mu\text{-}OR)_3(OR)_4(THF)_2]$ (R = Bu^t). Combined with one mole of MgR′$_2$ (R′ = n-hexyl), it is a catalyst for the pseudo-living polymerization of ethene.[570] Refluxing $[Ln_3(OBu^t)_9(Bu^tOH)_2]$ (Ln = La, Nd, Yb) in toluene gives[571] the oxo-centered $[Ln_3O(OBu^t)_{13}]$. Higher nuclearity clusters $[Ln_4O_3(OBu^t)_6]$ (Ln = Pr, Y) have been obtained from similar reactions. Alkylation of "Ln(O-Bu^t)$_3$" with AlMe$_3$ give related mixed alkyl/alkoxy bridged $[Ln(\mu\text{-}OBu^t)_3(\mu\text{-}Me)_3(AlMe_2)_3]$ (Ln = Pr, Nd, Y).[572] The alternative route to butoxides, using salt-elimination from LnCl$_3$ and the alkali-metal alkoxide has in some cases given chlorine-containing products, such as $Y_3(OBu^t)_8Cl(THF)_2$ and $Y_3(OBu^t)_7Cl_2(THF)_2$, although $Ln_3(OBu^t)_9(THF)_2$ (Ln = La, Y) have also been made.[573–575] Using an even bulkier ligand affords[576] the compound $Ce(OCBu^t_3)_3$ which is believed to be a monomer. It undergoes high-yield thermolysis in vacuo at 150 °C affording an alkoxy-bridged dimer $[(Bu^t_2CHO)_2Ce(\mu\text{-}OCHBu^t_2)_2Ce(OCHBu^t_2)_2]$ with four-coordinate cerium

$$2Ce(OCBu^t_3)_3 \rightarrow [Ce(OCHBu^t_2)_3]_2 + 6C_4H_8 \tag{16}$$

Reaction of $[Gd\{N(SiMe_3)_2\}_3]$ with $(Me_3Si)_3SiOH$ in THF gives hexane-soluble $[Gd\{OSi-(SiMe_3)_3\}_3(THF)_2]$, a compound that can also be made from GdCl$_3$ and $(Me_3Si)_3SiONa$. This has a trigonal bipyramidal structure with axial THF molecules, Gd—O(OR) = 2.142 Å and Gd—O (THF) = 2.314 and 2.448 Å.[577] $[Gd\{OSi(SiMe_3)_3\}_3(MeCN)_2]$ can be made similarly in acetonitrile. $[Gd\{OSi(SiMe_3)_3\}_3(THF)_2]$ reacts with DABCO (1,4-diazabicyclo[3.2.2]octane) forming $[Gd\{OSi(Si-Me_3)_3\}_3(DABCO)]$ which again has a trigonal bipyramidal structure with DABCO acting as a monodentate ligand instead of the hoped-for coordination polymer; here Gd—O(OR) = 2.161 Å and Gd—O(THF) = 2.520 Å. Reaction of $[Gd\{OSi(SiMe_3)_3\}_3(THF)_2]$ with 4,4′-bipyridyl gives $[Gd\{OSi(SiMe_3)_3\}_3(4,4\text{-bipy})_2]$, possibly a monomer, from which one mole of bipy can be leached in MeCN forming what is believed to be a polymer, $[Gd\{OSi(SiMe_3)_3\}_3(4,4\text{-bipy})_2]_n$. Reaction of $[La\{N(SiMe_3)_2\}_3]$ with $(Me_3Si)_3SiOH$ in THF gives hexane-soluble $La\{OSi(SiMe_3)_3\}_3(THF)_4$, believed to be [La{OSi(SiMe$_3$)$_3$}$_3$(THF)$_3$].THF. The homoleptic silyloxides [La{OSi(SiMe$_3$)$_3$}$_3$] cannot be made by direct exchange between [Ln{N(SiMe$_3$)$_2$}$_3$] with (Me$_3$Si)$_3$SiOH in a nonpolar solvent, however. La{OSi(SiMe$_3$)$_3$}$_3$(THF)$_n$ lose their THF on sublimation in vacuum. La{OSi(Si-Me$_3$)$_3$}$_3$(THF)$_n$ absorb CO$_2$ from the atmosphere forming carbonates. A number of Ln{OQPh$_3$}$_3$ (Ln = Y, La, Ce; Q = C, Si) have been synthesized. Some of these are definitely dimers, such as [La(OCPh$_3$)$_3$]$_2$ and [Ce(OSiPh$_3$)$_3$]$_2$, which are [La(OCPh$_3$)$_2$(μ-OCPh$_3$)]$_2$ and [Ce(OSiPh$_3$)$_2$(μ-OSiPh$_3$)]$_2$ respectively.[578] As expected, the bridging M—O distances are longer than the terminal ones; in [La(OCPh$_3$)$_2$(μ-OCPh$_3$)]$_2$, the La—O (terminal) are 2.175–2.184 Å whilst La—O (bridging) = 2.389–2.483 Å. All these compounds may conveniently be synthesized by the alcoholysis of [Ln{N(SiMe$_3$)$_2$}$_3$](Ln = La,Ce). ^{29}Si-NMR studies suggest that [Y(OSiPh$_3$)$_2$(μ-OSiPh$_3$)]$_2$ retains its dimeric structure in solution.[579] In this solid state the Y—O (terminal) distances are 2.058–2.062 Å whilst Y—O (bridging) = 3.211–3.288 Å. The shorter terminal bond lengths are again expected, and may be compared with values of 2.118–2.138 Å in five and six-coordinate Lewis base adducts. The alkoxide bridges can be cleaved by Lewis bases[580–582] (Py, THF, Bu$_3$P=O) forming adducts such as *fac*-Ln(OSiPh$_3$)$_3$(THF)$_3$ (Ln = Y, La, Ce) and Ln(OSiPh$_3$)$_3$(OPBu$_3$)$_2$ (trigonal bipyramidal, with axial phosphine oxides), as well as the ionic [K(DME)$_4$][Y(OSiPh$_3$)$_4$(DME)]. In Y(OSiPh$_3$)$_3$(THF)$_3$ Y—OR bond lengths fall in the range 2.118–2.138 Å whilst Y—O(THF) distances are 2.374–2.462 Å, whilst in Y(OSiPh$_3$)$_3$(OPBu$_3$)$_2$ Y—OR bond lengths are in the range 2.118–2.129 Å whilst Y—O (THF) distances are 3.261–3.266 Å. A comparison of the structures of Ln(OSiPh$_3$)$_3$(THF)$_3$ (Ln = La, Y) indicates that although these compounds are isostructural, there was a greater contraction in the Ln—O (THF) bond length than in the Ln—OR distance on passing from La to Y, interpreted in terms of a greater "malleability" in the weaker bonds to the ether. In a further study of lanthanide silyloxides, the structures of [Gd(OSi(SiMe$_3$)$_3$)$_3$(L)$_2$] (L = THF; 2L = H$_2$N(C$_2$H$_4$)NH$_2$) show them to have tbp coordination of gadolinium. [Gd(OSi(SiMe$_3$)$_3$)$_3$(THF)$_2$] and [La(OSi(Si-Me$_3$)$_3$)$_3$(THF)$_4$] both absorb CO$_2$ to give carbonates; they also lose THF in vacuo on sublimation at 205 °C to afford homoleptic silyloxides.[583] A chloro-bridged dimer, [Nd(OSiBu^t$_3$)$_2$(THF)(μ-Cl)]$_2$ has been characterized.[584] Silanol ligands with alkylamide groups enable the isolation[585] of volatile monomers like five-coordinate [Y{OSi(Bu^t)[(CH$_2$)$_3$NMe$_2$]$_2$}$_3$] where one amide is uncoordinated. An yttrium compound with a remarkable cyclic decameric structure, [Y(OCH$_2$CH$_2$OCH$_3$)$_3$]$_{10}$ has been made both by reaction of yttrium chips with 2-methoxyethanol and by alcoholysis of Y$_5$O(OPr^i)$_{13}$.Yttrium attains pentagonal bipyramidal seven-coordination by forming one terminal Y—O link and by linking to six bridging oxygens.[586] As they are easily made, "[Ln(OPr^i)$_3$]", which may be clusters (see below), have attracted attention as catalysts and as starting materials. Thus [Y(OPr^i)$_3$] catalyzes the ring-opening of epoxides with Me$_3$SiN$_3$.[587] [La(OPr^i)$_3$] is a very efficient catalyst for the transesterification of esters with alcohols.[588] [La(OPr^i)$_3$] reacts with anthracenebis (resorcinol) forming an insoluble 1:2 polycondensate which catalyzes enolization and aldol reactions of ketones such as cyclohexanone in pure water at normal pH.[589]

A number of oxo-centered clusters have attracted attention. Compounds previously formulated as Ln(OPr^i)$_3$ have been recognized[566,590–594] as Ln$_5$O(OPr^i)$_{13}$ (Ln = Y, Yb, Er, Nd, Gd, Eu, Pr) and more of these undoubtedly can be made. Crystallography has established the structures of most of these compounds, and shown them to be clusters [Ln$_5$(μ_5-O)(μ_3-OPr^i)$_4$(μ_2-OPr^i)$_4$(OPr^i)$_5$] containing a square-pyramidal arrangement of the lanthanides around a μ_5-oxo group. The μ_2 groups link basal metal atoms and the μ_3 groups link two basal metal atoms with an apical atom. Thus in [Y$_5$(μ_5-O)(μ_3-OPr^i)$_4$(μ_2-OPr^i)$_4$(OPr^i)$_5$], the Y—O bond distances follow the expected pattern Y-μ_3-OR > Y-μ_2-OR > Y-OR (they average 3.27, 3.25, and 2.01 Å respectively). The Y-μ_3-OR distances involving the apical Y are, at 2.18–2.32 Å, significantly shorter than those involving the basal yttriums at 2.37–2.45 Å; the Y-μ_5-O distances are at 2.35 Å, rather longer than those involving μ_3-OR. The metal–metal distances are relatively long (3.26–3.38 Å in the ytterbium compound; 3.30–3.47 Å in the yttrium compound) showing the absence of metal–metal bonding. The ^{89}Y-NMR spectrum of Y$_5$O(OPr^i)$_{13}$ shows two signals with an intensity ratio of 4:1 confirming the retention of the square pyramidal structure in solution.[590] Syntheses have been reported for the oxo-centered alkoxides [Ln$_5$O(OPr^i)$_{13}$] (Ln = Nd, Gd); they have similar structures, based on a square pyramidal M$_5$O core, to the known Er compound, a structure largely retained in solution. They react with [Al$_4$(OPr^i)$_{12}$] forming [LnAl$_3$(OPr^i)$_{12}$].[593] Reaction of EuCl$_3$ with KOPr^i followed by stoichiometric hydrolysis yields Eu$_5$O(OPr^i)$_{13}$, whereas reaction of Eu metal with HOPr^i in toluene gives the mixed-valence Eu$_4$(OPr^i)$_{10}$(HOPr^i)$_3$.[594] Praseodymium alkoxides have been investigated,[595] a number of oxo-centered species being isolated. Reaction of Pr metal with ROH affords [Pr$_5$O(OR)$_{13}$] (R = Pr^i, Am^i), [Pr$_6$O$_2$(ONp)$_8$] (Np = neopentyl), and [Pr(OC$_2$H$_4$NMe$_2$)$_3$]. Alcoholysis of [Pr{N(SiMe$_3$)$_2$}$_3$] gives [Pr$_4$O(ONp)$_{10}$], [Pr$_4$O$_2$(ONp)$_8$], [Pr$_3$(OR)$_9$(ROH)$_2$] (R = Bu^t, Am^t), and [Pr$_4$O$_2$(O-C$_2$H$_4$OMe)$_8$]. Reaction of Nd chips with isopropanol in the presence of Hg(OAc)$_2$ as the customary catalyst affords two neodymium alkoxides. [Nd(OPr^i)$_3$(Pr^{i}OH)] is tetrameric, with a structure similar to [Ti(OMe)$_3$]$_4$, involving six-coordinate neodymium. The second product is an oxo-centered compound, Nd$_5$O(OPr^i)$_{13}$(Pr^{i}OH)$_2$, which in contrast to the square pyramidal Ln$_5$O(OPr^i)$_{13}$ has a M$_5$O trigonal bipyramidal core. Nd—O distances vary from 2.121 Å for a terminal linkage to 2.719 Å for a Nd-μ_5-O bond.[596] Several mixed-metal alkoxides have been synthesized, including [Y$_4$PrO(OPr^i)$_{13}$] which has the familiar structure [Y$_4$Pr(μ_5-O)(μ_3-OR)$_4$(μ-OR)(OR)$_5$]. The structure of [Pr$_4$(μ_4-O)(μ_3, η^2-OR)$_4$(μ, η^1-OR)(OR) (OPMe$_3$)]$_2$ was also determined. [Y$_5$O(OPr^i)$_{13}$] reacts with HACAC forming [Y$_2$(μ-OAc)$_2$(ACAC)$_4$(H$_2$O)$_2$].[597] [Ln{N(SiMe$_3$)$_2$}$_3$] (Ln = Y, Lu) undergo alcoholysis with donor-functionalized alcohols HOCR$_2$CH$_2$do (do = OMe, R = Me, Et; do = NMe$_2$, R = Me) forming volatile, alkane soluble [Ln(OCR$_2$CH$_2$do)$_3$]. [Lu(OCMe$_2$CH$_2$OMe)$_3$] is dimeric; inadvertent hydrolysis yields the novel [Lu$_4$(O)(OH)(OCMe$_2$CH$_2$OMe)$_9$] whose Lu$_4$O$_{15}$ core has a butterfly rather than a tetrahedral geometry.[598] A cluster involving six metal atoms is Gd$_6$O(OCH$_2$-CH$_2$OCH$_3$)$_{16}$, obtained from the Hg^{II} catalyzed reaction of gadolinium with 2-methoxyethanol or by alcohol exchange with Gd$_5$O(OPr^i)$_{13}$. It has the structure [Gd$_6$(μ_4-O)(μ_3, η^2-OR)$_4$(μ, η^2-OR)$_6$(μ, η^1-OR)$_2$(OR)$_4$] (R = OCH$_2$CH$_2$OCH$_3$), with four gadoliniums surrounding the oxo ligand and two only bound to alkoxides. Gd—O bond lengths fall into the range 2.152–2.674 Å, showing the pattern Gd—OR < Gd-μ_4-O < Gd-μ-OR < Gd-μ_3-OR < Gd—OR (ether). The gadolinium atoms are seven and eight-coordinate.[599] A number of volatile fluoroalkoxides, some also involving sodium, have been described; including [Y{OCH(CF$_3$)$_2$}$_3$.L$_3$] (L = THF, Pr^{i}OH), [YNa$_3${OCH(CF$_3$)$_2$}$_6$(THF)$_3$], [YNa$_2${OCMe(CF$_3$)$_2$}$_5$(THF)$_3$] and [YNa$_2${OC(CF$_3$)Me$_2$}$_5$.THF]. The sodium is retained on sublimation, but the Lewis base is lost.[600,601] A number of volatile hexafluoro-*t*-butoxides have also been synthesized.[602,603] The Raman spectrum of [Eu(OCH(CF$_3$)$_2$)$_3$] indicates strong Eu$\cdots$F interactions; acid hydrolysis gives EuF$_3$.[604] A number of mixed metal alkoxo/diketonates (of obvious utility as possible materials for the synthesis of thin superconductor films) have been synthesized and the structure of [BaY$_2$(μ-OCH(CF$_3$)$_2$)$_4$(tmhd)$_4$] determined.[605]

There have been major developments in aryloxide chemistry. There is now known a wide range of aryloxides Ln(OAr)$_3$, some solvated, but some can be isolated as three-coordinate monomers, especially with 2,6-di-*t*-butylaryloxides. Tested syntheses have appeared[606] for [Ln(OC$_6$H$_3$Bu$^t_2$-2,6-Me-4)$_3$] (Ln = Y, La, Pr, Nd, Dy–Er, Yb) and [Ln(OC$_6$H$_3$Bu$^t_2$-2,6)$_3$] (Ln = Y, La, Sm). In these three-coordinate species, the possibility arises of the LnO$_3$ grouping being planar or pyramidal, as in the silylamides [Ln{N(SiMe$_3$)$_2$}$_3$]. Both possibilities seem to be realized. [Y(OC$_6$H$_3$Bu$^t_2$-2,6)$_3$] is trigonal planar[607] but [Ce(OC$_6$H$_2$Bu$^t_2$-2,6-Me-4)$_3$] is trigonal pyramidal.[608] Possibly the balance of small and variable Van der Waals forces is the determining factor. [Sm(OR)$_3$] (R = 2,6-Bu$^t_2$-4-MeC$_6$H$_2$)[609] is a catalyst for the Michael reaction of ketones with α,β-unsaturated ketones that also shows catalytic activity for tandem Aldol–Tischenko reaction of ketones and aldehydes to form 1,3-diol monoesters. Solvates and adducts are sometimes obtained, even with these bulky aryloxides, thus NdCl$_3$ reacts with three moles of RONa (R = 2,6-Bu$^t_2$-4-MeC$_6$H$_2$) forming four-coordinate [Nd(OR)$_3$(THF)];[610] using four moles of RONa, [Na(THF)$_6$][Nd(OR)$_4$] was obtained, again with tetrahedrally coordinated neodymium. The structure of four-coordinate [Sm(OR)$_3$(OPPh$_3$)] (R = 2,6-Bu$^t_2$-4-MeC$_6$H$_2$) has been determined.[611] Using the relatively unhindered 2,6-dimethylphenoxide ligand, reaction of YCl$_3$ with NaOC$_6$H$_3$Me$_2$-2,6 in THF has been found to afford six-coordinate [Y(OC$_6$H$_3$Me$_2$-2,6)$_3$(THF)$_3$], isolated as the *fac*-isomer. If this is crystallized from toluene, a dimer [Y(OAr)$_3$(THF)]$_2$ is formed, having the structure [(ArO)$_2$(THF)Y-(μ-OAr)$_2$Y(THF)(OAr)$_2$]. This equilibrium is completely reversible. In the dimer, yttrium is in square-pyramidal five-coordination. Bridging Y—O distances are as usual longer, at 3.275–3.277 Å, than the terminal Y—O distances of 2.046–2.075 Å.[612] Terbium metal reacts with phenols in refluxing isopropanol (probably via intermediate isopropoxides) forming *fac*-[Tb(OC$_6$H$_3$Me$_2$-2,6)$_3$(THF)$_3$] and [Tb(OC$_6$H$_3$Pr$^i_2$-2,6)$_3$(THF)$_2$], the latter having a trigonal bipyramidal structure with axial thf molecules.[613] Although terbium is a relatively unreactive lanthanide metal, the reaction proceeds well with mercury salts as catalysts. A polyether complex [La(OC$_6$H$_3$Me$_2$-2,6)$_3${MeO(CH$_2$OCH$_2$)$_4$Me}] is monomeric with eight-coordinate lanthanum.[614] The slightly bulkier 2,6-diisopropylphenolate ligand leads to unsolvated Ln(OC$_6$H$_3$Pr$^i_2$-2,6)$_3$ species which are in fact η^6-arene bridged dimers Ln$_2$(OC$_6$H$_3$-Pr$^i_2$-2,6)$_6$. These dissolve in THF to form conventionally bound trigonal bipyramidal THF adducts [Ln(OC$_6$H$_3$Pr$^i_2$-2,6)$_3$(THF)$_2$] (axial THF ligands)[615] (Ln = Pr, Nd, Sm, Gd, Er, Yb, Lu). These compounds form Lewis base adducts, including those with ammonia. The structures of [La$_2$(OC$_6$H$_3$Pr$^i_2$-2,6)$_6$(NH$_3$)$_2$], [La(OC$_6$H$_3$Pr$^i_2$-2,6)$_3$(NH$_3$)$_4$] and [La(OC$_6$H$_3$Pr$^i_2$-2,6)$_3$(THF)$_2$] have been determined;[616] [La$_2$(OC$_6$H$_3$Pr$^i_2$-2,6)$_6$] is bridged by two η^6-aryl groups. Like the other THF adducts, [Sm(OC$_6$H$_3$Pr$^i_2$-2,6)$_3$(THF)$_2$] is tbpY with axial THF molecules;[617] it is the synthon for [Sm(OC$_6$H$_3$Pr$^i_2$-2,6)$_3$(Py)$_2$], [Sm(OC$_6$H$_3$Pr$^i_2$-2,6)$_3$(Py)$_3$], K[Sm(OC$_6$H$_3$Pr$^i_2$-2,6)$_4$] and K[Sm(OC$_6$H$_3$Pr$^i_2$-2,6)$_4$(Py)].

Reaction of [Ln$_2$(OC$_6$H$_3$Pr$^i_2$-2,6)$_6$] with LiOC$_6$H$_3$Pr$^i_2$-2,6 or NaOC$_6$H$_3$Pr$^i_2$-2,6 yields [(THF)La(OC$_6$H$_3$-Pr$^i_2$-2,6)$_2$(μ-OC$_6$H$_3$Pr$^i_2$-2,6)$_2$Li(THF)] and [(THF)La(OC$_6$H$_3$Pr$^i_2$-2,6)$_2$(μ-OC$_6$H$_3$ Pr$^i_2$-2,6)$_2$Na(THF)$_2$]

respectively[618]; the corresponding reaction with $CsOC_6H_3Pr^i_2$-2,6 affords $[CsLa(OC_6H_3Pr^i_2$-2,6)_4]$. The latter contains alternating Cs^+ and $[La(OC_6H_3Pr^i_2$-2,6)_4]^-$ ions in a one-dimensional chain structure held together by Cs-arene π-interactions. Similar interactions are found in $[Cs_2La(OC_6H_3$-Pr^i_2-2,6)_5]$.[619] Similarly the aryloxide $K[Ln(OC_6H_3Pr^i_2$-2,6)_4]$ has aryloxide anion chains bridged by K-arene interactions;[620] this compound is obtained even when just three moles of KOAr are reacted with $LnCl_3$, another example of "alkali-metal retention." The phenolates $[Ln(OC_6H_3Ph_2$-2,6)_3]$ (Ln = La, Ce, Pr, Nd, Gd, Ho, Er, Lu, Y) have been synthesized from the reaction of $HOC_6H_3Ph_2$-2,6 and the lanthanide in the presence of mercury at 200 °C. All have monomeric structures with the lanthanide slightly out of the O_3 plane, but with some additional ring-metal interactions.[621] $[Ln(OC_6H_3Ph_2$- 2,6)_3(THF)_2]$.2 THF (dpp = 2,6-diphenylphenolate; Ln = La, Nd) have "conventional" five-coordinate trigonal bipyramidal coordination of the metal with one axial and one equatorial THF;[622,623] in contrast, $[Nd(OC_6H_3Ph_2$-2,6)_3(THF)]$ has pseudo-tbp coordination with three equatorial phenoxides, an apical THF and an apical position occupied by a phenyl group, and, as already remarked, unsolvated $[Nd(OC_6H_3Ph_2$-2,6)_3]$ also features Nd-ring interactions.

Crystallization of $[Yb(OC_6H_3Ph_2$-2,6)_3(THF)_2]$ from DME affords $[Yb(OC_6H_3Ph_2$-2,6)_3(DME)].0.5DME$; both this and the Nd analogue have a *sp* structure with an axial aryloxide ligand and two *cis*-aryloxides in the basal plane.[624] A number of anionic diphenylphenolates have been made.[625] $LnCl_3$ (Ln = Nd, Er) react with $Na(OC_6H_3Ph_2$-2,6).0.5THF in 1,3,5-tri-t-butylbenzene at 300 °C forming $[Na\{Ln(OC_6H_3Ph_2$-2,6)_4\}]$; on crystallization of $[Na\{Ln(OC_6H_3Ph_2$-2,6)_4]$ from DME or diglyme the species $[Na(diglyme)_2][Ln(OC_6H_3Ph_2$-2,6)_4]$ and $[Na(dme)_3][Ln(OC_6H_3Ph_2$-2,6)_4]$ which contain discrete anions are obtained. $[ClLn(OR)_3Na]$ (Ln = Lu, Y) and $[ClY(OR')_3Y(OR')_3Na]$ $(OR = 4\text{-}O\text{-}2,6\text{-}(CH_2NMe_2)_2C_6H_2$; $OR' = 2\text{-}OC_6H_4(CH_2NMe_2))$. Reaction of $[Y\{N(SiMe_3)_2\}_3]$ and $[Ba\{N(SiMe_3)_2\}_2]$ with Bu^tOH yields $[YBa_2(OBu^t)_7(Bu^tOH)]$, which has a triangular structure with two μ_3 and three μ_2 ligands.[626] "Unsolvated" $[NaLa(OC_6H_3Ph_2$-2,6)_4]$ contains $[La(OC_6H_3Ph_2$-2,6)_4]^-$ with sodium bound to three oxygens and interacting with three different phenyl groups.[619] Another route to Yb^{III} aryloxides involves oxidation of the Yb^{II} aryloxides $[Yb(OAr)_2(THF)_2]$ $(Ar = OC_6H_2Bu^t_2$-2,6-R-4; R = H, Me, Bu^t) with HgX_2 or CH_2X_2 (X = Cl, Br, I) affording $[Yb(OAr)_2X(THF)_2]$; $[Yb(OAr)_2I(THF)_2]$ (R = Me) has a square pyramidal structure with apical iodine and *trans*- aryloxides and THF molecules. Inadvertent hydrolysis of $[Yb(OAr)_2Cl(THF)_2]$ (R = H) affords the hydroxy-bridged dimer $[(ArO)_2(THF)Yb(\mu\text{-}OH)_2Yb(OAr)_2(THF)]$, also with five-coordinate ytterbium.[627] Other aryloxides have been synthesized using O-amino phenolate ligands to facilitate binding of anions and cations in the complexes.[628] Normally alkoxides, aryloxides, and amides are hydrolyzed by even traces of water, but the structure of a water adduct of an aryloxide, $[Pr(OC_6H_2(CH_2NMe_2)_3$-2,4,6)_3(H_2O)_2]$ has been reported; its structure shows two of the aryloxides to be bidentate, so that the praseodymium is seven-coordinate.[629]

3.2.2.6 Mixed Group 15 and 16 Donors

3.2.2.6.1 *MRI agents*

Magnetic resonance imaging (MRI) is probably the most important new application of lanthanide compounds to emerge in the last 20 years. Many hospitals now have MRI scanners and use contrast agents in examinations. The introduction of lanthanide-based contrast agents has revolutionized diagnostics, assisting doctors in distinguishing between normal and diseased tissue and thus improving prognosis. A book on MRI agents, mainly concerned with gadolinium complexes, has now appeared.[630] Various reviews, all relevant, have appeared since the previous volume,[631–639] some more detailed than others.[631–634] The literature relevant to gadolinium complexes with possible MRI applications is immense and expanding, so for comprehensive coverage the interested reader is referred to the above reviews. MRI relies on detecting the NMR signals of water molecules in the body as a function of space. Since 60% of the body is water, it is the obvious substance to examine. Spatial information is obtained by making the ^1H resonance frequency position-dependent. Thus, within a particular piece of tissue, otherwise-identical water protons resonate at slightly different frequencies dependent upon their position in the field, so that the resulting NMR signal is spatially encoded and a two-dimensional image is obtained. The signal intensity depends upon the relaxation times of the protons. In general, the shorter the relaxation times, the more intense the signal. The imaging agent enhances the contrast to distinguish between healthy and diseased tissue. MRI uses a paramagnetic contrast agent, which shortens the relaxation time (t_1) for the protons in water molecules in that tissue.

What makes a good MRI agent? The choice is dictated by a combination of several factors:

  (i)   high magnetic moment,
 (ii)   long electron spin relaxation time,
(iii)   osmolarity similar to serum,
 (iv)   low toxicity,
  (v)   solubility,
 (vi)   targeting tissue, and
(vii)   coordinated water molecules.

The Gd^{3+} ion is especially suitable for its magnetic properties on account of its large number of unpaired electrons ($S = 7/2$) and because its magnetic properties are isotropic. Its relatively long electron-spin relaxation time, at $\sim 10^{-9}$ s, is more suitable than other highly paramagnetic ions such as Dy^{3+}, Eu^{3+}, and Yb^{3+} ($\sim 10^{-13}$ s). Taken together, these factors are very favorable for nuclear spin relaxation. However, since the free Gd^{3+} ion is toxic (the LD_{50} is ~ 0.1 mmol kg^{-1}, which is less than the imaging dose, which is normally of the order of 5 g for a human), complexed Gd^{3+} is used, using a ligand that forms a very stable complex *in vivo*. Relaxation times are shorter the nearer the water molecules are to the Gd^{3+} ion, so that the complex ideally must have water molecules in the coordination sphere, the more the better, so that more solvent water molecules can readily be exchanged with coordinated water molecules; however, the use of multidentate ligands to ensure a high stability constant for the gadolinium complex (to minimize the amount of toxic, free Gd^{3+} ions present) tends to reduce the number of bound waters, and in practice most contrast agents have one coordinated water.

Complexes of polyaminocarboxylic acids such as $[Gd(DTPA)(H_2O)]^{2-}$ (gadopentetate dimeglumine; Magnevist) or $[Gd(DOTA)(H_2O)]^-$ (gadoterate meglumine; Dotarem) have been widely used; these meet most of the above criteria. Both neutral and charged complexes are used, the former having less osmotic effect and the latter being more hydrophilic, so that amide derivatives like $[Gd(DTPA-BMA)(H_2O)]$ (gadodiamide; Omniscan) have advantages. (DTPA = diethylenetriaminepentaacetate; DOTA = DOTA = 1,4,7,10-tetraaza-cyclododecane-N, N',N',N'''-tetraacetate; DTPA-BMA = dimethylamide of diethylenetriaminepentaacetic acid).

3.2.2.6.2   *EDTA complexes*

These have not been much studied of late, since although EDTA forms relatively stable complexes with Gd^{3+} (log K = 17.35) and the complex $[Gd(EDTA)(H_2O)_n]^{2-}$ ($n = 2$–3) will, owing to the lower denticity of EDTA compared to DTPA, have the advantage of more coordinated water molecules, nevertheless its tolerability in animal studies was poor and it was therefore superseded by DTPA, which forms a more stable gadolinium complex (log K for $[Gd(DTPA)(H_2O)]^{2-} = 22.46$).[640] The thermodynamics of complex formation by aminopolycarboxylic acids with lanthanides have been discussed.[641] The pattern of solid-state structures of lanthanide EDTA complexes previously determined is that there is a change in the number of coordinated waters from three to two near the end of the series, following the lanthanide contraction. $Na[Ln(EDTA)(H_2O)_3].5H_2O$ (Ln = La, Nd, Eu) are isostructural, with nine-coordinate lanthanides.[642] $[K[Yb(EDTA)(OH_2)_2].5H_2O$ contains eight-coordinate ytterbium.[643] Structures of several $M[Ln(EDTA)(H_2O)]_n]$ (M = alkali metal; Ln = lanthanide) show that the coordination number depends upon the ionic radii of both the lanthanide and the alkali metal.[644] The complex (guanidinium)$_2$ $[[Eu(EDTA)F(H_2O)_2]_2].2H_2O$ contains nine-coordinate europiums linked to two fluorine bridges.[645] UV/vis studies[646] on the $^7F_0 \rightarrow {}^5D_0$ transition in Eu^{3+} complexes of polyaminocarboxylates such as EDTA and PDTA indicate an equilibrium in solution between eight- and nine-coordinate complex species, as known for the aqua ion. ^{17}O-NMR studies were also reported on a number of eight-coordinate species $[Ln(PDTA)(H_2O)_2]^-$ (Ln = Tb, Dy, Er, Tm, Yb) and $[Er(EDTA)(H_2O)_2]^-$.

3.2.2.6.3   *Complexes of DTPA and its derivatives*

Many solid-state structures have been determined for complexes of DTPA and its derivatives with Gd^{3+} and other Ln^{3+} ions. In $Na_2[Gd(dtpa)(H_2O)]$, dtpa is octadentate and one water molecule is coordinated (**18**).[647]

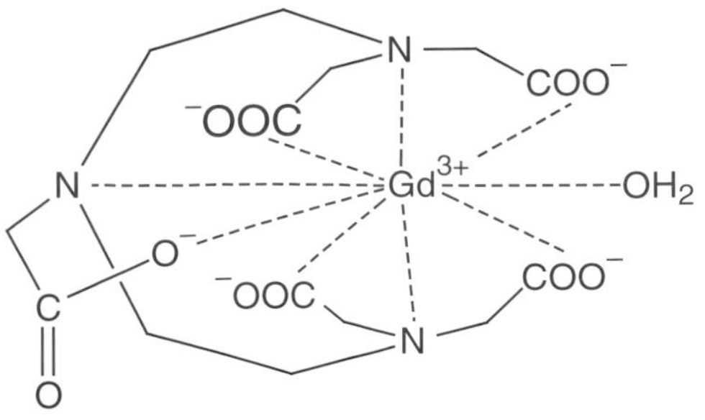

(18)

The Gd—O(H_2O) distance is 2.490 Å, Gd—O (carboxylate) distances fall in the range 2.363–2.437 Å and Gd—N 2.582 Å (central N) and 2.626–2.710 Å. Similar ions are present in the Ba[Gd(dtpa)-(H_2O)][648] and (CN_3H_6)$_2$[Gd(dtpa)(H_2O)][649] and in Ba[Nd(dtpa) (H_2O)].[650] On the other hand, a number of salts are known, like (NH_4)$_2$[Gd(dtpa)](H_2O), which have no water coordinated in the solid state. Instead, a hydrogen bonding network involving the ammonium ions causes Gd(dtpa) units to associate[651] into dimers, in contrast to the monomeric nature of Na_2[Gd(dtpa)(H_2O)].

Similarly Cs_4[Dy(dtpa)]$_2$ has an unprecedented dimeric structure in which Dy is nine-coordinate (as in other dtpa complexes) but with the ninth coordination position occupied by a bridging carboxylate, rather than a water molecule.[652] The structure and optical spectra of [(NH_2)$_3$]$_2$[Nd(dtpa)(H_2O)].$7H_2O$ have been reported.[653] The structure of K_2[Yb(dtpa)(H_2O)] has been determined at low temperatures; in addition to the bound water, there are six waters in the outer coordination sphere and another molecule hydrogen bonded to carboxylate oxygens and significantly near the metal ion, prompting a reassessment of the relaxivity data for the Gd analogue, taking account of second-sphere water molecules.[654] The structure found in salts like Na_2[Gd(dtpa)(H_2O)] is thought to represent the species found in solution data from a number of measurements on solutions containing [Ln(dtpa)(H_2O)$_n$]$^{2-}$ species, such as luminescence spectra of the Eu and Tb analogues indicate that one water molecule is coordinated.[655–658] A wide variety of complexes of amide derivatives of DTPA have been synthesized, in order to create neutral species; likewise a number of other octadentate pentacarboxylic acids and their derivatives have also been investigated. The structure of [Lu(bba-DTPA)(H_2O)] (bba-dtpa = bis(benzylamide) of dtpa) has been determined;[659] [Gd(bba-DTPA)(H_2O)] has a relaxivity comparable to other dtpa amide complexes. Complexes of the bis(phenylamide) of DTPA with Gd, Y, and Lu have been reported;[660] the gadolinium complex having a comparable proton relaxation enhancement to [Gd-DTPA]$^{2-}$. NMR indicates that the yttrium complex of the bis(butylamide) of DTPA exists as one eight-coordinate isomer in solution.[661] A ^{17}O-NMR study of amide derivatives of [Gd(DTPA)(H_2O)]$^{2-}$ has been carried out;[662] results indicating dissociatively activated water exchange. The potential MRI contrast agent, [Gd(DTPA-BMEA)(H_2O)] has very similar properties to the existing [Gd(DTPA-BMA)(H_2O)].[663] (DTPA-BMEA is compound (**19**)).

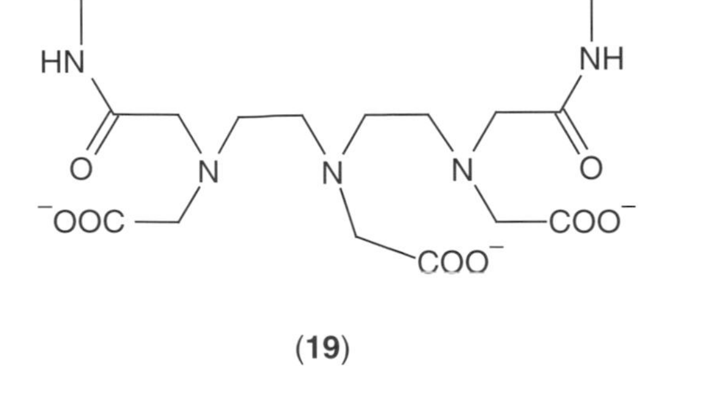

(19)

Stability constants have been determined for MS-325, a rationally designed contrast agent based on Gd(DTPA) with a sidechain containing a phosphate group and a lipophilic diphenylcyclohexyl moiety which gives a strong noncovalent interaction with human serin albumin. This gives MS-325 three advantages over [Gd(DTPA)]$^{2-}$, namely targeting the agent to blood; slowing down the tumbling time and hence improving the relaxation enhancement by up to an order of magnitude; and increasing the half-life of the drug *in vivo*.[664] MS-325 (**20**) has a high affinity for serum proteins and a greater proton relaxivity than Gd-DTPA itself.[665]

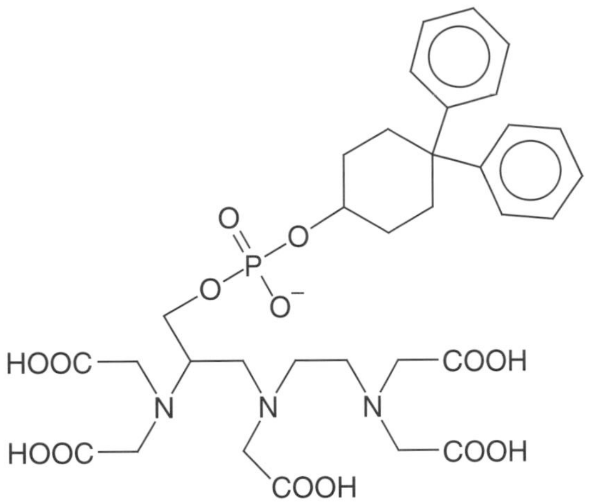

(**20**)

Five bis(amide) derivatives of DTPA form neutral Gd complexes[666] with similar relaxivity to [Gd(dtpa)(H$_2$O)]$^{2-}$. The structure of a gadolinium complex of a bis(amide) of DTPA (**21**), a potential contrast agent, has been determined.[667]

The Gd^{3+} complex of a diamide (**22**) has an octadentate ligand in a nine-coordinate tricapped trigonal prismatic molecule with one coordinated water molecule.[668]

Its relaxivity is comparable to that of [Gd(DTPA)(H$_2$O)]$^{2-}$. A study[669] of the complexes of DTPA, related ligands and diamides with Gd^{3+} and Ca^{2+} has shown that the amide groups in the diamides do not contribute to calcium complexation but do enhance Gd^{3+} complexation. [La(DTPA-dien)-(H$_2$O)]$_2$(CF$_3$SO$_3$)$_2$.18H$_2$O is a carboxylate-bridged dimer in the solid state whilst [Eu(DTPA-dien)]$_4$(CF$_3$SO$_3$)$_4$.6CF$_3$SO$_3$Na.20H$_2$O is a carboxylate-bridged tetramer; in solution, the europium tetramer breaks up and binds a water molecule.[670] The structures of other amide complexes have been determined.[671–677]

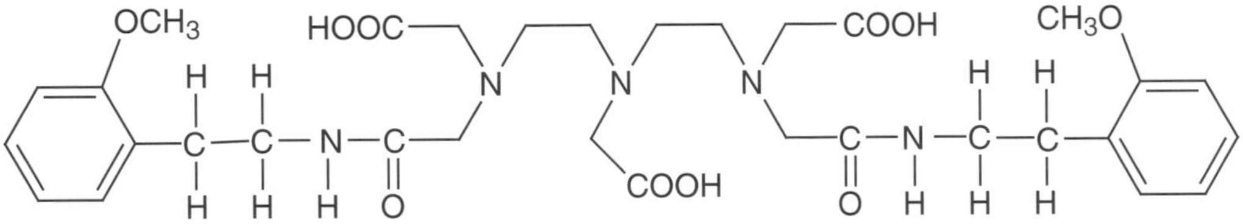

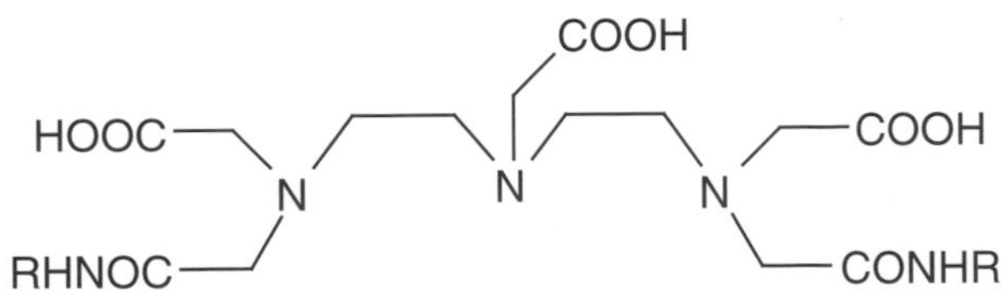

(**21**)

(**22**)

Relaxivity and other studies on other Gd DTPAbis(amide) complexes have been reported.[678–683] Dissociation kinetics of Ce and Gd complexes of bis(amide) ligands derived from DTPA have been studied.[684] Complexes of a new octadentate ligand H_5BOPTA (23) similar to DTPA have been made.

(4) 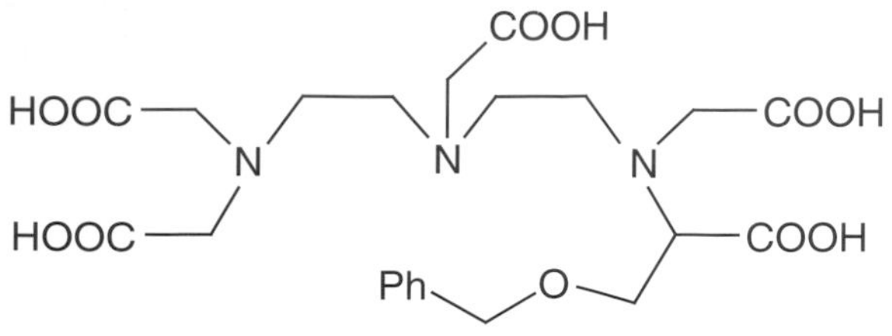 H_5BOPTA

(23)

$Na_2[Gd(BOPTA)(H_2O)]$ has nine-coordinate gadolinium, with one coordinated water molecule.[685] Spin-lattice relaxation data for the Gd^{3+} complex of 3,6,10-tri(carboxymethyl)-3,6,10-triazadodeca-nedioic acid (24) indicates that there is probably one water molecule coordinated.[686]

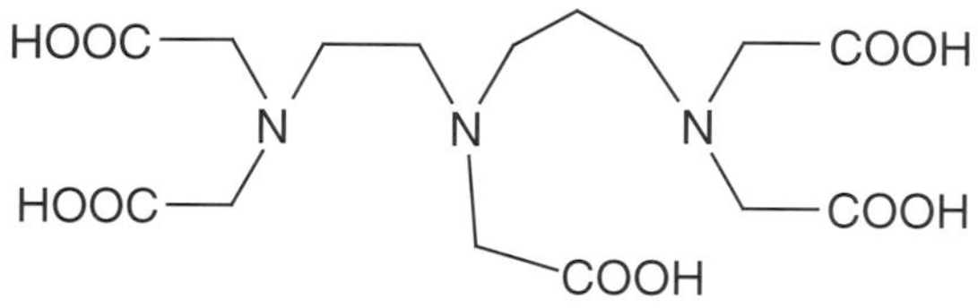

(24)

Very high relaxivities have been found for three Gd(DTPA-bisamide)alkyl copolymers.[687] A gadolinium complex of a substituted DTPA (25) was undergoing phase III clinical trials at the end of the twentieth century as a liver-specific contrast agent for MRI.[688]

Complexes of monoamide derivatives of dtpa with Ln^{3+} ions have been studied. Their stability constants are, as expected, less than those of dtpa itself.[689] Using $[Gd(DTPA)]^{2-}$ it has now been shown that these complexes get absorbed by the DNA of the cell they have been used to locate. If the gadolinium is subjected to thermal neutron treatment, short-range high-energy electrons are emitted that can kill the tumour cell whilst nearby healthy tissue is unaffected.[690]

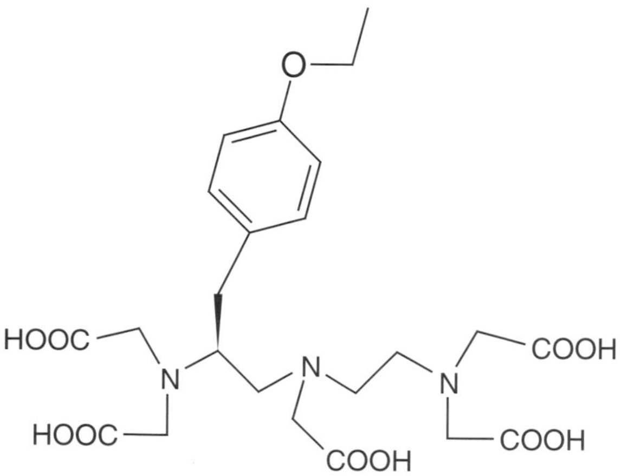

(25)

3.2.2.6.4 Complexes of DOTA and other complexing agents

The potentially octadentate ligand DOTA (1,4,7,10-tetra(carboxymethyl)-1,4,7,10-tetraazacyclo-dodecane, has been widely studied. It forms very stable lanthanide complexes with one water molecule coordinated, $[Ln(DOTA)(H_2O)]^-$ (26). Log K for $[Gd(DOTA)(H_2O)]^-$ is ~24.7, compared with the value of 22.46 for $[Gd(DTPA)(H_2O)]^{2-}$.

(26)

Crystal structures have been reported for a number of these complexes. The structure of Na[Gd-(DOTA)(H$_2$O)].4H$_2$O shows that gadolinium is nine-coordinate[691,692] in the solid state, with Gd—O(water) at 2.458 Å, and Gd—O distances in the range 2.362–2.370 Å and Gd—N distances between 2.648 Å and 2.679 Å. The Y,[692,693] Eu,[694] Lu,[695] and Ho[696] complexes have the same structure, whereas in the solid state, Na [La(DOTA)La(HDOTA)].10H$_2$O has a novel structure in which two [La(DOTA)]$^-$ units are joined by a carboxylate bridge.[697] Kinetics of formation and dissociation of [Eu(DOTA)]$^-$ and [Yb(DOTA)]$^-$ have been investigated.[698] NMR study of [Yb(DOTA)]$^-$ shows two conformations related by slow inter- and intramolecular exchange.[699] Conformation and coordination equilibria in DOTA complexes have been studied by ^{1}H-NMR.[700] An EXAFS study of [Gd(DOTA)(H$_2$O)]$^-$ and [Gd(DTPA)(H$_2$O)]$^{2-}$ in the solid state and solution permits comparison with existing solid-state diffraction data.[701] The stability constants of [Ce(DOTA)]$^-$ and [Yb(DOTA)]$^-$ are reported as $10^{24.6}$ and $10^{26.4}$ respectively.[702] Measurements of excited state lifetimes of europium(III) complexes with DOTA-derived ligands show that N—H and C—H vibrations allow a vibronic deactivation pathway of the Eu $^5D_0$ excited state; estimates of apparent hydration states can be made.[703] The solid-state structure of [Gd(DTMA)(H$_2$O)]$^{3+}$ (DTMA = DOTA tetrakis(methylamide) (27)) is a capped square antiprism.[704] In solution the complex has only limited stability (log $K = 12.8$). NMR studies of the La, Gd, Ho, and Yb complexes of the DOTA analogue (28) have been reported.[705]

Stability constants have been measured for the Gd and Y complexes of DOTA, DO3A (1,4,7,10-tetraazacyclododecane-1,4,7-triacetic acid) and HP-DO3A (10-(2-hydroxypropyl)-1,4,7,10-tetraazacyclododecane-1,4,7-triacetic acid); the Gd and Y complexes of HP-DO3A are isostructural.[706] (again capped square-antiprismatic nine-coordination). If one carboxylic acid sidechain in H$_4$DOTA is replaced by hydrogen, the resulting acid, H$_3$DO3A, forms neutral lanthanide complexes. In solution they are believed to have two water molecules coordinated so that the species present is [Gd(DO3A)(H$_2$O)$_2$]. Other derivatives have been made in which the fourth carboxylic acid group is replaced by hydroxyalkyl groups that are marketed and approved for use (—CH$_2$CH(CH$_3$)OH = Prohance; —CH(CH$_2$OH)CH(OH)CH$_2$OH = Gadovist). These compounds have a sidechain hydroxyalkyl group coordinated, in addition to four nitrogens, three carboxylate oxygens, and a water molecule (29).[706,707]

A temperature-dependent UV–visible study[708] on [Eu(DO3A)(H$_2$O)$_n$] shows the existence of a hydration equilibrium with $n = 1$, 2, strongly weighted towards [Eu(DO3A)(H$_2$O)$_2$]. Extrapolation to the Gd analogue indicated a water exchange rate in [Gd(DO3A)(H$_2$O)$_n$] to be twice as fast as in

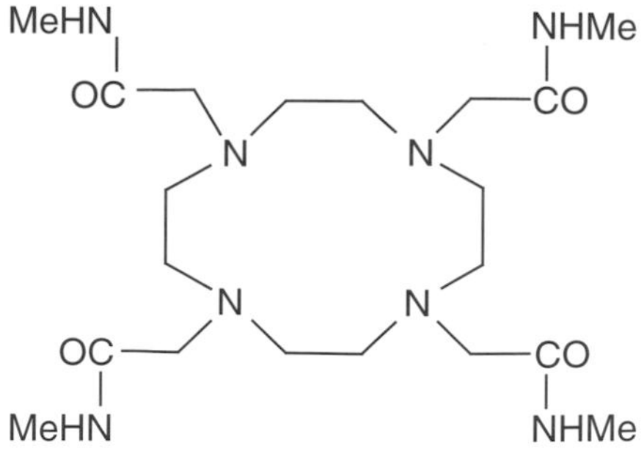

(DTMA)

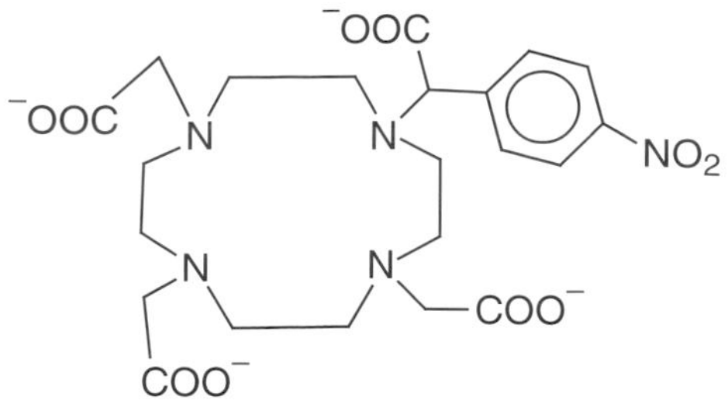

(27)   (28)

[Gd(DOTA)(H$_2$O)]$^-$ but still much slower than in [Gd(H$_2$O)$_8$]$^{3+}$(DO3A = 1,4,7-tris (carboxy-methyl)1,4-7,10-tetraazacyclododecane). The complex Pr[moe-do3a] (**30**) has been reported to be an *in vivo* NMR thermometer.[709]

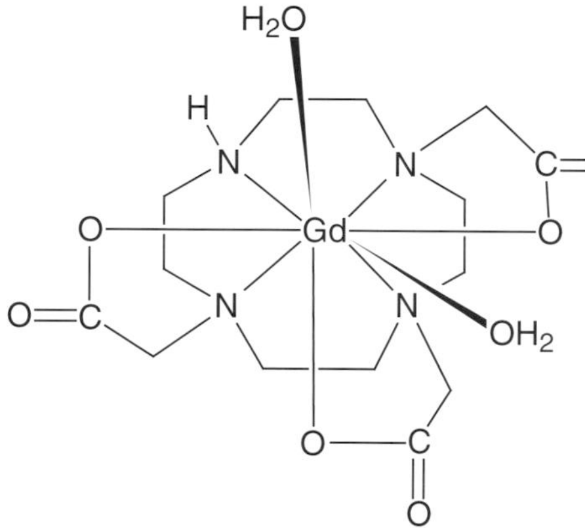

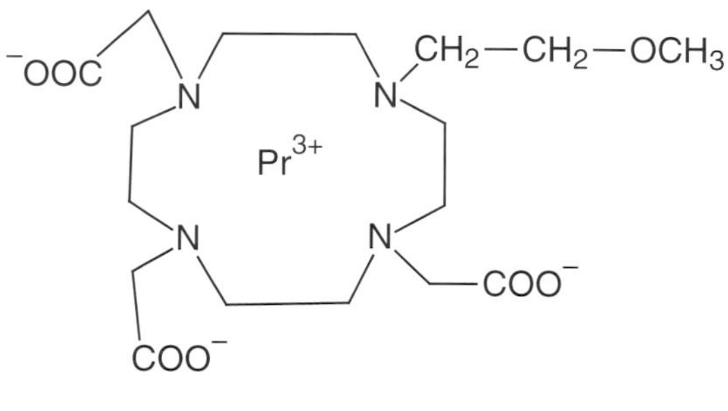

(29)   (30)

The thermodynamics of Gd^{3+} complexation by DOTA and DO3A, leading to determination of the thermodynamic parameters, has been studied.[710] The first gadolinium complex to bind two water molecules that appears to have a high relaxivity, and minimizes anion and protein binding, [GdaDO3A]$^{2-}$, has been described.[711] The gadolinium complex of BO(DO3A)$_2$$^{6-}$ (**31**), [BO{Gd-(DO3A)(H$_2$O)}$_2$], has been designed as a MRI agent with slow rotation and rapid water exchange[712] to enhance ^{1}H relaxivity.

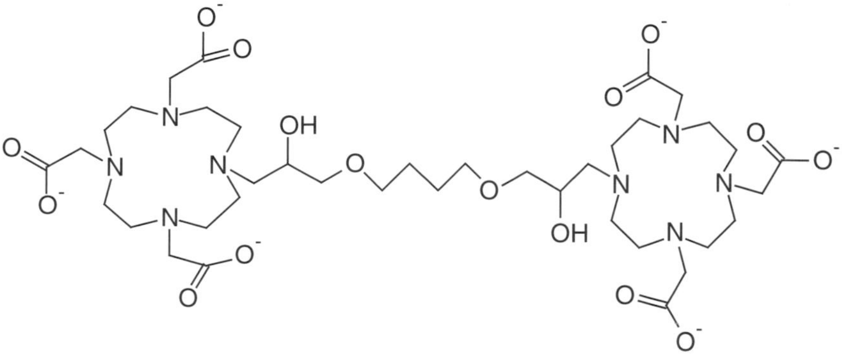

(31)

A DOTA-based contrast agent (**32**) has a high relaxivity owing to its capability to self-organize in micelles.[713] A gadolinium complex with a DOTA-derived ligand that acts as a "smart" MRI agent that reports on specific enzymatic activity has been reported.[714] A ligand based on a tetraaza macrocycle with four attached phosphinate groups forms stable eight-coordinate lanthanide complexes; although water does not enter the coordination sphere, the Gd complex is a promising outer-sphere MRI agent.[715] Diamides and other ligands affording neutral gadolinium complexes with concomitant advantages as MRI agents continue to be studied. Neutral Eu, Gd, and Yb complexes of a tetraaza macrocycle (**33**) with three phosphinate and one carboxamide side arms have been synthesized.[716]

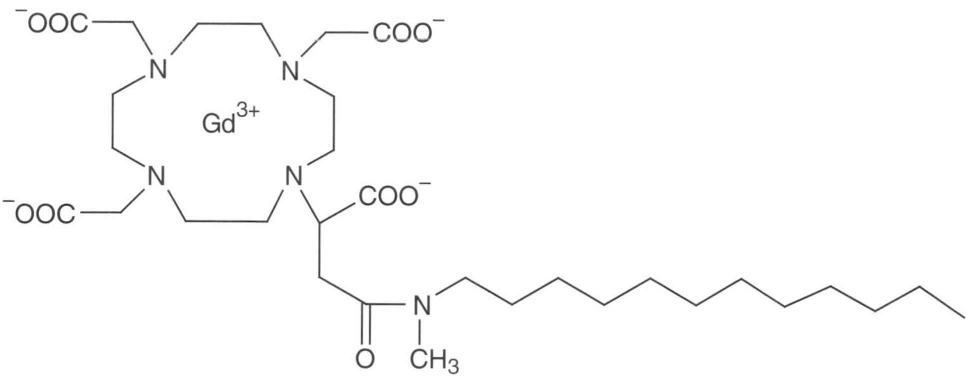

(**32**)

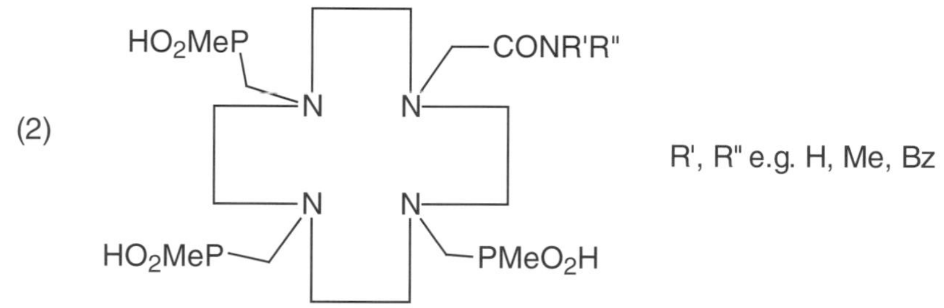

(2)

R', R" e.g. H, Me, Bz

(**33**)

Lanthanide complexes $[Ln(DO2A)(H_2O)_n]^+$ of the potentially hexadentate DO2A have been studied by measuring the lanthanide-induced shifts in the ^{17}O-NMR spectra.[717] Analysis of the contact contribution indicates a decrease in the hydration number from $n = 3$ (Ln = Ce-Eu) to $n = 2$ (Ln = Tb-Lu). Study of the $^7D_0 \rightarrow {}^5F_0$ transition in the UV–visible spectra of the Eu complex indicates an equilibrium in solution between eight and nine-coordinate species. (DO2A = 1,7-bis(carboxymethyl)1,4-7,10-tetraazacyclododecane).

Gadolinium complexes of *N*-tris(2-aminoethyl)amine-*N'*,*N'*,*N''*,*N''*,*N'''*,*N'''*-hexaacetic acid (H_6ttaha) (**34**) and *N*-(pyrid-2-yl-methyl)ethylenediamine-*N*,*N'*,*N''*-triacetic acid (H_3PEDTA) are potential NMR imaging agents.[718]

Two lanthanide complexes of triethylenetetramine-*N*,*N*,*N'*,*N'*,*N'''*,*N'''*-hexaacetic acid (H_6ttha), $[C(NH_2)_3]_2[La(ttha)].3H_2O$ and $[C(NH_2)_3]_2[Nd(ttha)].5H_2O$, have been characterized.[719] The former compound has 10-coordinate lanthanum, with four nitrogens and all six carboxylate groups bound to the metal, though the one protonated carboxylate has a La—O distance about 0.3 Å longer than the other La—O distances; the neodymium compound has nine-coordinate neodymium, with the —COOH group uncoordinated. The nine-coordinate anion in $[C(NH_2)_3]_2[Gd(Httha)].5H_2O$ has no coordinated waters and thus is a poor MRI agent[720] but can be used as a reference standard since its relaxivity is caused by "outer-sphere" (i.e., non-coordinated) water interactions.

The ttaha ligand is heptadentate (H_6ttaha = tris(2-aminoethyl)amine hexaacetic acid) but nine-coordination is attained in $[C(NH_2)_3]_3[Gd(ttaha)]\cdot 3H_2O$ in the solid state by coordination of two

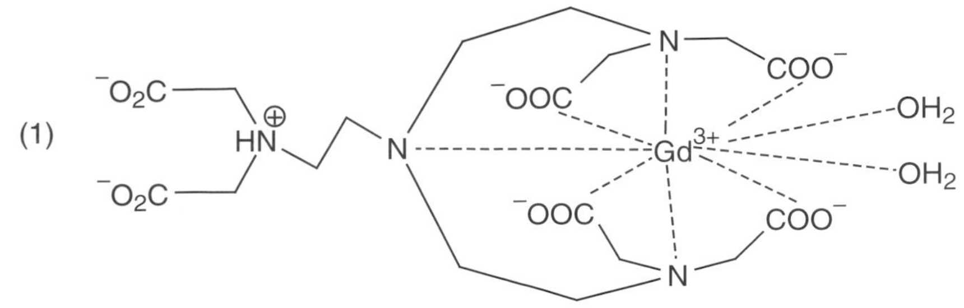

(34)

carboxylate oxygens from neighboring complexes. However in solution, when binding to Gd^{3+}, the ttaha has one leg free, acting as a heptadentate ligand; two water molecules also coordinate, giving it a high relaxivity compared to many MRI agents,[721] so that it acts as a MRI agent. K[La(Httha)(H$_2$O)].8H$_2$O (35) (H$_6$ttha = triethylenetetraamine hexaacetic acid) has lanthanum in a bicapped square antiprismatic geometry[722] whilst K$_3$[Yb(ttha)].5H$_2$O (36) has four nitrogens and five carboxylate oxygens bound to ytterbium.[723]

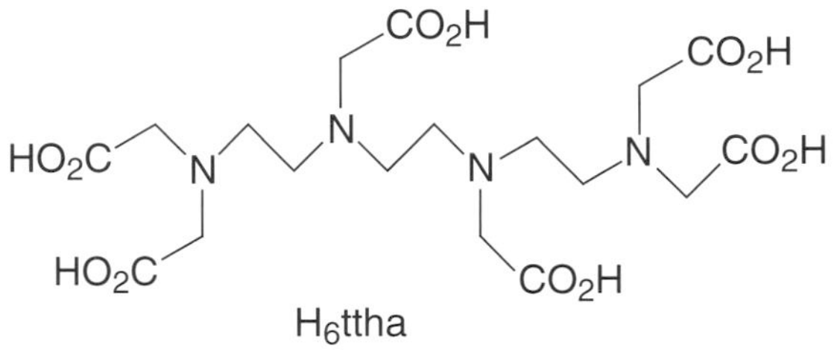

H$_6$ttha

(35)

H$_6$ttaha

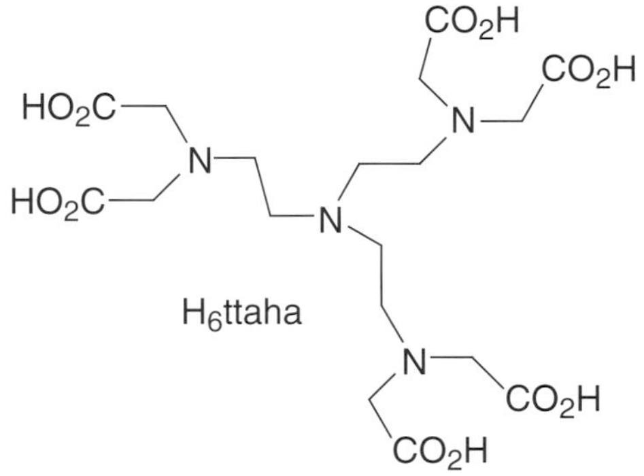

(36)

Solution dynamics of complexes of EGTA^{4-} (3,12-bis(carboxymethyl)-6,9-dioxa-3,12-diazate-tradecanedioate) have been studied; a change in coordination number from 10 to eight is believed to occur across the lanthanide series.[724] Binuclear La and Y complexes of 9,14-dioxo-1,4,7,10,13-pentaaza-1,4,7-cyclopentadecanetriacetic acid have nine-coordinate lanthanides.[725] Lanthanide polyamine carboxylates and complexes of macrocycles with pendant amide groups show promise as catalysts for RNA cleavage;[726,727] such complexes have considerable stability in aqueous solution.[728] Complexes of amides of calixarenes have been investigated, including study of relaxivity.[729,730] A 10-coordinate Gd complex with a relaxivity 3.5 times greater than [Gd(dtpa)]$^{2-}$ has been reported.[731] [Gd(DOTP)]$^{5-}$ (DOTP = 1,4,7,10-tetraazacyclododecane-N,N',N',N'''-tetrakis (methylenephosphonate) has been used as a relaxation agent in the study of human adult haemoglobin.[732]

Stability constants for [Ln(dotp)]$^{5-}$ complexes have been determined; log K ranges from 27.6 (La) to 29.6 (Lu) (dotp = 1,4,7,10-tetraazaccyclodododecane-1,4,7,10-tetrakis(methylenephosphonic acid).[733] In view of its use as a cation-shift reagent, it is notable that ^{23}Na-NMR studies[734] show NH$_4^+$ and K$^+$ compete effectively with Na$^+$ for the binding sites on [Tm(dotp)]$^{5-}$. Ca^{2+} and Mg^{2+} also complex with [Tm(dotp)]$^{5-}$.

Molecular mechanics calculations have been reported[735] for a number of gadolinium complexes, including those with EDTA, dtpa, dtpa–bma and do3a. Other molecular mechanics calculations using a simple force field, computable on a PC, have been reported for [Gd(edta)(OH$_2$)$_3$]$^-$ and a variety of Schiff base complexes.[736] NMR and EPR studies are also reported on a number of compounds of this type.[737] In the solid state, [La(pedta)(H$_2$O)].2H$_2$O (pedta = N-(pyrid-2-ylmethyl)ethylenediamine-N,N',N'-triacetate) adopts a polymeric structure, giving 10-coordinate La; in solution, luminescence results for the Eu complex indicate three coordinated water molecules.[738]

Complexes $[Ln(hedtra)(H_2O)_n]$ (Ln = most lanthanides; H_3hedtra = N-(2-hydroxyethyl)ethylene-diamine triacetic acid) have been synthesized. They fall into three series; $[M(hedtra)(H_2O)_2].3H_2O$ (M = Ho, Tm) have eight-coordinate square antiprismatic coordination in which the acid is coordinating through the hydroxo oxygen in addition to the two nitrogens and three carboxylate oxygen atoms.[739] A DOTA-based peptide complexed with ^{90}Y has been studied with a view to receptor-mediated radiotherapy. The structure of the yttrium complex of the model peptide DOTA-D-PheNH$_2$ shows it to have eight-coordinate yttrium.[740] Gadolinium-loaded nanoparticles have been examined as potential contrast agents.[741] Bis(amide) derivatives of ethylenedioxydiethylenedinitrilotetraacetic acid (H_4egta) form cationic gadolinium complexes.[742] The increase in positive charge results in a slower water exchange rate; the complexes have rather lower stability constants than $[Gd(egta)]^-$ and are thus not suitable for use as MRI agents *in vivo*. In a study of the luminescence properties of $[EuTETA]^-$ and $[EuDOTA]^-$, (TETA = 1,4,8,11-tetraazacyclotetradecane-1,4,8,11-tetraacetate), the structure of the $[EuTETA]^-$ ion has been determined.[743] Lanthanide complexes of a new potentially heptadentate macrocycle, 1,4,7,10-tetraazacyclodecane-1,4,7,10-tetramethyltrimethylenetris(phenylphosphinic acid) (H_3L), have been studied. Dimeric complexes $[LnL]_2$ have phosphinic acid bridges and eight-coordinate lanthanides; additionally a water molecule is present at a distance strongly dependent upon the Ln^{III} ion. The dimer is strong enough to resist coordination by donors like Ph_3PO.[744] $CoCO_3$, $Gd(OH)_3$, and H_4DCTA react to form a novel cluster $[Gd_2Co_2(\mu_4\text{-}O)(\mu\text{-}H_2O)(DCTA)_2(H_2O)_6].10H_2O$, which features four metal ions round a central oxide and with bridging waters and carboxylate groups.[745] A study of the kinetics of formation of DCTA complexes of Ln^{3+} ions indicates that they are first order in the reactants[746] (H_4DCTA = trans-1,2-diaminocyclohexane-N,N,N',N'-tetraacetic acid).

A nonadentate ligand based on 1,4,7-triazacyclononane has three amine nitrogen, three imine nitrogen, and three carboxylate oxygen donor atoms; it forms isostructural La, Sm, and Y complexes.[747] Tetrahydrofuran-2,3,4,5-tetracarboxylic acid (THFTCA) forms unexpectedly weak complexes with the uranyl ion and unexpectedly large sensitivity to the ionic radius of Ln^{3+} in complex formation. A study of the thermodynamics of complex formation by THFTCA with Ln^{3+} ions (Ln = La, Nd, Eu, Dy, Tm) and UO_2^{2+} indicates that these anomalies arise from the complexation entropy rather than the enthalpy.[748] Derivatives of 3,6,10-tri(carboxymethyl)-3,6,10-triazadodecanoic acid form Gd^{3+} complexes that are more stable than the Gd–DTPA complex and optimal water exchange rates, thus having potential as MRI contrast agents.[749] Reversible intramolecular binding of a sulfonamide sidechain in some gadolinium macrocyclic complexes has been used to effect pH-dependent relaxivity.[750] Other studies with complexes of ligands derived from tetraazamacrocycles include phosphonate ligands,[751] chiral amides,[752] di-[753] and triacids.[754,755] The role of water exchange in attaining maximum relaxivity in MRI agents has been examined.[756]

3.2.2.6.5 *Complexes of mixed Group 15 and 16 donors as spectroscopic probes*

In addition to their applications as MRI contrast agents, there is considerable interest in applications of lanthanide complexes of these complexing agents as spectroscopic probes.

The area has been well reviewed and selected examples are discussed here.[757,758] Ln^{III} (Ln = Eu, Tb, Y) complexes of a macrocycle with a pendant dansyl group ((37): R = CH_3, CF_3, OCH_3) have been synthesized; reversible intramolecular binding of the dansyl group is achieved by varying the pH (Scheme 6). This does not affect luminescence of the Tb and Eu complexes, as the chromophore triplet state is not populated, but protonation of the NMe_2 group does sensitize luminescence of the Eu complex.[759] The effect on the Eu^{3+} or Tb^{3+} luminescence, and its pH dependence, of changing ligand substituents and hence the energy of the ligand singlet or triplet states has been examined. The resulting complexes have been incorporated into thin film sol-gel matrices and evaluated as pH sensors.[760] Time-resolved near-IR luminescence of the Yb^{3+} and Nd^{3+} complexes of the Lehn cryptand ligand have been investigated. Luminescent lifetimes of the Yb complex have been used to determine the number of inner-sphere water molecules, but the Nd complex exhibits unexpectedly long luminescent lifetimes.[761] A terbium complex (38) is the first example of a molecular logic gate corresponding to a two-input INHIBIT function; the output, (a terbium emission line) is only observed when the "inputs," the presence of proteins and the absence of oxygen, are both satisfied.[762]

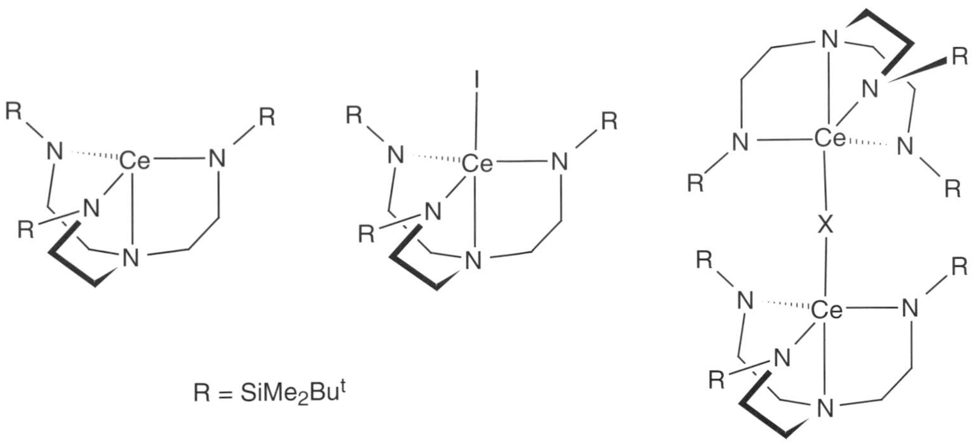

R = SiMe₂Buᵗ

Scheme 6

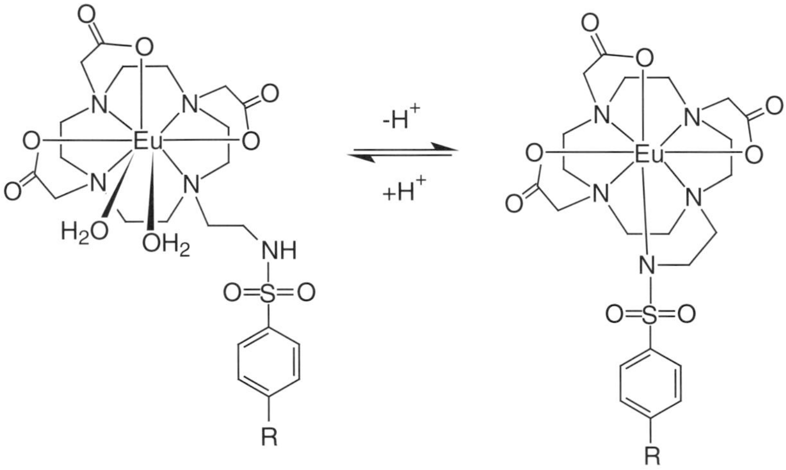

(**37**)

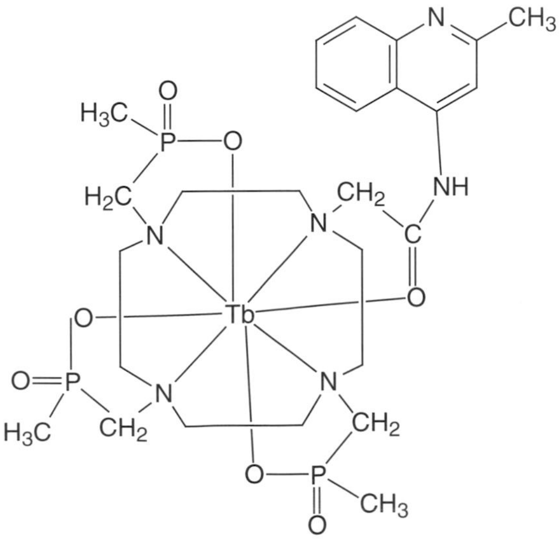

(**38**)

When Nd or Yb complexes of a similar macrocycle are covalently linked to a palladium porphyrin complex, this sensitizes near-IR emission from the lanthanide, enhanced in the absence of oxygen and in the presence of a nucleic acid.[763] Another application of luminescence accompanies the terbium complex shown in (39) whose luminescence is enhanced by binding to zinc, and can therefore signal for that metal.[764]

(39)

Terbium and europium complexes (40) with pH sensitive luminescence have been described.[765] At high pH, the amine group in the sidechain is deprotonated and able to coordinate to the metal.

A study has been made of the emission of some related Tb and Eu macrocylic complexes, immobilized in a sol-gel glass, which is made pH-dependent either by perturbing the energy of the aryl singlet or triplet state, or by modulating the degree of quenching of the lanthanide excited state.[766] The effect of bicarbonate chelation on the polarized luminescence from chiral europium (41) and terbium complexes has been followed.[767,768] The change in geometry (and chirality) at the metal on bicarbonate binding has a pronounced effect upon the emission polarization.

The luminescence of terbium (42) and europium macrocyclic complexes have been studied and found to depend upon pH. The phenanthridine sidechain acts as a photosensitizer and (42) exhibits luminescence quenched by molecular oxygen; corresponding Eu compounds exhibit halide-quenched emisssion.[769,770]

3.2.2.7 Complexes of Macrocycles Involving Group 15 and 16 Donors

3.2.2.7.1 *Porphyrins and phthalocyanines*

Compounds fall into a number of series, simple mono- derivatives, simple sandwich compounds (both homogeneous and heterogeneous) and triple-decker sandwiches. In the case of the sandwich compounds, both anionic species and neutral species are known, the latter containing the tervalent lanthanide bound to one one-electron oxidized tetrapyrrole ligand. Reactions of Ln (acac)$_3$ with H$_2$TPP and other porphyrins in boiling solvents like trichlorobenzene have been reported to give products like [Ln(acac)(porph)] although for the earlier lanthanides Ln(H)(porph)$_2$ seems more usual. On extended reflux, the double-decker sandwiches [Ln$_2$(porph)$_3$] tend to result. The initial synthesis gives a high yield but this is reduced considerably on chromatographic purification.[771] Thus refluxing [Eu(acac)$_3$] with H$_2$OEP in 1,2,4-trichlorobenzene for four hours leads to a mixture of [Eu(OEP)$_2$] and [Eu$_2$(OEP)$_3$] (along with a small amount of [Eu(OEP)(OAc)]), the former being a good deal more soluble in organic solvents, so that the mixture is separable by crystallization.[772] [Eu(OEP)$_2$], isomorphous with [Ce(OEP)$_2$] and similar compounds, has square antiprismatic coordination of Eu with staggered porphyrin rings, Eu—N distances being in the range 2.473(4)–2.556(4) Å. [Eu(OEP)$_2$] undergoes

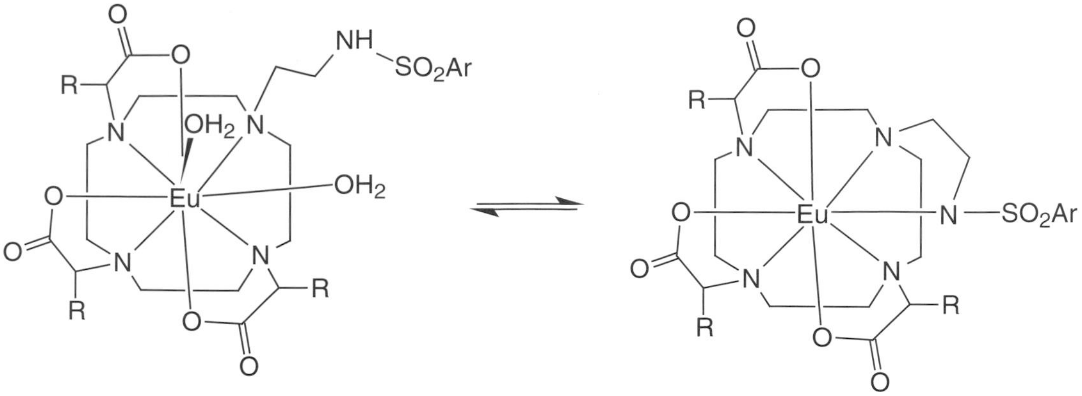

R e.g. CF$_3$, CH$_3$, OCH$_3$

Ar e.g. p-CF$_3$, p-CH$_3$, p-CH$_3$OPh

(40)

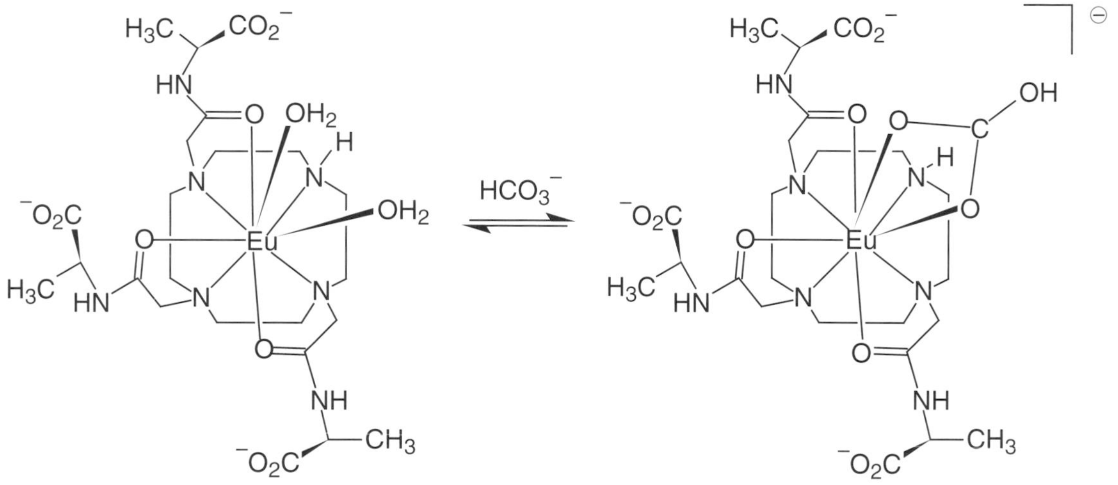

(41)

reduction to [Eu(OEP)$_2$]$^-$ using sodium naphthalenide. Similarly cerium triple-deckers Ce$_2$(OEP)$_3$ and double-deckers Ce(OEP)$_2$ have been reported.[773] The triple-decker is paramagnetic and relatively insoluble, whilst the double-decker is more soluble and diamagnetic. It can be oxidized to [Ce(OEP)$_2$]$^+$. Syntheses generally give a mixture of products, so that careful separations are often the order of the day. Thus refluxing freshly-prepared [Sm(acac)(TPP)] with Li$_2$OEP for three hours leads to a mixture of various complexes including [SmH(OEP)(TPP)], [SmH(TPP)$_2$], and [Sm$_2$(OEP)$_3$], separable by chromatography.[774] [SmH(OEP)(TPP)] has a sandwich structure with square antiprismatic coordination of Sm, with mean Sm—N (tpp) distances of 2.538(4) Å and Sm—N (oep) of 2.563(4) Å. Other, less forcing, methods have been described involving reaction of homolytic alkyls and alkoxides with H$_2$OEP.[775,776] [LuR$_3$] (R = CH(SiMe$_3$)$_2$) react with H$_2$OEP in toluene forming [Ln(OEP)R]. [Lu(OEP)(CH(SiMe$_3$)$_2$)] has a square pyramidal structure with a highly dished porphyrin skeleton; Lu—C is 2.374(8) Å and Lu—N bonds are in the range 3.236(7)–3.296(7) Å; the lutetium atom is 0.918 Å above the plane of the porphyrin ring. Steric effects are clearly important as this route does not work with the lanthanum or yttrium analogues, or with more demanding porphyrins such as tetratolylporphyrin or tetramesitylporphyrin. [Lu(OEP)(CH(SiMe$_3$)$_2$)] reacts with protic species, forming the aryloxide [Lu(OEP)(OAr)] (Ar = O-2,6-C$_6$H$_3$Bu$^t_2$). A more direct route is the reaction between [M(OAr)$_3$] (M = Lu, Y) and

(**42**)

H$_2$OEP, which yields [M(OEP)(OAr)]; these then can be reacted with LiCH(SiMe$_3$)$_2$ in a more convenient route to the alkyls [M(OEP)(CH(SiMe$_3$)$_2$)] (M = Lu, Y). These undergo hydrolysis forming the dimers [(OEP)M(μ-OH)$_2$M(OEP)]. [Lu(OEP)(OAr)] reacts with MeLi forming [(OEP)Y(μ-Me)$_2$Li(OEt$_2$)] which in turn reacts with AlMe$_3$ affording [(OEP)Y(μ-Me)$_2$AlMe$_2$]; this activates oxygen forming [(OEP)Y(μ-OMe)$_2$AlMe$_2$].[775,776] An up-to-date high-yield route[777] for later members of the [Ln(TPP)Cl] series is provided by reaction of [LnCl$_3$(THF)$_3$] (Ln = Ho, Er, Tm, Yb) with Li$_2$TPP(DME)$_2$ in refluxing toluene. Use of the hydrocarbon solvent forces precipitation of LiCl, reducing possibilities of –ate complexes; after removal of LiCl, [Ln(TPP)Cl(DME)] crystallizes from the cold reaction mixture. Structures have been reported for the seven-coordinate Ho and Yb compounds; the DME and Cl are coordinated to the metal on the same side of the porphyrin ring, which takes on a saddle-shaped distortion. In [Ln(TPP)Cl(DME)], Yb—N = 2.324(2) Å, Yb—Cl = 2.605 Å, and the Yb—O distances are 2.395 Å and 2.438 Å, with Yb 1.105(1) Å above the N$_4$ plane of the ring; corresponding values for the Ho compound are Ho—N = 2.357(2) Å, Ho—Cl = 2.603 Å, and the Yb—O distances are 2.459 Å and 2.473 Å, with Ho 1.154(3) Å above the N$_4$ plane. *In situ* prepared[778] [Ln{(N(SiMe$_3$)$_2$}$_3$·x(Li(THF)$_3$)] reacts with H$_2$(porph) (porph is, for example, TTP, TMPP) forming [Ln(TMPP)(OH$_2$)$_3$]Cl (Ln = Yb, Er, Y) and [Yb(TTP)(OH$_2$)$_2$(THF)]Cl. Using an excess of [Ln(NR$_2$)$_3$·x(Li(THF)$_3$)] the product depends upon the nature of both R and the porphyrin, giving either mono or dinuclear complexes.[779] Compounds studied structurally include [YCl(TTP)-{(CH$_3$OCH$_2$CH$_2$)$_2$O}]; [YCl(TPP)$_2$(H$_2$O)][YCl(TPP)(H$_2$O)(THF)].2THF.H$_2$O; [{Yb(TMPP)(μ-OH)}$_2$]; [{Yb(TMTP)}$_2$(μ-OH)(μ-OCH$_2$CH$_2$OCH$_3$)]; and [YbCl(TPP)(H$_2$O)(THF)] (TTP = tetratolylporphyrin; TMPP = tetrakis(p-methoxyphenyl)porphyrin). [Yb(N(SiMe$_3$)$_2$)$_3$·xLiCl(THF)$_3$] reacts with H$_2$TPP in refluxing bis(2-methoxyethyl)ether forming the oxalate bridged dimer [((TPP)Yb(MeO(CH$_2$)$_2$OMe))$_2$(μ-η^2:η^2-O$_2$CCO$_2$)].[780]

Syntheses of [Ln(TPP)Cl] (Ln = Eu, Gd, Tm) and [Tm(TPP)(OAc)] have been given.[781] Gadolinium is eight-coordinate in [Gd(TPP)(acac)(H$_2$O)$_2$]·6H$_2$O.3TCB.[782] A achiral gadolinium porphyrin complex extracts chiral amino acids from aqueous solution forming complexes that exhibit chirality-specific CD activities.[783] The structure of [Tb(β-Cl$_8$tpp)(O$_2$CMe)(Me$_2$SO)$_2$] shows a pronounced saddle distortion of the porphyrin ring.[784] The structures of two dilutetium cofacial porphyrin dimers, in which each lutetium is bound to two(μ-OH) groups as well as to four porphyrin nitrogens, have been determined.[785] Studies of [Ln(OEP)$_2$] (La, Nd, Eu) indicate that the electron hole is delocalized over both the porphyrin rings in the electronic and vibrational timescales;[786] similarly studies by a number of techniques on [Ln$_2$(OEP)$_3$] systems indicate strong intra-ligand interactions with electron hole again being delocalized over all three rings.[787] Proton NMR spectra of [Ln$_2$(OEP)$_3$] give evidence for inter-ring steric crowding and concomitant limited rotation of the OEP ligands.[788] NMR spectra of the asymmetric double-decker sandwich compounds [LnH(oep)(tpp)] (Ln = Dy, Lu) have been analyzed.[789] Synthesis of lipophilic lanthanide(III) bis(tetrapyridylporphyrinates) and their conversion into water-soluble N-methylated systems are reported.[790] Lanthanide porphyrazine complexes, [Ln(OPTAP)$_2$] (Ln = Ce, Eu, Lu) and a triple-decker [Eu$_2$(OPTAP)$_3$] have π-radical electronic structures whilst the sandwich structure of the cerium compound has been confirmed by X-ray diffraction.[791] An investigation into complexes of general type Ln_xPc_y shows that Ln_2Pc_3 mainly obtained for La and Nd, becoming less plentiful for Eu and Gd; later on in the series [LnHpc$_2$] is become the norm for Dy, Yb, and Lu.[792] Sandwich structures are reported for $(Bu_4N)[Ln^{III}(pc)_2]$ compounds (Ln = Nd, Gd, Ho, Lu). The skew angle between the two pc rings increases as the ionic radius decreases, apparently depending upon the π—π interaction between the two rings. Syntheses of the green [Ln^{III}(pc)$_2$] (Ln = La–Lu except Ce, Pm; pc = phthalocyaninate) by electrochemical oxidation of $(Bu_4N)[Ln^{III}(pc)_2]$ are reported;[793] one phthalocyanine ligand is present as a pc$^-$ π-radical, the other as pc^{2-}. The structure of α1-[Er(pc)$_2$] shows two equivalent, slightly saucer-shaped, pc rings and distorted square antiprismatic coordination of erbium,[794] whilst in [La(pc)$_2$]·CH$_2$Cl$_2$ one pc is planar, the other convex, again with square antiprismatic coordination.[795] Spectroscopic and structural information on [Na(18-crown-6)][Lupc$_2$], [NBu$_4$] [Lupc$_2$], and [LuHPc$_2$] has been published.[796] [Lu(pc)$_2$], presumably [Lu(pc)(pc·)] has a sandwich structure with staggered rings.[797,798] A study of the magnetic properties of the tert-butylphthalocyanine complexes [Lu(Bupc)$_2$] concludes that the green forms are H[Lu(Bupc)$_2$].[799] The lanthanide ion has square antiprismatic coordination in the isostructural series (PNP)[Lu(pc)$_2$] (Lu = La–Tm).[800] Detailed characterization has been reported of the compounds [Lu(2,3-nc)$_2$] and [Lu(2,3-nc)(pc)] (2,3-nc = 2,3-naphthalocyaninate) in which the unpaired electron is delocalized over the two macrocycles.[801] Bis(phthalocyaninato) lutetium complexes with long-chain alkyl groups substitutents on the pyrrole rings display three kinds of discotic mesophase and also three primary colors of electrochromism.[802] A new sandwich bis(phthalocyanine) complex of lutetium with eight hexylthio groups has been characterized.[803] The synthesis has been reported of the triple-decker sandwich, [Lu$_2$(1,2-naphthacyaninato)$_3$].[804] Reaction of Gd with 1,2-dicyanobenzene under a nitrogen atmosphere leads to a bicyclic gadolinium phthalocyanine with a trigonal prismatic GdN$_6$ coordination geometry.[805] Reaction of a metal salt with 1,2-cyanobenzene affords K[M(pc^{2-})$_2$] (M = La, Ce, Pr, Sm), convertible into other salts Bu$_4$N [M(pc^{2-})$_2$] (M = La, Ce, Pr, Sm) and Pe$_4$N [La(pc^{2-})$_2$]. The metallic bronze [M(pc$^-$)$_2$]I$_2$ (M = Sc, Y) have been made similarly in the presence of NH$_4$I; these react with Bu$_4$NOH forming Bu$_4$N[M(pc^{2-})$_2$]·xMeOH (M = Sc, Y; 0 < x < 1). Crystal structures show M—N distances that increase monotonically with ionic radius of the metal ion.[806] [Ln^{III}(pc)(tpp)] (Ln = La, Pr, Nd, Eu, Gd, Er, Lu, Y) have been synthesized;[807] their spectroscopic properties are consistent with a one-electron oxidized π-radical phthalocyanine group, Pc$^{·-}$. The structures of [La(pc)(tpp)] and [Gd(pc)(tpp)]$^+$SbCl$_6^-$ were determined. The two rings in these compounds adopt a staggered arrangement. In [La(pc)(tpp)], the average La—N distances are 2.520(3) Å for the porphyrin ring and 2.590(3) Å for the phthalocyanine ring, the corresponding values for [Gd(pc)(tpp)]$^+$SbCl$_6^-$ being 2.452(7) Å and 2.450(7) Å repectively. A one-pot synthesis has been reported[808] for [Ln(Nc)(TBPP)] (Ln = La, Pr, Nd, Sm–Tm) from the reaction of Ln-(acac)$_3$.nH$_2$O with the free naphthalocyanines and porphyrins in the presence of 1,8-diazabicyclo [5.4.0]undec-7-ene in n-octanol. Dilanthanide triple-decker porphyrin and phthalocyanine complexes have excited interest for some time, not least because of their potential in molecular information storage applications. Originally, methods for making homoleptic triple-decker porphyrin complexes relied on refluxing a lanthanide acetylacetonate with a porphyrin in 1,2,4-trichlorobenzene, forming a lanthanide acetylacetonate porphyrin complex, which was subsequently reacted with a lithiated phthalocyanine.[809] Previous synthetic routes have tended to yield mixtures of [(Pc)Ln(Por)Ln(Pc)], [(Pc)Ln(Pc)Ln(Por)], and [(Por)Ln(pc)Ln(Por)]. Sandwich compounds result from the action of Li$_2$Pc on [Ln(acac)(TPP)], of the types [(TPP)Ln(pc)Ln(TPP)] and [(pc)Ln(pc)Ln(TPP)] along with [(pc)Ln(TPP)].[810] New routes have been described to unsymmetrical [Ln(pc)(pc')] and [Ln(pc)(por)] systems.[811,812] Heteroleptic triple deckers [Eu$_2$(pc)$_2$(por)] and [Eu$_2$(pc)(por)$_2$] have also been synthesized.[813] [CeI((N(SiMe$_3$)$_2$)$_2$] with a porphyrin to form a half-sandwich [(Por)EuX] or [(Por)CeX'] (X = Cl, N(SiMe$_3$)$_2$; X' = I, N(SiMe$_3$)$_2$); this on reaction with Li$_2$Pc gives a double-decker ([(Por)Ln(Pc)]. [(Por')EuX] react with [Eu(pc)$_2$] to form a triple-decker; similarly, [(Por^1)CeX']) reacts with europium double deckers to give triple deckers [(Por^1)Ce(tBPc)Eu(Por^2)] and [(Por^1)Ce(tBPc)Eu(tBPc)].[814] In the triple-decker [(T4-OMePP)Nd(Pc)Nd(Pc)], the three rings are exactly parallel and perpendicular to the Nd—Nd axis. The Nd—Nd separation is 3.688(9) Å, the Nd—N (porphyrin) distance is 2.47(2) Å, the Nd—N(Pc) distance being 2.43(2), 2.57(2), and 2.74(2) Å, the latter being that involving the neodymium also bound to the porphyrin. The longest Nd—N distances are those involving the shared phthalocyanine ring. The neodymium sandwiched between the two phthalocyanines has square antiprismatic coordination, whilst the environment of the other neodymium is described as a distorted cube;[815] (T4-OMePP = tetrakis(4-methoxyphenyl)porphyrin). Reaction of [Ce(acac)$_3$(H$_2$O)$_2$] with H$_2$TPP, followed by further reaction with Li$_2$Pc (and correspondingly with other prophyrins and phthalocyanines) led to a mixture of [Ce^{IV}(TPP)(Pc)], [(TPP)Ce^{III}(Pc)Ce^{III}(TPP)], and [(Pc)Ce^{III}(TPP)Ce^{III}(Pc)]. The structures of the compounds show distorted cubic coordination of cerium.[816] In [(TPP)Ce^{III}(Pc(OMe)$_8$)Ce^{III}(TPP)], Ce—N (porphyrin) = 2.469(3) Å, and Ce—N (pc) 2.721(3) Å, with Ce—Ce 3.844(3) Å; in [(Pc)Ce^{III}(TP(MeO)P)-Ce^{III}(Pc)] Ce—N (porphyrin) = 2.727(2) Å, and Ce—N (pc) 2.473(2) Å, with Ce—Ce 3.664(2) Å. As seems to be general, the bigger bond lengths involve the "shared" ring. Pc(OMe)$_8$ = 2,3,9,10,12,13,17,18-octamethoxy(phthalocyaninate); TP(MeO)P = mesotetra (anisyl)porphyrinate. [Sm^{III}H(oep)(tpp)] has a sandwich structure, again with square antiprismatic coordination;[817] there is evidence to suggest that it is best represented by [Sm^{III}(oepH)(tpp)]. The first yttrium porphyrinogen complex has been prepared.[818] Remarkably, porphyrinogen complexes of neodymium and praseodymium fix N$_2$ and reduce it to the N$_2^{2-}$ anion in dimeric μ-N$_2$ complexes.[819] Ln(acac)$_3$ react with (tpyp)H$_2$ forming [Ln(tpyp)(acac)] (tpypH$_2$ = meso-tetrakis(4-pyridyl)porphyrin; Ln = Sm, Eu, Gd, Tb);[820] on methylation, they are converted into the cationic monoporphyrinates [Ln(tmepyp)(acac)].

3.2.2.7.2 *Texaphyrins*

A new type of complex is provided by the texaphyrins, compounds of "extended" porphyrins where the ring contains five donor nitrogens. These have attracted considerable interest because of their possible medicinal applications.[821–825] Two texaphyrin complexes are undergoing clinical trials; a gadolinium compound ((43); Gd-tex; XCYTRIN) is an effective radiation sensitizer for tumor cells. It assists the production of reactive oxygen-containing species, whilst the presence of the Gd^{3+} ion means that the cancerous lesions to which it localizes can be studied by MRI; it was undergoing Phase III clinical testing in 2002. The lutetium analogue ((44); Lu-Tex)[826,827] selectively absorbs light in the far-infrared (λ_{max} of 732 nm), localizes preferentially to cancerous tissue (like the Gd analogue), is water soluble (again like the Gd analogue), and generates singlet oxygen in high quantum yield. One form (LUTRIN^{TM}) is being developed for photodynamic therapy for breast cancer, where it is in Phase II trials, and another, (ANTRIN^{TM}) is being developed for photoangioplasty, where it has potential for treating arteriosclerosis, for example as an imaging agent for the study of retinal vascular disease,[828] since it photosensitizes human or bovine red blood cells on irradiation at 730 nm.[829]

One report indicates Gd-texaphyrin to be a tumor sensitive sensitizer detectable at MRI,[830] but another study casts doubt on its potential as a radiosensitizing agent.[831] The effect of axial ligation by nitrate and phosphate on the NMR spectra of paramagnetic lanthanide texaphyrins has been studied;[832] the structure of [Dy(tx)(Ph$_2$PO$_4$)$_2$] shows approximately pentagonal bipyramidal coordination (tx = texaphyrin). A ^{13}C-NMR study of [Ln(tx)(NO$_3$)$_2$] has been reported[833] and spectroscopic studies have been reported on texaphyrin complexes for a wide range of lanthanides, examining the influence of the metal ion and its spin state.[834] Phosphoramidite derivatives of dysprosium texaphyrin have been used to prepare ribozyme analogues.[835]

3.2.2.7.3 *Crown ethers and other macrocycles*

The ability of lanthanides to form stable crown ether complexes was first developed in the 1970s. Two types of complexes are known, those where the crown ether is directly bonded to the lanthanide and those where the crown ether hydrogen-bonds to a coordinated water molecule. The choice is determined by the match between the size of the central cavity in the crown ether ring and the size of the lanthanide ion. A striking comment on the developments in crown ether chemistry is the synthesis of the first structurally characterization of cationic lanthanide alkyl complexes.[155] [Ln(CH$_2$SiMe$_3$)$_3$(THF)$_2$] (Ln = Y, Lu) react with B(C$_6$X$_5$)$_3$ (X = H, F) in the presence of crown ethers forming [Ln(CH$_2$SiMe$_3$)$_2$(CE)(THF)$_n$]$^+$[B(C$_6$X$_5$)$_3$(CH$_2$SiMe$_3$)]$^-$ (CE = [12]-crown-4, n = 1; CE = [15]-crown-5, [18]-crown-6, n = 0). In all these complexes, the crown ethers utilize all their donor atoms. In THF, but in the absence of crown ether, [Lu(CH$_2$SiMe$_3$)$_3$(THF)$_2$] reacts with B(C$_6$F$_5$)$_3$ forming [Lu(CH$_2$SiMe$_3$)$_2$(THF)$_3$]$^+$[B(C$_6$X$_5$)$_3$(CH$_2$SiMe$_3$)]$^-$.

The first crown ether complexes to be made were usually with lanthanide nitrates; when crown ethers complex with lanthanide nitrates, the two ligands generally combine to saturate the coordination sphere, but since chloride is a monodentate ligand, coordinative saturation is not always achieved that way and indeed chlorides are frequently displaced from the coordination sphere by water. This account therefore begins with lanthanide chloride complexes. 12-crown 4 tends to give complexes of the type [Ln(H$_2$O)$_5$(12-C-4)]Cl$_3$·2H$_2$O and [LnCl$_2$(H$_2$O)$_2$(12-C-4)]Cl·2H$_2$O but coordinating ability is affected by the presence of water and other solvents.[836] The crown ether does not coordinate to the

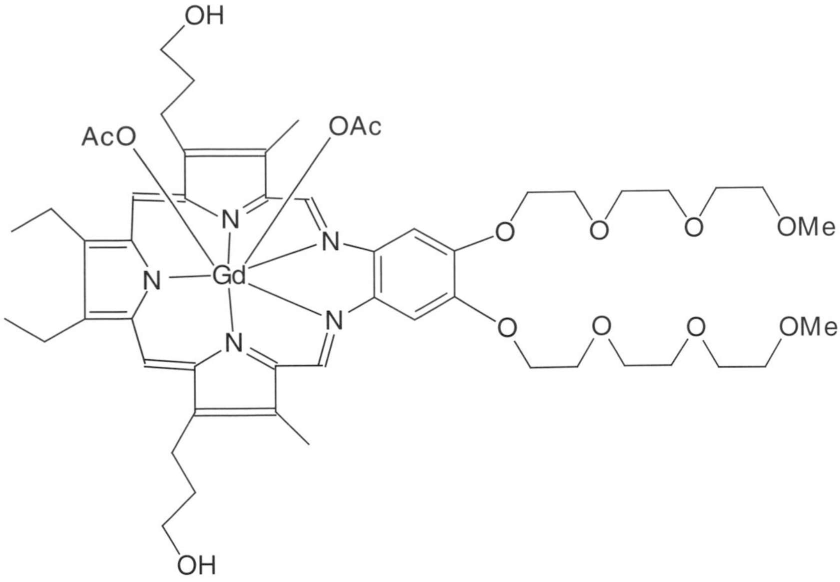

(43)

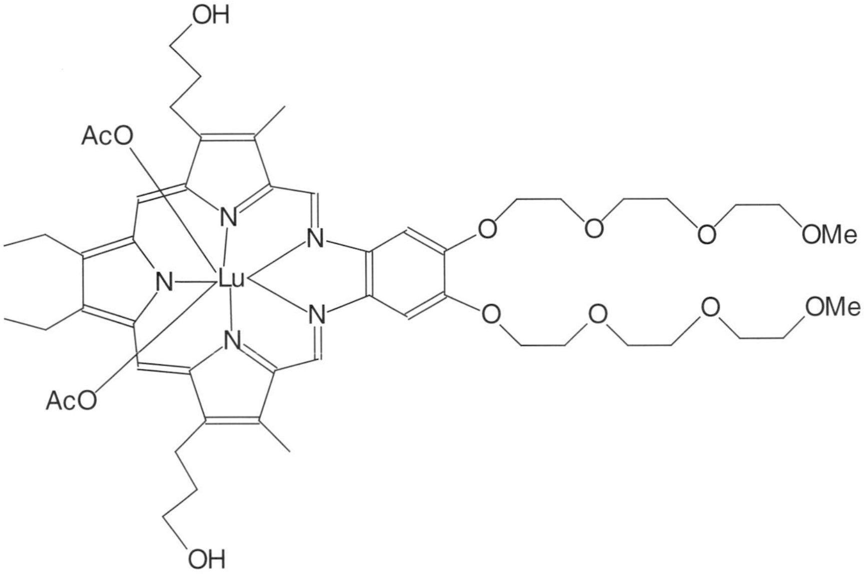

(44)

lanthanide at all in [Lu(H$_2$O)$_8$]·1$\frac{1}{2}$ (12-C-4).2H$_2$O, instead forming hydrogen bonds to coordinated water molecules.[837] With the larger lanthanum ion, [LaCl$_3$(12-crown-4)(MeOH)] is eight-coordinate, owing to a molecule of solvent coordinating.[838] By modifying conditions, different complexes can be obtained. Thus in MeCN or MeOH, normal syntheses from NdCl$_3$ and 15-crown-5 get complexes where [Nd(H$_2$O)$_9$]$^{3+}$ and [NdCl$_2$(H$_2$O)$_6$]$^+$ ions are hydrogen-bonded to the crown ether, without direct lanthanide-crown ether bonding. However by electrocrystallization, you get eight-coordinate [Nd(15-crown-5)Cl$_3$].[839] Similarly, in the eight-coordinate [Pr(15-crown-5)Cl$_3$], Pr—Cl distances are 2.707, 2.724, and 2.736 Å, averaging 2.722 Å.[840] Further structures determined include [Er(H$_2$O)$_8$]Cl$_3$.15-crown-5.[841] The structures of [LaCl$_3$(15-crown-5)] and [LaCl$_2$(phen)(OH$_2$)$_2$(μ-Cl)]$_2$(15-crown-5·MeCN] show lanthanum-crown ether coordination in the former but not the latter.[842] Ytterbium is not bound to the crown ether in [[Yb(OH$_2$)$_8$]Cl$_3$.15-crown-5].[843] 18-Crown-6 forms two series of complexes with lanthanide chlorides [LnCl(H$_2$O)$_2$(18-crown-6)]Cl$_2$.2H$_2$O (Ln = Pr-Tb) and [LnCl$_2$(H$_2$O)(18-crown-6)]Cl.2H$_2$O (Ln = La, Ce) although, depending upon conditions, [LnCl$_3$(18-crown-6)] can also be made. Its inability to form complexes with the chlorides of the smaller lanthanides is ascribed to a lack of flexibility found in polyethylene glycols, which can form such complexes.[844] New oxonium complexes (H$_9$O$_4$)[LaCl$_2$(H$_2$O)-(18-crown-6)]Cl$_2$ and (H$_3$O)[EuCl(H$_2$O)$_2$(18-crown-6)]Cl$_2$ have been reported.[845] Unusually, chloride is found in the first coordination sphere of lanthanum as well as nitrate in the serendipitously discovered mixed anion complexes [LaCl$_2$(NO$_3$)(12-crown-4)]$_2$ and [LaCl$_2$(NO$_3$)(18-crown-6)]; the structure of [La(OH$_2$)$_4$(NO$_3$)(12-crown-4)]Cl$_2$.MeCN has also been reported.[846]. Structures have been determined for the in-cavity complex [SmI$_3$(dibenzo-18-crown-6)][847] (tricapped trigonal prismatic) and the out-of-cavity complex [LaBr$_3$(12-crown-4)(acetone)][848] (distorted square anti-prismatic). The complex [La$_2$I$_2$(μ-OH)$_2$(dibenzo-18-crown-6)$_2$](I$_3$)I was isolated from the reaction of LaI$_3$ with the crown ether in MeCN; evidently a hydrolysis product, it contains nine-coordinate lanthanum[849] with the lanthanide sitting in the cavity. The out-of cavity hydroxide-bridged cationic complex [[Y(OH)(benzo-15-crown-5)(MeCN)]$_2$]I$_2$ results from the reaction of YI$_3$ and the crown ether in MeCN.[850] In the presence of SbCl$_5$, halide abstraction reactions of LnCl$_3$ (Ln = La, Y) form [Ln(MeCN)$_3$(18-crown-6)](SbCl$_6$)$_3$ and [Ln(MeCN)$_5$(12-crown-4)](SbCl$_6$)$_3$.[851] The crown ether dibenzo-18-crown-6 forms [Pr(dibenzo-18-crown-6)Cl$_2$(H$_2$O)]SbCl$_6$ containing nine-coordinate praseodymium.[852] [GdCl$_2$(dibenzo-18-crown-6)(MeCN)]SbCl$_6$ similarly has nine-coordinate gadolinium, with a roughly planar hexadentate crown ether giving a 2:6:1 coordination geometry;[853] in contrast, [Dy$_2$(dibenzo-18-crown-6)$_2$Cl$_4$]$_2$[Dy$_2$(MeCN)$_2$Cl$_8$] contains dimeric cations (and anions) linked by double chloride bridges.[854] Many nitrate complexes continue to be characterized. [La(NO$_3$)$_3$(12-crown-4)(H$_2$O)] and [Yb(NO$_3$)$_3$(12-crown-4)] have 11 and 10-coordinate lanthanides respectively.[855] Yttrium is not bound to the crown ether in [Y(NO$_3$)$_3$-(OH$_2$)$_3$.(Me$_2$-16-crown-5)(H$_2$O)].[856] The crown ether is nonbonded in [Ln(NO$_3$)$_3$(OH$_2$)$_3$](NO$_3$).2(15-crown-5) (Ln = Y, Gd, Lu).[857] Structural characterization has been achieved for [Pr(NO$_3$)$_3$(18-crown-6)] and [Ln(NO$_3$)$_3$(H$_2$O)$_3$].18-crown-6.[858] (Ln = Y, Eu, Tb-Lu). The complex [La(NO$_3$)$_3$(H$_2$O)$_2$(MeOH)(bipy)].15-crown-5 features an unbound crown ether and simultaneous coordination of both water and methanol.[859] Crown ethers are well known to encapsulate many lanthanide salts. When heavier lanthanide nitrates are recrystallized in the presence of 18-crown-6, [Ln(NO$_3$)$_3$(H$_2$O)$_3$] molecules are encapsulated into the crown ether (Ln = Sm, Eu, Yb).[388] The tosylates [Ln(tos)$_3$(H$_2$O)$_6$] cannot be incorporated.[860] 4-t-Butylbenzo-15-crown-5 (L) forms [La(NO$_3$)$_3$.L.0.5MeCN] in which [La(NO$_3$)$_3$.L] molecules have 11-coordinate lanthanum,[861] broadly similar in geometry to the benzo-15-crown-5 and cyclohexyl-15-crown-5 analogues. Europium is 10-coordinate[862] in [Eu(NO$_3$)$_2$(16-crown-5)(MeCN)]$^+$ (in contrast to 11-coordinate La in [La(NO$_3$)$_3$(16-crown-5)]; however, the smaller lutetium does not coordinate to 16-crown-5, for [Lu(NO$_3$)$_3$(H$_2$O)$_3$] molecules hydrogen-bond to the crown ether. Benzo-15-crown-5 has been used to extract europium in the presence of picrate with a diaphragm electrolyser.[863]

Comparative studies between crown ethers and the noncyclic glymes (linear polyethers) suggest that glyme complexes are favored — the chelate effect is thus more important than the macrocyclic effect. The more flexible glymes may accommodate water molecules and other small ligands (e.g., anions) more easily. Solubility effects may play a part too![864]

The factors affecting the complexes formed by LnX$_3$, Ln(NCS)$_3$, and Ln(NO$_3$)$_3$ with polyethylene glycols have been examined.[865] The chain length is the main factor controlling the coordination sphere and thus the number of additional inner sphere ligands present. Hydrogen bonding involving the terminal OH groups and water molecules generates supermolecular structures. Solubility of hydrated lanthanide chlorides in diethylene glycol and triethylene glycol (teg) has been studied;[866] the structure of nine-coordinate [Nd(teg)$_3$]Cl$_3$ has been determined. Three yttrium polyalcohol complexes have been synthesized. [Y{(HOCH$_2$)$_3$CMe}$_2$Cl$_2$]Cl.MeOH and [Y{(HOCH$_2$)$_3$CN(CH$_2$CH$_2$OH)$_2$}Cl$_2$.MeOH]Cl both have eight-coordinate yttrium; in [Y{(HOCH$_2$)$_3$CMe}$_2$(NO$_3$)(H$_2$O)](NO$_3$)$_2$ yttrium is nine-coordinate, with one bidentate nitrate.[867] Structures have been reported for [Pr(NO$_3$)$_3$(stilbeno-15-crown-5)] and [Nd(NO$_3$)$_3$(EO$_5$)] (EO$_5$ = tetraethyleneglycol);[868] the lanthanide is 11-coordinate in both. Unlike [Pr(NO$_3$)$_3$ (stilbeno-15-crown-5)], the Nd analogue is unstable and reacts with formic acid forming [Nd(NO$_3$)$_3$(EO$_5$)]. Lanthanides are nine-coordinate in [Ln(EO$_2$)$_3$](ClO$_4$)$_3$.3H$_2$O (Ln = Nd, Ho; EO$_2$ = diethylene glycol). Europium triflate complexes of homochiral polyether and polyethyleneglycol ligands have been characterized.[869]

Complexes of lanthanide triflates with polyethene glycol (HO(CH$_2$CH$_2$O)$_n$H; n = 2,3,4) and polyethene glycol dimethyl ether (MeO(CH$_2$CH$_2$O)$_n$Me; n = 2,3,4) are effective Lewis acid catalysts for the Diels–Alder reaction and for the allylation of aldehydes with allyltributyltin.[870] Structures have been reported for [La(OTf)$_3$(THF)(HO(CH$_2$CH$_2$O)$_4$H)], [DyOTf)$_2$(MeO(CH$_2$-CH$_2$O)$_4$Me)(H$_2$O)$_2$](OTf), and [Eu(OTf)$_3$(MeO(CHPhCHPhCH$_2$(OCH$_2$CH$_2$)$_2$OCHPhCH-

PhOMe)].[871,872] 2,2′-Bipyridyl complexes of Eu^{3+} and Tb^{3+} bound to (poly)ethylene glycol are strongly luminescent.[873] In [La(NO$_3$)$_3$(triethyleneglycol)(H$_2$O)], lanthanum is 11-coordinate[874] [CeCl(tetraglyme)(H$_2$O)$_3$]Cl$_2$.H$_2$O has nine-coordinate cations.[875] [PrCl$_3$(tetraethyleneglycol)]$_2$ has nine-coordinate praseodymium.[876] Pentaethylene glycol (EO$_5$) forms two series of nine-coordinate complexes with lanthanide chlorides, [LnCl$_2$(H$_2$O)(EO$_5$)]Cl.2H$_2$O (Ln = La-Pr) and [Ln(H$_2$O)$_3$(EO$_5$)]Cl$_3$.H$_2$O (Ln = Sm-Lu, Y).[877] The ability of polyethylene glycols to wrap themselves round a lanthanide ion irrespective of size, makes them more flexible coordinating agents than crown ethers. The serendipitous synthesis of the 10-coordinate [NdCl(NO$_3$)$_2$(tetraglyme)] has been reported.[878] Cerium is nine-coordinate in [Ce(NCS)$_3$(tetraethyleneglycol)(H$_2$O)].[879]

The study of lanthanide thiocyanate complexes of crown ethers and other macrocycles commenced during the mid-1990s. Nine-coordinate europium is found in [M(NCS)$_3$L] (M = La, Eu; L = (2,3,7,18-tetramethyl-3,6,14,17,23,24-hexaazatricyclo-[17.3.1.1^{8,12}]-tetracosa-1(23),2,6,8(24),9, 11,13,17,19,21-decaene-N^3,N^6,N^{14},N^{17},N^{23},N^{24}))[880,881] and in [M(NCS)$_3$(tdco)][882] (M = Eu, Yb; tdco = 1,7,10,16-tetraoxo-4,13-diazacyclooctadecane). Several thiocyanate complexes of the tetradentate 13-crown-4 have been isolated in [Ln(NCS)$_3$(13-crown-4)(H$_2$O)$_2$] (Ln = La, Pr-Tb, Er-Yb)[883] whilst lanthanum is 10-coordinate in [La(NCS)$_3$(18-crown-6)(DMF)].[884] In the compounds [Ln(dibenzyldiaza-18-crown-6)(NCS)$_3$] (Ln = Ce, Nd, Eu),[885,886] the lanthanide is bound to two nitrogens and four oxygens in the crown ether as well as three N-bonded thiocyanates. Thus in the cerium complex, Ce—N(NCS) distances are 2.496(3), 2.523(3), and 2.544(3) Å; Ce—O 2.565(2), 2.574(2), 2.603(2), and 2.635(2) Å; Ce—N(ring) distances are 2.782(3) Å and 2.810(3) Å. The CeN$_3$(NCS) moiety exists as a trigonal planar arrangement at the center of the macrocycle cavity with the trigonal plane perpendicular to the plane of the macrocycle. A number of complexes involve the lanthanide bonding to nitrogens in rings, either in aza crowns or in more classic macrocycles like cyclen. 1,4,7,10,13-Pentaazacyclopentadecane (L) forms nitrate complexes [Ln(L)(NO$_3$)$_3$] (Ln = Y, La-Yb except Ce, Pm, Ho); the structure of [La(L)(NO$_3$)$_3$] shows it to have 11-coordinate lanthanum.[887] As another example of structurally characterized complex of a N-donor macrocycle, [Nd([18]aneN$_6$)(NO$_3$)$_3$] has two monodentate and one bidentate nitrate giving 10-coordination.[888] Schiff base complexes can be synthesized by reaction of a lanthanide salt with a diamine and a suitable carbonyl derivative such as 2,6-diacetylpyridine.[889] A number of complexes of Schiff base and aminephenol ligands have been reported,[890-892] including a dinuclear complex where two Schiff base ligands bridge two lanthanums in a sandwich.[892] Polymeric complexes of the 1,5,9,13-tetraazacyclohexadecane ligand are reported.[893] Several studies of hexaaza macrocycle complexes have appeared,[894-896] typically reporting structures of 10- and 12-coordinate lanthanum complexes, whilst a nine-coordinate N$_6$ macrocycle Gd complex [GdL(H$_2$O)$_2$]$^{3+}$ studied as possible MRI agent,[897] and another series of complexes is believed to have a N$_3$O$_3$ macrocycle coordinating to the lanthanides.[898] A σ-bonded organometallic of a deprotonated macrocycle (HMAC = aza-18-crown-6) [Y{CH(SiMe$_3$)$_2$}(MAC)] has been synthesized.[899] [Er(CF$_3$SO$_3$)$_3$(cyclen)(MeCN)] has an eight-coordinate structure with monodentate triflates.[900] A number of complexes of a cyclen-based ligand system bearing four amide arms attached to the ring have been examined.[901] Three Eu^{III} complexes (R = Ph, 4-NO$_2$Ph, CH$_2$(4-NO$_2$Ph)) have been synthesized and their structures determined; the twist angle between the O$_4$ and N$_4$ planes is mainly determined by the flexibility of the pendant arms. In a study of the luminescence properties of Ln^{III} complexes (Ln = Sm, Eu, Tb), it was found that the efficiency of L → Ln intramolecular energy transfer is affected by both the gap between the ligand triplet state and the excited state and the donor–acceptor distance and the orientation of the chromophore. There is interest in lanthanide complexes capable of cleaving DNA continues. Dimeric Y and Nd macrocycle complexes are active catalysts for the degradation of double-stranded DNA, whereas the corresponding monomeric complexes are inactive. The mechanism probably involves random attack at single strands in which closed circular plasmid DNA is converted to a nicked intermediate, followed by attacks on this.[902] The hydrolysis of phosphate esters by yttrium complexes with bis-trispropane and related ligands has also been studied.[903]

Cryptands have attracted steady interest, partly because of their applications. Ln^{3+} ions are not complexed by them under conditions where Ln^{2+} ions, such as Eu^{2+} and Sm^{2+} are, and thermodynamic functions have been evaluated.[904] The application of europium cryptate complexes to bioassays has become significant;[905,906] this principle has been extended to diagnosis of mutations.[907] Such compounds have used cryptate ligands containing three bipy units,[908,909] in which the Ln^{3+} ion is contained within a ligand cavity; though bis(isoquinoline dioxide) within the cryptate has been used to ligate.[910-912] The Nd and Yb complexes of the cryptate ligands containing three bipy units have also been shown to luminesce in the near-IR.[913] Another application has been the use of lanthanide cryptates of the early lanthanides (La, Ce, Eu) as catalysts in the hydrolysis of phosphate monoesters,[914] diesters,[915] and trimesters.[916] The structure of [La(3.2.1)Cl$_2$]Cl. MeOH shows lanthanum to be nine-coordinate, though luminescence measurements on the europium complex of (3.2.1) show three water molecules to be coordinated in solution. The dissociation kinetics of the Eu complex of 4,7,13,16,21-pentaoxa-1,10-diaza-bicyclo[8.8.5]tricosane has been studied by monitoring its absorption spectrum.[917] The electrochemistry of Sm, Eu, and Yb complexes of the cryptates (3.2.2), (3.2.1), and (2.1.1) has been studied; stability constants have been determined from the redox potentials.[918] The Ln^{II} cryptate complexes were found to be more stable in DMF than the Ln^{III} analogues. A one-pot synthesis of yttrium and lanthanum cryptates has been described.[919] Cryptands have been described that have phenolic and amine donor sets and can accommodate one or two lanthanide ions.[920] A europium(III) complex of a N$_6$-donor cryptate that includes two bipy units has nine-coordination completed by three chlorides.[921] Its luminescence is modulated by pH, owing to protonation of the amine groups in the backbone. A combinatorial approach[922] has been applied to the formation of a yttrium-containing molecular capsule [{18-crown-6}({Y(H$_2$O)$_7^{3+}$}$_{1.33}$(p-sulfatocalix[4]arene^{4-})}$_2$]. The first *d–f* heteronuclear cryptate, involving nine-coordinate dysprosium and six-coordinate copper, has been reported.[923] The europium complex of a novel podand ligand has N$_3$O$_6$ coordination of europium.[924] Other structures of two 10-coordinate lanthanum complexes were reported.[925] Eight- and nine-coordinate gadolinium cryptates have been described.[926]

3.2.2.7.4 Calixarenes

The large Bu^t-calix[8]arene (LH$_6$) can accommodate a metal at both end, forming complexes [Ln$_2$(H$_6$L)(DMSO)$_5$] (Ln = La, Eu, Tm, Lu; L = p-t-Butylcalix[8]arene).[927,928] A calix[4]arene, p-t-butylcalix[4]arene (H$_4$L), forms a dimeric europium complex [Eu$_2$(HL)$_2$(DMF)$_4$].7DMF, in which the Eu^{3+} ion is seven-coordinate, bound to two DMF molecule, two bridging oxygens and three terminal oxygens from the macrocycle. The Eu—O (calixarene) bonds are very short, around 2.15 Å.[929] When an "expanded" calixarene containing another donor atom in the ring (H$_4$L′) was employed, a similar binuclear complex [Eu$_2$(HL′)$_2$(DMSO)$_2$] was obtained, with seven-coordination of europium again, being just bound to one coordinated DMSO molecule.[930] p-t-Butylcalix[8]arene (H$_8$L) forms a 1:1 complex with Eu^{III}. The calixarene is bidentate in [Eu(H$_6$L)(NO$_3$)(DMF)$_4$].3DMF.[931] p-Butylcalix[5]arene (H$_5$L) forms dimeric lanthanide complexes [Ln$_2$(H$_2$L)$_2$(DMSO)$_2$] (Ln = Eu, Gd, Tb).[932] The larger p-butylcalix[8]arene and p-nitrocalix[8]arene (H$_8$L) rings each incorporate two lanthanides, forming [Ln$_2$(H$_2$L)(DMF)$_5$] (Ln = Eu, Lu). The structures of europium complexes of both p-butylcalix[5]arene and p-nitrocalix[8]arene both contain eight-coordinate europium.[933] p-Chloro-N-benzylhexahomotriazacalix[3]arene forms a 1:1 complex with neodymium nitrate in which neodymium is bound to three bidentate nitrates and to three phenolic oxygens in the calixarene.[934] Much of the work with calixarenes has been prompted by their potential in separating lanthanides for uranium and other metals. A calixarene ligand with two amide substituents has been synthesized as an extractant for lanthanides; dimeric Sm and Eu and monomeric Lu complexes have been prepared;[935,936] similarly an erbium complex of a calixarene ligand with four amide substituents, [Er(L^1-2H)(picrate)] (L^1 = 5,11,17,23-Tetra-t-butyl-25,27-bis (diethylcarbamoylmethoxy)calix[4]arene) has been reported.[937] Syntheses and structures are reported for [Tm(L-2H)(A)], [Ce(L-2H)(A)(HOMe)$_2$].HA, [PrLA$_3$], and L.2HA. (L = 5,11,17,23-tetra-t-butyl-25,27-bis(diethylcarbamoylmethoxy)-26,28-dihydroxycalix[4]arene, HA = picric acid).[938] Hexahomotrioxacalix[3] arene macrocyles selectively bond Sc^{3+}, Y^{3+}, and lanthanides, forming μ-aryloxobridged dimers.[939] Calix[4]arenes substituted by acetamidophosphine oxide groups at the rim show selectivity, not just to trivalent ions but also to light lanthanides and actinides.[940] Calix[4]arene podands and barrelands incorporating bipy groups form lanthanide complexes; their Eu^{3+} and Tb^{3+} complexes have high metal luminescence quantum yields.[941] A calixarene with four phosphine oxides attached has been fixed to silica particles and the resulting system has been found to give very efficient extraction of Eu^{3+} and Ce^{3+} from simulated waste.[942] A calix[4]arene complex of Gd^{3+} binds noncovalently to human serum albumin and is a potential contrast agent.[943]

3.2.2.8 Group 16 Ligands Involving Sulfur, Selenium, and Tellurium

Complexes of lanthanides of neutral S-donors were unprecedented until 2002. Reaction of LaI$_3$(THF)$_3$ with 9S3 (9S3 = 1,4,7-trithiacyclononane) affords eight-coordinate [LaI$_3$(9S3)

(MeCN)$_2$]; the Ce analogue has been studied in solution by NMR. La—S distances are 3.064, 3.089, and 4.126 Å; La—I are 3.114, 3.177, and 3.186 Å and La—N 2.641 and 2.672 Å.[944] A number of studies have concerned bidentate thiolates, especially dithiocarbamates. Their adducts, especially with phen or bipy, seem more stable than the parent complexes, possibly because of coordinative saturation (and are discussed in Section 3.2.2.3.3) Another way in which this can be achieved is by forming an ionic complex. These points are illustrated by the first lanthanide chalcogenocarboxylate complexes [Sm(RCOS)$_3$(THF)$_2$] and [Na(THF)$_4$][Sm(RCSS)$_4$] (R = 4-MeC$_6$H$_4$)[945] and by monoalkyldithiocarbamates (RNH$_3$)[Ln(S$_2$CNHR)$_4$] (Ln = La-Nd, Sm-Gd; R = Me, Et), which have been synthesized for the lighter lanthanides.[946] The structure and luminescence of Na[Eu(S$_2$CNMe$_2$)$_4$] have reported, europium having dodecahedral eight-coordination. The two slightly different types of europium coordination sites in the lattice are reflected in the multiplicity of signals in the luminescence spectrum.[947] Europium is also eight-coordinate in (Ph$_4$P)[Eu(S$_2$P(OEt)$_2$)$_4$][948] and in [Me$_2$NH$_2$][Nd(S$_2$CNMe$_2$)$_4$],[949] a member of a series [Me$_2$NH$_2$][Ln(S$_2$CNMe$_2$)$_4$] (Ln = La, Pr-Nd, Sm–Ho). [Me$_2$NH$_2$][Ln(S$_2$CNMe$_2$)$_4$], [MeNH$_3$] [La(S$_2$CNHMe)$_4$], [EtNH$_3$][La(S$_2$CNHEt)$_4$], and [Ln(S$_2$CNMe$_2$)$_3$(DMSO)$_2$] exhibit wide antibacterial activity.[950]

The first monomeric lanthanide thiolates, [Ln(SBu^t)$_3$] (Ln = La, Ce, Pr, Nd, Eu, Yb, Y) have been made[951] from [Ln{(N(SiMe$_3$)$_2$)$_3$}] and HSBu^t. They are intensively reactive, doubtless owing to coordinative unsaturation, and the adducts [(BuS^t)$_2$(bipy)Ln(μ-SBu^t)$_2$Ln(bipy)(SBu^t)$_2$] (Ln = Y, Yb) have been isolated. Transmetallation of Sm with Hg(SC$_6$F$_5$)$_2$ affords Sm(SC$_6$F$_5$)$_3$; this has the solid-state structure [(THF)$_2$Sm(μ_2-SC$_6$F$_5$)(SC$_6$F$_5$)$_2$]$_2$. It undergoes thermal decomposition to SmF$_3$.[952] Use of SC$_6$F$_5$ as a ligand gives stabler and more hydrocarbon-soluble products, which have been isolated as Lewis base adducts. The large Ce^{3+} ion forms the seven-coordinate dimer [Ce(SC$_6$F$_5$)$_3$(THF)$_3$]$_2$ with thiolate bridges, whilst Ho and Er give monomeric [Ln(SC$_6$F$_5$)$_3$(THF)$_3$] (Ln = Ho, Er). Other adducts [Ln(SC$_6$F$_5$)$_3$(py)$_4$] (Ln = Sm, Yb) and [Er(SC$_6$F$_5$)$_3$(DME)$_2$] have also been isolated. Many of these compounds feature Ln···F interactions.[953] Lanthanide metals react with organic disulfides forming thiolates like [Yb(SAr)$_3$(Py)$_3$] (Ar = 2,4,6-triisopropylphenyl).[954] A series of thiophenolates and their selenium analogues have been synthesized and examined. [Ln(SPh)$_3$(Py)$_3$]$_2$ (Ln = Ho, Tm) have two thiolate bridges with seven-coordinate lanthanides; [Sm(SPh)$_3$(Py)$_2$]$_4$ has a linear arrangement of four seven-coordinate samariums with three, two, and three μ_2-bridging thiolates and [Sm(SPh)$_3$(THF)]$_{4n}$ is a polymer. [Ln(SePh)$_3$(THF)$_3$] (Ln = Tm, Ho, Er) have monomeric *fac*-octahedral structures; [Sm(SePh)$_3$(Py)$_3$]$_2$ has two selenothiolate bridges with seven-coordinate samarium; and [Ln$_3$(SePh)$_9$(THF)$_4$]$_n$ (La = Pr, Nd, Sm) are polymeric wiith three doubly bridging selenothiolates.[955] The thiolate [{(Et$_3$CS)$_2$Y(μ-SCEt$_3$)Py$_2$}$_2$] is dimeric with a planar Y$_2$S$_2$ core.[956] THF solutions of thiolates Ln(SPh)$_3$ react with S forming octanuclear clusters [Ln$_8$S$_6$(SPh)$_{12}$(THF)$_8$] (Ln = Ce, Pr, Nd, Sm, Gd, Tb, Dy, Ho, Er), their structure is based on a cube of lanthanides, edge bridged by mercaptides and face-bridged by sulfur;[957] analogous pyridine clusters [Ln$_8$S$_6$(SPh)$_{12}$(Py)$_8$] (Ln = Nd, Sm, Er) have also been isolated.[958] Clusters are also found in [Yb$_4$Se$_4$(SePh)$_4$(Py)$_8$] and in [Yb$_6$S$_6$(SPh)$_6$(Py)$_{10}$].[959] Ph$_2$Se$_2$ reacts with lanthanide amalgams[960] and pyridine forming Ln(SePh)$_3$Py$_3$ (Ln = Ho, Tm, Yb); these are dimeric with seven-coordinate lanthanides. In contrast, [Yb(SPh)$_3$Py$_3$] has a monomeric meridional structure. Thermolysis of the selenolates gives Ln$_2$Se$_3$ (Ln = Tm, Yb) and a mixture of MSe and MSe$_2$ (M = Ho). In addition to the mononuclear [Yb(SePh)$_3$(THF)$_3$], the tetranuclear [Yb$_4$(SePh)$_8$O$_2$(THF)$_6$] and ionic [Yb$_3$(SePh)$_6$-(DME)$_4$][Yb(SePh)$_4$(DME)] are reported.[961] Reaction of Sm with PhSeSePh in THF gives[962] a cluster [Sm$_8$Se$_6$(SePh)$_{12}$(THF)$_8$] whilst Sm(SePh)$_3$ reacts with S in DME affording [Sm$_7$S$_7$(SePh)$_6$-(DME)$_7$]$^+$[Hg$_3$ (SePh)$_7$]$^-$[Ln$_8$E$_6$(EPh)$_{12}$L$_8$] clusters (E = S, Se; Ln = lanthanide; L = Lewis base) can be prepared by reduction of Se—C bonds by low-valent Ln or by reaction of Ln(SePh)$_3$. Structures reported include [Sm$_8$E$_6$(SPh)$_{12}$(THF)$_8$] (E = S, Se) and [Sm$_8$Se$_6$ (SePh)$_{12}$(Py)$_8$]. They contain cubes of lanthanide ions with E^{2-} ions capping the faces and EPh bridging the edges of the cube. Reaction of Nd(SePh)$_3$ with Se to form [Nd$_8$Se$_6$(SePh)$_{12}$(Py)$_8$] shows that this series is not restricted to redox-active lanthanides.[963] La(SeSi(SiMe$_3$)$_3$)$_3$ is thought to be a three-coordinate monomer in toluene[964] but the Y analogue is postulated to be dimeric {Y(SeSi(SiMe$_3$)$_3$)$_2$(μ-SeSi(SiMe$_3$)$_3$}$_2$ (a surprise, in view of the smaller size of yttrium). Lewis base adducts such as Ln(SeSi(SiMe$_3$)$_3$)$_3$(THF)$_2$ (Ln = La, Sm, Yb) have been made. Tellurolate analogues Ln(TeSi(SiMe$_3$)$_3$)$_3$ (Ln = La, Ce, Y) have been prepared and the structure of the dmpe adduct [La(TeSi(SiMe$_3$)$_3$)$_3$(dmpe)$_2$] determined, the latter having a distorted pentagonal bypyramidal structure with two axial tellurolates. A number of pyridinethiolate (Spy; 2-S-NC$_5$H$_4$) complexes have been characterized. Ce(SPy)$_3$ reacts with Et$_4$P[SPy] forming [Et$_4$P][Ce(SPy)$_4$], in which cerium is eight-coordinate; [Et$_4$P][Ln(SPy)$_4$] (Ln = Ho, Tm) were also reported. [Yb(SPy)$_3$] crystallizes from pyridine as eight-coordinate [Yb(SPy)$_3$(py)$_2$].[965] The europium(III) pyridinethiolate [PEt$_4$][Eu(SPy)$_4$] have been synthesized[966] as well as some Eu^{II} compounds. Pyridine 2-thiolates [Ln(SC$_5$H$_4$N)$_2$(HMPA)$_3$]I (Ln = Pr, Nd, Sm, Eu, Er, Yb), formed by a cleavage reaction of 2,2'-dipyridyl disulfide in HMPA with iodine and Ln, have pentagonal bipyramidal coordination of the lanthanides.[967] A number of heterometallic Lanthanide-Group 12 chalcogen-olates have been reported[968] including [Py$_3$Eu(μ_2-SePh)$_2$(μ_3-SePh)Hg (SePh)$_2$], [(THF)$_4$Eu(μ_2-SePh)$_3$Zn(SePh)], [Sm(THF)$_7$][Zn$_4$(μ_2-SePh)$_6$(SePh)$_4$], and [Yb(THF)$_6$][Hg$_5$(μ_2-SePh)$_8$(SePh)$_4$].2THF. Reaction of later lanthanides with mixtures of I$_2$ and PhSeSePh in THF, followed by reaction with Se in pyridine affords [(THF)$_6$Ln$_4$I$_2$(SeSe)$_4$(μ_4-Se)].THF (Ln = Tm, Ho, Er, Yb). These are clusters containing a square array of Ln^{III} ions connected through a single (μ_4-Se) ligand. There are two I$^-$ ligands coordinating nonadjacent Ln^{III} ions on the side of the cluster opposite the (μ_4-Se), and the edges of the square are bridged by μ_2-SeSe groups. With a 1/1/1/1 Yb/I/Ph$_2$S$_2$/Se stoichiometry, the product is [Yb$_6$Se$_6$I$_6$(THF)$_{10}$] which contains a Yb$_4$Se$_4$ cubane fragment, with an additional Yb$_2$Se$_2$ layer capping one face of the cube. Upon thermolysis, the selenium rich compounds give iodine-free Ln$_2$Se$_3$.[969]

3.2.2.8.1 Group 17 ligands

The structure of LiKYF$_5$ has been determined.[970] NH$_4$LnF$_4$ (Ln = La-Dy) have been synthesized; the structure of NH$_4$DyF$_4$ was reported.[971] Pentagonal bipyramidal coordination of Y is found in two tetrafluorometallates.[972,973] NaMCl$_4$(M = Eu–Yb, Y) have the α-NiWO$_4$ structure at room temperature with six-coordinate lanthanides;[974] at high temperatures, they change to the seven-coordinate NaGdCl$_4$ structure. Na$_2$MCl$_5$ (M = Sm, Eu, Gd) adopt the Na$_2$PrCl$_5$ structure, again with seven-coordination.[975] Gadolinium is eight-coordinate in BaGdCl$_5$.[976] Na$_2$EuCl$_5$ reacts with Eu to form[977] the mixed valence compound NaEu$_2$Cl$_6$.Cs$_2$K[LnCl$_6$] (Ln = Eu, Tb) have the cubic elpasolite structure.[978]

Rb$_3$MCl$_6$.2H$_2$O (M = La–Nd) contain eight-coordinate lanthanides[979] in anionic trimers [Ln$_3$Cl$_{12}$(H$_2$O)$_6$]$^{3-}$. Cs$_3$LnCl$_6$.3H$_2$O (Ln = La–Nd) contain slightly distorted capped trigonal pris-matic coordination of Ln in edge-linked [LaCl$_2$Cl$_{4/2}$(H$_2$O)$_3$]$^-$$_\infty$ units; on dehydration, Cs$_3$LnCl$_6$ are formed, with two structural types.[980] On thermal decomposition,[981] (NH$_4$)$_3$YCl$_6$ first forms (NH$_4$)Y$_2$Cl$_7$ then YCl$_3$. MCl$_3$ (M = all Ln except Pm; Y) react with PyHCl in THF forming (PyH)$_3$MCl$_6$.THF, whose THF is lost in vacuo.[982] Reaction of RbCl and LnCl$_3$ at 850 °C leads to hexachlorometallates, and the structure of Rb$_3$[YCl$_6$] has been determined.[983] Li$_3$[YCl$_6$] is an ionic conductor.[984] (NHMe$_3$)$_4$[YbCl$_7$] has six-coordinate Yb.[985] Ba$_2$[EuCl$_7$] is isostructural with Ba$_2$[LnCl$_7$] (Ln = Gd-Lu, Y), but not with Ba$_2$[ScCl$_6$]Cl, containing capped trigonal prismatic [EuCl$_7$]$^{2-}$ ions.[986] [Cs$_4$YbCl$_7$] contains discrete [YbCl$_6$]$^{3-}$ octahedra.[987] Complex holmium chlor-ides synthesized[988] include Cs$_4$HoCl$_7$ and Cs$_3$HoCl$_6$. Enthalpies of formation have been deter-mined for LaX$_3$ as -258.5 ± 0.8 kcal mol^{-1} (X = Cl); -218.7 ± 1.7 kcal mol^{-1} (X = Br); and -166.9 ± 2.0 kcal mol^{-1} (X = I);[989] values have also been calculated for (NH$_4$)$_2$LaX$_5$,[989] (NH$_4$)Y$_2$Cl$_7$, and (NH$_4$)$_3$YX$_6$.[990] Structural and magnetic properties are reported for [Me$_2$NH$_2$][MCl$_4$(H$_2$O)$_2$] (M = Nd, Pr), which contain edge-connected [MCl$_{4/2}$Cl$_2$(H$_2$O)$_2$] trigon-dodecahedra.[991,992] Cs$_3$Lu$_2$Cl$_9$ is isostructural with Cs$_3$Tb$_2$Cl$_9$.[993]

Cs$_3$[Yb$_2$Cl$_9$] and Cs$_3$[Yb$_2$Br$_9$] have been synthesized and studied by high resolution inelastic neutron scattering.[994] Polymeric species such as [Ln$_2$Cl$_{11}$]$^{5-}$ are indicated in conductivity studies on LnCl$_3$ and M$_3$LnCl$_6$ (M = K, Rb, Cs; Ln = La, Ce, Pr, Nd).[995]

(NH$_4$)$_2$PrBr$_5$ is isostructural with K$_2$PrBr$_5$, where edge sharing between the PrBr$_5$ units gives capped trigonal prismatic seven-coordinate praseodymium.[996] Li$_3$LnBr$_6$ (Ln = Sm-Lu,Y) have been synthesized and the structure of Li$_3$ErBr$_6$ determined (Er-Br 2.767 Å).[997]

Rb$_2$Li[DyBr$_6$] has a tetragonally distorted elpasolite structure.[998] MCl$_3$ (M = Tb, Dy) react with PPh$_4$Cl in MeCN forming Ph$_4$P[MCl$_4$(MeCN)] which contain dimeric [(MeCN)Cl$_3$M(μ-Cl)$_2$MCl$_3$(MeCN)]$^{2-}$ anions.[999]

3.2.2.9 Complexes of the Ln^{2+} ion

3.2.2.9.1 Group 14 ligands

A number of simple alkyls and aryls have been made. The use of a bulky silicon-substituted *t*-butyl ligand permitted the isolation[1000,1001] of simple monomeric (bent) alkyls [Ln{C(SiMe$_3$)$_3$}$_2$] (Ln = Eu, Yb) which are sublimeable in vacuo. Other compounds include [Yb{C(SiMe$_3$)$_2$(Si-Me$_2$X)}] (X = CH=CH$_2$; CH$_2$CH$_2$OEt) and Grignard analogues [Yb{C(SiMe$_3$)$_2$(SiMe$_2$X)}I.OEt$_2$]

(X = Me, CH=CH$_2$, Ph, OMe), synthesized from RI and Yb. The alkyl ytterbium iodides have iodo-bridged dimeric structures, containing four-coordinate ytterbium when X = Me, but five-coordinate for X = OMe, due to chelation. When R = CH=CH$_2$, Ph, OMe (but not Me), the equilibrium can be displaced to the right on heating

$$2YbRI \rightleftharpoons YbR_2 + YbI_2 \qquad\qquad (17)$$

Reaction of powdered Ln (Ln = Eu, Yb) and Ph$_2$Hg in the presence of catalytic amounts of LnI$_3$ affords the compounds LnPh$_2$(THF)$_2$.[148,149] [YbPh$_2$(THF)$_2$] has been used as a synthon in the preparation of [Yb(GePh$_3$)$_2$(THF)$_4$].[1002] Structural characterization of lanthanide(II) aryls has been achieved by two groups. One approach has been to use a very bulky aryl ligand; compounds [Ln(Dpp)I(THF)$_3$] and [Ln(Dpp)$_2$(THF)$_2$] (Ln = Eu, Yb) have been made, with structures determined for [Yb(Dpp)I(THF)$_3$] and [Eu(Dpp)$_2$(THF)$_2$] (Dpp = 2,6-Ph$_2$C$_6$H$_3$).[1003] Similarly in [Yb(Dpp)$_2$(THF)$_2$] the geometry is a strongly distorted tetrahedron, with Yb—C bonds averaging 2.520 Å and Yb—O bonds averaging 2.412 Å.[1004] There are additionally two weak η^1-π-arene interactions (Yb—C 3.138 Å) involving α-carbons of the terphenyl groups. The pentafluorophenyl [Eu(C$_6$F$_5$)$_2$(THF)$_5$] has the pentagonal bipyramidal coordination geometry familiar for simple coordination compounds.[1005] Homoleptic Yb^{II} aluminates YbAl$_2$R$_8$ (R = Me, Et, Bu^i) have been made by a silylamide elimination reaction between [Yb(N(SiMe$_3$)$_2$)$_2$(THF)$_2$] and excess AlR$_3$. The methyl compound is an involatile and insoluble oligomer, but hexane-soluble YbAl$_2$Et$_8$ has a three-dimensional network based on [Yb(AlEt$_4$)]$^+$ and [Yb(AlEt$_4$)$_3$]$^-$ fragments linked by bridging α-carbons and secondary Yb···H—C agostic interactions.[1006]

3.2.2.9.2 *Group 15 ligands involving nitrogen*

[SmI$_2$(THF)$_2$] reacts with *N*-methylimidazole (*N*-Meim) forming [SmI$_2$(*N*-Meim)$_4$]. On recrystallization from hot THF, this loses a molecule of *N*-Meim and dimerizes, forming [{SmI(μ-I)-(*N*-Meim)$_3$}$_2$]. This contains six-coordinate Sm, with Sm—I = 3.237(1) Å (terminal) and 3.280(1)–3.307(1) Å (bridging); Sm—N distances are 2.621(7)–2.641(6) Å. On slow crystallization, this tends to oxidize to Sm^{III} complexes. Eu behaves similarly in forming [EuI$_2$(N-Meim)$_4$] and [{EuI(μ-I)(N-Meim)$_3$}$_2$].[1007] LnI$_2$ (Ln = Sm, Yb) reacts with substituted pyridines in THF forming [LnI$_2$(pyridine)$_4$]; the structures of [LnI$_2$(3,5-lutidine)$_4$] (Ln = Sm, Yb) and [YbI$_2$(4-*t*-Butpy)$_4$] were determined, all have *trans*- structures.[1008] Deacon *et al.* have reported[1009] a rich and extensive chemistry of the ytterbium(II) complex [Yb(NCS)$_2$(THF)$_2$], which undergoes a range of oxidative addition reactions affording complexes such as Yb(NCS)$_3$(THF)$_4$, Yb(NCS)$_3$(Odpp)(THF)$_3$ (HOdpp = 2,6-diphenylphenol), and Yb(NCS)$_2$(Cp)(THF)$_3$. The THF ligands in Yb(NCS)$_2$(THF)$_2$ can be replaced by DME forming eight-coordinate Yb(NCS)$_2$(DME)$_3$. Yb(NCS)$_2$(THF)$_2$ reacts with CCl$_3$CCl$_3$ forming Yb(NCS)$_2$Cl(THF)$_4$, which turns out to have the solid-state structure [YbCl$_2$(THF)$_5$]$^+$[Yb(NCS)$_4$(THF)$_3$]$^-$ in contrast to the monomer Yb(NCS)$_3$(THF)$_4$. Oxidation of Yb(NCS)$_2$(THF)$_2$ with TlO$_2$CPh initially affords a solvated of Yb(NCS)$_2$(O$_2$CPh) which then gives a mixture of Yb(NCS)$_3$(THF)$_4$ and dimeric [[Yb(NCS)(O$_2$CPh)$_2$(THF)$_2$]$_2$], the latter having a [Yb(μ-O$_2$CPh)$_4$Yb] core with both bi- and tridentate bridging benzoates. Ph$_2$CO reacts with Yb(NCS)$_2$(THF)$_2$ producing another dimer, [[Yb(NCS)$_2$(THF)$_3$]$_2$(μ-OC(Ph)$_2$C(Ph)$_2$O)], where two benzophenones have coupled together. In the area of amides, [SmI$_2$(THF)$_2$] reacts with KNPh$_2$ to form [Sm(NPh$_2$)$_2$(THF)$_4$] where the coordination geometry approximates to trigonal prismatic.[295] New syntheses are reported for [Ln{N(SiMe$_3$)$_2$}$_2$(THF)$_2$] (Ln = Sm, Yb).[1010] New amides [Ln{N(SiMe$_3$)(2,6-Pr$^i_2$C$_6$H$_3$)}$_2$(THF)$_2$] have been reported to have distorted tetrahedral structures[1011] [Yb{N(SiMe$_3$)(2,6-Pr$^i_2$C$_6$H$_3$)}$_2$(THF)$_2$] partially desolvates in vacuo forming [Yb{N(SiMe$_3$)(2,6-Pr$^i_2$C$_6$H$_3$)}$_2$(THF)]; [Yb{N(SiMe$_3$)$_2$}$_2$(THF)$_2$] desolvates completely to [Yb{N (SiMe$_3$)$_2$}$_2$] on heating in vacuo. The structures have been determined of [Sm(N(SiMe$_3$)$_2$)$_2$(THF)$_2$] and of the dimer [Sm$_2${N(SiMe$_2$)$_2$}(DME)$_2$(THF)$_2$(μ-I)$_2$], in the latter, Sm—O (THF) is 2.592 Å; Sm—O (DME) 2.685 Å; and Sm—I 3.847 Å. The former contains significant short-Sm···C distances, evidence for agostic interactions.[1012] The Sm^{II} amide [Sm(N(SiHMe$_2$)$_2$)$_2$(THF)$_x$] (x < 1) crystallizes from hexane as a remarkable trimeric [Sm[[μ-N(SiHMe$_2$)$_2$]$_2$Sm[N(SiHMe$_2$)$_2$](THF)$_2$], in which coordinative saturation involves multiple metal···SiH β-agostic interactions.[1013] [KSm{N(SiMe$_3$)$_2$}$_3$], which is oligomeric, has been reported.[1014] A large number of europium (II), samarium(II), and ytterbium(II) η^2-pyrazolides have been synthesized; [Ln{(Me$_2$pz)$_3$}(THF)] (Ln = La, Er) are considered to be[1015] a centrosymmetric dimer with four chelating and two bridging pyrazolides and two bridging THF ligands. Compounds [Ln{(Ph$_2$pz)$_3$}(THF)$_x$] (Ln = Sc, Y, Gd, Er, n = 2; Ln = Lu, n = 3) are also reported; [Ln{(Ph$_2$pz)$_3$}(THF)$_2$] are believed to be eight-coordinate. [Er{(Ph$_2$pz)$_3$}(DME)$_2$] has nine-coordinate erbium, with three chelating pyrazolates,[1016] one chelating and one η^1-dme; [Nd{(Bu$^t_2$pz)$_3$}(DME)] has a structure based on chains of Nd{(Bu$^t_2$pz)$_3$ units bridged by DME.[1016] [Yb(Ph$_2$pz)$_2$(DME)$_2$] has two chelating dimethoxyethane ligands and two η^2 3,5-diphenylpyrazolates.[1017] [Ln(η^2-Ph$_2$Pz)$_2$(OPPh$_3$)$_2$] (Ln = Er, Nd) are eight-coordinate; [Ln(η^2-Ph$_2$Pz)$_2$(THF)$_3$] (Ln = La, Nd) are nine-coordinate.[1018] Syntheses are reported of [Ln(Bu$^t_2$pz)$_3$(THF)$_2$] and [Ln(Bu$^t_2$pz)$_3$(Ph$_3$PO)$_2$] (Bu$^t_2$pz = 3,5 di-t-butylpyrazolato). The compound [Er(Bu$^t_2$pz)$_3$(THF)$_2$] has eight-coordinate erbium with chelating pyrazole ligands.[1019] Structures have been reported for [Ln(η^2-Ph$_2$pz)$_2$(DME)$_2$] (Ln = Sm, Eu), [Eu(η^2-Bu$^t_2$pz)$_2$(dme)$_2$], and for some analogous 7-azaindolides. (Me$_2$pz = 3,5-dimethylpyrazolide; Ph$_2$pz = 3,5-diphenylpyrazolide; Bu$^t_2$pz = 3,5-di-t-butylpyrazolide).[297] Reaction of Bu$^t_2$pzH with lanthanide metals and mercury at 220 °C affords [Eu(η^2-Bu$^t_2$pz)$_2$]; and [Yb$_2$(η^2-Bu$^t_2$pz)$_5$], the latter having the structure [(η^2-Bu$^t_2$pz)Yb^{II}(μ-η^2: η^2-Bu$^t_2$pz)$_2$Yb^{III}(η^2-Bu$^t_2$pz)$_2$].[1020] Several carbazolyl complexes have been synthesized; in all cases so far, it behaves as a monohapto-amide. [Eu(carbazolyl)$_2$(THF)$_4$] is *cis*-,[1021] as [Yb(carbazolyl)$_2$(THF)$_4$] is presumed to be. On dissolution in DME/THF, [Yb(carbazole)$_2$(THF)$_4$] forms all *cis*-[Yb(carbazolyl)$_2$(DME)(THF)$_2$]. This has Yb—N = 2.43(3)–2.45(2) Å; Yb—O (THF) = 2.41(2)–2.48(2) Å and Yb—O (DME) = 2.44(2)–2.46(1) Å.[1022] [SmI$_2$(THF)$_2$] reacts with potassium carbazolyl forming *cis*-[Sm(carbazolyl)$_2$(THF)$_4$]; this has Sm—N of 2.565(13) Å and Sm—O of 2.582(7) Å. *cis*-[Sm(carbazolyl)$_2$(THF)$_4$] reacts with *N*-methylimidazole to form *trans*-[Sm(carbazolyl)$_2$(N-Meim)$_4$], which has Sm—N (carbazolyl) = 2.591(3) Å and Sm—N (*N*-Meim) = 2.685(14) Å.[1023] Some Ln^{II} β-diketiminates have also been made, [Ln(L-L)$_2$(THF)$_2$] (Ln = Sm, Yb), [Ln(L-L)$_2$] and [Ln(L-L')$_2$] (L-L' = (R)NC(Ph)C(H)C(Bu^t)NR). LnCl$_3$ reacts with [{Me$_2$SiN(R)Li}$_2$] (R = Ph, Bu^t) forming[1024] chloride bridged dimers [{(R)NSiMe$_2$SiMe$_2$N(R)}Ln(μ-Cl)(THF)}$_2$] (Ln = Nd, Gd, Yb) which can be converted into trifluoracetates [{(Bu^t)NSiMe$_2$SiMe$_2$N(Bu^t)}Ln(μ-OCOCF$_3$)(THF)}$_2$]. The benzamidinate [(PhC(N(SiMe$_3$)$_2$)$_2$)Yb(THF)$_2$] has been made.[1025]

Polypyrazolylborates have been very extensively investigated. SmCl$_2$ reacts with one mole of KTp in THF forming [SmCl(Tp)$_2$(Hpz)] which has square antiprismatic coordination of samarium.[1026] [Sm(Tp)Cl(L)] (L = Hpz, N-Mepz) react with sodium β-diketonates forming [Sm(Tp)(ACAC)], [Sm(Tp)(tfac)] and [Sm(Tp)(hfac)]. [SmCl(Tp)(Hpz)] reacts with K[BH$_2$pz$_2$] forming the bicapped trigonal prismatic [SmCl(Tp)(BH$_2$pz$_2$)] whilst [SmCl(Tp)(Hpz)] is square-antiprismatic.[1027] [Ln(Tp)$_2$(THF)$_2$] and [Ln(Tp^{Me2})$_2$] (Ln = Eu, Sm and Yb) have been synthesized;[1027] the latter are very insoluble and are usually purified by sublimation in the cases of the Eu and Yb compounds.[318,1029] Using Tp^{Me2}, a very wide range of compounds has been synthesized. [Ln(Tp^{Me2})$_2$] (Ln = Sm, Yb, Eu) tend to be insoluble, but introduction of an additional 4-ethyl group increases solubility in [Ln(Tp^{Me2,4Et})$_2$] (Ln = Sm, Yb, Eu), whilst [Ln(Tp^{Ph2})$_2$] (Ln = Sm, Yb) and [Ln(Tp^{Th2})$_2$] (Ln = Sm, Yb, Eu) have also been made. These compounds tend to have six-coordinate trigonal antiprismatic coordination of the lanthanide.[320] The THF molecules in [Ln(Tp)$_2$(THF)$_2$] can be replaced by other donors, as seen in the structures[1030] of [Eu(Tp)$_2$(L)$_2$] (L = Ph$_2$SO, (Me$_2$N)$_2$C=O). [Sm(Tp^{Me2})$_2$] undergoes a range of one electron transfer reactions, whilst complexes of the sterically demanding Tp^{But,Me} ligand [Sm(Tp^{But,Me})R] are resistant to redistribution reactions.[1031] [Sm(Tp^{Me2})$_2$] has six-coordinate samarium. In its wide range of oxidations,[318,1032] it forms [Sm(Tp^{Me2})$_2$X] (X = Cl, Br, F) by halogen abstraction from organic compounds, and [Sm(Tp^{Me2})$_2$]I by iodine oxidation. Reaction with TlBPh$_4$ gives [Sm(Tp^{Me2})$_2$]BPh$_4$. [Sm(Tp^{Me2})$_2$] cleaves dichalcogenides REER forming [Sm(Tp^{Me2})$_2$(ER)] (R is, for example, Ph, E = S, Se).[1033] [Sm(Tp^{Me2})$_2$] reacts with azobenzene forming [Sm(Tp^{Me2})$_2$(N$_2$Ph$_2$)] where the Tp^{Me2} ligands remain tridentate.[1034] [Sm{Tp^{Me2}}$_2$] reacts with dioxygen forming the first lanthanide superoxo complex [Sm{Tp^{Me2}}$_2$(η^2-O$_2$)];[1035] isotopic substitution confirmed this assignment. X-ray diffraction shows that O$_2$ is bound side-on. The samarium(II) poly(pyrazolyl)borate [Sm(Tp^{Me2})$_2$] undergoes a one-electron oxidation with [Hg(C≡CPh)$_2$] forming monomeric seven-coordinate [Sm(Tp^{Me2})$_2$(C≡CPh)].[321] This undergoes a remarkable rearrangement at 105 °C in benzene solution with the exchange between a pyrazole ring and a alkynyl group, forming [Sm(Tp^{Me2})((HB(Me$_2$pz)$_2$(C≡CPh))(Me$_2$pz)]. The insolubility of [Ln(Tp^{Me2})$_2$] makes it difficult to obtain mono(Tp^{Me2}) complexes, as the equilibrium gets driven to the right, even with a 1:1 stoichiometry in reaction mixture

$$2[Ln(Tp^{Me2})I(THF)_x] \rightleftharpoons [Ln(Tp^{Me2})_2] + LnI_2(THF)_n \qquad (18)$$

This can be obviated by using a bulky hydrotris(3-*t*-butyl-5-methylpyrazolyl) borate ligand (Tp^{But,Me}), which affords monomeric ytterbium(II) compounds with just one pyrazolylborate bound to ytterbium such as [Yb(Tp^{But,Me})I(L)$_n$] (L = THF, Bu^{t}NC, n = 1; L = 3,5-lutidine, n = 2), [Yb(Tp^{But,Me})(N(SiMe$_3$)$_2$)], and [Yb(Tp^{But,Me})(CH(SiMe$_3$)$_2$)].[318,1036]

It is still possible to obtain bis complexes with Tp^{tBu,Me}, but [Ln(Tp^{tBu,Me})$_2$] (Ln = Sm, Yb) shows[1037] two different mode of ligand bonding; one is a conventional η^3-ligand, the other is bound via two nitrogen atoms and an agostic B—H···Sm interaction. The NMR spectrum of [Yb(Tp^{tBu,Me})$_2$] shows[182] Yb–HB coupling, confirming that the agostic interaction persists in solution. Use of a bulky ligand stabilizes the dimeric hydride [(Tp^{But,Me})Yb(μ-H)$_2$Yb(Tp^{But,Me})] which does have a rich chemistry,[1038,1039] forming an ene-diolate by CO insertion and reacting with alkynes. In [Eu(B(pz)$_4$)$_2$(THF)$_2$], (pz = pyrazolyl) europium has the expected eight-coordination.[334] Analogous [Ln(B(pz)$_4$)$_2$(THF)$_2$] (Ln = Sm, Yb) react with alkyl halides forming [Ln(B(pz)$_4$)$_3$] by oxidative disproportionation. A europium(II) poly(pyrazolyl)borate complex (**45**) exhibits an orange electroluminescence.[1040]

(**45**)

[SmI$_2$(THF)$_2$] reacts[1041] with a dipyrrolide dianion under N$_2$ forming a remarkable tetranuclear dinitrogen complex [((μ-Ph$_2$C(η^1:η^5-C$_4$H$_3$N)$_2$)Sm)$_4$(μ-η^1:η^1:η^2:η^2-N$_2$)].

3.2.2.9.3 *Group 15 ligands involving phosphorus*

A number of organophosphides of divalent lanthanides have been reported, with structures of [Yb(PPh$_2$)$_2$(THF)$_4$],[1042] [Sm(PPh$_2$)$_2$(N-Methylimidazole)$_4$],[1043] and [Yb(P(mesityl)$_2$)$_2$(THF)$_4$],[1043] all having *trans*-octahedral structures. Both [YbI$_2$(THF)$_2$] and [Yb{(N(SiMe$_3$)$_2$)$_2$(THF)$_2$] react with KPPh$_2$ forming [Yb(PPh$_2$)$_2$(THF)$_4$]; the THF can be displaced by N-methylimidazole forming *trans*-[Yb(PPh$_2$)$_2$(N-mim)$_4$].[1044] The structures of the compounds [Sm(P(mesityl)$_2$)$_2$(THF)$_4$] and [Sm(As(mesityl)$_2$)$_2$(THF)$_4$] have been reported.[1045] [Ln(P(H)(mes*))$_2$(THF)$_4$] (Ln = Eu, Yb; mes* = 2,4,6-Bu$^t_3$C$_6$H$_2$) are the first lanthanide complexes with primary phosphide ligands;[1046] they feature distorted octahedral coordination of the lanthanide. The three isostructural phosphides [(THF)$_2$Li(μ-PBu$^t_2$)$_2$M(μ-PBu$^t_2$)$_2$Li(THF)$_2$] (M = Sm, Eu, Yb) have tetrahedrally coordinated lanthanides.[1047] The structure of *trans*-[Eu(PPh$_2$)$_2$(N-mim)$_4$] has been determined.[1048] A monomeric four-coordinate phosphide complex, [Sm(P(CH(SiMe$_3$)$_2$)(C$_6$H$_4$-2-NMe$_2$)$_2$)] has been reported.[1049] [SmI$_2$(THF)$_2$] reacts with KP(SiMe$_3$)$_2$ forming Sm$_2${P(SiMe$_3$)$_2$}$_4$.(THF)$_3$, which has the unsymmetrical structure [((Me$_3$Si)$_2$P)Sm(μ-P(SiMe$_3$)$_2$)$_3$Sm(THF)$_3$].[1050] Sm^{II} and Yb^{II} complexes of chelating secondary phosphide ligands have also been reported.[1051]

3.2.2.9.4 *Group 16 ligands involving oxygen*

The rate of water exchange at the europium(II) aqua ion, believed to be [Eu(OH$_2$)$_8$]$^{2+}$, is the fastest ever measured for a non Jahn–Teller ion.[1052] YbI$_2$.H$_2$O has six-coordinate Yb (5 I, 1O).[1053] KEu(CH$_3$COO)$_3$ is the first ternary Eu^{II} carboxylate, with eight and nine-coordinate Eu.[1054]

[Ln(O$_3$SCF$_3$)$_3$] have been widely used for years; now [Ln(O$_3$SCF$_3$)$_2$] (Ln = Sm, Yb) have been prepared[1055] by reduction of the Ln^{III} analogues and used as pinacol coupling catalysts. A large number of complexes of *O*-donors, particularly of SmI$_2$ have been examined, because of the importance of SmI$_2$ as a one-electron reductant in organic synthesis, much of which has been carried out using hexamethylphosphoramide (HMPA) as a solvent.[1056–1060] The addition of HMPA to THF solutions of SmI$_2$ brings considerable rate enhancement; HMPA is a suspected carcinogen so certain alternative ligands have been investigated. The structures of a number of complexes, both of HMPA and other O-donors, have been investigated, shedding light on SmI$_2$-promoted reactions. In general these compounds of amide ligands have a coordination number of six, suggesting steric effects at work, in addition to pointing to strong electron-donating power of the ligands, as the CN of Sm^{II} complexes is usually in the range seven to nine. [LnI$_2$(THF)$_2$] (Ln = Sm, Yb[1061] and Eu[1062,1063]) are useful starting materials, conveniently made by reaction of Ln with ICH$_2$CH$_2$I. Under other circumstances, different complexes result. The initial product of crystallization of SmI$_2$ from THF is pentagonal bipyramidal [SmI$_2$(THF)$_5$];[1064] other seven-coordinate complexes obtained using mixtures of THF and dimethoxyethane are [SmI$_2$(THF)$_3$(DME)] and [SmI$_2$(THF)(DME)$_2$]. In [SmI$_2$(DME)$_2$(THF)] Sm—O (DME) is 2.618 Å; Sm—O (THF) is 2.530 Å; and Sm—I is 3.246 Å, whilst in [SmI$_2$(DME)(THF)$_3$] Sm—O (DME) is 2.641 Å; Sm—O (THF) is 2.571 Å; and Sm—I is 3.323 Å. YbI$_2$ reacts with 4.5 moles HMPA in THF forming [Yb(HMPA)$_4$(THF)$_2$]I$_2$; a similar reaction with SmI$_2$ affords [Sm(HMPA)$_4$I$_2$]. Excess HMPA gives [Sm(HMPA)$_6$]I$_2$. These compounds all have six-coordinate Ln^{II}. In SmI$_2$(HMPA)$_4$, average bond lengths of Sm—O are 2.500 Å and Sm—I are 3.390 Å, whereas in [Sm(HMPA)$_6$]I$_2$ the Sm—O distances average 2.53 Å.[1065] For comparison, in [SmI$_2$(Ph$_3$PO)$_4$], Sm—O is 3.27(1) Å.[1066] SmI$_2$(THF)$_2$ reacts with two moles of tetramethylurea (tmu) forming[1067] [SmI$_2$(tmu)$_2$(THF)$_2$] (tmu = -tetramethylurea) has an all *trans*-geometry, with Sm—O (tmu) of 2.446 Å, Sm—O (THF) of 2.528 Å, with Sm—I of 3.061 Å and 3.317 Å. Using excess tmu did not result in the isolation of a complex where the other THF ligands had been replaced. Another complex to be isolated is *trans*-[SmI$_2$(dmi)$_4$] (dmi = 1,3-dimethyl-2-imidazolidone, N,N-dimethylacetamide); here all the THF groups have been substituted. The average Sm—O distance here is 2.48 Å whilst the Sm—I distances are even longer, at 3.345 Å and 3.579 Å. *trans*-[SmI$_2$(dma)$_4$] (dma = dimethylacetamide) has similar Sm—O distances, averaging 2.45 Å and Sm—I 3.309 Å. Use of four moles (i.e., a deficit on the stoichiometric amount) of dimethylpropylene urea (dmpu) causes displacement of the iodides from SmI$_2$(THF)$_2$ forming [Sm(dmpu)$_6$]I$_2$ where the cation has a distorted trigonal antiprismatic geometry; Sm—O distances are 2.475–2.488 Å; however if just two moles of dmpu are used in the reaction, [SmI$_2$(dmpu)$_3$(THF)] results, in which the iodides are *trans*- and the dmpu ligands have a *mer*-arrangement.

The developing coordination chemistry of Sm^{II} has thrown up some interesting cases of isomerism. Reaction of SmI$_2$(THF)$_2$ with diglyme results in the isolation of *trans*-[SmI$_2${O(CH$_2$-CH$_2$OMe)$_2$}$_2$], whilst the *cis*-isomer was obtained as a by-product in the reaction of SmI$_2$ with *t*-BuOK in diglyme. These were the first examples to be isolated of geometric isomers in an eight-coordinate complex.[1068,1069] The Sm–O distances in *cis*-[SmI$_2${diglyme}$_2$], fall into the range 2.653–2.699 Å, averaging 2.68 Å, which is slightly shorter than the value for the *trans*-isomer. The Sm–I distances in the *cis*-isomer, however, at 3.332–3.333 Å, are significantly longer than those in *trans*-[SmI$_2${diglyme}$_2$] (3.265 Å). In another interesting study,[1070] reaction of SmI$_2$ with 1,2-diiodoethane at 50 °C yields [SmI$_2$(DME)$_3$]. Crystallization at −20 °C affords racemic [SmI$_2$(DME)$_3$], whilst crystallization from solution at ambient temperature yields a mixture of crystals of the two different enantiomers. Sm–I distances are 2.3550(8) Å and 2.3832(8) Å and Sm–O distances range from 2.656(7) Å to 2.681(6) Å. It was suggested that interactions between methyl groups prevent ready interconversion. The samarium(II) complex [(DME)$_2$BrSm(μ-Br)$_2$SmBr(DME)$_2$] has been synthesized.[1071] A number of solvated Yb^{II} and Sm^{II} complexes, mainly with diethylene glycol dimethylether(dime), have been characterized,[1072] notably [Yb(dime)$_3$]$^{2+}$, [Yb(dime)$_2$(MeCN)$_2$]$^{2+}$, [Yb(dime)(MeCN)$_5$]$^{2+}$, and [Yb(py)$_5$(MeCN)$_2$]$^{2+}$; these have tricapped trigonal prismatic, square antiprismatic, and pentagonal bipyramidal coordination respectively.

Reaction of Nd and Dy with I$_2$ at 1,500 °C followed by dissolution of the product affords [LnI$_2$(DME)$_3$] and [LnI$_2$(THF)$_5$] (Ln = Nd, Dy).[1073] [NdI$_2$(THF)$_5$] has the familiar pentagonal bipyramidal structure with axial iodines.[1074] [TmI$_2$(DME)$_2$(THF)] has a similar coordination geometry. [TmI$_2$(DME)$_2$] is the first molecular Tm^{II} compound;[1075] one DME is monodentate, the iodides occupy axial positions in a pentagonal bipyramidal geometry. Another Tm^{II} complex, this time with a calixarene, has been made.[1076] The first lanthanide(II) diketonate, [Eu(tmhd)$_2$(DME)$_2$], has been made, by reaction of [EuI$_2$(THF)$_2$] with Kthd in THF, evaporation and crystallization from DME. The Sm analogue, [Sm(tmhd)$_2$(DME)$_2$] has also been synthesized by a similar route; it decomposes to [Sm(tmhd)$_3$(DME)] on keeping a solution at $-34\,°C$ for two weeks (tmhd = Me$_3$CCOCHCOCMe$_3$; dme = dimethoxyethane).[1077] The Eu^{II} crown ether complex [Eu(benzo-15-crown-5)$_2$] (ClO$_4$)$_2$ has been synthesized by electrochemical reduction of a solution of Eu(ClO$_4$)$_3$ and benzo-15-crown-5 in MeOH/H$_2$O. It contains 10-coordinate europium, with Eu—O distances in the range 2.662(3)–2.728(4) Å. This compound exhibits strong luminescence, taking on a violet hue in daylight.[1078] The Eu^{II} EDTA complex Na$_3$[Eu(EDTA)]Cl.7H$_2$O is in fact polymeric in the solid state with europium bound to two nitrogens and to six carboxylate oxygens.[1079] [(NH$_2$)$_3$]$_3$[Eu^{II}(dtpa)(H$_2$O)].8H$_2$O is isostructural with the Sr analogue.[1080] In solution, [Eu^{II}(dtpa)(H$_2$O)]$^{3-}$ is less stable to oxidation than Eu^{2+}(aq).[1081]

The redox stability of complexes of two other ligands, ODDA^{2-} and ODDM^{4-} have been studied; both [Eu^{II}(ODDA)(H$_2$O)] and [Eu^{II}(ODDM)]$^{2-}$ are more stable[1082] and the ODDM complex has a significantly greater stability constant than [Eu^{II}(dtpa)(H$_2$O)]$^{3-}$ (log K values of 13.07 vs. 10.08 respectively) (ODDM^{4-} = 1,4,10,13-tetraoxa-7,16-diaza-cyclooctadecane-7,16-dimalonate; ODDA^{2-} = 1,4,10,13-tetraoxa-7,16-diazacyclooctadecane-7,16-diacetate). A number of Ln^{II} aryloxides have been reported including [Eu(OC$_6$H$_2$Bu$^t_2$-2,6-Me-4)$_2$(THF)$_3$].[1083] [Yb(OC$_6$H$_2$Bu$^t_2$-2,6-Me-4)$_2$(Et$_2$O)$_2$] and dimeric [Yb(OC$_6$H$_2$Bu$^t_2$-2,6-Me-4)$_2$][1084] [Sm{N(Si-Me$_3$)$_2$}$_2$(THF)$_3$] reacts with HOC$_6$H$_3$Bu$^t_2$-2,6-Me-4 forming[1085] five-coordinate [Sm(OC$_6$H$_3$Bu$^t_2$-2,6-Me-4)$_2$(THF)$_3$]; {KSm{N(SiMe$_3$)$_2$}$_3$} reacts with HOC$_6$H$_3$Bu$^t_2$-2,6-Me-4 forming [KSm(OC$_6$H$_3$Bu$^t_2$-2,6-Me-4)$_3$(THF)]$_\infty$ which has tetrahedral coordination of Sm^{II} and in which potassium ions act as bridges with K-arene interactions. [Sm{N(SiMe$_3$)$_2$}$_2$(THF)$_2$] reacts with HOC$_6$H$_3$Bu$^t_2$-2,6-Me-4 forming the five-coordinate aryloxide [Sm{OC$_6$H$_3$Bu$^t_2$-2,6-Me-4}$_2$(THF)$_3$] which is a convenient synthon for a number of Sm^{II} and Sm^{III} compounds.[1086]

Eu reacts with 2-methoxyethanol[1087] forming the hydrocarbon-soluble oligomer [Eu(OCH$_2$-CH$_2$OMe)$_2$]$_n$ ($n > 10$ in toluene). This reacts with 2,6-dimethylphenol or 2,6-diisopropylphenol forming the tetrametallic [[Eu(μ^3-η^2-OCH$_2$CH$_2$OMe)(η^2-OCH$_2$CH$_2$OMe)(OC$_6$H$_3$R$_2$-2,6)][H$^+$]$_4$ (R = Me, Pr^i). These have a tetrahedron of seven-coordinate europium atoms, each bound to one terminal bidentate alkoxide, one bridging bidentate alkoxide, a terminal aryloxide, and two bridging oxygens from other aryloxides. [Eu(OCH$_2$CH$_2$OMe)$_2$]$_n$ reacts with Me$_3$Al forming the hexametallic [Me$_3$Al(μ:η^2-OCH$_2$CH$_2$OMe)Eu(μ:η^2-OCH$_2$CH$_2$Me)$_2$(AlMe$_2$)$_2$. 2,6-Disubstituted phenols react with Eu in liquid NH$_3$ forming [Eu(OC$_6$H$_3$Bu$^t_2$-2,6)$_2$(NCMe)$_4$] and [Eu$_4$(μ-OC$_6$H$_3$Pr$^i_2$-2,6)$_4$(OC$_6$H$_3$Pr$^i_2$-2,6)$_2$(μ_3-OH)$_2$(NCMe)$_6$].[1088] An interesting variety of phenoxides with a range of europium polyhedra have been reported. Eu reacts with Pr^{i}OH forming[1089] arene-soluble [Eu(OPr^i)$_2$(THF)$_x$]$_n$ which is a synthon for a trimetallic dimethylphenoxide [Eu(OC$_6$H$_3$-2,6-Me$_2$)$_2$]$_3$ that reacts with isopropanol vapor forming a cluster H$_{10}$[Eu$_8$O$_8$(OC$_6$H$_3$-2,6-Me$_2$)$_2$(OPr^i)$_2$(THF)$_6$], containing a cubic arrangement of europiums. Other clusters have been formed by direct reaction between Eu and phenols, such as H$_x$[Eu$_8$O$_6$(OC$_6$H$_3$-2,6-Me$_2$)$_{12}$(OPr^i)$_8$] and H$_5$[Eu$_5$O$_5$(OC$_6$H$_3$-2,6-Pr$^i_2$)$_6$(MeCN)$_8$]. H$_{18}$[[Eu$_9$O$_8$(OC$_6$H$_3$-2,6-Me$_2$)$_{10}$(THF)$_{10}$] [Eu$_9$O$_9$-(OC$_6$H$_3$-2,6-Me$_2$)$_{10}$(THF)$_6$]] was also synthesized. 2,6-diphenylphenol (dppOH) reacts with Eu or Yb on heating in the presence of mercury[1090] forming [Ln(Odpp)$_2$]. These have the structures [Eu$_2$(Odpp)(μ-Odpp)$_3$] and [Yb$_2$(Odpp)(μ-Odpp)$_3$] whilst small amounts of the mixed-valence [Yb$_2$(μ-Odpp)$_3$]$^+$[Yb(Odpp)$_4$]$^-$ were also obtained. All three compounds have additional Ln–aryl interactions. Europium in liquid ammonia reacts with 2,6-dimethylphenol forming the asymmetric [(DME)(RO)Eu^{II}(μ-RO)$_3$Eu^{II}(DME)$_2$] (R = 2,6-Me$_2$C$_6$H$_3$).[1091] Reduction of [Yb(OC$_6$H$_3$Ph$_2$-2,6)$_3$-(THF)$_2$] leads to the isolation of [Yb(OC$_6$H$_3$Ph$_2$-2,6)$_2$(THF)$_3$] and [Yb(OC$_6$H$_3$Ph$_2$-2,6)$_2$(DME)$_2$]; the former has a tbp structure with axial aryloxides.[624]

Another divalent compound is[1092] tbp[Sm(OAr)$_2$(THF)$_3$].THF (Ar = 2,6-Bu$^t_2$-4-Me-C$_6$H$_2$). [Sm(OAr)$_2$(THF)$_4$] (Ar = 2,6-Bu$^t_2$C$_6$H$_3$) catalyzes[1093] the polymerization of phenyl isocyanate. Three-coordinate Yb^{II} exists in [{YbX(μ-X)}$_2$](X = OAr, OC(Bu^t)$_3$) and [{Yb(NR$_2$)(μ-X)}$_2$] (X = OAr, OC(Bu^t)$_3$) where Ar = 2,6-Bu$^t_2$-4-Me-C$_6$H$_2$; R = SiMe$_3$).[1094] The first Ln^{II} siloxide to be characterized[1095] structurally is [{Yb(OSiMe$_2$Bu^t)(η^2-DME)(μ-OSiMe$_2$Bu^t)}$_2$].

3.2.2.9.5 Group 16 ligands involving sulfur, selenium, and tellurium

Ytterbium reacts with organic disulfides forming [Yb(SAr)$_2$(Py)$_2$] (Ar = 2,4,6-triisopropyl-phenyl).[954] Transmetallation of Sm and Eu with Hg(SC$_6$F$_5$)$_2$ affords Sm(SC$_6$F$_5$)$_3$ and Eu(SC$_6$F$_5$)$_2$ respectively; these have the solid-state structures [(THF)$_2$Sm(μ_2-SC$_6$F$_5$)(SC$_6$F$_5$)$_2$]$_2$ and [(THF)$_2$Eu(μ_2-SC$_6$F$_5$)$_2$]$_n$, the latter being a one-dimensional polymer.[952] These undergo thermal decomposition to LnF_3. Some remarkable Ln^{II} thiolates have been reported, in the form of $[Ln(SAr)_2]$ ($Ar = 2,6$-$Trip_2C_6H_3$; $trip = 2,4,6$-$Pr^i_3C_6H_2$). Though formally two-coordinate (and though in the case of the europium compound, uncoordinated THF is present in the lattice), there are η^6-π-interactions present. Structures are also reported of six-coordinate $[Yb(SAr)_2(DME)_2]$ and the Yb^{III} compound $[YbI(SAr)_2(THF)_3]$.[1096] A sterically crowded Yb^{II} thiolate $[Yb(SAr^*)_2(THF)_4]$ ($Ar^* = 2,6$-$Trip_2C_6H_3$, where Trip is $2,4,6$-$^iPr_3C_6H_2$) has the *trans-* structure now becoming familiar for these Ln^{II} sytems.[1097] The overcrowding caused by the sterically encumbered Ar^* ligands is reflected in the large Yb—S—C angle of 151.16. A number of pyridinethiolate (Spy; 2-S-NC$_5$H$_4$) complexes have been characterized.[965] $[Yb(SPy)_2]$ crystallizes from pyridine as seven-coordinate $[Yb(SPy)_2(Py)_3]$. Europium pyridinethiolates $[Eu(SPy)_2(Py)_4]$ and $[Eu(SPy)_2(bipy)(THF)_2]$ have been synthesized.[966] Heterometallic chalcogenides $[(THF)_2Eu(\mu_2\text{-}SePh)_6Pb_2]$, $[Yb(THF)_6][Sn(SePh)_3]_2$, and $[Py_2Eu(2\text{-}S\text{-}NC_5H_4)_2Sn(2\text{-}S\text{-}NC_5H_4)_2]_n$ have been characterized.[1098] $Ln(SePh)_2$ reacts with selenium in DME forming heterovalent clusters $[Ln_4Se(SePh)_8(DME)_4]$ ($Ln = Sm, Yb, Nd^{III}/Yb^{II}, Sm^{III}/Yb^{II}$) which contain a square of lanthanide ions with a capping selenide.[1099] Lanthanide(II) selenolate and tellurolate complexes,[1100–1102] usually obtained as etherates, are possible precursors to lanthanide monochalcogenides. $M(TePh)_2$ ($M = Yb, Eu$) crystallize as one-dimensional polymers like $[(THF)_2Eu(TePh)_2]_\infty$ and $[(THF)_2Yb(TePh)_2.1/2 (THF)]_\infty$. $[Ln(ESi(SiMe_3)_3)_2 (tmeda)_2]$ ($Ln = Eu, Yb; E = Se, Te$) have been reported and the structures of $[Yb(SeSiMe_3)_3)_2(tmeda)_2]$ and $[\{Eu(SeSiMe_3)_3)_2(dmpe)_2\}_2]$ determined; the ytterbium tmeda complexes eliminate $E(Si(SiMe_3)_3)_2$ at $200\,°C$ affording YbSe and YbTe.

3.2.2.9.6 Group 17 ligands

Rb_4TmI_6, synthesized by heating a mixture of RbI, Tm, and HgI_2, has the K_4CdI_6 structure with trigonal antiprismatic coordination of Tm.[1103] $RbYbI_3$ has a structure based on edge-sharing $[YbI_6]$ octahedra.[1104] M_2EuI_6 ($M = Cs, Rb$) have isolated $[EuI_6]^{2-}$ ions,[1105] similarly Rb_2YbI_6 has octahedral $[YbI_6]^{2-}$ ions.[1106] Reduction of MX_3 ($M = Sm, Dy, Tm, Yb; X = Br, I$) with alkali metals, In, and Tl (A) leads generally to the ternary compounds AMX_3 and AM_2X_5; some similar compounds with divalent metals (Ca, Sr, Ba) have also been made.[1107] $LiDy_2Br_5$, prepared by reduction of $DyBr_3$ with Li metal at $700\,°C$, is isostructural with $LiDy_2Cl_5$, $LiYb_2Cl_5$, and $LiLn_2Br_5$ ($Ln = Sm, Tm$).[1108,1109]

3.2.2.10 Complexes of the Ln^{4+} ion

There have been fewer developments here than in the (+2) state.

3.2.2.10.1 Group 15 ligands

A highlight here is the synthesis of the first Ln^{IV} silylamide. Although it cannot be oxidized with Cl_2, $[Ce(N(SiMe_3)_2)_3]$ reacts with $TeCl_4$ forming $[CeCl(N(SiMe_3)_2)_3]$; this has a trigonal prismatic structure, with Ce $0.36\,Å$ out of the N_3 basal plane (and, interestingly a $0.05\,Å$ lengthening of the N—Si bond).[1110] A preliminary mention of the reaction of PPh_3Br_2 with $[Ce(N(SiMe_3)_2)_3]$ forming $[CeBr(N(SiMe_3)_2)_3]$ may also be noted. A previous Ce^{IV} amide was synthesized by iodine oxidation of a cerium(III) compound of a triamidoamine, affording a notable compound with a rare Ce^{IV}—I linkage (Scheme 9); analogous oxidation with X_2 ($X = Cl, Br$) yields mixed-valence dimers.[1111]

The reaction of 2,4,6-tri-*t*-butylpyridyl-1,3,5-triazine (L) with $(NH_4)_2[Ce(NO_3)_6]$ leads to a number of species,[1112] including 11-coordinate $[Ce(L)(NO_3)_4]$, $[Ce(NO_3)_5(OH_2)]^-$, and $[Ce(NO_3)_6(OEt)]^{2-}$ (all crystallographically characterized). Syntheses are reported for some Ce^{IV} bis (porphyrinates);[1113,1114] NMR shows that the two porphyrin rings do not rotate with respect to each other even at $140\,°C$. COSY NMR-spectral data have been reported for Ce^{IV} porphyrins.[1115] Ring oxidized Ce^{IV} phthalocyanines have been prepared[1116] by oxidation of $[Ce^{IV}(pc)_2]$ or $[Ce^{III}(pc)_2]^-$, and the structure of $[Ce^{IV}(pc)_2](BF_4)_{0.33}$ determined. The synthesis and structure of $[Ce^{IV}(pc)_2]$ is reported.[1117]

3.2.2.10.2   Group 16 ligands

Hydrated cerium(IV) triflate dehydrates on stirring with triflic anhydride.[1118] It is potentially a valuable source of other Ce^{IV} compounds. The crystal structure of $(NH_4)_4Ce(SO_4)_4.2H_2O$ has been reported,[1119] together with a DTA study involving $(NH_4)_2Ce(SO_4)_3$, $(NH_4)_2Ce(NO_3)_6$, and $Cs_2Ce(NO_3)_6.MgCe(NO_3)_6.8H_2O$ is isomorphous with the Th analogue, containing 12-coordinate $[Ce(NO_3)_6]^{2-}$ ions;[1120] $K_2Ce(NO_3)_6$ exists in two polymorphs,[1121] again with icosahedral $[Ce(NO_3)_6]^{2-}$. The iodates $Ce^{IV}HIO_6.4H_2O$ and $MCe^{IV}IO_6.nH_2O$ (M = alkali metal) have been reported.[1122] A number of diketonates are now better characterized, prompted by the possibility of using them as CVD materials and petrol additives, as well as a source of cerium oxide as an oxygen store for catalytic converters. Thus the coordination geometries in the air stable potential CVD precursors $[Ce(tmhd)_4]$ and $[Ce(pmhd)_4]$ (tmhd = 2,2,6,6-tetramethyl-3,5-heptanedionate; pmhd = 1-phenyl-5-methylhexane-1,3-dionate) are now known to be distorted dodecahedral and square antiprismatic respectively; the former sublimes unchanged whilst the latter is involatile.[1123] $[Ce\{Me_3CCOCHCOCMe_2(OMe)\}_4]$ is square antiprismatic.[1124] Alkoxides are among the more important (and best characterized) Ce^{IV} compounds. General methods are available to make many $Ce(OR)_4$ (R, e.g., Me, Et, OPr^i, n-C_8H_{17}).[1125]

$$(NH_4)_2Ce(NO_3)_6 + 4ROH + 4NH_3 \rightarrow Ce(OR)_4 + 6NH_4NO_3 \qquad (19)$$

$$(NH_4)_2Ce(NO_3)_6 + 4ROH + 6NaOMe \rightarrow Ce(OR)_4 + 6NaNO_3 + 2NH_3 + 6MeOH \qquad (20)$$

As usual, many of these alkoxides are oligomers, though coordinative saturation can be achieved by adduct formation. $Ce(OSiPh_3)_4$, prepared by alcoholysis of $Ce(OPr^i)_4$ in dimethoxyethane, is isolated as $[Ce(OSiPh_3)_4(DME)_x]$ ($0.5 < x < 1$). Crystals of $[Ce(OSiPh_3)_4(DME)]$ display octahedral coordination of cerium with Ce—O (Si) of 2.10–2.14 Å and two rather long Ce—O (ether) bonds at 2.58–2.59 Å. Although the solid is air- and moisture-stable, it undergoes immediate reaction with HACAC forming $Ce(ACAC)_4$.[1126] Thermal desolvation of $[Ce(OPr^i)_4 (HOPr^i)_2]$ affords $[Ce_4O(OPr^i)_{14}]$, which has the structure $[Ce_4(\mu_4\text{-}O)(\mu_3\text{-}OPr^i)_2(\mu\text{-}OPr^i)_8(OPr^i)_{14}$.[1127] Alcohol exchange between $[Ce_2(OPr^i)_8(Pr^iOH)_2]$ and hexafluoroisopropanol (Hhfip) affords $[Ce(hfip)_4(thf)_2(Pr^iOH)_x]$, convertible into the stable adducts $[Ce(hfip)_4L_2]$ ($L_2 = 2$ bipy; tmen; diglyme) {diglyme = 2,5,8-trioxanonane; tmen = N,N,N',N'-tetramethylethane-1,2-diamine}. Reaction with pmdien (pmdien = N,N,N',N'',N-pentamethyldiethylenetriamine) results in $[Ce(hfip)_3(OPr^i)(pmdien)]$ and $[Hpmdien]_2[Ce(hfip)_6]$, the latter having octahedrally coordinated Ce.[1128] $[Ce_2(OPr^i)_8(Pr^iOH)_2]$ reacts with Hthd (Hthd = 2,2,6,6,-tetramethylheptane-3,5-dione) and barium isoprooxide forming $[Ba_4Ce_2(\mu_6\text{-}O)(thd)_4(\mu_3\text{-}OPr^i)_8(OPr^i)_2]$.[1129] Reaction of $(NH_4)_2Ce(NO_3)_6$ with two to eight equivalents of $NaOCMe_3$ has given a range of *t*-butoxide species, represented by general formulae $Ce(OCMe_3)_a(NO_3)_b(solvent)_cNa_d$ ($a = 1–6$; $b = 0–3$; $c = 2, 4$; $d = 0, 2$) in addition to $NaCe_2(OCMe_3)_9$ and $Ce_3O(OCMe_3)_{10}$. $(NH_4)_2Ce(NO_3)_6$ reacts with three moles of $NaOCMe_3$ forming $Ce(OCMe_3)(NO_3)_3$, isolable as $Ce(OCMe_3)(NO_3)_3 (HOCMe_3)_2$. This reacts with one mole of $NaOCMe_3$ forming $Ce(OCMe_3)_2(NO_3)_2(HOCMe_3)_2$ or $Ce(OCMe_3)_2(NO_3)_2(THF)_2$. $Ce(OCMe_3)_2(NO_3)_2(HOCMe_3)_2$ has eight-coordinate cerium, or, if each bidentate nitrate is considered to occupy one coordination position, the coordination geometry at Ce approximates to distorted octahedral, with Ce—O (Bu^t) distances of 2.023 Å and 2.025 Å; the Ce—O (NO_3) distances average 2.56 Å and the Ce—O (alcohol) distances 2.521–2.529 Å. Further reaction of "$Ce(OCMe_3)_2(NO_3)_2$" with $NaOCMe_3$ yields $Ce(OCMe_3)_3(NO_3)$ and $Ce(OCMe_3)_4(THF)_2$. $Ce(OCMe_3)_4(THF)_2$ reacts with excess $NaOCMe_3$ to afford $Na_2Ce(OCMe_3)_6(DME)_2$ which has octahedral $[Ce(OCMe_3)_6]^{2-}$ ions surrounded by $Na(DME)^+$ ions, sodium being coordinated facially to three oxygens of the CeO_6 unit. The Ce—O distances vary, as two butoxides are terminal (Ce—O of 2.141 Å), two are doubly bridging (Ce—O of 3.230 Å) and two are triply bridging (Ce—O of 2.374 Å). $Ce(OCMe_3)_4(THF)_2$ reacts with 0.5 mole of $NaOCMe_3$ to form $NaCe_2(OCMe_3)_9$. Both $NaCe_2(OCMe_3)_9$ and $Ce(OCMe_3)_4(THF)_2$ slowly convert in solution into $Ce_3O(OCMe_3)_{10}$.[1130] The alkoxide group plays an important role in stabilizing the Ce^{IV} state, witness the existence of $[CeCp_2(OBu^t)_2]$ and $[CeCp_3(OBu^t)]$ when other Ce^{IV} organometallics cannot be isolated says something about the role of OR is supporting the (IV) state of cerium.[1131] An improved synthesis of $[Ce(NO_3)_4(Ph_3PO)_2]$, from $(NH_4)_2[Ce(NO_3)_6]$ and Ph_3PO in MeCN is reported; it was also stated, however, that the literature reaction of $(NH_4)_2[Ce(NO_3)_6]$ and Ph_3PO in propanone tends to lead to reduction and the formation of $[Ce(NO_3)_3(Ph_3PO)_3]$.[432] Newer Ce^{IV} THF complexes, $[CeClZ(THF)_5] [Ce(THF)Cl_5]$ (Z = Cl, NO_3) have been synthesized.[1132]

3.2.2.10.3 Group 17 ligands

Fluorination of cerium oxide with NH_4HF_2 affords $(NH_4)_4[CeF_8]$.[1133] β-$BaTbF_6$ has a structure[1134] based on infinite chains of edge-sharing $[TbF_8]^{4-}$ units. α-$BaTbF_6$ contains $[Tb_4F_{26}]^{10-}$ ions based on association of four square antiprisms by sharing corners and edges.[1135]

3.2.3 REFERENCES

1. Atwood, J. L.; Smith, K. D. *J. Chem. Soc., Dalton Trans.* **1974**, 921.
2. Anderson, T. J.; Neuman, M. A.; Melson, G. A. *Inorg. Chem.* **1973**, *12*, 927.
3. Shannon, R. D. *Acta Crystallogr., Sect. C* **1976**, *32*, 751.
4. Jensen, W. B. *J. Chem. Educ.* **1982**, *59*, 634.
5. Hart, F. A. Scandium, Yttrium and the Lanthanides. In *Comprehensive Coordination Chemistry*; Wilkinson, G., Gillard, R. D., McCleverty, J. A., Eds.; Pergamon: Oxford, UK, 1987; Vol. 3, p 1059.
6. Cotton, S. A. Scandium, Yttrium and the Lanthanides: Inorganic and Coordination Chemistry. In *Encyclopedia of Inorganic Chemistry*; King, R. B., Ed.; Wiley: New York, 1994, p 3595.
7. Cotton, S. A. *Polyhedron* **1999**, *18*, 1691.
8. Meehan, P. R.; Aris, D. R.; Willey, G. R. *Coord. Chem. Rev.* **1999**, *181*, 121.
9. Kobayashi, S., Ed. *Lanthanides: Chemistry and Use in Organic Synthesis.* Springer: Berlin, 1999.
10. Kobayashi, S. *Eur. J. Org. Chem.* **1999**, 15.
11. Putzer, M. A.; Wickleder, M. S. *Z. Anorg. Allg. Chem.* **1999**, *625*, 1777.
12. Schmann, H. *J. Organomet. Chem.* **1985**, *281*, 950.
13. Mu, Y.; Piers, W. E.; MacQuarrie, D. C.; Zaworotko, M. J.; Young, V. G. *Organometallics* **1996**, *15*, 2720.
14. Westerhausen, M.; Hartmann, M.; Schwarz, W. *Inorg. Chim. Acta* **1998**, *269*, 91.
15. Guttenberger, C.; Amberger, H.-D. *J. Organomet. Chem.* **1997**, *545–546*, 601.
16. Schaverien, C. J. *Adv. Organomet. Chem.* **1994**, *36*, 283.
17. Kohn, R. D.; Kociok-Kohn, G.; Schumann, H. Scandium, Yttrium and the Lanthanides: Organometallic Chemistry. In *Encyclopaedia of Inorganic Chemistry*; King, R. B., Ed.; Wiley: New York, 1994; p 3618.
18. Edelmann, F. T. In *Comprehensive Organometallic Chemistry*, 2nd ed.; Abel, E. W.; Stone, F. G. A.; Wilkinson, G., Eds.; Pergamon: Oxford, UK, 1995; Vol. 4, p 10.
19. Schumann, H.; Meese-Marktscheffel, J. A.; Essar, L. *Chem. Rev.* **1995**, *95*, 865.
20. Cotton, S. A. *Coord. Chem. Rev.* **1997**, *160*, 159.
21. Melson, G. A.; Stotz, R. W. *Coord. Chem. Rev.* **1971**, *7*, 133 and references therein.
22. Simon, M.; Meyer, G. *Z. Krist.* **1996**, *211*, 327.
23. Drew, M. G. B.; Iveson, P. B.; Hudson, M. J.; Liljenzin, J. O.; Spljuth, L.; Cordier, P.-Y.; Enarsson, A.; Hill, C.; Madic, C. *J. Chem. Soc., Dalton Trans.* **2000**, 821.
24. Ahrens, B.; Cotton, S. A.; Feeder, N.; Noy, O. E.; Raithby, P. R.; Teat, S. J. *J. Chem. Soc., Dalton Trans.* **2002**, 2027.
25. Drew, M. G. B.; Hudson, M. J.; Iveson, P. B.; Madic, C.; Russell, M. L. *J. Chem. Soc., Dalton Trans.* **2000**, 2711.
26. Mullica, D. F.; Kautz, J. A.; Farmer, J. M.; Sappenfield, E. L. *J. Mol. Struct.* **1999**, *479*, 31.
27. Anwander, R.; Runte, O.; Eppinger, J.; Gerstberger, G.; Herdtweck, E.; Spiegler, M. *J. Chem. Soc., Dalton Trans.* **1998**, 847.
28. Roussel, P.; Alcock, N. W.; Scott, P. *Chem. Commun.* **1998**, 801.
29. Edelmann, F. T.; Richter, J. *Eur. J. Solid State Inorg. Chem.* **1996**, *33*, 157.
30. Hagedorn, J. R.; Arnold, J. *Organometallics* **1996**, *15*, 984.
31. Sewchok, M. G.; Haushalter, R. C.; Merola, J. S. *Inorg. Chim. Acta* **1988**, *144*, 47.
32. Arnold, J.; Hoffmann, C. G. *J. Am. Chem. Soc.* **1990**, *112*, 8620.
33. Arnold, J.; Hoffmann, C. G.; Dawson, D. Y.; Hollander, F. J. *Organometallics* **1993**, *12*, 3645.
34. Radecka-Paryzek, W.; Luks, E. *Monatsh. Chem.* **1995**, *126*, 795.
35. Radecka-Paryzek, W.; Patroniak-Krzyminiewska, V. *Pol. J. Chem.* **1995**, *69*, 1.
36. Fruyzuk, M. D.; Giesbrecht, G.; Rettig, S. J. *Organometallics* **1996**, *15*, 3329.
37. Karsch, H. H.; Ferazin, G.; Kooijman, H.; Steigelman, O.; Schier, A.; Bissinger, P.; Hiller, W. *J. Organomet. Chem.* **1994**, *482*, 151.
38. Favier, F.; Pascal, J.-L. *C. R. Acad. Sci. Paris, Ser II* **1991**, *313*, 619.
39. Hamidi, M. E. M.; Pascal, J.-L. *Polyhedron* **1994**, *13*, 1787.
40. Sugita, Y.; Ohki, Y.; Suzuki Y.; Ouchi, A. *Bull. Chem. Soc. Jpn.* **1987**, *60*, 3441 and references therein.
41. Brodkin, J. S.; Foxman, B. M. *J. Chem. Soc., Chem. Commun.* **1991**, 1073.
42. Wickleder, M. S. *Z. Anorg. Allg. Chem.* **1999**, *625*, 1556.
43. Bergmann, H, ed., *Gmelin Handbook of Inorganic Chemistry*; Springer-Verlag, Berlin, 1974, *C2*, 226.
44. Bergmann, H, ed., *Gmelin Handbook of Inorganic Chemistry*; Springer-Verlag, Berlin, 1977, *C5*, 98.
45. Bergmann, H, ed., *Gmelin Handbook of Inorganic Chemistry*; Springer-Verlag, Berlin, 1978, *C6*, 94.
46. Jin, Z.; Liu, Y.; Zhang, S.; Yu, F.; Li, J. *Acta. Chim. Sin.* **1987**, *45*, 1048.
47. Tan, M.; Gan, X.; Tang, N.; Zhou, J.; Wang, X.; Zhu, Y. *Wuji Huaxue Xuebao* **1990**, *6*, 5 [Chem. Abs. **1991**, *115*, 221711].
48. Gan, X.; Tang, N.; Zhu, Y.; Zhai, Y.; Tan, M. *Zhonggo Xitu Xuebao* **1989**, *7*, 13 [Chem. Abs. **1990**, *112*, 150561; J. Chinese Rare Earth Society **1990**, *8*, 10].
49. Tan, M.; Gan, X.; Tang, N.; Zou, J.; Zhu, Y.; Wang, X. *Gaodeng Xuexiao Huaxue Xuebao* **1988**, *9*, 1217 [Chem. Abs. **1989**, *111*, 125632].
50. Castellani, C. B.; Carugo, O.; Giusti, M.; Sardone, N. *Eur. J. Solid State Inorg. Chem.* **1995**, *32*, 1089.
51. Waller, F. J.; Barrett, A. G. M.; Braddock, D. C.; Ramprasad, D.; McKinnell, R. M.; White, A. J. P.; Williams, D. J.; Ducray, R. *J. Org. Chem.* **1999**, *64*, 2910.

52. Lim, K. C.; Skelton, B. W.; White, A. H. *Aust. J. Chem.* **2000**, *53*, 875.
53. Harrowfield, J. M.; Skelton, B. W.; White, A. H. *Aust. J. Chem.* **1994**, *47*, 397.
54. Ohni, Y.; Suzuki, Y.; Takeguchii, T.; Ouchi, A. *Bull. Chem. Soc. Jpn.* **1988**, *61*, 393.
55. Willey, G. R.; Meehan, P. R.; Rudd, M. D.; Drew, M. G. B. *J. Chem. Soc., Dalton Trans.* **1995**, 3175.
56. Aquino, M. A. S.; Clegg, W.; Lin, Q.-T.; Sykes, A. G. *Acta Crystallogr., Sect. C* **1995**, *51*, 560.
57. Matsumoto, F.; Ohki, Y.; Suzuki, Y.; Ouchi, A. *Bull. Chem. Soc. Jpn.* **1989**, *62*, 2081.
58. Ilyushin, A. B.; Petrosyants, S. P. *Zh. Neorg. Khim.* **1994**, *39*, 1517 [*Russ. J. Inorg. Chem.* **1994**, *39*, 1449].
59. Ripert, V.; Hubert-Pfalzgraf, L. G.; Vaissermann, J. *Polyhedron* **1999**, *18*, 1845.
60. Ma, J.-F.; Jin, Z.-S.; Ni, J.-Z. *Polyhedron* **1995**, *14*, 563.
61. Kanno, H.; Yamaguchi, T.; Ohtaki, H. *J. Phys. Chem.* **1989**, *93*, 1695.
62. Yamaguchi, T.; Niihara, M.; Takamuru, T.; Wakita, H.; Kanno, H. *Chem. Phys. Lett.* **1997**, *274*, 485.
63. Henning, Th-J.; Jacobs, H.; *Z. Anorg. Allg. Chem.* **1992**, *616*, 71.
64. Willey, G. R.; Meehan, P. R.; Drew, M. G. B. *Polyhedron* **1995**, 3175.
65. Fawcett, J.; Platt, A. W. G.; Russell, D. R. *Polyhedron* **2002**, *21*, 287.
66. Deakin, L.; Levason, W.; Popham, M. C.; Reid, G.; Webster, M. *J. Chem. Soc., Dalton Trans.* **2000**, 2439.
67. Levason, W.; Patel, B.; Popham, M. C.; Reid, G.; Webster, M. *Polyhedron* **2001**, *20*, 2711.
68. Hill, N. J.; Levason, W.; Popham, M. C.; Reid, G.; Webster, M. *Polyhedron* **2002**, *21*, 1579.
69. Imamoto, T.; Nishiura, M.; Yamanoi, Y.; Tsutura, H.; Yamaguchi, K. *Chem. Lett.* **1996**, 875.
70. Imamoto, T.; Okano, N.; Nishiura, M.; Yamaguchi, K. *Kidorui* **1997**, *30*, 344.
71. Vincentini, G.; Ayala, J. D.; Matos, J. R. *Thermochim. Acta* **1991**, *191*, 317.
72. Ayala, J. D.; Vincentini, G.; Bombieri, G. *J. Alloys Compd.* **1995**, *225*, 357.
73. Gan, X.; Tan, N.; Zhang, W.; Tan, M. *Wuji Huaxue Xuebao* **1989**, *5*, 17 [Chem. Abs. **1990**, *113*, 143966].
74. Liu, W.; Tan, M. *Wuji Huaxue Xuebao* **1992**, *8*, 84 [Chem. Abs. **1993**, *118*, 138535].
75. Ripert, V.; Hubert-Pfalzgraf, L. G.; Vaissermann, J. *Polyhedron* **1999**, *18*, 1845.
76. Addison, C. C.; Greenwood, A. J.; Haley, M. J.; Logan, N. J. *Chem. Soc., Chem. Commun.* **1978**, 580.
77. Meyer, G.; Stockhouse, S. *Z. Kristallogr.* **1994**, *209*, 180.
78. Bradley, D. C.; Chudzynska, H.; Frigo, D. M.; Hammond, M. E.; Hursthouse, M. B.; Mazid, M. A. *Polyhedron* **1990**, *9*, 719.
79. Tuirevskaya, E. P.; Belokon, A. I.; Starikova, Z. A.; Yanovsky, A. I.; Kiruschenkov, E. I.; Turova, N. Ya. *Polyhedron* **2000**, *19*, 705.
80. Gromilov, S. A.; Lisovian, V. I.; Baldina, I. A.; Borisov, S. V. *Zh. Neorg. Khim.* **1988**, *33*, 1482 [*Russ. J. Inorg. Chem.* **1993**, *33*, 840].
81. Lisovian, V. I.; Gromilov, S. A. *Izv. Akad. Nauk. SSSR Ser. Khim.* **1987**, 2098.
82. Narbutt, J.; Krejzler, J. *Inorg. Chim. Acta* **1999**, *286*, 175.
83. Ezhov, Yu. S.; Komarov, S. A.; Sevast'yanov, V. G. *J. Struct. Chem.* **1998**, *39*, 514.
84. Fleeting, K. A.; Davies, H. O.; Jones, A. C.; O'Brien, P.; Leedham, T. J.; Crosbie, M. J.; Wright, P. J.; Williams, D. J. *Chem. Vap. Deposition* **1999**, *5*, 261.
85. Li, B. G.; Zhang, Y.; Gan, L. B.; Lin, T. Z.; Huang, C. H.; Xu, G. X. *J. Rare Earths* **1994**, *12*, 241.
86. Willey, G. R.; Meehan, P. R. *Inorg. Chim. Acta* **1999**, *284*, 71.
87. Strel'tsova, N. R.; Bel'skii, V. K.; Bulychev, B. M.; Kireeva, O. V. *Zh. Neorg. Khim.* **1992**, *37*, 1822 [*Russ. J. Inorg. Chem.* **1992**, *37*, 934].
88. Woodman, T. J.; Errington, W. *Transition Met. Chem.* **1998**, *23*, 387.
89. Willey, G. R.; Lakin, M. T.; Alcock, N. W. *J. Chem. Soc., Chem. Commun.* **1992**, 1619.
90. Willey, G. R.; Lakin, M. T.; Alcock, N. W. *J. Chem. Soc., Dalton Trans.* **1993**, 3407.
91. Willey, G. R.; Meehan, P. R.; Rudd, M. D.; Drew, M. G. B. *J. Chem. Soc., Dalton Trans.* **1995**, 811.
92. Willey, G. R.; Meehan, P. R.; Drew, M. G. B. *Polyhedron* **1996**, *15*, 1397.
93. Daitch, C. E.; Hampton, P. D.; Duesler, E. N. *Inorg. Chem.* **1995**, *34*, 5641.
94. Masuda, Y.; Zhang, Y.; Yan, C.; Li, B. *J. Alloys Compd.* **1998**, *275–277*, 873.
95. Zheng, Y.-W.; Wang, Z.-M.; Jia, J.-T.; Liao, C.-S.; Yan, C.-H. *Acta Crystallogr., Sect C* **1999**, *55*, 1418.
96. Zhang, Y.; Li, B.; Gao, S.; Jin, T.; Xu, G. X. *J. Rare Earths* **1995**, *13*, 1.
97. Fjellvåg, H.; Karen, P. *Acta Chem. Scand.* **1994**, *48*, 294.
98. Zasorin, E. Z.; Ivanov, A. A.; Ermanova, L. I.; Spiridonov, V. P. *Zh. Phys. Khim.* **1989**, 63 669.
99. Ezhov, Yu.; Komarov, S. A.; Sevastyanov, V. G. *Zh. Strukt. Khim.* **1997**, *38*, 489.
100. Haaland, A.; Martinsen, K.-J.; Shorokhov, D. J.; Girichev, G. V.; Sokolov, V. I. *J. Chem. Soc., Dalton Trans.* **1998**, 2787.
101. Champarnaud-Mesjard, J. C.; Frit, B. *Eur. J. Solid State Inorg. Chem.* **1992**, *29*, 161.
102. Dahlke, P.; Babel, D. *Z. Anorg. Allg. Chem.* **1994**, *620*, 1686.
103. Carlson, S.; Xu, Y.; Norrestam, R. *J. Solid State Chem.* **1998**, *135*, 116.
104. Lin, Y. B.; Keszler, D. A. *Mater. Res. Bull.* **1993**, *28*, 931.
105. Faget, H.; Grannec, J.; Tressaud, A.; Rodriguez, V.; Roisnel, T.; Flerov, I. N.; Gorev, M. V. Eur. *J. Solid State Inorg. Chem.* **1996**, *33*, 893.
106. Reber, C.; Gudel, H. U.; Meyer, G.; Schleid, T.; Daul, C. A. *Inorg. Chem.* **1989**, *28*, 3249.
107. Masselmann, S.; Meyer, G. *Z. Anorg. Allg. Chem.* **1998**, *624*, 551.
108. Meyer, G.; Ax, P.; Schleid, T.; Irmler, M. *Z. Anorg. Allg. Chem.* **1987**, *554*, 25.
109. Bohnsack, A.; Meyer, G.; Wickleder, M. *Z. Krist.* **1996**, *211*, 394.
110. Gutsol, A. F.; Kuznetsov, V. Ya.; Rys'kina, M. P.; Tikhomirova, E. L.; Kalinnikov, V. T. *Zh. Prikl. Kkim (St. Petersberg)* **1998**, *71*, 543 [*Chem. Abs.* **1998**, *129*, 156119].
111. Bohnsack, A.; Meyer, G. *Z. Anorg. Allg. Chem.* **1996**, *622*, 173.
112. Metallinou, M. M.; Nalbandain, L.; Papatheodorough, G. N.; Voigt, W.; Emons, H. H. *Inorg. Chem.* **1991**, *30*, 4260.
113. Corbett, J. D. *Pure Appl. Chem.* **1992**, *54*, 1395.
114. McCollum, B. C.; Dudis, D. S.; Lachgar, A.; Corbett, J. D. *Inorg. Chem.* **1990**, *29*, 2030.
115. Lachgar, A; Dudis, D. S.; Dorhout, P. K.; Corbett, J. D. *Inorg. Chem.* **1991**, *30*, 3321.
116. Dorhout, P. K.; Corbett, J. D. *Inorg. Chem.* **1991**, *30*, 3326.

117. Artelt, H. M.; Schleid, T.; Meyer, G. *Z. Anorg. Allg. Chem.* **1994**, *620*, 1521.
118. Hughbanks, T.; Corbett, J. D. *Inorg. Chem.* **1988**, *27*, 2022 and references therein.
119. Cotton, S. A. *Lanthanides and Actinides* **1991**, Macmillan: London.
120. Kaltsoyannis, N.; Scott, P. *The f Elements* **1999**, Oxford University Press: Oxford, U.K.
121. Aspinall, H. C. *Chemistry of the F-block Elements* **2001**, Gordon and Breach: London.
122. Bunzli, J. C. G.; Choppin, G. R. *Lanthanide Probes in Life, Chemical and Earth Sciences: Theory and Practice* **1990**, Elsevier: Amsterdam.
123. Morss, L. R.; Meyer, G. *Synthesis of Lanthanide and Actinide Compounds* **1991**, Kluwer: Dordrecht, The Netherlands.
124. Herrmann, W. A., Ed. Organolanthanoid Chemistry: Synthesis, Structure, Catalysis. In *Topics in Current Chemistry* Springer-Verlag: Berlin, 1996; Vol. 179.
125. Bochkarev, M. N.; Zakharov, L. N.; Kalinina, G. S. *Organoderivatives of Rare Earth Elements* **1995**, Kluwer: Dordrecht.
126. Kobayashi, S. *Lanthanides: Chemistry and Use in Organic Synthesis* **1999**, Springer Verlag: Berlin.
127. Imamoto, T. *Lanthanides in Organic Synthesis* **1994**, Academic Press: New York.
128. Evans, C. H. *Biochemistry of the Lanthanides* **1990**, Plenum: New York.
129. Evans, C. H., Ed. Episodes from the History of the Rare Earth Elements. In *Chemists and Chemistry*; Kluwer: Dordrecht, The Netherlands 1996; Vol. 15.
130. Hitchcock, P. B.; Lappert, M. F.; Smith, R. G.; Bartlett, R. A.; Power, P. P. *J. Chem. Soc., Chem. Commun.* **1988**, 1007.
131. Van Der Sluys, W. G.; Burns, C. J.; Sattelberger, A. P. *Organometallics* **1989**, *8*, 855.
132. Westerhausen, M.; Hartmann, M.; Schwarz, W. *Inorg. Chim. Acta* **1998**, *269*, 91.
133. Schaverien, C. J.; Orpen, A. G. *Inorg. Chem.* **1991**, *30*, 4968.
134. Guttenberger, C.; Amberger, H-D. *J. Organomet. Chem.* **1997**, *545–546*, 601.
135. Reddmann, H.; Guttenberger, C.; Amberger, H.-D. *J. Organomet. Chem.* **2000**, *602*, 65.
136. Hitchcock, P. B.; Lappert, M. F.; Smith, R. G. *J. Chem. Soc., Chem. Commun.* **1989**, 369.
137. Schaverien, C. J.; van Mechelen, J. B. *Organometallics* **1991**, *10*, 1704.
138. Westerhausen, M.; Hartmann, M.; Pfitzner, A.; Schwarz, W. *Z. Anorg. Allg. Chem.* **1995**, *621*, 837.
139. Niemeyer, M. *Acta. Crystallogr., Sect. E* **2001**, *57*, m553.
140. Niemeyer, M. *Z. Anorg. Allg. Chem.* **2000**, *626*, 1027.
141. Atwood, J. L.; Lappert, M. F.; Smith, R. G.; Zhang, H. *J. Chem. Soc., Chem. Commun.* **1988**, 1308.
142. Evans, W. J.; Shreeve, J. L.; Broomhall-Dillard, R. N. R.; Ziller, J. W. *J. Organomet. Chem.* **1995**, *501*, 7.
143. Schumann, H.; Müller, J.; Brunks, N.; Lauke, H.; Pickardt, J.; Schwarz, H.; Eckart, K. *Organometallics* **1984**, *3*, 69.
144. Schumann, H.; Lauke, H.; Hahn, E.; Pickardt, J. *J. Organomet. Chem.* **1984**, *263*, 29 and references therein.
145. Biagini, P.; Lugli, G.; Abis, L.; Millini, R. *J. Organomet. Chem.* **1994**, *474*, C16.
146. Evans, W. J.; Anwander, R.; Doedens, R. J.; Ziller, J. W. *Angew. Chem., Int. Ed. Engl.* **1994**, *33*, 1641.
147. Evans, W. J.; Anwander, R.; Ziller, J. W. *Organometallics* **1995**, *14*, 1107.
148. Bochkarev, L. N.; Stepantseva, T. A.; Zakharov, L. N.; Fukin, G. K.; Yanovsky, A. I.; Struchkov, Yu.T. *Organometallics* **1994**, *14*, 2127.
149. Bochkarev, L. N.; Kharamenkov, V. V.; Rad'kov, Yu.F.; Zakharov, L. N.; Struchkov, Yu.T. *J. Organomet. Chem.* **1992**, *429*, 27.
150. Jin, Z.; Zhang, Y.; Chen, W. *J. Organomet. Chem.* **1990**, *396*, 407.
151. Rabe, G. W.; Strissel, C. S.; Liable-Sands, L. M.; Concolino, T. E.; Rheingold, A. L. *Inorg. Chem.* **1999**, *38*, 3446.
152. Rabe, G. W.; Bérubé, C. D.; Yap, G. P. A. *Inorg. Chem.* **2001**, *40*, 2682.
153. Rabe, G. W.; Bérubé, C. D.; Yap, G. P. A. *Inorg. Chem.* **2001**, *40*, 4780.
154. Arndt, S.; Spaniol, T. P.; Okuda, *J. Chem. Commun.* **2002**, 896.
155. Hermann, W. Preparation of lanthanoid complexes with heterocyclic carbenes (Hoechst A.-G., Germany), German Patent Application 4447070 [*Chem. Abs.* **1996**, *125*, 143016].
156. Young, D. M.; Schimek, G. L.; Kolis, J. W. *Inorg. Chem.* **1996**, *35*, 7620.
157. Li, J.-S.; Neumuller, B.; Dehnicke, K. *Z. Anorg. Allg. Chem.* **2002**, *628*, 45.
158. Berthet, J.-C.; Rivière, C.; Miquel, Y.; Nierlich, M.; Madic, C.; Ephritikhine, M. *Eur. J. Inorg. Chem.* **2002**, 1439.
159. Trikkha, A. K. *Polyhedron* **1992**, *11*, 2273.
160. Evans, W. J.; Rabe, G. W.; Ziller, J. W. *Inorg. Chem.* **1994**, *33*, 3072.
161. Evans, W. J.; Shreeve, J. L.; Boyle, T. J.; Ziller, J. W. *J. Coord. Chem.* **1995**, *34*, 229.
162. Fréchette, M.; Butler, I. R.; Hynes, R.; Detellier, C. *Inorg. Chem.* **1992**, *31*, 1650.
163. Fréchette, M. *Can. J. Chem.* **1993**, *71*, 377.
164. Thomas, R. R.; Chebolu, V.; Sen, A. *J. Am. Chem. Soc.* **1986**, *108*, 4096.
165. Hu, J.-Y.; Shen, Q.; Lin, Z.-S. *Chinese Sci. Bull.* **1990**, *35*, 1090.
166. Shen, Q.; Hu, J.-Y.; Lin, Z.-S.; Sun, Y. *Zhonggo Xitu Xuebao (J. Chinese Rare Earth Society)* **1990**, *8*, 359.
167. Deacon, G. B.; Görtler, B.; Junk, P. C. ; E. Lork E.; Mews, R.; Petersen, J.; Zemva, B. *J. Chem. Soc., Dalton Trans.* **1998**, 3887.
168. Semenova, L. I.; Skelton, B. W.; White, A. H. *Aust. J. Chem.* **1999**, *52*, 551.
169. Semenova, L. I.; White, A. H. *Aust. J. Chem.* **1999**, *52*, 571.
170. Rivière, C.; Nierlich, M.; Ephritikhine, M.; Madic, C. *Inorg. Chem.* **2001**, *40*, 4428.
171. Rivière, C.; Lance, M.; Ephritikhine, M.; Nierlic, M. *Z. Kristallogr-New Cryst. Struct.* **2000**, *215*, 239.
172. Kepert, D. L.; Semenova, L. I.; Sobolev, A. N.; White, A. H. *Aust. J. Chem.* **1996**, *49*, 1005.
173. Boudalius, A. K.; Nastopoulos, V.; Perlepes, S. P.; Raptopoulou, C. P.; Terzis, A. *Trans. Met. Chem.* **2001**, *26*, 276.
174. Bower, J. F.; Cotton, S. A.; Fawcett, J.; Russell, D. R. *Acta Crystallogr., Sect. C* **2000**, *56*, e8.
175. Dong, N.; Zhu, L. G.; Wang, J. R. *Chinese Chem. Lett.* **1992**, *3*, 745.
176. Ji, Z. P.; Rogers, R. D. *J. Chem. Crystallogr.* **1994**, *24*, 415.
177. Zhu, W. X.; Yang, R.-N.; Zhao, J.-Z.; Luo, B.-S.; Chen, L.-R. *Jiegou Huaxue (Chinese J. Struct. Chem.)* **1990**, *9*, 286.
178. Jin, L.; Lu, S.; Lu, S. *Polyhedon* **1996**, *15*, 4069.
179. Pisarevskii, A. P.; Mitrofanova, N. D.; Frolovskaya, S. N.; Martynenko, L. I. *Russ. J. Coord. Chem.* **1996**, *21*, 832.

180. Zou, Y.-Q.; Li, X.; Li, Y.; Hu, H.-M. *Acta Crystallogr., Sect. C* **2001**, *57*, 1048.
181. Brodkin, J. S.; Foxman, B. M.; Clegg, W.; Cressey, J. T.; Harbron, D. R.; Hunt, P. A.; Straughan, B. P. *Chem. Mater.* **1996**, *8*, 242.
182. Su, C.; Tan, M.; Tang, N.; Gan, X.; Lu, W.; Wang, X. *J. Coord. Chem.* **1996**, *38*, 207.
183. Su, C.; Tang, N.; Tan, M.; Wu, K. *Polyhedron* **1996**, *15*, 233.
184. Kuz'mina, N. P.; Ivanov, R. A.; Ilyukhin, A. B.; Paramonov, S. E. *Russ. J. Coord. Chem.* **1999**, *25*, 635.
185. Varand, V. L.; Glinskaya, L. A.; Klevtsova, R. F.; Larionov, S. V. *J. Struct. Chem.* **2000**, *41*, 544.
186. Varand, V. L.; Klevtsova, R. F.; Glinskaya, L. A.; Larionov, S. V. *Russ. J. Coord. Chem.* **2000**, *26*, 869.
187. Mincheva, L. Kh.; Skogareva, L. S.; Razgonyaeva, G. A.; Sakharova, V. G.; Sergienko, V. S. *Russ. J. Coord. Chem.* **1997**, *42*, 1828.
188. Ji, Z. P.; Rogers, R. D. *J. Chem. Crystallogr.* **1994**, *24*, 797.
189. Zheng, Y.-Q.; Zhou L.-X.; Lin, J.-L. *Z. Anorg. Allg. Chem.* **2001**, *627*, 1643.
190. Mirochnik, A. G.; Bukvetskii, B. V.; Zhikhareva, P. A.; Karasev, V. E. *Russ. J. Coord. Chem.* **2001**, *27*, 443.
191. Zheng, Y.-Q.; Zhou, L.-X.; Lin, J.-L.; Zhang, S.-W. *Z. Kristallogr.- New Cryst. Struct.* **2001**, *216*, 357.
192. Antsyshkina, A. S.; Sadikov, G. G.; Rodnikova, M. N.; Mikhiaklichenko, A. I.; Nevzorova, L. V. *Russ. J. Coord. Chem.* **2002**, *47*, 361.
193. Wu, D. M.; Lin, X.; Lu, C.-Z.; Zhuang, H.-H. *Jiegou Huaxue (Chinese J. Struct. Chem.)* **2000**, *19*, 69.
194. Zheng, Y-Q.; Zhou, L-X.; Lin, J-L.; Zhang, S-W. *Z. Anorg. Allg. Chem.* **2001**, *627*, 2425.
195. Kepert, C. J.; Lu, W. M.; Semenova, L. I.; Skelton, B. W.; White, A. H. *Aust. J. Chem.* **1999**, *52*, 481.
196. Panagiotopoulos, A.; Zafiropoulos, T. F.; Perlepes, S. P.; Bakaklbassis, E.; Masson-Ramade, I.; Kahn, O.; Terzis, A.; Raptopoulou, C. P. *Inorg. Chem.* **1995**, *34*, 4918.
197. Legendziewicz, J.; Tsaryuk, V.; Zolin, V.; Lebedeva, E.; Borzechowska, M.; Karbowiak, M. *New J. Chem.* **2001**, *25*, 1031.
198. Varand, V. L.; Glinskaya, L. A.; Klevtsova, R. F.; Larionov, S. V. *J. Struct. Chem.* **1998**, *39*, 244.
199. Ivanov, R. A.; Korsakov, I. E.; Kuzmina, N. P.; Kaul, A. R. *Mendeleev Commun.* **2000**, 98.
200. Chen, Z.; Li, J.; Chen, F.; Proserpio, D. M. *Inorg. Chim. Acta* **1998**, *273*, 255.
201. George, C.; Purdy, A. P. *Acta Crystallogr., Sect. C* **1997**, *53*, 1381.
202. Commuzzi, C.; Di Bernardo, P.; Polese, P.; Portanova, R.; Tolazzi, M.; Zanonato, P. L. *Polyhedron* **2000**, *19*, 2427.
203. Kepert, C. J.; Lu, W.; Skelton, B. W.; White, A. H. *Aust. J. Chem.* **1994**, *47*, 365.
204. Semenova, L. I.; White, A. H. *Aust. J. Chem.* **1999**, *52*, 539.
205. Berthet, J.-C.; Rivière, C.; Miquel, Y.; Nierlich, M.; Madic, C.; Ephritikhine, M. *Eur. J. Inorg. Chem.* **2002**, 1439.
206. Semenova, L. I.; Sobolev, A. N.; Skelton, B. W.; White, A. H. *Aust. J. Chem.* **1999**, *52*, 519.
207. Mürner, H.-R.; Chassat, E.; Thummel, R. P.; Bünzli, J.-C. G. *J. Chem. Soc., Dalton Trans.* **2000**, 2809.
208. Mallet, C.; Thummel, R. P.; Hery, C. *Inorg. Chim. Acta* **1993**, *210*, 223.
209. Sinha, S. P. *Z. Naturforsch. Teil A* **1965**, *20*, 1661.
210. Fréchette, M.; Bensimon, C. *Inorg. Chem.* **1995**, *34*, 3520.
211. Drew, M. G. B.; Iveson, P. B.; Hudson, M. J.; Liljenzin, J. O.; Spljuth, L.; Cordier, P.-Y.; Enarsson, A.; Hill, C.; Madic, C. *J. Chem. Soc., Dalton Trans.* **2000**, 821; See also; Cotton, S. A.; Noy, O. E.; Liesener, F.; Raithby, P. R. *Inorg. Chim. Acta*, **2003**, *344*, 37.
212. Drew, M. G. B.; Hudson, M. J.; Iveson, P. B.; Russell, M. L.; Liljenzin, J.-O.; Sklberg, M.; Spjuth, L.; Madic, C. *J. Chem. Soc., Dalton Trans.* **1998**, 2973.
213. Grigoriev, M. S.; Den Auwer, C.; Madic, C. *Acta Crystallogr., Sect. C* **2001**, *57*, 1141.
214. Semenova, L. I.; White, A. H. *Aust. J. Chem.* **1999**, *52*, 507.
215. Leverd, P. C.; Charbonnel, M.-C.; Dognon, J.-P.; Lance, M.; Nierlich, M. *Acta Crystallogr., Sect. C* **1999**, *55*, 368.
216. Hayashi, K.; Nagao, N.; Jalielehvand, F.; Satou, N.; Fukuda, Y. *Kidorui* **1996**, *28*, 210.
217. Hayashi, K.; Nagao, N.; Harada, K.; Haga, M.; Fukuda, Y. *Chem. Lett.* **1998**, 1173.
218. Boudalius, A. K.; Nastopoulos, V.; Terzis, A.; Raptopoulou, C. P.; Perlepes, S. P. *Z. Naturforsch. Teil B* **2001**, *56*, 122.
219. Cotton, S. A.; Raithby, P. R. *Inorg. Chem. Commun.* **1999**, *2*, 86.
220. Hayashi, K.; Nagao, N.; Harada, K.; Haga, M.; Fukuda, Y. *Chem. Lett.* **1998**, 1173.
221. Nakao, A.; Hayashi, K.; Fukuda, Y. *Kidorui* **1999**, *34*, 156.
222. Nakao, A.; Fukuda, Y. *Kidorui* **2000**, *36*, 188.
223. Fukuda, Y.; Nakao, A.; Hayashi, K. *J. Chem. Soc., Dalton Trans.* **2002**, 527.
224. Wietzke, R.; Mazzanti, M.; Latour, J.-M.; Pécaut, J. *Inorg. Chem.* **1999**, *38*, 3581.
225. Drew, M. G. B.; Hudson, M. J.; Iveson, P. B.; Madic, C. *Acta Crystallogr,. Sect. C* **2000**, *56*, 434.
226. Drew, M. G. B.; Hudson, M. J.; Iveson, P. B.; Madic, C.; Russell, M. L. *J. Chem. Soc., Dalton Trans.* **2000**, 2711.
227. Drew, M. G. B.; Guillaneux, D.; Hudson, M. J.; Iveson, P. B.; Russell, M. L.; Madic, C. *Inorg. Chem. Commun.* **2001**, *4*, 12.
228. Iveson, P. B.; Rivière, C.; Guillaneux, D.; Nierlich, M.; Thuéry, P.; Ephritikhine, M.; Madic, C. *Chem. Commun.* **2001**, 1512.
229. Drew, M. G. B.; Guillaneux, D.; Hudson, M. J.; Iveson, P. B.; Russell, M. L.; Madic, C. *Inorg. Chem. Commun.* **2001**, *4*, 462.
230. Wang, S.; Luo, Q.; Zhou, X.; Zeng, Z. *Polyhedron* **1993**, *12*, 939.
231. Cui, Y. X.; Luo, Q.; Wang, S.; Wang, L.; Zhu, Y. *J. Chem. Soc., Dalton Trans.* **1994**, 2523.
232. Piguet, C.; Williams, A. F.; Bernardinelli, G.; Bünzli, J.-C. G. *Inorg. Chem.* **1993**, *32*, 874.
233. Piguet, C.; Williams, A. F.; Bernardinelli, G.; Bünzli, J.-C. G. *Inorg. Chem.* **1993**, *32*, 4139.
234. Petoud, S.; Bünzli, J.-C. G.; Renoud, F.; Piguet, C.; Schenk, K. J.; Hopfgartner, G. *Inorg. Chem.* **1997**, *36*, 5750.
235. Piguet, C.; Williams, A. F.; Bernardinelli, G.; Moret, E.; Bünzli, J.-C. G. *Helv. Chim. Acta* **1992**, *75*, 1697.
236. Piguet, C.; Williams, A. F.; Bernardinelli, G.; Moret, E.; Bünzli, J.-C. G. *J. Chem. Soc., Dalton Trans.* **1995**, 83.
237. Bernardinelli, G.; Piguet, C.; Williams, A. F. *Angew. Chem.,Int. Ed. Engl.* **1992**, *31*, 1629.
238. Bünzli, J.-C. G.; Bernardinelli, G.; Hofgartner, G.; Williams, A. F. *J. Am. Chem. Soc.* **1993**, *115*, 8197.
239. Renaud, F.; Piguet, C.; Bernardinelli, G.; Bünzli, J.-C. G.; Hofgartner, G. *Chem. Eur. J.* **1997**, *3*, 1646.
240. Renaud, F.; Piguet, C.; Bernardinelli, G.; Bünzli, J.-C. G.; Hofgartner, G. *Chem. Eur. J.* **1997**, *3*, 1660.
241. Martin, N.; Bünzli, J.-C. G.; McKee, V.; Piguet, C.; Hofgartner, G. *Inorg. Chem.* **1998**, *37*, 577.
242. Elhabiri, M.; Scopelliti, R.; Bünzli, J.-C. G.; Piguet, C. *J. Am. Chem. Soc.* **1999**, *121*, 510747.

243. Drew, M. G. B.; Hudson, M. J.; Madic, C.; Russell, M. L. *J. Chem. Soc., Dalton Trans.* **1999**, 2433.
244. Wietzke, R.; Mazzanti, M.; Latour, J.-M.; Pécaut, J.; Cordier, P.-Y.; Madic, C. *Inorg. Chem.* **1998**, *37*, 6690.
245. Mazzanti, M.; Wietzke, R.; Pécaut, J.; Latour, J.-M.; Maldivi, P.; Remy, M. *Inorg. Chem.* **2002**, *41*, 2389.
246. Wietzke, R.; Mazzanti, M.; Latour, J.-M.; Pécaut, J. *Chem. Commun.* **1999**, 209.
247. Su, C. Y.; Kang, B. S.; Mu, X. Q.; Sun, J.; Tong, Y. X.; Chen, Z. N. *Aust. J. Chem.* **1998**, *51*, 565.
248. Yamada, T.; Shinoda, S.; Tsukube, T. *Chem. Commun.* **2002**, 1218.
249. Morss, L. R.; Rogers, R. D. *Inorg. Chim. Acta* **1997**, *255*, 193.
250. Yashiro, M.; Ishikubo, A.; Takarada, T.; Komiyama, M. *Chem. Lett.* **1995**, *8*, 655.
251. Constable, E. C.; Elder, S. M.; Tocher, D. A. *Polyhedron* **1992**, *11*, 2599.
252. Constable, E. C.; Chotalia, R.; Tocher, D. A. *Chem. Commun.* **1992**, 771.
253. Westerhausen, M.; Hartmann, M.; Pfitzner, A.; Schwarz, W. *Z. Anorg. Allg. Chem.* **1995**, *621*, 837.
254. Rees Jr, W. S.; Just, O.; Van Derveer, D. S. *J. Mater. Chem.* **1999**, *9*, 249.
255. Herrmann, W. A.; Anwander, R.; Munck, F. C.; Scherer, W.; Dufaud, V.; Huber, N. W.; Artus, G. R. J. *Z. Naturforsch.* **1994**, *49b*, 1789.
256. Niemeyer, M. *Z. Anorg. Allg. Chem.* **2002**, *628*, 547.
257. Collin, J.; Giuseppone, N.; Jaber, N.; Domingois, A.; Maria, L.; Santos, I. *J. Organomet. Chem.* **2001**, *628*, 271.
258. Schuetz, S. A.; Day, V. W.; Sommer, R. D.; Rheingold, A. L.; Belot, J. A. *Inorg. Chem.* **2001**, *40*, 5292.
259. Jank, S.; Hanss, J.; Reddmann, H.; Amberger, H.-D.; Edelstein, N. M. *Z. Anorg. Allg. Chem.* **2002**, *628*, 1355.
260. Niemeyer, M. *Z. Anorg. Allg. Chem.* **2002**, *628*, 547.
261. Aspinall, H. C.; Bradley, D. C.; Hursthouse, M. B.; Sales, K. D.; Walker, N. P. C.; Hussain, B. *J. Chem. Soc., Dalton Trans.* **1989**, 623.
262. Berg, D. J.; Gendron, R. A. L. *Can. J. Chem.* **2000**, *78*, 454.
263. Karl, M.; Seybert, G.; Massa, W.; Agarwal, S.; Greiner, A.; Dehnicke, K. *Z. Anorg. Allg. Chem.* **1999**, *625*, 1405.
264. Berg, D. J.; Gendron, R. A. L. *Can. J. Chem.* **2000**, *78*, 454.
265. Herrmann, W. A.; Anwander, R.; Munck, F. C.; Scherer, W.; Dufaud, V.; Huber, N. W.; Artus, G. R. J. *Z. Naturforsch. Sect. B* **1994**, *49*, 1789.
266. Anwander, R. Runte O.; Eppinger, J.; Gerstberger, G.; Herdtweck, E.; Spiegler, M. *J. Chem. Soc. Dalton Trans.* **1998**, 847.
267. Rabe, G. W.; Yap, G. P. A. *Z. Kristallogr.-New Cryst. Struct.* **2000**, *215*, 457.
268. Anwander, R.; Eppinger, J.; Nagl, I.; Schere, W.; Tafipolsky, M.; Sirch, P. *Inorg. Chem.* **2000**, *39*, 4713.
269. Görlitzer, H. W.; Spiegler, M.; Anwander, R. *J. Chem. Soc., Dalton Trans.* **2000**, 4287.
270. Anwander, R.; Roesky, R. *J. Chem. Soc., Dalton Trans.* **1997**, 137.
271. Gerstberger, G.; Palm, C.; Anwander, R. *Chem. Eur. J.* **1999**, *5*, 997.
272. Deacon, G. B.; Fallon, G. D.; Forsyth, C. M.; Schumann, H.; Weimann, R. *Chem. Ber. Recl.* **1997**, *130*, 409.
273. Click, D. R.; Scott, B. L.; Watkin, J. G. *Chem. Commun.* **1999**, 633.
274. Click, D. R.; Scott, B. L.; Watkin, J. G. *Acta Crystallogr., Sect. C* **2000**, *56*, 1095.
275. Rees, W. S.; Just, O.; Schumann, H.; Weimann, R. *Angew. Chem., Int. Ed. Engl.* **1996**, *35*, 419.
276. Annand, J.; Aspinall, H. C. *J. Chem. Soc., Dalton Trans.* **2000**, 1867.
277. Pernin, C. G.; Ibers, J. A. *Inorg. Chem.* **2000**, *39*, 1222.
278. Berberich, H.; Roesky, P. W. *Angew. Chem., Int. Ed. Engl.* **1998**, *37*, 1569.
279. Annand, J.; Aspinall, H. C.; Steiner, A. *Inorg. Chem.* **1999**, *38*, 3941.
280. Aspinall, H. C.; Bickley, J. F.; Dwyer, J. L. M.; Greeves, N.; Kelly, R. V.; Steiner, A. *Organometallics* **2000**, *19*, 5416.
281. Essig, M. W.; Keogh, D. W.; Scott, B. L.; Watkin, J. G. *Polyhedron* **2001**, *20*, 373.
282. Schumann, H.; Winterfeld, J.; Eosenthal, E. C. E.; Hemling, H.; Esser, L. *Z. Anorg. Allg. Chem.* **1995**, *621*, 122.
283. Aspinall, H. C.; Tillotson, M. R. *Polyhedron* **1994**, *13*, 3229.
284. Wong, W.-K.; Zhang, L.; Xue, F.; Mak, T. C. W. *Polyhedron* **1996**, *15*, 345.
285. Wong, W.-K.; Zhang, L.; Xue, F.; Mak, T. C. W. *Polyhedron* **1997**, *16*, 2013.
286. Evans, W. J.; Anwander, R.; Ziller, J. W.; Khan, S. I. *Inorg. Chem.* **1995**, *34*, 5927.
287. Karl, M.; Seybert, G.; Massa, W.; Harms, K.; Agarwal, S.; Maleika, R.; Stelter, W.; Greiner, A.; Heitz, W.; Neumüller, B.; Dehnicke, K. *Z. Anorg. Allg. Chem.* **1999**, *625*, 1301.
288. Evans, W. J.; Anwander, R.; Ziller, J. W. *Inorg. Chem.* **1995**, *34*, 5927.
289. Karl, M.; Neumuller, B.; Seybert, G.; Massa, W.; Dehnicke, K. *Z. Anorg. Allg. Chem.* **1997**, *623*, 1203.
290. Edelmann, F. T.; Steiner, A.; Stalke, D.; Gilje, J. W.; Jagner, S.; Hakansson, M. *Polyhedron* **1994**, *13*, 539.
291. Deacon, G. B.; Forsyth, C. M.; Scott, N. M. *Eur. J. Inorg. Chem.* **2000**, 2501.
292. Evans, W. J.; Anwander, R.; Doedens, R. J.; Ziller, J. W. *Angew. Chem., Int. Ed. Engl.* **1994**, *33*, 1641.
293. Kraut, S.; Magull, J.; Schaller, U.; Karl, M.; Harms, K.; Dehnicke, K. *Z. Anorg. Allg. Chem.* **1998**, *624*, 1193.
294. Evans, W. J.; Ansari, M. A.; Ziller, J. W.; Khan, S. I. *Inorg. Chem.* **1996**, *35*, 5435.
295. Minhas, R. K.; Ma, Y.; Song, J.-I.; Gambarotta, S. *Inorg. Chem.* **1996**, *35*, 1866.
296. Just, O.; Rees, W. S. *Inorg. Chem.* **2001**, *40*, 1751.
297. Deacon, G. B.; Delbridge, E. E.; Skelton, B. W.; White, A. H. *Eur. J. Inorg. Chem.* **1999**, 751.
298. Deacon, G. B.; Delbridge, E. E.; Forsyth, C. M. *Angew. Chem., Int. Ed. Engl.* **1999**, *38*, 1766.
299. Deacon, G. B.; Gitlits, A.; Skelton, B. W.; White, A. H. *Chem. Commun.* **1999**, 1213.
300. Pfeiffer, D.; Ximba, B. J.; Liable-Sands, L. M.; Rheingold, A. L.; Heeg, M. J.; Coleman, D. M.; Schlegel, H. B.; Kuech, T. F.; Winter, C. H. *Inorg. Chem.* **1999**, *38*, 4539.
301. Hitchcock, P. B.; Lappert, M. F.; Tian, S. *J. Chem. Soc., Dalton Trans.* **1997**, 1945.
302. Zhou, Y.; Yap, G. P. A.; Richardson, D. S. *Organometallics* **1998**, *17*, 4387.
303. Edelmann, F. T.; Richter, J. *Eur. J. Solid State Inorg. Chem.* **1996**, *33*, 157.
304. Wedler, M.; Krösel, F.; Pieper, U.; Stalke, D.; Edelmann, F. T.; Amberger, H.-D. *Chem. Ber.* **1992**, *125*, 2171.
305. Duchateau, R.; van Wee, C. T.; Metsma, A.; van Duijnen, P. T.; Teuben, J. H. *Organometallics* **1996**, *15*, 2279.
306. Santos, I.; Marques, N. *New J. Chem.* **1995**, *19*, 551.
307. Marques, N.; Sella, A.; Takats, J. *Chem. Rev.* **2002**, *102*, 2137.
308. Apostolidis, C.; Rebizant, J.; Kanellakopulos, B.; von Ammon, R.; Dornberger, E.; Mueller, J.; Powietzka, B.; Nuber, B. *Polyhedron* **1997**, *16*, 1057.

309. Reger, D. L.; Lindeman, J. A.; Lebioda, L. *Inorg. Chim. Acta* **1987**, *139*, 71.
310. Sun, C. D.; Wong, W. T. *Inorg. Chim. Acta* **1997**, *255*, 355.
311. Onishi, M.; Nagoaka, N.; Hiraki, K.; Itoh, K. J. *Alloys Compd* **1987**, *139*, 71.
312. Lawrence, R. G.; Jones, C. J.; Kresinski, R. A. *Polyhedron* **1996**, *15*, 2011.
313. Lawrence, R. G.; Jones, C. J; Kresinski, R. A. *Inorg. Chim. Acta* **1999**, *285*, 283.
314. Lawrence, R. G.; Jones, C. J.; Kresinski, R. A. *J. Chem. Soc., Dalton Trans.* **1996**, 501.
315. Lawrence, R. G.; Hamor, T. A.; Jones, C. J.; Paxton, K.; Rowley, N. M. *J. Chem. Soc., Dalton Trans.* **2001**, 2121.
316. Long, D. P.; Chandrasekaran, A.; Day, R. O.; Bianconi, P. A.; Rheingold, A. L. *Inorg. Chem.* **2000**, *39*, 4476.
317. Liu, S. Y.; Maunder, G. H.; Sella, A.; Stephenson, M.; Tocher, D. A. *Inorg. Chem.* **1996**, *35*, 76.
318. Maunder, G. H.; Sella, A.; Tocher, D. A. *J. Chem. Soc., Chem. Commun.* **1994**, 885.
319. Clark, R. J. H.; Liu, S. Y.; Maunder, G. H.; Sella, A.; Elsegood, M. R. J. *J. Chem. Soc., Dalton Trans.* **1997**, 2241.
320. Hillier, A. C.; Zhang, X. W.; Maunder, G. H.; Liu, S. Y.; Eberspacher, T. A.; Metz, M. V.; McDonald, R.; Domingos, A.; Marques, N.; Day, V. W.; Sella, A.; Takats, J. *Inorg. Chem.* **2001**, *40*, 5106.
321. Lin, G.; McDonald, R.; Takats, J. *Organometallics* **2000**, *19*, 1814.
322. Hillier, A. C.; Liu, S.-Y.; Sella, A.; Elsegood, M. R. J. *Inorg. Chem.* **2000**, *39*, 2635.
323. Zhang, X.-W.; Loppnow, G. R.; McDonald, R.; Takats, J. *J. Am. Chem. Soc.* **1995**, *117*, 7828.
324. Deng, D.-L.; Zhang, Y.-H.; Dai, C.-Y.; Zeng, H.; Ye, C.-Q.; Hage, R. *Inorg. Chim. Acta* **2000**, *310*, 51.
325. Long, D. P.; Chandrasekaran, A.; Day, R. O.; Bianconi, P. A.; Rheingold, A. L. *Inorg. Chem.* **2000**, *39*, 4476.
326. Long, D. P.; Bianconi, P. A.; Rheingold, A. L. *J. Am. Chem. Soc.* **1996**, *118*, 12453.
327. Amoroso, A. J.; Jeffery, J. C.; Jones, P. L.; McCleverty, J. A.; Rees, L.; Rheingold, A. L.; Sun, Y.; Takats, J.; Trofimenko, S.; Ward, M. D.; Yap, G. P. A. *J. Chem. Soc., Chem. Commun.* **1995**, 1881.
328. Jones, P. L.; Amoroso, A. J.; Jeffery, J. C.; McCleverty, J. A.; Psillakis, E.; Rees, L. H.; Ward, M. D. *Inorg. Chem.* **1997**, *36*, 10.
329. Reeves, Z. R.; Mann, K. L. V.; Jeffery, J. C.; McCleverty, J. A.; Ward, M. D.; Barigelletti, F.; Armaroli, N. *J. Chem. Soc., Dalton Trans.* **1999**, 349.
330. Ward, M. D.; McCleverty, J. A.; Mann, K. L. V.; Jeffery, J. C.; Motson, G. R.; Hurst, J. *Acta Crystallogr., Sect. C* **1999**, *55*, 2055.
331. Amoroso, A. J.; Thompson, A. M. C.; Jeffery, J. C.; Jones, P. L.; McCleverty, J. A.; Ward, M. D. *J. Chem. Soc., Chem. Commun.* **1994**, 2751.
332. Rheingold, A. L.; Incarvito, C. D.; Trofimenko, S. *J. Chem. Soc., Dalton Trans.* **2000**, 1233.
333. Bell, Z. R.; Motson, G. R.; Jeffery, J. C.; McCleverty, J. A.; Ward, M. D. *Polyhedron* **2001**, *20*, 2045.
334. Domingos, A.; Marçalo, J.; Marques, N.; Pires De Matos, A.; Galvão, A.; Isolini, P. C.; Vicentini, G.; Zinner, K. *Polyhedron* **1995**, *14*, 3067.
335. Reger, D. L.; Chou, P. T.; Studer, S. L.; Knox, S. J.; Martinez, M. L.; Brewer, W. E. *Inorg. Chem.* **1991**, *30*, 2397.
336. Bardwell, D. A.; Jeffery, J. C.; Jones, P. L.; McCleverty, J. A.; Psillakis, E.; Reeves, Z.; Ward, M. D. *J. Chem. Soc., Dalton Trans.* **1997**, 2079.
337. Armaroli, N.; Accorsi, G.; Barigelletti, P.; Couchman, S. M.; Fleming, J. S.; Harden, N. C.; Jeffery, J. C.; Mann, K. L. V.; McCleverty, J. A.; Rees, L. H.; Starling, S. R.; Ward, M. D. *Inorg. Chem.* **1999**, *38*, 5769.
338. Onishi, M.; Yamaguchi, H.; Shimotsuma, H.; Hiraki, K.; Nagaoka, J.; Kawano, H. *Chem. Lett.* **1999**, 573.
339. Arai, H.; Suzuki, Y.; Matsumura, N.; Ouchi, A. *Bull. Chem. Soc. Jpn.* **1989**, *62*, 2530.
340. Li, J.; Huang, C.; Xu, Z.; Xu, G.; He, C.; Zheng, Q. *Chin. J. Inorg. Chem.* **1992**, *8*, 49.
341. Tateyama, Y.; Kuniyasu, Y.; Suzuki, Y.; Ouchi, Y. *Bull. Chem. Soc. Jpn.* **1988**, *61*, 2805.
342. Ouchi, A. *Bull. Chem. Soc. Jpn.* **1989**, *62*, 2431.
343. Matsumoto, F.; Takeuchi, T.; Ouchi, A. *Bull. Chem. Soc. Jpn.* **1989**, *62*, 2078.
344. Matsumoto, F.; Takeuchi, T.; Ouchi, A. *Bull. Chem. Soc. Jpn.* **1989**, *62*, 1809.
345. Matsumoto, F.; Takeuchi, T.; Ouchi, A. *Bull. Chem. Soc. Jpn.* **1990**, *63*, 620.
346. Rabe, G. W.; Ziller, J. W. *Inorg. Chem.* **1995**, *34*, 5378.
347. Westerhausen, M.; Hartmann, M.; Schwarz, W. *Inorg. Chim. Acta* **1998**, *269*, 91.
348. Westerhausen, M.; Schneiderbauer, S.; Hartmann, M.; Warchhold, M.; Nöth, H. *Z. Anorg. Allg. Chem* **2002**, *628*, 330.
349. Rabe, G. W.; Riede, J.; Schier, A. *Inorg. Chem.* **1996**, *35*, 2680.
350. Aspinall, H. C.; Moore, S. R.; Smith, A. K. *J. Chem. Soc., Dalton Trans.* **1992**, 153.
351. Hitchcock, P. B.; Lappert, M. F.; Mackinnon, I. A. *Chem. Commun.* **1988**, 1557.
352. Fryzuk, M. D.; Haddad, T. S.; Retting, S. J. *Organometallics* **1992**, *11*, 2967.
353. Rizkalla, E. N.; Choppin, G. R. Lanthanides and actin ideo hydration and hydrolysis. In *Handbook on the Physics and Chemistry of Rare Earths*; Gschneider, Jr., K. A.; Eyring, L.-R.; Choppin, G. R.; Lander, G. H., Eds.; North Holland: Amsterdam, 1994; Vol. 18, p 529.
354. Cossy, C.; Helm, L.; Merbach, A. E. *Inorg. Chem.* **1988**, *27*, 1973.
355. Helm, L.; Merbach, A. E. *J. Solid State Chem.* **1991**, *28*, 245.
356. Cossy, C.; Helm, L.; Powell, D. H.; Merbach, A. E. *New J. Chem.* **1995**, *19*, 27.
357. Kowall, T.; Foglia, F.; Helm, L.; Merbach, A. E. *J. Am. Chem. Soc.* **1995**, *117*, 3790.
358. Näslund, J.; Lindqvist-Reis, P.; Persson, I.; Sandström, M. *Inorg. Chem.* **2000**, *39*, 4006.
359. Allen, P. G.; Bucher, J. J.; Shuh, D. K.; Edelstein, N. M.; Craig, I. *Inorg. Chem.* **2000**, *39*, 595.
360. Kimura, T.; Kato, Y.; Choppin, G. R. In *Recent Progress in Actinides Separation Chemistry*; Yoshida, Z.; Kimura, T.; Meguro, Y., Eds.; World Scientific: Singapore, 1997; p 149.
361. Kimura, T.; Kato, Y. *J. Alloys Compd.* **1998**, *278*, 92.
362. Kimura, T.; Kato, Y. *Kidorui* **1995**, *26*, 134.
363. Kajinami, A.; Miwa, K.; Deki, S. *Kidorui* **1995**, *26*, 206.
364. Kanno, H.; Yokoyama, H. *Polyhedron* **1996**, *15*, 1437.
365. Galera, S.; Lluch, J. M.; Oliva, A.; Bertrán, J.; Foglia, F.; Helm, L.; Merbach, A. E. *New. J. Chem.* **1993**, *17*, 773.
366. David, F. H.; Fourest, B. *New J. Chem.* **1997**, *21*, 167.
367. Chatterjee, A.; Maslen, E. N.; Watson, K. J. *Acta Crystallogr., Sect. B* **1988**, *44*, 381.
368. Kurisaki, T.; Yamaguchi, T.; Wakita, H. *J. Alloys Compd.* **1993**, *192*, 293.

369. Kepert, C. J.; Skelton, B. W.; White, A. H. *Aust. J. Chem.* **1994**, *47*, 391.
370. Yamase, T.; Naruke, H.; Wéry, A. M. S. J.; Kaneko, M. *Chem. Lett.* **1998**, 1281.
371. Naruke, H.; Yamase, T.; Kaneko, M. *Bull. Chem. Soc. Jpn.* **1999**, *72*, 1775.
372. Nicolo, F.; Plancherel, D.; Chapuis, G.; Bünzli, J.-C. G. *Inorg. Chem.* **1988**, *27*, 3518.
373. Okhi, Y.; Suzuki, Y.; Takeuchi, T.; Ouchi, A. *Bull. Chem. Soc. Jpn.* **1988**, *61*, 393.
374. Faithfull, D. L.; Harrowfield, J. M.; Ogden, M. I.; Skelton, B. W.; Third, K.; White, A. H. *Aust. J. Chem.* **1992**, *45*, 583.
375. Zhang, H.; Wang, R.; Jin, T.; Zhou, Z.; Zhou, X. *J. Rare Earths* **1998**, *16*, 311.
376. Wickleder, M. S. *Z. Anorg. Allg. Chem.* **1999**, *625*, 1556.
377. Belin, C.; Gavier, F.; Pascal, J. L.; Tillard-Charbonnel, M. *Acta Crystallogr. Sect. C* **1996**, *52*, 1872.
378. Favier, F.; Pascal, J.-L.; Cunin, F.; Fitch, A. N.; Vaughan, G. *Inorg. Chem.* **1998**, *37*, 1776; Favier, F.; Pascal, J.-L.; Cunin, F.; Fitch, A. N.; Vaughan, G. *J. Solid State. Chem.* **1998**, *139*, 259.
379. Pascal, J. L.; El Haddad, M.; Rieck, H.; Favier, F. *Can. J. Chem.* **1994**, *72*, 2044.
380. Wickleder, M. S.; Schafer, W. *Z. Anorg. Allg. Chem.* **1999**, *625*, 309.
381. Wickleder, M. S. *Z. Anorg. Allg. Chem.* **1999**, *625*, 11.
382. Hamidi, M. El.M.; Hnach, M.; Zineddine, H. *J. Chim. Phys.- Chim. Biol.* **1997**, *94*, 1295.
383. Yanagihara, N.; Nakamura, S.; Nakayama, M. *Polyhedron* **1998**, *17*, 3625.
384. Aricó, E. M.; Zinner, L. B.; Apostolidis, C.; Dornberger, E.; Kanellakopulos, B.; Rebizant, J. *J. Alloys Compd.* **1997**, *249*, 111.
385. Junk, P. C.; Kepert, D. L.; Skelton, B. W.; White, A. H. *Aust. J. Chem.* **1994**, *52*, 497.
386. Kepert, C. J.; Skelton, B. W.; White, A. H. *Aust. J. Chem.* **1994**, *47*, 385.
387. Reuter, G.; Fink, H.; Seifert, H-J. *Z. Anorg. Allg. Chem.* **1994**, *620*, 665.
388. Junk, P. C.; Semenova, L. I.; Skelton, B. W.; White, A. H. *Aust. J. Chem.* **1999**, *52*, 531.
389. Lim, K. C.; Skelton, B. W.; White, A. H. *Aust. J. Chem.* **2000**, *53*, 867.
390. Meyer, G.; Gieseke-Vollmer, D. *Z. Anorg. Allg. Chem.* **1994**, *619*, 1603.
391. Lossin, A.; Meyer, G. *Z. Anorg. Allg. Chem.* **1994**, *620*, 428.
392. Lossin, A.; Meyer, G. *Z. Anorg. Allg. Chem.* **1994**, *619*, 1609.
393. Junk, P. C.; Kepert, C. J.; Lu, W. M.; Skelton, B. W.; White, A. H. *Aust. J. Chem.* **1999**, *52*, 437.
394. Lossin, A.; Mayer, G.; Fuchs, R.; Strähle, J. *Z. Naturforsch. Teil B* **1992**, *47*, 179.
395. Deiters, D.; Meyer, G. *Z. Anorg. Allg. Chem.* **1996**, *622*, 325.
396. Lossin, A.; Meyer, G.; Fuchs, R.; Straehle, J. *Z. Naturforsch Teil B.* **1992**, *47*, 179.
397. Binnemans, K.; Jongen, L.; Bromant, C.; Hinz, D.; Meyer, G. *Inorg. Chem.* **2000**, *39*, 5938.
398. Junk, P. C.; Kepert, C. J.; Lu, W. M.; Skelton, B. W.; White, A. H. *Aust. J. Chem.* **1999**, *52*, 459.
399. Petrochenkova, N. V.; Bukvetskii, B. V.; Mirochnik, A. G.; Karasev, V. E. *Russ. J. Coord. Chem.* **2002**, *28*, 67.
400. Brodkin, J. S.; Foxman, B. M.; Clegg, W.; Cressey, J. T.; Harbron, D. R.; Hunt, P. A.; Straughan, B. P. *Chem. Mater.* **1996**, *8*, 242.
401. Kemp, T. J.; Read, P. A.; Beatty, R. N. *Inorg. Chim. Acta* **1995**, *238*, 109.
402. Lin, Y. *J. Rare Earths* **1996**, *14*, 98.
403. Katti, K. V.; Singh, P. R.; Barnes, C. L. *Synth. React. Inorg. Met.-Org. Chem.* **1996**, *26*, 349 [*Chem. Abs.* **1996**, *124*, 248713].
404. Karipides, A. G.; Jai-nhuknan, J.; Cantrell, J. S. *Acta Crystallogr., Sect. C* **1996**, *52*, 2740.
405. Deiters, D.; Meyer, G. *Z. Anorg. Allg. Chem.* **1996**, *622*, 325.
406. Ma, J.-F.; Hu, N.-H.; Ni, J.-Z. *Polyhedron* **1996**, *15*, 1797.
407. Abram, U.; Dell'Amico, D. B.; Calderazzo, F.; Porta, C. D.; Englert, U.; Marchetti, F.; Merigo, A. *Chem. Commun.* **1999**, 2053.
408. Wickleder, M. S. *Z. Anorg. Allg. Chem.* **1998**, *624*, 1347.
409. Yamaguchi, T.; Nakamura, K.; Wakita, H.; Nomura, M. In *Recent Progress in Actinides Separation Chemistry*; Yoshida, Z.; Kimura, T.; Meguro, Y., Eds.; World Scientific: Singapore 1997; p 25.
410. Zalewicz, M.; Golinski, B. *J. Chem. Crystallogr.* **1999**, *28*, 879.
411. Becker, A.; Uhrland, W. *Z. Anorg. Allg. Chem.* **1999**, *625*, 217.
412. Becker, A.; Uhrland, W. *Z. Anorg. Allg. Chem.* **1999**, *625*, 1033.
413. Wickleder, M. S. *Z. Anorg. Allg. Chem.* **1999**, *625*, 1771.
414. Meyer, G.; Kutlu, I. *Z. Anorg. Allg. Chem.* **2000**, *626*, 975.
415. Chen, Y.; Ma, B.-Q.; Liu, Q.-D.; Li, J.-R.; Gao, S. *Inorg. Chem. Commun.* **2000**, *3*, 319.
416. Wickleder, M. S.; Müller, I.; Meyer, G. *Z. Anorg. Allg. Chem.* **2001**, *627*, 4.
417. Guillou, N.; Auffredic, J. P.; Louër, M.; Louër, D. *J. Solid State Chem.* **1994**, *106*, 295.
418. Zeng, G. F.; Guo, X.; Wang, C. Y.; Lin, Y. H.; Li, H. *Chinese J. Struct. Chem.* **1994**, *13*, 24.
419. Manck, E.; Meyer, G. *Eur. J. Solid State Inorg. Chem.* **1993**, *30*, 883.
420. Runde, W.; Neu, M. P.; Van Pelt, C; Scott, B. L. *Inorg. Chem.* **2000**, *39*, 1050.
421. Clark, D. L.; Donohoe, R. J.; Gordon, J. C.; Gordon, P. L.; Keogh, D. W.; Scott, B. L.; Tait, C. D.; Watkin, J. G. *J. Chem. Soc., Dalton Trans.* **2000**, 1975.
422. Bond, D. L.; Clark, D. L.; Donohoe, R. J.; Gordon, J. C.; Gordon, P. L.; Keogh, D. W.; Scott, B. L.; Tait, C. D.; Watkin, J. G. *Inorg. Chem.* **2000**, *39*, 3934.
423. Hill, N. J.; Levason, W.; Popham, M. C.; Reid, G.; Webster, M. *Polyhedron* **2002**, *21*, 445.
424. Hill, N. J.; Leung, L.-S.; Levason, W.; Webster, M. *Acta Crystallogr., Sect. C* **2002**, *58*, m 295.
425. Wang, H. K.; M. J. Zhong, H. K.; Jing, X. Y.; Wang, J. T.; Wang, R. J.; Wang, W. G. *Inorg. Chim. Acta.* **1989**, *163*, 19.
426. Deakin, L.; Levason, W.; Popham, M. C.; Reid, G.; Webster, M. *J. Chem. Soc., Dalton Trans.* **2000**, 2439.
427. Levason, W.; Newman, E. H.; Webster, M. *Polyhedron* **2000**, *19*, 2697.
428. Levason, W.; Newman, E. H.; Webster, M. *Acta Crystallogr Sect. C* **2000**, *56*, 1308.
429. Huang, C.; Li, G.; Zhou, Y.; Jin, T.; Xu, G. *Beijing Dax. Xue. Zir. Kex.* **1987**, 12 [*Chem. Abs.* **1987**, *106*, 167758].
430. Sakamoto, J.; Miyake, C. *Kidorui* **1993**, *22*, 154 [*Chem. Abs.* **995**, *122*, 278608].
431. Valle, G.; Casotto, G.; Zanonato, P. L.; Zarli, B. *Polyhedron* **1986**, *5*, 2093.

432. Lin, J.; Hey-Hawkins, E.; von Schnering, H. G. *Z. Naturforsch.* **1990**, *45a*, 1241.
433. Sakamoto, J.; Miyake, C. *Kidorui* **1993**, *22*, 154 [*Chem. Abs.* **1995**, *122*, 278608].
434. Bosson, M.; Levason, W.; Patel, T.; Popham, M. C.; Webster, M. *Polyhedron* **2001**, *20*, 2055.
435. Levason, W.; Patel, B.; Popham, M. C.; Reid, G.; Webster, M. *Polyhedron* **2001**, *20*, 2711.
436. Casellato, U.; Graziani, R.; Russo, U.; Zarli, B. *Inorg. Chim. Acta* **1989**, *166*, 9.
437. Long, D.-L.; Lu, H.-M.; Chen, J.-T.; Huang, J.-S. *Acta Crystallogr. Sect. C* **1999**, *55*, 1664.
438. Fawcett, J.; Platt, A. W. G.; Russell, D. R. *Inorg. Chim. Acta* **1998**, *274*, 177.
439. Platt, A. W. G.; Fawcett, J.; Hughes, R. S.; Russell, D. R. *Inorg. Chim. Acta* **1999**, *295*, 146.
440. Barr, D.; Brooker, A. T.; Drake, S. R.; Raithby, P. R.; Snaith, R.; Wright, D. S. *Angew. Chem., Int. Ed. Engl.* **1990**, *29*, 285.
441. Hou, Z.; Kobayasgi, K.; Yamazaki, H. *Chem. Lett.* **1991**, 265.
442. Petricek, S.; Demšar, A.; Golic, L.; Košmrlj, J. *Polyhedron* **2000**, *19*, 199.
443. Nishiura, M.; Tsuruta, H.; Yamaguchi, K.; Imamoto, T. *Kidorui* **1997**, *30*, 342.
444. Nishiura, M.; Yamanoi, Y.; Tsuruta, H.; Yamaguchi, K.; Imamoto, T. *Bull. Soc. Chim. Fr.* **1997**, *134*, 411.
445. Imamoto, T.; Yamanoni, Y.; Tsuruta, H.; Yamaguchi, K.; Yamazaki, M.; Inanaga, J. *Chem. Lett.* **1995**, 949.
446. Cabrera, A.; Rosas, N.; Alvarez, C.; Sharma, P.; Toscano, A.; Salmón, M.; Arias, J. L. *Polyhedron* **1996**, *15*, 2971.
447. Gusev, Yu. K.; Lychev, A. A. *Radiochemistry* **1996**, *38*, 388.
448. Asakura, K.; Imamoto, T. *Bull. Chem. Soc. Jpn.* **2001**, *74*, 731.
449. Evans, W. J.; Broomhall-Dillard, R. N. R.; Ziller, J. W. *Polyhedron* **1998**, *17*, 3361.
450. Semenova, L. I.; Skelton, B. W.; White, A. H. *Aust. J. Chem.* **1996**, *49*, 997.
451. Antsyshkina, A. S.; Adikov, G. G.; Rodnikova, M. N.; Mikhailichenko, A. I.; Balyknova, T. V. *Russ. J. Inorg. Chem.* **2002**, *47*, 367.
452. Cherkasova, T. G. *Zh. Neorg. Khim.* **1994**, *39*, 1316.
453. Evans, W. J.; Feldman, J. D.; Ziller, J. W. *J. Am. Chem. Soc.* **1996**, *118*, 4581.
454. Hubert-Pfalzgraf, L. G.; Machado, L.; Vaissermann, J. *Polyhedron* **1996**, *15*, 545.
455. Willey, G. R.; Woodman, T. J.; Drew, M. G. B. *Polyhedron* **1997**, *16*, 3385.
456. Deacon, G. B.; Tuong, T. D.; Wilkinson, D. L. *Inorg. Synth.* **1990**, *27*, 136.
457. Deacon, G. B.; Tuong, T. D.; Wilkinson, D. L. *Inorg. Synth.* **1990**, *28*, 286.
458. Wu, S.-H.; Ding, Z.-B.; Li, X.-J. *Polyhedron* **1994**, *13*, 2679.
459. Deacon, G. B.; Feng, T.; Junk, P. C.; Skelton, B. W.; Sobolev, A. N.; White, A. H. *Aust. J. Chem.* **1998**, *51*, 75.
460. Willey, G. R.; Woodman, T. J.; Errington, W. *J. Indian Chem. Soc.* **1998**, *75*, 435.
461. Jin, S.-H.; Dong, Z.-C.; Huang, J.-S.; Zhang, Q.-E.; Lu, J.-X. *Acta Crystallogr Sect. C* **1991**, *47*, 426.
462. Sobota, P.; Utko, J.; Szafert, S. *Inorg. Chem.* **1994**, *35*, 5203.
463. Willey, G. R.; Woodman, T. J.; Drew, M. G. B. *Polyhedron* **1997**, *16*, 3385.
464. Evans, W. J.; Shreeve, J. L.; Ziller, J. W.; Doedens, R. J. *Inorg. Chem.* **1995**, *34*, 576.
465. Willey, G. R.; Meehan, P. R.; Woodman, T. J.; Drew, M. G. B. *Polyhedron* **1997**, *16*, 623.
466. Woodman, T. J.; Errington, W.; Willey, G. R. *Acta Crystallogr., Sect. C* **1997**, *53*, 1801.
467. Anfang, S.; Dehnicke, K.; Magull, J. *Z. Naturforsch. B* **1996**, *51*, 531.
468. Deacon, G. B.; Feng, T.; Junk, P. C.; Meyer, G.; Scott, N. M.; Skelton, B. W.; White, A. H. *Aust. J. Chem.* **2000**, *53*, 853.
469. Anfang, S.; Karl, M.; Faza, N.; Massa, W.; Magull, J.; Dehnicke, K. *Z. Anorg. Allg. Chem.* **1997**, *623*, 1425.
470. Galliazzi, M. C. *Polymer* **1988**, *29*, 1516.
471. Karraker, D. G. *Inorg. Chim. Acta* **1987**, *139*, 189.
472. Xie, Z.; Chiu, K.; Wu, B.; Mak, T. C. W. *Inorg. Chem.* **1996**, *35*, 5957.
473. Niemeyer, M. *Acta Crystallogr., Sect. E* **2001**, *57*, m364.
474. Niemeyer, M. *Z. Anorg. Allg. Chem.* **1999**, *625*, 848.
475. Gradeef, P. S.; Yunlu, K.; Deming, T. J.; Olofson, J. M.; Ziller, J. W.; Evans, W. J. *Inorg. Chem.* **1989**, *28*, 2600.
476. Depaoli, G.; Ganis, P.; Zanonato, P. L.; Valle, G. *Polyhedron* **1993**, *12*, 1933.
477. Depaoli, G.; Ganis, P.; Zanonato, P. L.; Valle, G. *Polyhedron* **1993**, *12*, 671.
478. Willey, G. R.; Woodman, T. J.; Carpenter, D. J.; Errington, W. *J. Chem. Soc., Dalton Trans.* **1997**, 2677.
479. Nishiura, M.; Okano, N.; Imamoto, T. *Bull. Chem. Soc. Jpn.* **1999**, *72*, 1793.
480. Imamoto, T.; Okano, N.; Nishiura, M.; Yamaguchi, K. *Kidorui* **1997**, *30*, 344.
481. Alvarez, H. A.; Matos, J. R.; Isolani, P. C.; Vincentini, G.; Castellano, E. E.; Zukerman-Schpector, L. *J. Coord. Chem.* **1998**, *43*, 349.
482. Munhoz, C.; Isolani, P. C.; Vincentini, G.; Zukerman-Schpector, L. *J. Alloys. Compd.* **1998**, *275-277*, 782.
483. Dai, J.; Xu, Q. F.; Nukada, R.; Qian, P.; Wang, H. Z.; Mikuriya, M.; Munakata, M. *J. Coord. Chem.* **1998**, *43*, 13.
484. Carvallo, L. R. F.; Zinner, L. B.; Vincentini, G.; Bombieri, G.; Benetollo, F. *Inorg. Chem. Acta* **1992**, *191*, 49.
485. Evans, W. J.; Shreeve, J. L.; Ziller, J. W.; Doedens, R. J. *Inorg. Chem.* **1995**, *34*, 576.
486. Evans, W. J.; Shreeve, J. L.; Ziller, J. W.; Doedens, R. J. *Inorg. Chem.* **1993**, *32*, 245.
487. Barnhart, D. M.; Frankcom, T. M.; Gordon, P. L.; Sauer, N. N.; Thompson, J. A.; Watkin, J. G. *Inorg. Chem.* **1995**, *34*, 4863.
488. Barash, E. H.; Coan, P. S.; Lobkovsky, E. B.; Streib, W. E.; Caulton, K. G. *Inorg. Chem.* **1993**, *32*, 497.
489. Harlan, C. J.; Kareiva, A.; MacQueen, D. N.; Cook, R.; Barron, A. R. *Adv. Mater.* **1997**, *9*, 68.
490. Zaityseva, I. G.; Kuz'mina, N. P.; Martynenko, L. I.; Makhaev, V. D.; Borisov, A. P. *Zh. Neorg. Chem.* **1998**, *43*, 805.
491. Filotti, L.; Bugli, G.; Ensuque, A.; Bozon-Verduraz, F. *Bull. Soc. Chim. Fr.* **1996**, *133*, 1117.
492. Christidis, P. C.; Tossidis, I. A.; Paschalidis, D. G.; Tzavellas, L. C. *Acta Crystallogr., Sect. C* **1998**, *54*, 1233.
493. Trikkha, A. K.; Dilbagi, K. *J. Rare Earths* **1992**, *10*, 175.
494. Shankar, G.; Ramalingam, S. K. *Indian. J. Chem. Sect. A* **1988**, *27*, 61.
495. Amao, Y.; Okura, I.; Miyashita, T. *Chem. Lett.* **2000**, 1286.
496. Shen, C. *Jiegou Huaxue* **1997**, *16*, 371 [*Chem. Abs.* **1997**, *127*, 287181].
497. Lis, S.; Ptaziak, A. S.; Elbanowski, M. *Inorg. Chim. Acta* **1989**, *155*, 259.
498. Fritsch, E.; Mächler, E.; Arrouy, F.; Orama, O.; Berke, H.; Povey, I.; Willmott, P. R.; Locquet, J.-P. *Chem. Mater.* **1997**, *9*, 127.

499. Baxter, I.; Drake, S. R.; Hursthouse, M. B.; Malik, K. M. A.; McAleese, ;J.; Otway, D. J.; Plakatouras, J. C. *Inorg. Chem.* **1995**, *34*, 1384.
500. Darr, J. A.; Mingos, D. M. P.; Hibbs, D. E.; Hursthouse, M. B.; Malik, K. M. A. *Polyhedron* **1996**, *15*, 3225.
501. Luten, H. A.; Rees, W. S.; Goedken, V. L. *Chem. Vap. Deposition* **1996**, *2*, 149.
502. Rees, W. S.; Carris, M. W. *Inorg. Synth.* **1997**, *31*, 302.
503. Clegg, W.; Sage, I.; Oswald, I.; Brough, P.; Bourhill, G. *Acta Crystallogr,. Sect. C* **2000**, *56*, 1323.
504. Holz, R. C.; Thompson, L. C. *Inorg. Chem.* **1993**, *32*, 5251.
505. Holz, R. C.; Thompson, L. C. *Inorg. Chem.* **1988**, *27*, 4641.
506. Schumann, H.; Glanz, M.; Winterfeld, J.; Hemling, H.; Kuhn, N.; Kratz, T. *Chem. Ber.* **1994**, *127*, 2369.
507. Hong, Z.; Liang, C.; Li, R.; Li, W.; Zhao, D.; Fan, D.; Wang, D.; Chu, B.; Zang, F.; Hong, L.-S.; Lee, S.-T. *Adv. Mater.* **2001**, *13*, 1241.
508. Tao, Y.; Shao, X.; Zhao, G.; Jiu, X. *Huaxue Tongbao* **1997**, *128*, 39 [Chem. Abs. **1997**, *128*, 186071].
509. Kang, S.-J.; Jung, Y. S.; Sohn, Y. S. *Bull. Korean Chem. Soc.* **1997**, *18*, 75.
510. Lim, J. T.; Hong, S. T.; Lee, J. C.; Lee, I.-M. *Bull. Korean Chem. Soc.* **1996**, *17*, 1023.
511. Malandrino, G.; Fragalà, I. L.; Aime, S.; Dastrù, W.; Gobetto, R.; Benelli, C. *J. Chem. Soc., Dalton Trans.* **1998**, 1508.
512. Malandrino, G.; Benelli, C.; Castelli, F.; Fragalà, I. L. *Chem. Mater.* **1998**, *10*, 3434.
513. Malandrino, G.; Licata, R.; Castelli, F.; Fragalà, I. L.; Benelli, C. *Inorg. Chem.* **1995**, *34*, 6233.
514. Malandrino, G.; Incontro, O.; Castelli, F.; Fragalà, I. L.; Benelli, C. *Chem. Mater.* **1996**, *8*, 1292.
515. Kang, S.-J.; Jung, Y. S. *Bull. Korean Chem. Soc.* **1997**, *18*, 75 [Chem. Abs. **1997**, *126*, 194445].
516. Pollard, K. D.; Vittal, J. J.; Yap, G. P. A.; Puddephatt, R. J. *J. Chem. Soc., Dalton Trans.* **1998**, 1264.
517. Fragala, I. L.; Maladrino, G.; Benelli, C.; Castelli, F. *Chem. Mater.* **1998**, *10*, 3434.
518. Fragala, I. L.; Maladrino, G.; Incontro, O.; Castelli, F. *Chem. Mater.* **1996**, *8*, 1292.
519. Evans, W. J.; Giarikos, D. G.; Johnston, M. A.; Greci, M. A.; Ziller, J. W. *J. Chem. Soc., Dalton Trans.* **2002**, 520.
520. McAleese, J.; Darr, J. A.; Steele, B. C. H. *Chem. Vap. Deposition* **1996**, *2*, 244.
521. Werts, M. H. V.; Duin, M. A.; Hofstraat, J. W.; Verhoeven, J. W. *Chem. Commun.* **1999**, 799.
522. Iftikhar, K. *Polyhedron* **1996**, *15*, 1113.
523. Li, H.; Inoue, S.; Machida, K.; Adachi, G. *Chem. Mater.* **1999**, *11*, 3171.
524. Petrov, V. A.; Marshall, W. J.; Grushin, V. V. *Chem. Commun.* **2002**, 520.
525. Baxter, I.; Darr, J. A.; Hursthouse, M. B.; Malik, K. M. A.; Mcaleese, J.; Mingos, D. M. P. *Polyhedron* **1998**, *17*, 1329.
526. Guillon, H.; Daniele, S.; Hubert-Pfalzgraf, L. G.; Bavoux, C. *Inorg. Chim. Acta* **2000**, *304*, 99.
527. Gu, J. S.; Tong, J. Y.; Ma, M. H.; Tuck, D. G. *Chinese Chem. Lett.* **1995**, *6*, 259.
528. Tsukabe, H.; Shiba, H.; Uenishi, J. *J. Chem. Soc., Dalton Trans.* **1995**, 181.
529. Tsukube, H.; Shinoda, S.; Uenishi, J.; Kanatani, T.; Itoh, H.; Shiode, M.; Iwachido, T.; Yonemitsu, O. *Inorg. Chem.* **1998**, *37*, 1585.
530. Van Meervelt, L.; Froyen, A.; D'Olieslager, W.; Görller-Walrand, C.; Drisque, I.; King, G. S. D.; Maes, S.; Lenstra, A. T. H. *Bull. Soc. Chim. Belg.* **1996**, *105*, 377 [Chem. Abs. **1996**, *125*, 345999].
531. de Mello Donegá, C.; Junior, S. A.; de Sá, G. F. *J. Chem. Soc., Chem. Commun.* **1996**, 1199.
532. Kido, J.; Ikeda, W.; Kimura, M.; Nagai, K. *Kidorui* **1995**, *26*, 110.
533. Trikha, A. K.; Zinner, L. B.; Zinner, K.; Isolini, P. C. *Polyhedron* **1996**, *15*, 1651.
534. Batista, H. J.; de Andrade, A. V. M.; Longo, R. L.; Simas, A. M.; de Sa, G. F.; Ito, N. K.; Thompson, L. C. *Inorg. Chem.* **1998**, *37*, 3542.
535. Uekawa, M.; Miyamoto, Y.; Ikeda, H.; Kaifu, K.; Nayada, T. *Bull. Chem. Soc. Jpn.* **1998**, *71*, 2253.
536. Wang, M.-Z.; Jin, L.-P.; Cai, G.-L.; Liu, S.-X.; Luang, J.-H.; Qin, W.-P.; Huang, S-H. *J. Rare Earths* **1994**, *12*, 166.
537. Li, H.-Y.; Yang, Y.-S.; Huang, Y.-Q.; Hu, S.-Z. *Chinese J. Struct. Chem.* **1994**, *13*, 371.
538. Kuz'mina, N. P.; An'Tu, Z.; Pisarevskii, A. P.; Martynenko, L. I.; Struchkov, Yu.T. *Russian J. Coord. Chem.* **1994**, *20*, 665.
539. Plakatouras, J. C.; Baxter, I.; Hursthouse, M. B.; Malik, K. M. A.; McAleese, J.; Drake, S. R. *J. Chem. Soc., Chem. Commun.* **1994**, 2455.
540. Baxter, I.; Drake, S. R.; Hursthouse, M. B.; McAleese, J.; Malik, K. M. A.; Mingos, D. M. P.; Otway, D. J.; Plakatouras, J. C. *Polyhedron* **1998**, *17*, 3777.
541. Thompson, L. C.; Atchison, F. W.; Young, V. G. *J. Alloys Compd.* **1998**, *275–277*, 765.
542. Okada, K.; Uekawa, M.; Wang, Y. F.; Chen, T. M.; Nakaya, T. *Chem. Lett.* **1998**, 801.
543. McGhee, M. D.; Bergstedt, T.; Zhang, C.; Saab, A. P.; O'Regan, M. B.; Bazan, G. C.; Srdanov, V. I.; Heeger, A. J. *Adv. Mater.* **1999**, *11*, 1354.
544. Hasegawa, Y.; Sogabe, K.; Wada, Y.; Kitamura, T.; Nakashima, N.; Yanagida, S. *Chem. Lett.* **1999**, 35.
545. Tsukube, H.; Hosokubo, M.; Wada, M.; Shinoda, S.; Tamiaki, H. *J. Chem. Soc., Dalton Trans.* **1999**, 11.
546. Gleizes, A.; Julve, M.; Kuzmina, N.; Alikhanyan, A.; Lloret, F.; Malkerova, I.; Sanz, J. L.; Senocq, F. *Eur. J. Inorg. Chem.* **1998**, 1169.
547. Kuz'mina, N. P.; Kupriyanova, G. N.; Troyanov, S. I. *Russ. J. Coord. Chem.* **2000**, *26*, 367.
548. Sasaki, M.; Manseki, K.; Horiuchi, H.; Kumagai, M.; Sakamoto, M.; Saakiyama, H.; Nishida, Y.; Sakai, M.; Sadaoka, Y.; Ohba, M.; Okawa, H. *J. Chem. Soc., Dalton Trans.* **2000**, 259.
549. Guillon, H.; Hubert-Pfalzgraf, L. G.; Vaissermann, J. *Eur. J. Inorg. Chem.* **2000**, 1243.
550. Kido, T.; Nagasto, S.; Sunatsuki, Y.; Matsumoto, N. *Chem. Commun.* **2000**, 2113.
551. Rees, W. S.; Just, O.; Castro, S. L.; Matthews, J. S. *Inorg. Chem.* **2000**, *39*, 3736.
552. Belot, J. A.; Wang, A.; McNeely, R. J.; Liable-Sands, L.; Rheingold, A. L.; Marks, T. J. *Chem. Vap. Deposition* **1999**, *5*, 65.
553. Polyanskaya, T. M.; Romanenko, G. V.; Podberezskaya, N. V. *J. Struct. Chem.* **1997**, *38*, 637 [Chem. Abs. **1997**, *128*, 161181].
554. Baxter, I.; Darr, J. A.; Hursthouse, M. B.; Malik, K. M. A.; Mingos, D. M. P.; Plakatouras, J. C. *J. Chem. Crystallogr.* **1998**, *28*, 267.
555. Sweeting, L. M.; Rheingold, A. L. *J. Am. Chem. Soc.* **1987**, *109*, 2652.
556. Cotton, F. A.; Daniels, L. M.; Huang, P. *Inorg. Chem. Commun.* **2001**, *4*, 319.
557. Gao, L. H.; Whang, K. Z.; Huang, C. H.; Zhao, X. S.; Xia, X. H.; Li, T. K.; Xu, J. M. *Chem. Mater.* **1995**, *7*, 1047.
558. Whang, K. Z.; Huang, C. H.; Xu, G. X.; Wang, R. J. *Polyhedron* **1995**, *14*, 3669.
559. Yu, G.; Liu, Y.; Wu, X.; Zhu, D.; Li, H.; Jin, L.; Wang, M. *Chem. Mater.* **2000**, *12*, 2537.

560. Huang, C.; Zhu, X.; Guo, F.; Song, J.; Xu, Z.; Liao, C.; Jin, Z. *Beijing Daxue Xuebao, Ziran Kexueban* **1992**, *28*, 428 [Chem. Abs. **1994**, *120*, 93867].
561. Qian, D.; Nakahara, H.; Fukuda, K.; Yang, K. *Chem. Lett.* **1995**, 175.
562. Whang, K. Z.; Jiang, W.; Huang, C. H.; Xu, G. X.; Xu, L. G.; Li, T. K.; Zhao, X. S.; Xie, X. M. *Chem. Lett.* **1994**, 1761.
563. Xiao, Y. J.; Gao, X. X.; Huang, C. H.; Whang, K. Z. *Chem. Mater.* **1994**, *6*, 1910.
564. Mehrotra, R. C.; Singh, A.; Tripathi, U. M. *Chem. Rev.* **1991**, *91*, 1287.
565. Andersen, R. A.; Templeton, D. H.; Zalkin, A. *Inorg. Chem.* **1978**, *17*, 1962.
566. Westin, G.; Kritikos, M.; Wijk, M. *J. Solid State Chem.* **1998**, *141*, 168.
567. Barnhart, D. M.; Clark, D. L.; Huffman, J. C.; Vincent, R. L.; Watkin, J. G. *Inorg. Chem.* **1993**, *32*, 4077.
568. Barnhart, D. M.; Clark, D. L.; Gordon, J. C.; Huffman, J. C.; Watkin, J. G.; Zwick, B. D. *J. Am. Chem. Soc.* **1993**, *115*, 8461.
569. Bradley, D. C.; Chudzynska, H.; Hursthouse, M. B.; Motevalli, M. *Polyhedron* **1991**, *10*, 1049.
570. Gromada, J.; Chenal, T.; Mortreux, A.; Ziller, J. W.; Leising, F.; Carpentier, J.-F. *Chem. Commun.* **2000**, 2183.
571. Daniele, S.; Hubert-Pfalzgraf, L. G.; Hitchcock, P. B.; Lappert, M. F. *Inorg. Chem. Commun.* **2000**, *3*, 218.
572. Biagini, P.; Lugli, G.; Abis, L.; Millini, R. *J. Organomet. Chem.* **1994**, *474*, c16.
573. Evans, W. J.; Sollberger, M. S.; Hanusa, T. P. *J. Am. Chem. Soc.* **1988**, *110*, 1841.
574. Evans, W. J.; Sollberger, M. S. *Inorg. Chem.* **1988**, *27*, 4417.
575. Evans, W. J.; Olofson, J. M.; Ziller, J. W. *J. Am. Chem. Soc.* **1990**, *112*, 2308.
576. Stecher, H. A.; Sen, A.; Rheingold, A. L. *Inorg. Chem.* **1989**, *28*, 3280.
577. Kornev, A. N.; Chesnokova, T. A.; Zhelova, E. V.; Zakharov, L. N.; Fukin, G. N.; Kursky, Y. A.; Domrachev, G. A.; Lickiss, P. D. *J. Organomet. Chem.* **1999**, *587*, 113.
578. Evans, W. J.; Golden, R. E.; Ziller, J. W. *Inorg. Chem.* **1991**, *30*, 4963.
579. Coan, P. S.; McGeary, M. J.; Lobkovsky, E. B.; Ziller, J. W. *Inorg. Chem.* **1991**, *30*, 3570.
580. Gradeef, P. S.; Yunlu, K.; Deming, T. J.; Olofson, J. M.; Doedens, R. J.; Evans, W. J. *Inorg. Chem.* **1990**, *29*, 420.
581. McGeary, M. J.; Coan, P. S.; Folting, K.; Streib, W. E.; Caulton, K. G. *Inorg. Chem.* **1989**, *28*, 3282.
582. McGeary, M. J.; Coan, P. S.; Folting, K.; Streib, W. E.; Caulton, K. G. *Inorg. Chem.* **1991**, *30*, 1723.
583. Kornev, A. N.; Chesnokova, T. A.; Zhelova, E. V.; Zakharov, L. N.; Fukin, G. N.; Kursky, Y. A.; Domrachev, G. A.; Lickiss, P. D. *J. Organomet. Chem.* **1999**, *587*, 113.
584. Wedler, M.; Gilje, J. W.; Pieper, U.; Stalke, D.; Noltenmmeyer, M.; Edelmann, F. T. *Chem. Ber.* **1991**, *124*, 1163.
585. Shao, P.; Berg, D. J.; Bushnell, G. W. *Inorg. Chem.* **1994**, *33*, 3453.
586. Poncelet, O.; Hubert-Pfalzgraf, L. G.; Daran, J.-C.; Caulton, K. G. *Chem. Commun.* **1989**, 1846.
587. Meguro, M.; Asao, N.; Yamamoto, Y. *J. Chem. Soc., Chem. Commun.* **1995**, 1021.
588. Okano, T.; Miyamoto, K.; Kiji, J. *Chem. Lett.* **1995**, 246.
589. Dewa, T.; Saiki, T.; Aoyama, Y. *J. Am. Chem. Soc.* **2001**, *123*, 502.
590. Poncelet, O.; Sartain, W. J.; Hubert-Pfalzgraf, L. G.; Folting, K.; Caulton, K. G. *Inorg. Chem.* **1989**, *28*, 263.
591. Bradley, D. C.; Chudzynska, H.; Friogo, D. M.; Hammond, M. E.; Hursthouse, M. B.; Mazid, M. A. *Polyhedron* **1990**, *9*, 719.
592. Coan, P. S.; Hubert-Pfalzgraf, L. G.; Caulton, K. G. *Inorg. Chem.* **1992**, *31*, 1262.
593. Kritikos, M.; Moustiakimov, M.; Wijk, M.; Westin, G. *J. Chem. Soc., Dalton Trans.* **2001**, 1931.
594. Westin, G.; Moustiakimov, M.; Kritikos, M. *Inorg. Chem.* **2002**, *41*, 3249.
595. Hubert-Pfalzgraf, L. G.; Daniele, S.; Bennaceur, A.; Daran, J.-C.; Vaissermann, J. *Polyhedron* **1997**, *16*, 1223.
596. Helgesson, G.; Jagner, S.; Poncelet, O.; Hubert-Pfalzgraf, L. G. *Polyhedron* **1991**, *10*, 1559.
597. Poncelet, O.; Hubert-Pfalzgraf, L. G.; Daran, J. C. *Polyhedron* **1990**, *9*, 1305.
598. Anwander, R.; Munck, F. C.; Priermeyer, T.; Scherer, W.; Runte, O.; Herrmann, W. A. *Inorg. Chem.* **1997**, *36*, 3545.
599. Daniele, S.; Hubert-Pfalzgraf, L. G.; Daran, J.-C. *Polyhedron* **1996**, *15*, 1063.
600. Labrize, F.; Hubert-Pfalzgraf, L. G. *Polyhedron* **1995**, *14*, 881.
601. Laurent, F.; Huffman, J. C.; Folting, K.; Caulton, K. G. *Inorg. Chem.* **1995**, *34*, 3980.
602. Bradley, D. C.; Chudzynska, H.; Hursthouse, M. B.; Motevalli, M.; Wu, R. *Polyhedron* **1994**, *13*, 1.
603. Bradley, D. C.; Chudzynska, H.; Hursthouse, M. B.; Motevalli, M. *Polyhedron* **1994**, *13*, 7.
604. Poncelet, O.; Guilment, J.; Martin, D. J. *Sol-Gel. Sci. Technol.* **1998**, *13*, 129.
605. Labrize, F.; Hubert-Pfalzgraf, L. G.; Daran, J.-C.; Halaut, S.; Tobaly, P. *Polyhedron* **1996**, *15*, 2707.
606. Lappert, M. F.; Singh, A.; Smith, R. G. *Inorg. Synth.* **1990**, *27*, 164.
607. Hitchcock, P. B.; Lappert, M. F.; Smith, R. G. *Inorg. Chim. Acta* **1987**, *139*, 183.
608. Stecher, H. A.; Sen, A.; Rheingold, A. L. *Inorg. Chem.* **1988**, *27*, 1130.
609. Nishiura, M.; Kameoka, M.; Imamoto, T. *Kidorui* **2000**, *36*, 294.
610. Zhang, L.-L.; Yao, Y.-M.; Luo, Y.-J.; Shen, Q.; Sun, J. *Polyhedron* **2000**, *19*, 2243.
611. Yao, Y.-M.; Shen, Q.; Yu, K.-B. *Acta Crystallogr., Sect. C* **2000**, *56*, 1330.
612. Evans, W. J.; Olofson, J. M.; Ziller, J. W. *Inorg. Chem.* **1989**, *28*, 4308.
613. Evans, W. J.; Johnston, M. A.; Greci, M. A.; Ziller, J. W. *Polyhedron* **2001**, *20*, 277.
614. Aspinall, H. C.; Williams, M. *Inorg. Chem.* **1996**, *35*, 255.
615. Barnhart, D. M.; Clark, D. L.; Gordon, J. C.; Huffman, J. C.; Vincent, R. L.; Watkin, J. G.; Zwick, B. D. *Inorg. Chem.* **1994**, *33*, 3487.
616. Butcher, R. J.; Clark, D. L.; Grumbine, S. K.; Vincent-Hollis, R. L.; Scott, B. L.; Watkin, J. G. *Inorg. Chem.* **1995**, *34*, 5468.
617. Clark, D. L.; Gordon, J. C.; Watkin, J. G.; Huffman, J. C.; Zwick, B. D. *Polyhedron* **1996**, *15*, 2279.
618. Clark, D. L.; Gordon, J. C.; Huffman, J. C.; Vincent-Hollis, R. L.; Watkin, J. G.; Zwick, B. D. *Inorg. Chem.* **1994**, *33*, 5903.
619. Clark, D. L.; Deacon, G. B.; Feng, T.; Hollis, R. V.; Scott, B. L.; Skelton, B. W.; Watkin, J. G.; White, A. H. *J. Chem. Soc., Dalton Trans.* **1996**, 1729.
620. Clark, D. L.; Hollis, R. V.; Scott, B. L.; Watkin, J. G. *Inorg. Chem.* **1996**, *35*, 667.
621. Deacon, G. B.; Feng, T.; Forsyth, C. M.; Gitlits, A.; Hockless, D. C. R.; Shen, Q.; Skelton, B. W.; White, A. H. *J. Chem. Soc., Dalton Trans.* **2000**, 961.
622. Deacon, G. B.; Gatehouse, B. M.; Shen, Q.; Ward, G. N.; Tiekink, E. R. T. *Polyhedron* **1993**, *12*, 1289.

623. Deacon, G. B.; Feng, T.; Skelton, B. W.; White, A. H. *Aust. J. Chem.* **1995**, *48*, 741.
624. Deacon, G. B.; Feng, T.; Junk, P. C.; Skelton, B. W.; White, A. H. *Chem. Ber. Rec.* **1997**, *130*, 851.
625. Deacon, G. B.; Feng, T.; Junk, P. C.; Skelton, B. W.; White, A. H. *J. Chem. Soc., Dalton Trans.* **1997**, 1181.
626. Borup, B.; Streib, W. E.; Caulton, K. G. *Chem. Ber.* **1996**, *129*, 1003.
627. Deacon, G. B.; Meyer, G.; Stellfeldt, D.; Zelesny, G.; Skelton, B. W.; White, A. H. *Z. Anorg. Allg. Chem.* **2001**, *627*, 1652.
628. Hogerheide, M. P.; Ringelberg, S. N.; Grove, D. M.; Jastrzebski, J. T. B. H.; Boersma, J.; Smeets, A. L.; Spek, W. J. J.; van Koten, G. *Inorg. Chem.* **1996**, *35*, 1185.
629. Daniele, F. S.; Hubert-Pfalzgraf, L. G.; Vaissermann, J. *Polyhedron* **1995**, *14*, 327.
630. Merbach, A. E.; Toth, E., Eds.; *The Chemistry of Contrast Agents for Medical Magnetic Resonance Imaging*; Wiley: London, 2001.
631. Caravan, P.; Ellison, J. J.; McMurry, T. J.; Lauffer, R. B. *Chem. Rev.* **1999**, *99*, 2293.
632. Lauffer, R. B. *Chem. Rev.* **1987**, *87*, 901.
633. Tweedle, M. F. Relaxation Agents in NMR Imaging. In *Lanthanide Probes in Life, Chemical and Earth Sciences*; Bunzli, J.-C. G.; Choppin, G. R., Eds.; Elsevier: Amsterdam, 1989; p. 127.
634. Parker, D. Imaging and Targeting. In *Comprehensive Supramolecular Chemistry*; Atwood, J. L.,; Davies, J. E. D.; MacNicol, D. D.; Vogtle, F., Eds.; Pergamon: Oxford, UK, 1996; Vol. 10, 486.
635. Yam, V. W.-W.; Lo, K. K.-W. *Coord. Chem. Rev.* **1999**, *184*, 157.
636. Aime, S.; Botta, M.; Fasano, M.; Crich, S. G.; Terreno, E. *Coord. Chem. Rev.* **1999**, *185–186*, 321.
637. Watson, A. D. *J. Alloys Compd.* **1994**, *207/208*, 14.
638. Aime, S.; Botta, M.; Fasano, M.; Terreno, E. *Chem. Soc. Rev.* **1998**, *27*, 19.
639. Kumar, K. *J. Alloys. Compd.* **1997**, *249*, 163.
640. Weinmann, H. J.; Brasch, R. C.; Press, W.-R.; Wesbey, G. E. *Am. J. Roentgenol.* **1984**, *142*, 619, cited by Anelli, P. L.; Lattuda, L. In *The Chemistry of Contrast Agents for Medical Magnetic Resonance Imaging*; Merbach, A. E.; Toth, E., Eds.; Wiley: London, 2001; pp. 133–134.
641. Choppin, G. R. *Thermochim. Acta* **1993**, *227*, 1.
642. Nakamura, K.; Kurisaki, T.; Wakita, H.; Yamaguchi, T. *Acta Crystallogr., Sect. C* **1995**, *51*, 1559.
643. Sakagami, N.; Homma, J.; Konno, T.; Okamoto, K. *Acta Crystallogr., Sect. C* **1997**, *53*, 1376.
644. Sakagami, N.; Yamada, Y.; Konno, T.; Okamoto, K. *Inorg. Chim. Acta* **1999**, *288*, 7.
645. Mistryukov, V. E.; Mikhailov, Yu.N.; Chuklanova, E. B.; Sergeev, A. V.; Shchelokov, R. N. *Zh. Neorg. Khim.* **1998**, *43*, 1997 [Russ J. Inorg. Chem **1998**, *43*, 1862].
646. Graeppi, N.; Powell, D. H.; Laurenczy, G.; Zékany, L.; Merbach, A. E. *Inorg. Chim. Acta* **1995**, *235*, 311.
647. Gries, H.; Miklautz, H. *Physiol. Chem. Phys. Med. NMR* **1984**, *16*, 105.
648. Jin, T.; Zhao, S.; Xu, G.; Han, Y.; Shi, N.; Ma, Z. *Huaxue Xuebao* **1991**, *49*, 569.
649. Ruloff, R.; Gelbrich, T.; Hoyer, E.; Sieler, J.; Beyer, L. *Z. Naturforsch. B* **1998**, *53*, 955.
650. Stezowski, J. J.; Hoard, J. *Isr. J. Chem.* **1984**, *24*, 323.
651. Inoue, M. B.; Inoue, M.; Fernando, Q. *Inorg. Chim. Acta* **1995**, *232*, 203.
652. Sakagami, N.; Homma, J.; Konno, T.; Okamoto, K. *Acta Crystallogr., Sect. C* **1997**, *53*, 1378.
653. Mondry, A.; Starynowicz, P. *Polyhedron* **2000**, *19*, 771.
654. Hardcastle, K. I.; Botta, M.; Fasano, M.; Digilio, G. *Eur. J. Inorg. Chem.* **2000**, 971.
655. Horrocks, W. D.; Sudnick, D. R. *J. Am. Chem. Soc.* **1979**, *101*, 334.
656. Geraldes, C. F. G. C.; Sherry, A. D.; Cacheris, W. P.; Kuan, K. T.; Brown, R. D.; Koenig, S. H.; Spiller, M. *Magn. Reson. Med.* **1988**, *8*, 191.
657. Chang, C. A.; Brittain, H. G.; Telser, J.; Tweedle, M. F. *Inorg. Chem.* **1990**, *29*, 4468.
658. Beeby, A.; Clarkson, I. M.; Dickins, R. S.; Faulkner, S.; Parker, D.; Royle, L.; de Sousa, A. S.; Williams, J. A. G.; Woods, M. *J. Chem. Soc. Perkin Trans 2* **1999**, 493.
659. Aime, S.; Benetollo, F.; Bombieri, G.; Colla, S.; Fasano, M.; Paoletti, S. *Inorg. Chim. Acta* **1997**, *254*, 63.
660. Aime, S.; Fasano, M.; Paoletti, S.; Terreno, E. *Gazz. Chim. Ital.* **1995**, *125*, 125.
661. Geraldes, C. F. C.; Delgado, R.; Urbano, A. M.; Costa, J.; Jasanda, F.; Neveu, F. *J. Chem. Soc., Dalton Trans.* **1995**, 327.
662. Toth, E.; Burai, L.; Brucher, E.; Merbach, A. E. *J. Chem. Soc., Dalton Trans.* **1997**, 1587.
663. Tóth, E.; Connac, F.; Helm, L.; Adamli, K.; Merbach, A. E. *Eur. J. Inorg. Chem.* **1998**, 2017.
664. Aime, S.; Chiaussa, M.; Diglio, G.; Gianolio, E.; Terreno, E. *J. Biol. Inorg. Chem.* **1999**, *4*, 766.
665. Muller, R.; Radüchel, B.; Laurent, S.; Platzek, J.; Piérart, C.; Mareski, P.; Vander Elst, L. *Eur. J. Inorg. Chem.* **1999**, 1949.
666. Zhuo, R-X.; Wen, J.; Wang, L. *Chem. Res. Chin. Univ.* **1997**, *13*, 150 [Chem Abs. **1997**, *127*, 116674].
667. Wang, Y.-M.; Wang, Y.-J.; Sheu, R.-S.; Liu, G.-C.; Lin, W.-C.; Liao, J.-H. *Polyhedron* **1999**, *18*, 1147.
668. Bligh, S. W. A.; Chowdhury, A. H. M. S.; McPartlin, M.; Scowen, I. J.; Bulman, R. A. *Polyhedron* **1995**, *14*, 567.
669. Paul-Roth, C.; Raymond, K. N. *Inorg. Chem.* **1995**, *34*, 1408.
670. Franklin, J.; Raymond, K. N. *Inorg. Chem.* **1994**, *33*, 5794.
671. Ehnebom, L.; Fjaertoft Pedersen, B. *Acta Chem. Scand.* **1992**, *46*, 126.
672. Konings, M. S.; Dow, W. C.; Love, D. B.; Raymond, K. N.; Quay, S. C.; Rocklage, S. M. *Inorg. Chem.* **1990**, *29*, 1488.
673. Parker, D.; Pulukkody, K.; Smith, F. C.; Batsanov, A.; Howard, J. A. K. *J. Chem. Soc., Dalton Trans.* **1994**, 689.
674. Inoue, M. B.; Inoue, M.; Fernando, Q. *Acta Crystallogr., Sect. C* **1994**, *50*, 1037.
675. Inoue, M. B.; Inoue, M.; Munoz, I. C.; Bruck, M. A.; Fernando, Q. *Inorg. Chim. Acta* **1993**, *209*, 29.
676. Inoue, M. B.; Navarro, R. E.; Inoue, M.; Fernando, Q. *Inorg. Chem.* **1995**, *34*, 6074.
677. Inoue, M. B.; Oram, P.; Inoue, M.; Fernando, Q.; Alexander, A. L.; Unger, E. C. *Magn. Reson. Imaging* **1994**, *12*, 429.
678. Lammers, H.; Maton, F.; Pubanz, D.; van Laren, M. W.; van Bekkum, H.; Merbach, A. E.; Muller, R. N.; Peters, J. A. *Inorg. Chem.* **1997**, *36*, 2527.
679. Imura, H.; Choppin, G. R.; Cacheris, W. P.; de Learie, L. A.; Dunn, T. J.; White, D. H. *Inorg. Chim. Acta* **1997**, *258*, 227.
680. Aime, S.; Botta, M.; Fasano, M.; Paolettii, S.; Terreno, E. *Chem. Eur. J.* **1997**, *3*, 1499.
681. Bovens, E.; Hoefnagel, M. E.; Boers, E.; Lammers, H.; van Bekkum, H.; Peters, J. A. *Inorg. Chem.* **1996**, *35*, 7678.
682. Aukrust, A.; Raknes, A.; Sjøgren, C. E.; Sydnes, L. K. *Acta Chem. Scand.* **1997**, *51*, 918.
683. Tóth, E.; Burai, L.; Brücher, E.; Merbach, A. A. E. E. *J. Chem. Soc., Dalton Trans.* **1997**, 1587.
684. Chou, K. Y.; Kim, K. S.; Kim, J. C. *Polyhedron* **1994**, *13*, 567.

685. Uggeri, F.; Aime, S.; Anelli, P. L.; Botta, M.; Brochetta, M.; de Haen, C.; Ermondi, G.; Grandi, M.; Paoli, P. *Inorg. Chem.* **1995**, *34*, 633.
686. Wang, Y. M.; Lee, C. H.; Liu, G. C.; Sheu, R. S. *J. Chem. Soc., Dalton Trans.* **1998**, 4113.
687. Tóth, E.; Helm, L.; Kellar, K. E.; Merbach, A. E. *Chem. Eur. J.* **1999**, *5*, 1202.
688. Schmitt-Willich, H.; Brehm, M.; Ewers, C. L. J.; Michl, G.; Müller-Fahrnow, A.; Petrov, O.; Platzek, J.; Radüchel, B.; Sülzle, D. *Inorg. Chem.* **1999**, *38*, 1134.
689. Sarka, L.; Bányai, I.; Brücher, E.; Király, R.; Platzek, J.; Radüchel, B.; Schmitt-Willich, H. *J. Chem. Soc., Dalton Trans.* **2000**, 3699.
690. De Stasio, G.; Casalbore, P.; Pallini, R.; Gilbert, B.; Sanità, F.; Ciotti, M. T.; Rosi, G.; Festinesi, A.; Larocca, L. M.; Rinelli, A.; Perret, D.; Mogk, D. W.; Perfetti, P.; Minha, M. P.; Mercanti, D. *Cancer Research* **2001**, *61*, 4272.
691. Dubost, J. P.; Leger, J. M.; Langlois, M. H.; Meyer, D.; Schaefer, M. C. *C. R. Acad. Sci., Ser. II Univers.* **1991**, *312*, 349.
692. Chang, C. A.; Francesconi, L. C.; Malley, M. F.; Kumar, K.; Gougoutas, J. Z.; Tweedle, M. F.; Lee, D. W.; Wilson, J. *Inorg. Chem.* **1993**, *32*, 3501.
693. Parker, D.; Pulukkody, K.; Smith, F. C.; Batsanov, A.; Howard, J. A. K. *J. Chem. Soc., Dalton Trans.* **1994**, 689.
694. Spirlet, M. R.; Rebizant, J.; Desreux, J. F.; Loncin, M. F. *Inorg. Chem.* **1984**, *23*, 359.
695. Aime, S.; Barge, A.; Botta, M.; Fasano, M.; Ayala, J. D.; Bombieri, G. *Inorg. Chim. Acta* **1996**, *246*, 423.
696. Benetollo, F.; Bombieri, G.; Aime, S.; Botta, M. *Acta Crystallogr., Sect. C* **1999**, *55*, 353.
697. Aime, S.; Barge, A.; Benetollo, F.; Bombiieri, G.; Botta, M.; Uggeri, F. *Inorg. Chem.* **1997**, *36*, 4287.
698. Tóth, E.; Brücher, E.; Lazár, I.; Tóth, I. *Inorg. Chem.* **1994**, *33*, 4070.
699. Jacques, V.; Desreux, J. F. *Inorg. Chem.* **1994**, *33*, 4048.
700. Aime, S.; Botta, M.; Fasano, M.; Marques, M. P. M.; Geraldes, C. F. G. C.; Pubanz, D.; Merbach, A. E. *Inorg. Chem.* **1997**, *36*, 2059.
701. Bénazeth, S.; Purans, J.; Chalbot, M.-C.; Nguyen-van-Duong, M. K.; Nicholas, L.; Keller, F.; Gaudemer, A. *Inorg. Chem.* **1998**, *37*, 3667.
702. Burai, L.; Fabian, I.; Kiraly, R.; Szilagyi, E.; Bruchner, E. *J. Chem. Soc., Dalton Trans.* **1998**, 243.
703. Dickens, R. S.; Parker, D.; de Sousa, A. S.; Williams, J. A. G. *J. Chem. Soc., Chem. Commun.* **1996**, 696.
704. Alderighi, L.; Bianchi, A.; Calabi, L.; Dapporto, P.; Giorgi, C.; Losi, P.; Paleari, L.; Paoli, P.; Rossi, P.; Valtancoli, B.; Virtuani, M. *Eur. J. Inorg. Chem.* **1998**, 1581.
705. Aime, S.; Botta, M.; Ermondi, G.; Terreno, E.; Anelli, P. L.; Fedeli, F.; Uggeri, F. *Inorg. Chem.* **1996**, *35*, 2726.
706. Kumar, K.; Chang, C. A.; Francesconi, L. C.; Dischino, D. D.; Malley, M. F.; Gougoutas, J. Z.; Tweedle, M. F. *Inorg. Chem.* **1994**, *33*, 3567.
707. Platzek, J.; Blaszkiewicz, P.; Gries, H.; Luger, P.; Michl, G.; Mueller-Fahrnow, A.; Raduechel, B.; Sulzle, D. *Inorg. Chem.* **1997**, *36*, 6086.
708. Tóth, E.; Dhubhghaill, O. M. N.; Besson, G.; Helm, L.; Merbach, A. E. *Magn.Reson.Chem.* **1999**, *37*, 701.
709. Roth, K.; Bartholomae, G.; Bauer, H.; Frenzel, T.; Kossler, S.; Platzek, J.; Radüchel, B.; Weinmann, H.-J. *Angew. Chem., Int. Ed. Engl.* **1996**, *35*, 655.
710. Bianchi, A.; Calabi, L.; Ferrini, L.; Losi, P.; Uggeri, F.; Valtancoli, B. *Inorg. Chim. Acta* **1996**, *249*, 13.
711. Messeri, D.; Lowe, M. P.; Parker, D.; Botta, M. *Chem. Commun.* **2001**, 2742.
712. Toth, E.; Vauthey, S.; Pubanz, D.; Merbach, A. E. *Inorg. Chem.* **1996**, *35*, 3375.
713. André, J. P.; Tóth, E.; Fischer, H.; Seelig, A.; Mäcke, H. R.; Merbach, A. E. *Chem. Eur. J.* **1999**, *5*, 2977.
714. Moats, R. A.; Fraser, S. E.; Meade, T. J. *Angew. Chem., Int. Ed. Engl.* **1997**, *36*, 726.
715. Aime, S.; Batsanov, A. S.; Botta, M.; Howard, J. A. K.; Parker, D.; Senanayake, K.; Williams, G. *Inorg. Chem.* **1994**, *33*, 4696.
716. Aime, S.; Botta, M.; Parker, D.; Williams, J. A. G. *J. Chem. Soc., Dalton Trans.* **1995**, 2259.
717. Yerly, F.; Dunand, F. A.; Tóth, E.; Figuciria, A.; Kóvacs, Z.; Sherry, A. D.; Geraldes, C. F. G. C.; Merbach, A. E. *Eur. J. Inorg. Chem.* **2000**, 993.
718. Ruloff, R.; Arnold, K.; Beyer, L.; Dietze, F.; Gründer, W.; Wagner, M.; Hoyer, E. *Z. Anorg. Allg. Chem.* **1995**, *621*, 807.
719. Ruloff, R.; Prokop, P.; Sieler, J.; Hoyer, E.; Beyer, L. *Z. Naturforsch. Teil B* **1996**, *51*, 963.
720. Ruloff, R.; Gelbrich, T.; Sieler, J.; Hoyer, E.; Beyer, L. *Z. Naturforsch. Teil B* **1997**, *52*, 805.
721. Ruloff, R.; Muller, R. N.; Pubanz, D.; Merbach, A. E. *Inorg. Chim. Acta* **1998**, *275–276*, 15.
722. Wang, R.-Y.; Li, J.-R.; Jin, T.-Z.; Xu, G.-X.; Zhou, Z.-Y.; Zhou, X.-G. *Polyhedron* **1997**, *16*, 1361.
723. Wang, R.-Y.; Li, J.-R.; Jin, T.-Z.; Xu, G.-X.; Zhou, Z.-Y.; Zhou, X-G. *Polyhedron* **1997**, *16*, 2037.
724. Aime, S.; Barge, A.; Borel, A.; Botta, M.; Chemerisov, S.; Merbach, A. E.; Muller, U.; Pubanz, D. *Inorg. Chem.* **1997**, 36, 5104.
725. Inoue, M. B.; Inoue, M.; Fernando, Q. *Acta Crystallogr., Sect. C* **1994**, *50*, 1037.
726. Morrow, J. R.; Shelton, V. M. *New J. Chem.* **1994**, *18*, 371.
727. Amin, S.; Morrow, J. R.; Lake, C. H.; Churchill, M. R. *Angew. Chem., Int. Ed. Engl.* **1994**, *33*, 773.
728. Chin, K. O. A.; Morrow, J. R.; Lake, C. H.; Churchill, M. R. *Inorg. Chem.* **1994**, *33*, 656.
729. Georgiev, E. M.; Roundhill, D. M. *Inorg. Chim. Acta* **1997**, *258*, 93.
730. Beer, P. D.; Drew, M. G. B.; Ogden, M. I. *J. Chem. Soc., Dalton Trans.* **1997**, 1489.
731. Benson, M. T.; Cundari, T. F.; Saunders, L. C.; Sommerer, S. O. *Inorg. Chim. Acta* **1997**, *258*, 127.
732. Aime, S.; Ascenzi, P.; Comoglio, E.; Fasano, M.; Paolettii, S. *J. Am. Chem. Soc.* **1995**, *117*, 9365.
733. Sherry, A. D.; Ren, J.; Huskens, J.; Brucher, E.; Toth, E.; Geraldes, C. F. C. G.; Castro, M. M. C. A.; Cacheris, W. P. *Inorg. Chem.* **1996**, *35*, 4604.
734. Ren, J.; Sherry, A. D. *Inorg. Chim. Acta* **1996**, *246*, 331.
735. Reichert, D. E.; Hancock, R. D.; Welch, M. J. *Inorg. Chem.* **1996**, *35*, 7013.
736. Cundari, T. R.; Moody, E. W.; Sommerer, S. O. *Inorg. Chem.* **1995**, *34*, 5989.
737. Powell, D. H.; Dhubhghaill, O. M. N.; Pudanz, D.; Helm, L.; Lebedev, Y. S.; Schlaefer, W.; Merbach, A. E. *J. Am. Chem. Soc.* **1996**, *118*, 9333.
738. Ruloff, R.; Rainer, R.; Beyer, L. *Z. Anorg. Allg. Chem.* **1998**, *624*, 902.
739. Yamaguchi, K.; Inomata, Y.; Howell, F. S. *Kidorui* **1998**, *32*, 280 [Chem. Abs. **1998**, *129*, 239043].
740. Heppeler, A.; Froidevaux, S.; Macke, H. R.; Jermann, E.; Behe, M.; Powell, P.; Hennig, M. *Chem. Eur. J.* **1999**, *5*, 1974.
741. Reynolds, C. H.; Annan, N.; Beshah, K.; Huber, J. H.; Shaber, S. H.; Lenkinski, R. E.; Wortman, J. A. *J. Am. Chem. Soc.* **2000**, *122*, 8940.

742. Aime, S.; Barge, A.; Botta, M.; Frullano, L.; Merlo, U.; Hardcastle, K. I. *J. Chem. Soc., Dalton Trans.* **2000**, 3435.
743. Kang, J.-G.; Na, M.-K.; Yoon, S.-K.; Sohn, Y.; Kim, Y.-D.; Suh, I.-H. *Inorg. Chim. Acta.* **2000**, *310*, 56–64.
744. Rohovec, J.; Vojtíšek, P.; Hermann, P.; Ludvik, J.; Lukeš, I. *J. Chem. Soc., Dalton Trans.* **2000**, 141.
745. Ma, B.-Q.; Gao, S.; Bai, O.; Sun, H.-L.; Xu, G.-X. *J. Chem. Soc., Dalton Trans.* **2000**, 1003.
746. Szilágyi, E.; Brücher, E. *J. Chem. Soc., Dalton Trans.* **2000**, 2229.
747. Tei, L.; Baum, G.; Blake, A. J.; Fenske, D.; Schröder, M. *J. Chem. Soc., Dalton Trans.* **2000**, 2793.
748. Morss, L. R.; Nash, K. L.; Ensor, D. D. *J. Chem. Soc., Dalton Trans.* **2000**, 285.
749. Caravan, P.; Comuzzi, C.; Crooks, W.; McMurray, T. J.; Choppin, G. R.; Woulfe, S. R. *Inorg. Chem.* **2001**, *40*, 2170.
750. Parker, D.; Lowe, M. P.; Reany, O.; Aime, S.; Botta, M.; Castellano, G.; Gianolio, E. *J. Am. Chem. Soc.* **2001**, *123*, 7601.
751. Sherry, A. D. *J. Alloys Compd.* **1997**, *249*, 153.
752. Dickins, R. S.; Howard, J. A. K.; Lehmann, C. W.; Moloney, J.; Parker, D.; Peacock, R. D. *Angew. Chem., Int. Ed. Engl.* **1997**, *36*, 521.
753. Huskens, J.; Torres, D. A.; Kovacs, Z.; André, J. P.; Geraldes, C. F. G. C.; Sherry, A. D. *Inorg. Chem.* **1997**, *36*, 1495.
754. Lecomte, C.; Dahaoui-Gindrey, V.; Chollet, H.; Gros, C.; Miishra, A. K.; Barbette, F.; Pullumbi, P.; Guilard, R. *Inorg. Chem.* **1997**, *36*, 3827.
755. Spirlet, M.-R.; Rebizant, J.; Wang, X.; Jin, T.; Gilsoul, D.; Comblin, V.; Maton, F.; Muller, R. N.; Desreux, J. F. *J. Chem. Soc., Dalton Trans.* **1997**, 497.
756. Toth, E.; Pubanz, D.; Vauthey, S.; Helm, L.; Merbach, A. E. *Chem. Eur. J.* **1996**, *2*, 1607.
757. Parker, D.; Williams, J. A. G. *J. Chem. Soc., Dalton Trans.* **1996**, 3613.
758. Parker, D. *Coord. Chem. Rev.* **2000**, *205*, 109.
759. Lowe, M. P.; Parker, D. *Inorg. Chim. Acta* **2001**, *317*, 163.
760. Blair, S.; Lowe, M. P.; Mathieu, C. E.; Parker, D.; Senanayake, P. K.; Kataky, R. *Inorg. Chem.* **2001**, *40*, 5860.
761. Faulkner, S.; Beeby, A.; Carrié, M.-C.; Dadabhoy, A.; Kenright, A. M.; Sammes, P. G. *Inorg. Chem. Commun.* **2001**, *4*, 187.
762. Gunnlaugsson, T.; MacDonail, D. A.; Parker, D. *Chem. Commun.* **2000**, 93.
763. Beeby, A.; Dickins, R. S.; Fitzgerald, S.; Gowenlock, L. J.; Maupin, G. L.; Parker, D.; Riehl, J. P.; Siligiardi, G.; Williams, J. A. G. *Chem. Commun.* **2000**, 1183.
764. Reany, O.; Gunnlaugsson, T.; Parker, D. *Chem. Commun.* **2000**, 473.
765. Lowe, M. P.; Parker, D. *Chem. Commun.* **2000**, 797.
766. Blair, S.; Lowe, M. P.; Mathieu, C. E.; Parker, D.; Ksenanayake, P.; Kataky, R. *Inorg. Chem.* **2001**, *40*, 5860.
767. Dickins, R. S.; Gunnlaugsson, T.; Parker, D.; Peacock, R. D. *Chem. Commun.* **1998**, 1643.
768. Bruce, J. I.; Dickins, R. S.; Gunnlaugsson, T.; Lopinski, S.; Lowe, M. P.; Parker, D.; Peacock, R. D.; Perry, J. J. P.; Aime, S.; Botta, M. *J. Am. Chem. Soc.* **2000**, *122*, 9674.
769. Parker, D.; Senanayake, P. K.; Williams, J. A. G. *J. Chem. Soc., Perkin Trans. 2* **1998**, 2129.
770. Parker, D.; Williams, J. A. G. *Chem. Commun.* **1998**, 245.
771. Wong, C. P. *Inorg. Synth.* **1983**, *22*, 156.
772. Buchler, J. W.; De Cian, A.; Fischer, J.; Kihn-Botulinski, M.; Weiss, R. *Inorg. Chem.* **1988**, *27*, 339.
773. Buchler, J. W.; De Cian, A.; Fischer, J.; Kihn-Botulinski, M.; Paulus, H.; Weiss, R. *J. Am. Chem. Soc.* **1986**, *108*, 3652.
774. Spyroulias, G. A.; Coutsoleos, A. G.; Raptopoulou, C. P.; Terzis, A. *Inorg. Chem.* **1995**, *34*, 2476.
775. Schavieren, C. J.; Orpen, A. G. *Inorg. Chem.* **1991**, *30*, 4968.
776. Schavieren, C. J. *Chem. Commun.* **1991**, 458.
777. Foley, T. J.; Abboud, K. A.; Boncella, J. M. *Inorg. Chem.* **2002**, *41*, 1704.
778. Wong, W.-K.; Zhang, L.; Wong, W.-T.; Xue, F.; Mak, T. C. W. *J. Chem. Soc., Dalton Trans.* **1999**, 615.
779. Wong, W.-K.; Zhang, L.; Xue, F.; Mak, T. C. W. *J. Chem. Soc., Dalton Trans.* **1999**, 3053.
780. Wong, W.-K.; Zhang, L.; Xue, F.; Mak, T. C. W. *J. Chem. Soc., Dalton Trans.* **2000**, 2245.
781. Moussavi, M.; De Cian, A; Fischer, J.; Weiss, R. *Inorg. Chem.* **1988**, *27*, 1287.
782. Jiang, J. Z.; Hub, T. D.; Xie, J. L.; Zhang, J. Z. *Kidorui* **1998**, *32*, 278 [Chem. Abs. **1999**, *129*, 239042].
783. Tamiaki, H.; Matsumoto, N.; Tsukabe, H. *Tetrahedron Lett.* **1997**, *38*, 4235.
784. Spyroulias, G. A.; Despotoupoulos, A.; Raptopoulou, C. P.; Terzis, A.; Coutsolelos, A. G. *J. Chem. Soc., Chem. Commun.* **1997**, 782.
785. Lachgar, M.; Tabard, A.; Brandes, S.; Guilard, R.; Atmani, A.; De Cian, A.; Fischer, J.; Weiss, R. *Inorg. Chem.* **1997**, *36*, 4141.
786. Duchowski, J. K.; Bocian, D. F. *J. Am. Chem. Soc.* **1990**, *112*, 3312.
787. Duchowski, J. K.; Bocian, D. F. *J. Am. Chem. Soc.* **1990**, *112*, 8807.
788. Buchler, J. W.; Kihn-Botulinski, M.; Löffer, J.; Wicholas, M. *Inorg. Chem.* **1989**, *28*, 3770.
789. Babailov, S. P.; Coutsolelos, A. G.; Dikiy, A.; Spryoulias, G. A. *Eur. J. Inorg. Chem.* **2001**, 303.
790. Spyroulias, G. A.; de Montauzon, D.; Maisonat, A.; Poilblanc, R.; Coutsolelos, A. G. *Inorg. Chim. Acta* **1998**, *275–276*, 182.
791. Montalban, A. G.; Michel, S. L. J.; Baum, S. M.; Vesper, B. J.; White, A. J. P.; Williams, D. J.; Barrett, A. G. M.; Hoffman, B. M. *J. Chem. Soc., Dalton Trans.* **2001**, 3269.
792. M'Sadak, M.; Roncali, J.; Garnier, F. *J. Chim. Phys., Phys.-Chim. Biol.* **1986**, *83*, 211.
793. Koike, N.; Uekusa, H.; Ohashi, Y.; Harnoode, C.; Kitamura, F.; Ohsaka, T.; Tokuda, K. *Inorg. Chem.* **1996**, *35*, 5798.
794. Ostendorp, G.; Werner, J.-P.; Homberg, H. *Acta Crystallogr., Sect. C* **1995**, *51*, 1125.
795. Ostendorp, G.; Homberg, H. *Z. Naturforsch. B* **1995**, *50*, 1200.
796. Evans, W. J.; Sollberger, M. S.; Hanusa, T. P. *J. Am. Chem. Soc.* **1988**, *110*, 1841.
797. De Cian, A; Moussavi, M.; Fischer, J.; Weiss, R. *Inorg. Chem.* **1985**, *24*, 3162.
798. Ostendorp, G.; Homberg, H. *Z. Anorg. Allg. Chem.* **1996**, *622*, 1222.
799. Zelentsov, V. V. *Russ. J. Coord. Chem.* **1997**, *23*, 68.
800. Haghighi, M. S.; Franken, A.; Homberg, H. *Z. Naturforsch. B* **1994**, *49*, 812.
801. Guyon, F.; Pondaven, A.; Guenot, P.; L'Her, M. *Inorg. Chem.* **1994**, *33*, 4787.
802. Komatsu, T.; Ohta, K.; Fujimoto, T.; Yamamoto, I. *J. Mater. Chem.* **1994**, 533.
803. Guerek, A. G.; Ahsen, V.; Luneau, D.; Pecaut, J. *Inorg. Chem.* **2001**, *40*, 4793.
804. Guyon, F.; Pondaven, A.; L'Her, M. *J. Chem. Soc., Chem. Commun.* **1994**, 1125.

805. Janzak, J.; Kubiak, R. *Acta Crystallogr., Sect. C* **1995**, *51*, 2039.
806. Hückstädt, H.; Tutass, A.; Göldner, M.; Corneliessen, U.; Homberg, H. *Z. Anorg. Allg. Chem.* **2001**, *627*, 485.
807. Chabacch, D.; Tahiri, M. De Cian A.; Fischer, J.; Weiss, El Malouli Bibout M. *J. Am. Chem. Soc.* **1995**, *117*, 8548.
808. Jiang, J.; Liu, W.; Cheng, K.-L.; Poon, K.-W.; Ng, D. K. P. *Chem. Eur. J.* **2001**, 413.
809. Buchler, J. W.; De Cian, A; Fischer, J.; Kihn-Botulinski, M.; Paulus, H.; Weiss, R. *J. Am. Chem. Soc.* **1986**, *108*, 3652.
810. Jiang, J.; Lau, R. L. C.; Chan, T. W. D.; Mak, T. C. W.; Ng, D. K. P. *Inorg. Chim. Acta* **1997**, *255*, 59.
811. Jiang, J.; Liu, W.; Law, W.-F.; Lin, J.; Ng, D. K. P. *Inorg. Chim. Acta* **1998**, *268*, 141.
812. Jiang, J.; Choi, M. T. M.; Law, W.-F.; Chen, J.; Ng, D. K. P. *Polyhedron* **1998**, *17*, 3903.
813. Jiang, J.; Liu, W.; Law, W.-F.; Ng, D. K. P. *Inorg. Chim. Acta* **1998**, *268*, 49.
814. Gross, T.; Chevalier, F.; Lindsey, J. S. *Inorg. Chem.* **2001**, *40*, 4762.
815. Moussavi, M.; De Cian, A; Fischer, J.; Weiss, R. *Inorg. Chem.* **1986**, *25*, 2107.
816. Chabach, D.; Lachkar, M.; de Cian, A; Fischer, J.; Weiss, R. *New J. Chem.* **1992**, *16*, 431.
817. Spyroulias, G. A.; Coutsolelos, A. G.; Raptopoulou, C. P.; Terzis, A. *Inorg. Chem.* **1995**, *34*, 2476.
818. Jubb, J.; Gambarotta, S.; Duchateau, R.; Teuben, J. H. *J. Chem. Soc., Chem. Commun.* **1994**, 2641.
819. Campazzi, E.; Solari, E.; Floriani, C.; Scopelliti, R. *Chem. Commun.* **1998**, 2603.
820. Spyroulias, G. A.; Sioubara, M. P.; Coutsolelos, A. G. *Polyhedron* **1995**, *14*, 3563.
821. Sessler, J. L.; Tvermoes, N. A.; Davis, J.; Anzenbacher, Jr P.; Jursikova, K.; Sato, W.; Seidel, D.; Lynch, V.; Black, C. B.; Try, A.; Andrioletti, B.; Hemmi, G.; Mody, T. D.; Magda, D. J.; Kral, V. *Pure Appl. Chem.* **1999**, *71*, 2009.
822. Rosenthal, D. I.; Nurenberg, P.; Beccera, C. R.; Frenkel, E. P.; Carbonne, D. P.; Lum, B. L; Miller, R.; Engel, J.; You, S.; Miles, D.; Renschler, M. F. *Clin. Cancer Res.* **1999**, *5*, 739.
823. Young, S. W.; Quing, F.; Harriman, A.; Sessler, J. L.; Dow, W. C.; Mody, T. D.; Hemmi, G.; Hao, Y.; Miller, R. A. *Proc. Natl. Acad. Sci. USA* **1996**, *93*, 6610.
824. Timmerman, B.; Carde, P.; Koprowski, C.; Arwood, D.; Ford, J.; Mehta, M.; Tishler, R.; Larner, J.; Miller, R.; Koffler-Horovita, S.; Hoth, D.; Renschler, M. *Int. J. Radiat. Oncol. Biol. Phys.* **1998**, *42*, 198.
825. Sharman, W. M.; Allen, C. M.; van Lier, J. E. *Drug Design and Testing* **1999**, *4*, 507.
826. Adams, A. *Science* **1998**, *279*, 1307.
827. Rouhi, A. M. *Chem. Eng. News* Nov. 2, **1998**, 22.
828. Blumenkranz, M. S.; Woodburn, K. W.; Qing, F.; Verdooner, S.; Kessel, D.; Miller, R. *Am. J. Ophthalmol.* **2000**, *129*, 353.
829. Bilgin, M. D.; Al-Akhras, M.-A.; Khalil, M.; Hemmati, H.; Grossheimer, L. I. *Photochem. Photobiol.* **2000**, *72*, 121.
830. Viala, J.; Vanel, D.; Meignan, P.; Lartigau, E.; Carde, P.; Renschler, M. *Radiology* **1999**, *212*, 755.
831. Bernhard, E. J.; Mitchell, J. B.; Deen, D.; Cardell, M.; Rosenthal, D. I.; Martin, J. *Cancer Research* **2000**, *60*, 86.
832. Lisowski, J.; Sessler, J. L.; Lynch, V.; Mody, T. D. *J. Am. Chem. Soc.* **1995**, *117*, 2273.
833. Lisowski, J.; Sessler, J. L.; Mody, T. D. *Inorg. Chem.* **1995**, *34*, 4336.
834. Guldi, D. M.; Mody, T. D.; Gerasimchuk, N. N.; Magda, D.; Sessler, J. L. *J. Am. Chem. Soc.* **2000**, *122*, 8289.
835. Magda, D.; Crofts, S.; Lin, A.; Miles, D.; Wright, M.; Sessler, J. L. *J. Am. Chem. Soc.* **1997**, *119*, 2293.
836. Rogers, R. D.; Rollins, A. N.; Benning, M. M. *Inorg. Chem.* **1988**, *27*, 3826.
837. Rogers, R. D. *J. Coord. Chem.* **1988**, *16*, 415.
838. Mao, G. J.; Jin, Z.-S.; Ni, J.-Z. *Jiegou Huaxue* (Chinese J. Struct. Chem.), **1994**, *13*, 377.
839. Rogers, R. D.; Rollins, A. N.; Henry, R. F.; Murdoch, J. S.; Etzenhouser, R. D.; Huggins, S. E.; Nuñez, L. *Inorg. Chem.* **1991**, *30*, 4946.
840. Nuñez, L.; Rogers, R. D. *J. Crystallogr., Spec. Res.* **1992**, 22, 265.
841. Rogers, R. D.; Rollins, A. N. *J. Chem. Crystallogr.* **1994**, *24*, 531.
842. Mao, J.; Jin, ; Ni, J. *J. Coord. Chem.* **1995**, *230*, 195.
843. Hassaballa, H.; Steed, J. W.; Junk, P. C.; Elsegood, M. R. J. *Inorg. Chem.* **1998**, *37*, 4666.
844. Rogers, R. D.; Rollins, A. N.; Etzenhouser, R. D.; Voss, E. J.; Bauer, C. B. *Inorg. Chem.* **1993**, *32*, 3451.
845. Hassaballa, H.; Steed, J. W.; Junk, P. C.; J. Elsegood, M. R. *Inorg. Chem.* **1998**, *37*, 4666.
846. Rogers, R. D.; Rollins, A. N. *Inorg. Chim. Acta* **1995**, *230*, 177.
847. Runschke, C.; Meyer, G. *Z. Anorg. Allg. Chem.* **1997**, *623*, 983.
848. Runschke, C.; Meyer, G. *Z. Anorg. Allg. Chem.* **1997**, *623*, 1017.
849. Runschke, C.; Meyer, G. *Z. Anorg. Allg. Chem.* **1997**, *623*, 1493.
850. Runschke, C.; Meyer, G. *Z. Anorg. Allg. Chem.* **1998**, *623*, 1243.
851. Willey, G. R.; Lakin, M. T.; Alcock, N. W. *J. Chem. Soc., Dalton Trans.* **1993**, 3407.
852. Willey, G. R.; Meehan, P. R.; Salter, P. A.; Drew, M. G. B. *Polyhedon* **1996**, *15*, 4227.
853. Willey, G. R.; Meehan, P. R.; Rudd, M. D.; Clase, H. J.; Alcock, N. W. *Inorg. Chim. Acta* **1994**, *215*, 209.
854. Crisci, G.; Meyer, G. *Z. Anorg. Allg. Chem.* **1994**, *620*, 1023.
855. Mao, J.-G.; Jin, Z.-S.; Yu, F.-L. *Chinese J. Struct. Chem.* **1994**, *13*, 276.
856. Lu, T.; Peng, X.; Inoue, Y.; Ouchi, M.; Yu, K.; Ji, L. *J. Chem. Crystallogr.* **1998**, *28*, 197 [Chem. Abs. **1999**, *28*, 156017].
857. Rogers, R. D.; Kurihara, L. K. *Inorg. Chim. Acta* **1987**, *130*, 131.
858. Rogers, R. D.; Rollins, A. N. *J. Chem. Crystallogr.* **1994**, *24*, 321.
859. Li, Z. P.; Rogers, R. D. *J. Chem. Crystallogr.* **1994**, *24*, 415.
860. Jones, C.; Junk, P. C.; Smith, M. K.; Thomas, R. C. *Z. Anorg. Allg. Chem.* **2000**, *626*, 2491.
861. Mao, J.-G.; Wang, R.-Y.; Jin, Z.-S *Jiegou Huaxue (Chinese J. Struct. Chem.)* **1994**, *13*, 56.
862. Mao, J.-G.; Jin, Z.-S.; Ni, J.-Z.; Yu, L. *Polyhedron* **1994**, *13*, 313.
863. Fu, Z. Y.; Kong, F.; Qin, M.; Wang, B. H.; Zhao, B.; Miao, S. H. *J. Chem. Soc., Chem. Commun.* **1992**, 1753.
864. Rogers, R. D.; Etzenhouser, R. D.; Murdoch, J. S.; Reyes, E. *Inorg.Chem.* **1991**, *30*, 4946.
865. Rogers, R. D.; Zhang, J.; Bauer, C. B. *J. Alloys Compd.* **1997**, *249*, 41.
866. Ohno, H.; Saito, Y.; Yamase, T. *Chem. Lett.* **1997**, 213.
867. Chen, Q.; Chang, Y. D.; Zubieta, J. *Inorg. Chim. Acta* **1997**, *258*, 257.
868. Lu, T.; Ji, L.; Tan, M.; Liu, Y.; Yu, K. *Polyhedron* **1997**, *16*, 1149.
869. Aspinall, H. C.; Greeves, N.; Lee, W.-M.; McIver, E. G.; Smith, P. M. *Tetrahedron Lett.* **1997**, *38*, 4679.
870. Gu, J.; Hu, X.; Li, Q.; Chen, L. *Wuji Huaxue Xuebao* **1998**, *14*, 313 [Chem. Abs. **1999**, *129*, 283762].
871. Aspinall, H. C.; Greeves, N.; McIver, E. G. *J. Alloys Compd.* **1998**, *275–277*, 773.

872. Aspinall, H. C.; Dwyer, J. L. M.; Greeves, N.; McIver, E. G.; Woolley, J. C. *Organometallics* **1998**, *17*, 1884.
873. Bekiari, V.; Lianos, P. *Adv. Mater.* **1998**, *10*, 1455.
874. Erman, L. Y.; Mindrul, L. F.; Gal'perin, E. L.; Kurochkin, V. K.; Petunin, V. A. *Koord. Khim.* **1991**, *17*, 1286.
875. Rogers, R. D.; Henry, R. F. *J. Crystallogr., Spec. Res.* **1992**, *22*, 361.
876. Rogers, R. D.; Henry, R. F. *Acta Crystallogr., Sect. C* **1992**, *48*, 1099.
877. Rogers, R. D.; Rollins, A. N.; Etzenhouser, R. D.; Voss, E. J.; Bauer, C. B. *Inorg. Chem.* **1993**, *32*, 3451.
878. Moller, A.; Scott, N.; Meyer, G.; Deacon, G. B. *Z. Anorg. Allg. Chem.* **1999**, *625*, 181.
879. Ni, Z.; Lin, F.; Hu, C. *Gaodeng Xuexiao Huaxue* **1992**, *13*, 1349 [Chem. Abs. **1993**, *119*, 19228].
880. Benetollo, F.; Bombieri, G.; Samaria, K. M.; Vallarino, L. M. *J. Chem. Crystallogr.* **1996**, *28*, 9.
881. Benetollo, F.; Bombieri, G.; Vallarino, L. M. *Acta. Crystallogr., Sect. C* **1996**, *52*, 1190.
882. Benetollo, F.; Bombieri, G.; Depaoli, G.; Truter, M. R. *Inorg. Chim. Acta* **1996**, *245*, 223.
883. Lu, T.; Tan, M.; Su, H.; Liu, Y. *Polyhedron* **1993**, *12*, 1055.
884. Mao, J.-G.; Jin, Z.-S.; Ni, J.-Z. *Chinese J. Struct. Chem.* **1994**, *13*, 329.
885. Saleh, M. I.; Salhin, A.; Talipov, S.; Saad, B.; Fun, H-K.; Ibrahim, A. R. *Z. Krist.- New Cryst. Struct.* **1999**, *214*, 45.
886. Saleh, M. I.; Salhin, A.; Saad, B.; Fun, H.-K. *J. Mol. Struct* **1999**, *475*, 93.
887. Li, D.; Gan, X.; Tan, M.; Wang, X. *Polyhedron* **1997**, *23*, 3991.
888. Bu, X.-H.; Lu, S.-L.; Zhang, R.-H.; Wang, H.-G.; Yao, X.-K. *Polyhedron* **1997**, *16*, 3247.
889. Bombieri, G. *Inorg. Chim. Acta* **1987**, *139*, 21.
890. Yang, L.-W.; Liu, S.; Wong, E.; Rettig, S. J.; Orvig, C. *Inorg. Chem.* **1995**, *34*, 2164.
891. Yang, L.-W.; Liu, S.; Rettig, S. J.; Orvig, C. *Inorg. Chem.* **1995**, *34*, 4921.
892. Aguiari, A; Brianese, N.; Tamburini, S.; Vigato, P. A. *New J. Chem.* **1995**, *19*, 627.
893. De Maria Ramirez, F.; Sosa-Torres, M. E.; Castro, M.; Basurto-Uribe, E.; Zamorano-Ulloa, R.; del Rio-Portilla, F. *J. Coord. Chem.* **1997**, *41*, 303.
894. Benetollo, F.; Bombieri, G.; Vallarino, L. M. *Polyhedron* **1994**, *13*, 573.
895. Bligh, S. W. A.; Choi, N.; Evagoorou, E. G.; Li, W.-S.; McPartlin, M. *J. Chem. Soc., Chem. Commun.* **1994**, 2399.
896. Kasuga, K.; Moriguchi, T.; Yamada, K.; Hiroe, M.; Handa, M.; Sogabe, K. *Polyhedron* **1994**, *13*, 159.
897. Bligh, S. W. A.; Choi, N.; Evagorou, E. G.; McPartlin, M.; Cummins, W. J.; Kelly, J. D. *Polyhedron* **1992**, *11*, 2571.
898. Bandin, R.; Bastida, R.; de Bals, A.; Castro, P.; Fenton, D. E.; Macias, A.; Rodriguez, A.; Rodriguez, T. *J. Chem. Soc., Dalton Trans.* **1994**, 1185.
899. Lee, L.; Berg, D. J.; Einstein, F. W.; Batchelor, R. J. *Organometallics* **1997**, *16*, 1819.
900. Wang, X.-Q.; Yamg, W.-C.; Jin, T.-Z.; Xu, G.-X.; Ma, Z.-S.; Shi, N.-C. *Huaxue Xuebao* **1996**, *54*, 347 [Chem. Abs. **1996**, *125*, 47522].
901. Zucchi, G.; Scopelleti, R.; Bunzli, J.-C. G. *J. Chem. Soc., Dalton Trans.* **2001**, 1975.
902. Bligh, S. W. A; Choi, N.; Evegorou, E. G.; McPartlin, M.; White, K. N. *J. Chem. Soc., Dalton Trans.* **2001**, 3169.
903. Gómez-Tagle, P.; Yatsimirsky, A. K. *J. Chem. Soc., Dalton Trans.* **2001**, 2663.
904. Cassol, A.; Di Bernardo, P.; Pilloni, G.; Tolazzi, M.; Zanonato, P. L. *J. Chem. Soc., Dalton Trans.* **1995**, 2689.
905. Lopez, E.; Chypre, C.; Alpha, B.; Mathis, G. *Clin. Chem.* **1993**, *39*, 196.
906. Mathis, G. *Clin. Chem.* **1995**, *41*, 1391.
907. Lopez-Crapez, E.; Bazin, H.; Andre, E.; Noletti, J.; Grenier, J.; Mathis, G. *Nucleic Acids Research* **2001**, *29*(14), e70.
908. Bkouche-Waksman, I.; Guilhem, J.; Pascard, C.; Alpha, B.; Deschenaux, R.; Lehn, J.-M. *Helv. Chim. Acta* **1991**, *74*, 1163.
909. Arnaud, F. *Eur. J. Solid State Inorg. Chem.* **1991**, *28*, 229.
910. Saraidarov, T.; Reisfeld, R.; Pietraszkiewicz, M. *Chem. Phys. Lett* **2000**, *330*, 515.
911. Gawryszewska, P.; Jerzykiewicz, L. B.; Pietraszkiewicz, M.; Legendziewicz, J.; Riehl, J. P. *Inorg. Chem.* **2000**, *39*, 5365.
912. Gawryszewska, P.; Pietraszkiewicz, M.; Riehl, J. P.; Legendziewicz, J. *J. Alloys Compd.* **2000**, *300–301*, 283.
913. Faulkner, S.; Beeby, A.; Carrié, M.-C.; Dadabhoy, A.; M. Kenwright, A.; Sammes, P. G. *Inorg. Chem. Commun.* **2001**, *4*, 187.
914. Oh, S. J.; Song, K. H.; Park, J. W. *Chem. Commun.* **1995**, 575.
915. Oh, S. J.; Song, K. H.; Whang, D.; Yoon, T. H.; Moon, H.; Park, J. W. *Inorg. Chem.* **1996**, *35*, 3780.
916. Oh, S. J.; Yoon, C. W.; Park, J. W. *J. Chem. Soc., Perkin Trans. 2* **1996**, 329.
917. Oh, S. J.; Park, J. W. *J. Chem. Soc., Dalton Trans.* **1997**, 753.
918. Marolleau, I.; Gisselbrecht, J.-P.; Gross, M.; Arnaud-Neu, F.; Schwing-Weill, M.-J. *J. Chem. Soc., Dalton Trans.* **1989**, 367.
919. Avecilla, F.; Bastida, R.; de Blas, A.; Carrera, E.; Fenton, D. E.; Macias, A.; Platas, C.; Rodriguez, A.; Rodriguez-Blas, T. *Z. Naturforsch. Teil B* **1997**, *52*, 1273.
920. Platas, C.; Avecilla, F.; de Blas, A; Rodriguez-Blas, T.; Geraldes, C. F. G. C.; Tóth, E.; Merbach, A. E.; Bünzli, J.-C. G. *J. Chem. Soc., Dalton Trans.* **2000**, 611.
921. Bazzicalupi, C.; Bencini-Bianchi, A.; Giorgi, C.; Fusi, V.; Masotti, A.; Valtancoli, B.; Roque, A.; Pina, F. *Chem. Commun.* **2000**, 561.
922. Hardie, M. J.; Johnson, J. A.; Raston, C. L.; Webb, H. R. *Chem. Commun.* **2000**, 849.
923. Chen, Q.-Y.; Luo, Q.-H.; Wang, Z.-L.; Chen, J-T. *Chem. Commun.* **2000**, 1033.
924. Renaud, F.; Piguet, C.; Bernardinelli, G.; Hopfgartner, G.; Bunzli, J.-C. G. *Chem. Commun.* **1999**, 457.
925. Mao, J. G.; Jin, Z. S. *Polyhedron* **1994**, *13*, 319.
926. Drew, M. G. B.; Howarth, O. W.; Harding, C. J.; Martin, N.; Nelson, J. *J. Chem. Soc., Chem. Commun.* **1995**, 903.
927. Furphy, B. M.; Harrowfield, J. M.; Kepert, D. L.; Skelton, B. W.; White, A. H.; Wilner, F. R. *Inorg. Chem.* **1987**, *26*, 4231.
928. Harrowfield, J. M.; Ogden, M. I.; White, A. H. *Aust. J. Chem* **1991**, *44*, 1237.
929. Furphy, B. M.; Harrowfield, J. M.; Ogden, M. I.; Skelton, B. W.; White, A. H.; Wilner, F. R. *J. Chem. Soc., Dalton Trans.* **1989**, 2217.
930. Asfari, Z.; Harrowfield, J. M.; Ogden, M. I.; Vicens, J.; White, A. H. *Angew. Chem., Int. Ed. Engl.* **1991**, *30*, 1149.
931. Harrowfield, J. M.; Ogden, M. I.; Richmond, W. R.; White, A. H. *J. Chem. Soc., Dalton Trans.* **1991**, 2153.
932. Charbonnière, L. J.; Balsiger, C.; Schenk, K. J.; Bünzli, J.-C. G. *J. Chem. Soc., Dalton Trans.* **1998**, 505.
933. Bünzli, J.-C. G.; Ihringer, F.; Dumy, P.; Sager, C.; Rogers, R. D. *J. Chem. Soc., Dalton Trans.* **1998**, 497.
934. Thuéry, P.; Nierlich, M.; Vicens, J.; Takemura, H. *J. Chem. Soc., Dalton Trans.* **2000**, 279.
935. Beer, P. D.; Drew, M. G. B.; Grieve, A.; Kan, M.; Leeson, P. B.; Nicholson, G.; Ogden, M. I.; Williams, G. *J. Chem. Soc., Chem. Commun.* **1996**, 1117.
936. Beer, P. D.; Drew, M. G. B.; Kan, M.; Leeson, P. B.; Ogden, M. I.; Williams, G. *Inorg. Chem.* **1996**, *35*, 2202.

937. Beer, P. D.; Drew, M. G. B.; Leeson, P. B.; Ogden, M. I. *Inorg. Chim. Acta* **1996**, *246*, 133.
938. Beer, P. D.; Drew, M. G. B.; Grieve, A.; Ogden, M. I. *J. Chem. Soc., Dalton Trans.* **1995**, 3455.
939. Daitch, C. E.; Hampton, P. D.; Duesler, E. N.; Alam, T. M. *J. Am. Chem. Soc.* **1996**, *118*, 7769.
940. Dekmau, L. H.; Simon, N.; Schwing-Weill, M.-J.; Arnaud-Neu, F.; Dozol, J.-F.; Eymard, S.; Tournois, B.; Böhmer, V.; Grüttner, C.; Musigmann, C.; Tunayar, A. *Chem. Commun.* **1998**, 1627.
941. Ulrich, G.; Ziessel, R.; Manet, I.; Guardigli, M.; Sabbatini, N; Fraternali, F.; Wipff, G. *Chem. Eur. J.* **1997**, *3*, 1815 .
942. Matthews, S. E.; Parzuchowski, P.; Garcia-Carrera, A.; Grüttner, C.; Dozol, J-F.; Böhmer, V. *Chem. Commun.* **2001**, 417.
943. Bryant, L. H.; Yordanov, A. T.; Linnoila, J. J.; Brechbiel, M. W.; Frank, J. A. *Angew. Chem., Int. Ed. Engl.* **2000**, 39, 1641.
944. Karmazin, L.; Mazzanti, M.; Pécaut, J. *Chem. Commun.* **2002**, 654.
945. Kanda, T.; Ibi, M.; Mocizuki, K-I.; Kato, S. *Chem. Lett.* **1998**, 957.
946. Su, C.; Tang, N.; Tan, M.; Liu, W.; Gan, X. *Polyhedron* **1996**, *15*, 73.
947. Kobayashi, T.; Naruke, H.; Yamase, T. *Chem. Lett.* **1997**, 907.
948. Kobayashi, T.; Yamase, T. *Kidorui* **1997**, *30*, 166.
949. Su, C.; Tan, M.; Tang, N.; Gan, X.; Zhang, Z.; Xue, Q.; Yu, K. *Polyhedron* **1997**, *16*, 1643.
950. Su, C.; Zhou, Q.; Zhang, C.; Tan, B.; Kang, B. *Zhongshan Daxue Xuebao, Ziran Kexueban* **1997**, *36*, 131 [Chem. Abs. **1997**, *127*, 305231].
951. Aspinall, H. C.; Cunningham, S. A.; Maestro, P.; Macaudiere, P. *Inorg. Chem.* **1998**, *37*, 5396.
952. Melman, J. H.; Emge, T. J.; Brennan, J. G. *Inorg. Chem.* **2001**, *40*, 1078.
953. Mclman, J. H.; Rohde, C.; Emge, T. J.; Brennan, J. G. *Inorg. Chem.* **2002**, *41*, 28.
954. Mashima, K.; Nakayama, Y.; Fukumoto, H.; Kanehisa, N.; Kai, Y.; Nakamura, A. *J. Chem. Soc., Chem. Commun.* **1994**, 2523.
955. Lee, J.; Freedman, D.; Melman, J. H.; Brewer, M.; Sun, L.; Emge, T. J.; Long, F. H.; Brennan, J. G. *Inorg. Chem.* **1998**, *37*, 2512.
956. Purdy, A. P.; Berry, A. D.; George, C. F. *Inorg. Chem.* **1997**, *36*, 3370.
957. Melman, J. M.; Emge, T. J.; Brennan, J. G. *Chem. Commun.* **1997**, 2268.
958. Melman, J. H.; Emge, T. J.; Brennan, J. G. *Inorg. Chem.* **1999**, *38*, 2117.
959. Freedman, D.; Melman, J. H.; Emge, T. J.; Brennan, J. G. *Inorg. Chem.* **1998**, *37*, 4162.
960. Lee, J.; Brewer, M.; Berardini, M.; Brennan, J. G. *Inorg. Chem.* **1995**, *34*, 3215.
961. Geissinger, M.; Magull, J. *Z. Anorg. Allg. Chem.* **1995**, *621*, 2043.
962. Freeman, D.; Emge, T. J.; Brennan, J. G. *J. Am. Chem. Soc.* **1997**, *119*, 11112.
963. Freedman, D.; Emge, T. J.; Brennan, J. G. *Inorg. Chem.* **1999**, *38*, 4400.
964. Cary, D. R.; Ball, G. E.; Arnold, J. *J. Am. Chem. Soc.* **1995**, *117*, 3492.
965. Berardini, M.; Lee, J.; Freedman, D.; Lee, J.; Emge, T. J.; Brennan, J. G. *Inorg. Chem.* **1997**, *36*, 5772.
966. Berardini, M.; Brennan, J. G. *Inorg. Chem.* **1995**, *34*, 6179.
967. Mashima, K.; Shibahara, T.; Nakayama, Y.; Nakamura, A. *Inorg. Chem.* **1995**, *34*, 263.
968. Berardini, M.; Emge, T. J.; Brennan, J. G. *Inorg. Chem.* **1995**, *34*, 5327.
969. Kornienko, A.; Melman, J. H.; Hall, G.; Emge, T. J.; Brennan, J. G. *Inorg. Chem.* **2002**, *41*, 121.
970. Goyunov, A. V.; Popov, A. I.; Khaidukov, N. M.; Fedorov, P. P. *Mater. Res. Bull* **1992**, *27*, 213.
971. Plitzko, C.; Meyer, G. *Z. Anorg. Allg. Chem.* **1997**, *623*, 1393.
972. le Fur, Y.; Khaidukov, N. M.; Leonard, S. A. *Acta Crystallogr., Sect. C* **1992**, *48*, 978.
973. Goyunov, A. V.; Popov, A. I. *Zh. Neorg. Chem.* **1992**, *37*, 276.
974. Wickleder, M. S.; Meyer, G. *Z. Anorg. Allg. Chem* **1995**, *621*, 546.
975. Wickleder, M. S.; Meyer, G. *Z. Anorg. Allg. Chem.* **1995**, *621*, 740.
976. Masselmann, S.; Meyer, G. *Z. Anorg. Allg. Chem.* **1998**, *624*, 357.
977. Wickleder, M. S.; Meyer, G. *Z. Anorg. Allg. Chem.* **1996**, *622*, 593.
978. Villafuerte-Castrejón, M. E.; Estrada, M. R.; Gómez-Lara, J.; Duque, J.; Pomés, R. *J. Solid State Chem.* **1997**, *132*, 1.
979. Reuter, G.; Roffe, M.; Frenzen, G. *Z. Anorg. Allg. Chem.* **1995**, *621*, 630.
980. Reuter, G.; Frenzen, G. *J. Solid State Chem.* **1995**, *116*, 329.
981. Oppermann, H.; Huong, D. Q. *Z. Anorg. Allg. Chem.* **1995**, *621*, 665.
982. Rossmanith, K.; Unfried, P. *Monatsh. Chem.* **1995**, *126*, 687.
983. Mattfeld, H.; Meyer, G. *Z. Anorg. Allg. Chem.* **1992**, *618*, 13.
984. Steiner, H. J.; Lutz, H. D. *Z. Anorg. Allg. Chem.* **1992**, *613*, 26.
985. Czjek, M.; Fuess, H.; Pabst, I. *Z. Anorg. Allg. Chem.* **1992**, *617*, 105.
986. Masselmann, S.; Meyer, G. *Z. Kristallogr.- New Cryst. Struct.* **1998**, *213*, 690.
987. Reuter, G.; Sebastian, J.; Frenzen, G. *Acta Crystallogr., Sect. C* **1996**, *52*, 1859.
988. Roffe, M.; Seifert, H. J. *J. Alloys Compd.* **1997**, *257*, 128.
989. Oppermann, H.; Morgenstern, A.; Ehrlich, S. *Z. Naturforsch. Teil B* **1997**, *52*, 1062.
990. Ehrlich, S.; Oppermann, H.; Hennig, C. *Z. Naturforsch. Teil B* **1997**, *52*, 311.
991. Becker, A.; Uhrland, W. *Z. Anorg. Allg. Chem.* **1999**, *625*, 217.
992. Becker, A.; Uhrland, W. *Z. Anorg. Allg. Chem.* **1999**, *625*, 1033.
993. Schleid, T.; Meyer, G. *Z. Krist.* **1995**, *210*, 145.
994. Güdel, H. U.; Furrer, A.; Blank, H. *Inorg. Chem.* **1990**, *29*, 4081.
995. Gaune, P.; Gaune-Escard, M.; Rycerz, L.; Bogacz, A. *J. Alloys Compd.* **1996**, *235*, 143.
996. Bohnsack, A.; Meyer, G. *Z. Krist.* **1996**, *211*, 327.
997. Bohnsack, A.; Balzer, G.; Wickleder, M.; Gudel, H. U.; Meyer, G. *Z. Anorg. Allg. Chem.* **1997**, *623*, 1352.
998. Bohnsack, A.; Meyer, G. *Z. Krist.* **1997**, *212*, 2.
999. Crisci, G.; Meyer, G. *Z. Anorg. Allg. Chem.* **1998**, *624*, 927.
1000. Eaborn, C.; Hitchcock, P. B.; Izod, K.; Smith, J. D. *J. Am. Chem. Soc.* **1994**, *116*, 12071.
1001. Eaborn, C.; Hitchcock, P. B.; Izod, K.; Lu, Z.-R.; Smith, J. D. *Organometallics* **1996**, *15*, 4783.
1002. Makarov, V. M.; Bochkarev, L. N.; Dumkina, E. V.; Zhil'tsov, S. F. *Russ. J. General Chem.* **1999**, *69*, 88.
1003. Heckmann, G.; Niemeyer, M. *J. Am. Chem. Soc.* **2000**, *122*, 4227.
1004. Niemeyer, M. *Acta Crystallogr., Sect. E* **2001**, *57*, m578.
1005. Forsyth, C. M.; Deacon, G. B. *Organometallics* **2000**, *19*, 1205.

1006. Klimpel, M. G.; Anwander, R.; Tafipolsky, M.; Scherer, W. *Organometallics* **2001**, *20*, 3983.
1007. Evans, W. J.; Rabe, G. W.; Ziller, J. W. *Inorg. Chem.* **1994**, *33*, 3072.
1008. Maunder, G. H.; Sella, A. *Polyhedron* **1998**, *17*, 63.
1009. Deacon, G. B.; Forsyth, C. M.; Wilkinson, D. L. *Eur. Chem. J.* **2001**, *7*, 1784.
1010. Evans, W. J.; Drummond, D. K.; Zhang, H.; Atwood, J. L. In *Synthetic Methods of Organometallic and Inorganic Chemistry*; Edelman, F. T, Ed.; Georg. Thieme: Stuttgart, 1997; Vol. 6, p 28.
1011. Deacon, G. B.; Fallon, G. D.; Forsyth, C. M.; Schumann, H.; Weimann, R. *Chem. Ber. Recl.* **1997**, *130*, 409.
1012. Evans, W. J.; Drummond, D. K.; Zhang, H.; Atwood, J. L. *Inorg. Chem.* **1988**, *27*, 575.
1013. Nagl, I.; Scherer, W.; Tafipolsky, M.; Anwander, R. *Eur. J. Inorg. Chem.* **1999**, 1405.
1014. Evans, W. J.; Anwander, R.; Ansari, M. A.; Ziller, J. W. *Inorg. Chem.* **1995**, *34*, 5.
1015. Cosgriff, J. E.; Deacon, G. B.; Gatehouse, B. M.; Lee, P. R.; Schumann, H. *Z. Anorg. Allg. Chem.* **1996**, *622*, 1399.
1016. Cosgriff, J. E.; Deacon, G. B.; Fallon, G. D.; Gatehouse, B. M.; Schumann, H.; Weimann, R. *Chem. Ber.* **1996**, *129*, 953.
1017. Deacon, G. B.; Delbridge, E. E.; Skelton, B. W.; White, A. H. *Eur. J. Inorg. Chem.* **1998**, 543.
1018. Cosgriff, J. E.; Deacon, G. B.; Gatehouse, B. M. *Aust. J. Chem.* **1993**, *46*, 1881.
1019. Cosgriff, J. E.; Deacon, G. B.; Gatehouse, B. M.; Hemling, H.; Schumann, H. *Aust. J. Chem.* **1994**, *47*, 1223.
1020. Deacon, G. B.; Gitlits, A.; Skelton, B. W.; White, A. H. *Chem. Commun.* **1999**, 1213.
1021. Deacon, G. B; Forsyth, C. M.; Gatehouse, B. M; White, A. H. *Aust. J. Chem.* **1990**, *43*, 1.
1022. Abrahams, C. T.; Deacon, G. B.; Gatehouse, B. M.; Ward, G. N. *Acta Crystallogr., Sect. C* **1994**, *50*, 504.
1023. Evans, W. J.; Rabe, G. W.; Ziller, J. W. *Organometallics* **1994**, *13*, 1641.
1024. Shah, S. A. A.; Dorn, H.; Roesky, H. W.; Lubini, P.; Schmidt, H.-G. *Inorg. Chem.* **1997**, *36*, 1102.
1025. Wedler, M.; Noltemeyer, M.; Schmidt, H.-G.; Pieper, U.; Stalke, D.; Edelmann, F. T. *Angew. Chem., Int. Ed. Engl.* **1990**, *29*, 894.
1026. Onoshi, M.; Itoh, K.; Hiraki, K. *Nagashi Daigakubu Kogakubu Kenkyu Hohoku* **1997**, *27*, 167 [Chem. Abs. **1997**, *126*, 180400].
1027. Onishi, M.; Itoh, K.; Hiraki, K.; Oda, R.; Aoki, K. *Inorg. Chim. Acta* **1998**, *277*, 8.
1028. Moss, M. A. J.; Kresinski, R. A.; Jones, C. J.; Evans, W. J. *Polyhedron* **1993**, *12*, 1953.
1029. Takats, J.; Zhang, X. W.; Day, V. W.; Eberspacher, T. A. *Organometallics* **1993**, *12*, 4286.
1030. Carvalho, A.; Domingos, A.; Isolani, P. vC.; Marques, N.; de Matos, A. vP.; Vincentini, G. *Polyhedron* **2000**, *19*, 1707.
1031. Takats, J. *J. Alloys Compd.* **1997**, *249*, 52.
1032. Hillier, A. C.; Liu, S. Y.; Sella, A.; Elsegood, M. R. J. *Angew. Chem., Int. Ed. Engl.* **1999**, *38*, 2745.
1033. Hillier, A. C.; Liu, S. Y.; Sella, A.; Elsegood, M. R. J. *Inorg. Chem.* **2000**, *39*, 2635.
1034. Moss, M. A. J.; Kresinski, R. A.; Jones, C. J.; Evans, W. J. *Polyhedron* **1993**, *12*, 1953.
1035. Zhang, X.; Loppnow, G. R.; McDonald, R.; Takats, J. *J. Am. Chem. Soc.* **1995**, *117*, 7828.
1036. Hasinoff, L.; Takats, J.; Zhang, X. W.; Bond, A. H.; Rogers, R. D. *J. Am. Chem. Soc.* **1994**, *116*, 8833.
1037. Zhang, X.; McDonald, R.; Takats, J. *New J. Chem.* **1995**, *19*, 573.
1038. Ferrence, G. M.; McDonald, R.; Takats, J. *Angew. Chem., Int. Ed. Engl.* **1999**, *38*, 2233.
1039. Ferrence, G. M.; Takats, J. *J. Organomet. Chem.* **2002**, *647*, 84.
1040. Shipley, C. P.; Capecchi, S.; Salata, O. V.; Etchells, M.; Dobson, P. J.; Christou, V. *Adv. Mater.* **1999**, *11*, 533.
1041. Dube, T.; Concini, S.; Gambarotta, S.; Yap, G. P. A.; Vasapollo, G. *Angew. Chem., Int. Ed. Engl.* **2000**, *38*, 3657.
1042. Rabe, G. W.; Yap, G. P. A.; Rheingold, A. L. *Inorg. Chem.* **1995**, *34*, 4521.
1043. Atlan, S.; Nief, F.; Ricard, L. *Bull. Soc. Chim. Fr.* **1995**, *132*, 649.
1044. Rabe, G. W.; Riede, J.; Schier, A. *Main Group Chem.* **1996**, *1*, 273 [Chem. Abs. **1996**, *125*, 184109].
1045. Nief, F.; Ricard, L. *J. Organometallic Chem.* **1997**, *529*, 357.
1046. Rabe, G. W.; Guzei, I. A.; Rheingold, A. L. *Inorg. Chem.* **1997**, *36*, 4924.
1047. Rabe, G. W.; Yap, G. P. A.; Rheingold, A. L. *Inorg. Chem.* **1997**, *36*, 3212.
1048. Rabe, G. W.; Yap, G. P. A.; Rheingold, A. L. *Inorg. Chim. Acta* **1998**, *267*, 309.
1049. Clegg, W.; Izod, K.; Liddle, S. T. *J. Organomet. Chem.* **2000**, *613*, 128.
1050. Rabe, G. W.; Riede, J.; Schier, A. *Organometallics* **1996**, *15*, 439.
1051. Izod, K.; O'Shaughnessy, P.; Sheffield, J. M.; Clegg, W.; Liddle, S. T. *Inorg. Chem.* **2000**, *39*, 4741.
1052. Caravan, P.; Merbach, A. E. *J. Chem. Soc., Chem. Commun.* **1997**, 2147.
1053. Lasocha, W. *J. Solid State Chem.* **1995**, *114*, 308.
1054. Lossin, A.; Meyer, G. *Z. Anorg. Allg. Chem.* **1992**, *614*, 12.
1055. Hanamoto, T.; Sugimoto, Y.; Sugino, A.; Inaga, J. *Synlett.* **1994**, 377.
1056. Molander, G. A. *Chem. Rev.* **1992**, *92*, 29.
1057. Molander, G. A.; Harris, C. R. *Chem. Rev.* **1996**, *96*, 307.
1058. Imamoto, T. *Lanthanides in Organic Synthesis*; Academic Press: London 1994.
1059. Kagan, H. B.; Sasaki, M.; Collin, J. *Pure Appl. Chem.* **1988**, *60*, 1725.
1060. Kagan, H. B.; Namy, J. L. *Tetrahedron* **1986**, *42*, 6573.
1061. Namy, J. L.; Girard, P.; Kagan, H. B.; Caro, P. E. In *Synthetic Methods of Organometallic and Inorganic Chemistry*; Herrman, W. A., Ed.; Thieme Verlag: Stuttgart, 1997; Vol. 6, p 26.
1062. Watson, P. L.; Tulip, T. H.; Williams, I. In *Synthetic Methods of Organometallic and Inorganic Chemistry*; Herrman, W. A., Ed.; Thieme Verlag: Stuttgart, 1997; Vol. 6, p 27.
1063. Watson, P. L.; Tulip, T. H.; Williams, I. *Organometallics* **1990**, *9*, 1999.
1064. Evans, W. J.; Gummersheimer, T. S.; Ziller, J. W. *J. Am. Chem. Soc.* **1995**, *117*, 8999.
1065. Hou, Z.; Zhang, Y.; Wakatsuki, Y. *Bull. Chem. Soc. Jpn.* **1997**, *70*, 149.
1066. Sen, A.; Chebolu, V.; Holt, E. M. *Inorg. Chim. Acta* **1986**, *118*, 87.
1067. Nishiura, M.; Katagiri, K.; Imamoto, T. *Bull. Chem. Soc. Jpn.* **2001**, *74*, 1417.
1068. Chebolu, V. R.; Whittle, R.; Sen, A. *Inorg. Chem.* **1985**, *24*, 3082.
1069. Chebolu, V. R.; Whittle, R.; Sen, A. *Inorg. Chem.* **1987**, *26*, 1821.
1070. Hakonsson, M.; Vestergren, V.; Gustafsson, B.; Hilmersson, G. *Angew. Chem., Int. Ed. Engl.* **1999**, *38*, 2199.
1071. Mandel, A.; Magull, J. *Z. Anorg. Allg. Chem.* **1997**, *623*, 1542.
1072. White, J. P.; Deng, H.; Boyd, E. P.; Gallucci, J.; Shore, S. G. *Inorg. Chem.* **1994**, *33*, 1685.
1073. Bochkarev, M.; Fagin, A. A. *Chem. Eur. J.* **1999**, *5*, 2990.
1074. Bochkarev, M. N.; Fedushkin, I. L.; Dechert, S.; Fagin, A. A.; Schumann, H. *Angew. Chem., Int. Ed. Engl.* **2001**, *40*, 3176.

1075. Bochkarev, M. N.; Fedushkin, I. L.; Fagin, A. A.; Petrovskaya, T. V.; Ziller, J. W.; Broomhall-Dillard, R. N. R.; Evans, W. J. *Angew. Chem., Int. Ed. Engl.* **1997**, *36*, 133.
1076. Fedushkin, I. L.; Weydert, M.; Fagin, A. A.; Nefedov, S. E.; Eremenko, I. L.; Bochkarev, M. N.; Schumann, H. *Z. Naturforsch. Teil B* **1999**, *54*, 4661032.
1077. Evans, W. J.; Shreeve, J. L.; Ziller, J. W. *Inorg. Chem.* **1994**, *33*, 6435.
1078. Starynowicz, P.; Bukietynska, K. *Eur. J. Inorg. Chem.* **2002**, 1827.
1079. Starynowicz, P. *J. Alloys Compd.* **1998**, *269*, 67.
1080. Burai, L.; Tóth, E.; Seibig, S.; Scopelliti, R.; Merbach, A. E. *Chem. Eur. J.* **2000**, *6*, 3761.
1081. Seibig, S.; Tóth, E.; Merbach, A. E. *J. Am. Chem. Soc.* **2000**, *122*, 5822.
1082. Burai, L.; Tóth, E.; Seibig, S.; Scopelliti, R.; Merbach, A. E. *Chem. Eur. J.* **2000**, *6*, 3761.
1083. van den Hende, J. R.; Hitchcock, P. B.; Holmes, S. A.; Lappert, M. F.; Leung, W.-P.; Mak, T. C. W.; Lappert, M. F. *J. Chem. Soc., Dalton Trans.* **1995**, 1427.
1084. van den Hende, J. R.; Hitchcock, P. B.; Holmes, S. A.; Lappert, M. F. *J. Chem. Soc., Dalton Trans.* **1995**, 1435.
1085. Evans, W. J.; Anwander, R.; Ansari, M. A.; Ziller, J. W. *Inorg. Chem.* **1995**, *34*, 5.
1086. Hou, Z.; Fujita, A.; Yoshimura, T.; Jesorka, A.; Zhang, Y.; Yamazaki, H.; Wakatsuki, Y. *Inorg. Chem.* **1996**, *35*, 7190.
1087. Evans, W. J.; Greci, M. A.; Ziller, J. W. *Inorg. Chem.* **1998**, *37*, 5221.
1088. Evans, W. J.; Greci, M. A.; Ziller, J. W. *J. Chem. Soc. Dalton Trans.* **1997**, 3035.
1089. Evans, W. J.; Greci, M. A.; Ziller, J. W. *Inorg. Chem.* **2000**, *39*, 3213.
1090. Deacon, G. B.; Forsyth, C. M.; Junk, P. C.; Skelton, B. W.; White, A. H. *Chem. Eur. J.* **1999**, *5*, 1452.
1091. Evans, W. J.; McClelland, W. G.; Greci, M. A.; Ziller, J. W. *Eur. J. Solid State Inorg. Chem.* **1996**, *33*, 145.
1092. Qi, G.-Z; Shen, Q.; Lin, Y.-H. *Acta Crystallogr., Sect. C* **1994**, *50*, 1456.
1093. Yuan, F. G.; Shen, Q. *Chinese Chem. Lett.* **1997**, *8*, 639.
1094. van den Hende, J. R.; Hitchcock, P. B.; Lappert, M. F. *J. Chem. Soc., Chem. Commun.* **1994**, 1413.
1095. Duncalf, D. J.; Hitchcock, P. B.; Lawless, G. A. *J. Organomet. Chem.* **1996**, *506*, 347.
1096. Niemeyer, M. *Eur. J. Inorg. Chem.* **2001**, 1969.
1097. Niemeyer, M. *Acta Crystllogr., Sect. E* **2001**, *57*, m396.
1098. Lee, J.; Emge, T. J.; Brennan, J. G. *Inorg. Chem.* **1997**, *36*, 5064.
1099. Freedman, D.; Syan, S.; Emge, T. J.; Croft, M.; Brennan, J. G. *J. Am. Chem. Soc.* **1999**, *121*, 11713.
1100. Khansis, D. V.; Brewer, M.; Lee, J.; Emge, T. J.; Brennan, J. G. *J. Am. Chem. Soc.* **1994**, *116*, 7129.
1101. Cary, D. R.; Arnold, J. *Inorg. Chem.* **1994**, *33*, 1791.
1102. Strzelecki, A. R.; Likar, C. L.; Helsel, B. A.; Utz, T.; Lin, M. C.; Bianconi, P. A. *Inorg. Chem.* **1994**, *33*, 5188.
1103. Miller, G.; Smith, M.; Wang, M.; Wang, S. *J. Alloys Compd.* **1998**, *265*, 140.
1104. Wang, M. T.; Wang, S. H. *J. Rare Earths* **1997**, *15*, 246.
1105. Ling, C.; Wang, M.; Wang, S. *J. Alloys Compd.* **1997**, *256*, 112.
1106. Meitian, W.; Shihua, W. *J. Solid State Chem.* **1997**, *128*, 66.
1107. Schilling, G.; Meyer, G. *Z. Anorg. Allg. Chem.* **1996**, *622*, 759.
1108. Schilling, G.; Meyer, G. *Z. Krist.* **1996**, *211*, 255.
1109. Schleid, T.; Meyer, G. *Z. Krist.* **1995**, *210*, 144.
1110. Eisenstein, O.; Hitchcock, P. B.; Hulkes, A. G.; Lappert, M. F.; Maron, L. *Chem. Commun.* **2001**, 1560.
1111. Morton, C.; Alcock, N. W.; Lees, M. R.; Munslow, I. J.; Sanders, C. J.; Scott, P. *J. Am. Chem. Soc.* **1999**, *121*, 11255.
1112. Chan, G. Y. S.; Drew, M. G. B.; Hudson, M. J.; Isaacs, N. S.; Byers, P.; Madic, C. *Polyhedron* **1996**, *15*, 3385.
1113. Buchler, J. W.; Nawra, M. *Inorg. Chem.* **1994**, *33*, 2830.
1114. Buchler, J. W.; Eiermann, V.; Hanssum, H.; Heinz, G.; Rüterjans, H.; Schwarzkopf, M. *Chem. Ber.* **1994**, *127*, 589.
1115. Davoras, E. M.; Spyroulias, G. A.; Mikros, E.; Coutsolelos, A. G. *Inorg. Chem.* **1994**, *33*, 3430.
1116. Ostendorp, G.; Rotter, H. W.; Homberg, H. *Z. Naturforsch. Teil. B* **1996**, *51*, 567.
1117. Hagighi, M. S.; Homborg, H. *Z. Naturforsch. Teil. B* **1991**, *16*, 1641.
1118. Berthet, J. C.; Lance, M.; Nierlich, M.; Ephritikhine, M. *Eur. J. Inorg. Chem.* **2000**, 1969.
1119. Pokol, G.; Leskelae, T.; Niinsto, L. *J. Therm. Anal.* **1995**, *42*, 343.
1120. Guillou, N.; Louer, M.; Auffredic, J.-P.; Louer, D. *Eur. J. Solid State Inorg. Chem.* **1995**, *32*, 35.
1121. Guillou, N.; Louer, M.; Auffredic, J-P.; Louer, D. *Acta Crytsallogr., Sect. C* **1995**, *51*, 1029.
1122. Levason, W.; Oldroyd, R. D. *Polyhedron* **1996**, *15*, 409.
1123. Baxter, I.; Darr, J. A.; Hursthouse, M. B.; Malik, K. M. A.; Mingos, D. M. P.; Plakatouras, J. C. *J. Chem. Crystallogr.* **1998**, *28*, 267.
1124. Troyanov, S. I.; Moroz, S. A.; Pechurova, N. I.; Snezkho, N. I. *Koord. Khim.* **1993**, *18*, 1207.
1125. Gradeef, P. S.; Schreiber, F. G.; Mauermann, H. *J. Less Common Met.* **1986**, *126*, 335.
1126. Gradeef, P. S.; Yunlu, K.; Gleizes, A.; Galy, J. *Polyhedron* **1989**, *8*, 1001.
1127. Sirio, C.; Hubert-Pfalzgraf, L. G.; Bois, C. *Polyhedron* **1997**, *16*, 1129.
1128. Hubert-Pfalzgraf, L. G.; Abada, V.; Vaissermann, J. *J. Chem. Soc., Dalton Trans.* **1998**, 3437.
1129. Hubert-Pfalzgraf, L. G.; Sirio, C.; Bois, C. *Polyhedron* **1998**, *17*, 821.
1130. Evans, W. J.; Deming, T. J.; Olofson, J. M.; Ziller, J. W. *Inorg. Chem.* **1989**, *28*, 4027.
1131. Evans, W. J.; Deming, T. J.; Ziller, J. W. *Organometallics* **1989**, *8*, 1581.
1132. Evans, W. J.; Edinger, L. A.; Ziller, J. W. *Polyhedron* **1999**, *18*, 1475.
1133. Patwe, S. J.; Wani, B. N.; Rao, U. K.; Venkateswarlu, K. S. *Can. J. Chem.* **1989**, *67*, 1815.
1134. Largeau, E.; El-Ghozzi, M.; Metin, J.; Avignant, D. *Acta Crystallogr., Sect. C* **1997**, *53*, 530.
1135. Largeau, E.; Gaumet, V.; El-Ghozzi, M.; Avignant, D.; Cousseins, J. C. *J. Mater. Chem.* **1997**, *7*, 1881.

Comprehensive Coordination Chemistry II
ISBN (set): 0-08-0437486

Volume 3, (ISBN 0-08-0443257); pp 93–188

3.3
The Actinides

C. J. BURNS, M. P. NEU, and H. BOUKHALFA
Los Alamos National Laboratory, NM, USA
and
K. E. GUTOWSKI, N. J. BRIDGES, and R. D. ROGERS
The University of Alabama, Tuscaloosa, AL, USA

3.3.1   INTRODUCTION

3.3.1.1   Historical Development of Actinide Coordination Chemistry

Investigation of the coordination chemistry of the actinide elements began with the isolation of uranium from a pitchblende sample in 1789;[1] thorium was similarly isolated from thorite mineral samples in 1829.[2] The chemistry of these elements remained somewhat obscure until the discovery in 1895 by Becquerel that uranium undergoes radioactive decay. Interest in the chemistry (and more specifically the nuclear chemistry) of the earliest actinide elements blossomed (including protactinium),[3] as scientists sought to understand and systematize the chemical and radioactive properties of naturally occurring radioactive elements. The discovery of artificial radioactivity in 1934 proved to be the next watershed development in the history of the actinide elements, as the promise was offered of truly synthesizing new elements not previously found in nature. The development of man-made elements began in 1940 with the production of neptunium.[4]

Coordination chemistry (in the form of descriptive chemistry) played a key role in this discovery phase. The chemical properties of the elements (redox characteristics and stoichiometry of simple compounds such as oxides and halides) were often used to argue for their placement in the periodic table. It was for this reason that a series of compounds was first proposed in which the 5*f*-orbitals were being successively populated. The first actinides were considered to be members of a new *d*-transition series, until it became clear that their chemical properties did not mimic those of their supposed cogeners. Although thorium is chemically similar to group 4 elements, and uranium can bear some similarity to group 6 metals, neptunium did not share many similarities with group 7 elements, and plutonium bore no resemblance to osmium or other group 8 elements. Seaborg first developed the hypothesis that the actinide elements actually constituted a second "*f*-transition series," analogous to the lanthanide elements. With this recognition, the search began to populate all positions in the series. Table 1 provides a listing of the manmade actinide elements, along with their dates of first synthesis.

The discovery of nuclear fission in 1938 proved the next driver in the development of coordination chemistry. Uranium-235 and plutonium-239 both undergo fission with slow neutrons, and can support neutron chain reactions, making them suitable for weaponization in the context of the Manhattan project. This rapidly drove the development of large-scale separation chemistry, as methods were developed to separate and purify these elements. While the first recovery processes employed precipitation methods (e.g., the bismuth phosphate cycle for plutonium isolation),

Table 1   Man-made actinide elements.

Element	Date
Np	1940
Pu	1940–1
Am	1944–5
Cm	1944
Bk	1949
Cf	1950
Es	1952
Fm	1953
Md	1958
No	1958
Lw	1961

subsequent methods employed extraction of actinide ions from aqueous into nonaqueous solution by the use of organic extractants such as ethers, amines, or organophosphates. With increasing sophistication required in separation processes (e.g., separation of actinides from lanthanides), new classes of extractants have appeared including bifunctional and chelating ligands. Some emphasis will be given in this chapter to the use of these types of extractants in the separation of the elements, given the historical importance of this application.

Investigations of the nonaqueous coordination chemistry of the actinides began with the Manhattan project. Isotopic separation of uranium and plutonium generally involved distillation or centrifugation of volatile metal complexes. While the higher oxidation state fluoride complexes had favorable volatility, their corrosive properties led to a search for alternative classes of compounds. These investigations supported the development of several new classes of compounds, including volatile alkoxide and borohydride complexes. The expansion of these interests to include the organometallic chemistry of the elements began in 1956 (shortly after the discovery of ferrocene in 1951) with the preparation of the first cyclopentadienyl complexes of actinides.[5] Organometallic chemistry of the actinide elements lies outside the scope of this chapter, and no discussion will be provided of complexes containing carbon-based σ-bonding (i.e., alkyl, aryl) or π-bonding ligands. For further information on organometallic chemistry, the reader is referred to other recent reviews.[6]

Most recently, interest has grown in the chemistry of the early actinides in biologically and environmentally relevant conditions, as interest has turned to the remediation of contaminated sites, and the evaluation of long-term fate and transport in the environment. The first impact of this has been to stimulate research in the aqueous coordination chemistry of actinides under conditions dissimilar to process media (near-neutral pH, lower concentration, lower ionic strength). Research has also focused on the complexation of actinides by ligands that are derived from (or which mimic) metal complexation under biological conditions, such as catecholate groups or amino acids. These classes of ligands will be included in the context of the broader suite of multidentate ligands.

A comprehensive treatment of the early literature covering several ligand classes is available in the Gmelin series; emphasis in this chapter will be placed on referencing more recent developments. Where primary literature has not been cited, information has been drawn from Gmelin as a primary reference.[7]

The depictions of molecular structures presented were generated using the program Crystal-Maker 2,[8] using atomic coordinates obtained from the Cambridge Structural Database.[9] Several of the tables have been reproduced from the Actinide chapter by K. W. Bagnall in the previous edition of *Comprehensive Coordination Chemistry* (*CCC*, 1987); this served as the starting point for many ligand classes.

3.3.1.1.1 *Characteristics of the actinides*

It should be noted that one of the most significant characteristics of the actinides is their radio-activity; all isotopes are radioactive, although some have half-lives of greater than 1×10^5 years. Precautions must be taken in their handling, ranging from the use of special enclosures (HEPA-filtered exhaust hoods, negative-pressure gloveboxes) to the use of shielded facilities.

Similarities exist between the chemical characteristics of the actinides and those of the lanthanides. The metal ions are generally considered to be relatively "hard" Lewis acids, susceptible to complexation by hard (i.e., first row donor atom) ligands and to hydrolysis. Both actinide and lanthanide ions are affected by the "lanthanide contraction," resulting in a contraction of ionic radius and an increasing reluctance to exhibit higher oxidation states later in the series. Most species are paramagnetic, although the electron spin–nuclear spin relaxation times often permit observation of NMR spectra, and disfavor observation of ESR spectra except at low temperatures. The elements display more than one accessible oxidation state, and one-electron redox chemistry is common.

The actinide elements display much more diversity in their chemistry than their lanthanide counterparts, however. The greater radial extent and energetic availability of the 5*f*- and 6*d*-orbitals result in increased interaction with ligand-based orbitals. While the electronic structure of lanthanide complexes is dominated by spin-orbit coupling and electron–electron repulsion, that of actinide complexes is often significantly impacted by ligand-field effects, leading to complex optical spectra. Due to the energetic accessibility of metal valence electrons early in the series, the early actinides display a much wider range of attainable oxidation states (see Table 2), and the bonding in chemical compounds is often described to be somewhat more covalent than that in

Table 2 Known oxidation states of the actinide elements.

Th	Pa	U	Np	Pu	Am	Cm	Bk	Cf	Es	Fm	Md	No	Lw
					2			2	2	2	2	2	
	3	3	3	3	3	3	3	3	3	3	3	3	3
4	4	4	4	4	4	4	4	4					
	5	5	5	5	5								
		6	6	6	6								
			7	7	7								

lanthanide complexes.[10] Therefore, it may be said that the coordination chemistry of the actinides is richer and more varied than that for the lanthanide elements.

Some distinction may be made between the chemistry of the early actinide elements (Th–Pu) and that of the later actinide elements (Am–Lw). The early actinide metals are much more readily available, either from natural ores or as products of nuclear materials and fuel production. The later actinides are rarer, and are only available in extremely limited quantities from specialized sources. The early actinides have the greatest range of accessible oxidation states. Isotopes of the early actinides generally have longer half-lives, reducing the risk of self-radiolysis inherent to compounds of radioactive elements (and yielding more stable products). Finally, broader interest exists in the technological applications of the early actinides, due to their role both in energy production and nuclear weapons production. The more widespread use of these elements has also contributed to a more acute need to investigate their behavior in the environment. For all these reasons, the coordination chemistry of the early actinide elements is much more developed, and will be discussed separately.

3.3.1.1.2 Coordination numbers and geometries

Actinide ions display relatively large ionic radii, and therefore support higher coordination numbers; coordination numbers of 8–10 are common, and 12- and 14-coordinate metal centers have been observed (see Table 3). Because ionic radii decrease across the series, however, accessible coordination numbers often decrease across the series for a given oxidation state with the same ligand. Although metal–ligand orbital overlap in complexes of the actinides may exceed that in lanthanide compounds, the actinides still exhibit chemical behavior largely consistent with ionic bonding. As a consequence, the geometry of coordination complexes is not strongly driven by orbital considerations, but rather by steric considerations (ligand–ligand repulsions). Ligands are generally labile, and kinetic barriers for reactions are most often moderate.

3.3.2 EARLY ACTINIDE METALS—THORIUM TO PLUTONIUM

The chemistry of the early actinide metals has been most extensively studied for many reasons. Chief among these is the availability of materials for study. Thorium and uranium obtained from ores as described above have been available for chemical investigations for well over 100 years. In fact, all early actinide elements may be found in nature, although only thorium, protactinium, and uranium are present in sufficient quantities to justify extraction. The remaining early actinide elements, neptunium and plutonium, are produced in large quantities in nuclear reactors.

A second reason for the wealth of chemical investigations of the early actinide elements is the relative diversity of their chemistry. While the chemistry of the later actinides is most often restricted to that of the tri- and tetravalent oxidation states, compounds of the early actinides can be isolated in all oxidation states from +3 to +7. The accessibility of a range of oxidation states is the impetus for signficant chemical interest in the early actinides, but also vastly complicates investigation of these elements under some circumstances, such as aqueous redox behavior. In the case of plutonium, ions in four different oxidation states (+3, +4, +5, and +6) can exist simultaneously in comparable concentrations in the same solution.

Table 3 Coordination numbers and geometries of actinide compounds.

Coordination number	Complex	Coordination geometry	References
3	$[U(N\{SiMe_3\}_2)_3]$	Pyramidal	a
4	$U(O\text{-}2,6\text{-}Bu^t_2C_6H_3)_4$	Tetrahedral	b
	$[U(NPh_2)_4]$	Highly distorted tetrahedral	c
5	$[U(NEt_2)_4]_2$	Distorted trigonal pyramidal	d
6	$[UCl_6]^{2-}$	Octahedral	e
	$U(dbabh)_6$	Octahedral	f
	$U[H_2B(3,5\text{-}Me_2pz)_2]_3$	Trigonal prismatic	g
7	$[UCl(Me_3PO)_6]^{3+}$	Distorted monocapped octahedron	h
	$UI_3(THF)_4$	Pentagonal bipyramidal	i
	$[UO_2(NCS)_5]^{3-}$	Pentagonal bipyramidal	j
	$[PuF_7]^{2-}$ (in Rb_2PuF_7)	Capped trigonal prism	k
8	$[U(NCS)_8]^{4-}$ (NEt_4^+ salt)	Cube	l
	$[B(pz)_4]_2UCl_2$	Distorted square antiprism	m
	$[UCl_2(DMSO)_6]^{2+}$	Distorted dodecahedron	n
	$PuBr_3$	Bicapped trigonal prism	o
	$[UO_2(S_2CNEt_2)_3]^{-}$	Hexagonal bipyramidal	p
9	UCl_3	Tricapped trigonal prism	o
	$[Pu(H_2O)_9][CF_3SO_3]_3$	Tricapped trigonal prism	q
	$[Th(C_7H_5O_2)_4(DMF)]$ $C_7H_6O_2 =$ tropolone	Monocapped square antiprism	r
10	$[Th(NO_3)_4(Ph_3PO)_2]$	Best described as *trans*-octahedral	s
		with four bidentate NO_3 groups in the Equatorial plane	
	$[Th(NO_3)_3(Me_3PO)_4]^{+}$	1:5:4 geometry	t
	$[Th(C_2O_4)_4]^{4-}$ [in $K_4Th(C_2O_4)_4$]	Bicapped square antiprism	u
11	$[Th(NO_3)_4(H_2O)_3]\cdot2H_2O$	Best described as a monocapped trigonal prism with four bidentate NO_3 groups occupying four apices	v
12	$[Th(NO_3)_6]^{2-}$	Icosahedral	t
	$[^{py}Tp_2U][BPh_4]$	Icosahedral	w
14	$[U(BH_4)_4]$	Bicapped hexagonal antiprism	x
	$[U(BH_4)_4(THF)_2]$	Bicapped hexagonal antiprism	y

[a] Stewart, J. L.; Andersen, R. A. *Polyhedron* **1998**, *17*, 953. [b] Van Der Sluys, W. G.; Sattelberger, A. P.; Streib, W. E.; Huffman, J. C. *Polyhedron* **1989**, *8*, 1247. [c] Reynolds, J. G.; Zalkin, A.; Templeton, D. H.; Edelstein, N. M. *Inorg. Chem.* **1977**, *16*, 1090. [d] Reynolds, J. G.; Zalkin, A.; Templeton, D. H.; Edelstein, N. M.; Templeton, L. K. *Inorg. Chem.* **1976**, *15*, 2498. [e] Zachariasen, W. H. *Acta Crystallogr.* **1948**, *1*, 268. [f] Meyer, K.; Mindiola, D. J.; Baker, T. A.; Davis, W. M.; Cummins. C. C. *Angew. Chem., Int. Ed. Engl.* **2000**,*39*, 3063. [g] Carvalho, A.; Domingos, A.; Gaspar, P.; Marques, N.; Pires de Matos, A.; Santos, I. *Polyhedron* **1992**, *11*, 1481. [h] Bombieri, G.; Forsellini, E.; Brown, D.; Whittaker, B. *J. Chem. Soc., Dalton Trans.* **1976**, 735. [i] Clark, D. L.; Sattelberger, A. P.; Bott, S. G.; Vrtis, R. N. *Inorg. Chem.* **1989**, *28*, 1771. [j] Bombieri, G.; Forsellini, E.; Graziani, R.; Pappalardo, G. C. *Transition Met. Chem.* **1979**, *4*, 70. [k] Penneman, R. A.; Ryan, R. R.; Rosenzweig, A. *Struct. Bonding (Berlin)* **1973**, *13*, 1. [l] Countryman, R.; McDonald, W. S. *J. Inorg. Nucl. Chem.* **1971**, *33*, 2213. [m] Campello, M. P.; Domingos, A.; Galvão, A.; Pires de Matos, A.; Santos, I. *J. Organomet. Chem.* **1999**, *579*, 5. [n] Bombieri, G.; Bagnall, K. W. *J. Chem. Soc., Chem. Commun.* **1975**, 188. [o] Zachariasen, W. H. *Acta Crystallogr.* **1948**, *1*, 265. [p] Bowmann, K.; Dori, Z. *Chem. Commun.* **1968**, 636. [q] Matonic, J. H.; Scott, B. L.; Neu, M. P. *Inorg. Chem.* **2001**, *40*, 2638. [r] Day, V. W.; Hoard, J. L. *J. Am. Chem. Soc.* **1970**, *92*, 3626. [s] Mazur-ul-Haque; Caughlin, C. N.; Hart, F. A.; van Nice, R. *Inorg. Chem.* **1971**, *10*, 115. [t] Alcock, N. W.; Esperàs, S.; Bagnall, K. W.; Wang Hsian-Yun, *J. Chem. Soc., Dalton Trans.* **1978**, 638. [u] Akhtar, M. N.; Smith, A. J. *Chem. Commun.* **1969**, 705. [v] Ueki, T.; Zalkin, A.; Templeton, D. H. *Acta Crystallogr.* **1966**, *20*, 836. [w] Amoroso, A. J.; Jeffery, J. C.; Jones, P. L.; McCleverty, J. A.; Rees, L. Rheingold, A. L.; Sun, Y.; Takats, J.; Trofimenko, S.; Ward, M. D.; Yap, G. P. A. *Chem. Commun.* **1995**, 1881. [x] Bernstein, E. R.; Hamilton, W. C.; Keiderling, T. A.; La Placa, S. J.; Lippard, S. J.; Mayerle, J. J. *Inorg. Chem.* **1972**, *11*, 3009. [y] Rietz, R. R.; Zalkin, A.; Templeton, D. H.; Edelstein, N. M.; Templeton, L. K. *Inorg. Chem.* **1978**, *17*, 658.

The most practical reason for interest in the chemistry of the early actinides is their technological importance. In particular, the importance of uranium and plutonium in applications ranging from nuclear power to radiothermal generators for deep space missions. The production and use of nuclear materials in nuclear weapons from the 1940s has also driven the development of a great deal of the chemistry of these metals, from their separation and isolation to investigations of their behavior under biologically relevant conditions.

3.3.2.1 Trivalent Oxidation State

3.3.2.1.1 *General characteristics*

With the exception of thorium and protactinium, all of the early actinides possess a stable +3 ion in aqueous solution, although higher oxidation states are more stable under aerobic conditions. Trivalent compounds of the early actinides are structurally similar to those of their trivalent lanthanide counterparts, but their reaction chemistry can differ significantly, due to the enhanced ability of the actinides to act as reductants. Examples of trivalent coordination compounds of thorium and protactinium are rare. The early actinides possess large ionic radii (effective ionic radii = 1.00–1.06 Å in six-coordinate metal complexes),[11] and can therefore support large coordination numbers in chemical compounds; 12-coordinate metal centers are common, and coordination numbers as high as 14 have been observed.

3.3.2.1.2 *Simple donor ligands*

(i) Ligands containing anionic group 15 donor atoms

Amide ligands. The trivalent chemistry of the actinides with N-donor ligands is limited to sterically bulky ligands that provide kinetic stabilization against ligand exchange and polymerization. The bis(trimethylsilyl)amide ligand ($N(SiMe_3)_2^-$) supports a wide array of oxidation states of uranium.

Trivalent homoleptic complexes $An[N(SiMe_3)_2]_3$ have been generated for uranium, neptunium, and plutonium[12–14] by metathesis reactions (see Equations (1) and (2)). The molecular structure of $U[N(SiMe_3)_2]_3$ has been determined:[15]

$$UCl_3(THF)_x + 3\ NaN(SiMe_3)_2 \xrightarrow{\ THF\ } U[N(SiMe_3)_2]_3 \tag{1}$$

$$AnI_3(THF)_4 + 3\ NaN(SiMe_3)_2 \xrightarrow{\ THF\ } An[N(SiMe_3)_2]_3 \tag{2}$$

$$(An = U,\ Np,\ Pu)$$

The geometry about the uranium center is trigonal pyramidal, with a U—N distance of 3.320(4) Å, and a N—U—N angle of 116.24(7)°. The magnetic susceptibility shows that the complex has an effective moment comparable to those determined for trivalent metallocenes and halides ($\mu_{eff} = 3.354(4)$, $\theta = -13$ K at 5 kG), consistent with a $5f^3$ electronic configuration. A low energy $5f$ ionization band observed in the photoelectron spectroscopy is consistent with the electronic configuration.[16] The steric congestion about the metal center prohibits isolation of stable adducts.

A tris(amido)amine framework, consisting of the ligand $\{N[CH_2CH_2N(Si(Bu^t)Me_2)]_3\}^{3-}$, has been used to produce U^{III} derivatives. Initial attempts to reduce the complex $\{N[CH_2CH_2N(Si(Bu^t)Me_2)]_3\}UCl$ resulted in the formation of a mixed-valence complex $\{[N[CH_2CH_2N(Si(Bu^t)Me_2)]_3]U\}_2(\mu\text{-}Cl)$.[17] The bimetallic complex is thought to possess electronically distinct U^{III} and U^{IV} centers. The purple U^{III} species, $N[CH_2CH_2N(Si(Bu^t)Me_2)]_3U$, was originally isolated by fractional sublimation of the bimetallic U^{III}/U^{IV} complex.[18] This complex can be prepared directly by reduction of $N[CH_2CH_2N(Si(Bu^t)Me_2)]_3UI$ by potassium in pentane. A variety of adducts of this complex have been reported.[19] Reaction of the U^{III} complex with dinitrogen produces one of the most unusual adducts isolated in this system (see Equation (3)):

$$N[CH_2CH_2N(Si(t\text{-}Bu)Me_2)]_3U + N_2 \longrightarrow \{N[CH_2CH_2N(Si(t\text{-}Bu)Me_2)]U\}_2(\mu^2\text{-}\eta^2{:}\eta^2\text{-}N_2) \tag{3}$$

Despite the apparent reversibility of the N_2 addition in solution (based on ^{1}H-NMR experimental data), a molecular structure of the complex was obtained[18] (see Figure 1).

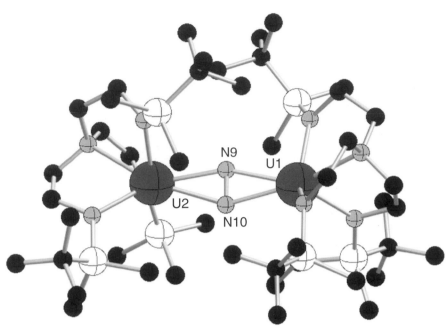

Figure 1 Crystal structure of {N[CH$_2$CH$_2$N(Si(Bu^t)Me$_2$]U}$_2$(μ^2-η^2: η^2-N$_2$) (Roussel and Scott *J. Am. Chem. Soc.* **1998**, *120*, 1070).

The N–N distance in the dinitrogen unit is essentially unperturbed. Metrical data, along with magnetic data, suggest that the complex is best formulated as a U^{III} species. The electronic structure of this complex has been investigated; the only significant U–N$_2$–U interaction was found to consist of U → N$_2$ π backbonding.[20]

An additional bulky amide ligand type, which supports novel coordination complexes of lower valent uranium has been developed. Complexes of the formula U(NRAr)$_3$(THF) (R = Bu^t, adamantyl; Ar = 3,5-Me$_2$C$_6$H$_3$) cannot be generated directly from trivalent halide precursors; instead, they are produced in the reduction of the uranium(IV) iodide complex by sodium amalgam.[21]

Aside from the neutral tris(amido)actinide complexes that have been prepared with sterically encumbering ligands as described, an alternate approach to the stabilization of trivalent actinide amides is the generation of anionic "ate"-type complexes. As an example, reaction of UI$_3$(THF)$_4$ with excess KHNAr (Ar = 2,6-Pr$^i_2$C$_6$H$_3$) produces the anionic complex [K(THF)$_2$]$_2$[U(NHAr)$_5$] which has been crystallographically characterized.[22]

Polypyrazolylborate ligands. Monoanionic poly(pyrazolyl)borate ligands (B(pz)$_4^-$, HB(pz)$_3^-$, H$_2$B(pz)$_2^-$, and substituted derivatives, pz = pyrazol-1-yl) commonly bind to *f*-elements in either a trihapto or dihapto geometry through nitrogen atoms in the pyrazolyl substituents. Most chemistry with trivalent actinides involves the substituted ligand HB(3,5-Me$_2$pz)$_3^-$. The complex U[HB(3,5-Me$_2$pz)$_3$]$_2$Cl has been generated both by metathesis reaction of UCl$_3$ with K[HB(3,5-Me$_2$pz)$_3$][23] and by reduction of the U^{IV} precursor U[HB(3,5-Me$_2$pz)$_3$]$_2$Cl$_3$ with sodium napthalenide.[24] The complex is somewhat unstable, and upon recrystallization can be oxidized to generate the tetravalent oxo complex {UCl[HB(3,5-Me$_2$pz)$_3$](μ-O)}$_4$.[25] The use of uranium triiodide has become increasingly common in the synthesis of trivalent complexes. Reaction of UI$_3$(THF)$_4$ with M[HB(3,5-Me$_2$pz)$_3$] (M = Na, K) in a 1:1 or 1:2 ratio results in the formation of the compounds U[HB(3,5-Me$_2$pz)$_3$]I$_2$(THF)$_2$ and U[HB(3,5-Me$_2$pz)$_3$]$_2$I respectively.[26,27] In the monoligand compound the pyrazolylborate ligand is tridentate, while the bis(ligand) compound demonstrates two different coordination modes for the two [HB(3,5-Me$_2$pz)$_3$] groups. One of the ligands is η^3-coordinated to the metal center, while in the second ligand, two of the pyrazolyl rings appear to coordinate in a "side-on" type of arrangement with the N—N bond of the ring within bonding distance to the uranium atom. Upon abstraction of the iodide ligand with TlBPh$_4$, however, this ligand reverts to a conventional tridentate geometry; the uranium center is seven-coordinate in {U[HB(3,5-Me$_2$pz)$_3$]$_2$(THF)}$^+$; the tetrahydrofuran ligand occupies the seventh site. A limited number of U^{III} complexes have been reported with other pyrazolylborate ligands. Uranium trichloride or triiodide react with bis(pyrazolyl)borate ligands to generate the species and

U[H$_2$B(3,5-Me$_2$pz)$_2$]$_3$ and U[H$_2$B(pz)$_2$]$_3$(THF).[28,29] The coordinated tetrahydrofuran may be removed from the latter to yield the base-free complex U[H$_2$B(pz)$_2$]$_3$. The solid state structure of U[H$_2$B(3,5-Me$_2$pz)$_2$]$_3$ reveals that the metal lies in a trigonal prismatic arrangement of six pyrazole nitrogen atoms, with the three rectangular faces of the trigonal prism capped by three B—H bonds. When a related ligand devoid of B—H bonds is employed, such as (Ph$_2$B(pz)$_2$), the resulting tris(ligand) complex U[Ph$_2$B(pz)$_2$]$_3$ contains a six-coordinate uranium center.[30] The lower coordination number may be reflected in the shorter U—N bond distances in the crystal structures. However, the bond distance is only very slightly shorter and may not be statistically significant (2.53(3) Å, vs. 2.59(3) Å or 2.58(3) Å in the 10- and nine-coordinate complexes, respectively). A mixed halide/bis(pyrazolyl)borate complex has been produced by the reaction of UI$_3$(THF)$_4$ with K[H$_2$B(3-Bu^t,5-Mepz)$_2$]. The complex UI$_2$[H$_2$B(3-Bu^t,5-Mepz)$_2$](THF)$_2$ reacts with triphenyl-phosphine oxide to yield the base adduct UI$_2$[H$_2$B(3-Bu^t,5-Mepz)$_2$](OPPh$_3$)$_2$.[30] Only one complex of a trivalent transuranic metal has been reported; reaction of PuCl$_3$ with M[HB(3,5-Me$_2$pz)$_3$] in refluxing THF generates the dimeric complex [PuCl(μ-Cl){HB(3,5-Me$_2$pz)$_3$}-(3,5-Me$_2$pzH)]$_2$.[31] A particularly interesting encapsulating ligand is found in the tris(3-(2-pyridyl)-pyrazol-1-yl]-borate ligand (^{py}Tp). Reaction of the potassium salt of this ligand with UI$_3$(THF)$_4$ forms the complex [^{py}Tp$_2$U]I, or in the presence of NaBPh$_4$, [^{py}Tp$_2$U][BPh$_4$].[32] The structure of this complex is found in Figure 2. The complex consists of a rare example of 12-coordinate uranium, where the metal lies in an icosahedral coordination environment. The pyrazolylborate groups are approximately staggered with respect to one another, and six pyridyl nitrogens form the equatorial belt.

(ii) Ligands containing neutral group 15 donor atoms

The chemistry of simple actinide complexes employing neutral group 15-atom donor complexes is extensive.

Ammonia. Ammonia adducts of trivalent uranium are rare. The trihalide complexes of uranium and plutonium are reported to form adducts when exposed to liquid or gaseous ammonia. Higher-coordinate complexes (e.g., UCl$_3$·7NH$_3$ are suggested to be stable at lower temperatures; above room temperature the complex loses ammonia to form UCl$_3$·3NH$_3$. Further ligand loss occurs above 45 °C to yield UCl$_3$·NH$_3$. Uranium tribromide has been reported to yield

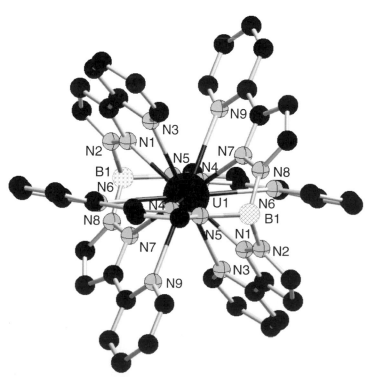

Figure 2 Crystal structure of [^{py}Tp$_2$U][BPh$_4$] (Amoroso, Jeffery *et al.*, *Chem. Commun.* **1995**, 1881).

adducts with either four or six molecules of ammonia, depending on the conditions of preparation. The products $PuCl_3 \cdot 8NH_3$ and $PuI_3 \cdot 9NH_3$ have been reported;[33] they appear to be similarly susceptible to loss of ammonia. There are no reported adducts of trivalent actinides with neutral amines.

Heterocyclic ligands. The advent of the use of trivalent actinide iodides has enabled the characterization of pyridine adducts of uranium, neptunium, and plutonium.[13,34] The complexes AnI_3py_4 (An = U, Np, Pu; py=pyridine) are generated from actinide metals and halide sources in coordinating solvents. They are readily soluble in organic solvents, and serve as convenient precursors to a variety of trivalent actinide species.[14,35] Several related adducts have been generated using neutral tris(N-heterocycle)amine ligands. Reaction of tris((2-pyridyl)methyl)-amine (tpa) with $[UI_3(THF)_4]$ in pyridine results in the isolation of $U(tpa)I_3(pyridine)$.[36] The analogous complex $[U(Mentb)_2]I_3$ complex was prepared by treating $[UI_3(THF)_4]$ with two equivalents of tris(N-methylbenzimidazol-2-ylmethyl)amine (Mentb). Crystallographic studies of the latter reveal that the uranium center is eight-coordinate; the two tetradentate tris(imidazolyl-methyl)amine groups fold around the metal center in a pseudo-D_3 symmetric manner. Solution NMR studies in pyridine show a large difference in the behavior of Mentb and tpa towards uranium binding; the bis(ligand) complex of Mentb is found to be more stable in solution than that of tpa. A related complex employing the tris[(2,2'-bipyridin-6-yl)methyl]amine (tbpa) ligand, $[UI_2(tbpa)][I] \cdot py$, has also been reported.[37] The complexation of uranium triiode by 2,2'-bipyr-idine (bipy) has been investigated in anhydrous pyridine solution.[38] At room temperature, both a 1:1 and 1:2 complex (U:bipy) are observed to form in solution; the 1:2 complex appears to be enthalpically favored. Addition of excess ligand permits observation of a 1:3 complex at low temperature. The "$U(bipy)_2I_3$" complex formed in solution behaves as a 1:1 electrolyte, suggesting a formulation $[U(bipy)_2I_2][I]$. The complex $UI_3(bipy)_2(py) \cdot py$ was isolated from solution and structurally characterized.

Nitriles. The simple acetonitrile adducts $UCl_3 \cdot MeCN$ and $NpCl_3 \cdot 4MeCN$ have been reported; the Np-237 Mössbauer spectrum of the latter has been reported.[39] As in the case of N-heterocyclic ligands, the isolation of nitrile adducts of trivalent uranium have been spurred by the availability of soluble iodide starting materials. The complex $UI_3(MeCN)_4$ has been prepared and characterized crystallographically as well as by magnetic susceptibility and solid-state absorption spectroscopy,[40] and the complexes $UCl_3(MeCN)(H_2O)_5$ and $NH_4[UBr_2(MeCN)_2(H_2O)_5]$-$[Br]_2$ have been isolated.[41,42]

Phosphines. A limited number of trivalent uranium borohydride phosphine complexes have been reported. The complexes $U(BH_4)_3(dmpe)_2$ (dmpe = bis(1,2-dimethylphosphino)ethane)[43] and $U(BH_4)_3(o\text{-}PPh_2(C_6H_4N))_2$[44] are prepared from the reaction of $U(BH_4)_3(THF)_x$ and the corresponding ligand, while $U(MeBH_3)_3(dmpe)_2$[45] is generated when the tetravalent precursor $U(BH_4)_4(dmpe)$ is heated in the presence of excess dmpe.

(iii) Ligands containing anionic group 16 donor atoms

Oxides. The binary oxide, Pu_2O_3, has been observed as an intermediate between Pu and PuO_2 and it has hexagonal and cubic forms. The hexagonal phase is of the La_2O_3 "type A," rare earth sesquioxide structure and contains seven-coordinate Pu^{III}. The analogous Np phase prob-ably exists as a bulk compound, but has not been as well studied. It has been observed in an XPS study on the oxidation of Np metal.[46] Ternary oxides are generally prepared from high tempera-ture reactions of binary oxides. One class has the general formula $PuMO_3$, where M = Al, V, Cr, Mn, and the perovskite structure, in which MO_6 octahedra are linked in a network and 12-coordinate Pu^{III} ions are located in the interstices between octahedra. Quaternary oxides of Pu^{III} are also known, such as Ba_2PuNbO_6 and Ba_2PuTaO_6.

Hydroxides. The hydrolysis and carbonate complexation of the actinides has been recently reviewed.[47] Plutonium(III) hydrolysis is not well known because Pu^{III} is readily oxidized to Pu^{IV} in aqueous solutions, particularly at near-neutral and basic pH. The first hydrolysis product, $Pu(OH)^{2+}$, has been identified in acid solution up to pH ~ 3 (where it is about 70% formed) before oxidation to Pu^{IV} prevents further study.[48] The first hydrolysis product of Np^{III} has been similarly studied.[46] The hydroxide solids, $Pu(OH)_3 \cdot xH_2O$ and $Np(OH)_3 \cdot xH_2O$, are prepared by precipitation and presumed to be isostructural with $Am(OH)_3$.

Carbonates. Trivalent actinide carbonates generally oxidize rapidly to An^{IV} species. Only the u^{III} and Np^{III} complexes of this type, generally prepared via reduction, have been studied in any detail. In aqueous Pu^{III} solutions, there is evidence for the stepwise formation of the carbonato complexes, $Pu(CO_3)^+$ and $Pu(CO_3)_2$. Additional carbonate and hydroxocarbonate complexes may form, but are immediately oxidized to Pu^{IV} species. Neptunium(III) carbonate, hypothesized to be $Np(CO_3)_3{}^{3-}$, has been prepared by electrochemical reduction of Np^{IV} carbonate.[49]

Nitrates and Phosphates. Trivalent nitrate species have been prepared in nitric acid solution, but they are unstable with respect to oxidation. A plutonium nitrate has been prepared, and by analogy with the lanthanides presumed to be $Pu(NO_3)_3$, although it was not characterized.[50] Neptunium and plutonium phosphates solution species are proposed to have the formula $An(H_2PO_4)_n{}^{3-n}$ ($n = 1$–4), but not spectroscopically or structurally characterized.[51] For Pu^{III}, the blue, hexagonal $PuPO_4 \cdot 0.5H_2O$ has been prepared by precipitation from acid solution and heated to yield the anhydrate. Additional binary, ternary, and quaternary phosphates have been prepared by Bamberger and others and have generally been characterized by chemical analysis, Raman, and X-ray powder diffraction.[52]

Sulfates. Sulfate complexes in solution, of the form $An(SO_4)_n{}^{3-2n}$ ($n = 1,2$), have been reported for Pu^{III}.[53] These anions can be precipitated as hydrates; and partially dehydrated solids can be obtained by addition of less polar solvents. There is some evidence for a Pu^{III} sulfate, $Pu_2(SO_4)_3 \cdot xH_2O$, but it is not as well characterized as the complex salts. Hydrated sulfato complexes of the type $MAn(SO_4)_2 \cdot xH_2O$, where An is U, Pu, and M is a monovalent cation, are known. The Pu^{III} sulfate, $KPu(SO_4)_2 \cdot H_2O$ and the dehydrate are isostructural with the Nd^{III} analogues. Similarly, the $NH_4Pu(SO_4)_2 \cdot 4H_2O$ is isomorphous with the corresponding Ce^{III} compound. The structures of U^{III} sulfates have been reconsidered with some new X-ray diffraction data. A crystal structure of $(NH_4)_2U(SO_4)_2 \cdot 4H_2O$ show that U is nine-coordinate with six oxygen atoms from four sulfate groups and the remaining three inner-sphere waters. A noncoordinated water is also present. The nonahydrate $(NH_4)_2U_2(SO_4)_4 \cdot 9H_2O$ likely contains nine and 12-coordinate U, in contrast with Am sulfate, $Am_2(SO_4)_3 \cdot 8H_2O$, which is comprised of eight coordinate Am^{III}.[54] Salts of other complex anions, such as $K_5An(SO_4)_4 \cdot 4H_2O$ are also known for Np^{III} and Pu^{III}.

Alkoxide Compounds. Despite the variety of higher valent actinide alkoxide complexes since the 1950s, successful preparations of trivalent actinide compounds employing alkoxide ligands have only appeared in the literature since the 1980s. Much of the attention regarding synthesis of trivalent actinide alkoxides has focused on the preparation of homoleptic uranium(III) aryloxide complexes. Among the earliest reports is that involving reaction of three equivalents of sodium phenoxide with $UCl_3(THF)_x$ in THF, from which a light red-brown solution was obtained;[55] the reaction did not result in the isolation of $U(OPh)_3$. It was later reported that alcoholysis of $U[N(SiMe_3)_2]_3$ with three equivalents of $HO-2,6-R_2C_6H_3$ ($R = Bu^t$, Pr^i) in hexane produced dark green ($R = Bu^t$) or dark purple ($R = Pr^i$) solutions from which homoleptic $[U(O-2,6-R_2C_6H_3)_3]_x$ ($R = Bu^t$, $x = 1$; $R = Pr^i$, $x = 2$) compounds were isolated[56] (see Equation (4)). The molecular structure of $[U(O-2,6-Pr^i_2C_6H_3)_3]_2$ demonstrates an unprecedented structure composed of a centrosymmetric bis η^6-arene-bridged dimer (see Figure 3). Based upon analysis of the infrared spectrum, it was suggested that $U(O-2,6-Bu^t_2C_6H_3)_3$ is monomeric. Similarly, hexane solutions of $An[N(SiMe_3)_2]_3$ ($An = Np,Pu$) react with three equivalents of $HO-2,6-Bu^t_2C_6H_3$ to form $An(O-2,6-Bu^t_2C_6H_3)_3$:[14]

$$U[N(SiMe_3)_2]_3 + 3\ HO\text{-}2,6\text{-}R_2C_6H_3 \longrightarrow [U(O\text{-}2,6\text{-}R_2C_6H_3)_3]_x + 3\ HN(SiMe_3)_2 \quad (4)$$

A number of adducts of uranium trisaryloxides are readily prepared. The THF adduct, $U(O-2,4,6-Me_3C_6H_2)_3(THF)_2$, is isolated from the reaction of $NaO-2,4,6-Me_3C_6H_3$ with $UCl_3(THF)_x$ in tetrahydrofuran.[57] Sattelberger and co-workers reported that the compound $U(O-2,6-Bu^t_2C_6H_3)_3$ readily coordinates a number of Lewis bases (THF, EtCN, Ph_3PO) to form isolable, and presumably tetrahedral 1:1 adducts, $LU(O-2,6-Bu^t_2C_6H_3)_3$.[56] Analysis of the ^{1}H-NMR spectra and infrared data suggest that both a 1:1 and 1:2 adduct are obtained upon coordination of $CNBu^t$ to $U(O-2,6-Bu^t_2C_6H_3)_3$.[58] Alternatively, a THF adduct can simply be prepared by allowing three equivalents of $KO-2,6-R_2C_6H_3$ ($R = Pr^i$, Bu^t, $x = 1$; $R = Me$, $x = 2$) to react with $UI_3(THF)_4$ in tetrahydrofuran to produce a dark red solution from which $U(O-2,6-R_2C_6H_3)_3(THF)_x$ is isolated[59] (see Equation (5)):

$$UI_3(THF)_4 + 3\ KO\text{-}2,6\text{-}R_2C_6H_3 \longrightarrow U(O\text{-}2,6\text{-}R_2C_6H_3)_3(THF)_x + 3\ KI \quad (5)$$

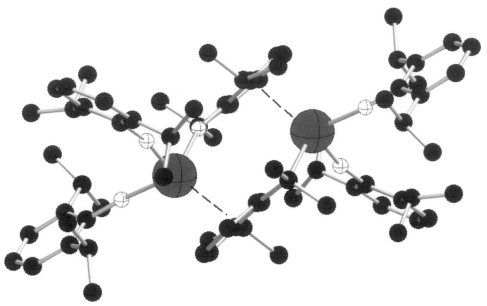

Figure 3 Crystal structure of [U(O-2,6-Pr$^i_2$C$_6$H$_3$)$_3$]$_2$ (Van Der Sluys, Burns *et al. Inorg. Chem.* **1998**, *110*, 5924).

An assessment of the relative binding constants of Lewis bases with different donor sites (triphenylphosphine oxide, *N,N*-di-iso-propylbenzamide, 4,4′-dimethoxybenzophenone) to the Pu^{III} trisaryloxide, Pu(O-2,6-Bu$^t_2$C$_6$H$_3$)$_3$, has been reported using variable temperature ^{1}H-NMR spectroscopy.[60]

Only one study has suggested the formation of an actinide(III) alkoxide (–OR) compound in which R is an alkyl. A recent investigation of the reactivity of Pu^{III} *iso*-propoxide, prepared *in situ* from the reaction of Pu[N(SiMe$_3$)$_2$]$_3$ and three equivalents HOPr^i, indicates that the trivalent alkoxide complex is an effective catalyst in the Meerwein–Ponndorf–Verley reduction of ketones by isopropanol.[61]

Triflate complexes. Another recent addition to this class of compounds is the isolation of a trivalent trifluoromethanesulfonate (OTf$^-$ = triflate) derivative of uranium, U(OTf)$_3$, from the reaction of UH$_3$ and triflic acid.[62] A Lewis base adduct of the complex was prepared to facilitate characterization; the complex [U(OTf)$_2$(OPPh$_3$)$_4$][OTf] has been crystallographically characterized, and possesses both a monodentate and a bidentate triflate ligand in the coordination sphere of the metal.

Sulfur donor ligands. A ligand related to the pyrazolylborate family has been employed to stabilize a U^{III} cation. The reaction of bis(2-mercapto-1-methylimidazolyl)borate, [H(R)-B-(timMe)$_2$]$^-$ (R = H, Ph) with UI$_3$(THF)$_4$ and Tl(BPh$_4$) generates the ionic species {U[H(R)B(timMe)$_2$]$_2$(THF)$_3$}{BPh$_4$}.[63] The uranium atom in these species lies within a distorted tricapped trigonal prism of ligands consisting of four sulfur atoms and two hydrogen atoms from the borate ligands and three THF oxygen atoms (Figure 4).

(iv) Ligands containing neutral group 16 donor atoms

The class of oxo-donor atom ligands is the most prevalent in actinide coordination chemistry, owing to their predominant use in separation chemistry of the *f*-elements.

Aqua species. A number of investigations of trivalent actinide ions in aqueous media have been directed at identifying the coordination environment of the metal center. Examination by luminescence, X-ray absorption (i.e., extended X-ray absorption fine structure (EXAFS)), and NMR spectrometry suggest that the early actinides are likely ligated by nine water molecules.[64–68] Confirmation of this assignment may be found in the crystal structure of [Pu(H$_2$O)$_9$][CF$_3$SO$_3$]$_3$, prepared by the dissolution of plutonium metal in triflic acid.[69] The plutonium ion in this complex is coordinated by nine water molecules arranged in an ideal tricapped trigonal prismatic geometry with Pu—O distances of 2.574(3) Å and 2.476(2) Å (Figure 5).

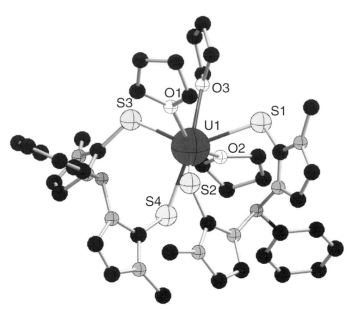

Figure 4 Crystal structure of {U[H(R)B(timMe)$_2$]$_2$(THF)$_2$}{BPh$_4$} (Maria, Domingos *et al. Inorg. Chem.* **2001**, *40*, 6863).

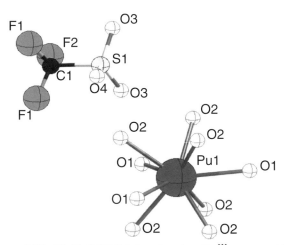

Figure 5 Crystal structure of [Pu(H$_2$O)$_9$][CF$_3$SO$_3$]$_3$, showing Pu^{III} aqua with ideal tricapped trigonal prismatic geometry (Matonic, Scott *et al. Inorg. Chem.* **2000**, *40*, 2638).

There are many examples of hydrates of solid actinide halide complexes (see Table 4). In some instances, the complexes are reported to be easily dehydrated, and it is therefore suggested that the metal ion is not coordinated by water. This is most often the case where X = F, and strong M—F—M bridge bonding in the solid precludes the formation of molecular hydrates. In other cases, complexes (e.g., AnCl$_3$·6H$_2$O, An = Pu, Am; AnBr$_3$·6H$_2$O, An = U, Np, Pu) and have been found to be isostructural with known lanthanide halide hydrates of the formula [LnX$_2$(H$_2$O)$_6$]X. The molecular structure of the anionic complex [NH$_4$][UCl$_4$(H$_2$O)$_4$] has been reported.[70]

Ethers, cyclic ethers. The complex UCl$_3$(THF)$_n$ was reported by Moody *et al.*[71] and subsequently used by a number of researchers as a precursor for entry into U^{III} chemistry. The molecular nature of the complex was not well characterized, however. Subsequently, the complex UI$_3$(THF)$_4$ was prepared via halide oxidation of uranium metal and structurally characterized.[13] The metal center is found to lie within a pentagonal bipyramidal coordination environment, with

Table 4   Representative hydrates of actinide(III) compounds.

$PuF_3 \cdot (0.4 \text{ to } 0.75)H_2O$	
$PuCl_3 \cdot 6H_2O$	
$M^{III}Br_3 \cdot 6H_2O$	$M^{III} = U, Np, Pu$
$M^{I}UCl_4 \cdot 5H_2O$	$M^{I} = Rb, NH_4$
$M_2(SO_4)_3 \cdot xH_2O$	$M^{III} = U, x = 8; Pu, x = 5 \text{ or } 7$
$NaNp(SO_4)_2 \cdot xH_2O$	
$M^{I}Pu(SO_4)_2 \cdot xH_2O$	$x = 1, 2, 4 \text{ or } 5 \text{ variously with}$ $M^{I} = Na, K, Rb, Cs, Tl, NH_4$
$(NH_4)_2U_2(SO_4)_4 \cdot 9H_2O$	
$Pu_2(SO_3)_3 \cdot xH_2O$	
$PuPO_4 \cdot 0.5H_2O$	
$HM^{III}[Fe^{II}(CN)_6] \cdot xH_2O$	$M^{III} = U, x = 9 \text{ to } 10.$
$Pu[Fe^{III}(CN)_6] \cdot ca. \ 7H_2O$	
$M^{III}_2(C_2O_4)_3 \cdot xH_2O$	$M^{III} = Np, x = ca. \ 11; Pu, x = 1, 2, 3, 6, 9 \text{ or } 10$

two axial and one equatorial iodide ligands. Neptunium and plutonium analogues of this complex have been reported;[14,34,35] these species now serve as the most common reagents for entry into trivalent actinide molecular chemistry.

Carbamides. A series of U[III] homoleptic complexes of the ligands antipyrine (2,3-dimethyl-1-phenylpyrazol-5-one, ap) and pyrimidone (4-dimethylaminoantipyrine, dma) have been prepared from complex halide precursors in the presence of the appropriate ligand. Reaction of $RbUCl_4 \cdot 5H_2O$ with ap results in the formation of the ionic species $[U(ap)_6]Cl_3$, whereas reaction of $NH_4UCl_4 \cdot 5H_2O$ and $NaBPh_4$ with these ligands yields the corresponding tetraphenylborate species, $[U(ap)_6](BPh_4)_3$ and $[U(dma)_6](BPh_4)_3$.[72,73]

Phosphine oxides. The complex $[U(OTf)_2(OPPh_3)_4][OTf]$ (OTf = triflate) has been prepared by the reaction of $U(OTf)_3$ with phosphine oxide. One of the inner sphere triflate ligands is monodentate, and the other is bidentate, supporting an overall pentagonal bipyramidal coordination environment about the uranium center.[62]

(v) Ligands containing group 17 donor atoms

The preparation and properties of halides of the actinides have been described fully.[7] As most of these complexes are solid state, rather than molecular in nature, only overview information on classes of compounds will be provided. Adducts of the halide complexes will be discussed in the context of compounds of the respective Lewis bases (*vide infra*).

Binary halides. Trihalide complexes of all elements Ac–Pu have been reported except for thorium and protactinium; trihalide complexes are among the few reported complexes of actinium. The trifluorides, MF_3 (M = Ac, U, Np, Pu), exist in a LaF_3-type structure. Most can be prepared by precipitation from solution or hydrofluorination of oxides, although the uranium fluoride is formed in reduction reactions, and is highly sensitive to hydrolysis. The trichlorides, MCl_3 (M = Ac, U, Np, Pu) and tribromides, MBr_3 (M = Ac, U, α-Np), adopt the UCl_3-type structure in which the nine-coordinate metal atom lies at the center of a tricapped trigonal prism. β-NpBr$_3$ and PuBr$_3$ have the eight-coordinate $PuBr_3$-type structure in which the coordination geometry is a bicapped trigonal prism and this is found also for the triiodides, MI_3 (M = U, Np, Pu). The complexes PaI_3 and ThI_3 have been reported, although their identification is more tentative.

Complex halides. Ternary fluoride complexes of trivalent uranium, neptunium, and plutonium are well known, and are formed by the reaction of binary halides and additional metal halides (alkali halides) in melts, solvents such as thionyl chloride, or by precipitation from aqueous solution. Fewer chloroactinates and bromoactinates are known. Complexes of the formula $MAnX_4$, M_2AnX_5, M_3AnX_6, $M'AnX_5$, and M'_2AnX_7 (M = alkali metal; M' = alkaline earth) are the most common, although more complex formulations (e.g., MAn_2Cl_7) have also been reported.

Common structural types have been reported among these groups of compounds. The complex $NaPuF_4$ is isostructural with $NaNdF_4$[74] and therefore consists of a tricapped trigonal prismatic arrangement of fluorine atoms about the metal center. Published reports on the complex $CsUCl_4$ differ in their assignment of the symmetry of the structure.[75,76] The compounds M_2AnX_5 contain metal centers in either a monocapped trigonal prismatic or distorted pentagonal bipyramidal coordination environment.[77,78] Diffraction data for $SrUCl_5$ is available, although the structure has not been unambiguously assigned.[76] Complexes of the formula M'_2AnX_7 are isostructural with related lanthanide complexes, and are therefore assumed to contain metal centers that lie within a monocapped trigonal prismatic arrangement of halide ions.[76,79] Complexes of the formula M_3AnCl_6 are known for uranium, neptunium, and plutonium; these compounds contain isolated $AnCl_6^-$ ions.[76]

3.3.2.1.3 Chelating ligands

(i) Multidentate donor ligands

Hydroxamate. Hydroxamate complexes of trivalent actinides can be prepared directly in aqueous solution and other polar solvents and extracted into organic solvents, but due to the high thermodynamic stability of the corresponding tetravalent actinide complexes they are rapidly oxidized. They can also be prepared in solution via electrochemical reduction of the tetravalent complexes. These complexes have been studied for their role in separating high and low valent actinides in nuclear fuel processing schemes.[80]

Catecholate. Am^{III} and Pu^{III} complexes of sulfonated and carboxylated catecholamide ligands (CAMS and CAMC), including potentially octadentate chelators have been studied for their potential utility in removing actinides from humans via chelation. The complex coordination is pH dependent, with a triscatecholate Pu^{III} complex forming above pH 12. The stoichiometry of the Am^{III} complex was not determined; however, its optical absorbance characteristics were determined.[81]

8-Hydroxyquinoline and derivatives. Trivalent plutonium complexes with 8-hydroxyquinoline (Oxine, Ox) of the formula $An(Ox)_3$ are prepared by precipitation from aqueous solution in the presence of sulyite or dithionite as a reducing agent (to retain An^{III}). Attempts to prepare analogous U^{III} and Np^{III} complexes result in immediate oxidation.

Oxalate. The trivalent oxalates have been widely used in actinide separation and purification. For this application the very low solubility and physical properties of $Pu_2(C_2O_4)_3 \cdot xH_2O$ are key. These solids are often precursors that are dehydrated and fired to produce oxides, such as PuO_2 and AmO_2.

Oxalate complexes in solution are mainly of the form, $Pu(C_2O_4)_n^{3-2n}$, $n = 2$–4. The intermediate $Pu(C_2O_4)_3^{3-}$ is relatively unimportant, as is the species $Pu(HC_2O_4)_4^-$, which predominates in the narrow pH range 1.7 to 2.2.[82]

Polyoxometallates. Polyoxometallates of the group 6 transition metals (iso- and hetero-polyoxoanions) form a special class of metallate ligands for the actinide elements. These species can incorporate other atoms as either primary or secondary (peripheral) heteroatoms. Primary heteroatoms are necessary to complete the polyoxoanion structure; secondary heteroatoms can be removed without disruption of the stable polyanion unit. The early actinides serve in both roles in known compounds.

The relative large ionic radii of actinide cations require polyoxometallate ligands that can generate high coordination numbers at the metal center. Complexes of three main classes of polyoxoanions have been described: decatungstometalates $[An^{IV}W_{10}O_{36}]^{8-}$, An = Th, U; dodeca-molybdometalates $[An^{IV}Mo_{12}O_{42}]^{8-}$, An = Th, U, or Np; and derivatives of the Keggin and Dawson structures, $An[XW_{11}O_{39}]_2^{n-}$ and $An[X_2W_{17}O_{61}]_2^{n-}$ (X = P, Si, B, As; An = Th, U, Np, Pu). Of these, only one has been reported to stabilize a trivalent actinide. Reaction of Pu^{III} with the anions $PW_{11}O_{39}^{7-}$, $P_2W_{17}O_{61}^{10-}$, $SiW_{11}O_{39}^{8-}$, $BW_{11}O_{39}^{9-}$, and $AsW_{11}O_{39}^{7-}$, result in the isolation of 1:2 (An:ligand) complexes as potassium or cesium salts.[83]

(ii) Macrocyclic ligands

Crown ethers. Metal and crown ether complexes have to fulfill two requirements to form stable complexes. First, the coordination sphere of the metal must be stabilized by the crown ether and any complexing anions. A trivalent uranium coordination sphere is typically satisfied

with a coordination number between seven or eight for crown ether inclusion complexes. Second, the oxidation state of the metal must be counterbalanced to neutralize the charge. This is easily accomplished with numerous ligands, ranging from coordinating species such as nitrate to those that are relatively noncoordinating like perchlorate. Through the use of EXAFS analysis using the uranium L_{III} absorption edge, as well as X-ray single crystal diffraction, a proposed solution of U^{III} forming an inclusion complex with dicyclohexyl-18-crown-6 (dch-18-crown-6) has been determined. The uranium is bound in a *trans* manner to two BH_4^- anions, yielding the monovalent cation, $U(BH_4)_2^+$. When the $U(BH_4)_2^+$ complexes with dch-18-crown-6, the proposed structure is similar to that of (UO_2)(dch-18-crown-6), to be discussed later in this chapter. The resulting geometry around the U^{III} is hexagonal bipyramidal with equatorial crown ether complexation. The anion for the complex is a trivalent uranium ion oxidized to the U^{IV} species and complexed with five chlorides and one BH_4^-, giving the divalent anion $[U^{IV}Cl_5BH_4]^{2-}$. The coordination environment of the anion is pseudo-octahedral.[84]

3.3.2.1.4 *Borohydride and aluminohydride ligands*

Borohydride compounds of trivalent actinides are limited to those of uranium. The initial reports of $U(BH_4)_3$ indicated it was prepared from thermal or photochemical decomposition of $U(BH_4)_4$[85–89] as in Equation (6):

$$2\ U(BH_4)_4 \longrightarrow 2\ U(BH_4)_3 + B_2H_6 + H_2 \tag{6}$$

Other methods for its preparation invariably result in the isolation of Lewis base adducts. The complexes $U(BH_4)_3(THF)_x$[71] and $U(BH_4)_3$(18-crown-6)[90] were prepared by the metathesis reaction of $LiBH_4$ with $UCl_3(THF)_x$ or UCl_3(18-crown-6) in THF. Although the stoichiometry of the THF adduct was not characterized in the initial report, the compound was later prepared from the reaction of UH_3 and BH_3 in THF, and characterized to be $U(BH_4)_3(THF)_3$.[91] The molecular structure of the compound reveals that it adopts an octahedral geometry about the metal center with the borohydride and THF ligands mutually facial; all borohydride ligands are tridentate.

The THF adduct of $U(BH_4)_3$ serves as a useful reagent in the synthesis of other base adducts. Reaction of $U(BH_4)_3(THF)_x$ with dmpe (1,2-dimethylphosphinoethane) results in the formation of $U(BH_4)_3$(dmpe)$_2$.[43] The uranium center in this complex has a pentagonal bipyramidal geometry (considering each BH_4 unit as one ligand). Two of the borohydride ligands are tridentate, while the third is bidentate, presumably owing to steric encumbrance at the metal center. Reduction in the coordination number reduces this strain; the five-coordinate complex $U(BH_4)_3(Ph_2Ppy)_2$, similarly prepared from the reaction of $U(BH_4)_3(THF)_x$ and 2-(diphenylphosphino)pyridine, possess three tridentate borohydride ligands.[44]

Trivalent borohydride Lewis base adduct complexes can also be prepared by reduction of their tetravalent analogues. Reduction of $U(BH_4)_4$ in the presence of phosphines is reported to yield the adducts $U(BH_4)_3L_2$ (L = PEt_3, PEt_2Ph).[92] Thermolysis of $U(MeBH_3)_4$(dmpe) in the presence of excess dmpe results in reduction of the metal center to generate the complex $U(MeBH_3)_3$(dmpe)$_2$.[45] In select cases, reduction leads to the formation of polymetallic complexes. Reduction of $U(BH_4)_4$ in the presence of crown ether ligands such as 18-crown-6 or dicyclohexyl-(18-crown-6) generates products of the overall stoichiometry $U_3(BH_4)_9$(crown)$_2$.[93] Subsequent investigation of the complex $U_3(BH_4)_9$(18-crown-6)$_2$ by EXAFS suggest a structural model in which $U(BH_4)_2^+$ cations are coordinated within the cavity of the crown ligands, while the other uranium resides within a $[U(BH_4)_5]^{2-}$ coordination environment.[94]

Trivalent cationic and anionic borohydride complexes have also been generated. Protonation of $U(BH_4)_3(THF)_3$ with $[NEt_3H][BPh_4]$ yields the cationic species $[U(BH_4)_2(THF)_5][BPh_4]$.[95] The uranium atom adopts a pentagonal bipyramidal geometry, with tridentate borohydride ligands in the apical positions. The anionic compound $[Na$(18-crown-6)$][U(BH_4)_4]$ has been isolated from the reaction of $U(BH_4)_3(THF)_3$ with $NaBH_4$ in the presence of 18-crown-6.[96]

Attempts have been made to isolate aluminohydride analogs in trivalent chemistry. Reaction of UCl_3 with three equivalents of $LiAlH_4$ yields a gray powder, proposed to be $U(AlH_4)_3$.[97] The complex is reported to decompose at temperatures above $-20\,°C$.

3.3.2.2 Tetravalent Oxidation State

3.3.2.2.1 *General characteristics*

All early actinides from thorium to plutonium possess a stable +4 ion in aqueous solution; this is the most stable oxidation state for thorium and generally for plutonium. The high charge on tetravalent actinide ions renders them susceptible to solvation, hydrolysis, and polymerization reactions. The ions are readily hydrolyzed, and therefore act as Brønsted acids in aqueous media, and as potent Lewis acids in much of their coordination chemistry (both aqueous and nonaqueous). Ionic radii are in general smaller than that for comparable trivalent metal cations (effective ionic radii = 0.96–1.06 Å in eight-coordinate metal complexes),[11] but are still sufficiently large to routinely support high coordination numbers.

3.3.2.2.2 *Simple donor ligands*

(i) *Ligands containing group 14 donor atoms*

The small steric size and propensity of cyanide groups to bridge metal centers have limited their use as ligands in molecular coordination chemistry of the actinides, where they are prone to form amorphous polymeric products. Limited metathesis studies have been conducted. Reaction of tetravalent halides with alkali metal cyanides in liquid ammonia is reported to give rise to a product of the formula $UX_3(CN)\cdot4NH_3$,[98] whereas use of the larger thorium ion yields unidentified products.

The neutral isocyanide ligands (CNR, R = alkyl, aryl) have been used extensively in organometallic actinide chemistry, where they are commonly observed to undergo insertion reactions into metal–carbon sigma bonds.[99,100] These ligands are relatively weak bases in coordination chemistry, however, and few complexes have been isolated. Lewis base adducts of a tetravalent halides, $AnX_4(CNc\text{-}C_6H_{11})_4$ (An = Th, X = I; An = U, X = Cl, Br, I), were generated by direct reaction of the constituents in organic solvent.[101] As is characteristic for isocyanide ligands acting principally as σ-donor ligands, the isocyanide νC—N band in the IR spectra of these complexes moves approximately $50\,\text{cm}^{-1}$ to higher frequency that that in free isocyanide ligand.

(ii) *Ligands containing anionic group 15 donor atoms*

Amide complexes. The first report of a tetravalent actinide amide complex was the isolation of $U(NEt_2)_4$.[102] In general, compounds of the formula $An(NR_2)_4$ (An = Th, U) are generated by reaction of metal tetrahalides and alkali metal amide salts in nonaqueous solvents. $U(NEt_2)_4$ has been found to exist as a dimer both in the solid state and in benzene solution,[103] with uranium centers bridged by two diethylamido groups. With even smaller alkyl substituents, larger aggregates are obtained; the molecular structure of $U(NMe_2)_4$ reveals it to be a trimer in the solid state.[104] Each metal lies within a roughly octahedral arrangement of amide ligands, with the central metal sharing an octahedral face with each of its neighboring uranium centers. The polymeric structures can be broken up by the addition of Lewis bases; addition of two equivalents of hexamethylphosphoramide (HMPA) to $U(NMe_2)_4$ leads to the isolation of the monomeric base adduct $U(NMe_2)_4(HMPA)_2$,[105] which exists as a mixture of *cis*- and *trans*-isomers in solution. Monomeric homoleptic anionic complexes can also be isolated; reaction of UCl_4 with excess lithium amide salts $LiNMe_2$ and $LiNEt_2$ in THF results in the isolation of the complexes $[Li(THF)]_2[U(NMe_2)_6]$ and $[Li(THF)][U(NEt_2)_5]$, respectively.[106]

Larger amide ligands give rise to monomeric products. $U(NPh_2)_4$ may be prepared either by reaction of UCl_4 with $LiNPh_2$, or by aminolysis reaction of $HNPh_2$ with $U(NEt_2)_4$, or the uranium metallacycle $\overline{U[N(SiMe_3)(SiMe_2CH_2)][N(SiMe_3)_2]_2}$.[107,108] The uranium atom lies within a severely distorted tetrahedron of nitrogen atoms. If the filtrate from the metathesis reaction is allowed to react slowly with air, a product of partial hydrolysis is isolated ($[Li(OEt_2)\{UO(NPh_2)_3\}_2]$), which contains terminal amide ligands and two μ^3-Li,U,U'-oxo ligands. Higher coordination numbers can be observed; reaction of $ThBr_4(THF)_x$ with four equivalents of $KNPh_2$ in THF results in the isolation of Th(NPh$_2$)$_4$(THF), while use of the smaller amide NMePh$^-$ yields the bis-tetrahydrofuran adduct Th(NMePh)$_4$(THF)$_2$.[109] A related five-coordinate amide complex K[Th(NMePh)$_5$] can be prepared by reaction of thorium tetrabromide with five equivalents of K(NMePh).

The smaller alkylamide complexes are susceptible to redistribution reactions. Reaction of UCl$_4$ and U(NEt$_2$)$_4$ in a 1:1 or 3:1 ratio generates the complexes U(NEt$_2$)$_2$Cl$_2$ and U(NEt$_2$)Cl$_3$(THF) in high yield.[110] In contrast, the complex U(NEt$_2$)$_3$Cl is not stable in solution, but exists in equilibrium with U(NEt$_2$)$_2$Cl$_2$ and U(NEt$_2$)$_4$.

The bis(trimethylsilyl)amido ligand has been used extensively in supporting the tetravalent chemistry of thorium and uranium. Tetravalent complexes of the formula ClAn[N(SiMe$_3$)$_2$]$_3$ (An = Th, U) have been prepared[111] from the 3:1 reaction of NaN(SiMe$_3$)$_2$ with AnCl$_4$ (Equation (7)), and the complex Cl$_2$U[N(SiMe$_3$)$_2$]$_2$(DME) can be generated from a 2:1 reaction of ligand:-halide salt.[112] Substituted complexes of the formula RAn[N(SiMe$_3$)$_2$]$_3$ (An = Th, U; R = Me, Et, Pr^i, Bu, BH$_4$) are formed by the reaction of ClAn[N(SiMe$_3$)$_2$]$_3$ with the appropriate lithium or magnesium reagents.[111,113] Unlike comparable cyclopentadienyl analogues, the methyl compound does not undergo ready insertion of CO, although a number of other insertion and protonation reactions have been reported, including insertion of ketones, aldehydes, isocyanides, and aliphatic nitriles.[113,114] The methyl ligand is further susceptible to removal by protic reagents such as secondary amines:

$$AnCl_4 + 3\ NaN(SiMe_3)_2 \xrightarrow{\ THF\ } An[N(SiMe_3)_2]_3Cl \tag{7}$$

$$(An = Th,\ U).$$

The hydride compounds HAn[N(SiMe$_3$)$_2$]$_3$ (An = Th, U) are the sole products of attempts to introduce an additional equivalent of the bis(trimethylsilyl)amide ligand.[115] Reaction of the uranium hydride complex with the Lewis acid B(C$_6$F$_5$)$_3$ results in loss of H$_2$ and formation of the zwitterionic product U$^+$[N(SiMe$_3$)$_2$]$_2$[N(SiMe$_3$)(SiMe$_2$CH$_2$B$^-$(C$_6$F$_5$)$_3$].[116]

Pyrolysis of the hydrides result in the loss of dihydrogen and the formation of an unusual metallocycle[117] (see Equation (8)):

$$[(Me_3Si)_2N]_3MH \xrightarrow[-H_2]{} [(Me_3Si)_2N]_2M \underset{\overset{|}{SiMe_3}}{\overset{\displaystyle \triangle}{\underset{N}{\diagup}} Si\diagdown} \tag{8}$$

$$M = U,\ Th$$

The metallacycles of uranium and thorium have been shown to undergo a large number of insertion and protonation reactions,[118–124] as shown in Figure 6. In some cases these reactions (such as reduction of carbonyl- containing organic compounds) have been found to be stereoselective.

Reactions of the metallacycle complexes with protic reagents are frequently used to generate derivatives. For example, reaction of An[N(SiMe$_3$)(SiMe$_2$CH$_2$)][N(SiMe$_3$)$_2$]$_2$ (An = Th, U) with excess aryl alcohols generates the complexes An(O-2,6-R$_2$C$_6$H$_3$)$_3$[N(SiMe$_3$)$_2$],[125,126] and reaction with a stoichiometric amount of aryl alcohol or aryl thiol yields the mixed ligand complexes An(E-2,6-R$_2$C$_6$H$_3$)[N(SiMe$_3$)$_2$]$_3$ (E = O, S).[126,127] Reaction of the thorium metallacyclic complex Th[N(SiMe$_3$)(SiMe$_2$CH$_2$)][N(SiMe$_3$)$_2$]$_2$ with smaller protic amines similarly results in incomplete transamination; reaction with four equivalents of HNMePh yields only the mixed amide complex Th(NMePh)$_2$[N(SiMe$_3$)$_2$]$_2$.[109]

The bis(trimethylsilyl)amide ligand is capable of supporting the formation of organoimido complexes at actinide centers. The tetravalent uranium dimer {U[N(SiMe$_3$)$_2$]$_2$(μ-N-p-C$_6$H$_4$Me)}$_2$ was prepared by reaction of ClU[N(SiMe$_3$)$_2$]$_3$ with LiHN(p-C$_6$H$_4$Me),[128] presumably by α-elimination of HN(SiMe$_3$)$_2$ from an intermediate amide complex. As observed for the related cyclopentadienyl compound, the arylimido ligand bridges the two metal centers in an asymmetric fashion, with U—N bond distances of 3.378(3) Å and 2.172(2) Å.

Tris(amido)amine ligands, N[CH$_2$CH$_2$NR]$_3$$^{3-}$ (R = trialkylsilyl), support unusual reactivity in the early actinides. Complexes of both thorium and uranium have been generated by metathesis reactions involving both the ligands N[CH$_2$CH$_2$N(SiMe$_3$)]$_3$$^{3-}$ and N[CH$_2$CH$_2$N(Si(Bu^t)Me$_2$)]$_3$$^{3-}$.

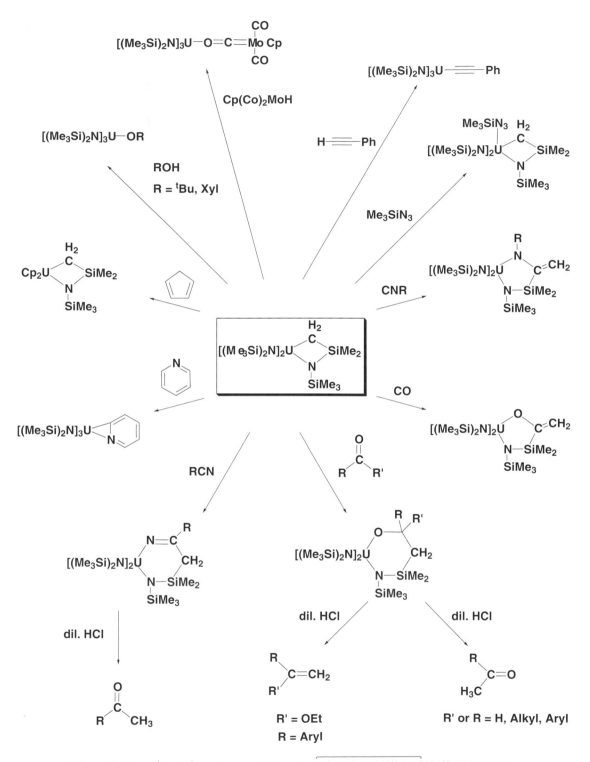

Figure 6   Reactions of uranium metallacycle $\overline{U_x[N(SiMe_3)(SiMe_2CH_2)]}[N(SiMe_3)_2]_2$.

The complexes $[\{N[CH_2CH_2NSiMe_3]_3\}AnCl]_2$ (An = Th, U) were first reported;[129] the molecular structure of the uranium complex demonstrated it was dimeric in the solid state. The chloride ligand may be substituted, and derivatives incorporating cyclopentadienyl, borohydride, alkoxide, amide, and diazabutadiene derivatives have been characterized.[130–132] Attempts to alkylate the complex $N[CH_2CH_2N(Si(Bu^t)Me_2)]_3UI$ with alkyllithium or alkylpotassium reagents resulted in the isolation of a metallacyclic product resulting from intramolecular activation of a methyl group, as show in Equation (9):[134]

$$N[CH_2CH_2N(Si(t\text{-}Bu)Me_2)]_3UI + KCH_2Ph \xrightarrow{-KI} PhCH_3 + \text{[structure]} \tag{9}$$

R = Si(*t*-Bu)Me$_2$

The U—C bond length in the metallacyclic unit is unusually long (2.752(11) Å), and is susceptible to protonation by alcohols, amines, and terminal alkynes; reaction with pyridine leads to the generation of a η^2-pyridyl complex.

The bulky amide ligand, NRAr- (R = Bu^t, adamantyl; Ar = 3,5-Me$_2$C$_6$H$_3$) has been used to synthesize monomeric complexes of U^{IV}. Complexes of the formula UI(NRAr)$_3$ may be prepared by the reaction of UI$_3$(THF)$_4$ with LiNRAr;[21] oxidation of the uranium center is presumed to be accompanied by sacrificial generation of U^0. A limited number of tetravalent derivatives of this ligand set have been reported, including the silyl complex [N(*t*-Bu)Ar]$_3$USi(SiMe$_3$)$_3$[135] and the bridging cyanoimide complex [N(Bu^t)Ar]$_3$U(μ-NCN)U[N(Bu^t)Ar]$_3$ (Ar = 3,5-Me$_2$C$_6$H$_3$).[136] Reaction of the trivalent complex with Mo[NPh(R′)]$_3$ (R′ = Bu^t, adamantyl) under dinitrogen results in the formation of (NRAr)$_3$U(μ-N$_2$)Mo[NPh(R′)]$_3$, which contains a linear Mo—N—N—U unit.[21] Both metals are seemingly best regarded as tetravalent. Reduction of UI(NRAr)$_3$ by KC$_8$ in toluene has been found to give rise to an interesting series of bridging arene complexes.[137,138] Reduction of IU[N(R)Ar]$_3$ (R = Bu^t, Ar = 3,5-Me$_2$C$_6$H$_3$) by KC$_8$ generates the complex (μ-C$_7$H$_8$){U[N(R)Ar]$_3$}$_2$. A related compound (μ-C$_7$H$_8$){U[N (R)Ar]$_3$}$_2$ (R = adamantyl), could also be generated in low yield by reaction of UI$_3$(THF)$_4$ with LiN(R)Ar(OEt$_2$) in toluene. These species serve as convenient precursors into U^{IV} derivatives. Reaction of (μ-C$_7$H$_8$){U[N(R)Ar]$_3$}$_2$ (R = Bu^t, Ar = 3,5-Me$_2$C$_6$H$_3$) with Ph$_2$S$_2$ generates a dimeric thiolate-bridged species, [U(μ-SPh)(SPh)[N(R)Ar]$_2$]$_2$, and reaction of the μ-arene complex with azobenzene yields the bridgine imido complex [U(μ-NPh)[N(R)Ar]$_2$]$_2$.[137]

Polymetallic species are formed in the aminolysis reactions of U(NEt$_2$)$_4$ with chelating diamines.[107,139] Reaction of U(NEt$_2$)$_4$ with *N,N′*-dimethylethylenediamine generates principally a trimeric product, U$_3$(MeNCH$_2$CH$_2$NMe)$_6$; a tetramer U$_4$(MeNCH$_2$CH$_2$NMe)$_8$ is obtained as a byproduct. Bimetallic complexes are similarly formed when bi- or polydentate ligands are used. The introduction by metathesis of the potentially tridentate ligand [(Pr^i)$_2$PCH$_2$CH$_2$]$_2$N$^-$ into the coordination sphere of thorium and uranium sheds light on the relative ionic radius of the two metals. Reaction of ThCl$_4$ with two equivalents of the amide salt results in the formation of [{[(Pr^i)$_2$PCH$_2$CH$_2$]$_2$N}$_2$ThCl(μ-Cl)]$_2$, in which each metal center is seven-coordinate. Each amide ligand is bidentate through the amide and one phosphine donor site; one pendant phosphine arm remains uncoordinated. In contrast, structural characterization of the analogous uranium complex reveals that eight-coordinate metal centers are in the dimeric complex; one amide ligand is tridentate, while the second is bidentate, with an uncomplexed phosphine arm.[140]

A series of cationic uranium amide complexes have recently been generated by protonation of neutral amide precursors with an acidic trialkylammonium salt of the weakly coordinating anion tetraphenylborate. The neutral species U(NEt$_2$)$_2$Cl$_2$ and U(NEt$_2$)Cl$_3$(THF) may be treated with NHEt$_3$BPh$_4$ to generate the cationic species [U(NEt$_2$)Cl$_2$(THF)$_2$][BPh$_4$] and [UCl$_3$(THF)$_2$][BPh$_4$], respectively. Reaction of NHEt$_3$BPh$_4$ with U(NEt$_2$)$_4$ generates the monocation [U(NEt$_2$)$_3$]$^+$; this can be further protonated in refluxing THF to generate the dication [U(NEt$_2$)$_2$(THF)$_3$]$^{2+}$. Similarly, reaction of [U(NEt$_2$)Cl$_2$(THF)$_2$][BPh$_4$] with additional ammonium salt at room temperature results in the formation of [UCl$_2$(THF)$_4$]$^{2+}$. The molecular structure of [U(NEt$_2$)$_3$(THF)$_3$][BPh$_4$] (Figure 7) and the pyridine adduct [U(NEt$_2$)$_2$(py)$_5$][BPh$_4$]$_2$[110] have been reported. The former contains a pseudooctahedral uranium atom, with a facial arrangement of amide ligands. The latter possesses *trans* amide ligands, with five pyridine ligands arrayed in the equatorial plane. Attempts to reduce U^{IV} cationic species such as [U(NEt$_2$)$_3$]$^+$ with sodium amalgam resulted only in the isolation of U(NEt$_2$)$_4$, suggesting disproportionation of the uranium-(III) intermediate.[141] The cationic species [U(NEt$_2$)$_3$][BPh$_4$] has been demonstrated to catalyze the dehydrocoupling of primary alkylamines and phenylsilane to generate the aminosilanes PhSiH$_{3-x}$-(NHR)$_x$ (x = 1–3).[142] In addition, the same species will catalyze the selective dimerization of terminal alkynes.[143]

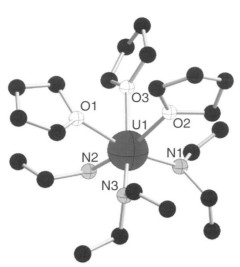

Figure 7   Crystal structure of [U(NEt$_2$)$_3$(THF)$_3$][BPh$_4$] (Berthet, Boisson *et al. J. Chem. Soc., Dalton Trans.*
1995, 3019).

Amidinate complexes. Amidinate ligands have been employed as ancillary ligands in the generation of compounds of tetravalent uranium and thorium. Reaction of Li[N(SiMe$_3$)$_2$] and Na[N(SiMe$_3$)$_2$] with *para*-substituted benzonitriles yields the benzamidinate ligands M[4-RC$_6$H$_4$C(NSiMe$_3$)$_2$] (M = Li, Na; R = H, Me, OMe, CF$_3$). Alternatively, more substituted Li[2,4,6-R$_3$C$_6$H$_2$C(NSiMe$_3$)$_2$] (R = CF$_3$, Me) is generated by the addition of aryllithium reagents to Me$_3$SiNCNSiMe$_3$. Amidinate ligands (L) have been used to generate complexes of the formula L$_2$AnCl$_2$ (An = Th, U) and L$_3$AnCl (for less sterically demanding substituents) by metathesis reactions.[144] Substitution of the halide precursors has been reported to generate methyl and borohydride derivatives.[145] The molecular structure of the complex [C$_6$H$_5$C(NSiMe$_3$)$_2$]$_3$UMe has been determined. The benzamidinate ligands coordinate to the metal center in a η^3-manner; the relatively long U—C σ bond of 2.498(5) Å is taken as an indication of steric crowding in the complex. Related amidinate and 1-aza- allyl ligands also have been shown to generate bis-(ligand)thorium dichloride complexes,[146] as well as an interesting mixed valence U^{III}/U^{IV} complex.[147]

Phosphide complexes. Rare examples of actinide phosphide complexes devoid of organo-metallic co-ligands (such as alkyl, cyclopentadienyl, or cyclooctatetraenyl) exist for tetravalent uranium and thorium. The reaction of ThCl$_4$ and four equivalents of the monoanionic and potentially tridentate ligand (PMe$_2$CH$_2$CH$_2$)$_2$P$^-$ (as the lithium salt) results in the formation of Th[P(CH$_2$CH$_2$PMe$_2$)$_2$]$_4$, a homoleptic phosphido complex.[148] The molecular structure of the complex indicates that each phosphine ligand has one coordinated and one uncomplexed phos-phine arm, yielding an eight-coordinate metal center. The compound is fluxional in solution at temperatures above −80 °C. The uranium analogue, U[P(CH$_2$CH$_2$PMe$_2$)$_2$]$_4$, has also been pre-pared,[149] and has been shown to have a similar structure. Although the compounds are isostruc-tural, they do not exhibit identical chemical behavior. Although the thorium complex will insert CO, the uranium compound will not.[149] The product of the thorium reaction is an unusual product of "double insertion" where the CO is incorporated into a diphospha-alkoxide.[150] The two phosphide phosphorus atoms become bonded to the inserted carbon atom, and the newly-generated P$_2$CO unit is η^3-bonded to the thorium center (see Figure 8).

Polypyrazolylborate ligands. The first report of an actinide complex employing a poly-(pyrazolyl)borate ligand was the preparation of complexes of the formula U[BH$_2$(pz)$_2$]$_4$, U[HB(pz)$_3$]$_4$, and U[HB(pz)$_3$]$_2$Cl$_2$ by reaction of UCl$_4$ with the potassium salt of the appropriate ligand.[151] On the basis of ^{13}C-NMR spectroscopy, the HB(pz)$_3$ ligands were assigned as bidentate in the complex U[HB(pz)$_3$]$_2$Cl$_2$, while the complex U[HB(pz)$_3$]$_4$ was speculated to have two bidentate and two tridentate ligands.[152]

The first report of metathesis reactions with thorium involved the preparation of the complexes Th[HB(pz)$_3$]$_{4-n}$X$_n$ (n = 2, X = Cl, Br; n = 1, X = Cl), Th[HB(3,5-Me$_2$-pz)$_3$]Cl$_2$, Th[B(pz)$_4$]$_2$Br$_2$, and adducts of the complexes Th[HB(pz)$_3$]Cl$_3$ and Th[HB(pz)$_3$]$_4$,[153] although subsequent reports have appeared of other derivatives, including Th[HB(3,5-Me$_2$-pz)$_3$]Cl$_3$.[154] The larger ionic radius of thorium enables higher coordination numbers; unlike the uranium complexes, the thorium

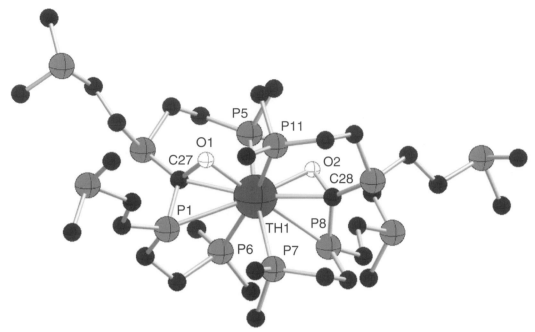

Figure 8 Phospha-alkoxide complex generated by insertion of carbon monoxide into a homoleptic thorium dialkylphosphide complex (Edwards, Hursthouse *et al. J. Chem. Soc., Chem. Commun.* **1994**, 1249).

derivatives Th[HB(pz)$_3$]$_2$X$_2$ (X = Cl, Br) were proposed to have tridentate pyrazolylborate ligands on the basis of spectroscopy.

Several routes have been identified to produce UI$_2$[HB(pz)$_3$]$_2$, including reaction of UI$_4$ with two equivalents of K[HB(pz)$_3$] in CH$_2$Cl$_2$[155] and oxidation of U[HB(pz)$_3$]$_2$I(THF)$_2$ with iodine.[156] The reaction of UI$_4$ with two equivalents of K[HB(pz)$_3$] in THF does not yield the same compound. The iodobutoxide complex U[HB(pz)$_3$]$_2$I(O(CH$_2$)$_4$I) was isolated, presumably generated by ring-opening of solvent.[155] The smaller size of the U^{IV} ion, combined with the larger steric size of the [HB(3,5-Me$_2$pz)$_3$] ligand inhibits formation of bis(ligand) complexes of the substituted poly(pyrazolyl)borate; reaction of UCl$_4$ with two equivalents of K[HB(3,5-Me$_2$pz)$_3$] leads to ligand degradation and the formation of UCl$_2$[HB(3,5-Me$_2$pz)$_3$](3,5-Me$_2$pz).[157]

The complex UCl$_3$[HB(3,5-Me$_2$pz)$_3$](THF) contains a relatively weakly coordinated solvent molecule; the base free complex can be isolated, and has been crystallographically characterized.[158] The THF is also readily replaced by a number of other coordinating bases, permitting comparisons of relative ligand affinity. The relative affinities of a series of bases for UCl$_3$[HB(3,5-Me$_2$pz)$_3$] was found to be: OPPh$_3$ > C$_6$H$_{11}$NC > PhCN > MeCN > OP(OEt)$_3$ > OP(O-Bu^n)$_3$ > C$_5$H$_5$N > THF.

Attempts to introduce a larger poly(pyrazolyl)borate ligand have demonstrated the steric limits of this system. Reaction of UCl$_4$ with one equivalent of the thallium salt of [HB(3-Mspz)$_3$]$^-$ (Ms = mesityl) generates only the product containing an isomerized ligand, UCl$_3$[[HB(3-Mspz)$_2$(5-Mspz)].[159]

A variety of metathesis reactions have been carried out with the bis(ligand) actinide species An[HB(pz)$_3$]$_2$Cl$_2$ to generate complexes containing oxygen, nitrogen, or sulfur donors.[160–163] Steric factors can be significant in these reactions. For example, reaction of bulky alkylamides with U[HB(pz)$_3$]$_2$Cl$_2$ generates only the monoamide complexes U[HB(pz)$_3$]$_2$Cl(NR$_2$).

In an attempt to reduce the steric constraints of the ancillary ligands, derivatives of the mono-(pyrazolylborate) complexes An[HB(3,5-Me$_2$pz)$_3$]Cl$_3$(THF) have also been prepared.[160,164–167] As before, the degree of substitution is often dependent on the size of the ligand introduced; tris(amide) derivatives such as An[HB(3,5-Me$_2$pz)$_3$](NR$_2$)$_3$ can be produced for R = Et, Ph, whereas for the larger ligand N(SiMe$_3$)$_2$$^-$, only a monoamide complex can be isolated. The monoalkoxide and monoaryloxide complexes of thorium have been reported to be unstable; uranium mono- and bis(phenoxide) complexes are only stable in the presence of a coordinating molecule of THF.[160]

The neptunium derivatives Np[HB(pz)$_3$]$_2$Cl$_2$ and Np[HB(3,5-Me$_2$pz)$_3$]Cl$_3$(THF) have been produced from NpCl$_4$.[168]

Reaction of uranium tetrachloride with two equivalents of the bulky ligand B(pz)$_4$$^-$ as the potassium salt yields the complex [B(pz)$_4$]$_2$UCl$_2$.[169] Although a limited number of derivatives of

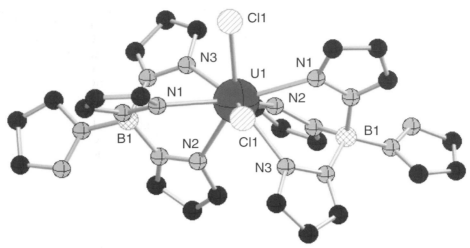

Figure 9   Crystal structure of [B(pz)₄]₂UCl₂ (Campello, Domingos *et al. J. Organomet. Chem.* **1999**, 579, 5–17).

this compound could be produced, in general the ligand set provided less thermal stability than comparable complexes of the "U[HB(pz)₃]₂" fragment. The complex [B(pz)₄]₂UCl₂ displays eight-coordinate geometry in the solid state, in an distorted square antiprismatic arrangement of ligands (Figure 9). The complex is fluxional in solution; ¹H-NMR spectra demonstrate that all coordinated pyrazolylborate rings are equivalent. For the derivatives [B(pz)₄]₂UCl(O-Buᵗ), [B(pz)₄]₂-UCl(O-2,4,6-Me₃C₆H₂), [B(pz)₄]₂U(S-Prⁱ)₂, and [B(pz)₄]₂U(O-Buᵗ)₂, it is possible to slow down the interconversion of the typical eight-coordinate polyhedra (square antiprism—dodecahedron—bicapped trigonal prism). At higher temperatures, it was possible for some of these compounds to reach a regime where all pyrazolyl groups were equivalent on the NMR timescale, indicating dissociative exchange of free and coordinated rings.

The potassium salt of the "podand" ligand tris[3-(2-pyridyl)-pyrazol-1-yl]borate (ᵖʸTp) reacts with thorium tetra(nitrate) to generate the complex (ᵖʸTp)Th(NO₃)₃.[170] The crystal structure of the complex reveals that the metal center is 12-coordinate, binding to the six nitrogen atoms of the podand ligand, and to two of the oxygen atoms of each nitrate. The molecule has three-fold symmetry, and the nitrates are located between the bidentate arms of the podand.

Thiocyanate and selenocyanate. A variety of tetravalent complexes of actinides are known incorporating the thiocyanate ligand, NCS⁻ (Table 5). The most prevalent member of this class is the anion An(NCS)₈⁴⁻. The tetraethylammonium complexes are known for An = Th, Pa, U, Np, and Pu. All possess a similar crystal structure, in which the metal ion lies within a cubic arrangement of thiocyanate ligands. The structure is dependent on counterion, however; the structure of the anion in Cs₄U(NCS)₈·H₂O reveals that the coordination environment about the uranium atom is square antiprismatic. When dehydrated, the complexes M₄An(NCS)₈ (M = Cs, Rb; An = Th, U) possess a dodecahedral metal environment in the solid state, but square antiprismatic geometry in acetone solution.[171] The analogous selenocyanate complexes (NEt₄)₄An(NCSe)₈ (An = U, Pa) have also been reported; they are isostructural with their isocyanate counterparts.

Table 5   Actinide(IV) thiocyanates and thiocyanato complexes.

[Th(NCS)₄(H₂O)₄]	
[U(NCS)₄(H₂O)₄]·(18-crown-6)₁.₅·3H₂O·MeCOBuⁱ	
Rb[Th(NCS)₅(H₂O)₃]	
Na₂[Th(NCS)₅(OH)(H₂O)ₓ]	$x = 2$ to 3
(NH₄)₃Th(NCS)₇·5H₂O	
K₄[U(NCS)₈]·xH₂O	$x = 0$ or 2
(NH₄)₄[M(NCS)₈]·xH₂O	M = Th or U, $x = 0$ and M = Th, $x = 2$
Rb₄[M(NCS)₈]·xH₂O	M = Th, $x = 0$, 2 or 3 and M = U, $x = 0$ or 1
Cs₄[M(NCS)₈]·xH₂O	M = Th, $x = 0$ or 2 and M = U, $x = 0$, 1 or 2
(NEt₄)₄[M(NCS)₈]	M = Th, Pa, U, Np and Pu

Neutral complexes of the thiocyanate ligands can be isolated as Lewis base adducts. Thorium and uranium form tetra(hydrate) complexes $An(NCS)_4(H_2O)_4$. The compound $U(NCS)_4(piperazine)$ has been reported.[172] The neutral adducts $Th(NCS)_4(L)_4$ and $U(NCS)_4(L)_3$ ($L = N,N$-diisopropylpropionamide) have been structurally characterized.[173] The smaller size of the uranium ion is reflected in the isolation of a seven-coordinate compound, while an eight-coordinate thorium compound is formed. If more sterically demanding carbamide ligands are employed ($L = N,N$-diisopropylbutyramide, $L' = N,N$-dicyclohexylacetamide), seven-coordinate thorium complexes, $Th(NCS)_4(L)_3$ and $Th(NCS)_2Cl_2(L')_3$, can be isolated.[174]

Sulfenamide. A homoleptic sulfenamido complex of uranium has been isolated from the reaction of $Li(Bu^tNSPh)$ with UCl_4 in the presence of PMe_3 in toluene.[175] The complex $U(PhS=N-Bu^t)_4$ possesses two η^2-coordinated sulfenamide ligands, and is nearly isostructural with the analogous zirconium complex.

(iii) Ligands containing neutral group 15 donor atoms

Ammonia. Ammonia adducts have been characterized for tetravalent halides of thorium, uranium, and plutonium. Adducts of all of the uranium halides have been reported,[7] as have adducts for thorium tetrachloride, -bromide, -iodide, and plutonium tetrachloride. Most lose ammonia at elevated temperatures. There is some ambiguity concerning the identity of these compounds as conflicting reports of the chemical composition exist. For example, $Cs_2[PuCl_6]$ is reported to react with ammonia to yield the simple adduct, $PuCl_4 \cdot xNH_3$, and yet it has been suggested that the related thorium chloride and bromide "ate" complexes $[ThX_6]^{2-}$ undergo ammonolysis in liquid ammonia to generate amide complexes of the composition $(NH_4)_2ThBr_2(NH_2)_2$.[176] Similar controversy exists in reports of possible ammonolysis of other tetravalent thorium complexes, such as the nitrate and the sulfate.

Amines, hydrazines, and hydroxylamines. Amine complexes are known for tetravalent complexes of the earliest actinides (Th, U), particularly for the halides, nitrates, and oxalates. The complexes are generated either in neat amine, or by addition of amine to the parent compound in a nonaqueous solvent. Some of the known simple amine compounds are presented in Table 6. The molecular structure of $ThCl_4(NMe_3)_3$ has been determined.[177] The coordination environment about the metal is a chloride capped octahedron. A very limited number of adducts exist in which a tetravalent actinide is coordinated by a hydrazine or hydroxylamine ligand; the parent compound is generally a halide or sulfate complex. Cationic metal hydrates coordinated with primary, secondary, or tertiary amines have also been isolated with acetylacetonate, nitrate, or oxalate as counterions.

Table 6 Representative actinide(IV) amine and hydrazine compounds.

$ThX_4 \cdot 4L$	$X = Cl$, Br; $L = RNH_2$, with $R = alkyl$, $PhCH_2$, aryl $X = Cl$; $L = RR'NH$, with $R = Ph$ and $R' = Me$, Et, $PhCH_2$, $PhX = Cl$; $L = R_2R'N$, with $R = Me$ or Et and $R' = Ph$
$[ThCl_4(R_3N)_x]$	$R = Me$, $x = 3$; $R = Et$, $x = 2$
$ThCl_4 \cdot 3MeC_6H_4NH_2$	(toluidine)
$ThCl_4 \cdot \beta\text{-}C_{10}H_7NH_2$	
$Th(acac)_4 \cdot PhNH_2$	
$Th(NO_3)_4 \cdot xL \cdot yH_2O$	$L = Bu^nNH_2$, Me_2NH, Et_3N
$Th(C_2O_4)_2 \cdot 4Bu^nNH_2 \cdot 2H_2O$	
$UCl_4 \cdot xL$	$L = RNH_2$: $x = 1$, $R = Et$, Pr^n; $x = 2$, $R = Me$, Et, Ph; $x = 3$, $R = Bu^t$; $x = 4$. $R = Pr^n$, $Bu^nL = R_2NH$: $x = 2$, $R = Et$; $x = 3$, $R = Me$, Pr^i, $Bu^tL = R_3N$: $x = 1$, $R = Et$; $x = 2$, $R = Me$
$UBr_4 \cdot 2Et_2NH$	
$U(OPh)_4 \cdot 2PhOH \cdot Et_3N$	
$M^{IV}F_4 \cdot xN_2H_4$	$M^{IV} = Th$, $x = 1$, 1.66; $M^{IV} = U$, $x = 1$, 1.5 or 2
$ThCl_4 \cdot 4PhNHNH_2$	
$UCl_4 \cdot xN_2H_4$	$x = 6$ or 7
$Th(SO_4)_2 \cdot xN_2H_4$	$x = 1.5$ or 2

The Actinides

Table 7   Actinide(IV) diamine compounds.

Ethylenediamine, en	
$Th(C_9H_6NO)4 \cdot C_9H_7NO \cdot en$	$C_9H_7NO = 8\text{-hydroxyquinoline}$
$\quad UCl_4 \cdot 4en$	
N, N, N', N'-Tetramethylethylenediamine, tmed	
$\quad U\{(CF_3)_2CHO\}_4 \cdot tmed$	
Diaminoalkanes	
$ThBr_4 \cdot xL \cdot yH_2O$	$x = 2, \ y = 5, \ L = 1,2\text{-diaminopropane and}$
	$1,4\text{-diaminobutane } x = 2, \ y = 2 \text{ and } x = 4, \ y = 6,$
	$L = 1,4\text{-diaminobutane} x = 4, \ y = 2,$
	$L = 1,2\text{-diaminopropane}$
Diaminoarenes	
$ThCl_4 \cdot 2L$	$L = 1,2\text{-}, \ 1,3\text{- or } 1,4\text{-diaminobenzene},$
	$4,4'\text{-diamino-biphenyl (benzidine)},$
	$4,4'\text{-diamino-3,3'-dimethyl-biphenyl (}o\text{-tolidine) or}$
	$4,4'\text{-diamino-3,3'-dimethoxybiphenyl (}o\text{-dianisidine})$
$UCl_4 \cdot 2L$	$L = 1,8\text{-diaminonaphthalene}$
$2UCl_4 \cdot 5L$	$L = 1,2\text{-diaminobenzene}$
$UBr_4 \cdot 4L$	$L = 1,2\text{-diaminobenzene}$
$[Th(NO_3)_2L_2](NO_3)_2$	$L = 1,2\text{-diaminobenzene}$

Given the apparent lability of amines coordinated to tetravalent actinide centers, amine complexes have often been stabilized by the introduction of a chelating diaminoalkane or diaminoarene. The most common derivatives are those of the parent actinide halides,[178] as shown in Table 7. Both the complexes $UCl_4(tmeda)_2$ (tmeda = N,N,N',N'-tetramethylethylenediamine)[179] and $ThCl_4\text{-}(tmeda)_2$[180] have been characterized crystallographically. Both complexes are eight-coordinate with bidentate tmeda ligands. The geometry about the metal center approximates a D_{2d} dodecahedron. The tmeda ligands are readily replaced by chelating diphosphine ligands (*vide infra*), indicating that the tetravalent actinides have a stronger affinity for softer phosphine donors.

These complexes can act as reagents in subsequent reactions,[178] although displacement of the tmeda ligand is observed. The tetravalent derivative $U(MeBH_3)_4(tmeda)$ has been reported, although it was produced by displacement of THF from the complex $U(MeBH_3)_4(THF)_x$.[181]

Heterocyclic ligands. N-heterocyclic adducts of simple tetravalent actinide salts exist for halides, nitrates, carboxylates, alkoxides, and perchlorate complexes of thorium, as well as halides and alkoxides of uranium. Most common among these are complexes with pyridine and its derivatives. Coordination number for the metal center range from six to eight (Table 8). The

Table 8   Some complexes of N-heterocyclic ligands with actinide(IV) compounds.

Pyridine, C_5H_5N (py) and substituted pyridines	
$ThCl_4 \cdot 2L$	$L = 2\text{-Me- or } 2\text{-}H_2N\text{-}C_5H_4N$
$UX_4 \cdot 2py$	$X = Cl, \ Br$
$ThX_4 \cdot 4L$	$X = Cl, \ Br, \ NCS; \ L = py, \ 2\text{-Me-}, \ 2,4\text{-Me}_2\text{- and } 2,6\text{-Me}_2\text{-pyridine}$
$UCl_4 \cdot 4L$	$L = (2\text{-}H_2N, \ 3\text{-HO})C_5H_3N$
$ThI_4 \cdot 6L$	$L = py, \ 2\text{-Me-}, \ 2,4\text{-Me}_2\text{- and } 2,6\text{-Me}_2\text{-pyridine}$
$Th(NO_3)_4 \cdot 2L$	$L = py, \ 2\text{-Me-}, \ 2\text{-}H_2N\text{-}, \ 2,4\text{-Me}_2\text{- and } 2,6\text{-Me}_2\text{-pyridine}$
$Th(ClCCO_2)_4 \cdot 2py$	
$[ThL_6](ClO_4)_4$	$L = 2\text{-}H_2N\text{-}, \ 2,4\text{-Me}_2\text{- and } 2,6\text{-Me}_2\text{-pyridine}$
$[ThL_8](ClO_4)_4$	$L = py, \ 2\text{-Me-}C_5H_4N$
$[M(Py)_8]_3[Cr(NCS)_6]_4$	$M = Th, \ U$
$UCl_2(C_7H_5O_2)_2 \cdot 2Py$	$C_7H_6O_2 = 2\text{-hydroxybenzaldehyde}$
Quinoline and	
isoquinoline, C_9H_7N	
$ThX_4 \cdot 4C_9H_7N$	$X = Cl, \ Br, \ NCS$
$Th(NCS)_4 \cdot 4\text{-}i\text{-}C_9H_7N$	
$ThI_4 \cdot 6C_9H_7N$	
$Th(NO_3)_4 \cdot 2L$	$L = C_9H_7 \ N \text{ or } i\text{-}C_9H_7N$
$[ThL_8](ClO_4)_4$	$L = C_9H_7 \ N \text{ or } i\text{-}C_9H_7N$

steric bulk of the ligand dictates the precise coordination number. Thorium will coordinate eight ligands in the complexes $[Th(L)_8](ClO_4)_4$ (L = pyridine, 2-Me-pyridine), whereas the bulkier L = 2,4-Me$_2$-pyridine and 2,6-Me$_2$-pyridine supports only six coordinate thorium in $[Th(L)_6](ClO_4)_4$.[182] Compounds of uranium and thorium halides and perchlorates have also been isolated with coordinating piperidine, quinoline, and isoquinoline ligands and their derivatives. As in the case of pyridine derivatives, the metal centers are most often eight-coordinate. Displacement of halides is possible to maintain this coordination environment. For example, the quinoline complex $ThI_4 \cdot C_9H_7N$ behaves as the 1:2 electrolyte $[ThI_2L_6]I_2$ in solution, suggesting that the metal center remains eight-coordinate.[183]

Bidentate heterocyclic ligands (e.g., 2,2'-bipyridine, bipy, or 1,10-phenanthroline, phen) are also commonly used as coordinating bases in tetravalent chemistry, although there are few structurally characterized examples. These ligands are also presumed to support metal coordination numbers up to eight. Some of these complexes are neutral, such as $ThX_4(bipy)_2$ (X = Cl, Br, NCS),[184] whereas some behave as salts in solution, indicating displacement of the counterion from the primary coordination sphere of the metal ion.[182] 2,2'-Bipyridine and 1,10-phenanthroline can also act as Brønsted bases in reactions with protic solvents. Reaction of UCl_4 and bipy or phen in alcohols such as ethanol results in the formation of products of partial alcoholysis such as $UCl_3(OEt)bipy_2$, accompanied by the formation of $(bipyH)_2UCl_6$.[185]

Other N-containing heterocycles that have been employed as coordinating bases include phenazine, phthalazine, pyrazine, triazine, imidazole, and piperizine, as well as pyridine-containing complexes such as terpyridine, dipyridylethanes, dipyridylketone, and dipyridylamine. In some cases it has been speculated that the products of ligands containing more than one nitrogen in the ring are polymeric, with ligands coordinated through both nitrogen atoms. In a more recent study of coordination piperazine compounds of uranium tetrahalides, perchlorates, and thiocyanates, however,[172] optical spectroscopy is consistent with six- and eight-coordinate coordination environments about the uranium centers in many of the derivatives reported, suggesting simpler coordination modes.

Nitriles. Nitrile complexes of uranium and thorium halides have been well studied, particularly complexes of acetonitrile (MeCN). Halide complexes $AnX_4(MeCN)_n$ were intially prepared either by reaction of the anhydrous halide with acetonitrile, or by electrochemical dissolution of a thorium anode in the presence of dissolved chlorine. Initial estimates of stoichiometry suggested $n = 2$ or 4, depending on the steric bulk of the base. It has been suggested from UV–visible spectroscopy that most complexes possess eight-coordinate metal centers, although with larger nitriles (e.g., Bu^{t}CN), complexes of the formula $UX_4(Bu^tCN)_3$ could be isolated. The molecular structure of $UCl_4(MeCN)_4$ has been determined,[186] confirming the coordination number of the metal center.

More recently, more well defined adducts of the tetrahalides with acetonitrile have been isolated by oxidation of the appropriate metal (uranium, thorium) by halide sources in the presence of the nitrile.[187] It has been reported that addition of more strongly coordinating bases to $UI_4(MeCN)_4$ in acetonitrile can generate Lewis base adducts (e.g., $UI_4(tmu)_2$, tmu = tetramethylurea). The complex appears to undergo some halide or nitrile dissociation in polar media, however; addition of $OPPh_3$ to a solution in THF yields only the THF ring-opened product $UI_2(OCH_2CH_2CH_2CH_2I)_2(OPPh_3)_2$.[188]

Aliphatic phosphines. As discussed in Section 3.3.2.2.2(ii), tetravalent actinide complexes possess a surprisingly high affinity for soft phosphine donor groups. In addition to the pendant phosphinoamine and phosphine complexes discussed previously, a number of other tetravalent complexes of uranium and thorium containing neutral phosphine ligands have been reported. The first report of a phosphine adduct of a tetrahalide was the bridging diphosphine complex $[UCl_4]_2(dppe)$.[189] All other reports of diphosphine complexes have involved chelation of a single metal center, and the species $ThX_4(dmpe)_2$ (X = Cl, I) and $UX_4(dmpe)_2$ (X = Cl, Br) have been characterized.[190] The disphosphine remains coordinated during metathesis reactions, and the derivatives $AnX_4(dmpe)_2$ (An = U, Th; X = Me, OPh) can be prepared by reaction with the appropriate lithium reagents.[190] The slightly larger benzyl group forces displacement of one of the chelating disphosphine ligands, resulting in the formation of $An(CH_2Ph)_4(dmpe)$ (An = U, Th).[191]

A single uranium halide adduct of a monodentate phosphine has been reported; reaction of UCl_4 with excess PMe_3 permits isolation of the trimethylphosphine adduct, $UCl_4(PMe_3)_3$.[190]

Arsines. The ligand *o*-phenylenebis(dimethylarsine) (diars) has been used to complex actinide halides. The complexes $PaCl_4(diars)$ and $UCl_4(diars)$ have been reported. Both are produced by the reduction of pentavalent precursors in solution upon addition of the arsine.[192]

(iv) Ligands containing anionic group 16 donor atoms

Oxides. Due to their importance as nuclear fuel material, actinide oxides have been intensively investigated. They are very complicated compounds, due to the formation of nonstoichiometric or polymorphic materials. The dioxides, AnO_2, have the fluorite structure, wherein the actinide has eight nearest neighbor oxygen atoms in a cubic geometry. They can be readily prepared by heating of the actinide hydroxide, oxalate, carbonate, peroxide, nitrate, and other oxyacid salts. For the elements beyond thorium, sub- and superstoichiometric oxides remain an area of research. This is mostly due to two main issues. The dominant disposal or repository form of nuclear waste is spent fuel rods, which are based on UO_2. It is very insoluble in its crystalline form; however, as the material ages it becomes brittle, and under common conditions undergoes phase transformations to hydroxides and oxidized forms that would significantly increase uranium solubility and potential for release into the environment. Secondly, there is concern that stored Pu will be slightly hydrated and highly self-irradiated Pu^0 and PuO_2 and may transform to PuO_{2+x}, which is significantly less stable.[193] Earlier reports suggested that PuO_{2+x}, like UO_{2+x}, contains interstitial oxygen in clusters of defect sites. There is more recent spectroscopic data, however, that suggest the presence of actinyl species.[194-196]

Ternary actinide(IV) oxides are numerous and varied. Some classic types are M_2AnO_3 (M = Na, K, Rb, Cs; An = Th–Pu), $BaAnO_3$(for Th–Pu), and Li_8PuO_6, which can be prepared by fusing the respective actinide and alkali or alkaline earth oxides to form double oxides. In addition, ternary thorium oxides have been reported with lanthanides ($(ThCe)O_2$, $(ThCe)O_{2-x}$ with $x < 0.25$), niobium ($Th_{0.25}NbO_3$), tantalum ($ThTa_2O_7$, $Th_2Ta_2O_9$), molybdenum ($Th(MoO_4)_2$, $ThMo_2O_8$), germanium ($ThGeO_4$), titanium and vanadium. Analogous U and Pu phases are known for most of these compounds. Superconducting properties have been observed in the δ-compound $Nd_{2-x}Th_x CuO_4$ at $x = 0.16$.

Hydroxides. Pure and mixed metal actinide hydroxides have been studied for their potential utility in nuclear fuel processing. At the other end of the nuclear cycle, the hydroxides are important in spent fuel aging and dissolution, and environmental contamination. Tetravalent actinides hydrolyze readily, with Th^{IV} more resistant and Pu^{IV} more likely to undergo hydrolysis than U^{IV} and Np^{IV}. All of these ions hydrolyze in a stepwise manner to yield monomeric products of formula $An(OH)_n^{4-n}$ with $n = 1, 2, 3$ and 4, in addition to a number of polymeric species. The most prevalent and well characterized are the mono- and tetra-hydroxides, $An(OH)^{3+}$ and $An(OH)_4$.[197-199] Characterization of isolated bis and tri-hydroxides is frustrated by the propensity of hydroxide to bridge actinide centers to yield polymers. For example, for thorium, other hydroxides include the dimers, $Th_2(OH)_2^{6+}$ and $Th_2(OH)_4^{4+}$, the tetramers, $Th_4(OH)_8^{8+}$ and $Th_4(OH)_{12}^{4+}$, and two hexamers, $Th_6(OH)_{14}^{10+}$ or $Th_6(OH)_{15}^{9+}$.[200-203] These polynuclear complexes are common in chloride and nitrate solutions. It is noteworthy that these polynuclear hydrolysis products have only been well defined for thorium (i.e., not for other tetravalent actinide ions). For U^{IV} there is limited evidence for polymeric species such as $U_6(OH)_{15}^{9+}$.[197] Characterization of additional distinct Pu^{IV} hydroxide species has been thwarted by the formation of the colloidal oxy/hydroxide. This form is very common in aqueous Pu chemistry and can form under widely varying conditions, including in concentrated electrolytes. It has a distinctive optical absorbance spectrum, can range in size from ten to hundreds of angstroms, and is generally described as hydroxylated nanoparticles of PuO_2.

The An^{IV} hydroxide solids are amorphous and their exact composition and structure are not known.[204] Ternary hydroxides have been characterized, mostly for Th. The structure of $Th(OH)_2$-$CrO_4·H_2O$ is built up of infinite chains, $[Th(OH)_2]_n^{2n+}$ containing two almost parallel rows of OH groups so that each thorium atom is in contact with four OH groups; the CrO_4^{2-} groups are so packed that each thorium atom is in contact with four oxygen atoms of four different CrO_4^{2-} groups, making up a square antiprismatic arrangement of oxygen atoms about each thorium atom. The structure of $Th(OH)_2SO_4·H_2O$ is similar.

Peroxides and other dichalcogenides. Peroxide ligands oxidize uranium and protactinium, so peroxo complexes of tetravalent early actinides are restricted to An = Th, Np, and Pu. The compounds $AnO_4·xH_2O$ precipitated from dilute acid solutions of neptunium(IV) and plutonium-(IV) by hydrogen peroxide appear to be actinide(IV) compounds, although the stoichiometry has not been well determined. Pu^{IV} peroxide evidences two crystalline forms, hexagonal and cubic face-centered.[82] The former contains 3-3.4 and the latter, 3 peroxo oxygens atoms per Pu. Soluble intermediates of the type $[Pu(\mu-O_2)_2Pu]^{4+}$ reportedly form at low hydrogen peroxide concentrations. The hydrated thorium peroxide sulfate, $Th(O_2)_4 (SO_4)·3H_2O$, is very stable. Several mixed-ligand thorium peroxo complexes have been isolated, including the sulfate $Th(O_2)(SO_4)·H_2O$, carboxylates $Th(O_2)(RCO_2)$, phenoxo compounds, and mixed composition

Table 9 Actinide(IV) carbonates and carbonato complexes.

$An(CO_3)_2 \cdot xH_2O$	$An = Th$, $x = 0.5$, 3–4; Pu, $x = ?$
$AnO(CO_3) \cdot xH_2O$	$An = Th$, $x = 2$, 8; U, $x = 0$; Pu, $x = 2$
$Th(OH)_2(CO_3) \cdot 2H_2O$	
$xAnO_2 \cdot AnO(CO_3) \cdot yH_2O$	$An = Th$, $x = 1$, $y = 1.5$ or 4; $x = 3$, $y = 1$; $x = 6$,
	$y = 0$, $An = Pu$, $x = 1$, $y = 0$ or 3
$M^I_2[Th(CO_3)_3] \cdot xH_2O$	$M^I = NH_4$, $x = 6$; CN_3H_6 (guanidinium), $x = 0$, 4
$(enH_2)[U(CO_3)_3(H_2O)_2] \cdot 2H_2O$	
$M^I_4[An(CO_3)_4] \cdot xH_2O$	Generally known for $An = Th\text{-}Pu$; $M^I = Na$, K, NH_4, CN_3H_6
$M^I_6[An(CO_3)_5] \cdot xH_2O$	Generally known for $An = Th\text{-}Pu$; $M^I = Na$, K, NH_4, CN_3H_6
	$An = Th$, U; $M = [Co(NH_3)_6]_2$, $x = 4$, 5
	$An = Th$; $M = Ca_3$, Ba_3, $x = 7$;
	$(CN_3H_6)_3(NH_4)_3$, $x = 3$ $An = U$; $M^I = (CN_3H_6)_4(NH_4)_2$, $x = 1$;
	$(CN_3H_6)_3(NH_4)_3$, $x = 2$
$M^I_8[Pu(CO_3)_6] \cdot xH_2O$	$M^I = Na$, K, NH_4, x unspecified
$M^I_{12}[Pu(CO_3)] \cdot xH_2O$	$M^I = Na$, K, NH_4, x unspecified
$Na[Th(OH)(CO_3)_2(H_2O)_3] \cdot 3H_2O$	
$M^I_2[Th(OH)_2(CO_3)_2(H_2O)_2] \cdot xH_2O$	$M^I = Na$, $x = 8$; K, $x = 3$
$K_3[Th(OH)(CO_3)_3(H_2O)_2] \cdot 3H_2O$	
$Na_5[Th(OH)(CO_3)_4(H_2O)] \cdot 8H_2O$	
$(CN_3H_6)_5[An(OH)_3(CO_3)_3] \cdot 5H_2O$	$An = Th$, U
$(enH_2)[U(OH)_2(CO_3)_2] \cdot 3H_2O$	
$(enH_2)_2[U_2(OH)_2(CO_3)_5)(H_2O)_4] \cdot 2H_2O$	
$(CN_3H_6)_5[Th(CO_3)_3F_3]$	
$M^I_2[U(HCO_3)_2F_4]$	$M^I = Na$, NH_4

(and potentially polymetallic) halides and nitrates, $Th(O_2)_{1.6}(A^-)_{0.5}(O^{2-})_{0.15} \cdot 2.5H_2O$ with $A = Cl$ or NO_3.

The reaction of uranium metal with polyselenides in molten potassium polyselenide generates an interesting molecular diselenide complex $K_4[U(\eta^2\text{-}Se_2)_4]$,[205] containing a discrete $U(\eta^2\text{-}Se_2)_4^-$ anion.

Carbonates. Actinide carbonates have been very thoroughly studied by a variety of solution and solid state techniques. These complexes are of interest because of their fundamental chemistry and environmental behavior, including aspects of actinide mineralology. In addition, separation schemes based on carbonate have been proposed. Coordination numbers are generally quite high, eight to ten; carbonate is bound to the metal center in a bidentate fashion and is often hydrogen-bonded to outer sphere waters or counter ions (see Table 9).

Aqueous carbonate complexes of An^{IV} ions, $An(CO_3)_n^{4-2n}$, $n = 1\text{-}5$, form stepwise with increasing solution pH and carbonate concentration.[46,197,202,206] As with other oxoanionic ligand systems, the stability of the carbonate complexes decreases across the series, such that the pentacarbonato complex is well studied for Th^{IV} and U^{IV}. The tetracarbonato complex is more important for Np^{IV} and Pu^{IV} in solution, although salts of the pentacarbonato anion are known across the series. Most studies of Th, U, Np, and Pu do indicate that mixed hydroxyocarbonate complexes, $An(OH)_x(CO_3)_y^{(2y+4-x)-}$, e.g, $Th(OH)_3(CO_3)^-$ for Th, are important in describing the aqueous solution behavior. For the lower order carbonates the actinide is presumably nine or ten-coordinate with waters and bidentate carbonate in the inner coordination sphere. For the penta- and hexacarbonato complexes there is no evidence that any water molecules remain bound to the actinide center.

Actinide(IV) carbonato solids of general formula $M_xAn(CO_3)_y \cdot nH_2O$ have been prepared for a variety of metal cations ($M = Na^+$, K^+, NH_4^+, $C(NH_2)_3^{2+}$, $y = 4, 5, 6, 8$). The only well-characterized actinide(IV) hexacarbonato compound is the mineral tuliokite $Na_6BaTh(CO_3)_6 \cdot 6H_2O$.[207] The three dimensional structure consists of alternating chains of barium and thorium icosahedra which share common polyhedral faces. The sodium atoms are found interspersed between the barium and thorium columns. The thorium chains contain discrete $Th(CO_3)_6^{8-}$ icosahedra, which have three mutually perpendicular planes formed by the *trans* carbonate ligands, giving virtual T_h symmetry.

The pentacarbonato salts of thorium(IV) and uranium(IV) are among the most well-studied actinide solids. They can be prepared directly by precipitation from carbonate solutions, or indirectly by the decomposition of oxolates or reduction of actinyl(V, VI) species. The salts of formula $M_6An(CO_3)_5 \cdot nH_2O$ ($An = Th$, U; $M_6 = Na_6$, K_6, Tl_6, $[Co(NH_3)_6]_2$, $[C(NH_2)_3]_3[(NH_4)]_3$,

[C(NH$_2$)$_3$]$_6$; $n = 4$–12) have all been reported.[208–210] These hydrated salts contain bidentate carbonate ligands and no water molecules bound directly to the central actinide. Structures from single crystal X-ray diffraction studies are known for salts of Th(CO$_3$)$_5$$^{6-}$. For example, triclinic Na$_6$Th(CO$_3$)$_5$·12H$_2$O contains Th^{IV} coordinated to 10 oxygen atoms of five bidentate carbonato ligands in an irregular geometry.[211–213] Use of the hydrogen-bond donating guanidinium cation provides a more regular geometric structure in [C(NH$_2$)$_3$]$_6$[Th(CO$_3$)$_5$], where the coordination geometry about the metal is hexagonal bipyramidal, comprised of three bidentate carbonate ligands in an approximately hexagonal plane and two *trans* bidentate carbonate ligands occupying pseudo-axial positions.[214] Uranium(IV) carbonates are readily oxidized in air to uranium(VI) complexes, and are therefore not as well characterized. The potassium salt, K$_6$U(CO$_3$)$_5$·6H$_2$O, can be prepared by dissolution of freshly prepared U^{IV} hydroxide in K$_2$CO$_3$ solution in the presence of CO$_2$, and the guanidinium salt can be prepared by addition of guanidinium carbonate to a warm U(SO$_4$)$_2$ solution, followed by cooling.[215]

Salts of An^{IV} carbonates of lower carbonate to actinide ratio have also been reported but are far less common and detailed structural information is generally not available. Simple, binary thorium(IV) carbonates of formula Th(CO$_3$)$_2$ and Th(CO$_3$)$_2$·nH$_2$O ($n = 0.5$ and 3.00) are reported to form during the pyrolysis of Th(C$_2$O$_4$)$_2$, or by heating thorium hydroxide under CO$_2$.[216] Solids of formula ThO(CO$_3$) and Th(OH)$_2$(CO$_3$)·2H$_2$O have also been reported, but not characterized.[216] An example of a mixed ligand carbonate is the carbonatothorofluorothorate(IV), (CN$_3$H$_6$)$_5$[Th(CO$_3$)$_3$F$_3$], which is nine-coordinate. Th^{IV} is bonded to three bidentate carbonate groups and three terminal fluorine atoms to give a monocapped square antiprismatic geometry in this complex.

Tetracarbonato U^{IV} salts, such as [C(NH$_2$)$_3$]$_4$[U(CO$_3$)$_4$] and [C(NH$_2$)$_3$]$_3$(NH$_4$)[U(CO$_3$)$_4$] have been reported,[215] and a tricarbonate is reportedly formed by addition of ethylenediammonium sulfate to uranium(IV) solutions of (NH$_4$)$_2$CO$_3$ or KHCO$_3$ followed by precipitation of [C$_2$H$_4$(NH$_3$)$_2$][U(CO$_3$)$_3$(H$_2$O)]·2H$_2$O.[215] Hydrolysis of this complex occurs with dissolution to give [C$_2$H$_4$(NH$_3$)$_2$]$_2$[U$_2$(OH)$_2$(CO$_3$)$_5$(H$_2$O)$_4$]·2H$_2$O or [C$_2$H$_4$(NH$_3$)$_2$][U(OH)$_2$(CO$_3$)$_2$(H$_2$O)$_2$]·H$_2$O.

Plutonium(IV) carbonato complexes can be similarly prepared, as demonstrated by the single-crystal X-ray diffraction structure reported for the precipitated sodium salt, [Na$_6$Pu(CO$_3$)$_5$]$_2$·Na$_2$CO$_3$·33H$_2$O.[217] Plutonium is coordinated by 10 carbonate oxygens in the anion shown in Figure 10. This type of complex can also be prepared by dissolving plutonium(IV) oxalate in the alkali metal carbonate solution and precipitating the solid by addition of alcohols. Depending on reaction conditions, green amorphous powders of compositions K$_4$Pu(CO$_3$)$_4$·nH$_2$O, K$_6$Pu(CO$_3$)$_5$·nH$_2$O, K$_8$Pu(CO$_3$)$_6$·nH$_2$O, and K$_{12}$Pu(CO$_3$)$_8$·nH$_2$O have all been reported.[218] Sodium salts of formula Na$_4$Pu(CO$_3$)$_4$·3H$_2$O, Na$_6$Pu(CO$_3$)$_5$·2H$_2$O, and Na$_6$Pu-(CO$_3$)$_5$·4H$_2$O have been claimed as light green crystalline compounds that appear to dehydrate in air.[218] Similarly, the (NH$_4$)$_4$Pu(CO$_3$)$_4$·4H$_2$O and [Co(NH$_3$)$_6$]$_2$Pu(CO$_3$)$_5$·5H$_2$O salts

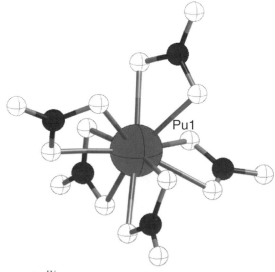

Figure 10 The pentacarbonato Pu^{IV} anion from the crystal structure of the hydrated sodium salt (Clark, Conradson *et al. Inorganic Chemistry* **1998**, 37, 2893–2899).

have been reported.[219] Although few of these compounds have been well characterized, their formulas are consistent with the known Th^{IV} phases, and they are presumably isostructural with them.

Carboxylates. Aminocarboxylate and other polycarboxylate actinide complexes are typically formed in aqueous solution by addittion of actinide(IV) salts to a solution of the polycarboxylic acid ligand. Most have been characterized by NMR and optical visible and infrared spectroscopy, and a very few have been fully structurally characterized using single-crystal X-ray diffraction. They can also form by reduction of the metal ion present in higher oxidation states, as demonstrated for Pu(VI) and Pu(V). Tetravalent actinide ions form 1:1 complexes An(IV)-L (L = NTA, HEDTA, EDTA, DTPA, citrate) in acidic solutions, but even hexadentate EDTA leaves coordination sites for hydrolysis, polymerization, or other additional complexation, often at higher pH. For example, mixed aminocarboxylate/hydroxide, An(IV)-L(OH)$_x$, and aminocarboxylate/carbonate complexes, have been prepared for Th(IV) and Pu(IV). Several additional types of mixed ligand Th(IV) complexes have been characterized, including ThLL′, where L = NTA, HEDTA, EDTA, CDTA or DTPA; L′ = halide, resorcinol (res), 2-methylresorcinol (2-Me-res), 5-methylresorcinol (5-Me-res) or 4-chlororesorcinol (4-Cl-res), salicylic acid (SA) or 5-sulfosalicylic acid (SSA), ethylenediamine, 1,2-propylenediamine, 1,3-propylenediamine, diethylenetriamine or triethylenetetramine, and purines, and pyrimidines (adenosine, guanosine, cytidine, uridine, adenine, etc.). Complexes with 2:1 aminocarboxylate to actinide ion ratio are isolated from solutions containing excess ligand and are more stable with respect to hydrolysis. The crystal structure of U(IV) and Th(IV) complexes (CN$_3$H$_6$)$_3$[AnEDTAF$_3$] reveal that the central actinide atom is surrounded by 3F atoms, 4O atoms, and 2N atoms, with EDTA in a gauche conformation. Malonic acid complexes with thorium(IV) to create a distorted square antiprism coordination environment. The malonic acid complexes through both acid groups in a 1,5 arrangement to create a six-membered ring. The ring is planar where it complexes with the thorium, but the alpha carbon deviates from the plane. This is illustrated in Figure 11. Addition of a water molecule raises the coordination number from eight to nine forming a monocapped square antiprism where the capped face is more planer than the uncapped faced.

Since malonic acid is a diacid, the formation of polymeric chains is possible. Uranium(IV) has been shown to form polymeric chains creating a three-dimensional lattice. For every uranium atom, three water molecules are complexed along with two tridentate malonic acids (η^2 is present and a μ^2 is also present). The polymeric chain is depicted in Figure 12.[220]

Nitrates. Aqueous nitrate complexes (see Table 10) of An^{IV} ions are very well studied mostly because of their importance in nuclear material processing, particularly in liquid/liquid extractions and ion exchange chromatography. The solution species, An(NO$_3$)$_n^{4-n}$, $n = 1$–6, have been extensively studied for Th^{IV} and Pu^{IV}. Numerous cationic resins have been developed that have strong affinity for the hexanitrato species, Pu(NO$_3$)$_6^{2-}$. Although later work suggests that this complex is not present at significant concentration in the absence of resins, even in concentrated nitrate solution. There is good evidence, including recent NMR and EXAFS data, that indicates the mono-, bis-, tetra, and the hexa- complexes are significant, but the tris- and penta-nitrato complexes are not.[221] X-ray absorbance data for the system suggest aquo ligation decreases in the inner sphere even before sequential planar, bidentate nitrates bind the metal center. The coordination numbers in these complexes is approximately 11–12 for the first coordination sphere, with average bond

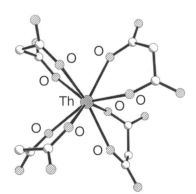

Figure 11 Crystal structure of (C$_4$H$_{12}$N$_2$)$_2$[Th(C$_3$H$_2$O$_4$)$_4$]H$_2$O (Zang, Collison *et al. Polyhedron* **2000**, *19*, 1757–1767).

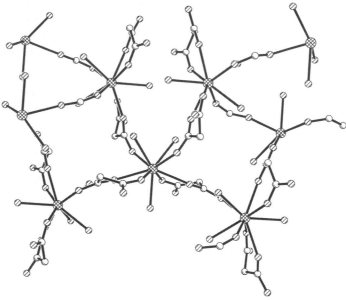

Figure 12   Crystal structure of [U(C$_3$H$_2$O$_4$)$_2$(H$_2$O)$_3$]$_n$ depicting the coordination environment of the U^{IV} in the polymeric chain (Zhang, Collison *et al. Polyhedron* **2000**, *19*, 1757–1767).

Table 10   Actinide(IV) nitrate complexes.

M$^I_2$[An(NO$_3$)$_6$]	Generally known for An = Th, U, Pu; M^I = NH$_4$, K, Rb, Cs, Tl, Et$_4$ N, Bu^n, Me$_2$(PhCH$_2$)$_2$ N, Me$_3$(PhCH$_2$)N, C$_5$H$_5$NH,...; An = U, Pu; M^I = C$_9$H$_7$NH (quinolinium) An = Th; M^I = NO, NO$_2$
M^{II}[An(NO$_3$)$_6$]·8H$_2$O	Generally known for An = Th, U, Pu; M^{II} = Mg, Zn, Co, Ni; An = Th; M^{II} = Mn,
(bipyH$_2$)[U(NO$_3$)$_6$]	
M^{I}Th(NO$_3$)$_5$·xH$_2$O	M^I = Na, x = 8.5; K, x = 6
K$_3$H$_3$An(NO$_3$)$_{10}$·xH$_2$O	An = Th, x = 4; U, x = 3

distances of 2.49Å (nitrate) and 2.38Å (water).[222] Stability and the relative importance of the mono- and bisnitrato complexes have recently been re-evaluated.[223]

Numerous additional solution studies on mixed ligand, nitrate complexes have been performed in the development and performance testing of extractants. Most notably these include tributyl-phosphate (TBP) and other phosphine oxides. As other examples, a variety of mixed amide, nitrate complexes have been proposed based upon NMR, IR and extraction behavior.[224–226] The composition and proposed structures of these types of species are described in the sections corresponding to the functionality of the extractant.

Actinide(IV) nitrates solids are readily formed in nitric acid by dissolution of hydroxides or carbonates followed by precipitation (Table 10). Depending upon the pH of solution, crystalline orthorhombic An(NO$_3$)$_4$·5H$_2$O(An = Th, Np, Pu), or for Th^{IV}, Th(NO$_3$)$_4$·4H$_2$O can be obtained.[227] For Pu, the tetranitrate pentahydrate can also be prepared by heating a Pu^{VI} nitrate salt.[228] The coordination geometry about the 11-coordinate thorium atom in [Th(NO$_3$)$_4$·5H$_2$O.([Th(NO$_3$)$_4$-(H$_2$O)$_3$)]·2H$_2$O is a monocapped trigonal prism in which four of the prism apices are occupied by bidentate nitrate groups. In the dimeric basic nitrate, [Th$_2$(OH)$_2$)(NO$_3$)$_6$(H$_2$O)$_6$)]·2H$_2$O, the thorium atoms are bridged by two OH groups, and each thorium atom is also coordinated to three bidentate nitrate groups and three water molecules. The geometry can be considered as a rather distorted dodecahedron in which the nitrate groups occupy three apices.

Anhydrous Th(NO$_3$)$_4$ is obtained by heating more complex nitrates under vacuum. Hexani-trato complexes are obtained from moderately concentrated (8 M to 14 M) nitric acid, in the presence of sulfamic acid to inhibit oxidation by nitrite in the case of uranium(IV). The nitrate groups in these compounds are bidentate and the structure of the anion is distorted icosahedral, such as in $[M(H_2O)_6][Th(NO_3)_6] \cdot 2H_2O$, where $M = Mg$, Zn, Co, Ni. The anions in similar Pu^{IV} salts, such as $M_2Pu(NO_3)_6 \cdot 2H_2O$, where $M = Rb$, Cs, NH_4, and pyridinium, are presumably isostructural.[229] Uranium(IV) does not form solid binary nitrates, but is apparent in ternary phases of the general formula $M_2[U(NO_3)_6]$, where $M = NH_4$, Rb, Cs, and $M[U(NO_3)_6] \cdot 8H_2O$, where $M = Mg$, Zn.

Phosphates. Because of their very low solubility, as exemplified by stable minerals and ore bodies, actinide phosphates have been proposed as potential radioactive waste forms.[230] Together with this property, the multiple protonation states and possible coordination modes make the solution An^{IV} phosphate species particularly challenging to characterize. Generally, complexes of the formula $An(H_3PO_4)_x(H_2PO_4)_y^{(4-y)+}$ ($x = 0$, 1; $y = 0$, 1, 2) have been proposed to form under acidic conditions, $An(HPO_4)_3(H_2PO_4)x^{(2+x)-}$ ($x = 1,2$), at neutral pH, and $An(HPO_4)_x^{4-2x}$ ($x = 1–3$) under basic conditions.

In the solid state, the major classes actinide(IV) phosphate are orthophosphates, hydrogenphosphates, pyrophosphates, metaphosphates, and polyphosphates.[231,232] In addition there are numerous ternary compounds, mixed valent uranium phosphates, halophosphates, organophosphates, and most recently, open framework and templated phases.[233]

Binary and ternary thorium compounds have been synthesized with varying ratios of metal, thorium and phosphate. Recently, Bernard *et al.* reported two distinct thorium types in $Th_4(PO_4)_4(P_2O_7)$, one eight-coordinate with oxygen from five phosphate and one diphosphate group around the thorium atom.[234] Ternary compounds of the general formula $M^ITh_2(PO_4)_3$ and $M^{II}Th(PO_4)_2$ with $M^I = $ alkali cation, Tl, Ag, Cu,[235,236] and $M^{II} = $ Ca, Sr, Cd, Pb,[237,238] have been studied. In the structure of $NaTh_2(PO_4)_3$, thorium is eight-coordinate, and the local coordination environment can be described as $[Th(\eta^2\text{-}PO_4)_2(\eta^1\text{-}PO_4)_4]$ in a pseudosquare bipyramidal configuration with bidentate phosphates in the axial positions and monodentate phosphates in the equatorial positions. The Th^{IV} ion in $KTh_2(PO_4)_3$ is nine-coordinate with a local coordination environment described as $[Th(\eta^2\text{-}PO_4)_2(\eta^1\text{-}PO_4)_5]$ containing both bridging and bidentate phosphate groups. In $Na_2Th(PO_4)_3$ two different thorium atoms are identified with eight and ten neighboring oxygen atoms.[239]

Few uranium(IV) phosphates have been fully characterized. They generally include the uranium atom in seven-coordinate, distorted pentagonal bipyramidal; eight coordinate, square antiprismatic; or nine-coordinate irregular geometries. Hydrogen phosphates, $U(HPO_4)_2 \cdot xH_2O$ can be prepared by precipitation from phosphoric acid solutions. Among them the bis- and tetrahydrates are the best characterized but single-crytal data are still lacking.[240] The simplest binary phase is the triclinic metaphosphate, $U(PO_3)_4$, with eight-coordinate square antiprisms of UO_8 connected by $(P_4O_{12})^{4-}$ rings.[241,242] The ortho phosphate UP_2O_7 can be prepared by thermal decomposition of the uranyl hydrogen phosphates. The mixed valent orthophosphate, $U(UO_2)(PO_4)_2$, can be prepared either via a solid state reaction, combining UO_2 and ammonium phosphates or by reducing uranyl chloride with hydrazine, followed by addition of concentrate phosphoric acid.[243] In the structure seven-coordinate U^{IV} alternate with uranyl within $PaCl_5$ type chains, which are connected by phosphate groups to form a three-dimensional network. The coordination environment about the metal center is similar to that found in $U_2O(PO_4)_2$, which is thought to be the correct formula for compounds previously believed to be $(UO)_2P_2O_7$.[244] $U_2O_3P_2O_7$ and $U_3O_5P_2O_7$ have been synthesized containing uranium in the oxidation state $+IV$ and $+VI$ in a ratio 1:1 and 2:1, respectively.[245]

The pyrophosphate of uranium(IV) has been obtained and the structure determined to belong to the ZrP_2O_7-type structure.[246] Octahedral sites in the zirconium phosphates can accommodate U^{IV}, as exemplified by the Na dizirconium tris(phosphate) structural family ([NZP]). An end member in this study was monoclinic $KU_2Zr(PO_4)_3$, which contains nine-coordinate U^{IV}.[247] Compounds of the general formula $MU_2(PO_4)_3$ have been reported for $M = $ Li, Na, and K, where U^{IV} is nine coordinate; similar compounds could not be obtained with the larger Rb and Cs ions.[248] Recent examples of three-dimensional structures exist for the halophosphate phases, $UXPO_4 \cdot 2H_2O$, $x = Cl$, Br.[249] In these compounds, all four phosphate oxygens are bound to uranium atoms, and the U^{IV} is in a distorted pentagonal bipyramidal geometry.

Fewer, but still numerous, Pu phosphates have been characterized.[82] Plutonium metaphosphate, orthorhombic $Pu(PO_3)_4$, can be crystallized from solutions of PuO_2 in metahosphoric acid.[250] The hydrogen phosphate, $Pu(HPO_4)_2 \cdot xH_2O$ is prepared by precipitation from phosphoric acid solutions and can be used as precursor for other phosphates. Red $Pu_2H\text{-}(PO_4)_3 \cdot xH_2O$ is made by heating the hydrodrogen phosphate; decomposition above $100\,°C$ reportedly yields $Pu_3(PO_4)_4 \cdot xH_2O$. Anhydrous Pu pyrophosphate PuP_2O_7, has been prepared by the thermal decomposition of plutonium

Table 11 Actinide(IV) sulfato complexes.

$M^I_2[M^{IV}(SO_4)_3]\cdot xH_2O$	$M^{IV} = Th$; $M^I = Na$, $x = 6$; K, $x = 4$;
	NH$_4$, $x = 0$ or 5; Rb, $x = 0$ or 2; Cs, $x = 2$; Tl, $x = 4$
	$M^{IV} = U$; $M^I = K$ or Cs, $x = 2$; NH$_4$ or Rb, $x = 0$
$M^I_4[M^{IV}(SO_4)_4]\cdot xH_2O$	$M^{IV} = Th$; $M^I = Na$, $x = 4$; K, $x = 2$; NH$_4$, $x = 0$ or 2; Cs, $x = 1$
	$M^{IV} = U$; $M^I = Na$, $x = 6$; K, $x = 2$; NH$_4$, $x = 0$ or 3;
	Rb, $x = 2$; enH, $x = 2$; $M^{IV} = Np$; $M^I = K$, $x = 3$
	$M^{IV} = Pu$; $M^I = K$ or NH$_4$, $x = 2$; Rb, $x = 0$, 1 or 2; Cs, $x = 0$
$[Co(NH_3)_6]Na[Np(SO_4)_4]\cdot 8H_2O$	
$M^I_6[M^{IV}(SO_4)_5]\cdot xH_2O$	$M^{IV} = Th$; $M^I = NH_4$ or Cs, $x = 3$
	$M^{IV} = U$; $M^I = NH_4$, $x = 4$
	$M^{IV} = Pu$; $M^I = Na$, $x = 1$; K, $x = 0$; NH$_4$ $x = 2$ to 4
$M^I_8[M^{IV}(SO_4)_6]\cdot xH_2O$	$M^{IV} = Th$; $M^I = NH_4$, $x = 2$ $M^{IV} = U$; $M^I = NH_4$, $x = 3$
$Na_6[U_2(SO_4)_7]\cdot 4H_2O$	
$U(SO_4)(C_2O_4)\cdot xH_2O$	$x = 0$, 1, 2 or 3
$U_2(SO_4)(C_2O_4)_3\cdot xH_2O$	$x = 0$, 2, 4, 8 or 12
$M^I_6U_2(SO_4)_4(C_2O_4)_3\cdot xH_2O$	$M^I = NH_4$ or Rb, $x = 0$, 2 or 4
$Rb_4U_2(SO_4)_3(C_2O_4)_3\cdot xH_2O$	$x = 0$, 4 or 6

oxolatophosphates.[52,251] Pyro- and metaphosphates of Np^{IV} and two double orthophosphates NaNp$_2$(PO$_4$)$_3$ and Na$_2$Np(PO$_4$)$_2$ have been prepared and determined to be isostructural with the analogous Th^{IV} and U^{IV} compounds.[252]

Alkyl phosphates, U{O$_2$P(OR)$_2$}$_4$ (R = Me, Et or Bu) and U{O$_2$PH(OR)}$_4$ (R = Me, Et, PR′ or Bu″), have been reported, as has the phenyl derivative, U(O$_3$PPh)$_2$. Plutonium monobutyl phosphate was reportedly prepared by addition of monobutyl phosphate to Pu^{IV} in nitric acid solution.

Sulfates and sulfites. Sulfate has high affinity for tetravalent actinides and forms complexes of the type An(SO$_4$)$_n^{4-2n}$ ($n = 1, 2$) in solution, with the tetrasulfato being the most important (predominant at sulfate concentrations greater than 0.2 M), and the trissulfato not detected under most conditions. These anions can be precipitated as hydrates and subsequently dehydrated at 400 °C. Representative An^{IV} sulfate complexes are shown in Table 11. For example, hydrated thorium sulfate, Th(SO$_4$)$_2\cdot n$H$_2$O ($n = 9, 8, 6, 4$), is easily crystallized from thorium and sulfuric acid. Analogous U^{IV} and Pu^{IV} compounds are well known; the red tetrahydrate Pu^{IV} phase is noteworthy because of its very high purity.[82] The octahydrate loses four waters at relatively low temperature, and can be fully dehydrated. The common bicapped square antiprismatic geometry is adopted by the An^{IV} centers in the tetra- and octahydrates.[253] For uranium, the basic salt, UOSO$_4\cdot$2H$_2$O, is formed in sulfate solution at neutral pH. Ternary salts have been characterized, such as the green Pu^{IV} compounds, M$_4$An(SO$_4$)$_4\cdot x$H$_2$O where and M is K or NH$_4$. The pentasulfato complex has not been identified in solution, but the potassium and other monovalent salts, M$_6$An(SO$_4$)$_5\cdot x$H$_2$O have been characterized.

Fluorosulfates, U(SO$_3$F)$_4$, U(SO$_3$F)$_2$, and MU(SO$_3$F)$_6$, (M = Mg, Zn) have been obtained by treating U(MeCO$_2$)$_4$, with HSO$_3$F.[254] The compound U(SO$_3$F)$_4$ appears to involve two mono- and two bi-dentate SO$_3$F groups. The structure of the anion in K$_4$Th(SO$_4$)$_4\cdot$2H$_2$O consists of chains of thorium atoms linked by pairs of bridging sulfate groups, and the coordination geometry about the thorium atom is a tricapped trigonal prism.

A simple sulfite is known for Th^{IV}, Th(SO$_3$)$_2\cdot$4H$_2$O. Salts of hydrated complexes are known for thorium(1 V) and uranium(IV) (see Table 12), both of which form a series of hydrated salts of what

Table 12 Actinide(IV) sulfites and sulfito complexes.

$Th(SO_3)_2\cdot 4H_2O$	
$M^I_2Th(SO_3)_3\cdot xH_2O$	$M^I = Na$, $x = 5$; K, $x = 7.5$; NH$_4$, $x = 4$; CN$_3$H$_6$ (guanidinium), $x = 12$
$M^I_4Th(SO_3)_4\cdot xH_2O$	$M^I = Na$, $x = 3$ or 6; NH$_4$, $x = 5$
$Na_{2n}U(SO_3)_{n+2}\cdot xH_2O$	$n = 3$, 4, 5 and 6; x, unspecified
$Na_{2n}M^{IV}(SO_3)_n(C_2O_4)_2\cdot xH_2O$	$M^{IV} = Th$; $n = 3$, 4, 5 or 7, $x = 5$ to 6; $M^{IV} = Th$; $n = 3$, 4, 5 or 7, $x = 5$ to 6; $n = 9$, $x = 6$
$Ba_6Th(SO_3)_6(C_2O_4)_2\cdot 7H_2O$	$n = 3$, $x = 5$; $n = 4$, $x = 4$; $n = 5$, $x = 7.5$; $n = 6$, $x = 7$ to 8

appear to be sulfitooxalato complex anions, but definitive characterization is needed. They are obtained by dissolving thorium oxalate in concentrated aqueous sodium sulfite.

Perchlorates and iodates. Thorium perchlorate forms upon dissolution of thorium hydroxide in perchloric acid and crystallizes as $Th(ClO_4)_4 \cdot 4H_2O$. The precipitation of tetravalent actinides as iodates has long been used to separate these elements from lanthanides at low pH. One of the earliest forms that ^{239}Pu was isolated in was that of $Pu(IO_3)_4$.[255] The structure and most properties of $Pu(IO_3)_4$ are currently unknown, but a remarkable feature is that it is insoluble in 6M HNO_3.

Alkoxides. In 1954, Bradley and co-workers reported the synthesis of the thorium tetra-kisalkoxide compound $Th(O^iPr^i)_4$ (see Equation (10)); other rational reaction routes yield impure products.[256] Additional $Th(OR)_4$ compounds ($R = Me$, Et, Bu, Bu^t, pentyl, CH_2CMe_3, OCH-MeEt, OCHEt₂, CMe₂Et, CMeEt₂, CMe₂Pr, CMe_2Pr^i, CEt₃, CMeEtPr, $CMeEtPr^i$) are prepared from alcoholysis of $Th(OPr^i)_4$[256,257] (see Equation (11)). Subsequent studies confirmed that alcoholysis of $[(Me_3Si)_2N]_2Th(CH_2SiMe_2NSiMe_3)$ with $HOPr^i$ generates a homoleptic compound of empirical formula $Th(OPr^i)_4$.[258] Addition of either excess pyridine to the fresh reaction mixture of $[(Me_3Si)_2N]_2Th(CH_2SiMe_2NSiMe_3)$ and $HOPr^i$ or a stoichiometric amount of pyridine to the metathesis reaction of $UBr_4(THF)_4$ and four equivalents of $KOPr^i$ permits isolation of the tetramer $Th_4(OPr^i)_{16}(py)_2$. A similar reaction between the metallacycle and pentan-3-ol in the presence of pyridine yields the dimer $Th_2(OCHEt_2)_8(py)_2$. In solution studies, treatment of the metallacycle $[(Me_3Si)_2N]_2Th(CH_2SiMe_2NSiMe_3)$ with four equivalents of the bulkier alcohol $HOCH(Pr^i)_2$ yields the dimer $[Th(OCHPr^i_2)_4]_2$, which exists in equilibrium with monomer $Th(OCHPr^i_2)_4$.[259] Addition of Lewis bases dimethoxyethane or quinuclidine to this dimer allows for the isolation of $[Th(OCHPr^i_2)_4(DME)]$ or $[Th(OCHPr^i_2)_4(C_7H_{13}N)]$, respectively. Similar reaction products employing tertiary alkoxide ligands were investigated. The metathesis reaction of $ThI_4(THF)_4$ with four equivalents of $KOBu^t$ in the presence of pyridine generates $Th(OBu^t)_4(py)_2$, and alcoholysis of $[(Me_3Si)_2N]_2Th(CH_2SiMe_2NSiMe_3)$ with $HOBu^t$ provides the dimer $Th_2(OBu^t)_8(HOBu^t)$.[260] The coordinated alcohol of the latter compound is deprotonated by $Na[N(SiMe_3)_2]$ to yield $NaTh_2(OBu^t)_9$, while addition of a stoichiometric amount of water to $Th_2(OBu^t)_8(HOBu^t)$ under reflux conditions in toluene yields the cluster $Th_3O(OBu^t)_{10}$ where one alkoxy group and the oxo are both triply bridging the three thorium centers. Lewis base adducts of homoleptic alkoxides may be isolated, such as the complex $Th(OBu^t)_4(py)_2$. (see Figure 13):

$$ThCl_4 + 4\ NaO^iPr \longrightarrow Th(O^iPr)_4 + 4\ NaCl \qquad (10)$$

$$Th(O^iPr)_4 + 4\ ROH \longrightarrow Th(OR)_4 + 4\ HO^iPr \qquad (11)$$

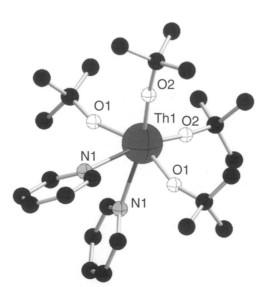

Figure 13 Crystal structure of $Th(OBu^t)_4(py)_2$ (Clark and Watkin *Inorg. Chem.* **1993**, *32*, 1766–1772).

Gilman and co-workers reported the synthesis of uranium tetrakisalkoxide complexes $U(OR)_4$ ($R' = Et$, $R = Me$, Et; $R' = H_2$, $R = Bu^t$) from alcoholysis and metathesis reactions[102] (see Equations (12) and (13)):

$$U(NR'_2)_4 + 4\ ROH \longrightarrow U(OR)_4 + 4\ HNR'_2 \qquad (12)$$

$$UCl_4 + 4\ LiOR \longrightarrow U(OR)_4 + 4\ LiCl \qquad (13)$$

Additional uranium tetrakisalkoxides ($U(OPr)_4$, $U(OPr^i)_4$) were prepared via metathesis routes carried out in dimethylcellosolve. In one report, it was suggested that Gilman's initial report of $U(OBu^t)_4$ actually represented an oxidized uranium *t*-butoxide species.[261] Cotton and co-workers later published the structure of $UO_3(OBu^t)_{10}$, a product isolated from Gilman's reported procedure for the synthesis of $U(OBu^t)_4$.[262] If, however, this reaction mixture is maintained at $\leq -10\ ^{\circ}C$, then the complex $KU_2(OBu^t)_9$ is isolated; this is readily oxidized to $U_2(OBu^t)_9$ in solution.[263,264] A high-yield synthesis of the neutral species $U_2(OBu^t)_8(HOBu^t)$ was reported from reaction of *t*-butanol with either $[(Me_3Si)_2N]_2U[N(SiMe_3)SiMe_2CH_2]$ or $U(NEt_2)_4$.[263] Treatment of $U_2(OBu^t)_8(HOBu^t)$ with $KOBu^t$ or KH further yields $KU_2(OBu^t)_9$. Both $U_2(OBu^t)_8(HOBu^t)$ and $KU_2(OBu^t)_9$ react with O_2 to form $U_2(OBu^t)_9$, or with H_2O to form $U_3O(OBu^t)_{10}$. A spectroscopic study has been carried out on the $[U_2(OBu^t)_9]^-$ dimeric anion in the presence of different cations (H^+, K^+, and TBA^+); it reveals the sensitivity of the $5f$–$5f$ spectra to the coordination sphere of the anion.[265]

Alternative routes to homoleptic U^{IV} alkoxide complexes have been described, including electrochemical generation of $U(OCH_2CH_3)_4$ from uranium metal in ethanol,[266] alcoholysis (ROH; $R = Et$, Pr^i, Bu^t) of $U(\eta\text{-}C_3H_5)_4$ at $-30\ ^{\circ}C$,[267] and generation of $U[OCH(Bu^t)_2]_4$ via metathesis reactions. The latter yields the addition compound $LiU(Me)[OCH(Bu^t)_2]_4$ in the presence of LiMe.[268] Analogous fluoroalkoxide compounds $U(OC(CF_3)_3)_4(THF)_2$ and $U(OCH(CF_3)_2)_4(THF)_2$ have been prepared from the reactions of UCl_4 and respective sodium alkoxide in tetrahydrofuran.[12]

The preparation of homoleptic neptunium tetrakisalkoxides, $Np(OR)_4$ ($R = Me$, Et) has been reported[269] (Equation (14)). $Pu(OPr^i)_4\text{-}HOPr^i$ was purified from a reaction mixture of $(C_5H_6N)_2PuCl_6$, $HOPr^i$, and NH_3; the authors further suggested that alcoholysis of $Pu(OPr^i)_4$ with $HOBu^t$ produced $Pu(OBu^t)_4$.[270]

$$NpCl_4 + LiOR \longrightarrow Np(OR)_4 + 4\ LiCl \qquad (14)$$

The molecular and electronic structures of a variety of uranium(IV) aryloxide compounds have been described. Sattelberger and co-workers reported the first structural characterization of a homoleptic tetrakisaryloxide compound, $U(O\text{-}2,6\text{-}Bu^t_2C_6H_3)_4$, prepared from alcoholysis of the metallacycle $[(Me_3Si)_2N]_2U(CH_2SiMe_2N(SiMe_3)_2]$ with $HO\text{-}2,6\text{-}Bu^t_2C_6H_3$ in refluxing toluene.[125,126] Subsequent studies show that $U(O\text{-}2,6\text{-}Bu^t_2C_6H_3)_4$ can also be generated from the metathesis reaction between $UI_4(CH_3CN)_4$ or UCl_4 and $KO\text{-}2,6\text{-}Bu^t_2C_6H_3$ in tetrahydrofuran at room temperature or from oxidation of $U(O\text{-}2,6\text{-}Bu^t_2C_6H_3)_3$ by molecular oxygen.[271,272] An investigation of the electronic structure of this highly symmetric, $5f^2$ compound using low temperature absorption spectroscopy was reported.[273] The syntheses of $Th(O\text{-}2,6\text{-}R_2C_6H_3)_4$ ($R = Me$, Pr^i) and $U(O\text{-}2,6\text{-}Pr^i_2C_6H_3)_4$ using an aminolysis reaction in toluene were also described, but a metathesis route using the $ThI_4(THF)_4$ precursor is necessary to generate the analogous thorium *t*-butoxide substituted derivative, $Th(O\text{-}2,6\text{-}Bu^t_2C_6H_3)_4$.[59,126]

Lewis base adducts of thorium(IV) and uranium(IV) aryloxides are readily prepared. Initial reports of phenoxide compounds of uranium(IV) describe NH_3 derivatives from the reaction of UCl_4 or $UOCl_2$ with appropriate phenols in the presence of ammonia[274] (see Equations (15) to (17)):

$$UCl_4 + 4\ ROH + (n+4)\ NH_3 \longrightarrow U(OR)_4 \cdot nNH_3 + 4\ NH_4Cl$$

$R = Ph$, $2\text{-}ClC_6H_4$, $3\text{-}ClC_6H_4$, $4\text{-}ClC_6H_4$, $2\text{-}MeC_6H_4$
$4\text{-}MeC_6H_4$, $\alpha\text{-}C_{10}H_7$, $\beta\text{-}C_{10}H_7$

$$(15)$$

$$UCl_4 + 6\ ROH + 5\ NH_3 \longrightarrow U(OR)_4 \cdot 2ROH \cdot NH_3 + 4\ NH_4Cl \tag{16}$$

$$R = Ph,\ 4\text{-}ClC_6H_4,\ 2\text{-}MeC_6H_4,\ 4\text{-}MeC_6H_4$$

$$UOCl_2 + 6\ ROH + 3\ NH_3 \longrightarrow UO(OR)_2 \cdot 4ROH \cdot NH_3 + 2\ NH_4Cl \tag{17}$$

$$R = Ph,\ 4\text{-}ClC_6H_4$$

Alkoxide/phosphine uranium(IV) complexes $Th(OPh)_4(dmpe)_2$ and $U(OPh)_4(dmpe)_2$ are isolated as toluene solvates from the alcohol exchange of HOPh with $M(Me)_4(dmpe)_2$ (M = Th, U).[190] Analogous Lewis base adducts of thorium tetrakis(aryloxide) complexes $(Th(O\text{-}2,6\text{-}R_2C_6H_3)_4(THF)_2,\ (R = Me,\ Pr^i)\ Th(O\text{-}2,6\text{-}Me_2C_6H_3)_4(py)_2$, and $Th(O\text{-}4\text{-}Bu^tC_6H_4)_4(py)_3)$ have been reported.[59,126] Coordination of the less sterically demanding but more electron poor aryloxide ligand, $O\text{-}2,6\text{-}Cl_2C_6H_3$, produces $U(O\text{-}2,6\text{-}Cl_2C_6H_3)_4(THF)_2$.[275]

A convenient preparation of the mono- and bisalkoxide uranium derivatives $U(BH_4)_3(OCHR_2)(THF)_2$ and $U(BH_4)_2(OCHR_2)_2(THF)_2$ (R = $CHMe_2$, $CHPh_2$, C_6H_{11}) involves the reduction of an appropriate ketone with $U(BH_4)_4$ in tetrahydrofuran.[276] The monoalkoxides are alternatively prepared from the reaction of the ketones with four equivalents of UCl_4 in the presence of $LiBH_4$, treatment of $U(BH_4)_4$ with the $B(OCHR_2)_nH_{3-n}$ (R = OPr^i, OCy) formed from the reaction of excess ketone employed in the reaction with liberated BH_3, the redistribution reaction of $U(BH_4)_2(OR)_2(THF)_2$ with $U(BH_4)_4$ (R = Pr^i, Cy), or the addition of the respective alcohols, $HOPr^i$, $HOCHPh_2$, or HOCy, to $U(BH_4)_4$. Similar products were obtained from reactions between $U(BH_4)_4$ and ketones 2-methylcyclohexanone, 4-t-butylcyclohexanone and norcamphor.

A comparison between the electronic influence of the tri-t-butylmethoxide ligand (tritox = $(OC(Bu^t)_3)$) and sterically analogous cyclopentadienyl ligand on a uranium(IV) metal center has been conducted.[277,278] The tetravalent uranium complexes $(tritox)UCl_3(THF)_x$ and $(tritox)_2UCl_2(THF)_2$ have been isolated; these species serve as precursors in the isolation of a series of mixed ligand compounds: $((tritox)_2(C_5H_5)UCl$; $(tritox)_2UR_2$, R = BH_4, $CH(COMe_3)_2$, $\eta\text{-}C_3H_5$, CH_2Ph; $(tritox)U(BH_4)_3(THF)$, $(tritox)_2U(BH_4)_2$, $(tritox)_3U(BH_4))$.

Actinide(IV) alkoxide complexes have been reported which are coordinated by a variety of other bulky ligand sets. Uranium(IV) amido compounds are reagents for the preparation of homoleptic uranium(IV) alkoxides as well as mixed alkoxide/amido species. A variety of mixed aryloxide–diethylamide derivatives have been prepared including $(U(NEt_2)(O\text{-}2,6\text{-}Bu^t_2C_6H_3)_3$ and $U(NEt_2)(O\text{-}2,6\text{-}R_2C_6H_3)_3$, R = Pr^i, Bu^t).[279,280] The previously described metallacycle $[(Me_3Si)_2N]_2\underline{M}CH_2Si(Me)_2NSiMe_3$ (M = Th, U) is a useful starting material for the preparation of both homoleptic and mixed alkoxide/amide actinides compounds, including the compounds $Th(O\text{-}2,6\text{-}Me_2C_6H_3)[N(SiMe_3)_2]_3$, $Th(O\text{-}2,6\text{-}Bu^t_2C_6H_3)[N(SiMe_3)_2]_3$, $Th(O\text{-}2,6\text{-}Bu^t_2C_6H_3)_2\text{-}[N\text{-}(SiMe_3)_2]_2$, $Th(O\text{-}2,6\text{-}Bu^t_2C_6H_3)_3[N(SiMe_3)_2]$, $U(O\text{-}2,6\text{-}Pr^i_2C_6H_3)[N(SiMe_3)_2]_3$, $U(O\text{-}2,6\text{-}Bu^t_2C_6H_3)[N(SiMe_3)_2)]_3$, as well as the products $Th_4(OPr^i)_{16}(py)_2$, $Th_2(OCHEt_2)_8$, $[Th(OCHPr^i)_4]_2$, $M_2(OBu^t)_8(HOBu^t)$ (M = Th, U) and $U(OR)_4$ (M = Th, R = $2,6\text{-}Bu^t_2C_6H_3$; M = U, R = $2,6\text{-}Bu^t_2C_6H_3$ or $2,6\text{-}Pr^i_2C_6H_3$) (*vide supra*).[119] Substituted triamidoamine uranium(IV) compounds $U(N(CH_2CH_2NSiMe_3)_3)(OR)$ (R = ^tBu, $^tC_4F_9$, Ph, $2,6\text{-}^tBu_2\text{-}4\text{-}MeC_6H_2$) and three *ate* derivatives $[U(N(CH_2CH_2NSiMe_3)_3(OR)(OR')Li(THF)_n]$ (R, R' = Bu^t, Ph) are prepared via reactions of the (triamidoamine)uranium chloride compound with an appropriate alkali metal alkoxide.[132]

Appropriate chalcogenide sources allow for the one-electron oxidation of $U(O\text{-}2,6\text{-}Bu^t_2C_6H_3)_3$ to chalcogenide-bridged uranium(IV) dimers $(\mu\text{-}X)[U(O\text{-}2,6\text{-}Bu^t_2C_6H_3)_3]_2$ (X = O, oxidant = N_2O, NO, Me_3NO, pyNO; X = S, oxidant = COS, Ph_3PS).[272]

Mixed halide/alkoxide ligand compounds have also been reported. Derivatization of $Th(OCH\text{-}(Pr^i)_2)_4(py)_2$ with Me_3SiI yields $ThI(OCH(Pr^i)_2)_3(py)_2$.[281] The compound $UI_2(OPr^i)_2(HOPr^i)$ is prepared by treatment of U metal with iodine in the presence of $HOPr^i$; the product of solvent loss, $U_2I_4(OPr^i)_4(HOPr^i)$, is isolated under reduced pressure.[282] The instability of $UI_4(MeCN)$ to tetrahydrofuran solvent allows for ring-opening of THF and recrystallization of $UI_2(OCH_2CH_2CH_2CH_2I)_2(Ph_3PO)_2$ following addition of triphenylphosphine oxide.[188] Other routes to mixed aryloxide–halide species include oxidation of $U(O\text{-}2,6\text{-}Bu^t_2C_6H_3)_3$ by a variety of halogenating agents. Compounds of the formula $XU(O-2,6-Bu^t_2C_6H_3)_3$ $(X = F, Cl, Br, I;$ oxidant = $AgBF_4$, $AgPF_6$, $C_6H_5CH_2Cl$, PCl_5, $AgBr$, CBr_4, PBr_5, I_2, HCl_3, $C_2I_4)$ and $X_2U(O-2,6-Bu^t_2C_6H_3)_2$ $(X = I;$ oxidant = $CI_4)$ have been prepared.[271,272] Lappert and co-workers reported the synthesis of mixed ligand compounds $UCl_2(O-2,6-Bu^t_2C_6H_3)_2$, and $[Li(THF)_3UCl_2(O-2,6-Bu^t_2C_6H_3)_2(\mu\text{-}Cl)]$.[279] In one report, $Th(OR)_4$ $(R = Pr^i, Bu^t)$ was allowed to react with various quantities of acetyl chloride, resulting in the formation of mixed halide/alkoxide compounds and in the case of the Bu^t compounds, alkoxide/halide/acetate derivatives.[283] The compound $UCl_2(Et_2)_2\text{-}xHOEt_2$ was isolated from a reaction of uranium metal with ethanol in CCl_4.[266] In an attempt to oxidize $Np(OEt)_4$ in the presence of bromine, $NpBr(OEt)_3$ and $NpBr_2(OEt)_2$ were generated.[269]

Thiolates. The first reported reaction route to homoleptic thiolate compounds (reaction of uranium tetrakisdiethylamide with four equivalents of either ethanethiol or butanethiol) appeared in 1956[102] (Equation (18)); this reaction was subsequently reinvestigated.[284] The homoleptic thiolate complexes are reported to be insoluble, but the addition of Lewis bases permits isolation of monomeric products; the complex $U(SPr^i)_4(hmpa)_2$ (hmpa = hexamethylphosphoramide) has been crystallographically characterized.[284] Protonation of $U(SPr^i)_4(hmpa)_2$ with $[NEt_3H][BPh_4]_2$ in the presence of hmpa generates $[U(SPr^i)_2\text{-}(hmpa)_2][BPh_4]$, and iodinolysis of $U(SPr^i)_4$ in pyridine yields the iodo derivatives $[U(SPr^i)_{4-n}\text{-}I_n(py)_x]$ $(n = 1\text{–}3)$. The complex $[U(SPr^i)_2I_2(py)_3]$ has been characterized by single-crystal X-ray diffraction.[284] A uranium-sulfur cluster, $U_3S(SBu^t)_{10}$, is isolated from the reaction of uranium tetrakisdiethylamide and t-butylthiol, a reaction expected to afford $U(SBu^t)_4$.[284,285] The Lewis base adduct, $U(SBu^t)_4(py)_3$, is obtained from this same reaction in the presence of pyridine, and the pyridine adduct is then cleanly converted to $U_3S(SBu^t)_{10}$ in refluxing benzene. Other synthetic routes employing reaction of either UCl_4 or $U(BH_4)_4$ with NaSR lead to the formation of the red ionic complexex $[Na(THF)_3]_2[U(SR)_6]$ $(R = Bu, Pr^i, Bu^t, Ph)$. It has been suggested that protonation of $[Na(THF)_3]_2[U(SBu)_6]$ with NEt_3HBPh_4 forms the green compound $U(SBu)_4$ first reported by Gilman.[102,286] Treatment of $U(BH_4)_4$ with HSBu also allows for the preparation of $U(SBu)_4$:[284]

$$U[NEt_2]_4 + 4HSR \longrightarrow U(SR)_4 + 4 HNEt_2$$

$$R = Et, {}^iPr, Bu$$

(18)

The synthesis and characterization of uranium(IV) phenylthiolates has also been investigated. In contrast with the reactions of $U(NEt)_4$ with alkylthiols to form either uranium(VI) tetrakisthiolates $U(SR)_4$ $(R = Et, Bu, Pr^i)$[102,286] or the cluster $[U_3S(S^tBu)_{10}]$,[285] the reaction of phenylthiol with uranium(IV) tetrakisdiethylamide affords the red ionic product $[NEt_2H_2][U(SPh)_6]$.[287] The reaction mixture of UCl_4 with NaSPh, CuSPh, and PPh_3 yields red $[(Ph_3P)Cu(\mu\text{-}SPh)_3U(\mu\text{-}SPh)_3Cu(PPh_3)]$, which has a core uranium environment analogous to that found in $[Na(THF)_3]_2[U(SR)_6]$.[284,287,288] Homoleptic uranium(IV) tetrakisphenylthiolates are synthesized from reaction of either $U(BH_4)_4$ or $U(SBu)_4$ with phenylthiol, thiol exchange of $U(SBu)_4$ with HSPh, or oxidation of uranium metal with RSSR $(R = Et, Pr^i, Ph)$.[284]

Thorium and uranium thiolates coordinated by additional bulky ligands can be prepared. Reaction of the uranium metallacycle $[\{(Me_3Si)_2N\}_2\overline{U(CH_2SiMe_2NSiMe_3}]$ with one equivalent of 2,6-dimethylthiophenol allows for the isolation of monothiolate $U(S-2,6-Me_2C_6H_3)[N-(SiMe_3)_2]_3$.[127]

The reactivity of select uranium(IV) thiolate compounds has been investigated. The product $(SPr^i)_2C{=}S$ was identified from reaction of carbon disulfide with $U(SPr^i)_4$.[284]

Triflates. Actinide triflate complexes have been investigated both as promising reagents for further synthesis, and as potent Lewis acids. The initial reports of triflate complexes of tetravalent actinides were thorium species $Th[N(SiMe_3)_2](OTf)_3$ and $Th[N(SiMe_3)_2]_3(OTf)$, generated by protonation of the correspondent thorium metallacycle by triflic acid.[289] Subsequently, routes have been devised for the generation of the homoleptic compound $U(OTf)_4$ by treatment of the trivalent triflate with triflic acid, or by reaction of UCl_4 with TfOH.[62] The tetravalent triflate reacts with triphenylphosphine oxide to generate the complex $U(OTf)_4(OPPh_3)_2$.

(v) Ligands containing neutral group 16 donor atoms

Aqua species. The coordination number of tetravalent actinide ions Th^{4+} and U^{4+} has been examined in aqueous solution.[290] These studies suggest the metal ions have 10 ± 1 water molecules in their primary coordination sphere, at distances of 2.45 Å (Th) or 2.42 Å (U).

Early literature contains a large number of hydrates of tetravalent actinides, but as in the case of the trivalent species, it is difficult to ascertain whether these constitute complexes with water in the inner coordination sphere of the metal ion. It has been suggested that ease of removal of one water of hydration indicates it resides principally in the lattice. As an illustration of this, the reported actinide sulfate hydrates, $An(SO_4)_4 \cdot 8H_2O$, (An = Th, U, Pu) readily lose four molecules of water at temperatures $<100\,°C$. The fact that four molecules of water remain in the inner coordination sphere has been confirmed by single crystal X-ray diffraction.[253] Coordination numbers as high as 10 (bicapped square antiprismatic arrangement of atoms about the metal center) have been reported for complexes with multidentate anions such as the hydroxyacetate $U(OHCH_2CO_2)_4(H_2O)_2$[291] and the pyridine-2,6-dicarboxylate $Th\{C_5H_3N\text{-}2,6\text{-}(CO_2)_2\}_2(H_2O)_4$.[292] Aqua complexes can also display relatively high coordination numbers, particularly when the metal coordination sphere does not contain other strongly coordinating ligands. Examples of this may be found in the compounds $\{Th(\mu\text{-}OH)(H_2O)_5(pic)\}_2[pic]_4$ (pic = picrate)[293] $[ThCl_2\text{-}(H_2O)_7]Cl_2 \cdot 18\text{-crown-6},$[294] and $[UBr(H_2O)_8]Br_3 \cdot H_2O$[295] in which the inner coordination sphere of the metal is heavily hydrated.

Ethers, cyclic ethers. As previously discussed, the actinide centers are often regarded to universally prefer hard Lewis base ligands, and yet in some instances (e.g., replacment of ethers by phosphines), it has been found that ethers do not serve as strong ligands. In nearly all isolated structures of tetravalent actinides with ethers, the coordinating base is either a cyclic ether (the constrained angle about oxygen increases the σ-orbital character in lone pairs and thereby the donor strength) or a bidentate di-ether such as 1,2-dimethoxyethane (dme), although the diethyl and dimethyl ether complexes of tetrahalides have been reported.

Several structures of actinide etherate complexes have appeared in recent years. Thorium complexes such as $ThBr_4(THF)_4,$[296] $ThCl_4(H_2O)(THF)_3,$[297] and in $ThBr_4(dme)_2$[298] display distorted dodecahedral geometry. The smaller size of tetravalent uranium reduces the metal coordination number in $UCl_4(THF)_3$.[299] One bimetallic halide etherate has been reported; the complex $[UCl_3(\mu\text{-}Cl)(THF)_2]_2$ has been structurally characterized.[300] The molecular structure of $[NBu_4][UCl_5(THF)]$ has also been reported,[301] the metal center in this complex is pseudooctahedral.

Alcohols. Alcohols are among the most common "solvate" ligands in actinide chemistry (Table 13); historically the hydrated chloride complexes were reacted with alcohol in benzene, and the water of hydration removed by azeotropic distillation of the benzene. More recent examples result from the crystallization of anhydrous halides from alcoholic solvent. Similarly, solvates of alkoxide complexes result from metathesis or solvolysis reations in alcohol. The molecular structures of the halides $AnCl_4(Pr^iOH)_4$ (An = Th, U) have been reported,[302] the coordination geometry about the metal is a distorted dodecahedron.

As in the presence of aqua complexes, the presence of crown ethers facilitates the crystallization of actinide halide solvate complexes; in this manner $ThCl_4(EtOH)_3(H_2O) \cdot 18\text{-crown-6}$ and $ThCl_4(MeOH)_2(H_2O)_2 \cdot 15\text{-crown-5}$ have been isolated.[303,304]

Ketones, aldehydes, esters. Complexes of ketones, aldehydes, and esters have been made with uranium and thorium halide complexes by isolating the product from a ligand-containing solution. Ethyl- and n-propyl acetates also react with uranium tetrachloride to yield mixed halide–acetate salts with ester as an additional coordinating base.

Carbamides. Carbamides (along with the closely related ureas, *vide infra*) constitute one of the most numerous ligand sets available for actinide coordination ($RCONR'_2$). Despite

Table 13 Representative alcohol and phenol adducts of actinide(IV) compounds.

$MCl_4 \cdot 4ROH$	M = Th, R = Me, Et, Pr^n, Pr^i, Bu^n, Bu^i M = U, R = Me, Et, Pr^n, Pr^i
$ThCl_4 \cdot C_7H_7OH$	$C_7H_7OH = o$- or m-cresol
$Th(C_9H_6NO)_4 \cdot EtOH$	
$Th(C_9H_6NO)_4 \cdot 2ROH$	R = 2,4-$(O_2N)_2C_6H_3$, 2,4,6-$(O_2N)_3C_6H_2$
$Th(C_9H_6NO)(OMe)(Cl_3CCO_2)_2 \cdot MeOH$	
$U(CF_3COCHCOPh)_4 \cdot Bu^nOH$	
$Np(OEt)_4 \cdot EtOH$	
$Pu(OPr^i)_4 \cdot Pr^iOH$	

Table 14 Structurally characterized monodentate amides of Th^{IV} and U^{IV}.

Compound	References
ThBr$_4$(CH$_3$CON(Pr^i)$_2$)$_2$	a
Th(NCS)$_2$Cl$_2$(CH$_3$CON(Cy)$_2$)$_3$	b
Th(NCS)$_4$(Pr^{i}CON(Pr^i)$_2$)$_3$	c
[UCl$_3$(HCONMe$_2$)$_5$]$_2$[UCl$_6$]	d
[Th(NO$_3$)$_2$(2-pyridonato)$_6$](NO$_3$)$_2$	d

^{a} Al-Daher, Bagnall *et al. J. Less-Common Met.* **1986**, *122*, 167. ^{b} Bagnall, Benetollo *et al. Polyhedron* **1992**, *11*, 1765. ^{c} Charpin, Lance *et al. Acta Crystallogr., Sect. C* **1988**, *44*, 257. ^{d} Goodgame, Newnham *et al. Polyhedron* **1990**, *9*, 491.

the presence of another heteroatom, coordination to the metal center generally occurs through the oxygen atom. Initial investigations of coordination chemistry emphasized the effect of contraction in metal ion radius in tetravalent halides across the actinide series (An = Th, U, Np, Pu) on the coordination number of the metal ion.[305] Complexes of the stoichiometry AnCl$_4$·xL can be prepared from the reactions of AnCl$_6^{2-}$ and the appropriate ligand; for the larger amides CH$_3$CON(Pr^i)$_2$ and CH$_3$CH$_2$CON(Pr^i)$_2$ the neptunium and plutonium chlorides form adducts where x = 2, whereas for the smaller amide HCON(Me)$_2$, x = 2.5. Similar correlations between steric size of a coordinating amide ligand and actinide coordination number have been noted for uranium and thorium nitrates,[306,307] carboxylates,[308] and thiocyanates.[309] The complexes are stable in protic (and potentially coordinating) solvents such as alcohols and carboxylic acids, but are decomposed in water. Several monoamide derivatives of plutonium(IV) nitrate have been isolated; these are reported to form the adducts Pu(NO$_3$)$_4$(L)$_3$ (L = amide).[310]

A number of complexes of monodentate amides have been structurally characterized with varying numbers of ligands, including those in Table 14. A limited number of complexes have also been reported for related lactam and antipyrine ligands.

Ureas. Urea adducts (and those of the closely related *N*-alkylated derivatives) may be prepared from nonaqueous solvents; alternatively, preparation in aqueous alcoholic solution leads to the formation of hydrates. In contrast to the carbamides discussed above, there is relatively little variability in the coordination number of reported urea adducts of tetravalent actinides. Most complexes are either six- or seven-coordinate; higher coordination numbers are observed for the larger thorium ion (Table 15).

The six-coordinate complexes are octahedral with the neutral ligands occupying *trans* positions. The eight-coordinate complex Th(NCS)$_4$[OC(NMe$_2$)$_2$]$_4$ is best described as a slightly distorted dodecahedron.

Table 15 Urea adducts of tetravalent uranium and thorium.

Compound	References
UBr$_4$[OC(NMePh)$_2$]$_2$	a
UCl$_4$[OC(NMePh)$_2$]$_2$	a
Th(NO$_3$)$_4$[OC(NMe$_2$)$_2$]$_2$	b
UCl$_4$[OC(NMe$_2$)$_2$]$_2$	c
UBr$_4$[OC(NMe$_2$)$_2$]$_2$	c
UI$_4$[OC(NMe$_2$)$_2$]$_2$	d
ThBr$_4$[OC(NEt$_2$)$_2$]$_3$	e
ThCl$_4$[OC(NMePh)$_2$]$_3$	f
Th(SO$_4$)$_2$(H$_2$O)[OC(NH$_2$)$_2$]$_4$	g
Th(NCS)$_4$[OC(NMe$_2$)$_2$]$_4$	h

^{a} De Wet, J. F. and M. R. Caira *J. Chem. Soc., Dalton Trans.* **1986**, 2035. ^{b} Al-Daher, A. G. M., K. W. Bagnall, *et al. J. Chem. Soc., Dalton Trans.* **1986**, 615. ^{c} Du Preez, J. G. H., B. Zeelie, *et al. Inorg. Chim. Acta* **1986**, *122*, 119. ^{d} Du Preez, J. G. H., B. Zeelie, *et al. Inorg. Chim. Acta* **1987**, *129*, 289. ^{e} Al-Daher, A. G. M., K. W. Bagnall, *et al. J. Less-Common Met.* **1986**, *122*, 167. ^{f} Bagnall, K. W., A. G. M. Al-Daher, *et al. Inorg. Chim. Acta* **1986**, *115*, 229. ^{g} Habash, J., R. L. Beddoes, *et al. Acta Crystallogr., Sect. C* **1991**, *47*, 1595. ^{h} Rickard, C. E. F. and D. C. Woollard *Aust. J. Chem.* **1980**, *33*, 1161.

N-oxides, phosphine oxides, arsine oxides, and related ligands. The vast majority of compounds associated with this class of ligands contain phosphine oxide or similar ligands; compounds reported in the literature with coordinated phosphine oxide ligands are too numerous to list (to illustrate, Table 16 presents just simple "binary compounds" of the formula $AnX_4 \cdot nL$). Organophosphate esters are the principal extractant in a number of actinide liquid–liquid separation and purification schemes, and so the structural chemistry of the P=O bond has drawn a great deal of attention. The prototypical species for this class of compounds is $AnX_4(Ph_3PO)_2$ (An = Th, Pa, U, Np, Pu; X = Cl, Br). Both *cis-* and *trans-*octahedral coordination geometries have been identified; for the most part, complexes are isostructural for a given halide. A number of more recent structurally characterized examples are presented in Table 17.

By comparison, many fewer complexes of *N*-oxides and arsine oxides have been discussed. Given the relatively oxidizing nature of *N*-oxide compounds, it is difficult to stabilize tetravalent actinide complexes with these ligands except in the case of thorium, where no higher oxidation state is available. In this case, base adducts are known for a number of different simple inorganic salts of thorium (Table 18). Most complexes are those of pyridine *N*-oxide and related heterocyclic *N*-oxides. For more strongly complexing anions (e.g., NCS-, lighter halides), the complexes do not behave as electrolytes in solution and form adducts with two or three neutral base ligands, indicating that no anion dissociation takes place. For less strongly coordinating ligands (nitrate, iodide, perchlorate, etc.), larger numbers of *N*-oxide ligands

Table 16 Complexes of simple actinide(IV) compounds with *P*-oxides.

$M^{IV}Cl_4 \cdot xR_3PO$	R = Me, $x=2$, M^{IV} = Th, U, Np; $x=3$, M^{IV} = Th, U; $x=6$, M^{IV} = Th, Pa, U, Np, Pu
	R = Et, $x=2$, M^{IV} = Th, U
	R = Bu^n, $x=1.5(+6H_2O)$ 2, 3.5, 4, 5, 8, M^{IV} = U
	R = Bu^{n}O, $x=2, 3$, M^{IV} = U
	R = Ph, $x=2$, M^{IV} = Th, Pa, U, Np, Pu; $x=3$, M^{IV} = Th R = Me$_2$N, $x=2$, M^{IV} = Th, Pa, U, Np, Pu
	$R_3 = Et_2Ph, EtPh_2$, $x=2$, M^{IV} = Th, U
$M^{IV}Br_4 \cdot xR_3PO$	R = Me, $x=2$, M^{IV} = U; $x=6$, M^{IV} = Th, U
	R = Et, Bu^n, $x=2$, M^{IV} = U
	R = Ph, $x=2$, M^{IV} = Th, Pa, U, Np, Pu; $x=3$, M^{IV} = Th R = Me$_2$N, $x=2$, M^{IV} = Th, Pa, U, Np, Pu; $x=3$, M^{IV} = Th R_3 = (Me$_2$N)$_2$Ph, $x=2$, M^{IV} = U
$M^{IV}(NO_3)_4 \cdot xR_3PO$	R = Me, $x=3.33, 2.67$, M^{IV} = Th; $x=3, 4, 5$, M^{IV} = Th, U; $x=3$, M^{IV} = Np R = MeO, $x=3, 4$, M^{IV} = Th
	R = Et, $x=2.67$, M^{IV} = Th
	R = Pr^n, $x=2.67$, M^{IV} = Th, U, Np
	R = Bu^n, $x=2$, M^{IV} = U; $x=4$, M^{IV} = Th
	R = Bu^{n}O, $x=2$, M^{IV} = U; $x=2, 3.33$, M^{IV} = Th R = Bu^{i}O, $x=3$, M^{IV} = Th
	R = i-C$_5$H$_{11}$, $x=2$, M^{IV} = U
	R = n-C$_8$H$_{17}$, $x=2, 3$, M^{IV} = Th
	R = Ph, $x=2$, M^{IV} = Th, U, Np, Pu
	R = Me$_2$N, $x=2$, M^{IV} = Th, U, Np; $x=2.67, 3$, M^{IV} = Th; $x=4$, M^{IV} = Th, U
	R_3 = (MeN)$_2$Ph, $x=2$, M^{IV} = U
	R_3 = Bu^n(Bu^{n}O)$_2$P, $x=2.21, 2.67$, M^{IV} = Th
	R_3 = (MeO)(PhO)$_2$, $x=4$, M^{IV} = Th
$M^{IV}(ClO_4)_4 \cdot xR_3PO$	R = Me, $x=6$, M^{IV} = U; R = Et, $x=4$, M^{IV} = Th; R = Pr^n, $x=4$–5, M^{IV} = U; R = Bu^n, $x=5$ (+3H$_2$O), M^{IV} = Th; R = Ph, $x=4, 5$, M^{IV} = Th; $x=5$–6, M^{IV} = U R = Me$_2$N, $x=6$, M^{IV} = Th, $x=5, 6$, M^{IV} = U
$M^{IV}(NCS)_4 \cdot xR_3PO$	R = Me, $x=4$, M^{IV} = U, Np, Pu; $x=6$, M^{IV} = Th
	R = Ph, $x=4$, M^{IV} = Th, U, Np
	R = Me$_2$N, $x=4$, M^{IV} = Th, U, Np, Pu
U(NCSe)$_4 \cdot 4R_3PO$	R = Bu^n, Me$_2$N
Th(CF$_3$COCHCOR)$_4 \cdot$L	R = Me, L = Ph$_3$PO, (Bu^{n}O)$_3$PO, (n-C$_8$H$_{17}$)$_3$PO
	R = Me, L = Ph$_3$PO, (Bu^{n}O)$_3$PO, (n-C$_8$H$_{17}$)$_3$PO, (C$_3$H$_2$F$_5$O)$_3$PO, (Bu^{n}O)$_3$PO, (PhO)$_3$PO R = 2-C$_4$H$_3$S, L = BuPO, Bu^n(Bu^{n}O)$_2$PO, (n-C$_8$H$_{17}$)$_3$PO, Ph$_3$PO, (Bu^{n}O)$_3$PO
U(CF$_3$COCHCOR)$_4 \cdot$2L	R = Me, CF$_3$, L = (Bu^{n}O)$_3$PO

Table 17 Structurally characterized An^{IV} P-oxides.

Compound	References
$UCl_4\{(Me_2N)_3PO\}_2$	a
$UCl_4\{(pyrrolidinyl)_3PO\}_2$	b
$UBr_4(Ph_3PO)_2$	c
$UBr_4(Ph_3PO)_2$	d
$UBr_4\{(pyrrolidinyl)_3PO\}_2$	e
$[UCl(Me_3PO)_6]Cl_3$	f
$U(NCS)_4(Me_3PO)_4$	g
$U(NCS)_4\{(Me_2N)_3PO\}_2$	h
$[Th(NO_3)_3(Me_3PO)_4]_2[(Th(NO_3)_6]$	i
$\{Th(NO_3)_3[(Me_2N)_3PO]_4\}_2(Th(NO_3)_6)$	j
$(Ph_4P)[Th(NO_3)_5(Me_3PO)_2]$	i
$ThCl_4(Ph_3PO)_3$	k
$[UBr_2\{(pyrrolidinyl)_3PO\}_4][BPh_4]_2$	l
$[UI_2\{(pyrrolidinyl)_3PO\}_4][BPh_4]_2$	l
$U(NO_3)_4\{(pyrrolidinyl)_3PO\}_2$	m
$U(NO_3)_4(Ph_3PO)_2$	m
$[U(S-i\text{-}Pr)_2\{(Me_2N)_3PO\}_4][BPh_4]_2$	n

[a] De Wet, J. F. and S. F. Darlow *Inorg. Nucl. Chem. Lett.* **1971**, *7*, 1041. [b] De Wet, J. F. and M. R. Caira *J. Chem. Soc., Dalton Trans.* **1986**, 2035. [c] Bombieri, G., F. Benetollo, *et al. Journal of the Chemical Society, Dalton Transactions: Inorganic Chemistry* **1983**, 343–348. [d] Bombieri, G., D. Brown, *et al. J. Chem. Soc.,Dalton Trans.* **1975**, 1873. [e] Du Preez, J. G. H., H. E. Rohwer, *et al. Inorg. Chim. Acta* **1991**. *189*, 67. [f] Bombieri, G., E. Forsellini, *et al. J. Chem. Soc., Dalton Trans.* **1976**, 735. [g] Rickard, C. E. F. and D. C. Woollard *Aust. J. Chem.* **1979**, *32*, 2182. [h] Kepert, D. L., J. M. Patrick, *et al. J. Chem. Soc., Dalton Trans.* **1983**, 385. [i] Alcock, N. W., S. Esperas, *et al. J. Chem. Soc., Dalton Trans.* **1978**, 638. [j] English, R. P., J. G. H. du Preez, *et al. S. Af. J. Chem.* **1979**, *32*, 119. [k] Van den Bossche, G., J. Rebizant, *et al. Acta Crystallogr., Sect. C* **1988**, *44*, 994. [l] Du Preez, J. G. H., L. Gouws, *et al. J. Chem. Soc., Dalton Trans.* **1991**, 2585. [m] Dillen, J. L. M., C. A. Strydom, *et al. Acta Crystallogr., Sect. C* **1988**, *44*, 1921. [n] Leverd, P. C., M. Lance, *et al. J. Chem. Soc., Dalton Trans.* **1995**, 237.

Table 18 Representative complexes of actinide(IV) compounds with *N*-oxides.

L = pyridine *N*-oxides, $R^1R^2C_5H_3NO$	
$ThCl_4 \cdot 2L$	$R^1 = H$ and $R^2 = H$ (also $+2H_2O$), 2-, 3- or 4-Me; $R^1 = 2$-Me, $R^2 = 6$-Me
$UCl_4 \cdot 2L$	$R^1 = R^2 = H$; $R^1 = H$, $R^2 = 2$-, 3- or 4-CO_2H
$ThBr_4 \cdot 2L$	$R^1 = H$ and $R^2 = H$, 2-, 3- or 4-Me
$ThI_4 \cdot 4L$	$R^1 = H$ and $R^2 = H$, 2-, 3- or 4-Me; $R^1 = 2$-Me, $R^2 = 6$-Me (all $[ThI_2L_4]I_2$)
$Th(NCS)_4 \cdot xL$	$R^1 = R^2 = H$, $x = 4$; $R^1 = H$, $R^2 = 2$-, 3- or 4-Me, $x = 4$; $R^1 = 2$-Me, $R^2 = 6$-Me, $x = 2$
$Th(NO_3)_4 \cdot xL$	$R^1 = R^2 = H$, $x = 2$ ($+MeCO_2Et$), 8; $R^1 = H$, $R^2 = 2$-Me, $x = 3$; $R^1 = 2$-Me, $R^2 = 6$-Me, $x = 3, 4$
$Th(ClO_4)_4 \cdot xL \cdot yH_2O$	$R^1 = H$; $R^2 = H$, $x = 8$, 9, $y = 0$; $R^2 = 2$-Me, 4-Me, $x = 8$, $y = 0$; $R^2 = 3$-Me, $x = 8$, $y = 1$ or 3; $R^2 = 4$-Cl, 4-NO_2, 4-MeO, $x = 8$, $y = 0$; $R^1 = 2$-Me, $R^2 = 6$-Me, $x = 6$ or 8, $y = 0$
L = quinoline *N*-oxide	
$ThX_4 \cdot yL$	$X = Cl$, $y = 2$ ($+2H_2O$); $X = NCS$, $y = 4$; $X = NO_3$, $y = 3$; $X = ClO_4$, $y = 6$
$Th(OH)_2(NO_3)_2 \cdot 2L$	
L = isoquinoline *N*-oxide	
$Th(NCS)_4 \cdot 4L$	
L = 2,2′-bipyridyl *N*-oxide or 1,10-phenanthroline *N*-oxide	
$ThX_4 \cdot 2L$	$X = Cl$, Br, NCS, NO_3
$ThX_4 \cdot 3L$	$X = I$, ClO_4
L = 2,2′-bipyridyl *N,N′*-dioxide	
$ThX_4 \cdot L$	$X = NCS$, NO_3
$ThX_4 \cdot 3L$	$X = Cl$, Br
$ThX_4 \cdot 4L$	$X = I$, ClO_4
L = 1,10-phenanthroline *N,N′*-dioxide	
$ThX_4 \cdot 2L$	$X = Cl$, Br, NCS, NO_3, ClO_4
$ThI_4 \cdot 3L$	

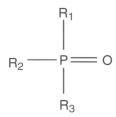

Figure 14 General diagram of phosphine oxide.

coordinate (up to eight or nine), indicating that no inner-sphere anion coordination to the metal takes place.

The few known arsine oxide complexes are very similar in behavior to the corresponding phosphine oxide analogues. The only structurally characterized examples of this class are UCl_4 $(Et_3AsO)_2$[311] and $UBr_4(Ph_3AsO)_2$;[312] both are *trans* octahedral.

Phosphine oxides have the general structure shown in Figure 14. The commercially available compound Cyanex 923 (TRPO), available through Cytec Inc., Canada, is a mixture of four trialkyl phosphine oxides, with substituents as indicated in Table 19. TRPO is very favorable as an extractant due to its hydrophobicity, solubility in organics, and stability with respect to hydrolysis.[313] TRPO has been investigated as an extractant for Th^{IV} from nitric acid solutions into xylene. Extractant dependency analysis shows that the uptake of Th^{IV} into the organic phase increases (slope of two) with increasing extractant concentration. Hence, the expected stoichiometry for the extraction is given by Equation (19):

$$Th^{4+}_{(a)} + 4\ NO_3^-{}_{(a)} + 2\ TRPO_{(o)} \longrightarrow Th(NO_3)_4 \bullet 2TRPO_{(o)} \qquad (19)$$

As a comparison, the extraction of Th^{IV} from xylene by trioctylphosphine oxide (TOPO), Cyanex 921 (similarly available through Cytec Inc.) (Table 19), has also been studied and extractant dependency indicates a 1:2 metal:extractant complex ratio like that seen for TRPO.

IR data for the Th^{IV}/TRPO complex shows that the phosphoryl stretching frequency moves down in energy from $1{,}146\ cm^{-1}$ to a lower value of $1{,}095\ cm^{-1}$, indicating a direct interaction between the phosphoryl oxygen and the Th^{IV} metal in the extracted complex.[314]

Plutonium(IV) has also been extracted by a series of phosphine oxides including TRPO, bis(2,4,4-trimethylpentyl)octylphosphine oxide, commercially known as Cyanex 925 and TOPO.

Cyanex 925 with Pu^{IV} shows the best extractive ability in nitrate media due to better complexing ability of the nitrate anion to the metal. At high acid concentrations, nitric acid competes with $Pu(NO_3)_4$ for the binding site of Cyanex 925, causing the uptake of Pu^{IV} to fall off. At high HCl and $HClO_4$ concentrations, Pu^{IV} uptake increases dramatically due to the easier formation of neutral Pu^{IV} complexes and hydration considerations.

Extractant dependencies from both HCl and HNO_3 into various organic solvents show a slope of two for the uptake of Pu^{IV} by Cyanex 925. The resultant complexation stoichiometry in nitrate media is as shown in Equation (20):[315]

$$Pu^{4+}_{(a)} + 4\ X^-{}_{(a)} + 2\ [Cyanex\ 925]_{(o)} \longrightarrow Pu(X)_4 \bullet 2[Cyanex\ 925]_{(o)}$$
$$(20)$$
$$(X = NO_3,\ Cl)$$

Table 19 R-group substituents for various type of phosphine oxide extractants.

Extractant	Alkyl chain (R) length
Cyanex 923 (TRPO) (4 components)	$R_1 = R_2 = R_3 = C_8H_{17}$; $R_1 = C_8H_{17}$, $R_2 = R_3 = C_6H_{13}$; $R_1 = C_6H_{13}$, $R_2 = R_3 = C_8H_{17}$; $R_1 = R_2 = R_3 = C_6H_{13}$
Cyanex 921 (TOPO)	$R_1 = R_2 = R_3 = C_8H_{17}$
Cyanex 925	$R_1 = C_8H_{17}$, $R_2 = R_3 = 2,4,4$-trimethyl pentyl

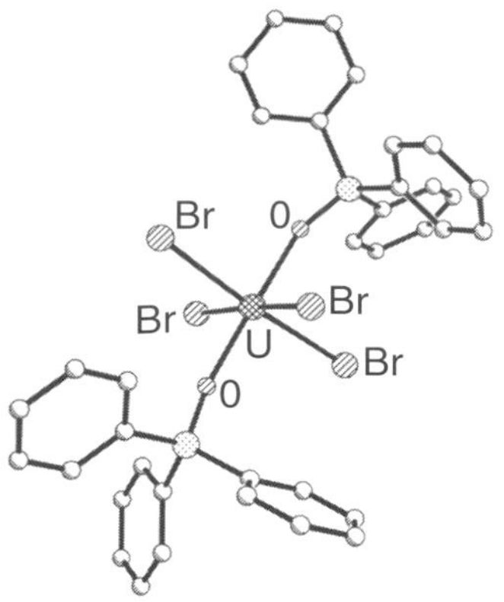

Figure 15   Crystal structure of UBr$_4$(C$_{18}$H$_{15}$PO)$_2$ Bombieri, Benetollo *et al. J. Chem. Soc., Dalton-Trans.* **1983**, 343).

A representative crystal structure of U^{IV} with two triphenylphosphine oxide molecules and four bromide ions is shown in Figure 15. Coordination of the triphenylphosphine oxide ligands occurs via the oxygen of the P=O group in *trans* positions. The bromide atoms coordinate equatorially, making the uranium six-coordinate with an octahedral geometry.[316]

Sulfoxides. Sulfoxides, particularly dimethylsulfoxide (DMSO), can act as oxygen-donor ligands to the electropositive actinides Th, U, Np, and Pu (Table 20). These complexes are most often prepared by reaction of the actinide salt (halides, nitrates, perchlorates) with the ligand in nonaqueous media, although some complexes can be prepared by anion exchange starting with the parent halide sulfoxide adduct. Adducts of the formula AnI$_3$L$_4$ (An = U, Np, Pu; L = DMSO) are prepared by oxidation of the corresponding metal by iodine in DMSO.[14]

Dimethylsulfoxide acts as a strong donor, leading to the formation of ionic compounds in the presence of excess base. The molecular structure of UCl$_4$·3DMSO reveals it to be [UCl$_2$(DMSO)$_6$][UCl$_6$];[317] the uranium atom in the cation lies at the center of a distorted dodecahedron. Other reported complexes have a large number of associated sulfoxide molecules. These species are likely also to be ionic, as indicated by their IR spectra. With bulky sulfoxides as ancillary ligands, complexes with lower coordination numbers can be isolated; the complex UCl$_4$(Bu$^i_2$SO)$_2$ possesses *trans*-octahedral geometry.[312]

Thioethers. As in the case of ethers, complexes contain either cyclic thioethers (e.g., tetrahydrothiophene, THT) or bidentate dithioethers such as MeSCH$_2$CH$_2$SMe. There is one report of a dimethyldithioethane complex of a halide, UCl$_4$ (MeSCH$_2$CH$_2$SMe)$_2$. The only structurally characterized complexes are those of borohydride derivatives U(MeBH$_3$)$_4$ (MeSCH$_2$CH$_2$SMe)[181] and [U(MeBH$_3$)$_4$(μ-THT)]$_2$.[318]

Thioureas. A few thorium complexes of the formula ThCl$_4$(L)$_x$ have been reported (L = SC(NH$_2$)$_2$, x = 2,8; L = SC(NH$_2$)(NHp-ClC$_6$H$_4$), x = 4; and L = S(cyclo-CNHCH$_2$CH$_2$NH), x = 8).

Phosphine sulfides, selenides. Phosphine sulfides and selenides do not often act as Lewis bases towards tetravalent actinides, owing to their propensity for chalcogen transfer. Only the thorium complex ThCl$_4$(SePPh$_3$)$_2$ has been reported.[319]

(vi) Ligands containing group 17 donor atoms

Binary halides. Hydrous fluoride complexes are generally obtained by precipitation of the metal ion from aqueous solution, while anhydrous fluorides are produced from halogenation of oxides, or by thermal decomposition of salts (e.g., NH$_4$UF$_5$). The anhydrous tetrafluoride complexes of thorium, protactinium, uranium, neptunium, and plutonium are known; all are halogen-bridged polymers in the solid state. The complexes are isostructural; the metal center lies within a distorted square antiprismatic arrangement of fluoride ions. Hydrous fluorides can display alternate geometries; in the compound Np$_3$F$_{12}$·H$_2$O, there are three distinct neptunium sites, none of which is coordinated by the water of hydration.[320]

Table 20 Complexes of actinide(IV) compounds with sulfoxides.

$M^{IV}Cl_4 \cdot xR_2SO$	R = Me; $x = 3$, M^{IV} = Th, Pa, U, Np, Pu; $x = 5$, M^{IV} = Th, Pa, U, Np; $x = 6$, M^{IV} = Th; $x = 7$, M^{IV} = Th, U, Np, Pu
	R = Et; $x = 2.5$, M^{IV} = Np, Pu; $x = 3$, M^{IV} = Th, U, Np; $x = 4$, M^{IV} = Th
	R = Pr^n; $x = 3$, M^{IV} = U
	R = Bu^n; $x = 2$, M^{IV} = Th; $x = 4$, M^{IV} = U
	R = Bu^i, $x = 2$, M^{IV} = U; $x = 3$, M^{IV} = Th
	R = Bu^t; $x = 2$, M^{IV} = U
	R = n-C$_5$H$_{11}$, n-C$_6$H$_{13}$; n-C$_7$H$_{15}$, n-C$_8$H$_{17}$, $x = 2$, M^{IV} = Th
	R = Ph, $x = 2$ (+2H$_2$O), M^{IV} = U; $x = 3$, M^{IV} = U, Np; $x = 4$, M^{IV} = Th, U, Np
	R = α-C$_{10}$H$_7$; $x = 3$, M^{IV} = Th, U, Np
$[UCl_3(Me_2SO)_5]ClO_4$	
$[UCl_2(Me_2SO)_6](ClO_4)_2$	
$M^{IV}Br_4 \cdot xR_2SO$	R = Me; $x = 6, 8$, M^{IV} = Th, U
	R = Et; $x = 5, 6$, M^{IV} = U
	R = Pr^n; $x = 7$, M^{IV} = U
	R = Bu^n; $x = 8$, M^{IV} = U
	R = Bu^i; $x = 4$, M^{IV} = U
	R = Bu^t; $x = 2$, M^{IV} = U
	R = Ph; $x = 4$, M^{IV} = Th, U
$ThI_4 \cdot 6Ph_2SO$	
$Th(NCS)_4 \cdot 4Ph_2SO$	
$M^{IV}(NO_3)_4 \cdot xR_2SO$	R = Me; $x = 3$, M^{IV} = Th, U, Np, Pu; $x = 6$, M^{IV} = Th, Np, Pu
	R = Et; $x = 3$, M^{IV} = Th, U, Np
	R = Bu^n, n-C$_5$H$_{11}$, n-C$_6$H$_{13}$, n-C$_7$H$_{15}$, n-C$_8$H$_{17}$, $x = 2$, M^{IV} = Th
	R = n-C$_8$H$_{17}$; $x = 3$, M^{IV} = Th
	R = PhCH$_2$; $x = 4$, M^{IV} = Th
	R = Ph; $x = 3, 4$, M^{IV} = Th, U, Np, Pu
	R = α-C$_{10}$H$_7$; $x = 3$, M^{IV} = Th, Np
$Th(ClO_4)_4 \cdot xR_2SO$	R = Me, $x = 6, 12$; R − Ph, $x = 6$
$Th(R^1COCHCOR^2)_4 \cdot BuSO$	$R^1 = CF_3$, $R^2 = Me, CF_3$; $R^1 = R^2 = C_2F_5$
$ThX_4 \cdot yMe_2SO$	X = HCO$_2$, OSC$_7$H$_5$ (thiotroponate), $y = 1$; X = C$_9$H$_6$NO (8-hydroxyquinolinate), $y = 2$
$M^{IV}(trop)_4 \cdot Me_2SO$	Htrop = tropolone; M^{IV} = Th, U
$ThX_2 \cdot yMe_2SO \cdot nH_2O$	X = SO$_4$, $y = 4$, n = 3 or 9; X = C$_2$O$_4$, $y = 2$, n = 1
L = thianthrene 5-oxide, $M^{IV}Cl_4 \cdot 4L$	C$_{12}$H$_8$OS$_2$ M^{IV} = Th, U
L = tetrahydrothiophene-S-oxide	
$ThX_4 \cdot yL$	X = Cl, Br, $y = 4$; X = I, $y = 8$; X = NCS, $y = 2$; X = NO$_3$, $y = 6$; X = ClO$_4$, $y = 10$

Tetrachloride and tetrabromide complexes are known for thorium, protactinium, uranium, neptunium, and plutonium. These are similarly produced by halide-based oxidation of metals or hydrides, or by halogenation of oxides. A common structural type is reported for most compounds. The reported structure of thorium tetrachloride reveals that the coordination geometry about the metal is dodecahedral.[321] The compounds are generally volatile and can be sublimed. The gas-phase electron diffraction structure of UCl$_4$[322] suggests that the molecule is tetrahedral, with a U–Cl distance of 2.51 Å.

The iodide complexes are somewhat less stable, and well-characterized examples exist only for thorium, protactinium, and uranium. The thorium and uranium derivatives can be coveniently prepared by the reaction of iodine and metal, while protactinium tetraiodide is generated by reduction of PaI$_5$. The molecular structure of ThI$_4$ has been examined;[323] the metal lies within a distorted square antiprism of iodide ions.

Complex halides. A large number of complex halides of tetravalent actinides have been prepared, particularly for the fluoride complexes.[324] The most common formulations are shown in Table 21.

Prototype structures within these classes of compounds have been determined. The structure of LiUF$_5$[325] is also representative of the compounds of thorium, protactinium, neptunium, and

Table 21 Classes of complex halides of tetravalent actinides.

$M^IM^{IV}F_5$	$M^I = Li$, $M^{IV} = Th$, Pa, U, Np, Pu
	$M^I = Na$, K, Rb, $M^{IV} = U$
	$M^I = Cs$, $M^{IV} = Th$, U, Pu
	$M^I = NH_4$, $M^{IV} = Th$, U, Np, Pu
$M^IM^{IV}_2F_9$	$M^I = Li$, $M^{IV} = Th$; $M^I = Na$, $M^{IV} = Th$, U
	$M^I = K$, $M^{IV} = Th$, U, Np, Pu; $M^I = Rb$, $M^{IV} = Th$, U;
	$M^I = Cs$, $M^{IV} = U$
$M^IM^{IV}_3F_{13}$	$M^I = Li$, $M^{IV} = Th$, U, Np, Pu;
	$M^I = K$, $M^{IV} = Th$, U; $M^I = Rb$, Cs, $M^{IV} = Th$;
	$M^I = NH_4$, $M^{IV} = U$, Np, Pu
$M^I_2M^{IV}F_6$	$M^I = Na$, $M^{IV} = Th$, U, Np, Pu
	$M^I = K$, $M^{IV} = Th$, U, Np
	$M^I = Rb$, $M^{IV} = Th$, Pa, U, Np, Pu
	$M^I = Cs$, $M^{IV} = Th$, U, Pu
	$M^I = NH_4$, $M^{IV} = U$, Np, Pu
	$M^I = Et_4 N$, $M^{IV} = Pa$, U
$M^{II}M^{IV}F_6$	$M^{II} = Ca$, Sr, $M^{IV} = Th$, U, Np, Pu
	$M^{II} = Ba$, Pb, $M^{IV} = Th$, U, Np
	$M^{II} = Cd$, Eu, $M^{IV} = Th$
	$M^{II} = Co$, $M^{IV} = U$, Np ($+3H_2O$)
$M^I_2M^{IV}F_7$	$M^I = Li$, $M^{IV} = Th$, U; $M^I = Na$, K, $M^{IV} = Th$, Pa, U;
	$M^I = Rb$, Cs, $M^{IV} = Th$, U; $M^I = NH_4$, $M^{IV} = Th$
$M^I_4M^{IV}F_8$	$M^I = Li$, $M^{IV} = U$, Np, Pu
	$M^I = NH_4$, $M^{IV} = Th$, Pa, U, Np, Pu
$M^I_7M^{IV}_6F_{31}$	$M^I = Na$, K, $M^{IV} = Th$, Pa, U, Np, Pu
	$M^I = Rb$, $M^{IV} = Th$, Pa, U, Np, Pu
	$M^I = NH_4$, $M^{IV} = U$, Np, Pu
$M^I_2 [M^{IV}Cl_6]$	$M^{IV} = Th$, U; $M^I = Li–Cs$, Me_4N, Et_4N
	$M^{IV} = Pa$, Np; $M^I = Cs$, Me_4N, Et_4N
	$M^{IV} = Pu$; $M^I = Na–Cs$, Me_4N, Et_4N
$M^{II}[UCl_6]$	$M^{II} = Ca$, Sr, Br
$M^I_2 [M^{IV}Br_6]$	$M^{IV} = Th$, Pa; $M^I = Me_4N$, Et_4N
	$M^{IV} = U$; $M^I = Na–Cs$, Me_4N, Et_4N
	$M^{IV} = Np$; $M^I = Cs$, Et_4N
	$M^{IV} = Pu$; $M^I = Et_4N$
$M^I_2 [MI_6]$	$M^{IV} = Th$, U; $M^I = Et_4N$, Me_3PhN
	$M^{IV} = Pa$; $M^I = Et_4N$, Me_3PhN, Me_3PhAs

plutonium. The uranium atom in this structure is surrounded by nine fluorides in a tricapped trigonal prismatic array, with adjacent prisms sharing edges and corners. This local coordination environment persists in other complex fluorides such as KAn_2F_9 ($An = Th–Pu$)[326] and MAn_3F_{13} ($M = NH_4$, Rb; $An = Th$, U, Np).[327] Compounds of the type M_2AnF_6 ($M = Rb$; $An = U$, Np, Pu) contain chains of AnF_8 dodecahedra, whereas in the complexes of the lighter alkali metals ($M = Na$, K; $An = U$), the UF_9 polyhedra are tricapped trigonal prisms.[326,328] The compound $CoAnF_6 \cdot 3H_2O$ ($An = U$, Np) consists of chains of units $[AnF_8(H_2O)]$ (capped square antiprisms sharing two fluorides).[329] Compounds of the formula $(NH_4)_4AnF_8$ generally contain distinct dodecahedral $[AnF_8]^{4-}$ ions ($An = Th$, U, Pa, Np, Pu);[330] the exception is the thorium compound, in which ThF_9 tricapped trigonal prisms share edges to form chains.[331]

More recently, a number of novel uranium fluoride complexes have been produced in hydrothermal syntheses in the presence of organic structure-directing agents[332–336] (see Table 22). Reactions of UO_2 with aqueous orthophosphoric acid, aqueous hydrofluoric acid, and organic templating agents such as alkanediamines generate a variety of solid-state structures incorporating negatively charged sheets of uranium fluoride polyhedra separated by alkylammonium counterions and occluded water molecules. In most of the layered structures, the layers are constructed from equivalent UF_9 tricapped trigonal prisms that share three edges and two corners, whereas the complex $[HN(CH_2CH_2NH_3)_3]U_5F_{24}$ contains both UF_8 bicapped trigonal prisms and UF_9 tricapped trigonal prisms. The ammonium ions can subsequently be exchanged for a wide range of group 1, group 2, and transition metals. A subsequent study investigated the role of water stoichiometry in determining the structure of the product in the UO_2/2-methyl-piperazine/HF(aq.)/H_2O system.

Table 22 Uranium(IV) fluoride complexes produced by hydrothermal syntheses.

Compound	References
$(H_3N(CH_2)_3NH_3)U_2F_{10}\cdot 2H_2O$	a
$(H_3N(CH_2)_4NH_3)U_2F_{10}\cdot 3H_2O$	a
$(H_3N(CH_2)_6NH_3)U_2F_{10}\cdot 2H_2O$	a
$[HN(CH_2CH_2NH_3)_3]U_5F_{24}$	b
$(C_5N_2H_{14})_2(H_3O)U_2F_{13}$	c
$(C_5N_2H_{14})_2U_2F_{12}\cdot H_2O$	c
$(C_5N_2H_{14})(H_3O)U_2F_{11}$	c
$(C_4N_2H_{12})_2U_2F_{12}\cdot H_2O$	d
$(C_5H_{14}N_2)U_2F_{10}(H_2O)$	e
$(NH_4)_7U_6F_{31}$	f
$(NH_4)U_3F_{13}$	f

[a] Francis, R. J. and O. H. D. Halasyamani *Angew. Chem., Int. Ed. Engl.* **1998**, *37*, 2214. [b] Francis, R. J., P. S. Halasyamani, *et al. Chemistry of Materials* **1998**, *10*, 3131–3139. [c] Francis, R. J., P. S. Halasyamani, J. S. Bee and D. O'Hare *J. Am. Chem. Soc.* **1999**, *121*, 1609. [d] Walker, S. M., P. S. Halasyamani, S. Allen and D. O'Hare *J. Am. Chem. Soc.* **1999**, *121*, 10513. [e] Almond, P. M., L. Deakin, A. Mar and T. E. Albrecht-Schmitt. *Inorg. Chem.* **2001**, *40*, 886. [f] Cahill, C. L. and P. C. Burns *Inorg. Chem.* **2001**, *40*, 1347–51.

The reaction conditions appear to control the dimensionality of the complexes; an increase in solution acidity results in an increase in bridging between uranium centers. The complex $(C_5N_2H_{14})_2(H_3O)U_2F_{13}$ is a molecular, or "zero-dimensional," phase consisting of dimeric $[U_2F_{13}]^{5-}$ units separated by 2-methyl-piperazine and hydronium cations. Each dimer consists of face-sharing trigonal prisms, wherein uranium cations are bonded to eight fluorine atoms in a distorted bicapped trigonal prismatic coordination. $(C_5N_2H_{14})_2U_2F_{12}\cdot H_2O$ has a one-dimensional structure and contains uranium fluoride chains formed from edge-sharing polyhedra with eight-coordinate uranium. $(C_5N_2H_{14})(H_3O)U_2F_{11}$ consists of anionic sheets of nine-coordinate uranium cations that are separated by protonated 2-methyl-piperazine and occluded hydronium cations. The related aqua complex $(C_5H_{14}N_2)U_2F_{10}(H_2O)$[337] has also been prepared under hydrothermal conditions, and possesses a uranium fluoride chain structure. The complexes $(NH_4)_7U_6F_{31}$ and $(NH_4)U_3F_{13}$ are prepared in the presence of DABCO as the templating base;[336] the ammonium counterion is presumed to arise from decomposition of the organic base. $(NH_4)_7U_6F_{31}$ consists of chains of nine-coordinate uranium fluoride polyhedra, while $(NH_4)U_3F_{13}$ consists of a three-dimensional network.

The heavier halides display a significantly reduced propensity to form bridging structures, and the dominant class of complexes is that containing the $[AnX_6]^{2-}$ unit, although evidence exists for short chain-like ions such as $[Th_2Cl_{10}]^{2-}$, $[Th_3Cl_{14}]^{2-}$, and $[Th_3Cl_{10}]^{2+}$ in thorium-rich molten salts such as $ThCl_4$, A_2ThCl_6, and A_3ThCl_7 (A = Li, Na, K, Cs).[338] Crystallographically characterized examples of complexes $A_2[AnX_6]$ (An = U, Np; X = Cl, Br, I) reveal octahedral coordination about the metal center.[339–342]

Crystallization of complex halides from solution containing crown ethers results not in the complexation of the actinide by the ether oxygen atoms, but rather in the isolation of $[AnX_6]^{2-}$ salts in which the crown appears to act as a crystallization aid.[304,343–345]

Oxohalides. A number of oxohalides of the formula $AnOX_2$ have been reported; the structure of $PaOCl_2$ consists of an infinite polymer in which the Pa atoms are seven-, eight-, and nine-coordinate.[346]

3.3.2.2.3 Chelating ligands

(i) Multidentate donor ligands

Hydroxamates, cupferron, and related ligands. As anionic oxygen donor ligands, hydroxamates have a strong affinity for the oxophilic tetravalent actinides, with solution complex formation constants generally greatest for Pu^{IV} and decreasing as follows: $Pu^{IV} > Np^{IV} > U^{IV} > U^{VI}$. A significant effort has been made to prepare ligands with high specific affinity and selectivity for actinides that could be used for mammalian chelation therapy or as a specific extractant. A biomimetic approach to such ligand design, based on naturally occurring hydroxamate and catecholate siderophores and hydroxypyridinoate moieties, has been the most vital. The actinide complexes reported include synthetic and biogenic bi-, hexa- and octadentate hydroxamates.

Proposed therapeutic removal of actinides has evolved from substituting a nontoxic metal for the metal bound in blood and tissue, to chelating the Pu with general or specific chelating agents.[347–352]

Simple bidentate hydroxamates commonly bind actinides via replacement of the hydroxamate proton by the metal to form a five-membered chelate ring. The known $M^{IV}L_4$ complexes are usually prepared by treating an aqueous solution of the metal with an excess of the hydroxamic acid.[353,354] The complexes $Th[(CH_3)_2CHN(O)O(O)R]_4$ ($R=C(CH_3)_3$ (**1**), or $CH_2C(CH_3)_3$ (**2**)) have been prepared directly in aqueous solution. The U^{IV} analog of (**1**) was prepared similarly, but it is unstable and undergoes an internal oxygen transfer reaction to form a bis(hydroxamato)uranyl complex. These complexes have been characterized using single-crystal X-ray diffraction and optical absorbance spectroscopy. Complex (**1**) has approximately S_4 symmetry and the eight-coordinate polyhedron is nearly cubic, whereas the structure of (**2**) shows an eight-coordinate metal, with D_{2d} trigonal-faced dodecahedral geometry (Figure 16).[355] Somewhat surprisingly, one hydroxamic acid, (PhCO)NHOH, has been reported to behave as a neutral ligand in a postulated 10-coordinate complex, $Th(NO_3)_4((PhCO)NHOH)_2$; however, the stoichiometry of the complex has not been confirmed nor has the complex been fully characterized. Several complexes of *N*-phenyl-benzoylhydroxamic acid (HL^1) and cupferron (*N*-nitrosophenlyhydroxylamine, (HL^2) have been reported.[356] The complexes $Th(L^1)_4 \cdot 4H_2O$ and $Th(L^1)_3 \cdot 2H_2O$ have been prepared by reacting an aqueous solution of thorium nitrate with an excess of the ligand. $Th(L^2)_4 \cdot H_2O$ was obtained similarly from combination of thorium nitrate and cupferron in $H_2O/MeOH$. Several other ternary complexes, including $Th(L^2)_4Ph_3PO$, $Th(L^2)_4py$, and $Th(L^2)_4dmf$, were prepared by treating solutions of $Th(L^2)_4 \cdot H_2O$ in $CHCl_3$ with an excess of the ancillary ligand. Some U^{IV} complexes of cupferronate and neocupferronate were prepared and characterized by optical absorbance, vibrational, and electron spin resonance spectroscopy.[357,358] However, their solution and solid-state structures have not been determined.

The complexation of actinides with multidentate hydroxamate ligands comprise naturally-occurring siderophores and synthetic ligands designed based on these Fe^{III} chelators. For example, a series of ligands based on desferrioxamines have been synthesized and their metal complexes characterized. The structure of a Pu^{IV}–desferrioxamine E complex was determined from X-ray diffraction analysis[359] (Figure 17). Other types of ferrioxamine complexes of Th^{IV} and Pu^{IV} have been characterized in solution by NMR, potentiometry, and optical absorbance spectroscopy, including desferrioxamine B (DFO), octadentate derivatives [*N*-(2,3-dihydroxy-4-carboxybenzoyl)desferrioxamine B (DFOCAMC), *N*-(1,2-dihydro-1-hydroxy-2-oxopyridin-6-yl)carbonyl)desferrioxamine B (DFO-1,2-HOPO), and *N*-(2,3-dihydroxy-4-(methylamido)benzoyl)- desferrioxamine B (DFOMTA)].[360]

The complexation of thorium(IV) and plutonium(IV) with a tetrahydroxamate ligand based on the cyclohexane-1,2-diyldinitrilotetraacetate complexon, with hydroxamate instead of carboxylate groups has been reported. The speciation appears to be pH dependent. Up to pH 9 the complexes

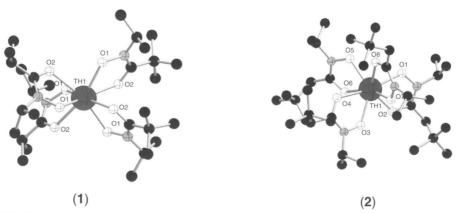

(1)   (2)

Figure 16 Crystal structures of $Th[(CH_3)_2CHN(O)O(O)R]_4$ ($R = C(CH_3)_3$ (**1**) or $CH_2C(CH_3)_3$ (**2**)). (Smith and Raymond *J. Am. Chem. Soc.* **1981**, *103*, 3341–3349).

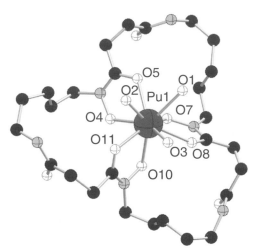

Figure 17 Plutonium(IV) coordination sphere in the crystal structure of Pu^{IV} complexed by the siderophore desferrioxamine E (Neu, Matonic *et al. Angew. Chem., Int. Ed. Engl.* **2000**, *39*, 1442–1444).

are monomeric; then dimeric complexes, M_2L_2, have been suggested based on the magnetic properties of these complexes and modeling calculations.[361]

Catecholate. Actinide(IV) complexes formed by catechol and the related compounds resorcinol, phloroglucinol, orcinol, and pyrogallol include the mono, bis, tris, and tetra complexes as well as polymeric compounds (Table 23). Thorium dichloride catecholate, and the corresponding resorcinolate, phloroglucinolate, and orcinolate have been obtained by evaporating an ether solution of the components to dryness and heating the residue until the evolution of hydrogen chloride ceased. When thorium tetrachloride is added to an excess of the molten catechol using this preparation, the product is $H_2[Th(C_6H_4O_2)_3]$.[362] More common complexes are tetrakis(catecholato)uranate(IV) and -thorate(IV) complexes $Na_4[M(C_6H_4O_2)_4]\cdot21H_2O$, M = Th, U, which are obtained from basic aqueous solutions of the metal chlorides. The complexes show D_{2d} molecular symmetry (structure determined by single-crystal X-ray diffraction, see Figure 18).[363] The geometry of the anion is a trigonal faced dodecahedron and the oxygen atoms of the water molecules form a hydrogen-bonded network through the crystal. The other compounds, thorium(IV) bis derivatives of 2,2′-dihydroxybiphenyl or dinaphthyl and 1,8-dihydroxynaphthalene, ThL_2, are precipitated from methanolic solutions of the tetrachloride and the diol in the presence of base. These complexes have not been structurally characterized.[362]

Tiron complexes of Th^{IV} and other actinides have been prepared, generally in aqueous solution[364] The EXAFS data have been modeled to include binding of the sulfonate group to Th^{IV} at low pH. This preferred complexation of a sulfonate over a catecholate, even at low pH, is unexpected. Bidentate catechol ligation of thorium Th(tiron)x, ($x \geq 2$), has been proposed at very high excess Tiron.

Table 23 Actinide(IV) catecholates and related compounds.

$ThCl_2(L)$	L = 1,2-dihydroxybenzene, $C_6H_6O_2$; Resorcinol, 1,3-dihydroxybenzene, and hydroquinone, 1,4-dihydroxybenzene, $C_6H_6O_2$; Orcinol, 2,5-dihydroxytoluene, $C_7H_8O_2$; Phloroglucinol, 1,3,5-trihydroxybenzene, $C_6H_6O_3$; Resorcinol, 1,3-dihydroxybenzene, and hydroquinone, 1,4-dihydroxybenzene, $C_6H_6O_2$
$Th(L)_2$	L = $C_{12}H_8O_2$, $C_{20}H_{12}O_2$
$M^I_2[Th(C_6H_3O_3)_2]\cdot7H_2O$	M^I = Na, K
$M^I[U(C_6H_4O_2)_2(OH)]\cdot xH_2O$	M^I = pyH, $x = 4$; $C_2N_4H_5$ (dicyandiamidinium), $x = 20$
$M^I_2[Th(C_6H_4O_2)_3]\cdot xH_2O$	M^I = H, $x = 0$; NH_4, $x = 5$
$2(NH_4)_2[U(C_6H_4O_2)_3]\cdot C_6H_6O_2\cdot8H_2O$	
$Na_4[M^{IV}(C_6H_4O_2)_4]\cdot21H_2O$	M^{IV} = Th, U
$(NH_4)_4H_2[Th(C_6H_4O_2)_4]$	[or $(NH_4)_2[Th(C_6H_4O_2)_3]\cdot C_6H_6O_2$]
$(NH_4)_2[Th_3(C_6H_4O_2)_6(OH)_2]\cdot10H_2O$	
$M^I_2[Th_3(C_6H_4O_2)_7]\cdot20H_2O$	M^I = Na, K
$M^I_2H_2[U_2(C_6H_4O_2)_7]\cdot xH_2O$	M^I = K, $x = 3$; NH_4, $x = 6$; CN_3H_6 (guanidinium), $x = 14$

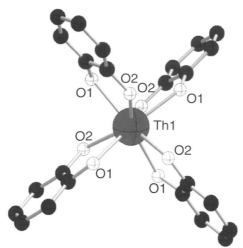

Figure 18   Crystal structures of $Na_4[M(C_6H_4O_2)_4]\cdot21H_2O$, M = Th, U (Sofen, Abu-Dari *et al. J. Am. Chem. Soc.* **1978**, *100*, 7882–7887).

The similarities between Pu^{IV} and Fe^{III} (charge to ionic radius ratios, formation of highly insoluble hydroxides) have stimulated the design of specific Pu^{IV} sequestering agents modeled after iron(III) chelators.[81,347,350,365] This approach led to the design of potentially octadentate catecholamide ligands including both catechol only and mixed functional catechol and hydroxypyridinone ligands. These ligands have been studied for intended application in mammalian actinide decorporation.[366–368] Plutonium and americium complexes have been reported for a series of sulfonated and carboxylated catechols. The stoichiometry of the complexes formed depends on pH. Above pH 12, the Pu^{IV} complex is tetrakis(catecholate) and at neutral pH it is tris(catecholate).[81,365] Tetravalent actinide complexes can also be prepared indirectly. For example, the reduction of Np^V by catecholate and hydroxypyridinoate ligands yields a Np^{IV} species as determined using X-ray absorbance spectroscopy.[350]

Pyoverdin. Pyoverdin complexes of tetravalent actinides have been investigated due to the potential of this class of ligand to solubilize and sequester these metals (as they do for Fe^{III}). At near-neutral pH pyoverdine forms a 1:1 Pu:pyoverdine complex with Pu^{IV}. The stoichiometry changes to 1:2 when excess ligand is present. Thorium(IV), U^{IV}, and U^{VI} complexes have also been reported. Their optical absorbance spectroscopic properties, but no structural studies, are reported. The selectivity of pyoverdin for common actinides in the order $Th^{IV} > U^{IV} > U^{VI}$ has been proposed.[369,370]

Pyridonate. Tetravalent Th^{IV} and U^{IV} complexes of 1 oxy-2-pyridonate, $Th(C_5H_4NO_2)_4H_2O$, and $U(C_5H_4NO_2)_4CHCl_3$, have been prepared by slowly adding a basic aqueous solution of excess ligand to solutions of the metal tetrachlorides. The crystal structure of the thorium complex, $Th(C_5H_4NO_2)_4H_2O$, shows a nine-coordinate, neutral complex of low symmetry. Four bidentate ligands and one water molecule are bonded to thorium to form a D_{3h} tricapped trigonal prismatic coordination geometry.[371] The related compound $Th(C_5H_4NO_2)_4\cdot MeOH$ was prepared by refluxing a methanolic solution of thorium nitrate with excess *O*-hydroxypyridine-*N*-oxide. The complex has the same general coordination geometry as the aqueous complex, with methoxide in the inner coordination sphere.[372] Multifunctional ligands containing one to four hydroxypyridinone binding units have been researched for their potential use in actinide separations and and chelation therapy. For example, the octadentate, mixed hydroxypyridinone (HOPO) ligand, 3,4,3-LI-(1,2-Me-3,2-HOPO), when administered orally, removes actinides from animals more efficiently than any injected ligand studied previous.[349,367,373]

1-hydroxy-6-*N*-octylcarboxamide-2(1H)-pyridinone (octyl-1,2-HOPO) has been shown to be a highly selective extractant for tetravalent plutonium from acidic solutions. The structure of octyl-1,2-HOPO is illustrated in Figure 19. The general equilibrium for the extraction from nitric acid solutions is given in Equation (21):

$$Pu^{4+}{}_{(a)} + m\ HHOPO_{(o)} + (4-m)\ NO_3{}^-{}_{(a)} \longrightarrow [Pu(HOPO)_m(NO_3)_{4-m}]_{(o)} + m\ H^+{}_{(a)} \quad (21)$$

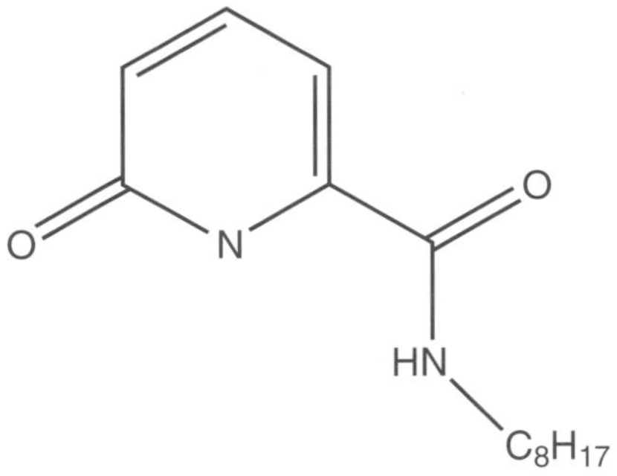

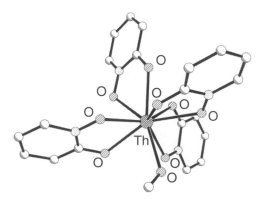

Figure 19 Octyl-1,2-HOPO.

In Equation (21) the ligand acts in a deprotonated bidentate manner, where m can range from 0 to 4. While the hydroxypyridinonate ligands are themselves a class of compounds, their ability to extract Pu^{IV} is directly related to their protonation constants. Octyl-1,2-HOPO has the lowest protonation constants among all hydroxypyridinonates thus making it the best agent for extraction from acid solutions, particularly at low acid concentrations.

Extractant dependency at low concentrations indicates independent behavior, even though high distributions are obtained. At high concentrations, slope analysis gives a value of four as discussed in the equilibrium above. This could possibly be explained by the presence of $Pu(HOPO)(NO_3)_3$ at low concentration and at $Pu(HOPO)_4$ at higher concentrations.[226]

From an X-ray single crystal diffraction study of Th^{IV} with a 1-hydroxy-2(1H)-pyridinone, a coordination number of nine can be seen. This coordination number is due to four of the bidentate ligands chelating to thorium with the additional complexation of one methanol molecule, as illustrated in Figure 20.[374]

8-Hydroxyquinoline and derivatives. The complex of Th^{IV} with 8-hydroxyquinoline(Ox), $Th(Ox)_4 \cdot HOx$ was prepared by precipitation from aqueous solution.[375] The IR spectra of the Th complexes have absorbance frequencies corresponding to a N-H$\cdots$O bond, similar to those observed in the spectrum of $UO_2(Ox)_2HOx$. This vibrational band is not observed in $Th(Ox)_4$, suggesting that HOx is bound to the metal through the phenolic O in this case.[376] Several other complexes of Th^{IV} with 8-hydroxyquinoline derivatives have also been prepared similarly, including those with 7-nitroso-8-hydroxyquinoline-5-sulfonic acid[377] and 5-chloro-7-nitro-8-hydroxyquinoline.[378] When $Th(Ox)_4 \cdot HOx$ is dissolved in DMSO, the $Th(Ox)_4 \cdot 2(DMSO)$ complex forms, in which only one DMSO is coordinated to the metal center. The complexes were characterized in solution by vibrational spectroscopy. The molecular structure of the complex determined from X-ray diffraction is shown in Figure 21. The oxine groups are arranged in a distorted square antiprismatic configuration about the metal ion, with the coordinated DMSO in a capping position.[379] Similarly, in the complex $Th(Ox)_4DMF$, the four 8-quinolinolato ligands are bidentate, and a DMF ligand, bonded through the oxygen, completes the coordination sphere. In this case the coordination polyhedron of the thorium atom is best described as a slightly distorted tricapped trigonal prism.[380] Analogous heavier actinide complexes of Np and Pu with 8-hydroxyquinoline

Figure 20 Crystal structure of $Th(C_5H_4NO_2)_4CH_3OH$ (Casellato, Vigato *et al. Inorganica Chimica Acta* **1983**, *69*, 77–82).

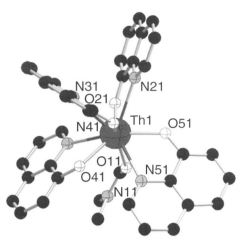

Figure 21   Crystal structure of Th(oxine)$_4$(DMSO) (Singer, Studd *et al. Chem. Commun.* **1970**, 342).

Table 24   Tetravalent actinide carbamate complexes.

M^{IV}(O$_2$CNR$_2$)$_4$	M^{IV} = U, Th, R = Me, Et
M^{IV}(OSCNR$_2$)$_4$	M^{IV} = U, Th, R = Me, Et
M^{IV}(S$_2$CNEt$_2$)$_4$	M^{IV} = U, Th
U(Se$_2$CNEt$_2$)$_4$	
[U(O$_2$CR)$_2$L$_2$]	(R = NEt$_2$, Me; L = tris(pyrazolyl)borate)
Cp$_2$U(XYCNEt$_2$)$_2$	(X, Y = O or S and X = O, Y = S).

and some of its 2-, 5-, 7-Me- and halogen-substituted derivatives have been reported and assigned the stoichiometries Np(Ox)$_4$ and Pu(Ox)$_4$.[381]

Carbamate. Thorium(IV) and uranium(IV) carbamate complexes M(R$_2$NCX$_2$) are usually obtained by the insertion of CX$_2$ (X = O, S, Se) or COS into the metal–nitrogen bonds of thorium(IV) and uranium(IV) dialkylamides M(NR$_2$)$_4$.[382] The complexes are precipitated from *n*-hexane solutions of the M^{IV}-tetrakis dialkylamide by addition of excess CX$_2$. A much simpler route is by reaction of UCl$_4$ with R$_2$NH and carbon dioxide in benzene (R = Et) or toluene (R = Me). The carbamates precipitate on addition of *n*-heptane after concentrating the solution.[383] Complexes in Table 24 have been reported, based mostly on elemental analysis, IR and NMR spectroscopy. The complex U(Et$_2$NCO$_2$)$_4$, prepared by reaction of UCl$_4$ with Et$_2$NH and CO$_2$, is a monomer in benzene and the ^{1}H-NMR spectra of this compound indicate that the alkyl groups are equivalent. A by-product of the preparation of U(Et$_2$NCO$_2$)$_4$ from the tetrachloride and the amine is a product of composition U$_4$O$_2$(Et$_2$NCO$_2$)$_{12}$. This is a tetramer in which there are two inequivalent uranium(IV) sites. One U^{IV} is coordinated in a distorted tricapped trigonal prism, and the geometry of the other does not fit any type of regular polyhedron.[383] These compounds are very sensitive to oxygen and water. Related Th^{IV} and U^{IV} thiocarbamates M(R$_2$NCXY) (X and Y are O or S) are obtained similarly from the dialkylamides M(NR$_2$)$_4$. For example, M(OSCNEt$_2$)$_4$ and M(S$_2$CNEt$_2$)$_4$ have been characterized in direct analogy with the carbamates.[382] Related pyrazolylborate complexes of the form, U(O$_2$CR)$_2$L$_2$ (R = NEt$_2$, Me; L = tris(pyrazolyl)borate) have also been prepared and characterized by elemental analysis and NMR, IR, and reflectance spectroscopies.[384] Cp$_2$U(XYCNEt$_2$)$_2$ (X, Y = O or S and X = O, Y = S) were prepared by treating Cp$_2$U(NEt$_2$)$_2$ with CS$_2$, COS, and CO$_2$. The compounds, Cp$_2$U(S$_2$CNEt$_2$)$_2$ and Cp$_2$U(OSCNEt$_2$)$_2$ are monomeric in benzene; for Cp$_2$U(O$_2$CNEt$_2$)$_2$ poly-meric behavior is indicated. Spectroscopic data are consistent with a bidentate coordination of the carbamate ligands in all cases. The coordination geometry around the U center is pseudoocta-hedral with cyclopentadienyl groups occupying mutually *cis* positions.[385–387]

Oxalate. A large number of oxalato and mixed oxalato complexes (Tables 25 and 26) have been reported. The hydrated oxalates, M(C$_2$O$_4$)$_2$·xH$_2$O (x = 0, 1, 2 or 6) are precipitated from aqueous media. The Th^{IV} and U^{IV} compounds are isomorphous. The neptunium(IV) compound consists of [Np(C$_2$O$_4$)$_2$]$_n$ layers, in which all oxalato ions are tridentate chelate-bridged, and the coordination polyhedron of the neptunium atom is a distorted cube comprised of eight oxygen atoms from four

Table 25 Actinide(IV) oxalate and oxalato complexes.

$M(C_2O_4)_2 \cdot xH_2O$	$M = Th$, $x = 0, 1, 2, 4, 6$; U, $x = O, 1, 2, 3, 5, 6$; Np, Pu, $x = 2, 6$
$UO(C_2O_4) \cdot xH_2O$	$x = 0, 4, 6$
$[Np(C_2O_4)_3 2H_2O]_n \cdot H_2O^a$	
$M^I_2 M^{IV}(C_2O_4)_3 \cdot xH_2O$	$M^{IV} = Th$, $M^I = CN_3H_6$, $x = 6, 8$; NH_4, x unspecified $M^I_2 = [(PhCH_2,)N(C_9H_7)]^+H^+$, $M^{IV} = Th$, U (NC_9H_7), = quinoline)
$H_2Ca[U_2(C_2O_4)_6] \cdot 24H_2O$	
$M^I_4[M^{IV}(C_2O_4)_4] \cdot xH_2O$	$M^{IV} = Th$; $M^I = Na$, $x = 0, 5.5, 6$; K, $x = 0, 4$; NH_4, $x = 0, 3, 4.7,6.5,7$; Me_2NH_2, $x = 0, 2, 9$; $Bu^n_2NH_2$, $x = 0, 4$; CN_3H_6, $x = 2$
	$M^I_4 = (CN_3H_6)_3,(NH_4)$, $x = 3$
	$M^{IV} = U$; $M^I = K$, $x = 0, 1,2, 4, 4.5, 5$; NH_4, $x = 0, 3, 5, 6,7$; Cs, $x = 3$; CN_3H_6, $x = 0, 2$
	$M^{IV} = Np$; $M^I = Na$, $x = 3$; K, $x. = 4$; NH_4, x unspecified
	$M^{IV} = Pu$; $M^I = Na$, $x = 5$; K, $x = 4$
$Ba_2U(C_2O_4)_4 \cdot 8H_2O^b$	
$K_2MnU(C_2O_4)_4 \cdot 9H_2O$	
$M^{II}_2[M^{IV}(C_2O_4)_4] \cdot xH_2O$	$M^{IV} = Th$, $M^{II} = Ba$, $x = 11$; enH_2, $x = 2.5$
	$M^{IV} = U$, $M^{II} = Ca$, $x = 0, 1, 4,6, 10$; Sr, $x = 0,4,6$; Ba, $x = 0, 6, 6.5, 7, 8, 9$; Cd, $x = 0, 6, 7$; Pb, $x = 0, 6, 8$; $[Pt(NH_3)_4]$, $x = 3$
$M^{III}_4[M^{IV}(C_2O_4)_4]_3 \cdot xH_2O$	$M^{IV} = Th$, $M^{III} = [Co(en)3]$, $x = 22$; $[Co(tn)_3]$, $x = 3$; $tn = H_2N(CH_2)_3NH_2$
	$M^{IV} = U$, $M^{III} = La$, $x = 22$; $M^{III} = Cr(urea)_6]$, $x = 6$ to 11
	$M^{IV} = Pu$, $M^{III} = [Cr(urea)_6]$, x unspecified
$[Pt(NH_3)_6][U(C_2O_4)_4]_3 \cdot 3H_2O$	
$M^I_6[M^{IV}(C_2O_4)_5] \cdot xH_2O$	$M^{IV} = Th$, $M^I = NH_4$, $x = 3, 7.5$; $M^{IV} = Pu$, $M^I = K$, $x = 4$; NH_4, x unspecified
$M^{III}_2[M^{IV}(C_2O_4)_5] \cdot xH_2O$	$M^{IV} = Th$, $M^{III} = [Co(NH_3)_6]$, $x = 3$; $[Cr(NH_3)_6]$, $x = 20$; $[Cr(urea)_6]$, $x = 0.5$
$M_2Np_2(C_2O_4)_5 \cdot nH_2O$	$(M = H, Na, K, and NH_4)$
$M^I_2[M^{IV}_2(C_2O_4)_5] \cdot xH_2O$	$M^{IV} = Th$, $M^I = H$, $x = 9^a$; NH_4, $x = 2, 7$
	$M^{IV} = U$, $M^I = H$, $x = 0, 4, 8$; Na, K, $x = 8$; NH_4, $x = 0,2,4, 8$; CN_3H_6, $x = 0, 1, 4$
$M^I_8[Th(C_2O_4)_6] \cdot xH_2O$	$M^I = Et_3NH$, $x = 0,3$; $Bu^n_2NH_2$, $x = 0$
$H_2CaU_2(C_2O_4)_6 \cdot 24H_2O$	
$M^I_6Th_2(C_2O_4)_7 \cdot xH_2O$	$M^I = Et_2NH_2$, $x = 0, 6$; $Pr^n_2NH_2$, $x = 0, 8$; CN_3H_6, $x = 5, 8, 12.5$ to 13.7

[a] Charushnikova, I. A., N. N. Krot, *et al. Radiokhimiya* **1998**, *40*, 538. [b] Spirlet, M. R., J. Rebizant, *et al. Acta Crystallographica, Section C: Crystal Structure Communications* **1987**, *C43*, 19–21.

oxalate ligands. The hydrated basic oxalate, $UO(C_2O_4) \cdot 6H_2O$, precipitates on photoreduction of $UO_2(HCO_2)_2$ in the presence of oxalic acid. Other hydrates are known; some authors describe them as hydroxo compounds [e.g., $U(OH)_2(C_2O_4) \cdot 5H_2O$], but this requires confirmation.

A few salts of the trisoxalato actinide(IV) anions are known, such as the acid benzylquinolinium compounds (Table 25), but the more usual complexes are the tetraoxalato and pentaoxalato species. The coordination geometry of the 10-coordinate thorium atom in the anion of $K_4[Th(C_2O_4)_4] \cdot 4H_2O$[388] is a slightly irregular bicapped square antiprism with an oxalate bridged structure that is cross-linked into a three-dimensional framework by hydrogen bonding (Figure 22). The geometry in both crystal modifications of $K_4[U(C_2O_4)_4] \cdot 4H_2O$[389] is the same as in the thorium compound. In one phase the three bidentate C_2O_4 groups and a tetradentate bridging C_2O_4 group link the metal atoms in a one dimensional polymeric array; the other phase is isostructural with the thorium compound. The uranium atom in $Ba_2U(C_2O_4)_4 \cdot 8H_2O$ is nine-coordinate, bound by four oxalates and one water molecule. The coordination geometry about the U atom is between tricapped trigonal prism and mono capped square antiprism.[390] Ba atoms interact with the oxalate O atoms, making the oxalates appear as quadridentate ligands that bridges U and Ba atoms. Additional An^{IV} oxalato complexes with molar ratios 1:5 or 1:6 metal ion to oxalate have been reported, but little is known about their coordination geometry and they could be mixtures of other known oxalato compounds.

Table 26 Actinide(IV) mixed oxalate and oxalato complexes.

$UF_2(C_2O_4)\cdot1.5H_2O$	
$UX_2(C_2O_4)_3\cdot yH_2O$	$X=F, y=0; X=Cl, y=0, 2, 4$ or 12
$M^I_4M^{IV}F_4(C_2O_4)_3\cdot xH_2O$	$M^{IV}=Th, MI=K, x=0$
	$\quad M^{IV}=U, M^I=NH_4, x=4$
$K_2(Pu(C_2O_4)_2(CO_3)\cdot ca.1.5H_2O$	
$K_4U(C_2O_4)_4$	
$M^I_4M^{IV}(C_2O_4)_x(CO_3)_{4-x}\cdot yH_2O$	$M^{IV}=Th, M^I=K, x=1, y=4.6$
	$\quad M=NH_4, x=2, y=0.5$
	$\quad M^I_4=(CN_3H_6)_3(NH_4), x=1, y=1.5,$ or 2–3.5, and $x=2, y=3$
	$\quad M^{IV}=U, M^I=(CN_3H_6)_3(NH_4), x=1\ y=2$
	$\quad M^{IV}=Pu, M^I=Na, K, x=1, y$ unspecified
	$\quad M^I=Na, x=2, y=3$
$(NH_4)_4Th_2(C_2O_4)(CO_3)_5\cdot10H_2O$	
$K_6M^{IV}(C_2O_4)(CO_3)_{5-x}\cdot yH_2O$	$M^{IV}=Th, x=1, y=6$-8 and $x=2, y=0, 1$ or 4
	$\quad M^{IV}=Pu, x=2, y$ unspecified
$M^I_6Th_2(C_2O_4)_x(CO_3)_{7-x}\cdot yH_2O$	$M^I=K, x=3, y=6$
	$\quad M^I=CN_3H_6, x=2, y=4$ or 8 and $x=3, y=14$
$Na_8Th(C_2O_4)_x(CO_3)_{6-x}\cdot yH_2O$	$x=1, y=10$ to 11 and $x=2, y=9$ to 10.5 or 11
$M^I_8Th_2(C_2O_4)_x(CO_3)_{8-x}\cdot yH_2O$	$M^I=K, x=3, y=13$ or 16
	$\quad M^I=CN_3H_6, x=1, y=6$ and $x=3, y=o$
$Na_{10}Th(OH)_2(C_2O_4)_3(CO_3)\cdot xH_2O$	$x=10, 11, 11.5$ or 16
$M^I_{10}M^{IV}_2(C_2O_4)_x(CO_3)_{9-x}\cdot yH_2O$	$M^{IV}=Th, M^I=K, x=2, y=8, 12$ or 14 and $x=4,$
	$\quad y=5$ or 7
	$\quad M^I=CN_3H_6, x=1, y=8$
	$\quad M^{IV}=U, M^I_{10}=(CN_3H_6)_8(NH_4)_2, x=1, y=4$ or 8
	$\quad M^I_{10}=[Cr(urea)_6]_3(NH_4), x=1, y=6$
$Na_{12}Th(C_2O_4)_2(CO_3)_6\cdot13H_2O$	
$K_2Th_2(OH)_2(C_2O_4)(CO_3)_3\cdot xH_2O$	$x=0, 1$ or 2
$K_5Th_2(OH)(C_2O_4)_2(CO_3)_4\cdot2H_2O$	
$Na_4[M^{IV}_2(OH)_{2x}(C_2O_4)(CO_3)_{5-x}]\cdot yH_2O$	$M^{IV}=Th, x=1, y=4$ and $x=3, y=2$
	$\quad M^{IV}=U, x=2, y=4$
$Na_{10}Th(OH)_2(C_2O_4)_3(CO_3)\cdot xH_2O$	$x=8$-9
$(NH_4)_4U_2(C_2O_4)_3(HCO_2)_3\cdot H_2O\cdot2HCO_2H$	
$K[U(C_2O_4)_2(NCS)(H_2O)_3]$	
$Cs[U(C_2O_4)(NCS)_2(H_2O)x]$	$x=0$ or 2
$K_4Th(C_2O_4)_2(HPO_4)_2\cdot6H_2O$	
$K_4[Th(C_2O_4)_2(C_4H_4O_6)_2]\cdot3H_2O$	$C_4H_6O_6=$ tartaric acid
$K_4[Th(C_2O_4)_2(C_6H_5O_7)_2]\cdot3H_2O$	$C_6H_8O_7=$ citric acid
$K_2MnU(C_2O_4)_4\cdot9H_2O$	

Analysis of crystals of $M_2Np_2(C_2O_4)_5\cdot nH_2O$ ($M=H$, Na, K, and NH_4) by electronic absorption spectroscopy in the long wave region of the spectrum showed that the coordination polyhedron of neptunium(IV) in these compounds differs from that in previously studied crystal compounds of Np^{IV}. The crystal structure of $H_2Np_2(C_2O_4)_5\cdot9H_2O$ (Figure 23) shows that Np^{4+} cations and $C_2O_4^{2-}$ anions form an openwork skeleton with channels extending along z-axis of the crystal. Oxonium cations and H_2O molecules are located in the channels. Two independent neptunium(IV) atoms are surrounded by oxygen atoms of five oxalate ions and four water molecules (CN 12); the coordination polyhedron is a distorted hexagonal analogue of cubooctahedron.[391]

Mixed oxalates and oxalato complexes (Table 24) also require further investigation. The sulfito and sulfato oxalates have been mentioned earlier and an equally large number of carbonato-oxalato species have been recorded,[388,392] some of which may well be mixtures. In addition to the compounds listed in Table 24, products of the rather unlikely compositions $K_7[U(OH)(C_2O_4)_2-(CO_3)_3]\cdot6H_2O$ and $K_{16}[U_2(OH)_2(C_2O_4)_3(CO_3)_8]\cdot10H_2O$ have been reported.

Polymeric $K_2MnU(C_2O_4)_4\cdot9H_2O$[393] has been prepared by the reaction of the tetraoxalato uranate compound, $K_4U(C_2O_4)_4$, with Mn^{II} in aqueous solution. The U ion is linked to four Mn^{II} ions via each of its oxalate ligands. The U^{IV} ion is nine-coordinate, bonded to four oxalate ligands and one water molecule.

β-Diketonates. β-Diketones chelate with metal ions, including actinides, to form neutral species via the deprotonated enolate anions as illustrated in Figure 24.[394] A very wide array of homoleptic complexes of the general formula $An(R^1COCR^2COR^3)_4$ have been reported for

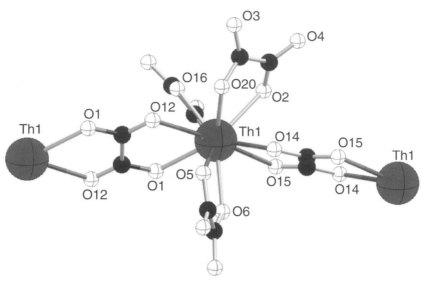

Figure 22 Oxalate bridged three-dimensional structure of $K_4[Th(C_2O_4)_4]\cdot 4H_2O$; oxalates bridge in each plane, either directly or via hydrogen-bonding (Akhtar and Smith Acta *Crystallographica, Section B* **1975**, *31*, 1361).

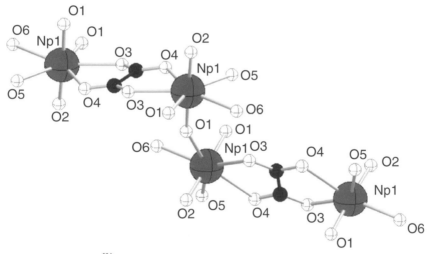

Figure 23 Oxalato bridge Np^{IV} centers in the crystal structure of $H_2Np_2(C_2O_4)_5\cdot H_2O$ (Bykhovskii, Kuz'mina *et al. Radiokhimiya* **1988**, *30*, 37–41).

An = Th, U, Np, and Pu. In complexing with metal ions, the β-diketones form planar six-member chelate rings with elimination of the enol proton. The simpler β-diketones, such as acetylacetone (HAA), are fairly water soluble, but form complexes that may be soluble in organic solvents. This is especially true for the An^{IV} ions which form strong complexes with HAA and can be effectively sequestered to the organic phase, making HAA a potentially useful extractant (See Table 27). The four stability constants in Table 27 for tetravalent actinides imply that four HAA ligands coordinate with each metal ion in the formation of the extracted neutral ML_4 complexes.[395]

Like with trivalent actinides (*vide infra*), 2-thenoyltrifluoroacetone (HTTA) is also effective at complexing with tetravalent actinides. Extractant dependency studies have shown than Th^{IV} displays a 1:4 extraction stoichiometry (Th:extractant) with HTTA.[396]

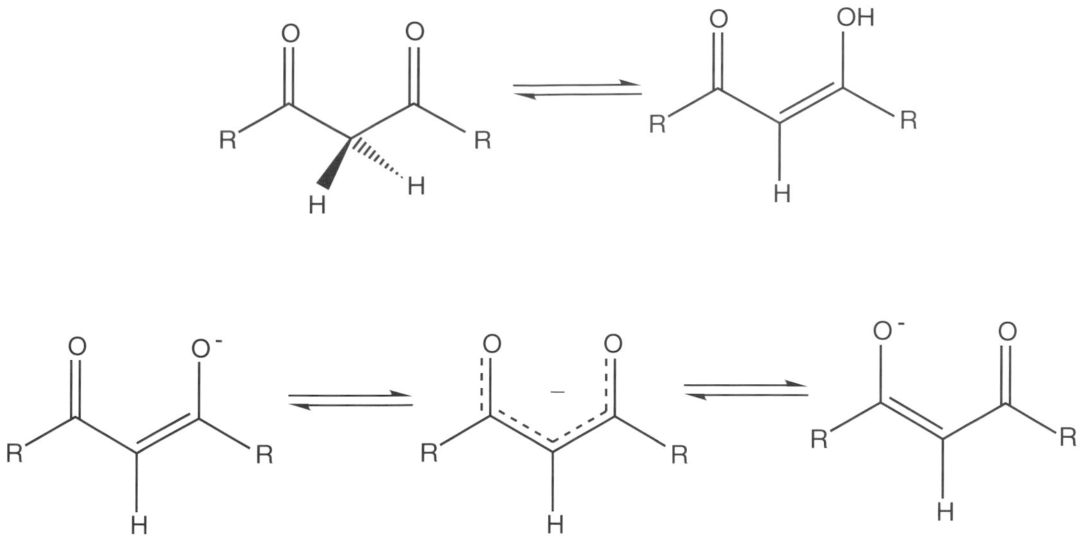

Figure 24 Bond tautomerism in β-diketone and β-diketonate.

Table 27 Stability constants for acetylacetonate complexes and distribution constants (from benzene or chloroform) in perchlorate media.[a]

An^{z+}	Th^{4+}	U^{4+}	Np^{4+}	Pu^{4+}
log K_1	8.00	9.02	8.58	10.5
log K_2	7.48	8.26	8.65	9.2
log K_3	6.00	6.52	6.71	8.4
log K_4	5.30	5.60	6.28	5.91
log K_{D4} (benzene)	2.52	3.64	3.45	2.54
log K_{D4} (chloroform)	2.55	4.0		2.6

[a] Adapted from Ahrland, S. In *The Chemistry of the Actinide Elements*; J. Katz, G. Seaborg, L. Morss, Eds.; Chapman and Hall: New York, 1986; Vol. 2, p 1480.

In perchlorate media, HTTA extracts Th^{IV} according to the extraction equilibrium equation (Equation (22)).This arguably makes HTTA a potentially useful extractant for Th^{IV} by itself:[397]

$$Th^{4+} + 2\,HTTA + 2\,ClO_4^- \longrightarrow Th(TTA)_2(ClO_4)_2 + 2\,H^+ \tag{22}$$

In many cases, synergists are added to HTTA extraction systems to enhance the separation of actinide ions. One example is the addition of the crown ethers (CE) dibenzo-18-crown-6, dicyclo-hexyl-18-crown-6, dibenzyl 24-crown-8, and benzyl-15-crown-5. These crown ethers have been shown to synergistically enhance extraction into benzene and the increase follows $Eu^{3+} > UO_2^{2+} > Th^{4+}$. The extraction equilibrium for crown ether/HTTA systems for the separation of Th^{IV} is shown in Equation (23). The binding of the crown ether in the extracted complex seems to be a function of crown ether basicity and steric effects:[396]

$$Th^{4+}_{(a)} + 4\,HTTA_{(o)} + CE_{(o)} \longrightarrow Th(TTA)_4 \bullet CE_{(o)} + 4\,H^+_{(a)} \tag{23}$$

Bis(1-phenyl-3-methyl-4-acylpyrazol-5-one) derivatives of the type H_2BP_n, where $n = 3, 4, 5, 6, 7, 8, 10,$ and 22, will extract the tetravalent actinides U^{4+}, Np^{4+}, and Pu^{4+}. As with the trivalent actinides, the H_2BP_n proved a better extractant than 1-phenyl-3-methyl-4-benzoylpyrazolone-5 (HPBMP), and the highest extractability occurred with the H_2BP_7 and H_2BP_8 ligands. Dependency studies indicate that 1:2 (An:L) complexes are formed for U^{IV}, Np^{IV}, and Pu^{IV} upon extraction from nitrate media into chloroform. Perchlorate solutions caused precipitates to form for various n values(4–6), probably due to ion pair formation in greater than 5 M $HClO_4$ solutions.[398]

An oxa-derivative of HPBMP, 3-phenyl-4-acetyl-5-isoxazolone (HPAI), has been studied as an attractive extractant for Th^{IV}. The structure is illustrated in Figure 25. HPAI, like other

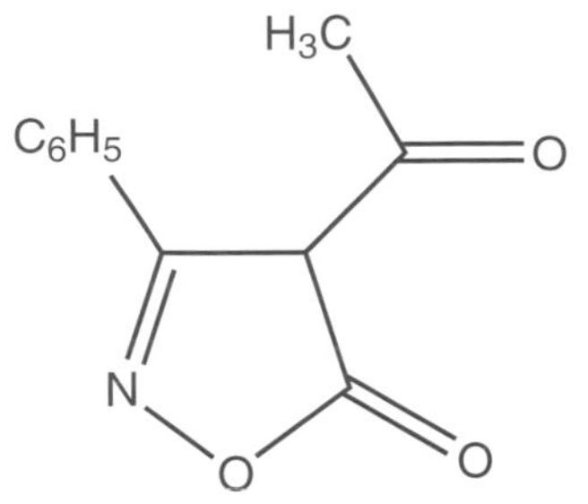

Figure 25 3-Phenyl-4-acetyl-5-isoxazolone (HPAI).

β-diketones, acts as a bidentate ligand in its enolic form. Extractant dependency indicates that four HPAI molecules are involved in the extraction of Th^{IV} from nitrate media into 4-methyl-2-pentanone. HPAI shows higher extractability than both HPMBP and HTTA due to the lower pK_a value of the ligand. IR spectrophotometric measurements indicate deprotonation of the enolic hydroxy group, allowing the charged oxygen to chelate with the metal. This is confirmed by C=O stretch shifts and the presence of typical 400–500 cm^{-1} metal/ligand bands, suggesting that the carbonyl oxygen is involved in the chelation. The lack of bands between 3,100 cm^{-1} and 3,600 cm^{-1} confirm that no nitrogen interactions are occurring with the metal. Additionally, there is no coordination of water to the metal complex.[399]

CMPO. CMPO, or octyl(phenyl)-*N,N*-diisobutylcarbamoylmethylphosphine oxide (see Figure 26), was developed by Horwitz and co-workers as an efficient actinide extractant for use in the TRUEX process in the remediation of acidic nuclear waste solutions. Derivatives of carbamoylphosphine oxides (CMPO) have been studied in nuclear fuel processing schemes involved in transmutation concepts.[400]

In general, bifunctional carbamoylmethylphosphonates (CMP) and carbamoylmethylphosphine oxides (CMPO) readily form complexes with actinide ions in aqueous and nonaqueous solutions. Complexes isolated in the solid state contain ligands chelated to the central metal ion, and the bidentate chelate interaction has been confirmed by single crystal X-ray structure determinations with uranyl and Th^{IV}.[401–403] However, spectroscopic studies of several complexes suggest that the ligands may only bind in a monodentate mode in solution, and this characteristic probably plays a role in determining the solvent extraction performance.[404] Although data for actinide complexes are sparse, trifunctional CMP and CMPO-like ligands containing two P=O donor groups and an amide or ester group also have been studied as actinide chelators.[405,40] In these cases, the ligands generally form bidentate chelates where a six-membered ring results and the third donor group acts as a bridging connector to another metal/ligand unit.

While the TRUEX process has been optimized for the removal of trivalent actinides, particularly Am^{III}, from nuclear waste solutions, CMPO has the ability to complex with and extract tetravalent actinides as well. Th^{IV}, Np^{IV}, and Pu^{IV} are all effectively extracted from hydrochloric acid solutions into tetrachloroethylene, even at moderate HCl concentrations, with extractability following the trend Pu^{IV} > Np^{IV} > Th^{IV} under all experimental conditions. Additionally, Pu^{IV} shows the highest extraction efficiency of all actinides by CMPO into TBP–dodecane

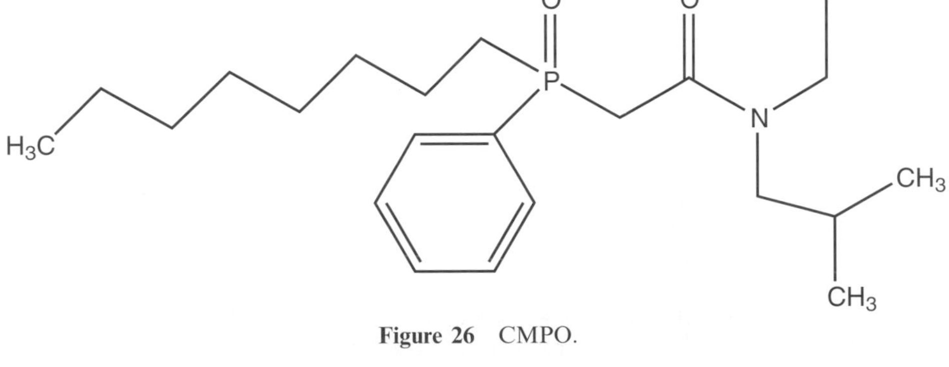

Figure 26 CMPO.

at HNO_3 concentrations up to 6 M. Extractant dependency studies show that CPMO complexes with Th^{IV} in a 3:1 ligand:metal stoichiometry to form the extracted species (see Equation (24)):

$$Th^{4+}{}_{(a)} + 4\ Cl^-{}_{(a)} + 3\ CMPO_{(o)} \longrightarrow ThCl_4 \cdot 3CMPO_{(o)} \qquad (24)$$

Interestingly, extractant dependency shows two different extracted species with Pu^{IV} depending on the acid from which it is extracted. Slope analysis for Pu^{IV} from HNO_3 solutions indicate the formation of a 2:1 CMPO:Pu^{IV} complex, while in HCl a 3:1 complex is observed (see Equations (25) and (26)):

$$Pu^{4+}{}_{(a)} + 4\ NO_3{}^-{}_{(a)} + 2\ CMPO_{(o)} \longrightarrow Pu(NO_3)_4 \cdot 2CMPO_{(o)} \qquad (25)$$

$$Pu^{4+}{}_{(a)} + 4\ Cl^-{}_{(a)} + 3\ CMPO_{(o)} \longrightarrow PuCl_4 \cdot 3CMPO_{(o)} \qquad (26)$$

It is proposed that coordination of CMPO with Pu^{IV} is similar to that in Am^{III}, meaning that monodentate coordination through the phosphoryl oxygen is observed for the nitrate complexes, and bidentate coordination through both the phosphoryl and carbonyl oxygen atoms occurs for the chloride complexes, yielding a coordination number of 10 for Pu^{IV}, which is interesting since it must change its extractant dependency to maintain the same coordination number in both types of complexes.[407]

Polydentate P,P- and N,P-dioxides. The coordination chemistry of polydentate phosphine oxides with actinide ions is of interest since several of these ligands show unique solvent extraction properties.[408,409]

Polyfunctional phosphinopyridine N,P-dioxides, (phosphinomethyl)pyridine N,P-dioxides and bis(phosphinomethyl)pyridine N,P,P-trioxides have been prepared, and selected coordination chemistry with actinide ions has been explored. The phosphinopyridine N,P-dioxides form bidentate chelates with uranyl and Th^{IV}, and in the solid state these complexes display six-membered chelate rings that appear to be relatively sterically congested.[410,411] The solvent extraction properties of these ligands are not unique since they resemble the performance of trialkylphosphine oxides.[412]

The coordination chemistry of the (phosphinomethyl)pyridine N,P-dioxides and bis(phosphino-methyl)pyridine N,P,P-trioxides shows that seven-membered chelate ring structures are quite stable when formed with trivalent and tetravalent actinide ions. For example, crystal structure determinations for 2:1 complexes between the trifunctional ligand, 2,6-[Ph$_2$P(O)CH$_2$]$_2$C$_5$H$_3$NO and Pu(NO$_3$)$_4$ and Th(NO$_3$)$_4$ show that two ligands bond in a tridentate fashion to the actinide ions. Two bidentate nitrate ions also appear in the inner coordination sphere, but two are displaced to the outer sphere.[413,414] The structures also show that the metal ions are "encased" in a lipophilic envelope generated by the ligands, and as a result, the complexes are soluble in organic solvents. The bifunctional ligand 2-[Ph$_2$P(O)CH$_2$]C$_5$H$_4$NO and Pu(NO$_3$)$_4$ produce a 2:1 complex [Pu(L)$_2$(NO$_3$)$_3$][Pu(NO$_3$)$_6$]$_{0.5}$ when combined in a 1:1 ratio. The two bifunctional ligands chelate to the Pu^{IV} ion along with three bidentate nitrate ions resulting in a coordination number of 10. Interestingly, an expected 4:1 ligand/metal chelate structure, related to that found with lanthanide ions, was not isolated. Solvent extraction studies with chloroform and dodecane soluble derivatives of these two ligands show performance closely parallel with CMPO ligands in the same solvents.[415–417]

Diphosphonic acids. Phosphorus-based extractants with the structure shown in Figure 27 are known as phosphonic acids. They are highly acidic and tend to form protonated complexes. Diphosphonic acids have been studied for the extraction of tetravalent actinides such as Th^{IV}. P,P'-di(2-ethylhexyl)methanediphosphonic acid (H$_2$DEH[MDP]) (see Figure 28) shows limited

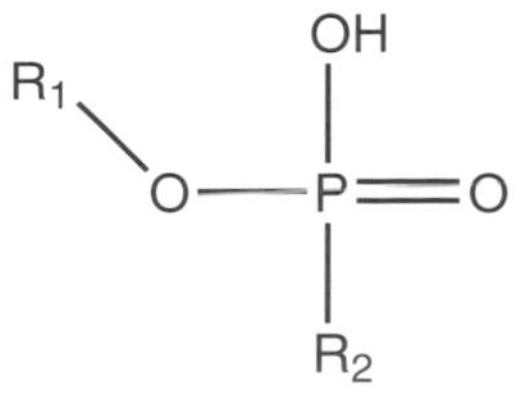

Figure 27   General diagram of a phosphonic acid.

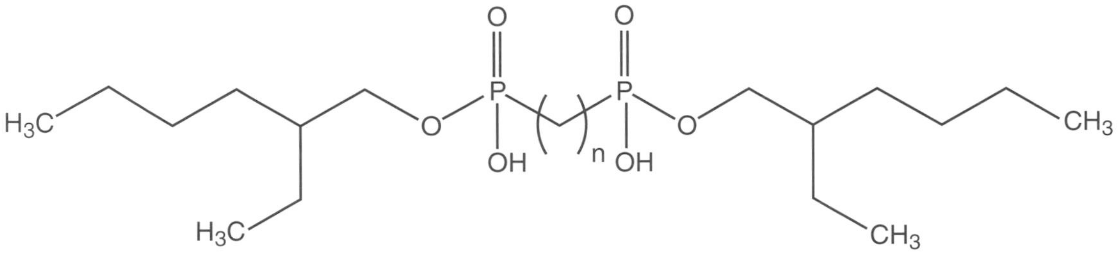

Figure 28 *P,P'*-di(2-ethylhexyl)alkanediphosphonic acids (*n* − 1: H$_2$DEH[MDP]; *n* = 2: H$_2$DEH[EDP]; *n* = 4: H$_2$DEH[BuDP]).

acid dependency for Th^{IV} that allows H$_2$DEH[MDP] to behave like a neutral extractant, even at high acid concentrations, due to the competition between nitric acid and the metal for the phosphoryl donor site.

Interestingly, extractant dependency studies with Th^{IV} show a very small slope over the entire extractant concentration range, indicating that its extraction is independent of both variables—nitric acid concentration and extractant concentration. This is indicative of a low solubility of the metal/extractant complex in both phases, perhaps due to a phenomenon observed in uranium/dialkylpyrophosphoric acid extractions, where the actinide is part of a highly polymerized complex present in the organic layer. This colloidal species is probably formed via oxo-bridges and can be precipitated at high Th^{IV} concentrations.[418]

P,P'-di(2-ethylhexyl)ethanediphosphonic acid (H$_2$DEH[EDP]) extraction with Th^{IV} into *o*-xylene shows no acid or extractant dependency. At low Th^{IV} concentrations, extraction occurs via bonding with the phosphoryl oxygens, giving the protonated complexes. At high Th^{IV} concentrations, complexation leads to the release of H$^+$ ions. Furthermore, the lack of acid and extractant dependency leads to the conclusion that the extracted complexes are polymeric in nature under all conditions.[419]

Interestingly *P,P'*-di(2-ethylhexyl)butanediphosphonic acid (H$_2$DEH[BuDP]) shows a strong extractant dependency with Th^{IV}, especially at higher nitric acid concentrations. At lower acidities, a zero dependency is observed, indicating the formation of a polymeric species. At higher acidities, this behavior is not observed.[420] Unlike Am^{III}, where complexation stoichiometry depends on extractant concentration, extractant dependency studies show a slope of two for Th^{IV} with all three extractants; H$_2$DEH[MDP], H$_2$DEH[EDP], and H$_2$DEH[BDP]. Considering the observed acid dependency having a slope of three, it is likely that Th^{IV} is extracted by a mechanism involving Th(NO$_3$)$_3$(L)(HL) species, where Th(NO$_3$)$_4$ only becomes important at high acid concentrations, where L is one of the three diphosphonic acids.[421]

Diamides. Malonamides are a relatively new class of extractants that have chelating abilities with tetravalent actinides as well as with the lanthanides. Malonamides are nonphosphorus containing extractants and are completely incinerable since they contain only carbon, hydrogen, oxygen, and nitrogen, thus following the "CHON" principle. Malonamides are amide-substituted malonic acids and have the general structure seen in Figure 29.

The R groups in Figure 29 can be hydrogenic, aliphatic, or aromatic, and the extracting properties of malonamides can be fine-tuned by varying the identity of these substituents. The R$_1$ chain is usually a methyl or ethyl chain to decrease the steric hindrance that can occur when complexing. R$_2$ can be an aliphatic or aromatic carbon chain. R$_3$ is usually a long carbon chain to aid in the solubility of the malonamide in an organic solvent.

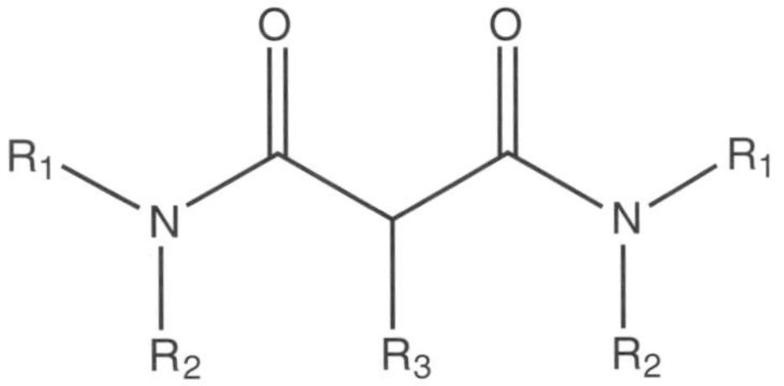

Figure 29 General diagram of a malonamide.

Several studies have looked at the extraction of Th^{IV} and Pu^{IV} ions by malonamides of varying structural character. Nigond *et al.* investigated the extraction of Pu^{IV} from nitric acid media with the malonamide *N,N'*-dimethyl-*N,N'*-dibutyltetradecylmalonamide (DMDBTDMA).[422] UV–vis experiments indicate the presence of two extracted species that are formed according to the following equilibria shown in Equation (27) (*n*=1 or 2):

$$Pu^{4+}_{(a)} + 4\ NO_3^{-}_{(a)} + n\ DMDBTDMA_{(o)} \longrightarrow Pu(NO_3)_4 \bullet (DMDBTDMA)_{n(o)} \quad (27)$$

The complexes that are formed are nonionic, and coordination to the Pu^{IV} metal occurs in a bidentate mode through the carbonyl oxygens of the malonamide ligand. IR spectroscopy indicates C_{2v} geometry of the extracted complex due to nitrate stretching bands at 1530–1540 cm^{-1} and 1280 cm^{-1}. The extracted species from complexation with DMDBTDMA are different than those that would be obtained with monamides, where the anionic complex $Pu(NO_3)_6H_2(amide)_x$ would be observed in the organic phase. Monamides are weaker complexants for Pu^{IV} than are malonamides, due to nitrate/metal competition at high acid concentrations.[422]

Nair *et al.* studied the extraction of Pu^{IV} by *N,N'*-methyl-*N,N'*-butylmalonamide(MBMA),[423] *N,N',N',N'*-tetra-butyl-malonamide (TBMA), and its more sterically-hindered analogue, *N,N',N',N'*-tetra-isobutyl-malonamide(TiBMA).[423] Extractant dependency studies yield a slope of two for the malonamide ligands complexing with Pu^{IV} in extraction to the organic phase.[424]

Studies with Pu^{IV} polymer have shown that efficient extraction is possible by pentaalkylpropane diamides over a large range of nitric acid concentrations (1–5 M). The extractive ability of the diamide is found to depend on the age of the plutonium polymer. When the polymer is over six months old, better extraction is observed, although the mechanism is not clearly understood.[425]

The oxygen-based diglycolamide (see Figure 30), *N,N'*-dimethyl-*N,N'*-dihexyl-3-oxapentanedia-mide (DMDHOPDA) is also an effective extractant for Th^{IV} with HTTA as a synergist, and experimental data indicates that two extracted species may be present. As a result, limits were set on the experimental conditions for the extraction of only one of the two species ($-2.7 < \log[DMDHOPDA] < -2$), resulting in an extraction stoichiometry consistent with the extraction of $Th(TTA)(DMDHOPDA)(ClO_4)_3$ into the organic phase. Without the synergist, the coordination environment around thorium is filled by the addition of another diglycolamide and the perchloric anion for charge balance to generate $Th(DMDHOPA)_2(ClO_4)_4$.[397,426]

The sulfur-based thiodiglycolamides, as seen in Figure 31, N,N'-dimethyl-N,N'-dihexyl-3-thio-pentanediamide (DMDHTPDA) and N,N'-dihexyl-3-thiopentanediamide (DHTPDA), both extract thorium(IV) with HTTA as a synergist in the same manner. The extraction stoichiometry

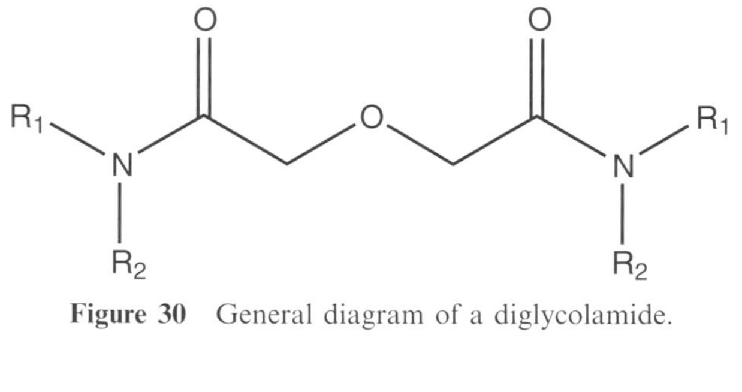

Figure 30   General diagram of a diglycolamide.

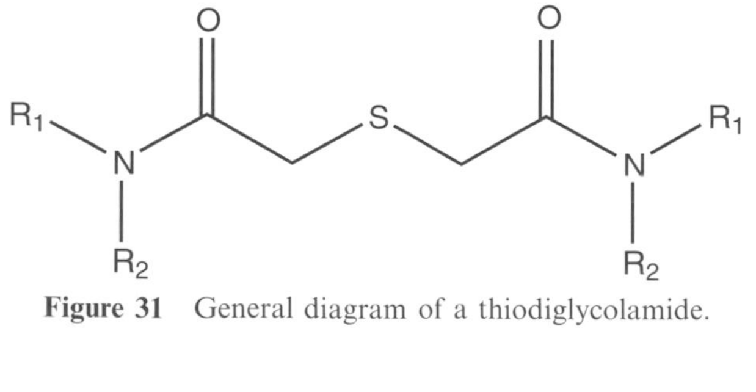

Figure 31   General diagram of a thiodiglycolamide.

for both DMDHTPDA and DHTPHA (L) is given by Equation (28). Extraction by the ligand alone is negligible, indicating a synergistic mechanism with HTTA:

$$Th^{4+}{}_{(a)} + 3\,HTTA_{(o)} + L_{(o)} + ClO_4{}^-{}_{(o)} \longrightarrow Th(TTA)_3(L)(ClO_4)_{(o)} + 3\,H^+{}_{(a)} \qquad (28)$$

Polyoxometallates. As previously discussed, several classes of polyoxometallates can serve as ligands in the complexation of tetravalent actinide ions. The first of these is the decatungsto-metallates, $[An^{IV}W_{10}O_{36}]^{8-}$, An = Th, U.[427,428] The molecular structure of the uranium complex has been determined.[427,429] The actinide ion in this complex is eight-coordinate, ligated by two tetradentate W_5O_{18} groups (lacunary derivatives of the W_9O_{19} structure). The overall symmetry of the anion is close to D_{4d}, with U—O bond lengths of 2.29–3.32 Å. Although six distinct oxygen chemical environments exist in the structure, only three signals are observed in the ^{17}O-NMR spectra.[430] The complexes are not stable outside the pH range 5.5–8.5.

Among the first polyoxometallate complexes to be prepared were those of the dodecamolyb-dometallate family, $[AnMo_{12}O_{42}]^{8-}$ (An = tetravalent Th, U, Np). The thorium complex was first prepared,[431,432] followed later by uranium and neptunium analogues.[433,434] The structure of the complexes contains an icosahedrally coordinated actinide surrounded by six face-sharing Mo_2O_9 units linked by corner sharing.[435] A variety of other characterization data on these complexes have been reported.[436–439] The uranium compound appears to undergo reversible oxidation to form a U^V complex.[438] The complexes $[AnMo_{12}O_{42}]^{8-}$ (An = Th, U) can themselves further act as ligands toward other metal cations. Weak complexes of $AnMo_{12}O_{42}$ with varying stoichiometries form in aqueous solution with M = divalent (Mn, Fe, Co, Ni, Zn, Cd, Cu), trivalent (Y, Er, Yb), and tetravalent (Th) cations.[440,441] In the crystallographically characterized examples $(NH_4)_2$-$[UMo_{12}O_{42}(Er-(H_2O)_5)_2]\cdot nH_2O$ and $(NH_4)_3[UThMo_{12}O_{42}]$,[442,443] $[UMo_{12}O_{42}]^{8-}$ serves as a tridentate ligand towards the other metal centers.

The complexes $Th[XMo_{12}O_{40}]^{n-}$ (X = P, Si) have been proposed principally from analytical data.

A more extensive set of actinide complexes is formed with tungstates of the Keggin and Dawson structure, $An[XW_{11}O_{39}]_2{}^{n-}$ and $An[X_2W_{17}O_{61}]_2{}^{n-}$ (X = P, Si, B, As; An = Th, U, Np, Pu).[438,444–449] These ligands form very stable complexes of tetravalent lanthanides and actinides. A review of complexes of f-elements with this class of polyoxometalates provides references to a range of characterization data.[450] The lacunary heteropolyanions act as tetradentate ligands toward the actinide center, generating an eight-coordinate metal center in an approximate square antiprismatic geometry.[451] Although the stability of molybdenum analogs is markedly decreased, a few mixed-metal analogs have been isolated, including $K_{10}[An(PMo_2W_9O_{39})_2]\cdot22H_2O$ and $K_{16}[An(P_2MoW_{16}O_{61})_2]\cdot28H_2O$ (An = Th, U).[452,453]

Other. Complexes of tetravalent uranium have been synthesized using the anion $\{(C_5H_5)Co[PO(OEt)_2]_3\}^-$, or a Kläui ligand, as the ancillary group. The complexes $LUCl_3(THF)$[454] and L_2UCl_2,[455] $L = \{(C_5H_5)Co[PO(OEt)_2]_3\}^-$, have been prepared by metathesis reactions employing uranium tetrachloride. The molecular structure of the complexes indicate that the cobalt tris(phosphate) complex is tridentate, coordinating the uranium center through the three P=O groups (see Figure 32).

(ii) Schiff base-derived ligands

Schiff bases are macrocyclic or macro-acyclic ligands that typically contain both nitrogen and oxygen donors and are often polydentate in coordinating ability. However, the identity of the donor can be varied between sulfur, phosphorus, nitrogen, and oxygen to change the donor properties, and hence the coordination abilities, of the ligand. Schiff bases are sometimes synthe-sized as compartmental ligands where binding at one site influences a change in conformation in another site on the molecule for cooperative complexation with two or more metal ions metals. Schiff bases have been traditionally prepared by the condensation reaction between a formyl- or carbonyl-containing derivative with primary amine groups in the presence of certain metal ions, such as alkaline earth cations, that act as templating agents. However, tailoring of the Schiff base often requires modifications to this very simplistic synthetic procedure.[456]

Thorium(IV) has been reacted with a pentadentate compartmental ligand for the first crystal structure reported on a complex with Th^{IV} as a binucleating metal ion. The crystal structure of the complex has been solved to reveal two $Mg[Th_2L_3]_2\cdot6H_2O$ units in the unit cell. Each $Th_2L_3{}^-$ anion, where all the oxygen atoms are deprotonated, comprises a dinuclear unit, where the two

The Actinides

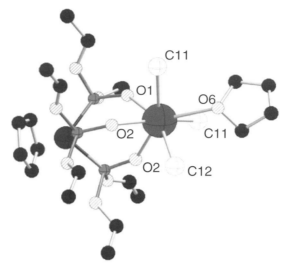

Figure 32 Crystal structure of LUCl₃(THF) (L = {(C₅H₅)Co[PO(OEt)₂]₃}) (Wedler, Gilje *et al. J. Organomet. Chem.* **1991**, *411*, 271).

thorium atoms are coordinated in a bridging fashion with the three central oxygen atoms from three separate ligands (i.e., each central oxygen donates to both thorium atoms). Each ligand then donates an oxygen and nitrogen to one thorium atom and the other oxygen and nitrogen to the second thorium atom. The thorium atoms each have a coordination number of nine and adopt a slightly distorted tricapped trigonal prismatic conformation.[457]

Examples of bidentate Schiff bases as extractants for Th^{IV} have also been illustrated. The ligands *N*-salicylidene-*p*-toluidene (HSalTol) and *N*-salicylidene-*p*-phenetidine (HSalPhen) can nearly quantitatively extract Th^{IV} from chloride media into benzene at a pH of 7. The deprotonated form of both ligands (designated HSB) is proposed to take place in complexing with Th^{IV} according to the following extraction equilibrium obtained from slope analysis (see Equation (29)):

$$Th^{4+}{}_{(a)} + HSB_{(o)} + 3\ Cl^-{}_{(a)} \longrightarrow Th(SB)Cl_{3(o)} + H^+{}_{(a)} \quad (29)$$

The maximum in the extraction at pH 7 indicates that solubilization of the extracted complex due to hydrolysis or ligand dissolution at high basicity.[458]

(iii) Macrocyclic ligands

N-Heterocyclic ligands. Porphyrins (see Figure 33) have been shown to complex well with late transition metals and have recently been shown to complex with actinides of varying oxidation state. Porphyrins are good as complexing agents, but have poor selectivity. The coordination of the actinide with a porphyrin is controlled by the oxidation state of the actinide, the cavity size of the porphin, and the molar ratio between the metal and the porphyrin. There are cases of the metal being completely contained within the cavity, adjacent to the cavity, or being sandwiched between multiple porphyrins.[459]

5,10,15,20-Tetraphenylporphyrin (H₂TPP) complexes with thorium to yield the product Th(TPP)₂. As indicated by X-ray single crystal diffraction data, a 2:1 sandwich style coordination is present, which creates a coordination number of eight around the thorium. The phenyl groups which are attached to the porphyrin cause some distortion in the square anitprismatic geometry, causing the ligands to be offset by about a 30° angle, as seen in Figure 34.

2,3,7,8,12,13,17,18-Octaethylporphyrin (H₂OEP) and thorium complex to give a crystal structure similar to that seen with H₂TPP and thorium. In Th(OEP)₂, the coordination environment around the thorium is an ideal square antiprism. The replacement of phenyl groups with ethyl chains removes any steric hindrance that was present in Th(TPP)₂. This is illustrated in Figure 35.[460]

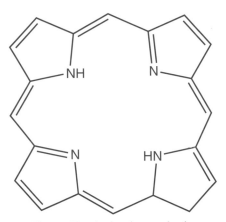

Figure 33 A simple porphyrin.

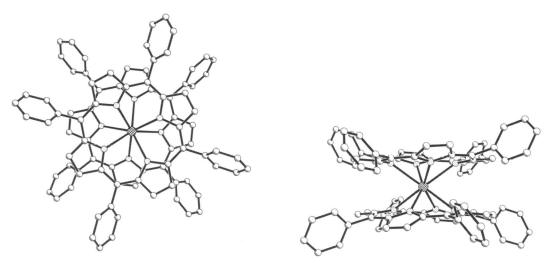

Figure 34 Crystal structure of [Th(C$_{44}$H$_{28}$N$_4$)$_2$]·C$_7$H$_8$ (top and side views) (Girolami, Gorlin *et al. Journal of Coordination Chemistry* **1994**, *32*, 173–212).

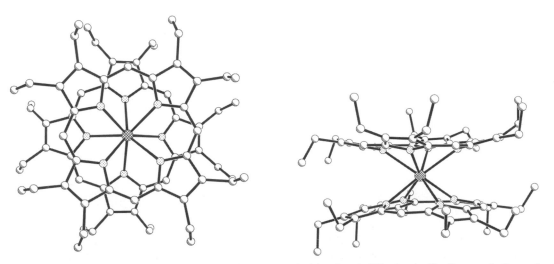

Figure 35 Crystal structure of Th(C$_{36}$H$_{44}$N$_4$)$_2$ (top and side views) (Girolami, Gorlin *et al. Journal of Coordination Chemistry* **1994**, *32*, 173–212).

A study of H_2OEP ligand complexes with U^{IV} and Th^{IV} in the presence of a coordinating solvent such as THF, benzonitrile, and pyridine give complexes of the type $[M^{IV}(OEP)Cl_2L_n]$, where L is the solvent-type ligand. From NMR and IR data, the proposed structure (Figure 36) is similar to the $Th(OEP)_2$ structure, except two solvent molecules are bound to the metal along with two chlorides anions taking the place of one of the porphinato ligands. This would give the metal a coordination number of eight.[461]

When H_2TPP is treated with a five-fold excess of anhydrous UCl_4 and 2,6-lutidine in benzonitrile, the resulting structure is reported to be $U(TPP)Cl_2$. Upon crystallization from THF, a solvent adduct is formed of the type $U(TPP)Cl_2(THF)$. In Figure 36, a 4:3 piano stool coordination geometry of the solvent adduct structure is observed with the uranium being complexed above the cavity of the porphryin, due to the TPP cavity being too small to form a uranium inclusion complex. Bonding is also improved via the "saucer-shape" of the porphyrin ring. The chlorides maintain charge balance, while the THF increases the coordination up from six to seven. The coordination around the uranium is not a traditional coordination arrangement.[462]

While not common, porphyrins can complex in a manner so as to create a trimeric metalloporphyrin as in the case with $[(TPP)Th(OH)_2]_3 \cdot H_2O$. The thorium atoms lie within a square antiprismatic coordination environment with the hydroxides bridging between thorium atoms. The bridging oxygens of the hydroxide group are in an ideal trigonal prism with respect to one another. This environment around the thorium atoms can be seen in Figure 37, where all water molecules and hydrogen and carbon atoms in the porphyrin rings have been removed for clarity.[463]

Examples of $(\eta^5\text{-}C_4N)$ coordination in pyrrole-derived macrocyles may be found in the reaction products of uranium halides with the tetraanion of the macrocycle $\{[(\text{-CH}_2\text{-})_5]_4\text{-calix}[4]\text{tetrapyr-role}\}$.[464] As described in Equation (30), the reaction of $UI_3(THF)_4$ with the potassium salt of the tetrapyrrolide in THF generates a dinuclear U^{IV} complex, $\{[\{[(\text{-CH}_2\text{-})_5]_4\text{-calix}[4]\text{tetrapyrrole}\}\text{-}UK(THF)_3]_2(\mu\text{-O})\} \cdot 2THF$; the oxo group is proposed to come from deoxygenation of a THF molecule:

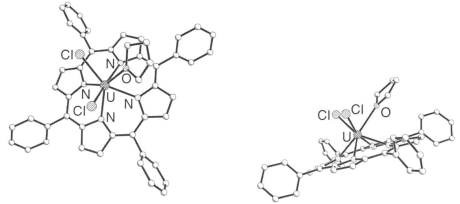

Figure 36   Crystal structure of $U(C_{44}H_{28}N_4)Cl_2(C_4H_8O)$ (top and side views) (Girolami, Milam *et al.* *Inorganic Chemistry* **1987**, *26*, 343–344).

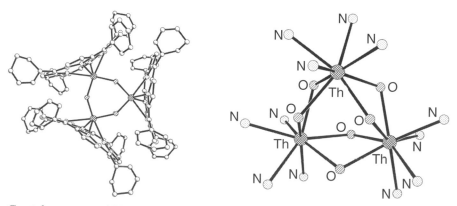

Figure 37   Crystal structure of $[(C_{44}H_{28}N_4)_3Th(OH)_2]_3 \cdot 2H_2O \cdot 3C_7H_6$ (Kadish, Liu *et al.* *J. Am. Chem. Soc.* **1988**, *110*, 6455–6462).

$$UIC_3(THF)_4 \ + \ [K(THF)_4]\{calix\text{-}[4]\text{-}tetrapyrrole\} \longrightarrow$$

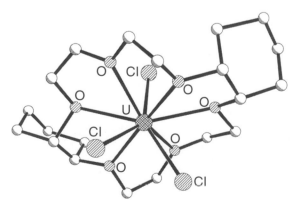

(30)

Reaction of $UI_3(THF)_4$ with the corresponding lithium tetrapyrrolide salt in a 1:2 ratio generates instead $[\{[(\text{-}CH_2\text{-})_5]_4\text{-}calix[4]tetrapyrrole\}ULi(THF)_2]_2 \cdot hexane$, in which the β-carbon of one of the pyrrole rings has undergone a metallation reaction. Reaction of the potassium salt with $UI_3(DME)_4$ avoids the complication of THF activation, and the simple trivalent uranate complex, $[\{[(\text{-}CH_2\text{-})_5]_4\text{-}calix[4]tetrapyrrole\}U(DME)][K(DME)]$, is generated. The geometry about the metal center in these compounds is qualitatively similar to a metallocene complex. The ligand adopts a σ/π bonding mode, in which two of the four pyrrole rings in the macrocycle are η^5-bonded to the uranium, and the other two rings are σ-coordinated only through the pyrrole nitrogen. The U—N (σ) bond lengths for the tetravalent derivatives range from 3.39 Å to 2.47 Å; these distances are slightly longer in the trivalent derivative ($\sim$2.53 Å). The π-coordination of the pyrrole ring yields somewhat longer U—N bond distances ($\sim$2.65 Å in tetravalent compounds, 2.74 Å in the trivalent compound), and U—$C_{pyrrole}$ bond distances that range from 2.68 Å to 2.88 Å.

Reaction of $UI_3(THF)_4$ with $[Li(THF)]_4\{[(\text{-}CH_2\text{-})_5]_4\text{-}calix[4]tetrapyrrole\}$ in a 1:2 ratio generates the dinuclear complex $[Li(THF)_4]_2[U_2I_4\{[(\text{-}CH_2\text{-})_5]_4\text{-}calix[4]\text{-}tetrapyrrole\}].$[465] Partial reduction of UCl_4, followed by reaction with one half of an equivalent of the lithium salt is reported to generate the mixed-valence compound $[Li(THF)_2](\mu\text{-}Cl)_2\{U_2[(\text{-}CH_2\text{-})_5]_4\text{-}calix[4]tetrapyrrole\}Cl_2 \cdot THF$. Both of these complexes display alternate $\sigma/\eta^5,\pi$-coordination to opposite pairs of pyrrole ligands in a single tetrapyrrole group. The bridging nature of the macrocyclic ligand brings the uranium centers into relatively close proximity (3.4560(8) Å and 3.365(6) Å, respectively); magnetic susceptibility measurements on the U^{III}/U^{III} dimer suggests weak antiferromagnetic coupling occurs between metal centers.

Crown ethers. X-ray single crystal diffraction was used to determine the crystal structure of $U^{IV}Cl_3$(dicyclohexyl-18-crown-6) as seen in Figure 38. The coordination geometry around the uranium is distorted tricapped trigonal prism where the two planar triangles are offset from one another. This geometry is formed through the nonplanar oxygens of the crown ether and three chlorides bound to the uranium. Distortion of the crown ether is required in order to achieve complexation with the uranium ion.

Figure 38 Crystal structure of $[UCl_3(C_{20}H_{40}O_6)]_2 \cdot UCl_6 \cdot (C_3H_8)_2$ depicting the coordination of the U^{IV} in one of the crown complexes (de Villardi, Charpin *et al. J. Chem. Soc., Chem. Commun.* **1978**, 90–92).

3.3.2.2.4 Borohydride ligands

Borohydride complexes of the tetravalent actinides are more common and members of the series $An(BH_4)_4$ exist for $An = Th$, Pa, U, Np, and Pu. The initial method employed for the preparation of $An(BH_4)_4$ involved reaction of $AnCl_4$ or AnF_4 with $Al(BH_4)_3$ or $Li(BH_4)$,[85,466,467] (see Equation (31)):

$$AnF_4 + 2\,Al(BH_4)_3 \longrightarrow An(BH_4)_3 + 2\,Al(BH_4)F_2 \qquad (31)$$

Other synthetic routes have been reported,[468,469] including metathesis reactions in ethereal solvents.[470] Given the difference in ionic radii of the metal ions, it is not surprising that not all $An(BH_4)_4$ compounds are isomorphous. Two different polymeric morphologies of $U(BH_4)_4$ have been identified.[471–474] In the most common form,[471,472] the uranium atom is coordinated by six borohydride ligands in a pseudooctahedral fashion. Two *cis*-borohydride groups are tridentate, while the other four are bidentate, and bridge two uranium atoms. The overall polymeric chain is helical. Another form has been identified in which the two tridentate borohydride groups reside in *trans*-positions of the octahedron, while equatorial bidentate BH_4 groups bridge metal centers to create a polymeric sheet structure. $Th(BH_4)_4$ and $Pa(BH_4)_4$ are reported to be isostructural with the major form of $U(BH_4)_4$.[466,475] In contrast, the neptunium and plutonium compounds are monomeric, with a pseudotetrahedral arrangement of tridentate borohydride groups surrounding the metal center.[475]

Substituted analogues $An(MeBH_3)_4$ ($An = Th$, Pa, U, Np) have been prepared either by reaction of $An(BH_4)_4$ with BMe_3 (see Equation (32)),[476] or by metathesis routes employing $LiBH_3Me$ (see Equation (33)):[477–479]

$$AnCl_4 + 4\,LiBH_3Me \longrightarrow An(MeBH_3)_4 + 4\,LiCl \qquad (32)$$

$$(An = Th, Pa, U, Np)$$

$$PaCl_5 + LiBH_3Me \longrightarrow Pa(MeBH_3)_4 \qquad (33)$$

As in the case of trivalent borohydride complexes, a number of base adducts have been prepared and characterized. In the case of adducts of $U(BH_4)_4$, the size of the base can control the dimensionality of the resulting product. The 1:1 adducts with small dialkylethers (e.g., $[U(BH_4)_4(OMe_2)]_n$, $[U(BH_4)_4(OEt_2)]_n)$[480] form chains in the solid state, in which bidentate borohydride groups bridge pseudooctahedral uranium centers; the remaining borohydride groups are tridentate, and the remaining coordination site is occupied by the ether ligand. Use of the slightly larger Pr^n_2O ligand results in the formation of an unusual dimer formulated as $(Pr^n_2O)_2(\eta^3\text{-}BH_4)_3U(\mu\text{-}\eta^2,\eta^2\text{-}BH_4)U(\eta^3\text{-}BH_4)_4$[481] (see Figure 39). Use of the methylborohydride group inhibits the formation of polymeric products, due to its inability to act as a bridging bidentate ligand. Therefore, the diethylether and THF adducts of $Th(MeBH_3)_4$ are found to be dimeric, with two

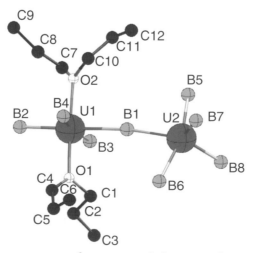

Figure 39 Crystal structure of $(n\text{-}Pr_2O)_2$ $(\eta^3\text{-}BH_4)_3U(\mu\text{-}\eta^2,\eta^2\text{-}BH_4)U(\eta^3\text{-}BH_4)_4$ (Zalkin, Rietz *et al. Inorg. Chem.* **1978**, *17*, 661).

bridging methylborohydride ligands.[482] The complex $(MeBH_3)_3Th(\mu\text{-}MeBH_3)_2Th(MeBH_3)_3(OEt_2)$ only exhibits ether coordination to one end of the dimer, presumably due to steric factors.

Tetrahydrofuran forms 2:1 adducts with $U(BH_4)_4$ and $U(MeBH_3)_4$. In the solid state the complexes exists as a pseudooctahedral monomer with *trans*-THF ligands and tridentate borohydride groups.[318,483,484] The tetrahydrothiophene analog of $U(MeBH_3)_4$ is not isostructural. The complex $[U(MeBH_3)_4(THT)]_2$ is a dimer with metal centers bridged by the sulfur atoms of the tetrahydrothiophene groups.[318] The complex $U(BH_4)_4(OPPh_3)_2$ has also been reported.[485,486]

Coordination of $U(MeBH_3)_4$ by the bidentate ligands $Me_2PCH_2CH_2PMe_2$, $MeOCH_2CH_2OMe$, $Me_2NCH_2CH_2NMe_2$, and $MeSCH_2CH_2SMe$ produces monomeric, octahedral adducts.[45,181]

Few cationic or anionic derivatives are known. Addition of $LiBH_4$ to $Th(BH_4)_4$ is reported to generate the "ate" complexes $Li[Th(BH_4)_5]$ and $Li_2[Th(BH_4)_6]$.[470]

3.3.2.3 Pentavalent Oxidation State

3.3.2.3.1 General characteristics

Protactinium, uranium, neptunium, and plutonium all can be generated in the pentavalent oxidation state in aqueous media, although hydrolysis results in the formation of the dioxo species AnO_2^+ for all but protactinium. The NpO_2^+ ion is most stable in aqueous solutions; in contrast, UO_2^+ and PuO_2^+ disproportionate readily. The actinyl ions display a linear O—An—O unit, and coordination chemistry is restricted to that of the equatorial plane, or "belly band" about the metal center. The lower charge-to-surface area ratio of these ions makes them much weaker acids, thereby reducing hydrolysis. Complexes of pentavalent early actinides not containing the AnO_2^+ unit can be isolated from nonaqueous media, either by oxidation of lower valent precursors or from reactions employing precursors such as the pentavalent halides.

3.3.2.3.2 Simple donor ligands

(i) Ligands containing anionic group 15 donor atoms

Amide complexes. Pentavalent amide and related N-donor complexes of the actinides are relatively rare in comparison to analogous alkoxide complexes (*vide infra*). In most cases, these species are prepared by oxidation of tetravalent precursors. The complexes $[Li(THF)]_2$-$[U(NMe_2)_6]$ and $[Li(THF)][U(NEt_2)_5]$, prepared by the reaction of UCl_4 with excess lithium amide salts $LiNMe_2$ and $LiNEt_2$ in THF,[106] can be oxidized by either $TlBPh_4$ or AgI to generate the uranium(V) species $[U(NMe_2)_6]^-$ and $U(NEt_2)_5$, with concomitant formation of Tl^0 or Ag^0.[104,106] Determination of the molecular weight of $U(NEt_2)_5$ indicates that it is a monomer in benzene solution.

In several instances, the generation of a uranium(V) complex is the result of fortuitous oxidation during reaction. A hexakis(amido)uranate complex, $[Li(THF)_x][U(dbabh)_6]^-$ (Hdbabh = 2,3:5,6-dibenzo-7-azabicyclo[2.2.1]hepta-2,5-diene) is generated in an unusual redox reaction employing $UI_3(THF)_4$ as a starting material (see Equation (34)):[487]

$$UI_3(THF)_4 + 7\ Li(OEt_2)(dbabh) \xrightarrow[-3\ LiI,\ -anthracene,\ -"Li_3N"]{THF,\ -100\,°C,\ warmed\ to\ 20\,°C} \left[\left(U\!-\!N=\!\!\left\langle\!\!\!\!\!\!\!\right.\right)_6\right]^- [Li(thf)_x]^+$$

$$\tag{34}$$

$$UCl_4 + 2\ MeC_6H_4C(NSiMe_3)(NSiMe_3)_2 \xrightarrow{O_2} [4\text{-}MeC_6H_4C(NSiMe_3)_2]_2UCl_3 \tag{35}$$

The complex $UCl_2\{N[CH_2CH_2P(Pr^i)_2]_2\}_3$ has also been reported.[488] This complex was produced adventitiously in the reaction of UCl_4 with $LiN[CH_2CH_2P(Pr^i)_2]_2$, presumably by oxidation of U^{IV} by traces of oxygen.

Amidinate complexes. In other cases, isolation of U^V comes about as the result of aerobic oxidation. The interesting pentavalent benzamidinate derivative $[4\text{-}MeC_6H_4C(NSiMe_3)_2]_2UCl_3$ was produced by adventitious aerobic oxidation during reaction of UCl_4 with the corresponding silylated benzimidine (Equation 35).[489]

(ii) Ligands containing neutral group 15 donor atoms

Ammonia and amines. Complexes of pentavalent actinides with ammonia or amine adducts are rare. The only reported members of this series are adducts of the electron-poor alkoxide complex, $U(OCH_2CF_3)_5$. The ammonia adduct, $U(OCH_2CF_3)_5\cdot(6\text{--}12)NH_3$, was proposed as the product of the reaction between UCl_5 and CF_3CH_2OH in the presence of excess ammonia. The amine adducts $U(OCH_2CF_3)_5\cdot xR_2NH$ ($x = 3$, $R = Me$; $x = 2$, $R = Pr^n$) and $U(OCH_2CF_3)_5\cdot2NMe_3$ are prepared by reaction of the alkoxide complex with excess amine in ether, followed by removal of the solvent under reduced pressure, and vacuum distillation of the products. All are reported to be green liquids.

Heterocyclic ligands. Complexes of UCl_5 with a variety of N-heterocyclic ligands, including pyridine, 2-mercaptopyridine, quinoline, isoquinoline, 2,2'-bipyridine, pyrazole, and substituted pyrazoles, pyrazine, pthalazine, and phenazine have been reported. These complexes are generally prepared by reaction of the ligand with UCl_5 or its trichloroacrylolyl chloride compound, $UCl_5\cdot C_3Cl_4O$. The majority of the complexes in this series are not well characterized. In addition, ambiguity exists in several cases regarding the ligand to metal ratio, which also brings into question the coordination number of the uranium species. In at least one case ($UCl_5\cdot bipy$), the complex is a 1:1 electrolyte in solution, and is therefore probably best formulated as $[UCl_4(bipy)]Cl$.[490] Adducts have been reported to form between N-heterocycles 1,10-phenanthroline, and phenazine and the ion $UOCl_5{}^{2-}$;[491,492] it is likely that these are ionic species in solution as well.

A more thorough study has been conducted of the chemistry of UF_5 with the heterocyclic bases 2-fluoropyridine (F-py) and 2,2-bipyridine (bipy).[493] While the reaction of F-py with β-UF_5 appears to lead to reduction of the metal center, reaction with bipy in acetonitrile generated the compounds $UF_5(bipy)$ and $[(bipy)_2H][UF_6]$ (obtained in the presence of excess bipy). The complex $UF_5(bipy)$ has been characterized by single crystal X-ray diffraction. Two different morphologies may be isolated from solution, depending on the temperature of the reaction. In both forms, the coordination geometry about the uranium center is a distorted fluoride monocapped trigonal prism (see Figure 40).

Nitriles. Reaction of UF_5 in acetonitrile with either Me_3SiCl or UCl_5 gives rise to the mixed halide nitrile adduct, $UCl_2F_3(MeCN)$.[494] Acetonitrile adducts of the pentabromide complexes $AnBr_5(MeCN)_x$ ($An = Pa$, $x = 3$; $An = U$, $x = 2\text{--}3$) have also been reported, as has the complex $Pa_2O(NO_3)_8\cdot2MeCN$.[495]

Phosphines, arsines. It has been reported that the adducts $UCl_5(PPh_3)$ and $UCl_5(dppe)$ can be prepared by reaction of UCl_5 with the corresponding phosphine,[490,496] although subsequent

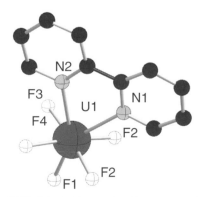

Figure 40   Crystal structure of $UF_5(bipy)$ (Arnaudet, Bougon *et al. Inorg. Chem.* **1994**, *33*, 4510).

papers have called these formulations into question.[497] One example of a crystallographically characterized pentavalent phosphine complex has appeared; the complex $UCl_2\{[(Pr^i)_2PCH_2CH_2]_2N\}_3$ has been reported.[488]

The diars complex $PaCl_5(diars)_x$ ($x = 1-2$) has been reported (diars = o-phenylenebis(dimethylarsine)).[192]

Thiocyanate and selenocyanate. Thiocyanate complexes of dioxoneptunium(V) have been prepared. The species $Cs_4[NpO_2(NCS)_5]$ and $NpO_2(NCS)(urea)_4$ have been reported; the latter complex has been structurally characterized.[498]

(iii) Ligands containing anionic group 16 donor atoms

Oxides. The most common pentavalent actinide oxides are the monoclinic Pa_2O_5 and Np_2O_5.[499] There are also mixed valent oxides, such as U_3O_8 and some evidence that the superstoichiometric oxide PuO_{2+x} contains Pu^V.[500] Ternary oxides, M_3AnO_4 and M_7AnO_6 exist for $An = Pa$, U, Np, and Pu, where M is generally an alkali metal and $Ba_2U_2O_7$ has been reported.

Hydroxides. The hydrolysis of Np^V has been studied more than that of any other pentavalent actinide because it is the most stable oxidation state for Np and it is an actinide ion of significant concern for environmental migration. Pentavalent uranium disproportionates in aqueous solution at pH values where hydrolysis would occur. Hydrolysis products for Pa^V, Pu^V, and Am^V are very similar to, but much less stable than those of Np, so only Np hydroxides will be described in detail. Neptunyl hydrolyzes at about pH 9, to form the stepwise products, $NpO_2(OH)$ and $NpO_2(OH)_2^{2+}$, which have been identified by optical absorbance and Raman spectroscopy.[501–503] In addition to the hydroxide these complexes likely have two or three inner-sphere waters in the equatorial plane and pentagonal bipyramidal coordination geometry.

The monohydroxide hydrate solid is amorphous and has not been fully structurally characterized. Attempts to increase the crystallinity have produced Np_2O_5. Mixed hydroxo carbonato complexes, such as $NpO_2(OH)(CO_3)_2^{4-}$ or $NpO_2(OH)_2(CO_3)^{4-}$ have been proposed to explain the solubility behavior of Np^V solids in basic carbonate solution but they have not been characterized.

Single crystal structure X-ray diffraction analyses and structural classification of synthetic and natural mineral phases have revealed interesting actinide coordination chemistry.[504,505] This approach has led to the identification of U^V in $CaU(UO_2)_2(CO_3)O_4(OH)(H_2O)_7$, the mineral wyartite.[506] The structure contains three unique U positions. Two of these are uranyl ions with the typical pentagonal-bipyramidal coordination. The third is also seven-coordinate, but does not contain '-yl' oxygens; and polyhedral geometry and electroneutrality requirements indicate that this site contains U^V.

Carbonates and Carboxylates. Pentavalent actinide carbonato complexes are generally prepared by addition of alkali metal carbonate solutions to acidic solutions of the An^V ion. For example, the mono-, bis- or triscarbonato Np complexes, $NpO_2(CO_3)^-$, $NpO_2(CO_3)_2^{3-}$, and $NpO_2(CO_3)_3^{5-}$, can be isolated by varying the carbonate concentration.[507] The triscarbonato complexes of Np^V and Pu^V can also be prepared electrochemically, and the U^V complexes has only been prepared electrochemically from $UO_2(CO_3)_3^{4-}$.[508] These complexes all have the general actinyl carbonate structure with the axial AnO_2 and the oxygen atoms of the aquo and bidentate carbonato ligands arrayed about the equatorial plane to form a pentagonal or hexagonal bipyramidal coordination polyhedron. The triscarbonato complex is isostructural with the hexavalent analog, with a longer actinyl distance of 1.85 Å (vs. ~1.75 Å for Np^{VI}) and very similar carbonate bond distances.[509] Interestingly, the Raman frequencies of the actinyl decrease linearly with increasing atomic number of the actinide.[510] Mixed hydroxo carbonates, $NpO_2(CO_3)_2(OH)^{2-}$ and $NpO_2(CO_3)_2(OH)_2^{3-}$, have been studied in solution.[502]

Solids corresponding to nearly all of the solution species (the U^V is one exception) have been prepared as microcrystalline powders via precipitation.[511–515] The structures of compounds $MNpO_2(CO_3)$ and $M_3AnO_2(CO_3)_2$, where M is an alkali metal or ammonium, have been described in detail.[47,515,516] These compounds show interesting structural changes due to the alkali metal cation present (size), the size similarity of hydrated ions such as K and NpO_2^+, and the extent of hydration.

For example, for $MNpO_2(CO_3)$ where $M = Cs^+$, Rb^+, NH_4^+, K^+, Na^+, and Li^+, a hexagonal-to-orthorhombic phase change is observed within the $NpO_2(CO_3)$ layer at the potassium–sodium boundary. The solids both contain actinyl carbonate layers and the hexagonal and orthorhombic sheets are related by displacement of the chains of actinyl units through half a translation along the crystallographic *a*-axis. The orthorhombic structure appears to allow for the closer contacts necessary for the smaller sodium and lithium cations. The potassium monocarbonate appears to swell along the *c*-axis with (reversible) hydration, suggesting the pentavalent actinides have a more complex structure than the actinyl(VI) carbonate layers and may be represented by the general formula of $KanO_2(CO_3) \cdot nH_2O$ with intercalated water molecules (see Figure 41).

The biscarbonato solid $M_3NpO_2(CO_3)_2$ maintains the same orthorhombic layered structure as seen in $MAnO_2(CO_3)$, except that one half of the AnO_2^+ ions in the anionic carbonate layer have been replaced by alkali metal cations. One can envision that M^+ and AnO_2^+ cations form alternating chains within the familiar hexagonal sheet and give rise to the approximate composition $[M_{0.5}(AnO_2)_{0.5}(CO_3)]$ within the layer. The cation and anion layers are now oriented such that an alkali metal cation, M^+, lies directly above and below the linear AnO_2^+ ion of adjacent sheets. The anionic carbonate layer and the cationic potassium layers line up such that they are parallel to the crystallographic *c*-axis, and this allows for an M—O=An interaction between layers. In this way, a second infinite chain of O=An=O—M—O=An=O units is formed, resulting in a maximally ordered structure.

The observations by Volkov *et al.*[517] that alkali cations can occupy the same sites as the AnO_2^+ ion explains the structure of $M_3AnO_2(CO_3)_2$, and may explain the presence of nonstoichiometric solids such $M_4AnO_2(CO_3)_{2.5} \cdot nH_2O$ (see Figure 42). This solid could easily arise from further replacement of AnO_2^+ ions in the layers by alkali metal cations, M^+. In this way it was proposed that solids of intermediate compositions $M_{(3+2x)}AnO_2(CO_3)_{(2+x)} \cdot nH_2O$, where $0 = x = 0.5$, with cations and waters exchanging into the solid, could exist while still preserving the basic structural features.[517]

Neptunium(V) complexes with polycarboxylic acid ligand are the most described among other actinides in the pentavalent oxidation state. In solution of EDTA at pH 5–6 NpO_2Y^{3-} is formed, its thermodynamic stability and the complex extractability have been reported. With citric acid at pH 4–5 the compounds NpO_2Cit2- and NpO_2Hcit are formed. The formation $[Co(NH_3)_6][NpO_2L] \cdot 3H_2O$, and $(NpO_2)_2H_2L \cdot 5H_2O$ through the reaction of NpO_2^+ with $EDTA(H_4L)$ in aqueous solution has been reported.

Nitrates. No inner-sphere An^V nitrate solution complexes have been characterized. The solid nitrates of Np^V and Pa^V, $NpO_2(NO_3) \cdot xH_2O$ ($x=1,5$), $RbNpO_2(NO_3)_2 \cdot H_2O$, and $PaO(NO_3)_3 \cdot xH_2O$ ($x=1–4$) can be precipitated from aqueous solution at high nitrate concentrations.[518] The hexanitratoprotactinates, $MPa(NO_3)_6$, where M is a alkali metal or quaternary amine, have been prepared by treating the chloro complex salts $MPaCl_6$ with liquid N_2O_5. The acid, $MPa(NO_3)_6$ is also known. In these compounds the protactinium(V) is presumably 12-coordinate by comparison with the tetravalent Np and Th nitrates. Neptunyl(V) nitrates have been starting materials for the preparation of "cation—cation" complexes, where actinyl ion interactions such as Np^V–Np^V and Np^V–U^{VI} are thought to be significant. An interesting solid in this class is the orthorhombic $Cs_4(NpO_2)_3Cl_6(NO_3)_2 \cdot H_2O$, in which pentagonal bipyramidal Np polyhedra are linked to form layers of composition $[(NpO_2)_2Cl_4(NpO_2)(Cl)(H_2O)]_n^{4n-}$.[519] Both

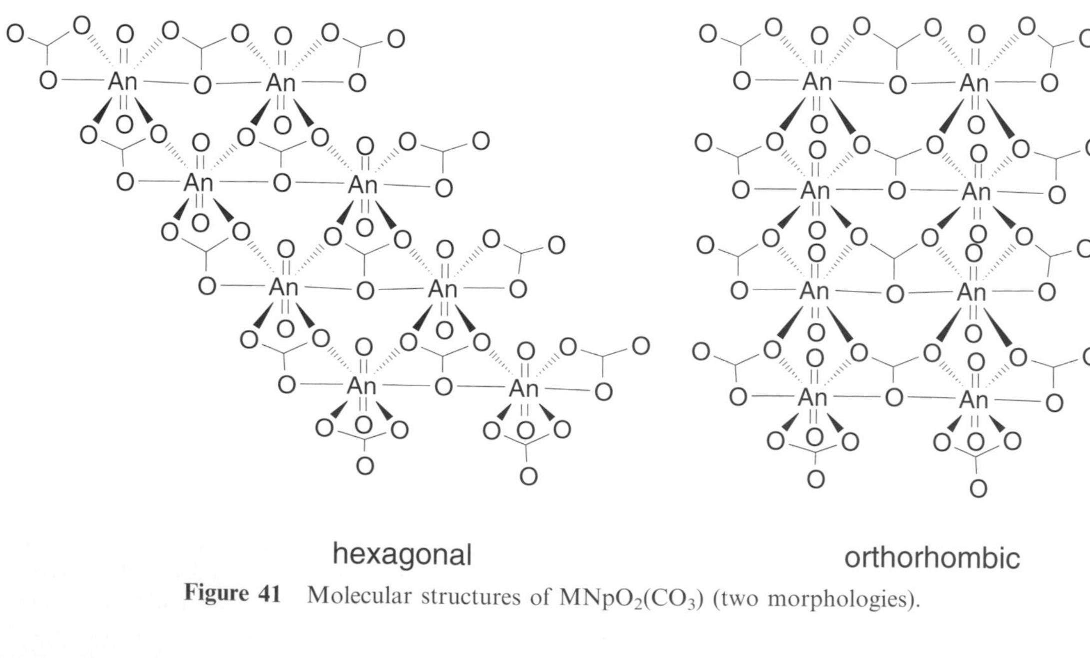

hexagonal   orthorhombic

Figure 41 Molecular structures of $MNpO_2(CO_3)$ (two morphologies).

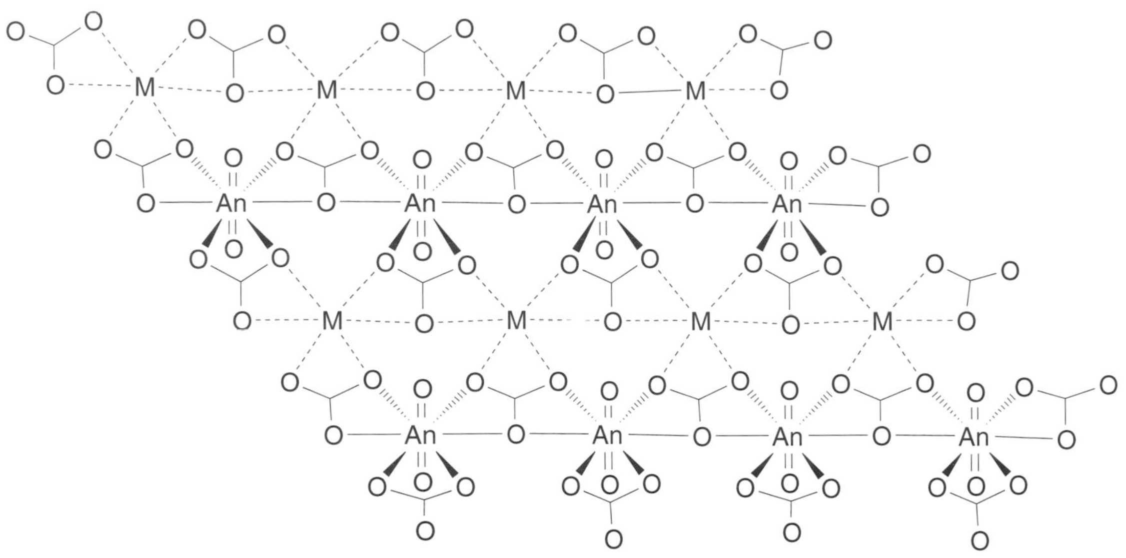

Figure 42 Molecular structure of $M_3AnO_2(CO_3)_2$.

this compound and its mixed-valent Np^V/Np^{VI} decomposition product, which also contains neptunyl oligomers, have structures and bond distances that suggest actinyl–actinyl interactions.[520] While these structural features in the solid state can be alternatively attributed to packing forces, the numerous reports of increased extraction of one actinyl with the addition of another actinyl suggest the interaction may be significant.[521] For example, the extraction of Np^V with CMPO from nitric acid increases with the addition of U^{VI}.[522] A pentavalent actinide nitrite complex, NpO_2NO_2, was reported in a study of Np complexation by a variety of inorganic ligands but it has not been characterized.[523]

Phosphates and arsenates. Neptunyl and plutonyl phosphate complexes have been prepared from An^{IV} phosphoric acid solutions. There is good evidence for $NpO_2HPO_4^-$ and $PuO_2HPO_4^-$, but their structures were not reported.[524,525] Additional complexes undoubtedly are formed, but their stoichiometries are not certain.[46] Protactinium phosphate solids, such as $PaO_2(H_2PO_4)_3\cdot 2H_2O$, have been reported, but without structural information. Simlarly Pa^V and Np^V arsenato complexes, such as $H_3PaO_2(PhAsO_3)_2$ $NpO_2H_2PO_4$ have been reported, but no structural information is available for them.

Sulfates and selenates. Two types of neptunyl sulfates have been well characterized. The simple binary salt, $(NpO_2)_2SO_4\cdot xH_2O$, where $x=2$, 4.5, and 6, can be precipitated from neptunyl sulfuric acid solutions. And bis(sulfato) complexes, $[Co(NH_3)_6]NpO_2(SO_4)_2\cdot 3H_2O$ and $[Co-(NH_3)_6]NpO_2(SO_4)_2\cdot M_2SO_4$ xH_2O, where M = Na, K can be isolated by adding Np^V to the preparation of $Co(NH_3)_6(SO_4)_2$.[526] Interestingly, Am^V analogues of these sulfates have been reported, but those for Pu^V have not. Protactinium oxosulfate, $H_3PaO(SO_4)_3$ can be be precipitated from H_2SO_4/HF. Similarly, $H_3PaO(SeO_4)_3$ can be precipitated from H_2SeO_4/HF solutions of Pa^V.

Perchlorates and iodates. Hydrated neptunium(V) iodate and a salt of a complex anion, $[Co(NH_3)_6](NpO_2)_2(IO_3)_5\cdot 4H_2O$, have been reported based on elemental analyses and powder diffraction data. The structure of $NpO_2(IO_3)$ was determined by single crystal X-ray diffraction.[527] Its structure consists of neptunyl(V) cations linked to one another by both $NpO_2^+-NpO_2^+$ bonds and bridging iodate anions creating a pentagonal bipyramidal NpO_7 unit. Oxygen atoms from the iodate anions occupy three of the equatorial sites in the NpO_7 units. Both oxo atoms of the neptunyl(V) units are involved in coordinating adjacent Np^V centers, leading to the creation of a two-dimensional neptunium oxide sheet. A perrhenate complex of Pa^V, $PaO(ReO_4)_3\cdot xH_2O$, has also been reported.

Alkoxide complexes. In contrast to the propensity of many uranium(V) species to disproportionate to uranium(IV) and uranium(VI), homoleptic uranium(V) alkoxide compounds are quite stable toward disproportionation. Gilman and co-workers reported the synthesis of dark brown uranium(V) pentakis(ethoxide) from a metathesis reaction between UCl_4 and four equivalents of sodium ethoxide.[528] In this early report, it was noted that better yields were obtained "when no great care was taken to exclude air from the reaction," and in the presence of oxygen, the product yield was 80%. The mechanism shown in Equations (36) and (37) was suggested for this reaction.

$$UCl_4 + 4\,Na(OCH_2CH_3) \longrightarrow U(OCH_2CH_3)_4 \tag{36}$$

$$5\,U(OCH_2CH_3)_4 + O_2 \longrightarrow 4\,U(OCH_2CH_3)_5 + UO_2 \tag{37}$$

Molecular weight determinations were consistent with a dimeric structure, $[U(OCH_2CH_3)_5]_2$, and the compound can be distilled at 123 °C (0.001 Torr). Species that are thermally stable, distillable or sublimable are desirable for use in the separation of metal isotopes. The number of ensuing reports describing various synthetic routes to $[U(OCH_2CH_3)_5]_2$ are evidence of that motive.[528–536] Some of these methods are described by Equations (38) to (43):

$$U(OCH_2CH_3)_4 + 1/2\,Br_2 \longrightarrow U(OCH_2CH_3)_4Br \xrightarrow[-NaBr]{Na(OCH_2CH_3)} U(OCH_2CH_3)_5 \tag{38}$$

$$UCl_5 + 5\,NaOCH_2CH_3 \longrightarrow U(OCH_2CH_3)_5 + 5\,NaCl \tag{39}$$

$$UCl_5 + 5\,HOCH_2CH_3 + 5\,NH_3 \longrightarrow U(OCH_2CH_3)_5 + 5\,NH_4Cl \tag{40}$$

$$NaU(OCH_2CH_3)_6 + HCl \longrightarrow U(OCH_2CH_3)_5 + NaCl + HOCH_2CH_3 \tag{41}$$

$$(C_5H_6N)_2U(O)Cl_5 + 5\,NH_3 + 3\,HOCH_2CH_3 \longrightarrow OU(OCH_2CH_3)_3 + 5\,NH_4Cl + 2\,C_5H_5N \tag{42}$$

$$UF_5 + 5\,NaOCH_2CH_3 \longrightarrow U(OCH_2CH_3)_5 + 5\,NaF \tag{43}$$

Uranium(V) homoleptic pentakisalkoxides, mixed alkoxides, oxo/alkoxides ($UO(OR)_3$), or solvate derivatives (OR, R = Me, Pr, Pr^i, Bu, Bu^i, Bu^s, Bu^t, CH_2CF_3, $CH_2CH{=}CH_2$, $(CH_2)_4CH_3$, $CH_2CH_2Pr^i$, $CH_2CHMeEt$, CH_2Bu^t, $CHEt_2$, $CHMePr$, $CHMePr^i$, CMe_2Et, CMe_2Pr, CMe_2Pr^i, $CMeEt_2$, $CMeEtPr^i$, CEt_3) were prepared using these reaction routes or simple alcohol exchange in refluxing benzene.[262,529,537,538] Displaced ethanol is removed azeotropically with benzene.

Molecular weight determinations of uranium(V) pentakisalkoxide complexes[537–540] suggest that most are dimeric, except for polymeric $[U(OMe)_5]_x$ and a few species incorporating sterically bulky alkoxide ligands. Spectroscopic data (absorption, ^1H NMR) supports the prediction that $[U^V(OCH_2CH_3)_5]_x$ exists as a dimer at room temperature, and ^1H-NMR analysis suggests that $[U^V(OPr^i)_5]_x$ exists as a monomer–dimer equilibrium at room temperature.[541–544] The structure of $[U(OPr^i)_5]_2$ dimer was later confirmed and further elucidated by single crystal X-ray diffraction analysis. The compound has an edge-sharing bioctahedral structure.[262]

Attempts to prepare other homoleptic $An(OCH_2CH_3)_5$ compounds (An = Pa, Np) have been reported. Protactinium(V) pentakisethoxide was prepared from the metathesis reaction between $PaCl_5$ and $NaOCH_2CH_3$ in ethanol, and the compound was formulated as $[Pa(OCH_2CH_3)_5]_x$ ($x > 5$) based upon analysis of the infrared spectrum and molecular weight determination in benzene.[545] Another study showed that oxidation of $Np(OCH_2CH_3)_4$ with bromine and $NaOCH_2CH_3$ in CCl_4 produced $NpBr(OCH_2CH_3)_4$.[269] Further addition of $NaOCH_2CH_3$ to a solution of $NpBr(OCH_2CH_3)_4$ in tetrahydrofuran only resulted in reduction to an unidentified Np^{IV} species, based on absorption spectra of the solution.

Complex salts have also been prepared. The $M[U(OCH_2CH_3)_6]_x$ salts (M = Na, $x = 1$; Ca, $x = 2$; Al, $x = 3$) were prepared by allowing $U(OCH_2CH_3)_5$ to react with respective metal alkoxides in a 1:1 ratio.[530,543] $NaU(OCH_2CH_3)_6$ decomposed with heat, but $Ca[U(OCH_2CH_3)_6]_2$ was purified by sublimation and $Al[U(OCH_2CH_3)_5]_3$ can be distilled.

Lewis base adducts of $U(OCH_2CH_3)_5$ have been prepared with acetonitrile, THF, pyridine, and SO_2,[543] and adducts of $U(OCH_2CF_3)_5$ were prepared with a number of aliphatic amines (NMe_3, $NPrH_2$, NPr^iH_2, NPr_2H, NMe_2H, ethylenimine).[529,530] Later reports of the synthesis of polyfluoroalkoxides ethanol adducts $U[OC(CF_3)_3]_4(OCH_2CH_3)(HOCH_2CH_3)$, and $U[OCH(CF_3)_2]_4(OCH_2CH_3)(HOCH_2CH_3)$ from the reaction between the respective fluorinated alcohol with $U(OCH_2CH_3)$[543,546] determined these complexes to be monomeric.

A variety of mixed ligand/alkoxide uranium(V) products are also isolable. Substitution compounds ($U(OR)_4L$, $U(OR)_3L_2$, $U(OR)_2L_3$) were prepared from the reactions of $U(OCH_2CH_3)_5$ with HCl, β-ketoesters (2,2,2,-trifluoroaceto acetate, methyl acetate, ethyl acetate), acetyl chlorides

(MeCO$_2$R, R = Et, Pr^i, Pentyl^t) or β-diketones (acetylacetone, benzoylacetone).[530,547,548] Other mixed halogen/alkoxide uranate products have been reported.[549] Anhydrous ethanol was allowed to react with hexahalogenouranates, MU^{V}X$_6$ (M = N(CH$_2$CH$_3$)$_4$, As(C$_6$H$_5$)$_4$; X = Cl, Br) at room temperature to yield MU(OCH$_2$CH$_3$)$_2$X$_4$. The reaction of HF with U(OCH$_2$CH$_3$)$_5$ is suggested to form U(OCH$_2$CH$_3$)$_2$F$_3$.[543]

The mixed valence dinuclear species U$_2$(OBu^t)$_9$ was obtained from unstable K[U$_2$(OBu^t)$_9$]. The dimer crystallizes as a face-sharing bioctahedron.[262] Theoretical studies have been carried out to understand the lack of metal–metal bonding in these dinuclear uranium alkoxide structures.[550]

Syntheses of a variety of uranium(V) species employing phenoxide ligands have been described. In the preparation of uranium(V) aryloxide compounds via alcohol exchange, both products of partial alcohol replacement (U(OC$_6$H$_5$)$_4$(OCH$_2$CH$_3$) and U(OC$_6$H$_5$)$_3$(OCH$_2$CH$_3$)$_2$)[551] and complete exchange (U(OPh)$_5$)[543] have been reported from reactions with U(OCH$_2$CH$_3$)$_5$, depending on stoichiometry and reaction conditions. The synthesis and characterization of analogous uranium(V) perfluorophenoxide, U(OC$_6$F$_5$)$_5$(HOCH$_2$CH$_3$), have also been presented.[543] The metathesis reaction between CsUCl$_6$ and NaOC$_6$H$_5$, followed by extraction with *N,N*-dimethylformamide led to an "ate" product of composition close to U(OC$_6$H$_5$)$_4$Cl·2 DMF.[551] A unique U^V/U^{VI} mixed valence uranium phenoxide aggregate, {[U^V(OC$_6$H$_5$)$_3$(THF)]$_2$[U^{VI}O$_2$(THF)]$_2$}-(μ-OC$_6$H$_5$)$_4$(μ-O)$_2$, was synthesized by the reaction of NaOC$_6$H$_5$ with UCl$_3$·xTHF in tetrahydrofuran.[55] The structure of the complex consists of two seven-coordinate uranium(V) metal centers and two five-coordinate uranyl groups (U^{VI}O$_2$$^{2+}$) bridged by phenoxide and oxo ligands.

Thiolate complexes. Uranium(V) thiolate compounds have also been prepared. It was reported that addition of H$_5$C$_6$SSC$_6$H$_5$ to UCl$_5$-Cl$_2$C=CClCOCl allowed for the formation of a uranium(V) arylsulfide compound, [UCl$_4$(SPh)]$_2$, as characterized by elemental analysis.[490,552] In another report, *p*-thiocresol was allowed to react with [U^V(OCH$_2$CH$_3$)$_5$]$_x$ in benzene under reflux to obtain U(SC$_6$H$_4$CH$_3$)$_4$(OCH$_2$H$_3$) in 74% yield.[551] Reactions of [U^V(OCH$_2$CH$_3$)$_5$]$_x$ were carried out with a series of thiosalicylic, thiolactic, and thiobenzoic acids, as well as alkyl thioglycolates in variable stoichiometric ratios to form substitution compounds that were characterized by elemental analysis.[553]

(iv) Ligands containing neutral group 16 donor atoms

The pentavalent oxidation state is accessible for the early actinides uranium, protactinium, neptunium, and plutonium. Pentavalent species with neutral Group 16 bases can include either adducts of AnX$_5$ or complexes incorporating oxo-containing cations, AnO^{3+} or AnO$_2$$^+$.

Aqua species. Ready hydrolysis ensures that all aqua species of pentavalent actinide species include oxo or hydroxide ligands. Representative aqua species are presented in Table 28. Early reports of hydrates were unable to differentiate between coordinated water and water included in the lattice of a complex. There are several structurally characterized examples in this class. Examples include the complex [(NpO$_2$)$_2$ (SO$_4$)(H$_2$O)], in which the water is bound to one of the two neptunium centers to complete a coordination number of eight,[554] and NpO$_2$-ClO$_4$·4H$_2$O,[555] which is shown to be an ionic complex with four water molecules in equatorial positions of the pentagonal bipyramidal geometry. Other structurally characterized neptunyl hydrates include (NpO$_2$)$_2$ (NO$_3$)$_2$·5H$_2$O,[556] NpO$_2$Cl·H$_2$O,[554] and the tri- and tetrahydrates of neptunyl malonate, (NpO$_2$)$_2$C$_3$H$_2$O$_4$·xH$_2$O (x = 3, 4).[556]

Ethers. The only reported ether compounds of pentavalent actinides are the species UCl$_5$·ether, where ether = THF or R$_2$O (R = Me, Et, Pr^i, Bu^n, and *i*-C$_5$H$_{11}$). Dioxane is suggested to form either both 1:1 and 1:3 (U:L) adducts.[557]

Ketones, aldehydes, esters. Adducts of UCl$_5$ with a number of ketone derivatives of polycyclic aromatics have been reported. The complexes UCl$_5$·L (L = anthr-10-one, 9-methyleneanthr-10-one, 1,9-benzoanthr-10-one, and 9-benzylideneanthr-10-one) are likely six coordinate. The ligands anthr-10-one and 1,9-benzoanthr-10-one also appear to form 1:2 (An:L) complexes. The trichloroacryloyl chloride complex UCl$_5$·Cl$_2$C=CClCOCl has been identified as the initial product of the reaction of UO$_3$ with hexachloropropene; this species subsequently thermally decomposes to yield UCl$_5$. Under the common reaction conditions the UCl$_5$ thus generated spontaneously converts to UCl$_4$.

The ester complexes U(OR)X$_4$·MeCO$_2$R (R = Et, X = Cl or Br; R = Pr^i, X = Cl) were generated by the reaction of U(OR)$_5$ with acyl halides MeCOX.

Carbamides. The only reported carbamide complex is U(OC$_6$H$_5$)$_4$Cl·2DMF.[551]

Table 28 Some hydrates of protactinium(V), uranium(V), neptunium(V), and plutonium(V) compounds.

$PaF_5 \cdot xH_2O$	$x = 1, 2$
$(Et_4N)_2(UOF_5) \cdot 2H_2O$	
$NpOF_3 \cdot 2H_2O$	
$NpO_2(ClO_4) \cdot xH_2O$	$x = 3, 7$
$M^IM^VO_2(CO_3) \cdot xH_2O$	$M^I = Na$, $M^V = Np$, $x = 0.5, 1, 2, 3, 3.5$ or 4; $M^V = Pu$, x unspecified
	$M^I = K, Rb$, $M^V = Np, Pu$
	$M^I = NH_4$, $M^V = Np, Pu$ ($x = 3$)
$(NH_4)_2PuO_2(CO_3)(OH) \cdot xH_2O$	
$K_3M^VO_2(CO_3)_2 \cdot xH_2O$	$M^V = Np, Pu$ ($x \leq 2$)
$Pa(C_2O_4)_2(OH) \cdot 6H_2O$	
$PaO(C_2O_4)(OH) \cdot xH_2O$	$2 < x < 4$
$NpO_2(HC_2O_4) \cdot 2H_2O$	
$(NpO_2)_2C_2O_4 \cdot H_2O$	
$M^INpO_2(C_2O_4) \cdot xH_2O$	$M^I = Na$, $x = 1, 3$; K, $x = 2$; NH_4, $x = 2.2, 3$; Cs, $x = 2, 3$
$MNpO_2(C_2O_4)_2 \cdot xH_2O$	$M^I = Na, K, NH_4$
$MNpO_2(C_2O_4)_3 \cdot xH_2O$	$M^I = Na, K, NH_4, Cs$
$PaO(NO_3)_3 \cdot xH_2O$	$1 < x < 4$
$NpO_2(NO_3) \cdot xH_2O$	$x = 1, 5$
$RbNpO_2(NO_3)_2 \cdot H_2O$	
$PaO(H_2PO_4)_3 \cdot 2H_2O$	
$NH_4PuO_2HPO_4 \cdot 4H_2O$	
$[Co(NH_3)_6]NpO_2(C_2O_4)_2nH_2O$	
($n = 3, 4$)	
$[Co(NH_3)_6]NpO_2(SO_4)_2 \cdot 3H_2O$	
$[Co(NH_3)_6]NpO_2(SO_4)_2 \cdot$	$MSO_4 \cdot xH_2OM^I = Na, K$
$[Co(NH_3)_6](NpO_2)_2(IO_3)_5 \cdot 4H_2O$	
$PaO(ReO_4)_3 \cdot xH_2O$	

Urea. The complex $NpO_2(NCS)(urea)_4$ has been structurally characterized;[498] the four urea molecules lie in the equatorial plane of the pentagonal bipyramidal geometry.

Phosphine oxide. Many reported actinide-phosphine oxide adducts are those of the actinide halide complexes (see Table 29). Most are obtained by direct reaction of the two constituents in nonaqueous media. $UCl_5(OPPh_3)$ has been structurally characterized;[558] the coordination environment is best described as approximately octahedral.

Sulfoxide. The complex $Pa(tropolonate)_4Cl \cdot DMSO$ has been reported.[559]

Phosphine sulfide, phosphine selenide. Considering the aforementioned instability of P=S and P=Se bonds, it is not surprising that the only reported complexes in this class involve Pa^V, the most stable actinide of that oxidation state. The complexes PaX_5L ($X = Cl, Br$; $L = Ph_3PS$, $(Ph_2PS)_2CH_2$, Ph_3PSe, and $(Ph_2PSe)_2CH_2$ have been identified.[560]

(v) Ligands containing Group 17 ligands

Binary halides. A number of homoleptic halides of pentavalent protactinium, uranium, and neptunium have been reported. In particular, the fluoride complexes AnF_5 are prepared by high

Table 29 Complexes of actinide(V) compounds with *P*-oxides.

$MX_5 \cdot R_3PO$	$R = n\text{-}C_8H_{17}$, $M = U$, $X = Cl$
	$R = Me_2 N$, $M = U$, $X = Br$
	$R = Ph$, $M = Pa, U$, $X = Cl, Br$
$PaCl_5 \cdot Ph_2(PhCH_2)PO$	
$MX_5 \cdot 2Ph_3PO$	$M = Pa$, $X = F, Cl, Br$; $M = U$, $X = F, Cl$
$UCl_2F_3 \cdot 2Ph_3PO$	
$PaBr_5 \cdot (Ph_2PO)_2CH_2$	
$Pa(OEt)_2X_3 \cdot Ph_3PO$	$X = Cl, Br$
$PaOX_3 \cdot 2Ph_3PO$	$X = Cl, Br$
$NpO_2NO_3 \cdot 3(n\text{-}C_8H_{17})_3PO$	
$[NpO_2\{(n\text{-}C_8H_{17})_3PO\}_4]ClO_4 \cdot H_2O$	

temperature oxidation of the tetravalent fluorides with fluorine or other potent fluorinating agents (e.g., KrF_2), or by reduction of the hexafluorides with iodine or PF_3. The synthesis of NpF_5 by oxidation of $[NpF_6]^-$ with BrF_3 in anhydrous HF has also been reported.[561] Two crystallographic forms (α-, β-) exist for UF_5; the neptunium compound is reported to be isostructural with α-UF_5,[562] whereas PaF_5 has the same structure as β-UF_5.[563] In the alpha form, the uranium lies within an octahedral coordination environment. In the beta form, the uranium center is eight-coordinate, with a geometry intermediate between dodecahedral and square antiprismatic.[564]

Heavier pentahalides have also been reported for protactinium and uranium. Protactinium is reported to form $PaCl_5$, $PaBr_5$, and PaI_5; the chloride and bromide derivatives have been structurally characterized.[565,566] UCl_5 exists in two morphologies (α-, β-); UBr_5 appears to be isostructural with β-UCl_5.

Several mixed halides $AnX_nX'_{5-n}$ have also been reported.

Complex halides. The major classes of complex halides are $M[AnX_6]$, $M_2[AnX_7]$, and $M_3[AnX_8]$ (M = alkali metal). The complexes are generally prepared either by reaction of the appropriate pentahalide complex with the alkali metal halide salt, or by *in situ* reaction, where the pentahalide is produced directly in the presence of MX. The uranium complexes $M[UX_6]$ (X = F, Cl); the metal ion is octahedrally coordinated,[567–570] as are the actinide centers in $[NO][AnF_6]$ (An = Np, Pu).[571] For the larger protactinium ion, higher coordination numbers can be accomodated, and the complex $RbPaF_6$, the metal center is eight-coordinate.[572] Similarly, the compounds $MPaF_7$ display a nine-coordinate metal center,[573] whereas the smaller metal ions in Rb_2AnF_7 (An = Np, Pu) can only accommodate seven fluoride ions in their coordination sphere. The octafluoro complexes $Na_3[AnX_8]$ (An = Pa, U, Np) have been reported; each the metal center lies in a cubic environment.[574]

Oxohalides. Binary oxohalides of the formulae $AnOX_3$, AnO_2X, An_2OX_8, $An_2O_3X_4$, and An_3O_7X have been reported, most commonly for protactinium, but also in some cases for uranium and neptunium. The complexes are polymeric in the solid state; the structure of $PaOBr_3$ demonstrates that the metal center is in a pentagonal bipyramidal geometry of three oxygen and four bromine atoms.[575] The structure of $NpO_2Cl \cdot H_2O$ has been determined.[576]

A limited number of complex oxohalide complexes have been characterized. Complexes of the classes $(NR_4)_2[AnOX_5]$, $MAnO_2X_2$, and $M_3[UO_2Cl_4]$ (M = alkali metal) have been reported. Several members of the latter class have been structurally characterized and display a pseudo-octahedral geometry about the metal center.[577]

3.3.2.3.3 Chelating ligand

(i) Multidentate donor ligands

Hydroxamate. Similar to trivalent hydroxamates, pentavalent actinide hydroxamate complexes are generally unstable relative to tetravalent and/or hexavalent complexes. Pu^{VI} or Pu^V hydroxamate complexes can be prepared; however, at near-neutral and basic pH they rapidly reduce to Pu^{IV} complexes.[80]

8-Hydroxyquinoline and derivatives. A Np^V complex of 8-hydroxyquinoline $NpO_2(Ox)(H_2O)_2$ has been prepared by precipitation from aqueous solution.[578] The $NpO_2(Ox) \cdot DMSO$ is prepared by replacing the two water molecules by DMSO. The bis(quinoline) complex, $NpO_2(Ox)_2^-$ is formed by increasing the pH of a $NpO_2(Ox)(H_2O)_2$ solution. This anion precipitates as $[(C_6H_5)_4As][NpO_2(Ox)_2 \cdot H_2O]$ with addition of tetraphenylarsonium chloride.

Oxalate. Pentavalent actinide oxalates are limited to a few Np^V hydrated oxalates and oxalato complex salts (see Table 26). The structure of a simple neptunyl oxalate complex $(NpO_2)_2$-$(C_2O_4) \cdot 6H_2O$ has been determined by X-ray diffraction. The complex was precipitated from aqueous solutions by the reaction of NpO_2NO_3 and $(NH_4)_2C_2O_4$. The structure of the complex consists of electroneutral layers of $(NpO_2)_2$ $(C_2O_4) \cdot 4H_2O$ between which coordinated aqua ligands and waters of crystallization are located. Neptunyl(V) acts as a monodentate ligand for the adjacent Np and oxalate coordinates each Np in a bidentate mode.[579]

Diamides. Sasaki and Choppin used the diglycolamide, N,N'-dimethyl-N,N'-dihexyl-3-oxapentanediamide (DMDHOPDA), to successfully extract Np^V from an aqueous $NaClO_4$ solution into nitrobenzene or toluene (less effective). This is significant since Np^V possesses weak complexation ability with organic extractants. NpO_2^+ was characterized in solution by absorption spectroscopy. Extraction dependency studies show nonintegral slopes, indicative of $NpO_2(DMDHOPDA)^+$ and $NpO_2(DMDHOPDA)^{2+}$ species being present in the extracted phase; pH studies also show an acid dependence on its extraction (90% complete at a pH of 3).[580]

Polyoxometallates. The relatively high negative charge of the Dawson and Keggin anions $[XW_{11}O_{39}]^{n-}$ and $[X_2W_{17}O_{61}]^{n-}$ render them capable of supporting actinides in higher oxidation states. Chemical or electrochemical oxidation of the U^{IV} species $U[PW_{11}O_{39}]_2^{10-}$, $U[P_2W_{17}O_{61}]_2^{16-}$, and $U[SiW_{11}O_{39}]_2^{12-}$ results in the formation of stable pentavalent complexes without the formation of actinyl-type species.[438,581] Other examples of the complexation of pentavalent actinides with polyoxometallates indicate somewhat weaker bonding. Absorption spectroscopy suggests the complexation of Np^V by $[P_2W_{17}O_{61}]^{10-}$ in solution, but this complexation is supressed by the addition of excess Na^+ or K^+, suggesting that the pentavalent actinide may be only weakly coordinated to the outer sphere of the oxoanion as the actinyl species NpO_2^+.[582,583]

(ii) Macrocyclic ligands

N-Heterocyclic ligands. The porphyrin [22]hexaphyrin(1.0.1.0.0.0) has been complexed with NpO_2^+ to form an inclusion complex. This complex was studied by single crystal X-ray diffraction and the crystal structure is shown in Figure 43. In this complex the neptunium and the nitrogen atoms of the porphyrin are almost perfectly co-planar. The coordination around the neptunium is nearly an ideal hexagonal bipyramidal with the two oxo-ligands being *trans* to the equatorial plane.[584]

The plutonium(V) analog of the neptunium(V) [22]hexaphyrin(1.0.1.0.0.0) was confirmed by UV–vis spectroscopy. In this complex, plutonium(VI) is exposed to the ligand and is reduced to plutonium(V). It is postulated that the coordination is the same as the neptunium(V) complex.[585]

Crown ethers. In neptunium(V) and neptunium(VI) crown ether complexes, the two axial oxygen ligands help in forming more stable complexes. The reaction of neptunium(VI) with 18-crown-6 in the presence of 1 M HX (X=ClO_4 or CF_3SO) results in the reduction of neptunium(VI) to neptunium(V), thus allowing better metal complexation. The geometry around Np^V in the complex is hexagonal bipyramidal. Disorder in the crystal leads to two possible spatial locations for the crown ether: 30° offset from one another and each refined with 50/50 occupancy. This does not affect the coordination environment of the metal.[586]

3.3.2.4 Hexavalent Oxidation State

3.3.2.4.1 General characteristics

Stable hexavalent ions of the type AnO_2^{2+} (An = U, Np, Pu) can be generated in aqueous media. The uranyl ion (UO_2^{2+}) is the most stable form of uranium in aqueous media under aerobic conditions. In general, aqueous Np^{VI} and Pu^{VI} compounds can be prepared from Np^V and Pu^{IV}

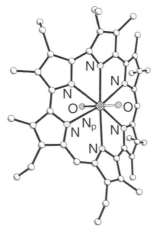

Figure 43 Crystal structure of $[NpO_2(C_{44}H_{51}N_6)]\cdot(HN(C_2H_5)_3)$ (Sessler, Seidel, *et al. Angew. Chem., Int. Ed. Engl.* **2001**, *40*, 591).

or Pu^V electrochemically by adding chemical oxidants such as ozone. Resulting Pu^{VI} complexes are generally more stable than Np^{VI} complexes, but both classes are far less stable than their U^{VI} analogues. For example, hydrated Np^{VI} and Pu^{VI} carbonates will decompose to Np^V carbonates and Pu^{IV} oxyhydroxides, respectively, in days to weeks, depending on the initial preparation. The hexavalent actinyl ions also display a linear O—An—O unit, with other ligands coordinating roughly in the plane bisecting and perpendicular to this unit. Consistent with the lower thermodynamic stability of NpO$_2^{2+}$ and PuO$_2^{2+}$, these ions possess weaker An—O bonds. Non-actinyl coordination complexes of hexavalent actinides are rare, and are limited to a few ligand types (generally those containing halide or alkoxide ligands, although one example of a stable homoleptic amide complex of U^{VI} has been reported).

3.3.2.4.2 *Simple donor ligands*

(i) *Ligands containing group 14 donor atoms*

The most recent class of Group 14 donor ligands to be employed in actinide chemistry is that of *N*-heterocyclic carbenes. These ligands act as σ-donor bases toward a number of metals in coordination chemistry. Reaction of UO$_2$Cl$_2$(THF)$_3$ with 1,3-dimesitylimidazole-2-ylidene and its 4,5-dichlorosubstituted derivative generate 1:2 (uranium:carbene) adducts, UO$_2$Cl$_2$(L)$_2$.[587] Crystallographic characterization reveals an octahedral metal center, with *trans*-oxo, chloro, and carbene ligands. The uranium–carbon bond distances in these species are long at 2.626(7) Å and 2.609(4) Å, consistent with the formulation of the C—U bond as a dative interaction.

A rare example of U–C interaction in hexavalent actinide chemistry is found in the isolation of a bis(iminophosphorano)methanide complex of the uranyl ion[588] Reaction of UO$_2$Cl$_2$(THF)$_3$ with Na[CH(Ph$_2$P=NSiMe$_3$)$_2$] generates the dimer {UO$_2$(μ-Cl)[CH(Ph$_2$P=NSiMe$_3$)$_2$]}$_2$. The U–C distance is 2.691(8) Å; the length indicates a very weak interaction, although it falls within the sum of the van der Waals radii of the two atoms.

An adduct of uranyl chloride with cyclohexylisocyanide has been reported.[101]

(ii) *Ligands containing anionic group 15 donor atoms*

Amide complexes.
Hexavalent complexes containing N-donor ligands comprise two major classes: homoleptic amide complexes, and species derived from the uranyl ion (UO$_2^{2+}$). Only two homoleptic amide complexes have been reported. The complex U(NMe$_2$)$_6$ is reportedly produced in the oxidation of [U(NMe$_2$)$_6$]$^-$ by silver iodide;[104,106] the complex is unstable in solution, and could be characterized only by its NMR spectrum. It decomposes to form U(NMe$_2$)$_5$ and unidentified side products. A more stable hexavalent amide complex is produced by the oxidation of [Li(THF)$_x$][U(dbabh)$_6$]$^-$ (Hdbabh = 2,3:5,6-dibenzo-7-azabicyclo[2.2.1]hepta-2,5-diene) by any of several oxidants (air, Cp$_2$Fe$^+$, Ag$^+$, I$_2$).[487] The complex U(dbabh)$_6$ has been characterized crystallographically (see Figure 44). The coordination environment of the metal center is nearly octahedral, with U–N distances ranging from 2.178(6) Å to 2.208(5) Å. Density functional theory (DFT) calculations suggest that the HOMO is triply degenerate with U—N π-bonding character. Although the orbitals are largely nitrogen-based, they contain contributions from U 6*d* and 5*f*-orbitals.

A series of uranyl complexes containing the bis(trimethyl)silylamide ligand have been reported.[12,589,590] The complex UO$_2$[N(SiMe$_3$)$_2$]$_2$(THF)$_2$ is produced in the reaction of UO$_2$Cl$_2$ with two equivalents of K[N(SiMe$_3$)$_2$]. The anionic species [Na(THF)$_2$]$_2${UO$_2$[N(SiMe$_3$)$_2$]$_4$} is produced in the reaction of UO$_2$Cl$_2$ and excess Na[N(SiMe$_3$)$_2$]. The intervening members of the series, [M(THF)$_2$]{UO$_2$[N(SiMe$_3$)$_2$]$_3$} (M = Na, K) have been more recently reported, and serve as rare examples of uranyl complexes with only three ligands in the equatorial plane. While the sodium derivative can only be produced by Lewis-acid abstraction or protonation of N(SiMe$_3$)$_2^-$ from the "ate" complex [Na(THF)$_2$]$_2${UO$_2$[N(SiMe$_3$)$_2$]$_4$}, the potassium derivative may be directly synthesized by reaction of [UO$_2$Cl$_2$(THF)$_2$]$_2$ with K[N(SiMe$_3$)$_2$]. A base adduct, UO$_2$[N(SiMe$_3$)$_2$]$_2$(OPPh$_3$)$_2$, has also been reported.[591] Attempts to introduce more electron-rich ligands into the coordination sphere of the uranyl ion are found to result in reduction of the metal center. Reaction of [K(18-crown-6)]$_2$[UO$_2$Cl$_4$] with the tris(amido)amine ligand Li$_3$[N(CH$_2$CH$_2$NSi(Bu^t)Me$_2$)$_3$] produces the mixed valence oxo- and imido-containing species [K(18-crown-6)(Et$_2$O)][UO(2-CH$_2$CH$_2$N(CH$_2$CH$_2$NSi-(Bu^t)Me$_2$)$_2$)]$_2$

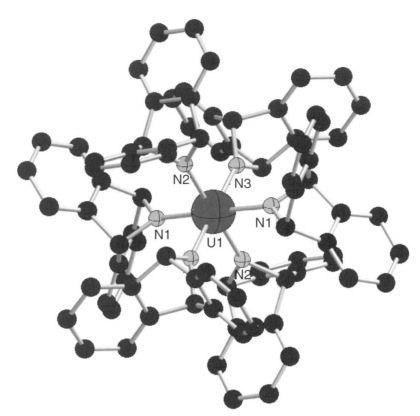

Figure 44 Crystal structure of U(dbabh)$_6$ (Meyer, Mindiola *et al. Angew. Chem., Int. Ed. Engl.* **2000**, *39*, 3063–3066).

as the major isolated uranium product in moderate yield.[592] The multidentate triamidoamine ligand coordinates to uranium through the capping amine and two of the three pendant amido ligands, while the third pendant amido donor has been activated to generate a bridging imido ligand by loss of the silyl substituent. One of the uranyl oxo groups is retained as a terminal ligand to complete the coordination sphere for each uranium center. The oxo and imido nitrogen may be regarded as the axial ligands of a capped trigonal bipyramidal geometry about the metal center, while the two amido ligands and the other imido donor occupy equatorial coordination sites. The central amine of the tripodal set serves as the capping ligand.

Amidinate ligands. The benzamidinate ligand has also been found to coordinate the uranyl ion. The uranyl complex, [C$_6$H$_5$C(NSiMe$_3$)$_2$]$_2$UO$_2$, was prepared by a metathesis reaction with UO$_2$Cl$_2$.[593]

Pyrazolylborate. Only one reported complex of a tris(pyrazolyl)borate complex has been reported. The chelating ligand tris[3-(2-pyridyl)-pyrazol-1-yl]borate (^{py}Tp) reacts in a 1:1 ratio with uranyl nitrate to yield the complex UO$_2$(^{py}Tp)(OEt) after recrystallization from EtOH/CH$_2$Cl$_2$.[170] The ethoxide ligand is presumably derived from the alcohol of recrystallization. The three bidentate arms of the pyrazolylborate ligand are incapable of spanning the equatorial plane of the uranyl unit; instead two arms provide four ligands of the pentagonal bipyramidal plane. The third arm of the ligand is pendant, with the pyridyl and pyrazolyl rings adopting a *trans*-coplanar arrangement (see Figure 45).

Phosphoraniminato complexes. A very different class of species with U—N bonds is derived from the uranyl ion by replacement of one or both of the oxo groups by a phosphoran iminato group (PR$_3$-N^{2-}). The first example of such a complex, [PPh$_4$][UOCl$_4$\{NP(*m*-Tol)$_3$\}] (Tol = tolyl) was generated in modest yield by elimination of Me$_3$SiCl from the mono-oxo complex [UOCl$_5$]$^-$ (Equation (44)):[594]

$$[\text{PPh}_4][\text{UOCl}_5] + \text{Me}_3\text{SiNP}(\textit{m}\text{-Tol})_3 \longrightarrow [\text{PPh}_4][\text{OUCl}_4(\text{NP}(\textit{m}\text{-Tol})_3)] + \text{Me}_3\text{SiCl} \quad (44)$$

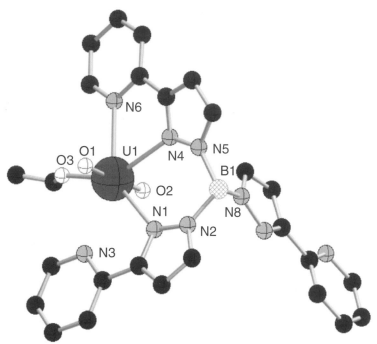

Figure 45 Crystal structure of UO$_2$(pyTp)(OEt) (pyTp=tris[3-(2-pyridyl)-pyrazol-1-yl]borate) (Amoroso, Jeffery *et al. Polyhedron* **1995**, *15*, 2023–2027).

The procedure was subsequently extended to include other members of the class [UOCl$_4${NPR$_3$}]$^-$.[595] The resulting complexes are stable in the absence of moisture; in the presence of water they undergo rapid hydrolysis to generate uranyl salts. The molecular structure of the compound [PPh$_4$][UOCl$_4${NP(*m*-Tol)$_3$}] reveals that the O—U—N unit retains the linearity inherent to the actinyl species (179.0(6)°), and that the U—O bond is within the range expected for uranyl complexes (1.759(13) Å). The U—N bond is short (1.901(14) Å), suggesting a significant degree of metal–ligand multiple bonding. Bis(iminato) complexes UCl$_4${NPR$_3$}$_2$ can be isolated from the same reaction mixture by controlling the conditions under which the products are isolated. They appear to be formed concurrently with UO$_2$Cl$_4^{2-}$, although isolated complexes of [UOCl$_4${NPR$_3$}]$^-$ appear to be stable to redistribution. In the vibrational spectra of these species the antisymmetric U—N—P stretching frequencies in the oxo-iminato complexes appear to lie at higher frequencies than those in the bis(iminato) species. This has been cited as supporting evidence for the existence of an inverse *trans* influence in actinide complexes.[596] More recently, a sulfiliminato analogue, [PPh$_4$][UOCl$_4${NSPh$_2$}], has been produced by the reaction of [PPh$_4$][UOCl$_5$] with Me$_3$Si{NSPh$_2$}.[597]

Additional support for the existence of non-oxo uranyl analogues may be found in the reactions of uranium atoms with small molecules (N$_2$, NO, CO, etc.) in argon matrices. Although not stable outside of the stabilizing matrix, vibrational spectroscopy is consistent with the formation of other linear triatomic species such as NUN, NUO, and CUO.[598–601]

Cyanate and thiocyanate. Cyanate salts of the uranyl ion are thought to be formed in reactions of uranyl and Et$_4$N(NCO). The complex (Et$_4$N)$_2$[UO$_2$(NCO)$_4$(H$_2$O)] has been reported,[602] and the partially hydrolyzed salt K$_3$[(U$_2$O$_5$)(NCO)$_5$]·H$_2$O has also been reported. Although no complexes have been isolated, anions of the formula [UO$_2$(NCO)$_4$]$^{2-}$ and [UO$_2$(NCO)Cl$_3$]$^{2-}$ are thought to be formed in solutions containing UO$_2$Cl$_4^{2-}$ and R$_4$N[Ag(NCO)$_2$], on the basis of IR data.[603]

Neutral complexes of the formula UO$_2$(NCS)$_2$(H$_2$O)$_x$ have been reported, as have monoanionic species containing the [UO$_2$(NCS)$_3$L$_x$]$^-$ anion and the complex anion [UO$_2$(NCS)$_5$]$^{3-}$. The complex Cs$_3$[UO$_2$(NCS)$_5$] has been crystallographically characterized;[604] as expected, the metal lies within a pentagonal bipyramidal coordination sphere.

It has been found that addition of crown ethers facilitates the crystallization of anionic derivatives. In this manner, the anion [UO$_2$(NCS)$_4$(H$_2$O)]$^-$ has been isolated both as the ammonium and potassium salts.[605]

Azide. There are few reports of actinide azide complexes. One compound isolated by the reaction of uranyl with tetraalkylammonium azide salts has an extended structure in the solid state.[606] The structure of catena-(tetraethylammonium bis(μ_2-azido-*N,N*)azidodioxouranium incorporates a chain of uranyl ions bridged in the equatorial plane by two azide group; each uranyl is further ligated by a terminal azide to complete a pentagonal bipyramidal coordination environment about uranium. If a tetramethylammonium counterion is used, the large aggregate [NMe$_4$]$_8$[(UO$_2$)$_6$(μ_3-O)$_2$(μ_2-N$_3$)$_8$(N$_3$)$_8$] is isolated (see Figure 46).

(iii) Ligands containing neutral group 15 donor atoms

All neutral base adducts of hexavalent actinides are those of the dioxo, or actinyl ions (AnO$_2^{2+}$). Although at least one report has appeared which indicates the stability of base adducts of UF$_6$.[607] However, subsequent researchers have demonstrated that reduction occurs in these systems.[493]

Ammonia and amines. A number of ammonia and amine adducts of uranyl complexes have been reported (see Tables 30 and 31), isolated from reaction of the uranyl salt with amine either in aqueous or nonaqueous media. The molecular structure of the acetylacetonate derivative UO$_2$(CF$_3$COCHCOCF$_3$)(NH$_3$) is prototypical of the class. The uranyl *trans*-dioxo geometry is preserved, and the ammonia nitrogen is one of the five coordinating atoms in the "belly band" of the ion.[608] Occasionally ammonia can react with the coordinating ligand; reaction with electron-rich acetylacetonate ligands leads to condensation to form β-ketoimine complexes. Several proposed hydrazine adducts of the uranyl ion have also been reported,[609] although little characterization data is available. Several simple salts of uranyl with chelating diamine ligands have also been reported (see Table 32).

Heterocyclic ligands. Complexes of actinyl complexes with coordinating *N*-heterocycles in the equatorial plane are proposed to have metal coordination numbers ranging from six to eight, depending on the anion and the size of the base (*cf.* uranyl complexes presented in Table 33). Several compounds containing pyridine or bidentate heterocyclic ligands have been crystallographically characterized, and are found to contain seven- or eight-coordinate actinide(VI) ions. The structure of UO$_2$(NO$_3$)$_2$(py)$_2$ has been determined.[610] The complex consists of an eight-coordinate (considering nitrate as bidentate) uranium, with mutually *trans*-pyridine and nitrate ligands. A *cis* geometry of two nitrate ligands is enforced by the use of the chelating ligand 1,10-phenanthroline (phen) in the complex UO$_2$(NO$_3$)$_2$(phen);[611] this ligand also coordinates to the uranyl species [UO$_2$(O$_2$CCH$_3$)(phen)(μ-OH)]$_2$.[611] Monopyridine adducts AnO$_2$(acac)$_2$(py)

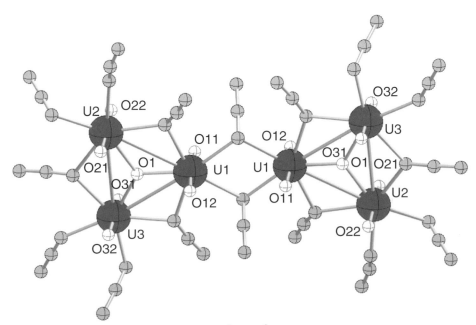

Figure 46 Crystal structure of [NMe$_4$]$_8$[(UO$_2$)$_6$(μ^3-O)$_2$(μ^2-N$_3$)$_8$] (Charpin, Lance *et al. Acta Crystallogr., Sect. C* **1986**, *42*, 1691).

Table 30 Ammonia complexes of dioxouranium(VI) compounds.

$UO_2F_2 \cdot x NH_3$	$x = 2$, 3 or 4
$UO_2Cl_2 \cdot x NH_3$	$x = 0.5$, 1, 2, 3, 4, 5 or 10
$UO_2Y_2 \cdot x NH_3$	$Y = Br$, I, $x = 2$, 3 or 4
$UO_2Y_2 \cdot x NH_3 \cdot Et_2O$	$Y = Cl$, Br, I, NO_3 and $x = 2$; $Y = Cl$, NO_3 and $x = 3$
$UO_2(NO_3)_2 \cdot x NH_3$	$x = 2$ or 4
$UO_2(NO_3)_2 \cdot 3NH_3 \cdot 2H_2O$	
$UO_2(NH) \cdot x NH_3$	$x = 1$ or 2
$UO_2Y_2 \cdot x NH_3$	$Y = MeCO_2$, $0.5SO_4$ and $x = 2$, 3 or 4;
	$Y = HCO_2$, $0.5C_2O_4$, and $x = 2$
$UO_2(OH)Y \cdot NH_3$	$Y = HCO_2$, $0.5C_2O_4$
$UO_2(C_{10}H_8NO)_2 \cdot NH_3$	$C_{10}H_9NO = $ 7-methyl-8-hydroxyquinoline
$UO_2(RCOCHCOR)_2 \cdot NH_3$	$R = CF_3$, Ph

Table 31 Amine complexes of dioxouranium(VI) compounds.

$UO_2Cl_2 \cdot 2RNH_2$	$R = PhNH_2$, $H_2NC_6H_4NO_2$, 2-, 3-, 4-$MeC_6H_4NH_2$,
	2-, 3-$MeOC_6H_4NH_2$, 4-$EtOC_6H_4NH_2$, $H_2NC_6H_3$-3,
	4-Me_2, 1-, 2-$H_2NC_{10}H_7$
$UO_2Cl_2 \cdot 2$ to $3RNH_2$	$R = Me$, Et, Pr^n, Bu^n
$UO_2Cl_2 \cdot R_2NH$	$R_2 = Ph_2$, $(PhCH_2)_2$, (Ph)(Et), (Ph)($PhCH_2$)
$UO_2Cl_2 \cdot 2L$	$L = Ph_3 N$, $PhCH = NPh$, $Et_2N(C_2H_4OH)$
$UO_2(NO_3)_2 \cdot 2L$	$L = MeNH_2$, $PhNH_2$, 2-$H_2NC_6H_4OH$, Et_2NH, Et_2N
$UO_2(HCO_2)_2 \cdot 2MeNH_2$	
$UO_2(MeCO_2)_2 \cdot 2RNH_2$	$R = Me$, Ph
$UO_2(EtCO_2)_2 \cdot 2MeNH_2$	
$UO_2SO_4 \cdot 2PhNH_2 \cdot 3H_2O$	
$UO_2C_2O_4 \cdot 2PhNH_2 \cdot 2H_2O$	
$UO_2(C_7H_3NO_4) \cdot 2C_3H_7N$	$C_3H_7N = $ allylamine; $C_7H_5NO_4 = $ pyridine-2,6-dicarboxylic acid
$UO_2A_2 \cdot L$	$HA = H(acac)$ with $L = Me_3 N$ and Et_2N; $PhCOCH_2COPh$,
	$C_7H_6O_2$ (tropolone), C_9H_7NO (8-hydroxyquinoline) with $L = PhNH_2$
$[UO_2(S_2CNEt_2)_2(Et_2NH)]$	

Table 32 Complexes of ethylenediamine and other polydentate nitrogen ligands with dioxouranium(VI) compounds.

Ethylenediamine, en	
$UO_2(NO_3)_2 \cdot x$en	$x = 1$ or 2
$UO_2(HPO_4) \cdot$en	
$L = $ 1,2-diaminopropane	
$UO_2(NO_3)_2 \cdot L$	
$L = $ hexamethylenetetramine, $(CH_2)_6N_4$	
$UO_2X_2 \cdot L$	$X = NCS$, $MeCO_2$
Diaminoarenes	
$UO_2Cl_2 \cdot x L$	$x = 1$ or 2, $L = $ 1,2-diaminobenzene
$UO_2Cl_2 \cdot 2L$	$L = $ 1,3- or 1,4-diaminobenzene
$UO_2(NO_3)_2 \cdot 2L \cdot H_2O$	$L = $ 1,2-diaminobenzene
$UO_2(NO_3)_2 \cdot x L \cdot y H_2O$	$x = 3$, $y = 0$, $L = $ 3,4-diaminotoluene;
	$x = y = 2$, $L = $ 2,3-diaminotoluene
$UO_2SO_4 \cdot L \cdot THF$	$L = $ 4,4'-diaminobiphenyl (benzidine)

($An = U$, Np; acac = acetylacetonate) provide examples of seven-coordinate uranium.[612] In the presence of weakly coordinating anions, it is possible to displace the anion from the coordination sphere of the metal (as is proposed for the complex $[UO_2(py)_5][ClO_4]_2$), but formation of ionic salts is not always the outcome. The base-free complex $UO_2(OTf)_2$ (OTf = triflate) is generated from the reaction of UO_3 with triflic acid or its anhydride. Given the weakly coordinating nature of the triflate ion, it might be anticipated that it could readily be replaced by N-heterocyclic ligands. Recrystallization of $UO_2(OTf)_2$ from pyridine, however, merely results in the base adduct

Table 33 Some complexes of N-heterocyclic ligands with dioxouranium(VI) compounds.

Pyridine and substituted pyridines	
$UO_2X_2 \cdot yL$	$X = Cl$, $y = 1$ ($+H_2O$) or 2, $L = py$; $y = 2$, $L = $ 2-Me-, 3-Me- and 2-H_2N-C_5H_4N; $y = 4$, $L = py$
	$X = NO_3$, $y = 1$ ($+H_2O$ or Et_2O) or 2, $L = py$
	$X = OPh$, O(2- or 4-MeC_6H_4), O(4-ClC_6H_4), $y = 1$, $L = py$
	$X = O$(2-MeC_6H_4), $y = 3$, $L = py$
	$X = MeCOCHCOMe$, $MeCOCHCOPh$, $PhCOCHCOPh$, $y = 1$, $L = py$
	$X = MeCOCHCOMe$, $y = 1$, $L = $ 3- or 4-H_2N-, 4-HO-, 3- or 4-$MeCO$-, 3- or 4-NC-, 3-Cl, or 4-Me-C_5H_4N
	$X = S_2CNEt_2$, $y = 1$, $L = py$, 4-Bu^t-, 4-Ph-, 4-$Ph(CH_2)_3$-, or 2-$(CHEt_2)$-C_5H_4N
	$X = C_9H_6NO$, $y = 1$, $L = py$; $C_9H_7NO = $ 8-hydroxyquinoline
$UO_2X \cdot ypy$	$X = SO_4$, $y = 2$ or 3; Cr_2O_7, $y = 2$; dibasic Schiff base anions, $y = 1$
$[UO_2(py)_5](ClO_4)_2$	
Piperidine, $C_5H_{11}CN$	
$UO_2L_2 \cdot C_5H_{11}N$	$HL = CF_3COCH_2CO$(2-C_4H_3S) or n-$C_3F_7COCH_2COBu^t$
Quinolines	
$UO_2X_2 \cdot L$	$X = NO_3$, $L = $ quinoline ($+H_2O$) or 2-methylquinoline ($+H_2O$)
	$X = MeCOCHCOMe$, $L = $ quinoline
	$X = C_9H_6NO$ (8-hydroxyquinolinate), $L = $ quinoline
$UO_2SO_4 \cdot L$	$L = $ 2-methylquinoline
$UO_2Cr_2O_7 \cdot 2L$	$L = $ isoquinoline
$UO_2X_2 \cdot ybipy$	$X = Cl$, NO_3, $MeCO_2$, 2-, 3- or 4-$H_2NC_6H_4CO_2$, $C_8H_4F_3S$ (tta), $y = 1$
	$X = NCS$, $NCSe$, NO_3, $N(CN)_2$, $y = 2$
$UO_2X_2 \cdot yC_{10}H_8N_2$	$X = Cl$, $y = 1$; $X = NO_3$, $y = 1.5$
$UO_2Cl_2 \cdot bipy \cdot xH_2O$	$x = 1$ or 2
$UO_2(OH)X \cdot bipy$	$X = NO_3$, $MeCO_2$
$UO_2X \cdot ybipy$	$X = SO_4$, $y = 1$; $X = Cr_2O_7$, $y = 2$
$UO_2SO_4 \cdot C_{10}H_8N_2$	
$UO_2X \cdot bipy \cdot yH_2O$	$X = SO_4$, $y = 4$; $X = CrO_4$, C_2O_4, $y = 1$

$UO_2(OTf)_2(py)_3$.[613] It is only in the structure of the hydrolysis product, $[\{UO_2(py)_4\}_2(\mu\text{-O})][OTf]_2$, that displacement of the triflate anions is observed.

N-heterocyclic adducts have also been reported for piperidine, quinoline, isoquinoline, imidizoles, dipyridylamines, 2,2-bipyridine, and 1,10-phenanthroline.

Nitriles. Acetonitrile adducts have been reported for several uranyl salts, including the chloride and nitrate. Although lower metal coordination numbers (e.g., five) have been suggested in the formulation of some of these compounds, it is likely that the species contain six- or seven-coordinate metal ions. As illustration of this, the molecular structure of $UO_2Cl_2(MeCN)_2(H_2O)$[614] possesses a seven-coordinate uranium center, with mutually *trans*-acetonitrile and chloride ligands. The adduct $\{(C_{10}H_{21})_4N\}NpO_2Cl_3 \cdot MeCN$ has also been reported.[615]

(iv) Ligands containing anionic group 16 donor atoms

Oxides. The binary actinide oxides include AnO_3, An_3O_8, An_4O_{9-y}, and $AnO_{2\pm x}$ with all of the phases well characterized in the solid state for U. An additional oxide, An_3O_7 has been reported;[616] the distinct stoichiometric tetragonal phase is not prevalent and many reported occurances may in fact correspond to mixtures of the other more stable oxides.[617] In addition to being characterized by X-ray diffraction studies, X-ray photoelectron spectroscopy, and optical and X-ray absorbance techniques have also been used to distinguish the phases.[196,618,619]

Although there have been numerous reports of analogous phases that may contain Np^{VI} and Pu^{VI}, and there is evidence for some of them in the oxidation of mixed U, Np, Pu oxide fuels, most have not been structurally characterized or otherwise confirmed. For example, neptunium trioxide has been reported, and is thought to be the solid formed from neptunyl hydroxide precipitates (formulated as either $NpO_3 \cdot 2H_2O$ or $NpO_2(OH)_2 \cdot H_2O$).[620,621] Plutonium trioxide has only been observed in the gas phase when PuO_2 or $(U, Pu)O_2$ are oxidized.[622] Reported preparations of Np_3O_8, yielded instead Np_2O_5, or mixtures of Np_2O_5 and NpO_2.[499,623] The superstoichiometric $PuO_{2\pm x}$ was reported to contain Pu^{VI};[193] however, more recent X-ray absorbance studies suggests that both the hydrated solid and solution suspension contain Pu^V.[500]

Uranium trioxide has been very well studied in part because of its applications in the nuclear fuel cycle. It occurs naturally in pitchblende and is generally prepared by thermal decomposition of oxide hydrates and uranyl salts or oxidation of lower oxides or halides.[624] The trioxide has been isolated in six well-defined stoichiometric modifications as well as a substoichiometric $UO_{2.9}$. The alpha phase comprises a three-dimensional network of chains of bicapped hexagonal prisms of UO_8 polyhedra and it contains structural features observed in numerous mineral phases. The U—O bond length along the chains are 2.083 Å and the bonds in the pentagonal chains are 2.07 to 2.7Å. This coordination geometry is also observed in molecular hydroxides. The uranyl type bond length is 2.08 Å and the bonds in the hexagonal plane range from 2.03 Å to 2.80 Å in length. The trioxide decomposes into lower oxides prior to melting or subliming.

Uranium trioxide is a key precursor to UF_4 and UF_6, which are used in the isotopic enrichment of nuclear fuels.[625–627] It is also used in the production of UO_2 fuel,[628] and microspheres of UO_3 can themselves be used as nuclear fuel. Fabrication of UO_3 microspheres has been accomplished using sol-gel or internal gelation processes.[629–632] Finally, UO_3 is also a support for catalytic oxidative destructive of organics.[633,634]

Triuranium octaoxide, U_3O_8, is also found in pitchblende. The common centered orthorhombic structure contains staggered chains of UO_7, similar to molecular uranyl species. XPS studies of U_3O_8 have indicated the presence of two oxidation states, U^{IV} and U^{VI}, in a 1:2 ratio, respectively.[635] The relatively flexible structure and mixed valency allows for super- and substoichiometric phase formation, depending on the temperature and O_2 partial pressure during preparation. The preparation of U_3O_8 has been accomplished by thermal decomposition of urananites, oxide hydrates and uranyl salts. In addition to the varying U/O ratios, U_3O_8 has been found to exist in at least five crystalline forms.

Industrially, U_3O_8 has been shown to be active in the decomposition of organics, including benzene and butanes,[636,637] and as supports for methane steam reforming catalysts.[638] In the nuclear fuel industry, U_3O_8 is an oxidation product of UO_2 (SIMFUEL), and thus a major component of spent fuel rods.[639,640] The density of U_3O_8 is significantly less than that of UO_2 and as a result, the production of U_3O_8 in nuclear fuel can lead to the destruction of the UO_2 pellet by pulverization. Triuranium octaoxide is used in the initial production of UO_2 pellets for fuel,[641,642] in the manufacturing of MOX (mixed oxide) pellets,[643] as well as a dispersive nuclear fuel itself.[644]

Two oxygen-rich UO_{2+x} phases, U_3O_7, and U_4O_9 contain U^{VI} units with unusual coordination geometries. The beta form of triuranium heptaoxide, formed by the low temperature oxidation of UO_2, has an anion excess defect structure. The tetragonal, alpha-U_3O_7 is naturally occurring, although rare.[645] The cubic phase U_4O_9 has different coordination geometries of uranium; uranium polyhedra from eight- to eleven-coordinate are found in interesting combinations within the unit cell, including UO_{11} polyhedra and UO_8 square antiprisms (see Figure 47).[646]

The uranyl oxide hydrates are important corrosion products of uraninite and UO_2 in spent nuclear fuel under oxidizing conditions. However, the systematics of the structures had not been well described until the studies reported by Miller *et al.*[505] With the exception of the synthetic $UO_2(OH)_2$ polymorphs, all hydrate crystal structures are based on sheets of edge-sharing uranyl pentagonal or square bipyramids. Only four structural unit chains are required to construct the uranyl oxide hydrate sheets (as well as the structurally similar U_3O_8 sheets). One chain is made up

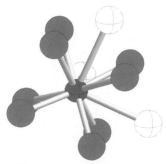

Figure 47 An unusual 11-coordinate U^{IV} within the extended structure of U_4O_9 with a pseudotricapped cubic geometry, where the longer, "capping" oxygens are shown in white (Bevan, Grey *et al. J. Solid State Chem.* **1986**, *61(1)*, 1–7).

of hexagonally coordinated uranyl ions sharing opposing edges. A second chain, composed of pentagonal bipyramids sharing edges and alternating with trigonal vacancies, is present in all other UOH sheets. Another zigzag chain consists of edge-sharing pentagonal bipyramids forming a zigzag chain. And the remaining structural unit is a discontinuous "chain" of rhombic bipyramids. This chain occurs in sheets which contain only four-coordinate uranyl ion and those containing both four- and five-coordinate uranyl ions. Burns, Miller, and Ewing have similarly analyzed a large number of U^{VI} minerals and describe their systematic structural and thermodynamic properties using a number of uranyl polyhedra.[505]

There are a tremendous number of ternary oxides, generally alkali and alkaline earth salts of the form M_2AnO_4 and $MAnO_4$ with many known for U, Np, and Pu. In addition, there are large families of uranates, with the sodium and ammonium salts being the most common. In nuclear fuels production the "yellow cake" contains ammonium diuranate which is actually a mixture of compounds ranging from $(NH_4)_2UO_4$ to $(NH_4)_2U_8O_{25}$, and having the approximate composition $(NH_4)_2U_2O_7$.[647] Also characterized are Li_4AnO_4 for Np, Pu, M_2UO_4, M_2UO_5, $M_2U_2O_7$, $Li_2U_3O_{10}$, $Li_2U_4O_{12}$, $Cs_2Np_3O_{10}$, Li_6AnO_6, for Np, Pu, and $M_2Np_2O_7$, many characterized by Cordfunke and co-workers.[648–654] The phase Li_6AnO_6 and its transition to $Li_2U_3O_{10}$ has been re-evaluated.[655] The structure of the anion in $Li_2U_3O_{10}$ ($Li_2(UO_2)_3O_4$) consists of octahedral $(UO_2)O_4$ and pentagonal bipyramidal $(UO_2)O_5$ groups linked by oxygen bridges.[656] Burns and co-workers have characterized numerous hydrated ternary oxy/hydroxides, particularly those containing Na, NH_4, Ca, and Pb, related to minerals.[657–660]

Hydroxides. The hydrolysis of hexavalent actinides has been studied extensively and recently reviewed.[46,197,661,662] For uranyl, the primary solution and solid state species have been well characterized and include polymers and low dimensional solids. An impressive number of potentiometric titration studies established several solution species that have been confirmed by fluorescence, X-ray absorbance, and X-ray diffraction studies. The products often have interesting structural features, including linked uranyl oxo/hydroxo/aqua polyhedra. Within these structures U centers are generally seven-coordinate, with pentagonal bipyramidal geometry, but also may be six- or eight-coordinate with rhombic and hexagonal bipyramidal geometries.

At approximately micromolar and lower solution concentrations, the initial hydrolysis product is $UO_2(OH)^+$, which likely has four waters in addition to the hydroxide in the equatorial plane and a pentagonal bipyramidal geometry.[663,664] The hydrolysis proceeds stepwise to form the bishydroxo, $UO_2(OH)_2$, which has very low solubility. At higher uranium concentrations the first hydrolysis product is the dimer, $(UO_2)_2(OH)_2^{2+}$, with two bridging hydroxides and pseudopentagonal bipyramidal coordination of the dioxo, hydroxo, and aquo oxygens.[197,665] Further hydrolysis products include the trimeric uranyl hydroxide complexes $(UO_2)_3(OH)_5^+$ and $(UO_2)_3(OH)_4^{2+}$.[197] A triply bridging oxygen has been identified in an adamantane-like trimer of this type, the species $[(UO_2)_3(\mu_3\text{-}O)(\mu_2\text{-}OH)_3]^+$, formed in the sol-gel process.[666] There is good evidence for additional trimeric species with the formulas $(UO_2)_3(OH)_7^-$, $(UO_2)_3(OH)_8^{2-}$, and $(UO_2)_3(OH)_{10}^{4-}$.[667]

Solids precipitated from these systems have been formulated based on stoichiometry as $UO_2(OH)_2$ or various hydrated $UO_3 \cdot xH_2O$ phases, but their structural formulas can be very complicated. Several researchers have recently analyzed these hydroxides and related weathered minerals, such as schoepite $([(UO_2)_8O_2(OH)_{12}](H_2O)_{12})$, metaschoepite, and becquerelite $(Ca[(UO_2)_6O_4(OH)_6] \cdot 8H_2O)$, and have described useful structural classifications.[505,668–673] The schoepite structure, for example, consists of $(UO_2)_8O_2(OH)_{12}$ sheets of edge- and corner-sharing uranyl pentagonal bipyramids that are hydrogen-bonded to each other through interstitial waters.[674] There are also a large number and variety of mixed ligand hydroxide species, particularly for carbonate and other oxoanion complexes.

There are a number of structurally interesting mixed-ligand uranyl hydroxides. For example, the basic compound of composition $Zn(UO_2)_2SO_4(OH)_4 \cdot 1.5H_2O$, has a structure based on chains of $UO_2(OH)_3O_2$ pentagonal bipyramids containing tridentate bridging OH^- groups. Species of this type have also been studied in solution, but the complexity of the system has precluded structural characterization.[673] There are many hydrated binary and ternary uranium oxides, such as the uraninites, that contain uranyl hydroxide complexes within their structure.

There is much less known about the hydrolysis of the transuranic ions. Data indicate they hydrolyze at higher pH, approximately 5–6 for Pu^{VI}, and form the stepwise hydroxide products $AnO_2(OH)^+$ and $AnO_2(OH)_2$. Tetrahydroxide species have been suggested, but not yet well characterized. Polymerization is less pronounced than for uranyl, with dimers, $(AnO_2)_2(OH)_2^{2+}$, indicated in Pu^{VI} and Np^{VI} solutions of approximately 0.1 mM and higher concentrations. Only one trimeric species, $(NpO_2)_3(OH)_5^+$, is suggested for Np^{VI}; and none have been identified for Pu^{VI}.

Peroxides. Peroxide has been used on a large scale in the production of "yellow cake" in uranium processing. There is a simple U^{VI} peroxide, $UO_4 \cdot 2H_2O$, which is probably of the form $[(UO_2)(O_2)(H_2O)]$. Hydrated salts of peroxo complex anions, such as $Na_4AnO_2(O_2)_3 \cdot 9H_2O$, have been prepared. The anion has approximate D_{3h} symmetry, common to many actinyl oxoanion complexes, with three coplanar peroxide groups surrounding the linear uranyl group.[675] Ternary uranyl peroxide complexes containing triphenyl phosphine oxides and related ligands $[U(O)(O_2)L_2)] \cdot H_2O$, $L = Ph_3PO$, Ph_3AsO, $pyNO$ have been prepared.[676] Uranyl complexes with bridging peroxide, such as $M_6[(UO_2)_2(\mu\text{-}O_2)(C_2O_4)]$, $M = Na$, K, NH_4, have been prepared by combining the peroxide, $UO_4 \cdot 2H_2O$, with oxalates, carbonates, and other salts.

Carbonates. Hexavalent actinide carbonates have been very thoroughly studied by a variety of solution and solid state techniques. These complexes are of interest not only because of their fundamental chemistry and environmental behavior, but also because of extensive industrial applications, such as in uranium mining and nuclear fuel production and reprocessing. Uranyl carbonates are very soluble, very stable, and can be readily precipitated to produce powders suitable for industrial scale transformations.

A general feature seen in all actinyl carbonate structures is that the linear triatomic AnO_2 unit forms the axis of a hexagonal bipyramidal coordination polyhedron and the oxygen atoms of the carbonato ligand are arrayed about the equator. In solution, carbonato complexation is stepwise with increasing pH and carbonate concentration to yield the mono-, bis-, and triscarbonato species, $AnO_2(CO_3)$, $AnO_2(CO_3)_2^{2-}$, $AnO_2(CO_3)_3^{4-}$ (for $An = U$, Np, and Pu). For uranyl there is also a great deal of evidence for additional polymeric species and $(UO_2)_2(CO_3)(OH)_3^-$, $(UO_2)_3O(OH)_2(HCO_3)^+$, and $(UO_2)_{11}(CO_3)_6(OH)_{12}^{2-}$ under conditions of high metal ion concentration or high ionic strength.[197] In the solid state, $AnO_2(CO_3)$, $M_6(AnO_2)_3(CO_3)_6$, and $M_4AnO_2(CO_3)_3$ are well characterized for uranium. The analogous neptunium and plutonium solids are not as well described, and only the triscarbonato complex is known for americium. (A very high carbonate concentration has been used to stabilize this high oxidation state for Am). Bicarbonate complexes are, at most, minor species. Those that had been reported previously are now generally believed to be ternary carbonato, hydroxo species, or the known pure carbonates.

The triscarbonato species have hexagonal bipyramidal geometry and can be readily precipitated to form salts with monovalent cations. Single crystal X-ray diffraction studies have been reported for a large number of this type of uranyl carbonate and a few of the neptunyl analogues. The anion in the $M_4AnO_2(CO_3)_3$ and $M'_2AnO_2(CO_3)_3$ salts, where M is an alkali metal or other monovalent cation and M' is a alkali earth or other divalent cation, essentially has the same coordination geometry, with approximately D_{3h} symmetry, where three bidentate carbonate ligands lie in the plane perpendicular to the actinyl axis (see Figure 48). Structural parameters vary little among the many compounds, with An=O bond distances of 1.7–1.9 Å, and An—O carbonate bond lengths of 2.4–2.6 Å. Not only are the triscarbonato uranyl complexes among the most studied synthetic compounds, they are also the basis for numerous naturally occurring minerals, such as andersonite ($Na_2CaUO_2(CO_3)_4 \cdot nH_2O$), bayleyite ($Mg_2UO_2(CO_3)_4 \cdot nH_2O$), and liebigite ($Ca_2UO_2(CO_3)_3 \cdot 10H_2O$), to name a few.[505,677] Uranyl carbonate units can be linked together via counter ions or hydrogen bonds or to form polyhedra clusters and sheets as exemplified in new calcium uranyl carbonates.[678]

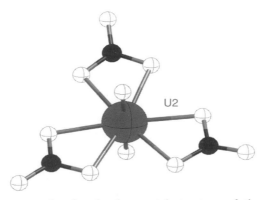

Figure 48 The triscarbonato uranyl anion in the crystal structure of the hydrated guanadinium salt, showing the hexagonal bipyrimadal coordination of U^{VI} that is common to most uranyl oxo anion complexes (Anderson, Bombieri *et al. J. Chem. Soc., Dalton Trans.* **1972**, 2059).

Analogous Np^{VI} and Pu^{VI} carbonates can be assumed to be isostructural with the well-characterized uranyl compounds, albeit with slightly shorter bond distances to reflect the actinide contraction. For the triscarbonato complexes of Np^{VI} this has been confirmed by the structures of $K_4NpO_2(CO_3)_3$ by single-crystal structure determinationof $[(CH_3)_4N]NpO_2(CO_3)_3$.[679,680]

For uranyl, the biscarbonato species, which is likely seven-coordinate, is in equilibrium with the hexakiscarbonato trimer, $(UO_2)_3(CO_3)_6^{6-}$, and is therefore prevalent only at relatively low uranium concentrations.[681] The trimer, $(UO_2)_3(CO_3)_6^{6-}$, was inferred from NMR data and other solution methods until being structurally characterized using EXAFS and single crystal X-ray diffraction (Figure 49).[682] The anion in $[C(NH_2)_3]_6[(UO_2)_3(CO_3)_6] \cdot 6.5H_2O$ possesses nearly ideal D_{3h} symmetry in which the three uranyl axis are perpendicular to the plane defined by the six carbonates and the three uranium atoms.[682] A number of solids previously believed to contain the monomeric biscarbonato uranium(VI) complex may in fact also be salts of this anion. Similar to the hexavalent hydroxides, polymer formation, or at least stability of polymers, appears to decrease dramatically across the series. Mixed actinyl trimers, such as $[(UO_2)_2(NpO_2)(CO_3)_6]$, have been identified from NMR and optical absorbance studies for both Np and Pu;[683,684] however, $(NpO_2)_3(CO_3)_6^{6-}$ and $(PuO_2)_3(CO_3)_6^{6-}$, have not been isolated in the solid state.

The solid state structure of rutherfordine, UO_2CO_3, has been determined from crystals of both the natural mineral and synthetic samples (see Figure 50). It has a layered structure in which the local coordination environment of the uranyl ion is hexagonal bipyramidal, with the uranyl units perpendicular to the orthorhombic plane. Each uranium atom forms six equatorial bonds with the oxygen atoms from two bidentate and two monodentate carbonates. The neptunyl and plutonyl analogs are isostructural with UO_2CO_3 based on Rietveldt analysis of powder X-ray diffraction data.

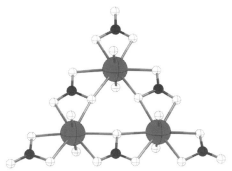

Figure 49 The pseudo D_{3h} symmetric trimeric uranyl carbonate anion in the crystal structure of $[C(NH_2)_3]_6[(UO_2)_3 (CO_3)_6] \cdot 6.5H_2O$ (Allen, Bucher *et al. Inorg. Chem.* **1995**, *34*, 4797–4807).

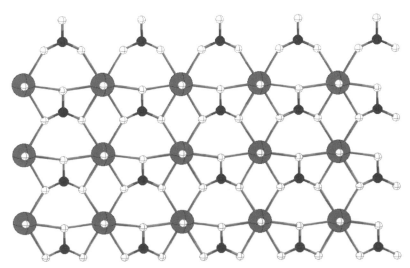

Figure 50 Uranyl carbonate layers in the crystal structure of the mineral rutherfordine, UO_2CO_3 (Finch, Cooper *et al. Cannadian Mineralogist* **1999**, *37*, 929–938).

In addition to the pure carbonates and hydroxy carbonates, there are mixed-ligand uranyl carbonates, such as in $K_4[UO_2(O_2)(CO_3)_2]$ and $K_3[UO_2F_3(CO_3)]$.

Nitrates. The actinyl nitrates are weak complexes formed in the presence of excess nitrate. For uranyl, the mono-, bis-, and trisnitrato actinyl species, $AnO_2(NO_3)^+$, $AnO_2(NO_3)_2$, $AnO_2(NO_3)_3^-$ have been inferred from spectrophotometric solution studies. For neptunyl only the mono- and bis-, and for plutonyl only the mono-, species are significant. The coordination geometries of these complexes are presumably the same as the related carbonates, although some of the species may be protonated in the nitrate system. Indeed, a protonated tris species $HUO_2(NO_3)_3$ is reported; but as a minor species and not in most studies. Mixed tributylphosphate (TBP), nitrate complexes have been widely studied, including an EXAFS study of the structural changes as the actinyl species are reduced to An^{IV}.[685]

The hydrated solids, $AnO_2(NO_3)_2 \cdot xH_2O$, are easily obtained and are very common An^{VI} starting materials. The orthorhombic uranyl nitrate hexahydrate is prepared from dilute nitric acid solutions, and the trihydrate from concentrated acid. Analogous neptunyl and plutonyl trihydrate and hexahydrate solids have also been studied. Uranyl nitrates are used on an industrial scale in the nuclear fuel cycle; for example, for extraction by TBP and other separations processes. Trisnitrato actinyl salts $MAnO_2(NO_3)_3 \cdot xH_2O$ where M is a monovalent cation are well known.[468,686] The anions in this class of compounds are isostructural. The complex Np^{VI} anion in $RbNpO_2(NO_3)_3$, for example, consists of a hexagonal bipyramidal arrangement of oxygen atoms about the Np atom with six oxygen atoms from the bidentate nitrate groups in the equatorial plane. Nitrites of the formula AnO_2NO_2 have been prepared and characterized, but quantitative structural data are lacking.

Phosphates. Most actinyl phosphates have been prepared in aqueous solution at low pH and very high phosphate concentrations. These conditions lead to a number of species in equilibria, with most containing one or two partially protonated phosphates. The products are relatively insoluble making structural and spectroscopic characterization of the species even more challenging. The major species identified are $UO_2(PO_4)^-$, $UO_2(HPO_4)$, $UO_2(H_2PO_4)^+$, $UO_2(H_2PO_4)_2$, $UO_2(H_3PO_4)^{2+}$, and $UO_2(H_3PO_4)(H_2PO_4)^+$. A detailed discussion of the complex formation in the uranium–$H_kP_mO_n$ system and its chemical thermodynamics has been reported.[197] Since Np^{VI} and Pu^{VI} are far less stable at low pH and generally have less affinity for oxoanionic ligands than U^{VI}, the species that have been characterized contain less protonated forms of the ligand and the bis ligand complexes are less stable. For Np^{VI}, $NpO_2H_2PO_4^-$, $NpO_2(HPO_4)$, and $NpO_2(HPO_4)_2^{2-}$ have been reported. Only the first two are known for Pu^{VI}.[687,688]

Uranium(VI) phosphates have been widely investigated and can be divided in several structure types: orthophosphates $M(UO_2)_n(PO_4)_m \cdot xH_2O$, hydrogenphosphates $M(UO_2)_n(H_kPO_4)_m \cdot xH_2O$, pyrophosphates $U_mO_nP_2O_7$, metaphosphates $(UO_2)_n(PO_3)_m \cdot xH_2O$, and polyphosphates $(UO_2)_n(P_aO_b)_m \cdot xH_2O$.[689] They have commonly been prepared from dissolution of uranium metal in phosphoric or mixed acid solutions, by addition of phosphate to aqueous solutions of the nitrates, or by decomposition.[689,690]

More recently hydrothermal conditions and amine or ammonium structure-directing agents have been used to prepare $[NHEt_3][(UO_2)_2(PO_4)HPO_4]$ and $[NPr_4][(UO_2)_3(PO_4)HPO_4)_2]$ (Figure 51).[233] The structures of these compounds contain infinite chains of edge-sharing UO_7 pentagonal bipyramids cross linked by bridging PO_4 tetrahedra to form two-dimensional anionic sheets. This same method yielded the first uranyl phosphate with a three-dimensional open framework structure, $[Et_2NH_2]_2(UO_2)_5(PO_4)_4$ (Figure 52).[691]

Several salts of formula, $M(UO_2)_n(PO_4)_m \cdot xH_2O$, where $M = H^+$, M^+ or M^{2+}, have also been characterized. Some of the latter compounds are identical with natural minerals. In fact the Ca salt, the very common mineral autunite $Ca[(UO_2)(PO_4)_2] \cdot 11H_2O$, has recently been redetermined.[692] This same group has prepared the Cs, Rb, and K salts by hydrothermal methods. The structures consist of sheets of phosphate tetrahedra and uranyl pentagonal bipyramids, $[(UO_2)(PO_4)]^-$. These sheets are connected by a uranyl pentagonal bipyramid in the interlayer that shares corners with two phosphate tetrahedra on each of two adjacent sheets and whose fifth equatorial ligand is water.[693] The hydrogen uranyl phosphates readily exchange the hydrogen with alkali or alkaline earth metals.

The tetrahydrate, $H(UO_2)(PO_4) \cdot 4H_2O$, is reported to form three different polymorphic modifications at room temperature.[694] The geometry about the uranyl ion in $K_4(UO_2)(PO_4)_2$ is tetragonal bipyramidal with four oxygen atoms in the equatorial plane from four tetrahedral phosphate groups, making up a $[(UO_2)(PO_4)_2]_n^{4n-}$ layer.[695] The neutral compound, $(UO_2)_3(PO_4)_2 \cdot xH_2O$, has been synthesized as mono-, tetra-, and hexahydrate. Orthorhombic $(UO_2)_3(PO_4)_2 \cdot 4.8H_2O$ was prepared by addition of 0.5 M uranyl nitrate to 0.36 M H_3PO_4 at 60 °C and pH 1.[696]

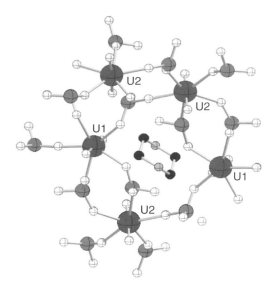

Figure 51 Phosphate coordination geometries and cocrystallized template from the structure-directed preparation of a layered uranyl phosphate (Francis, Drewitt *et al. Chem. Commun.* **1998**, 279–280).

A few uranyl metaphosphates have been reported. $UO_2(PO_3)_2$ is formed in 85% H_3PO_4 at 300–350 °C or by thermal decomposition of $UO_2(H_2PO_4)_2$ at 800–850 °C. In addition, there are uranyl phosphites, $UO_2HPO_3 \cdot xH_2O$, $x = 0,3$, and hypophosphites $UO_2(H_2PO_2)_2 \cdot xH_2O$. A new phosphite with a three-dimensional structure was recently prepared using hydrothermal methods.[697]

The trihydrate of uranyl dihydrogenphosphates, $UO_2(H_2PO_4)_2 \cdot 3H_2O$, has been obtained from a suspension of $HUO_2PO_4 \cdot 4H_2O$ in 85% phosphoric acid after stirring for several days. The monohydrate is also known. It has been reported that $UO_2HPO_4 \cdot 3H_2O$ behaves as a solid ionic conductor because of H^+ mobility across the hydrogen bonds network in the compound.[698] Pyrophosphates, $(UO_2)_2P_2O_7$ and $UO_2H_2P_2O_7$ have been studied.[699]

Uranium(VI) polyphosphates, $(UO_2)_2(P_3O_{10})_2 \cdot xH_2O$, were obtained by precipitation of a uranyl solution with $Na_5P_3O_{10}$ $M_2UO_2P_2O_7$, with M = alkali metal, have been synthesized by heating

(a)   **(b)**

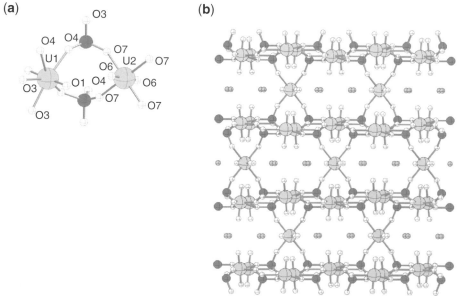

Figure 52 Coordination environment (a) about uranyl and within and (b) between uranyl phosphate layers in the first organically templated open-framework uranium phosphate, $[Et_2NH_2]_2[(UO_2)_5(PO_4)_4]$ (Danis, Runde *et al. Chem. Commun.* **2001**, 2378–2379).

uranyl nitrate in the presence of alkali metal pyrophosphates and ammonium dihydrogenphosphate. The new mixed valence $U(UO_2)(PO_4)_2$ has been synthesized and characterized spectroscopically showing the absence of pyrophosphate and the existence of the dioxocation unit, UO_2^{2+}, as one of the two independent U atoms. Bidentate phosphates are connecting the chains generating a three-dimensional network.[700]

Several Np^{VI} and Pu^{VI} phosphates have been prepared well-characterized. They have not been characterized by single-crystal X-ray diffraction; thus their coordination chemistry has not been fully described.

Arsenates. Similar to the phosphates, the actinyl arsenate AsO_3, AsO_4, and As_2O_7 units bind to the actinide center to produce open framework and extended structures. The arsenates have not been studied in solution, as expected given their low solubilities. Examples of recently characterized arsenates include $(UO_2)_3(AsO_4)_2$, $(UO_2)_2As_2O_7$, and $UO_2(AsO_3)_2$.[689,699,701]

Molybdates. Common types of molybdates are $M_2(UO_2)(MoO_4)_2 \cdot xH_2O$ and $M_6(UO_2)(MoO_4)_4 \cdot xH_2O$, where M is a monovalent cation (Na, K, Rb, Cs, NH_4.).[702,703] Structures of these compounds include linked MoO_4 tetrahedra and uranyl square bipyramids and/or uranyl pentagonal bipyramids. Some related and characterized molybdates include alkyl amine salts, $(C_6H_{14}N_2)_3(UO_2)_5(MoO_4)_8 \cdot 4H_2O$ and $(C_2H_{10}N_2)[(UO_2)(MoO_4)_2]$, which contain sheets of molybdate linked uranyl bipyramids, and the amine cations in the interlayers.[704] A silver salt $(Ag_6(UO_2)_3(MoO_4)_5)$ has been reported; it is one of the dozens of known inorganic uranyl compounds containing sheets of polyhedra that contain trimers of uranyl pentagonal bipyramids that are connected only by the sharing of vertices with other polyhedra.[705] Transuranic analogues formulated as NpO_2-$MoO_{42} \cdot xH_2O$ and $PuO_2MoO_{42} \cdot xH_2O$ and Pu(M = Np, Pu) have also been prepared.[706,707]

The common six- and seven-coordinate uranyl-based polyhedra are also observed in the extended structures of uranyl molybdates containing Mo_2O_7 units. For example, the mineral iriginite, $[(UO_2)Mo_2O_7(H_2O)_2](H_2O)$, and the hydrothermally prepared $[(UO_2)Mo_2O_7(H_2O)_2]$, which is formed under more basic conditions, contain electroneutral sheets of $[(UO_2)Mo_2O_7(H_2O)_2]$.[705]

Sulfates and sulfites. Mono- and bis-sulfate complexes of actinyl ions, AnO_2SO_4 and $AnO_2(SO_4)_2^{2-}$, are generally prepared from acidic solutions. The geometry about the actinide metal center is pentagonal bipyramidal from actinyl, sulfato, and aquo oxygen atoms. A tris(sulfato) complex has been reported, but it is very weak if it does exist. Ternary hydroxo, sulfato complexes have been reported for uranyl, but they have not been structurally characterized.

These solution species can be precipitated to prepare uranyl sulfate hydrates. The anhydrate and additional hydrates are formed after subsequent dehydration or other treatments to complete the series, $UO_2SO_4 \cdot xH_2O$, where $x = 0, 0.5, 1, 2, 2.5, 3, 3.5$ and $UO_2SO_4)_2 \cdot xH_2O$, where $x = 0, 4, 8$. The same types of compounds have been prepared for Np^{VI} and Pu^{VI}, but they are not as numerous or as well characterized. Like the carbonates, the sulfates have been used in leaching uranium from ore. After acidic dissolution, addition of ammonia gives a precipitate thought to be $(NH_4)_2(UO_2)_2SO_4(OH)_4 \cdot nH_2O$. While the solution ternary hydroxo, sulfato complexes are not yet well known, the related solid, $M_2(UO_2)_6(SO_4)_3(OH)_{10} \cdot 8H_2O$ (M = divalent cation), the mineral zippeite, has been known since the early nineteenth century.[708] New uranyl sulfates with extended structures containing both inner- and outer-sphere sulfates continue to be discovered.[709,710] Novel hydrothermal synthetic methods, such as the organic templating that has produced interesting uranyl fluorides and phosphates has also been applied to prepare uranyl sulfates, such as $[N_4H_{12}](UO_2)_6(H_2O)(SO_4)_7$.[711] The fluorosulfate, $UO_2(SO_3F)_2$, is obtained by treating $UO_2(MeCO_2)_2$ with HSO_3F.[254] A large number of ternary U^{VI} sulfates of the general formula $M(UO_2)_m(SO_4)_n \cdot xH_2O$, where M = alkali metals, ammonia or transition metals (Mn, Cd, Hg), have been reported. A layered structure is observed for $(NH_4)_2UO_2(SO_4)_2 \cdot 2H_2O$ with local pentagonal bipyramidal coordination around the uranium atom, and bridging sulfate groups joining the uranyl polyhedra.[712] In $K_4UO_2(SO_4)_3$ each uranium in the pentagonal bipyramid is coordinated to five oxygen atoms from four sulfate groups in the equatorial plane.[713] Analogous sulfates have been isolated for Np^{VI} and characterized by IR and X-ray diffraction.[714]

Similar to the sulfates, mono- and bis(sulfites) have been prepared in solution and precipitated as the solids, UO_2SO_3 and $UO_2SO_3 \cdot 4.5H_2O$. The analogous Np and Pu compounds have not been structurally characterized.

Selenates, selenites, and tellurates. Uranyl selenates and selenites have been prepared and can be expected to have the same coordination features as the sulfates and sulfites. In addition to the selenate, UO_2SeO_4, the ternary selenite $(UO_2)_2(OH)_2(SeO_3)$, has been reported to form in uranyl, selenous acid solutions.[715,716] A uranyl tellurate, UO_2TeO_4, has been reported, but no tellurites.[197]

Perchlorates, iodates. Actinyl(VI) iodates are generally prepared via hydrothermal syntheses, rather than solid state reactions owing to the thermal disproportionation of iodate at high-temperature. In the absence of additional cations, $UO_2(IO_3)_2$, $AnO_2(IO_3)_2(H_2O)$ (An = U, Np), and $AnO_2(IO_3)_2 \cdot H_2O$ (An = Np, Pu) form under both mild and supercritical hydrothermal conditions.[717,718] $UO_2(IO_3)_2$ is one-dimensional, and contains chains of edge-sharing UO_8 hexagonal pyramids. $AnO_2(IO_3)_2(H_2O)$ (An = U, Np) and $AnO_2(IO_3)_2 \cdot H_2O$ (An = Np, Pu) are both layered and contain AnO_7 pentagonal bipyramids linked by iodate anions. The structure of $AnO_2(IO_3)_2 \cdot H_2O$ (An = Np, Pu) is also polar owing to the alignment of the stereochemically active lone-pair of electrons on the iodate anions. The incorporation of alkali metal, alkaline-earth metal, and main group cations into these hydrothermal syntheses allows for the isolation of structurally complex compounds exemplified by $M_2[(UO_2)_3(IO_3)_4O_2]$ (M = K, Rb, Tl) and $M[(UO_2)_2(IO_3)_2O_2]$ (M = Sr, Ba, Pb).[717] The former compounds contain one-dimensional chains of edge-sharing UO_6 tetragonal bipyramids and UO_7 pentagonal bipyramids. The latter compounds contain one-dimensional ribbons of distorted UO_7 pentagonal bipyramids that share edges. In both cases, the edges of the one-dimensional chains are terminated by iodate anions. Finally, UO_2^{2+} can be used to stabilize the new iodate anion, tetraoxoiodate(V), IO_4^{3-}, in $Ag_4(UO_2)_4(IO_3)_2(IO_4)_2O_2$.[717] The uranium oxide substructure of $Ag_4(UO_2)_4(IO_3)_2(IO_4)_2O_2$ contains ribbons of edge-sharing UO_8 hexagonal bipyramids and UO_7 pentagonal bipyramids.

Crystals of uranyl perchlorate, $UO_2(ClO_4)_2 \cdot xH_2O$, have been obtained with six and seven hydration water molecules. The uranyl is coordinated with five water molecules in the equatorial plane with a distance of U—O(aqua) of 2.45 Å. The unit cells contain two $[ClO_4]^-$ and one or two molecules of hydration water held together by hydrogen bonding.[719] Because perchlorates and perchloric acid solutions (similar to nitrates) are very common starting materials in actinyl chemistry there are numerous types of perchlorate containing structures. Examples are provided by the diperchlorates of uranyl complexes of neutral donor ligands, as in the phosphoramide complex.[720]

Carboxylates. Uranyl complexes formed with numerous aminocarboxylate ligands, such as IMDA, NTA, HEDTA, EDTA, CDTA, or DTPA, have been prepared in water and other very polar solvents and characterized predominantly by optical and NMR spectroscopy. Mixed ligand complexes, MLL' have also been reported, with the secondary ligands including resorcinols, salicylic acids, amines, and nucleosides, (e.g. adenosine, guanosine, adenine, etc.) EDTA and mixed EDTA, hydroxide complexes of Pu(VI) have been reported; but these complexes slowly reduce to corresponding Pu(IV) compounds. In complexes of malonic acid derivatives with the uranyl ion, bidentate bonding is unaffected by the substitution of the alpha carbon. X-ray single crystal diffraction of the complex $(C_4H_{12}N_2) \cdot [UO_2(C_4H_4O_4)_2(H_2O)] \cdot 2H_2O$ shows that alkyl substitution on the alpha carbon of malonic acid has no affect on the coordination environment around the uranium. In all cases, pentagonal bipyramidal coordination is observed around the uranium with the ligand forming a six-membered ring. To achieve the desired coordination around the uranium, one water molecule is complexed in the inner sphere. The six-membered rings are not planar in nature; in fact, the alpha carbons stray away from planarity in a *trans* orientation to one another.[721]

In some cases, malonic acid has been shown to bridge between two uranyl ions. In the structure of $(C_{10}H_{26}N_2)[(UO_2)_3(C_7H_{10}O_4)_5] \cdot 2H_2O$, the two bridging malonic acids clearly create pentagonal bipyramidal geometry around the uranium. The malonic acid bonds in two possible ways: either as a bidentate ligand or as a tridentate ligand where one acid molecule has 1,3-bridging between two uranyl moeities and another is mondentate to the uranyl group.[721]

Malonic acid has the ability to form several different polymeric complexes with uranyl, many of which have been studied by X-ray single crystal diffraction. A zig-zag polymer is shown in Figure 53 having two different coordination environments around the uranium atoms. In this polymeric chain, two closely-neighboring uraniums are bridged by two tetradentate malonic acid ligands. One of the carboxylic acids has μ^2 bonding forming an oxo-bridge between the two uraniums. The other carboxylic acid forms a monodentate bond to one of the uraniums. Also bound to each uranium is a tridentate malonic acid molecule with one side having an η^2 bond through one of the carboxylic acids and the other having monodentate coordination to the third uranium of the asymmetric group. The two uraniums which are doubly bridged have a hexagonal bipyramidal coordination environment while the third uranium has a pentagonal bipyramidal environment.[721]

Another version of the previous polymeric chain has been studied with X-ray crystallography, where the only difference lies in the uranyl with the pentagonal bipyramidal coordination.

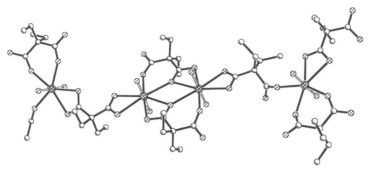

Figure 53 Crystal structure of $(C_{10}H_{26}N_2)[(UO_2)_3(C_7H_{10}O_4)_5]\cdot 2H_2O$ depicting the polymeric coordination of the U^{VI} (Zhang, Collison *et al. Polyhedron* **2002**, *21*, 81–96).

In another case, the bidentate malonic acid is replaced with two *trans*-water molecules, leaving the coordination environment the same, but causing the *trans*-oxo ligands to rotate 90°.[722]

Alkoxides. Exposure of the pentavalent alkoxide complex $U(OCH_2CH_3)_5$ to oxygen was first reported to produce a small amount of a volatile and thermally unstable red liquid byproduct.[723] Elemental analysis of the compound was consistent with the stoichiometry $U(OCH_2CH_3)_6$, and molecular weight determinations suggested a monomeric structure in benzene. Improved yields of $U(OCH_2CH_3)_6$ have been achieved using other reaction routes (see Equations (45) to (47)):

$$NaU(OCH_2CH_3)_5 + 1/2\ Br_2 \longrightarrow U(OCH_2CH_3)_6 + NaBr \quad (3.45) \qquad (45)$$

$$NaU(OCH_2CH_3)_6 + 1/2\ Pb(CO_2CH_3)_4 \longrightarrow U(OCH_2CH_3)_6 + NaCO_2CH_3 + 1/2\ Pb(CO_2CH_3)_2 \qquad (46)$$

$$U(OCH_2CH_3)_5 + NaOCH_2CH_3 \longrightarrow NaU(OCH_2CH_3)_6$$

$$2\ NaU(OCH_2CH_3)_6 + (C_6H_5CO_2)_2 \longrightarrow 2\ U(OCH_2CH_3)_6 + 2\ NaO_2CC_6H_5 \qquad (47)$$

Reduction of $U(OCH_2CH_3)_6$ to the more stable $U^V(OCH_2CH_3)_5$ occurs in the presence of $U^{IV}(OCH_2CH_3)_4$, ethanethiol, diethylamine, or ethyl cyanoacetate in ether solution.

Other homoleptic uranium(VI) hexakisalkoxides are prepared from alcohol exchange reactions of $U(OCH_2CH_3)_6$ with other alcohols (OR: R = Me, Pr, Pr^i, Bu).[723]

Bradley reported that homoleptic uranium hexakis(alkoxide) complexes coordinated by secondary and tertiary alkoxides ($U(OR)_6$: R = Pr^i, Bu^s, Bu^t) were produced from thermal disproportionation of $UO(OR)_4$ (*vide supra*).[724] $U(OMe)_6$ was initially prepared from oxidation of $U^V(OMe)_5$ in the presence of benzoyl peroxide.[723] Interest in a more convenient synthetic route to $U(OMe)_6$ was stimulated by its potential use in uranium isotope separation, which can be achieved with a CO_2 laser.[725–727] Facile syntheses of $U(OMe)_6$ were reported by different groups (see Equations (48) to (51)):[725,726,728]

$$UF_6 + 6\ NaOCH_3 \xrightarrow{\ -78\,°C,\ CH_2Cl_2\ } U(OCH_3)_6 + 6\ NaF \qquad (48)$$

$$UF_6 + 2\ CH_3Si(OCH_3)_3 \xrightarrow{\ -78\,°C,\ CH_2Cl_2\ } U(OCH_3)_6 + 2\ CH_3SiF_3 \qquad (49)$$

$$UCl_4 + 6\ LiOCH_3 \longrightarrow Li_2U(OCH_3)_6 + 4\ LiCl$$

$$2\ NaU(OCH_2CH_3)_6 + (C_6H_5CO_2)_2 \longrightarrow 2\ U(OCH_2CH_3)_6 + 2\ NaO_2CC_6H_5 \qquad (50)$$

$$UF_6 + 6\ Si(OCH_3)_4 \longrightarrow U(OCH_3)_6 + 6\ FSi(OCH_3)_3 \qquad (51)$$

Chemical, vibrational spectroscopic, and infrared multiphoton photochemical properties of $U(OMe)_6$ were reported[725] and the electronic structure and bonding of this compound was investigated by using a combination of He I/He II photoelectron spectroscopy and discrete variational DV-Xα molecular orbital calculations.[729]

The reaction of U(OCH(CH$_3$)$_2$)$_6$ with lithium, magnesium, or aluminum alkyls does not generate uranate(VI) compounds containing uranium–carbon bonds, but rather addition complexes.[544] The structures of these products, (MeLi)$_3$-U(OPr^i)$_6$, (R$_2$Mg)$_3$-U(OPr^i)$_6$ (R = Me, CH$_2$Bu^t, CH$_2$SiMe$_3$), and (Me$_3$Al)$_6$-U(OPr^i)$_6$, were suggested based upon ^{1}H-NMR spectroscopic analysis. There was no evidence that an anionic complex could be prepared from treatment of U(OCH$_2$CH$_3$)$_6$ with NaOCH$_2$CH$_3$.[723]

Mixed uranium(VI) halide/alkoxide products have been described. The compound U(OCH$_2$CH$_3$)$_5$Cl is prepared by reaction of one equivalent of HCl with U(OCH$_2$CH$_3$)$_6$. A thermally unstable mixed halide/alkoxide compound, U(OMe)F$_5$, is prepared from the reaction of UF$_6$ with MeOH in CFCl$_3$ at $-90\,^\circ$C.[730] The methoxyfluorouranium(VI) compounds, U(OCH$_3$)$_n$F$_{6-n}$, $n = 1$–5, are obtained from the reaction of UF$_6$ with Me$_3$Si(OMe) or U(OMe)$_6$.[725] Characterization by ^{1}H- and ^{19}F-NMR spectroscopy suggests that these monomers have six-coordinate uranium centers, and undergo rapid intermolecular ligand exchange.

The first uranyl alkoxide complex, uranyl di-*iso*-amyloxide, was reported in 1952.[731] Conflicting reports ensued regarding the composition of uranyl bisalkoxide compounds. Gilman and co-workers reported that the yellow brown, ether and alcohol soluble uranyl ethoxide, UO$_2$(OCH$_2$CH$_3$)$_2$-3HOCH$_2$CH$_3$, was prepared from a metathesis reaction between anhydrous uranyl chloride and sodium ethoxide, and the bright red ethanol soluble uranyl *t*-butoxide analogue, UO$_2$(OBu^t)$_2$-4HOBu^t, was formed upon oxidation of U^{IV}(OBu^t)$_4$.[732] Bradley and co-workers, however, were unable to reproduce the preparation of UO$_2$(OCH$_2$CH$_3$)$_2$-3HOCH$_2$CH$_3$ using Gilman's synthetic procedure. A compound of stoichiometry UO$_2$(OCH$_2$CH$_3$)$_2$-2HOCH$_2$CH$_3$ was instead prepared from the metathesis reaction between UO$_2$Cl$_2$ and LiOCH$_2$CH$_3$ or ethanol exchange with uranyl methoxide.[724]

In a later report, UO$_2$(OMe)$_2$-MeOH was prepared from solvolysis of uranyl nitrate in alcohol solutions in the presence of tridecylamine as the deprotonating base, although no yield was reported.[733] Alcohol exchange was identified between UO$_2$(OMe)$_2$ and primary alcohols (OR: R = Et, Pr, Bu^i, amyl), generating a new route for the production of uranyl bisalkoxides.[724] However, when UO$_2$(OMe)$_2$ was allowed to react with secondary or tertiary alcohols (OR: R = Pr^i, Bu^s, Bu^t), the exchange was accompanied by redistribution resulting in loss of a uranyl oxo ligand to form UO(OR)$_4$-HOR (along with an insoluble residue), which further disproportionated to U(OR)$_6$ upon heating. The compound assignments were based solely upon elemental analysis.

Subsequent studies in nonalcoholic solvent confirm that redistributive exchange of uranyl oxo and ancillary alkoxide ligands can occur. Reaction of two equivalents of potassium *t*-butoxide with uranyl chloride results in the formation of the trimetallic species [UO$_2$(OBu^t)$_2$][UO(O-Bu^t)$_4$]$_2$.[734] Apparently, redistributive exchange is precluded by steric saturation of the uranyl alkoxide through coordination of a strong Lewis base. The complex UO$_2$(OBu^t)$_2$(Ph$_3$PO)$_2$ is prepared from the reaction of KOBu^t and UO$_2$Cl$_2$(Ph$_3$PO)$_2$ in tetrahydrofuran.[591]

Electron-poor donor alkoxide ligands, such as aryloxides or fluoroalkoxides, also inhibit ligand redistribution. The metathesis reaction of UO$_2$(NO$_3$)$_2$·2THF with sodium nonafluoro-*t*-butoxide yielded the yellow diamagnetic complex UO$_2$[OC(CF$_3$)$_3$]$_2$·2THF.[12] In order to assess the relative influence of steric versus electronic effects on formation of oxo-alkoxide bridged species, a series of binary uranyl alkoxide complexes were studied.[735] The reaction of uranyl chloride with primary alkoxide potassium neopentoxide gives as the sole isolable species a uranium complex generated by replacement of each oxo ligand by two alkoxide groups, U(OCH$_2$CH$_3$)$_6$. A reaction pathway for the redistributive exchange of alkoxide and oxo ligands was proposed. Sterically bulky alkoxide ligands inhibit redistributive exchange. Metathesis reactions of KOR (R = CHPh$_2$, CH(Bu^t)Ph) with uranyl chloride allowed for the isolation of simple monomeric uranyl species, UO$_2$(OCHPh$_2$)$_2$(THF)$_2$ and UO$_2$[OCH(Bu^t)Ph]$_2$(THF)$_2$. Larger aggregates can be isolated utilizing bulky alkoxide ligands. The tetrameric aggregate, [UO$_2$(OCH(Pr^i)$_2$)$_2$]$_4$, was isolated from the reaction of KOCH(Pr^i)$_2$ with uranyl chloride. Mixed ligand uranyl alkoxide compounds have been reported. A chloride alkoxide compound, U$_2$O$_5$Cl(OPr^i)-0.5CH$_3$COOPr^i, was reported in which Bradley's proposed uranyl isopropoxide species, U$_2$O$_5$(OPr^i)$_2$-2HOPr^i (*vide supra*) was reacted with one equivalent of acetyl chloride in refluxing benzene.[724,736] A series of air and moisture stable monothiocarbamate uranyl alkoxide products, [R$_2$NH$_2$][UO$_2$(R$_2$NCOS)$_2$(OR')] (R = Me, Et, Pr; R' = Me, Et), were prepared from the reaction of uranyl chloride trihydrate with a solution in which carbonyl sulfide was bubbled through dipropylamine in ethanol.[737–739] Uranyl phenoxide compounds have also been prepared. In an initial study of binary uranyl aryloxides, compounds UO$_2$(OR)$_2$ (R = Ph, 2-NO$_2$C$_6$H$_4$, 2-ClC$_6$H$_4$) were prepared from refluxing uranyl chloride or uranyl acetate with an excess of the appropriate phenol in xylene.[740] The authors suggested that these species were polymeric, but could form monomeric Lewis base adducts in the presence of the appropriate ligands (pyridine, piperidine, phenanthroline). They also reported that the salts, $M_2UO_2(OC_6H_5)_4$ (M = Na, K) were isolated from a refluxing solution of $UO_2(OC_6H_5)_2$ and two equivalents of MOR in xylene. Subsequently, a series of uranyl aryloxide complexes $(UO_2(O-2,6-Bu^t_2C_6H_3)_2(THF)_2$, $UO_2(O-2,6-Ph_2C_6H_3)_2(THF)_2$, $[UO_2(O-2,6-Cl_2C_6H_3)_2(THF)_2]_2$, and $[UO_2 (O-2,6-Me_2C_6H_3)Cl(THF)_2]_2)$ were prepared via metathesis reactions between uranyl chloride and the appropriate o-substituted phenoxides.[741] The formation of monomeric or dimeric compounds appears to be dictated purely by the steric requirements of the ligands. Alcoholysis of uranyl amide complexes with substituted phenols leads to the formation of neutral $(UO_2(O-2,6-Pr^i_2C_6H_3)_2(py)_3)$ and anionic $[Na(THF)_3]_2[UO_2(O-2,6-Me_2C_6H_3)_4]$ uranyl phenoxide species.[589] A series of pyridine uranyl phenoxide adducts, $UO_2(OR)_2 \cdot n\ C_5H_5N$, were prepared (OR: R = Ph, $n = 1$; R = p-ClC$_6$H$_4$, $n = 1$; R = p-MeC$_6$H$_4$, $n = 1$; R = o-MeC$_6$H$_4$, $n = 3$).[274] A series of pyridinium "ate" species $[pyH]_2[UO_2(OR)_4]$ were also reported.

Thiolates, selenates. Few thiolate complexes of uranyl exist. Those complexes that have been prepared are stabilized by the use of ligands with pendant heteroatom bases that can coordinate to the metal center.[742,743] Reaction of $UO_2(NO_3)_2$ with pyridine-2-thiol and 3-trimethyl-silyl-pyridine-2-thiol results in the formation of the complexes $[HSC_5H_4][UO_2(NO_3)_2(SC_5H_4N)]$ and $[(C_5H_3NS-3-SiMe_3)_2H][UO_2(NO_3)_2(C_5H_3NS-3-SiMe_3)]$, wherein each uranyl contains a 2-mercaptopyridine ligand in the equatorial ligand plane.[744] The latter contains a protonated disulfide counterion derived from coupling of two pyridylthiol anions. When reaction of uranyl nitrate with 2-mercaptopyrimidine or 2-mercapto-4-methylpyrimidine is carried out in the presence of atmospheric oxygen and triethylamine, the binuclear complexes $[HNEt_3]_2[(UO_2)_2(O_2)(SC_4N_2H_3)_4]$ and $[HNEt_3][H(UO_2)_2(O_2)(SC_4N_2H_2Me)_4] \cdot Me_2CO \cdot 0.5Et_3N$ are isolated; these species are unique examples of peroxo-bridged diuranyl compounds.[745] The analogous reaction with 2-mercaptopyridine yields the tetranuclear bridging oxo complex $[HNEt_3]_2[(UO_2)_4(O)_2(SC_5NH_4)_6] \cdot Me_2CO$.[745]

Triflates. The synthesis of $UO_2(O_3SCF_3)_2(H_2O)_n$ was first reported in 1994,[746] and involved the treatment of UO_3 with triflic acid in water. The molecular structure of the complex $UO_2(O_3SCF_3)_2(H_2O)_3 \cdot 2(15\text{-crown-}5)$ was subsequently reported,[747] as was the solution spectrum of $UO_2(O_3SCF_3)_2(MeCN)_3$. The metal coordination environment in both complexes are best regarded as pentagonal bipyramidal uranyl units. Anhydrous and solvent free uranyl triflate has since been prepared by several routes,[613] but most conveniently by reaction of UO_3 with neat triflic acid or triflic anhydride at elevated temperatures. As discussed above, recrystallization of $UO_2(OTf)_2$ from pyridine generates the base adduct $UO_2(OTf)_2(py)_3$.

(v) Ligands containing neutral group 16 donor atoms

The dominant species in the +6 oxidation state is the actinyl ion, AnO_2^{2+}; where the most stable member of the series is the uranyl ion, UO_2^{2+}. The majority of well-characterized coordination compounds are of this species, with the exception of some halide and hydroxide complexes.

Aqua species. Among isolated solids, hydrates of actinyl species are common (see Table 34), although it is not always clear that water is coordinated directly to the metal center. The equatorial plane of the actinyl ion can accommodate between one and five coordinated water molecules, depending both on the extent of inner-sphere anion coordination and the size of the metal ion. A total coordination number of seven for the metal ion is very common; pentagonal bipyramidal geometries are found in such complexes as $(enH_2)[UO_2F_4(H_2O)]$[748] and $UO_2(acac)_2 \cdot (H_2O)$.[749] EXAFS investigations indicate that the aquated uranyl ion in solution also has the expected pentagonal bipyramidal structure $[UO_2(H_2O)_5]^{2+}$.[746] An experimental study of the exchange between $[UO_2(H_2O)_5]^{2+}$ and bulk water (observing ^{17}O-NMR signals) suggests that the mechanism for exchange is dissociative.[750] Crown ethers have been found to effect the inclusion of uranyl hydrates. Crystallization of a variety of uranyl salts from acids (or from biphasic liquid clathrate systems[133]) in the presence of crowns results in the formation of these inclusion compounds. Some of the structurally characterized examples in this class are listed in Table 35. Little structural variability is observed in these complexes; all display metal coordination numbers of seven or eight.

Ethers. Complexes with dialkylether ligands (as well as tetrahydrofuran) have been reported for uranyl chlorides, nitrates, thiocyanides, perchlorates, and *beta*-diketonate ligands. The formulation of the complexes would suggest uranium coordination numbers ranging from six

Table 34 Some hydrates of dioxoactinide(VI) compounds.

$UO_2X_2 \cdot yH_2O$	$X = Cl$, $y = 1$, 3; $X = Br$, $y = 3$; $X = NCS$, $y = 1$, 3; $X = acac$, $y = 1$
$M^{VI}O_2(NO_3)_2 \cdot 6H_2O$	$M^{VI} = U$, Np, Pu
$M^{VI}O_2(ClO_4)_2 \cdot xH_2O$	$M^{VI} = U$, $x = 3$, 5, 7; $M^{VI} = Pu$, $x = 6$
$M^{VI}O_2(IO_3)_2 \cdot xH_2O$	$M^{VI} = U$, $x = 1$, 2; $M^{VI} = Np$, $x = 2$
$UO_2(RCO_2)_2 \cdot xH_2O$	$R = H$, $x = 1$; $R = Me$, Et, Pr^n, Pr^i, Bu^n, $x = 2$
$M^{VI}O_2(C_5H_4N\text{-}3\text{-}CO_2)_2 \cdot 2H_2O$	$M^{VI} = U$, Np, Pu
$M^{VI}O_2C_2O_4 \cdot 3H_2O$	$M^{VI} = U$, Np, Pu
$UO_2(H_2PO_4)_2 \cdot 3H_2O$	
$(UO_2)_3(PO_4)_2 \cdot 3H_2O$	
$(enH_2)[UO_2F_4(H_2O)]$	
$M^IPuO_2F_3 \cdot H_2O$	$M^I = Na$, K, Rb, Cs, NH$_4$
$M^{II}UO_2F_4 \cdot 4H_2O$	$M^{II} = Zn$, Cd, Cu, Mn, Co, Ni
$Rb_2UO_2Cl_4 \cdot 2H_2O$	
$MU_2O_5Cl_4 \cdot 2H_2O$	$M^I = Rb$, Cs
$M^I[UO_2(NCS)_3(H_2O)_2]$	$M^I = K$, Rb, NH$_4$
$M^{II}[UO_2(MeCO_2)_3]_2 \cdot xH_2O$	$M^{II} = Mg$, $x = 6$, 7, 8, 12; $M^{II} = Ca$, $x = 6$; $M^{II} = Sr$, $x = 2$, 6; $M^{II} = Ba$, $x = 2$, 3, 6, 10
$M[UO_2(CO_3)_3] \cdot xH_2O$	$M^{II} = Mg$, $x = 16$–18, 20; $M^{II} = Ca$, $x = 4$; $M^{II} = Sr$, $x = 9$; $M^{II} = Ba$, $x = 5$, 6
$M^IM^{VI}O_2PO_4 \cdot xH_2O$	$M^{VI} = U$, $M^I = H$, Na, NH$_4$, $x = 3$; $M^I = Li$, Na, K, $x = 4$; $M^{VI} = Np$, $M^I = H$, Li, $x = 4$; $M^I = Na$, K, NH$_4$, $x = 3$
$M^{II}(M^{VI}O_2PO_4)_2 \cdot xH_2O$	$M^{VI} = U$, $M^{II} = Ca$, Sr, $x = 6$; Cu, $x = 8$; $M^{VI} = Np$, $M^{II} = Ca$, Sr, Ba, $x = 6$
$M^IM^{VI}O_2AsO_4 \cdot xH_2O$	$M^{VI} = U$, $M^I = H$, $x = 4$; $M^I = NH_4$, $x = 3$; $M^{VI} = Np$, $M^I = H$, Li, $x = 4$; $M^I = Na$, $x = 3.5$; $M^I = K$, NH$_4$, $x = 3$
$M^{II}(M^{VI}O_2AsO_4)_2 \cdot xH_2O$	$M^{VI} = U$, $M^{II} = Mg$, Zn, Ni, Co, $x = 8$; $M^{VI} = Np$, $M^{II} = Mg$, $x = 8$, 10; $M^{II} = Ca$, $x = 6$, 10; $M^{II} = Sr$, $x = 8$; $M^{II} = Ba$, $x = 7$
$Na_4NpO_2(O_2)_3 \cdot 9H_2O$	

to eight. The rigid directionality and hence accessibility of the electron pairs (donor strength) of tetrahydrofuran make it particularly well suited to facilitate isolation of uranyl complexes. The molecular structures of several THF adducts have been determined (see Table 36). The complex $UO_2Cl_2(THF)_3$ is isolated from the dehydration of $UO_2Cl_2(H_2O)_x$ by Me_3SiCl in THF; it readily loses THF under reduced pressure to generate the known compound $[UO_2Cl(\mu\text{-}Cl)(THF)_2]_2$. This species is particularly useful as an anhydrous reagent for subsequent nonaqueous synthetic chemistry.

Table 35 Structurally characterized $U^{VI}O_2$ crown ether inclusion compounds.

Compound	References
$UO_2(NO_3)_2(H_2O)_2 \cdot 2$(dibenzo-18-crown-6)	a
$UO_2(NO_3)_2(H_2O)_2 \cdot 2$(15-crown-5)	b
$UO_2(O_2CCH_3)_2(H_2O)_2 \cdot 2$(dibenzo-18-crown-6)	c
$UO_2Cl_2(H_2O)_2$(12-crown-4-O)$\cdot$(12-crown-4)	d
$UO_2(NO_3)_2(H_2O)_2 \cdot 2$(15-benzo-crown-5)	e
$[H_3O][UO_2Cl_3(H_2O)_2] \cdot$(15-crown-5)	f
$UO_2Cl_2(H_2O)_3 \cdot$(18-crown-6)	g
$UO_2(O_3SCF_3)_2(H_2O)_3 \cdot 2$(15-benzo-crown-5)	h
$UO_2Cl_2(H_2O)_3 \cdot$(15-crown-5)	f
$[UO_2(H_2O)_5][ClO_4]_2 \cdot 3$(15-crown-5)	g
$[UO_2(H_2O)_5][ClO_4]_2 \cdot 2$(18-crown-6)	g
$[UO_2(H_2O)_5][O_3SCF_3]_2 \cdot$(18-crown-6)	i

[a] Xinmin, G., T. Ning, *et al. J. Coord. Chem.* **1989**, *20*, 21. [b] Gutberlet, T., W. Dreissig, *et al. Acta Crystallogr., Sec. C.* **1989**, *45*, 1146. [c] Mikhailov, Y. N., A. S. Kanishcheva, *et al. Zh. Neorg. Khim.* **1997**, *42*, 1980. [d] Rogers, R. D., M. M. Benning, *et al. Chem. Commun* **1989**, 1586. [e] Rogers, R. D., A. H. Bond, *et al. J. Crystallogr.Spectrosc. Res.* **1992**, *22*, 365. [f] Hassaballa, H., J. W. Steed, *et al. Chem. Commun.* **1998**, 577. [g] Rogers, R. D., L. K. Kurihara, *et al. J. Inclusion Phenom. Macrocyclic Chem.* **1987**, *5*, 645. [h] Thuery, P., M. Nierlich, *et al. Acta Crystallogr., Sect. C* **1995**, *51*, 1300. [i] Deshayes, L., N. Keller, *et al. Acta Crystallogr., Sect., C* **1994**, *50*, 1541.

Table 36 Structurally characterized $U^{VI}O_2$ THF adducts.

Compound	References
$UO_2(NO_3)_2(THF)_2$	a
$UO_2(CF_3COCHCOCF_3)_2(THF)$	b
$UO_2Cl_2(THF)_3$	c
$[UO_2Cl(\mu\text{-}Cl)(THF)_2]_2$	d
$UO_2Br_2(THF)_3$	e
$[UO_2Cl(THF)_4][UCl_5(THF)]$	f

[a] Reynolds, J. G.; Zalkin, A.; *et al. Inorg.Chem.* **1977**, *16*, 3357. [b] Kramer, G. M.; Dines, M. B.; *et al. Inorg. Chem.* **1980**, *19*, 1340. [c] Wilkerson, M. P.; Burns, C. J.; *et al. Inorg. Chem.* **1999**, *38*, 4156. [d] Rogers, R. D.; Green, L. M.; *et al. Lanth. Actin. Res.* **1986**, *1*, 185. [e] Rebizant, J.; Van den Bossche, G.; *et al. Acta Crystallogr., Sect. C* **1987**, *43*, 1298. [f] Noltemeyer, M.; Gilje, J. W.; *et al. ActaCrystallogr., Sect. C* **1992**, *48*, 1665.

Alcohols. Alcohol adducts of uranyl species of the formula $UO_2X_2\cdot yROH$ (X = Cl, NO_3, $MeCO_2$, alkoxides, *beta*-diketonates, tropolonate, etc.; $y = 1$–3) have been reported. In some instances, adducts are formed in alcohol solvent, while in others they are prepared by removal of water by azeotropic distillation. An interesting example of an alcohol adduct is the compound formed between uranyl chloride and the polyethyleneglycol (PEG), hexaethyleneglycol, UO_2Cl_2-$(H_2O)(PEG)$.[751] The molecular structure of this and a handful of other structurally characterized complexes in this class[752,753] demonstrate that the coordination geometry about the uranium atom is pentagonal bipyramidal.

Ketones, aldehydes, esters. The majority of representatives in this class of compounds contain cyclic or acyclic ketones as ligands (see Table 37). Complexes are known for uranyl halides, nitrates, cyanates, *beta*-diketonates, etc. The formulation of the compounds would suggest typical six- to eight-coordinate uranium. Confirmation of this is found in the structure of the acetic acid solvate $UO_2Cl_2[OC(CH=CHPh)_2]_2\cdot 2MeCO_2H$. The uranium center is six-coordinate, with all of the identical ligands mutually *trans*. Fewer compounds are known of adehydes and organic esters, most of the formula $UO_2X_2(L)_2$ (X = chloride, nitrate, L = aldehyde, ester), although mixed ligand (carbonyl-containing base, water) species have also been reported.

Carbamide and related ligands. Given the somewhat "harder" acid nature of actinyl ions relative to lower oxidation states, carbonyl-containing compounds constitute strong donor groups, and carbamides are perhaps the most widely studied type of ligand in this class (see Table 38); monamides are widely explored in extraction chemistry. This class of compounds also includes closely related ligands such as lactams, lactones, and antipyrines. In particular, the use of antipyrine (atp) for precipitating uranyl from solution in the presence of thiocyanate (presumably as $UO_2(SCN)_2(atp)_3$) was used as an assay for uranium in minerals. Adducts are most often prepared by direct reaction of actinyl salt and ligand in nonaqueous media. The stoichiometry of the adducts is controlled by the size of the coordinating ligand; stoichiometries between 1:1 and 1:5 (metal:ligand) are observed. The molecular structures of a number of amide complexes have been determined; most are monomeric, and possess uranium coordination numbers of seven or eight (see Table 39).

Oxoanions promote the formation of dimeric or polymeric products. The local coordination environment in these species is that of pentagonal bipyramidal uranyl groups with two neutral ligands and three oxo groups from two different bridging sulfate, chromate, or acetate groups.[754]

Table 37 Some ketone complexes of dioxouranium(VI) compounds.

L = ketones RCOR'	
$[UO_2Cl_2L_2]\cdot 2MeCO_2H$	R = R' = PhCH=CH
$UO_2(NO_3)_2\cdot xL\cdot yH_2O$	R = R' = Me, $x = 1$, $y = 2$ or 3; $x = 2$, $y = 0$
	R = Me, R' = Et, $x = 1$, $y = 3$; $x = 2$, $y = 0$
	R = Me, R' = Bu^i, $x = 1$, $y = 2$; $x = 2$, $y = 0$
$UO_2X_2\cdot L$	R = R' = Me, X = NCO, acac, trop
$UO_2SO_4\cdot L\cdot 2H_2O$	R = R' = Me
L = cyclic ketones	
$UO_2(NO_3)_2\cdot 2L$	L = cyclohexanone
$UO_2(trop)_2\cdot L$	L = cyclopentanone, cyclohexanone

Table 38 Amide complexes of dioxouranium(VI) compounds.

$[UO_2F_2(L)]_n$	$L = HCONMe_2(DMF)$, $MeCONMe_2(DMA)$, $MeCONH_2$
$UO_2Cl_2 \cdot xL$	$X = 1$, $L = Me_2NCO(R)CONMe_2$ with $R = CMe_2$ or $CH_2C(Me)_2CH_2$
	$x = 1.5$, $L = Me_2NCO(CH_2)_nCONMe_2$ with $n = 1$ or 3
	$x = 2$, $L = MeCONH_2(+H_2O)$; $MeCONHR$ with $R = Pr^i$, $p\text{-}H_2NC_6H_4$ or $p\text{-}EtOC_6H_4$; $HCONMe_2$, $MeCONR_2$ with $R = Me$, Pr^n, Pr^i or $n\text{-}C_8H_{17}$; $RCONMe_2$ with $R = Pr^i$, Me_2CHCH_2 or Me_3C
	$x = 3$, $L = MeCONH(p\text{-}EtOC_6H_4)$, $MeCONHEt$, $HCONMe_2$
$UO_2Br_2 \cdot xL$	$x = 2$, $L = DMA$
	$x = 3$, $L = DMF$
	$x = 4$, $L = MeCONH(p\text{-}EtOC_6H_4)$
$[UO_2(DMA)_6][UBr_6]$	
$UO_2I_2 \cdot 4DMF$	
$Cl(L)_2UO_2\{\mu_2\text{-}(O_2)\}UO_2(L)_2Cl$	$L = DMF$, DMA
$UO_2(NO_3)_2 \cdot xL$	$x = 1$, $L = RNCOCH_2CONR^2_2$ with $R^1 = Bu^n$ or Bu^i and $R^2 = Me(CH_2Ph)$
	$x = 2$, $L = DMF$, $HCON(Me)(CH_2Ph)$, $MeCONH_2$, $MeCONH(p\text{-}EtOC_6H_4)$, $MeCONR_2$ (with $R = Et$, Pr^i, $n\text{-}C_8H_{17}$, $n\text{-}C_{10}H_{21}$, $n\text{-}C_{12}H_{25}$ or Ph), $MeCONEt(MeC_6H_4)$, $RCONMe_2$ (with $R = Pr^i$, Bu^n, Me_2CHCH_2 or Me_3C), $Pr^nCONBu^n_2$, $Me_3CCONBu^n_2$
	$x = 3$, $L = MeCONHPh$
$[UO_2L_5](ClO_4)_2$	$L = DMF$, DMA or $MeCONEt_2$
$UO_2(NCS)_2 \cdot 2L \cdot H_2O$	$L = MeCONH_2$
$UO_2(MeCO_2)_2 \cdot xL$	$L = DMF$, $x = 1$ or 2; $L = DMA$, $MeCONPr^i_2$, $x = 1$
$UO_2(MeCCO_2)_2 \cdot L$	$L = MeCONPr^i_2$
$UO_2(C_2O_4) \cdot xL$	$x = 1$, $L = MeCONH_2$
	$x = 2$, $L = MeCONH(p\text{-}EtOC_6H_4)$
	$x = 3$, $L = DMF$
$UO_2SO_3 \cdot xL \cdot yH_2O$	$L = HCONH_2$, $x = 1$, $y = 2$; $L = MeCONH_2$, $x = 1$, $y = 1.5$ or $x = 1.5$, $y = 0.5$; $L = DMF$, $x = 1$, y unspecified
$UO_2SO_4 \cdot xL \cdot yH_2O$	$L = MeCONH_2$, $x = 1$, $y = 2$; $x = 2$, $y = 0$ or $x = 3$, $y = 1$; $L = DMF$, $x = 2$, $y = 0$
$UO_2CrO_4 \cdot 2L$	$L = MeCONH_2$, DMF
$UO_2(HPO_3) \cdot L$	$L = H_2NCOCH_2CONH_2$
$UO_2(HPO_3) \cdot 2L$	$L = MeCONH_2$
$[UO_2(HPO_3)(H_2O)L]$	$L = MeCONH_2$, DMF
$UO_2(Et_2NCS)_2 \cdot xL$	$x = 1$, $L = DMF$; $x = 2$, $L = HCON(CH_2Ph)(Me)$
$UO_2L_2 \cdot DMF$	$L = CF_3COCHCO(2\text{-}C_4H_3S)$
$UO_2L \cdot DMF$	$L = OC_6H_4CH = NCH_2CH_2N = CHC_6H_4O$

Extraction of U^{VI} from nitrate and thicyanate solutions into various organic solvents has been performed using three different monamides: *N,N*-methylbutyldecanamide (MBDA), *N,N*-dibutyl-decanamide (DBDA), and *N,N*-dihexyldecanamide (DHDA). Distributions from nitrate vary as a function of organic diluent and are greatest for aromatic diluents. A similar trend is not observed for thiocyanate. The extraction equilibria from NO_3^- and SCN^- media from slope analysis are as follows (see Equations (52) and (53); L = ligand):

$$UO_2^{2+}{}_{(a)} + 2\,NO_3^-{}_{(a)} + nL_{(o)} \longrightarrow UO_2(NO_3)_2 \cdot nL_{(o)} \tag{52}$$

($n = 2$ for dodecane, 1.6 for other organics)

$$UO_2^{2+}{}_{(a)} + 2\,SCN^-{}_{(a)} + 3L_{(o)} \longrightarrow UO_2(SCN)_2 \cdot 3L_{(o)} \tag{53}$$

(all organics)

Table 39 Structurally characterized U^{VI} carbamides.

Compound	References
$UO_2(NO_3)_2(N,N$-dibutyldodecanamide$)_2$	a
$UO_2(NO_3)_2(N,N$-dibutyl-3,3-dimethylbutanamide$)_2$	b
$UO_2Cl_2(N$-butylformamide$)_3$	c
$UO_2(NO_3)_2$(hexahydro-2H-azepin-2-one$)_2$	d
$UO_2(NO_3)_2(N$-ethylcaprolactam$)_2$	e
$UO_2Cl(O_2CCH_3)(N,N$-dimethylformamide$)_2$	f
$[UO_2(N,N$-diethylacetamide$)_5][BF_4]_2$	g
$[UO_2(\mu\text{-}O_2CCH_3)_2(N,N$-dimethylacetamide$)_2]_2$	h
$[UO_2(\mu\text{-}CrO_4)(N,N$-diethylacetamide$)_2]_2$	I
$[UO_2(\mu\text{-}CrO_4)(N,N$-dimethylformamide$)_2]_x$	j
$[UO_2(\mu\text{-}SO_4)(N,N$-diethylacetamide$)(H_2O)]_x$	k
$[UO_2(\mu\text{-}SO_4)(N,N$-dimethylformamide$)_2]_x$	l
$[UO_2(\mu\text{-}SO_4)(H_2O)$(acetamide$)_2]_2$	m
$[UO_2(\mu\text{-}SO_4)(H_2O)$(acetamide$)_3]_2$	m

[a] Charpin, P., Lance, M.; *et al. Acta Crystallogr., Sec. C.* **1986**, *42*, 560. [b] Charpin, P.; Lance, M.; *et al. Acta Crystallogr., Sec. C.* **1987**, *43*, 231. [c] Charpin, P.; Lance, M.; *et al. Acta Crystallogr., Sect. C* **1988**, *44*, 257. [d] Cao, Z.; Wang, H.; *et al. ActaCrystallogr., Sec. C.* **1993**, *49*, 1942. [e] Cao, Z. B.; Wang, H. Z.; *et al. Chin. Chem. Lett.* **1992**, *3*, 211. [f] Zhang, D. C.; Zhu, Z. Y.; *et al. Acta Chim. Sinica (Chin.)* **1989**, *47*, 588. [g] Deshayes, L.; Keller, N.; *et al. Acta Crystallogr., Sect. C.* **1992**, *48*, 1660. [h] Mistryukov, V. E.; Mikhailov, Y. N.; *et al. Sov. J. Coord. Chem.* **1983**, *9*, 163. [i] Mikhailov, Y. N.; Gorbunova, Y. E.; *et al. Russ. J. Inorg. Chem.* **1998**, *43*, 885–889. [j] Mikhailov, Y. N.; Orlova, I. M.; *et al. Sov. J. Coord. Chem.* **1976**, *2*, 1298. [k] Mikhailov, Y. N.; Gorbunova, Y. E.; *et al. Zh. Neorg. Khim.* **1997**, *42*, 1300. [l] Thuéry, P.; Keller, N.; *et al. Acta Crystallogr., Sect. C* **1995**, *51*, 1526. [m] Serezhkina, L. B.; Vlatov, V. A.; *et al. Zh. Neorg. Khim.* **1989**, *34*, 1251.

Extraction of uranyl with all three amides was always much higher from the SCN^- solution than from the NO_3^- solution, possibly indicating a more stable U^{VI}/SCN^- complex. IR spectra and ^{1}H-NMR data confirm the bonding of the amide carbonyl directly to the metal ion in all three ligands. Lack of OH stretching modes in the IR spectra eliminates the presence of water in the extracted complexes. IR data also confirms direct bonding on NO_3^- and SCN^- to the metal. These analyses confirm the slope analysis observations.[755]

The monoamides N,N-di(ethyl-2-hexyl)butanamide (DOBA) and N,N-di(ethyl-2-hexyl)-*i*-butanamide (DOiBA) extract U^{VI}, as well as the lower oxidation state An^{IV}. DOiBA tends to have lower extractive ability due to the bulky isobutyl groups attached to the nitrogen. Uranium(VI) extracts better than Pu^{IV} for both monoamides and always forms 2:1 U^{VI}/monoamide complexes. Hence, the observed extracted complexes at low acidity are $UO_2(NO_3)_2(DOBA)_2$ and $UO_2(NO_3)_2(DOiBA)_2$. IR analysis indicates direct carbonyl bonding to the metal and C_{2v} geometry in the final extracted complexes. Metal nitrate anions coordinated with protonated ligands are the likely species extracted in high acid conditions.[756]

Ureas. The coordination environment of complexes with urea (and related) ligands is very similar to that of carbamide ligands (see Table 40). Nearly all representatives of this class consist structurally of pentagonal bipyramidal uranyl units. Adducts with three or four neutral ligands are monomeric; those with two urea ligands are dimeric (edge sharing units bridged by anionic ligands). Polymeric sulfate complexes have been characterized, $UO_2(SO_4)\cdot2$urea and $UO_2(SO_4)\cdot3$urea,[757,758] with pentagonal bipyramidal uranyl units linked by either tri- or bidentate sulfate groups, respectively.

One structure of a homoleptic uranyl urea compound has been reported, $[UO_2(N,N',N'$-tetramethylurea$)_4][B_{12}H_{12}]$,[759] in which four urea groups lie in the equatorial plane of an octahedral uranium.

Nitroalkanes. A limited number of solvates of uranyl salts have been reported to crystallize from nitromethane or nitrobenzene solutions, including the following formulations $UO_2(NO_3)_2\cdot2RNO_2$ and $UO_2(ClO_4)_2\cdot2MeNO_2$.

Amine N-oxides, phosphine oxides, arsine oxides, and related ligands. The prototypical system for extraction of the uranyl ion from aqueous solution into organic solvent is tributylphosphate in hydrocarbons such as kerosene. This has stimulated interest in understanding the coordination chemistry of actinyl ions with P=O (and related) functional groups in order to optimize extraction efficiency or discrimination among actinides to be separated. Of all classes of neutral group 16-atom donor ligands, phosphine oxide adducts are the most common examples of complexes of transuranic elements (Np, Pu).

Base adducts are most often prepared by direct reaction of actinyl salt and ligand in nonaqueous media. The donor strength of this class of ligands is evidenced by the fact that phosphine

Table 40 Urea complexes of U^{VI}.

Compound	References
[UO$_2$(urea)$_4$(H$_2$O)](NO$_3$)$_2$	a
UO$_2$(SO$_4$)(N,N'-ethylenecarbamide)$_2$(H$_2$O)	b
UO$_2$(SO$_4$)(N,N-dimethylurea)$_3$	b
UO$_2$(SO$_4$)(urea)$_4$	c
UO$_2$(HPO$_3$)(N,N-dimethylurea)(H$_2$O)	d
[UO$_2$F(μ-F)(urea)$_2$]$_2$	e
[UO$_2$(μ-OH)(urea)$_3$]$_2$I$_4$	f
[UO$_2$(O$_2$CMe)(urea)$_3$][UO$_2$(O$_2$CMe)$_3$]	g
UO$_2$(NO$_3$)$_2$(N,N,N',N'-tetramethylurea)$_2$	h
UO$_2$(NO$_3$)$_2$(1,3-bis(n-butyl)imidazolidin-2-one)$_2$	i
UO$_2$(NO$_3$)$_2$(urea)$_2$	j

[a] Dalley, N. K.; Mueller, M. H.; *et al. Inorg.Chem.* **1972**, *11*, 1840. [b] Mikhailov, Y. N.; Gorbunova, Y. E.; *et al. Zh. Neorg. Khim.* **1999**, *44*, 415. [c] Serezhkin, V. N.; Soldatkina, M. A.; *et al. Sov.Radiochem. (Engl. Transl.)* **1981**, *23*, 551. [d] Mistryukov, V. E.; Kanishcheva, A. S.; *et al. Sov. J. Coord. Chem. (Engl. Transl.)* **1982**, *8*, 860. [e] Mikhailov, Y. N.; Ivanov, S. B. O. I. M.; *et al. Sov. J. Coord. Chem.(Engl. Transl.)* **1976**, *2*, 1212. [f] Mikhailov, Y. N. K. V. G.; Kovaleva, E. S. *J.Struct. Chem. (Engl. Transl.)* **1968**, *9*, 620. [g] Mistryukov, V. E.; Mikhailov, Y. N.; *et al. Sov. J. Coord. Chem. (Engl. Transl.)* **1983**, *9*, 163. [h] Van Vuuren, C. P. J.; Van Rooyen, P. H.; *Inorg. Chim. Acta* **1988**, *142*, 151. [i] Cao, Z.; Qi, T.; *et al. Acta Crystallogr., Sect. C* **1999**, *55*, 1270. [j] Alcock, N. W.; Kemp, T. J.; *et al. Acta Crystallogr., Sect. C* **1990**, *46*, 981.

oxide adducts will form in solution with a stoichiometric amount of ligand even in such potentially coordinating solvents as acetonitrile, ethanol, or 1,2-dimethoxyethane.[760] The reported stoichiometry of complexes prepared ranges from 1:1 to 5:1 (L:An) (see Table 41), with higher ligand to metal ratios resulting in the case of weakly coordinating anions (such as perchlorate) where ionic complexes [AnO$_2$L$_x$][ClO$_4$]$_2$ might reasonably be expected to form. The complexes are usually identified by the intense IR-active νP=O stretching frequency. The frequency of free trialkylphosphine oxides generally falls between 1150–1200 cm^{-1}; reduction in this frequency by ~ 100 cm^{-1} is common upon coordination.

The coordination environment about the metal atoms is very frequently octahedral; higher coordination numbers such as eight arise from coordination of bidentate counterions (acetate, nitrate). The most common geometry for octahedral complexes is *trans*, although a *cis* isomer of UO$_2$Cl$_2$(OPPh$_3$)$_2$ has been isolated,[761] and a *cis* geometry is observed for the derivative UO$_2$-(OBu^t)$_2$(OPPh$_3$)$_2$.[591]

Commercially available Cyanex 923, or TRPO (see Table 19), has been used for the successful extraction of U^{VI} ions from nitric acid solutions into xylene. Extractant dependency gives a slope of two for hexavalent uranium, similar to the behavior observed for trioctylphosphine oxide (T-OPO). The extraction stoichiometry for TRPO with UO$_2^{2+}$ is given by Equation (54):

$$UO_2^{2+}{}_{(a)} + 2\,NO_3^{-}{}_{(a)} + 2\,TRPO_{(o)} \longrightarrow UO_2(NO_3)_2 \bullet 2TRPO_{(o)} \tag{54}$$

The shift in the phosphoryl stretching frequency for complexes of TRPO with U^{VI} to lower values (1146 cm^{-1} to 1071 cm^{-1}) indicate strong donation of the phosphoryl oxygen lone pair to the metal center, and comparison with Th^{IV} indicate a stronger U^{VI} complex.[314]

Tributylphosphate (TBP) is used as an extractant in the PUREX process for the selective extraction of both Pu^{IV} and U^{VI} ions (see Figure 54). As discussed in Section 3.3.2.2.3.1.9, TBP is used as a synergist with CMPO in the TRUEX process for the treatment of nuclear wastes.

The equilibrium for the extraction of UO$_2^{2+}$ by TBP into various organic solvent from nitric acid media has been thoroughly characterized; the extracted species is proposed to be UO$_2$(NO$_3$)$_2$·2TBP.[762]

Den Auwer *et al.*[763] have studied the complexation of the uranyl ion by a series of trialkyl and triaryl phosphates: tri-*i*-butylphosphate(T*i*BP), tri-*n*-butylphosphate(TBP), trimethylphosphate(TMP), and triphenylphosphate(TPhP) by EXAFS. EXAFS was used to help better understand the coordination environment of the extracted complexes, as well as the extractive ability of the organophosphorus compounds.

In the experiment uranyl nitrate was dissolved into excess TBP, T*i*BP, TMP, and TPhP and the spectra were taken both in the solid state at 77 K and also at room temperature as a liquid using the uranium L$_{III}$ edge. From the experiments it was determined that all four ligands coordinated

Table 41 Structurally characterized examples of actinyl phosphine oxide complexes.

Compound	References
$UO_2(S_2PR_2)_2(OPMe_3)$	a
$UO_2\{CF_3COCHCO(2\text{-}C_4H_3S)_2\}\{OP(n\text{-}C_8H_{17})_3\}$	b
$UO_2(CF_3COCHCOCF_3)_2\{OP(OMe)_3\}$	c
$UO_2(S_2CMe)_2(OPPh_3)$	d
$UO_2(CF_3COCHCOPh)_2\{OP(NMe_2)_3\}$	e
$UO_2(NO_3)_2[OP(OEt)_2(4,6\text{-piperidino-}1,3,5\text{-trazine-O})]_2$	f
$[UO_2(O_2CMe)_2(OPPh_3)]_2$	g
$UO_2(NO_3)_2\{OP(OMe)_2(endo\text{-}8\text{-camphanyl})\}_2$	h
$UO_2Cl_2[OP(NMe_2)_3]_2$	i
$UO_2Cl_2(OPPh_3)_2$	j
	k
$NpO_2Cl_2(OPPh_3)_2$	l
$AnO_2(NO_3)_2(OPPh_3)_2$ (An = U, Np)	l
$UO_2(NO_3)_2[OP(OEt)_3]_2$	m
$UO_2(NO_3)_2[OPPh_2(NHEt)]_2$	n
$UO_2(NO_3)_2[OP(NC_5H_{10})_3]_2$	n
$UO_2Cl(O_2CCCl_3)(OPPh_3)_2$	o
$UO_2(NO_3)_2[OP(Oi\text{-}Pr)_2(CH_2SO_2(c\text{-}C_6H_{11}))]_2$	p
$UO_2(NO_3)_2[OPPh_2(CH_2COPh)]_2$	q
$UO_2(NO_3)_2[OPPh_2(CH_2SO_2NMe_2)]_2$	r
$UO_2(\eta^1\text{-}O_2CC(=CH_2)Cl)(\eta^2\text{-}O_2CC(=CH_2)Cl)(OPPh_3)_2$	s
$UO_2(NO_3)_2[OP(NMe_2)_2(NHCOCCl_3)]_2$	t
$\{UO_2[OP(NMe_2)_3]_4\}(ClO_4)_2$	u
$\{UO_2[OP(NMe_2)_3]_4\}(I_3)_2$	v

[a] Storey, A. E.; Zonnevijlle, F.; *et al. Inorg. Chim. Acta* **1983**, *75*, 103. [b] Lu, T. H.; Lee, T. J.; *et al. Inorg. Nucl. Chem. Lett.* **1977**, *13*, 363. [c] Taylor, J. C.; Waugh, A. B. *J. Chem. Soc., Dalton Trans.* **1977**, 1630. [d] Bombieri, G.; Croatto, U.; *et al. J. Chem. Soc., Dalton Trans.* **1972**, 560. [e] Charpin, P.; Lance, M ; *et al. Acta Crystallogr., Sect. C* **1986**, *42*, 987. [f] Conary, G. S.; Duesler, E. N.; *et al. Inorg. Chim. Acta* **1988**, *145*, 149. [g] Panattoni, C.; Graziani, R.; *et al. Inorg. Chem.* **1969**, *8*, 320. [h] Henderson, W.; Leach, M. T.; *et al. Polyhedron* **1998**, *17*, 3747. [i] Julien, R.; Rodier, N.; *et al. Acta Crystallogr., Sect. B.* **1977**, *33*, 2411. [j] Bombieri, G.; Forsellini, E.; *et al. J. Chem. Soc., Dalton Trans.* **1978**, 677. [k] Akona, S. B.; Fawcett, J.; *et al. Acta Crystallogr., Sect. C* **1991**, *47*, 45. [l] Alcock, N. W.; Roberts, M. M.; *et al. J. Chem. Soc., Dalton Trans.* **1982**, 25. [m] Kanellakopulos, B.; Dornberger, E.; *et al. Z. Anorg. Allg. Chem.* **1993**, *619*, 593. [n] de Aquino, A. R.; Bombieri, G.; *et al. Inorg. Chim. Acta* **2000**, *306*, 101. [o] Alcock, N. W.; Flanders, D. J.; *et al. Acta Crystallogr., Sect. C* **1986**, *42*, 634. [p] Karthikeyan, S.; Ryan, R. R.; *et al. Inorg. Chem.* **1989**, *28*, 2783. [q] Babecki, R.; Platt, A. W. G.; *et al. Polyhedron* **1989**, *8*, 1357. [r] Cromer, D. T., Ryan, R. R.; *et al. Inorg. Chim. Acta* **1990**, *172*, 165. [s] Saunders, G. D.; Foxon, S. P.; *et al. Chem. Commun.* **2000**, 273. [t] Amirkhanov, V. M.; Sieler, J.; *et al. Z. Naturforsch., Teil B.* **1997**, *52*, 1194. [u] Nassimbeni, L. R.; Rodgers, A. L. *Acta Crystallogr., Sect. C.* **1976**, *5*, 301. [v] Caira, M. R.; de Wet, J. F.; *et al. Inorg. Chim. Acta* **1983**, *77*, L73.

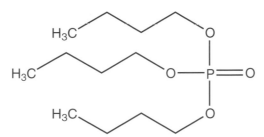

Figure 54 Tributylphosphate (TBP).

similarly to the uranium center. Table 42 shows that the distance between the U–O(P) does not differ much with a change in the alkyl group. The change that is observed may possibly be attributed to the cone angle of the ligand, with the less sterically hindered group being the methyl groups and the most sterically hindered being the phenyl groups. With nitrate and trialkylphosphate ligands being bidentate and monodentate, respectively, the resulting coordination around the uranium is hexagonal bipyramidal with the complex formula of $[UO_2(NO_3)_2X_2]$ where X is TBP, T*i*BP, TMP or TPhP.[763]

The EXAFS data are in agreement with structures solved by X-ray single crystal diffraction. It has been shown that uranyl and neptunyl complexes with TPhP are isostructural. The hexagonal bipyramidal coordination geometry around the metal center can be seen in Figure 55. The observed complexes are $UO_2(NO_3)_2(TPhP_2)$ and $NpO_2(NO_3)_2(TPhP_2)$.

Table 42 Best fit parameters for uranium to nearest neighbor (r, Å) from adjusted, filtered EXAFS spectra. (Den Auwer, Charbonnel, *et al.*, *Polyhedron*, **1998**, *17*, 4507.)

	Crystallographic value 77 K	EXAFS values				
		TBP	TiBP		TMP	TPhP
		295 K *(l)*	77 K *(s)*	298 K *(s)*	298 K *(l)*	298 *(l)*
U=O	1.757	1.77(1)	1.78(1)	1.78(1)	1.78(1)	1.77(1)
U—O(P)	3.372	2.41(1)	3.37(2)	3.38(2)	3.36(2)	3.39(1)
U—O(N)	2.509	2.54(1)	2.53(1)	2.54(1)	2.53(2)	2.53(1)
U—O'(N)	2.510	2.54(1)	2.53(1)	2.54(1)	2.53(2)	2.53(1)
U···N	2.960	2.99(1)	3.00(1)	2.98(2)	2.96(8)	2.93(3)

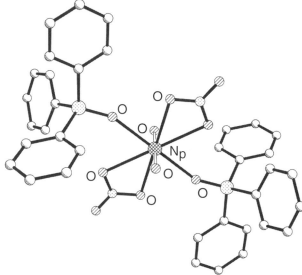

Figure 55 Crystal structure of $NpO_2(NO_3)_2(C_{18}H_{15}PO)_2$ (Alcock, Roberts *et al. J. Chem. Soc., Dalton Trans.* **1982**, 25).

If the anion bound to the neptunyl is changed from the bidentate singly-charged nitrate ion to the monodentate singly-charged chloride anion, the coordination geometry around the neptunium changes to square bipyramidal. This type of coordination sphere for an actinyl complex is not as favored; most actinyl complexes have coordination numbers of seven or eight. The difference in the coordination sphere around the neptunium can be seen in Figure 56.[764]

Monophosphoric acid-based extractants have the general structure shown in Figure 57. The compound di-(2-ethylhexyl)-phosphoric acid (DEHPA) has been used to extract U^{VI} from various acid solutions, including hydrochloric, nitric, and sulfuric acids, into kerosene. At low acidities, the extraction mechanism for all three types of acid favors the formation of a U^{VI}/DEHPA polymer complex, dictated by the equilibrium shown in Equation (55) ($L = (C_8H_{17}O)_2PO_2^-$). For both HNO_3 and HCl at higher acidities, the U^{VI} cation is extracted by a mechanism similar to that with nonionic reagents to form neutral complexes of the type $UO_2X_2(HL)_2$:

$$2\ UO_2^{2+}{}_{(a)} + 3\ (HL)_{2(o)} \longrightarrow (UO_2)_2L_6H_{2(o)} + 4\ H^+{}_{(a)} \qquad (55)$$

The partitioning of U^{VI} from HCl increases with increasing metal concentration up to around 0.05 M, indicating that two DEHPA molecules complex with the uranyl ion, similar to sulfuric and nitric acid solutions. This suggests the formation of other extracted species at high U^{IV} loadings.

Addition of a synergist to the DEHPA/HCl extraction system such as tributylphosphate (TBP) has a two-sided effect in improving extraction at high acidity and hindering extraction at low acidity. At low acidity, as in the analogous sulfuric acid system, the presence of TBP disrupts the creation of polymeric complexes between U^{VI} and DEHPA.[765-767]

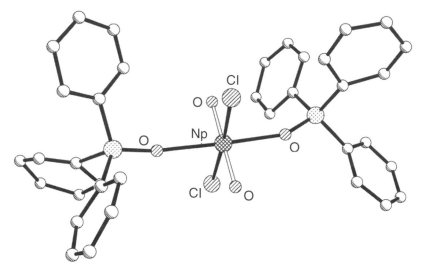

Figure 56 Crystal structure of NpO$_2$Cl$_2$(C$_{18}$H$_{15}$PO)$_2$ (Alcock, Roberts *et al. J. Chem. Soc., Dalton Trans.* **1982**, 25).

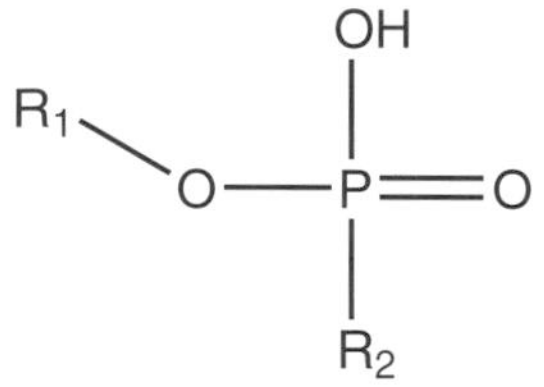

Figure 57 General diagram of a phosphoric acid.

Synergism in the sulfuric acid/U^{VI} system by the addition of both DEHPA and TBP results in a 1:2:2 (U:DEHPA:TBP) extraction stoichiometry with deprotonation of the phosphoric acid. Synergism is observed at high acidities, leading to better distribution coefficients, but at lower acidities, an antagonistic effect is observed due to the reason discussed above. Absorption spectroscopy data suggests the above equilibrium, where P—O stretching frequencies for DEHPA are shifted to lower values and those for TBP are unchanged, confirming that the extracted complex involves hydrogen bonding to the phosphoryl group of DEHPA.[768]

A similar study has looked at the extraction of both U^{VI} and Pu^{VI} by di-(2-ethylhexyl)phosphoric acid from perchloric acid into dodecane, toluene, and chloroform. Extractant dependency studies show two molecules of DEHPA complex with one U^{VI} or Pu^{VI} ion in the extracted species, accompanied by the liberation of two protons. This agrees with the ion exchange mechanism reported by Sato for U^{VI} from HCl media at low acid concentrations.

Addition of tri-*n*-octylphosphine oxide (TOPO), a proposed synergist, to the system works to improve the uptake of both actinide ions. Two plausible complexation stoichiometries have been reported for the synergistic effect of TOPO, suggesting the formation of two different extracted species (see Equations (56) and (57)):[769]

$$MO_2(ClO_4)_2 \cdot 2TOPO_{(o)} + H_2L_{2(o)} \longrightarrow MO_2(ClO_4)(HL_2) \cdot TOPO_{(o)} + TOPO_{(o)} + H^+_{(a)} + ClO_4^-{}_{(a)} \tag{56}$$

$$MO_2(ClO_4)_2 \cdot 2TOPO_{(o)} + 2 H_2L_{2(o)} \longrightarrow MO_2(HL_2)_2 \cdot TOPO_{(o)} + TOPO_{(o)} + 2 H^+_{(a)} + 2 ClO_4^-{}_{(a)} \tag{57}$$

Sulfur-containing derivatives of phosphoric acids, such as di-2-ethylhexyl-dithiophosphoric acid (HEhdtp), have been used to extract U^{VI}, presumably for the softer nature of the sulfur atom. Various noncoordinating solvents, including cyclohexane, chloroform, and carbon tetrachloride, have been used for extractions from perchlorate solutions. Analysis of distribution date indicates that the extraction of U^{VI} into these oxygen-free solvents occurs by way of an ion-exchange mechanism, similar to previous studies involving benzene. Slope analysis studies in all three solvents give a value of two, suggesting that $UO_2(Ehdtp)_2$ is formed. Studies conducted as a function of pH show that no type of polymeric species is formed in the extracted or aqueous phases. Additionally, analysis of the results in the chlorinated solvents tend to suggest that the polarity of the solvent is not a major factor in the extraction of U^{VI}.

It is observed that shorter alkyl chain dithiophosphoric acid extractants show poorer extracting ability than the HEhdtp, which can be explained as a solubility phenomenon, causing the branched alkyl chain, 2-ethylhexyl, to favor the organic phase.[836]

A relatively large number of N-oxide complexes with the uranyl ion have been reported employing both the trimethylamine N-oxide and pyridine N-oxide ligands (see Table 43). The latter has been particularly used in the isolation of beta-diketonate complexes of the general type, $UO_2L_2(pyNO)$. Common formulations for monomers include $UO_2L_2(N\text{-oxide})_x$ ($x = 2$ or 3) for monodentate L, $UO_2L_2(N\text{-oxide})$ for bidentate L, and $[UO_2(N\text{-oxide})_x][L]_2$ ($x = 4$ or 5) for weakly coordinating anions. Structurally characterized examples of this class of adducts include $UO_2(Et_2NCS_2)_2(ONMe_3)$,[770] and the compounds derived from N-oxides of polycyclic heteroaromatic ligands: $UO_2(NO_3)_2(2,2'\text{-bipyridine-}N,N'\text{-dioxide-O,O})$ and $[UO_2(NO_3)(2,2'\text{-bipyridine-}N,N'\text{-dioxide-O,O})_2](NO_3)$.[771]

There are fewer reported examples of arsine oxide adducts of actinyl species. Most appear to have structures similar to those of their phosphorus analogs. As an example, the complexes $UO_2(O_2CMe)_2(OAsPh_3)_2$ and $[UO_2(O_2CMe)_2(OAsPh_3)]_2$ are isomorphous with the phosphine oxide complexes. The molecular structure of $UO_2(NO_3)_2(OAsPh_3)_2$ has been determined;[772] the complex possesses *trans*-nitrate and arsine oxide ligands in the equatorial plane. Analysis of the metrical data indicates some shortening of the U—O (As) bond lengths with respect to the U—O (P) bond lengths in $UO_2(NO_3)_2(OPPh_3)_2$.

Sulfoxides. Table 44 presents some of the reported derivatives of actinyl ions coordinated by sulfoxide and related ligands. These ligands coordinate through the oxide oxygen atom in the equatorial plane of the actinyl ion. Coordination numbers about the metal can range from six to eight, as is typical for actinyl species. One of the most common geometries of complexes is

Table 43 Complexes of N-oxides with dioxouranium(VI) compounds.

$UO_2X_2 \cdot yL$	$L = Me_3NO$; $X = NO_3$, $y = 1$ or 4
	$X = ClO_4$, $y = 4$
	$X = Et_2NCS_2$ or Et_2NCSe_2, $y = 1$
	$L = C_5H_5NO$; $X = Cl$, $y = 2$ or 3
	$X = NCS$, $y = 3$
	$X = NO_3$, $y = 2$ or 3
	$X = ClO_4$, $y = 5$
	$X = acac$, trop or 2-OC_6H_4CHO, $y = 1$
$[UO_2(Me_3NO)_4][BPh_4]_2$	
$UO_2X \cdot yL$	$L = C_5H_5NO$; $X = SO_4$, $y = 2$
	$X = NC_5H_3\text{-}2,6\text{-}(CO_2)_2$, $y = 2$
	$X = O(CH_2CO_2)_2$, $y = 1$ (polymer) or 2
	$L = 2\text{-MeC}_5H_4NO$; $X = ONC_5H_3\text{-}2,6\text{-}(CO_2)_2$, $y = 2$

Table 44 Complexes of S-oxides with dioxouranium(VI) compounds.

$UO_2X_2 \cdot yR_2SO$	$y = 1$; $R = Me$, $X = F$ (polymer), $MeCO_2$, trop
	$y = 2$; $R = Me$, Et, Bu^n, $n\text{-}C_6H_{13}$, $X = NO_3$
	$R = n\text{-}C_8H_{17}$, $X = Cl$
	$R = Ph$, $X = Cl$, Br, NCS, NO_3, $MeCO_2$
	$y = 3$; $R = Me$, $X = Cl$, Br, NCS, NCSe
	$R = Ph$, $X = Cl$, Br
	$y = 4$; $R = Me$, $X = Br$, NO_3, ClO_4
	$R = PhCH_2$, $X = ClO_4$
	$R = Ph$, $X = I$, ClO_4
	$y = 4.5$; $R = Me$, $X = Br$
	$y = 5$; $R = Me$, $X = NO_3$, ClO_4
	$R = Ph$, $X = ClO_4$
$UO_2X \cdot 2Me_2SO$	$X = SO_3(+0.5H_2O)$, SO_4, CrO_4, $ONC_5H_3\text{-}2,6\text{-}(CO_2)_2$ (polymer)
$3UO_2(C_2O_4) \cdot 5Me_2SO$	
$(UO_2)_2(O_2)Cl_2 \cdot 4Me_2SO$	

trans-$UO_2L_2(OSR_2)_2$, as exemplified by the molecular structures of $UO_2(NO_3)_2(OSMe_2)_2$[773] and $UO_2(NO_3)_2(p$-tolyl-*n*-butylsulfoxide$)_2$.[774] Weakly coordinating anions promote the formation of homoleptic ionic sulfoxide complexes such as $[UO_2(OSMe_2)_5](ClO_4)_2$[775] and $[UO_2(OSMe_2)_5](BF_4)_2$.[776] Finally, sulfoxides have frequently been used in coordinating uranyl *beta*-diketonate complexes, generating 1:1 adducts such as $UO_2(PhCOCHCOPh)_2[OS(Me)CH_2Ph]$[777] and $UO_2(PhCOCHCOPh)_2(OSPh_2)$.[778]

Thioureas. A modest number of thiourea complexes of the uranyl ion have been reported (see Table 45), although none have been structurally characterized to date.

(vi) Ligands containing group 17 ligands

Binary halides. The stability of hexavalent halide complexes of the actinides is restricted to the lighter halides. The hexafluoride complexes of uranium, neptunium, and plutonium are well known, and have been studied extensively in the development of volatility processes for isotope separation. Uranium hexafluoride is a colorless, readily sublimable solid with a high vapor pressure (120 torr) at room temperature. It can be produced by a large number of routes, generally involving fluorination of lower-valent compounds, or oxidation of the tetravalent precursor $[UF_6]^{2-}$. A review of the preparation and properties of UF_6 has been published.[779] NpF_6 is an orange solid with similarly low melting point and high volatility. Although somewhat less stable than the uranium and neptunium analogues, reddish-brown PuF_6 can be prepared, and has been studied thoroughly. All hexafluoride complexes are octahedral in both the solid state and gas phase.

One binary chloride complex has been isolated. UCl_6 may be prepared by chlorination of lower-valent chlorides, chlorination of UF_6 with BCl_3, or by the disproportionation of UCl_5 at 102–150 °C under high vacuum. The complex is volatile, although it decomposes at relatively low temperatures (178 °C).

Complex halides. A large number of complex fluorides of the formula $M(UF_7)$ (M = alkali metal) and $M_2(UF_8)$ have been produced by reaction of MF with UF_6 or by thermal decomposition of $M(UF_7)$, respectively.[780,781] High temperature fluorination of M_3UF_7 is reported to yield M_3UF_9 (M = K, Rb).[782]

Oxohalides. Oxohalide complexes of hexavalent actinides have been prepared for U, Np, and Pu. Complexes of the formula $AnOF_4$ have been reported, for An = U, Np, and Pu; these are prepared by controlled hydrolysis of the hexafluoride complexes. The complexes contain actinides in a pentagonal bipyramidal coordination environment. The most common form[783–785] has the oxo ligand and one fluoride ligand in the axial positions (an alternate phase of UOF_4 has also been identified).[786] Dioxo complexes are somewhat more common, owing to the thermodynamic stability of the actinyl (AnO_2^{2+}) unit. The complexes AnO_2X_2 (An = U, X = F, Cl, Br; An = Np, X = F; An = Pu, X = F, Cl) have been reported. The compound $PuO_2Cl_2 \cdot 6H_2O$ is reported to be unstable, and decomposes slowly to a Pu^{IV} species. One common means of preparation of the actinyl halides is dissolution of AnO_3 in the corresponding acid HX.

Complex oxohalides are derived from the binary species, including the classes $M_2AnO_2X_4 \cdot 2H_2$ (An = U, Np, Pu; M = alkali metal or ammonium; X = F, Cl, Br), $MAnO_2F_3 \cdot xH_2O$ (An = U, M = ammonium; An = Pu, M = alkali metal, ammonium), $M_3[AnO_2F_5]$ (An = U, Np, Pu; M = alkali metal), $M_2AnO_2Cl_4$ (An = U, Pu; M = alkali metal or ammonium), MUO_2Cl_4 or $MUO_2F_4 \cdot 4H_2O$ (M = group 2 or group 12 dication), $M(AnO_2)_2F_5 \cdot xH_2O$ or $M(UO_2)_2Cl_5$ (An = U, Pu; M = alkali metal or ammonium), $M_3(UO_3)_3F_7 \cdot xH_2O$ or $M_3(UO_3)_3Cl_7$,

Table 45 Complexes of thioureas with dioxouranium(VI) compounds.

$L = SC(NH_2)_2$	
$UO_2X_2 \cdot yL$	X = Cl, y = 2; X = NO_3, y = 2 or 4; X = $MeCO_2$, y = 1, 2 or 4
$UO_2SO_4 \cdot yL$	y = 1 or 2
$L = SC(NHMe)_2$	
$UO_2(NCS)_2 \cdot 3L$	
$L = SC(NH_2)(NHPh)$	
$UO_2(MeCO_2)_2 \cdot 2L$	
$L = SC(NMe_2)_2$	
$UO_2X_2 \cdot yL$	X = Cl, NO_3, y = 2; X = NCS, y = 3
$UO_2(NCS)(NO_3) \cdot 2L$	

$M_2U_2O_5Cl_4 \cdot xH_2O$ (M = alkali metal), and KUO_3Cl. The hydrated compounds are obtained by crystallization from HCl; the anhydrous compounds are generated from molten salts. The *trans*-dioxo geometry of the actinyl ions is preserved in these species and the metal ions are generally six- or seven-coordinate. The molecular structure of organic salts of $[UO_2Br_4]^{2-}$ have been examined;[342,787] these confirm the pseudooctahedral coordination environment of the metal center. As in the case of hexahalogenates of uranium, crown ether ligands have been found to promote the crystallization of $[UO_2Cl_4]^{2-}$ in different salts.[344,345,788–790]

As in the case of U^{IV}, hydrothermal syntheses using structure-directing organic agents have been reported to yield unusual new classes of complex uranyl fluorides.[334–336,791] Reaction of uranyl acetate, uranyl nitrate, or UO_3 in aqueous HF solutions with organic bases (piperazine, pyridine, pyrazole, DABCO) generates a range of structures in which "UO_2F_5" pentagonal bipyramidal units share vertices and/or edges to form structures of variable dimensionality, from molecular complexes such as $(C_4N_2H_{12})UO_2F_4 \cdot 3H_2O$, to chains (e.g., $(C_5H_6N)UO_2F_3$, $(C_3H_5N)UO_2F_3$) to sheets (e.g., $(C_4N_2H_{12})_2(U_2O_4F_5)_4 \cdot 11H_2O$, $(C_6H_{14}N_2)(UO_2)_2F_6$, (C_5H_6N)-$U_2O_4F_5$, $(C_3H_5N)U_2O_4F_5 \cdot 1.75H_2O$), and three-dimensional structures (e.g., $(C_4N_2H_{12})U_2O_4F_6$). A mixed-valence U^{IV}/U^{VI} complex, $(C_6H_{14}N_2)_2(UO_2)_2\ F_5UF_7 \cdot H_2O$, containing a chain-like structure has also been reported.[336]

3.3.2.4.3 *Chelating ligands*

(i) *Multidentate donor ligands*

Hydroxamates, cupferron and related ligands. Most of the hexavalent actinide complexes reported involve the complexation of uranium(VI) by cupferron and its derivatives to form 1:2 and 1:3 U:L complexes. Several complexes of the form $(NH_4)_2[UO_2L_2L']$, $(NH_4)_2[UO_2L_2X_2])$ ($L' = CO_3^{2-}$, $C_2O_4^{2-}$, X = F, Cl; HL = cupferron) and $NH_4UO_2(C_6H_5N_2O_2)_3$ have been prepared by reacting excess ligand with the metal ion.[792,793] Uranyl complexes of *N*-phenyl-benzoylhy-droxamic acid (HL^1), cupferron N-nitrosophenlyhydroxylamine, (HL^2) $UO_2(L^1)_2A$ (A = MeOH, Ph_3PO, DMF, py) and $UO_2(L^1)_2DMF \cdot H_2O$, have been reported. The complexes were character-ized by elemental analysis and optical absorbance spectroscopy, and some were also characterized by X-ray diffraction. The structure of $UO_2(L^1)_2MeOH$ shows a seven-coordinate pentagonal bipyramidal geometry around the uranium center (see Figure 58). The inner coordination sphere of U is composed of two bidentate hydroxamato ligands, one methanol in the equatorial plane, and the axial uranyl oxygens. The methanol molecule is easily replaced by other neutral, more basic monodentate ligands, such as Ph_3PO, DMSO, DMF, or pyridine.[356] A similar structure of $UO_2(L)_2EtOH$ (L = *p*-isopropylbenzophenylhydroaxamic acid) has been reported.[794] Neocupferron forms 3:1 complexes with dioxouranium(VI) of the type $M[UO_2-(C_{10}H_7O_2N_2)_3]$-$\cdot xH_2O$ and $M'[UO_2(C_{10}H_7O_2N_2)_3]_2 \cdot xH_2O$, where M and M' are univalent and divalent cations.[795] Additional complexes with similar stoichiomerties and structures have been reported and

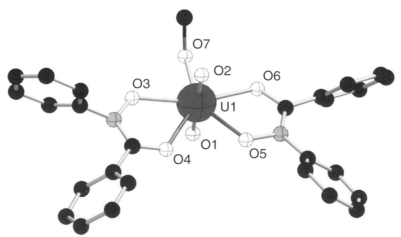

Figure 58 Structure of $UO_2(L^1)_2MeOH$ (Casellato, Vigato *et al. Inorg. Chim. Acta* **1984**, *81*, 47–54).

characterized in solution by conductance, DTA, NMR, and IR and UV spectra.[795–797] In the presence of acetohydroxamic acid and desferrioxamines, Pu^{VI} and Np^{VI} reduce to the Pu^{IV} and Np^{IV} complexes, respectively, at pH-dependent rates.[80]

Catecholate. Pyrocatecholates of composition $UO_2(1,2-C_6H_4O_2) \cdot xH_2O$ ($x = 1$ or 3) and (pyH)-$H[UO_2(1,2-C_6H_4O_2)(OH)] \cdot H_2O$, as well as the resorcinol compound, $UO_2-(1,3-C_6H_4O_2)$, have been reported. The uranyl complexes formed in aqueous solution with 4,5-dihydroxy-3,5-benzenedisulfonate (Tiron) have been postulated to be trimeric, with the stoichiometry 3:3 UO_2^{2+}:tiron based on EXAFS studies.[364] Mixed catecholate–hydroxypyridinonate ligands are described in the pyridonate section.

Pyridonate. Uranyl complexes with tetradentate ligands composed of two hydroxypyridonate groups linked by an amine $[UO_2(L^1)_2 \cdot DMF]$, $[UO_2L^3 \cdot DMSO]$, $[UO_2L^4 \cdot DMSO]DM$-DMSO]DMSO$\cdot H_2O \cdot 0.5C_6H_{12}$ and $[UO_2L^5 \cdot DMSO] \cdot DMSO$ have been prepared and structurally characterized (see Figure 59).[798] These uranyl complexes have been prepared by refluxing a methanolic solution containing equivalent amounts of $UO_2(ClO_4)_2$ and the hydroxypyridinone ligand. The linking amines are propaneamine, 1,3-diaminopropane, 1,4-diaminobutane, 1,5-diaminopentane respectively. In these complexes uranyl is bound by four oxygens from the hydroxypyridinone ligands and one solvent oxygen to generate a pentagonal bipyramidal coordination polyhedron as shown in Figure 60.

The extractant 1-hydroxy-6-N-octylcarboxamide-2(1H)-pyridinone (octyl-1,2-HOPO) has an appreciable affinity for Pu^{VI}, though much less than for Pu^{IV} under identical conditions. The equilibrium for the extraction from nitric acid media is given by Equation (58). Deprotonated HOPO generally coordinates the metal in a bidentate fashion, but in this case, m can range from 0 to 2. Extraction is greatest at low acid concentrations of around 0.001 M. Extractant dependency gives a slope of 1.3, which equates to a value of 2 for m in the above extraction stoichiometry:[226]

$$PuO_2^{2+}{}_{(a)} + m\ HHOPO_{(o)} + 2\text{-}m\ NO_3^-{}_{(a)} \longrightarrow [PuO_2(HOPO)_m(NO_3)_{2\text{-}m}]_{(o)} + m\ H^+{}_{(a)}$$

$$(58)$$

A series of multidentate ligands containing catecholate or hydroxypyridinonate metal binding groups for removal of actinides *in vivo* have been developed. Tetradentate ligands with two bidentate groups per chelator molecule attached to linear 4- or 5-carbon backbones were the most promising of a series of ligands studied, with respect to metal removal from mammals.[350]

8-Hydroxyquinoline and derivatives. Uranyl complexes of 8-hydroxyquinoline (Ox = oxine) and its derivatives have been prepared in aqueous solution.[375,377,799–803] The solid state IR spectra of $[UO_2(Ox)_n(HOx)]$ shows that Ox is bound to the metal through the phenolic O, the proton

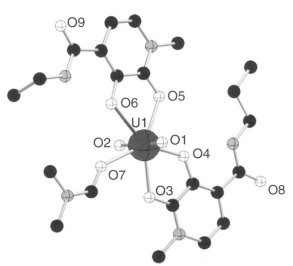

Figure 59 Crystal structure of $[UO_2(L^1)_2 \cdot DMF]$, $[UO_2L^3 \cdot DMSO]$ (Xu and Raymond *Inorganic Chemistry* **1999**, *38*, 308–315).

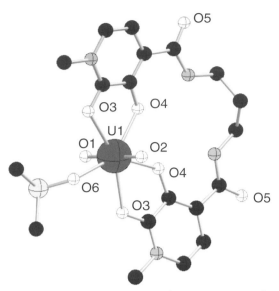

Figure 60   Crystal structure of [UO$_2$(L^1)$_2$·DMF], [UO$_2$L^3·DMSO].

forming an intramolecular H-bond between the N atom of the adducted molecule and the O atom of a neighboring chelate ring.[376] This bonding was confirmed by the structure of the complex UO$_2$(Ox)$_2$HOx·L obtained by X-ray diffraction studies (Figure 61).[804] The crystal structure shows that the three hydroxyquinoline ligands are in the plane perpendicular to the uranyl unit. Two of the hydroxyquinoline ligands are bidentate and the third is monodentate; its nitrogen is linked to one of the phenolic oxygens. Similarly, the uranyl complex formed with 5,7-dihalo-8-hydroxyquinoline precipitated from an aqueous acetone solution has been formulated as UO$_2$(5,7X,Ox)$_2$OC(CH$_3$)$_2$ based upon IR analysis. The presence of the acetone molecule instead of a third Ox is attributed to steric hindrance. Salts of triquinoline complexes, M[UO$_2$(Ox)$_3$] (M═Na$^+$, NR$_4^+$, Ph$_4$As$^+$) can be prepared by increasing the pH of the solution containing UO$_2$(Ox)$_2$HOx. Analysis of the anions by ^{1}H NMR shows that the three ligands are equivalently bound to the metal, presumably to yield hexagonal bipyramidal geometry.[805] Chelates of Np^{VI}

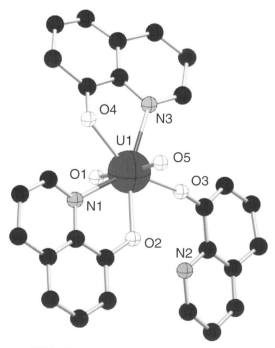

Figure 61   The structure of UO$_2$(Ox)$_2$HOx·CHCl$_3$ (Hall, Rae *et al. Acta. Crist.* **1987**, *22*, 258).

and Pu^{VI} could not be prepared, as the hexavalent ions are reduced to a lower oxidation state at pH values where metal-ligand complexes would form.[578] This is expected given the redox sensitivity of the ligand and the high thermodynamic stability of the resultant An^{IV} complexes.

Carbamate. Carbamate-containing derivatives of uranyl nitrate, uranyl oxalate, uranyl acetate, and other uranyl complexes of simple oxoanions ($[UO_2(NO_3)_2L_2]$, $[UO_2(tropolonate)_2L]$, $[UO_2(acetate)(OH)L]$, $[UO_2(oxalate)L]$, and $[UO_2(phthalate)L]_n$) have been prepared by addition of ethyl carbamate to uranyl salts. Bis(ethylcarbamate) dinitratodioxouranium(VI) $[UO_2(NO_3)_2L_2]$, for example, shows an irregular hexagonal bipyramidal geometry in which the linear uranyl group is perpendicular to the equatorial plane formed by four oxygen atoms of two nitrate groups and the two amidic oxygen atoms from the ethyl carbamate ligand (Figure 62).[806]

A related uranyl complex with the disulfide ligand, has been prepared and characterized by X-ray diffraction. The structure of $[(n\text{-}C_3H_7)_2NH_2]_2[UO_2((n\text{-}C_3H_7)_2NCOS)_2(S_2)]$ consists of $[(n\text{-}C_3H_7)2NH_2]^+$ cations and $[UO_2((n\text{-}C_3H_7)_2NCOS)_2(S_2)]^{2-}$ anions with the uranium atom at the center of an irregular hexagonal bipyramid. The equatorial coordination plane contains the disulfide (S_2^{2-}) group bonded in a "side-on" fashion and two oxygen and two sulfur donor atoms from the mono-thiocarbamate ligands. The nitrogen atom in the dipropylammonium cation is hydrogen bonded to the uranyl oxygen atoms (see Figure 63).[807]

A dinuclear mixed ligand uranyl complex with dioxime, carbonate, and oxalate ligands $(C_2N_2H_{10})_2[(UO_2)_2(CO_3)(C_2O_4)_2(C_3H_4N_2O_2)]\cdot H_2O$ has been reported and characterized by X-ray diffraction.[808] The uranyl ions in the dimer are six-coordinate, characterized by the presence of one three membered, one four membered, and one five membered chelate ring in the equtorial plane.

Several uranyl dithiocarbamates $UO_2(R_2NCS_2)_2\cdot L$ have been prepared via precipitation from solutions containing potassium R-dithiocarbamate and uranyl acetate.[809] Uranyl diethylthiocarbamate complexes with triphenylphosphine, triphenylarsine oxide, or trimethylamine N-oxide (L) $UO_2(Et_2NCS_2)_2\cdot L$, $L = (C_6H_5)_3AsO$, $(C_6H_5)_3PO$, and $(CH_3)_3NO$ have been prepared by the reaction of L with $K[UO_2(Et_2NCS_2)_3]\cdot H_2O$.[810,811] Structural characterization reveals the uranium metal center is commonly seven-coordinate, at the center of a distorted pentagonal bipyramid, with the linear uranyl unit perpendicular to a plane containing four sulfur atoms of two carbamate groups and one oxygen atom from an ancillary ligand (see Figure 64).

Similarly, many tris(R-carbamate)dioxouranium(VI) complexes (exemplified by $K[UO_2(Et_2NCS_2)_3]\cdot H_2O$) have been reported. The complexes are prepared by reacting uranyl acetate with the R-carbamate salt obtained by addition of R-amine to an aqueous solution containing equimolar amounts of carbon disulfide and potassium hydroxide.[810,812,813]

Oxalate. A large number of actinide(VI) oxalate and mixed ligand oxalate complexes have been reported (see Table 46). Different coordination modes of uranyl by the oxalate group have

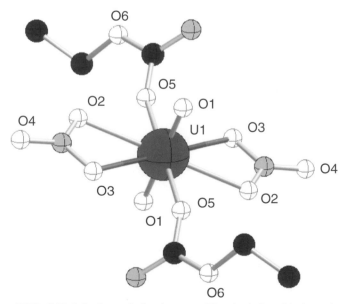

Figure 62 Structure of $UO_2(NO_3)_2L_2$ L = ethylcarbamate (Graziani, Bombieri *et al. J. Chem. Soc., Dalton Trans.* **1973**, 451–454).

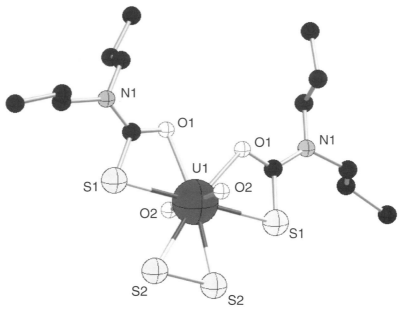

Figure 63 Molecular structure of [(*n*-C₃H₇)₂NH₂]₂[UO₂((*n*-C₃H₇)₂NCOS)₂(S₂)] (Perry, Zalkin *et al. Inorg. Chem.* **1982**, *21*, 237–240).

been reported. The typical coordination modes of the oxalato anions are bidentate, as in $UO_2C_2O_4 \cdot 3H_2O$, and tetradentate bridging as in $[(NpO_2)_2C_2O_4 \cdot 4H_2O]_n$ (see Figure 65). Bidentate chelating and bidentate bridging coordination modes are also observed as in $M^I_6[(UO_2)_2(C_2O_4)_5]$. The trioxalatouranyl complexes $M^I_4[UO_2(C_2O_4)_3]$ are characterized by the standard bidentate chelating (with the formation of a five membered ring) and bidentate chelating having a four-membered chelating ring. As expected, the coordination sphere about the metal center is very similar to that observed in carbonato complexes. The complex $UO_2C_2O_4 \cdot 3H_2O$ has a pentagonal bipyramidal coordination geometry comprised of four oxygen atoms from two independent C_2O_4 groups and one from the water molecule in the equatorial plane. Each C_2O_4 group is tetradentate, bridging two UO_2^{2+} ions.

Many complexes of the type $M^I[UO_2(C_2O_4)_2]$ and hydrates $M_n[MO_2(C_2O_4)_2L]xH_2O$ are also known (see Table 46). In the complex $K_2[UO_2(C_2O_4)_2\{CO(NH_2)_2\}] \cdot H_2O$, the basic structural unit is $[UO_2(C_2O_4)_2\{CO(NH_2)_2\}]^{2-}$, in which U has pentagonal bipyramidal UO_7 coordination.[814] The

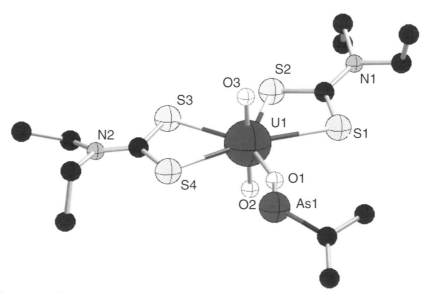

Figure 64 Structure of UO₂(Et₂NCS₂)₂·L, L = (C₆H₅)₃AsO, (C₆H₅)₃PO, and (CH₃)₃NO (Graziani, Zarli *et al. Inorg. Chem.* **1970**, *9*, 2116–1124; Forsellini, Bombieri *et al. Inorg. Nucl. Chem. Lett.* **1972**, *8*, 461–463).

Table 46 Actinide(VI) oxalate complexes.

Compound	Key	References
$MO_2C_2O_4 \cdot 3H_2O$	$M^{VI} = $ U, Np, Pu,	a
$UO_2C_2O_4 \cdot xH_2O$	$x = 0$ or 1	
$M[UO_2(C_2O_4)F_3] \cdot H_2O$	M = Na, K, Rb, Cs,	b
$Cs_n[UO_2(C_2O_4)(SeO_4)]$	$n = 1$ or 23	
$Rb[UO_2(C_2O_4)SO_4] \cdot H_2O$		d
$NH_4[NpO_2(C_2O_4)]2 \cdot 7H_2O$		e
$M_2[UO_2(C_2O_4)_2]$	M = Li, Na, K, Rb, Cs,	
	NH$_4$, Ti, CN$_3$H$_6$	
$M[UO_2(C_2O_4)_2]$	M = Sr, Ba	
$M^{II}[UO_2(C_2O_4)_2] \cdot nH_2O$	(M = K, Rb and Cs)	f
$M_2^I[UO_2(C_2O_4)_2(H_2O)x] \cdot yH_2O$		
$[UO_2(C_2O_4)_2H_2O]_n \cdot H_2O$		g
$Na[NpO_2(C_2O_4)_2 \cdot H_2O]_n$		h
$Cs_2[NpO_2(C_2O_4)_2] \cdot 2H_2O$		I
$K_2[UO_2(C_2O_4)_2\{CO(NH_2)_2\} \cdot H_2O$		
$Cs_4[UO_2(C_2O_4)_2(SeO_4)]2.7H_2O$		c
$M[UO_2(C_2O_4)_2F] \cdot H_2O$	M = NH$_4$, K, Rb, Cs	b
$K_2[UO_2(C_2O_4)_2\{CO(NH_2)_2\} \cdot H_2O$		j
$NH_4[UO_2(C_2O_4)_2(NH_2O)] \cdot H_2O$		k
$CH(NH_2)_3[(UO_2(C_2O_4)_2(CH_3NHO)] \cdot H_2O$		l
$NH_2NH_3[UO_2(C_2O_4)_2] \cdot H_2O$		m
$NH_2NH_3[UO_2(C_2O_4)_2$		n
$\quad C(CH_3)_2NO] \cdot H_2O$		
$NH_2NH_3[UO_2(C_2O_4)_2$		o
$\quad CCH_3CHNOHNO] \cdot H_2O$		
$Co(NH_3)_6[(NpO_2(C_2O_4)_2]_n \cdot H_2O$		p
$Co(NH_3)_6[((NpO_2(C_2O_4)_2)_2] \cdot 6H_2O$		q
$(C_2N_2H_{10})_2[(UO_2)_2(CO_3)$		r
$\quad (C_2O_4)_2(C_3H_4N_2O_2)] \cdot H_2O$		
$M_4^I[UO_2(C_2O_4)_3]$	$M^I = $ Na, K, Rb,	
	Cs, NH$_4$, Ti	
$[(NpO_2)_2C_2O_4 \cdot 4H_2O]n$		
$K_2[(UO_2)_2(C_2O_4)_3] \cdot 4H_2O$		
$Cs_2[(NpO_2)_2(C_2O_4)_3]$		
$M_6^I[(UO_2)_2(C_2O_4)_5]$	$M^I = $ Na, K, Rb, Cs,	
	NH$_4$, CN$_3$H$_6$	
$K_6[(NpO_2)_2(C_2O_4)_5] \cdot 2-4H_2O$		
$(N_2H_5)[(UO_2)_2(C_2O_4)_5] \cdot 2H_2O$		s
$M[(UO_2)_2(C_2O_4)_5] \cdot 2H_2O$	M = NH$_4$, C(NH$_2$)$_3$	t
$C_2H_4(NH_3)_2[(UO_2)_2(C_2O_4)_3(i-PrNHO)_2] \cdot H_2O$		t, u
$C_2H_4(NH_3)_2[(UO_2)_2(C_2O_4)_2$		t, u
$\quad ((CH_3)_2NO)_2] \cdot H_2O$		
$(NH_4)_6[(UO_2)_2(C_2O_4)(SeO_4)_4] \cdot 2H_2O$		v
$(NH_4)_4[((UO_2(C_2O_4)H_2O)_2(SeO_4)] \cdot H_2O$		w
$C_2H_2(NH_3)_2[(UO_2(C_2O_4))_2$		r
$\quad (CH_3C_2HN_2O_2)O_3] \cdot H_2O$		

[a] Mefod'eva, M. S.; Grigor'ev, M. S.; *et al. Sov. Radiochem.* **1981**, *23*, 565. [b] Nguyen Q.-D.; Bkoucke-Waksman, I.; *et al. Bull. Soc. Chim. Fr.* **1984**, 129–132. [c] Mikhailov, Y. N.; Gorbunova, Y. E.; *et al. Zhurnal Neorganicheskoi Khimii* **2000** *45*, 1825–1829. [d] Mistryukov, V. E.; Mikhailov, Y. N.; *et al. Zhurnal Neorganicheskoi Khimii* **1993**, *38*, 1514–16. [e] Grigor'ev, M. S.; Bessonov, A. A.; *et al. Radiokhimiya* **1991**, *33*, 46. [f] Dahale, N. D.; Chawla, K. L.; *et al. Journal of Thermal Analysis and Calorimetry* **2000**, *61*, 107–117. [g] Mikhailov, Y. N.; Gorbunova, Y. E.; *et al. Zh. Neorg. Khim.* **1999**, *44*, 1448. [h] Tomilin, S. V.; Volkov, Y. F.; *et al. Radiokhimiya* **1984**, *26*, 734–9. [i] Mefod'eva, M. P.; Grigor'ev, M. S.; *et al. Radiokhimiya* **1981**, *23*, 697–703. [j] Mikhailov, Y. N.; Gorbunova, Y. E.; *et al. Zhurnal Neorganicheskoi Khimii* **2002**, *47*, 936–939. [k] Shchelokov, R. N.; Orlova, I M.; *et al. Koord. Khim.* **1984**, 1644. [l] Shchelokov, R. N.; Mikhailov, Y. N.; *et al. Zhurnal Neorganicheskoi Khimii* **1987**, *32*, 1173–1179. [m] Poojary, M. D.; Patil, S. K.; *Proc. Indian Acad. Sci., Chem. Sci.* **1987**, *99*, 311. [n] Beirakhov, A. G.; Orlova, I. M.; *et al. Zhurnal Neorganicheskoi Khimii* **1990**, *35*, 3139–3144. [o] Beirakhov, A. G.; Orlova, I. M.; *et al. Zhurnal Neorganicheskoi Khimii* **1991**, *36*, 647–653. [p] Grigor'ev, M. S.; Baturin, N. A.; *et al. Radiokhimiya* **1991**, *33*, 19. [q] Beirakhov, A. G., I. M. Orlova, et al. *Zhurnal Neorganicheskoi Khimii* **1999**, *44*, 1492–1498. [r] Govindarajan, S.; Patil, S. K.; *et al. Inorg. Chim. Acta.* **1986**, 103. [s] Chumaevsky, N. A.; Minaeva, N. A.; *et al. Zh. Neorg. Khim.* **1998**, *43*, 789. [t] Shchelokov, R. N.; Mikhailov, Y. N.; *et al. Zhurnal Neorganicheskoi Khimii* **1986**, *31*, 2050–2054. [u] Shchelokov, R. N.; Mikhailov, Y. N.; *et al. Zhurnal Neorganicheskoi Khimii* **1986**, *31*, 2339–2344. [v] Mikhailov, Y. N.; Gorbunova, Y. E.; *et al. Zhurnal Neorganicheskoi Khimii* **1999**, *44*, 1448–1453. [w] Mikhailov, Y. N.; Gorbunova, Y. E.; *et al. Zhurnal Neorganicheskoi Khimii* **1996**, *41*, 2058–2062.

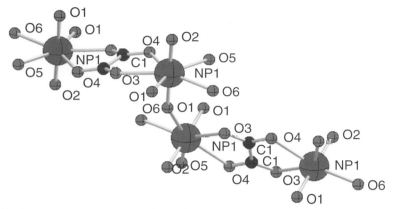

Figure 65 $(NpO_2)_2(C_2O_4)\cdot 4H_2O$ (Grigor'ev, Charushnikova *et al. Zh. Neorg. Khim.* **1996**, *41*, 539).

U-containing complexes are connected in a three-dimensional framework by potassium ions and a system of hydrogen bonds with the hydrogen atoms of urea molecules. The complex $Rb_2[UO_2(SO_4)C_2O_4]\cdot H_2O$, prepared by reacting $UO_2C_2O_4\cdot 3H_2O$ with Rb_2SO_4, has similar structural features about the uranium center with pentagonal bipyramidal (UO_7) coordination geometry.[815, 816] For the complex $Cs_4[UO_2(C_2O_4)_2(SO_4)]$ (see Figure 66) prepared in an analogous fashion from the oxalate trihydrate and Cs_2SO_4, the uranium also has pentagonal bipyramidal coordination geometry, with the axial uranyl oxygen atoms perpendicular to the equatorial plane composed of four oxygen atoms from two bidentate oxalates and one sulfate oxygen.[817] The coordination polyhedron of uranium in the complex $[(UO_2)_2(C_2O_4)(SeO_4)_4]$ is approximately pentagonal bipyramidal UO_7, with the axial uranyl unit and five equatorial oxygen atoms from one bidentate oxalate group and three selenate ions (see Figure 67).[819]

The coordination polyhedron of uranium in the polymeric anion $[UO_2(C_2O_4)_2]_n^{2n-}$ in $((NH_4)_2[UO_2(C_2O_4)_2])_n$ is also pentagonal bipyramidal, with axial uranyl oxygen atoms. In the infinite chains $[(UO_2)(C_2O_4)_2]_n^{2n-}$ one oxalate ligand is coordinated to uranyl via all four oxygen atoms, bridging two uranium centers; the other is bidentate to one uranium atom and unidentate to another.[818]

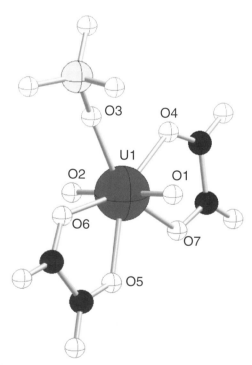

Figure 66 The structure of the complex $Cs_4UO_2(C_2O_4)_2(SO_4)$ (Mikhailov, Gorbunova *et al. Zhurnal Neorganicheskoi Khimii* **2000**, *45*, 1825–1829).

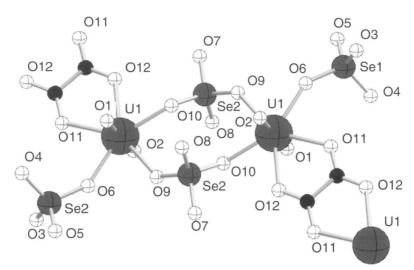

Figure 67 Structure of [(UO₂)₂(C₂O₄)(SeO₄)₄] (Alcock *J. Chem. Soc., Dalton. Trans.* **1973**, 1614; Mikhailov, Gorbunova *et al. Zhurnal Neorganicheskoi Khimii* **2000**, *45*, 1999–2002).

The oxalate groups in the anion of $(NH_4)_4[UO_2(C_2O_4)_3]$ are all bidentate, giving rise to distorted hexagonal bipyramidal coordination geometry,[818] whereas in the polymeric anion of $(NH_4)_4[(UO_2)_2(C_2O_4)_3]$ the coordination geometry is pentagonal bipyramidal, with one quadri-dentate C_2O_4 group coordinated to two uranium atoms and the other bidentate to one and unidentate to a second uranium atom, forming infinite double chains $[(C_2O_4)UO_2(C_2O_4)UO_2$ $(C_2O_4)]_n^{2n-1}$.[818] In $K_6[(UO_2)_2(C_2O_4)_5]\cdot H_2O$, the coordination geometry is again pentagonal bipyr-amidal, with two oxygen atoms each from two bidentate C_2O_4 groups and one from the bridging C_2O_4 group in the equatorial plane.[820]

A single crystal X-ray diffraction study of a mixed-ligand uranyl complex with a bidging carbonate group has been reported. The complex $(C_2N_2H_{10})_2[(UO_2)_2(CO_3)(C_2O_4)_2(C_3H_4-N_2O_2)]\cdot H_2O$ has been prepared by reacting $[UO_2(C_2O_4)(H_2O)]^-$ ion with α-dioxime in the presence of carbonate. The uranyl moieties in the dimer have a coordination number of eight and are characterized by the presence of one three-membered, one four-membered, and one five-memberd chelate ring in the equatorial plane (see Figure 68).[808,821,822]

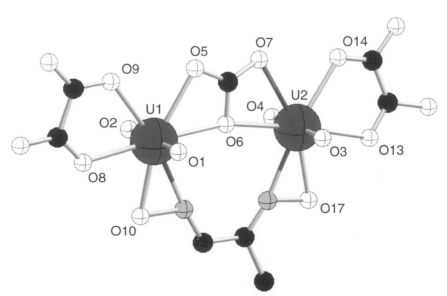

Figure 68 Structure of $(C_2N_2H_{10})_2[(UO_2)_2(CO_3)(C_2O_4)_2(C_3H_4N_2O_2)]\cdot H_2O$ (Beirakhov, Orlova *et al. Zh. Neorg. Khim.* **1998**, *44*, 1414–1419).

A number of basic oxalates, such as $(UO_2)_2C_2O_4(OH)_2 \cdot 2H_2O$, and salts of basic oxalato complex ions of the types $M^I_3UO_2(C_2O_4)_2(OH)$ and $M^I_5(UO_2)_2(C_2O_4)_4(OH)$ have also been reported, as well as salts of a wide range of peroxo-, halogeno-, sulfato-, selenito-, selenato-, thiocyanato-, and carbonato-oxalato complex anions. Only the preparation and stoichiometries of these complexes have been reported.

β-Diketonates. In a few cases, the β-diketone has been reported to act as a neutral ligand to the uranyl ion, but by far the most common type of compound is UO_2L_2, where L is the deprotonated β-diketonate ligand. Many members of this class of compounds have been characterized with a range of substituents. In addition, a number of mixed ligand complexes have been reported, such as the mixed halide species $[UO_2(CH_3\text{-}COCHCOCH_3)F(H_2O)_2] \cdot 3H_2O$ and $K_2[UO_2(CH_3COCHCOCH_3)F_3]$.[823]

The extraction of U^{VI} by a synergistic mixture of 2-thenoyltrifluoroacetone (HTTA) and tributylphosphate (TBP) from nitric acid media into benzene has been studied. Previous literature reports have described the synergistically extracted species as $UO_2(NO_3)(TTA) \cdot TBP$. However, Patil *et al.*[824] have used extraction studies to show that the nitrate anion is not present in the extracted complex. Rather, the only species involved in the synergistic extraction is $UO_2\text{-}(TTA)_2 \cdot TBP$.[824]

HTTA has been shown to extract tri- and tetravalent actinides with crown ethers as synergists according to a "size-fitting" effect. The first crystal structures of UO_2^{2+} with HTTA and two different crown ethers have been reported: $[UO_2(TTA)_2H_2O]_2(\text{benzo-15-crown-5})$ and $[UO_2(TTA)_2(\mu\text{-}H_2O)]_2(H_2O)_2(\text{dibenzo-18-crown-6})$.

Both compounds were formed by the reaction of $[UO_2(TTA)_2 \cdot 2H_2O]$ with either benzo-15-crown-5 or dibenzo-18-crown-6 in chloroform. In the stable benzo-15-crown-5 structure, two $[UO_2(TTA)_2 \cdot H_2O]$ units are coordinated to the crown ether via bridging due to the hydrogen bonding interaction of the hydrogen on a water molecule and various crown ether oxygens. Interestingly, no direct crown ether/uranyl coordination exists in the molecule; rather the hexavalent uranium atom is surrounded by seven oxygen atoms (four from two TTA ligands, one from water, and the two axial uranyl oxygens) to give pentagonal bipyramidal geometry.

In the dibenzo-18-crown-6 complex, a $[UO_2(TTA)_2 \cdot H_2O]$ group has the uranyl bound to two TTA units and a water molecule, once again giving pentagonal bipyramidal geometry around the uranium. Two of these seven-coordinate units form hydrogen bonded dimers via one hydrogen from the water molecule and the uranyl oxygen to give the $[UO_2(TTA)_2(\mu\text{-}H_2O)]_2$ complex. The other hydrogen from the water molecule is hydrogen bonded to a second water molecule (second coordination sphere), which is in turn weakly coordinated to two crown ether oxygen atoms (third coordination sphere).[825]

Bis(1-phenyl-3-methyl-4-acylpyrazol-5-one) derivatives of the type H_2BP_n, where n equals 3, 4, 5, 6, 7, 8, 10, and 22, will also extract the hexavalent actinide UO_2^{2+} from both perchlorate and nitrate media into chloroform. As with all other actinide ions, H_2BP_n extracted better than HPBMP and the highest extractability occurred with the H_2BP_7 and H_2BP_8 ligands. The extracted species for the uranyl ion was found to vary according to polymethylene chain length:

$$UO_2^{2+}{}_{(a)} + 2\,H_2BPn_{(o)} \longrightarrow UO_2(HBPn)_{2(o)} + 2\,H^+{}_{(a)} \tag{59}$$

$$(n = 3, 4)$$

$$UO_2^{2+}{}_{(a)} + H_2BPn_{(o)} \longrightarrow UO_2(BPn)_{(o)} + 2\,H^+{}_{(a)} \tag{60}$$

$$(n = 5\text{-}8, 10, 22)$$

The longer chain length extractants form 1:1 complexes, allowing the two bifunctional groups of BP_n to coordinate in a bidentate manner. The shorter chain lengths do not allow a bidentate coordination, thus forcing two ligands to coordinate in a monodentate fashion with the metal.[398]

3-phenyl-4-acetyl-5-isoxazolone (HPAI) extractant dependency indicates that two HPAI molecules are involved in the extraction of U^{VI} from nitrate media into 4-methyl-2-pentanone in an ionic mechanism. IR spectrophotometric measurements indicate chelate interaction similar to those in Th^{IV}. Deprotonation of the enolic hyroxyl group allows the charged oxygen atoms to chelate with the metal. This is confirmed by shift of the C=O stretching frequency in the IR spectrum and the presence of typical 400–500 cm^{-1} metal/ligand bands, suggesting that the carbonyl oxygen is involved in the chelation. The lack of features between $3,100 \, \text{cm}^{-1}$ and $3,600 \, \text{cm}^{-1}$ confirm that no nitrogen interactions are occurring with the metal. Additionally, there is no coordination of water to the metal complex.[399]

Amino acids. The interaction of uranyl with a number of amino acids have been investigated, motivated by the interest in *in vivo* actinide chemistry and potential actinide–nucleic acid interations. Most of the investigations focused on the determination of the stability constants of the resulting complexes. The nature of amino acid structures suggests potentially strong bond formation between uranyl and carboxylate oxygens. A stronger binding of Th^{4+} over UO_2^{2+} is expected. Most of the reports have investigated the aqueous formation of the Th^{4+} and UO_2^{2+} complexes by addition of the amino acid to a solution of the metal ion prepared from its most soluble salts (e.g., $UO_2(NO_3)_2 \cdot 6H_2O$). Potentiometric, spectrophotometric titrations, calorimetry and polarography techniques have been applied.

Uranyl forms 1:1 complexes with amino acids in acidic aqueous solution. Complexes of up to a 1:3 U:amino acid ratio may form, when the amino acid is not sterically limited and can form favorable chelate rings. The complexes generally have a hexagonal bipyramidal structure, in which dioxo uranyl is perpendicular to the equatorial plane that contains bidentate coordinated carboxylates (Figure 69).[826] Complexes of UO_2^{2+}, and Th^{4+} with the α-amino acids (H_2L) serine, cysteine, methionine, threonine, substituted glycines,[827,828] succinate, aspertate, glutamate,[829] glycylglycine, L(+)asparagines, D,L-β-phenylalanine, D,L -α-alanine, and α-amino isobutyric acid,[830] alanine, phenylalanine, valine, leucine, and isoleucine[827] have been reported. Most of those complexes contain the amino acids in the zwitterionic form binding the metal through the ionized carboxyl group. Amino acids like L-serine and L-threonine which have carboxyl, hydroxy, and amino groups have a potential to bind uranyl in a tridentate fashion; however, mostly bidentate carboxylate binding has been observed. The exception is the uranyl complex with 4-amino-3-hydroxybutyric acid in which the hydroxy group is on the fourth carbon, making it possible to form two chelate rings, one involving both the carboxylate and the hydroxy groups.[831]

Mixed ligand uranyl amino acid complexes, containing malonic, diglycolic, glutraric, maleic, glycolic thioglycolic acids, and the simple amino acids β-alanine and glycine have been reported. These complexes are prepared by addition of the ligand to solutions containing 1:1 β-alanine-UO_2 complex.[832] The synthesis of mixed ligand fluoro complexes of the type $A_3[UO_2(R)_2F_5] \cdot nH_2O$ [n = 2 or 3, A = K or NH_4^+, R = glycine, alanine, cysteine] have been reported. The complexes were characterized by a combination of chemical analysis, solution conductance measurements, and spectroscopic studies. The complexation of uranyl by glycine and alanine occurs in the zwitterionic form of the amino acid; whereas, cysteine is reported to be present as a uninegative ligand. In all of these mixed-ligand complexes the amino acids bind uranyl in unidentate fashion through one carboxylate oxygen atom.[823]

CMPO. As with tetravalent actinides, CMPO has the ability to strongly extract hexavalent actinides along with trivalent americium. Extractant dependency studies from hydrochloric and

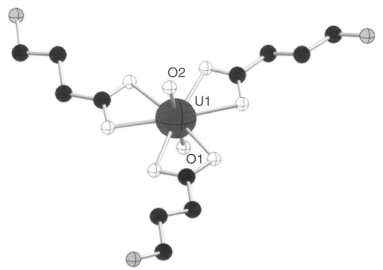

Figure 69 Molecular structure of $[UO_2(\gamma\text{-aminobutanoic acid})_3](NO_3)_2 \cdot 2H_2O[UO_2(L')_3](NO_3)_2 \cdot H_2O$ (Bismondo, Casellato *et al. Inorganica Chimica Acta* **1985**, *110*, 205–210).

nitric acid solutions show a 2:1 coordination stoichiometry for the CMPO–uranyl extracted species:

$$UO_2^{2+}{}_{(a)} + 2\,X^-{}_{(a)} + 2\,CMPO_{(o)} \longrightarrow UO_2X_2 \bullet 2CMPO_{(o)}$$

$$(X = Cl, NO_3)$$

(61)

It is assumed CMPO coordination with uranyl is the same as that for the trivalent and tetravalent actinides. Monodentate coordination occurs via the phosphoryl oxygen for the nitrate complexes and bidentate coordination occurs through both the phosphoryl and carbonyl oxygen atoms for the chloride complex, leading to a coordination number of eight for both kinds of complexes.[407]

A crystal structure of a UO_2^{2+} complex with octyl(phenyl)-N,N-diisobutylcarbamoylmethyl-phosphine oxide and nitrate has been reported (see Figure 70).[1015] The stoichiometry of the complex is $UO_2(NO_3)_2 \cdot CMPO$ and shows bidentate coordination through both the phosphoryl and carbonyl oxygen atoms on CMPO. While it is difficult to compare species in solution with those observed in the solid state, the structure is interesting since it shows a 1:1 complex of CMPO and uranyl, as well as bidentate CMPO coordination in a nitrate complex. This is in contrast to the solution-phase complex proposed by Horwitz and co-workers in 1987.[407]

Diamides. Studies on the complexation of the uranyl (UO_2^{2+}) ion with amide-based extractants are very common in the literature. Malonamides, in general, are very effective extractants for U^{VI}. The extraction of U^{VI} by the malonamide, N,N'-dimethyl-N,N'-dibutyltetradecylmalonamide (DMDBTDMA), has been characterized to gain useful insights into metal/diamide complexation. UV–vis analysis indicates only one extracted, nonacidic species in the complexation of uranyl with DMDBTDMA as shown in Equation (62). UV–vis and IR spectra indicate that the source of nitrate in the complex is from nitrate salt rather than the nitric acid. As in the Pu^{IV} system, the position of the nitrate bands in IR spectroscopy indicates a C_{2v} geometry. The malonamide extracted species are nonionic in contrast to monoamide counterparts where ion-pairs are common at high acidities:[422]

$$UO_2^{2+}{}_{(a)} + 2\,NO_3^-{}_{(a)} + DMDBTDMA_{(o)} \longrightarrow UO_2(NO_3)_2 \bullet DMDBTDMA_{(o)} \quad (62)$$

The extraction of U^{VI} in HNO_3 into toluene by N,N,N',N'-tetrabutylmalonamide (TBMA) indicates a 3:1 ligand/metal coordination in the extracted species. IR stretching frequencies at $1{,}606\,cm^{-1}$ and $1{,}574\,cm^{-1}$ indicate the bidentate nature of the malonamide ligand, and the absence frequencies at $746\,cm^{-1}$, $1{,}031\,cm^{-1}$, and $1{,}267\,cm^{-1}$ indicate that the nitrate anion is not directly coordinated to the uranyl ion.[833]

A crystal structure of a UO_2^{2+}/NO_3/malonamide complex has been reported by Lumetta *et al.*[834] Uranyl nitrate has been crystallized with N,N,N',N'-tetramethylmalonamide (TMMA)

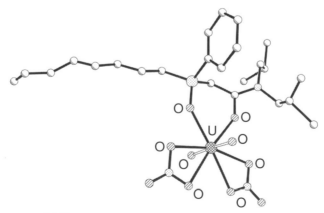

Figure 70 Crystal structure of $UO_2(NO_3)_2(C_{24}H_{42}PO_2N)$ (Cherfa, Pécaut *et al. Z. Kristallogr.-New Crys. Struct.* **1999**, *214*. 523–525).

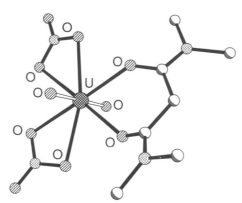

Figure 71 Crystal structure of $UO_2(NO_3)_2(C_7H_{14}N_2O_2)$ (Lumetta, McNamara *et al. Inorganica Chimica Acta* **2000**, *309*, 103–108).

acting as a bidentate ligand through both carbonyl oxygens, giving the formula $UO_2(NO_3)_2(TMMA)$ (see Figure 71). The nitrate groups are bidentate, and the nitrate malonamide ligands all coordinate equatorially to the linear uranyl fragment, giving the hexagonal bipyramidal geometry.[834]

The oxygen-containing diglycolamides, DMDHOPDA and DHOPDA, as well as the sulfur containing ones, DMDHTPDA and DHTPDA, are all very effective U^{VI} extractants when synergistically combined with thenoyltrifluoroacetone (HTTA). The stoichiometry for the extraction by all four diglycolamides (L) can be described by Equation (63). All four ligands probably serve as bidentate β-diketonate groups in the extracted species:[397,426]

$$UO_2{}^{2+}{}_{(a)} + 2\,HTTA_{(o)} + L_{(o)} \longrightarrow UO_2(TTA)_2(L)_{(o)} + 2\,H^+{}_{(a)} \qquad (63)$$

Adipicamides, like malonamides, are diamides with a butylene group bridging the carbonyl groups of the amides (see Figure 72). N,N,N',N'-tetrabutyladipicamide (TBAA) was shown to extract U^{VI} and Th^{IV} from nitric acid, where U^{VI} extraction decreases at high acid concentrations due to proton/metal competition for the TBAA coordination site. The extracted complex as determined by the slope in an extraction dependence is suggested to be $UO_2(NO_3)_2 \cdot TBAA$.[835]

Diphosphonic acids. P,P'-di(2-ethylhexyl) methanediphosphonic acid ($H_2DEH[MDP]$) is effective at extracting hexavalent actinides such as U^{VI} from *o*-xylene. Similar to Th^{IV}, extraction of U^{VI} shows minimal acid dependency due to the competition between the metal and nitric acid for the binding site. Extractant dependency studies for U^{VI} show interesting behavior at low acid concentrations where no extractant dependency is observed, possibly due to the coexistence of species having differing stoichiometries, protonation, and aqueous stabilities which have not been characterized.[418]

The extraction of U(VI) into oxylene by P,P'-di(2-ethylhexyl) ethanediphosphonic acid ($H_2DEH[EDP]$) is slightly more efficient than the methylene counterpart while showing no acid dependency. Extractant dependency analysis indicates a complexation equilibrium that is more complicated than Am(III), although it is believed that both occur via a similar mechanism under some experimental conditions: the hydrated U(VI) metal is transferred into the cavity of an aggregated micelle with release of H^+ ions.[419]

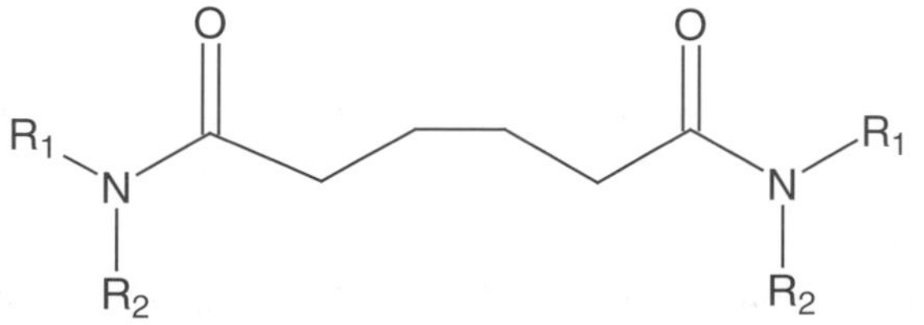

Figure 72 General diagram of an adipicamide.

With *P,P′*-di(2-ethylhexyl) butanediphosphonic acid (H$_2$DEH[BuDP]), U(VI) data analysis points to the formation of two separate extracted complexes with different stoichiometries whose mechanisms are not consistent with extraction via a micelle mechanism.[420]

Similar extractions using H$_2$DEH[MDP], H$_2$DEH[EDP], and H$_2$DEH[BuDP] in 1-decanol show differences in the distribution ratios due to the depolymerizing nature of the solvent, causing the extractants to exist as monomers in solution. Extractant dependency studies yield a slope of nearly two for the U^{VI} ion, suggesting that UO$_2$L·H$_2$L or UO$_2$(HL)$_2$ species are present, where L refers to the doubly deprotonated acid. However, the lower value of 1.5 for the H$_2$DEH[MDP] slope indicates that UO$_2$NO$_3$(HL)·H$_2$L may also be present in the organic phase.[420]

Polyoxometalates. Complexation of higher valent actinides by polyoxometalate complexes is weaker than that for tri- and tetravalent cations, due to the low charge and steric constraints imposed by the *trans*-dioxo geometry of the prevalent actinyl geometry (e.g., AnO$_2^{2+}$). In one study, the stability constants of 1:1 complexes of the uranyl ion with several heteropolymolybdates (CrMo$_6$O$_{24}$H$_6^{3-}$, IMo$_6$O$_{24}^{5-}$, TeMo$_6$O$_{24}^{6-}$, and MnMo$_9$O$_{32}^{6-}$) and isopolymetalates (V$_{10}$O$_{28}^{6-}$ and Mo$_7$O$_{24}^{6-}$) were determined; these fell in the range 2–4.[837] Conflicting evidence exists for the complexation of uranyl by the Keggin and Dawson ions; some reports suggest stability constants on the order of log $\beta = 1$,[838] while other studies reported no evidence of complexation at lower acid strengths.[839] NpO$_2^{2+}$ appears to form weak complexes with P$_2$W$_{17}$O$_{61}^{10-}$, except in higher ionic strength solutions.[582]

Other. The complex [Pt$_2$(PPh$_3$)$_4$(μ^3-S)$_2$UO$_2$(η^2-NO$_3$)$_2$] was prepared by direct reaction of the constituent metal species;[840] the molecular structure has been determined.

(ii) Schiff base ligands

The most commonly used actinide in Schiff base coordination studies is the uranyl ion. Many comprehensive reviews pulling together numerous examples of U^{VI} (primarily) coordination with these ligands have been published and the reader is referred to these for a more thorough discussion.[456,841–843] A few current representative examples will be presented here.

Recently, the novel design of compartmental Schiff bases has led to the incorporation of crown ether moieties into the macrocyclic structure. One such ligand containing an N$_3$O$_2$ Schiff base and a O$_2$O$_n$ ($n = 3$ [H$_2$L$_A$] or 4 [H$_2$L$_B$]) crown-ether functionality (see Figure 73) have been used to

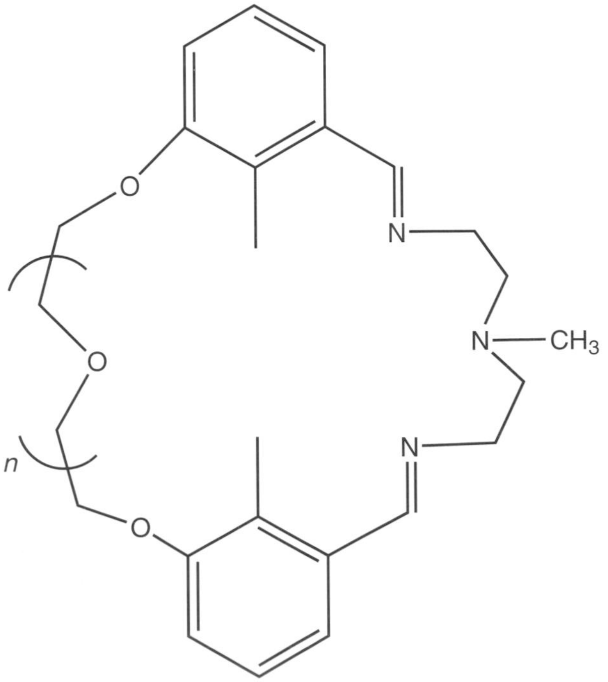

Figure 73 Diagram of compartmental Schiff base ligand (n = 1 or 2).

complex U^{VI}. The bi-compartmental nature of this ligand allows for metal ion recognition and complexation at two independent sites within the molecule.

Previous studies have shown that the compartments are very selective for certain metal ions, with alkaline earth metals and 4f-metals preferring the "hard" crown ether-like chamber and transition metals preferring the "soft" Schiff base chamber. Due to the soft nature of the uranyl ion and its transition metal-like character, it is always observed to prefer coordination to the Schiff base site on the ligand. As a result, uranyl is seven-coordinate in a pentagonal bipyramidal geometry with the three nitrogen and two deprotonated oxygen Schiff base donors. This leaves the potential for the formation of heterodinuclear complexes with metal ions occupying the crown ether-like chamber.

While many Schiff bases are known to contain mixed nitrogen and oxygen donors, pure nitrogen donor ligands are also known such as the pyrrole-derived ligands, three of which are illustrated in Figure 74. Each of the three ligands coordinates with the uranyl ion as a hexadentate ligand as indicated by X-ray crystallography (see Figure 75). The complexes of the first two ligands (non-phenyl) are both nonplanar, the second even more so than the first. The third complex with the phenyl-containing ligand is completely planar, probably due to the steric constraints imposed by the phenyl rings. While changes in the ring size and shape cause distinct changes in the overall geometry of the complex itself, it has little influence on the inner-sphere coordination adopted by the uranyl ion. In all three cases, the uranium atom adopts nearly ideal hexagonal bipyramidal geometry via coordination with all six equatorial nitrogen atoms. In complexes with the first two ligands, the U—N$_{pyrrole}$ bond distances are very similar, as are the U—N$_{imine}$ bond distances. The major difference occurs for the third complex where the U—N$_{pyrrole}$ and U—N$_{imine}$ bond lengths are shorter and longer, respectively.[844]

(iii) Macrocyclic ligands

N-Heterocyclic ligands. The oxidized form of [24]hexaphyrin(1.0.1.0.0.0) was created when exposed to uranium(VI), and formed a [22]hexaphyrin(1.0.1.0.0.0)/UO$_2^{+2}$ inclusion complex. The crystal structure, as solved by X-ray single crystal diffraction, is shown in Figure 76. The nitrogen atoms from the porphyrin have some deviation from the least squares plane. This gives a slightly disordered hexagonal bipyramidal with the oxo ligands *trans* to one another.[584]

Calixarenes. Calixarenes are macrocyclic donor ligands that have the ability to bind with a wide range of metal ions, including actinides, due to the synthesis and availability of a series of various ring sizes. The most basic structure of a calixarene is a cyclic arrangement of phenol units linked by methylene groups, where the phenols can act as anionic ligands for the actinide center. A generalized example of a calix[4]arene ligand is shown in Figure 77.

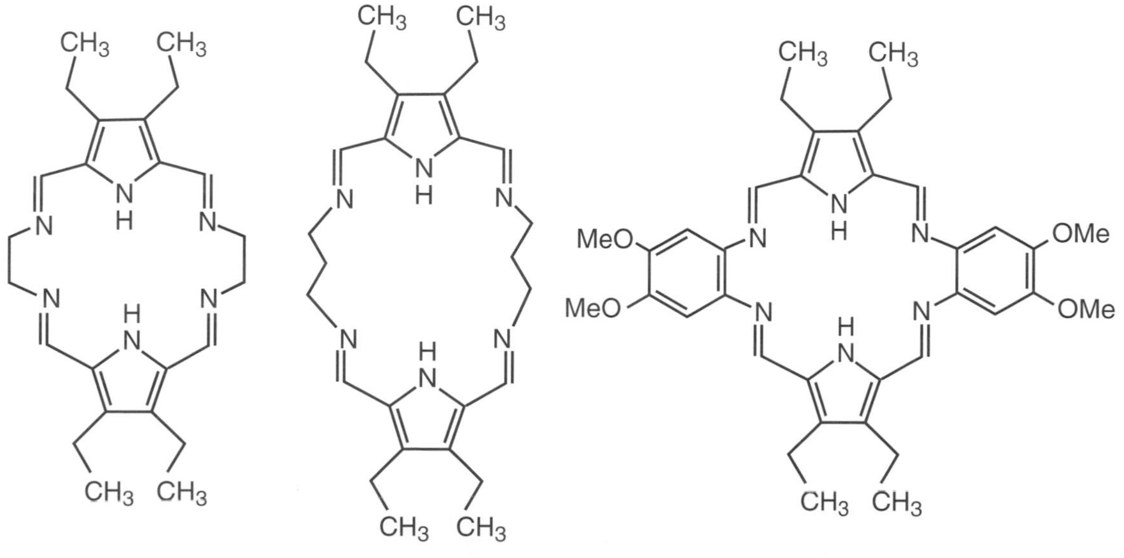

Figure 74 Diagrams of pyrrole-derived Schiff bases.

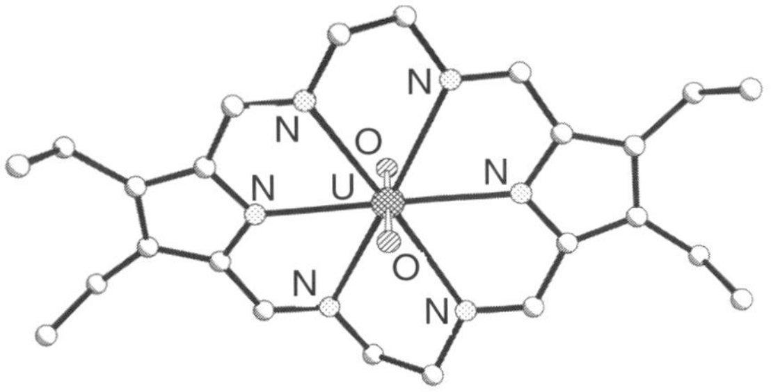

Figure 75 Crystal structure of UO$_2$(C$_{24}$H$_{32}$N$_6$) (Sessler, Mody *et al. Inorganica Chimica Acta* **1996**, *246*, 23–30).

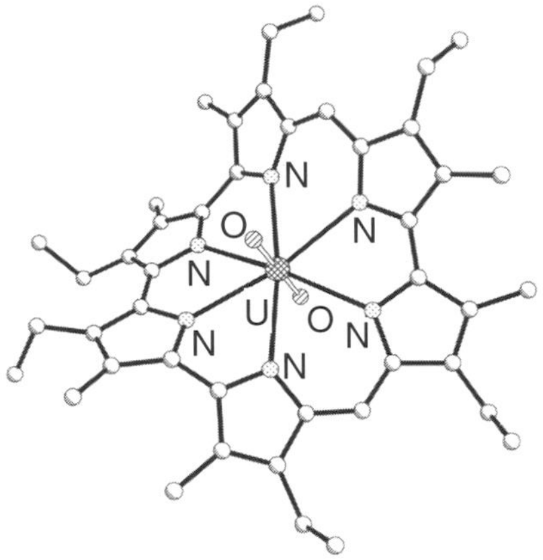

Figure 76 Crystal structure of UO$_2$(C$_{44}$H$_{51}$N$_6$) (Sessler, Seidel *et al. Angew. Chem., Int. Ed., Engl.* **2001**, *40*, 591–594).

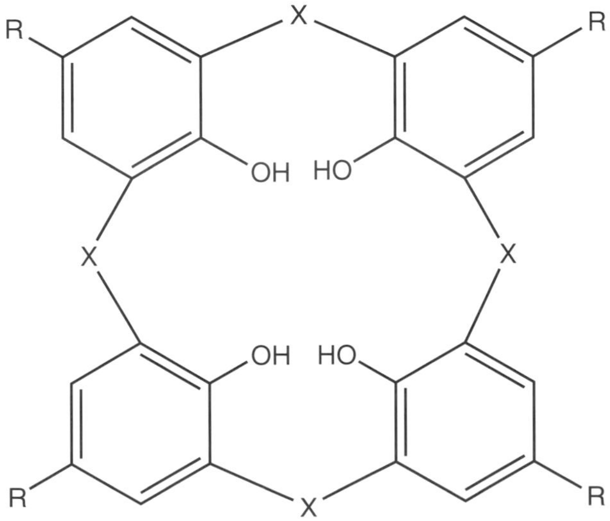

Figure 77 General diagram of a calixarene.

Oxa-calixarenes have -CH$_2$-O-H$_2$C- units replacing the methylene bridges connecting phenol groups. Aza- and thia- crown derivatives, with nitrogen and sulfur replacing the oxygens, are also common. A recent review by Thuéry *et al.*[845] gives an in-depth analysis of the coordination chemistry of phenolic calixarenes. As of 2003, the only actinides whose coordination with calixarenes has been observed or investigated are U^{V} and U^{VI}, with only a single known example of Th^{IV}. All complexation of calixarenes with actinides is known to occur only via the phenolic oxygen atoms, with the common mode for UO_2^{2+} being through four deprotonated phenolic-OH groups, yielding an anionic complex. However, U^{VI} is known to accept as few as three donors and as many as five, but never six, in its complexation with calixarenes. In bonding to any metal, including actinides, calixarene coordination is controlled by the number and geometry of the donor oxygen atoms in the macrocyclic ring.

Basic calixarenes of the type calix[n]arene have been complexed with U^{VI} with n ranging from 3 to 9, including 12, and can be either inclusion complexes or exclusion complexes (see Table 47). Inclusion complex have three or more phenolic-oxygen bonds with the uranium, essentially enclosing the metal entirely, and the exclusion complexes have three or less, leading to a less than perfect encircling of the metal.[845]

While calixarenes have great potential as being extractants for actinides, particularly in the hexavalent state, most of their coordination chemistry is known by isolation and structural characterization of discrete complexes. The compound p-t-butylhexahomotrirooxacalix[3]arene forms a complex with the uranyl ion to generate an unusually low coordination number. Two similar complexes were obtained: $[(UO_2^{2+})(L^{3-})(HNEt_3^+)]\cdot3H_2O$ and $[(UO_2^{2+})(L_3^-)(HDABCO^+)]\cdot3CH_3OH$, where HL_3 is the calixarene and DABCO is diazabicyclo[2.2.2]octane. A crystal structure of $[(UO_2^{2+})(L^{3-})(HNEt_3^+)]\cdot3H_2O$ is shown in Figure 78. The linear uranyl fragment is coordinated to all three deprotonated phenolic-oxygen atoms in its equatorial plane and is at the center of the calixarene. The ether oxygens do not participate in bonding, which is the case for nearly all calixarene/actinide complexes. The uranyl is not completely co-planar with its three bound oxygens and adopts a pseudotrigonal coordination geometry. The entire calixarene ligand adopts a "cone-like" conformation around the metal.[846]

The first reported crystal structure of a calixarene/actinide complex was that of the bis(homooxa)-p-t-butylcalix[4]arene ligand coordinated with the uranyl ion.[847] The uranyl ion forms a 1:1 inclusion complex with the calixarene, unsymmetrically binding the ligand via the four phenolic-oxygen atoms, giving a chemical formula of $[(UO_2^{2+})(L^{4-})(HNEt_3^+)_2]\cdot2H_2O$, where H_4L is the calixarene ligand. The interaction of the metal with the etherial oxygen is weak, if at all present, due to the lone pair of electrons pointing away from the uranium. As with other calixarenes, the geometry of the ligand itself is cone-shaped around the metal.[848]

p-t-Butylcalix[5]arene bonds with uranyl to give the inclusion complex $[(UO_2^{2+})(HL^{4-})(HNEt_3^+)_2]\cdot2MeOH$, where H_5L is the neutral form of the ligand. The linear uranyl fragment sits in the cavity of the calixarene ring and bonds equatorially with the five phenolic oxygens, four of which are in their anionic forms from deprotonation. U—O bond lengths for three of the five bonds fall between 2.25 Å and 3.30 Å, with the fourth and fifth being significantly longer at 2.571(7) Å and 2.836(8) Å. The bonding environment around the uranium from the phenolic oxygens is a pentagonal one, with an overall configuration of pentagonal bipyramidal. From a side profile, the calixarene takes its usual cone configuration around the metal center (see Figure 79).[849]

p-t-Butylcalix[6]arene complexes with uranyl are unique in that they have not been observed to form inclusion complexes with the metal. Only exclusion complexes are observed with 2:1, 2:2, and 3:3 metal:ligand stoichiometries. Similarly, calix[7]arenes only have the ability to complex with uranyl via one or both of its trimeric or tetrameric subunits.[850]

Table 47 Coordination geometry of various uranyl calixarene complexes.[a]

Ligand	Coordination of complex
Hexahomotrioxacalix[3]arene	Pseudo-trigonal
Calix[4]arene	No "internal" complex formed
Dihomooxacalix[4]arene	Square planar
Tetrahomodioxacalix[4]arene	Square planar
Octahomotetraoxacalix[4]arene	Pentagonal
Calix[5]arene	Pentagonal
Calix[6]arene	No "internal" complex formed
Tetrahomodioxacalix[6]arene	Square planar
Calix[7]arene	Square planar; direct UO_2^{2+}–UO_2^{2+} bonding
Calix[8]arene	Pentagonal
Octahomotetraoxacalix[8]arene	Pentagonal
Calix[9]arene	Pentagonal
Calix[12]arene	Pentagonal; two dimers

[a] Table taken from Thuéry, P.; Nierlich, M.; Vicens, J.; Masci, B. *J. Chem. Soc., Dalton Trans.*, **2001**, 867–874.

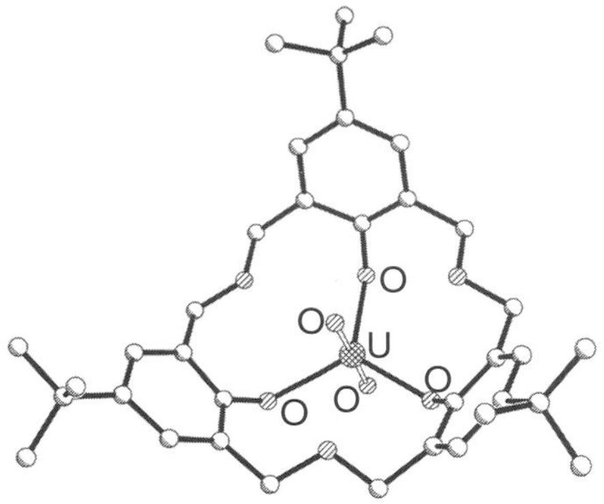

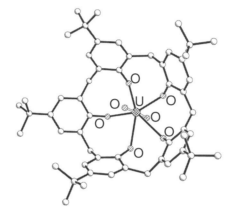

Figure 78   Crystal structure of [(UO₂)(C₃₆H₄₅O₆)(HN(C₂H₅)₃)]·3H₂O (Thuéry, Nierlich *et al. J. Chem. Soc., Dalton Trans.* **1999**, 3151–3152).

Figure 79   Crystal structure of [(UO₂)(C₅₅H₆₆O₅)](HN(C₂H₅)₃)₂·(CH₃OH)₂ (Thuéry, P. and Nierlich, M. *J. Incl. Phenom. Mol. Recogn. Chem.* **1997**, *27*, 13).

A bimetallc inclusion complex is formed when $UO_2(NO_3)_2$ reacts with p-*t*-butylcalix[8]arene. The resulting complex stoichiometry is $[(UO_2^{2+})_2(H_4L^{4-})(OH^-)]^-·2HNEt_3^+·OH^-·2NEt_3·3H_2O·4\text{-}CH_3CN$ where the fully protonated ligand is given by H_8L. Two linear UO_2^{2+} fragments lie within the calixarene ring and each are equatorially bonded to four phenolic oxygens, only two of which remain prontonated. A bridging OH^- group gives each uranyl ion a distorted pentagonal arrangement with the bonded oxygens, typical for uranyl coordination. The uranium adopts a pentagonal bipyramidal arrangement, slightly distorted due to the equatorial environment being slightly nonplanar as seen in Figure 80. U—O bond lengths are observed to vary with the protonation, or lack thereof, of the oxygen.[851]

The largest calixarene that has been used in complexation with UO_2^{2+} is *t*-butylcalix[12]arene. Two bimetallic uranyl units (four uranyl groups) are enclosed within the ring, giving it a stoichiometry of $[HNEt_3]_2[\{(UO_2^{2+})_2(NO_3)_2(py)\}_2(H_4L)]$, where H_4L^{8-} is the deprotonated form of the ligand. The bimetallic uranyl units are assembled with bridging nitrate anions. Each bimetallic array has both uranium atoms bound to five phenolic oxygens, four of which are in a deprotonated form. The remaining two phenols in the calixarene are unbound. In each bimetallic unit, one uranium is bound to three phenolic oxygens and the other uranium bound to two, with the third coordination site being occupied by a pyridine molecule. The resulting geometry around the uranyl is a pentagonal arrangement as illustrated in Figure 81.[852]

Crown ethers. U^{VI}-crown ether chemistry has been extensively studied due to the wide variey of complexes that can be formed. Uranyl can form both inclusion and exclusion complexes with both hetero- and homocrown ethers. Exclusion complexes can form bonds to other metals in the crown ether cavity directly or via hydrogen bonds to crown ether oxygens. The nature of complexation is dependent on the chemical environment.

Both EXAFS spectroscopy and X-ray single crystal diffraction have been used to study uranyl complexes with 18-crown-6 and dicylohexyl-18-crown-6 (dch-18-crown-6) in the presence of the

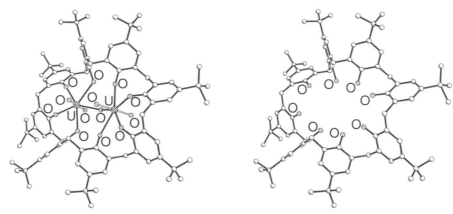

Figure 80 Crystal structure of $(HN(C_2H_5)_3)_2[(UO_2)_2(C_{88}H_{108}O_8)(OH)]\cdot(N(C_2H_5)_3)_2\cdot(H_2O)_3\cdot(CH_3CN)_4$ and of the calixarene ligand with the metal complex removed for clarity (Thuéry, Keller *et al. Acta Crystallogr.*, *Section C* **1995**, *C51*, 1570–1574).

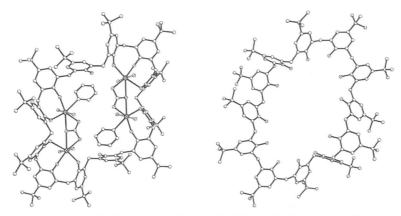

Figure 81 Crystal structure of $[HN(C_2H_5)_3]_2[\{(UO_2)_2(NO_3)(C_5H_5N)\}(C_{132}H_{152}O_{12})]$ and of the ligand with the uranium complexes removed (Leverd, Dumazet-Bonnamour *et al. Chem. Commun.* **2000**, 493–494).

trifluoromethanesulfonate (OTf⁻) anion. The EXAFS data was collected using the uranium L_{III} absorption edge in both solution and solid state for $UO_2(18\text{-crown-}6)(OTf)_2$ and $UO_2(dch\text{-}18\text{-}crown\text{-}6)(OTf)_2$. A comparison of EXAFS and X-ray diffraction models is provided in Table 48. It can be concluded from the data that the coordination in the solid state is probably very similar to coordination in solution. Both techniques are consistent with proposed hexagonal bipyramidal coordination geometry around the uranium. No significant deviations are observed in the planarity of the equatorial coordination region, indicating a nearly ideal geometry (Figure 82). Deviations from ideality are indicated by the sum of the O—U—O bond angles (ideal 360°); in $UO_2(18\text{-crown-}6)$ the sum is 363.3° while in the $UO_2(dch\text{-}18\text{-}crown\text{-}6)$ the sum is 366.8°.[853]

Table 48 A comparison of EXAFS and X-ray diffraction data showing distance(R) to nearest neighbors (Peshayes, Keller *et al. Polyhedron*, **1994**, *13*, 1725).

		R_{EXAFS} (Å)		$R_{X\text{-ray}}$
Complex		*Solution*	*Solid*	*Solid*
$UO_2(18\text{-crown-}6)(OTf)_2$	U—O_{ax}	1.77	1.77	1.64(5)
	U—O_{eq}	2.58	2.57	2.50(5)
	U—C_{eq}	3.49	3.45	3.49(8)
$UO_2(dch\text{-}18\text{-}crown\text{-}6)(Tf)_2$	U—O_{ax}	1.77	1.76	1.78(5)
	U—O_{eq}	2.59	2.63	2.58(8)
	U—C_{eq}	3.49	3.53	3.47(9)

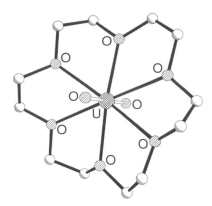

Figure 82   Crystal structure of [UO$_2$(C$_{12}$H$_{24}$O$_6$)]·(CF$_3$SO$_3$)$_2$ (Deshayes, Keller *et al. Polyhedron* **1994**, *13*, 1725).

Azacrowns have been studied as possible extractants for the uranyl cation. Complexation of actinyls by azacrowns (possessing slightly softer nitrogen donor atoms) is less favored then the harder oxygen donors of traditional crown ethers. Crowns can be completely substituted, as in 18-azacrown-6, or only partly substituted as in diaza-18-crown-6. While substitution of a softer donor atom does effect the strength of the ligand, the presence of a weakly-complexing anion such as OTf$^-$ offers limited competition for complexation to the uranyl. The coordination environment is not affected by the presence of nitrogen donors; the uranyl still adopts a hexagonal bipyramidal coordination geometry. The nitrogen donors also tend to effect the planarity of the equatorial region of a complex. In the UO$_2$(diaza-18-crown-6), the sum of the equatorial donor angles is 361.2°; this represents the closest geometry to ideality of any of the substituted or unsubstituted 18-crown-6 ligands. The complex UO$_2$(18-azacrown-6) has the greatest deviation with a donor angle sum of 378.8°.[854,855]

Uranium(VI)/crown ether complexes can form polymeric chains in the presence of anions such as sulfate. Each uranyl is bound to two water molecules and one oxygen from three different sulfate anions. Each sulfate anion therefore serves as a μ^3-bridging ligand to three uraniums. The crown ethers are not complexed to the polymer chain but instead surround it in a sheet-like fashion. An example of this polymeric network with 12-crown-4 can be seen in Figure 83. The inorganic polymer chains are separated from each other by the formation of an organic crown ether layer, resulting in organic/inorganic layering. Hydrogen bonding occurs between each crown ether and an adjacent water molecule to form the organic layer; this water in turn hydrogen bonds to the inorganic chain.

The complex [UO$_2$(SO$_4$)(OH$_2$)$_3$]·0.5(18-crown-6) exhibits another means of forming polymers wherein the uranyl-sulfate bridging forms a zig-zag-like polymeric pattern. For each uranyl unit, one of the three equatorial water ligands is hydrogen bonded to two bridging sulfate ions via oxygen atoms on the sulfate. The remaining two water molecules are hydrogen bonded to uncomplexed 18-crown-6 molecules. These crown ethers exist on both sides of the polymer in an alternating fashion.

The complex [(H$_3$O)(18-crown-6)]$_2$[(UO$_2$(NO$_3$)$_2$)$_2$C$_2$O$_4$] consists of stacked [(UO$_2$-(NO$_3$)$_2$)$_2$C$_2$O$_4$]$^{2-}$ anions surrounded by stacked [(H$_3$O)(18-crown-6)]$^+$ cations. In the anions,

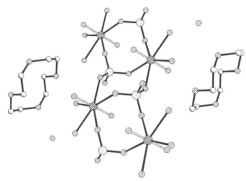

Figure 83   Crystal structure of {[UO$_2$(SO$_4$)(H$_2$O)$_2$]·0.5(C$_8$H$_{16}$O$_4$)(H$_2$O)}$_n$ depicting the coordination of the U^{VI} in a sulfate polymeric chain (Rogers, Bond *et al. Inorg. Chem.* **1991**, *30*, 2671).

two bidentate nitrate ions coordinate to a single uranyl and two of these uranyl/nitrate units are bridged by an oxalate moiety where two oxygens are bound to each uranyl. This arrangement creates a hexagonal bipyramidal geometry around each uranium. In the cation, the hyrdronium ion sits in the cavity of the crown ether.[856]

Polymeric and bridged complexes are not the only form of exclusionary crown complexes with uranyl. In complexes of the uranate ions $UO_2X_4^{2-}$ (where $X = Cl$ or Br), sandwiching of the metal unit between crown ether-complexed counterions can occur. In cationic 12-crown-4 complexes with lithium, 15-crown-5 complexes with sodium, and 18-crown-6 complexes with potassium, the negatively charged uranyl complex will bridge between two crown ether/group 1 metal cations. The uranium center in these complexes lies in a psuedooctahedral coordination environment. Structures differ in the nature of the alkali metal-uranium bridging groups. In some cases, the two metal centers are bridged by two halide ligands as seen in Figure 84. In this structure, the two equatorial bromides bond to the potassium in an 18-crown-6/potassium complex. The uranyl oxo groups remain uncoordinated to the alkali metal. In the $[K(18\text{-}crown\text{-}6)]_2[UO_2Br_4]$ complex the $O{=}U{\cdots}K$ angle is 63°, and 56° in the chloro complex, indicating that actinyl moeiety is tilted toward the potassium atoms, possibly to bring the uranyl oxo group into proximity with the acidic alkali metal.

In the complex $[Na(15\text{-}crown\text{-}5)]_2[UO_2Cl_4]$, some of the "sandwiches" display a comparable geometry, in which the the $O{=}U{\cdots}Na$ angle is observed to be 87°. A second type of bonding was also observed in this species, however, in which a single chloride coordinates to each sodium atom in the crown ether/sodium complex. The $O{=}U{\cdots}Na$ angle is much more acute (31°), with the axial oxygens tilted toward the sodium for a weak interaction. The size of the halide ligand impacts the coordination mode; in the analogous bromide complex the sodium and uranium centers are bridged by one bromide anion and a uranyl oxo group, leaving the other two bromide atoms unbound and a $Br{-}U{\cdots}Na$ angle of 90° (see Figure 85). The size of the cation also influences the coordination geometry; in the complex $[Li(12\text{-}crown\text{-}2)]_2[UO_2Cl_4]$, bridging occurs only via the uranyl oxo groups.[691]

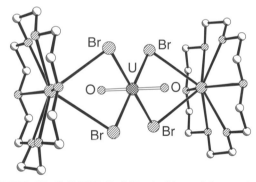

Figure 84 Crystal structure of $[K(C_{12}H_{24}O_6)]_2[UO_2Br_4]$ (Danis, Lin *et al. Inorganic Chemistry* **2001**, *40*, 3389–3394).

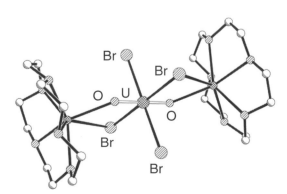

Figure 85 Crystal structure of $[Na(C_{10}H_{20}O_5)]_2[UO_2Br_4]$ (Danis, Lin *et al. Inorganic Chemistry* **2001**, *40*, 3389–3394.)

3.3.2.5 Heptavalent Oxidation State

3.3.2.5.1 *General characteristics*

Heptavalent neptunium and plutonium can be prepared in highly alkaline aqueous media via electrochemical or chemical oxidation of An^{IV}, An^V, or An^{VI} species, with Np^{VII} being more easily obtained and isolated than Pu^{VII}.[857–863] The complexes formed have been characterized in solution primarily by optical absorbance and vibrational spectroscopy, and more recently by NMR and EXAFS, and in the solid state by EXAFS and X-ray diffraction. Most research in this area was conducted at the Russian Academy of Sciences.

3.3.2.5.2 *Simple donor ligands*

(i) *Ligands containing group 16 donor atoms*

Hydroxide and aqua complexes. The coordination environment of Np^{VII} in solution remains topical because the species stabilized under highly alkaline conditions could be technologically useful in nuclear waste processing. In addition it provides an opportunity to examine structure/bonding relationships in f^0 systems with unusual coordination geometries. Solution studies report various coordination geometries, including a dioxo moiety coordinated equatorially by hydroxo or aquo ligands,[858,864] and a square planar tetraoxo complex with two additional axial hydroxo ligands.[865] The tetraoxo Np^{VII} coordination is common in the solid state. Analogies with other high-valent actinide ions, including An^{VI} species suggest that the hydroxy neptunyl ion $(NpO_2)(OH)_x(H_2O)_y$ is the prevalent species in solution. In contrast, the oxo anions of hexa- and heptavalent transition metals exhibit tetrahedral MO_4 coordination. Recent EXAFS and computational studies provide evidence that Np^{VII} has a tetraoxo first coordination sphere, slightly distorted from square-planar geometry, with a Np—O bond distance of 1.87 Å, with two hydroxy ligands at a distance of 3.3 Å (see Figure 86). This unusual coordination environment appears to result from a competition for ligand electron density among *d*- and *f*-orbitals of various symmetries of the Np atom, with a slight distortion from square-planar-based geometry favored by increased *f*-orbital participation in bonding with oxygen *p*-orbitals.[866,867] Interestingly, the Np^{VI}/Np^{VII} redox couple is reversible, indicating rapid electron exchange between the two forms. Considering the stoichiometry and conformations of the complexes, slow electron exchange and irreversible electrode potentials may be expected. However, the $NpO_4(OH)_2^{3-}$ species may be considered a deprotonated form of $NpO_2(OH)_4^-$; and proton-transfer reactions are often rapid. This reasoning may explain the relative instability of Pu^{VII} species in solution and suggests that dioxo Np^{VII} species may be observed under less basic conditions.[867]

A large number of hydrated solid compounds, such as $M(NpO_4) \cdot xH_2O$, $M_3[(NpO_4)(OH)_2]_2 \cdot xH_2O$ or $M(NpO_4)(OH)_2 \cdot 5H_2O$, have been prepared by oxidation of Np^{VI} followed by precipitation of complex Np^{VII} anions from aqueous alkaline solutions using alkali (M = Cs, K, Na, Li), alkaline earth (M = Ca, Sr, Ba), or hydrogen-bond donating cations, such as cobalt hexamine and cobalt ethylene diamine. Many of these complexes were initially reported to contain the MO_5^{3-} anion, as in the formulation $K_3NpO_5 \cdot xH_2O$; however, structural characterization has revealed the coordination geometry of the Np^{VII} to be tetragonal bipyramidal, with two hydroxide ligands in the axial positions, and the formula to be $K_3(NpO_4)(OH)_2 \cdot 2.2H_2O$.[868] Single crystal X-ray structures have been determined for compounds such as $Na_3NpO_4(OH)_2 \cdot nH_2O$[869] and $Co(NH_3)_6NpO_4(OH)_2 \cdot 2H_2O$[864,870] which also contain $NpO_4(OH)_2^{3-}$ in this highly unusual coordination geometry. Crystalline samples of analogous Pu^{VII} compounds

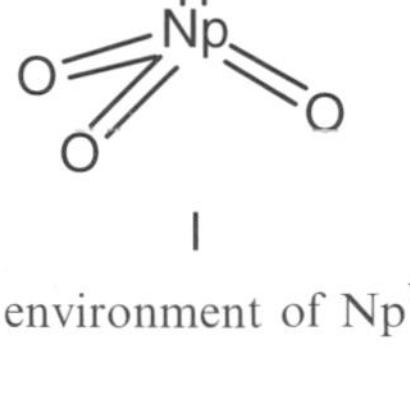

Figure 86 Coordination environment of Np^{VII} in aqueous solution.

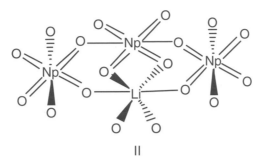

Figure 87 Neptunium coordination environment in $Li[Co(NH_3)_6][(Np_2O_8)(OH)_2]$.

have been prepared.[871] Similar, nonhydrated Np^{VII} and Pu^{VII} salts can be prepared by heating alkali metal peroxides with Np and Pu dioxides above 250 °C.

The ternary oxides Li_5NpO_6 and Li_5PuO_6 have been obtained by heating the actinide dioxide with lithium oxide in oxygen at ~400 °C.[872] The AnO_6^{5-} anion appears to be octahedral. The IR spectrum of $Li_3(NpO_2)(OH)_6$ has been reported for a sample that was obained by dissolving $NpO_2(OH)_3 \cdot 3H_2O$ in aqueous lithium hydroxide followed by solvent evaporation. The compound $Li[Co(NH_3)_6][(Np_2O_8)(OH)_2] \cdot 2H_2O$, was isolated from a lithium hydroxide solution of neptunium-(VII) by adding $[Co(NH_3)_6]Cl_3$. In the structure there are two independent neptunium centers, each coordinated by an octahedron of oxygen atoms; the octahedra share corners to form a chain (see Figure 87).[873] A carbonate complex has also been proposed, but not yet characterized.

3.3.2.5.3 *Chelating ligands*

There is some evidence that $NpO_4(OH)_2^{3-}$ can be extracted into organic solvents by phenol and pyrazolone ligands that replace hydroxide in the inner coordination sphere. For example, extraction by bis(2-hydroxy-5-octylbenzyl)amine or 2-hydroxy-5-tert-butylphenyl disulfide was confirmed by ^{13}C NMR. Optical absorption data indicated that Np^{VII} reduction to Np^{VI} does occur, but that Np^{VI} is extracted more slowly and with smaller distribution coefficients than is Np^{VII}.[874,875]

3.3.3 THE LATER ACTINIDE METALS—TRANSPLUTONIUM ELEMENTS

Americium and curium are by far the most studied transplutonium elements, being available in milligram to gram quantities. The inorganic and organic complexes of Am and Cm, in addition to other chemistry of these elements, are described in detail in the element-specific chapters on Am and Cm chemistry by Runde and Schulz[1043] and Lumetta, Thompson, Penneman, and Eller,[1044] respectively, in "The Chemistry of the Actinide and Transactinide Elements," to be published in 2003. A recommended review of the chemical thermodynamic data of Am compounds has been published by the Nuclear Energy Agency and the Organisation for Economic Co-operation and Development.[876] This reference reports critically evaluated stability constants for most characterized Am solids and solution species (and, by analogy, most expected Cm compounds). Myasoedov and Kremliakova[877] have reviewed Russian literature up to the mid-1980s on americium and curium chemistry, including separations.[877] A subsequent review has described research in this area up to 1994.[878]

The coordination chemistry of the heavier transplutonium elements relevant to separations is often the first type of chemistry examined for these metals. Indeed, a great deal of actinide coordination chemistry was founded in the separations developed in the 1950s and 1960s. For elements prepared an "atom-at-a-time" this information is first used to determine preferred oxidation state(s) and group classification of new heavy and super heavy elements. There are recent reviews that describe this chemistry for transactinides.[879–882] This approach remains very useful, as demonstrated by recent reports of the chemistry of element 108.[883] Being highly radio-active, the elements beyond plutonium have autoradiolytic properties that strongly affects their oxidation state and coordination chemistry, particularly in solution. These reactions and their corresponding rates are well described for the early actinides by Newton.[884–886]

Most transplutonium chemistry was performed either in initial chemical characterization by groups in the USA and Russia in the 1950s and 1960s or in the continuing effort to develop lanthanide/actinide and actinide/actinide separations in these countries, as well as in France, India, and Japan. The majority of the coordination chemistry is related to the development and testing of new and improved liquid/liquid extraction processes. Worldwide, aside from purely academic investigations, scientists and engineers are motivated to find new ways to remove minor amounts of neptunium, plutonium, and americium from various stored wastes or reactor fuel so that final disposal is easier and less expensive and so energy production is efficient and economically affordable. In addition, aqueous complexes, such as hydroxides, carbonates, phosphates, and sulfates, are being studied in the context of potential release from nuclear waste and nuclear fuel repositories.

Extraction processes (TRUEX, PUREX, Talspeak, DIAMEX, PARC, etc.) generally involve complexation of transplutonium elements by alkyl phosphines, phosphine oxides, phosphoric acids, carbamoyl phosphonates, diamides, and thiophosphinates in aqueous/organic extractions, within derivatized solid supports, or on coated particles. There are excellent reviews of the processes and significant complexes by Mathur *et al.*[409] and selected chapters in "The Chemistry of the Actinide and Transactinide Elements" to be published in 2003.[1043,1044] Work on the separation for nuclear waste management in the United States, France, and Russia have been reviewed.[887–889]

In the last two decades the coordination chemistry of these elements has benefited greatly from advances in time-resolved laser fluorescence (TRLIFS) and synchrotron-based X-ray absorbance spectroscopies (XANES and XAFS). The advantageous luminescence properties of Am^{III} and Cm^{III} allow the study of dilute solutions ($\sim 10^{-5}$ M for Am and $\sim 10^{-9}$ M for Cm) and complex matrices and require relatively small masses of material.[890–894] Similarly, EXAFS has been used to study the coordination of americium in organic chelator complexes,[895] and inorganic complexes with $P_5W_{30}O_{110}^{15-}$[896] and chloride.[897]

By analogy with lighter actinides and from experimental data, the coordination geometries of transplutonium ions in a range of oxidation states can be generalized. Americium(VI) and (V) generally have coordination numbers ranging from six to nine and most often adopt coordination environments of pentagonal bipyramidal and hexagonal bipyramidal for actinyl species, and cubic, octahedral, square antiprismatic, tricapped trigonal prismatic, or irregular geometries for nonactinyl species. The trivalent and tetravalent transplutonium compounds can have coordination numbers as high as twelve, and the most common regular geometries being octahedral, tricapped trigonal prismatic, and dodecahedral.

3.3.3.1 Divalent Oxidation State (Am, Cm, Cf, Es, Fm, Md, No)

3.3.3.1.1 *General characteristics*

Of the divalent transplutonium actinides, only Am compounds have been prepared in any significant quantity.

3.3.3.1.2 *Simple donor ligands*

(i) Ligands containing group 16 donor atoms

Oxides and aqueous. Although AmO has been reported, characterization data are inconsistent. It likely that the monoxide can only be synthesized under high pressure from Am metal and Am_2O_3, in a manner analogous to some lanthanide oxides. Cm^{II} is unknown other than as a transient aqueous species and a species coprecipitated from melts, and possibly in CmO.

(ii) Ligands containing group 17 donor atoms

While solid structures are generally rare for divalent americium compounds, the black halides $AmCl_2$, $AmBr_2$,[898] and AmI_2[899] have been prepared by reacting metallic americium with the corresponding mercuric halides at 300–400 °C. Interestingly, all three compounds crystallize in different lattices: orthorhombic $AmCl_2$, tetragonal $AmBr_2$, and monoclinic AmI_2. Ternary Am^{II} halides may be prepared from americium trihalides with americium metal and lithium metal as reductants.[900] The californium(II) and einsteinium(II) dihalides have been prepared by hydrogen reduction of the trihalides at high temperatures. Californium bromide has also been obtained by heating Cf_2O_3 in HBr, and there is also tracer level evidence for the formation of $FmCl_2$.

3.3.3.2 Trivalent Oxidation State (Am–Lr)

3.3.3.2.1 General characteristics

The coordination chemistry of trivalent transplutonium elements is very similar to the trivalent lanthanides of similar ionic radii. Coordination numbers of six, and eight to ten, and coordination polyhedra of octahedron, square antiprism, tricapped trigonal prism and mono-capped square anti-prism are common. In agreement with other actinide(III) ions the stability of transplutonium complexes complexes with monovalent inorganic ligands follows the order: $F^- > H_2PO_4^- > SCN^- > NO_3^- > Cl^- > ClO_4^-$. In some cases, the stability of the trivalent actinide complex is slightly greater than that of the corresponding lanthanide complex, due to a combination of bonding and solvation differences.[51,901,902] As discussed widely, this difference in stability can be used to effectively separate Am^{III} from lanthanide elements.

3.3.3.2.2 Simple donor ligands

(i) Ligands containing group 15 donor atoms

The Am^{III} thiocyanates have been studied intensively because of the separation of lanthanide and actinide elements in thiocyanate media. Three complexes of general formula $Am(SCN)_n^{3-n}$ ($n = 1$–3) have been identified from spectroscopic and solvent extraction data.

(ii) Ligands containing group 16 donor atoms

Oxides. Binary oxides of the formula An_2O_3 have been prepared and well characterized for Am, Cm, Bk, and Cf, while the heavier transplutonium oxides have generally only been prepared on the scale of micrograms or less. Three crystal modifications (similar to lanthanides) have been reported for both Am_2O_3 and Cm_2O_3, two of which have been found for Bk_2O_3 (A and C types) and for Am_2O_3 (A and B types). These oxides are generally prepared by heating oxyanion complex precipitates, such as $An_2(C_2O_4)_3$, or AnO_2 compounds at temperatures greater than 600 °C. For both Am and Cm, An_2O_3 transforms to the hexagonal phase at room temperature within about three years due to self-irradiation.[903] Additional phases are predicted based on phase diagrams of related lanthanide systems. The ternary oxide $LiAmO_2$ is obtained by heating AmO_2 with Li_2O in hydrogen at 600 °C. Other ternary oxides include $M(AnO_2)_2$ (M = Sr or Ba) and $MAlO_3$, reported for both Am and Cm, and the Cm oxides, Cm_2O_2Sb and Cm_2O_2Bi,[904] $BaCmO_3$,[905] and Cm_2CuO_4.[906] The latter is of interest by its analogy to M_2CuO_4 (M = La, Pr–Eu), which are parent compounds for high-temperature superconductors. Although Cm_2CuO_4 is isostructural with the M_2CuO_4 (M = Pr–Gd) series, its Th-doped analogue is not superconducting, unlike analogous Pr–Eu doped materials.

Hydroxide, aqua, and hydrates. From the similar absorption spectra of Am^{3+} in aqueous solution, $AmCl_3$, and in $LaCl_3$, and the linear relationship between the decay rate of the americium fluorescence and the number of inner-sphere water molecules, it has been concluded that Am^{III} is coordinated by nine inner-sphere water molecules.[907–909] Similarly, the hydration number for the Cm^{III} ion has been estimated to be nine on the basis of fluorescence life-times.[910,911] EXAFS studies of aqueous Am^{3+} and Cm^{3+}, however, have suggested coordination numbers closer to 10.[64] EXAFS investigation of Cf^{3+} in aqueous solution indicates a coordination number of 8.5 (± 1.5), with a Cf—O distances of 2.41 ± 0.02 Å.[912] This coordination number was confirmed for Am in the solid state by isolation of single crystals of the triflate salt of nonaqua complex, which contains a tricapped, trigonal prismatic cation that is isostructural with the analogous Pu^{III} compound.[69]

The trivalent transplutonium ions have stepwise hydrolysis products of the type $An(OH)_n^{3-n}$, where $n = 1,2,3$, with $n = 4$ species postulated for Am^{III}. The Cm species have been studied using time-resolved laser fluorescence spectroscopy,[913] the Am species using optical spectroscopy, and the heavier actinide hydroxides less directly using precipitation methods. In the solid state, the trishydroxides, $An(OH)_3$ (An = Am, Cm, Bk, and Cf) can be prepared by aging aqueous hydroxide precipitates.[914] The Am hydroxide can also be prepared by hydration of Am_2O_3 with steam at 225 °C.[915]

Hydrates of the oxoanion and halide complexes are numerous and are mostly isostructural with the hydrated lanthanide chlorides. In the hydrated salicylate, $AmL_3 \cdot H_2O$, one water molecule is in the inner coordination sphere of the Am^{III} cation. The hydrated xenate(VIII), $Am_4(XeO_6)_3 \cdot 40H_2O$, has also been reported.

Oxoanions. Trivalent transplutonium nitrates can be isolated by evaporation of nitric acid solutions of the ions. For example, curium trinitrate, $^{244}Cm(NO_3)_3$, has been characterized. The phosphates are much more numerous and complicated, due to the multiple protonation states of the ligand. The solution complexes, $AnHPO_4^+$ and $An(H_2PO_4)_n^{3-n}$ (n = 1–4) have been used to interpret cation exchange, solvent extraction and spectroscopic data.[916] However, some of those data could be reinterpreted as solvation changes with concomitant changes in ionic strength, and not discrete inner-sphere phosphate complexes. The phosphate solids that have been isolated by precipitation include hydrates, such as $AnPO_4 \cdot 0.5H_2O$ (An = Am, Cm)[917–919] and dehydrated $AnPO_4$. For Am^{III}, the anhydrous compound has also been obtained by reacting AmO_2 with stoichiometric amounts of $(NH_4)_2HPO_4$ at 600–1000 °C. Hobart *et al.* reported the Raman spectra of the phosphate salts, as well as those of the salts of a number of other oxyanion complexes.[920]

Sulfate complexes in solution, of the form $An(SO_4)_2^{3-2n}$ (n = 1,2) have been reported. These anions can be precipitated as hydrates $An_2(SO_4)_3 \cdot 8H_2O$[921] and partially dehydrated solids can be obtained by addition of less polar solvents.[933] Anhydrous sulfates, such as $An_2(SO_4)_3$ (An = Am, Cm, or Cf) are also prepared by heating the hydrate to a temperature of 500–600 °C in air.[922] A number of double sulfates of Am^{III} with formulas $MAm(SO_4)_2 \cdot xH_2O$ (M = K, Na, Rb, Cs, Tl; x = 0, 1, 2, 4), $K_3Am(SO_4)_3 \cdot xH_2O$, and $M_8Am_2(SO_4)_7$ (M = K, Cs, Tl) have been prepared by adding metal sulfate to Am^{3+} in sulfuric acid solutions. The oxosulfates, $An_2O_2SO_4$ (An = Cm or Cf), have been reported; the curium compound is obtained by heating Cm^{III}-loaded resin (sulfonate form) in a stream of oxygen at 900 °C; it has a body-centered orthorhombic structure.[923]

Tabuteau and Pages[924,925] investigated the Am–molybdate and Am–tungstate systems. By reacting stoichiometric amounts of AmO_2 and MoO_3 or WO_3 at 1,080 °C, the monoclinic $Am_2(MoO_4)_3$ and $Am_2(WO_4)_3$ are prepared; with potassium present and at lower temperature, ternary phases, $KAm(MoO_4)_2$ and $K_5Am(MoO_4)_4$, are isolated. Higher order tungstate and heteropolyanionic complexes have been studied, including for their use as solution precursors to solid state materials. Shirokova *et al.*[926] reported the complexation of Am^{III} with *N,N*-dimethylacetamide and the Keggin-type heteropolyanion $PW_{12}O_{40}^{3-}$. Complexes of Am^{III} and Cm^{III} with $W_{10}O_{36}^{12-}$, $PW_{11}O_{39}^{7-}$ and $SiW_{11}O_{39}^{8-}$ have also been prepared.[927–930] In contrast, Williams *et al.* reported that Am^{III} can be integrated into the Preyssler anion, $AmP_5W_{30}O_{110}^{12-}$.[896]

Carbonates and carboxylates. The trivalent transplutonium formate, carbonate, and oxalate complexes have been relatively well studied. The oxalato complexes, particularly $Am_2(C_2O_4)_3 \cdot 10H_2O$, have been used extensively for separations and other processing. The carbonates have been studied primarily in the context of waste management and environmental risk. Americium formate can prepared by evaporating a solution of $Am(OH)_3$ in concentrated formic acid. The binary Am^{III} carbonate, $Am_2(CO_3)_3 \cdot 4H_2O$ precipitates from a CO_2-saturated solution of $NaHCO_3$.[931,932] The analogous Cm^{III} solid forms after addition of K_2CO_3 to Cm^{III} solution.[933] The ternary compounds $NaAm(CO_3)_2 \cdot 4H_2O$ and $Na_3Am(CO_3)_3 \cdot 4H_2O$ can also be precipitated from bicarbonate solutions.[859] In analogy to neodymium and europium analogs, orthorhombic $AmOHCO_3$ was characterized by X-ray powder diffraction data.[931,934] A hexagonal form has been reported, but not confirmed.[935] The anhydrous carbonates, $An_2(CO_3)_3$ (M = Am or Cm) are formed by the radiolytic decomposition of the oxalate or by heating the anhydrous oxalates.

The hydrated oxalate, $Am_2(C_2O_4)_3 \cdot xH_2O$ (x = 7, 9, 10 or 11), is precipitated from aqueous solutions containing americium(III) by oxalic acid and the anhydrous oxalate is obtained by heating the decahydrate above 340 °C.[936] The corresponding Bk and Cf solids are precipitated from acid solutions by oxalic acid. Both the Am and Cm oxalates are used for calcination to the oxides. For example, oxalate precipitation has been used to process large amounts of ^{244}Cm, with subsequent metathesis with 0.5 M hydroxide to form $Cm(OH)_3$.[937,938] The ternary oxalate complexes of general formula $MAm(C_2O_4)_2 \cdot xH_2O$ have been prepared from Am^{III} oxalate and MC_2O_4 (M = NH₄, Na, K, Cs) in neutral solution.[939] It has been demonstrated that a substantial

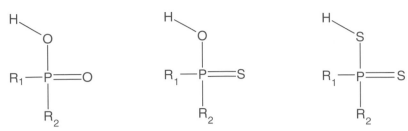

Figure 88 General diagrams of phosphinic acid, monothiophosphinic acid and dithiophosphinic acid, respectively.

separation of americium from lanthanum can be obtained by fractional precipitation of americium and lanthanum oxalates; about 50% of the lanthanum can be rejected at each stage with only about 4% of the americium.

Phosphoric, phosphinic, and phosphonic acid. Phosphinic acid-based extractants contain the P(O)OH acid functionality as well as two R groups attached to the phosphorus, where R can be hydrogenic, aliphatic, or aromatic. Sulfur containing derivatives are known as monothiophosphinic acids, where the phosphonyl oxygen is replaced with a sulfur, and dithiophosphinic acids, where both oxygens are replaced with sulfur (Figure 88).

In the extraction of trivalent lanthanides and actinides with phosphinic acids, the hard donor nature of oxygen makes it difficult to effectively distinguish between the hard cations Ln^{III} and An^{III}, thus making separation of one from the other difficult. The replacement of oxygen with softer donor atoms such as sulfur make Ln^{III} and An^{III} more distinguishable, hence the development of mono- and dithiophosphinic acids as more effective extractants in the separation of these species. In most studies, the dithiophosphinic acids have been proven to be the most effective agents for selectively separating these ions. To this end, Am^{III} and Cm^{III} complexes with bis(2,4,4-trimethylpentyl)phosphinic acid (HC272), bis(2,4,4-trimethylpentyl)monothiophosphinic acid (HC302), and bis(2,4,4-trimethylpentyl)dithiophosphinic acid (HC301) (extracted into *n*-dodecane), all available through Cytec, Inc., Canada, have been studied by visible absorption spectroscopy and X-ray absorption fine structure spectroscopy (XAFS) to determine the origin of this selectivity.

XAFS spectroscopy attempts to model experimental data to propose a coordination environment of a metal complex. XAFS modeling proposed that for HC272 there is only oxygen donation to the Am^{III} and Cm^{III} inner coordination sphere, while for HC301, only sulfur donation is observed. Am^{III} and Cm^{III} HC302 complexes have both oxygen and sulfur bound to the metal. These coordination models make sense in light of the chemical structure of the extractants.

When an excess of HC272 is present in the organic phase, it is proposed to coordinate to trivalent actinides in a fashion similar to most oxygen-containing diphosphinic acids, yielding a coordination stoichiometry of $M[H(C272)_2]_3$. The bonding consists of three $H(C272)_2^-$ hydrogen-bonded dimers coordinated in a bidentate mode to the metal as seen in Figure 89, allowing for excellent extraction into a non polar organic phase. XAFS modeling also indicates that the hexacoordinate complexes of HC272 arrange in a distorted octahedral (O_h) geometry, comparable to most An^{III} ions in highly ionic coordination complexes.[940]

Coordination studies of monothiophosphinic acids with actinides are scarce in the literature. Slope analysis from extractant dependency studies using trivalent curium indicate a 3:1 stoichiometry for the coordination of HC302 with the metal.[941] XAFS studies indicate that all M–HC302 bonds are shorter than those typically observed with R_2POS^- complexes and are more like bond lengths seen in hexacoordinate $R_2PO_2^-$ and $R_2PS_2^-$ complexes. These studies also show that the extracted complex consists of a trivalent actinide cation coordinated with two monodentate $C302^-$ molecules through oxygen, one bidentate $C302^-$ molecule through both oxygen and sulfur, and one water molecule, all of which are bound in the inner sphere as seen in Figure 90. Although the coordination number of between four and five shown above is unusually low for actinide cations, it is consistent with all experimental and modeling data.

XAFS data for complexes of Cm^{III} with the HC301 extractant indicate only sulfur donation to the metal in the inner sphere of coordination. HC301 forms 3:1 complexes with the trivalent actinides and are coordinated in a bidentate mode as seen in Figure 91. Data indicates a hexacoordinate structure that resembles D_3 symmetry in lanthanide dithiophosphinic acid complexes.

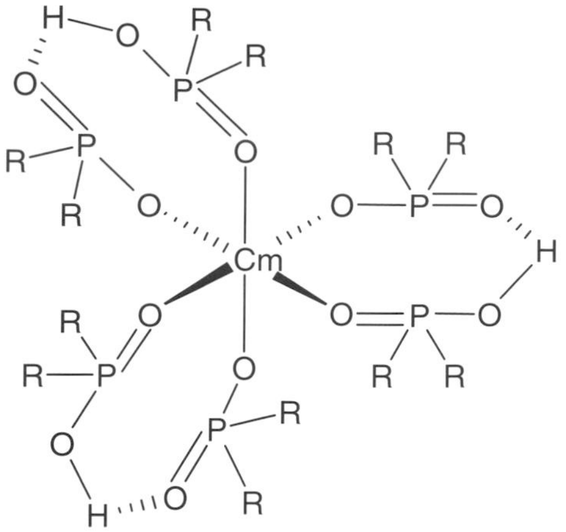

316 *The Actinides*

Figure 89 Predicted complexation of HC272 with Cm^{III} from XAFS (Jensen and Bond, *J. Am. Chem. Soc.* **2002**, *124*, 9870–9877).

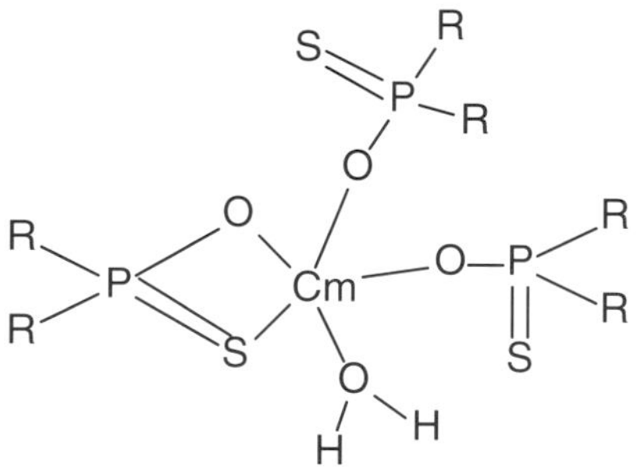

Figure 90 Proposed diagram of HC302 complexation of Cm^{III} from XAFS.

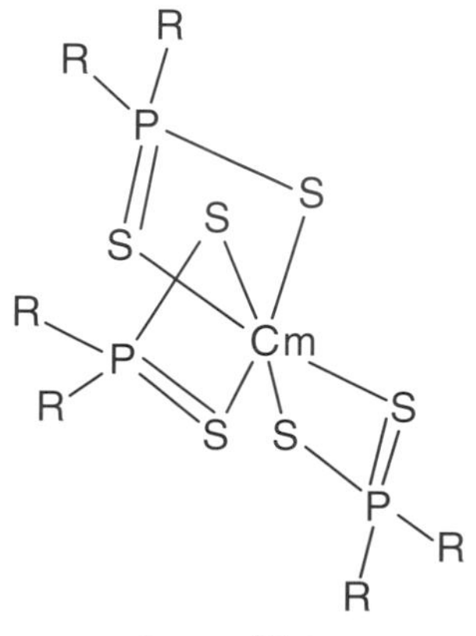

Figure 91 Predicted complexation of HC301 with Cm^{III} from XAFS (Jensen and Bond, *J. Am. Chem. Soc.* **2002**, *124*, 9870–9877).

The fact that dithiophosphinic acids show the greatest selectivity for An^{III} over Ln^{III} as compared to phosphinic and monothiophosphic acids is due to an increased covalency in the An—S bonds but not necessarily a shorter An—S bond. In addition, trivalent actinides show a thermodynamic preference to form bonds with soft donor atoms. The structural differences observed in the complexes of An^{III} with the three ligands studied are due to differences in the hard and soft nature of the oxygen and sulfur atoms, respectively, and their hydrogen-bonding ability.[940]

A study by Zhu *et al.* using purified HC301 (>99% bis(2,4,4-trimethylpentyl)dithiophosphinic acid) in kerosene seems to indicate a different reaction stoichiometry with trivalent americium.[943] Previous studies indicated that the extractant is mainly found in a dimeric form at higher extractant concentration ranges.[942] Slope analyses in the more recent study, however, show a pH dependence using nitric acid of about three and a log extractant dependence of about two,[943] suggesting an alternate extraction stoichiometry for Am^{III} with HC301 in kerosene as shown in Equation (64):

$$Am^{3+}{}_{(a)} + 2\,(HC301)_{2(o)} \longrightarrow Am(C301)_3(HC301)_{(o)} + 3\,H^+{}_{(a)} \qquad (64)$$

Despite this stoichiometry ambiguity, both studies suggest a high selectivity for trivalent actinides over lanthanides by dithiophosphinic acids.[943]

A group of aromatic dithiophosphinic acids, $R_2PS(SH)$, where $R = C_6H_5$, ClC_6H_4, FC_6H_4, and $CH_3C_6H_4$, have been made and used with various synergists to extract trivalent actinides from nitric acid media into toluene. Extraction of Am^{III} requires the presence of a synergist such as tributyl phosphate (TBP), and extraction ratios increase in the order $(C_6H_5)_2PS(SH)$ $<(FC_6H_4)_2PS(SH) <(ClC_6H_4)_2PS(SH)$.

Slope analyses to determine the extraction stoichiometries of Am^{III} for both the C_6H_5 and ClC_6H_4 extractants were performed to determine pH, TBP, and extractant dependencies. For Am^{III}, many nonintegral slopes were obtained, indicating that mixed complexes with varying stoichiometries are being extracted. It is believed that the extractants may act similarly to Cyanex 301 and form complexes of the type $AmA_3(HA)_xTBP_y$, but the values for x and y are still undetermined and further work needs to be done. Trioctylphosphine oxide was found to be the best synergist for Am^{III} extraction.[944]

Phosphine oxides. Few molecular complexes of trivalent transplutonium elements have been reported. Several studies examine the extraction chemistry of Am^{3+}, Cm^{3+}, and Bk^{3+} with a combination of β-diketones and tri-*n*-alkyl phosphine oxides and trialkylphosphates. From these, compounds reported to be of the formula $AnL_3(R_3PO)_x$ (An = Am, Cm; R = *n*-octyl, Bu^{n}O) were isolated, where L = $CF_3COCHCOR$ (R = Me, CF_3, Bu^t).[945–947] The stoichiometry of the complexes (An:P=O) was not always reported. The complex $Am(CF_3COCHCOCF_3)_3[OP(OBu^n)_3]_2$ is reported to be volatile at 175 °C.[947]

(iii) Ligands containing group 17 donor atoms

The trivalent transplutonium halides have been extensively studied. Several reviews deal specifically with actinide halides.[948–951] In aqueous solution the mono- and bis-complexes have been characterized, with the formation of the latter decreasing down the halide series. For example, AmF^{2+} and AmF_2^+ have been studied, but only a very weak monochloride complex $AmCl^{2+}$ has been reported. These species are reported to have coordination numbers as high as 11, although recent EXAFS studies show that the hydration number decreases with increasing halide concentration (and ionic strength) at concentrations below which the halo complexes form. These data suggest the coordination numbers of the mixed aquo halo complexes are probably seven to ten.

In the solid state, the complete series AnX_3 (X = F, Cl, Br, I) are known for Am and Cm and most are known for Bk–Es. A variety of preparative methods have been reported and reviewed, from single step element combinations at high temperature, to treatment of oxides with anhydrous HX, to complex salt precipitation followed by multistep decompositions. The trifluorides are prevalent and can be prepared by dehydrating precipitates from HF or other concentrated fluoride solutions, or by treating oxides or hydroxides with $HF_{(g)}$. Most adopt the 11-coordinate LaF_3 structure. One form of BkF_3 and the known form of CfF_3 and EsF_3 have the eight-coordinate, high temperature LaF_3-type structure. With a larger anion, the chlorides have the nine-coordinate hexagonal UCl_3-type structure across the series to Cf, for which one form has this nine-coordinate structure and a second modification has the eight-coordinate $PuBr_3$ structure. The chlorides can be prepared by treating the oxides or oxychlorides with anhydrous HCl at elevated temperature. The tribromides (Am, Cm, Bk) and α-AmI_3 are isostructural with the $PuBr_3$. However, β-AmI_3 and the other triiodides (Cm, Bk, Cf) have the orthorhombic six-coordinate BiI_3 structure (Am is the only dimorphic actinide triiodide.).[952]

318 *The Actinides*

The oxyhalides AmOCl, AmOI, and CmOCl are isostructural PbClF type (hexagonal), and contain nine coordinate An^{III} surrounded by four oxygens and five halides. Americium(III) oxyboride has also been prepared, but the structure is unconfirmed.

Hydrates of the oxoanion and halide complexes are numerous for Am^{III} and Cm^{III} and a few have also been reported for Bk^{III} and Cf^{III}. The hexahydrate trichlorides and tribromides, $AnCl_3 \cdot 6H_2O$ (An = Am or Bk) and $AnBr_3 \cdot 6H_2O$ (An = Am or Cf), are isostructural with the hydrated lanthanide chlorides and involve mixed ligand complexes, such as $[AnCl_2(H_2O)_6]^+$.

Many other trihalide adducts have been prepared. For example, americium chloride has been treated with a number of salts to yield $AmCl_3 \cdot MCl$ where MCl is LiCl, CsCl, $(C_4H_9)_4NCl$, or $(C_2H_5)_4NCl$.

Ternary complexes, such as $MAmX_4$, M_2AmX_5, KAm_2F_7, and M_2AmX_6 and M_3AmX_3, where M = an alkali eath metal, have been prepared and characterized for Am and some are also known for heavier transplutonium ions. The fluoro complexes $MAmX_4$ (M=phosphonium) are isostructural with the analogous Pu^{III} compounds. Octahedral anions, such as $AmCl_6^{3-}$ and $AmBr_6^{3-}$ are present in M_2AmX_5 and in triphenylphosphonium salts in anhydrous ethanol.[953]

3.3.3.2.3 Chelating ligands

CMPO. Carbamoylmethylphosphine oxide (CMPO) acts as a neutral extractant in complexing with Am^{III} in both nitric acid (low to high concentrations) and hydrochloric acid (moderate to high concentrations). Extractant dependency studies have shown the stoichiometric relationship between the CMPO ligand and the actinide to be 3:1 in forming the extracted species in both nitric and hydrochloric acid systems as indicated in Equation (65):

$$Am^{3+}_{(a)} + 3\ X^-_{(a)} + 3\ CMPO \bullet (HX)_{n(o)} \longrightarrow Am(X)_3 \bullet (CMPO)_3(HX)_{m(o)} + (3n-m)\ HX_{(a)}$$

$$(X = NO_3, Cl;\ n = 0\text{-}2;\ m = 0\text{-}3)$$

(65)

In the TRUEX process, CMPO is typically used with tributylphosphate (TBP) as a phase modifier and some organic solvent such as dodecane, carbon tetrachloride, tetrachloroehtylene, or paraffinic hydrocarbons. Distribution studies indicate that the Am^{III}/nitrate/CMPO complex is considerably more extractable than the analogous chloride complex at low/moderate acid concentrations due to the soft nature and larger hydration energy of the chloride ion. While CMPO extracts tetra- and hexavalent actinides in addition to trivalent actinides, selective partitioning of Am^{III} over the higher oxidation states can be achieved by suitable selection of acid concentrations.[954] Complexation of CMPO with Am^{III} occurs in one of two ways. In nitrato complexes, monodentate coordination between the electronegative phosphoryl oxygen and the metal center occurs. In chloro complexes, bidentate coordination occurs with the metal center via the phosphoryl oxygen, as well as the carbonyl oxygen. This leads to a coordination number of nine for both types of complexes.[955]

Comparative studies have shown that changing the substituents attached to the phosphoryl group alters the basicity of the donor group and strengthens or weakens the extractive ability of the molecule. Substitution with two alkoxy groups, as in dihexyl-N,N-diethylcarbamoylmethylphosphine oxide, decreases the basicity of the donor group and lowers the Am^{III} extraction. Substitution with two phenyl groups also shows an inductive effect, but less so than an alkoxy substitution.[956]

The effectiveness of CMPO tends to drop off at high acid concentration due to the competition between actinide cations and protons at the basic bonding site on CMPO. CMPO extraction is also very dependent upon the diluent, making generalized statements about CMPO extraction very difficult.[957]

β-Diketonate. The hydrated americium β-diketonate compounds, $AmL_3 \cdot xH_2O$, HL = (HL = $MeCOCH_2COMe$, CF_3COCH_2COPh, $CF_3COCH_2CO(2\text{-}C_4H_3S)$, and $Bu^tCOCHCO\text{-}Bu^t$), are precipitated from aqueous or ethanolic Am^{III} solutions. The latter is isomorphous with the analogous Ln^{III} chelates of similar ionic radii, in which the central metal atoms are seven coordinate. The Bk^{III} compounds, $Bk(CF_3COCHCOR)_3$ (R = CF_3, Bu^t and 2-C_4H_3S), have been obtained by a solvent extraction method. The salt $CsAn(HFAA)_4 \cdot xH_2O$, where HFAA = hexafluoroacetylacetonate, has been studied in detail for An = Am, Cm, Bk, Cf, and Es.[958] This compound forms readily when HFAA is added to ethanol solutions of the trivalent actinides in the presence of cesium ion. The coordination environment about the metal center in the anion of these salts is dodecahedral. Base adducts of fluorinated diketonates have been isolated, including those of tributyl phosphate or trioctylphosphine oxide.[946]

2-Thenoyltrifluoroacetone (HTTA) (Figure 92) has been used as an extraction agent with some soft donor ligands for Am^{III} in the presence of Eu^{III}. The soft donor ligands used include triphenylamine (Ph_3N), triphenylarsine (Ph_3As), and triphenylphosphine (Ph_3P). The addition of Ph_3As and Ph_3P showed a synergistic enhancement of the extraction into benzene, with the latter showing a greater effect. In both cases, the ligand and synergist combinations showed no preference for Am^{III} over Eu^{III}, and the soft donor ligands gave weaker distribution ratios than previous studies with hard donor ligands. Extractant dependency indicates that three HTTA molecules are involved in the complexation.[959]

The extraction of trivalent actinides with β-diketones in the presence of crown ethers has also been investigated. Trifluoroacetylacetone(HTFA), HTTA, benzoyltrifluoroacetone(HBFA), and 2-napthoyltrifluoroacetone (HNFA) have all been used with 18-crown-6 to remove Am^{III} and Cm^{III} from perchlorate media into 1,2-dichloroethane. It has been shown that the mechanism of extraction for all of the β-diketones (HA) is the formation of a cationic complex via a synergistic ion-pair extraction (SIPE) as shown in Equation (66):

$$M^{3+}_{(a)} + 2\,HA_{(o)} + \text{18-crown-6}_{(o)} + ClO_4^{-}_{(a)} \longrightarrow [MA_2(\text{18-crown-6})][ClO_4]_{(o)} + 2\,H^+_{(a)}$$

$$(M = Am, Cm)$$

$$(66)$$

The preference of the SIPE for the trivalent actinides is due to the stabilization of the crown ether/ M^{III} bonds due to a "size-fitting effect" where the diameter of the metal ion matches that of the crown ether cavity, leading to significantly greater extraction efficiency due to this synergistic effect.[960]

Another widely studied β-diketone for the extraction of trivalent actinides has been 1-phenyl-3-methyl-4-benzoylpyrazolone-5 (HPMBP), illustrated in Figure 93. The extraction of Am^{III}, Cm^{III}, Bk^{III}, and Cf^{III} by HPMBP may be expressed in two ways. First, the extraction of Cm^{III} has been observed to follow the extraction equilibrium shown in Equation (67):

$$Cm^{3+}_{(a)} + 3\,HPMBP_{(o)} \longrightarrow Cm(PMBP)_{3(o)} + 3\,H^+_{(a)} \tag{67}$$

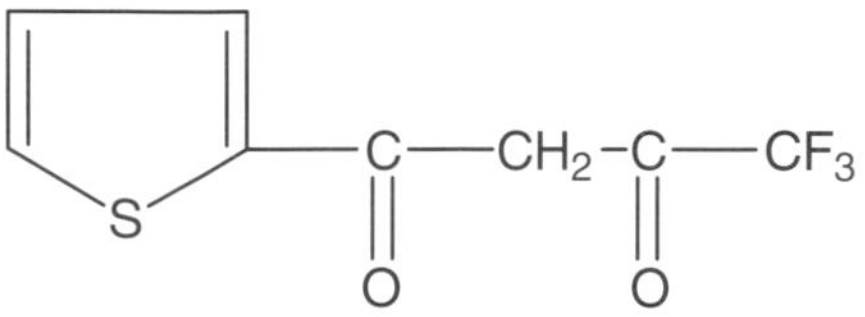

Figure 92   Thenoyltrifluoroacetone (HTTA).

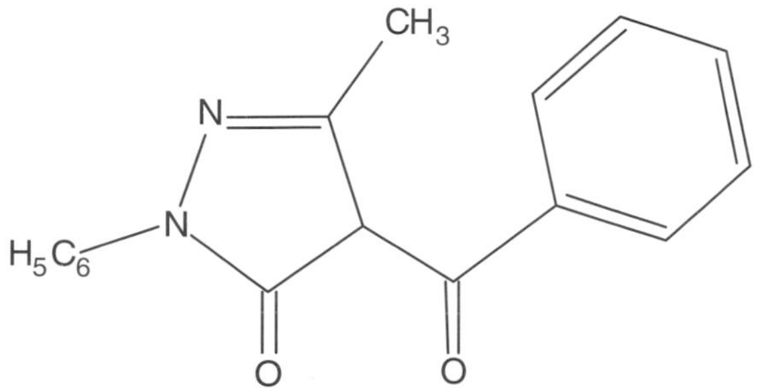

Figure 93   1-phenyl-3-methyl-4-benzoylpyrazolone-5 (HPMBP).

In contrast, Am^{III}, Bk^{III}, and Cf^{III} extraction equilibria are characterized by the formation of self-adduct species as shown in Equation (68):

$$M^{3+}_{(a)} + (3+n)\ HPMBP_{(o)} \longrightarrow M(PMBP)_3 \bullet n(HPMBP)_{(o)} + 3\ H^+_{(a)} \tag{68}$$

$$(M = Am, Bk, Cf)$$

In the above equilibrium, extractant dependency studies have indicated that $n = 1$ for Am^{III} and Bk^{III} and $n = 2$ for Cf^{III}. These stoichiometries have been observed for extractions into chloroform, and the self-adduct formation with trivalent actinides has been possibly connected with "tetrads" where Cm is one of the minima. Extractions into xylene, however, leads to the formation of self-adducts with all four actinides due to better distribution coefficients in xylene over chloroform. The formation of self-adducts is due to ligand concentration, ionization constant of the ligand, basicity of the bound ligand, solvent identity, and oxidation state of the metal ion.

The addition of a neutral donor synergist such as tri-*n*-octylphosphine(TOPO) oxide to the above HPBMP extractions leads to the equilibrium shown in Equation (69). The equation holds true for all four trivalent actinides and m can take the value of one or two:[961]

$$M^{3+}_{(a)} + 3\ HPMBP_{(o)} + mTOPO_{(o)} \longrightarrow M(PMBP)_3 \bullet m(TOPO)_{(o)} + 3\ H^+_{(a)} \tag{69}$$

Derivatives of HPBMP of the type bis(1-phenyl-3-methyl-4-acylpyrazol-5-one), hereafter referred to as H_2BP_n, where $n = 3$, 4, 5, 6, 7, 8, 10, and 22 have been studied as quadridentate extractants for trivalent actinides(from both perchlorate and nitrate media into chloroform (Figure 94). These derivatives were found to extract better than HPBMP due to their multidentate nature and high hydrophobic character. For Am^{III}, Cm^{III}, and Cf^{III}, the highest degree of extraction was obtained with the H_2BP_7 and H_2BP_8 ligands. Extraction dependency for n equals 5–8, 10, and 22 indicates that the extraction equilibrium involves one singly and one doubly deprotonated ligand in the extracted species.

For H_2BP_3 and H_2BP_4, similar dependency studies indicate that another equilibrium mechanism may come into play where one H_2BP_n ligand is coordinated with the species being $M(BP_n)(OH)$ or $M(HBP_n)(OH)_2$.[398]

Diamides. Malonamides have excellent chelating abilities with trivalent actinides such as Am^{III}, as well as with the lanthanides. The extracting ability of malonamides tends to increase with the decreasing basicity of the molecule; this phenomenon is due to lower proton/metal competition with the less basic malonamide. Spjuth *et al.*[957] studied the uptake of Am^{III} from HNO_3 solutions by malonamides that had been tailored to vary the basicity of the ligand (see Table 49).

The coordination of the malonamides with Am^{III} is believed to be bidentate in nature via the carbonyl oxygens. The oxygen in alkoxy R_3 substituents is not believed to take part in binding to the Am^{III} metal. As nitric acid concentration increases, the average number of HNO_3 molecules associated with each malonamide in the organic phase increases, with a maximum of four occurring for BUDOPx (N,N'-dimethyl-N,N'-dibutylundecylpropoxy malonamide), but differing depending on R-groups.[957]

In malonamides, one nitrogen substituent is usually a methyl group to minimize the steric interactions that may hinder complexation. In order to make the malonamide soluble in organics, the other nitrogen substituent is typically a long alkyl chain.[962] However, the presence of a phenyl

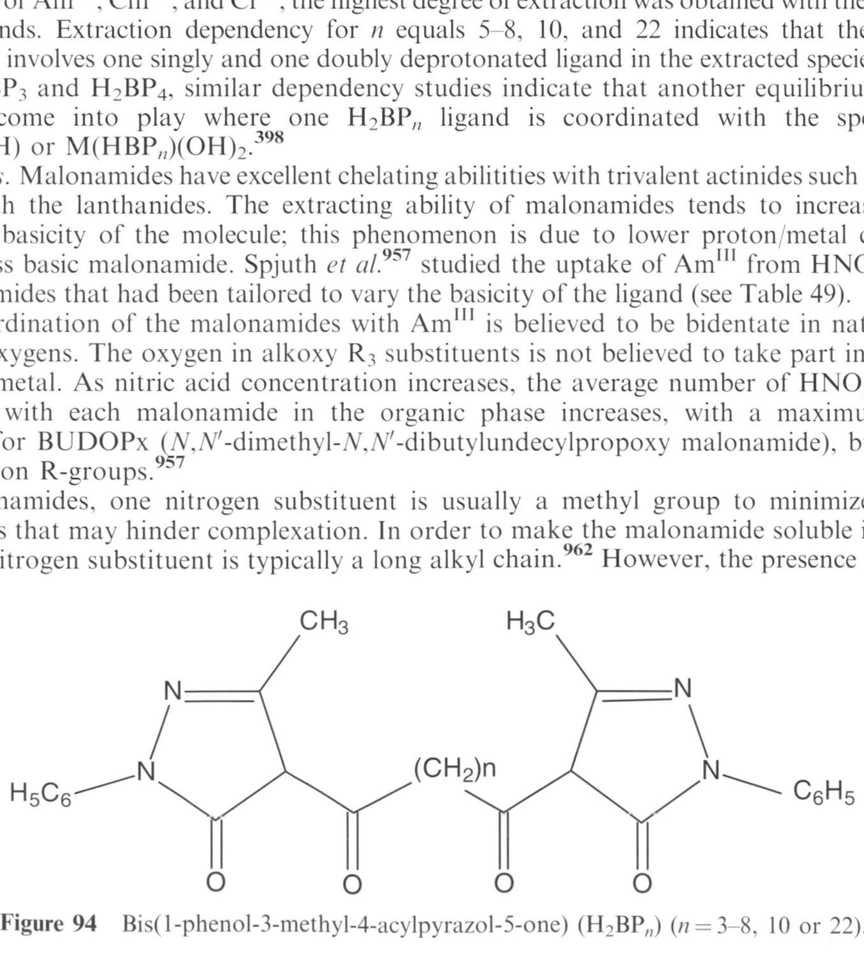

Figure 94   Bis(1-phenol-3-methyl-4-acylpyrazol-5-one) (H_2BP_n) ($n = 3$–8, 10 or 22).

Table 49 Basicity of substituted malonamides (reproduced from Spjuth, L., *et al. Solvent Extr. Ion Exch.* **2000**, *18*, 1.).

R_1	R_2	R_3	Abbreviation	Basicity
methyl	butyl	$C_{14}H_{29}$	BTD	Most basic
methyl	butyl	$C_{18}H_{37}$	BOD[a]	
methyl	cyclohexyl	$C_{14}H_{29}$	CHTD	
methyl	butyl	$C_3H_6OC_{11}H_{23}$	BUDOPx	
methyl	butyl	$C_2H_4OC_{12}H_{25}$	BDDEx	
methyl	octyl	$C_2H_4OC_6H_{13}$	OHEx	
methyl	phenyl	$C_{14}H_{29}$	PHTD	
methyl	4-chlorophenyl	$C_{14}H_{29}$	CLPHTD	Least basic

[a] Basicity data not available; assumed to be similar to BTD and CHTD

group allows for increased electron withdrawing ability and lower basicity, thus making phenyl-substituted malonamides better complexing reagents for Am^{III} than the alkyl-substituted analogs.[957] Addition of a long alkyl or alkoxy chain (14 carbons or more) at the malonamide methylene carbon leads to even better solubility in the organic phase, leading to enhanced extractive power.[962] Studies where the methylene substituent is a hydrogen atom, such as in N,N'-dimethyl-N,N'-dibutylmalonamide and N,N'-dimethyl-N,N'-dioctlymalonamide, show poorer extractive ability than those with long alkyl/alkoxy chains, probably due to the a lower solubility of the metal complex in the organic phase.[429,963]

Mechanisms for the extraction of Am^{III} or Cm^{III} can be either coordination or ion-pair, depending on acid concentrations as suggested by a study using the CHTD (N,N'-dimethyl-N,N'-dicyclohexyltetradecyl) malonamide and PHTD (N,N'-dimethyl-N,N'-diphenyltetradecyl malonamide malonamides). At low nitric acid concentrations, a coordination mechanism is suspected due to lower competition between protons and the actinide for the carbonyl oxygens (L = malonamide ligand, M = Am, Cm). Extractant dependency studies indicate that $n = 2-4$ malonamide molecules in the extracted species. At moderate and high nitric acid concentrations, two ion-pair mechanisms dominate, respectively as shown in Equations (70) and (71). Under these conditions, the malonamide is likely protonated at one or both carbonyl oxygens:[964]

$$HL^+_{(o)} + M(NO_3)_4^-_{(a)} \longrightarrow HL^+M(NO_3)_4^-_{(o)} \tag{70}$$

$$H_2L^{2+}_{(o)} + M(NO_3)_5^{2-}_{(a)} \longrightarrow H_2L^{2+}M(NO_3)_5^{2-}_{(o)} \tag{71}$$

Sasaki and Choppin have looked at diglycolamides and thiodiglycolamides with thenoyltrifluoroacetate(HTTA) in perchlorate solutions as more effective extractants for trivalent actinides. The diamides examined were N,N'-dimethyl-N,N'-dihexyl-3-oxapentanediamide (DMDHOPDA), N,N'-dihexyl-3-oxapentanediamide (DHOPDA), N,N'-dimethyl-N,N'-dihexyl-3-thiopentanediamide (DMDHTPDA), and N,N'-dihexyl-3-thiopentanediamide (DHTPDA). When trivalent americium was present only with the diamide, poor extraction was seen, probably due to hydrogen bonding and the soft donor nature of the sulfur atoms. Addition of HTTA, however, showed a synergistic effect and drastically increased the complexation and extraction ability of the diamide. Extraction stoichiometry from dependency studies for Am^{III} is given by Equation (72), where L is a diamide:[397,426,965]

$$Am^{3+}_{(a)} + 3\,HTTA_{(o)} + L_{(o)} \longrightarrow Am(TTA)_3(L)_{(o)} + 3\,H^+_{(a)} \tag{72}$$

Coordination of the diamide with the metal ion can occur in both a bidentate mode through both carbonyl oxygens or one carbonyl oxygen and the bridging oxygen/sulfur, or in a tridentate mode via both carbonyl oxygens and the bridging atom.[966]

Diphosphinic acids. The diphosphonic acids such as the compound P,P'-di(2-ethylhexyl) methanediphosphonic acid (H$_2$DEH[MDP]) have been investigated as extractants for Am^{III} from nitric acid into *o*-xylene. The acid dependence for Am^{III} extraction by H$_2$DEH[MDP] shows unusual behavior compared to other actinide cations. Am^{III} shows a peak maximum in the nitric acid range 0.1 to 1 M (0.1F H$_2$DEH[MDP]). At lower acidities, a positively charged complex is formed, resulting in a 1:1 Am^{III}/extractant complex which favors the aqueous phase, giving low distribution ratios. In the aqueous phase, the positive charge on the complex is balanced by negatively charged nitrate species. At higher acidities, 1:2 complexes are predicted, allowing neutralization of the trivalent charge by the extractant, eliminating the need for nitrate anions. Extractant dependency studies confirm these varying complexation stoichiometries.[418] The observed 2:1 stoichiometry at high acidities is due to the existence of the extractant in a dimeric form induced by the solvent.[421]

The related compounds, *P,P'*-di(2-ethylhexyl)ethanediphosphonic acid ($H_2DEH[EDP]$) and *P,P'*-di(2-ethylhexyl)butanediphosphonic acid ($H_2DEH[BuDP]$) have also been studied for the extraction of Am^{III}. Studies have shown that the increasing length of the alkyl bridge in the diphosphonic acids has the effect of decreasing the stability of the metal/extractant complexes formed due to their bidentate nature in forming chelating rings.[967] Hence, the extraction efficiency for Am^{III} by this series of acidic extractants follows the order $H_2DEH[MDP] > H_2DEH[EDP] > H_2DEH[BuDP]$. An interesting phenomenon observed for $H_2DEH[EDP]$ is the high degree of aggregation that the compound undergoes in the *o*-xylene diluent. It is believed that it forms inverted micelles with an aggregation number of six, leading to an extraction mechanism where the Am^{III} is hydrated and transferred into the hydrophilic interior of the aggregated micelle, followed by release of H^+ ions. The interaction of Am^{III} here is weaker than that in $H_2DEH[MDP]$, probably due to the hydration of the metal and its tendency to behave as a monodentate ligand.[419]

The longer length of the alkyl bridge in $H_2DEH[BuDP]$ causes it to behave even more as a monodentate extractant and lowering its extractant efficiency. It has a lower aggregation state than its $H_2DEH[EDP]$, causing Am^{III} complexes to form containing two trimers of the extractant (consistent with the observed extractant dependency of two).[420]

Similar studies in the depolymerizing diluent 1-decanol, where the diluent competes for hydrogen bonding with the phosphoryl functional sites, causes the extractants to exist as monomers in solution. Additionally, 1-decanol has the effect of suppressing metal extraction, and extractant dependency studies indicate slopes of one (1:1 metal/extractant complexes) to three (1:3 metal/extractant complexes) for Am^{III} at the lowest and highest extractant concentrations, respectively. The suppression of metal ion extraction by 1-decanol is presumably due to the hydrogen bonding described above, thereby increasing competition for the chelating site.[421]

The acidic extractant *P,P'*-di(ethylhexyl)benzene-1,2-diphosphonic acid ($H_2DEH[1,2-BzDP]$), shown in Figure 95, has been investigated as an extractant for Am^{III} into *o*-xylene due to the rigidity of functional groups on the benzene ring and the low likelihood for aggregation. $H_2DEH[1,2-BzDP]$ is unstable at room temperature, but was found to be sufficiently stable in *o*-xylene with refrigeration. The extractability of Am^{III} with this ligand (compared with analogues with alkane backbones) follows the order $H_2DEH[1,2-BzDP] > H_2DEH[MDP] > H_2DEH[EDP] > H_2DEH[BuDP]$ over a wide range of nitric acid concentrations. Extractant dependency studies indicate a complexation stoichiometry of 1:3 for metal to extractant.

The entrapment of Am^{III} by three extractant molecules leads to the conclusion that the $H_2DEH[1,2-BzDP]$ exists as a monomer in *o*-xylene, unlike $H_2DEH[MDP]$ where dimers yield a complex stoichiometry of two.[967]

Other. Hydroxyquinoline complexes have also been prepared from aqueous solution and have the form AmL_3, where HL is 8-hydroxyquinoline, 5-chloro- or 5,7-dichloro-8-hydroxyquinoline. There is

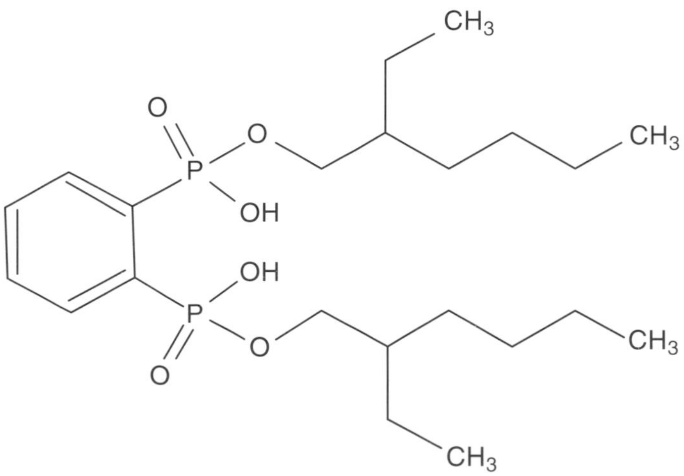

Figure 95 $H_2DEH[1,2-BzDP]$.

solvent extraction evidence for the Cf^{III} 5,7-dichloro-8-hydroxyquinoline complex. The citrate complexes have been prepared in solution and their simple salts $Am(C_6H_5O_7) \cdot xH_2O$ have been characterized. The citrate complex can be crystallized by the use of an hydrogen bond donor such as cobalt hexamine to form the solid $[Co(NH_3)_6][Am-(C_6H_5O_7)_2] \cdot xH_2O$.[968] An interesting structure determined using single crystal X-ray diffraction is that of the salicylate complex $Am(C_7H_5O_3)_3 \cdot H_2O$. In this molecule each Am^{III} is linked to six different salicylate groups and is surrounded by nine oxygen atoms, eight from the salicylate groups and one from the water molecule; two salicylate groups are bidentate, one via its carboxylate group and the other via its carboxylate and phenolic groups, and the other four are monodentate via the carboxylate group (Figure 96). Danford *et al.*[969] precipitated the dipivaloylmethane complex $Am(C_{11}H_{19}O_2)_3$ by adding aqueous Am^{III} sulfate to a solution of dipivaloylmethane and NaOH in 70% aqueous ethanol.

3.3.3.2.4 Macrocyclic ligands

The bis(phthalocyanine) (Pc) complex, $Am(Pc)_2^-$, has been obtained by heating AmI_3 with *o*-phthalodinitrile in l-chloronaphthalene, or from americium(III) acetate and *o*-phthalodinitrile; it is probably a sandwich compound similar to those obtained with the tripositive lanthanides.[970]

3.3.3.3 Tetravalent Oxidation State (Am, Cm, Bk, Cf)

3.3.3.3.1 General characteristics

Americium(IV) is stable in concentrated H_3PO_4, $K_4P_2O_7$, phosphotungstate, and fluoride (NH_4F, KF) solutions, and is otherwise reduced to Am^{III}. It can be prepared by dissolving $Am(OH)_4$ in concentrated NH_4F solutions and it is not reduced by water.[971] In contrast to americium, the oxidation of Cm^{III} to Cm^{IV} is achieved only with the strongest oxidizing agents, and only two reports claim evidence for an oxidation state greater than $+4$.[972,973]

Other than the CmF_4/MF system, the only claims for chemically generated Cm^{IV} in solution are the reports that red solutions result when aqueous Cm^{III} solutions are mixed with potassium peroxydisulfate and heteropolyanions such as $[P_2W_{17}O_{61}]^{10-}$.[446,447] The polytungstate Cm^{IV} complexes, $CmW_{10}O_{36}{}^{8-}$, $Cm(SiW_{11}O_{39})^{12-}$, and $Cm(PW_{11}O_{39})_2{}^{10-}$, display chemiluminescence upon reduction to Cm^{III}.[927,974] Chemiluminscence has also been observed during dissolution of the Cm^{IV} double oxide Li_xCmO_y in mineral acids.[975]

3.3.3.3.2 Simple donor ligands

(i) Ligands containing group 16 donor atoms

Oxides. All the reported dioxides, AnO_2, (An = Am, Cm, Bk and Cf) possess the fluorite structure. Interestingly, the lattice parameters determined for AmO_2 have not been consistent.[976–978] Morss and co-workers[979] reported a neutron diffraction and magnetic susceptibility study of Cm dioxide prepared by calcination of Cm^{III} oxalate. Based on the lattice parameter, the stoichiometry of this material was reported to be $CmO_{1.99 \pm 0.01}$, indicating that the material essentially contained only Cm^{IV}. Nevertheless, the effective paramagnetic moment was found to be (3.36 ± 0.06) μ_B, a value which had previously been attributed to the presence of Cm^{III}. These data possibly suggest $AnO_{2 \pm x}$ phases should be considered for Am and Cm, as they have been for Pu. The oxides can be prepared by heating a variety of oxoanion complexes (e.g., nitrates, oxalates, etc.) in air or oxygen above 600 °C. Other binary oxides are not known.

Ternary oxides of the types M_2AmO_3 (M = Li, Na), $MAmO_3$, and Li_2AnO_6 (An = Am, Cm) have been recorded. They are obtained by heating the dioxides with the alkali or alkaline earth metal oxide at high temperatures under vacuum or in nitrogen. The ternary oxides $BaCmO_3$[905] and Cm_2CuO_4[906] have recently been reported.

Hydroxides. Attempts to characterize $Am(OH)_4$ have not yet been successful. A precipitate reported to be $Am(OH)_4$ has been obtained by heating $Am(OH)_3$ at 90 °C in 0.2 M NaOH with

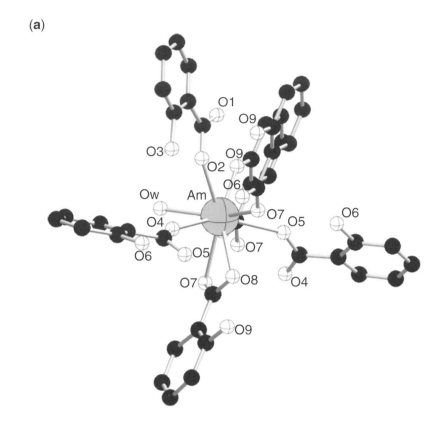

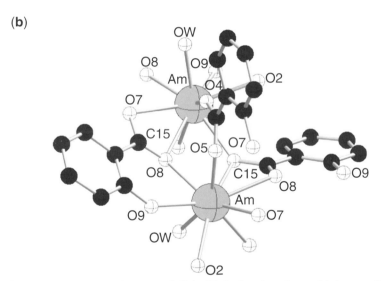

Figure 96 Single crystal structure of $Am(C_7H_5O_3)_3 \cdot H_2O$ showing the multiple coordination modes of the salicylate ligand and 10-coordinate Am^{III} center (The monodentate salicylate has been removed from view (b) for clarity) (Burns and Baldwin *Inorg. Chem.* **1997**, *16*, 289–294).

NaOCl or in 7 M KOH with peroxydisulfate.[980] The dissolution of this precipitate in sulfuric or nitric acid leads to a mixture of Am^{III}, Am^V, and Am^{VI}. $AmO_2(OH)_2$ has been suggested to precipitate in slightly basic concentrated NaCl solutions under inert atmosphere, but it also has not yet been characterized.[981,982]

Carbonates and carboxylates. There is only one Am^{IV} carbonato complex reported. From combined spectroscopy and cyclic voltammetry data in bicarbonate/carbonate solutions,[983] it was concluded that $Am(CO_3)_5{}^{6-}$ is the limiting carbonate complex of Am^{IV}. Its formation is consistent with the stability constant expected based on those of the analogous pentacarbonato species of lighter actinides. No solid Am^{IV} carbonates are known, but the limiting complex is likely isostructural with the analogous Pu^{IV} carbonate shown above.

Other. The only tetravalent transplutonium silicate known, $^{241}AmSiO_4$, is obtained by reacting $Am(OH)_4$ with excess SiO_2 in $NaHCO_3$ solution at 230 °C. This solid is patented to be used in manufactured alpha sources.

Polyoxometallates have been used recently to stablize Am^{IV}. Chartier *et al.* reported spectroscopic evidence for the formation of $AmP_2W_{17}O_{61}{}^{6-}$ and $Am(P_2W_{17}O_{61})_2{}^{16-}$ via their absorbance bands at 789 nm and 560 nm, respectively.[928] However, formation of a red Cm^{IV} complex in phosphotungstate solution was achieved by the use of peroxydisulfate as the oxidant.[447] Kosyakov *et al.* demonstrated that, in such solutions, the Cm^{IV} is reduced much more rapidly than can be accounted for by radiolytic effects, while Am^{IV} in such solutions is much more stable, being reduced at a rate attributable to radiolytic effects alone.[446]

(ii) Ligands containing group 17 donor atoms

The tetrafluorides AnF_4 (An = Am, Cm, Bk and Cf) are obtained by fluorination of lower-valent oxides or halides under elevated temperature and pressure.[562,984,985] All have the eight-coordinate UF_4 structure. The optical absorbance spectrum of Am^{IV} in concentrated fluoride solution resembles very closely that of solid AmF_4, suggesting the solution species has a similar coordination environment. CmF_4 is prepared by fluorine oxidation of the trifluoride; the curium ion in the solid is eight-coordinate, and lies within a square coordination environment.[562,948,984] Magnetic susceptibility measurements suggest a fluoride-deficient structure, CmF_{4-x}.[986,987] The orthorhombic fluoro complex Rb_2AmF_6 is formed in concentrated aqueous fluoride solutions with $RbAmO_2F_2$ or $Am(OH)_4$ and is isostructural with Rb_2UF_6, containing chains of fluoride dodecahedra.[948,988] The related $LiAnF_5$ is isostructural with $LiUF_5$. The Bk^{IV} complex, Cs_2BkCl_6, has been prepared and characterized, but surprisingly, it is not isostructural with the analogous Pu^{IV} and Ce^{IV} compounds. Compounds of composition $M_7An_6F_{31}$ (M = Na, K and An = Am, Cm, Bk) are also known.[989,990] The compounds were prepared by direct fluorination of evaporated salt mixtures of MX and CmX_3 at about 300 °C. The basic coordination polyhedron is a square antiprism.[948] In tetragonal $LiCmF_5$, the curium coordination is tricapped trigonal prismatic.[948] Hexafluorides and oxyfluorides CmF_6 and $CmOF_3$ (as well as $NpOF_3$, NpF_7, PuO_3F, AmF_5, AmF_6, and EsF_4) have been reported using thermochromatographic techniques,[973] but there has been no independent confirmation of these species.

3.3.3.3.3 *Chelating ligands*

β-Diketonates. The berkelium(IV) complexes BkL_4 (HL = $CF_3COCH_2COBu^t$ and $CF_3COCH_2CO(2\text{-}C_4H_3S)$) are formed by solvent extraction of aqueous Bk^{IV} solutions with the β-diketone.

Other. The first Am^{IV} compound with an organic ligand was prepared by reacting AmI_3 at 200 °C with phthalodinitrile in 1-chloronapthene to yield the dark violet phthalocyanine compound $Am(C_{32}H_{16}N_8)_2$.[991] There is evidence that americium also forms the monophthalocyaninato complex.

3.3.3.4 **Pentavalent Oxidation State (Am)**

3.3.3.4.1 *General characteristics*

There is tracer scale evidence for the formation of the Cf^V during ozonization of ^{249}Bk and subsequent decay to ^{249}Cf. However, the only stable isolated pentavalent transplutonium compounds are Am^V species. Americium(V) can be prepared in solution by oxidizing Am^{III} in near-neutral and alkaline aqueous solution using oxidants such as as ozone, hypochlorite and peroxydisulfate or alternatively by electrochemically or chemically reducing Am^{VI}.[992,993] It can also be isolated from mixtures of Am^{III}, Am^V, and Am^{VI} by extracting $AmO_2{}^+$ using thenoyltrifluoroacetone in isobutanol.[994] Americium(V) can be isolated in the solid state as sodium Am^V carbonate by heating a 2 M Na_2CO_3 Am^{VI} solution up to 60 °C or in solution by dissolution of An^{VI} hydroxides in mineral acids.[995,996] More exotic methods include the dissolution of solid Li_3AmO_4 in dilute perchloric acid or the electrolytic oxidation of Am^{III} in iodate solutions.[997]

3.3.3.4.2 Simple donor ligands

(i) Ligands containing group 16 donor atoms

Oxide. Simple binary oxides are unknown. However, ternary oxides, such as Na_3AmO_4, Li_3AmO_4, and Li_7AmO_6, are obtained by heating AmO_2 with the alkali metal oxide. These compounds have been characterized by X-ray diffraction.

Aquo and hydroxides. By analogy with lighter actinides and from recent experimental data, Am^V is thought to be coordinated by four or five water molecules in the equatorial plane, in addition to the two axial actinyl oxygens. EXAFS spectral data are consistent with this coordination geometry as demonstrated by Williams *et al.*[896] In addition, Shilov and Yusov analyzed reported variations in the $Am^{V/VI}$ potentials and the stability constants of the actinyl(V) oxalate complexes and proposed the same coordination number and geometry.[998] Solubility studies indicate the hydrolysis products of AmO_2^+ are $AmO_2(OH)$ and $AmO_2(OH)_2^-$, with structures and constants similar to those known for Np^V.[999] Tananaev suggests the formation of $AmO_2(OH)_3^{2-}$ and $AmO_2(OH)_4^{3-}$ in highly alkaline media based on absorbance spectroscopy measurements of Am^V in LiOH solutions.[1000] Species isolated under these conditions include the ternary Am^V hydroxides, $MAmO_2(OH)_2 \cdot nH_2O$ at 0.1–0.5 M $[OH^-]$ and $M_2AmO_2(OH)_3 \cdot nH_2O$ (M = Na, K, Rb, Cs) at 0.5–2.0 M $[OH]^-$, which have been characterized by a variety of solution techniques, including vibrational spectroscopy.[993,1001]

Carbonates and carboxylates. A number of carbonates of general formula, $MAmO_2CO_3$ (M = K, Na, Rb, Cs or NH_4) have been synthesized by precipitation of Am^V in dilute bicarbonate solutions of the corresponding cation.[511,513,932,1002,1003] The use of a large excess of alkali carbonate yields the $K_3AmO_2(CO_3)_2$ and $K_5AmO_2(CO_3)_3$ solids, which are almost certainly isostructural with the potassium neptunyl(V) carbonates described in the pentavalent carbonate section.[512,1004] The solids can alternatively be prepared by electrochemical reduction of americium(VI) in carbonate solutions. The Am=O and Am—O_{eq} bond distances are calculated from X-ray powder diffraction data to be 1.935 Å and 2.568 Å respectively. Both distances are significantly longer than those in the Np^V compounds, i.e., 1.75 Å for Np=O and 2.46 Å for Np—O_{eq} in aqueous $NpO_2(CO_3)_n^{1-2n}$ complexes.[509] The acetato complex salt, $Cs_2AmO_2(CH_3CO_2)_3$, has been prepared by precipitation from solution and is isostructural with the analogous neptunium(V) and plutonium(V) compounds. Hydrated oxalato complex salts $MAmO_2C_2O_4 \cdot xH_2O$ (M = K, Cs) were prepared by precipitation. The vibrational spectra of acetate and carbonates complexes have been reported.[510]

Oxoanion complexes. Fedoseev and Budentseva[929] claimed the preparation of the Am sulfates, $(AmO_2)_2(SO_4) \cdot xH_2O$, from evaporation of a Am^V-containing sulfuric acid solution, and two double salts, $CsAmO_2SO_4 \cdot xH_2O$ from evaporation of a solution containing $(AmO_2)_2(SO_4)$ and Cs_2SO_4 in a 3:1 ratio. They also report that $Co(NH_3)_6AmO_2(SO_4)_2 \cdot 2H_2O$ can be prepared by simply including Am^V in the chemicals used to prepare $Co(NH_3)_6(SO_4)_2$.[929] Am^V sulfate also crystallizes from an ozonated solution of $Am(OH)_3$ after addition of sulfuric acid and subsequent evaporation. A simple chromate salt, $(AmO_2)_x(CrO_4)_y \cdot H_2O$, has been reportedly prepared by slow evaporation of a Am^V/chromic acid solution.[1005] Optical spectroscopic data confirm the presence of Am^V in the solid, but the composition is unknown.

(ii) Ligands containing group 17 donor atoms

The ternary Am^V fluorides, $KAmO_2F_2$ and $RbAmO_2F_2$, precipitate from concentrated aqueous fluoride solutions of Am^V and consist of $AmO_2F_2^-$ layers held together by K^+ or Rb^+ ions from the rhombohedral $KAmO_2F_2$, where americium is eight-coordinate with two axial oxygen atoms and six fluorides in the equatorial plane.[513] In contact with acidic RbF solution, $RbAmO_2F_2$ reduces overnight to Rb_2AmF_6.[988] The chloride $Cs_3AmO_2Cl_4$ precipitates with ethanol from 6 M HCl containing Am^V hydroxide and CsCl[1006] and is isostructural with the analogous Np^V compound.

3.3.3.5 Hexavalent Oxidation State(Am)

3.3.3.5.1 *General characteristics*

Some theoretical work suggests that Cm^{VI} may be even more stable than Am^{VI}, and the lack of success in preparing Cm^{VI} may result from the low stability of Cm^{V} and the high Cm^{IV}/Cm^{III} potential.[1007,1008] One report claims the synthesis of Cm^{VI} by beta decay of $^{242}AmO_2{}^+$.[972]

Many of the classic partitioning processes rely on the formation of Am^{VI} to facilitate separation from trivalent lanthanides or heavier trivalent actinides. Americium(VI) can be prepared in basic aqueous solutions from Am^{III} using powerful oxidants, such as peroxydisulfate, and from Am^{V} using weaker oxidants, such as Ce^{IV}.[1009] It can be precipitated from solution as a carbonate by electrolytic or ozone oxidation of concentrated carbonate solutions of Am^{III} or Am^{V}, or solubilized by dissolution of sodium americyl(VI) acetate. These oxidations and the resulting coordination compounds have been used for relatively large scale processing. For examples, Stephanou *et al.*[1010] found that Cm^{III} could be separated from Am by oxidizing the latter to Am^{VI} with potassium persulfate, precipitating CmF_3, and retaining soluble Am^{VI}. Proctor *et al.*[1011,1012] used precipitation of cerium peroxide to separate gram quantities of americium from cerium, and precipitation of lanthanide trifluorides to accomplish lanthanide/actinide separation. Solution species and aqueous precipitates of Am^{VI} and Am^{V} generally have pentagonal or hexagonal bipyrimadal coordination geometries with mono- or bidentate ligands in the plane perpendicular to the americyl moeity.

3.3.3.5.1.1 Ligands containing group 16 donor atoms

Oxides. AmO_3 is unknown. Ternary oxides of the type M_2AmO_4 (M = K, Rb or Cs), M_6AmO_6(M = Li or Na), and Ba_3AmO_6, prepared by heating AmO_2 with the metal hydroxide or oxide in oxygen, have been reported.[997,1013,1014,1016]

Aqua, hydroxides. In aqueous solution, Am^{VI} likely has five water molecules coordinated in the equatorial plane based on EXAFS studies of lighter hexavalent actinides. Spectroscopic data indicate the formation of Am^{VI} hydrolysis species of general formula $AmO_2(OH)_n{}^{2-n}$, where $n = 1$, 2, (and minor species, $n = 3$ and 4), which can be generated from ozone oxidation of basic Am^{III} or Am^{V} solutions or $Am(OH)_3$ solid.[1017] In alkali hydroxide solutions, the Am^{VI} gradually reduces to form a light-tan solid which, when dissolved in mineral acid, yields Am^{V}. It is claimed that Am^{VI} disproportionates into Am^{VII} and Am^{V} in greater than 10 M NaOH solutions.[1018] The stoichiometric solid, $AmO_2(OH)_2$, has been suggested to precipitate in slightly basic concentrated NaCl solutions under inert atmosphere, but its amorphous nature of the solid phase precluded diffraction characterization.[981,982,1019]

Oxoanion complexes. The nitrato complexes $MAmO_2(NO_3)_3$ (M = Rb or Cs) are precipitated from nitric acid solutions of americium(VI) and have been characterized by powder X-ray diffraction and IR spectroscopy. Hydrated phosphato and arsenato complex salts of the form $MAmO_2PO_4 \cdot xH_2O$ (M = K, Rb, Cs or NH_4) are precipitated from americium(VI) solutions at pH 3.5–4 and are similar to analogous U^{VI} and Np^{VI} compounds. Lawaldt *et al.*[1020] also report the precipitation of Am^{VI} arsenates via complexation of Am^{VI} with *N,N*-dimethylacetamide and the Keggin-type heteropolyanion $PW_{12}O_{40}{}^{3-}$ to prepare Am^{VI} phosphates that are converted to arsenates.[1020] The obtained compounds were isostructural with the analogous Am^{VI} phosphates. The hydrated sulfato complex salt, $Co(NH_3)_6(HSO_4)_2(AmO_2(SO_4)_3) \cdot nH_2O$ is prepared by addition of cobalt hexamine to an aqueous Am^{VI} sulfate solution. It is isostructural with the analogous U^{VI} and Np^{VI} compounds.[1021] A reported chromate complex, prepared from Am^{V}, chromic acid solution and reported to be $AmO_2CrO_4 \cdot H_2O$ is likely an Am^{V} (and not Am^{VI}) complex, based on optical absorbance spectroscopy.[1005] Fedoseev *et al* report the synthesis of $AmO_2Mo_2O_7 \cdot 3H_2O$ at 100 °C; however, no phase characterization is provided.[1022]

Carbonates and carboxylates. By analogy with U^{VI} carbonates, one may expect mono-, bis-, and triscarbonato complexes in solution and in the solid state, containing bidentate carbonates in the equatorial plane perpendicular to the americyl oxygens. The Raman shifts have been used to distinguish these species, particularly the mono- and biscarbonate.[510,1023] However, only the limiting triscarbonate complex, $AmO_2(CO_3)_3{}^{4-}$, has been very well studied in solution[876,983] and isolated in the solid state as $M_4AmO_2(CO_3)_3$ (M = Cs or NH_4).[1024] Crystalline cubic sodium americyl acetate, $NaAmO_2(OOCCH_3)_3$, can be prepared by addition of sodium actetate to an acidic Am^{VI} solution.[1025,1026] The vibrational spectrum has been measured and the structure was inferred by comparison with the structures of lighter actinyl carbonates.[1027–1029] Oxalate salts of Am^{VI} have also been prepared similarly by precipitation from aqueous solution and structurally characterized by powder X-ray diffraction. These complexes are isolated in the solid state as $MAmO_2C_2O_4 \cdot xH_2O$, (M = K or Cs).[1030]

Other. The pyridine- and *N*-oxopyridine-2-carboxylates, AmO_2L_2, have been obtained from aqueous solutions of americium(VI) and the acid HL; the *N*-oxide 2-carboxylate is formed as the dihydrate.[1031] di-(2-ethylhexylphosphoric acid (HDEHP) solutions have been used to selectively extract Am^{VI} from Cm^{III} in such systems rapid reduction of Am^{VI} to lower oxidation states is a problem.[1032]

Ligands containing group 17 donor atoms. The binary Am^{VI} fluoride, AmO_2F_2, has been prepared by reacting solid sodium Am^{VI} acetate with anhydrous HF containing a small amount of F_2 at $-196\,^\circ C$.[1033] The compound is isostructural with other actinyl(VI) fluorides. The complex chloride, $Cs_2AmO_2Cl_4$, is obtained by the unusual oxidation of $Cs_3AmO_2Cl_4$ in concentrated HCl.[1034] Its cubic form appears to transform to a monoclinic form when washed with small volumes of concentrated HCl.[1006] It is suggested that the cubic form is likely to be a mixed oxidation state compound of formula $Cs_7(AmO_2)(AmO_2)_2Cl_{12}$.[1035] Conflicting claims have been put forth concerning the existence of AmF_6. More recently, Gibson and Haire[1036] have reported that they were not able to confirm the existence of AmF_6, despite exhaustive efforts. Interestingly, Am^{VI} hexachloride appears to be sufficiently stable to permit X-ray crystallographic studies.

3.3.3.6 Heptavalent Oxidation State

Green solutions believed to be Am^{VII} are prepared by oxidation of Am^{VI} in concentrated aqueous basic solution by either ozone or the $O^{\cdot}$ radical.[1037,1038] In contrast to Np^{VII}, and similar to Pu^{VII}, Am^{VII} is unstable and reduces to the hexavalent state within minutes. A review on the chemistry of heptavalent transplutonium elements can be found in the *Handbook of the Physics and Chemistry of the Actinides*.[1039]

3.3.4 OTHER SOURCES

Several reviews were invaluable in surveying advances in actinide coordination chemistry, including the following: *Gmelin Handbook of Inorganic Chemistry* published in 1988 and the supplements on thorium and uranium;[7,216] the chapters on thorium and uranium in the *Kirk-Othmer Encyclopedia of Chemical Technology*, 4th editon., 1997 by D. L. Clark, D. W. Keogh, M. P. Neu, and W. Runde;[1040,1041] *The Chemistry of the Actinide Elements*, 2nd edition, edited by Katz, J. J., Seaborg, G. T., Morss, L. R., 1986;[1045] preprints of the element-specific chapters on Am and Cm chemistry by Runde and Schulz and Lumetta, Thompson, Penneman, and Eller, respectively,[1043,1044] in *The Chemistry of the Actinide and Transactinide Elements*, to be published in 2003; *Handbook on the Physics and Chemistry of the Actinides* edited by A. J. Freeman and C. Keller, 1985;[1046] The Nuclear Energy Agency and the Organisation for Economic Co-operation and Development critical reviews of the chemical thermodynamic data of U, Am, and Np and Pu (U volume edited by I. Grenthe, J. Fuger, R. J. M. Konigs, R. J. Lemire, A. B. Muller, C. Nguyen-Trung, H. Wanner;[197] Am volume edited by Silva, R. J. Bidoglio, G., Rand, M. H., Robouch, P. B., Wanner, H., Puigdomenech, I.,[876] Np and Pu volume edited by Lemire, R. J., Fuger, J., Nitsche, H., Potter, P., Rand, M. H., Rydberg, J., Spahiu, K., Sullivan, J. C., Ullman, W. J., Vitorge, P., Wanner, H.;[46] *Actinide partitioning—a review in Solvent Extraction and Ion Exchange*, 2001 by Mathur, J. N., Murali, M. S., Nash, K. L.;[409] *The Crystal Chemistry of Uranium in Reviews in Mineralogy*, 1999 by Peter Burns.[1043]

ACKNOWLEDGMENTS

We thank Dr. Marianne Wilkerson for significant contributions to the sections on alkoxide and thiolate chemistry. We thank Dr. Jeff Golden for technical input and assistance with all aspects of manuscript preparation. Dr. Brian Scott assisted with structural database searching and provided crystallographic data from structures determined at Los Alamos National Laboratory, and Halo

Golden assisted with the preparation of tables. Finally, we are grateful to Drs. Wolfgang Runde and P. Gary Eller for providing their reviews and references on Am and Cm chemistry.

3.3.5 REFERENCES

1. Klaproth, M. H. *Chem. Ann. (Crell) II* **1789**, 387.
2. Berzelius, J. J. *K. Sven. Vetenskapsakad. Handl.* **1829**, *9*, 1.
3. Fajans, K.; Göhring, O. *Naturwissenschaften* **1913**, *1*, 339.
4. McMillan, E.; Abelson, P. *Phys. Rev.* **1940**, *57*, 1185.
5. Reynolds, L. T.; Wilkinson, G. *J. Inorg. Nucl. Chem.* **1956**, *2*, 246.
6. Edelmann, F. Scandium, yttrium, and the 4f and 5f Elements, Excluding their Zero Oxidation State Complexes. In *Comprehensive Organometallic Chemistry II*; Abel, E. W.; Stone, F. G. A.; Wilkinson, G., Eds.; Elsevier Science: Oxford, UK 1995, Vol 8, pp 2–192.
7. Gmelin, L. Gmelin's Handbuch der Anorganischen Chemie, Verlag Chemie: Weinheim, Germany; 1955, Vol. 44 (Th), Vol. 55 (U); 1973, Vol. 71 (Np, Pu, transuranium elements); Springer-Verlag: Berlin; 1975, Th Suppl. Vol. A1–E4, U Suppl. Vol. A1–E2.
8. Palmer, D. C. 2.1 ed. **1988**, Cambridge University Technical Services LTD: Cambridge (software, referenced as requested by company).
9. Allen, F. H. *Acta Cryst.* **2002**, *B58*, 380.
10. Burns, C. J.; Bursten, B. E. *Comm. Inorg. Chem.* **1989**, *9*, 61.
11. Shannon, R. D.; Prewitt, C. T. *Acta Crystallogr., Sect. B* **1969**, *25*, 925.
12. Andersen, R. A. *Inorg. Chem.* **1979**, *18*, 1507.
13. Clark, D. L.; Sattelberger, A. P.; Bott, S. G.; Vrtis, R. N. *Inorg. Chem.* **1989**, *28*, 1771.
14. Zwick, B. D.; Sattelberger, A. P.; Avens, L. R. Transuranium Organometallic Elements: The Next Generation. In *Transuranium Elements: A Half Century*; American Chemical Society: Washington, D. C., 1992, Chapter 25.
15. Stewart, J. L.; Andersen, R. A. *Polyhedron* **1998**, *17*, 953.
16. Green, J. C.; Payne, M.; Seddon, E. A.; Andersen, R.A. *J. Chem. Soc., Dalton Trans.* **1982**, 887.
17. Roussel, P.; Hitchcock, P. B.; Tinker, N. D.; Scott, P. *Chem. Commun.* **1996**, 2053.
18. Roussel, P.; Scott, P. *J. Am. Chem. Soc.* **1998**, *120*, 1070.
19. Roussel, P.; Boaretto, R.; Kingsley, A. J.; Alcock, N. W.; Scott, P. *J. Chem. Soc., Dalton Trans.* **2002**, 1423.
20. Kaltsoyannis, N.; Scott, P. *Chem. Commun.* **1998**, 1665.
21. Odom, A. L.; Arnold, P. L.; Cummins, C. C. *J. Am. Chem. Soc.* **1998**, *120*, 5836.
22. Nelson, J. E.; Clark, D. L.; Burns, C. J.; Sattelberger, A. P. *Inorg. Chem.* **1992**, *31*, 1973.
23. Santos, I.; Marques, N.; Pires de Matos, A. *J. Less-Comm. Met.* **1986**, *122*, 215.
24. Santos, I.; Marques, N.; Pires de Matos, A. *Inorg. Chim. Acta* **1985**, *95*, 149.
25. Domingos, A.; Marques, N.; Pires de Matos, A.; Santos, I.; Silva, M. *Polyhedron* **1992**, *11*, 2021.
26. McDonald, R.; Sun, Y.; Takats, J.; Day, V. W.; Eberspracher, T. A. *J. Alloys Compd.* **1994**, *213*, 8.
27. Sun, Y.; McDonald, R.; Takats, J.; Day, V. W.; Eberspracher, T. A. *Inorg. Chem.* **1994**, *33*, 4433.
28. Carvalho, A.; Domingos, A.; Gaspar, P.; Marques, N.; Pires de Matos, A.; Santos, I. *Polyhedron* **1992**, *11*, 1481.
29. Sun, Y.; Takats, J.; Eberspracher, T.; Day, V. *Inorg. Chim. Acta* **1995**, *229*, 315.
30. Maria, L.; Campello, M. P.; Domingos, A.; Santos, I.; Andersen, R. *J. Chem. Soc, Dalton Trans.* **1999**, 2015.
31. Apostolidis, C.; Carvalho, A.; Domingos, A.; Kanellakopulos, B.; Maier, R.; Marques, N.; Pires de Matos, A.; Rebizant, J. *Polyhedron* **1998**, *18*, 263.
32. Amoroso, A. J.; Jeffery, J. C.; Jones, P. L.; McCleverty, J. A.; Rees, L. R. A. L.; Sun, Y.; Takats, J.; Trofimenko, S.; Ward, M. D.; Yap, G. P. A. *Chem. Commun.* **1995**, 1881.
33. Cleveland, J. M.; Bryan, G. H.; Sironen, R. J. *Inorg. Chim. Acta* **1972**, *6*, 54.
34. Karraker, D. G. *Inorg. Chim. Acta* **1987**, *139*, 189.
35. Avens, L. R.; Bott, S. G.; Clark, D. L.; Sattelberger, A. P.; Watkin, J. G.; Zwick, B. D. *Inorg. Chem.* **1994**, *33*, 2248.
36. Wietzke, R.; Mazzanti, M.; LaTour, J. M.; Pecaut, J. *J. Chem. Soc, Dalton Trans.* **2000**, 4167.
37. Wietzke, R.; Mazzanti, M.; LaTour, J. M.; Pecaut, J. *J. Chem. Soc, Dalton Trans.* **1998**, 4087.
38. Riviere, C.; Nierlich, M.; Ephritikhine, M.; Madic, C. *Inorg. Chem.* **2001**, *40*, 4428.
39. Karraker, D. G.; Stone, J. A. *Inorg. Chem.* **1980**, *19*, 3545.
40. Drozdzynski, J.; du Preez, J. G. H. *Inorg. Chim. Acta* **1994**, *218*, 203.
41. Zych, E.; Drozdzynski, J. *Inorg. Chim. Acta* **1986**, *115*, 219.
42. Zych, E.; Starynowicz, P.; Lis, T.; Drozdzynski, J. *Polyhedron* **1993**, *12*, 1661.
43. Wasserman, H. J.; Moody, D. C.; Ryan, R. R. *J. Chem. Soc., Chem. Commun.* **1984**, 532.
44. Wasserman, H. J.; Moody, D. C.; Paine, R. T.; Ryan, R. R.; Salazar, K. V. *J. Chem. Soc., Chem. Commun.* **1984**, 533.
45. Brennan, J.; Shinomoto, R.; Zalkin, A.; Edelstein, N. *Inorg. Chem.* **1984**, *23*, 4143.
46. Lemire, R. J.; Fuger, J.; Nitsche, H.; Potter, P.; Rand, M. H.; Rydberg, J.; Spahiu, K.; Sullivan, J. C.; Ullman, W. J.; Vitorge, P.; Wanner, H. *Chemical Thermodynamics of Neptunium and Plutonium*; Elsevier: New York, 2001; Vol. 4.
47. Clark, D. L.; Hobart, D. E.; Neu, M. P. *Chem. Rev.* **1995**, *95*, 25.
48. Kraus, K. A.; Dam, J. R.; Seaborg, G. T.; Katz, J. J.; Manning, W. M., Eds.; McGraw-Hill: New York, 1949; IV–14B, pp 466.
49. Fedoseev, A. M.; Peretrukhin, V. F.; Krot, N. N. *Proc. Academy of Science, USSR Physical Chemistry Section* **1979**, *244*, 139.
50. Fuks, L.; Siekierski, S. *J. Radioanal. Nucl. Chem.* **1987**, *108*, 139.
51. Moskvin, A. I. *Soviet Radiochemistry* **1971**, *13*, 688.
52. Bamberger, C. E. Solid Inorganic Phosphates of the Transuranium Elements. In *Handbook on the Physics and Chemistry of the Actinides*; Vol. 3 Freeman, A. J.; Keller, C. Eds.; Elsevier: Amsterdam, 1985; pp 289–303.

53. Nash, K. L.; Cleveland, J. M. Stability Constants, Enthalpies, and Entropies of Plutonium(III) and Plutonium(IV) Sulfate Complexes. In *Plutonium Chemistry;* Carnall, W. T., Choppis, G. R., Eds.; American Chemical Society: Washington, DC 1983, Vol. 216, pp 251–262.

54. Bullock, J. I.; Ladd, M. F. C.; Povey, D. C.; Storey, A. E. *Inorganica Chimica Acta* **1980**, *43*, 101.

55. Zozulin, A. J.; Moody, D. C.; Ryan, R. R. *Inorg. Chem.* **1982**, *21*, 3083.

56. Van Der Sluys, W. G.; Burns, C. J.; Huffman, J. C.; Sattelberger, A. P. *Inorg. Chem.* **1988**, *110*, 5924.

57. Van de Weghe, P.; Collin, J.; Santos, I. *Inorganica Chimica Acta* **1994**, *222*, 91.

58. Van Der Sluys, W. G.; Sattelberger, A. P. *Inorg. Chem.* **1989**, *28*, 2496.

59. Clark, D. L.; Sattelberger, A. P.; Van Der Sluys, W. G.; Watkin, J. G. *J. Alloys Compd.* **1992**, *180*, 303.

60. Burns, C. J.; Sattelberger, A. P. Organometallic and Nonaqueous Coordination Chemistry. In *Advances in Plutonium Chemistry*; American Nuclear Society, La Grange Park, IL 2002.

61. Warner, B. P.; D'Alessio, J. A.; Morgan, A. N., III; Burns, C. J.; Schake, A. R.; Watkin, J. G. *Inorg. Chim. Acta* **2000**, *309*, 45.

62. Berthet, J. C.; Lance, M.; Nierlich, M.; Ephritikhine, M. *Eur. J. Inorg. Chem.* **1999**, 2005.

63. Maria, L.; Domingos, A.; Santos, I. *Inorg. Chem.* **2001**, *40*, 6863.

64. Allen, P. G.; Bucher, J. J.; Shuh, D. K.; Edelstein, N. M.; Craig, I. *Inorg. Chem.* **2000**, *39*, 595.

65. Farkas, I.; Grenthe, I.; Bányai, I. *J. Phys. Chem. A* **2000**, *104*, 1201.

66. Conradson, S. D. *Appl. Spectrosc.* **1998**, *52*, 252A.

67. Kim, J. I.; Klenze, R.; Wimmer, H. *Eur. J. Solid State Inorg. Chem.* **1991**, *28*, 347.

68. Fuger, J.; Khodakovosky, I. L.; Sergeyeva, E. I.; Medvedev, V. A.; Navratil, J. D. *The Chemical thermodynamics of Actinide Elements and Compounds*; IAEA: Vienna, 1992; Part 12.

69. Matonic, J. H.; Scott, B. L.; Neu, M. P. *Inorg. Chem.* **2001**, *40*, 2638.

70. Karbowiak, M.; Drozdynski, J.; Janczak, J. *Polyhedron* **1996**, *15*, 241.

71. Moody, D. C.; Odom, J. D. *J. Inorg. Nucl. Chem.* **1979**, *41*, 533.

72. Barnard, R.; Bullock, J. I.; Gellatly, B. J.; Larkworthy, L. F. *J. Chem. Soc., Dalton Trans.* **1972**, 1932.

73. Bullock, J. I.; Storey, A. E.; Thompson, P. *J. Chem. Soc.,Dalton Trans.* **1979**, 1040.

74. Burns, J. H. *Inorg. Chem.* **1965**, *4*, 881.

75. Suglobova, I. G.; Chirkst, D. E. *Koord. Khim.* **1981**, *7*, 97.

76. Karbowiak, M.; Drozdzynski, J. *J. Alloys Compd.* **1998**, *275–77*, 848.

77. Krämer, K.; Keller, L.; Fischer, P.; Jung, B.; Edelstein, N. N.; Güdel, H. U.; Meyer, G. *J. Solid State Chem.* **1993**, *103*, 152.

78. Krämer, K.; Güdel, H. U.; Meyer, G.; Heuer, T.; Edelstein, N.; Jung, B.; Keller, L.; Fischer, P.; Zych, E.; Drozdynski, J. *Z. Anorg. Allg. Chem.* **1994**, *620*, 1339.

79. Karbowiak, M.; Hanusa, J.; Drozdynski, J.; Hermarowicz, K. *J. Solid State Chem.* **1996**, *121*, 312.

80. May, I.; Taylor, R. J.; Denniss, I. S. B.; Geoff.; Wallwork, A. L.; Hill, N. J.; Rawson, J. M.; Less, R. *J. Alloys Compd.* **1998**, 275.

81. Kappel, M. J.; Nitsche, H.; Raymond, K. N. *Inorg. Chem.* **1985**, *24*, 605.

82. Cleveland, J. M. *The Chemistry of Plutonium*; American Nuclear Society: La Grange Park, IL, USA, 1979.

83. Saprykin, A. S.; Spitsyn, V. I.; Orlova, M. M. *Radiokhimiya* **1978**, *20*, 247.

84. Dejean, A.; Charpin, P.; Folcher, G.; Rigny, P.; Navaza, A.; Tsoucaris, G. *Polyhedron* **1987**, *6*, 189.

85. Schlesinger, H. I.; Brown, H. C. *J. Am. Chem. Soc.* **1953**, *75*, 219.

86. Ghiassee, N.; Clay, P. G.; Walton, G. N. *J. Inorg. Nucl. Chem.* **1981**, *43*, 2909.

87. Paine, R. T.; Schonberg, P. R.; Light, R. W.; Danen, W. C.; Freund, S. M. *J. Inorg. Nucl. Chem.* **1979**, *41*, 1577.

88. Ghiassee, N.; Clay, P. G.; Walton, G. N. *Inorg. Nucl. Chem. Lett.* **1978**, *14*, 117.

89. Ghiassee, N.; Clay, P. G.; Walton, G. N. *Inorg. Nucl. Chem. Lett.* **1980**, *16*, 149.

90. Moody, D. C.; Penneman, R. A.; Salazar, K. V. *Inorg.Chem.* **1979**, *18*, 208.

91. Männ, D.; Nöth, H. *Z. Anorg. Allg. Chem.* **1986**, *543*, 66.

92. Ban, B.; Folcher, G.; Marquet-Ellis, H.; Rigny, P. *Nouv. J. Chim.* **1985**, *9*, 51.

93. Dejean-Meyer, A.; Folcher, G.; Marquet-Ellis, H. *J. Chim. Phys.* **1983**, *80*, 579.

94. Dejean, A.; Chaprin, P.; Folcher, G. R. P.; Navaza, A.; Tsoucaris, G. *Polyhedron* **1987**, *6*, 189.

95. Arliguie, T.; Lance, M.; Nierlich, M.; Vigner, J.; Ephritikhine, M. *J. Chem. Soc., Chem. Commun.* **1994**, 847.

96. Baudry, D.; Bulot, E.; Charpin, P.; Ephritikhine, M.; Lance, M.; Nierlich, M.; Vigner, J. *J. Organomet. Chem.* **1989**, *371*, 163.

97. Le Maréchal, J. F.; Ephritikhine, M.; Folcher, G. *J. Organomet. Chem.* **1986**, *309*, C1.

98. Bagnall, K. W.; Baptista, J. O. *J. Inorg. Nucl. Chem.* **1970**, *32*, 2283.

99. Dormond, A.; El Bouadili, A. A.; Moise, C. *J. Chem. Soc., Chem. Commun.* **1984**, 749.

100. Zanella, P.; Brianese, N.; Casellato, U.; Ossola, F.; Porchia, M.; Rossetto, G.; Graziani, R. *J. Chem. Soc., Dalton Trans./* **1987**, 2039.

101. Lux, F.; Bufe, U. E. *Angew. Chem., Int. Ed. Engl.* **1971**, *10*, 274.

102. Jones, R. G.; Karmas, G.; Martin, J. G. A.; Gilman, H. *J. Am. Chem. Soc.* **1956**, *78*, 4285.

103. Reynolds, J. G.; Zalkin, A.; Templeton, D. H.; Edelstein, N. M.; Templeton, L. K. *Inorg. Chem.* **1976**, *15*, 2498.

104. Boisson, C. Dissertation *University of Orsay* **1996**, University of Orsay: Paris.

105. Berthet, J. C.; Ephritikhine, M. *Coord. Chem. Rev.* **1998**, *178*, 83.

106. Berthet, J. C.; Ephritikhine, M. *J. Chem. Soc., Chem. Commun.* **1993**, 1566.

107. Reynolds, J. G.; Zalkin, A.; Templeton, D. H. *Inorg.Chem.* **1977**, *16*, 3357.

108. Turman, S. E.; Van der Sluys, W. G. *Polyhedron* **1992**, *11*, 3139.

109. Barnhart, D. M.; Clark, D. L.; Grumbine, S. K.; Watkin, J. G. *Inorg. Chem.* **1995**, *34*, 1695.

110. Berthet, J. C.; Boisson, C.; Lance, M.; Vigner, J.; Nierlich, M.; Ephritikhine, M. *J. Chem. Soc., Dalton Trans.* **1995**, 3019.

111. Turner, H. W.; Andersen, R. A.; Zalkin, A.; Templeton, D. H. *Inorg. Chem.* **1979**, *18*, 1221.

112. McCullough, L. G.; Turner, H. W.; Andersen, R. A.; Zalkin, A.; Templeton, D. H. *Inorg. Chem.* **1981**, *20*, 2869.

113. Dormond, A.; Aaliti, A.; Moise, C. *J. Org. Chem.* **1988**, *53*, 1034.

114. Dormond, A.; El Bouadili, A. A.; Moise, C. *J. Org. Chem.* **1987**, *52*, 688.

115. Turner, H. W.; Simpson, S. J.; Andersen, R. A. *J. Am. Chem. Soc.* **1979**, *101*, 2782.
116. Muller, M.; Williams, V. C.; Doerrer, L. H.; Leech, M. A.; Mason, S. A.; Green, M. L. H.; Prout, K. *Inorg. Chem.* **1998**, *37*, 1315.
117. Simpson, S.; Andersen, R. A. *Inorg. Chem.* **1981**, *20*, 2991.
118. Simpson, S. J.; Turner, H. W.; Andersen, R. A. *Inorg. Chem* **1981**, *20*, 2991.
119. Dormond, A.; El Bouadili, A. A.; Moise, C. *J. Chem. Soc., Chem. Commun.* **1985**, 914.
120. Dormond, A.; Aaliti, A.; Moise, C. *Tetrahedron Lett.* **1986**, *27*, 1497.
121. Dormond, A.; El Bouadili, A. A.; Moise, C. *J. Less-Common Met.* **1986**, *122*, 159.
122. Dormond, A.; Aaliti, A.; El Bouadili, A.; Moise, C. *J. Organomet. Chem.* **1987**, *329*, 187.
123. Dormond, A.; El Bouadili, A. A.; Moise, C. *J. Org. Chem.* **1989**, *54*, 3747.
124. Baudry, D.; Dormond, A.; Visseaux, M.; Monnot, C.; Chardot, H.; Lin, Y.; Bakhmutov, V. *New J. Chem.* **1995**, *19*, 921.
125. Van Der Sluys, W. G.; Sattelberger, A. P.; Streib, W. E.; Huffman, J. C. *Polyhedron* **1989**, *8*, 1247.
126. Berg, J. M.; Clark, D. L.; Huffman, J. C.; Morris, D. E.; Sattelberger, A. P.; Smith, W. E.; Van Der Sluys, W. G.; Watkin, J. G. *J. Am. Chem. Soc.* **1992**, *114*, 10811.
127. Clark, D. L.; Miller, M. M.; Watkin, J. G. *Inorg. Chem.* **1993**, *32*, 772.
128. Stewart, J. L.; Andersen, R. A. *New J. Chem.* **1995**, *19*, 587.
129. Scott, P.; Hitchcock, P. B. *Polyhedron* **1994**, *13*, 1651.
130. Scott, P.; Hitchcock, P. B. *J. Chem. Soc., Dalton Trans* **1995**, *4*, 603.
131. Roussel, P.; Hitchcock, P. B.; Tinker, N. D.; Scott, P. *Inorg. Chem.* **1997**, *36*, 5716.
132. Roussel, P.; Alcock, N. W.; Boaretto, R.; Kingsley, A. J.; Munslow, I. J.; Sanders, C. J.; Scott, P. *Inorg. Chem.* **1999**, *38*, 3651.
133. Hassaballa, H.; Steed, J. W.; Junk, P. C. *Chem. Commun.* **1998**, 577.
134. Boaretto, R.; Roussel, P.; Kingsley, A. J.; Munslow, I. J.; Sanders, C. J.; Alcock, N. W.; Scott, P. *Chem. Commun.* **1999**, 1701.
135. Diaconescu, P. L.; Odom, A. L.; Agapie, T.; Cummins, C. C. *Organometallics* **2001**, *20*, 4993.
136. Mindiola, D. J.; Tsai, Y. C.; Hara, R.; Chen, Q.; Meyer, K.; Cummins, C. C. *Chem. Commun* **2001**, 125.
137. Diaconescu, P. L.; Arnold, P. L.; Baker, T. A.; Mindiola, D. J.; Cummins, C. C. *J. Am. Chem. Soc.* **2000**, *122*, 6108.
138. Diaconescu, P. L.; Cummins, C. C. *J. Am. Chem. Soc.* **2002**, *124*, 7660.
139. Reynolds, J. G.; Zalkin, A.; Templeton, D. H.; Edelstein, N. M. *Inorg. Chem.* **1977**, *16*, 599.
140. Coles, S. J.; Danopoulos, A. A.; Edwards, P. G.; Hursthouse, M. B.; Read, P. W. *J. Chem. Soc., Dalton Trans.* **1995**, 3401.
141. Boisson, C.; Berthet, J. C.; Ephritikhine, M.; Lance, M.; Nierlich, M. *J. Organomet. Chem.* **1997**, *533*, 7.
142. Wang, J. X.; Dash, A. K.; Berthet, J. C.; Ephritikhine, M.; Eisen, M. *J. Organomet. Chem.* **2000**, *610*, 49.
143. Dash, A. K.; Wang, J. X.; Berthet, J. C.; Ephritikhine, M.; Eisen, M. S. *J. Organomet. Chem.* **2000**, *604*, 83.
144. Wedler, M.; Knoesel, F.; Noltemeyer, M.; Edelmann, F. T.; Behrens, U. *J. Organomet. Chem.* **1990**, *388*, 21.
145. Wedler, M.; Knoesel, F.; Edelmann, F. T.; Behrens, U. *Chem. Ber.* **1992**, *125*, 1313.
146. Hitchcock, P. B.; Hu, J.; Lappert, M. F.; Tian, S. *J. Organomet. Chem.* **1997**, *536*, 473.
147. Hitchcock, P. B.; Lappert, M. F.; Liu, D. S. *J. Organomet. Chem.* **1995**, *488*, 241.
148. Edwards, P. G.; Harman, M.; Hursthouse, M. B.; Parry, J. S. *J. Chem. Soc, Chem. Commun.* **1992**, 1469.
149. Edwards, P. G.; Parry, J. S.; Read, P. W. *Organometallics* **1995**, *14*, 3649.
150. Edwards, P. G.; Hursthouse, M. B.; Abdul Malik, K. M.; Parry, J. S. *J. Chem. Soc., Chem. Commun.* **1994**, 1249.
151. Bagnall, K. W.; Du Preez, J. G. H.; Warren, R. F. *J. Chem. Soc., Dalton Trans.* **1975**, 140.
152. Bagnall, K. W.; Edwards, J.; Heatley, F. Uranium (IV) poly(pyrazol-l-yl)borate complexes—carbon-13 NMR spectra. *Transplutonium 1975, Proc. 4*[th] *Int. Transplutonium Elem. Symp. Baden-Baden Sept. 1975* Muller, W.; Lindner, R., Eds., North-Holland: Amsterdam, **1976**, 119.
153. Bagnall, K. W.; Beheshti, A.; Heatley, F. *J. Less-Comm. Met.* **1978**, *61*, 171.
154. Ball, R. G. E. F.; Matisons, J. G.; Takats, J.; Marques, N.; Marçalo, J.; Pires de Matos, A.; Bagnall, K. W. *Inorg. Chim. Acta* **1987**, *132*, 137.
155. Collin, J.; Pires de Matos, A.; Santos, I. *J. Organomet. Chem.* **1993**, *463*, 103.
156. Campello, M. P. C.; Domingos, A.; Santos, I. *J. Organomet. Chem.* **1994**, *484*, 37.
157. Marques, N.; Marçalo, J.; Pires de Matos, A.; Bagnall, K. W.; Takats, J. *Inorg. Chim. Acta* **1987**, *139*, 79.
158. Domingos, A.; Marques, N.; Pires de Matos, A. *Polyhedron* **1990**, *9*, 69.
159. Silva, M.; Domingos, A.; Pires de Matos, A.; Marques, N.; Trofimenko, S. *J. Chem. Soc., Dalton Trans.* **2000**, 4628.
160. Domingos, A.; Pires de Matos, A.; Santos, I. *J. Less-Common Met.* **1989**, *149*, 279.
161. Santos, I.; Marques, N.; Pires de Matos, A. *Inorg. Chim. Acta* **1987**, *139*, 87.
162. Domingos, Â.; Pires de Matos, A.; Santos, I. *Polyhedron* **1992**, *11*, 1601.
163. Domingos, A.; Marçalo, J.; Pires de Matos, A. *Polyhedron* **1992**, *11*, 909.
164. Marques, N.; Marçalo, J.; Pires de Matos, A.; Santos, I.; Bagnall, K. W. *Inorg. Chim. Acta* **1987**, *139*, 309.
165. Leal, J. P.; Marques, N.; Pires de Matos, A.; Calhorda, M. J.; Galvao, A. M.; Simoes, J. A. M. *Organometallics* **1992**, *11*, 1632.
166. Marçalo, J.; Marques, N.; Pires de Matos, A.; Bagnall, K. W. *J. Less-Comm. Met.* **1986**, *122*, 219.
167. Domingos, A.; Marçalo, J.; Marques, N.; Pires de Matos, A. *Polyhedron* **1992**, *11*, 501.
168. Apostolidis, C.; Kanellakopulos, B.; Maier, R.; Marques, N.; Pires de Matos, A.; Santos, I. *Proceedings of the 20e Journées des Actinides Prague* **1990**.
169. Campello, M. P.; Domingos, A.; Galvão, A.; Pires de Matos, A.; Santos, I. *J. Organomet. Chem.* **1999**, *579*, 5.
170. Amoroso, A. J.; Jeffery, J. C.; Jones, P. L.; McCleverty, J. A.; Ward, M. D. *Polyhedron* **1995**, *15*, 2023.
171. Grey, I. E.; Smith, P. W. *Aust. J. Chem.* **1969**, *22*, 311.
172. Manhas, B. S.; Pal, S.; Trikha, A. K. *Polyhedron* **1993**, *12*, 241.
173. Al-Daher, A. G. M.; Bagnall, K. W.; Benetollo, F.; Polo, A.; Bombicri, G. *J. Less-Common Met.* **1986**, *122*, 167.
174. Bagnall, K. W.; Benetollo, F.; Forsellini, E.; Bombieri, G. *Polyhedron* **1992**, *11*, 1765.
175. Danopoulos, A. A.; Hankin, D. A.; Cafferkey, S. M.; Hursthouse, M. B. *J. Chem. Soc., Dalton Trans.* **2000**, 1613.
176. Watt, G. W.; Baugh, D. W. *J. Inorg. Nucl. Chem. Lett.* **1974**, *10*, 1025.

177. Drew, M. G. B.; Willey, G. R. *J. Chem. Soc., Dalton Trans.* **1984**, 727.
178. Edwards, P. G.; Weydert, M.; Petrie, M. A.; Andersen, R. A. *J. Alloys Compd.* **1994**, *213*, 11.
179. Edwards, P. G.; Andersen, R. A.; Zalkin, A. *Acta Crystallogr., Sect. C* **1983**, *42*, 1480.
180. Rabinovich, D.; Schimek, G. L.; Pennington, W. T.; Nielsen, J. B.; Abney, K. D. *Acta Crystallogr., Sect. C* **1997**, *53*, 191.
181. Shinomoto, R.; Zalkin, A.; Edelstein, N. M.; Zhang, D. *Inorg. Chem.* **1987**, *26*, 2868.
182. Agarwal, R. K. S. A. K.; Srivastava, M.; Bhakru, N.; Srivastava, T. N. *J. Inorg. Nucl. Chem.* **1980**, *42*, 1775.
183. Srivastava, A. K.; Agarwal, R. K.; Srivastava, M.; Kapoor, V.; Srivastava, T. N. *J. Inorg. Nucl. Chem.* **1981**, *43*, 1393.
184. Kumar, N.; Tuck, D. G. *Can. J. Chem.* **1982**, *60*, 2579.
185. Gans, P.; Smith, B. C. *J. Chem. Soc. Abstracts* **1964** (Nov.), 4177–9.
186. van den Bossche, G.; Rebizant, J.; Spirlet, M. R.; Goffart, J. *Acta Crystallogr., Sect. C* **1986**, *42*, 1478.
187. du Preez, J. G. H.; Zeelie, B. *Inorg. Chim. Acta* **1986**, *118*, L25.
188. Avens, L. R.; Barnhart, D. M.; Burns, C. J.; McKee, S. D. *Inorg. Chem.* **1996**, *35*, 537.
189. Selbin, J.; Ortego, J. D. *J. Inorg. Nucl. Chem.* **1967**, *29*, 1449.
190. Edwards, P. G.; Andersen, R. A.; Zalkin, A. *J. Am. Chem.Soc.* **1981**, *103*, 7792.
191. Edwards, P. G.; Andersen, R. A.; Zalkin, A. *Organometallics* **1984**, *3*, 293.
192. Maddock, A. G.; Pires de Matos, A. *Radiochim. Acta* **1973**, *19*, 163.
193. Haschke, J. M.; Allen, T. H. *J. Alloys Compd.* **2002**, *336*, 124.
194. Allen, G. C.; Tempest, P. A.; Tyler, J. W. *Nature* **1982**, *295*, 48.
195. Allen, G. C.; Tucker, P. M.; Tyler, J. W. *J. Phys. Chem.* **1982**, *86*, 224.
196. Allen, G. C.; Tempest, P. A.; Garner, C. D.; Ross, I.; Jones, D. J. *J. Phys. Chem.* **1985**, *89*, 1334.
197. Grenthe, I.; Fuger, J.; Konigs, R. J. M.; Lemire, R. J.; Muller, A. B.; Nguyen-Trung, C.; Wanner, H. *Chemical Thermodynamics of Uranium*; Elsevier: New York, 1992; Vol. 1.
198. Lierse, C. *Institut für Radiochemie*, Report RCM 02286 (1986); Technische Universität München: Germany.
199. Pazukhin, E. M.; Kudryavtsev, E. G. *Radiokhimiya* **1990**, *32*, 18.
200. Milic, N. B.; Suranji, T. M. *Can. J. Chem.* **1982**, *60*, 1298.
201. Ryan, J. L.; Rai, D. *Inorg. Chem.* **1987**, *26*, 4140.
202. Bruno, J.; Grenthe, I.; Robouch, P. *Inorg. Chim. Acta* **1989**, *158*, 221.
203. Engkvist, I.; Albinsson, Y. *Radiochim. Acta.* **1992**, *58/59*, 109.
204. Rai, D.; Felmy, A. R.; Ryan, J. L. *Inorg. Chem.* **1990**, *29*, 260.
205. Sutorik, A. C.; Kanatzidis, M. G. *J. Am. Chem. Soc.* **1991**, *113*, 7754.
206. Ciavatta, L.; Ferri, D.; Grenthe, I.; Salvatore, F.; Spahiu, K. *Inorg. Chem.* **1983**, *22*, 2088.
207. Yamnova, N. A.; Pushcharovskii, D. Y.; Voloshin, A. V. *Doklady Akademii Nauk SSSR* **1990**, *310*, 99.
208. Dervin, J.; Faucherre, J. *Bull. Soc. Chim. France* **1973**, *3*, 2930.
209. Dervin, J.; Faucherre, J.; Herpin, P. *Bull. Soc. Chim. France* **1973**, *7*, 2634.
210. Chernyaev, I. I.; Golovnya, V. A.; Molodkin, A. K. *Russ. J. Inorg. Chem.* **1958**, *3*, 100.
211. Voliotis, P. S.; Rimsky, E. A. *Acta Crystallogr.* **1975**, *B31*, 2615.
212. Voliotis, S.; Fromage, F.; Faucherre, J.; Dervin, J. *Rev. Chim. Minérale* **1977**, *14*, 441.
213. Voliotis, P. S. *Acta Crystallogr.* **1979**, *B35*, 2899.
214. March, R. E.; Herbstein, R. H. *Acta Crystallogr.* **1988**, *B44*, 77.
215. Golovnya, V. A.; Bolotova, G. T. *Russ. J. Inorg. Chem.* **1961**, *6*, 1256.
216. Bagnall, K. W In *Gmelin's Handbook of Inorganic Chemistry, Supplement Volume C7*; Springer–Verlag: Berlin, 1988; p 1.
217. Clark, D. L.; Conradson, S. D.; Keogh, D. W.; Palmer, P. D.; Scott, B. L.; Tait, C. D. *Inorg. Chem.* **1998**, *37*, 2893.
218. Gel'man, A. D.; Zaitsev, L. M. *Zh. Neorgan. Khim.* **1958**, *3*.
219. Ueno, K.; Hoshi, M. *J. Inorg. Nucl. Chem.* **1970**, *32*, 381.
220. Zhang, Y.-J.; Collison, D.; Livens, F. R.; Powell, A. K.; Wocadlo, S.; Eccles, H. *Polyhedron* **2000**, *19*, 1757.
221. Veirs, D. K.; Smith, C. A.; Berg, J. M.; Zwick, B. D.; Marsh, S. F.; Allen, P.; Conradson, S. D. *J. Alloys Compds.* **1994**, *213/214*, 328.
222. Allen, P. G.; Veirs, D. K.; Conradson, S. D.; Smith, C. A.; Marsh, S. F. *Inorg. Chem.* **1996**, *35*, 2841.
223. Berg, J. M.; Veirs, D. K.; Vaughn, R. B.; Cisneros, M. A.; Smith, C. A. *J. Radioanal. Nucl. Chem.* **1998**, *235*, 25.
224. Preston, J. S.; du Preez, A. C. *Solvent Extr. Ion Exch.* **1995**, *13*, 391.
225. Berthon, C.; Chachaty, C. *Solvent Extr. Ion Exch.* **1995**, *13*, 781.
226. Romanovski, V. V.; White, D. J.; Xu, J.; Hoffman, D. C.; Raymond, K. N. *Solvent Extr. Ion Exch.* **1999**, *17*, 55.
227. Oetting, F. L.; Rand, M. H.; Ackermann, R. J. *The Chemical Thermodynamics of Actinide Elements and Compounds: Part 1, The Actinide Elements*; IAEA: Vienna, STI/PUB/424/1, 1976.
228. Staritzky, E. *Anal. Chem.* **1956**, *28*, 2021.
229. Ryan, J. L. *J. Phys. Chem.* **1961**, *65*, 1099.
230. Boatner, L. A.; Sales, B. C. Monazite. In *Radioactive Waste Forms for the Future;* Lutze, W., Ewing, R. C., Eds.; North-Holland: Amsterdam 1988.
231. Brandel, V.; Dacheux, N.; Genet, M. *J. Solid State Chem* **1996**, *121*, 467.
232. Kobets, L. V.; Umreiko, D. S. *Chem. Rev.* **1983**, 509.
233. Francis, R. J.; Drewitt, M. J.; Halasyamani, P. S.; Ranganathachar, C.; O'Hare, D.; Clegg, W.; Teat, S. J. *Chem. Commun* **1998**, 279.
234. Baglan, N.; Fourest, B.; Guillaumont, R.; Blain, G.; Le Du, J.-F.; Genet, M. *New J. Chem.* **1994**, *18(7)*, 809.
235. Benard, P.; Brandel, V.; Dacheux, N.; Jaulmes, S.; Launay, S.; Lindecker, C.; Genet, M.; Louer, D.; Quarton, M. *Chem. Mater.* **1996**, *8*, 181.
236. Louer, M.; Brochu, R.; Louer, D. *Acta Crystallogr.* **1995**, *B51*, 908.
237. Merigou, C.; Genet, M.; Ouillon, N.; Chopin, T. *New J. Chem.* **1995**, *19*, 275.
238. Matkovic, B.; Prodic, B.; Sljukic, M. *Croat. Chem. Acta* **1968**, *40*, 147.
239. Quarton, M.; Zouiri, M.; Freundlich, W. *C. R. Acad. Sci., Ser. 2* **1984**, *299*, 785.
240. Voinova, L. M. *Radiochemistry (Moscow)* **1998**, *40*, 299.
241. Masse, R.; Grenier, J. C. *Fr. Bull. Soc. Fr. Mineral Cryst.* **1972**, *95(1)*, 136.

242. Linde, S. A.; Gorbunovaz, Y. E.; Lavrov, A. V. *Zh. Neorg. Khim.* **1983**, *28(6)*, 1391.
243. Benard, P.; Loueur, D.; Dacheux, N.; Brandel, V.; Genet, M. *Chem. Mater.* **1994**, *6*, 1049.
244. Benard, P.; Loueur, D.; Dacheux, N.; Brandel, V.; Genet, M. *An. Quim. Int. Ed.* **1996**, *92(2)*, 79.
245. Schaekers, J. M.; Greybe, W. G. *J. Appl. Crystallogr..* **1973**, *6(Pt. 3)*, 249.
246. Cabeza, A.; Aranda, M. A. G.; Cantero, F. M.; Lozano, D.; Martinez-Lara, M.; Bruque, S. *J. Solid State Chem.* **1996**, 181.
247. Hawkins, H. T.; Spearing, D. R.; Veirs, D. K.; Danis, J. A.; Smith, D. M.; Tait, C. D.; Runde, W. H.; Spilde, M. N.; Scheetz, B. E. *Chem. Mater.* **1999**, *11*, 2851.
248. Burnaeva, A. A.; Volkov, Y. F.; Kryukova, A. I.; Skiba, O. V.; Spiryakov, V. I.; Korshunov, I. A.; Samoilova, T. K. *Radiokhim* **1987**, *29(1)*, 3.
249. Benard, P.; Loueur, M.; Loueur, D.; Dacheux, N.; Brandel, V.; Genet, M. *J. Solid State Chem.* **1997**, *132*, 315.
250. Douglas, R. M. *Acta Crystallogr..* **1962**, *15*, 505.
251. Bjorklund, C. W. *J. Am. Chem. Soc.* **1957**, *79*, 6347.
252. Nectoux, F.; Tabuteau, A. *Radiochem. Radioanal. Lett.* **1981**, *49*, 43.
253. Kierkegaard, P. *Acta Chem. Scand.* **1956**, *10*, 599.
254. Paul, R. C.; Singh, S.; Verma, R. D. *J. Fluorine Chem.* **1980**, *16*, 153.
255. Seaborg, G. T.; Wahl, A. C. *J. Am. Chem. Soc.* **1948**, *70*, 1128.
256. Bradley, D. C.; Saad, M. A.; Wardlaw, W. *J. Chem. Soc.* **1954**, 1091.
257. Bradley, D. C.; Chatterjee, A. K.; Wardlaw, W. *J. Chem. Soc.* **1956**, 2260.
258. Barnhart, D. M.; Clark, D. L.; Gordon, J. C.; Huffman, J. C.; Watkin, J. G. *Inorg. Chem.* **1994**, *33*, 3939.
259. Clark, D. L.; Huffman, J. C.; Watkin, J. G. *J. Chem. Soc., Chem. Commun.* **1992**, 266.
260. Clark, D. L.; Watkin, J. G. *Inorg. Chem.* **1993**, *32*, 1766.
261. Bradley, D. C.; Kapoor, R. N.; Smith, B. C. *J. Inorg. Nucl. Chem.* **1962**, *24*, 863.
262. Cotton, F. A.; Marler, D. O.; Schwotzer, W. *Inorg. Chim. Acta* **1984**, *85*, L31.
263. Van Der Sluys, W. G.; Sattelberger, A. P.; McElfresh, M. W. *Polyhedron* **1990**, *9*, 1843.
264. Arliguie, T.; Baudry, D.; Ephritikhine, M.; Nierlich, M.; Lance, M.; Vigner, J. *J. Chem. Soc., Dalton Trans.* **1992**, 1019.
265. Berg, J. M.; Sattelberger, A. P.; Morris, D. E.; Van Der Sluys, W. G.; Fleig, P. *Inorg. Chem.* **1993**, *32*, 647.
266. Vilhena, M. T.; Domingos, A.M. T. S.; Pires de Matos, A. *Inorg. Chim. Acta* **1984**, *95*, 11.
267. Brunelli, M.; Perego, G.; Lugli, G.; Mazzei, A. *J. Chem. Soc., Dalton Trans.* **1979**, 861.
268. Stewart, J. L.; Andersen, R. A. *J. Chem. Soc., Chem. Commun.* **1987**, 1846.
269. Samulski, E. T.; Karraker, D. G. *J. Inorg. Nucl. Chem.* **1967**, *29*, 993.
270. Bradley, D. C.; Harder, B.; Hudswell, F. *J. Chem. Soc.* **1957**, 3318.
271. McKee, S. D.; Burns, C. J.; Avens, L. R. *Inorg. Chem.* **1998**, *37*, 4040.
272. Avens, L. R.; Barnhart, D. M.; Burns, C. J.; McKee, S. D.; Smith, W. H. *Inorg. Chem.* **1994**, *33*, 4245.
273. Berg, J. M. *J. Alloys Compd.* **1994**, *213*, 497.
274. Funk, H.; Andrä, K. *Z. Anorg. Allg. Chem.* **1968**, *361*, 199.
275. Wilkerson, M. P.; Burns, C. J.; Paine, R. T.; Scott, B. L. *J. Chem. Crystallogr.* **2000**, *30*, 7.
276. Adam, R.; Villiers, C.; Ephritikhine, M.; Lance, M.; Nierlich, M.; Vigner, J. *New J. Chem.* **1993**, *17*, 455.
277. Baudin, C.; Ephritikhine, M. *J. Organomet. Chem.* **1989**, *364*, C1.
278. Baudin, C.; Baudry, D.; Ephritikhine, M.; Lance, M.; Navaza, A.; Nierlich, M.; Vigner, J. *J. Organomet. Chem.* **1991**, *415*, 59.
279. Blake, P. C.; Lappert, M. F.; Taylor, R. G.; Atwood, J. L.; Zhang, H. *Inorg. Chim. Acta* **1987**, *139*, 13.
280. Hitchcock, P. B.; Lappert, M. F.; Singh, A.; Taylor, R. G.; Brown, D. *J. Chem. Soc., Chem. Commun.* **1983**, 561.
281. Barnhart, D. M.; Clark, D. L.; Gordon, J. C.; Huffman, J. C.; Watkin, J. G.; Zwick, B. D. *Inorg. Chem.* **1995**, *34*, 5416.
282. Van Der Sluys, W. G.; Huffman, J. C.; Ehler, D. S.; Sauer, N. N. *Inorg. Chem.* **1992**, *31*, 1316.
283. Mehrotra, R. C.; Misra, R. A. *Indian J. Chem.* **1968**, *6*, 669.
284. Leverd, P. C.; Lance, M.; Vigner, J.; Nierlich, M.; Ephritikhine, M. *J. Chem. Soc., Dalton Trans.* **1995**, 237.
285. Leverd, P. C.; Arliguie, T.; Ephritikhine, M.; Nierlich, M.; Lance, M.; Vigner, J. *New J. Chem.* **1993**, *17*, 769.
286. Leverd, P. C.; Lance, M.; Nierlich, M.; Vigner, J.; Ephritikhine, M. *J. Chem. Soc., Dalton Trans.* **1993**, 2251.
287. Leverd, P. C.; Lance, M.; Nierlich, M.; Vigner, J.; Ephritikhine, M. *J. Chem. Soc., Dalton Trans.* **1994**, 3563.
288. Arliguie, T.; Baudry, D.; Berthet, J. C.; Ephritikhine, M.; Le Maréchal, J. F. *New J. Chem.* **1991**, *15*, 569.
289. Butcher, R. J.; Clark, D. L.; Grumbine, S. K.; Watkin, J. G. *Organometallics* **1995**, *14*, 2799.
290. Moll, H.; Denecke, M. A.; Jalilehvand, F.; Sandström, M.; Grenthe, I. *Inorg. Chem.* **1999**, *38*, 1795.
291. Alcock, N. W.; Kemp, T. J.; Sostero, S.; Traverso, O. *J. Chem. Soc., Dalton Trans.* **1980**, 1182.
292. Degetto, S.; Baracco, L.; Graziani, R.; Celon, E. *Transition Met. Chem.* **1978**, *3*, 351.
293. Harrowfield, J. M.; Peachey, B. J.; Skelton, B. W.; White, A. W. *Aust. J. Chem.* **1995**, *48*, 1349.
294. Rogers, R. D. *Lanth. Actin. Res.* **1989**, *3*, 71.
295. Rabinovich, D.; Schimek, G. L.; Pennington, W. T.; Nielsen, J. B.; Abney, K. D. *Acta Crystallogr., Sect. C* **1999**, *54*, 1740.
296. Clark, D. L.; Frankcom, T. M.; Miller, M. M.; Watkin, J. G. *Inorg Chem.* **1992**, *31*, 1628.
297. Spry, M. P.; Errington, W.; Willey, G. R. *Acta Crystallogr., Sect. C* **1997**, *53*, 1386.
298. Rabinovich, D.; Scott, B. L.; Nielsen, J. B.; Abney, K. D. *J. Chem. Crystallogr.* **1999**, *29*, 243.
299. Van der Sluys, W. G.; Berg, J. M.; Barnhart, D.; Sauer, N. N. *Inorg. Chim. Acta* **1993**, *204*, 251.
300. Rebizant, J.; Spirlet, M. R.; Apostolidis, C.; van den Bossche, G.; Kanellakopulos, B. *Acta Crystallogr., Sect. C* **1991**, *47*, 864.
301. Maury, O.; Ephritikhine, M.; Nierlich, M.; Lance, M.; Samuel, E. *Inorg. Chim. Acta* **1998**, *279*, 210.
302. Gordon, P. L.; Thompson, J. A.; Watkin, J. G.; Burns, C. J.; Sauer, N. N.; Scott, B. L. *Acta Crystallogr., Sect. C* **1999**, *55*, 1275.
303. Rogers, R. D.; Kurihara, L. K.; Benning, M. M. *J. Chem. Soc., Dalton Trans.* **1988**, 13.
304. Rogers, R. D.; Benning, M. M. *Acta Crystallogr., Sect. C* **1988**, *44*, 641.
305. Bagnall, K. W.; Payne, G. F.; Brown, D. *J. Less-Common Met.* **1985**, *109*, 31.
306. Al-Daher, A. G. M.; Bagnall, K. W.; Payne, G. F. *J. Less-Common Met.* **1986**, *115*, 287.

307. Bagnall, K. W.; Lopez, O. V. *J. Chem. Soc., Dalton Trans.* **1975**, 1409.
308. Bagnall, K. W.; Lopez, O. V. *J. Chem. Soc., Dalton Trans.* **1976**, 1109.
309. Bagnall, K. W.; Li, X. F.; Pao, P. J.; Al-Daher, A. G. M. *Can. J. Chem.* **1983**, *61*, 708.
310. Ruikar, P. B.; Nagar, M. S. *Polyhedron* **1995**, *14*, 3125.
311. Sommerville, P.; Laing, M. *Acta Crystallogr., Sect. B* **1976**, *32*, 1551.
312. De Wet, J. F.; Caira, M. R. *J. Chem. Soc., Dalton Trans.* **1986**, 2035.
313. Gupta, B.; Malik, P.; Deep, A. *J. Radioanal. Nucl. Chem.* **2002**, *251*, 451.
314. Sahu, S. K.; Reddy, M. L. P.; Ramamohan, T. R.; Chakravortty, V. *Radiochim. Acta* **2000**, *88*, 33.
315. Murali, M. S.; Michael, K. M.; Jambunathan, U.; Mathur, J. N. *J. Radioanal. Nucl. Chem.* **2002**, *251*, 387.
316. Bombieri, G.; Benetollo, F.; Bagnall, K. W.; Plews, M. J.; Brown, D. *J. Chem. Soc., Dalton Trans.* **1983**, 343.
317. Bombieri, G.; Bagnall, K. W. *J. Chem. Soc., Chem. Commun.* **1975**, 188.
318. Shinomoto, R.; Zalkin, A.; Edelstein, N. M. *Inorg. Chim. Acta* **1987**, *139*, 91.
319. Malhotra, K. C.; Mahajan, V. P.; Mehrotra, G.; Chaudhry, S. C. *Chem. Ind. (London)* **1978**, 921.
320. Cousson, A.; Abazli, H.; Pages, M.; Gasperin, M. *Acta Crystallogr., Sect. C* **1983**, *39*, 425.
321. Mucker, K.; Smith, G. S.; Johnson, Q.; Elson, R. E. *Acta Crystallogr., Sect. C* **1969**, *25*, 2362.
322. Haaland, A.; Martinsen, K. J.; Swang, O.; Volden, H. V.; Booij, A. S.; Konings, R. J. M. *J. Chem. Soc., Dalton Trans.* **1995**, 185.
323. Zalkin, A.; Forrester, J. D.; Templeton, D. H. *Inorg. Chem.* **1964**, *3*, 639.
324. Brown, D. Halides, Halates, Perhalates, Thiocyanates, Selenocyanates, Cyanates, and Cyanides. In *Comprehensive Inorganic Chemistry;* Bailar, J. C., Emeleus, H. J., Nyholm, R. N., Trotman-Dickenson, A. F., Eds.; Pergamon: Oxford, UK, 1973; Vol. 5, p 151.
325. Brunton, G. *Acta Crystallogr.* **1964**, *21*, 814.
326. Brunton, G. *Acta Crystallogr., Sect. B* **1969**, *25*, 1919.
327. Abazli, H.; Cousson, A.; Tabuteau, A.; Pages, M.; Gasperin, M. *Acta Crystallogr., Sect. B* **1980**, *36*, 2765.
328. Cousson, A.; Tabuteau, A.; Pages, M.; Gasperin, M. *Acta Crystallogr., Sect B* **1979**, *35*, 1198.
329. Abazli, H.; Cousson, A.; Jove, J.; Pages, M.; Gasparin, M. *J. Less-Common Met.* **1984**, *96*, 23.
330. Rosenzweig, A.; Cromer, D. T. *Acta Crystallogr., Sect B* **1970**, *26*, 38.
331. Zachariasen, W. H. *J. Am. Chem. Soc* **1948**, *70*, 2147.
332. Francis, R. J.; Halasyamani, O. H. D. *Angew. Chem., Int. Ed. Engl.* **1998**, *37*, 2214.
333. Francis, R. J.; Halasyamani, P. S.; Bee, J. S.; O'Hare, D. *J. Am. Chem. Soc.* **1999**, *121*, 1609.
334. Halasyamani, P. S.; Walker, S. M.; O'Hare, D. *J. Am. Chem. Soc.* **1999**, *121*, 7414.
335. Walker, S. M.; Halasyamani, P. S.; Allen, S.; O'Hare, D. *J. Am. Chem. Soc.* **1999**, *121*, 10513.
336. Cahill, C. L.; Burns, P. C. *Inorg. Chem.* **2001**, *40*, 1347.
337. Almond, P. M.; Deakin, L.; Mar, A.; Albrecht-Schmitt, T. E. *Inorg. Chem.* **2001**, *40*, 886.
338. Photiadis, G. M.; Paptheodorou, G. N. *J. Chem. Soc., Dalton Trans.* **1999**, 3541.
339. Magette, M.; Fuger, J. *Inorg. Nucl. Chem. Lett.* **1977**, *13*, 529.
340. Conradi, E.; Bohrer, R.; Weber, R.; Muller, U. *Z. Kristallogr.* **1987**, *181*, 187.
341. Casellato, U.; Graziani, R. *Z. Kristallogr.-New Cryst. Struct.* **1998**, *213*, 361.
342. Conradi, E.; Bohrer, R.; Muller, U. *Chem. Ber.* **1986**, *119*, 2582.
343. Wang, W. J.; Lin, J.; Shen, H.; Zheng, P.; Wang, M.; Wang, B. *Radiochim. Acta* **1986**, *40*, 199.
344. Rogers, R. D.; Benning, M. M. *J. Inclusion Phenom. Macrocyclic Chem.* **1991**, *11*, 121.
345. Rogers, R. D.; Kurihara, L. K.; Benning, M. M. *J. Inclusion Phenom. Macrocyclic Chem.* **1987**, *5*, 645.
346. Dodge, R. P.; Smith, G. S.; Johnson, Q.; Elson, R. E. *Acta Crystallogr., Sect. B* **1968**, *24*, 304.
347. Zhao, P.; Romanovski, V. V.; Whisenhunt, D. W., Jr.; Hoffman, D. C.; Mohs, T. R.; Xu, J.; Raymond, K. N. *Solvent Extr. Ion Exch.* **1999**, *17(5)*, 1327.
348. Paquet, F.; Montegue, B.; Ansoborlo, E.; Henge-Napoli, M. H.; Houpert, P.; Durbin, P. W.; Raymond, K. N. I. J. O. R. B. *Int. J. Radiat. Biol.* **2000**, *76(1)*, 113.
349. Xu, J.; Durbin, P. W.; Kullgren, B.; Ebbe, S. N.; Uhlir, L. C.; Raymond, K. N. *J. Med. Chem.* **2002**, *45(18)*, 3963.
350. Durbin, P. W.; Kullgren, B.; Ebbe, S. N.; Xu, J.; Raymond, K. N. *Health Physics Field* **1998**, *78*, 511.
351. O'Boyle, N. C.; Nicholson, G. P.; Piper, T. J.; Taylor, D. M.; Williams, D. R.; Williams, G. *Appl. Radiat. Isot.* **1997**, *48*, 183.
352. Durbin, P. W.; B. Kullgren, X. J.; Raymond, K. N. *Int. J. Radiat. Biol.* **2000**, *76*, 113.
353. Elving, P. J.; Olson, E. C. *J. Am. Chem. Soc.* **1956**, *78*, 420.
354. Horton, W. S. *J. Am. Chem. Soc.* **1956**, *78*, 897.
355. Smith, W. L.; Raymond, K. N. *J. Am. Chem. Soc.* **1981**, *103*, 3341.
356. Casellato, U.; Vigato, P. A.; Tamburini, S.; Graziani, R.; Vidali, M. *Inorg. Chim. Acta* **1984**, *81*, 47.
357. Yoshimura, T.; Miyake, C.; Imoto, S. *Technol. Rep. Osaka Univ.* **1972**, *22*, 791.
358. Yoshimura, T.; Miyake, C.; Imoto, S. *J. Inorg. Nucl. Chem.* **1975**, *37*, 739.
359. Neu, M. P.; Matonic, J. H.; Ruggiero, C. E.; Scott, B. L. *Angew. Chem., Int. Ed. Engl.* **2000**, *39*, 1442.
360. Whisenhunt, D. W., Jr.; Neu, M. P.; Hou, Z.; Xu, J.; Hoffman, D. C.; Raymond, K. N. *Inorg. Chem* **1996**, *35*, 4128.
361. Santos, M. A.; Rodrigues, E.; Gaspar, M. *J. Chem. Soc., Dalton Trans.* **2000**, 4398.
362. Von, K. A. *Z. Anorg. Allg. Chem.* **1968**, *361*, 254.
363. Sofen, S. R.; Abu-Dari, K.; Freyberg, D. P.; Raymond, K. N. *J. Am. Chem. Soc.* **1978**, *100*, 7882.
364. Sylwester, E. R.; Allen, P. G.; Dharmawardana, U. R.; Sutton, M. *Inorg. Chem.* **2001**, *40*, 2835.
365. Raymond, K. N.; Freeman, G. E.; Kappel, M. J. *Inorg. Chim. Acta* **1984**, *94*, 193.
366. Durbin, P. W.; Jones, E. S.; Raymond, K. N.; Weitl, F. L. *Radiat. Res.* **1980**, *81*, 170.
367. Durbin, P. W.; White, D. L.; Jeung, N.; Weitl, F. L.; Uhlir, L. C.; Jones, E. S.; Bruenger, F. W.; Raymond, K. N. *Health Phys.* **1989**, *56*, 839.
368. Uhlir, L. C.; Durbin, P. W.; Jeung, N.; Raymond, K. N. *J. Med. Chem.* **1993**, *36*, 504.
369. Bouby, M.; Billard, I.; MacCordick, J. *J. Alloys Compd.* **1998**, *271–273*, 206.
370. Bouby, M.; Billard, I.; Maccordick, H. J. *Czechoslovak J. Phys.* **1999**, *49*, 147.
371. Riley, P. E.; Abu-Dari, K.; Raymond, K. N. *Inorg. Chem.* **1983**, *22*, 3940.
372. Casellato, U.; Vigato, P. A.; Tamburini, S.; Vidali, M.; Graziani, R. *Inorg. Chim. Acta* **1983**, *69*, 77.

373. Durbin, P. W.; Kullgren, B.; Ebbe, S. N.; Xu, J.; Raymond, K. N. *Health Phys.* **2000**, *78*, 511.
374. Casellato, U.; Vigato, P. A.; Tamburini, S. *Inorg. Chim. Acta* **1983**, *69*, 77.
375. Frere, F. J. *J. Am. Chem. Soc.* **1933**, *55*, 4362.
376. Engelter, C.; Knight, C. L.; Thornton, D. A. *Spectrosc. Lett.* **1989**, *22*, 1161.
377. Mahmoud, M. R.; Awad, A.; Hammam, A. M.; Saber, H. *Indian J. Chem., Sect. A* **1980**, *19A*, 1131.
378. Unak, P.; Ozkayalar, T.; Ozdemir, D.; Yurt, F. *J. Radioanal. Nucl. Chem.* **1995**, *196*, 323.
379. Singer, N.; Studd, B. F.; Swallow, A. G. *Chem. Commun.* **1970**, 342.
380. Barton, R. J.; Dabeka, R. W.; Shengzhi, H.; Mihichuk, L. M.; Pizzey, M.; Robertson, B. E.; Wallace, W. J. *Acta. Crist.* **1983**, *C39*, 714.
381. Keller, C. *J. Inorg. Nucl. Chem.* **1965**, *27*, 321.
382. Bagnall, K. W.; Yanir, E. *J. Inorg. Nucl. Chem.* **1974**, *36*, 777.
383. Calderazzo, F.; Dell'Amico, G.; Pasquali, M.; Perego, G. *Inorg. Chem.* **1978**, *17*, 474.
384. Velasquez, O. *Revista Colombiana de Quimica* **1984**, *13*, 27.
385. Arduini, A. L.; Edelstein, N. M.; Jamerson, J. D.; Reynolds, J. G.; Schmid, K.; Takats, J. *Inorg. Chem.* **1981**, *20*, 2470.
386. Arduini, A. L.; Jamerson, J. D.; Takats, J. *Inorg. Chem.* **1981**, *20*, 2474.
387. Arduini, A. L.; Takats, J. *Inorg. Chem.* **1981**, *20*, 2480.
388. Akhtar, M. N.; Smith, A. J. *Acta Crystallogr., Sect. B* **1975**, *31*, 1361.
389. Favas, M. C.; Kepert, D. L.; Patrick, J. M.; White, A. H. *J.Chem. Soc., Dalton Trans* **1983**, 571.
390. Spirlet, M. R.; Rebizant, J.; Kanellakopulos, B.; Dornberger, E. *Acta Crystallogr., Sect. C: Cryst. Struct. Commun.* **1987**, *C43*, 19.
391. Bykhovskii, D. N.; Kuz'mina, M. A.; Maksimov, V. F.; Novikov, G. S.; Smirnov, A. N.; Solntseva, L. V. *Radiokhimiya* **1988**, *30*, 37.
392. Molodkin, A. K.; Skotnikova, G. A. *Russ. J. Inorg. Chem.* **1964**, *3*, 308.
393. Mortl, K. P.; Sutter, J.-P.; Golhen, S.; Ouahab, L.; Kahn, O. *Inorg. Chem.* **2000**, *39*, 1626.
394. Wai, C. M.; Lin, Y.; Ji, M.; Toews, K. L.; Smart, N. G. In *Metal-Ion Separation and Preconcentration: Progress and Opportunities*; Bond, A. H., Dietz, M. L., Rogers, R. D., Eds.; Oxford University Press: Washington, D.C., 1999, pp 390–400.
395. Ahrland, S. In *The Chemistry of the Actinide Elements*; Katz, J. J., Seaborg, G. T., Morss, L. R., Eds.; Chapman and Hall: New York, 1986; Vol. 2, 1480–1546.
396. Mathur, J. N.; Choppin, G. R. *Solvent Extr. Ion Exch.* **1993**, *11*, 1.
397. Sasaki, Y.; Choppin, G. R. *J. Radioanal. Nucl. Chem.* **1996**, *207*, 383.
398. Takeishi, H.; Kitatsuji, Y.; Kimura, T.; Meguro, Y.; Yoshida, Z.; Kihara, S. *Anal. Chim. Acta* **2001**, *431*, 69.
399. Jyothi, A.; Rao, G. N. *Polyhedron* **1989**, *8*, 1111.
400. Choppin, G. R.; Morgenstern, A. *J. Radioanal. Nucl. Chem* **2000**, *243*, 45.
401. Bowen, S. M.; Duesler, E. N.; Paine, R. T. *Inorg. Chem.* **1982**, *21*, 261.
402. Bowen, S. M.; Duesler, E. N.; Paine, R. T. *Inorg. Chem.* **1983**, *22*, 286.
403. Caudle, L. J.; Duesler, E. N.; Paine, R. T. *Inorg. Chim. Acta* **1985**, *110*, 91.
404. Kalina, D. G. *Solv. Extract. Ion Exch.* **1984**, *2*, 381.
405. McCabe, D. J.; Duesler, E. N.; Paine, R. T. *Inorg. Chem.* **1985**, *24*, 4626.
406. Conary, G. S.; McCabe, D. J.; Meline, R. L.; Duesler, E. N.; Paine, R. T. *Inorg. Chim. Acta* **1993**, *203*, 11.
407. Horwitz, E. P.; Diamond, H.; Martin, K. A. *Solvent Extr. Ion Exch.* **1987**, *5*, 447.
408. Nash, K. L. *J. Alloys Compd.* **1997**, *249*, 33.
409. Mathur, J. N.; Murali, M. S.; Nash, K. L. *Solvent Extr. Ion Exch.* **2001**, *19*, 357.
410. McCabe, D. J.; Russell, A. A.; Karthikeyan, S.; Paine, R. T.; Ryan, R. R.; Smith, B. *Inorg. Chem.* **1987**, *26*, 1230.
411. Russell, R. R.; Meline, R. L.; Duesler, E. N.; Paine, R. T. *Inorg. Chim. Acta* **1995**, *231*, 1.
412. Blaha, S. L.; McCabe, D. J.; Paine, R. T.; Thomas, K. W. *Radiochim. Acta* **1989**, *46*, 123.
413. Rapko, B. M.; Duesler, E. N.; Smith, P. H.; Paine, R. T.; Ryan, R. R. *Inorg. Chem.* **1993**, *32*, 2164.
414. Bond, E. M.; Duesler, E. N.; Paine, R. T.; Neu, M. P.; Matonic, J. H.; Scott, B. L. *Inorg. Chem.* **2000**, *39*, 4152.
415. Bond, E. M.; Engelhardt, U.; Deere, T. P.; Rapko, B. M.; Paine, R. T.; FitzPatrick, J. R. *Solv. Extract. Ion Exch.* **1997**, 381.
416. Bond, E. M.; Engelhardt, U.; Deere, T. P.; Rapko, B. M.; Paine, R. T.; FitzPatrick, J. R. *Solv. Extract. Ion Exch.* **1998**, 967.
417. Nash, K. L.; Lavallette, C.; Borkowski, M.; Paine, R. T.; Gan, X. *Inorg Chem.* **2002**, *41*, 5849.
418. Chiarizia, R.; Horwitz, E. P.; Rickert, P. G.; Herlinger, A. W. *Solvent Extr. Ion Exch.* **1996**, *14*, 773.
419. Chiarizia, R.; Herlinger, A. W.; Horwitz, E. P. *Solvent Extr. Ion Exch.* **1997**, *15*, 417.
420. Chiarizia, R.; Herlinger, A. W.; Cheng, Y. D.; Ferraro, J. R.; Rickert, P. G.; Horwitz, E. P. *Solvent Extr. Ion Exch.* **1998**, *16*, 505.
421. Chiarizia, R.; McAlister, D. R.; Herlinger, A. W. *Solvent Extr. Ion Exch.* **2001**, *19*, 415.
422. Nigond, L.; Musikas, C.; Cuillerdier, C. *Solvent Extr. Ion Exch.* **1994**, *12*, 297.
423. Nair, G. M.; Prabhu, D. R.; Mahajan, G. R. *J. Radioanal. Nucl. Chem.* **1994**, *186*, 47.
424. Nair, G. M.; Prabhu, D. R.; Mahajan, G. R.; Shukla, J. P. *Solvent Extr. Ion Exch.* **1993**, *11*, 831.
425. Cuillerdier, C.; Musikas, C.; Hoel, P.; Nigond, L.; Vitart, X. *Sep. Sci. Technol.* **1991**, *26*, 1229.
426. Sasaki, Y.; Choppin, G. R. *J. Radioanal. Nucl. Chem.* **1997**, *222*, 271.
427. Golubev, A. M.; Kazanskii, L. P.; Torchenkova, E. A.; Simonov, V. I.; Spitsyn, V. I. *Dokl. Chem.* **1975**, *221*, 198.
428. Kazanskii, L. P.; Golubev, A. M.; Baburina, I. I.; Torchenkova, E. A.; Spitsyn, V. I. *Bull. Acad. Sci. USSR, Div. Chem. Sci.* **1978**, 1956.
429. Golubev, A. M.; Muradyan, L. A.; Kazanskii, L. P.; Torchenkova, E. A.; Simonov, V. I.; Spitsyn, V. I. *Sov. J. Coord. Chem.* **1977**, *3*, 715.
430. Kazanskii, L. P.; Fedotov, M. A.; Spitsyn, V. I. *Dokl. Phys. Chem.* **1977**, *233*, 250.
431. Barbieri, G. A. *Atti Accad. Naz. Lincei* **1913**, *22*, 781.
432. Barbieri, G. A. *Atti Accad. Naz. Lincei* **1914**, *23*, 805.
433. Baidala, P.; Smurova, V. S.; Spitsyn, V. I. *Dokl. Chem.* **1971**, *197*, 202.

434. Torchenkova, E. A.; Golubev, A. M.; Saprykin, A. S.; Krot, N. N.; Spitsyn, V. I. *Dokl. Chem.* **1974**, *216*, 430.
435. Tat'yania, I. V.; Chernaya, T. S.; Torchenkova, E. A.; Simonov, V. I.; Spitsyn, V. I. *Dockl. Chem* **1979**, *247*, 1162.
436. Kazanskii, L. P.; Torchenkova, E. A.; Spitsyn, V. I. *Dokl. Phys. Chem.* **1973**, *209*, 208.
437. Tat'yanina, I. V.; Torchenkova, E. A.; Kazanskii, L. P.; Spitsyn, V. I. *Dokl. Phys. Chem.* **1977**, *234*, 597.
438. Termes, S. C.; Pope, M. T. *Transit. Met. Chem.* **1978**, *3*, 103.
439. Golubev, A. M.; Kazanskii, L. P.; Chuvaev, V. F.; Torchenkova, E. A.; Spitsyn, V. I. *Dokl. Chem.* **1973**, *209*, 326.
440. Spitsyn, V. I.; Orlova, M. M.; Saprykina, O. P.; Saprykin, A. S.; Krot, N. N. *Russ. J. Inorg. Chem.* **1977**, *22*, 1355.
441. Spitsyn, V. I.; Torchenkova, E. A.; Kazanskii, L. P. *Z. Chem.* **1974**, *14*, 1.
442. Molchanov, V. N.; Tat'yanina, I. V.; Torchenkova, E. A.; Kazanskii, L. P. *J. Chem. Soc., Chem. Commun.* **1981**, 93.
443. Tat'yanina, I. V.; Fomicheva, E. B.; Molchanov, V. N.; Zavodnok, V. E.; Bel'sky, V. K.; Torchenkova, E. A. *Sov. Phys., Crystallogr.* **1982**, *27*, 142.
444. Botar, A. V.; Weakley, T. J. R. *Rev. Roum. Chim.* **1973**, *18*, 1166.
445. Marcu, G.; Rusu, M.; Botar, A. V. *Rev. Roum. Chim.* **1974**, *19*, 827.
446. Kosyakov, V. N.; Timofeev, G. A.; Erin, E. A.; Andreev, V. I.; Kopytov, V. V.; Simakin, G. A. *Sov. Radiochem.* **1977**, *19*, 418.
447. Saprykin, A. S.; Spitsyn, V. I.; Krot, N. N. *Dokl. Chem.* **1976**, *226*, 114.
448. Saprykin, A. S.; Spitsyn, V. I.; Orlova, M. M.; Zhuravleva, O. P.; Krot, N. N. *Sov. Radiochem.* **1978**, *20*, 207.
449. Tourné, C.; Tourné, G. *Rev. Chim. Minéral* **1977**, *14*, 83.
450. Yusov, A. B.; Shilov, V. P. *Radiokhimiya* **1999**, *41*, 3.
451. Tourné, C.; Tourné, G. *Acta Crystallogr., Sect. B* **1980**, *36*, 2012.
452. Marcu, G.; Rusu, M.; Botar, A. V. *Stud. Univ. Babes-Bolyai, Chem.* **1986**, *31*, 76.
453. Marcu, G.; Rusu, M.; Botar, A. V. *Rev. Roum. Chim.*. **1989**, *34*, 207.
454. Wedler, M.; Gilje, J. W.; Noltemeyer, M.; Edelmann, F. T. *J. Organomet. Chem.* **1991**, *411*, 271.
455. Baudry, D.; Ephritikhine, M.; Kläui, W.; Lance, M.; Nierlich, M. *Inorg. Chem.* **1991**, *30*, 2333.
456. Brianese, N.; Casellato, U.; Tamburini, S.; Tomasin, P.; Vigato, P. A. *Inorg. Chim. Acta* **1998**, *272*, 235.
457. Casellato, U.; Guerriero, P.; Tamburini, S.; Vigato, P. A. *Inorg. Chim. Acta* **1987**, *139*, 61.
458. Panda, C. R.; Chakravortty, V.; Dash, K. C. *Indian J. Chem., Sect. A* **1985**, *24A*, 807.
459. Sessler, J. L.; Vivian, A. E.; Seidel, D.; Burrell, A. K.; Hoehner, M.; Mody, T. D.; Gebauer, A.; Weghorn, S. J.; Lynch, V. *Coord. Chem. Rev.* **2001**, *216–217*, 411.
460. Girolami, G. S.; Gorlin, P. A.; Milam, S. N.; Suslick, K. S.; Wilson, S. R. *J. Coord. Chem.* **1994**, *32*, 173.
461. Dormond, A.; Belkalem, B.; Guilard, R. *Polyhedron* **1984**, *3*, 107.
462. Girolami, G. S.; Milam, S. N.; Suslick, K. S. *Inorg. Chem.* **1987**, *26*, 343.
463. Kadish, K. M.; Liu, Y. H.; Anderson, J. E.; Charpin, P.; Chevrier, G.; Lance, M.; Nierlich, M.; Vigner, D.; Dormond, A.; Belkalem, B.; Guilard, R. *J. Am. Chem. Soc.* **1988**, *110*, 6455.
464. Korobkov, I.; Gambarotta, S.; Yap, G. P. A. *Organometallics* **2001**, *20*, 2552.
465. Korobkov, I.; Gambarotta, S.; Yap, G. P. A.; Thompson, L.; Hay, P. J. *Organometallics* **2001**, *20*, 5440.
466. Hoekstra, H. R.; Katz, J. J. *J. Am. Chem. Soc.* **1949**, *71*, 2488.
467. Banks, R. H.; Edelstein, N. M.; Rietz, R. R.; Templeton, D. H.; Zalkin, A. *J. Am. Chem. Soc.* **1978**, *100*, 1957.
468. Volkov, V. V.; Myakishev, K. G. *Radiokhim.* **1976**, *18*, 512.
469. Volkov, V. V.; Myakishev, K. G. *Radiokhim.* **1980**, *22*, 745.
470. Ehemann, M.; Nöth, H. *Z. Anorg. Allg. Chem.* **1971**, *386*, 87.
471. Bernstein, E. R.; Hamilton, W. C.; Keiderling, T. A.; La Placa, S. J.; Lippard, S. J.; Mayerle, J. J. *Inorg. Chem.* **1972**, *11*, 3009.
472. Bernstein, E. R.; Keiderling, T. A.; Lippard, S. J.; Mayerle, J. J. *J. Am. Chem. Soc.* **1972**, *94*, 2552.
473. Charpin, P.; Marquet-Ellis, H.; Folcher, G. *J. Inorg. Nucl. Chem.* **1979**, *41*, 1143.
474. Charpin, P.; Nierlich, M.; Vigner, D.; Lance, M.; Baudry, D. *Acta Crystallogr., Sect. C* **1987**, *43*, 1465.
475. Banks, R. H.; Edelstein, N. M.; Spencer, B.; Templeton, D. H.; Zalkin, A. *J. Am. Chem. Soc.* **1980**, *102*, 620.
476. Schlesinger, H. I.; Brown, H. C.; Horvitz, L.; Bond, A. C.; Tuck, L. D.; Walker, A. O. *J. Am. Chem. Soc.* **1953**, *75*, 222.
477. Shinomoto, R.; Gamp, E.; Edelstein, N. M.; Templeton, D. H.; Zalkin, A. *Inorg. Chem.* **1983**, *22*, 2351.
478. Gamp, E.; Shinomoto, R.; Edelstein, N. M.; McGarvey, B. R. *Inorg. Chem.* **1987**, *26*, 2177.
479. Kot, W. K.; Edelstein, N. M. *New J. Chem.* **1995**, *19*, 641.
480. Rietz, R. R.; Zalkin, A.; Templeton, D. H.; Edelstein, N. M. *Inorg. Chem.* **1978**, *17*, 653.
481. Zalkin, A.; Rietz, R. R.; Templeton, D. H.; Edelstein, N. M. *Inorg. Chem.* **1978**, *17*, 661.
482. Shinomoto, R.; Brennan, J. G.; Edelstein, N. M.; Zalkin, A. *Inorg. Chem.* **1985**, *24*, 2896.
483. Rietz, R. R.; Edelstein, N. M.; Ruben, H. W.; Templeton, D. H.; Zalkin, A. *Inorg. Chem.* **1978**, *17*, 658.
484. Charpin, P.; Lance, M.; Nierlich, M.; Vigner, D.; Musikas, C. *Acta Crystallogr., Sec. C.* **1987**, *43*, 231.
485. Charpin, P.; Nierlich, M.; Chevrier, G.; Vigner, D.; Lance, M.; Baudry, D. *Acta Crystallogr., Sect. C.* **1987**, *43*, 1255.
486. Charpin, P.; Lance, M.; Soulié, E.; Vigner, D.; Marquet-Ellis, H. *Acta Crystallogr., Sect. C.* **1985**, *41*, 1723.
487. Meyer, K.; Mindiola, D. J.; Baker, T. A.; Davis, W. M.; Cummins, C. C. *Angew. Chem., Int. Ed. Engl.* **2000**, *39*, 3063.
488. Coles, S. J.; Edwards, P. G.; Hursthouse, M. B.; Read, P. W. *J. Chem. Soc., Chem. Commun.* **1994**, 1967.
489. Wedler, M.; Noltemeyer, M.; Edelmann, F. T. *Angew. Chem. Int. Ed. Engl.* **1992**, *31*, 72.
490. Selbin, J.; Ahmad, N.; Pribble, M. J. *J. Inorg. Nucl. Chem.* **1970**, *32*, 3249.
491. Selbin, J.; Ballhausen, C. J.; Durrett, D. G. *Inorg. Chem.* **1972**, *11*, 510.
492. Selbin, J.; Durrett, D. G.; Sherrill, H. J.; Newkome, G. R.; Collins, M. *J. Inorg. Nucl. Chem.* **1973**, *35*, 3467.
493. Arnaudet, L.; Bougon, R.; Buu, B.; Lance, M.; Nierlich, M.; Vigner, J. *Inorg. Chem.* **1994**, *33*, 4510.
494. Berry, J. A.; Holloway, J. H.; Brown, D. *Inorg. Nucl. Chem. Lett.* **1981**, *35*, 3467.
495. Brown, D.; Jones, P. J. *J. Chem. Soc., A, Phys., Theoret.* **1966**, 733.
496. Selbin, J. N. A.; Pribble, M. J. *J. Chem. Soc., Chem. Commun.* **1969**, 759.
497. Fryzuk, M. D.; Haddad, T. S.; Berg, D. J. *Coord. Chem. Rev.* **1990**, *99*, 137.
498. Andreev, G. B.; Fedoseev, A. M.; Budantseva, N. A.; Antipin, M. Y. *Dokl. Akad. Nauk. SSSR* **2000**, *375*, 778.
499. Fahey, J. A.; Turcotte, R. P.; Chikalla, T. D. *J. Inorg. Nucl. Chem.* **1976**, *38(3)*, 495.
500. Conradson, S. D., Unpublished results.

501. Madic, C.; Begun, G. M.; Hobart, D. E.; Hahn, R. L. *Inorg. Chem.* **1984**, *23*, 1914.
502. Sullivan, J. C.; Choppin, G. R.; Rao, L. F. *Radiochim. Acta* **1991**, *54*, 17.
503. Neck, V.; Runde, W.; Kim, J. I.; Kanellakopulos, B. *Radiochim. Acta* **1994**, *65*, 29.
504. Burns, P. C. *Can. Mineral.* **1998**, *36*, 1061.
505. Burns, P. C.; Miller, M. L.; Ewing, R. C. *Can. Mineral.* **1996**, *34*, 845.
506. Burns, P. C.; Finch, R. J. *Am. Mineral.* **1999**, *84*, 1456.
507. Simakin, G. A.; Volkov, Y. F.; Visyashcheva, G. I.; Kapshukov, I. I.; Baklanova, P. F.; Yakovlev, G. N. *Radiokhimiya* **1974**, *16*, 859.
508. Bennett, D. A.; Hoffman, D. C.; Nitsche, H.; Russo, R. E.; Torres, R. A.; Baisden, P. A.; Andrews, J. E.; Palmer, C. E. A.; Silva, R. J. *Radiochim. Acta* **1992**, *56*, 15.
509. Clark, D. L.; Conradson, S. D.; Ekberg, S. A.; Hess, N. J.; Neu, M. P.; Palmer, P. D.; Runde, W.; Tait, C. D. *J. Am. Chem. Soc.* **1996**, *118*, 2089.
510. Madic, C.; Hobart, D. E.; Begun, G. M. *Inorg. Chem.* **1983**, *22*, 1494.
511. Volkov, Y. F.; Kapshukov, I. I.; Visyashcheva, G. I.; Osipov, S. V.; Yakovlev, G. N. "X-ray diffraction of neptunium(V), plutonium(V), and americium(V) monocarbonates with alkali metals," Nauch.-Issled. Inst. At. Reakt., Dimitrovgrad,USSR. FIELD URL **1974**.
512. Ellinger, R. H.; Zachariasen, W. H. *J. Phys. Chem.* **1954**, *58*, 405.
513. Nigon, J. P.; Penneman, R. A.; Staritzki, E.; Keenan, T. K.; Asprey, L. B. *J. Phys. Chem.* **1954**, *58*, 403.
514. Gorbeko-Germanov, D. S.; Klimov, V. C. *Russ. J. Inorg. Chem.* **1966**, *11*, 280.
515. Volkov, Y. F.; Tomilin, S. V.; Visyashcheva, G. I.; Kapshukov, I. I.; Mefod'eva, M. P.; Krot, N. N.; Rykov, A. G. *Radiokhimiya* **1981**, *23*, 690.
516. Volkov, Y. V.; Kapshukov, I. I. *Radiokhimiya* **1984**, *26*, 361.
517. Volkov, Y. F.; Visyashcheva, G. I.; Tomilin, S. V.; Kapshukov, I. I.; Rykov, A. G. *Radiokhimiya* **1981**, *23*, 254.
518. Katz, J. J.; Seaborg, G. T.; Morss, L. R. *The Chemistry of the Actinide Elements*; Chapman and Hall: London 1986.
519. Tomilin, S. V.; Volkov, Y. F.; Melkaya, R. F.; Spiryakov, V. I.; Kapshukov, I. I. *Radiokhimiya* **1986**, *28*, 695.
520. Volkov, Y. F.; Melkaya, R. F.; Spiryakov, V. I.; Tomilin, S. V.; Kapshukov, I. I. *Radiokhimiya* **1986**, *28*, 311.
521. Sullivan, J. C.; Choppin, G. R. *Radiochim. Acta* **1961**, *54*, 17.
522. Nagasaki, S.; Kinoshita, K.; Enokida, Y.; Suzuki, A. *J. Nucl. Sci. Technol.* **1992**, *29*, 1100.
523. Rao, P. R. V.; Gudi, N. M.; Bagawde, S. V.; Patil, S. K. *J. Inorg. Nucl. Chem.* **1979**, *41*, 235.
524. Moskvin, A. I.; Poznyakov, A. N. *Russ. J. Inorg. Chem.* **1979**, *24*, 1357.
525. Morgenstern, A.; Kim, J. I. *Radiochim. Acta* **1996**, *72*, 73.
526. Budantseva, N. A.; Fedoseev, A. M.; Grigor'ev, M. S.; Potemkina, T. I.; Afonas'eva, T. V.; Krot, N. N. *Soviet Radiochemistry* **1989**, *30*, 578.
527. Albrecht-Schmitt, T. E.; Almond, P. M.; Sykora, R. E. *Inorg. Chem.* **2003**, .
528. Jones, R. G.; Bindschadler, E.; Karmas, G.; Yoeman, F. A.; Gilman, H. *J. Am. Chem. Soc.* **1956**, *78*, 4287.
529. Jones, R. G.; Bindschadler, E.; Karmas, G.; Martin, G. A., Jr.; Thirtle, J. R.; Yoeman, F. A.; Gilman, H. *J. Am. Chem. Soc.* **1956**, *78*, 4289.
530. Jones, R. G.; Bindschadler, E.; Blume, D.; Karmas, G.; Martin, G. A., Jr.; Thirtle, J. R.; Gilman, H. *J. Am. Chem. Soc.* **1956**, *78*, 6027.
531. Bradley, D. C.; Chakravarti, B. N.; Chatterjee, A. K. *J. Inorg. Nucl. Chem.* **1957**, *3*, 367.
532. Traverso, O.; Portanova, R.; Carassiti, V. *Inorg. Nucl. Chem. Lett.* **1974**, *10*, 771.
533. Sostero, S.; Traverso, O.; Bartocci, C.; Di Bernardo, P.; Magon, L.; Carassiti, V. *Inorg. Chim. Acta* **1976**, *19*, 229.
534. Halstead, G. W.; Eller, P. G.; Asprey, L. B.; Salazar, K. V. *Inorg. Chem.* **1978**, *17*, 2967.
535. Sanyal, D. K.; Sharp, D. W. A.; Winfield, J. M. *J. Fluorine Chem.* **1980**, *16*, 585.
536. Halstead, G. W.; Eller, P. G. *Inorg. Synth.* **1982**, *21*, 162.
537. Bradley, D. C.; Chatterjee, A. K. *J. Inorg. Nucl. Chem.* **1957**, *4*, 279.
538. Bradley, D. C.; Kapoor, R. N.; Smith, B. C. *J. Chem. Soc.* **1963**, 204.
539. Bradley, D. C. *Nature* **1958**, *182*, 1211.
540. Bradley, D. C.; Holloway, H. *Can. J. Chem.* **1962**, *40*, 1176.
541. Karraker, D. G. *Inorg. Chem.* **1964**, *3*, 1618.
542. Karraker, D. G.; Siddall, T. H., III; Stewart, W. E. *J. Inorg. Nucl. Chem.* **1969**, *31*, 711.
543. Eller, P. G.; Vergamini, P. J. *Inorg. Chem.* **1983**, *22*, 3184.
544. Sigurdson, E. R.; Wilkinson, G. *J. Chem. Soc., Dalton Trans.* **1977**, 812.
545. Maddock, A. G.; Pires de Matos, A. *Radiochim. Acta* **1972**, *18*, 71.
546. Larson, E. M.; Eller, P. G.; Larson, A. C. *Lanthanide and Actinide Res.* **1986**, *1*, 307.
547. Bhandari, A. M.; Kapoor, R. N. *Can. J. Chem.* **1966**, *44*, 1468.
548. Bhandari, A. M.; Kapoor, R. N. *Aust. J. Chem.* **1967**, *20*, 233.
549. Brown, D.; Hurtgen, C. *J. Chem. Soc., Dalton Trans.* **1979**, 1709.
550. Cayton, R. H.; Novo-Gradac, K. J.; Bursten, B. E. *Inorg. Chem.* **1991**, *30*, 2265.
551. Bagnall, K. W.; Bhandari, A. M.; Brown, D. *J. Inorg. Nucl. Chem.* **1975**, *37*, 1815.
552. Selbin, J.; Ahmad, N.; Pribble, M. *J. Chem. Soc., Chem. Commun.* **1969**, 759.
553. Dubey, S.; Bhandari, A. M.; Misra, S. N.; Kapoor, R. N. *Ind. J. Chem.* **1970**, *8*, 97.
554. Grigor'ev, M. S.; Baturin, N. A.; Budantseva, N. A.; Fedoseev, A. M. *Radiokhimiya* **1993**, *35*, 29.
555. Grigor'ev, M. S.; Baturin, N. A.; Bessonov, A. A.; Krot, N. N. *Sov. Radiochem.* **1995**, *37*, 12.
556. Grigor'ev, M. S.; Charushnikova, I. A.; Krot, N. N.; Yanovskii, A. I.; Struchkov, Y. T. *Z. Neorg. Khim. (Engl. Transl.)* **1994**, *39*, 167.
557. Ortego, J. D.; Tew, W. P. *J. Coord. Chem.* **1972**, *2*, 13.
558. Bombieri, G.; Brown, D.; Mealli, C. *J. Chem. Soc., Dalton Trans.* **1976**, 2025.
559. Brown, D.; Rickard, C. E. F. *J. Chem. Soc. A: Inorganic, Physical, Theoretical* **1970**, 3373.
560. Brown, D. *Adv. Inorg. Chem. Radiochem.* **1969**, *12*, 1.
561. Malm, J. G.; Williams, C. W.; Soderholm, L.; Morss, L. R. *J. Alloys Compd.* **1993**, *194*, 133.
562. Asprey, L. B.; Haire, R. G. *Inorg. Nucl. Chem. Lett.* **1973**, *9*, 1121.
563. Brown, D.; Barry, J. A.; Holloway, J. H. UK Report AERE–R10415 *Atomic Energy Res. Establ.,* 1982.

564. Ryan, R. R.; Penneman, R. A.; Asprey, L. B.; Paine, R. T. *Acta Crystallogr., Sect. B* **1976**, *32*, 3311.
565. Dodge, R. P.; Smith, G. S.; Johnson, Q.; Elson, R. E. *Acta Crystallogr.* **1967**, *22*, 85.
566. Smith, G. S.; Johnson, Q.; Elson, R. E. *Acta Crystallogr.* **1967**, *22*, 300.
567. Eastman, M. P.; Eller, P. G.; Halstead, G. W. *J. Inorg. Nucl. Chem.* **1981**, *43*, 2839.
568. de Wet, J. F.; Caira, M. R.; Gellatly, B. J. *Acta Crystallogr., Sect. B.* **1978**, *34*, 1121.
569. Taylor, J. C.; Waugh, A. B. *Polyhedron* **1983**, *2*, 211.
570. Rybakov, V. B.; Aslanov, L. A.; Kolesnichenko, V. L. *Koord. Khim.* **2000**, *26*, 633.
571. Eller, P. G.; Malm, J. G.; Swanson, B. I.; Morss, L. R. *J. Alloys Compd.* **1998**, *269*, 50.
572. Burns, J. H.; Levy, H. A.; Keller, J. O. L. *Acta Crystallogr., Sect. B.* **1968**, *24*, 1675.
573. Brown, D.; Kettle, S. F. A.; Smith, A. J. *J. Chem. Soc. A* **1967**, 1429.
574. Brown, D.; Easey, J. F.; Rickard, C. E. F. *J. Chem. Soc. A* **1969**, 1161.
575. Brown, D.; Petcher, T.; Smith, A. J. *Nature* **1968**, *217*, 738.
576. Grigor'ev, M. S.; Bessonov, A. A.; Krot, N. N.; Yanovskii, A. I.; Struchkov, Y. T. *Sov. Radiochem.* **1993**, *35*, 382.
577. Vodovatov, V. A.; Ladygin, I. N.; Lychev, A. A.; Mashirov, L. G.; Suglobov, D. N. *Sov. Radiochem.* **1975**, *17*, 771.
578. Keller, C.; Eberle, S. H. *Radiochim. Acta* **1967**, *8*, 65.
579. Grigor'ev, M. S.; Charushnikova, I. A.; Krot, N. N.; Struchkov, Y. T. *Zh. Neorg. Khim.* **1996**, *41*, 539.
580. Sasaki, Y.; Tachimori, S. *Solvent Extr. Ion Exch.* **2002**, *20*, 21.
581. Maslov, L. P.; Sirotinkina, L. V.; Rykov, A. G. *Radiokhimiya* **1985**, *27*, 732.
582. Shilov, V. P. *Radiokhimiya* **1980**, *22*, 727.
583. Erin, E. A.; Kopytov, V. V.; Rykov, A. G.; Vasil'ev, V. Y. *Radiokhimiya* **1984**, *26*, 98.
584. Sessler, J. L.; Seidel, D.; Vivian, A. E.; Lynch, V.; Scott, B. L.; Keogh, D. W. *Angew. Chem., Int. Ed. Engl.* **2001**, *40*, 591.
585. Sessler, J. L.; Gorden, A. E. V.; Seidel, D.; Hannah, S.; Lynch, V.; Gordon, P. L.; Donohoe, R. J.; Tait, C. D.; Keogh, D. W. *Inorg. Chim. Acta* **2002**, *341*, 54.
586. Clark, D. L.; Keogh, D. W.; Palmer, P. D.; Scott, B. L.; Tait, C. D. *Angew. Chem., Int. Ed. Engl.* **1998**, *37*, 164.
587. Oldham, W. J.; Oldham, S. M.; Scott, B. L.; Abney, K. D.; Smith, W. H.; Costa, D. A. *Chem. Commun.* **2001**, 1348.
588. Sarsfield, M. J.; Helliwell, M.; Collison, D. *Chem. Commun.* **2002**, 2264.
589. Barnhart, D. M.; Burns, C. J.; Sauer, N. N.; Watkin, J. G. *Inorg. Chem.* **1995**, *34*, 4079.
590. Burns, C. J.; Clark, D. L.; Donohoe, R. D.; Duval, P. B.; Scott, B. L.; Tait, C. D. *Inorg. Chem.* **2000**, *39*, 3464.
591. Burns, C. J.; Smith, D. C.; Sattelberger, A. P.; Gray, H. B. *Inorg. Chem.* **1992**, *31*, 3724.
592. Duval, P. B.; Burns, C. J.; Buschmann, W. E.; Clark, D. L.; Morris, D. E.; Scott, B. L. *Inorg. Chem.* **2001**, *40*, 5491.
593. Wedler, M.; Roesky, H. W.; Edelmann, F. *J. Organomet. Chem.* **1988**, *345*, C1.
594. Brown, D. R.; Denning, R. G.; Jones, R. H. *J. Chem. Soc., Chem. Commun.* **1994**, 2601.
595. Brown, D. R.; Denning, R. G. *Inorg. Chem.* **1996**, *35*, 6158.
596. Denning, R. G. *Struct. Bonding* **1992**, *79*, 215.
597. Williams, V. C.; Müller, M.; Leech, M. A.; Denning, R. G.; Green, M. L. H. *Inorg. Chem.* **2000**, *39*, 2538.
598. Hunt, R. D.; Andrews, L. *J. Chem. Phys.* **1993**, *98*, 3690.
599. Hunt, R. D.; Yustein, J. T.; Andrews, L. *J. Chem. Phys.* **1993**, *98*, 6070.
600. Kushto, G. P.; Souter, P. F.; Andrews, L.; Neurock, M. *J. Chem. Phys.* **1997**, *106*, 5894.
601. Tague, T. J. Jr.; Andrews, L.; Hunt, R. D. *J. Phys. Chem.* **1993**, *97*, 10920.
602. Bailey, R. A.; Michelsen, T. W. *J. Inorg. Nucl. Chem.* **1972**, *34*, 2935.
603. Sles, V. G.; Skoblo, A. I.; Suglobov, D. N. *Sov. Radiochem.* **1974**, *16*, 504.
604. Alcock, N. W.; Roberts, M. W.; Brown, D. *Acta Crystallogr., Sect. B* **1982**, *38*, 2870.
605. Wang, M.; Zheng, P. J.; Zhang, J. Z.; Chen, Z.; Shen, J. M.; Yang, Y. H. *Acta Crystallogr., Sect. C* **1987**, *43*, 873.
606. Charpin, P.; Lance, M.; Nierlich, M.; Vigner, D.; Livet, J.; Musikas, C. *Acta Crystallogr., Sect. C* **1986**, *42*, 1691.
607. Muetterties, E. L. *Advances in the Chemistry of the Coordination Compounds* **1961**, Macmillan: New York.
608. Johnson, D. A.; Taylor, J. C.; Waugh, A. B. *J. Inorg. Nucl. Chem.* **1979**, *41*, 827.
609. Srivastava, A. K.; Agarwal, R. K.; Kapur, V.; Sharma, S.; Jain, P. C. *Transition Met. Chem.* **1982**, *7*, 41.
610. Pennington, W. T.; Alcock, N. W.; Flanders, D. J. *Acta Crystallogr., Sect. C* **1988**, *44*, 1664.
611. Alcock, N. W.; Flanders, D. J.; Pennington, M.; Brown, D. *Acta Crystallogr. C* **1988**, *44*, 247.
612. Alcock, N. W.; Flanders, D. J.; Pennington, M.; Brown, D. *Acta Crystallogr. C* **1987**, *43*, 1476.
613. Berthet, J. C.; Lance, M.; Nierlich, M.; Ephritikhine, M. *Eur. J. Inorg. Chem.* **2000**, 1969.
614. Hall, T. J.; Mertz, C. J.; Bachrach, S. M.; Hipple, W. G.; Rogers, R. D. *J. Crystallogr. Spectrosc. Res.* **1989**, *19*, 499.
615. Vodovatov, V. A.; Mashirov, L. G.; Suglobov, D. N. *Radiokhimiya* **1973**, *15*, 446.
616. Masaki, N. *J. Nucl. Mater.* **1981**, *101*, 229.
617. Janeczek, J.; Ewing, R. C.; Thomas, L. E. *J. Nucl. Mater.* **1993**, *207*, 177.
618. Allen, G. C.; Holmes, N. R. *Can. J. Applied Spectrosc.* **1993**, *38*, 124.
619. Allen, G. C.; Holmes, N. R. *Applied Spectrosc.* **1994**, *48*, 525.
620. Moskvin, A. I. *Sov. Radiochem.* **1971**, 700.
621. Kato, Y.; Kimura, T.; Yoshida, Z.; Nitani, N. *Radiochim. Acta* **1996**, *74*, 21.
622. Ronchi, C.; Capone, F.; Colle, J. Y.; Hiernaut, J. P. *J. Nucl. Mater.* **2000**, *280(1)*, 111.
623. Belyaev, Y. I.; Solntsev, V. M.; Kapshukov, I. I.; Sudakov, L. V.; Chistyakov, V. M. *Radiokhimiya* **1974**, *16(5)*, 747.
624. Kim, E. H.; Choi, C. S.; Park, J. H.; Chang, I. S. *Yoop Hakhoechi* **1993**, *30*, 289.
625. Girgis, B. S.; Rofail, N. H. *Radiochim. Acta* **1992**, *57*, 41.
626. Cartmell, H. R.; Ellis, J. F. Process and apparatus for the manufacture of uranium hexafluoride from recycled uranium trioxide. Fr. Demande 90-3100. *Chem Abstr* **1990**, *114*, 84812.
627. Pashley, J. H. *Radiochim. Acta* **1978**, *25*, 135.
628. Ozawa, T. Manufacture of uranium dioxide reactor fuel pellets. *Jpn. Kokai Tokkyo Koho* JP, 87–114291; *Chem. Abstr.* **1989**, *110*, 181394.
629. Tel, H.; Eral, M.; Altas, Y. *J. Nucl. Mater.* **1998**, *256(1)*, 18.
630. Lee, J.; Yamagishi, S.; Itoh, A.; Ogawa, T. *Nippon Genshiryoku Kenkyusho, [Rep.] Jaeri M* **1993**, .
631. Bishay, A. F.; Abdel, H. A. S.; Hammad, F. H.; Abadir, M. F.; Elaslaby, A. M. *J. Therm. Anal* **1989**, *35*, 1405.
632. Yamagishi, S.; Takahashi, Y. *J. Nucl. Sci. Technol* **1986**, *23*, 711.

633. Cortes, C. V.; Kremenic, G.; Gonzalez, T. L. *React. Kinet. Catal. Lett* **1988**, *36*, 235.
634. Mori, S.; Uchiyama, M. *Sekiyu Gakkai Shi* **1976**, *19*, 758.
635. Liu, S.; Guo, K.; Hu, Y.; Wang, Q.; Gu, D.; Shen, Z. *Fenxi Huaxue* **1994**, *22*, 984.
636. Taylor, S. H.; Hudson, I.; Hutchings, G. J. Catalytic Oxidation of Organic Compounds Pct. Int. Appl WO 96-GB705 19960325, 1996; *Chem. Abstr.* 1996, *125*, 307812.
637. Hutchings, G. J.; Heneghan, C. S.; Hudson, I. D.; Taylor, S. H. *ACS Symp. Ser* **1996**, *638*, 58.
638. Gordeeva, L. G.; Aristov, Y. I.; Moroz, E. M.; Rudina, N. A.; Zaikovskii, V. I.; Tanashev, Y. Y.; Parmon, V. N. *J. Nucl. Mater* **1995**, *218*, 202.
639. You, G. S.; Kim, K. S.; Min, D. K.; Ro, S. G.; Kim, E. K. *J. Korean Nucl. Soc* **1995**, *27*, 67.
640. Choi, J. W.; McEachern, R. J.; Taylor, P.; Wood, D. D. *J. Nucl. Mater* **1996**, *230*, 250.
641. Kim, B. G.; Song, K. W.; Lee, J. W.; Bae, K. K.; Yang, M. S.; Park, H. S. *Yoop Hakhoechi* **1995**, *32*, 471.
642. Suryanarayana, S.; Kumar, N.; Bamankar, Y. R.; Vaidya, V. N.; Sood, D. D. *J. Nucl. Mater* **1996**, *230*, 140.
643. Tokai, K.; Ooe, A. Manufacture of mixed oxide (MOX) pellets containg uranium oxide and plutonium oxide for fuel rods for power generation. JP 94-225519, *Chem Abstr.* **1996**, *124*, 272906.
644. Hofman, G. L.; Snelgrove, J. L. *Mater. Sci. Technol* **1994**, *104*, 45.
645. George, E.; Pagel, M.; Dusausoy, Y.; Gautier, J. M. *Uranium* **1986**, *1986*, 69.
646. Bevan, D. J. M.; Grey, I. E.; Willis, B. T. M. *J. Solid State Chem.* **1986**, *61(1)*, 1.
647. Cordfunke, E. H. P. *J. Inorg. Nucl. Chem.* **1962**, *24*, 303.
648. Cordfunke, E. H. P.; Ouweltjes, W. *J. Chem. Thermodyn.* **1981**, *13*, 193.
649. Cordfunke, E. H. P. *J. Nucl. Mater.* **1985**, *130*, 82.
650. Cordfunke, E. H. P.; Ijdo, D. J. W. *J. Phys. Chem. Solids* **1988**, *49*, 551.
651. Cordfunke, E. H. P.; Ijdo, D. J. W. *J. Solid State Chem.* **1994**, *109*, 272.
652. Cordfunke, E. H. P.; Gruppelaar, H.; Franken, W. M. P.; Abrahams, K.; Blankenvoorde, P. J. A. M.; Bultman, J. H.; Dodd, D. H.; Kloosterman, J. L.; Koning, A. J.; Transmutation of Nuclear Waste. Status Report TAS program 1994: Recycling and Transmutation of Actinides and Fission Products; Netherlands Energy Res. Foundation, Petten, The Netherlands, 1995.
653. Cordfunke, E. H. P.; Booij, A. S.; Smit-Groen, V.; van Vlaanderen, P.; Ijdo, D. J. W. *J. Solid State Chem.* **1997**, *131*, 341.
654. Cordfunke, E. H. P.; Booij, A. S.; Huntelaar, M. E. *J. Chem. Thermodyn.* **1999**, *31*, 1337.
655. Cordfunke, E. H. P.; Ouweltjes, W.; Prins, G.; Van Vlaanderen, P. *J. Chem. Thermodyn.* **1983**, *15*, 1103.
656. Spitsyn, V. I.; Kovba, L. M.; Tabachenko, V. V.; Tabachenko, N. V.; Mikhailov, Y. N. *Bull. Acad. Sci. USSR, Div. Chem. Sci.* **1982**, *31*, 711.
657. Glatz, R. E.; Li, Y.; Hughes, K.-A.; Cahill, C. L.; Burns, P. C. *Can. Mineral.t* **2002**, *40*, 217.
658. Burns, P. C.; Deely, K. M. *Can Mineral.* **2002**, *40*, 1579.
659. Li, Y.; Burns, P. C. *J. Nucl. Mater.* **2001**, *299*, 219.
660. Li, Y.; Burns, P. C. *Can. Mineral.* **2000**, *38*, 1433.
661. Sergeyeva, E. I.; Devina, O. A.; Khodakovsky, I. L.; Vernadsky, J. *J. Alloys Compd.* **1994**, *213/214*, 125.
662. Ahrland, S. Hydrolysis of the Actinide Ions. In *Handbook on the Physics and Chemistry of the Actinides* Vol. 6; Freeman, J. J., Keller, C., Eds.; Elsevier 1991; 471–510.
663. Moulin, C.; Decambox, P.; Moulin, V.; Decaillon, J. G. *Anal. Chem.* **1995**, *67(2)*, 348.
664. Aberg, M.; Ferri, D.; Glaser, J.; Grenthe, I. *Inorg. Chem.* **1983**, *22*, 3986.
665. Kato, Y.; Meinrath, G.; Kimura, T.; Yoshida, A. *Radiochim. Acta.* **1994**, *64(2)*, 107.
666. King, C. M.; King, R. B.; Garber, A. R. *Mater. Res. Soc. Symp. Proc.* **1990**, *180*, 1083.
667. Palmer, D. A.; Nguyen-Trung, C. *J. Solution Chem.* **1995**, *24(12)*, 1281.
668. Miller, M. L.; Finch, R. J.; Burns, P. C.; Ewing, R. C. *Mater. Res. Soc. Symp. Proc* **1996**, *412*, 369.
669. Finch, R. J.; Hawthorne, F. C. *Can. Mineral.* **1998**, *36*, 831.
670. Sowder, A. G.; Clark, S. B.; Fjeld, R. A. *Environ. Sci. Technol.* **1999**, *33*, 3552.
671. Weller, M. T.; Light, M. E.; Gelbrich, T. *Acta Crystallogr., Sect. B: Struct. Sci.* **2000**, *B56*, 577.
672. Allen, P. G.; Shuh, D. K.; Bucher, J. J.; Edelstein, N. M.; Palmer, C. E. A.; Marquez, L. N. *Mater. Res. Soc. Symp. Proc.* **1997**, *432*, 139.
673. Moll, H.; Reich, T.; Hennig, C.; Rossberg, A.; Szabo, Z.; Grenthe, I. *Radiochim. Acta* **2000**, *88*, 559.
674. Finch, R. J.; Cooper, M. A.; Hawthorne, F. C.; Ewing, R. C. *Can. Mineral.* **1996**, *34*, 1071.
675. Alcock, N. W. *J. Chem. Soc. A* **1968**, 1588.
676. Bhattacharjee, M.; Chaudhuri, M. K.; Purkayastha, R. N. D. *J. Chem. Soc., Dalton Trans.* **1990**, 2883.
677. Frondel, J. W.; Fleischer, M.; Jones, R. S. *Glossary of Uranium and Thorium-Bearing Minerals*; 4th ed.; US Geological Survey Bulletin 1250, 1967.
678. Li, Y.; Burns, P. C. *J. Solid State Chem.* **2002**, *166*, 219.
679. Grigor'ev, M. S.; Charushnikova, I. A.; Krot, N. N.; Yanovsky, A. I.; Struchkov, Y. T. *Radiokhimiya* **1997**, *39*, 419.
680. Musikas, C.; Burns, J. H. Structure and Bonding in Compounds Containing the Neptunyl (1^+) and Neptunyl (2^+) Ions. In *Transplutonium 1975, Proc. 4th Int. Transplutionium Elem. Symp., Baden Baden, Sept. 1975*, Mueller, W.; Lindner, R.; Eds.; North-Holland: Amsterdam, 1976, 237.
681. Bidoglio, G.; Cavalli, P.; Grenthe, I.; Omenetto, N.; Qi, P.; Tanet, G. *Talanta* **1991**, *38*, 433.
682. Allen, P. G.; Bucher, J. J.; Clark, D. L.; Edelstein, N. M.; Ekberg, S. A.; Gohdes, J. W.; Hudson, E. A.; Kaltsoyannis, N.; Lukens, W. W.; Neu, M. N.; Palmer, P. D.; Reich, T.; Shuh, D. K.; Tait, C. D.; Zwick, B. D. *Inorg. Chem.* **1995**, *34*, 4797.
683. Robouch, P.; Vitorge, P. *Inorg. Chim. Acta* **1987**, *140*, 239.
684. Grenthe, I.; Riglet, C.; Vitorge, P. *Inorg. Chem.* **1986**, *25*, 1679.
685. Den Auwer, C.; Revel, R.; Charbonnel, M. C.; Presson, M. T.; Conradson, S. D.; Simoni, E.; Le Du, J. F.; Madic, C. *J. Synchrotron Radiation* **1999**, *6*, 101.
686. Caville, C. *J. Raman Spectrosc.* **1976**, *4*, 395.
687. Weger, H. T.; Okajima, S.; Cunnane, J. C.; Reed, D. T. *Mater. Res. Soc. Symp. Proc.* **1993**, *294*, 739.
688. Mathur, J. N.; Choppin, G. R. *Radiochim. Acta* **1994**, *64*, 175.
689. Weigel, F. The Carbonates, Phosphates, and Arsenates of the Hexavalent and Pentavalent Actinides. In *Handbook on the Chemistry and Physics of the Actinides*, Vol. 3: Freeman, A. J.; Keller, C., Eds.; Elsevier: Amsterdam, 1985.

690. Weigel, F. In *Kirk-Othmer Encyclopedia of Chemical Technology*, 3rd ed.; Kroschwitz, J. I. Ed., Wiley: New York, 1983, 502–543.
691. Danis, J. A.; Lin, M. R.; Scott, B. L.; Eichhorn, B. W.; Runde, W. H. *Inorg. Chem.* **2001**, *40*, 3389.
692. Locock, A. J.; Burns, P. C. *Am. Mineral.* **2003**, *88*, 240.
693. Locock, A. J.; Burns, P. C. *J. Solid State Chem.* **2002**, *167*, 226.
694. Shilton, M. G.; Howe, A. T. *J. Solid State Chem.* **1980**, *34(2)*, 137.
695. Linde, S. A.; Gorbunova, Y. E.; Lavrov, A. V. *Russ. J. Inorg. Chem.* **1980**, *25*, 1105.
696. Sidorenko, G. A.; Zhil'tsova, I. G.; Moroz, I. K.; Valueva, A. *Dokl. Akad. Nauk SSSR* **1975**, *222(2)*, 444.
697. Doran, M.; Walker, S. M.; O'Hare, D. *Chem. Commun.* **2001**, 1988.
698. Morosin, B. *Acta Crystallogr., Sect. B: Struct. Sci.* **1978**, *34*, 327.
699. Barten, H. *Thermochim. Acta* **1988**, *124*, 339.
700. Dacheux, N.; Brandel, V.; Genet, M. *New J. Chem.* **1995**, *19(1)*, 15.
701. Cordfunke, E. H. P.; Muis, R. P.; Ouweltjes, W.; Flowtow, H. E.; O'Hare, P. A. G. *J. Chem. Thermodyn.* **1982**, *14*, 313.
702. Krivovichev, S. V.; Burns, P. C. *Can. Mineral.* **2001**, *39*, 207.
703. Andreev, G. B.; Antipin, M. Y.; Fedoseev, A. M.; Budantseva, N. A. *Russ. J. Coord. Chem.* **2001**, *27*, 208.
704. Krivovichev, S. V.; Burns, P. C. *J. Solid State Chem.* **2003**, *170*, 106.
705. Krivovichev, S. V.; Burns, P. C. *Can. Mineral.* **2002**, *40*, 1571.
706. Fedoseev, A. M.; Budantseva, N. A.; Yusov, A. B.; Grigor'ev, M. S.; Potyemkina, T. I. *Radiokhimiya* **1990**, *32*, 14.
707. Fedoseev, A. M.; Budantseva, N. A.; Shirokova, I. B.; Andreev, G. B.; Yurik, T. K.; Krupa, J. C. *Zh. Neorg. Khim.* **2001**, *46*, 45.
708. Frondel, C. l.; Ito, J.; Honea, R. M.; Weeks, A. M. *Can. Mineral.* **1976**, *12*, 429.
709. Hayden, L. A.; Burns, P. C. *Can. Mineral.* **2002**, *40*, 211.
710. Hayden, L. A.; Burns, P. C. *J. Solid State Chem.* **2002**, *163*, 313.
711. Doran, M.; Norquist, A. J.; O'Hare, D. *Chem. Commun.* **2002**, 2946.
712. Niinisto, L.; Toivonen, J.; Valkonen, J. *J. Acta. Chem. Scand., Ser. A* **1977**, *33*, 621.
713. Mikhailov, Y. N.; Kokh, L. A.; Kutznetsov, V. G.; Grevtseva, T. G.; Sokol, S. K.; Ellert, G. V. *Sov. J. Coord. Chem.* **1977**, *3*, 388.
714. Hellmann, H. Np (VI) Sulfates, *Technical Report INIS-mf-9276*, University of Munich: Munich, Germany, **1983**.
715. Khandelwal, B. L.; Verma, V. P. *Indian J. Chem.* **1975**, *13*, 967.
716. Verma, V. P.; Khandelwal, B. L. *Indian J. Chem.* **1973**, *11*, 602.
717. Bean, A. C.; Campana, C. F.; Kwon, O.; Albrecht-Schmitt, T. E. *J. Am. Chem. Soc.* **2001**, *123*, 8806.
718. Runde, W.; Bean, A. C.; Albrecht-Schmitt, T. E.; Scott, B. L. *Chem. Commun.* **2003**, 478.
719. Alcock, N. W.; Esperas, S. *J. Chem. Soc., Dalton Trans.* **1977**, *9*, 893.
720. Bokolo, K.; Courtois, A.; Delpuech, J. J.; Elkaim, E.; Protas, J.; Rinaldi, D.; Rodehueser, L.; Rubini, P. *J. Am. Chem. Soc.* **1984**, *106*, 6333.
721. Zhang, Y.; Collison, D.; Livens, F. R.; Helliwell, M.; Heatley, F.; Powell, A. K.; Wocadlo, S.; Eccles, H. *Polyhedron* **2002**, *21*, 81.
722. Zhang, Y.; Livens, F. R.; Collison, D.; Helliwell, M.; Heatley, F.; Powell, A. K.; Wocadlo, S.; Eccles, H. *Polyhedron* **2002**, *21*, 69.
723. Jones, R. G.; Bindschadler, E.; Blume, D.; Karmas, G.; Martin, G. A. Jr.; Thirtle, J. R.; Yeoman, F. A.; Gilman, H. *J. Am. Chem. Soc.* **1956**, *78*, 6030.
724. Bradley, D. C.; Chatterjee, A. K. *J. Inorg. Nucl. Chem.* **1959**, *12*, 71.
725. Cuellar, E. A.; Miller, S. S.; Marks, T. J.; Weitz, E. *J. Am. Chem. Soc.* **1983**, *105*, 4580.
726. Jacob, E. *Angew. Chem., Int. Ed. Engl.* **1982**, *21*, 142.
727. Miller, S. S.; DeFord, D. D.; Marks, T. J.; Weitz, E. *J. Am. Chem. Soc.* **1979**, *101*, 1036.
728. Cuellar, E. A.; Marks, T. J. *Inorg. Chem.* **1981**, *20*, 2129.
729. Bursten, B. E.; Casarin, M.; Ellis, D. E.; Fragalà, I.; Marks, T. J. *Inorg. Chem.* **1986**, *25*, 1257.
730. Vergamini, P. J. *J. Chem. Soc., Chem. Commun.* **1979**, 54.
731. Albers, H.; Deutsch, M.; Krastinat, W.; von Osten, H. *Chem. Ber.* **1952**, *85*, 267.
732. Jones, R. G.; Bindschadler, E.; Martin, G. A., Jr.; Thirtle, J. R.; Gilman, H. *J. Am. Chem. Soc.* **1957**, *79*, 4921.
733. Vdovenko, V. M.; Ladygin, I. N.; Suglobov, I. G.; Suglobov, D. N. *Radiokhimiya* **1969**, *11*, 236.
734. Burns, C. J.; Sattelberger, A. P. *Inorg. Chem.* **1988**, *27*, 3692.
735. Wilkerson, M. P.; Burns, C. J.; Dewey, H. J.; Martin, J. M.; Morris, D. E.; Paine, R. T.; Scott, B. L. *Inorg. Chem.* **2000**, *39*, 5277.
736. Solanki, A. K.; Bhandari, A. M. *Radiochem. Radioanal. Lett.* **1980**, *43*, 279.
737. Perry, D. L.; Templeton, D. H.; Zalkin, A. *Inorg. Chem.* **1978**, *17*, 3699.
738. Perry, D. L.; Templeton, D. H.; Zalkin, A. *Inorg. Chem.* **1979**, *18*, 879.
739. Perry, D. L. *Inorg. Chim. Acta* **1981**, *48*, 117.
740. Malhotra, K. C.; Sharma, M.; Sharma, N. *Indian J. Chem.* **1985**, *24A*, 790.
741. Wilkerson, M. P.; Burns, C. J.; Morris, D. E.; Paine, R. T.; Scott, B. L. *Inorg. Chem.* **2002**, *41*, 3110.
742. Casellato, U.; Vigato, P. A.; Tamburini, S.; Graziani, R.; Vidali, M. *Inorg. Chim. Acta* **1983**, *72*, 141.
743. Baghlaf, A. O.; Ishaq, M.; Ahmed, O. A. S.; Al-Julani, M. A. *Polyhedron* **1985**, *4*, 853.
744. Rose, D. J.; Chen, Q.; Zubieta, J. *Inorg. Chim. Acta* **1998**, *268*, 163.
745. Rose, D.; Chang, Y. D.; Chen, Q.; Zubieta, J. *Inorg.Chem.* **1994**, *33*, 5167.
746. Deshayes, L.; Keller, N.; Lance, M.; Navaza, A.; Nierlich, M.; Vigner, J. *Polyhedron* **1994**, *13*, 1725.
747. Thuery, P.; Nierlich, M.; Keller, N.; Lance, M.; Vigner, J. D. *Acta Crystallogr., Sect. C* **1995**, *51*, 1300.
748. Ivanov, S. B.; Davidovich, R. L.; Mikhailov, Y. N.; Shchelokov, R. N. *Koord. Khim.* **1982**, *8*, 211.
749. Frasson, E.; Bombieri, G.; Panattoni, C. *Coord. Chem. Rev.* **1966**, *1*, 145.
750. Farkas, I.; Bányai, I.; Szabó, Z.; Wahlgren, U.; Grenthe, I. *Inorg. Chem.* **2000**, *39*, 799.
751. Rogers, R. D.; Benning, M. M.; Etzenhouser, R. D.; Rollins, A. N. *J. Coord. Chem.* **1992**, *26*, 299.
752. Clemente, D. A.; Bandoli, G.; Vidali, M.; Vigato, P. A.; Portanova, R.; Magon, L. *J. Cryst. Mol. Struct.* **1973**, *3*, 221.
753. Mackinnon, P. I.; Taylor, J. C. *Polyhedron* **1983**, *2*, 217.

754. Mikhailov, Y. N.; Gorbunova, Y. E.; Demchenko, E. A.; Serezhkina, L. B.; Serezkhin, V. N. *Russ. J. Inorg. Chem.* **1998**, *43*, 885.
755. Vasudevan, T.; Murali, M. S.; Nagar, M. S.; Mathur, J. N. *Solvent Extr. Ion Exch.* **2002**, *20*, 665.
756. Condamines, N.; Musikas, C. *Solvent Extr. Ion Exch.* **1992**, *10*, 69.
757. Soldatkina, M. A.; Serezhkin, V. N.; Trunov, V. K. *J. Struct. Chem. (Engl. Transl.)* **1981**, *22*, 915.
758. Serezhkin, V. N.; Soldatkina, M. A.; Trunov, V. K. *Sov. Radiochem.* **1981**, *23*, 551.
759. Kuznetsov, I. Y.; Solntsev, K. A.; Kuznetsov, N. T.; Mikhailov, Y. N.; Orlova, A. M.; Alikhanova, Z. M.; Sergeev, A. V. *Koord. Khim.* **1986**, *12*, 1387.
760. Day, J. P.; Venanzi, L. M. *J. Chem. Soc. A* **1966**, 1363.
761. Akona, S. B.; Fawcett, J.; Holloway, J. H.; Russell, D. R.; Leban, I. *Acta Crystallogr., Sect. C* **1991**, *47*, 45.
762. Mathur, J. N.; Choppin, G. R. *Solvent Extr. Ion Exch.* **1998**, *16*, 459.
763. Den Auwer, C.; Charbonnel, M. C.; Presson, M. T.; Madic, C.; Guillaumont, R. *Polyhedron* **1998**, *17*, 4507.
764. Alcock, N. W.; Roberts, M. M.; Brown, D. *J. Chem. Soc., Dalton Trans.* **1982**, 25.
765. Sato, T. *J. Inorg. Nucl. Chem.* **1962**, *24*, 699.
766. Sato, T. *J. Inorg. Nucl. Chem.* **1963**, *25*, 109.
767. Sato, T. *J. Inorg. Nucl. Chem.* **1965**, *27*, 1853.
768. Sato, T. *J. Inorg. Nucl. Chem.* **1964**, *26*, 311.
769. Mapara, P. M.; Chetty, K. V.; Swarup, R.; Ramakrishna, V. V. *Radiochim. Acta* **1995**, *69*, 221.
770. Forsellini, E.; Bombieri, G.; Graziani, R.; Zarli, B. *Inorg. Nucl. Chem. Lett.* **1972**, *8*, 461.
771. Alcock, N. W.; Roberts, M. M. *Acta Crystallogr., Sect. C* **1987**, *43*, 476.
772. Panattoni, C.; Graziani, R.; Croatto, U.; Zarli, B.; Bombieri, G. *Inorg. Chim. Acta* **1968**, *2*, 43.
773. Sassmannshausen, M.; Lutz, H. D.; Zazhogin, A. *Z. Kristallogr -New Cryst. Struct* **2000**, *215*, 427.
774. Guo, S. S.; Zhang, D.; Wang, H. Z.; Yu, K. B. *Chin. J. Struct. Chem.* **1998**, *17*, 9.
775. Harrowfield, J. M. B.; Kepert, D. L.; Patrick, J. M.; White, A. H.; Lincoln, S. F. *J. Chem. Soc., Dalton Trans.* **1983**, 393.
776. Deshayes, L.; Keller, N.; Lance, M.; Nierlich, M.; Vigner, D. *Acta Crystallogr., Sect. C.* **1992**, *48*, 1660.
777. Kannan, S.; Venugopal, V.; Pilai, M. R. A.; Droege, P. A.; Barnes, C. L. *Polyhedron* **1996**, *15*, 97.
778. Kannan, S.; Raj, S. S. S.; Fun, H. K. *Acta Crystallogr., Sect. C* **2000**, *56*, e545.
779. Bacher, W.; Jacob, E. *Chemikerzeitung* **1982**, *106*, 117.
780. Wilson, W. W.; Christe, K. O. *Inorg. Chem.* **1982**, *21*, 2091.
781. Bougon, R.; Charpin, P.; Desmoulin, J. P.; Malm, J. G. *Inorg. Chem.* **1976**, *15*, 2532.
782. Iwasaki, M.; Ishikawa, N.; Ohwada, K.; Fujino, T. *Inorg. Chim. Acta* **1981**, *54*, L193.
783. Paine, R. T.; Ryan, R. R.; Asprey, L. B. *Inorg. Chem.* **1975**, *14*, 1113.
784. Peacock, R. D.; Edelstein, N. *J. Inorg. Nucl. Chem.* **1975**, *38*, 771.
785. Burns, R. C.; O'Donnell, T. A. *Inorg. Nucl. Chem. Lett.* **1977**, *13*, 657.
786. Taylor, J. C.; Wilson, P. W. *J. Chem. Soc., Chem. Commun.* **1974**, 232.
787. Bohrer, R.; Conradi, E.; Muller, U. *Z. Anorg. Allg. Chem.* **1988**, *558*, 119.
788. Zheng, P.; Wang, M.; Wang, B.; Wang, W. *Chin. J. Struct. Chem.* **1986**, *5*, 146.
789. Wang, W. J.; Chen, B.; Zheng, P.; Wang, B.; Wang, M. *Inorg. Chim. Acta* **1986**, *117*, 81.
790. Rogers, R. D.; Bond, A. H.; Hipple, W. G.; Rollins, A. N.; Henry, R. F. *Inorg. Chem.* **1991**, *30*, 2671.
791. Talley, C. E.; Bean, A. C.; Albrecht-Schmitt, T. E. *Inorg. Chem.* **2000**, *39*, 5174.
792. Klygin, A. E.; Kolyada, N. S. *Zh. Neorg. Khim.* **1961**, *6*, 216.
793. Kundu, P. C.; Roy, P. S.; Banerjeee, R. K. *J. Inorg. Nucl. Chem.* **1980**, *42*, 851.
794. Hojjatie, M.; Muralidharan, S.; Bag, S. P.; Panda, G. C.; Freiser, H. *Iran J. Chem. Chem. Eng.* **1995**, *15*, 81.
795. Kundu, P. C.; Bera, A. K. *Indian J. Chem., Sect. A* **1978**, *16A*, 865.
796. Kundu, P. C.; Bera, A. K. *Indian J. Chem., Sect. A* **1979**, *18A*, 62.
797. Kundu, P. C.; Bera, A. K. *Indian J. Chem., Sect. A* **1982**, *21A*, 1132.
798. Xu, J.; Raymong, K. N. *Inorg. Chem* **1999**, *38*, 308.
799. Fleck, H. R. *Analyst* **1937**, *62*, 378.
800. Claassen, A.; Visser, J. *Recl. Trav. Chim. Recueil des Traveux Chimiques des Pays-Bas et de la Belgique Pay-Bas* **1946**, *65*, 211.
801. Avinashi, B. K.; Banerji, S. K. *J. Indian Chem. Soc.* **1970**, *47*, 453.
802. Rudometkina, T. F.; Ivanov, V. M.; Busev, A. I. *Zh. Anal. Khim.* **1977**, *32*, 669.
803. El-Ansary, A. L.; Ali, A. A. *Indian J. Chem., Sect. A* **1986**, *25A*, 939.
804. Hall, D.; Rae, A. D.; Waters, T. N. *Acta. Crist.* **1967**, *22*, 258.
805. Baker, B.; Sawyer, D. T. *Inorg. Chem.* **1969**, *8*, 1160.
806. Graziani, R.; Bombieri, G.; Forsellini, E.; Degetto, S.; Marangoni, G. *J. Chem. Soc., Dalton Trans.* **1973**, 451.
807. Perry, D. L.; Zalkin, A.; Ruben, H.; Templeton, D. H. *Inorg. Chem.* **1982**, *21*, 237.
808. Beirakhov, A. G.; Orlova, I. M.; Gorbunova, Y. E.; Mikhailov, Y. N.; Schchelokov, R. N. *Zh. Neorg. Khim.* **1998**, *44*, 1414.
809. Jones, R. G.; Bindschadler, G. A.; Martin, G. A.; Thirtle, J. R.; Gilman, H. *J. Am. Chem. Soc.* **1957**, *79*, 4921.
810. Graziani, R.; Zarli, B.; Cassol, A.; Bombieri, G.; Forsellini, E.; Tondello, E. *Inorg. Chem.* **1970**, *9*, 2116.
811. Forsellini, E.; Bombieri, G.; Graziani, R.; Zarli, B. *Inorg. Nucl. Chem. Lett.* **1972**, *8*, 461.
812. Pennington, M.; Alcock, D. B. *Inorg. Chim. Acta* **1987**, *139*, 49.
813. Alcock, D. B.; Pennington, M. *J. Chem. Soc. Dalton. Trans.* **1989**, 471.
814. Mikhailov, Y. N.; Gorbunova, Y. E.; Artem'eva, M. Y.; Serezhkina, L. B.; Serezhkin, V. N. *Zh. Neorg. Khim.* **2002**, *47*, 936.
815. Serezhkina, L. B.; Losev, V. Y.; Mikhailov, Y. N.; Serezhkin, V. N. *Radiokhimiya* **1994**, *36*, 3.
816. Mistryukov, V. E.; Mikhailov, Y. N.; Kanishcheva, A. S.; Serezhkina, L. B.; Serezhkin, V. N. *Zh. Neorg. Khim.* **1993**, *38*, 1514.
817. Mikhailov, Y. N.; Gorbunova, Y. E.; Shishkina, O. V.; Serezhkina, L. B.; Caceres, D. *Zh. Neorg. Khim* **2000**, *45*, 1885.
818. Alcock, N. W. *J. Chem. Soc. Dalton. Trans.* **1973**, 1614.
819. Mikhailov, Y. N.; Gorbunova, Y. E.; Shishkina, O. V.; Serezhkina, L. B.; Serezhkin, V. N. *Zh. Neorg. Khim.* **1999**, *44*, 1448.

820. Legros, J. P.; Jeannin, Y. *Acta Crystallogr., Sect. B* **1976**, *32*, 2497.
821. Beirakhov, A. G.; Orlova, I. M.; Ashurov, Z. P.; Lobanova, G. M.; Mikhailov, G. M.; Shchelokov, R. N. *Zh. Neorg. Khim.* **1991**, *36*, 647.
822. Beirakhov, A. G.; Orlova, I. M.; Gorbunova, Y. E.; Mikhailov, Y. N.; Shchelokov, R. N. *Zh. Neorg. Khim.* **1999**, *44*, 1492.
823. Chaudhuri, M. K.; Srinivas, P.; Khathing, D. T. *Polyhedron* **1993**, *12*, 227.
824. Patil, S. K.; Bhandiwad, V.; Kusumakumari, M.; Swarup, R. *J. Inorg. Nucl. Chem.* **1981**, *43*, 1647.
825. Kannan, S.; Shanmugasundara Raj, S.; Fun, H.-K. *Polyhedron* **2001**, *20*, 2145.
826. Bismondo, A.; Casellato, U.; Sitran, S.; Graziani, R. *Inorg. Chim. Acta* **1985**, *110*, 205.
827. Nourmand, M.; Meissami, N. *J. Chem. Soc., Dalton Trans.: Inorg. Chem. (1972–1999)* **1983**, 1529.
828. Nourmand, M.; Bayat, I.; Yousefi, S. *Polyhedron* **1982**, *1*, 827.
829. Feldman, I.; Koval, L. *Inorg. Chem.* **1963**, *2*, 145.
830. Rangaraj, K.; Ramanujam, V. V. *J. Inorg. Nucl. Chem.* **1977**, *39*, 489.
831. Ramanujam, V. V.; Krishnan, C. N.; Rengaraj, K.; Sivasankar, B. *J. Indian Chem. Soc.* **1983**, *60*, 726.
832. Selvaraj, P. V.; Santappa, M. *J. Inorg. Nucl. Chem.* **1977**, *39*, 119.
833. Wang, Y. S.; Sun, G. X.; Bao, B. R. *J. Radioanal. Nucl. Chem.* **1997**, *224*, 151.
834. Lumetta, G. J.; McNamara, B. K.; Rapko, B. M.; Sell, R. L.; Rogers, R. D.; Broker, G.; Hutchison, J. E. *Inorg. Chim. Acta* **2000**, *309*, 103.
835. Wang, Y.-S.; Sun, G.-X.; Xie, D.-F.; Bao, B.-R.; Cao, W.-G. *J. Radioanal. Nucl. Chem.* **1996**, *214*, 67.
836. Curtui, M.; Haiduc, I. *J. Radioanal. Nucl. Chem.* **1984**, *86*, 281.
837. Saito, A.; Choppin, G. R. *J. Alloys Comp.* **1998**, *271–3*, 751.
838. Adnet, J. M.; Madic, C.; Bourges, J. *Proceedings of the 22nd Journees des Actinides: Meribel* **1992**, 15–16.
839. Bion, L.; Moisy, P.; Madic, C. *Radiochim. Acta* **1995**, *69*, 251.
840. Fong, S. W. A.; Yap, W. T.; Vittal, J. J.; Henderson, W.; Hor, T. S. A. *J. Chem. Soc., Dalton Trans.* **2002**, 1826.
841. Casellato, U.; Vidali, M.; Vigato, P. A. *Inorg. Chim. Acta* **1976**, *18*, 77.
842. Fenton, D. E.; Casellato, U.; Vigato, P. A.; Vidali, M. *Inorg. Chim. Acta* **1984**, *95*, 187.
843. Vigato, P. A.; Fenton, D. E. *Inorg. Chim. Acta* **1987**, *139*, 39.
844. Sessler, J. L.; Mody, T. D.; Dulay, M. T.; Espinoza, R.; Lynch, V. *Inorg. Chim. Acta* **1996**, *246*, 23.
845. Thuéry, P.; Nierlich, M.; Harrowfield, J.; Ogden, M. Phenoxide Complexes of the f-Elements [with respect to callixarenes] *Calixarenes 2001* **2001**, 561–582. Kluwer Academic: Dordrecht, The Netherlands, 561.
846. Thuéry, P.; Nierlich, M.; Masci, B.; Asfari, Z.; Vicens, J. *J. Chem. Soc., Dalton Trans.* **1999**, 3151.
847. Thuéry, P.; Keller, N.; Lance, M.; Vigner, J.-D.; Nierlich, M. *New J. Chem.* **1995**, *19*, 619.
848. Harrowfield, J. M.; Ogden, M. I.; White, A. H. *J. Chem. Soc., Dalton Trans.* **1991**, 979.
849. Thuéry, P.; Nierlich, M. *J. Inclusion Phenom. Mol. Recognit. Chem.* **1997**, *27*, 13.
850. Thuéry, P.; Nierlich, M.; Vicens, J.; Masci, B.; Takemura, H. *Eur. J. Inorg. Chem.* **2001**, 637.
851. Thuéry, P.; Keller, N.; Lance, M.; Vigner, J.-D.; Nierlich, M. *Acta Crystallogr., Sect. C: Cryst. Struct. Commun.* **1995**, *C51*, 1570.
852. Leverd, P. C.; Dumazet-Bonnamour, I.; Lamartine, R.; Nierlich, M. *Chem. Commun.* **2000**, 493.
853. Deshayes, L.; Keller, N.; Lance, M.; Navaza, A.; Nierlich, M.; Vigner, J. D. *Polyhedron* **1994**, *13*, 1725.
854. Nierlich, M.; Sabattie, J.-M.; Keller, N.; Lance, M.; Vigner, J.-D. *Acta Crystallogr., Sect. C: Cryst. Struct. Commun.* **1994**, *C50*, 52.
855. Thuéry, P.; Keller, N.; Lance, M.; Sabattié, J.-M.; Vigner, J.-D.; Nierlich, M. *Acta Crystallogr., Sect. C: Cryst. Struct. Commun.* **1995**, *C51*, 801.
856. Rogers, R. D.; Bond, A. H.; Hipple, W. G.; Rollins, A. N.; Henry, R. F. *Inorg. Chem.* **1991**, *30*, 2671.
857. Komkov, Y. A.; Krot, N. N.; Gel'man, A. D. *Radiokhimiya* **1968**, *10*, 625.
858. Spitsyn, V. I.; Gel'man, A. D.; Krot, N. N.; Mefod'eva, M. P.; Zakharova, F. A.; Komkov, Y. A.; Shilov, V. P.; Smirnova, I. V. *J. Inorg. Nucl. Chem.* **1969**, *31*, 2733.
859. Keller, C.; Fang, D. *Radiochim. Acta* **1969**, *11*, 123.
860. Musante, Y.; Ganivet, M. *J. Electroanal. Chem. Interfacial Electrochem.* **1974**, *57*, 225.
861. Varlashkin, P. G.; Begun, G. M.; Peterson, J. R. *Radiochim. Acta* **1984**, *35*, 211.
862. Tananaev, I. G. *Radiokhimiya* **1992**, *34*, 108.
863. Gelis, A. V.; Vanysek, P.; Jensen, M. P.; Nash, K. L. *Radiochim. Acta* **2001**, *89*, 565.
864. Clark, D. L.; Conradson, S. D.; Neu, M. P.; Palmer, P. D.; Runde, W.; Tait, C. D. *J. Am. Chem. Soc.* **1997**, *119*, 5259.
865. Appelman, E. H.; Kostka, A. G.; Sullivan, J. C. *Inorg. Chem.* **1988**, *27*, 2002.
866. Williams, C. W.; Blaudeau, J. P.; Sullivan, J. C.; Antonio, M. R.; Bursten, B.; Soderholm, L. *J. Am. Chem. Soc.* **2001**, *123*, 4346.
867. Bolvin, H.; Wahlgren, U.; Moll, H.; Reich, T.; Geipel, G.; Fanghaenel, T.; Grenthe, I. *J. Phys. Chem. A* **2001**, *105*, 11441.
868. Tomilin, S. V.; Volkov, Y. F.; Visyashcheva, G. I.; Kapshukov, I. I. *Radiokhimiya* **1983**, *25*, 58.
869. Grigor'ev, M. S.; Glazunov, M. P.; Krot, N. N.; Gavrish, A. A.; Shakh, G. E. *Radiokhimiya* **1979**, *21*, 665.
870. Nakamoto, T.; Nakada, M.; Masaki, N. M.; Saeki, M.; Yamashita, T.; Krot, N. N. *J. Radioanal. Nucl. Chem.* **1999**, *239*, 257.
871. Zakharova, F. A.; Orlova, M. M.; Gel'man, A. D. *Radiokhimiya* **1972**, *14*, 123.
872. Keller, C.; Seiffert, H. *Angew. Chem., Int. Ed. Engl.* **1969**, *8*, 279.
873. Burns, J. H.; Baldwin, W. H.; Stokely, J. R. *Inorg. Chem.* **1973**, *12*, 466.
874. Karalova, Z. I.; Lavrinovich, E. A.; Myasoedov, B. F. *J. Radioanal. Nucl. Chem.* **1992**, *159*, 259.
875. Karalova, Z. K.; Lavrinovich, E. A.; Ivanova, S. A.; Myasoedov, B. F.; Fedorov, L. A.; Sokolovskii, S. A. *Radiokhimiya* **1992**, *34*, 132.
876. Silva, R. J.; Bidoglio, G.; Rand, M. H.; Robouch, P. B.; Wanner, H.; Puigdomenech, I. *Chemical Thermodynamics of Americium* **1995**, Elsevier: New York.
877. Myasoedov, B. F.; Kremliakova, N. Y. *Americium and Curium Chemistry and Technology* **1985**, Reidel: New York.
878. Myasoedov, B. F. *J. Alloys Compd.* **1994**, *213–214*, 290.

879. Hoffman, D. C.; Lee, D. M. *J. Chem. Educ.* **1999**, *76(3)*, 332.
880. Kratz, J. V. Chemical Properties of the Transactinide Elements. In *Heavy Elements and Related New Phenomena*; Greiner, W., Gupta, R. K., Eds.; World Scientific: Singapore, 1999, Chapter 4.
881. Schaedel, M. *Radiochim. Acta* **2001**, *89(11–12)*, 721.
882. Tuerler, A. *Czech. J. Phys.* **1999**, *49*, 581.
883. Dullmann, C. E.; Bruchle, W.; Dressler, R.; Eberhardt, K.; Eichler, B.; Eichler, R.; Gaggeler, H. W.; Ginter, T. N.; Glaus, F.; Gregorich, K. E.; Hoffman, D. C.; Jager, E.; Jost, D. T.; Kirbach, U. W.; Lee, D. M.; Nitsche, H.; Patin, J. B.; Pershina, V.; Piguet, D.; Qin, Z.; Schadel, M.; Schausten, B.; Schimpf, E.; Schott, H. J.; Soverna, S.; Sudowe, R.; Thorle, P.; Timokhin, S. N.; Trautmann, N.; Turler, A.; Vahle, A.; Wirth, G.; Yakushev, A. B.; Zielinski, P. M. *Nature* **2002**, *V418*, 859.
884. Newton, T. W. *J. Inorg. Nucl. Chem.* **1976**, *38*, 1565.
885. Newton, T. W. *Kinetics of the Oxidation–Reduction Reactions of Uranium, Neptunium, Plutonium, and Americium in Aqueous Solutions*; Los Alamos Sci. Lab., Los Alamos, NM, 1975.
886. Fulton, R. B.; Newton, T. W. *J. Phys. Chem.* **1970**, *74*, 1661.
887. Choppin, G. R; Overview of Chemical Separation Methods and Technologies. In *Chemical Separation Technologies and Related Methods of Nuclear Waste Management*; Choppin, G. R., Khankhasayev, M. K., Eds.; Kluwer Academic: Dordrecht, The Netherlands, 1999, pp 1–16.
888. Jarvinen, G. D. Technology Needs for Actinide and Technetium Separations Based on Solvent Extraction, Ion Exchange, and Other Processes. In *Chemical Separation Technologies and Related Methods of Nuclear Waste Management*; Choppin, G. R., Khankhasayev, M. K., Ed., Kluwer Academic: Dordrecht, The Netherlands, 1999, pp 53–70.
889. Musikas, C. Review of Possible Technologies for Actinide Separation Using Other Extractants than TBP. In *Chemical Separation Technologies and Related Methods of Nuclear Waste Management*; Choppin, G. R., Khankhasayev, M. K., Ed., Kluwer Academic: Dordrecht, The Netherlands, 1999, pp 99–122.
890. Thouvenot, P.; Hubert, S.; Moulin, C.; Decambox, P.; Mauchien, P. *Radiochim. Acta* **1993**, *61*, 15.
891. Runde, W.; Van Pelt, C.; Allen, P. G. *J. Alloys Compd.* **2000**, *303*, 182.
892. Brundage, R. T. *J. Alloys Compd.* **1994**, *213*, 199.
893. Fanghaenel, T.; Weger, H. T.; Koennecke, T.; Neck, V.; Paviet-Hartmann, P.; Steinle, E.; Kim, J. I. *Radiochim. Acta* **1998**, *82*, 47.
894. Wimmer, H.; Kim, J. I.; Klenze, R. *Radiochim. Acta* **1992**, *58–59*, 165.
895. Yaita, T.; Tachimori, S.; Edelstein, N. M.; Bucher, J. J.; Rao, L.; Shuh, D. K.; Allen, P. G. *J. Synchrotron Radiation* **2001**, *8*, 663.
896. Williams, C. W.; Antonio, M. R.; Soderholm, L. *J. Alloys Compd.* **2000**, *303*, 509.
897. Allen, P. G.; Bucher, J. J.; Shuh, D. K.; Edelstein, N. M.; Craig, I. *Inorg. Chem.* **2000**, *39*, 505.
898. Baybarz, R. D. *J. Inorg. Nucl. Chem.* **1973**, *35*, 483.
899. Baybarz, R. D.; Asprey, L. B. *J. Inorg. Nucl. Chem.* **1972**, *34*, 3427.
900. Meyer, G. *J. Less Common Metals* **1983**, *93*, 371.
901. Moskvin, A. I. *Radiokhimiya* **1967**, *9*, 718.
902. Moskvin, A. I. *Radiokhimiya* **1973**, *15*, 504.
903. Hurtgen, C.; Fuger, J. *Inorg. Nucl Chem. Lett.* **1977**, *13*, 1186.
904. Charvillat, J. P.; Zachariasen, W. H. *Inorg. Nucl Chem. Lett.* **1977**, *13*, 161.
905. Fuger, J.; Haire, R. G.; Peterson, J. R. *J. Alloys Compd.* **1993**, *200*, 181.
906. Soderholm, L.; Antonio, M. R.; Williams, C.; Wasserman, S. R. *Anal. Chem.* **1999**, *71*, 4622.
907. Carnall, W. T. *J. Less Comm. Metals* **1989**, *156*, 221.
908. Barthelemy, P.; Choppin, G. R. *Inorg. Chem.* **1989**, *28*, 3354.
909. Kimura, T. K. Y. T. H.; Choppin, G. R. *J. Alloys Compd.* **1998**, *271/274*, 719.
910. Kimura, T.; Choppin, G. R. *J. Alloys Compd.* **1994**, *213/214*, 313.
911. Kimura, T.; Choppin, G. R.; Kato, Y.; Yoshida, Z. *Radiochim. Acta* **1996**, *72*, 61.
912. Revel, R.; Den Auwer, C.; Madic, C.; David, F.; Fourest, B.; Le Du, J. F.; Morss, L. R. *Inorg. Chem.* **1999**, *38*, 4139.
913. Fanghänel, T.; Kim, J. I.; Paviet, P.; Klenze, R.; Hauser, W. *Radiochim. Acta* **1994**, *66/67*, 81.
914. Haire, R. G.; Lloyd, M. H.; Milligan, W. O.; Beasley, M. L. *J. Inorg. Nucl. Chem.* **1977**, *39*, 843.
915. Morss, L. R.; Williams, C. W. *Radiochim. Acta* **1994**, *66*, 99.
916. Lebedev, I. A.; Frenkel, V. Y.; Kulyako, Y. M.; Myasoedov, B. F. *Radiokhim.* **1979**, *21*, 809.
917. Weigel, F.; Haug, H. *Radiochim. Acta* **1965**, *4*, 227.
918. Kazantsev, G. N.; Skiba, O. V.; Burnaevà, A. A.; Kolesnikov, V. P.; Volkov, Y. F.; Kryukova, A. I.; Korshunov, I. A. *Radiokhim.* **1982**, *24*, 88.
919. Rai, D.; Felmy, A. R.; Fulton, R. W. *Radiochim. Acta* **1992**, *56*, 7.
920. Hobart, D. E.; Begun, G. M.; Haire, R. G.; Hellwege, H. E. *J. Raman Spectrosc.* **1983**, *14*, 59.
921. Burns, J. H.; Baybarz, R. D. *Inorg. Chem.* **1972**, *11*, 2233.
922. Hall, G. R.; Markin, T. L. *J. Inorg. Nucl. Chem.* **1957**, *4*, 137.
923. Hale, W. H., Jr.; Mosley, W. C. *J. Inorg. Nucl. Chem.* **1973**, *35*, 165.
924. Tabuteau, A.; Pages, M. *J. Solid State Chem.* **1978**, *26*, 153.
925. Tabuteau, A.; Pages, M.; Freundlich, W. *Radiochem. Radioanal. Lett.* **1972**, *12*, 139.
926. Shirokova, I. B.; Grigor'ev, M. S.; Makarenkov, V. I.; DenAuwer, C.; Fedoseev, A. M.; Budantseva, N. A.; Bessonov, A. A. *Russ. J. Coord. Chem.* **2001**, *27*, 729.
927. Yusov, A. B. *Actinides* **1989**, *89*, 240–241.
928. Chartier, D.; Donnet, L.; Adnet, J. M. *Radiochim. Acta* **1999**, *85*, 25.
929. Fedoseev, A. M.; Budentseva, N. A. *Sov. Radiochem.* **1989**, *31*, 525.
930. Yusov, A. B.; Fedoseev, A. M. *Radiokhim.* **1990**, *32*, 73.
931. Meinrath, G.; Kim, J. I. *Eur. J. Inorg. Solid State Chem.* **1991**, *28*, 383.
932. Kim, J. I.; Klenze, R.; Wimmer, H.; Runde, W.; Hauser, W. *J. Alloys Compd.* **1994**, *213/214*, 333.
933. Dedov, V. D.; Volkov, V. V.; Gvozdev, B. A.; Ermakov, V. A.; Lebedev, I. A.; Razbitnoi, V. M.; Trukhlyaev, P. S.; Chuburkov, Y. T.; Yakovlev, G. N. *Radiokhim.* **1965**, *7*, 453.
934. Runde, W.; Meinrath, G.; Kim, J. I. *Radiochim. Acta* **1992**, *58*, 93.

344 *The Actinides*

935. Standifer, E. M.; Nitsche, H. *Lanthanide and Actinide Research* **1988**, *2*, 383.
936. Weigel, F.; ter Meer, N. *Inorg. Nucl. Chem. Lett.* **1967**, *3*, 403.
937. Scherer, V.; Fochler, M. *J. Inorg. Nucl. Chem.* **1968**, *30*, 1433.
938. Bibler, N. E. *Inorg. Nucl. Chem. Lett.* **1972**, *8*, 153.
939. Zubarev, V. G.; Krot, N. N. *Sov. Radiochem.* **1983**, *25*, 601.
940. Jensen, M. P.; Bond, A. H. *J. Am. Chem. Soc.* **2002**, *124*, 9870.
941. Jensen, M. P.; Bond, A. H. *Radiochim. Acta* **2002**, *90*, 205.
942. Ritcey, G. M.; Ashbrook, A. W. *Solvent Extraction: Principles and Applications to Process Metallurgy* **1984**, Elsevier: Amsterdam.
943. Zhu, Y.; Chen, J.; Jiao, R. *Solvent Extr. Ion Exch.* **1996**, *14*, 61.
944. Modolo, G.; Odoj, R. *Solvent Extr. Ion Exch.* **1999**, *17*, 33.
945. Fedoseev, E. V.; Ivanova, L. A.; Travnikov, S. S.; Davydov, A. V.; Myasoedov, B. F. *Sov. Radiochem.* **1983**, *25*, 343.
946. Davydov, A. V.; Myasoedov, B. F.; Travnikov, S. S.; Fedoseev, E. V. *Sov. Radiochem.* **1978**, *20*, 217.
947. Davydov, A. V.; Myasoedov, B. F.; Travnikov, S. S. *Dokl. Chem. (Engl. Transl.)* **1975**, *220/5*, 672.
948. Penneman, R. A.; Ryan, R. R.; Rosenzweig, A. *Struct. Bond.* **1973**, *13*, 1.
949. Brown, D.; Fletcher, S.; Holah, D. G. *J. Chem. Soc. A* **1968**, 1889.
950. Katz, J. J.; Sheft, I. *Adv. Inorg. Chem. Radiochem.* **1960**, *2*, 195.
951. Bagnall, K. W. *Coord. Chem. Rev.* **1967**, *2*, 145.
952. Haire, R. G.; Benedict, U.; Young, J. P.; Peterson, J. R.; Begun, G. M. *J. Phys. C: Solid State Phys.* **1985**, *18*, 4595.
953. Marcus, Y.; Bomse, M. *Israel J. Chem.* **1970**, *8*, 901.
954. Schulz, W. W.; Horwitz, E. P. *Sep. Sci. Technol.* **1988**, *23*, 1191.
955. Horwitz, E. P.; Diamond, H.; Martin, K. A.; Chiarizia, R. *Solvent Extr. Ion Exch.* **1987**, *5*, 419.
956. Chiarizia, R.; Horwitz, E. P. *Solvent Extr. Ion Exch.* **1992**, *10*, 101.
957. Spjuth, L.; Liljenzin, J. O.; Hudson, M. J.; Drew, M. G. B.; Iveson, P. B.; Madic, C. *Solvent Extr. Ion Exch.* **2000**, *18*, 1.
958. Nugent, L. J.; Burnett, J. L.; Baybarz, R. D.; Werner, G. K.; Tanner, J. P.; Tarrant, J. R.; Keller, O. L. *J. Phys. Chem.* **1969**, *73*, 1540.
959. El-Reefy, S. A.; Dessouky, N. A.; Aly, H. F. *Solvent Extr. Ion Exch.* **1993**, *11*, 19.
960. Meguro, Y.; Kitatsuji, Y.; Kimura, T.; Yoshida, Z. *J. Alloys Compd.* **1998**, *271–273*, 790.
961. Mathur, J. N.; Khopkar, P. K. *Polyhedron* **1984**, *3*, 1125.
962. Spjuth, L.; Liljenzin, J. O.; SkÜlberg, M.; Hudson, M. J.; Chan, G. Y. S.; Drew, M. G. B.; Feaviour, M.; Iveson, P. B.; Madic, C. *Radiochim. Acta* **1997**, *78*, 39.
963. Musikas, C.; Hubert, H. *Solvent Extr. Ion Exch.* **1987**, *5*, 877.
964. Chan, G. Y. S.; Drew, M. G. B.; Hudson, M. J.; Iveson, P. B.; Liljenzin, J.-O.; SkÜlberg, M.; Spjuth, L.; Madic, C. *J. Chem. Soc., Dalton Trans.* **1997**, 649.
965. Sasaki, Y.; Adachi, T.; Choppin, G. R. *J. Alloys Compd.* **1998**, *271–273*, 799.
966. Sasaki, Y.; Choppin, G. R. *Anal. Sci.s* **1996**, *12*, 225.
967. Otu, E. O.; Chiarizia, R.; Rickert, P. G.; Nash, K. L. *Solvent Extr. Ion Exch.* **2002**, *20*, 607.
968. Bouhlassa, S. *Chem. Abstract* **1983**, *98*, 82730.
969. Danford, M. D.; Burns, J. H.; Higgins, C. E.; Stokeley, J. R. J.; Baldwin, W. H. *Inorg. Chem.* **1970**, *9*, 1953.
970. Moskalev, P. N.; Shapkin, G. N. *Radiokhimiya* **1977**, *19*, 356.
971. Asprey, L. B.; Penneman, R. A. *Inorg. Chem.* **1962**, *1*, 134.
972. Peretrukhin, V. F.; Enin, E. A.; Dzyubenko, V. I.; Kopytov, V. V.; Polyukhov, V. G.; Vasil'ev, V. Y.; Timofeev, G. A.; Rykov, A. G.; Krot, N. N.; Spitsyn, V. I. *Dokl. Akad. Nauk SSSR* **1978**, *242*, 1359.
973. Fargeas, M.; Fremont-Lamouranne, R.; Legoux, Y.; Merini, J. *J. Less Common Metals* **1986**, *121*, 439.
974. Perminov, V. B.; Krot, N. N. *Radiokhim* **1986**, *28*, 72.
975. Yusov, A. B.; Fedoseev, A. M. *J. Radioanal. Nucl. Chem. Art.* **1991**, *147*, 201.
976. Zachariasen, W. H. *Acta Crystallogr.* **1949**, *2*, 288.
977. Zachariasen, W. H. *Phys. Rev.* **1949**, *73*, 1104.
978. Akimoto, Y. *J. Inorg. Nucl. Chem.* **1967**, *29*, 2650.
979. Morss, L. R.; Richardson, J. W.; Williams, C. W.; Lander, G. H.; Lawson, A. C.; Edelstein, N. M.; Shalimoff, G. V. *J. Less Common Metals* **1989**, *156*, 273.
980. Penneman, R. A.; Coleman, J. S.; Keenan, T. K. *J. Inorg. Nucl. Chem.* **1961**, *17*, 138.
981. Magirius, S.; Carnall, W. T.; Kim, J. I. *Radiochim. Acta* **1985**, *38*, 29.
982. Stadler, S.; Kim, J. I. *Radiochim. Acta* **1988**, *44*, 39.
983. Bourges, J. Y.; Guillaume, B.; Koehly, G.; Hobart, D. E.; Peterson, J. R. *Inorg. Chem.* **1983**, *22*, 1179.
984. Haug, H. O.; Baybarz, R. D. *Inorg. Nucl. Chem. Lett.* **1975**, *11*, 847.
985. Keenan, T. K.; Asprey, L. B. *Inorg. Chem.* **1969**, *8*, 235.
986. Haire, R. G.; Nave, S. E.; Huray, P. G. *Proceedings of the 12th Journée des Actinides Orsay* **1982**.
987. Nave, S. F.; Haire, R. G.; Huray, P. G. *Phys. Rev. B* **1983**, *28*, 2317.
988. Kruse, F. H.; Asprey, L. B. *Inorg. Chem.* **1962**, *1*, 137.
989. Keenan, T. K. *Inorg. Nucl. Chem. Lett.* **1966**, *2*, 153.
990. Keenan, T. K. *Inorg. Nucl. Chem. Lett.* **1967**, *3*, 391.
991. Lux, F. *Lanthanide and Actinide Phthalocyaninato Complexes Proc. 10th Rare Earth Res. conf.* May 1973, Carefree, AZ **1973**, *2*, 871.
992. Keenan, T. K. *Inorg. Chem.* **1965**, *4*, 1500.
993. Tananaev, I. G. *Radiokhimiya* **1991**, *33*, 24.
994. Hara, M. *Bull. Chem. Soc. Jpn.* **1970**, *43*, 89.
995. Coleman, J. S.; Keenan, T. K.; Jones, L. H.; Carnall, W. T.; Penneman, R. A. *Inorg. Chem.* **1963**, *2*, 58.
996. Coleman, J. S. *Inorg. Chem.* **1963**, *2*, 53.
997. Keller, C. *The Chemistry of the Transuranium Elements* **1971**, Verlag Chemie: Weinheim, Germany.
998. Shilov, V. P.; Yusov, A. B. *Radiochem. (Moscow)5* **1999**, *41*, 445.
999. Runde, W.; Neu, M. P.; Clark, D. L. *Geochim. Cosmochim. Acta* **1996**, *60*, 2065.

1000. Tananaev, I. G. *Radiokhim.* **1990**, *32*, 53.
1001. Tananaev, I. G. *Radiokhim.* **1990**, *32*, 4.
1002. Volkov, Y. F.; Kapshukov, I. I.; Visyashcheva, G. I.; Yakovlev, G. N. *Radiokhimiya* **1974**, *16*, 863.
1003. Volkov, Y. F.; Kapshukov, I. I.; Visyashcheva, G. I.; Yakovlev, G. N. *Radiokhimiya* **1974**, *16*, 868.
1004. Volkov, Y. F.; Visyashcheva, G. I.; Tomilin, S. V.; Kapshukov, I. I.; Rykov, A. G. *Radiokhimiya* **1981**, *23*, 248.
1005. Fedoseev, A. M.; Budantseva, N. A.; Grigor'ev, M. S.; Perminov, V. P. *Radiokhim.* **1991**, *33*, 7.
1006. Bagnall, K. W.; Laidler, J. B.; Stewart, M. A. A. *J. Chem. Soc. A* **1968**, 133.
1007. Ionova, G. V.; Spitsyn, V. I. *Dok. Acad. Sci. USSR* **1978**, *241*, 590.
1008. Spitsyn, V. I.; Ionova, G. V. *Radiokhim.* **1978**, *20*, 328.
1009. Penneman, R. A.; Asprey, L. B. A Review of Americium and Curium Chemistry. In *Proc. First Int. Conf. on the Peaceful Uses of Atomic Energy*, Geneva, Switzerland *1955*, **1956**, pp 355–362.
1010. Stephanou, S. E.; Penneman, R. A. *J. Am. Chem. Soc.* **1952**, *74*, 3701.
1011. Proctor, S. G.; Connor, W. V. *J. Inorg. Nucl. Chem.* **1970**, *32*, 3699.
1012. Proctor, S. G. *J. Less Common Metals* **1976**, *44*, 195.
1013. Hoekstra, H.; Gebert, E. *Inorg. Nucl Chem. Lett.* **1978**, *14*, 189.
1014. Keller, C.; Schmutz, H. *Z. Naturforsch. B* **1964**, *19*, 1080.
1015. Cherfa, S.; Pecaut, J.; Nierlreh, M. *Zeitschn Kristallogn-New Cryst. Struct.* **1999**, *214*, 523–5.
1016. Morss, L. R. Complex Oxide Systems of the Actinides. In *Actinides in Perspective*; Edelstein, N., Pergamon, Oxford: 1982, pp 381–407.
1017. Cohen, D. *Inorg. Nucl. Chem.* **1972**, *8*, 533.
1018. Nikolaevskii, V. B.; Shilov, V. P.; Krot, N. N.; Peretrukhin, V. F. *Radiokhimiya* **1975**, *17*, 426.
1019. Giffaut, E.; Vitorge, P. *Proc. Mater. Res. Soc.* **1993**, *294*, 747.
1020. Lawaldt, D.; Marquart, R.; Werner, G. D.; Wigel, F. *J. Less Common Metals* **1982**, *85*, 37.
1021. Ueno, K.; Hoshi, M. *J. Inorg. Nucl. Chem.* **1971**, *33*, 1765, 2631.
1022. Fedoseev, A. M.; Budantseva, N. A. *Radiokhim.* **1990**, *32*, 14.
1023. Basile, L. J.; Ferrarro, J. R.; Mitchell, M. L.; Sullivan, J. C. *Appl. Spectrosc.* **1978**, *32*, 535.
1024. Fedoseev, A. M.; Perminov, V. F. *Sov. Radiochem.* **1983**, *25*, 522.
1025. Asprey, L. B.; Stephanou, S. E.; Penneman, R. A. *J. Am. Chem. Soc.* **1950**, *72*, 1425.
1026. Asprey, L. B.; Stephanou, S. E.; Penneman, R. A. *J. Am. Chem. Soc.* **1951**, *73*, 5715.
1027. Jones, L. H.; Penneman, R. A. *J. Chem. Phys.* **1953**, *21*, 542.
1028. Jones, L. L. *J. Chem. Phys* **1953**, *21*, 1591.
1029. Jones, L. H. *J. Chem. Phys.* **1955**, *23*, 2105.
1030. Zubarev, V. G.; Krot, N. N. *Sov. Radiochem.* **1982**, *24*, 264.
1031. Eberle, S. H.; Robel, W. *Inorg. Nucl Chem. Lett.* **1970**, *6*, 359.
1032. Musikas, C.; Germain, M.; Bathellier, A. *ACS Symp. Ser.* **1980**, *117*, 157.
1033. Keenan, T. K. *Inorg. Nucl. Chem. Lett.* **1968**, *4*, 381.
1034. Bagnall, K. W.; Laidler, J. B.; Stewart, M. A. A. *Chem. Commun.* **1967**, *1*, 24.
1035. Melkaya, R. F.; Volkov, Y. F.; Sokolov, E. I.; Kapshukov, I. I.; Rykov, A. G. *Dokl. Chem./ (Engl. Transl.)* **1982**, / *262*/7, 42.
1036. Gibson, J. K.; Haire, R. G. *J. Nucl. Mater.* **1992**, *195*, 156.
1037. Shilov, V. P.; Gogolev, A. V.; Pikaev, A. K. *High Energy Chemistry (Translation of Khimiya Vysokikh Energii)* **1998**, *32*, 354.
1038. Krot, N. N.; Shillov, V. P.; Nikolaevskii, V. B.; Nikaev, A.; Gel'man, A. D.; Spitsyn, V. I. *Dokl. Acad. Sci. USSR* **1974**, *217*, 525.
1039. Mikheev, N. B.; Myasoedov B. F., Lower and Higher Oxidation States of Transplutonium Elements in Solutions and Melts. In *Handbook on the Physics and Chemistry of the Actinides:* Vol. 3; Freeman, A. J. Keller, C. Eds.; Elsevier: Amsterdam 1985, pp 347–386.
1040. Clark, D. L.; Keogh, D. W.; Neu, M. P.; Runde, W. *Thorium and Thorium Compounds*. In *Kirk-Othmer Encyclopedia of Chemical Technology*, 4th ed.; Kroschwitz, J. I., Ed., Wiley: New York, 1997, 639–94.
1041. Clark, D. L.; Keogh, D. W.; Neu, M. P.; Runde, W. *Uranium and Uranium Compounds*. In *Kirk-Othmer Encyclopedia of Chemical Technology*, 4th ed.; Kruschwitz, J. I., Ed., Wiley: New York, 1997, 69–88.
1042. Burns, P. C. *Rev. Mineral.* **1999**, *38*, 23.
1043. Runde, W.; Schulz, W. W. Americium. In *The Chemistry of the Actinide and Transactinide Elements*, 3rd edn, Katz, J. J; Morss, L. R.; Edetstein, N. M.; Fuger, J., Eds.; Kluwer Academic: Amsterdam, 2003, in press.
1044. Lumetta, G. J.; Thompson, M. C.; Penneman, R. A.; Eller, P. G. Curium. In *The Chemistry of the Actinide and Transactinide Elements*, 3rd edn. Katz, J. J.; Morss, L. R.; Edelstein, N. M.; and Fuger, J., Eds.; Kluwer Academic: Amsterdam, 2003, in press.
1045. *The Chemistry of the Actinide Elements*, 2nd edn., Katz, J. J.; Seaborg, G. T.; Morss, L. R., Eds.; Chapman and Hall: London, 1986, all chapters.
1046. *Handbook on the Physics and Chemistry of the Actinides*, Freeman, A. J.; Keller, C., Eds.; Elsevier: Amsterdam, 1985, all volumes.

Comprehensive Coordination Chemistry II
ISBN (set): 0-08-0437486

Volume 3, (ISBN 0-08-0443257); pp 189–345

3.4
Aluminum and Gallium

G. H. ROBINSON
The University of Georgia, Athens, GA, USA

3.4.1 ALUMINUM

3.4.1.1 Introduction

The coordination chemistry of aluminum is as rich as it is varied. The striking range and diversity of coordination modes of aluminum atoms spans both traditional inorganic chemistry and contemporary organometallic chemistry. Indeed, the coordination chemistry of aluminum goes beyond that which may be expected for an ns^2p^1 valence configuration. The fact that aluminum is the most abundant terrestrial metal only adds to the allure of this main group metal. The history of aluminum is equally fascinating. As an element that once held the crown jewels of France and was valued as a precious metal, to a critical component in various industrial and catalytic processes, aluminum has, in many regards, done it all. The coordination modes of aluminum virtually spans the gamut of structural motifs from low coordinate three-coordinate (trigonal planar and T-shaped) and normal four-coordinate (tetrahedral) to high coordinate five-coordinate (trigonal bipyramidal and square pyramidal) and six-coordinate (octahedral). Even seven-coordinate (pentagonal bipyramidal) has been reported. This contribution will emphasize this wide diversity as a function of the type of compound. While the coordination chemistry of aluminum was discussed in *Comprehensive Coordination Chemistry* (*CCC*, 1987) this review does not seek to repeat that accomplishment. Rather, this review will endeavor to concentrate more on the discoveries in the intervening years with more of an emphasis on the organometallic chemistry of aluminum and gallium. To this end, some historical background is in order.

3.4.1.2 Group 14 Ligands

Of the group 14-based ligands, the most important by far are the carbon-based congeners. The organometallic chemistry of aluminum is quite overwhelming. The Lewis acidity of aluminum alkyls and aryls is the dominant feature in chemistry. The organometallic Al—C bond has proven particularly important in a variety of industrial and catalytic processes. Reports of organoaluminum compounds date back to the eighteenth century. The direct synthesis of aluminum alkyls was a significant accomplishment in the development of this field (Equation (1)).

$$3/2\ H_2\ +\ Al\ +\ 3\ \diagdown\!\!/\ \longrightarrow\ R_3Al \tag{1}$$

The literature reveals a number of perhaps more convenient routes to aluminum alkyls. These include a simple oxidation–reduction reaction of aluminum metal with dialkylmercury compounds (Equation (2)).

$$2\ Al\ +\ 3\ R_2Hg\ \longrightarrow\ 2\ R_3Al\ +\ 3\ Hg \tag{2}$$

Other preparative routes to aluminum alkyls include reaction of aluminum halides with organometallic reagents such as lithium alkyls (Equation (3)).

$$AlX_3\ +\ 3\ RLi\ \longrightarrow\ R_3Al\ +\ 3\ LiX \tag{3}$$

or Grignard reagents (Equation (4)).

$$AlX_3\ +\ 3\ RMgX\ \longrightarrow\ R_3Al\ +\ 3\ MgX_2 \tag{4}$$

These routes are often more desirable than those involving organomercury reagents due to toxicity concerns.

The simple aluminum alkyls (R = Me, Et, Pr^i, etc.) are colorless, mobile, pyrophoric liquids. The pyrophoric nature of these substances may be traced to the considerable Al–O bond strength compared to the Al–C bond strength. Even though in Equations (1)–(4) the aluminum alkyls are depicted as R_3Al monomers with the aluminum atoms ostensibly in three-coordinate trigonal planar environments, in fact these substances are dimeric, R_6Al_2, with the aluminum atoms in four-coordinate tetrahedral environments. The bridging carbon atoms in these organoaluminum dimers are engaged in electron deficient, three center-two electron (3c-2e), bonding schemes. Although some debate initially ensued concerning the nature of the bonding in Me_6Al_2,[1,2] single crystal X-ray diffraction data[3] provided unambiguous data confirming the dimeric electron-deficient

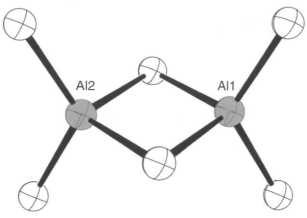

Figure 1 Solid-state structure of Me_6Al_2.

nature of these compounds. The Al–C–Al bond angle in Al_2Me_6 (Figure 1) is ~75°. The Al–C_{br} bond distance of 2.12 Å is considerably longer than the Al–C_{ter} bond distance of 1.95 Å. These factors result in a general weakness of the Al–C–Al bridges. Indeed, upon reaction the dimer readily cleaves in a symmetrical fashion across the electron deficient bridge providing R_3Al units.

A classic synthetic route to alkylaluminum halides is reaction of aluminum metal with alkyl halides. This procedure affords alkylaluminum sesquihalides, equimolar mixtures of dialkylaluminum halides and alkylaluminum dihalides (Equation (5)).

$$
2 \quad \underset{R}{\overset{R}{\diagdown}}Al\underset{X}{\overset{X}{\diagdown}}Al\overset{R}{\underset{X}{\diagup}} \quad \rightleftharpoons \quad \underset{R}{\overset{R}{\diagdown}}Al\underset{X}{\overset{X}{\diagdown}}Al\overset{R}{\underset{R}{\diagup}} \quad + \quad \underset{X}{\overset{R}{\diagdown}}Al\underset{X}{\overset{X}{\diagdown}}Al\overset{R}{\underset{X}{\diagup}} \qquad (5)
$$

Of course, these compounds are also dimeric, containing electron-deficient bonds with the aluminum atoms in four-coordinate tetrahedral environments. It is significant that the Al–X–Al angle in alkylaluminum halides is considerably widened from that observed in Me_6Al_2 (75°) to ~90°.

Alkylaluminum halides are important as they are often utilized to prepare other organoaluminum products. In particular, an industrial preparation of trimethylaluminum involves the sodium metal reduction of dimethylaluminum chloride, $[Me_2AlCl]_2$ (Equation (6)).

$$
3 \quad \underset{Me}{\overset{Me}{\diagdown}}Al\underset{Cl}{\overset{Cl}{\diagdown}}Al\overset{Me}{\underset{Me}{\diagup}} \quad + \; 6\,Na \quad \longrightarrow \quad \underset{Me}{\overset{Me}{\diagdown}}Al\underset{Me}{\overset{Me}{\diagdown}}Al\overset{Me}{\underset{Me}{\diagup}} \quad + \quad 2\,Al \quad + \quad 6\,NaCl \qquad (6)
$$

Although triphenylaluminum (**2**) exists as a dimer (Figure 2)[4,5] (with the aluminum atoms in four-coordinate tetrahedral environments) in the solid state with bridging η^1-phenyl groups—and is thus more accurately referred to as di-μ-phenyl-bis(diphenylaluminum)—sterically demanding carbon-based ligands can substantially affect the coordination environment of aluminum. It should be noted that the dimethylphenylaluminum derivatives of triphenylaluminum, di-μ-phenyl-bis(dimethylaluminum), $[Me_2PhAl]_2$, and tetra-*o*-tolyl-bis(μ-*o*-tolyl)dialuminum—the ortho-ligated toluene derivative—exists as a tetrahedral dimer about bridging η^1-phenyl groups.[6]

The role of sterically demanding ligand systems and their effect on the coordination of aluminum is conveniently illustrated when comparing the phenyl ligand with the mesityl ligand. Trimesitylaluminum, Mes_3Al[7] (Mes = 1,3,5-trimethylphenyl), is prepared by reaction of dimesitylmercury with aluminum metal (Equation (7)).

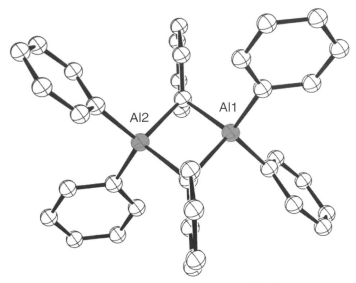

Figure 2   Solid-state structure of Ph_6Al_2.

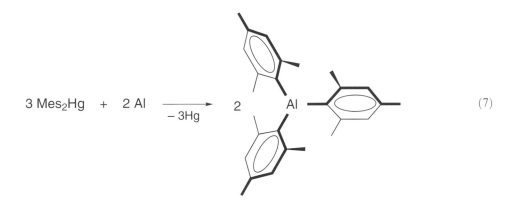

$$3\ Mes_2Hg\ +\ 2\ Al \xrightarrow[-3Hg]{}\ 2 \quad\quad\quad\quad\quad\quad\quad (7)$$

The steric demands of the mesityl ligands in Mes_3Al are critical in the aluminum atom assuming a virtually perfect three-coordinate trigonal planar coordination. The Al–C bond distance was shown to be 1.995(8) Å while the C–Al–C bond angle was 120°. The mesityl ligands are arranged in a propeller fashion about the metal center at dihedral angles of 55° (Figure 3).

The tetrahydrofuran adduct of trimesitylaluminum, $Mes_3Al\cdot THF$ was subsequently reported.[8] While the coordination of the aluminum atom in $Mes_3Al\cdot THF$ is distorted four-coordinate tetrahedral, the most meaningful comparison concerns the orientation of the three mesityl rings. In $Mes_3Al\cdot THF$ the mesityl ligands are no longer equivalent with dihedral angles of 56° relative to the AlC_3 basal plane. Rather, the mesityl rings now reside at angles of 96.6(2)°, 45.4(2)°, and 20.4(3)°. The tetrahedral environment of the aluminum atom is distorted as evidenced by the fact that two of the C–Al–C bond angles are approximately 120° while one is much smaller (and closer to that which is expected for a tetrahedral atoms) at 108.6°.

A particularly intriguing organoaluminum compound involving carbon-based ligands involves a recently reported carbene complex. Reaction of trimethylaluminum with 1,3-diisopropyl-4,5-dimethyl-imidazol-2-ylidene was carried out to afford the first organo-group 13 metal-carbenes, Me_3M:carbene (M = Al, Ga).[9] The ability of Lewis acids such as trimethylaluminum to form stable adduct complexes with suitable Lewis bases such as amines, phosphines, and oxygen containing compounds is obvious and well documented. Nonetheless, the concept of a "carbon-based" Lewis acid center interacting with a Lewis acid such as Me_3Al had received little attention. This compound is noteworthy in that while the four-coordinate tetrahedral coordination of the aluminum atom is not in itself unusual, the fact that the compound is monomeric is significant. Prior to the discovery of this compound, if an aluminum atom was involved in four-coordinate tetrahedral bonding to four carbon atoms, the resulting compound was almost always dimeric.

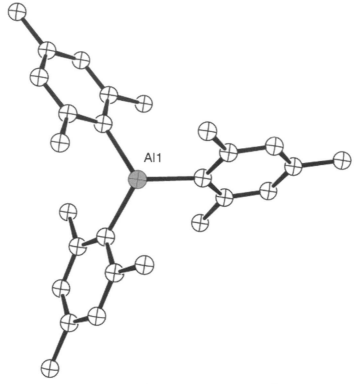

Figure 3 Solid-state structure of $(Mes)_3Al$.

The independent $Al-C_{(methyl)}$ bond distances in Me_3Al:carbene of $1.940(5)\,Å$ and $2.062(7)\,Å$, compares to $2.124(6)\,Å$ for the $Al-C_{(carbene)}$ bond distance. These distances are further placed in perspective when comparing them with the values reported for Al–C bond distances in Me_6Al_2 and Ph_6Al_2.

Certainly no discussion of the coordination chemistry of aluminum with carbon-based ligands would be complete without a discussion of the cyclopentadienyl ligand. This ligand, arguably the most important throughout the whole of organometallic chemistry, has an ever increasing chemistry with aluminum. Cyclopentadienyl(dimethyl)aluminum, a volatile solid isolated from reaction of trimethylaluminum with cyclopentadiene, displays different structures depending upon the physical state. For example, in the gas-phase it is monomeric with the cyclopentadiene (Cp) ligand interacting in a η^2 fashion with the aluminum atom basically being three-coordinate trigonal bipyramidal. However, in the solid state[10] the compound assumes a polymeric nature with each Me_2Al unit being bridged by an η^1-Cp ring. The closely related dicyclopentadienyl-(methyl)aluminum displays a dramatically different structural motif.[11] In the solid state this compound has unambiguously been shown to be a monomer with the Cp ligand interacting in a η^2 fashion—effectively resulting in the aluminum atom being five-coordinate square pyramidal. In this compound the molecule is monomeric with the aromatic rings residing in a somewhat asymmetric η^2 orientation.

The methyl(pentamethylcyclopentadienyl)aluminum chloride dimer, $[(C_5Me_5)MeAlCl]_2$, is pre-pared from reaction of dimethylaluminum chloride with (pentamethylcyclopentadienyl)lithium in toluene.[12] The solid-state structure of this compound reveals that the $C_5Me_5^-$ ligand interacts with the aluminum centers in an η^3 fashion across μ-chloro bridges.

A particularly interesting recently reported cyclopentadienylaluminum compound is the $[C_5Me_5)_2Al]^+$ cation, isolated from reaction of $(C_5Me_5)_2AlMe$ with $B(C_6F_5)_3$ in methylene chlor-ide (Equation (8)).

$$(C_5Me_5)_2AlMe \quad + \quad B(C_6F_5)_3 \quad \longrightarrow \quad [(C_5Me_5)_2Al][B(C_6F_5)_3Me] \qquad (8)$$

The authors note that this compound may be stored for months at $-17\,°C$ without appreciable decomposition. The $[C_5Me_5)_2Al]^+$ cation (Figure 4) sports a perfectly linear (ring-centroid)-

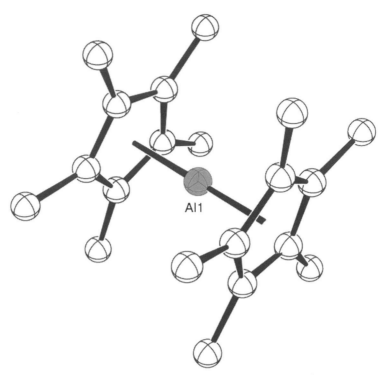

Figure 4   Solid-state structure of $[C_5Me_5)_2Al]^+$.

Al-(ring centroid) angle of 180°. The $C_5Me_5^-$ ligands are staggered relative to each other at a value of 36°.

Similar compounds containing the bis(pentamethylcyclopentadienyl)aluminum cation have been reported by other workers.[13]

3.4.1.3   Group 15 Ligands

3.4.1.3.1   Nitrogen ligands

Reactions between nitrogen species and aluminum compounds may be traced back to the 1800s. The most fundamental reaction in this regard is that of Me_3Al with ammonia (Equation (9)):

$$Me_3Al \quad + \quad NH_3 \quad \longrightarrow \quad (AlN)_n \quad + \quad 3\,CH_4 \qquad\qquad (9)$$

This reaction was initially studied by Wiberg[14] as a means to approach aluminum nitride. The thermodynamic driving force in this reaction is methane elimination (even though with each successive methane molecule that is eliminated, the subsequent elimination becomes more difficult). The reaction of trimethylaluminum with dimethylamine is another classic reaction which, after initially forming the Lewis acid–base adduct, $Me_3Al:N(H)Me_2$, forms the $[Me_2Al\text{-}NMe_2]_2$ dimer upon heating.[15] The formation of dimers and trimers with extensive Al–N association was quickly recognized[16] as a hallmark of the reactions of Me_3Al with simple amines.

The reaction of Me_3Al with methylamine proved to be very interesting.[17] Two products were suggested by NMR. After considerable effort the solid-state crystal structures of the reaction products confirmed *cis*- and *trans*-stereoisomers of $[Me_2Al\text{-}N(H)Me]_3$.[18] Both isomers contained nonplanar Al_3N_3 rings: a chair conformation was observed for the *cis*-$[Me_2Al\text{-}N(H)Me]_3$ isomer, while a boat confirmation was shown for the *trans*-isomer.

These reactions were subsequently found to be much more complicated than initially reported. Indeed, large clusters containing from eight to twelve aluminum atoms were ultimately isolated and characterized by single crystal X-ray diffraction.[19,20] The coordination of the aluminum atoms in all of these compounds, interesting as they are, was generally unremarkable as four-coordinate tetrahedral.

Sterically demanding amines have afforded a rich chemistry with organoaluminum compounds. The steric demands of a given amine are most prominently manifest in the coordination about the aluminum center. The 1980s proved to a rich decade for this type of work. Three prominent sterically demanding amines are aniline (**1**), 2,4,6-trimethylaniline (**2**), and 2,6-diisopropylaniline (**3**) (although the phenyl ligand is not generally considered to be sterically demanding, it is included in this group for comparative purposes).

(1)   **(2)**   **(3)**

One of the most interesting reactions involves trimethylaluminum with 2,6-diisopropylaniline.[21] The initial product is an aluminum–nitrogen dimer, however, upon further heating additional alkane elimination occurs resulting in the Al–N trimer (Equation (10)):

$$\text{Me}_3\text{Al} \quad + \quad \text{H}_2\text{NC}_6\text{H}_3(i\text{-Pr}_2) \quad \longrightarrow \quad [\text{Me}_2\text{Al-N(H)C}_6\text{H}_3(i\text{-Pr}_2)]_2$$

(10)

X-ray structural data confirmed the trimeric nature of $[\text{MeAl-NC}_6\text{H}_3(i\text{-Pr})_2]_3$ (Figure 5). A number of points are noteworthy regarding this compound. The neutral compound resided about a three-fold axis with a planar Al–N six-membered ring with the phenyl rings of the amine nearly orthogonal with the central plane. The bond angles at the aluminum and nitrogen atoms are 115.3(5)° and 124.7(5)°, respectively. While the overall structure of this compound bears a striking resemblance to borazine, an argument for true delocalization and aromaticity in this compound is problematic. In particular among other factors, the inter-ring Al–N bond distances are inequivalent.

Another interesting product is obtained from the condensation reaction of trimethylaluminum with mesitylamine. Similar to the previous compound, reaction of trimethylaluminum with mesitylamine initially yields a characteristic Al_2N_2 dimer. Further heating gives the aluminum-nitrogen tetramer $[\text{MeAlNC}_6\text{H}_2\text{Me}_3]_4$.[22] This unique "Al–N cube" may be viewed as the "fusing" of two Al_2N_2 dimers. The coordination sphere of each aluminum atom is completed by one methyl group and three nitrogen atoms. Similarly, the coordination sphere of each nitrogen atom is tetrahedral being completed by one mesityl group and three aluminum atoms of the cube. Thus, each atom residing in the cube is tetrahedral. The mean Al–N distance in this tetramer (1.948(7) Å) and the Al–C bond distance (1.949(3) Å) are unremarkable. Although Al–N tetramers are reasonably rare, the literature does reveal others involving various amines and LiAlH$_4$.[23]

It is informative to consider the dynamics that ultimately lead to an Al–N trimer rather than an Al–N tetramer. In both cases above the aluminum source was the same, trimethylaluminum. Thus, it is reasonable to examine the amine. The more sterically demanding amine, $\text{H}_2\text{NC}_6\text{H}_3\text{Pr}^i_2$, with the isopropyl groups give the trimer while the amine with less steric constraints around the

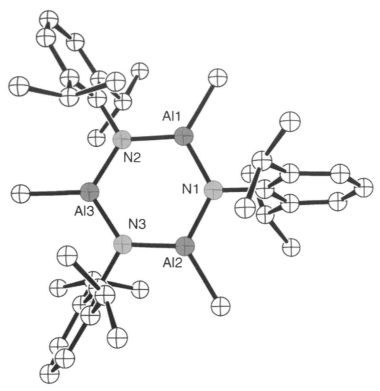

Figure 5   Solid-state structure of [MeAl-NC$_6$H$_3$(Pr^i)$_2$]$_3$.

nitrogen center, mesitylamine, gives the tetramer. We will later see that the steric demands of ligands on the aluminum (and gallium) centers will have an even more profound influence on the nature of the resulting compound.

Reaction of trimethylaluminum with the smallest aryl-based amine, aniline, is also intriguing. Although the initial product was not completely characterized, it was suggested to have the approximate formulation of [Me$_2$AlN(H)Ph]$_n$. Subsequent heating of this product gave, in low yield, [MeAlN(Ph)]$_6$.[24] Inclusion of solvent molecules into the crystal lattice made the structural solution problematic, but the structure of the molecule was unambiguously determined. The molecule resides about a S_6 axis. Along the sides of the hexamer are planar Al$_2$N$_2$ four-membered rings. Once can conceptualize that this hexamer is the combining of two Al$_3$N$_3$ units. The Al–N bond distances in this compound fall into two distinct categories: those in the Al$_3$N$_3$ rings (1.912(6) Å) and those within the Al$_2$N$_2$ four-membered rings (1.951(6) Å).

Due to the flexibility of the pendant amine groups, open-chain amines have demonstrated a varied chemistry in the coordination chemistry of organoaluminum species. For example, reaction of trimethylaluminum with diethylenetriamine results in a complex wherein the two open-chain amines are "bridged" by a series of four organoaluminum moieties.[25] It is noteworthy that the two middle aluminum atoms were found to be five-coordinate square pyramidal. This was the first example of a compound containing two five-coordinate aluminum atoms in square pyramidal environments. The only example of a six-coordinate aluminum alkyl was isolated from reaction of trimethylaluminum with N(CH$_2$CH$_2$OH)$_3$.[26] The molecule contained an Al$_4$O$_6$ core with two six-coordinate (distorted) octahedral aluminum atoms. The Al–C bond distance to the octahedral aluminum atoms 1.99(1) Å. Nitrogen-based crown ethers, azacrowns, are useful complexing agents for transition metals. The two most important azacrowns are cyclam [14]aneN$_4$ (**4**) and cyclen [12]aneN$_4$ (**5**). The driving force in the reaction of Me$_3$Al with such macrocyclic amines is a combination of the propensity to form Lewis acid–base adducts coupled with the thermodynamic advantage of Al–C/N–H bond cleavage and alkane elimination. Reaction of Me$_3$Al with [14]aneN$_4$ involves exhaustive alkane elimination and results in [Me$_3$Al]$_2$[14]ane-N$_4$[AlMe]$_2$.[27] Particularly noteworthy is the fact that the molecule resides about an Al$_2$N$_4$ four-membered ring while Me$_3$Al units occupy the other nitrogen sites (see Figure 6). The shorter Al–N bond distances are associated with the Al$_2$N$_2$ ring while the longer bond distances are associated with the terminal trimethlaluminum adducts.

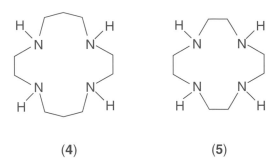

(4)   (5)

After the reporting of this compound a number of studies followed concerning [14]aneN$_4$ and various organoaluminum species were carried out. A particularly interesting product was isolated from reaction of trimethylaluminum with [14]aneN$_4$ in the presence of ZrCl$_4$, [Al·[14]aneN$_4$AlMe$_2$][Me$_2$AlCl]$_2$.[28] The most striking feature of the cation is the octahedral coordination of the central aluminum atom. Indeed, the [14]aneN$_4$ azacrown ether has been drawn back or pinned back by the dimethylaluminum unit, to further expose the nitrogen atoms. The coordination sphere of the central aluminum atom is completed by the four nitrogen atoms of the azacrown and by two μ-chlorine atoms.

3.4.1.3.2 *Phosphorus and arsenic ligands*

The sterically demanding phosphine tris(trimethylsilyl)phosphine, P(SiMe$_3$)$_3$, proved to be very important in the development of the chemistry of the Al–P bond. A number of aluminum-phosphorus adducts were reported in the last decade. In particular, Cl$_3$Al-P(SiMe$_3$)$_3$ toluene and Br$_3$Al·P(SiMe$_3$)$_3$ toluene, with Al–P bond distances of 2.392(4) Å and 2.391(6) Å, respectively, were reported.[29] These virtually identical Al–P bond distances are extremely short. The aluminum atoms in both complexes may be described as four-coordinate tetrahedral.

In many respects the organometallic chemistry of alkylaluminum halides with lithiated phosphines is very similar to that of the amines. The reaction chemistry with organoaluminum

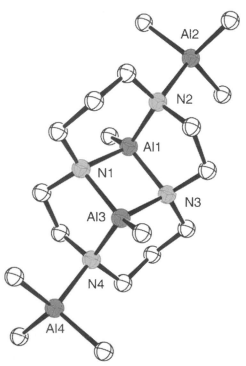

Figure 6 Solid-state structure of [Me$_3$Al]$_2$[14]aneN$_4$[AlMe]$_2$.

moieties is driven by initial adduct formation followed by Al–X and Li–P bond cleavage (and salt elimination). The compounds themselves are frequently Al_2P_2 four-membered ring centered dimers. Reaction of alkylaluminum halides with $P(SiMe_3)_3$ proved to be interesting. For example, reaction of $EtAlCl_2$ with $P(SiMe_3)_3$ yields the adduct $EtCl_2Al \cdot P(SiMe_3)_3$(Al–P: 2.435(3) Å). In the same study, the dimeric compound $[Et_2AlP(SiMe_3)_2]_2$ (**6**) was prepared from reaction of $EtAlCl_2$ with $LiP(SiMe_3)_2$ at $-78\,°C$.[30] This molecule contains a planar Al_2P_2 four-membered ring with the Al–P bond distance being 2.460(1) Å. Isolated from a similar synthetic scheme, this same laboratory reported the first example of an aluminum–phosphorus–arsenic mixed-pnicogen ring compound in $[Et_2Al\{Me_3Si\}_2PAs\{SiMe_3\}_2AlEt_2]$.[31]

Me₃Si   SiMe₃

Et   P   Et

Al   Al

Et   P   Et

Me₃Si   SiMe₃

(6)

The pentamethylcyclopentadienyl ligand, C_5Me_5, has played a significant role in the development of the chemistry concerning the Al–As bond. Reaction of $[(C_5Me_5)Al]_4$ (*vide infra*) with $[Bu^tAs]_4$ in toluene gives yellow crystals of $As_2[Al(C_5Me_5)]_3$.[32] While the gross structural features of this compound will be discussed in more detail later in this chapter, at this point the As_2Al_3 core will be examined (Figure 7). The two arsenic atoms are centered above and below the Al_3 ring at a distance of 2.48 Å. This Al–As bond distance is shorter than that reported in $[Et_2Al\{Me_3Si\}_2PAs\{SiMe_3\}_2AlEt_2]$ of 2.299(1) Å and 2.494(1) Å

Another noteworthy aluminum–arsenic compound is the trimeric $[Me_2AlAsPh_2]_3$, isolated from reaction of trimethylaluminum with diphenylarsine, Ph_2AsH.[33] This compound was one of the first Al–As six-membered ring compounds to be structurally characterized by single crystal X-ray diffraction. As shown in Figure 8, the Al–As ring is in a chair conformation with approximate tetrahedral environment about both Al and As atoms. The Al–As bond distances in this compound range from 2.512(3) Å to 2.542(3) Å.

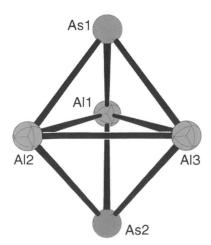

Figure 7   Solid-state structure of As_2Al_3 core of $As_2[AlCp^*]_3$.

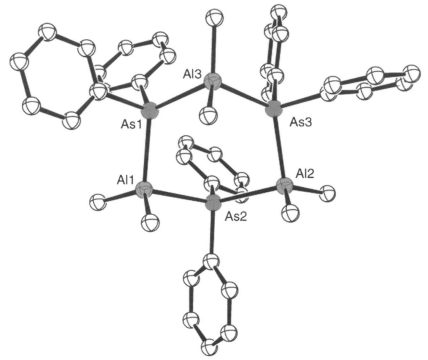

Figure 8 Solid-state structure of [Me$_2$Al-AsPh$_2$]$_3$.

3.4.1.4 Group 16 Ligands

3.4.1.4.1 Oxygen ligands

By far the most important group 16 compounds with aluminum are those of oxygen. This is due not only to the fact that the Al O bond is the thermodynamic driving force behind much of the chemistry of aluminum, but also due to the fact that Al–O compounds have found great utility in various industrial and catalytic processes. Indeed, one of the most important recent developments in this area may be found in a class of compounds knows as aluminoxanes. Aluminoxanes, methylaluminoxane (MAO) in particular, are very active cocatalysts in Ziegler–Natta systems. Two of the most common Al–O compounds are aluminum hydroxide, Al(OH)$_3$, and aluminum oxide, Al$_2$O$_3$.

The substantial bond strength of the Al–O bond is a major driving force in the chemistry of aluminum. This is evidenced by the ability of aluminum metal to form the ubiquitous Al$_2$O$_3$ oxide. Indeed, the pyrophoric nature of aluminum alkyls is traced to the great affinity between aluminum and oxygen. Certainly the simplest oxygen-based ligand is dioxygen itself. It is significant, therefore, that the literature reveals few discrete organoaluminum–dioxygen species. One notable example may be found in the reaction of potassium superoxide with trimethylaluminum in the presence of dibenzo-18-crown-6 (Equation (11)). The major point of interest

$$KO_2 \ + \ 2\,Me_3Al \ + \ \text{dibenzo-18-crown-6} \ \longrightarrow \ [K\cdot\text{dibenzo-18-crown-6}][Me_3Al\{O_2\}AlMe_3] \quad (11)$$

in the ionic [K·dibenzo-18-crown-6][Me$_3$Al{O$_2$}AlMe$_3$][34] compound is the dioxygen-based [Me$_3$Al{O$_2$}AlMe$_3$]$^-$ anion (Figure 9). An X-ray crystal structure of this compound confirms a most unusual bonding mode for oxygen—the two Me$_3$Al units are bridged by one of the oxygen atoms in an η^1 fashion. The rather long O–O bond distance of 1.47(2) Å was supported by the IR spectrum in which the stretch was observed at 851 cm^{-1}. The Al–O bond distances were 1.852(9) and 1.868(9) Å, while the Al–O–Al bond angle was 128.3(7)°. The value of 128.3(7)° for the Al–O–Al is comparable to the Al–N–Al bond angle observed for K[Al$_2$Me$_6$N$_3$].[35] Of course, the coordination of the aluminum atoms in this interesting anion is four-coordinate tetrahedral.

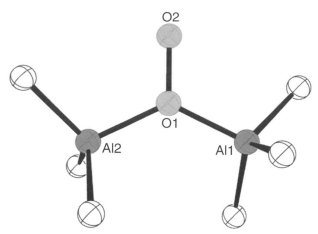

Figure 9 Solid-state structure Me$_3$Al{O$_2$}AlMe$_3$]$^-$ anion.

In fact, the major point of interest in this anion is not the coordination of the aluminum atoms but rather the unusual coordination of the oxygen atoms.

3.4.1.4.2 Crown ethers

An extensive chemistry has developed in the past two decades concerning the interaction of organoaluminum species with macrocyclic polyethers—crown ethers. Again, this chemistry is driven by the considerable Al–O bond strength. Characteristic of these compounds is a distortion of the crown ether by the organoaluminum species, essentially pulling the crown ether "inside-out". While 1,4-para-dioxane certainly does not technically qualify as a crown ether, it does share some similarities. To this end, this discussion begins with the simple Lewis acid–Lewis base complex bis(trimethylaluminum)-*p*-dioxane, Me$_3$Al(dioxane)AlMe$_3$ (**7**).[36] As illustrated in the compound below, the dioxane resides in a chair configuration as the two Me$_3$Al units bond to the two oxygen atoms. The Al–O bond distance in (**7**) is 2.02(2) Å.

/
―Al―
/
   O
 O
―Al―
/

(7)

Crown ethers (**8**)–(**11**), facilitated by the seminal discoveries of Pedersen,[37,38] have found great utility as phase transfer catalysts and as alkali metal complexing agents. This fact notwithstanding, the past two decades has witnessed the development of a rich crown ether chemistry involving organometallic compounds of aluminum. Distinct from the coordination mode observed for alkali metal ions (wherein the metal ion resides inside the macrocyclic cavity), aluminum alkyls typically form neutral Lewis acid–Lewis base complexes and reside along the macrocyclic perimeter (leaving the cavity empty). Alkylaluminum halides, given the appropriate crown ether, can reside within the crown ether cavity resulting in high coordination number (five or six) organoaluminum-crown ether cations.

The bis(trimethylaluminum)·12-crown-4 complex, [AlMe$_3$]$_2$12-crown-4, is a logical starting point. Prepared by reaction of trimethylaluminum with 12-crown-4,[39] the complex forms colorless crystals. The four oxygen atoms in 12-crown-4 were observed to be coplanar. The Al–O bond distance of 1.977(3) Å is remarkable as it compares to 2.02(2) Å reported for the dioxane-trimethylaluminum compound. Indeed, the overall conformation of this compound is quite similar to that observed in this compound as well.

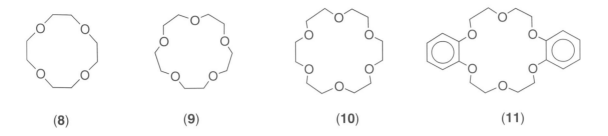

(8)   (9)   (10)   (11)

The product from the reaction of ethylaluminum dichloride with 12-crown-4 gives an unusual complex cation with an $AlCl_2$ fragment being complexed by the crown ether, $[Cl_2Al \cdot 12\text{-crown-}4]$ $[Cl_3AlEt]$.[40] While the Al–O bond distances are unremarkable, the aluminum resides in an octahedral environment with the crown ether being pulled back, allowing more of an "on edge" coordination mode for the crown ether (Figure 10). The orientation of the crown ether is noteworthy in this complex as it has been distorted and completely "drawn back" thereby more fully exposing the oxygen atoms. The aluminum atom, with its coordination sphere completed by the two chlorine atoms in equatorial positions, thus resides in an octahedral environment. The fact that the aluminum atom is octahedral in the smallest crown ether, 12-crown-4, is all the more impressive when one considers that octahedral coordination is also observed for the much larger 18-crown-6.

Octahedral coordination has also been observed for the larger 18-crown-6 with organoaluminum moieties. In the ionic complex $[Cl_2Al \cdot 18\text{-crown-}6][Cl_3AlEt]$[40] the aluminum atom is also found in an octahedral environment in the cation with four of the oxygen atoms of the crown ether bonding to the aluminum atom. With both 12-crown-4 and 18-crown-6 the generation of the Cl_2Al^+ cation from the respective alkylaluminum dihalide was cited as being critical in the preparation of these compounds.

A number of points are noteworthy with respect to the metrical values in the $[Cl_2Al \cdot 12\text{-crown-}4]^+$ and $[Cl_2Al \cdot 18\text{-crown-}6]^+$ cations. Regarding $[Cl_2Al \cdot 12\text{-crown-}4]^+$, the Al–Cl bond distances (2.200(8) Å and 2.202(5) Å) were considered rather long, while the Al–O bond distances (mean of 1.96(2) Å) fall within the expected range of aluminum–oxygen donor–acceptor bond distances. The larger 18-crown-6 displayed a wide range of Al–O bond distances (1.946(5) Å to 2.065(4) Å) while the Al–Cl bond distances (2.148(3) Å and 2.210(2) Å) were comparable to those observed with 12-crown-4. A particularly interesting complex, $[Cl_2Al \cdot benzo\text{-}15\text{-crown-}5][Cl_3AlEt]$,[41] results from reaction of benzo-15-crown-5 with ethylaluminum dichloride in toluene (Figure 11). As the coordination of aluminum in the anion is unremarkable four-coordinate tetrahedral, most of the

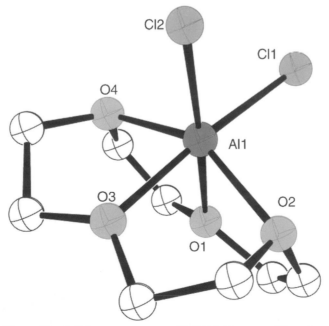

Figure 10   Solid-state structure of $[Cl_2Al \cdot 12\text{-crown-}4]^+$ cation.

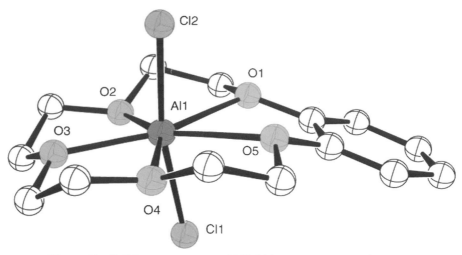

Figure 11   Solid-state structure of $[Cl_2Al \cdot benzo-15-crown-5]^+$ cation.

interest is directed toward the cation. In the $[Cl_2Al \cdot benzo-15-crown-5]$ cation the aluminum atom resides in an extremely rare seven-coordinate pentagonal bipyramidal geometry. The Al–O bond distances for the oxygen atoms adjacent to the aromatic group (2.28(1) Å and 2.30(1) Å) are considerably longer than those to the other three oxygen atoms (2.03(1) Å, 2.06(1) Å, and 2.08(1) Å). Consequently, the aluminum atom is located "off-center" in the crown ether cavity even as the metal atom is coplanar with the five oxygen atoms. Thus, the coordination sphere of aluminum consists of five equatorial oxygen atoms and two axial chlorine atoms (Al–Cl: 2.202(5) Å and 2.197(7) Å). It is intriguing that the aluminum atoms in complexes with 12-crown-4 and 18-crown-6 assumed octahedral structures, yet the rare pentagonal bipyramidal is found with benzo-15-crown-5.

With the larger crown ethers neutral trimethylaluminum compounds have been obtained. For example, the first reported organoaluminum-crown ether complexes were $[AlMe_3]_2 \cdot dibenzo-18-$crown-6 and $[AlMe_3]_4 \cdot 15-crown-5$ (Figures 12 and 13).[42] These compounds were prepared by

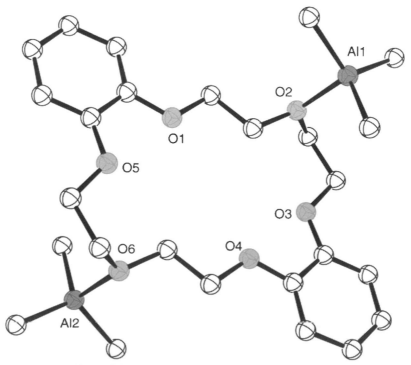

Figure 12   Solid-state structure of $[AlMe_3]_2 \cdot dibenzo-18-crown-6$.

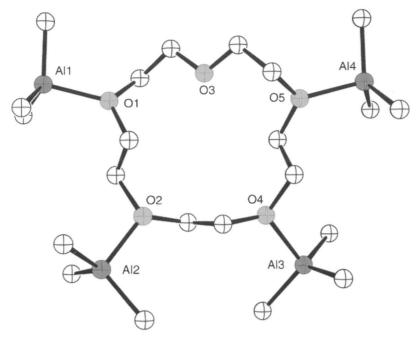

Figure 13 Solid-state structure of [AlMe₃]₄·15-crown-5.

reaction of excess trimethylaluminum with the respective crown ether. While it seems logical that the last remaining oxygen atom in [AlMe₃]₄·15-crown-5 could not be complexed by an Me₃Al unit (as it is forced toward the interior of the macrocyclic cavity), it is unexpected that with the larger dibenzo-18-crown-6 none of the remaining four oxygen atoms could be attacked by an Me₃Al unit. However, the authors suggested that the benzo rings imparted sufficient steric hindrance to discourage Me₃Al coordination to neighboring (four) oxygen atoms. In both compounds, the Me₃Al units served to "pull" the oxygen atoms along the macrocyclic perimeter affording a rather elongated and "flattened" orientation for the crown ether. Moreover, principally due to the absence of benzo groups, 15-crown-5 was deemed to be more flexible than dibenzo-18-crown-6. The six oxygen atoms of dibenzo-18-crown-6 assumed a chair configuration thereby allowing substantial Al–O interaction (by the two trimethylaluminum units), as evidenced by the Al–O bond distance of 1.967(3) Å in [AlMe₃]₂·dibenzo-18-crown-6. It is ironic that 15-crown-5 being more flexible, yet the mean Al–O bond distance in [AlMe₃]₄·15-crown-5 is considerably longer at 2.005(6) Å. The coordination of the aluminum atoms in these complexes were unremarkable four-coordinate tetrahedral.

Lastly, the mixed tetraoxo-diaza derivative of 18-crown-6, diaza-18-crown-6, has also been utilized to stabilize aluminum atoms in high coordination environments. In particular, the [(EtAl)₂·diaza-18-crown-6]²⁺ cation displays the aluminum center in a rare square pyramidal environment.[43]

3.4.1.4.3 *Sulfur, selenium, and tellurium ligands*

The chemistry of sulfur-based ligands with aluminum is striking in its range and diversity. While the organoaluminum chemistry of sulfur bears some resemblance to that of oxygen, there are notable differences. Perhaps most notable is the fact that sulfur is larger, softer, and more polarizable than oxygen. This has a direct bearing on the manner in which the sulfur center interacts with aluminum. The corresponding chemistry with selenium and tellurium ligands has not been developed to a comparable extent.

The aluminum–sulfur bond has not been explored to an extent comparable to that of the corresponding aluminum–oxygen bond. However, there does exist an interesting coordination chemistry of organoaluminum species involving sulfur-containing ligands. An unusual Al–S linear oligomer was reported for [Me₂AlSMe]ₙ,[44] with an Al–S bond distance of 2.348 Å. There is data that suggest that this substance exists in the gas phase as a cyclic Al₂S₂ dimer with the methyl groups oriented in a trans conformation.[45] Another Al–S compound, K[Al$_2$Me$_6$SCN], was synthesized containing the thiocyanide ligand and was characterized with an Al–S bond distance of 2.489(2) Å.[46] The coordination of aluminum in both of these compounds may be described as four-coordinate tetrahedral.

3.4.1.4.4 *Sulfur-based crown ethers*

The two most important thiacrown ethers are [14]aneS$_4$ (**12**) and [12]aneS$_4$ (**13**) (the thia equivalents of the aza-based crown ethers [14]aneN$_4$ and [12]aneN$_4$, respectively). Unlike oxygen-based crown ethers, sulfur-based crown ethers, thiacrown ethers, have a demonstrated ability to complex transition metals as opposed to alkali and alkaline earth metals. Nonetheless, interesting thiacrown ether complexes have been isolated with organoaluminum moieties.

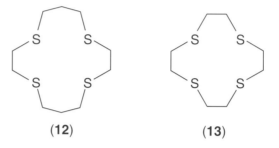

(**12**)   (**13**)

Reaction of trimethylaluminum with [14]aneS$_4$ gives [Me$_3$Al]$_4$[14]aneS$_4$ (Figure 14).[47] The conformation of the thiacrown ether was surprising as it assumed an "*exo*-dentate" geometry. Specifically, instead of the sulfur atoms residing along the macrocyclic cavity (as is the case for neutral oxygen-based crown ethers), the sulfur atoms have been pulled on the outside. Also noteworthy is the Al–S bond distance of 2.512(2) Å and 2.531(2) Å. These bond distances are considerably longer than those cited for [Me$_2$AlSMe]$_n$ (2.348 Å) and K[Al$_2$Me$_6$SCN] (2.489(2) Å).

Perhaps the most interesting organoaluminum-thiacrown ether complex is the [Me$_3$Al]·[12]aneS$_4$ complex (Figure 15).[48] Although the reaction was performed with a four-fold excess of trimethylaluminum with [12]aneS$_4$ only the 1:1 crystalline compound was isolated. Upon examination of the coordination of the [Me$_3$Al]·[12]aneS$_4$ monomer, the coordination of the

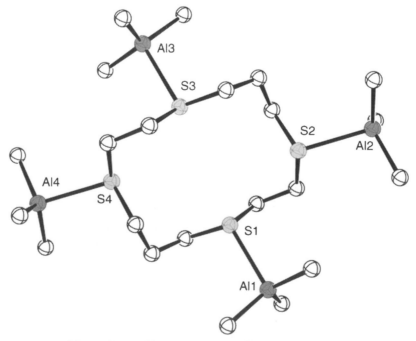

Figure 14   Solid-state structure of [Me$_3$Al]$_4$[14]aneS$_4$.

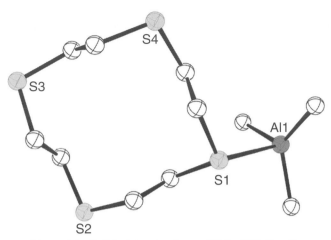

Figure 15   Solid-state structure of [Me$_3$Al][12]aneS$_4$.

aluminum atom appears decidedly nontetrahedral. Indeed, the aluminum atom appears to be coplanar with the three carbon atoms of the methyl groups. Thus, the "immediate" coordination of the aluminum atom goes beyond "distorted tetrahedral" and may be described as "trigonal pyramidal" with an extremely long Al–S bond distance of 2.718(3) Å. Indeed, the special coordination of the local environment about the aluminum atom suggested a more expansive view was in order. Upon examination of the unit cell it became clear that the coordination of the aluminum atom is not four-coordinate, but rather it is best described as five-coordinate as each aluminum atom has a secondary interaction with the sulfur atom of a neighboring [12]aneS$_4$ complex. The secondary Al–S contact is 3.052(3) Å. It is this secondary Al–S interaction, coupled with the primary Al–S interaction (bond), which causes the planarity of the Me$_3$Al unit. Thus, the coordination of the aluminum atom(s) in the "extended" [Me$_3$Al]·[12]aneS$_4$ complex is best described as five-coordinate trigonal bipyamidal. Essentially, a planar Me$_3$Al unit bridges two [12]aneS$_4$ moieties.

The literature reveals only a few examples of compounds that contain a direct Al–Se bond and fewer still of compounds that contain an Al–Te bond.

3.4.1.5   Group 17 Ligands

3.4.1.5.1   *Hydride ligands*

The fact that hydrogen can exist as either a cation (i.e., HCl) or an anion (i.e., NaH) belies its station as the simplest element. The chemistry of the H$^-$ hydride resembles that of the halides. Relative to a singular compound, the chemistry of aluminum hydride is embodied in the ubiquitous lithium aluminum hydride, LiAlH$_4$. This notwithstanding, relatively few compounds exists wherein a single hydrogen atom serves as a bridge between two organoaluminum moieties. One such compound results from the reaction of sodium hydride with trimethylaluminum, in the presence of 15-crown-5. This reaction yields the unusual [Me$_3$Al{H}AlMe$_3$]$^-$ anion.[49] Unlike the "bent" superoxide-based [Me$_3$Al{O$_2$}AlMe$_3$]$^-$ anion (previously discussed), the X-ray structure of the [Me$_3$Al{H}AlMe$_3$]$^-$ anion (Figure 16) unexpectedly reveals a perfectly linear, 180°, Al–H–Al linkage with an Al–H bond distance of 1.65 Å. The aluminum–hydride bond was comparable to that observed in the dimethylaluminum hydride dimer, [Me$_2$AlH]$_2$. The fact that the coordination of the aluminum atom in both [Me$_3$Al{O$_2$}AlMe$_3$]$^-$ and [Me$_3$Al{H}AlMe$_3$]$^-$ is tetrahedral does not diminish the remarkable nature of these organoaluminum anions.

3.4.1.5.2   *Halide ligands*

The coordination of aluminum with halogen-based ligands is generally straight forward. The halogen serves as a simple monodentate ligand with a 1−charge. The corresponding coordination of the aluminum atom is simple four-coordinate tetrahedral. In particular, in the simple alkylaluminum

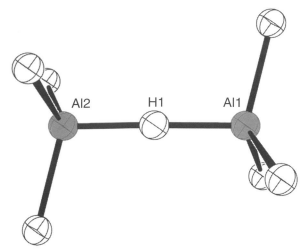

Figure 16 Solid-state structure of $[Me_3Al\{H\}AlMe_3]^-$ anion.

dihalides or dialkylaluminum halides the compounds exist as electron deficient dimers with μ-bridging halides (in much the same way as simple dimeric aluminum alkyls).

3.4.1.6 Compounds Containing Al–Al Bonds

3.4.1.6.1 *Neutral compounds containing the Al–Al bond*

The history of compounds containing Al–Al bonds is as colorful as it is interesting. Reports of organometallic alanes, compounds containing the iconic Al–Al bond, may be found as early as 1966.[50–54] However, these early reports are now viewed with considerable skepticism as neither spectroscopic nor compelling structural data were presented. As a point of origin, the Al–Al bond distance in aluminum metal has been reported as 2.348 Å. The first organometallic compound unambiguously shown to contain an Al–Al bond, tetrakis[bis(trimethylsilyl)methyl]dialane, $[(Me_3Si)_2HC]_2AlAl[CH(SiMe_3)_2]_2$, was reported in 1988.[55] This yellow crystalline compound was isolated from the potassium reduction of chloro-bis[bis(trimethylsilyl)methyl]aluminum (Equation (12)) (Figure 17):

$$2\ [(Me_3Si)_2HC]_2AlCl\ +\ 2\ K\ \longrightarrow\ \begin{array}{c}(Me_3Si)_2HC\\ \\(Me_3Si)_2HC\end{array}\!\!Al\!-\!Al\!\!\begin{array}{c}CH(SiMe_3)_2\\ \\CH(SiMe_3)_2\end{array} \qquad (12)$$

The Al–Al bond distance of 2.660(1) Å observed in $[(Me_3Si)_2HC]_2AlAl[CH(SiMe_3)_2]_2$ is a benchmark in organometallic chemistry as it stands as the first structural confirmation of a compound containing an Al–Al bond. The coordination of the aluminum atoms is also interesting as the core of the molecule is a planar C_2Al–AlC_2 core. It is interesting that the trigonal planar AlC_2 fragments are coplanar.

The "valence isomer of a dialane," $(\eta^5\text{-}C_5Me_5)Al$–$Al(C_6F_5)_3$, was prepared by treatment of $[Al(\eta^5\text{-}C_5Me_5)]_4$ with $Al(C_6F_5)_3$.[56] This compound is notable as it has an Al–Al bond wherein the two aluminum atoms reside in distinctly different coordination environments. Specifically, one aluminum atom $[(C_6F_5)_3Al\text{-}]$ is four-coordinate tetrahedral while the other one $[(\eta^5\text{-}C_5\,Me_5)Al\text{-}]$ is basically two-coordinate interacting in a η^5 fashion with the pentamethylcyclopentadienyl ligand. The Al Al ring centroid bond angle deviates from linearity at 170.1(3)°. The Al—Al bond distance was shown to be 2.591(3) Å.

Reactivity of $[(C_5Me_5)Al]_4$ has proven particularly interesting. Reaction of $[(C_5Me_5)Al]_4$ with $[Bu^tAs]_4$ gives a compound with a polyhedral As_2Al_3 framework, $As_2[(C_5Me_5)Al]_3$,[32] (along with 2-methylpropane and isobutene). This novel $As_2[(C_5Me_5)Al]_3$ compound was isolated as yellow

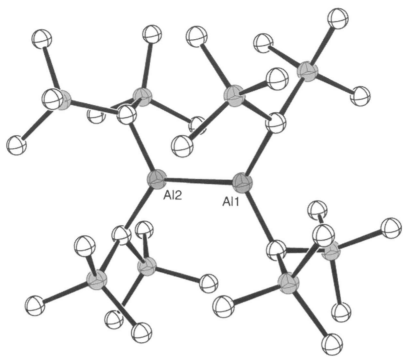

Figure 17 Solid-state structure of [(Me₃Si)₂HC]₂Al–Al[CH(SiMe₃)₂]₂.

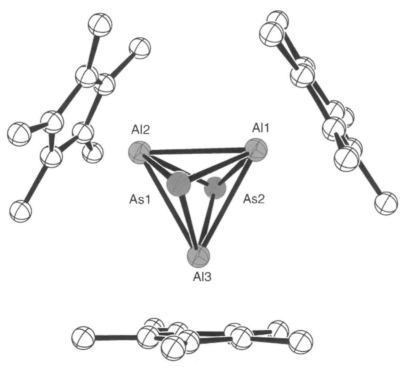

Figure 18 Solid-state structure of As₂[Cp*Al]₃.

crystals (Figure 18). While the short Al–As bond has been previously discussed herein, the Al₃ three-membered ring is noteworthy. The Al–Al bond distance in As₂[Cp*Al]₃ is 2.83 Å. This bond distance is slightly longer than those reported for R₂Al-AlR₂ (R = CH(SiMe₃)₂, 2.66 Å) and [(C₅Me₅)Al]₄ (2.77 Å). The authors suggest that there are only twelve electrons available for the nine bonds in the As₂Al₃ framework. Consequently, this results in an electron deficient situation. The bonding in the As₂Al₃ polyhedral, therefore, is suggested to be similar to that in the *closo*-boranes.

The gas-phase generation of aluminum(I) chloride, AlCl, in the presence of bis(pentamethyl-cyclopentadienyl)magnesium yields the tetramer $[(C_5Me_5)Al]_4$ (Equation (13)).[57]

$$4 \ (AlCl) \cdot (OEt_2)_x \ + \ 2 \ [Mg(C_5Me_5)_2]$$
$$\longrightarrow \ [(C_5Me_5)Al]_4 \ + \ 2 \ (MgCl_2 \cdot (Et_2O) \ + \ (4x - 4) \ Et_2O \tag{13}$$

This most novel compound contains an Al_4 tetrahedra core (each aluminum atom bonds to three other aluminum atoms) with pentamethylcyclopentadienyl ligands beyond the metallic center (Figure 19). The coordination of the aluminum atoms is technically tetrahedral as each aluminum atom bonds in a π-fashion to the pentamethylcyclopentadienyl ligand. The mean Al–Al bond distance in $[(C_5Me_5)Al]_4$ of 2.773(4) Å is expectedly longer than that observed for $[(Me_3Si)_2HC]_2AlAl[CH(SiMe_3)_2]_2$ (2.660(1) Å). The ^{27}Al-NMR spectrum (70.4 MHz, external standard $[Al(H_2O)_6]^{3+}$) of $[(C_5Me_5)Al]_4$ in benzene displayed a sharp signal at $\delta = -80.8$ ($\omega_{1/2} = 170$ Hz). This compound was also noteworthy in that it was the first molecular aluminum(I) compound stable under normal conditions (structurally characterized by single crystal X-ray diffraction). The relative weakness of the Al–Al bonds in $[(C_5Me_5)Al]_4$ was supported by quantum chemical calculations[58] and by the fact that monomeric $(C_5Me_5)Al$ units[59] could be obtained (both in solution and in the gas phase) by simply heating the $[(C_5Me_5)Al]_4$.

It should be noted that a second compound containing an Al_4-tetrahedra core was subsequently reported by the same research group.[60] In this study, reaction of $(AlI \cdot NEt_3)_4$ with donor-free Bu^t_3SiNa in toluene gives the tetramer $[(Bu^t_3Si)Al]_4$. The Al–Al bond distance in $[(Bu^t_3Si)Al]_4$ (2.604 Å) is shorter than the corresponding metal–metal distances reported for $[(C_5Me_5)Al]_4$ (0.17 Å shorter) and $[(Me_3Si)_2HC]_2AlAl[CH(SiMe_3)_2]_2$ (0.06 Å). Unlike the case for $[(C_5Me_5)Al]_4$ which yielded a very pronounced ^{27}Al-NMR signal, the $[(Bu^t_3Si)Al]_4$ tetramer did not readily yield an ^{27}Al-NMR spectrum due, in part, to a "different HOMO-LUMO gap" (as compared to $[(C_5Me_5)Al]_4$).

Even as we are often intrigued by compounds possessing short bonds, it is also important to examine the other extreme: those compounds with exceedingly long, in this case Al–Al, bonds. Reaction of AlX_3 (X = Cl or Br) with $Na[SiBu^t_3]$ yields $[Bu^t_3Si]_2AlAl[SiBu^t_3]_2$.[61] At a distance of 2.751(2) Å the central Si_2Al–$AlSi_2$ core of $[Bu^t_3Si]_2AlAl[SiBu^t_3]_2$, with as D_{2d} symmetry, has the longest Al–Al bond distance on record. In notable contrast, the next section will discuss compounds containing a measure of π-bonding.

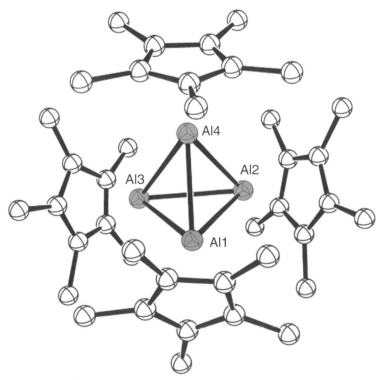

Figure 19   Solid-state structure of $[C_5Me_5Al]_4$.

In terms of cluster compounds containing more than four aluminum atoms, recent advances have proven quite encouraging. A novel aluminum cluster, $K_2[Bu^iAl]_{12}$, was obtained from the potassium metal reduction of diisobutylaluminum bromide.[62] This most unusual cluster has twelve aluminum atoms in a virtually perfect icosahedral geometry (Figure 20). This product was obtained in low yield as deeply red-colored crystals from a brown reaction mixture. The mean Al–Al bond distance of 2.660 Å in $K_2[Bu^iAl]_{12}$ is virtually identical to the Al–Al bond distance reported for the first organometallic alane, $[(Me_3Si)_2HC]_2AlAl[CH(SiMe_3)_2]_2$, 2.660(1) Å. This is somewhat surprising in that in the cluster there is more steric repulsion in such a cluster. Logic would suggest just the opposite: the small dialane dimer would have the shorter metal–metal interaction instead of the larger metallic cluster.

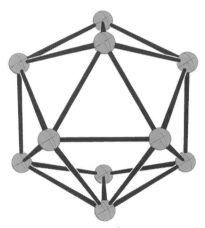

Figure 20 Solid-state structure of $K_2[Bu^iAl]_{12}$ core showing the Al_{12} cluster.

Other interesting aluminum cluster compounds have prominently utilized the Al(I) species. In particular, reaction of $LiN(SiMe_3)_2$ with a solution of Al(I) provided $Al_{77}R_{20}{}^{2-}$.[63] This compound remains the largest metalloid cluster yet structurally characterized. The authors viewed this cluster "as an intermediate on the way to aluminum metal." Schnöckel *et al.* subsequently reported that the $Al_{77}R_{20}{}^{2-}$ cluster is actually made up of smaller substituents including $Al_7R_6{}^-$ [64] and $Al_{12}R_8{}^-$ [65] Another interesting aluminum cluster, containing an Al_{14} core, results from a variation of the procedure established to prepare the $Al_{77}R_{20}{}^{2-}$ cluster.[66] The fact that these clusters contain more metal–metal bonds than metal–ligand bonds contributes to the authors employing the term "metalloid clusters" to distinguish them from traditional metallic clusters.

3.4.1.6.2 *Radical anions: A degree of multiple bonding in the Al–Al bond*

Soon after the experimental realization of compounds containing Al–Al bonds the concept of multiple bonding between two aluminum atoms began to gain attention. Beginning with the iconic compound of Uhl,[55] Pörschke *et al.*[67] allowed this compound to interact with lithium metal at $-30\,°C$, resulting in a black–violet solution. Crystallization of the product was achieved by the addition of TMEDA to complex the lithium ion leaving the $[(Me_3Si)_2HC]_2Al–Al[CH(SiMe_3)_2]_2{}^-$ radical anion at $0\,°C$. Most importantly, an X-ray crystal structure of the radical anion revealed 2.53 Å for the Al–Al bond (Figure 21). This represents a significant shortening of the Al–Al bond from the neutral alane (2.660(1) Å) distance. This is consistent with a measure of multiple bonding between the two metal atoms. The environment about the two aluminum atoms in the radical anion remain unchanged from that of the neutral species: three-coordinate trigonal planar.

3.4.2 GALLIUM

3.4.2.1 Introduction

In striking contrast to the ubiquitous nature of aluminum, gallium may legitimately be considered to be a rare element. Indeed, some of the so-called "rare earth metals" are more terrestrially abundant

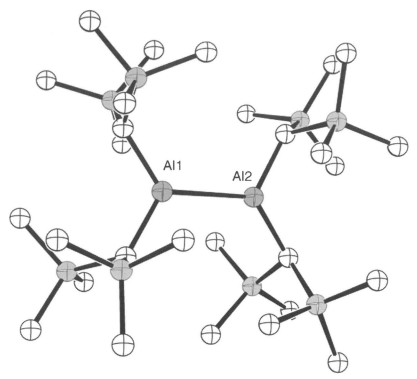

Figure 21   Solid-state structure of $[(Me_3Si)_2HC]_2Al–Al[CH(SiMe_3)_2]_2^{-}$.

than gallium. However, the history of gallium is just as interesting and engaging as that of aluminum. Paul-E'mile Lecoq de Boisbaudran is credited with discovering the element that would become known as gallium in 1875. He isolated little more than a single gram of this element from several hundred kilograms of the appropriate zinc blende ore. A particularly amusing historical anecdote concerns Dmitri Mendeleev and Lecoq de Boisbaudran. In his genius, Mendeleev had "predicted" the discovery of eka-aluminum, gallium, five years before Lecoq de Boisbaudran's actual discovery. Upon Lecoq de Boisbaudran's initial reporting of some of the physical properties of this new element Mendeleev wrote to him suggesting that he double check his value for the density of this new element as it was at odds with the value Mendeleev had predicted five years earlier. Upon closer examination of the density of gallium, Lecoq de Boisbaudran found that the experimental value for the density was indeed the value that Mendeleev had predicted.

There are significant differences between aluminum and gallium that directly affects the coordination chemistry exhibited by the two elements. One of the most intriguing points concern the atomic radius of gallium compared with that of aluminum. In striking contrast to the periodic trend of atomic radii increasing as one descends a given group, the atomic radius of gallium is observed to be slightly smaller (1.26 Å) than aluminum (1.48 Å). While size of the central atom is a prominent factor in coordination chemistry, it is difficult to quantitatively ascertain this effect relative to aluminum and gallium. Perhaps a more significant difference, as demonstrated by trimethylaluminum, is that aluminum often forms electron deficient bonding to obtain an octet of electrons. In notable contrast, gallium is perfectly at ease with only six electrons.

3.4.2.2   Group 14 Ligands

Similar to aluminum, the most important group 14 ligands for gallium are carbon based. The first organometallic compound of gallium, triethylgallium monoetherrate, $Et_3Ga \cdot OEt_2$, was reported in 1932 from reaction of ethylmagnesium bromide with gallium bromide in diethyl ether (Equation (14)):

$$3\ EtMgBr\ +\ GaBr_3\ \longrightarrow\ 3\ MgBr_2\ +\ Et_3Ga(Et_2O) \qquad (14)$$

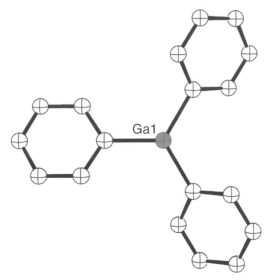

Figure 22 Solid-state structure of Ph$_3$Ga.

These workers also stated that the ether-free triethylgallium derivative, Et$_3$Ga, could be prepared by a redox reaction between gallium metal and diethylmercury. Trimethylgallium, like triethylgallium, is a monomer with the gallium atom residing in a trigonal planar environment. Indeed, the gallium atom in the simplest organometallic compound, trimethylgallium, Me$_3$Ga,[68] has been shown by gas phase electron diffraction to reside in a virtually idealized trigonal planar geometry.

A wealth of interesting chemistry concerns sterically demanding carbon-based ligands bonding to gallium. In this regard, the discussion must begin with the interactions of the phenyl, C$_6$H$_5$-, ligand with gallium even though this ligand is not normally considered to be sterically demanding. Triphenylgallium, Ph$_3$Ga,[69] is a convenient point of entry for this discussion. As supported by the solid-state crystal structure, the gallium atom in triphenylgallium is, on first glance, shown to reside in an unremarkable three-coordinate trigonal planar environment with Ga–C bond distances of 1.946(7) Å (Figure 22). A clue that the reality of the situation may be a bit more complicated is first hinted in the orientation of the phenyl rings. The three phenyl rings are observed to reside at dihedral angles of 0°, 13°, and 32° relative to the GaC$_3$ plane. Upon closer examination of the unit cell of this compound one observes that this arrangement of the phenyl rings allows for a significant secondary interaction of the gallium center with the *meta*-carbon atoms of other Ph$_3$Ga units. Thus—although not recognized or reported in the original article—the coordination of the gallium atom in Ph$_3$Ga may be best described as five-coordinate trigonal bipyramidal.

The synthesis and molecular structure of trimesitylgallium in 1986 marked the beginning of an exciting period in the organometallic chemistry of gallium. Trimesitylgallium was prepared by reaction of the Grignard reagent mesitylmagnesium bromide with gallium chloride (Equation (15)):[70]

$$3 \text{ MesMgBr } + \text{ GaCl}_3 \longrightarrow \text{ [Mes}_3\text{Ga]} \qquad (15)$$

The solid-state structure of this compound (Figure 23) reveals that the aromatic rings of Mes_3Ga are configured in a propeller arrangement at angles of 55.9° (relative to the GaC_3 basal plane). Indeed, the orientation of the mesityl groups provide substantial protection of the metal center rendering a virtually idealized trigonal planar geometry (C–Ga–C angle: 120°) about the deeply protected gallium center.

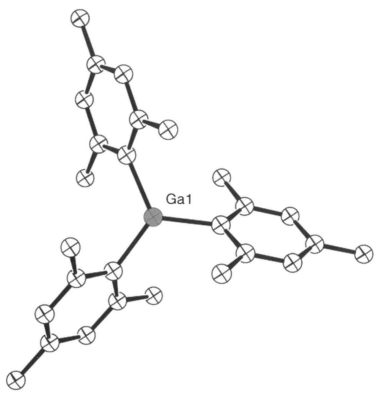

Figure 23   Solid-state structure of Mes_3Ga.

One of the most sterically demanding ligand systems used with gallium is the class of aryl-based *ortho*-substituted phenyl derivatives known as *m*-terphenyls.[71] Reaction of 2,6-dimesitylphenyl-lithium with gallium chloride forms bis(2,6-dimesitylphenyl)gallium chloride, $(Mes_2C_6H_3)_2GaCl$ (Figure 24).[72] Although the Ga–C bond length (1.956 Å and 2.000 Å) and Ga–Cl bond length (2.177(5) Å) were expectedly somewhat longer than normal, this compound was significant as this was the first example of a main group metal accommodating two such large sterically demanding ligands. Perhaps most significant, however, is the coordination about the gallium center. The steric bulk of the two ligands is such that the C–Ga–C bond angle has been significantly widened from 120° expected for trigonal planar (observed for Mes_3Ga) 153.5(5)°. Quite distinct from the trigonal planar coordination observed for gallium in Mes_3Ga, the gallium coordination in $(Mes_2C_6H_3)_2GaCl$ is T-shaped. Indeed, the 153.5(5)° C–Ga–C bond angle in $(Mes_2C_6H_3)_2GaCl$ is significantly greater than the corresponding C–Ga–C bond angle of 135.6(2)° for bis(2,4,6-tri-*tert*-butylphenyl)gallium chloride, $(Bu^t_3C_6H_2)_2GaCl$,[73] or the 134.3(3)° C–Ga–C bond angle for bis(diphenylphenyl)gallium iodide, $(C_6H_5)_2C_6H_3GaI$.[74] The significance of the T-shaped coordination for gallium lies in the fact that this generally obscure geometry is normally reserved for interhalogen compounds like ClF_3 and BrF_3. In such compounds the T-shaped geometry is predicated by the presence of two lone pairs of electrons in the equatorial plane on the central halogen atom. It is noteworthy, therefore, that the T-shaped geometry in $(Mes_2C_6H_3)_2GaCl$ results entirely from the interaction between the two sterically demanding ligands. It should be noted that the corresponding isostructural bis(2,6-dimesitylphenyl)gallium bromide, $(Mes_2C_6H_3)_2$-$GaBr$,[72] has been prepared and shown to be isostructural (C–Ga–C: 153.2°) with $(Mes_2C_6H_3)_2GaCl$.

While the organogallium chemistry of the cyclopentadienyl-based ligands (i.e., the pentamethyl derivative) will be discussed in detail later, it should be noted that a novel "ferrocenylgallane," $[(\eta^5\text{-}C_5H_5)Fe(\eta^5\text{-}C_5H_4)][Me_2Ga]_2[(\eta^5\text{-}C_5H_5)Fe(\eta^5\text{-}C_5H_4)]$, has been synthesized from reaction of

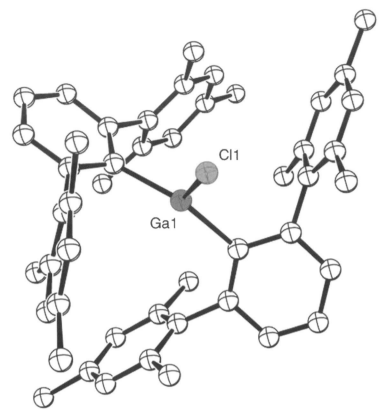

Figure 24 Solid-state structure of $(Mes_2C_6H_3)_2GaCl$.

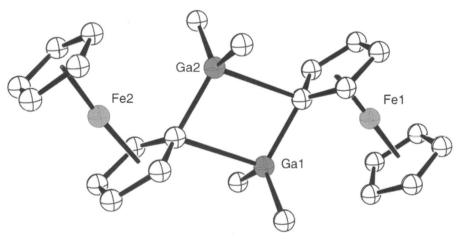

Figure 25 Solid-state structure of $[(\eta^5\text{-}C_5H_5)Fe(\eta^5\text{-}C_5H_4)][Me_2Ga]_2[(\eta^5\text{-}C_5H_5)Fe(\eta^5\text{-}C_5H_4)]$.

(chloromercurio)ferrocene, $[(\eta^5\text{-}C_5H_5)Fe(\eta^5\text{-}C_5H_4)HgCl]$, with trimethylgallium in toluene.[75] This ferrocenylgallane essentially consists of two ferrocene units bridged by two dimethylgallium units (Figure 25). The molecule resides about a center of symmetry located at the center of a planar, if asymmetric, Ga–C–Ga–C four-membered ring. While the ferrocenyl moieties are largely undistorted, the Ga–C bond distance of 2.587(5) Å was considered to be quite long. The coordination of the gallium atoms is distorted tetrahedral. However, upon closer examination, one can see that the coordination of the gallium centers may also be considered trigonal pyramidal wherein the trigonal plane consists of a Me_2Ga unit and one carbon atom of the ferrocenyl unit to give Me_2GaC_{Cp}. The fourth coordination site is completed by the axial approach of the carbon (Cp) approach of the other ferrocenyl unit.

3.4.2.3   Group 15 Ligands

3.4.2.3.1   *Nitrogen ligands*

The organogallium chemistry of nitrogen ligands is generally quite similar to that of aluminum with nitrogen ligands. Specifically, Lewis acid–Lewis base adducts are initially formed with primary amines. Further reaction leading to dimers or higher oligomers is driven by alkane elimination. In general, gallium is capable of all of the coordination modes displayed by aluminum earlier in this chapter. Thus, the coordination modes of gallium with various amines can range from three-coordinate trigonal planar to six-coordinate octahedral.

Although the coordination of gallium with nitrogen-based crown ethers, azacrown ethers, is not as well developed as that of aluminum, reports have demonstrated that gallium behaves in a fashion similar to that of its lighter congener with [14]aneN$_4$.[76]

3.4.2.3.2   *Phosphorus, arsenic, and antimony ligands*

An informative reaction in this regard involves that of trimethylgallium with the sterically demanding phosphine, tris(trimethylsilyl)phosphine (Me$_3$Si)$_3$P (Equation (16)):

$$Me_3Ga \ + \ P(SiMe_3)_3 \ \longrightarrow \ 1/2 \ [Ga_2P_2 \text{ ring structure}] \ + \ SiMe_4 \qquad (16)$$

This reaction, aided by evolution of tetramethylsilane, affords the organogallium dimer [Me$_2$Ga-P(SiMe$_3$)$_2$]$_2$.[77] The X-ray structure of this compound, while revealing the gallium atoms in four-coordinate tetrahedral environments, also highlights the planar Ga$_2$P$_2$(Ga–P: 2.456(1) Å; P–Ga–P: 88.0°; Ga–P–Ga: 90.0(1)) four-membered core of the molecule. Such "III–V" compounds were of interest as they often served as single-source molecular precursors to various materials. Indeed, the corresponding indium analog was shown to give indium phosphide upon pyrolysis.[78] Like nitrogen, the most common Ga–P structural motif is the Ga$_2$P$_2$ four-membered ring dimer. Nonetheless, Ga$_3$P$_3$ six-membered rings have also been reported. For example, reaction of trimethylgallium with diphenylphosphine results in [Me$_2$GaPPh$_3$]$_3$.[79] The Ga$_3$P$_3$ ring is in a chair conformation with Ga–P bond distances of 2.433(1) Å. The coordination of the gallium (and phosphorus) atoms is four-coordinate tetrahedral.

A striking gallium–phosphorus compound containing a P–P bond was isolated from reaction of the Lewis acid–Lewis base adduct Me$_3$GaPMe$_3$ with P(SiMe$_3$)$_3$, [P(SiMe$_3$)(Me$_2$Ga)$_2$]PP([Ga-Me$_2$)$_2$P(SiMe$_3$)$_2$].[80] The adduct, possessing C_{3v} symmetry was allowed to react with an excess of tris(trimethylsilyl)phosphine to give [P(SiMe$_3$)(Me$_2$Ga)$_2$]PP([GaMe$_2$)$_2$P(SiMe$_3$)$_2$]. While the coordination of the four gallium atoms in this complex is generally unremarkable as four-coordinate tetrahedral, the most noteworthy feature is the P–P bond of 2.25(3) Å. This complex represents a rare example of a phosphinogallane containing a P–P bond.

The organometallic coordination of gallium with arsenic ligands is quite similar to that of phosphorus. In particular, the predominant structural motif in gallium–arsenic compounds would be Ga–As dimers with a Ga$_2$As$_2$ four-membered ring core. The coordination of the arsenic and gallium atoms in such compounds would be tetrahedral. Typical examples of such compounds include [Me$_2$GaAs(Bu^t)$_2$]$_2$,[81] and [Ph$_2$GaAs(CH$_2$SiMe$_3$)$_2$]$_2$.[82] Occasionally, a Ga–As trimer with a Ga$_3$As$_3$ six-membered ring has been isolated. For example, although [Me$_2$GaAs(Pr^i)$_2$]$_3$ is a trimer (with the Ga atoms in four-coordinate tetrahedral environments) it is surprising that the ring was reported to have a distorted boat confirmation.[83]

The literature reveals a paucity of compounds containing the Ga–Sb bond. However, antimony seems to behave in a fashion similar to that of its lighter congeners. For example it can readily for Lewis acid–base adducts, (Bu^t)$_3$GaSb(Et$_3$).[84] Trimers such as [Me$_2$GaSb (SiMe$_3$)$_2$]$_3$ have also been reported.[85] The coordination of gallium in both of these compounds is unremarkable four-coordinate tetrahedral.

3.4.3.4 Group 16 Ligands

3.4.3.4.1 *Crown ethers*

The coordination chemistry of gallium with crown ethers is not developed to the same extent as that of aluminum. Indeed, the literature reveals only bis(trimethylgallium)(dibenzo-18-crown-6, $[GaMe_3]_2 \cdot$ dibenzo-18-crown-6,[86] the gallium analog of the previously reported aluminum complex, $[AlMe_3]_2 \cdot$ dibenzo-18-crown-6. The coordination of the gallium atoms in $[GaMe_3]_2 \cdot$ dibenzo-18-crown-6 is of course tetrahedral. Indeed, diaza-18-crown-6 has been shown to stabilize a gallium center in a five-coordinate trigonal bipyramidal environment.[87]

3.4.2.5 Group 17 Ligands

Of the gallium compounds concerning group 17 ligands, the gallium halides are may be considered the "work horses" of gallium chemistry as they are often the starting reagents. The gallium halides are differentiated from their aluminum analogs in their respective structures: the aluminum halides are dimeric with electron deficient Al–X–Al bridges (with the aluminum atoms being four-coordinate tetrahedral), while the gallium halides are monomeric, with the gallium atoms being three-coordinate trigonal planar. The first structurally characterized monomeric organogallium dihalides involved compounds of the type $sMesGaX_2$ (X = Cl, Br; sMes = supermesityl, $Bu^t_3C_6H_2$).[88] The coordination of the gallium atoms in $sMesGaX_2$ is three-coordinate trigonal planar. The monomeric nature of these compounds is particularly significant when one considers (Figure 26) that organogallium dihalides with considerably more sterically demanding ligands have been shown to be dimeric. In particular, even when the sterically demanding 2,6-dimesitylphenyl ligand is employed, the organogallium dichloride dimer (with μ-Cl bridges), $[(Mes_2C_6H_3)GaCl_2]_2$,[89] is isolated in the solid state. Thus, the monomeric nature of $sMesGaX_2$ is all the more remarkable considering the fact that dimers are found for much more sterically demanding ligands.

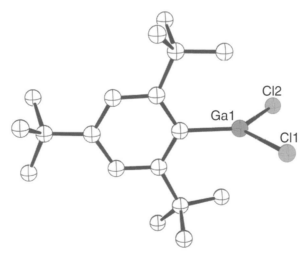

Figure 26 Solid-state structure of $sMesGaCl_2$.

3.4.2.5.1 *Two-coordinate gallium centers*

As should be evident at this point, depending upon the steric demands of the ligand, gallium is equally disposed to be three-coordinate trigonal planar or four-coordinate tetrahedral. With macrocyclic ligands such as crown ethers gallium can achieve five-coordinate square pyramidal or trigonal bipyramidal geometries. Only in the last few years have reports appeared describing gallium with novel two-coordinate motifs.

Reaction of solvent-free $Li\{(NDippCMe)_2CH\}$ (Dipp = $C_6H_3Pr^i_2$-2,6), "GaI," and potassium metal in toluene gave yellow crystals of $Ga\{(NDippCMe)_2CH\}$.[90] This striking compound features a two-coordinate gallium center in an extremely rare "V-shaped" (N–Ga–N: 87.53(5)°) structure (Figure 27). Moreover, the metal center was described as a six-electron gallium(I) center: electronically analogous to a singlet carbene carbon system. The authors suggested that the steric demands of this ligand are approximately similar to some of the sterically demanding *m*-terphenyl ligands. Equally amazing about this compound is the presence of a "lone pair" of electrons on the gallium center. This would suggest possibly significant Lewis base chemistry.

Another example of a two-coordinate gallium center is found in $[(Pr^i_3C_6H_2)_2C_6H_3]GaFe(CO)_4$ (Figure 28), isolated from reaction of $[(Pr^i_3C_6H_2)_2C_6H_3]GaCl_2$ with $Na_2[Fe(CO)_4]$.[91] Although this compound was described by the authors as a *ferrogallyne*, a compound containing an iron–gallium triple bond (*vide infra*), the issue at hand is that the gallium atom in $[(Pr^i_3C_6H_2)_2C_6H_3]GaFe(CO)_4$ is unambiguously two-coordinate with a C–Ga–Fe bond angle of 179.2(1)°. The Ga–Fe bond reported for $[(Pr^i_3C_6H_2)_2C_6H_3]GaFe(CO)_4$ of 2.2248(7) Å is among the shortest on record.

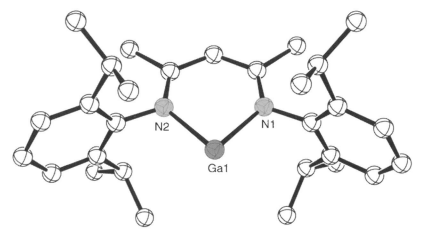

Figure 27   Solid-state structure of Ga carbene.

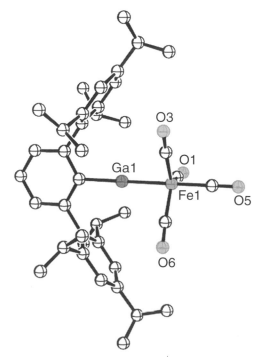

Figure 28   Solid-state structure of $[(Pr^i_3C_6H_2)_2C_6H_3]GaFe(CO)_4$.

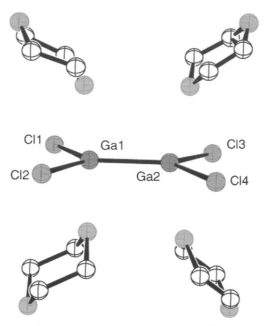

Figure 29 Solid-state structure of $Cl_2Ga-GaCl_2(dioxane)_4$.

3.4.2.6 Compounds Containing Ga–Ga Bonds

3.4.2.6.1 *Neutral compounds containing the Ga–Ga bond*

The first inorganic compound shown to contain a Ga–Ga bond is traced to the structure of bis[dibromo(1,4-dioxane)gallium], $Br_2GaGaBr_2(dioxane)_2$ (**14**). X-ray structural data on $Br_2GaGaBr_2(dioxane)_2$[92] confirmed the existence of a Ga–Ga bond in this solvent-stabilized species with the gallium atoms officially being in the (II) oxidation state (and not a "mixed" $Ga^{(I)}[Ga^{(III)}]$ system). The Ga–Ga bond in $Br_2GaGaBr_2(dioxane)_2$ was determined to be 2.395(6) Å. Similarly, the chloro derivative bis[dichloro(1,4-dioxane)gallium], $Cl_2GaGaCl_2(dioxane)_2$ (Ga–Ga: 2.406(1) Å; Ga–Cl: 2.406(1) Å; Ga–O: 2.021(5) Å), was reported to be isostructural with the bromine congener.[93] The coordination of $X_2GaGaX_2(dioxane)_2$ (X = Br, Cl) is four-coordinate tetrahedral in both cases.

(**14**)

Almost two decades after the reporting of the structure of $X_2GaGaX_2(dioxane)_2$ another modification of a dioxane-stabilized gallium(II) halide was reported. Room temperature (instead of 0 °C as in the original preparation) crystallization of Ga_2Cl_4 from a dioxane solution affords $Cl_2GaGaCl_2(dioxane)_4$ (Figure 29).[94] A number of issues are noteworthy concerning this compound. First of all, this compound is significant as it is a rare example of a dimeric compound containing a Ga–Ga bond wherein both gallium atoms are five-coordinate. For example, unlike the previous modification, in this case the coordination of both gallium atoms is five-coordinate. The coordination sphere of the gallium atoms in $Cl_2GaGaCl_2(dioxane)_4$ is completed by two chlorine atoms, two dioxane units, and a gallium atom. The coordination is virtually idealized trigonal bipyramidal (O–Ga–O: 179.10(10)°). While the Ga–O bond distance of 2.4087(19) Å in $Cl_2GaGaCl_2(dioxane)_4$ is considerably longer than that reported for $Cl_2GaGaCl_2$ (dioxane)$_2$ (2.021(5) Å), the Ga–Cl bond distance of 2.1721(7) Å in $Cl_2GaGaCl_2(dioxane)_4$ is substantially shorter than the $Cl_2GaGaCl_2(dioxane)_2$ value (2.406(1) Å). Perhaps the most significant difference between the bis(dioxane) and quadro(dioxane) gallium(II) chloride modifications is found in the

Ga–Ga bond distances: $Cl_2GaGaCl_2(dioxane)_2$, Ga–Ga: 2.406(1) Å; $Cl_2GaGaCl_2(dioxane)_4$, Ga–Ga: 2.3825(9) Å. It is most surprising that the Ga–Ga bond distance is shorter for the compound wherein the coordination number is higher. Logic would predict just the opposite!

The first organometallic compound containing a Ga–Ga bond, tetrakis[bis(trimethylsilyl) methyl]digallane, $[(Me_3Si)_2HC]_2GaGa[CH(SiMe_3)_2]_2$ (Figure 30), was reported in 1989.[95] This compound was prepared from reaction of the dioxane stabilized gallium(II) bromide, $Br_2Ga-GaBr_2(dioxane)_2$, with four equivalents of bis(trimethylsilyl)methyllithium, $LiCH(SiMe_3)_2$. The Ga–Ga bond distance in $[(Me_3Si)_2HC]_2GaGa[CH(SiMe_3)_2]_2$, isolated as yellow crystals from *n*-pentane, was determined to be 2.541(1) Å. This compound, like its aluminum analog, was shown to have a planar $C_2M–MC_2$ unit. However, the metal–metal bond distance in $[(Me_3Si)_2HC]_2Ga-Ga[CH(SiMe_3)_2]_2$ is 1.2 Å shorter than that observed for the aluminum analog. In addition, this compound exhibits a UV–vis absorption at 370 nm which was assigned to the Ga–Ga bond. It should be noted that even though the gallium(II) bromide bis(dioxane) starting compound contained a Ga–Ga bond (2.395(6) Å), it was conserved (and lengthened) in the organometallic compound.

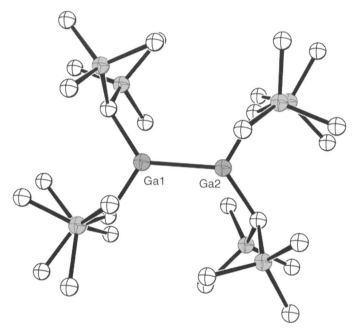

Figure 30   Solid-state structure of $[(Me_3Si)_2HC]_2Ga–Ga[CH(SiMe_3)_2]_2$.

The chemistry of molecules containing "gallium chains", strings of more than two gallium atoms, has not been extensively developed. To date, only a few such compounds have been reported. Reaction of the obscure "GaI" with phosphines resulted in a most unexpected product. This gallium subhalide was prepared by the ultrasonic irradiation of gallium metal and I_2. In the presence of triethylphosphine, "GaI" in toluene at $-78\,^\circ C$ results in $[Et_3P-GaI_2]_2Ga(I)PEt_3$.[29] The most striking point concerning this compound is the fact that it contains the first reported example of a "gallium chain" of three gallium atoms, -Ga–Ga–Ga-. It is noteworthy that the Ga–Ga bonds in this compound were shown to be reasonably short and asymmetric at distances of 2.451(1) Å and 2.460(1) Å. Moreover, this compound has mixed valences. Specifically, the center, bridging gallium atom was considered Ga(I) while the two terminal metal atoms were considered Ga(II). The Ga–Ga–Ga bond angle was shown to be 121.9(1)°.

A few years later another compound containing a "gallium chain" was reported. Interestingly, this case also involved phosphines. In this instance, reaction of $[(Pr^i_3C_6H_2)_2C_6H_3]_2GaCl_2$ with $P(SiMe_3)_3$ was shown to give the unusual organometallic compound $[(Pr^i_3C_6H_2)_2C_6H_3]Ga\{H_2P-Ga(H)PH_2\}Ga[C_6H_3(C_6H_2Pr^i_3)]$ (**15**).[96] Owing to the unusual nature of this compound characterization assumed added significance. To this end, this compound was characterized by multinuclear NMR, complete elemental analyses (C, H, Ga, and P), IR spectroscopy, and single crystal X-ray diffraction. The compound represented the first report of an organometallic compound containing a gallium chain, -Ga–Ga–Ga-. Surprisingly, yet consistent with the first gallium chain compound, $[Et_3PGaI_2]_2Ga(I)PEt_3$, the metallic chain is quite asymmetric with Ga–P

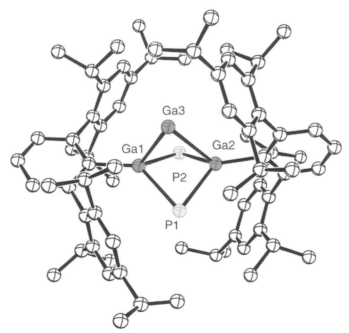

(15)

distances of 2.5145(13) Å and 2.7778(14) Å. The Ga–Ga–Ga bond angle in $[(Pr^i_3C_6H_2)_2C_6H_3]$-$Ga\{H_2PGa(H)PH_2\}Ga[C_6H_3(C_6H_2Pr^i_3)]$ is particularly acute at 69.68(4)°. This value is more than 50° less than the corresponding bond angle in $[Et_3PGaI_2]_2Ga(I)PEt_3$. Indeed, this compound may be considered to have a Ga_3P_2 core (Figure 31).

The cluster chemistry of gallium is a fertile, if still emerging, area of study. In most of the gallium clusters isolated sterically demanding ligands have been utilized. In a rather circuitous reaction involving the ultrasonication of gallium metal with iodine, both insoluble gallium subhalides and toluene-soluble "Ga_2I_3" were isolated. Addition of tris(trimethylsilyl)silyllithium·(THF)$_3$ to this complicated reaction yields an interesting ionic complex in which the anion contains a Ga_4Si trigonal bipyramidal core.[97] It was ambiguous whether there was Ga–Ga bonding in the equatorial plane.

Reaction of $Ga_2Br_4(dioxane)_2$ with a fourfold excess of $LiC(SiMe_3)_3$ results in another interesting gallium cluster: $[\{(Me_3Si)_3C\}Ga]_4$, a compound with a gallium tetrahedral core (**16**).[98] Each of the gallium atoms reside at the corners of an almost idealized pyramid. The mean Ga–Ga bond distance in the pyramid is 2.688 Å. This compound was reported to be thermally stable, decomposing only above 255 °C. Moreover, it was reported to be air-stable for months without significant decomposition. This compound is overall quite similar to the previously discussed Al_4 tetrahedral pyramid.

Figure 31 Solid-state structure of $[(Pr^i_3C_6H_2)_2C_6H_3]Ga\{H_2PGa(H)PH_2\}Ga[C_6H_3(C_6H_2Pr^i_3)]$.

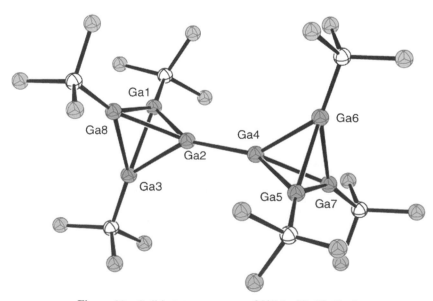

(16)

In 2001 a striking compound was prepared wherein two tetrahedra of gallium atoms are bridged by a single Ga–Ga bond. At the heart of this synthesis is the fabrication of gallium(I) bromide, GaBr. Reaction of trimethylsilyllithium (dissolved in toluene at $-78\,^{\circ}$C) with a GaBr solution was carried out. After workup a black residue was reported to remain. One of the products isolated from this residue was the neutral octagallane [{(Me$_3$Si)$_3$C}$_6$Ga$_8$].[99] The X-ray crystal structure of [{(Me$_3$Si)$_3$C}$_6$Ga$_8$] showing the Ga$_8$ core is shown in Figure 32. All angles within the triangular faces of the tetrahdra are virtually idealized 60°. It is surprising that the Ga–Ga bond distances in [{(Me$_3$Si)$_3$C}$_6$Ga$_8$] vary within a narrow range (2.605 Å to 2.648 Å). Perhaps even more surprising is the fact that the Ga–Ga bond distances in [{(Me$_3$Si)$_3$C}$_6$Ga$_8$] are significantly shorter than the corresponding distances reported for [{(Me$_3$Si)$_3$C}Ga]$_4$(Ga–Ga$_{mean}$: 2.688 Å). Indeed, the Ga–Ga bond connecting the two tetrahedra is 2.6143(11) Å. This was the first example of "two tetrahedral R$_3$M$_4$ units linked by a single metal–metal bond" for clusters containing one element.

A hexameric aggregate of (pentamethylcyclopentadienyl)gallium(I) was recently reported.[100] These workers grew a single crystal of this compound by "cooling a molten sample of the pure, freshly condensed material." The structure of the compound reveals a Ga$_6$ core inside a pentamethyl-cyclopentadienyl perimeter. The authors note that the Ga$_6$ unit "is not strictly octahedral but compressed along a C$_3$ axis to give two distinct Ga$_3$ units". While the C$_5$Me$_5^-$ ligands interact with the gallium atoms in an η_5 fashion, the authors argue that the "orientation of the C$_5$Me$_5^-$ ligands with respect to the M$_6$ core is consistent with a second order Jahn–Teller effect. It is important to note that other gallium clusters have been reported. For example, clusters containing nine, Ga$_9$(CMe$_3$)$_9$,[101] and twelve, [Ga$_{12}$(Flu)$_{10}$]$^{2-}$ (Flu = fluorenyl),[102] gallium atoms have recently been prepared and characterized.

Figure 32 Solid-state structure of [{(Me$_3$Si)$_3$C}$_6$Ga$_8$].

3.4.2.6.2 *Radical anions and multiple bond character*

The concept of a Ga–Ga bond with multiple bond character has only recently been brought to the fore. Perhaps the most compelling studies are those that provide a direct "gallane" to "gallene" comparison. The first gallane, $[(Me_3Si)_2HC]_2GaGa[CH(SiMe_3)_2]_2$ (Ga–Ga: 2.541(1) Å), was reduced with ethyllithium to give the radical anion $[(Me_3Si)_2HC]_2GaGa[CH(SiMe_3)_2]_2{}^-$ (Equation (17)):[103]

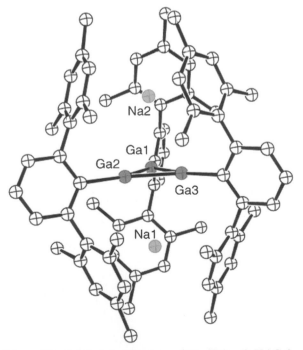

The Ga–Ga bond in the radical anion was determined to be 3.401(1) Å, a decrease of 0.14 Å from the neutral gallane. While the data strongly supports a measure of *p*-bonding among the gallium atoms, the coordination about the metal centers is unchanged from the neutral gallane (three-coordinate trigonal planar). Others have obtained similar results in different gallane to gallene comprisons.[104]

3.4.2.6.3 *Cyclogallenes and metalloaromaticity*

One of the more exciting developments in the coordination chemistry of gallium in the past few years has been the realization of metalloaromaticity. Metalloaromaticity, by definition, is traditional aromaticity exhibited by a metallic ring system rather than a carbon ring system. The first metalloaromatic compound was prepared by the sodium metal reduction of $(Mes_2C_6H_3)GaCl_2$ to give $Na_2[(Mes_2C_6H_3)Ga]_3$ (Figure 33).[105] As shown, the gallium atoms are three-coordinate in virtually idealized trigonal planar environments. The Ga–Ga–Ga bond angles within the ring are 60.01(1)°, while the Ga–Ga bond distance is 2.441(1) Å. The potassium-based cyclogallene, $K_2[(Mes_2C_6H_3)Ga]_3$, has also been reported.[106] In these compounds, the sodium atoms are not engaging in any meaningful metal–metal bonding with the gallium atoms (sodium–gallium approach: 3.1 Å). The sodium atoms appear to be assisted by subtle interactions with the π-cloud of the *m*-terphenyl ligands. Various computational quantum chemistry calculations, in addition to agreement with Schleyer's NICS (Nucleus Independent Chemical Shift),[107] have confirmed the metalloaromatic natures of these compounds.[108,109]

Figure 33 Solid-state structure of $Na_2[(Mes_2C_6H_3)Ga]_3$.

3.4.2.6.4 *Ga–Ga triple bonds*

Sodium metal reduction of $[(Pr^i_3C_6H_2)_2C_6H_3]GaCl_2$ does not result in a compound containing three-coordinate gallium atoms (like the cyclogallenes), rather, a most unexpected compound containing two-coordinate gallium atoms is isolated, $Na_2[\{(Pr^i_3C_6H_2)_2C_6H_3\}Ga\equiv Ga\{C_6H_3(C_6H_2Pr^i_3)_2\}]$ (see Figure 34).[110] The two-coordinate nature of each gallium atom simply consists of one sterically demanding ligand and the other gallium atom. Again, the sodium atoms do not appear to be engaging the gallium atoms. The Ga–Ga bond distance of 2.319(3) Å is noteworthy as being very short. Even though the bond angles about the two gallium atoms are decidedly nonlinear at angles of 128.5(4)° and 133.5(4)°, the authors referred to this compound as a *gallyne*—the first example of a gallium–gallium triple bond. While this description of the bonding was initially challenged,[111] the compelling nature of the compound is well documented.[112,113] Subsequent computational quantum chemistry calculations, including bond order analysis, provided a firm basis for the triple bond description.[114,115] Review articles have been published on the concept of triple bonding between two gallium atoms.[116–118]

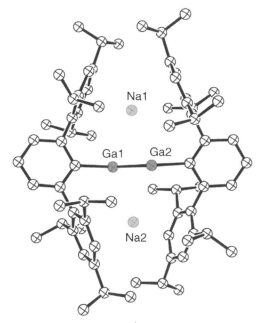

Figure 34   Solid-state structure of $Na_2[\{(Pr^i_3C_6H_2)_2C_6H_3\}Ga\equiv Ga\{C_6H_3(C_6H_2Pr^i_3)\}_2]$.

ACKNOWLEDGMENTS

The author is indebted to a number of gifted co-workers and students for their contributions over the years. Some of their names may be found in the list of references. Special thanks are extended to Jason K. Vohs for his expertise in generating many of the graphics for this chapter.

3.4.3 REFERENCES

1. Byram, S. K.; Fawcett, J. K.; Nyburg, S. C.; O'Brien, R. J. *Chem. Commun.* **1970**, 16–17.
2. Cotton, F. A. *Inorg. Chem.* **1970**, 9, 2804.
3. Vranka, R. G.; Amma, E. L. *J. Am. Chem. Soc.* **1967**, 89, 3121.
4. Malone, J. F.; McDonald, W. S. *J. Chem. Soc., Dalton Trans.* **1972**, 2646–2648.
5. Malone, J. F.; McDonald, W. S. *Chem. Commun.* **1967**, 444–445.
6. Malone, J. F.; McDonald, W. S. *J. Chem. Soc., Dalton Trans.* **1972**, 2649–2652.
7. Jerius, J. J.; Hahn, J. M.; Rahman, A. F. M. M.; Mols, O.; Ilsley, W. H.; Oliver, J. P. *Organometallics* **1986**, 5, 1812–1814.
8. Mel, V. S. J. D.; Oliver, J. P. *Organometallics* **1989**, 8, 827–830.
9. Li, X.-W.; Su, J.; Robinson, G. H. *Chem. Commun.* **1996**, 2683–2684.
10. Tecle, B.; Corfield, P. W. R.; Oliver, J. P. *Inorg. Chem.* **1982**, 21, 458.
11. Fisher, J. D.; Wei, M.-Y.; Willett, R.; Shapiro, P. J. *Organometallics* **1994**, 13, 3324–3329.

12. Schonberg, P. R.; Paine, R. T.; Campana, C. F.; Duesler, E. N. *Organometallics* **1982**, *1*, 799.
13. Dohmeier, C.; Schnockel, H.; Robl, C.; Schneider, U.; Ahlrichs, R. *Angew. Chem., Int. Ed. Engl.* **1993**, *32*, 1655.
14. Wiberg, W. *F.I.A.T. Review of German Science; Inorganic Chemistry Part II* **1939-45**, 159.
15. Davidson, N.; Brown, H. C. *J. Am. Chem. Soc.* **1942**, *64*, 316.
16. Laubengayer, A. W.; Smith, J. D.; Ehrlich, G. G. *J. Am. Chem. Soc.* **1961**, *83*, 542.
17. McLaughlin, G. M.; Sim, G. A.; Smith, J. D. *J. Chem. Soc., Dalton Trans.* **1972**, 2197.
18. Alford, J. K.; Gosling, A. K.; Smith, J. D. *J. Chem. Soc., Dalton Trans.* **1972**, 2203.
19. Hitchcock, P. B.; Smith, J. D.; Thomas, K. M. *J. Chem. Soc., Dalton Trans.* **1976**, 1433.
20. Amirkhalili, S.; Hitchcock, P. B.; Smith, J. D. *J. Chem. Soc., Dalton Trans.* **1979**, 1206.
21. Waggoner, K. M.; Hope, H.; Power, P. P. *Angew. Chem., Int. Ed. Engl.* **1988**, *27*, 1699–1700.
22. Waggoner, K. M.; Power, P. P. *J. Am. Chem. Soc.* **1991**, *113*, 3385.
23. Piero, G. D.; Cesari, M.; Dozzi, G.; Mazzei, A. *J. Organomet. Chem.* **1977**, *129*, 281.
24. Al-Wassil, A.-A.; Hitchcock, P. B.; Sarisaban, S.; Smith, J. D.; Wilson, C. L. *J. Chem. Soc., Dalton Trans.* **1985**, 1929.
25. Robinson, G. H.; Sangokoya, S. A. *J. Am. Chem. Soc.* **1987**, *109*, 6852–6853.
26. Healey, M. D.; Barron, A. R. *J. Am. Chem. Soc.* **1989**, *111*, 398–399.
27. Robinson, G. H.; Rae, A. D.; Campana, C. F.; Byram, S. K. *Organometallics* **1987**, *6*, 1227–1230.
28. Robinson, G. H.; Self, M. F.; Sangokoya, S. A.; Pennington, W. T. *J. Am. Chem. Soc.* **1989**, *111*, 1520–1522.
29. Schnepf, A.; Doriat, C.; Möllhausen, E.; Schnöckel, H. *Chem. Commun.* **1997**, 2111–2112.
30. Wells, R. L.; McPhail, A. T.; Self, M. F.; Laske, J. A. *Organometallics* **1993**, *12*, 3333.
31. Cooke, J. A. L.; Wells, R. L.; White, P. S. *Organometallics* **1995**, *14*, 3562.
32. Hänisch, C. K. F. v.; Üffing, C.; Junker, M. A.; Ecker, A.; Kneisel, B. O.; Schöckel, H. *Angew. Chem., Int. Ed. Engl.* **1996**, *35*, 2875–2877.
33. Cooke, J. A. L.; Purdy, A. P.; Wells, R. L. *Organometallics* **1996**, *15*, 84–90.
34. Hrncir, D. C.; Rogers, R. D.; Atwood, J. L. *J. Am. Chem. Soc.* **1981**, *103*, 4277–4278.
35. Atwood, J. L.; Newberry, W. R. *J. Organomet. Chem.* **1974**, *65*, 145.
36. Atwood, J. L.; Stucky, G. D. *J. Am. Chem. Soc.* **1967**, *89*, 5361.
37. Pedersen, C. J. *J. Am. Chem. Soc.* **1967**, *89*, 2495.
38. Pedersen, C. J. *J. Am. Chem. Soc.* **1967**, *89*, 7017.
39. Robinson, G. H.; Bott, S. G.; Elgamal, H.; Hunter, W. E.; Atwood, J. L. *Journal of Inclusion Phenomena* **1985**, *3*, 65–89.
40. Atwood, J. L.; Elgamal, H.; Robinson, G. H.; Bott, S. G.; Weeks, J. A.; Hunter, W. E. *J. Incl. Phenom.* **1984**, *2*, 367–376.
41. Bott, S. G.; Elgamal, H.; Atwood, J. L. *J. Am. Chem. Soc.* **1985**, *107*, 1796–1797.
42. Atwood, J. L.; Hrncir, D. C.; Shakir, R.; Dalton, M. S.; Priester, R. D.; Rogers, R. D. *Organometallics* **1982**, *1*, 1021–1025.
43. Self, M. F.; Pennington, W. T.; Laske, J. A.; Robinson, G. H. *Organometallics* **1991**, *10*, 36–38.
44. Bauerand, D. J.; Stucky, G. D. *J. Am. Chem. Soc.* **1969**, *91*, 5462.
45. Haaland, A.; Stoikkeland, O.; Weidlein, J. *J. Organomet. Chem.* **1975**, *94*, 353.
46. Shakir, R.; Zaworotko, M. J.; Atwood, J. L. *J. Organomet. Chem.* **1979**, *171*, 9.
47. Robinson, G. H.; Zhang, H.; Atwood, J. L. *Organometallics* **1987**, *6*, 887–889.
48. Robinson, G. H.; Sangokoya, S. A. *J. Am. Chem. Soc.* **1988**, *110*, 1494–1497.
49. Atwood, J. L.; Hrncir, D. C.; Rogers, R. D.; Howard, J. A. K. *J. Am. Chem. Soc.* **1981**, *103*, 6787–6788.
50. Schram, E. P. *Inorg. Chem.* **1966**, *5*, 1291–1294.
51. Schram, E. P.; Hall, R. E.; Glore, J. D. *J. Am. Chem. Soc.* **1969**, *91*, 6643.
52. Miller, M. A.; Schram, E. P. *Organometallics* **1985**, *4*, 1362–1364.
53. Hoberg, H.; Krause, S. *Angew. Chem., Int. Ed. Engl.* **1976**, *15*, 694.
54. Hoberg, H.; Krause, S. *Angew. Chem., Int. Ed. Engl.* **1978**, *17*, 949–950.
55. Uhl, W. *Z. Naturforsch.* **1988**, *43b*, 1113–1118.
56. Gorden, J. D.; Macdonald, C. L. B.; Cowley, A. H. *Chem. Commun.* **2001**, 75–76.
57. Dohmeier, C.; Robl, C.; Tacke, M.; Schnöckel, H. *Angew. Chem., Int. Ed. Engl.* **1991**, *30*, 564–565.
58. Gauss, J.; Schneider, U.; Ahlrichs, R.; Dohmeier, C.; Schnöckel, H. *J. Am. Chem. Soc.* **1993**, *115*, 2402.
59. Haaland, A.; Martinsen, K.-G.; Shlykov, S. A.; Volden, H. V.; Dohmeier, C.; Schnöckel, H. *Organometallics* **1995**, *14*, 3116.
60. Purath, A.; Dohmeier, C.; Ecker, A.; Schnöckel, H.; Amelunxen, K.; Passler, T.; Wiberg, N. *Organometallics* **1998**, *17*, 1894–1896.
61. Wiberg, N.; Amelunxen, K.; Blank, T.; Nöth, H.; Knizek, J. *Organometallics* **1998**, *17*, 5431–5433.
62. Hiller, W.; Klinkhammer, K. W.; Uhl, W.; Wagner, J. *Angew. Chem., Int. Ed. Engl.* **1991**, *30*, 179–180.
63. Ecker, A.; Weckert, E.; Schnöckel, H. *Nature* **1997**, *387*, 379.
64. Purath, A.; Köppe, R.; Schnöckel, H. *Angew. Chem., Int. Ed. Engl.* **1999**, *38*, 2969.
65. Purath, A.; Schnöckel, H. *Chem. Commun.* **1999**, 1933.
66. Köhnlein, H.; Stösser, G.; Baum, E.; Möllhausen, E.; Huniar, U.; Schnöckel, H. *Angew. Chem. Int., Ed. Engl.* **2000**, *39*, 799–801.
67. Pluta, C.; Pörschke, K.-R.; Kruger, C.; Hildenbrand, K. *Angew. Chem., Int. Ed. Engl.* **1993**, *32*, 388–390.
68. Beagley, B.; Schmidling, D. G.; Steer, I. A. *J. Mol. Struct.* **1974**, *21*, 437.
69. Malone, J. F.; McDonald, W. S. *J. Chem. Soc. (A)* **1970**, 3362–3367.
70. Beachley, O. T.; Churchill, M. R.; Pazik, J. C.; Ziller, J. W. *Organometallics* **1986**, *5*, 1814–1817.
71. Du, C.-J. F.; Hart, H.; Ng, K.-K. *J. Org. Chem.* **1986**, *51*, 3162–3165.
72. Li, X.-W.; Pennington, W. T.; Robinson, G. H. *Organometallics* **1995**, *14*, 2109–2111.
73. Meller, A.; Pusch, S.; Pohl, E.; Häming, L.; Herbst-Irmer, R. *Chem. Ber.* **1993**, *126*, 2255–2257.
74. Crittendon, R. C.; Beck, B. C.; Su, J.; Li, X.-W.; Robinson, G. H. *Organometallics* **1999**, *18*, 156–160.
75. Lee, B.; Pennington, W. T.; Laske, J. A.; Robinson, G. H. *Organometallics* **1990**, *9*, 2864–2865.
76. Lee, B.; Pennington, W. T.; Robinson, G. H.; Rogers, R. D. *J. Organomet. Chem.* **1990**, *396*, 269.
77. Dillingham, M. D. B.; Burns, J. A.; Byers-Hill, J.; Gripper, K. D.; Pennington, W. T.; Robinson, G. H. *Inorg. Chim. Acta* **1994**, *216*, 267–269.

78. Stuczynski, S. M.; Opila, R. L.; Marsh, P.; Brennan, J. G.; Steigerwald, M. L. *Chem. Mater.* **1991**, *3*, 379.
79. Robinson, G. H.; Burns, J. A.; Pennington, W. T. *Main Group Chem.* **1995**, *1*, 153–158.
80. Burns, J. A.; Dillingham, M. D. B.; Hill, J. B.; Gripper, K. D.; Pennington, W. T.; Robinson, G. H. *Organometallics* **1994**, *13*, 1514–1517.
81. Arif, A. M.; Benac, B. L.; Cowley, A. H.; Geerts, R.; Jones, R. A.; Kidd, K. B.; Power, J. M.; Schwab, S. T. *Chem. Commun.* **1986**, 1543.
82. Wells, R. L.; Purdy, A. P.; McPhail, A. T.; Pitt, C. G. *J. Organomet. Chem.* **1986**, *308*, 281.
83. Cowley, A. H.; Jones, R. A.; Mardones, M. A.; Nunn, C. M. *Organometallics* **1991**, *10*, 1635.
84. Schulz, S.; Nieger, M. *J. Chem. Soc., Dalton Trans.* **2000**, 639–642.
85. Schulz, S.; Nieger, M. *J. Organomet. Chem.* **1998**, *570*, 275.
86. Robinson, G. H.; Hunter, W. E.; Bott, S. G.; Atwood, J. L. *J. Organomet. Chem.* **1987**, *326*, 9–16.
87. Lee, B.; Pennington, W. T.; Robinson, G. H. *Organometallics* **1990**, *9*, 1709–1711.
88. Schulz, S.; Pusch, S.; Pohl, E.; Dielkus, S.; Herbst-Irmer, R.; Meller, A.; Roesky, H. W. *Inorg. Chem.* **1993**, *32*, 3343–3346.
89. Crittendon, R. C.; Li, X.-W.; Su, J.; Robinson, G. H. *Organometallics* **1997**, *16*, 2443–2447.
90. Hartman, N. J.; Eichler, B. E.; Power, P. P. *Chem. Commun.* **2000**, 1991–1992.
91. Su, J.; Li, X.-W.; Crittendon, R. C.; Campana, C. F.; Robinson, G. H. *Organometallics* **1997**, *16*, 4511–4513.
92. Small, R. W. H.; Worrall, I. J. *Acta Cryst. Sec. B* **1982**, *38*, 250–251.
93. Beamish, J. C.; Small, R. W. H.; Worrall, I. J. *Inorg. Chem.* **1979**, *18*, 220.
94. Wei, P.; Li, X.-W.; Robinson, G. H. *Chem. Commun.* **1999**, 1287–1288.
95. Uhl, W.; Layh, M.; Hildenbrand, T. *J. Organomet. Chem.* **1989**, *364*, 289–300.
96. Li, X.-W.; Wei, P.; Beck, B. C.; Xie, Y.; Schaefer, H. F.; Su, J.; Robinson, G. H. *Chem. Commun.* **2000**, 453–454.
97. Linti, G.; Köster, W.; Piotrowski, H.; Rodig, A. *Angew. Chem., Int. Ed. Engl.* **1998**, *37*, 2209–2211.
98. Uhl, W.; Hiller, W.; Layh, M.; Schwarz, W. *Angew. Chem., Int. Ed. Engl.* **1992**, *31*, 1364–1366.
99. Schnepf, A.; Köppe, R.; Schnöckel, H. *Angew. Chem., Int. Ed. Engl.* **2001**, *40*, 1241–1243.
100. Loos, D.; Baum, E.; Ecker, A.; Schnöckel, H.; Down, A. J. *Angew. Chem., Int. Ed. Engl.* **1997**, *36*, 860–862.
101. Uhl, W.; Cuypers, L.; Harms, L.; Kaim, W.; Wanner, M.; Winter, R.; Lich, R.; Saak, W. *Angew. Chem., Int. Ed. Engl.* **2001**, *40*, 566–568.
102. Schnepf, A.; Stößer, G.; Köppe, R.; Schnöckel, H. *Angew. Chem., Int. Ed. Engl.* **2000**, *39*, 1637–1639.
103. Uhl, W.; Schütz, W.; Kaim, W.; Waldhör, E. *J. Organomet. Chem.* **1995**, *501*, 79–85.
104. He, X.; Barlett, R. A.; Olmstead, M. M.; Ruhlandt-Senge, K.; Sturgeon, B. E.; Power, P. P. *Angew. Chem., Int. Ed. Engl.* **1993**, *32*, 717–719.
105. Li, X.-W.; Pennington, W. T.; Robinson, G. H. *J. Am. Chem. Soc.* **1995**, *117*, 7578–7579.
106. Li, X.-W.; Xie, Y.; Schreiner, P. R.; Gripper, K. D.; Crittendon, R. C.; Campana, C. F.; Schaefer, H. F.; Robinson, G. H. *Organometallics* **1996**, *15*, 3798–3803.
107. Schleyer, P. v. R.; Maerker, C.; Dransfeld, A.; Jiao, H.; Hommes, N. J. R. v. E. *J. Am. Chem. Soc.* **1996**, *118*, 6317–6318.
108. Xie, Y.; Schreiner, P. R.; Schaefer, H. F.; Li, X.-W.; Robinson, G. H. *J. Am. Chem. Soc.* **1996**, *118*, 10635–10639.
109. Xie, Y.; Schreiner, P. R.; Schaefer, H. F.; Li, X.-W.; Robinson, G. H. *Organometallics* **1998**, *17*, 114–122.
110. Su, J.; Li, X.-W.; Crittendon, R. C.; Robinson, G. H. *J. Am. Chem. Soc.* **1997**, *119*, 5471–5472.
111. Cotton, F. A.; Cowley, A. H.; Feng, X. *J. Am. Chem. Soc.* **1998**, *120*, 1795–1799.
112. Dagani, R. *Chem. Eng. News* **1997**, *75(June.16)*, 9–10.
113. Dagani, R. *Chem. Eng. News* **1998**, *76(March 16)*, 31–35.
114. Xie, Y.; Grev, R. S.; Gu, J.; Schaefer, H. F.; Schleyer, P. v. R.; Su, J.; Li, X.-W.; Robinson, G. H. *J. Am. Chem. Soc.* **1998**, *120*, 3773–3780.
115. Xie, Y.; Schaefer, H. F.; Robinson, G. H. *Chem. Phys. Letts.* **2000**, *317*, 174–180.
116. Robinson, G. H. *Acc. Chem. Res.* **1999**, *32*, 773–782.
117. Robinson, G. H. *Chem. Comm.* **2000**, 2175–2181.
118. Robinson, G. H. *Adv. Organomet. Chem.* **2001**, *47*, 283–294.

Comprehensive Coordination Chemistry II
ISBN (set): 0-08-0437486

Volume 3, (ISBN 0-08-0443257); pp 347–382

3.5
Indium and Thallium

H. V. RASIKA DIAS
The University of Texas at Arlington, USA

3.5.1 INDIUM

3.5.1.1 Introduction

The metallic element indium is the second heaviest member of the group 13 family. Indium has the electronic configuration of $[Kr]4d^{10}5s^25p^1$, and forms compounds in the oxidation states I, II, and III. Coordination compounds with indium in the trivalent state are the most common. In this chapter, primarily the developments in indium coordination chemistry since the early 1980s will be surveyed. *Comprehensive Coordination Chemistry-I* (*CCC*, 1987) is an excellent reference source for pre-1980 work.[1] In general, organometallic compounds are outside the scope of this chapter. Still, there is a huge body of literature that covers various aspects of indium coordination chemistry. Fortunately, there are several treatises pertinent to the coordination, organometallic, and general chemistry of indium.[1–21] For categories where there is a large amount of more recent work, and for early background material, the reader will be directed to some of these sources for more detailed coverage of the topic.

3.5.1.2 Indium (III)

3.5.1.2.1 Group 14 ligands

(i) Carbon ligands

The vast majority of molecules that belong to this category are organoindium compounds. Review articles on the chemistry of such compounds are available.[11–14] Although the carbon monoxide complexes of indium are unknown, the isoelectronic cyanide ligand forms thermally stable adducts with indium(III). During the attempted synthesis of indium oxycyanide by the action of cyanogen on indium oxyiodide, the monoclinic form of $In(CN)_3$ was obtained in low yield as a by-product.[22] Recently, a new form of indium(III) cyanide has been prepared in excellent yield by a low-temperature solution method, using $InCl_3$ and Me_3SiCN as starting materials.[23] X-ray crystallographic data show that $In(CN)_3$ has a cubic structure with an octahedrally coordinated indium atom surrounded by an average of three carbon and three nitrogen atoms. This material readily, and reversibly, incorporates Kr gas into the empty cavities to form $In(CN)_3 \cdot Kr$.[23]

(ii) Silicon, germanium, tin, and lead ligands

A few silyl complexes of indium(III) are known. The homoleptic trimethylsilyl derivative $In(SiMe_3)_3$ was reported in 1969.[1] It is a highly thermally, light- and oxygen-sensitive compound. Compounds with higher thermal stability have been obtained using sterically more demanding silyl ligands. For example, $\{(Me_3Si)_3Si\}_2In(\mu\text{-}Cl)_2Li(THF)_2$ (**1**) has been prepared by treating $InCl_3$ with $\{(Me_3Si)_3Si\}Li(THF)_3$.[24] It features a tetrahedral indium center with an unusually large Si—In—Si bond angle (139.9(2)°). The synthesis of $(t\text{-}Bu_2PhSi)_3In$ (**2**),[25] $\{(Me_5C_5)_2MeSi\}_2InMe$,[26] and the silyl indium halides $(t\text{-}Bu_3Si)_nInX_{3-n}$ (X = halide, n = 1, 2)[27–29] and $t\text{-}Bu_2PhSiInCl_2$[25] have also been reported. The synthesis of $(t\text{-}Bu_2PhSi)_3In$ and $(t\text{-}Bu_3Si)_3In$ involves a metathesis process between indium(III) halides and the sodium salt of the corresponding silyl ligand. The $\{(Me_5C_5)_2Si(Me)\}_2InMe$[26] has been obtained in high yield by treating the silylene $(Me_5C_5)_2Si$ with $InMe_3$. Some of the indium halide derivatives form adducts with oxygen- and nitrogen-containing donors.[25,27,28] For example, the dichlorides react with THF to form $t\text{-}Bu_2PhSiInCl_2(THF)$ and $t\text{-}Bu_3SiInCl_2(THF)$. The monochloride compound $(t\text{-}Bu_3Si)_2InCl$ reacts with $AlCl_3$ to give an ionic indium species $[(t\text{-}Bu_3Si)_2In][AlCl_4]$.

(1)

(2)

Compounds with In—Ge (e.g., $(Et_3Ge)_3In$)[1] and In—Sn bonds are rare. A series of stannyl compounds of the type $Ph_3SnInX_2(TMEDA)$ with apparently five-coordinate indium centers have been obtained from the reaction between InX in toluene/TMEDA (X = Cl, Br, I) and Ph_3SnX.[30] The treatment of $Ph_3SnInCl_2(TMEDA)$ with Et_4NCl leads to $[Et_4N][Ph_3SnInCl_3]$. The compound $L_2InSnPh_3$ (L = 2-[(dimethylamino)methyl]phenyl) can be synthesized using the chloro derivative L_2InCl and the sodium salt of $SnPh_3^-$.[31] There are no reports of coordination compounds with In—Pb bonds.

3.5.1.2.2 *Group 15 ligands*

(i) *Nitrogen ligands*

(a) *Neutral monodentate nitrogen ligands.* Neutral nitrogen donors form a variety of adducts with indium in the trivalent state. Indium salts, in particular those with weakly coordinating anions such as BF_4^-, NO_3^-, or ClO_4^-, form cationic species like $[In(en)_3]^{3+}$, $[In(py)_6]^{3+}$, $[In(bipy)_3]^{3+}$, and $[In(phen)_3]^{3+}$.[1,3,4] The formation of adducts containing acetonitrile donors, e.g., $[In(NCMe)_6](BF_4)_3$, is also established.[32] The cation $[In(NH_3)_6]^{3+}$ is present in liquid ammonia.[1] However, X-ray crystal structural data are not available. An ammonia adduct $InF_2(NH_2)(NH_3)$ has been prepared by reacting ammonium fluoride and indium nitride in supercritical ammonia.[33] The solid-state structure consists of octahedral indium moieties linked by fluoride and amide ligands. In addition, each indium atom is coordinated to one terminal F and one terminal NH_3 molecule.

(b) *Azide, NCO, and NCS ligands.* Indium nitride is an important semiconductor material.[34] Relatively milder routes (ideally below 600 °C) are preferred for the generation of indium nitride, due to its low thermal stability. Thus there is a constant need for new precursor material that generates InN under low-temperature conditions. One impetus for studying indium complexes of nitrogen-ligand compounds such as azido and amido derivatives is their potential utility in InN-related applications.

The isolation of several indium(III) adducts containing azide groups has been reported. These include Cl_2InN_3, Br_2InN_3, $Cl_2InN_3(py)_2$, $Cl_2InN_3(THF)_2$, $[(py)_2Na][(py)_2In(N_3)_4]$, $(py)_3In(N_3)_3$, $(2,2',2''-terpyridine)In(N_3)_3$, and $(2,2',2''-terpyridine)In(N_3)_2(O_2C(CH_2)_2CH_2OH)$.[35–39] Syntheses of $[(py)_2Na][(py)_2In(N_3)_4]$, $(py)_3In(N_3)_3$, $(2,2',2''-terpyridine)In(N_3)_3$, and $(2,2',2''-terpyridine)In-(N_3)_2(O_2C(CH_2)_2CH_2OH)$ involve the use of $InCl_3$ and sodium azide in the initial step.[36,38,39] Haloindium azides Cl_2InN_3, Br_2InN_3, $Cl_2InN_3(py)_2$, and $Cl_2InN_3(THF)_2$ have been synthesized, starting with the appropriate indium(III) halide and Me_3SiN_3.[35] They are reported to have relatively high thermal stability. $In(N_3)_3$, in contrast, is an explosive solid; Lewis-base adducts like $(py)_3In(N_3)_3$ and $(2,2',2''-terpyridine)In(N_3)_3$ are relatively less dangerous.

The pyridine adduct $(py)_3In(N_3)_3$ (3) is monomeric in the solid state. The indium atom adopts *mer*-octahedral geometry. In pyridine, the IR absorption bands corresponding to azide stretch appear at 2,084, 2,068, and 2,055 cm^{-1}. $Cl_2InN_3(THF)_2$ (4) forms dimers in the solid state, with a planar In_2N_2 core. Azido groups occupy the bridging sites. The X-ray crystal structures of $[(py)_2Na][(py)_2In(N_3)_4]$ and $(2,2',2''-terpyridine)In(N_3)_2(O_2C(CH_2)_2CH_2OH)$ have also been reported.

(3)

(4)

A few cyanate complexes of indium are reported. These include In(NCO)$_3$(py)$_3$, In(NCO)$_3$(DMSO)$_3$, and anionic species like [In(NCO)$_3$]$^-$.[40,41] Indium(III) adducts containing different ligand combinations of cyanate, fluoride, and water have been investigated using NMR spectroscopy.[42] The thiocyanate (or more correctly, isothiocyanate, considering the common mode of bonding with indium(III)) derivatives are relatively more common. ^{115}In NMR spectroscopy was used to study the reactions of indium(III) halides with halide and pseudohalide ions, and to observe NCS$^-$ and NO$_2^-$ complexes of indium(III).[43] The detection of N-bonded [In(NCS)$_6$]$^{3-}$ and [In(NO$_2$)$_6$]$^{3-}$, and the unique four- to six-coordination equilibrium, were observed between these and the tetracoordinated anions. The X-ray crystal structure of [Bu$_4$N]$_3$[In(NCS)$_6$] reveals that the six isothiocyanate ligands coordinate to indium octahedrally through the nitrogen atoms.[44] A calorimetric study of the coordination behavior of isothiocyanate ions in DMF has indicated the formation of [InNCS(DMF)$_5$]$^{2+}$, [In(NCS)$_2$(DMF)$_4$]$^+$, [In(NCS)$_3$(DMF)$_3$], [In(NCS)$_4$]$^-$, and [In(NCS)$_5$]$^{2-}$.[45] Indium(III) isothiocyanate has been synthesized from InCl$_3$ and KSCN, and used in the preparation of ionic salts containing [In(NCS)$_4$(bipy)]$^-$ and [In(NCS)$_4$(py)$_2$]$^-$ anions, as well as compounds with indium-transition-metal bonds such as [In(NCS){W(CO)$_3$(Cp)}$_2$].[46,47] The solid-state structural data of some of these ionic isothiocyanate compounds are available.[47]

(c) Amido and imido ligands. A convenient route to indium(III) amide has been reported. The reaction of indium(III) iodide with three equivalents of KNH$_2$ in anhydrous liquid ammonia affords In(NH$_2$)$_3$, which is insoluble in NH$_3$ but dissolves in NH$_3$ solutions containing KNH$_2$ to produce K$_x$In(NH$_2$)$_{3+x}$.[48] Related sodium indium amide may be obtained using a similar route. The compound Li$_3$In(NH$_2$)$_6$ can be synthesized from a mixture of InI$_3$, LiI, and KNH$_2$. Upon thermolysis, In(NH$_2$)$_3$, K$_x$In(NH$_2$)$_{3+x}$, and Na$_x$In(NH$_2$)$_{3+x}$ give InN, whereas Li$_3$In(NH$_2$)$_6$ affords Li$_3$InN$_2$.

H$_2$InNH$_2$ has been generated in argon matrices and characterized using IR spectroscopy.[49,50] Univalent and divalent indium amide derivatives are also observed under these conditions.

Many indium(III) adducts derived from primary or secondary amido ligands have been reported.[51] Syntheses of essentially all nonorganoindium amido complexes involve a salt-elimination process. The compound (THF)$_3$Li(μ-Cl)Cl$_2$InN(SiMe$_3$)(Dipp) (5) (Dipp = 2,6-(*i*-Pr)C$_6$H$_3$) represents a rare dihaloindium amide. It is obtained by the reaction of InCl$_3$ with LiN(SiMe$_3$)(Dipp) in tetrahydrofuran.[52] Although this reaction leads to the formation of an In—N bond, the LiCl elimination is incomplete. The phosphoranylidiniminodium(III) adduct [Cl$_2$(DMF)-In(NPPh$_3$)]$_2$ also has different ligands, in addition to nitrogen-based donors bonded to the indium atom. It is a dimeric molecule with pentacoordinate indium sites and NPPh$_3$ bridges.[53] The bromo derivative BrIn(tmp)$_2$ (tmp = 2,2,6,6-tetramethylpiperidinato) is reported to be monomeric in solution and in the gas phase.[54]

The compounds In[N(SiMe$_3$)$_2$]$_3$,[55] In(tmp)$_3$,[54] In[N(H)(2,4,6-(*t*-Bu)$_3$C$_6$H$_2$)]$_3$,[56] In(NEt$_2$)$_3$,[57] In(NCy$_2$)$_3$,[58] In[N(SiMe$_3$)Ph]$_3$, In[N(SiMe$_3$)*t*-Bu]$_3$, and In[N(SiHMe$_2$)*t*-Bu]$_3$ have been obtained as solvent- or halide-free indium(III) adducts via a metathesis route.[59] The use of smaller amido groups may lead to solvent-coordinated products or "ate" complexes. For example, the diethyl ether-coordinated compound (Et$_2$O)In[N(SiMe$_3$)Ph]$_3$ (6) was obtained initially during the synthesis of In[N(SiMe$_3$)Ph]$_3$ using InCl$_3$ and LiN(SiMe$_3$)Ph in Et$_2$O. However, the coordinated ether

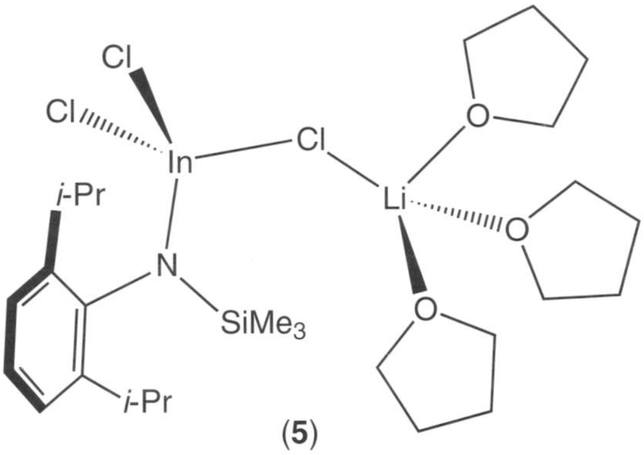

(5)

can be removed easily by dissolving the adduct in CH_2Cl_2, followed by the removal of solvent under reduced pressure. The attempted synthesis of $In[N(SiMe_3)Me]_3$ in Et_2O formed $Li\{In[N-(SiMe_3)Me]_4\}$.[59] However, the tris-amido adduct $(py)In[N(SiMe_3)Me]_3$ may be obtained by performing the reaction in pyridine. The neutral indium(III) complex $(py)In(NPh_2)_3$ **(7)** can be synthesized using $InCl_3$ and $LiNPPh_2$ in pyridine.[59] $[Li(THF)_4][ClIn(NPh_2)_3]$ **(8)** is obtained if the reaction is carried out in THF. The pyridine coordinated $(py)_2In[N(H)(2,6-(i-Pr)_2C_6H_3)]_3$ **(9)** and the *p*-(dimethylamino)pyridine adducts $(p-Me_2Npy)In[N(SiHMe_2)t-Bu]_3$ and $(p-Me_2N-py)In[N(SiMe_3)Me]_3$ have also been reported.[56,59] The reaction of $InCl_3$ with 3 or 4 equivalents of $LiNCy_2$ (Cy = cyclohexayl) affords only the neutral, trigonal-planar In(III) derivative $In(NCy_2)_3$.[58] The use of 4 equivalents of $LiN(CH_2Ph)_2$, however, leads to the ionic product $[Li(THF)_4][In\{N(CH_2Ph)_2\}_4]$.

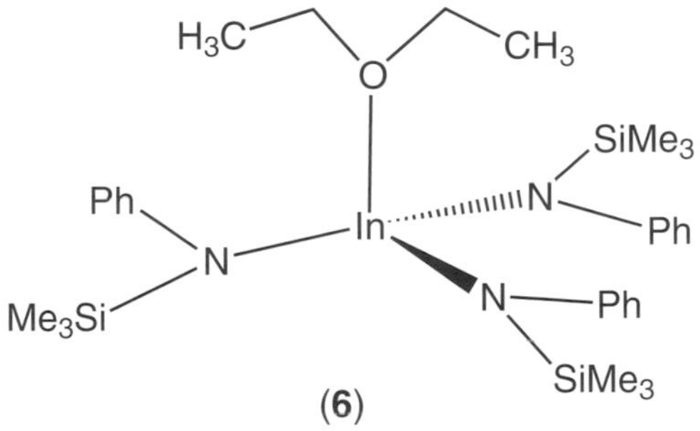

(6)

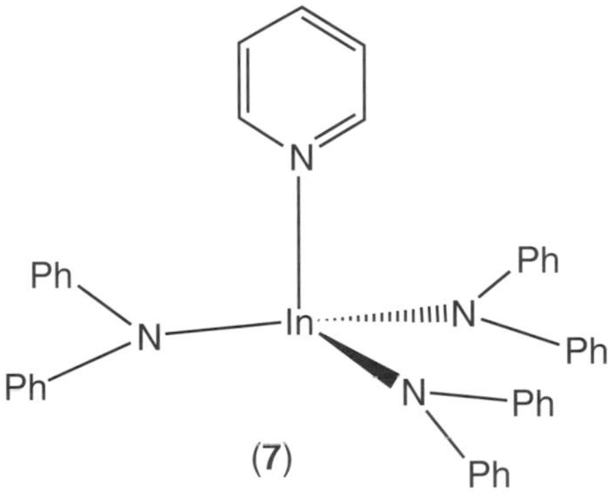

(7)

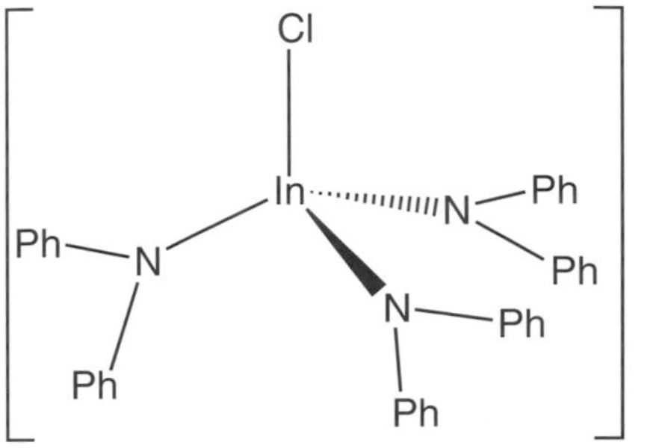

(8)

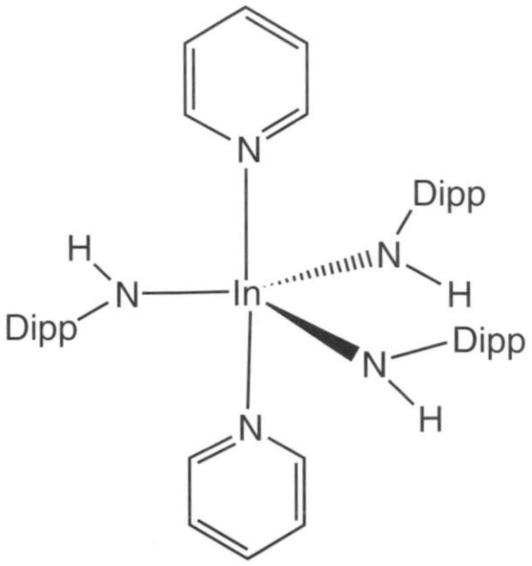

(9)

Indium and Thallium

The solid-state structures of In[N(SiMe$_3$)$_2$]$_3$,[60] In(tmp)$_3$, and In[N(H)2,4,6-(t-Bu)$_3$C$_6$H$_2$]$_3$ show that they are monomeric molecules with planar, three-coordinate indium centers. In[N(SiMe$_3$)$_2$]$_3$ reacts with CsF in toluene to produce [Cs(toluene)$_3$][FIn[N(SiMe$_3$)$_2$]$_3$.[61] The solid-state structure shows an essentially linear Cs–F–In moiety (174°). The four-coordinate indium(III) complexes (Et$_2$O)In[N(SiMe$_3$)Ph]$_3$ (6), (py)In(NPh$_2$)$_3$ (7), and (p-Me$_2$Npy)In[N(SiHMe$_2$)t-Bu]$_3$ have severely distorted tetrahedral metal sites (closer to the planar In(amido–N)$_3$kernel).[59] The compound [Li(THF)$_4$][ClIn(NPh$_2$)$_3$] (8) features a four-coordinate indium atom with the expected tetrahedral geometry. The X-ray crystal structure of (py)$_2$In[N(H)(2,6-(i-Pr)$_2$C$_6$H$_3$)]$_3$ (9) shows a five-coordinate, trigonal-bipyramidal indium center in which the axial sites are occupied by the two pyridine molecules.[56] Crystalline (p-Me$_2$Npy)Li{In[N(SiMe$_3$)Me]$_4$} has been obtained by treating Li{In[N(SiMe$_3$)Me]$_4$} with p-(dimethylamino)pyridine, and characterized using X-ray crystallography.[59]

One of the main interests in indium amides has been their potential utility as single-source precursors for indium nitride materials. They also serve as starting materials in the synthesis of various other indium compounds. For instance, amides such as In(NEt$_2$)$_3$ and In[N(SiMe$_3$)t-Bu]$_3$ react with alcohols or thiols to produce indium(III) alkoxides or thiolates, respectively.

The reaction of InX$_3$ (X = Cl, Br, I) with two equivalents of LiN(H)(t-Bu) led to imido derivatives [In$_4$X$_4$(t-BuN)$_4$] (10) with In$_4$N$_4$ heterocubane structures. The reaction involving InBr$_3$ also produced a minor by-product [In$_3$Br$_4$(t-BuN)(t-BuNH)$_3$] (11).[62]

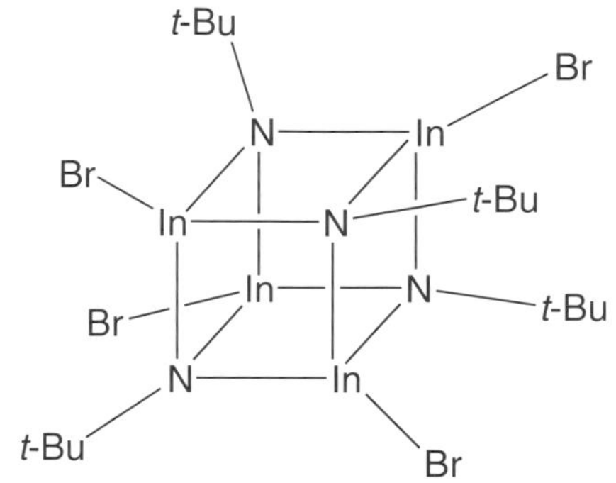

(10) 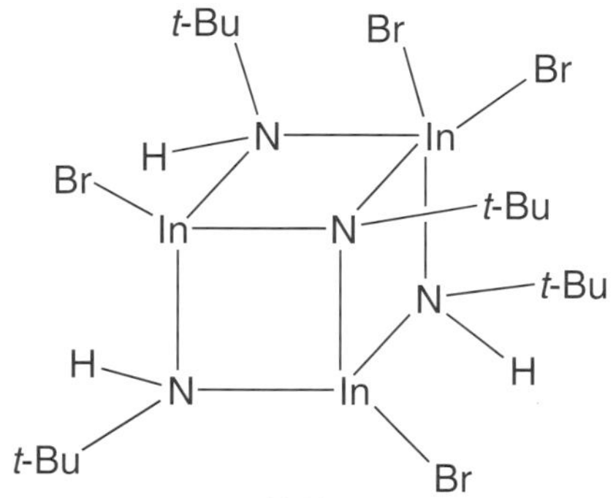   (11)

(d) Multidentate ligands. The dimeric [ClIn(N(Me)SiMe$_2$)$_2$NMe]$_2$ and monomeric LiIn[(NSiMe$_3$)$_2$SiMe$_2$]$_2$ complexes feature bidentate amido donors.[63,64] The synthesis involves the treatment of Li$_2$(N(Me)SiMe$_2$)$_2$NMe and Li$_2$(NSiMe$_3$)$_2$SiMe$_2$ with InCl$_3$ at 1:1 or 2:1 molar ratio, respectively. The reaction of Li$_2$(N(Me)SiMe$_2$)$_2$NMe with InCl$_3$ at 4:1 molar ratio affords an indium (III) compound [Li{In(HN(Me)SiMe$_2$NMe)$_2$(MeNSiMe$_2$NMe)}]$_2$, with a Li$_2$In$_2$Si$_2$N$_4$ adamantane-like core.[63]

Indium complexes of bidentate nitrogen-ligand systems such as triazenide, amidinate, aminotroponiminate, and β-ketiminate have been reported. These ligands feature unsaturated ligand backbones. The reaction of InCl$_3$ with either one or two equivalents of 1,3-diphenyltriazene in the presence of triethylamine gives an ionic product [HNEt$_3$][InCl$_2$(PhNNNPh)$_2$], rather than the expected neutral species InCl$_2$(PhNNNPh) or InCl(PhNNNPh)$_2$.[65,66] Interestingly, [HNEt$_3$]-[InCl$_2$(PhNNNPh)$_2$] reacts with a variety of Lewis bases to produce neutral indium(III) complexes [InCl$_2$(PhNNNPh)L$_2$] (L = pyridine, 3,5-dimethylpyridine, PEt$_3$; L$_2$ = 2,2′-bipyridine, 1,10-phenanthroline, Me$_2$PCH$_2$CH$_2$PMe$_2$, or Et$_2$PCH$_2$CH$_2$PEt$_2$). The [InCl$_2$(PhNNNPh)$_2$]$^-$ anion has a six-coordinate indium center with a highly distorted octahedral geometry. The tris(1,3-diphenyltriazenido)indium(III)In(PhNNNPh)$_3$ complex was prepared via an alkyl-group-elimination route (usually the preferred method for the synthesis of organoindium derivatives) using InMe$_3$ and H(PhNNNPh).[65]

The amidinate complex {In[CyNC(H)NCy]$_2$Cl}$_2$ (12) may be obtained by treating InCl$_3$ with two equivalents of Li[CyNC(H)NCy], or by reacting Me$_2$InCl with two equivalents of H[CyNC(H)NCy].[67] It has a dimeric, lantern-type structure with an unusual square-pyramidal geometry at the indium atoms.[67] In this molecule, four formamidinate ligands bridge the two indium atoms, while the chlorides occupy the apical sites. Bulkier substituents on the ligand backbone afford indium(III) adducts in which the amidinate serves as a chelating donor. Syntheses of In[CyNC(Me)NCy]$_2$Cl, In[CyNC(t-Bu)NCy]$_2$Cl (13), and In[Me$_3$SiNC(t-Bu)NSiMe$_3$]$_2$Cl, as well as the tris(amidinate) derivative In[CyNC(Me)NCy]₃, were reported.[68] The crystal-structure determination of In[CyNC(*t*-Bu)NCy]₂Cl (**13**) reveals that the indium atom adopts a distorted trigonal-bipyramidal geometry, with the chloride occupying one of the equatorial positions. The dichloroindium compound [Me₃SiNC(Ph)NSiMe₃]InCl₂ was obtained via a trimethylsilyl chloride elimination route.[69,70]

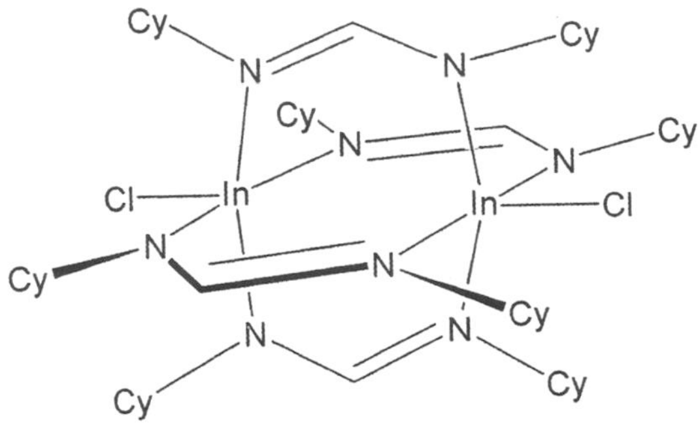

(12) 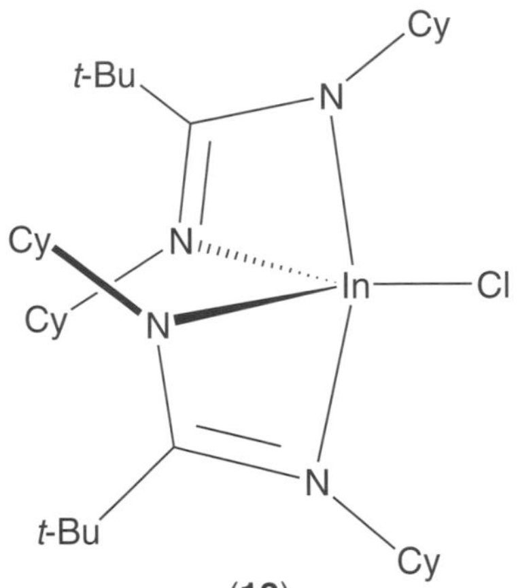   (13)

Unlike the triazenide or amidinate ligands that form four-membered metallacycles, the amino-troponiminates coordinate to metal ions forming five-membered metallacycles.[71] The dichloro In(III) adduct [(*i*-Pr)₂ATI]InCl₂ (**14**) ([(Me)₂ATI] = *N*-*i*-propyl-2-(*i*-propylamino)troponiminate) has been synthesized using InCl₃ and [(*i*-Pr)₂ATI]Li.[72] It has a tetrahedral indium center. The bis(aminotroponiminate) adduct [(Me)₂ATI]₂InCl (**15**) was obtained via an oxidative ligand-transfer process involving [(Me)₂ATI]₂Sn and InCl.[71,73] Its solid-state structure shows a slightly distorted trigonal-bipyramidal geometry at indium, with the chloride ion occupying one of the equatorial sites. This molecule shows fluxional behavior in solution at room temperature. Indium (III) complexes containing bridged aminotroponiminato ligands have also been synthesized. The indium atom in [(*i*-Pr)₂TP]InCl([(*i*-Pr)₂TP] = 1,3-di[2-(isopropylamino)troponiminate]propane) is five-coordinate and the geometry may be described as a distorted tetragonal pyramid.[74] The chloride ion occupies the apical position. The chloride may be replaced by a *t*-BuO⁻ group using *t*-BuOK to obtain [(*i*-Pr)₂TP]In(OBu').

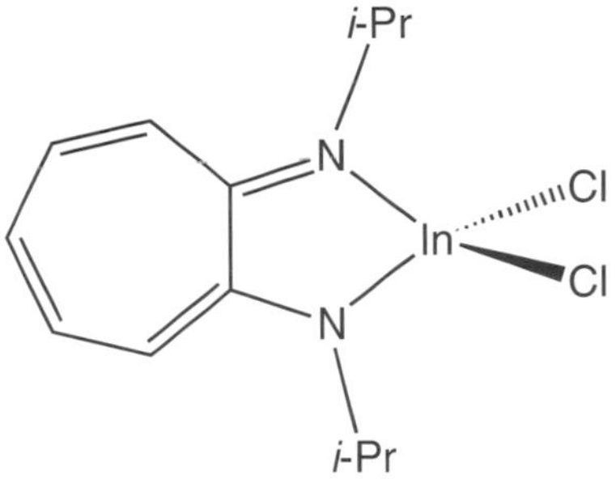

(14) 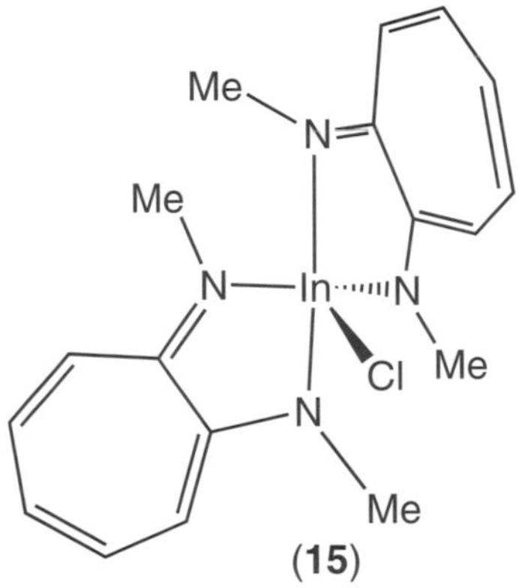   (15)

The β-ketiminate ligands typically form six-membered metallacycles. Synthesis of [HC{(Me)C(2,6-(*i*-Pr)₂C₆H₃)H}₂]InX₂ (X = Cl (**16**) or I) via a salt-elimination process has been reported.[75] It features the expected tetrahedral metal sites. An alkyl-elimination method involving Me₃In and 2-(benzylamino)pyridine starting material has been used in the synthesis of the six-coordinate, tris-ligand complex In[N(CH₂Ph)C₅H₄N]₃.[76] The 3-(2-pyridyl)pyrazolate ligand has been used in the isolation of a dichloroindium(III) derivative.[77] In[{(3-Py)Pz}InCl₂(DME)]₂, the 3-(2-pyridyl)pyrazolate ligand serves as a bridging as well as a chelating ligand for InCl₂(DME) fragments. Each indium atom has an N₃Cl₂O coordination sphere.

(16)   (17)

The dibenzo-tetraaza macrocycle 5,7,12,14-tetramethyldibenzo[b,i][1,6,9,10]-tetraazacyclotetra-decine (H$_2$TMTAA) is considered to be the intermediate between saturated cyclam-type ligands and aromatic porphyrin systems.[78] Treatment of Li$_2$TMTAA with indium(III) chloride yields ClIn(TMTAA) (**17**).[78,79] The In–Cl fragment occupies a site above the N$_4$ plane of the saddle-shaped macrocycle, with indium adopting a square-pyramidal geometry.[78] The chloride can be substituted by -N(SiMe$_3$)$_2$ and -OSiMe$_3$ groups to obtain (TMTAA)InN(SiMe$_3$)$_2$ and (TMTAA)InOSiMe$_3$, respectively.

Poly(pyrazolyl)borate ligands play an important role in indium coordination chemistry.[80] Indium(III) adducts of bis-, tris-, and tetrakis(pyrazolyl)borates have been synthesized and the structures and chemistry have been investigated. The reaction of [H$_2$B(Pz)$_2$]K (Pz = pyrazolyl) with In(NO)$_3$ or InCl$_3$ at 3:1 molar ratio affords [H$_2$B(Pz)$_2$]$_3$In.[81,82] The solid-state data show that the octahedrally coordinated indium center is surrounded by three puckered bis(pyrazolyl)borate ligands.[81] It is also possible to synthesize [H$_2$B(Pz)$_2$]$_2$InCl and [H$_2$B(Pz)$_2$]InCl$_2$ using InCl$_3$ and [H$_2$B(Pz)$_2$]K in appropriate proportions.[82] The reaction of [H$_2$B(Pz)$_2$]$_2$InCl with CH$_3$CO$_2$Na, or [H$_2$B(Pz)$_2$]$_3$In with CH$_3$CO$_2$H, leads to the acetate derivative [H$_2$B(Pz)$_2$]$_2$In(O$_2$CCH$_3$).[82]

In the absence of adverse steric constraints, the tris(pyrazolyl)borates have a tendency to produce six-coordinate In(III) complexes. [HB(3,5-(Me)$_2$Pz)$_3$]InCl$_2$(THF),[83] [HB(3,5-(Me)$_2$-Pz)$_3$]InCl$_2$(NCCH$_3$),[84] [HB(3,5-(Me)$_2$Pz)$_3$]InCl$_2$(3,5-(Me)$_2$PzH),[83] and [HB(3,5-(Me)$_2$Pz)$_3$]InI$_2$(3,5-(Me)$_2$PzH)[85] have been isolated from reaction mixtures involving [HB(3,5-(Me)$_2$Pz)$_3$]K and InCl$_3$ or InI$_3$. Cationic, six-coordinate species containing {[HB(3,5-(Me)$_2$Pz)$_3$]$_2$In}$^+$ and {[HB(Pz)$_3$]$_2$In}$^+$ moieties, and the related tris(pyrazolyl)gallate adduct {[MeGa(Pz)$_3$]$_2$In}[InI$_4$], are known.[83,86] They have been obtained as by-products resulting from the disproportionation of In(I) reagents. Alternative routes to {[HB(Pz)$_3$]$_2$In}$^+$ involve halide abstraction from [HB(Pz)$_3$]$_2$InCl using AgBF$_4$, or methyl-group abstraction from [HB(Pz)$_3$]$_2$InMe using [HNEt$_3$][BPh$_4$], or treating InCl$_3$ with one equivalent of [HB(Pz)$_3$]K.[83] In a toluene/dichloro-methane solution, [HB(Pz)$_3$]$_2$InCl exists as a mixture of ionic {[HB(Pz)$_3$]$_2$In}Cl and covalent [HB(Pz)$_3$]$_2$InCl forms. The compound {[HB(Pz)$_3$]$_2$In}{[HB(Pz)$_3$] InCl$_3$} (**18**) is an interesting example where there is a six-coordinate anion and a cation in the same molecule. It was obtained using InCl$_3$ and [HB(Pz)$_3$]K at 2:3 molar ratio in a THF/H$_2$O solvent system.[83] X-ray data show that both indium centers have octahedral geometry.

Tetrakis(pyrazolyl)borate complexes of In(III)[83] [B(Pz)$_4$]$_3$In, [B(Pz)$_4$]$_2$InCl, and mixed-ligand complexes such as [HB(3,5-(Me)$_2$Pz)$_3$]InCl[H$_2$B(Pz)$_2$], [HB(3,5-(Me)$_2$Pz)$_3$]InCl[H$_2$B(3,5-(Me)$_2$Pz)$_2$], [HB(3,5-(Me)$_2$Pz)$_3$]InCl[HB(Pz)$_3$], and [HB(3,5-(Me)$_2$Pz)$_3$]InCl[B(Pz)$_4$], have been synthesized.[87] All these adducts are believed to feature six-coordinate indium sites. The crystal structure of [HB(3,5-(Me)$_2$Pz)$_3$]InCl[H$_2$B(Pz)$_2$] shows that the indium center has a distorted octahedral geometry.

Ligand-substitution chemistry at the indium center has been investigated. Most of these reactions concern the halide substitution of [HB(3,5-(Me)$_2$Pz)$_3$]InCl$_2$(THF) by C-, N-, O-, and S-based ligands.[80] For example, [HB(3,5-(Me)$_2$Pz)$_3$]InCl$_2$(THF) reacts with K$_2$S$_5$ to form the In(III) polysulfide complex [HB(3,5-(Me)$_2$Pz)$_3$]In(S$_4$)(3,5-(Me)$_2$PzH).[88] The related [HB(3,5-(t-Bu)$_2$Pz)$_3$]In(S$_4$) is also known, although it was synthesized by treating an In(I) adduct with

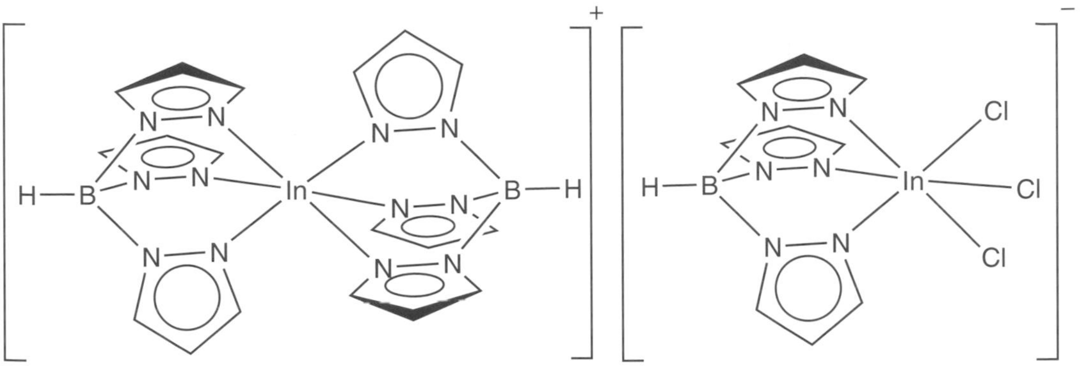

(18)

sulfur.[89] Metal adducts of formally monovalent indium are obtained by reacting [HB(3,5-(Me)$_2$-Pz)$_3$]InCl$_2$(THF) with [Fe(CO)$_4$]$^{2-}$ and [W(CO)$_5$]$^{2-}$.[90] Several products resulting from the hydrolysis of tris(pyrazolyl)boratoindium(III) complexes have also been isolated.[85]

The mixed-ligand complex {[CpCo{P(O)(OMe)$_2$}$_3$]In[HB(Pz)$_3$]}X (X = InCl$_4$ **(19)** or PF$_6$) contains two different tripodal ligands (N$_3$ and O$_3$ type).[91,92] It was synthesized starting from a 1:1:2 mixture of [CpCo{P(O)(OMe)$_2$}$_3$]Ag, [HB(Pz)$_3$]Tl, and InCl$_3$. Interestingly, no homoleptic products (e.g., {[HB(Pz)$_3$]$_2$In}$^+$) result from this mixture. However, heating a mixture of {[CpCo{P(O)(OMe)$_2$}$_3$]$_2$In}PF$_6$ and {[HB(Pz)$_3$]$_2$In}PF$_6$ for two days in water/MeOH produces {[CpCo{P(O)(OMe)$_2$}$_3$]In[HB(Pz)$_3$]}PF$_6$. The indium atom features octahedral geometry with a *fac*-N$_3$O$_3$ coordination sphere. The related {[CpCo{P(O)(OMe)$_2$}$_3$]In[HB(3,5-(Me)$_2$Pz)$_3$]}[AgCl$_2$] has been prepared from [HB(3,5-(Me)$_2$Pz)$_3$]InCl$_2$(NCCH$_3$) and [CpCo{P(O)(OMe)$_2$}$_3$]Ag. The synthesis of {[HB(Pz)$_3$]$_2$In}PF$_6$ using [HB(Pz)$_3$]Tl as a ligand-transfer agent is also described.[91]

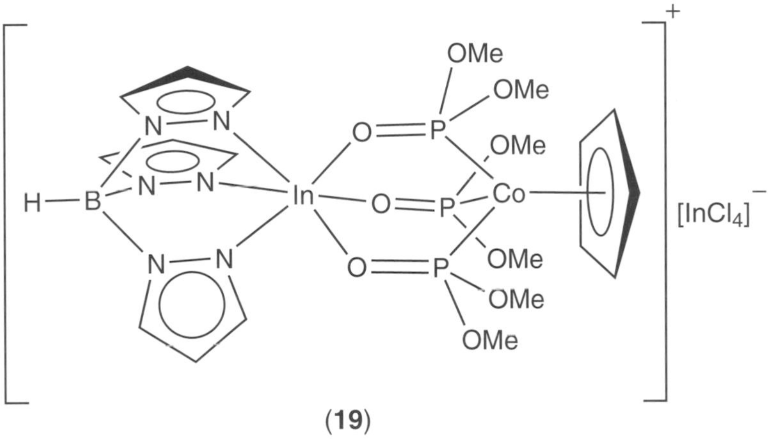

(19)

Indium(III) porphyrin complexes have been investigated by many groups.[93–113] They are of interest for applications related to photodynamic therapy, radiolabeled indium chemistry, light-emitting devices, photovoltaic cells, metal-catalyzed oxidation, and sensors. The indium(III) chloro derivatives of *meso*-tetraphenylporphine (TPPH$_2$), *meso*-tetrakis(*p*-methoxyphenyl)porphine (TMPPH$_2$), and *meso*-tetrakis(*p*-tolyl)porphine were reported in the early 1970s.[114] Synthesis typically involves the treatment of an indium(III) halide with the free ligand in an acetic acid/sodium acetate mixture. The indium(III) prophyrin complexes with axial acetate groups, such as In(TPP)(OAc), In(TPYP)(OAc) (TPYP = *meso*-tetra(4-pyridyl)porphyrinato), and In(TMPP)(OAc), can be obtained directly and in high yield by using In$_2$O$_3$ instead of InCl$_3$ in the above mixture.[101] Various other porphyrin-ligand systems, in particular OEP^{2-} (2,3,7,8,12,13,17,18-octaethylporphyrinato), have also been used to complex In(III) ions. Indium(III) porphyrin complexes with axial halide, N-donor, and O-donor groups (as well as alkyl, aryl, or transition-metal substituents) have been investigated.[96,101,105,107,108,113,115–119]

A series of indium(III) porphyrin complexes (porphyrin = TPP, OEP, T*p*-CF$_3$PP) containing axial tetrazolato and triazolato ligands have been prepared by the cycloaddition reactions of azidoindium(III) porphyrin complexes with nitriles and alkynes. Based on the mode of tetrazole or triazole linkage, two different isomers are possible (kinetic and the thermodynamic product). Structural and spectroscopic data reveal that the coordination mode of the tetrazolato or triazolato group depends on the alkyl or aryl substituent on the azolate moiety.[105,120] The crystal structure of (OEP)InL (L = 5-methyltetrazolate (**20**), 4-phenyltetrazolate (**21**)) reveals that the phenyl substituent occupies the tetrazolate ring 4-position, whereas the methyl group prefers the 5-position. The most important factor that determines the bonding mode of the crystallized product seems to be steric.[120] Redox properties and the reactivity towards donors such as pyridine or *N*-methyl imidazole have also been described.[108,118]

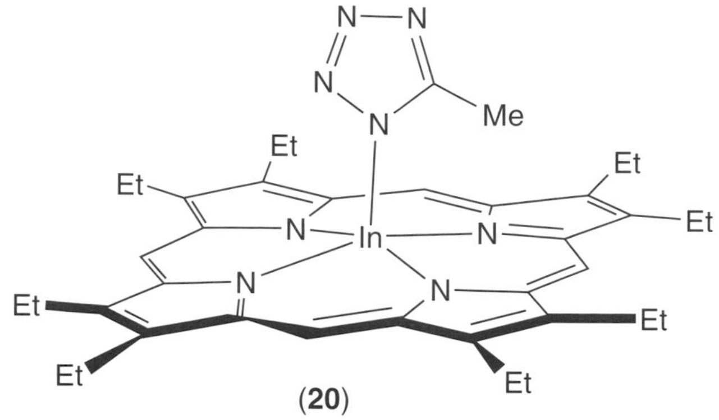

(20)

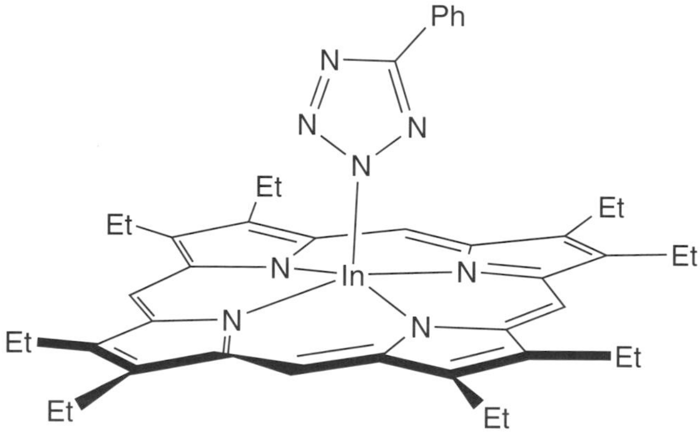

(21)

The indium center adopts a square-pyramidal geometry when coordinated to the porphyrin ligand and a monodentate donor (e.g., (TPP)InCl (**22**) or 10-(4′-*N*-pyridyl)-5,15,20-triphenylpor-phyrinatoindium(III) chloride).[115,116] The indium atom occupies a site above the plane formed by the four porphyrin nitrogen atoms. Solid-state structures of six-coordinate indium derivatives containing chelating bidentate ligands are known. In(TPYP)(OAc) (**23**) has an asymmetric bidentate acetato group, whereas in In(TMPP)(OAc) the two In–O(acetato) distances are equal.[101] Synthesis of [indium(III)(octaethyloxophlorin)]₂ featuring an oxophlorin ligand has been reported.[121] The X-ray structure shows a centrosymmetric dimer with the two metalla-oxophlorin units linked by In—O bonding.

Phthalocyanine derivatives of indium(III) have attracted even more interest.[94,122–144] Some indium phthalocyanine adducts show interesting nonlinear optical properties.[123,124,134] For example, *tert*-butyl-substituted chloro(phthalocyaninato)indium(III) (*t*-Bu)₄PcInCl (**24**) is one of the best substances for optical-limiting applications.[134] Optical limiters limit the intensity of transmitted light once the input intensity exceeds a threshold value. This ability is useful for the protection of sensitive objects, such as human eyes or light sensors from high-intensity light beams.

Synthesis of indium phthalocyanine complexes usually involves the assembly of a ring system in the presence of an indium source. For example, (*t*-Bu)₄PcInCl(chloro(tetra-(*tert*-butyl)phthalo-cyaninato)indium(III)) or (*n*-C₅H₁₁)₈PcInCl(chloro(octa-(*n*-pentyl)phthalocyaninato)indium(III))

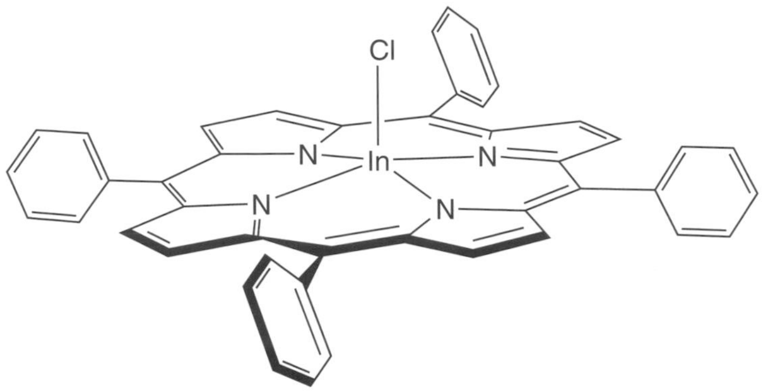

(22)

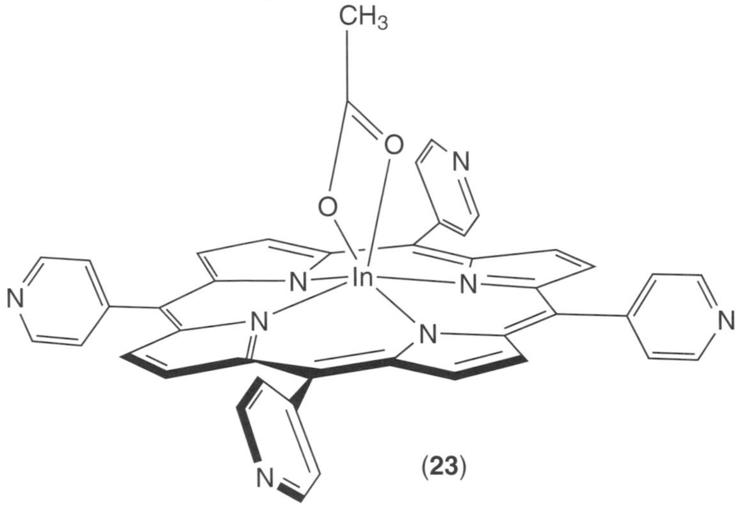

(23)

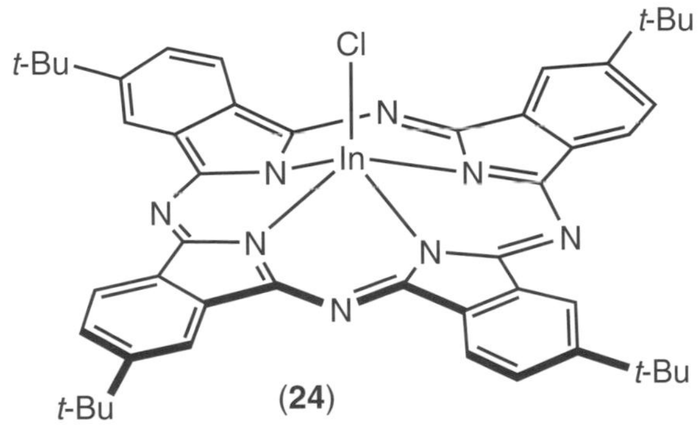

(24)

may be synthesized directly from a mixture of 4-*tert*-butylphthalonitrile or 4,5-bis(*n*-pentyl)phthalo-nitrile and InCl$_3$ in dry quinoline containing a catalytic amount of DBU.[130] Indium(III) adducts of various ring-substituted napthalocyanines (e.g., (25)) can also be obtained by using a similar route, starting with napthalenedicarbonitrile or more reactive diiminoisoindolines and InCl$_3$.[122,145] The use of indium metal and indium alloys has also been described.[128,139–141] The compound PcInI-(iodo(phthalocyaninato)indium(III)) was synthesized from the reaction of indium powder with 1,2-dicyanobenzene under a stream of iodine.[128]

Indium(III) phthalocyanine complexes show a rich structural diversity. Compounds like PcInI are monomeric, with a five-coordinate, square-pyramidal geometry at indium. The indium atom is located out of the N$_4$ plane, and the phthalocyaninato ring forms a dome shape.[128] In general,

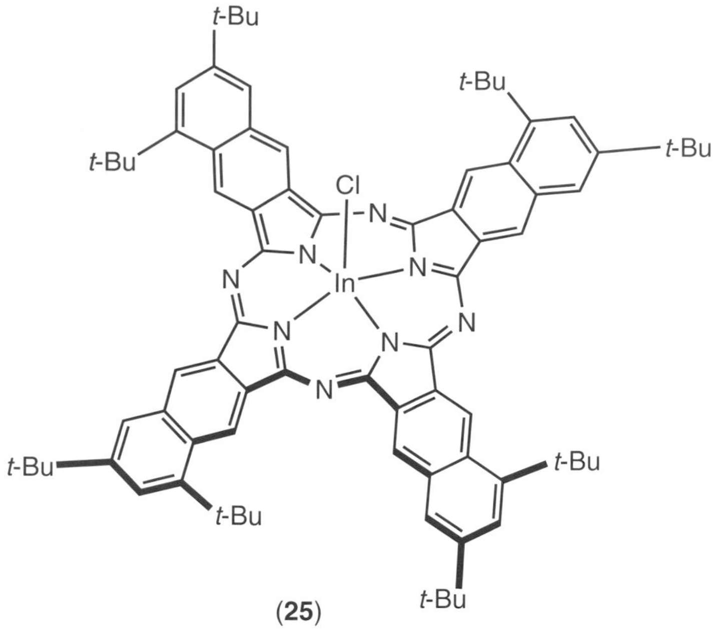

(25)

this type of ring distortion is observed when the metal atom is too large (about 0.7 Å or larger) to fit into the phthalocyaninato core.[126] Tetra *n*-butylammonium salts of *cis*-[PcInX$_2$]$^-$ (X = F (26), Cl, CN, HCO$_2$) compounds have been synthesized starting from PcInCl or *cis*-[PcIn(OH)$_2$]$^-$.[131] The X-ray crystal structure of [*n*-Bu$_4$][PcInF$_2$] (26) shows that the indium center is six-coordinate and the fluorides occupy *cis*-sites. Compounds containing the anion [PcInX$_2$]$^-$ (X = NCO, NO$_2$) are also known.[136,138] The carbonato derivative [*n*-Bu$_4$][PcInCO$_3$] features a *cis*-chelating ligand.[132]

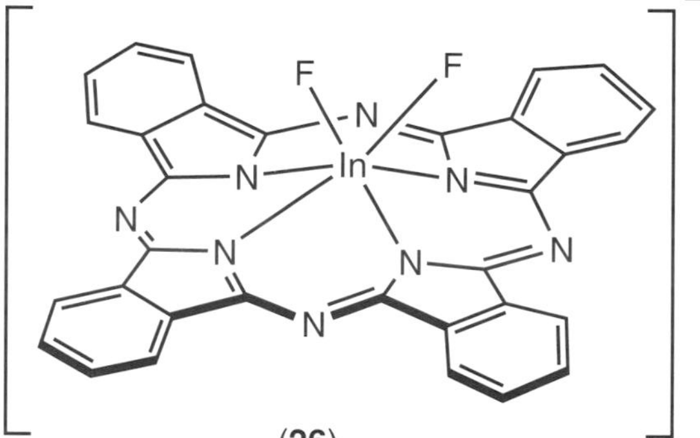

(26)

There are sandwich-type complexes of In(III) featuring phthalocyanine ligands. The neutral [Pc$_2$In] (27) is an interesting paramagnetic compound in which the In(III) is coordinated to a Pc^{2-} and to the radical anion Pc^{-}.[139] It was obtained by a direct reaction of InMg alloy and 1,2-dicyanobenzene.[139] The magnetic susceptibility measurement exhibits Curie-Weiss behavior. Structural data show that both halves of the sandwich are equivalent. The indium site is eight-coordinate and shows distorted square antiprismatic geometry. Compounds with anionic [Pc$_2$In]$^-$ moieties have also been prepared.[126]

The iodine-doped, paramagnetic compound [Pc$_2$In](I$_3$)$_{2/3}$ can be synthesized directly from In–Tl alloy and 1,2-dicyanobenzene under a stream of iodine.[146] The solid-state structure features one-dimensional stacks of [Pc$_2$In] columns and I$_3$ chains. A triple-decker indium(III) phthalocyanine complex [Pc$_3$In$_2$], which is diamagnetic, has been obtained by reacting In–Sn alloy with 1,2-dicyanobenzene at 210 °C.[141] Indium atoms are six-coordinate, and are located between the phthalocyanine rings. An indium(III) derivative of a bicyclic phthalocyanine system has also been synthesized and structurally characterized.[140]

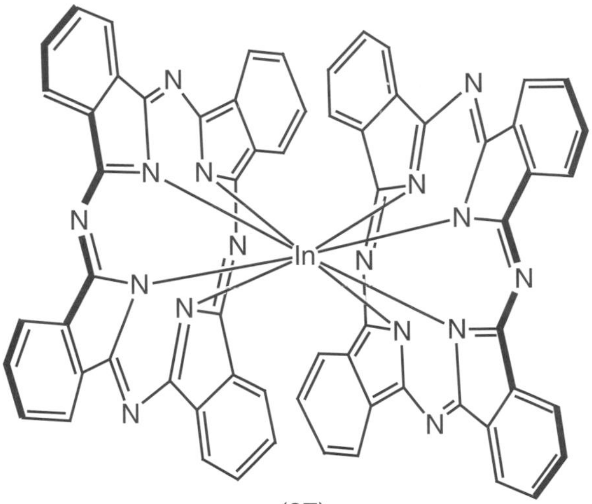

(27)

(ii) Phosphorus, arsenic, antimony, and bismuth ligands

Synthesis of ionic indium(III) compounds such as $[In(PPh_3)_4](ClO_4)_3$, $[In(AsPh_3)_4]ClO_4$, or $[In(diphos)_2]ClO_4$ have been reported.[1,3,4]

The preparation of nanometer-size isolated particles of group III–V materials such as InP is very challenging. It has been shown that indium(III) phosphides serve as useful single-source precursors to such material. The decomposition of $(t\text{-}Bu_2P)_3In$ (which has been obtained by reacting $InCl_3$ with three equivalents of $t\text{-}Bu_2PLi)^{[147]}$ in refluxing 4-ethylpyridine leads to nanometer-size InP.[148] This material shows clear quantum confinement effects. Nanocrystalline InP has been obtained from a mixture of $InCl_3$ and $(Me_3Si)_3P$ via dechlorosilylation followed by thermolysis.[149] An intermediate product of this process, believed to be an oligomeric $[Cl_2InP(Si-Me_3)_2]_n$, has been isolated as a yellow powder.[150] A low-temperature route to indium phosphide involving $InCl_3$, yellow phosphorus, and KBH_4 in ethylenediamine is also reported.[151] InAs and InSb may also be obtained via a similar technique.[152,153] The reaction of $InCl_3$ with $t\text{-}Bu_2PSiMe_3$ leads to $[(t\text{-}Bu_2P)_2InCl]_2$ (**28**), which contains an In_2P_2 ring.[154] Each tetrahedrally coordinated indium atom is linked by two bridging $t\text{-}Bu_2P$ groups, a terminal $t\text{-}Bu_2P$ group, and a terminal Cl atom.[154] Reaction of InCl with the potassium salt of phospholyl anion $[K(18\text{-crown-6})][PC_4Me_4]$ has resulted in the indium(III) adduct $[(\eta^1\text{-}PC_4Me_4)_2In(\mu\text{-Cl})_2K(18\text{-crown-6})]$ (**29**).[155] This is believed to be a product of a disproportionation reaction.[155] A compound with an In_4P_4 core has been observed in $[\{Cp(CO)_3Mo\}_4In_4(PSiMe_3)_4]$.[156] The $InCl_3$ reacts with 4 equivalents of $LiPPh_2$ in tetrahydrofuran to give $[Li(THF)_4][In(PPh_2)_4]$.[157] It is not possible to synthesize the analogous "ate" complex with bulkier $t\text{-}Bu_2P^-$ ligands.

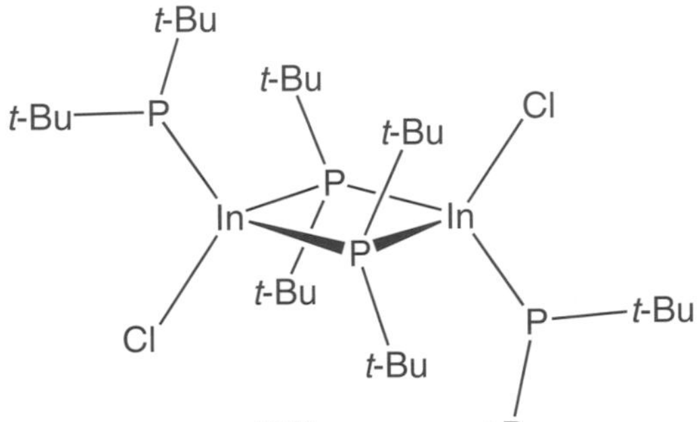

(28)

The synthesis of $(t\text{-}Bu_2As)_3In$ via a salt-elimination process has also been reported.[147] The stibido indium(III) complex $[(t\text{-}Bu_2Sb)_2InCl]_2$ (**30**) was obtained by the interaction of $InCl_3$ with $[t\text{-}Bu_2SbSiMe_3]_2$.[158] The X-ray crystal structure reveals that it is a dimer with stibido groups, rather than chlorides, acting as bridges.

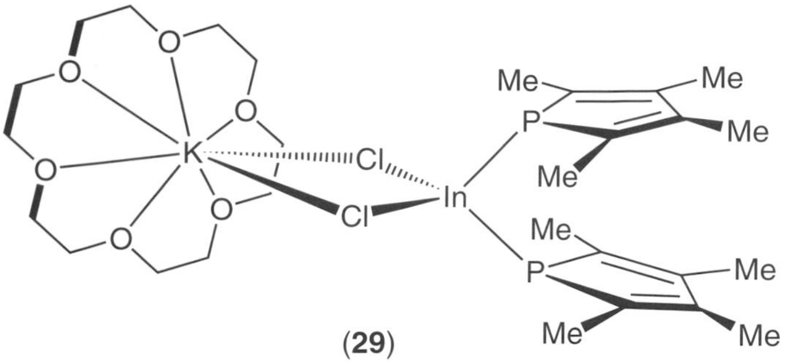

(29)

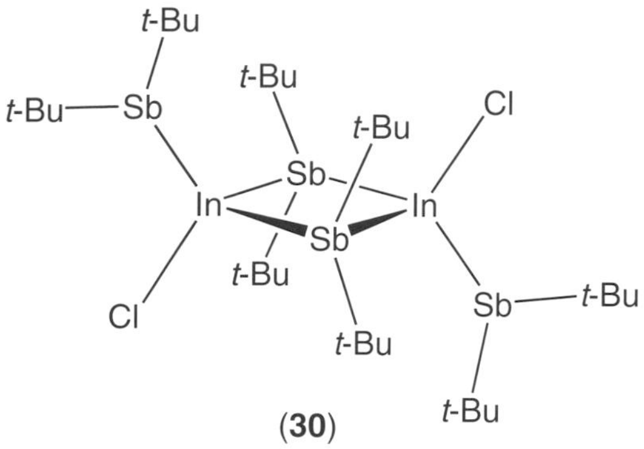

(30)

Reaction of indium(III) chloride with three equivalents of LiN(PPh$_2$)$_2$ in tetrahydrofuran affords the phosphazenide complex In[(PPh$_2$)$_2$N]$_3$ (31) as a yellow solid.[159] The solid-state structure shows that the six-coordinate indium atom is surrounded by three chelate rings in a propeller-like conformation.

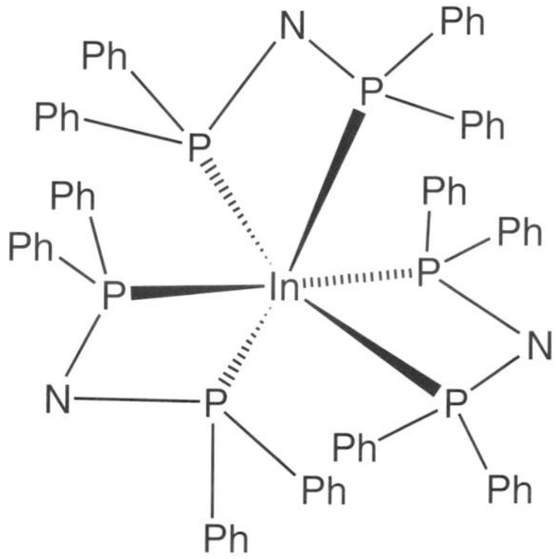

(31)

3.5.1.2.3 Group 16 ligands

(i) Oxygen ligands

(a) Neutral oxygen ligands. Ionic compounds of indium(III) with neutral oxygen are easily obtained by using indium salts of weakly coordinating anions. The existence of compounds with In^{3+} coordinated to water, dimethyl formamide, dimethyl sulfoxide, acetone, hexamethylphosphoramide, and OP(OMe)$_3$ has been established by various methods.[1,160–162] X-ray crystallographic data demonstrate the presence of octahedrally coordinated indium in [In(H$_2$O)$_6$]$^{3+}$ and [In(DMSO)$_6$]$^{3+}$ ions.[163–165] All the DMSO ligands are O–bonded. The structure of the hydrated indium(III) ions in aqueous perchlorate and nitrate solutions has been investigated by large-angle

X-ray scattering and extended X-ray absorption fine-structure techniques.[166] Data indicate that the indium ion is coordinated by six water molecules, and the In—O bond distance in the first hydration sphere is 2.131(7) Å. Changes in concentration or the anion had no influence on this distance. This In—O distance is very similar to the In—O distances observed for $[In(H_2O)_6]^{3+}$ ion in the solid state.[164,167]

(b) Hydroxide and oxide ligands. Hydrolysis of aqueous indium(III) ions occurs easily, leading to indium hydroxo species, and finally to insoluble $In(OH)_3$.[1,168,169] This is one of the challenges associated with designing indium complexes for radiopharmaceutical applications.[170] Indium(III) hydroxide can also be prepared at 0 °C by sonicating an aqueous $InCl_3$ solution.[171] It was obtained as needle-shaped, nanosized material. There are few well-authenticated indium adducts containing hydroxo ligands.[172–174] Indium(III) bromide reacts with 1,4-triazacyclononane (L) to give $LInBr_3$.[172] The hydrolysis of this adduct in alkaline aqueous solution leads to the first well-authenticated In(III) μ-hydroxo complex. The dithionate salt $[L_4In_4(\mu\text{-}OH)_6](S_2O_6)_3$ contains $[L_4In_4(\mu\text{-}OH)_6]^{6+}$ cations, with an admantane-like $In_4(OH)_6$ skeleton. The hydrolysis of $LInBr_3$ in sodium acetate affords a neutral, oxo-bridged dimer $L_2In_2(\mu\text{-}O)(MeCO_2)_4$. The compound $[In_2LCl_4(\mu\text{-}OH)_2]$ (L = bis[3-(2-pyridyl)pyrazol-1-yl]methane) is also a rare In(III) complex with bridging hydroxide ligands.[173] It has two pseudo-octahedral indium centers with *cis,cis,cis*-$N_2O_2Cl_2$ coordination environments.

The 1,3,5-triamino-1,3,5-trideoxy-*cis*-inositol (H_3taci) ligand (capable of N- or O bonding) shows an interesting group trend in which the ligand adopts O_6, O_3N_3, and N_6 modes of coordination for trivalent Al, Ga, and Tl ions, respectively.[175] The product obtained from the In(III) system is somewhat different.[176] $In(NO_3)_3$ reacted with H_3taci in MeOH to give $[In_6O(taci)_4](NO_3)_4\cdot8H_2O$, which was characterized by NMR and mass spectroscopy and by X-ray crystallography. The crystal structure shows that the central O^{2-} ion is surrounded by six indium atoms in an octahedral arrangement. This In_6O moiety is bonded to four hexadentate taci ligands, resulting in a larger octahedral In_6O_{13} core.[176]

(c) Alkoxide ligands. There is a significant interest in indium alkoxides because of their potential use as CVD precursors for indium oxide films.[177,178] Doped or undoped indium oxide thin films are used as heat insulators, transparent and conductive ceramics, solar-cell windows, and display panels. A series of tris(alkoxide) compounds $In(OR)_3$ (R = Me, Et, *i*-Pr, *n*-Bu, *s*-Bu, *t*-Bu, pentyl) was reported in the mid-1970s.[179] The isopropyl derivative was obtained from a reaction of $InCl_3$ with *i*-PrONa, and was used in the preparation of other alkoxides. Subsequent work suggested that this (*i*-PrO)$_3$In may not be a simple homoleptic alkoxide. For example, an oxo-centered cluster $In_5(\mu_5\text{-}O)(\mu_3\text{-}O\text{-}i\text{-}Pr)_4(\mu_2\text{-}O\text{-}i\text{-}Pr)_4(O\text{-}i\text{-}Pr)_5$ can be obtained using the same starting materials under similar conditions.[180,181]

More recently it has been shown that indium amides serve as convenient starting materials for obtaining alkoxide derivatives.[178] For example, $In[N(SiMe_3)t\text{-}Bu]_3$ reacts with *t*-BuOH, EtMe$_2$COH, Et$_2$MeCOH, and *i*-PrMe$_2$COH to yield dimeric $[In(\mu\text{-}OR)(OR)_2]_2$ (R = *t*-Bu (**32**), EtMe$_2$C, Et$_2$MeC, and *i*-PrMe$_2$C) alkoxides. Indium alkoxides with less bulky substituents have also been prepared, e.g., (*i*-PrO)$_3$In and (Et$_2$HCO)$_3$In (**33**). The (Et$_2$HCO)$_3$In is a tetramer in the solid state, whereas the insoluble isopropoxide analogue is believed to be a polymeric compound. The indium(III) amides $In(tmp)_3$ and $In(NEt_2)_3$ may also be used as starting materials for preparing alkoxides. Among this group of alkoxides, $[In(\mu\text{-}OCMe_2Et)(OCMe_2Et)_2]_2$ is reported to be the best precursor candidate for the deposition of indium oxide films.

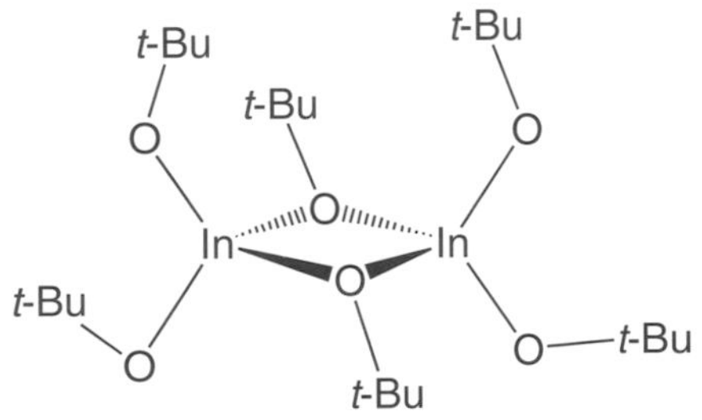

(**32**)

Acidic alkoxides (pK_a(O–H) of less than 10–11) also react with $In[N(SiMe_3)t\text{-}Bu]_3$ to afford alkoxides.[178] However, *t*-BuNH$_2$ resulting from the amido ligand decomposition often incorporates into the product. For example, an $In[N(SiMe_3)t\text{-}Bu]_3$ and 2,6-(*i*-Pr)$_2$C$_6$H$_3$OH mixture produces $[2,6\text{-}(i\text{-}Pr)_2C_6H_3O]_3In(t\text{-}BuNH_2)_2$.

(33)

The reactivity of some of the indium(III) alkoxides has been investigated. Monomeric molecules can be obtained by treating indium alkoxide aggregates with good Lewis bases, such as *p*-(dimethylamino)pyridine. Accordingly, compounds with *p*-(dimethylamino)pyridine donors, In(OCEtMe$_2$)$_3$(*p*-Me$_2$Npy) and In(OCMe$_3$)$_3$(*p*-Me$_2$Npy)$_2$, were synthesized and structurally characterized. They show four- and five-coordinate indium centers, respectively. The *p*-Me$_2$Npy ligands occupy axial positions of the five-coordinate, trigonal-bipyramidal system. The β-diketonate derivative [CH{(*t*-Bu)CO}$_2$]$_2$In(μ-OCMe$_3$)$_2$In(OCMe$_3$)$_2$ has also been obtained by the reaction of In(OCMe$_3$)$_3$ with [(*t*-Bu)CO]$_2$CH$_2$.

The synthesis and chemistry of indium fluoroalkoxides have also been reported.[182] For less acidic alkoxides, In[N(SiMe$_3$)*t*-Bu]$_3$ again serves as a good starting point. [In{μ-OCMe$_2$-(CF$_3$)}{OCMe$_2$(CF$_3$)}$_2$]$_2$ was obtained by treating the In(III)amide In[N(SiMe$_3$)*t*-Bu]$_3$ with HOCMe$_2$(CF$_3$). More acidic alcohols produce *t*-BuNH$_2$-incorporated products, such as In{OC-Me(CF$_3$)$_2$}$_3$(*t*-BuNH$_2$), In{OCH(CF$_3$)$_2$}$_3$(*t*-BuNH$_2$)$_3$, and [(*t*-BuNH$_3$][In{OCH(CF$_3$)$_2$}$_4$ (*t*-BuNH$_2$)]. Reactions involving In(tmp)$_3$ and In(NEt$_2$)$_3$ amides are less complicated, because they do not contain hydrolysable N–Si bonds. Synthesis and structures of trigonal bipyramidal [H$_2$NEt$_2$] [In{OCH(CF$_3$)$_2$}$_4$(HNEt$_2$)] and *mer*-octahedral In[OCMe(CF$_3$)$_2$]$_3$(py)$_3$ (**34**) have been reported as well. Chiral indium alkoxides were obtained by reacting Li$_2$(*S*)-BINOLate ((*S*)-BINOL = (*S*)-(−)-2,2′-dihydroxy-1,1′-binaphthyl) with InCl$_3$ in tetrahydrofuran.[183] Mixed-metal products were obtained, with three (*S*)-BINOLate ligands forming a distorted octahedral coordination sphere at indium.

(d) Multidentate oxygen ligands. Acetylacetonate complexes of indium(III) are also of interest as potential CVD precursors for the deposition of indium oxide materials. A number of homoleptic β-diketonates have been synthesized, including [CH{(Me)CO}$_2$]$_3$In, [CH{(*t*-Bu)CO}$_2$]$_3$In and [CH{(CF$_3$)CO}$_2$]$_3$In.[184–191] [CH{(CF$_3$)CO}$_2$]$_3$In can be synthesized from the free ligand and In(NO$_3$)$_3$, or [CH{(CF$_3$)CO}$_2$]Na and InCl$_3$.[191,192] Indium metal also reacts with bis(diketonato)-copper(II) derivatives (R^{1}COCHCOR^2)$_2$Cu (R^1, R^2 = Me, Me; or Ph, Ph; or Me, Ph; or Me, *t*-Bu) to afford tris(diketonato)indium(III) adducts.[189] The use of the parent diketone and In(OH)$_3$ to prepare indium(III) diketonates has been described as well.

X-ray structural data are available for some of these adducts.[184,193] The structure of [CH{(CF$_3$)CO}$_2$]$_3$In has been determined by gas-phase electron diffraction.[192] The coordination geometry at indium is described as distorted octahedral. The compounds [CH{(Me) CO}$_2$]$_3$In and [CH{(*t*-Bu)CO}$_2$]$_3$In have been used as precursors for indium oxide materials.[194–197]

A liquid–liquid extraction study using several β-diketones and Al^{3+} and In^{3+} ions reveals that the metal-ion extraction ability of β-diketone ligands depends on the distance between the two oxygen atoms and the interligand distance.[190] The ligands MeCOCH$_2$COMe and PhCOCH$_2$-COMe extract both smaller Al^{3+} and larger In^{3+} ions well. The PhCOC(Ph)HCOMe does not extract In^{3+}, because the phenyl group at the α-position prevents PhCOC(Ph)HCOMe from widening its bite size to accommodate the larger indium ion. Ligands with bulky terminal substituents, e.g., PhCOCH$_2$COPh, allow only the larger In^{3+} ion to be readily extracted.

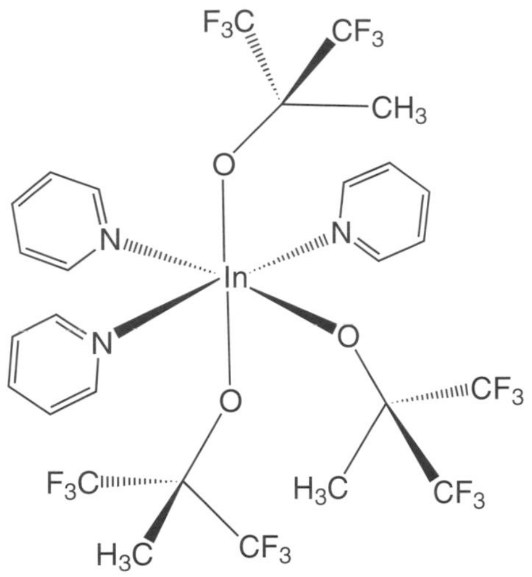

(34)

Mono-, bis-, and tris-monothio-β-diketonate complexes of In(III) have been prepared.[186,198–201] The crystal structure of In(PhCSC(H)COPh)₃ (**35**) shows that the indium has a distorted octahedral coordination with a *fac* arrangement of the S and O ligand atoms.

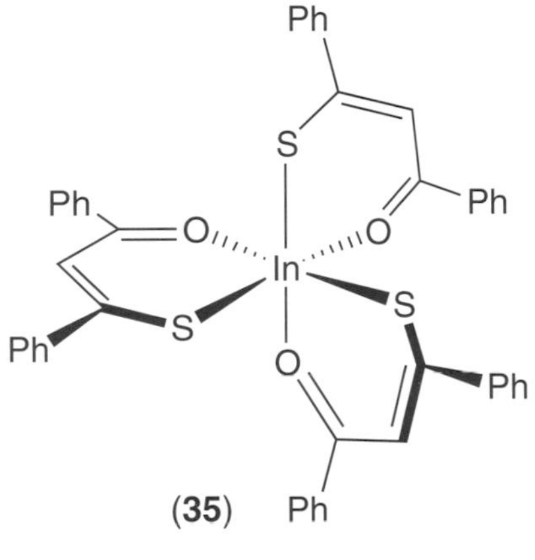

(35) Ph

Tris(tropolonato)indium(III) may be prepared in water from tropolone and indium(III) nitrate.[202] The indium center shows octahedral coordination. This tropolonato adduct is lipid-soluble and may be of radiopharmaceutical interest. The stability of adducts formed between trioctylphosphine oxide and tris(tropolonato)indium(III) has been investigated experimentally and computationally.[203] Unlike the gallium analogue, the indium complex can accommodate trioctylphosphine oxide while increasing its coordination number to seven. The thallium analogue can also take in one trioctylphosphine oxide. Indium derivatives of other chelating oxygen donors such as quinones,[204–207] γ-pyrone,[208] and pyridinone [209–217] have also been reported.

Indium halides react with *t*-butyl substituted orthoquinones to give either catecholate or semiquinonate complexes. For example, InI₃ reacts with two equivalents of (TBSQ)Na (TBSQ = 3,5-di-*tert*-butyl-1,2-benzosemiquinonate) to afford a paramagnetic product (TBSQ)₂InI.[206] The EPR data of (TBSQ)₂InI show characteristic signals attributable to semiquinonate (an anion radical) ligands bonded to an indium(III) center (with $A_{In} = 7.2$ G). Hyperfine constants for coupling to ^{115}In ($I = 9/2$) are about 5–7 G for similar In(III) adducts (for comparison, the monovalent derivatives show higher coupling constant values in the 9–10 G range).[218] Although the (TBSQ)₂InI species is present in solution, attempts to obtain a crystalline product by coordinating 4-methylpyridine molecules to indium led to the formation of the catecholate (TBC) derivative [(TBC)InI(4-Mepy)₂]₂ (**36**). Crystalline indium(III) complexes containing semiquinonate ligands are known. For example, solid (TBSQ)InI₂(4-Mepy)₂ can be obtained starting either with indium(II) iodide or InI₃. The compound (TBSQ)InBr₂(4-Mepy)₂ (**37**) has been obtained starting with InBr$_3$. The crystal structures of (TBSQ)InX$_2$(4-Mepy)$_2$ (X = Br, I) show that indium has a pseudooctahedral O$_2$N$_2$X$_2$ coordination sphere.[205,219]

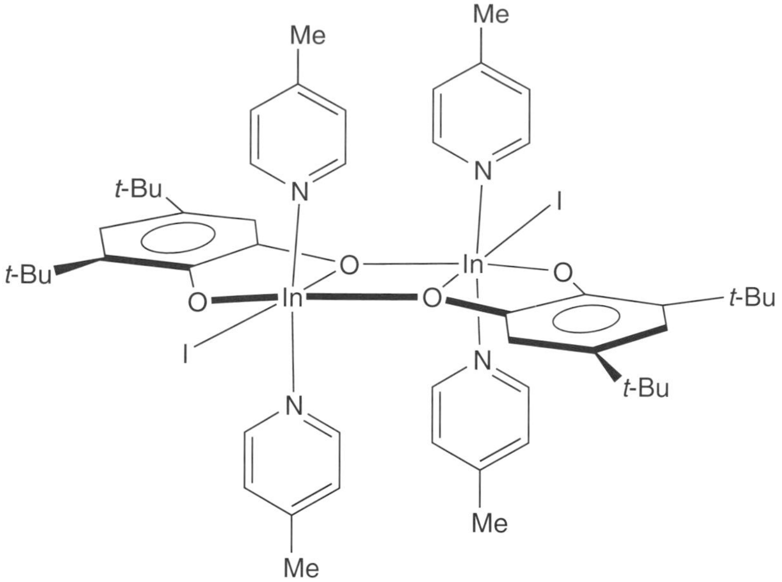

(36)

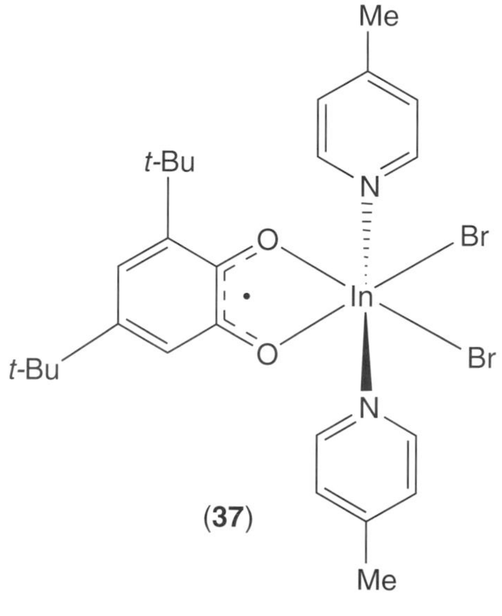

(37)

The chemistry of quinolinolato derivatives of In(III) has been investigated.[220–223] The crystal structure of tris(8-quinolinolato)indium(III) reveals that the indium has a pseudo *mer*-octahedral N$_3$O$_3$ coordination sphere. The ^{111}In analogue of this compound is useful for radiolabeling applications of white blood cells or platelets.[170,224,225]

Indium carboxylates and related compounds such as acetate, formate, and oxalate derivatives have been synthesized and investigated by various methods.[1,3,4,226] The structures of the In(III) oxalate complexes [In$_2$(C$_2$O$_4$)$_3$(H$_2$O)$_4$]·2H$_2$O, NH$_4$[In(C$_2$O$_4$)$_2$]·2H$_2$O, and Na[In(C$_2$O$_4$)$_2$]·2H$_2$O have been studied using X-ray crystallography.[227] Thermal decomposition of NH$_4$[In(C$_2$O$_4$)$_2$]·2H$_2$O leads to indium oxide.[228] The thermal behavior of mixed indium–thallium salts has also been examined.[229,230] Thermal decomposition of indium(III) formate provides another route to indium oxide materials.[231] It is possible to synthesize indium oxalato complexes with fluoride ligands. The preparation of In(C$_2$O$_4$)F and [In(C$_2$O$_4$)F$_2$]$^-$ has been reported.[232]

(e) Polyoxyanion ligands. Indium(III) compounds of polyoxyanions such as nitrate, sulfate, phosphate, and chlorate are well known.[1,4] Some of the work on indium phosphates in the late 1990s centered on developing different three-dimensional structures using organic templates.[233–242] Indium phosphonate derivatives have also been investigated as building blocks for preparing materials with well-defined internal spaces.[243–246]

Several periodato complexes of indium(III) have been obtained by using $In(NO_3)_3$ and H_5IO_6. At pH < 1, a crystalline product $H_{11}I_2InO_{14}$ forms.[247] At higher pH, insoluble, amorphous material of composition $In_5(IO_6)_3 \cdot nH_2O$ and $H_3In_4(IO_6)_3 \cdot nH_2O$ was produced.

(ii) Sulfur, selenium, tellurium ligands

(a) Neutral sulfur ligands. Neutral, sulfur-based donors, such as thioether and thiourea, form adducts with indium ions. The tridentate 1,4,7-trithiacyclononane reacts with $InCl_3$ to form a 1:1 adduct.[172] Structural data are not available. Cationic, neutral, and anionic indium(III) complexes of thiourea have been described.[4,248,249]

(b) Thiolate ligands. Alkyl and aryl thiolate derivatives of indium(III) can be synthesized by several methods. For example, $In(SEt)_3$, $In(S-n-Bu)_3$, and $In(SCMe_2Et)_3$ have been synthesized by an electrochemical method using an indium anode and the appropriate thiol in an acetonitrile medium.[250] This method also allows the synthesis of aryl thiolate complexes like $In(SPh)_3$, $In(SC_6F_5)_3$, $In(S-p-tolyl)_3$, and $In(SC_{10}H_7)_3$, as well as low-valent indium thiolates. A metathesis route (involving $InCl_3$ and $NaSPh$)[251] and an oxidative addition pathway (using indium metal and $PhSSPh$)[252] to $In(SPh)_3$ are also available. $In[S(2,4,6-(t-Bu)_3C_6H_2)]_3$ (38) has been synthesized by an amine-elimination process[253] using $In[N(SiMe_3)_2]_3$ and three equivalents of 2,4,6-$(t-Bu)_3C_6H_2SH$ in toluene. It is a rare monomeric indium thiolate, and has a trigonal-planar indium center. The steric bulk of the ligand obviously prevents aggregation. $In[N(SiMe_3)t-Bu]_3$ serves as a good starting point for $In(S-t-Bu)_3$ and $In(S-i-Pr)_3$.[254] The compound $In(S-t-Bu)_3$ is believed to be a dimer, whereas $In(S-i-Pr)_3$ is a polymeric solid.

(38)

$In[S(2,4,6-(i-Pr)_3C_6H_2)]_3$ is a yellow oil at room temperature.[255] Solid samples containing $In[S(2,4,6-(i-Pr)_3C_6H_2)]_3$ may be obtained by coordinating THF or acetonitrile to the indium atom. $In[S(2,4,6-(i-Pr)_3C_6H_2)]_3(THF)$ and $In[S(2,4,6-(i-Pr)_3C_6H_2)]_3(CH_3CN)_2$ have been structurally characterized. The thiolate ligands of the indium(III) complexes $[2-(MeO)-5-(Me)C_6H_3-S]_3In$ and $[o-(Me_2NCH_2)C_6H_4S]_3In$ have additional O- and N-donor sites.[255] An electrochemical route has been utilized to synthesize $In[2-(Ph_2P)C_6H_4S]_3$, $In[2-(Ph_2P)-6-(Me_3Si)C_6H_3S]_2[2-(Ph_2PO)-6-(Me_3Si)C_6H_3S]$, $In[2-(Ph_2PO)-6-(Me_3Si)C_6H_3S]_3$, and $[NMe_4][In\{PhP(C_6H_4S-2)_2\}_2]$.[256] The additional P- and/or O donors present on these thiolate ligands coordinate intramolecularly to indium forming octahedral structures with InS_3P_3, InS_3P_2O, and InS_4P_2 cores, respectively. The compound $In[2-(Ph_2P)C_6H_4S]_3$ (39) shows the *mer*-conformation. Phosphorus atoms of $[In\{PhP(C_6H_4S-2)_2\}_2]^-$ anion occupy *cis*-sites of an octahedron.

The Lewis-acidic property at the indium center also allows the synthesis of $In(S-t-Bu)_3(py)$, $In(S-t-Bu)_3(p-Me_2Npy)_2$,[254] $In(SPh)_3(py)_2$,[257] and $In[S(2,4,6-(CF_3)_3C_6H_2)]_3(Et_2O)$.[258] X-ray crystallographic data are available. The compounds $In(S-t-Bu)_3(p-Me_2Npy)_2$ and $In(SPh)_3(py)_2$ have trigonal-bipyramidal structures, with the apical sites occupied by pyridines. The In(S-*t*-Bu)₃(py) and In[S(2,4,6-(CF₃)₃C₆H₂)]₃(Et₂O) have four-coordinate, tetrahedral indium atoms.

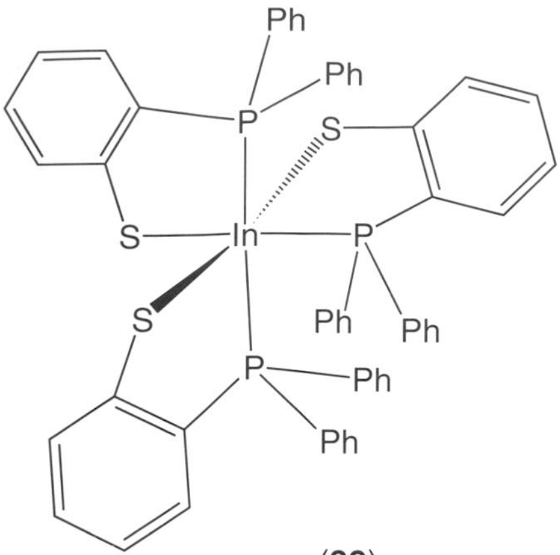

(39)

Synthesis of indium(III) thiolates containing halide donors have also been reported. The oxidative addition reactions of InX (X = Cl, Br, I) with PhSSPh yields (PhS)₂InX.[4,259] Related selenolates can also be synthesized, using PhSeSePh instead of PhSSPh.[259]

Ionic indium(III) thiolate compounds [Ph₄P][In(S-*t*-Bu)₄] and [Ph₄P][In(SCH₂CH₂S)₂] were obtained using InCl₃ and the appropriate thiolate *t*-BuSK or NaSCH₂CH₂SNa in the presence of a PPh₄⁺ salt. The synthesis of [XIn(SPh)₃]⁻ (X = Cl, Br, I) salts was achieved by treating In(SPh)₃ with tetraalkyl ammonium salts.[251] Anionic moieties have four-coordinate, tetrahedral indium atoms. The first indium–copper cluster [Ph₄P][Cu₆In₃(SEt)₁₆] has been prepared, using [Cu(CH₃CN)₄]PF₆ and [Ph₄P][In(SEt)₄]. Its crystal structure shows an adamantanoid Cu₆In₃S₁₃ framework (**40**).[260]

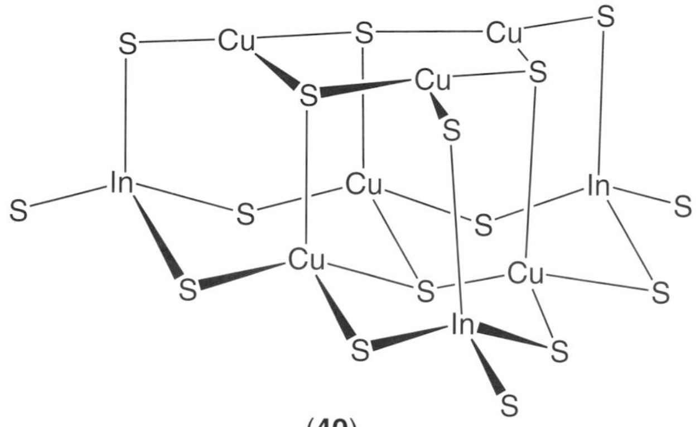

(40)

One of the primary interests in indium thiolates concerns their potential utility as precursors for chemical vapor deposition of indium sulfide and related materials. Thermal decomposition of In[S(2,4,6-(*i*-Pr)₃C₆H₂)]₃, [2-(MeO)-5-(Me)C₆H₃S]₃In, and [*o*-(Me₂NCH₂)C₆H₄S]₃In leads to In₂S₃.[255] Indium thiocarboxylates are also useful in this regard.[261] A sonochemical method for In₂S₃ involves the sonication of InCl₃ and thioacetamide in an aqueous solution at room temperature. At 0 °C, In₂O₃ was the major product.[262] Compounds like [Ph₄P][Cu₆In₃(SEt)₁₆] are of interest as potential sources of InCuS₂ materials. (Ph₃P)₂AgIn{SC(O)R}₄ (R = Me or Ph), which is derived from thiocarboxylate ligands, serves as an excellent precursor for AgInS₂ and AgIn₅S₈ materials. The indium(III) in (Ph₃P)₂AgIn{SC(O)Ph}₄ adopts a distorted octahedral coordination geometry.[263]

(c) Chelating anionic sulfur ligands. The chemistry of indium(III) complexes containing bidentate sulfur donors such as dithiocarbamates [S₂CNR₂]⁻, dithiophosphates [S₂P(OR)₂]⁻, dithiophosphinates [S₂PR₂]⁻, and dithioarsinates [S₂AsR₂]⁻ has been investigated.[264–283] The synthesis of In[S₂CNEt₂]₃ using a weakly acidic solution of an indium salt and sodium diethyldithiocarbamate

was reported many years ago, in the early 1940s.[284] Since then, many different tris(dialkyldithio-carbamates) complexes of indium, including $In[S_2CNR_2]_3$ (R = Me, Et, *n*-Pr, *i*-Pr, *n*-Bu, *i*-Bu), have been synthesized and characterized structurally, spectroscopically, and by thermodynamic methods.[267,269,273–275,277,279,280,284] Their use in the preparation of indium sulfide material, however, is more recent.[265,275,285] For example, the tris(dialkyldithiocarbamates) $In[S_2CN(Me)n\text{-}Bu]_3$ and $In[S_2CN(Me)n\text{-}hexyl]_3$ serve as excellent precursors for depositing In_2S_3 films under CVD conditions.[285] Indium(III) derivatives containing internally functionalized dithaiocarbamate ligands, $[S_2CN(CH_2CH_2)_2O]^-$, (**41**), and $[S_2CN(CH_2CH_2)_2NMe]^-$, are also known.[264,265]

(**41**)

A common route to synthesis involves the use of $InCl_3$ and the sodium salt of dialkyldithio-carbamate.[264,284] This method also allows the isolation of mixed-ligand adducts, e.g., $Cl_2In\text{-}[S_2CN(CH_2CH_2)_2O]$, $ClIn[S_2CN(CH_2CH_2)_2O]_2$, $O(CH_2CH_2S)_2In[S_2CN(CH_2CH_2)_2O]$, and $O(CH_2CH_2S)_2In[S_2CN(CH_2CH_2)_2NMe]$.[264] A related, but much simpler method uses an acidic solution of In(III) (generated from indium metal and HCl) and dialkylammonium dialkyldithio-carbamate ($[R_2NH_2][S_2CNR_2]$, synthesized from R_2NH and excess CS_2 in acetone).[269] Upon treatment of this mixture with NaOH, $In[S_2CNR_2]_3$ (R = Me, Et, *n*-Pr, *i*-Bu) precipitates as a white solid. $In[S_2CNMe_2]_3$ has been obtained from a reaction of $Me_2NC(S)SS(S)CNMe_2$ with indium metal in refluxing xylene. The diethyl analogue could not be obtained by this method.[269] The same reagents react in 4-methylpyridine at room temperature to afford $In[S_2CNMe_2]_3$.[267,274] Electrochemical methods that use a sacrificial indium anode and $Me_2NC(S)SS(S)CNMe_2$ are also reported.[269,276,277] Reaction of InX (X = Cl, Br, I) with $Et_2NC(S)SS(S)CNEt_2$ gives $In[S_2CNEt_2]_3$.[269] This reaction is believed to go through an $XIn[S_2CNEt_2]_2$ intermediate.

Tris(dialkyldithiocarbamato)indium(III) compounds $In[S_2CNR_2]_3$ exist as discrete molecules with distorted octahedral indium sites.[264,280] The related $In[S_2COEt]_3$ has a similar structure.[286] The five-coordinate $ClIn[S_2CN(i\text{-}Pr)_2]_3$ displays square-pyramidal geometry.[273] Compounds such as $In[S_2CO\text{-}i\text{-}Pr]_3$ serve as precursors for indium sulfide films.[287]

Dialkylmonothiocarbamato derivatives of indium(III) can be prepared using either the sodium or lithium salt of the carbamate ligand and $InCl_3$. Compounds $In[SOCNR_2]_3$ (R = Et, *i*-Pr (**42**)) have been prepared, structurally characterized, and used successfully as single-source precursors for the deposition of β-In_2S_3 films by low-pressure MOCVD at temperatures of 300–500 °C.[288–290] These monomeric compounds feature a distorted trigonal-prismatic geometry, with *mer*-O_3S_3 conformation at indium.

Indium(III) dithiophosphate $[S_2P(OR)_2]^-$,[270,272,278] dithiophosphinate $[S_2PR_2]^-$,[266,268,281–283,291] and dithioarsinate $[S_2AsR_2]^-$ complexes[271] contain somewhat similar sulfur-based chelating ligands. The compound $In[S_2P(i\text{-}Bu)_2]_3$ can be prepared by treating $InCl_3$ with $Na[S_2P (i\text{-}Bu)_2]$.[266] The ammonium salt of the ligand has been used in the synthesis of $In[S_2P(OR)_2]_3$ (R = Et, *n*-Pr, *i*-Pr, etc.).[272,278] The structurally characterized complexes are all monomeric, and contain six-coordinate indium atoms with distorted octahedral geometry.[266,270–272,281,283,291] More descriptive, and perhaps more proper, ways of describing the deviations from ideal octahedral geometry (often observed with these four-membered chelates) were discussed.[271] An X-ray

(42)

crystallographic study reveals that In[S$_2$P(*i*-Bu)$_2$] and Ga[S$_2$P(*i*-Bu)$_2$]$_3$ are not isostructural, which is rare for closely related systems of indium and gallium with S/S or O/O chelates.[266]

The imidobis(diphenylphosphinechalcogenide) ligands [Ph$_2$P(X)NP(X)Ph$_2$]$^-$ (X = O, S, Se) also form complexes with indium(III). They feature six-membered, phosphazene metallacycles. The dithioindium adduct [Ph$_2$P(S)NP(S)Ph$_2$]$_3$In,[292] and the related [Ph$_2$P(O)NP(O)Ph$_2$]$_3$In,[292,293] [Ph$_2$P(Se)NP(Se)Ph$_2$]$_3$In,[294] and [Ph$_2$P(S)NP(O)Ph$_2$]$_3$In[295] have been reported, including their X-ray crystal structural data. Indium adducts have distorted octahedral geometry. The structure of the mixed-donor tris(chelate) complex [Ph$_2$P(S)NP(O)Ph$_2$]$_3$ In corresponds to the *fac*-isomer. It was prepared by reacting [Ph$_2$P(S)NP(O)Ph$_2$]K with InCl$_3$ in a 3:1 molar ratio. Interestingly, the attempted synthesis of [Ph$_2$P(S)NP(Se)Ph$_2$]$_3$In by following a similar route leads only to the bis ligand adduct [Ph$_2$P(S)NP(Se)Ph$_2$]$_2$InCl. It is monomeric, and has a five-coordinate, distorted trigonal-bipyramidal indium center. The Cl and the Se atoms occupy the equatorial sites. Five-coordinate indium(III) adducts containing symmetric imidophosphinate ligands have also been reported. These include [*i*-Pr$_2$P(S)NP(S)*i*-Pr$_2$]$_2$InCl, [*i*-Pr$_2$P(Se)NP(Se)*i*-Pr$_2$]$_2$InCl, and [Ph$_2$P(Se)NP(Se)Ph$_2$]$_2$InCl.[296] They were obtained by the 2:1 stoichiometric reaction of the potassium or the sodium salt of the ligand with InCl$_3$. They all have distorted trigonal-bipyramidal geometry with equatorially bound chlorides.

Indium(III) complexes of dianionic sulfur ligands are mostly those derived from toluene-3,4-dithiolate (TDT^{2-}), 1,2-dicyanoethylene-1,2-dithiolate (MNT^{2-}), 1,2-ethanedithiol (EDT^{2-}), or 1,1-dicyanoethylene-2,2-dithiolate (*i*-MNT^{2-}). Various adducts (e.g., four-coordinate [In(MNT)$_2$]$^-$, five-coordinate [XIn(*i*-MNT)$_2$]$^{2-}$ (X = Cl, Br, or I), or six-coordinate [In(*i*-MNT)$_3$]$^{3-}$) have been reported.[1,4]

A number of indium thiolato complexes containing additional nitrogen-donor sites have been described in the literature. These include pyridine-2-thionate derivatives In(pyS)$_3$ (**43**),[282,297] In(3-CF$_3$pyS)$_3$,[298] In(3-Me$_3$SipyS)$_3$,[297] and pyrimidine-2-thionates In(RpymS)$_3$ (R = H; 4,6-Me$_2$; 5-Et-4,6-Me$_2$; 4,6-(Me, CF$_3$)).[299] The compounds In(pyS)$_3$ and In(3-Me$_3$SipyS)$_3$ were prepared using In(NO$_3$)$_3$, pyridine-2-thiole derivative and Et$_3$N in ethanol. Under anaerobic conditions in ethanol and with the use of InCl$_3$ and H(pyS), an alkoxy-bridged dimer [In(pyS)$_2$ (OEt)]$_2$ could be isolated.[297] Electrochemical oxidation of a sacrificial indium anode in a nonaqueous solution containing the precursor ligand is the method used in the synthesis of In(RpymS)$_3$. It is considered to be the preferred synthetic route to most of these compounds.[298] It is believed that the electrochemical reactions proceed via indium(I) derivatives.[298] Hydrogen gas forms at the cathode.

Solution NMR spectroscopic data of these tris(ligand) adducts point to the existence of *fac*-S$_3$N$_3$ isomers in solution. The same structure is retained in the solid state for In(pyS)$_3$, In(3-CF$_3$pyS)$_3$, In(3-Me$_3$SipyS)$_3$, and In(pymS)$_3$.[298] However, the compound In(5-Et-4,6-Me$_2$pymS)$_3$ adopts a *mer* conformation in the solid state.[299] The indium(III) complexes of

1-hydroxypyridine-2-thione (HPT) have also been synthesized by an electrochemical method.[300] The In(PT)$_3$ prefers the *fac* arrangement of ligands in chloroform, but crystallizes in the *mer* conformation.

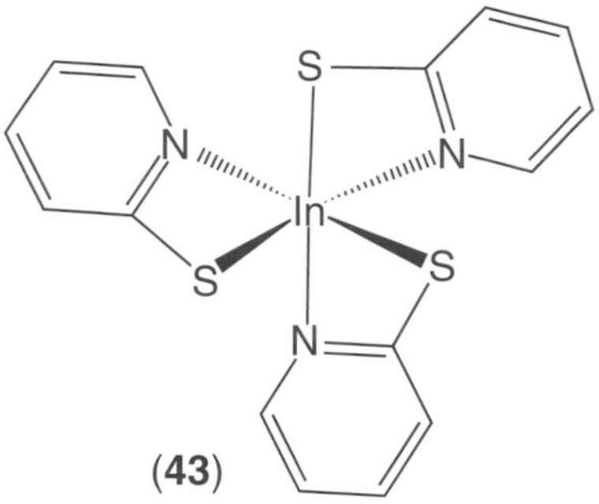

(43)

(d) Selenium and tellurium ligands. Group III/V material involving heavier chalcogens is also of interest.[301–303] Thus, as in the case of lighter thiolates, indium(III) selenolates and tellurolates have been investigated as possible single-source precursors for group III/V materials. However, compared to indium(III) thiolates, relatively little is known about the structures and properties of the heavier analogues. Neutral homoleptic complexes [In(SePh)$_3$]$_n$,[257,304] In[Se(2,4,6-(*t*-Bu)$_3$C$_6$H$_2$)]$_3$ (44),[305] In[SeC(SiMe$_3$)$_3$]$_3$,[305] In[SeSi(SiMe$_3$)$_3$]$_3$,[305] and In[TeSi(SiMe$_3$)$_3$]$_3$ have been synthesized.[305]

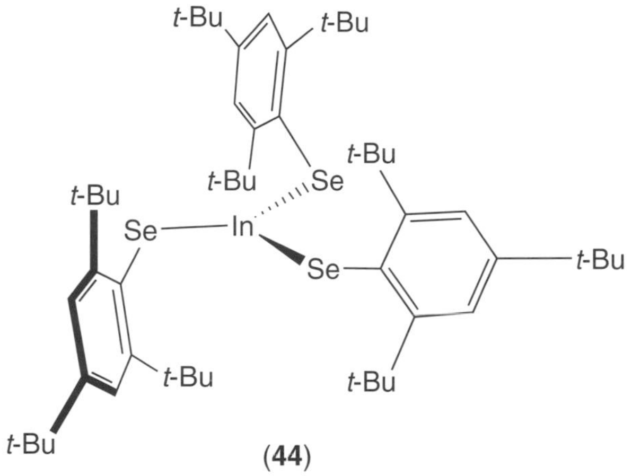

(44)

The diselenide PhSeSePh reacts with indium metal in refluxing toluene to give In(SePh)$_3$. The iodo derivative InI(SePh)$_2$ can be obtained by adding iodine to the mixture.[252] InCl$_3$ reacts with three equivalents of (DME)LiSeC(SiMe$_3$)$_3$ to afford In[SeC(SiMe$_3$)$_3$]$_3$, whereas with (THF)$_2$LiSe-Si(SiMe$_3$)$_3$ a THF adduct (THF)In[SeSi(SiMe$_3$)$_3$]$_3$ was obtained.[305] The THF-free compound In[SeSi(SiMe$_3$)$_3$]$_3$ and the related tellurium derivative In[TeSi(SiMe$_3$)$_3$]$_3$ can be synthesized by treating Cp$_3$In with HSeSi(SiMe$_3$)$_3$ and HTeSi(SiMe$_3$)$_3$, respectively. The synthesis of In[Se(2,4,6-(*t*-Bu)$_3$C$_6$H$_2$)]$_3$ involves an alkane-elimination process between HSe(2,4,6-(*t*-Bu)$_3$C$_6$H$_2$) and Et$_3$In.[253] The indium hydride complex InH$_3$(PCy$_3$) and PhMMPh (M = S, Se, Te) in DME were utilized in the synthesis of In(MPh)$_3$(PCy$_3$).[306]

The Lewis acidity of the indium center is apparent in the formation of adduct compounds like In(SePh)$_3$(PPh$_3$)$_2$, In(SePh)$_3$(PCy$_3$), In(TePh)$_3$(PCy$_3$), In(SePh)$_3$(py)$_2$, In(SePh)$_3$(2,2'-bipy), In(SePh)$_3$(phen), In[SeSi(SiMe$_3$)$_3$]$_3$(THF), In[SeSi(SiMe$_3$)$_3$]$_3$(py), In[SeSi(SiMe$_3$)$_3$]$_3$(TMEDA), In[SeSi(SiMe$_3$)$_3$]$_3$(DMPE), and {In[SeSi(SiMe$_3$)$_3$]$_3$}$_2$(μ-DMPE) (DMPE = 1,2-bis(dimethylphos-phino)ethane).[252,305] Although In[TeSi(SiMe$_3$)$_3$]$_3$ also forms adducts with Lewis bases, attempts to isolate adduct complexes have resulted in significant decomposition to the indium-free products Te[Si(SiMe$_3$)$_3$]$_2$ and [TeSi(SiMe$_3$)$_3$]$_2$. Note, however, that the In(TePh)$_3$(PCy$_3$) has been isolated as a thermally stable solid and characterized using X-ray crystallography.[306]

The indium selenolate [PPh$_4$][In(SePh)$_4$] can be prepared by treating InCl$_3$ with NaSePh, followed by the addition of [PPh$_4$]Cl.[307] Interestingly, if the product resulting from InCl$_3$ and

NaSePh is added to a flask containing $NaBH_4$ and elemental sulfur (not Se), a hydroselenido derivative $[In(SeH)(SePh)_3]^-$ can be isolated. Compounds like $[PPh_4][In(SeH)(SePh)_3]$ with hydroselenido ligands are rare.

The treatment of neutral $In(SePh)_3$ with $[PPh_4]Br$ also results in a selenolate adduct compound $[PPh_4][BrIn(SePh)_3]$.[257] A copper–indium complex with bridging selenolates $(Ph_3P)_2Cu[In(\mu-SeEt)_2(SeEt)_2]$ was also reported.[308] The action of selenium on $[HB(3,5-(t-Bu)_2Pz)_3]In$ or $[(t-Bu_3Si)_2In]_2$ affords $[HB(3,5-(t-Bu)_2Pz)_3]InSe$ or $(t-Bu_3SiIn)_4Se_4$ (45), respectively.[29] The tellurium does not react with $[HB(3,5-(t-Bu)_2Pz)_3]In$.[309]

(45)

The indium selenolate complex $[2,4,6-(t-Bu)_3C_6H_2Se]_3In$ (44) shows trigonal planar coordination of the indium atom.[253] The compound $In[SeSi(SiMe_3)_3]_3$ is also monomeric.[305] The indium atom appears to have close contacts with hydrogen atoms of methyl groups. Although the indium is four-coordinate in $\{In[SeSi(SiMe_3)_3]_3\}_2(\mu-DMPE)$ (46), it adopts a flattened tetrahedral geometry.[305] The anions $[In(SePh)_4]^-$ and $[In(SeH)(SePh)_3]^-$ (47) and the phosphine adducts $In(MPh)_3(PCy_3)$ (M = S, Se, Te) feature the expected tetrahedral coordination.[306,307] The IR stretching frequency corresponding to the Se–H stretch appears at 2,241 cm^{-1}. The neutral $In(SePh)_3$ (48) and (49), which lacks bulky substituents, is polymeric. However, it shows two crystalline modifications, a monoclinic form (48) that has six-coordinate, octahedral indium centers,[257] and a triclinic version (49) featuring five-coordinate, trigonal-bipyramidal indium sites.[304]

(46)   (47)

The In(III) adducts $[Ph_2P(Se)NP(Se)Ph_2]_3In$ and $[i-Pr_2P(Se)NP(Se)i-Pr_2]_2InCl$, containing chelating ligands, were described earlier. Dialkylselenocarbamate derivatives of indium, like $In[Se_2CN(Me)n-hexyl]_3$, are useful CVD precursor compounds for the deposition of In_2Se_3 films.[310] $In[Se_2CN(Me)n-hexyl]_3$ has been synthesized using $InCl_3$, CSe_2, and N-methylhexylamine. Ternary material $CuInSe_2$ has been prepared from a stoichiometric mixture of $In[Se_2CN(Me)n-hexyl]_3$ and $Cu[Se_2CN(Me)n-hexyl]_2$.[311,312] Pyrolysis of $In(SePh)_3$ affords hexagonal films of In_2Se_3. $In(SePh)_3$ has also been used in the preparation of III/V material using a spray MOCVD technique.[313]

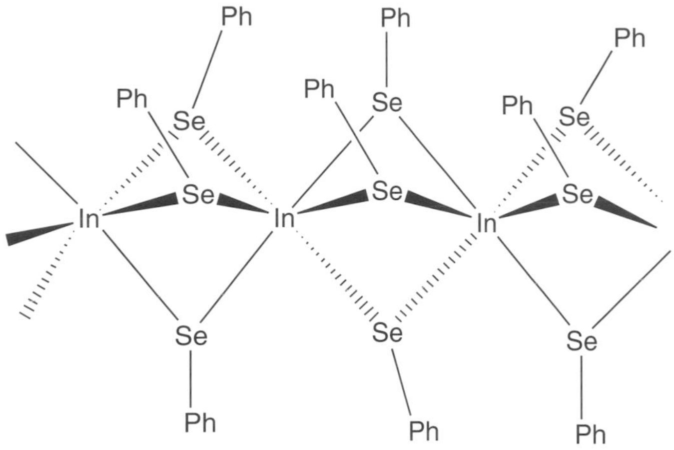

(48)

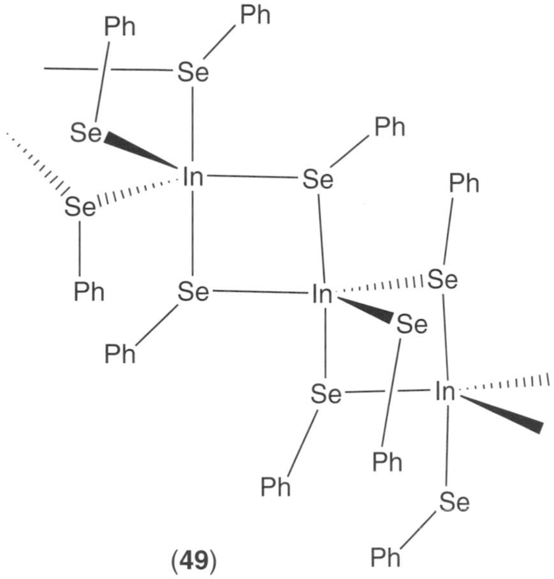

(49)

The pyridineselenolate ([SePy]⁻) and the 3-(trimethylsilyl) pyridineselenolate ([3-Me₃SipySe]⁻) ligands form air-stable, homoleptic In(III) compounds In(SePy)₃ and In(3-Me₃SipySe)₃.[314,315] The In(SePy)₃ has been synthesized by an electrochemical or a thermal method using indium metal and 2,2′-dipyridyldiselenide. Arrangements of the donor atoms of In(SePy)₃ (50) and In(3-Me₃SipySe)₃ around indium correspond to the *fac*-isomer. The same structures are maintained in solution, as indicated by the presence of single peak at δ 399 in the ⁷⁷Se NMR spectrum.[298] Compound In(SePy)₃ decomposes at 220 °C to afford In₂Se₃.[314]

3.5.1.2.4 Group 17 ligands

Indium(III) fluoride, chloride, bromide, and iodide are commercially available compounds. They are ionic compounds with six-coordinate metal sites.[2] The dimeric In₂I₆ (β-form) is also known.[316,317] Thermal decomposition of (NH₄)₃InF₆ is one of the routes to InF₃.[7] InF₃-based glass materials are important in optics-related applications. Unlike the Tl(III) derivative, which hydrolyses in water, InF₃ is insoluble in water. Hydrates of InF₃ are obtained by the evaporation of HF solutions of InF₃. The other trihalides of indium (InX₃; X = Cl, Br, I) are hygroscopic compounds and can be synthesized directly from the elements. InI₃ may be obtained easily by reacting indium with I₂ in diethyl ether.[318] These halides are widely used as starting materials for the synthesis of various other indium compounds.

(50)   **(51)**

Indium(III) halides form adducts with a variety of neutral and anionic donors of group 15, 16, or 17 elements. This area has been explored actively for many years. The interest in these adducts ranges from learning the effects of d^{10} configuration on the structure and stability of complexes, through their possible use as precursors for MOCVD processes (e.g., InN, InP material), to potential catalytic applications. Types of coordination compound formed by indium(III) halides (InX_3) include InX_3L, InX_3L_2, InX_3L_3, $[InX_2L_2]^+$, $[InX_2L_4]^+$, $[InX_4L_2]^-$, $[InX_5L_2]^{2-}$, $[InX_4]^-$, $[InX_5]^{2-}$, and $[InX_6]^{3-}$ (L = neutral donor). Note that not all these types are reported for all the halides. Based on the reported data, chloride derivatives appear to be the most diverse.

Donor–acceptor complexes involving indium(III) have been investigated using computational methods.[319,320] A recent theoretical study of $MX_3–D$ (M = Al, Ga, In; X = F, Cl, Br, I; D = YH_3, YX_3, X^-; Y = N, P, As), using self-consistent field and non-Hartree–Fock/density functional (B3LYP) methods with effective core potentials, reveals that the donor–acceptor strength decreases in the order F > Cl > Br > I and Al > Ga < In for all the donors D.[320] The study also finds that for all indium(III)(and Al and Ga) halides, the donor strength follows the order $X^- > NH_3 > H_2O > PH_3 > AsH_3 > PX_3$.

(i) Halides and carbon ligands

One of a rare group of donor-acceptor adducts concerns the nucleophilic carbene complexes of indium (III) halides.[321] Carbene complexes of boron,[322] aluminum,[323] gallium,[324] and thallium[325] have also been reported. The 1:1 adducts $Cl_3In[C\{N(i\text{-}Pr)CMe\}_2]$, $Br_3In[C\{N (i\text{-}Pr)CMe\}_2]$ (**51**), and 1:2 adducts $Cl_3In[C\{N(i\text{-}Pr)CMe\}_2]_2$, $Br_3In[C\{N(i\text{-}Pr)CMe\}_2]_2$ (**52**) can be synthesized by treating the appropriate InX_3 (X = Cl, Br) with either one or two equivalents of "stable" carbene $C\{N(i\text{-}Pr)CMe\}_2$.[321] The reaction at 1:5 metal halide-to-carbene ratio produced only 1:2 adducts, suggesting that the 1:3 complexes are sterically not viable. The conductivity and the ^{115}In NMR spectra suggest that the solid-state structures are retained in methylene chloride solutions, and are not in equilibrium with ionic structures of the type $[InX_2L_2][InX_4]$. For example, no signals corresponding to $[InCl_4]^-$ and $[InBr_4]^-$ ions were observed in ^{115}In NMR spectra at δ 430 and 176, respectively.

The X-ray data reveal that $Br_3In[C\{N(i\text{-}Pr)CMe\}_2]$ (**51**) is monomeric and tetrahedral. Both the 1:2 adducts $Cl_3In[C\{N(i\text{-}Pr)CMe\}_2]_2$ and $Br_3In[C\{N(i\text{-}Pr)CMe\}_2]_2$ (**52**) show essentially trigonal-bipyramidal indium sites but unusual halide ion coordination, with halide ions occupying one equatorial and two axial sites. Most 1:2 adducts between indium(III) halides and neutral donors show three equatorial halides. Ionic compounds featuring $\{Cl_4In[C\{N(i\text{-}Pr)CMe\}_2]\}^-$ and $\{Br_4In[C\{N(i\text{-}Pr)CMe\}_2]\}^-$ (**53**) anions have been obtained by treating a 1:1 mixture of InX_3 (X = Cl, Br) and $[C\{N(i\text{-}Pr)CMe\}_2]$ with half an equivalent of water.[321] The resulting ionic compounds $\{H[C\{N(i\text{-}Pr)CMe\}_2]\}\{X_4In[C\{N(i\text{-}Pr)CMe\}_2]\}$ are fluxional in solution at room temperature. Only one set of heterocyclic resonances has been observed in ^1H and ^{13}C NMR spectra for the coordinated carbene and the imidazolium cation. The anions show trigonal-bipyramidal geometry at indium, with the carbene occupying an equatorial site.

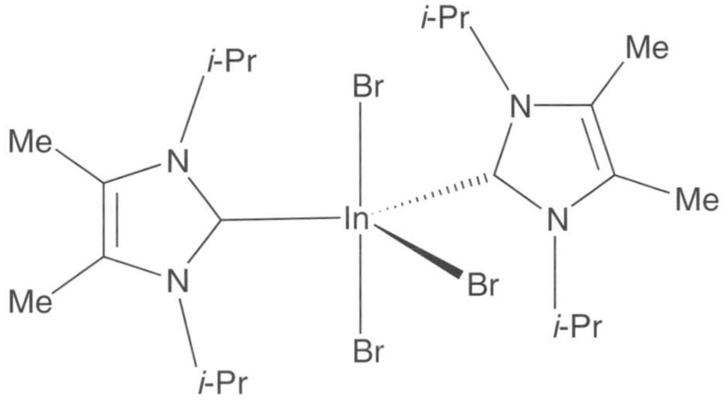

(52)

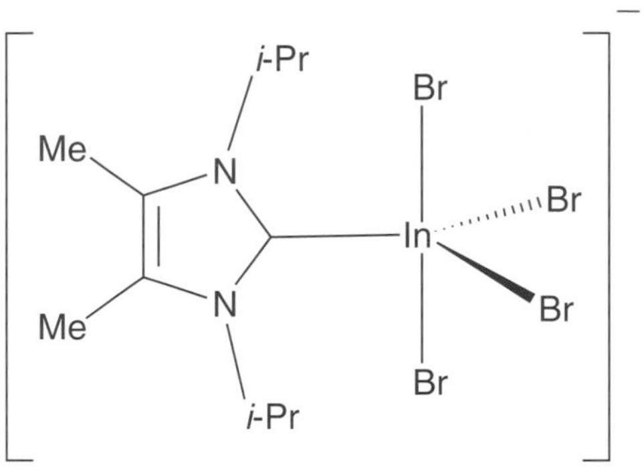

(53)

Indium halide compounds with ylide donors have been reported.[326,327] The reaction of InBr with CH_2Br_2 leads to Br_2InCH_2Br, which upon treatment with PPh_3 produces $Br_3InCH_2PPh_3$. Several other adducts of the type Br_3InCH_2L (L = NEt_3, $AsPh_3$, $SbPh_3$, $SC(NMe_2)_2$) are also known.[326,327] Some of these compounds have been analyzed by X-ray crystallography, semi-empirical quantum-mechanical methods, mass spectroscopy, and by thermogravimetric methods.

(ii) Halides and group 15 and group 16 ligands

Data on group 15, 16, and 17 donor adducts of indium(III) halides are more numerous. Compounds of the type InX_3L include $InCl_3(OCMe_2)$, $InCl_3(OCPh_2)$, $InCl_3(OPCl_3)$,[7] $InI_3(py)$,[7] $InI_3(L)$ (L = PPh_3, $P(i-Pr)_3$, $P(SiMe_3)_3$, $PHPh_2$, $PH(t-Bu)_2$, $AsPh_3$).[328–332] Structurally character-ized compounds of this type reveal the expected tetrahedral geometry at the indium. The $InI_3[P-(i-Pr)_3]$ can be synthesized by stirring a mixture of indium powder and $I_2P(i-Pr)_3$ (2:3 molar ratio) in Et_2O for 7 days.[330] Interestingly, the reaction involving $I_2P(n-Pr)_3$ which contains the *n*-propyl substituents leads to a divalent indium iodide product. Reaction of $(Et_2O)InI_3$ with PPh_3 or $AsPh_3$ affords $InI_3(L)$ (L = PPh_3 or $AsPh_3$) along with five-coordinate adducts $InI_3(L)_2$.[329]

The five-coordinate compounds are well represented. Some example of InX_3L_2 type include $InCl_3(THF)_2$,[333] $InCl_3(NMe_3)_2$,[334] $InCl_3\{OC(NMe_2)_2\}_2$,[335] $InCl_3\{SC(NMe_2)_2\}_2$,[335] $InCl_3\{SC[N(Me)-CH]_2\}_2$,[336] $InCl_3(PMe_3)_2$,[337] $InBr_3(THF)_2$,[338] $InBr_3\{SC[N(Me)CH]_2\}_2$,[336] $InBr_3(PPhMe_2)_2$,[339] $InI_3(PPhMe_2)_2$,[339] $InI_3(PPh_3)_2$,[332] and $InI_3(AsPh_3)_2$.[329] Solid-state structures (e.g., $InCl_3(NMe_3)_2$, $InCl_3-(PMe_3)_2$, and $InCl_3\{OC(NMe_2)_2\}_2$ (54))[334,335,337] consist of trigonal-bipyramidal indium sites with halides occupying the sites at the equatorial belt. There are exceptions, as in $InCl_3\{SC(NMe_2)_2\}_2$ (55),[335] $InCl_3\{SC[N(Me)CH]_2\}_2$,[336] $InBr_3\{SC[N(Me)CH]_2\}_2$,[336] and in the bis(carbene) adduct.[321] These adducts show structures with one halide occupying an equatorial site and the remaining two at axial positions. The compounds $InCl_3\{SC[N(Me)CH]_2\}_2$ and $InBr_3\{SC[N(Me)CH]_2\}_2$ have been synthesized by treating $InCl_3 \cdot 4H_2O$ and $InBr_3$ with a slight excess of 1,3-dimethyl-2(3H)-imidazolethione in hot $CH_3CN/EtOH$.[336]

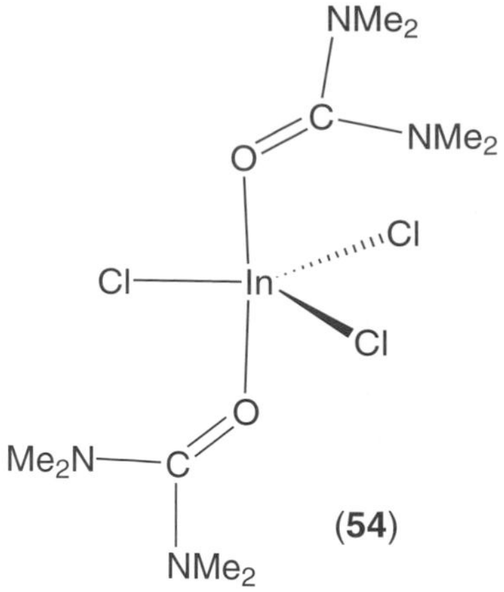

(54)

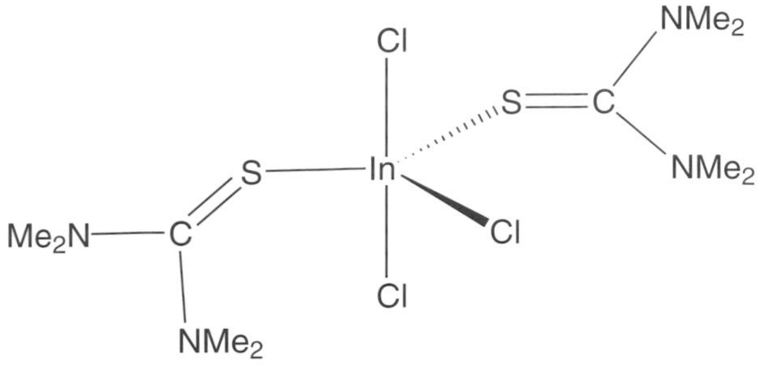

(55)

Compounds InF$_3$(bipy)(H$_2$O),[340] InCl$_3$(H$_2$O)$_3$,[341] InCl$_3$(THF)$_3$,[342] InCl$_3$(PhMe$_2$PO)$_3$, InCl$_3$(Me$_2$SO)$_3$,[343] InCl$_3$(Me$_3$PO)$_3$,[343] InBr$_3$(Me$_2$SO)$_3$,[343] InCl$_3$(Me$_3$PO)$_3$,[343] and InI$_3$(4-Me-py)$_3$[344] represent InX$_3$L$_3$-type molecules. Structural data show octahedral indium sites. However, both *mer*- and *fac*-configurations have been observed. The compounds InCl$_3$(PhMe$_2$PO)$_3$, InCl$_3$(Me$_2$SO)$_3$ (56), and InBr$_3$(Me$_2$SO)$_3$ show the *fac*-configuration,[343] whereas InF$_3$-(bipy)(H$_2$O),[340] InCl$_3$(Me$_3$PO)$_3$ (57), and InI$_3$(4-Mepy)$_3$ are *mer*-octahedral.[343,344]

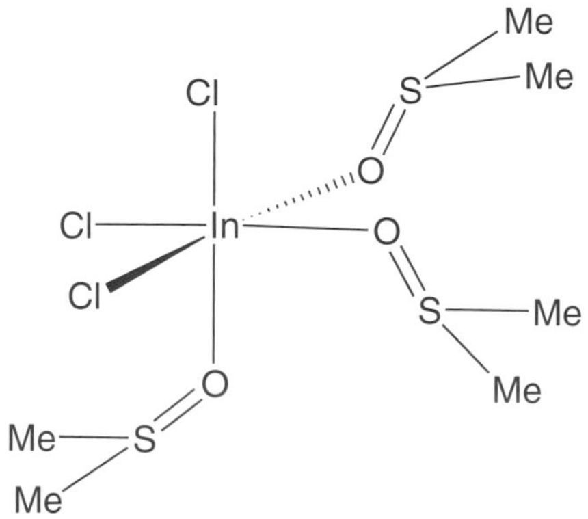

(56)

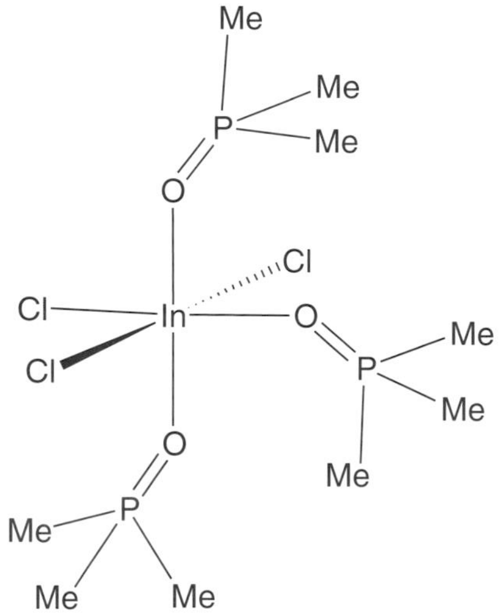

(57)

A group of aqua complexes of In(III) have been obtained as a part of supramolecular assemblies.[345] A macrocyclic cavitand cucurbituril has been used to facilitate the crystallization process. Compounds featuring [InCl$_2$(H$_2$O)$_4$]$^+$, [InCl$_4$(H$_2$O)$_2$]$^-$, and [In(H$_2$O)$_6$]$^{3+}$ ions have been isolated and characterized by X-ray crystallography. The cation [InCl$_2$(H$_2$O)$_4$]$^+$ has the *trans* arrangement of chlorides. The anion [InCl$_4$(H$_2$O)$_2$]$^-$ shows both *cis* and *trans* isomers; *cis*-[InCl$_4$(H$_2$O)$_2$]$^-$ has also been obtained using the [S$_4$N$_3$]$^+$ cation.[346] Molecules with [InCl$_5$(H$_2$O)$_2$]$^{2-}$ and [InBr$_5$(H$_2$O)$_2$]$^{2-}$ ions are also known.[347,348]

A study of indium(III)–iodine bond lengths as a function of coordination number of the indium shows a systematic change. The symmetry or the charge of the adduct has only a minor effect. For well-authenticated four-, five-, and six-coordinate systems, the average In—I distances are about 2.68, 2.73, and 2.83 Å, respectively.[344]

Overall, many factors—such as the halide ion, steric and electronic properties of the neutral donor, solvent, crystal packing forces, etc.—seem to control the nature of the product. Even minor variations lead to major structural changes.[343] For example, InCl$_3$(Me$_2$SO)$_3$ (56) shows *fac*-octahedral configuration, whereas InCl$_3$(Me$_3$PO)$_3$ (57) adopts the *mer* conformation. The InCl$_3$(Me$_3$PS)$_2$ is a five-coordinate complex. The compound InCl$_3$(Ph$_2$MePO)$_3$ is covalent, while the bromide analogue $[InBr_2(Ph_2MePO)_4][InBr_4]$ is ionic. The chloride of the bulkier Ph_3PO is also ionic, $[InCl_2(Ph_3PO)_4][InCl_4]$.

(iii) Halides and multidentate ligands

Cyclic tetraamine ligands such as cyclams (**58**) and cyclens (**59**) form adducts with indium halides.[349,350] Synthesis of $[InX_2(cyclam)][InX_4]_3$ (X = Cl, Br, I) and the 1:1 $InBr_3$ adducts of cross-bridged cyclam (1,4,8,11-tetraazabicyclo[6.6.2]hexadecane) and cross-bridged cyclen (1,4,7,10-tetraazabicyclo[5.5.2]tctradecane) have been reported. It is reported that, compared to cyclam and cyclen ligands, the cross-bridged ligands afford more kinetically inert metal complexes. Such adducts are of interest as potential indium-111-based pharmaceutical agents. The crystal structure of $InBr_3$(1,4,7,10-tetraazabicyclo[5.5.2]tetradecane) shows that it consists of $[InBr_2(1,4,7,10\text{-tetraazabicyclo}[5.5.2]\text{tetradecane})]^+$ (**60**) cations and bromide ions. The indium site is hexacoordinate and has a distorted octahedral geometry, with two bromides occupying *cis* sites.[350] Indium complexes of smaller ring systems are also known. Indium(III) chloride and bromide react with 1,4,7-triazacyclononane ([9]aneN_3) and 1,4,7-trimethyl-1,4,7-triazacyclononane (Me_3[9]aneN_3) to produce 1:1 complexes.[172,338] The X-ray crystal structure of $InBr_3(Me_3[9]aneN_3)$ (**61**) reveals *fac*-coordination of the macrocycle. A similar structure is observed for the $InBr_3$ adduct of 1,3,5-trimethyl-1,3,5-triazacyclohexane.[338]

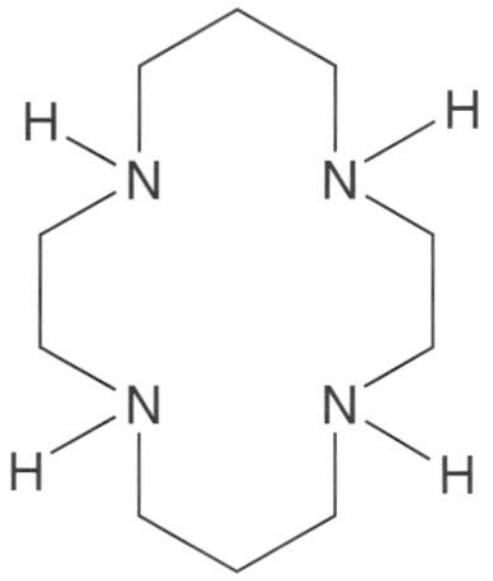

cyclam

(**58**)

cyclen

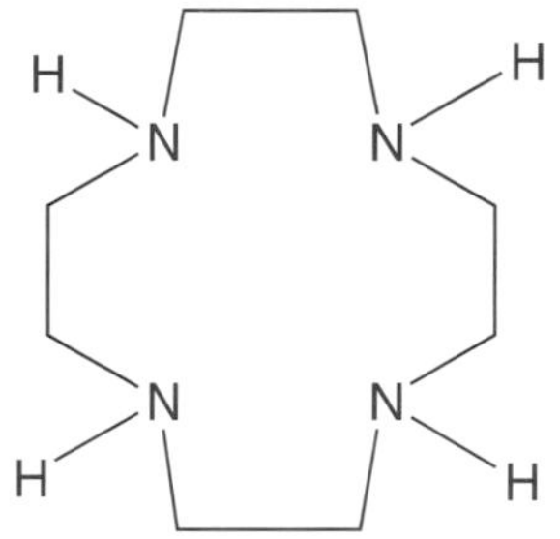

(**59**)

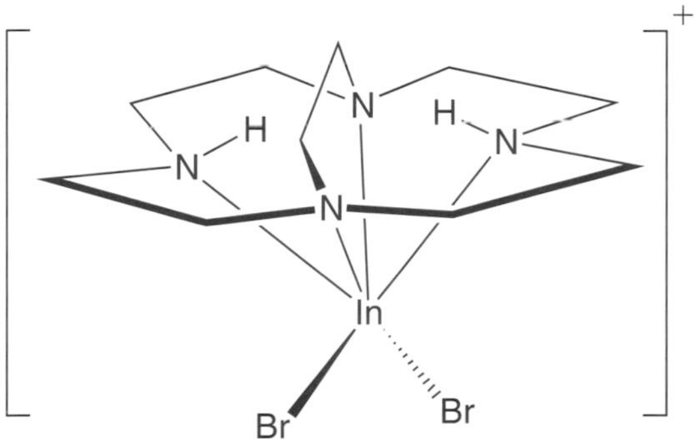

(**60**)

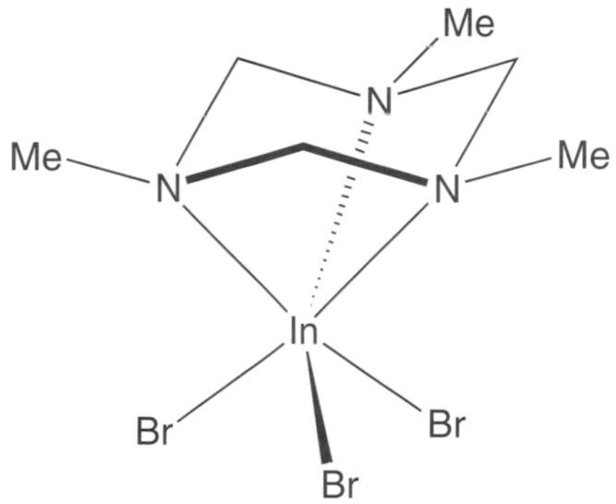

(**61**)

Reaction of 1,2-bis-(diphenylphosphanyl)benzene (DP) with an equimolar quantity of $InCl_3$ affords $[(DP)_2InCl_2][InCl_4]$.[351] The ionic structure was confirmed by X-ray crystallography. The indium atom in the cation adopts an octahedral geometry, and the chlorides occupy *trans* positions. The $InBr_3$ and InI_3 reactions lead to neutral, five-coordinate $(DP)InX_3$ compounds. The reaction of DP ligand with $InBr_3$ and InI_3 at 1:2 molar ratio, however, produces ionic $[(DP)InX_2][InX_4]$ (X = Br or I).[351] The chelating, potentially tridentate phosphine ligand, bis[(2-diphenylphosphanyl)phenyl]phenylphosphane (TP), also reacts with $InCl_3$ and InI_3 forming ionic species $[(TP)InX_2][InX_4]$ (X = Cl or I).[351,352] However, the indium centers are four-coordinate, and the phosphine ligand (although it has three P-donor sites) acts only as a bidentate donor. The indium atoms in these cations adopt essentially tetrahedral geometry.

The pyridine-2,6-bis(acetyloxime) acts as a tridentate chelator for In(III).[353,354] The reaction of 2,6-$(HONCMe)_2C_5H_3N$ with $InCl_3$ in MeOH yields seven-coordinate, distorted pentagonal-bipyramidal $InCl_3[2,6-(HONCMe)_2C_5H_3N](MeOH)$. The related, but bulkier, Schiff-base ligand 2,6-$(PhNCMe)_2C_5H_3N$ forms a six-coordinate adduct $InCl_3[2,6-(PhNCMe)_2C_5H_3N]$ (62). The MeOH can be replaced with Cl^- or water to obtain $\{InCl_4[2,6-(HONCMe)_2C_5H_3N]\}^-$ or $InCl_3[2,6-(HON-CMe)_2C_5H_3N](OH_2)$, respectively. These seven-coordinate adducts feature indium atoms with pentagonal-bipyramidal geometry. The chloride groups may be replaced by monoanionic dialkyl-thiocarbamates and pyridine-2-thiolate (PyS) ligands, or by dianionic oxalato (oxa) groups.[354] Compound In(oxa)Cl[2,6-$(HONCMe)_2C_5H_3N](OH_2)$ (63) is seven-coordinate, and has a pentagonal-bipyramidal indium center as well. The oxalato and amine oxime ligands form the pentagon.

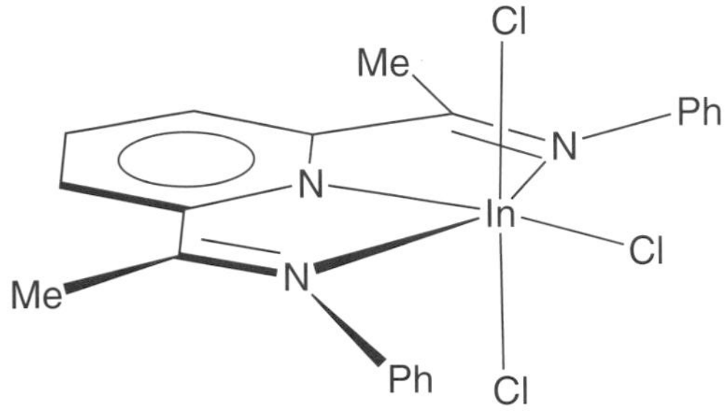

(62)

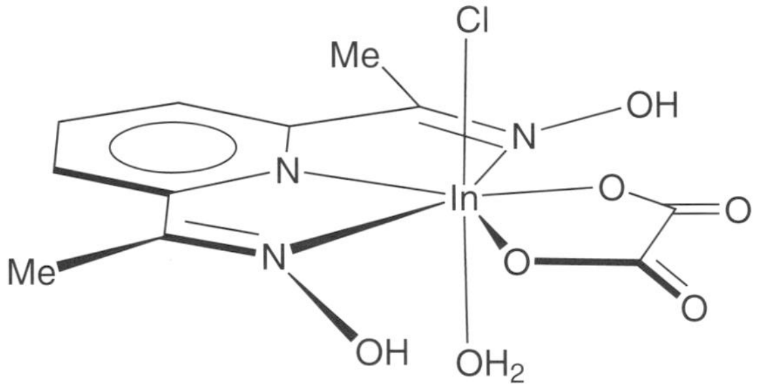

(63)

A few crown ether complexes of indium(III) have been reported. These include [InI_2(dibenzo-24-crown-8)(H_2O)](InI_4),[355] [InX_2(dibenzo-18-crown-6)][InX_4] (X = Cl, Br, I),[349] [InI_2(18-crown-6)][InI_4],[356] and [In(12-crown-4)_2][SbCl_6].[357] Crown ether-containing solids of aqua $InCl_3$ adducts were noted.[358,359] Synthesis of [InI_2(dibenzo-24-crown-8) (H_2O)][InI_4] involves the treatment of two equivalents of InI_3 with dibenzo-24-crown-8 in acetonitrile. The solid-state structural data show the InI_2 moiety located off-center within the crown ether cavity, with indium ions forming four bonds to ether oxygens and one to a water molecule. The compound [InI_2(18-crown-6)][InI_4] (64) has been prepared by the reaction of $InI_3 \cdot OEt_2$ with 18-crown-6. Again, the typical InI_2^+ threading through the ring is observed. The [In(12-crown-4)_2]^+ cation, which contains a smaller crown ether, features an eight-coordinate indium sandwiched between the two crown ethers.

(iv) Anionic complexes with halide ligands

Anionic indium(III) species containing only halides (e.g., [InX_4]^-, [InX_5]^{2-}, [InX_6]^{3-}) have been well known for many years.[1,3,4] The tetrahalo ion [InX_4]^- (X = Cl, Br, I) is a common counter-ion for cationic indium(III) compounds. It adopts tetrahedral geometry. The structure of the anion in [Et_4N]_2[InCl_5] (65) is particularly interesting, because it does not show the expected trigonal-bipyramidal geometry for a five-coordinate species.[360,361] It has a distorted square-pyramidal

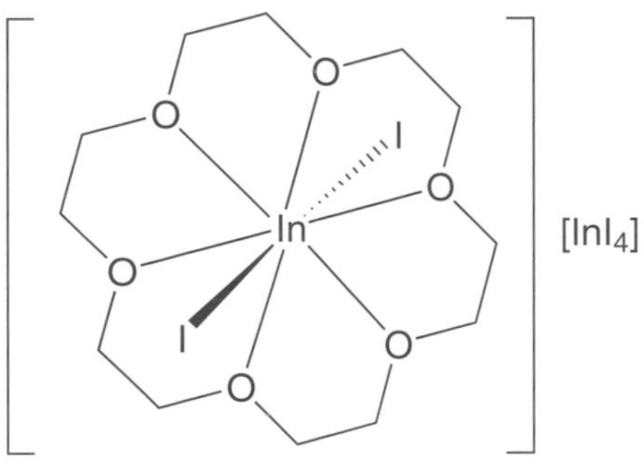

(64)

geometry, which is rare for a main-group compound.[362] The thallium salt $[Et_4N]_2[TlCl_5]$ is reported to be isomorphous with the indium analogue. The pentacoordinate anion in $[PPh_4]_2[InCl_5]\cdot CH_3CN$ (**66**), however, adopts trigonal-bipyramidal geometry.[363] It should be noted that the energies are not much different for the trigonal-bipyramidal and square-pyramidal geometries.[7] The pentacoordinate $[InBr_5]^{2-}$ and hexacoordinate $[InBr_6]^{3-}$ ions are known.[364] $[InBr_6]^{3-}$ displays the expected octahedral arrangement of bromides around indium(III). The pentabromoindate anion in $[4\text{-ClC}_5H_4NH]_2[InBr_5]$ adopts a rare, square-pyramidal geometry.[364]

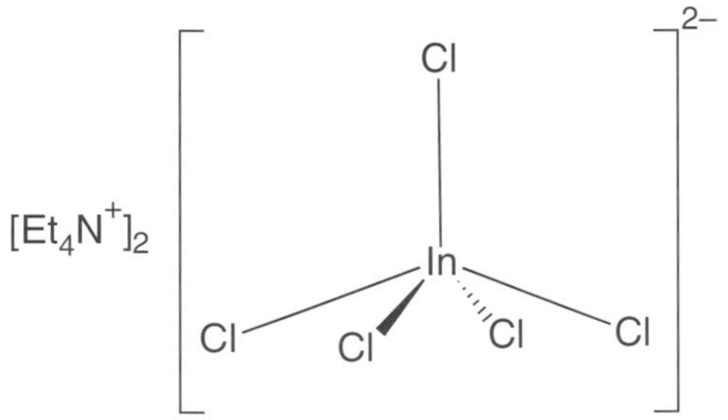

(65)

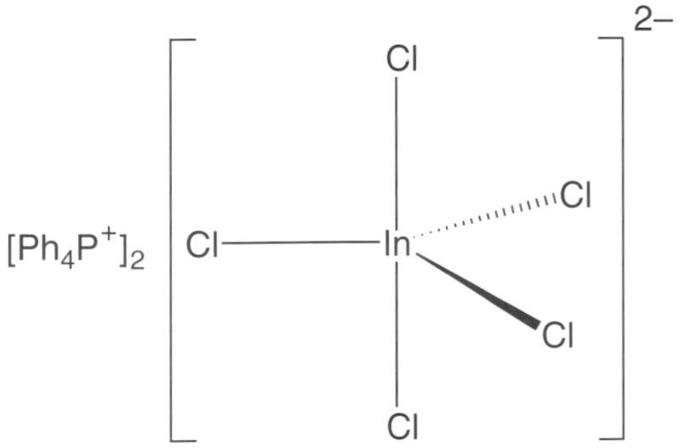

(66)

3.5.1.3.5 *Hydride ligands*

Indane or its aggregates are believed to be too unstable to exist as thermally stable species at room temperature.[365–367] Theoretical calculations predict that In_2H_6 is thermodynamically unstable in both the gas phase and as a solid.[367] However, since 1998 there have been some notable developments involving InH$_3$ complexes.[368] It is possible to synthesize tertiary amine complexes of InH$_3$ such as Me$_3$NInH$_3$, (quinuclidine)InH$_3$, and [N(CH$_2$)$_3$N]InH$_3$. These adducts decompose at room temperature, resulting in indium metal, H$_2$, and free amine. The compound Me$_3$NInH$_3$, however, can be decomposed in the presence of ammonia to obtain InN.[368] The (quinuclidine)InH$_3$ and LiBr afford an interesting indium aggregate [(quinuclidine)$_2$H][In{InBr$_2$-(quinuclidine)}$_4$], featuring a tetrahedron of indium atoms around an indium center.[368]

Phosphine adducts of InH$_3$ (prepared using LiInH$_4$ and Me$_3$NHCl, and then treating the resulting Me$_3$NInH$_3$ adduct with a phosphine) show better thermal stability in the solid state. Among the known InH$_3$ adducts of phosphines, the 1:1 and 1:2 Cy$_3$P adducts ((**67**), (**68**)) are the most stable.[306,369] Imidazol-2-ylidene (nucleophilic carbene) complexes of indane, [{(MeCN-(*i*-Pr)}$_2$C]InH$_3$ (**69**), and [{(HCN(Mes)}$_2$C]InH$_3$ (**70**) have been synthesized.[370,371] Solid samples of [{(HCN(Mes)}$_2$C]InH$_3$ are stable up to 115 °C.

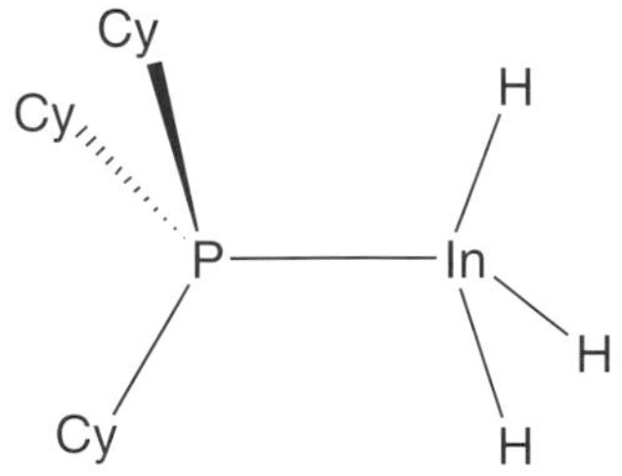

(67)

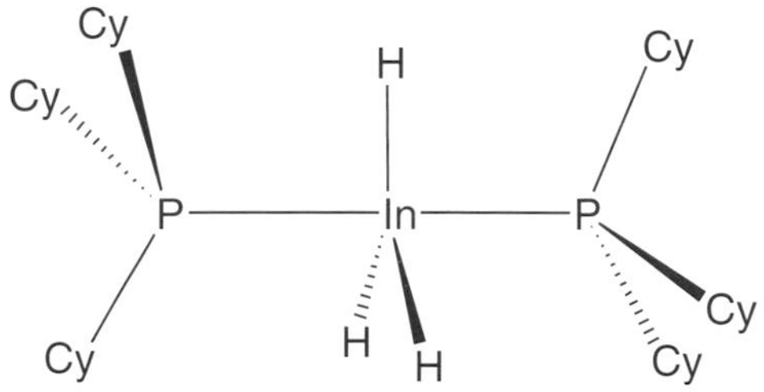

(68)

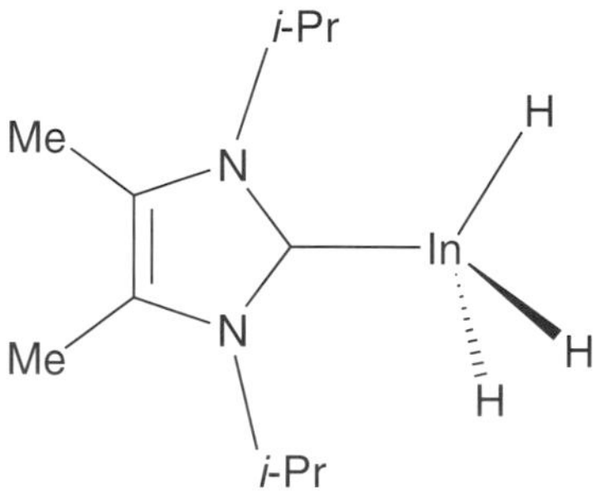

(69)

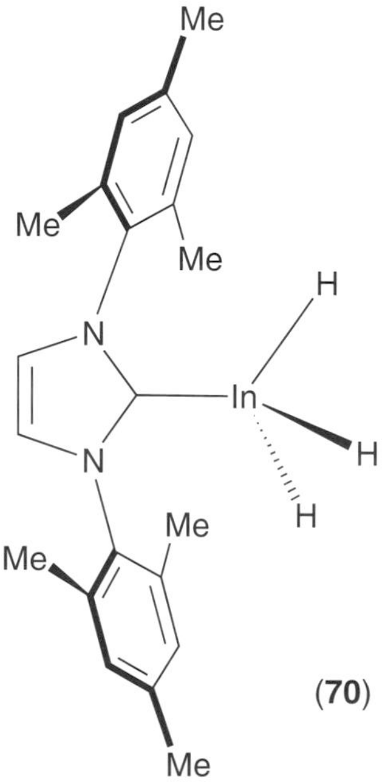

(70)

Some of these InH$_3$ adducts are useful starting materials for the preparation of other In–H compounds. The carbene adduct [{(HCN(Mes)}$_2$C]InH$_3$ reacts with quinuclidine·HCl to produce [{(HCN(Mes)}$_2$C]InH$_2$Cl. *In situ*-generated Me$_3$NInH$_3$ reacts with LiPCy$_2$ to give the trimeric phosphido–indium hydride complex (H$_2$InPCy$_2$)$_3$.[368] Review articles on the chemistry of group 13 hydrides, and more recent work involving donor stabilized InH$_3$, are available.[365,366,368]

There are several reports of organoindium compounds containing In–H bonds. These include [Li(THF)$_2$][{Me$_3$Si)$_3$C}$_2$In$_2$H$_5$],[372] K[H{In(CH$_2$CMe$_3$)$_3$}$_2$],[373] K$_3$[K(Me$_2$SiO)$_7$] [HIn(CH$_2$CMe$_3$)$_3$]$_4$,[374] HIn{2-Me$_2$NCH$_2$(C$_6$H$_4$)}$_2$,[375] Me$_2$InB$_3$H$_8$,[376] and [Li(TMEDA)$_2$][H(InMe$_3$)$_2$].[377] The anionic indium species MInH$_4$ (M = Li–Cs)[378] and Li[InH$_{4-n}$Ph$_n$] (where *n* = 1, 2)[379] are also known.

In addition to the coordination compounds, InH$_3$ and a variety of molecules containing In—H bonds have been generated in solid argon matrices.[49,50,380–382] These include HInCl$_2$, H$_2$InCl,

H_2InNH_2, and H_2InPH_2. The IR data of these species are reported. There is also a growing interest in the use of indium hydrides in organic synthesis.[365,383,384] For example, indium hydride species generated from a mixture of $InCl_3$ and $NaBH_4$ are shown to be promising alternatives to Bu_3SnH systems.

3.5.1.2.6 *Mixed-donor-atom ligands*

Mixed-donor-atom ligands play an important role in indium(III) coordination chemistry. One important application concerns their use in the synthesis of indium-111- (γ-emitter, half-life = 67.9 h) based radiopharmaceutical agents.[170] Choosing the ideal metal–ligand combination for this purpose is challenging. The metal adduct formation step should be fast, and resulting indium must be kinetically and thermodynamically stable. Hydrolysis reactions leading to indium hydroxo derivatives or $In(OH)_3$ are a concern. Furthermore, the complex should be stable enough to prevent exchange of indium from the radiopharmaceutical to transferrin (a plasma protein with a high affinity for indium(III), log $K_1 = 18.74$).[385] A large variety of multidentate ligands containing various combinations of N, O, and/or S donors have been used in the preparation of indium adducts for possible pharmaceutical use.[170,386–398] Multidentate ligands with neutral nitrogen and anionic sulfur donors,[399] and 6-coordinate metal sites appear to be the best.[400] The most popular ligand for indium-111, however, is diethylenetriaminepentaacetic acid (DTPA) (71).[170]

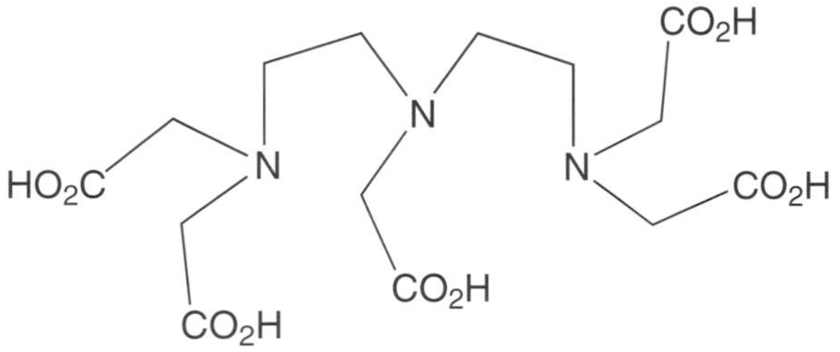

(71)

The diaqua(2,6-diacetaylpyridinedisemicarbazone)indium(III) cation (72) represents the first example featuring a pentagonal bipyramidal indium site.[401] It was isolated as [In(H$_2$DAPSC)(H$_2$O)$_2$](NO$_3$)$_2$(OH) by reacting hydrated indium nitrate and 2,6-diacetylpyridine bis(semicarbazone) (H$_2$DAPSC) in a water–ethanol solution.

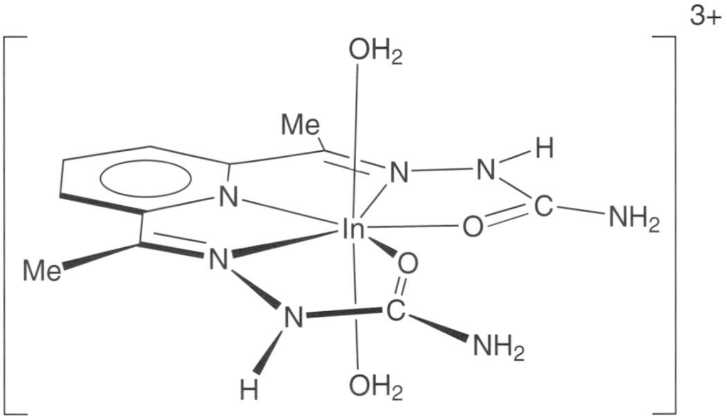

(72)

The chloroindium(III) complex of 1,4,7-triazacyclononanetriacetic acid also features pentagonal-bipyramidal geometry.[402] The chloride and one of the tertiary nitrogens occupy the axial sites. The compound {O[In(HDAPTSC)(OH)]$_2$} (73) (where H$_2$DAPTSC = 2,6-diacetylpyridine bis(thiosemicarbazone)) has two distorted pentagonal-bipyramidal units bridged by an oxo group.[174] The indium atoms also have terminal hydroxo ligands. Semicarbazone and thiosemi-carbazone complexes of indium have been reviewed.[403]

(73)

Transition-metal complexes of indium and indium adducts are available.[404] Structures, bonding, reactivity, and their materials-related applications are of current interest in the early 2000s.

3.5.1.3   Indium (II)

3.5.1.3.1   *Group 14 ligands*

Several well-defined In(II) compounds are known, and a few of these contain bulky silyl substituents such as supersilyl (t-Bu_3Si–) and hypersilyl ($[Me_3Si]_3Si$–) groups. The ruby-red colored $\{[(Me_3Si)_3Si]_2In\}_2$ (**74**) has been isolated in low yield from a reaction between $InCl_3$ and $(Me_3Si)_3$ SiLi in 1:3 molar ratio.[405] The major products of this reaction are $[(Me_3Si)_3Si]_2$, indium metal, and LiCl. The In—In bond distance is 2.868(1) Å. The reaction between t-Bu_3SiNa and $(Me_5C_5)In$ or $InCl_3$ or $InBr$ leads to deep violet $[(t$-$Bu_3Si)_2In]_2$ (**75**).[406-408] The red-violet $[(t$-$Bu_2PhSi)_2In]_2$ (**76**) can also be synthesized using similar routes.[29] The solid-state structures of $[(t$-$Bu_3Si)_2In]_2$ and $[(t$-$Bu_2PhSi)_2In]_2$ show trigonal-planar indium sites, orthogonal $InInSi_2$ planes, and a relatively long In—In distance of 2.922(1) and 2.938(1) Å, respectively.[29,407] The divalent $[(t$-$Bu_3Si)_2In]_2$ may be used in the synthesis of both subvalent and trivalent indium compounds. For example, thermolysis of $[(t$-$Bu_3Si)_2In]_2$ in boiling heptanes affords $(t$-$Bu_3Si)_8In_{12}$.[408] The indium(III) complexes $(t$-$Bu_3Si)InF_2$ and $(t$-$Bu_3Si)InBr_2$ were obtained by treating $[(t$-$Bu_3Si)_2In]_2$ with AgF_2 or HBr, respectively.[28] The action of selenium on $[(t$-$Bu_3Si)_2In]_2$ yields a heterocubane $(t$-Bu_3Si-$In)_4Se_4$.[29]

A few well-characterized organoindium(II) compounds are also known. These include $\{[(Me_3Si)_2CH]_2In\}_2$ (In–In = 2.828(1) Å), $\{[2,4,6$-$(CF_3)_3C_6H_2]_2In\}_2$ (In–In = 2.744(2) Å), $\{[2,4,6$-$(i$-$Pr)_3$-$C_6H_2]_2In\}_2$ (In–In = 2.775(2) Å), $\{[2,6$-$(Me_2NCH_2)_2C_6H_3](Cl)In\}_2$ (In–In = 2.7162(8) Å), $\{[(Me_3Si)_2C(Ph)C(Me_3Si)N]InBr\}_2$ (In–In = 2.728(4) Å).[409-412]   Perfluoroiodo   organics R_fI ($R_f = n$-C_yF_{2y+1} ($y = 1, 2, 3, 4, 6$), i-C_3F_7, C_6F_5) and C_6F_5Br react with indium metal in polyethers or THF to generate oxidative addition products of the general formula R_fInX (X = Cl, Br) involving In(II).[413]

In addition to In—C linkages, compounds $\{[2,6$-$(Me_2NCH_2)_2C_6H_3](Cl)In\}_2$ and $\{[Me_3Si)_2C(Ph)C$-$(Me_3Si)N]InBr\}_2$ have In—halide and In—N bonds. The coordination numbers at the indium are 4 and 5, respectively. The preparation of $\{[2,6$-$(Me_2NCH_2)_2C_6H_3](Cl)In\}_2$ involves the use of an In(III) precursor, whereas $\{[(Me_3Si)_2C(Ph)C(Me_3Si)N]InBr\}_2$ was a result of a disproportionation reaction involving In(I). Triindylindane $\{[2,4,6$-$(i$-$Pr)_3C_6H_2]_2In\}_3In$ has been prepared.[414] A collection of M—M bond distances (M = group 13 element) of metallanes is available.[410,412]

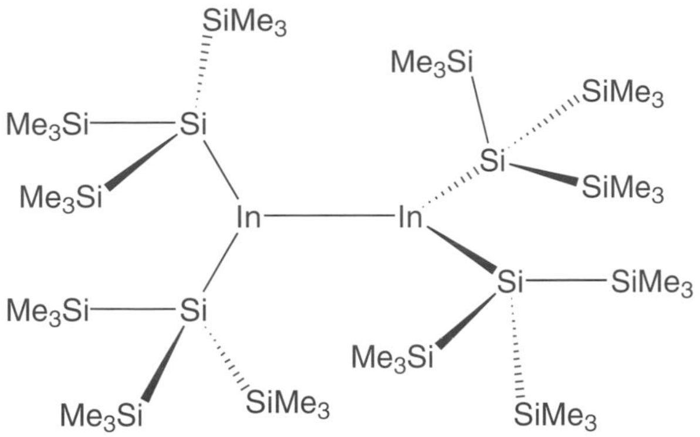

(74)

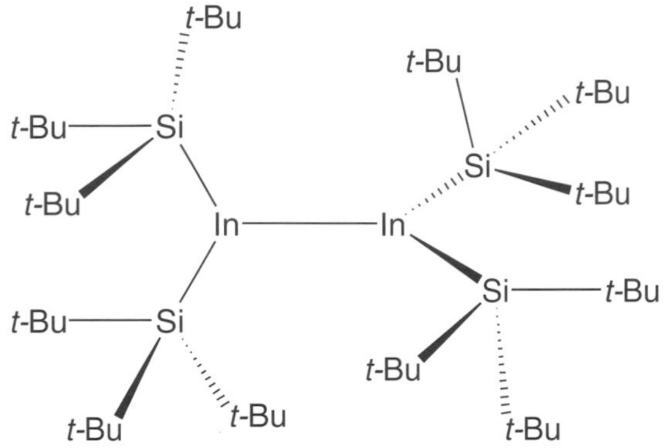

(75)

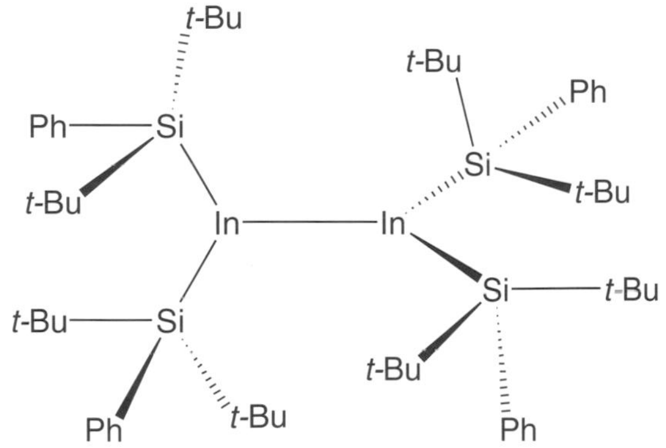

(76)

3.5.1.3.2 Group 15 ligands

The first well-characterized amide of In(II) was isolated using the cyclic silazane [(t-BuNSiMe)$_2$-(t-BuN)$_2$]H$_2$ ligand system.[415] The reduction of [(t-BuNSiMe)$_2$(t-BuN)$_2$]InCl using sodium naphthalene provides a convenient entry route to indium(II) species {[(t-BuNSiMe)$_2$-(t-BuN)$_2$]In}$_2$ (77). The related bis(tert-butylamido)cyclodiphosphazane [(t-BuNP)$_2$(t-BuN)$_2$]$^{2-}$ ligand system is also useful in this regard. However, {[(t-BuNP)$_2$(t-BuN)$_2$]In}$_2$ (78) has been obtained, as a redox disproportionation product, starting with InCl and the dilithium salt of [(t-BuNP)$_2$(t-BuN)$_2$]$^{2-}$.[416] In contrast to indium, the related thallium(I) derivative does not undergo disproportionation. The two indium(II) amido complexes {[(t-BuNSiMe)$_2$(t-BuN)$_2$]In}$_2$ and {[(t-BuNP)$_2$(t-BuN)$_2$]In}$_2$ are isostructural, and contain indium cages linked by unsupported In—In bonds with bond distances of 2.768(1) and 2.7720(4) Å, respectively. The indium atoms are four-coordinate, with each atom bonded to three nitrogens and an indium atom.

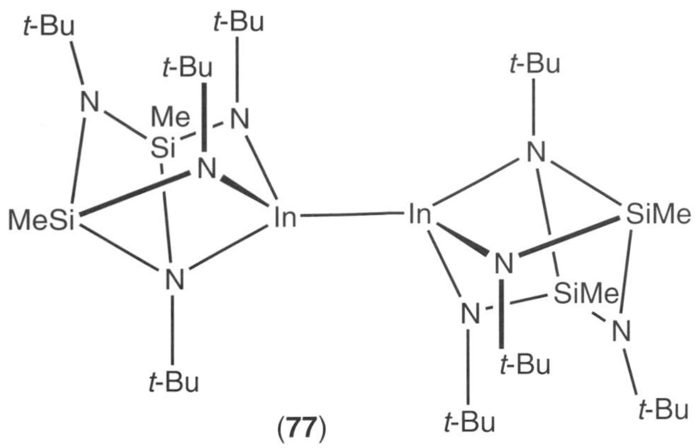

(77)

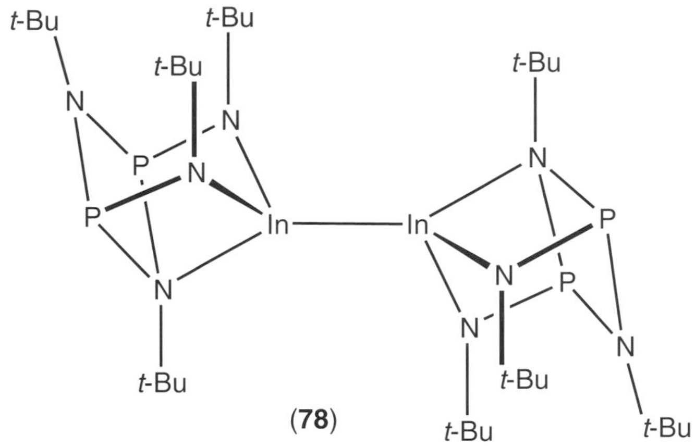

(78)

A mixed-valent In(I)/In(II) amide [MeC(CH$_2$NSiMe$_3$)$_3$In$_2$]$_2$ (**79**) can be obtained via a transmetal-ation reaction between the lithium salt of the ligand and InCl.[417] The thallium analogue is also known. The key feature is the In$_2$$^{4+}$ fragment with an In—In bond distance of 2.8067(9) Å. Inter-estingly, this bond distance is even longer than the Tl—Tl distance in [MeC(CH$_2$NSiMe$_3$)$_3$Tl$_2$]$_2$, (2.734(2) Å), but lies in the range found for covalent In—In bonds. The In—In bond is well shielded by the tripodal, *N*-SiMe$_3$-substituted ligand. This is reflected in the lack of reactivity towards isocyanides and heteroallenes, which were found to insert into In—In bonds of related compounds.

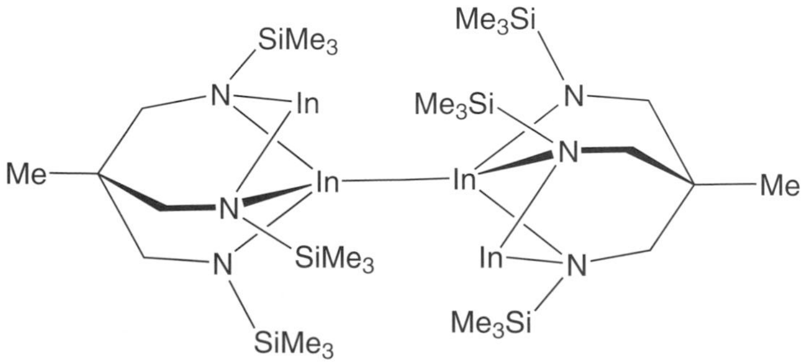

(79)

The redox disproportionation tendency of monovalent In(I) systems is further exemplified by the formation of [{C$_{10}$H$_6$(Me$_3$SiN)$_2$}In(THF)]$_2$ (**80**) during the attempted metal exchange of [C$_{10}$H$_6$(Me$_3$SiN)$_2$]Li$_2$(THF)$_4$ with InCl in tetrahydrofuran.[418] The In—In distance of 2.7237(6) Å in this molecule is one of the shortest established for a diindane. The related thallium analogue shows metal–ligand (rather than metal–metal) redox chemistry.

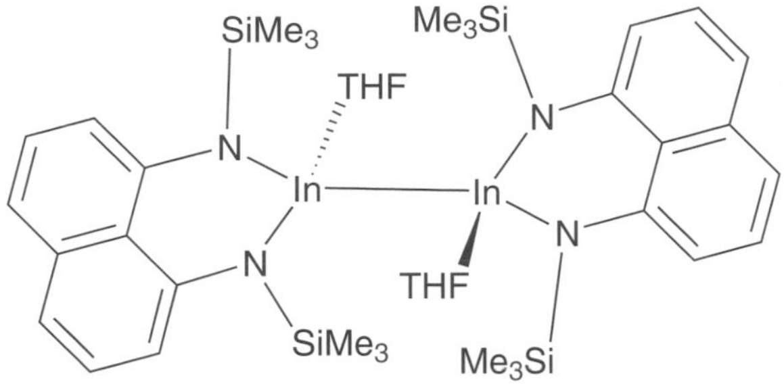

(80)

The compound {*syn*-In[PhP(CH$_2$SiMe$_2$NSiMe$_2$CH$_2$)$_2$PPh]}$_2$ **(81)** represents an example in which an In(II) species has been stabilized by a macrocyclic ligand.[419] Reduction of [PhP(CH$_2$Si-Me$_2$NSiMe$_2$CH$_2$)$_2$PPh]InCl with KC$_8$ in diethyl ether yields the dimeric In(II) complex as a colorless solid. The indium atoms exhibit distorted trigonal-bipyramidal geometries. The In—In bond length of 2.7618(12) Å is in the normal range.

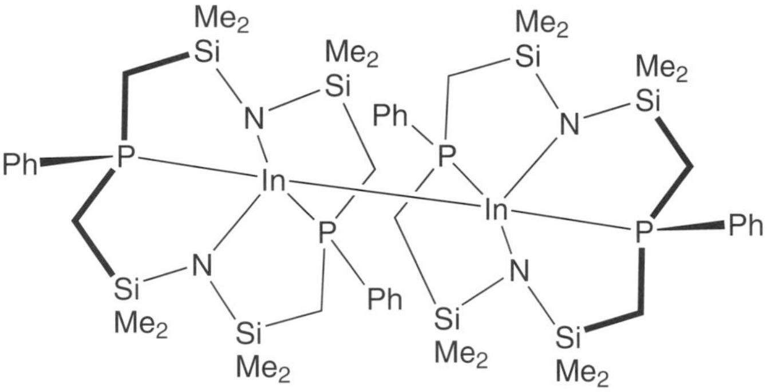

(81)

An interesting route to indium(II) compounds was discovered during an investigation of the oxidizing power of R$_3$PI$_2$ (R = Ph, *i*-Pr, *n*-Pr) with indium metal powder.[330] The action of (*n*-Pr)$_3$PI$_2$ on indium metal in diethyl ether leads to colorless, solid [(*n*-Pr)$_3$PInI$_2$]$_2$ **(82)**. The use of (*i*-Pr)$_3$PI$_2$ or Ph$_3$PI$_2$ with indium metal leads only to indium(III) iodides. In [(*n*-Pr)$_3$PInI$_2$]$_2$, the indium atoms have tetrahedral geometry, and the In—In distance is 2.745(3) Å.

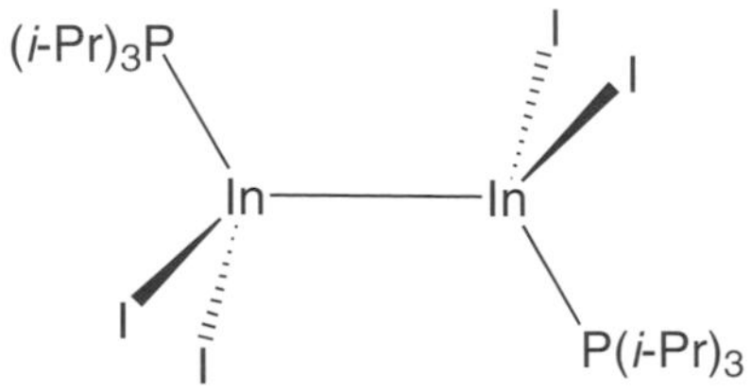

(82)

Close to nanometer size, molecular, group III/V compound [In$_3$(In$_2$)$_3$(PhP)$_4$(Ph$_2$P$_2$)$_3$Cl$_7$-(PEt$_3$)$_3$], featuring an unusual 19-atom cage, has been obtained from the reaction of InCl$_3$ with Et$_3$P and PhP(SiMe$_3$)$_2$. The 19-atom polyhedron, which has a diameter of about 0.7 nm, is built up by three formally trivalent indium atoms, six formally divalent indium atoms, and ten phosphorus atoms.[420]

3.5.1.3.3 Group 16 ligands

During an investigation involving InCl and Ph_3PAuCl in tetrahydrofuran, an indium(II) species was isolated as a colorless solid.[421] It was identified by X-ray crystallography as the tetrakis-THF adduct of the $InCl_2$ dimer (83). The indium atoms adopt trigonal-bipyramidal geometry. The dioxane complex of the lighter member gallium, $[(dioxane)Cl_2Ga]_2$, has tetrahedrally coordinated metal atoms. It is also possible to prepare $[(THF)_2Cl_2In]_2$ (83) by the reaction of In metal and $InCl_3$ in xylene, followed by the addition of tetrahydrofuran. The Raman spectrum of $[(THF)_2Cl_2In]_2$ shows an absorption at $180\,cm^{-1}$, suggesting that the In—In bond is retained in solution. The gold-containing indium adduct $(dppe)_2Au_3In_3(THF)_6$, containing two divalent indium atoms, may be obtained in the presence of 1,2-bis(diphenylphosphino)ethane(dppe) ligand.[421]

(83)

An electrochemical route to indium(II) thiolates has been described.[250] The electrochemical oxidation of anodic indium in acetonitrile and certain thiols leads to $In_2(SR)_4$ ($R = C_5H_{11}$, napthalide) derivatives. Thiols with different substituents produce In(I) or In(III) products. The corresponding oxidation of thallium metal gives only Tl(I) thiolates.

3.5.1.3.4 Group 17 ligands

Some of the earliest work on In(II) compounds centered around the identity of InX_2 (X = halide).[1,2,218] Structures of the type $X_2In–InX_2$ and $In[InX_4]$ fit the observed diamagnetic property of these halides. The structures of bromides and iodides have been confirmed to be of the latter type.[422,423] They involve indium(I) cations together with $[InX_4]^-$ anions.

There are several mixed-valent bromides involving In(II) as well (e.g., $In_2Br_3 = In^I_2[In^{II}_2Br_6]$, $In_5Br_7 = In^I_3[In^{II}_2Br_6]Br$).[2,424] The $[In^{II}_2Br_6]^{2-}$ unit features an In—In bond. The exact nature, or even the existence, of binary compounds of $InCl_2$ stoichiometry is less clear, and the early literature provides conflicting results.[218,425,426] However, there are various indium subchlorides known, including In_2Cl_3, In_5Cl_9, and In_7Cl_9.[2,427] They are formulated as mixed-valent compounds of In(I) and In(III) ions. Evidence has been found, during the electrochemical oxidation of indium metal in liquid ammonia solutions of NH_4X (X = Cl, Br, I), for the formation of In(II) species at the anode.[428] In the NH_4I-containing mixture, Raman data confirm the presence of In_2I_4. However, the isolation of neutral or anionic In(II) derivatives has not been successful, since the disproportionation reaction occurs on removal of solvent to give indium metal, In(I), and In(III) derivatives.

Preparation of a series of $In_2X_4\cdot 2L$ and $In_2X_4\cdot 4L$ (X = Cl, Br, or I; L = O, N, P, or S donors) has been reported.[2,3,218,429,430] These neutral In(II) halide complexes may be synthesized by treating the dihalide with various donors at low temperatures, or by starting with InX and InX_3 compounds. The presence of In—In bonds in $In_2X_4\cdot 2L$ and $In_2X_4\cdot 4L$ adducts is supported by Raman spectroscopic data. The reaction between InX (X = Cl, Br, I) and InY_3 (Br, I) in toluene-CH_2Cl_2–TMEDA solution at low temperature ($<-20\,°C$) produces an In(II) species $XYIn–InY_2$.[431] The crystal structure of $In_2Br_3I\cdot TMEDA$ (TMEDA = N,N,N',N'-tetramethylethanediamine) has been obtained.[432] It exhibits five-coordinate indium atoms with distorted trigonal-bipyramidal geometry, joined by an In—In bond 2.775(2) Å in length. The corresponding reaction with $InCl_3$ does not produce analogous chloro derivatives. A structural study of $In_2X_4\cdot(1,2\text{-dioxane})_2$ (X = Cl and Br) using NQR spectroscopy has been reported.[426] Synthesis of $In_2X_4\cdot(1,2\text{-dioxane})_2$ involves the use of In metal and InX_3 starting materials. The isolation of $[(n\text{-Pr})_3PInI_2]_2$ and $[(THF)_2Cl_2In]_2$ during the attempted syntheses of various other products has been described in earlier sections.

Anionic derivatives $In_2X_6{}^{2-}$ (X = Cl, Br, I) can also be prepared and isolated as stable solid materials.[2] For example, the reaction between In_2X_4 and Bu_4NX leads to anionic systems. Vibrational spectroscopic data support the existence of In—In bonded species.[4] The reaction between InCl and PPh_4Cl in acetonitrile has resulted in $[PPh_4]_2[In_2Cl_6]$, which contains centrosymmetric $[In_2Cl_6]^{2-}$ (84) ions having an In—In bond of length 2.727(1) Å.[363] The ternary In(II) bromide complex $KInBr_3$ (formulated as $K_2[In_2Br_6]$) has been prepared from $InBr_3$, KBr, and In.[433] The X-ray powder diffraction data indicate that the $[In_2Br_6]^{2-}$ ion has an eclipsed, ethane-like structure.

(84)

3.5.1.3.5 Hydride ligands

There are no reports of well-characterized InH_2 or its adducts. However, the indium(II) hydride species $HInNH_2$ and $HInPH_2$ have been generated by thermal and photoactivated reactions of indium metal atoms with NH_3 or PH_3 in solid argon matrices.[381,382] They have been identified by their IR spectra.

3.5.1.4 Indium(I)

3.5.1.4.1 Group 14 ligands

A dark green compound $(t\text{-Bu}_3Si)_6In_8$ (85) with an In_8 cluster framework has been isolated from a reaction between $t\text{-Bu}_3SiNa$ and $(Me_5C_5)In$.[406] This species is thermally stable up to 100 °C in solutions with the exclusion of light. The indium atom arrangement may be described as a doubly capped octahedron. The indium(II) species $[(t\text{-Bu}_3Si)_2In]_2$ is one of the by-products of this reaction. An even larger cluster compound $(t\text{-Bu}_3Si)_8In_{12}$ (86) results from the thermolysis of $[(t\text{-Bu}_3Si)_2In]_2$ in boiling heptane for 22 hours.[408] This black-violet, crystalline solid has a polyhedral framework of indium atoms, with the shape of a stretched ellipsoid rather than the icosahedron, and a longitudinal diameter of 750 pm. Thus these molecules reach nanoparticle sizes. The In—In bond distances range from 2.80 to 3.30 Å.

Synthesis of $InC_6H_3\text{-}2,6\text{-Trip}_2$ (where Trip = $2,4,6\text{-}(i\text{-Pr})_3C_6H_2$)[434] and $[(Me_3Si)_3CIn]_4$[435–437], and the matrix isolation of MeIn[438], are particularly noteworthy developments in organoindium(I) chemistry. X-ray data of $InC_6H_3\text{-}2,6\text{-Trip}_2$ and $[(Me_3Si)_3CIn]_4$ show the presence of a one-coordinate indium center and a tetrahedral In_4 core, respectively. They show interesting metal-ion coordination and oxidation chemistry.[409,434,436] Among indium(I) compounds, some of the earliest (e.g., $In(C_5H_5)$ was isolated in 1957) and the most widely known complexes are those containing cyclopentadienyl ligands.[13,218]

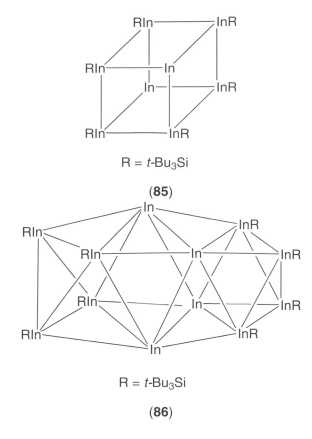

R = *t*-Bu$_3$Si

(85)

R = *t*-Bu$_3$Si

(86)

3.5.1.4.2   *Group 15 ligands*

As indicated in an earlier section, the amido indium(I/II) species [MeC(CH$_2$NSiMe$_3$)$_3$In$_2$]$_2$ is known.[417] The solid-state structure shows a longer In—N distance, with the monovalent indium atoms (d_{average}: In(I)–N = 2.43 Å, In(II)–N = 2.17 Å).

Tris(pyrazolyl)borate ligands play an important role in indium(I) chemistry.[439] This is perhaps not surprising, considering the close similarity between the ubiquitous cyclopentadienyl and tris(pyrazolyl)borate systems,[440] and the well-established utility of cyclopentadienyl ligands in indium(I) chemistry. The first thermally stable tris(pyrazolyl)boratoindium(I) adduct [HB(3-(Ph)Pz)$_3$]In (**87**) was reported in 1994.[441] It is an air-stable, monomeric compound. Shortly thereafter, the In(I) compounds of alkylated tris(pyrazolyl)borates were reported. These included [HB(3-(*t*-Bu)Pz)$_3$]In,[442] [HB(3,5-(*t*-Bu)$_2$Pz)$_3$]In,[443] and the fluoralkylated-ligand compound [HB(3,5-(CF$_3$)$_2$Pz)$_3$]In (**88**).[444,445] The alkali-metal derivatives of the ligand and InI or InCl have been utilized in the preparation of [HB(3-(Ph)Pz)$_3$]In and [HB(3,5-(*t*-Bu)$_2$Pz)$_3$]In. The thallium(I) and silver(I) complexes were used as ligand-transfer agents in the synthesis of [HB(3-(*t*-Bu)Pz)$_3$]In and [HB(3,5-(CF$_3$)$_2$Pz)$_3$]In.

Unlike some of the cyclopentadienylindium(I) compounds (e.g., (Me$_5$C$_5$)In is hexameric in the solid state), none of these tris(pyrazolyl)borates show any intermolecular In$\cdots$In interactions. It should be noted, however, that [HB(3-(Ph)Pz)$_3$]In, [HB(3-(*t*-Bu)Pz)$_3$]In, [HB(3,5-(*t*-Bu)$_2$Pz)$_3$]In, and [HB(3,5-(CF$_3$)$_2$Pz)$_3$]In contain much more sterically demanding ligands. The disproportionation of In(I) to In metal and In(III) has been observed in the presence of [HB(3,5-(CH$_3$)$_2$Pz)$_3$]$^-$.[86] The compounds [HB(3-(Ph)Pz)$_3$]In, [HB(3-(*t*-Bu)Pz)$_3$]In, and [HB(3,5-(CF$_3$)$_2$Pz)$_3$]In show C$_{3v}$ molecular symmetry. [HB(3,5-(*t*-Bu)$_2$Pz)$_3$]In exhibits a structure in which the [HB(3,5-(*t*-Bu)$_2$Pz)$_3$]$^-$ ligand adopts a highly twisted configuration, due to the intraligand repulsive interactions of the *tert*-butyl substituents.

Main-group and transition-metal adducts such as [HB(5-(*t*-Bu)Pz)(3-(*t*-Bu)Pz)$_2$](5-(*t*-Bu)PzH)In-InI$_3$ (**89**),[446] [HB(3,5-(CH$_3$)$_2$Pz)$_3$]InFe(CO)$_4$ (**90**), and [HB(3,5-(CH$_3$)$_2$Pz)$_3$]InW(CO)$_5$ are known.[90] The iron and tungsten adducts, however, have been obtained using In(III) precursors. The terminal selenido complex [HB(3,5-(*t*-Bu)$_2$Pz)$_3$]InSe may be obtained by the addition of selenium

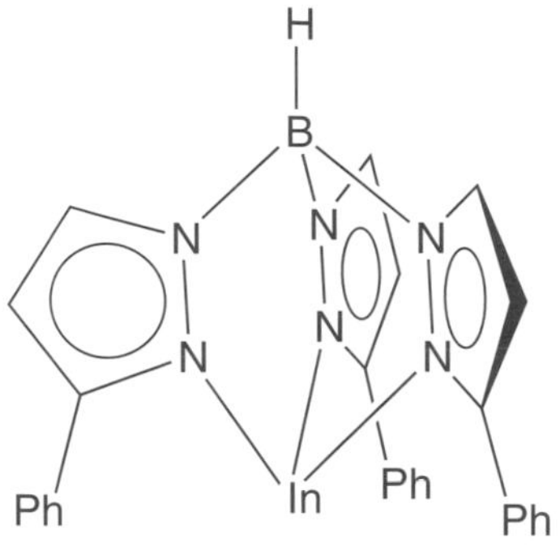

(87)

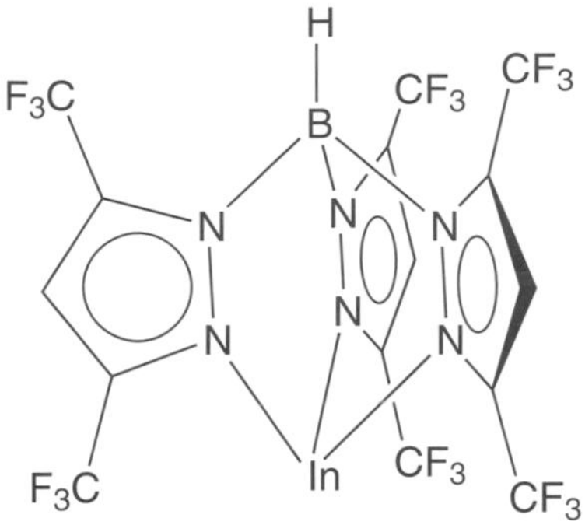

(88)

to the In(I) adduct [HB(3,5-(t-Bu)$_2$Pz)$_3$]In.[447] The indium(I) alkyl complex [{(Me$_3$Si)$_3$C)In]$_4$, however, does not afford similar adducts with selenium. [HB(3,5-(t-Bu)$_2$Pz)$_3$]In(S$_4$) and [HB(3,5-(t-Bu)$_2$Pz)$_3$]InI$_2$ have also been prepared from In(I) starting materials.[89] The zinc alkyl derivative [HB(3-(t-Bu)Pz)$_3$]ZnEt may be obtained by treating [HB(3-(t-Bu)Pz)$_3$]In with Et$_2$Zn.[442]

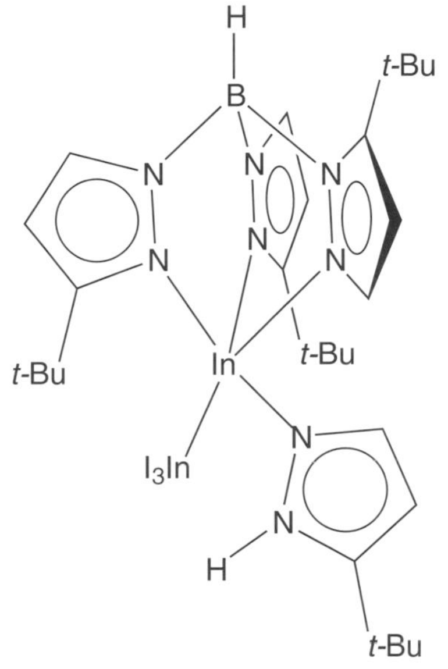

(89)

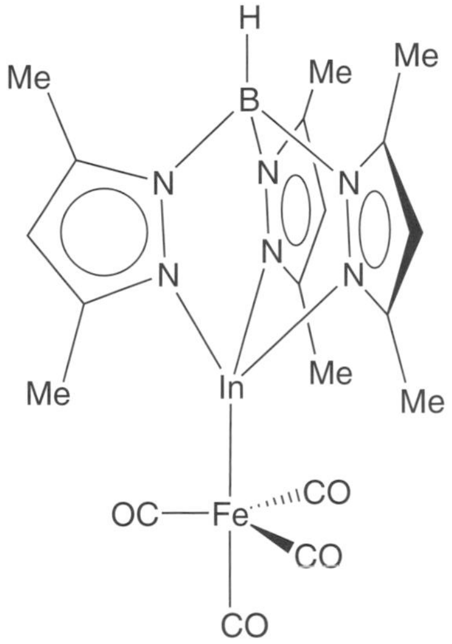

(90)

Co-condensation of indium vapor and t-BuC≡P at 77 K affords two phosphacyclopentadienyl derivatives of In(I), In[η^5-P$_2$C$_3$(t-Bu)$_3$] (**91**) and In[η^5-P$_3$C$_2$(t-Bu)$_2$] (**92**).[448,449] These compounds are volatile solids. X-ray data of In[η^5-P$_2$C$_3$(t-Bu)$_3$] show a monomeric structure, whereas In[η^5-P$_3$C$_2$-(t-Bu)$_2$] adopts a structure in which half-sandwich units are linked into chains by weak interactions between the indium atoms and the neighboring triphospholyl rings. In[η^5-P$_3$C$_2$(t-Bu)$_2$] may also be synthesized using InI and K[P$_3$C$_2$(t-Bu)$_2$]. Unlike the tris(pyrazolyl)boratoindium(I) derivatives, the indium(I) center in In[η^5-P$_3$C$_2$(t-Bu)$_2$] does not act as a Lewis-base site.

3.5.1.4.3 Group 16 ligands

An interesting indium(I) alkoxide [(2,4,6-(CF$_3$)$_3$C$_6$H$_2$O)In]$_2$ (**93**), featuring two-coordinate, bent indium sites, has been reported.[450] This compound is synthesized from CpIn and 2,4,6-(CF$_3$)$_3$C$_6$H$_2$OH in hexane, and features a planar In$_2$O$_2$ core. The reaction of Sn(μ-t-BuO)$_3$Tl or

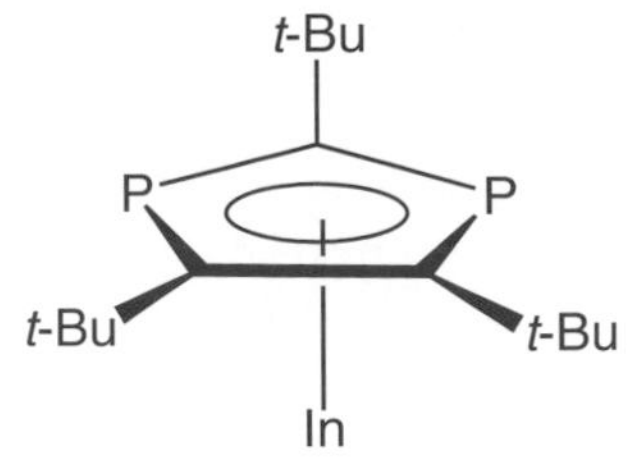

(91)

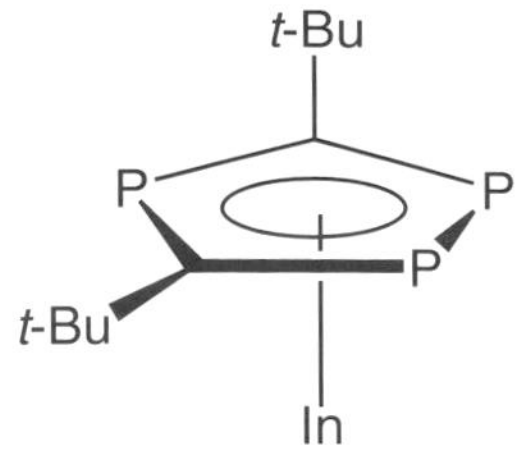

(92)

[Sn(*t*-BuO)$_3$Na]$_2$ with InBr yields an alkoxide Sn(μ-*t*-BuO)$_3$In (94) featuring both Sn(II) and In(I) sites.[451] A study of Lewis-base properties at the two low-valent, main-group metal sites reveals that transition-metal carbonyl fragments coordinate to the indium(I) site first and then to the tin atom(II).

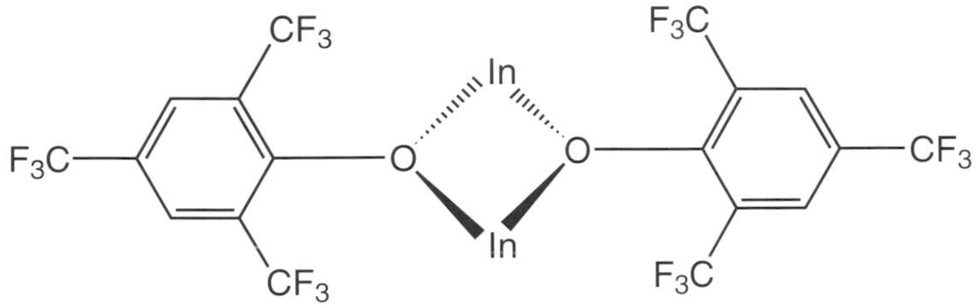

(93)

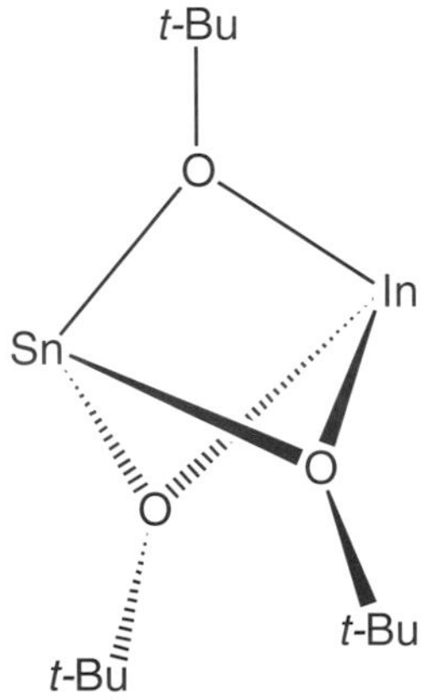

(94)

When indium metal is treated with 3,5-di-*tert*-butyl-1,2-benzoquinone (TBQ) in a 1:3 mole ratio, a paramagnetic, green solid In(TBSQ)(TBQ)$_2$ (where TBSQ = 3,5-di-*tert*-butyl-1,2-benzo-semiquinonate) forms. The same product may be obtained from InI$_3$ and three equivalents of NaTBSQ.[452] The ESR results confirm that the indium atom in this product is in the +1 state (and also confirm the presence of the semiquinone moiety).[218] The paramagnetic indium(I) compound In(TBSQ) has also been reported. It can be stabilized by coordination to 1,10-phenanthroline.[218]

The reaction of In(I) with an aqueous thiosulphate solution leads to In$_2$S$_2$O$_3$·2(H$_2$O), In$_2$S$_2$O$_3$·In-OH·2(H$_2$O), and In$_2$S$_2$O$_3$·2InNO$_3$·2(H$_2$O). These compounds have been characterized by IR spectroscopy, X-ray diffraction, and by thermal analytical methods.[453] A study of In/In(NO$_3$)$_3$ disproportionation to InNO$_3$ has been carried out, and shows that the nitrate ion exhibits a much stronger stabilizing influence on In(I) than Cl$^-$, SO$_4^{2-}$, Br$^-$, or ClO$_4^-$.[454]

The electrochemical oxidation of anodic indium in acetonitrile solutions of aromatic diols $R(OH)_2$ (diol = 1,2-dihydroxybenzene, 2,3-dihydroxynapthalene, 2.2'-dihydroxybiphenyl, 1,2-dihydroxy-tetrabromobenzene) leads to indium(I) derivatives of In[OR(OH)].[455] The deprotonation of –OH by Et_3N produces the anionic derivatives.

Related sulfur-donor compounds may also be prepared using electrochemical methods.[456] The electrochemical oxidation of anodic indium in the presence of alkanediols ($R(SH)_2$) produces In[SR(SH)] (R = various hydrocarbon bridges such as $1,2-C_2H_4$, $1,3-C_3H_6$, $1,4-C_4H_8$). Interestingly, under these conditions, thallium always produces $Tl_2(S_2R)$. The syntheses of alkene- and arenethiolato derivatives of indium InSR (R = Et, Bu), $In_2(SR)_4$ (R = C_5H_{11}, 2-$C_{10}H_7$), and $In(SR)_3$ (R = Ph, C_6F_5) have also been reported.[250] Again, thallium behaves somewhat differently under similar conditions. In general, electrochemical technique is simple, and leads to products in high yield. Furthermore, products are often low-valent derivatives. Some of the advantages and unique features of electrochemical synthesis have been described.[457]

Indium(I) complexes of Se and Te donors are known.[305] The reaction between InCl and $(Me_3Si)_3SiELi(THF)_2$ (E = Se or Te) and InCl leads to the metathesis product $(Me_3Si)_3SiEIn$. Chalcogenolysis with CpIn also affords the same In(I) selenolate or tellurolate. These indium(I) compounds are light-sensitive compounds. No solid-state structural data are available.

3.5.1.4.4 Group 17 ligands

Binary fluoride, chloride, bromide, and iodide derivatives of indium(I) are known. Among these InI is the most stable, whereas InF is known as an unstable gaseous species.[7] Many of these In(I) halides serve as useful sources of In(I) for chemical reactions.[218] Although they have a high tendency to disproportionate in solutions containing water or other bases, and are insoluble in common solvents, a few methods are now available to deliver In(I) in soluble form. For example, InX (X = Cl, Br, I) forms relatively stable solutions in mixtures of aromatic solvents and nitrogen donors like TMEDA at low temperatures.[218] The structures of these adducts are not well established. The effects of temperature on the stability, and the concentration of TMEDA on the solubility, of indium(I) species have been described. Interestingly, in ammonia InBr and InI disproportionate easily to produce indium(III) halide complexes. Solutions of indium(I) may also be prepared by the treatment of In/Hg with silver trifluoromethanesulfonate in dry acetonitrile.[458] In the absence of O_2, these solutions are stable for over five days. These mixtures serve as useful In(I) sources for synthetic applications.[459] The $InBF_4$ salt has been synthesized from the reaction of indium metal with anhydrous HF and BF_3.[460]

Several mixed-valent halide compounds (e.g., $In_2Cl_3 = In^I_3[In^{III}Cl_6]$, $In_7Cl_9 = In^I_6[In^{III}Cl_9]$, $In_5Cl_9 = In^I_3[In^{III}_2Cl_9]$, $InBr_2 = In^I[In^{III}Br_4]$, $In_2Br_3 = In^I_2[In^{II}_2Br_6]$, $In_4Br_7 = In^I_5In^{III}_3Br_{14}$, $In_5Br_7 = In^I_3[In^{II}_2Br_6]Br$, $In_7Br_9 = In^I_6In^{III}Br_9$), and $InI_2 = In^I[In^{III}I_4]$) are known.[2,427,461–463] Heating InCl with $InCl_3$ or $SnCl_2$ produces In_7Cl_6 via a redox process. The In_7Br_9 has a structure similar to that of the chloride analogue.[462] The solid-state structure of the bromide In_4Br_7 reveals unusual coordination numbers for In(I) of 10 and 12.[461] Evidence for the formation of $[InBr_2]^-$, along with InBr, has been reported when excess indium dissolves in HBr acid.[464]

The mixed-valent gold–indium cluster $[Au_3In_3Cl_6(dppe)_2]$ is composed of In(I) and In(II) atoms. Synthetic details were presented in an earlier section. Insertion and redox reactions of In(I) halides have been described. For example, oxidative addition reactions of InX (X = Cl, Br, I) with PhSSPh yields indium(III) thiolates of the type $(PhS)_2InX$.[4,259] Unexpected In(II) or In(III) products are common in reactions involving indium(I) halides, which result from facile disproportionation processes. Several such cases were described with divalent indium compounds.

3.5.2 THALLIUM

3.5.2.1 Introduction

Thallium is the largest and the heaviest element of the group 13 family. The metal and its compounds are dangerously toxic, even at low levels. It is a cumulative poison, and the lethal dose is considered to be about 13 mg of thallium for each kg of human body weight (or less than 1g of a thallium compound in a single ingestion). The ability of free Tl(I) ions in aqueous solutions to mimic potassium is one of the main causes of its high toxicity.[465] However, thallium-201 compounds are still useful for the diagnosis of myocardial perfusion.[466]

Thallium has the electronic configuration of $[Xe]4f^{14}5d^{10}6s^26p^1$, and forms compounds in the oxidation states I, II, and III. Thallium(II) derivatives are relatively rare. In general the bonding between Tl(III) and a donor is more covalent in nature, whereas Tl(I) compounds show more ionic behavior. Interactions between the Tl(I) atoms of neighboring molecules are common. The theoretical explanation has been controversial.[467–470] The coordination chemistry of Tl(III) is complicated by the highly oxidizing power of thallium(III) in both aqueous and nonaqueous solutions.

In this review, primarily the developments in thallium coordination chemistry since the early 1980s will be surveyed. *CCC* (1987) is an excellent reference source for work prior to that time.[1] In general, organometallic compounds are outside the scope of this article. There are several useful review and reference articles that describe thallium chemistry, including organothallium work.[1,6–8,12,14,16–19,471–475] Thallium is an important element in the field of high-temperature superconducting materials.[476]

3.5.2.2 Thallium (III)

3.5.2.2.1 Group 14 ligands

(i) Carbon ligands

A detailed, multimethod study of hydrated Tl(III) cyanide species in aqueous solution reveals that Tl(III) forms very strong complexes with cyanide ions (even stronger than halide–Tl(III) interactions).[477,478] Formation of a series of Tl(III) complexes $Tl(CN)_n^{3-n}$ ($n = 1$–4) has been established, and the solution structures and stability constants were reported. The mono- and dicyano complexes $[Tl(CN)(OH_2)_5]^{2+}$ and $[Tl(CN)_2(OH_2)_4]^+$ show six-coordinate thallium centers, whereas $Tl(CN)_3(OH_2)$ and $[Tl(CN)_4]^-$ have four-coordinate Tl(III) ions.

The equilibria, dynamics, and structures of $[Tl(edta)X]^{2-}$ (X = halide or pseudohalide) have been investigated.[479] The one-bond ^{205}Tl–^{13}C coupling constant for $[Tl(edta)CN]^{2-}$ is found to be 10,479 Hz. This indicates a strong Tl—C bond. The solid state structure of $Na_2[Tl(edta) CN]$ shows that the seven-coordinate thallium atom sits in the "edta-pocket" formed by the two nitrogen and four oxygen atoms with an axial bond to the cyanide ion (Tl–C = 2.14(3) Å).[479] Thallium(III) porphyrin complexes containing cyanide ligands are known.[480–482] The cyano group occupies an axial coordination site. The $^1J(^{205}Tl$–$^{13}C)$ coupling constants are large and typically >5,000 Hz.[480,482] The X-ray crystal structure of *meso*-tetra(4–pyridyl)porphyrinatothallium(III)cyanide, Tl(TPYP)(CN), shows that thallium–carbon (cyanide) distance is 2.12(2) Å. The characteristic IR band for υ_{CN} appears at 2,163 cm^{-1}.[482]

It is possible to prepare oligonuclear Pt–Tl compounds by using thallium(III) cyano complexes and $[Pt(CN)_4]^{2-}$.[483] The relative oxidation states of the metal atoms were estimated from their ^{195}Pt and ^{205}Tl NMR data, confirming that the $[(NC)_5Pt$–$Tl(CN)_n]^{n-}$ ($n = 1$–3) adducts can be considered as metastable intermediates in a two-electron process leading to Tl(I) and Pt(IV) final products.[483] These Pt-Tl bonded products show remarkably large one–bond ^{195}Pt-^{205}Tl spin–spin coupling constants, ranging from 25 to 71 kHz.[484] They have also been studied computationally.[485,486] Systems with such short metal–metal distances between relatively heavy atoms (e.g., Au, Pt, Tl, Pb) display interesting electronic properties.[487,488]

There are no Tl(III) carbonyl complexes isoelectronic with $[Tl(CN)_2]^+$. Theoretical studies predict that it would be difficult to observe $[Tl(CO)_2]^{3+}$ experimentally. Organothallium(III) compounds are well known, in particular those involving one or more noncarbon substituents of the type R_2TlX or $RTlX_2$.[473,474] Those compounds will not be considered here.

(ii) Silicon, germanium, tin, and lead ligands

Silyl thallium halides (t-$Bu_3Si)_nTlX_{3-n}$ (X = halide, $n = 1, 2$) have been reported. These compounds display relatively low thermal stability.[28,29,489] (t-$Bu_3Si)_2TlCl$ has been obtained by the addition of Me_3SiCl to a mixture containing $TlCl_3$ and t-Bu_3SiNa in a 1:3 molar ratio.[489] The reaction between $TlCl_3$ and t-Bu_3SiNa at a 1:2 molar ratio has produced thallium clusters $[(t$-$Bu)_3Si]_4Tl_3Cl$ and $[(t$-$Bu)_3Si]_6Tl_6Cl_2$, containing covalently linked Tl atoms.[489]

Tris(trimethylsilyl)thallium is prepared by the reaction between $Hg(SiMe_3)_2$ and trimethyl thallium.[2] Apart from the pre-1984 work, there are no significant new developments in Tl—Sn, Tl—Ge, or Tl—Pb bonded compounds.[1,3]

3.5.2.2.2 *Group 15 ligands*

(i) *Nitrogen ligands*

Simple complexes of thallium(III) with neutral nitrogen donors like ammonia or amines are not known in aqueous solution. The hydrolysis in aqueous solutions leading to hydroxo complexes $Tl(OH)_n^{3-n}$ ($n = 1, 2$) is one of the complications.[490] Mixed hydroxo complexes of Tl(III) containing ethylenediamine (en) are known.[491,492] Additional ligands on thallium can prevent the hydrolysis tendency. Complexes of the formula $[Tl(en)_3]X_3$, $[TlX_2(en)_2][TlX_4]$, and $[TlX_2(en)_2]X_3$ (X = Cl or Br) have been synthesized and identified based on IR spectroscopic or conductivity and molecular weight data.[493,494] More recently, the formation of $[Tl(en)_n]^{3+}$ ($n = 1$–3) complexes in a pyridine solution has been established by NMR spectroscopy.[495] The compound $[Tl(en)_3][ClO_4]_3$ has been crystallized and characterized using X-ray crystallography. The thallium ion features a distorted octahedral geometry, with nitrogen atoms of the three chelating ethylenediamine ligands forming the coordination sphere. The Tl(III) coordination chemistry involving diethylenetriamine (dien) and N,N,N',N'-tetrakis(2-aminoethyl)ethane-1,2-diamine (penten)[496] has also been investigated, including the solid-state structures of $[Tl(dien)_2][ClO_4]_3$ and $[Tl(NO_3)(penten)](NO_3)_2$. Thallium(III) nitrate and 1,4,7-triazacyclononane (L) at 1:4 ratio produce $[L_2Tl](NO_3)_3$.[497] The N,N',N''-trimethyl-1,4,7-triazacyclononane (L') derivatives of Tl(I) may also be synthesized from $TlNO_3$.

The ligand 1,3,5-triamino-1,3,5-trideoxy-*cis*-inositol (taci) is an interesting one, since it can provide four different coordination sites with variable softness and size. It features both nitrogen- and oxygen-donor sites. The metal coordination chemistry of taci with group 13 elements has been investigated.[175] Single-crystal X-ray analysis revealed a TlN_6 coordination sphere for Tl $(taci)_2(NO_3)_3 \cdot 2H_2O$ (95). Interestingly, the aluminum(III) in $Al(taci)_2Br_3 \cdot 7H_2O$ shows AlO_6 bonding, whereas gallium(III) in $Ga(taci)_2(NO_3)_3 \cdot 3H_2O$ adopts a GaN_3O_3 coordination sphere. The indium(III) shows an unusual structure, with a $(\mu_6$-$O)In_6$ unit.[176]

(95)

Reacting $TlCl_3$ with the silylated amidine $PhC(NSiMe_3)N(SiMe_3)_2$ gives the ionic derivative $[PhC(NHSiMe_3)_2][PhC(NSiMe_3)_2TlCl_3]$ (96).[498] The anion exhibits a five-coordinate thallium site in which the thallium atom is surrounded by three chlorine atoms and by two nitrogen atoms of the amidinato ligand. The axial sites of the trigonal bipyramid are occupied by one nitrogen and one chloride atom.

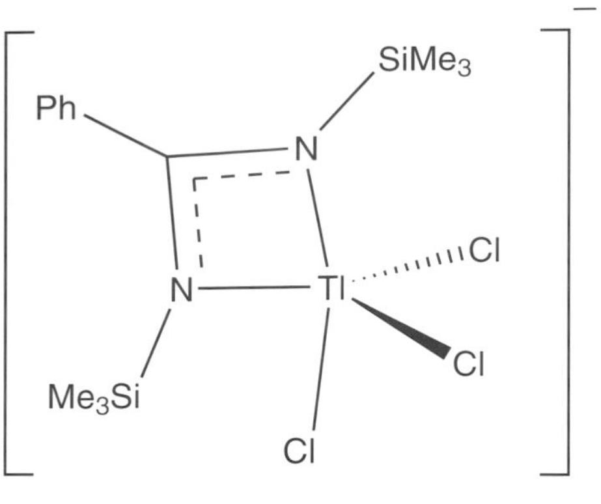

(96)

A dinuclear thallium(III) complex, tris[di(4,4'-phenyltriazenido)phenylmethane]dithallium(III), is formed by the reaction of TlNO₃ with di(4,4'-phenyltriazeno)phenylmethane in the presence of NaOH and air.[499] Each complex contains three doubly deprotonated bis(triazenido) ions $(PhN_3C_6H_4CH_2C_6H_4N_3Ph)^{2-}$ and two six-coordinate Tl^{3+} ions with trigonal-prismatic coordination of six N atoms.

Thallium(III) complexes of porphyrins are common.[97–99,117,480–482,500–515] Most studies involve 2,3,7,8,12,13,17,18–octaethylporphyrin (H₂OEP) and 5,10,15,20-tetraphenylporphyrin (H₂TPP) ligand systems. Synthesis, structures, spectroscopic data, and electrochemistry have been investigated. The typical coordination number at thallium(III) is either 5 or 6, and the coordination geometry may be described as a square-based pyramid formed by the porphyrin, in which the apical site is occupied either by a monodentate ligand (e.g., Cl, CN) or by a bidentate group (e.g., acetate). The X-ray crystal structure of Tl(TPYP)(CN)[482] (*meso*-tetra(4-pyridyl)porphyrinatothallium(III) cyanide) or Tl(TPP)CN[516] shows distorted square-pyramidal geometry at thallium. The thallium center in Tl(TPYP)(OAc) is six-coordinate, but the acetate group coordinates in asymmetric fashion, with two different Tl–O distances.[507] In contrast, the X-ray structure of Tl(III) *meso*-tetraphenylporphyrin acetate shows that the acetate group is coordinated as a symmetrically bonded bidentate ligand.[517] The N-methyltetraphenylporphyrin thallium(III) complex Tl(N-Me-TPP)(OAc)₂ has two *cis*-chelating acetate groups and an eight-coordinate thallium atom with a square-based antiprism geometry.[500] An unusual 4:3 tetragonal base–trigonal base, piano-stool, seven-coordinate geometry has been observed in Tl(N-Me-TPP)(O₂CCF₃)₂.[518] The compound Tl(TPP)(OSO₂CF₃)(THF) (97) features a six-coordinate thallium site, but a rare transoid geometry.[503]

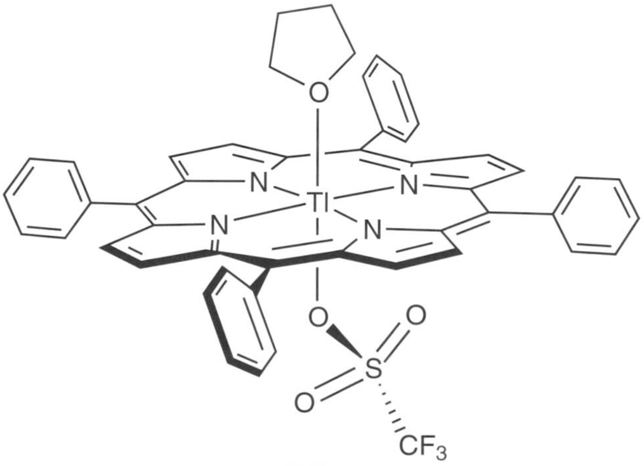

(97)

Heterometallic homo- and heteroleptic porphyrinate dimers with metal–thallium bonds have been described. These include (OEP)Rh–Tl(OEP), (TPP)Rh–Tl(OEP), (OEP)Rh–Tl(TPP), and (TPP)Rh–Tl(TPP).[519] The UV–visible spectroscopy confirms the presence of a strong π–π interaction between the macrocycles in each metal derivative.

Several phthalocyanato thallium(III) compounds have been synthesized and characterized.[520] Oxidation of dithalliumphthalocyaninate with excess iodine affords a blue-green iodophthalo-cyaninatothallium(III) derivative.[521] The thallium center has a tetragonal-pyramidal arrangement, with the iodo group occupying the axial site. The thallium atom is displaced out of the N_4 plane towards iodide by about 0.959 Å. Treatment of this iodo derivative (Pc)TlI with n-Bu$_4$NX (X = Cl, ONO, NCO) yields [n-Bu$_4$N][cis-(Pc)TlX$_2$] adducts.[522] Optical, vibrational data and the solid-state structures of [n-Bu$_4$N][cis-(Pc)Tl(ONO)$_2$] and [n-Bu$_4$N][cis-(Pc)TlCl$_2$] have been reported. The Tl(III) adducts [n-Bu$_4$N][cis-(Pc)Tl(ONO)$_2$] (98) and [n-Bu$_4$N][cis-(Pc)TlCl$_2$] feature eight- and six-coordinate thallium sites, respectively.

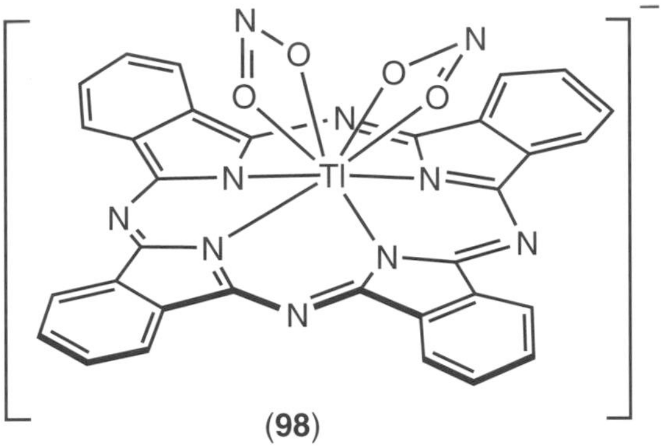

(98)

Complete lithiation of the tripodal amine HC{SiMe$_2$NH(p-Tol)}$_3$ with n-BuLi, followed by the reaction with TlCl$_3$, affords the Tl(III) amide HC{SiMe$_2$NTl(p-Tol)}$_3$ (99). The same product may be obtained by treating the HC{SiMe$_2$NLi(p-Tol)}$_3$ with three equivalents of TlCl. If the lithiation and transmetalation are performed in one step, an interesting mixed-valent amide [HC{SiMe$_2$N(p-Tol)}$_3$(n-BuTl)Tl] (100) forms.[523] The mixed-valent amide {CH$_2$[CH$_2$NSiMe$_3$]$_2$}$_2$Tl^{III}Tl^I, which features a bidentate ligand, has been prepared by treating two equivalents of the Tl(I) amide CH$_2$[CH$_2$N(Tl)SiMe$_3$]$_2$ with one equivalent of TlCl$_3$.[524]

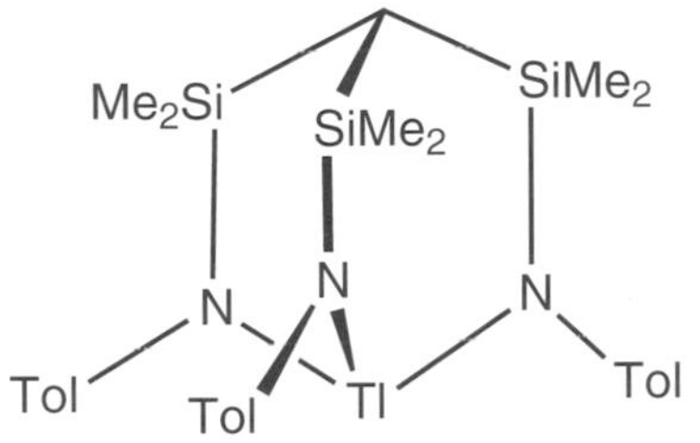

(99)

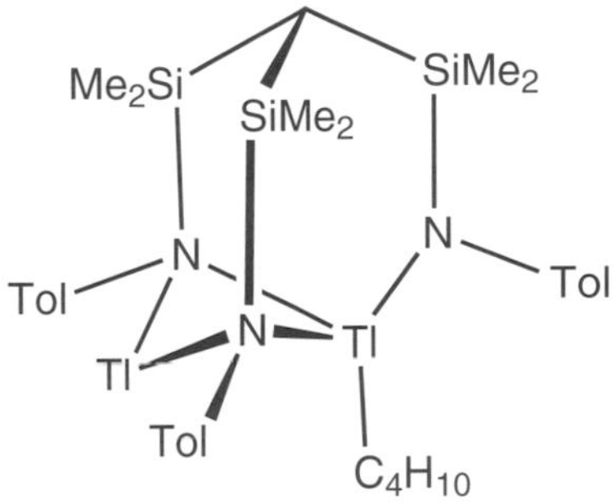

(100)

Thallium(III) adducts with a mixed-donor surrounding can be found in seven-coordinate [Tl(edta)OH]$^{2-}$, [Tl(edta)CN]$^{2-}$,[479] [Tl(bipy)$_3$(DMSO)]$^{3+}$,[525] and eight-coordinate [Tl(bipy)$_2$(NO$_3$)$_3$], [Tl(tpen)(NO$_3$)](ClO$_4$)$_2$ adducts (edta = ethylenediaminetetraacetate tpen = N,N,N',N'-tetrakis-(2-pyridylmethyl)ethylenediamine).[525] The thallium(III) ion has also been encapsulated in an iminophenolate cryptand ligand.[526] The NMR spectroscopic data suggest that the thallium(III) is held strongly within the host.

(ii) Phosphorus, arsenic, antimony, and bismuth ligands

There appear to be no notable nonorganothallium compounds that belong in this category. The highly oxidizing thallium(III) ion will most likely present problems during the synthesis of thallium(III) complexes of these soft donors. However, there are a few organothallium derivatives,[527,528] and Tl(III) adducts of sulfur ligands are known.

3.5.2.2.3   *Group 16 ligands*

(i)   *Oxygen ligands*

Solid-state data show that hexaaquathallium(III) ion has six water molecules coordinated to Tl(III) in a regular octahedral fashion.[529] However, in solution the water molecules are quite labile.[530,531] These solutions are acidic and the resulting hydroxides are fairly stable. Thallium(III) hydroxo species have also been investigated.[471] The solid-state structure of the first hydroxothallate, $Ba_2[Tl(OH)_6]OH$, has been reported.[532] It was synthesized by reacting $Tl_2O_3 \cdot xH_2O$ with NaOH in the presence of barium hydroxide. The structure and vibrational spectra of the DMSO-solvated Tl(III) ion were studied in a DMSO solution and in the solid state. The X-ray crystal structure of $[Tl(DMSO)_6](ClO_4)_3$ has been reported.[533]

Although Tl(III) adducts of β-diketones are not available, the homoleptic tropolonato derivative has been synthesized. The tris(tropolonato)thallium(III) can bind to trioctylphosphine oxide forming seven-coordinate complexes.[203] The indium(III) analogue behaves similarly.

Thallium(III) salts of polyoxyanions, including those of nitrate, halogenates, sulfate, phosphate, and acetate have been prepared, often starting from the thallium(III) oxide, and some solid-state structures have been investigated.[471,534–537] Thallium(III) acetate is a useful reagent in organic synthesis. Mixed-ligand complexes of thallium(III) containing donors in addition to oxygen have also been investigated.[538,539]

(ii)   *Sulfur, selenium, tellurium ligands*

Thallium(III) complexes in which the coordination sphere is made up exclusively of heavier, group 16 donors are rare.[540–545] The high oxidizing power of Tl(III) poses difficulties during the synthesis of such complexes. There are several early reports concerning the synthesis and structures of tris(*N,N*-dimethyldithiocarbamato)thallium(III) and tris(*N,N*-diethyldithiocarbamato) thallium(III) adducts, and compounds of the type $[Tl(dithiolene)_2]^-$ (where dithiolene = 1,2-$S_2C_2H_2$, 4,5-$S_2C_6H_2(CH_3)_2$, 1,2-S_2C_2-$(CN)_2$).[540,541,546] Thallium adducts of the type $[Tl(dithiolene)_3]^{3-}$ are also known.[547–549] Tris chelate adducts of Tl(III) containing 1,3-dithiole-2-thione-4,5-dithiolate (dmit) and 1,2-dithiole-3-thione-4,5-dithiolate (dmt) ligands can be synthesized using the alkali-metal salts of the ligand and $[Ph_4As][TlCl_4]$ as the thallium source. $[Ph_4As]_3[Tl(dmit)_3]$ and $[Ph_4As]_3[Tl(dmt)_3]$ have been isolated as red-brown and red crystalline solids, respectively.[547]

The Tl(III) adduct of dicyclohexyldithiophosphinic acid $[Tl\{S_2PCy_2\}_3]$ has been reported.[550] The reaction between diphenylthallium(III) compounds $TlPh_2X$ (X = Br, OH or S_2PCy_2) and dicyclohexyldithiophosphinic acid, $HS(S)PCy_2$, has resulted in the loss of one or more phenyl groups, leading to $Tl[S_2PCy_2]_3$ (**101**) as one of the products. The thallium atom is coordinated to three pairs of S atoms from two very anisobidentate ligands and one symmetrically bonded dithiophosphinate.

Anionic, tridentate tris(mercaptoimidazolyl)borates $[Tm^R]^-$ systems are useful for isolation of Tl(III) complexes.[542,543] Six-coordinate, sandwich complexes $\{[Tm^{Ph}]_2Tl\}ClO_4$, $\{[Tm^{Me}]_2 Tl\}I$ (**102**), and $\{[Tm^{Me}]_2Tl\}TlI_4$ have been synthesized and structurally characterized. The compound $\{[Tm^{Me}]_2Tl\}$ I has been isolated as a by-product during the synthesis of a tris(mercaptoimidazolyl)boratozinc complex using a thallium(I) starting material $[Tm^{Me}]Tl$.[542,543] Attempted synthesis of

(101)

$\{[Tm^{Me}]_2Tl\}^+$ using $[Tm^{Me}]Na$ and $Tl(NO_3)_3$ has been a failure, due to oxidation reductions between the Tl(III) ion and the ligand borohydride function. However, $[Tm^{Me}]_2Tl\}TlI_4$ can be synthesized in good yield by treating the $[Tm^{Me}]Tl$ with excess iodine. It is believed that, unlike the free ligand, the Tl(I) complexed ligand is more resistant to oxidation by Tl(III). The compound $\{[Tm^{Ph}]_2Tl\}ClO_4$ has been synthesized directly from $\{[Tm^{Ph}]Li$ and $Tl(ClO_4)_3 \cdot xH_2O$. The solid-state structures of these tris(mercaptoimidazolyl)borothallium(III) complexes show that the thallium ion is coordinated to six sulfur donors in a regular octahedral fashion. The Tl–S distances are 2.69 Å.

(102)

Electrochemical oxidation of thallium in the presence of 1-hydroxopyridine-2-thione (HPT) affords the thallium(I) adduct along with $Tl(PT)_3$ **(103)**.[300] The thallium atom is bonded to three oxygens and three sulfur atoms in an octahedral fashion, forming the *mer*-isomer. The ^{205}Tl NMR spectrum of $Tl(PT)_3$ exhibits the resonance due to thallium(III) at δ 2,968. The related thallium(I) adduct Tl(PT) displays the signal due to thallium(I) at δ 2,697.

3.5.2.2.4 *Group 17 ligands*

Compared with the trihalides of lighter, group 13 elements, the corresponding thallium(III) derivatives are less stable. TlF_3 hydrolyses easily in water, producing $Tl(OH)_3$.[7,471] Thus, hydrates of TlF_3 are not available. True double salts of the type $MTlF_4$ (M = Li or Na) and $MTlF_6$ (M = Na, K, Rb, and Cs) compounds containing the TlF_6^{3-} anion have been reported. TlI_3 is isomorphous with CsI_3, and is best described as Tl(I) triiodide. The $Tl^I I_3$ formulation is also consistent with the standard reduction potentials.[7] However, when treated with excess I^- TlI_4^- forms, in which the Tl(III) state becomes stable. There are many complexes containing the TlI_4^- ion.[551–559] The chlorides and bromides of Tl(III) show the most diverse coordination chemistry. Thallium(III) halides form complexes with a variety of N, O, and halide donors.[560–582] The Tl(III) adducts containing donors like P or S are rare, because of the oxidation reactions of the P- or S-donor center by the Tl(III) ions.

The nucleophilic carbene complexes of $TlCl_3$ and $TlBr_3$ have been reported.[325] These adducts are unique in that the divalent carbon site not only survives the oxidizing power of Tl(III), but also acts as a donor to Tl(III). The adducts $Cl_3Tl[C\{N(Mes)CH\}_2]$, $Cl_3Tl[C\{N(Mes)CBr\}_2]$, and $Br_3Tl[C\{N(Mes)CH\}_2]$ can be synthesized by treating TlX_3 with "stable" carbenes $C\{N(Mes)CH\}_2$ or $C\{N(Mes)CBr\}_2$. Although these complexes are moisture-sensitive, they possess high thermal stability. The bis(carbene) adduct $Cl_3Tl[C\{N(Mes)CH\}_2]C\{N(Me)C(Me)\}_2$ has also been prepared. The molecular structure of $Cl_3Tl[C\{N(Mes)CH\}_2]$ **(104)** has been reported. The thallium center adopts a distorted tetrahedral environment. The Tl–C(carbene) distance is 2.179(9) Å. Heating a mesitylene solution of $Cl_3Tl[C\{N(Mes)CH\}_2]$ and $Br_3Tl[C\{N(Mes)CH\}_2]$ to 163 °C leads to decomposition, resulting in the thallium(I) halide and the oxidized carbene product $[XC\{N(Mes)CH\}_2]X$.

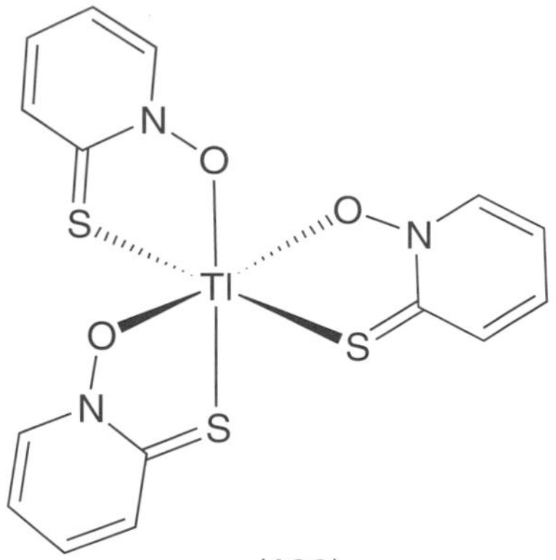

(103)

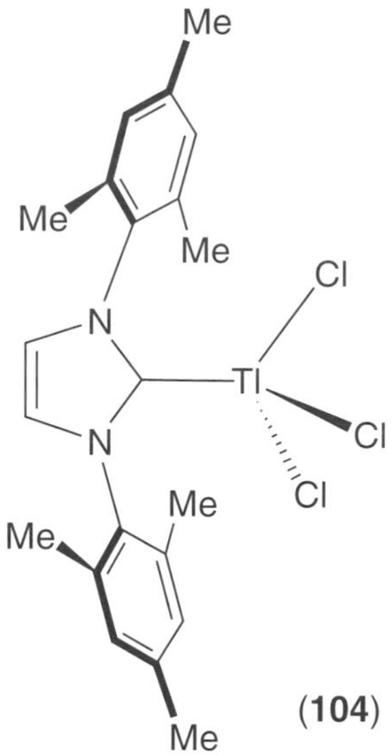

(104)

TlCl$_3$(py)$_3$ and TlBr$_3$(py)$_3$ feature octahedral Tl(III) centers with *mer*-geometry.[583,584] TlBr$_3$(OPPh$_3$)$_2$, [TlBr$_3$(μ-C$_4$H$_8$O$_2$)] and TlBr$_3$(py)$_2$ have trigonal-bipyramidal thallium centers in the solid state.[585–587] A series of Tl(III) compounds of the type TiClBrI(L)$_2$ (L = various pyridine N-oxides, HMPA, OPPh$_3$) containing three different halides on a single thallium atom have been prepared and characterized.[571] These compounds have been prepared by the action of IBr on TlCl in acetonitrile with the ligand present. The solid-state structure of TlClBrI(OPPh$_3$)$_2$ **(105)** has been reported. The thallium atom displays distorted trigonal-bipyramidal geometry, with the halide ions occupying the sites of the equatorial plane. Complexes of mixed halides such as TlBrCl$_2$, TlBrI$_2$, TlCl$_2$I, and TlCl$_2$Br have also been synthesized.[569,570]

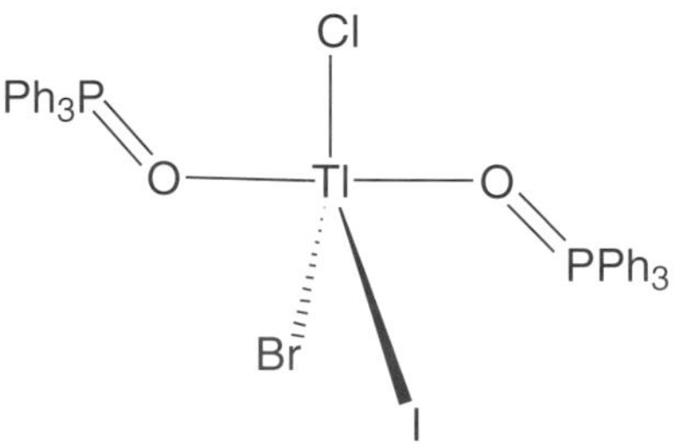

(105)

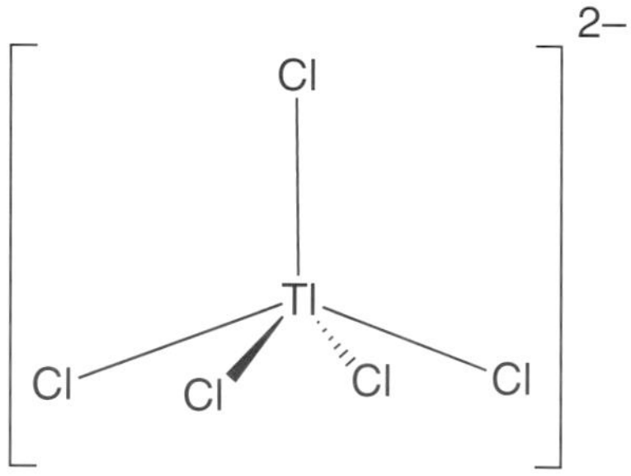

(106)

Thallium(III) chloride reacts with 1,4,7-triazacyclononane and 1,4,7-trimethyl-1,4,7-triazacyclo-nonane to produce 1:1 complexes.[497] It is also possible to prepare the InBr$_3$ adduct of 1,4,7-triazacyclononane. As noted earlier, the hydrolysis of this compound leads to the first well-authenticated In(III) μ-hydroxo complex.

Chlorothallate(III) complexes of various solid state structures are known.[588–599] Compounds with octahedral [TlCl$_6$]$^{3-}$ and tetrahedral [TlCl$_4$]$^-$ anions in the solid state have been well documented.[588] Pentanediammonium and 4-chloropyridinium salts of chlorothallates(III) contain distorted square-pyramidal [TlCl$_5$]$^{2-}$ **(106)** anions.[589,590] The presence of anions of the type [Tl$_2$Cl$_9$]$^{3-}$,[600,601] and [Tl$_2$Cl$_{10}$]$^{4-}$ has also been established.[590,593]

Bromothallate(III) complexes also show variable coordination numbers and structural diversity for the thallium(III) ion.[588,591,602–605] X-ray data, supported by Raman analysis, showed that the [TlBr$_5$]$^{2-}$ ion of 1,1,4,4-tetramethylpiperazinium and N,N'-dimethyltriethylenediammonium salts adopts a trigonal-bipyramidal geometry.[604] Compounds derived from 4,4'-dimethyl-2,2'-bipyridi-nium cation contain unusual bromothallate units, with four short Tl—Br bonds and one long Tl—Br interaction. The N-methyl-1,3-propanediammonium salt of [TlBr$_5$]$^{2-}$ is known. The X-ray structure reveals a distorted square pyramid with one long, additional Tl—Br contact.[606] Salts with octahedral $[TlBr_6]^{3-}$ and tetrahedral $[TlBr_4]^-$ anions are well known.[591] The anion $[Tl_2Br_9]^{3-}$ has also been reported.[603]

3.5.2.3.5 Hydride ligands

The chemistry of group 13 hydrides has been reviewed.[5] Thallium hydrides are the least stable among group 13 hydrides. Stability and the properties of Tl(III) hydrides have been analyzed by computational methods. Results suggest that Tl_2H_6 is thermodynamically unstable in both the gas phase and as a solid. Despite some claims in early literature, it is unlikely that TlH_3 aggregates may exist in the uncoordinated state. $LiTlH_4$ can be synthesized, but it decomposes rapidly at $0\,^\circ C$. $TlBH_4$ is a thermally stable compound.[2]

3.5.2.3 Thallium (II)

Only a very few well-authenticated molecules of divalent thallium are known.[2,3] The silyl derivatives of Tl(II) of the type R_2Tl—TlR_2 where $R = Si(SiMe_3)_3$ (**107**), $Si(t\text{-}Bu)_3$ (**108**), and $Si(t\text{-}Bu)_2Ph$ (**109**) have been synthesized and structurally characterized.[28,29,407,607] Dark red $\{[(Me_3Si)_3Si]_2Tl\}_2$ has been obtained from a reaction between $TlN(SiMe_3)_2$ and $(Me_3Si)_3SiRb$ in a toluene/pentane mixture.[607] Interestingly, the use of $(Me_3Si)_3SiM$ (M = Li, Na, K, Rb, Cs) and TlX (X = Cl, I) does not lead to Tl-Si bonded compounds, but results only in the formation of elemental Tl, MX, and $[(Me_3Si)_3Si]_2$. The thallium(II) derivative $\{[(Me_3Si)_3Si]_2Tl\}_2$ slowly decomposes in solution. However, solid samples are stable even in air for a short period. The crystal structure shows an approximately C_2-symmetric Tl_2Si_4 framework with a Tl—Tl bond length of 2.9142(5) Å and a $TlTlSi_2$ dihedral angle of 78.1°. The synthesis of $\{Tl[Si(t\text{-}Bu)_3]_2\}_2$ involves the use of TlBr and an alkali-metal salt $NaSi(t\text{-}Bu)_3$.[407] Thallium(II) radicals $(t\text{-}Bu)_3SiTl\cdot$ were suggested as being present in benzene solutions at room temperature to account for the unusually dark green color, the band-rich EPR signal, and some of the decomposition products. The compound $\{Tl[Si\text{-}(t\text{-}Bu)_2Ph]_2\}_2$ has been synthesized by treating TlBr with $NaSi(t\text{-}Bu)_2Ph$ in tetrahydrofuran.[29] It is a dark-blue-colored compound. According to X-ray crystal structure analysis, the thallium atoms in $\{Tl[Si(t\text{-}Bu)_3]_2\}_2$ and $\{Tl[Si(t\text{-}Bu)_2Ph]_2\}_2$ are planar, and coordinated with two Si atoms and one Tl atom. The Tl–Tl bond distances are 2.996(2) and 2.881(2) Å, respectively. The $TlTlSi_2$ dihedral angles are 89.6° and 82.2°, respectively. Larger substituents on the Tl lead to greater dihedral angles. For the $\{M[Si(t\text{-}Bu)_3]_2\}_2$ series (M = Al, In, Tl), the λ_{max} value of the visible absorption shifts with increasing atomic number and with increasing angle between the Si_2EE planes to a longer wavelength.[29] Selenium reacts with $\{Tl[Si(t\text{-}Bu)_3]_2\}_2$ to give hetorocubane $(t\text{-}Bu_3SiTl)_4Se_4$ (see (**45**) for a related structure).[29]

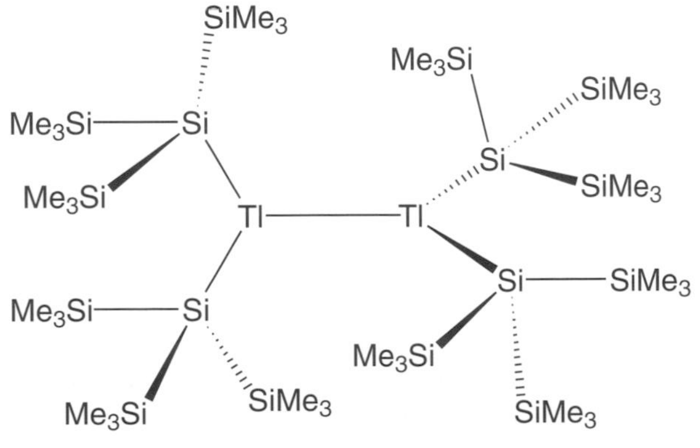

(**107**)

Molecular properties of organothallium compounds such as Me_2Tl have been calculated. The Tl–Me dissociation energy shows that Me_2Tl is unstable with respect to the disproportionation into either Me_3Tl and MeTl or $2Me_3Tl$ and Tl.[608] A paramagnetic Tl(II) complex $[NBu_4]_2[Tl\{Pt(C_6F_5)_4\}_2]$ containing a linear Pt—Tl—Pt core has been reported.[609]

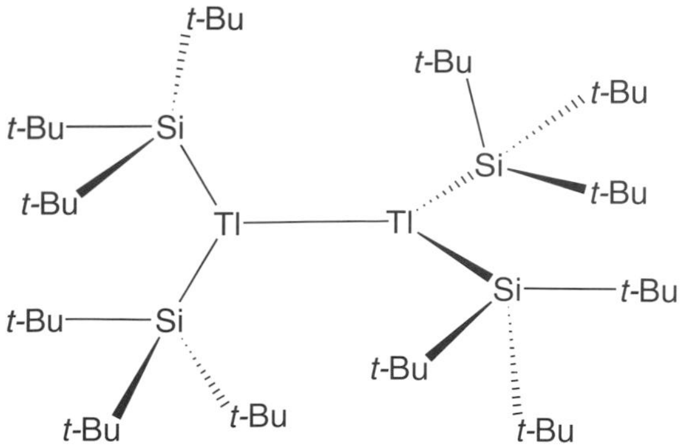

Indium and Thallium

(108)

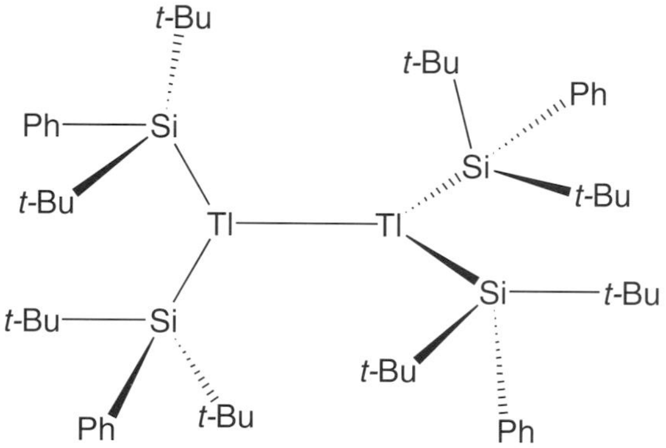

(109)

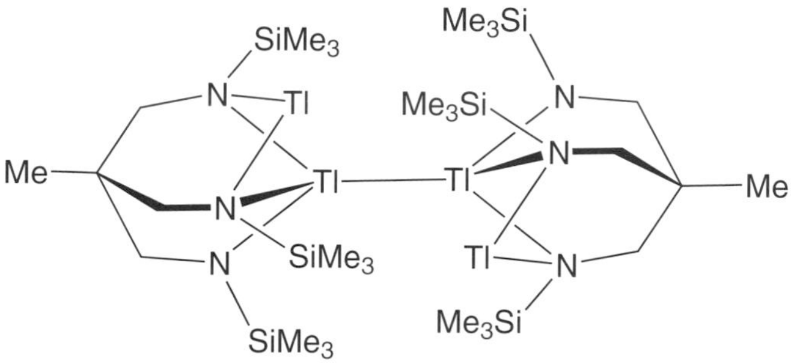

(110)

The diamagnetic, mixed-valent Tl(I)/Tl(II) amide $[MeC(CH_2NSiMe_3)_3Tl_2]_2$ has been obtained from the reaction between TlCl and $MeC\{CH_2N(Li)SiMe_3\}_3(dioxane)_3$ in tetrahydrofuran.[417] The compound $[MeC(CH_2NSiMe_3)_3Tl_2]_2$ is a red solid. It is also possible to prepare the related indium analogue, which is yellow in color, using a similar procedure. The key feature of $[MeC(CH_2NSi-Me_3)_3Tl_2]_2$ is the metal–metal-bonded Tl_2^{4+} fragment, which is shielded by the ligand framework. The Tl—Tl distance of 2.734(2) Å is relatively short compared to corresponding bond distances for silylated Tl(II) derivatives[29,607] described above.

The halides of compositions $TlCl_2$ and $TlBr_2$ are in fact $Tl^I[Tl^{III}X_4]$ species. Monovalent Tl is the most stable oxidation state for thallium in halide systems.

3.5.2.4 Thallium (I)

The monovalent thallium ion, with its relatively large ionic radius (1.50 Å for a 6-coordinate ion), has only weak electrostatic interactions with its ligands. The valence-shell electronic configuration of $d^{10}s^2$ with a lone pair makes the covalent interactions weak as well. Overall, the thallium ion is weakly solvated in most solvents, and crystallizes even without any coordinated solvent molecules. Thallium(I) compounds are the most widely explored group among thallium derivatives. The Tl^+ state is also the most stable ion in aqueous solutions.

3.5.2.4.1 Group 14 ligands

The thallium(I) ion forms salts with the cyanide ion.[1] However, the solution chemistry of TlCN is not well developed. In contrast, the cyanides of Tl(III) have been investigated in some detail.

The organothallium(I) compound $[(Me_3Si)_3CTl]_4$ features a distorted tetrahedron of Tl atoms in the solid state.[610] It is much less thermally stable than the analogous indium complex. An interesting, monomeric arylthallium(I) compound TlC_6H_3-2,6-Trip$_2$ (where Trip = 2,4,6-$(i$-Pr$)_3C_6H_2$) with a singly coordinated thallium atom has been described.[611] Cyclopentadienyl complexes of Tl(I) are well known.[612] Recently, the synthesis of an interesting Tl(I) derivative (η^5-$C_{60}Ph_5$)Tl involving C_{60} was reported.[613] The crystal structure reveals that the Tl(I) atom is bonded to C_{60} in η^5-fashion, and it lies deeply buried in a cavity created by the five phenyl groups.

Trithallane $[(t$-Bu$)_3Si]_4Tl_3Cl$ (111) and hexathallane $[(t$-Bu$)_3Si]_6Tl_6Cl_2$ (112) have been obtained during an attempt to synthesize $[(t$-Bu$)_3Si]_2TlCl$ from $(t$-Bu$)_3SiNa$ and $TiCl_3$.[489] These cluster compounds show high sensitivity to light, air, and moisture. A possible reaction pathway for the formation of $[(t$-Bu$)_3Si]_4Tl_3Cl$ and $[(t$-Bu$)_3Si]_6Tl_6Cl_2$ is also presented. The compound $[(t$-Bu$)_3Si]_4Tl_3Cl$, which is red in color, has a planar, four-membered Tl_3Cl core. The black hexathallane contains two four-membered Tl_3Cl rings which are linked by Tl—Tl and Tl—Cl interactions.

(111)   (112)

3.5.2.4.2 Group 15 ligands

(i) Nitrogen ligands

(a) Neutral nitrogen ligands. Little is known about thallium(I) adducts of neutral monodentate nitrogen donors like ammonia. Solution equilibria involving Tl(I) and ammonia have been investigated.[614–616] Data indicate the formation of mono- and diamminethallium(I) complexes. Studies involving pyridine and thallium(I) ions in aqueous solutions of NH_4ClO_4 suggest that, compared to NH_3, pyridine forms more stable adducts with Tl(I).[617] Thallium(I) methylamine interactions have been investigated using ^{205}Tl NMR spectroscopy.[618] The cyanomanganese carbonyls *trans*-[Mn(CN)(CO)(dppm)$_2$], *cis*- and *trans*-[Mn(CN)(CO){P(OR)$_3$}(dppm)] (R = Ph, Et; dppm = Ph$_2$PCH$_2$CH$_2$PPh$_2$), upon treatment with TlPF$_6$, form Tl—N-bonded complexes.[619] The formation of products with core geometries of the type Tl(μ-NC)Mn, {Tl(μ-NC)Mn}$_2$, and Tl{(μ-NC)Mn}$_2$ is observed.

Compared to neutral monodentate nitrogen donors, multidentate systems fare better in forming isolable Tl(I) adducts. Bidentate nitrogen donors like 2,2'-bipyridine (bipy) and 1,10-phenanthroline (phen) form TlL_2^+ cations.[1] The thallium–iron carbonyl compounds $[L_2Tl\{Fe(CO)_4\}_2]^-$ contain bidentate nitrogen donors L_2 such as bipy, en, phen, and tmda.[620] The Tl(I) encapsulated compound $[TlRh_4(\mu\text{-}2,6\text{-pyridinedithiolate})_2(cod)_2]^+$ shows an unusual see-saw coordination environment of the thallium atom, and pyridine–thallium coordination.[621]

The coordination chemistry of tris(2-pyridylmethyl)amine (TPA) with thallium(I) has been investigated. The reaction of $TlNO_3$ with TPA in aqueous acetonitrile results in $[Tl(TPA)]NO_3$.[622] Crystals of this compound were found to be $[Tl(TPA)]_2[H_3O][NO_3]_3$. The solid consists of two different $[Tl(TPA)]^+$ cations, one four-coordinate, while the second contains a seven-coordinate Tl site due to bonding to the three nitrate ions. Tris(pyrazolyl)methane[623] ligands also form Tl(I) complexes readily. Treatment of $HC(3,5\text{-}Me_2Pz)_3$ with $TlPF_6$ in tetrahydrofuran results in the immediate precipitation of $\{[HC(3,5\text{-}Me_2Pz)_3]_2Tl\}[PF_6]$ (113).[623] The mono ligand adduct $\{[HC(3,5\text{-}Me_2Pz)_3]Tl\}[PF_6]$ can be synthesized in acetone by using a mixture of 1:1 ligand/Tl(I) molar. The $\{[HC(3,5\text{-}Me_2Pz)_3]_2Tl\}[PF_6]$ complex has an octahedral structure with a stereochemically inactive lone pair. The coordination geometry at thallium in $\{[HC(3,5\text{-}Me_2Pz)_3]Tl\}[PF_6]$ is trigonal pyramidal. Tris(pyrazolyl)methane ligands are closely related to the anionic tris(pyrazolyl)borates.

Thallium nitrate reacts with N,N',N''-trimethyl-1,4,7-triazacyclononane (L) in the presence of $NaPF_6$ to yield the colorless solid $TlL[PF_6]$.[497] Crystals of $TlL[PF_6]$ consists of discrete TlL^+ cations and PF_6^- anions. The thallium(I) lone pair is stereochemically active in the solid.[497] Monomeric, four-coordinate Tl(I) complexes of mono-pendant-arm 1,4,7-triazacyclononane ligands have also been synthesized and characterized by X-ray crystallography.[624] The aminocryptand $N\{CH_2CH_2N(H)CH_2C_6H_4CH_2N(H)CH_2CH_2\}N$ has a large enough cavity to hold two metal ions in close proximity. The dilithium adduct of $N\{CH_2CH_2N(H)CH_2C_6H_4CH_2N(H)CH_2CH_2\}N$ (114) can be synthesized by treating the cryptate with CF_3SO_3Tl.[625] The thallium encapsulation, and the fact that it holds two Tl(I) ions closer to each other, have been established by NMR spectroscopy and

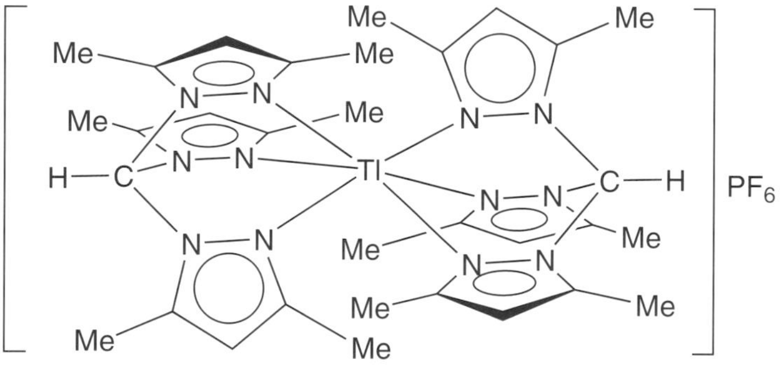

(113)

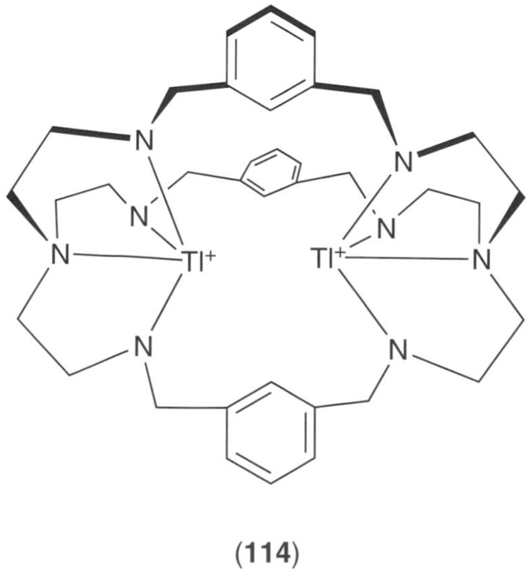

(114)

by solid-state structural studies. The Tl$\cdots$Tl distance of 4.3755(4) Å is longer than that observed in most dimeric or quasi-dimeric structures. However, the NMR data show that the two ^{205,203}Tl nuclei are coupled to each other through space with $J_{(Tl,Tl)} \gg 17$ Hz. This is the largest recorded through-space coupling between Tl atoms, indicating strong Tl$\cdots$Tl interaction in solution.

  (b) Amido ligands, monoanionic. The bis(trimethylsilyl)amido derivative of thallium (Me$_3$Si)$_2$NTl (**115**) has been synthesized by treating (Me$_3$Si)$_2$NTl with TlCl in toluene.[626] It is monomeric in benzene and in the gas phase.[627] It has a cyclic dimeric structure in the solid state, with intermolecular Tl$\cdots$Tl interactions. Related (2,6-(*i*-Pr)$_2$C$_6$H$_3$)(Me$_3$Si)NTl can be synthesized using a similar procedure.[628] It is a tetramer in the solid state. This amide shows weak Tl$\cdots$Tl and Tl—arene interactions.

  The bis(8-quinolinyl)amido (BQA) complex of Tl(I) has been synthesized by a transmetalation process involving the lithium derivative of the ligand and TlOTf.[629] The [BQA]Tl (**116**) exists as a monomeric species in solution. This compound serves as a good ligand-transfer agent for the preparation of group 10 metal adducts of the [BQA] ligand.

  (c) Diamido ligands. Difunctional thallium amides can also be synthesized. The reaction of CH$_2$[CH$_2$N{Li(dioxane)}SiMe$_2$R]$_2$ (R = Me or *t*-Bu) with TlCl leads to CH$_2$[CH$_2$N(Tl)SiMe$_2$R]$_2$ (**117**).[630] The thallium amide with the larger *t*-Bu group shows no significant Tl$\cdots$Tl contacts. Mixed-valence amides like {CH$_2$[CH$_2$NSiMe$_3$]$_2$}$_2$Tl^{III} Tl^I (**118**) and the related indium analogue {CH$_2$[CH$_2$NSiMe$_3$]$_2$}$_2$In^{III} Tl^I have been prepared.[524] The trivalent metal ion occupies the center of the tetrahedral coordination sphere of the amide nitrogens. The lithium ions of the diamide [(2-C$_5$H$_4$N)C(CH$_3$)(CH$_2$N(Li)SiMe$_3$)$_2$]$_2$ may be substituted in a stepwise manner to obtain a mixed lithium/thallium amide [(2-C$_5$H$_4$N)C(CH$_3$)(CH$_2$N(Li)SiMe$_3$)(CH$_2$N(Tl)SiMe$_3$)]$_2$ and the Tl(I) diamide [(2-C$_5$H$_4$N)C(CH$_3$)(CH$_2$N(Tl)SiMe$_3$)$_2$]$_2$.[631] Similar substitution of the lithium by a thallium ion has been achieved in [C$_{10}$H$_6${NLi(THF)$_2$SiMe$_3$}$_2$] to obtain [C$_{10}$H$_6${N[Li(THF)$_2$]Si-Me$_3$}{N(Tl)SiMe$_3$}] and [C$_{10}$H$_6${N(Tl)SiMe$_3$}$_2$] (**119**).[632]

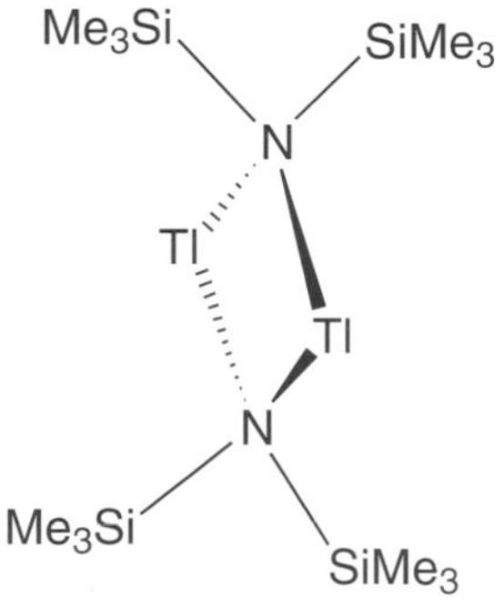

(115)

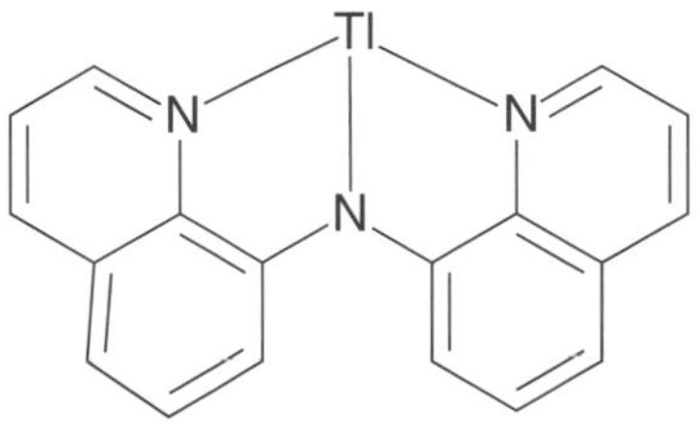

(116)

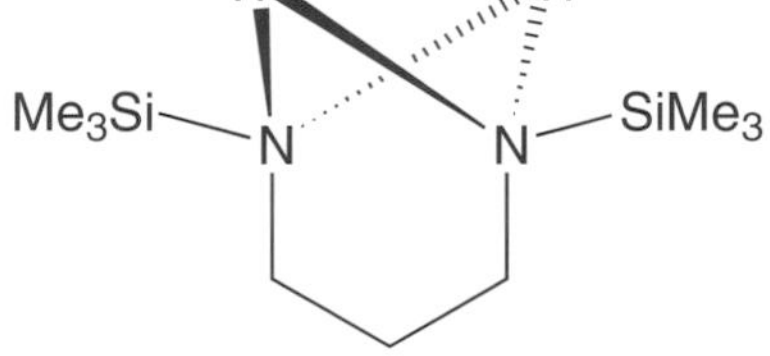

(117)

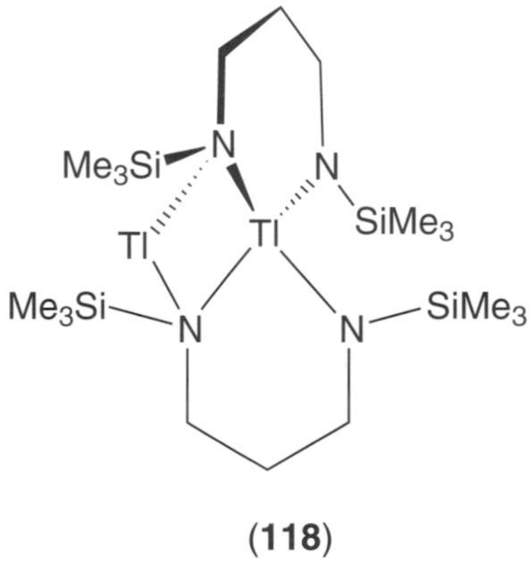

(118)

  The thallium amido [C$_{10}$H$_6${N(Tl)Si(R)Me$_2$}$_2$] (R = Me, *t*-Bu) derivatives, upon heating in dioxane to 90 °C, undergo metal–ligand redox chemistry leading to 4,9-diaminoperylenequinone-3,10-diimine derivatives.[418] The 4,9-diaminoperylenequinone-3,10-diimine is known; however, its synthesis is not an easy task. Related oxygen analogues are employed in photodynamic therapy and show cancerostatic and antiviral activity. The thallium route may provide an alternative, more convenient pathway for such organic compounds. Interestingly, the diindium analog $[C_{10}H_6\{N(In)SiMe_3\}_2]$ shows metal–metal redox chemistry leading to an In(II) complex $[\{C_{10}H_6(Me_3SiN)_2\}In(THF)]_2$.[418]

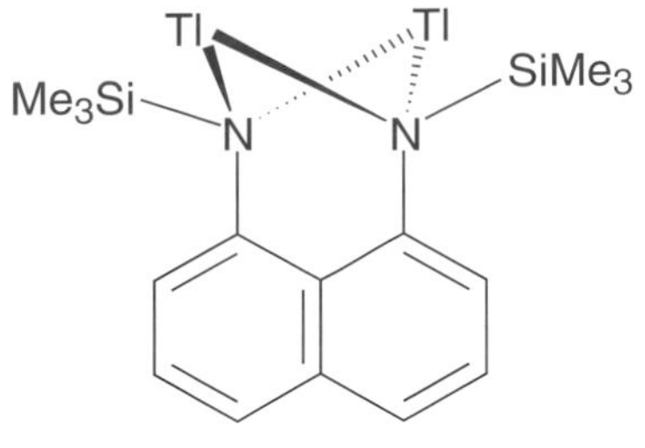

(119)

The thallium(I) complex of the bis(*tert*-butylamido)cyclodiphosphazane $[(t\text{-}BuNP)_2(t\text{-}BuN)_2]^{2-}$ ligand system is known. $[(t\text{-}BuNP)_2(t\text{-}BuNTl)_2]$ **(120)** has a dinuclear heterocubane structure.[416] The related $[(t\text{-}BuNSiMe)_2(t\text{-}BuNTl)_2]$ has also been synthesized and structurally characterized.[633]

(d) Triamido ligands. A few triamido derivatives of thallium are known.[523,634,635] They show different chemistry. They include $MeSi(Me_3CNTl)_3$,[636] $MeC[CH_2N(Tl)SiMe_3]_3$ **(121)**,[634] $(C_6H_5)C[CH_2N (Tl)SiMe_3]_3$,[634] and $MeSi[SiMe_2N(Tl)t\text{-}Bu]_3$ **(122)**. The compound $MeSi(Me_3CNTl)_3$ shows a dimeric structure in the solid state, with Tl$\cdots$Tl interactions. In fact, most of the thallium amides show Tl$\cdots$Tl interactions in the absence of stronger interactions with other functional groups in the molecule. The X-ray crystal structures of $(C_6H_5)C[CH_2N(Tl)SiMe_3]_3$ and $CH_3C[CH_2N(Tl)SiMe_3]_3$ demonstrate the relative importance of metal–arene vs. metal–metal interactions in thallium amide chemistry. In the tripodal Tl(I) amide with a phenyl group at the apical position of the ligand backbone, the competition between Tl$\cdots$Tl interaction and Tl$\cdots$arene interaction leads to an infinite chain structure in the solid state. The related $CH_3C[CH_2N(Tl)SiMe_3]_3$ is dimeric in the solid state.

During the synthesis of $HC[SiMe_2N(Tl)t\text{-}Bu]_3$, competing redox processes lead to the precipitation of thallium metal and the formation of $[HC\{SiMe_2N(H)t\text{-}Bu\}\{SiMe_2N(Tl)t\text{-}Bu\}_2]$ **(123)**.[637]

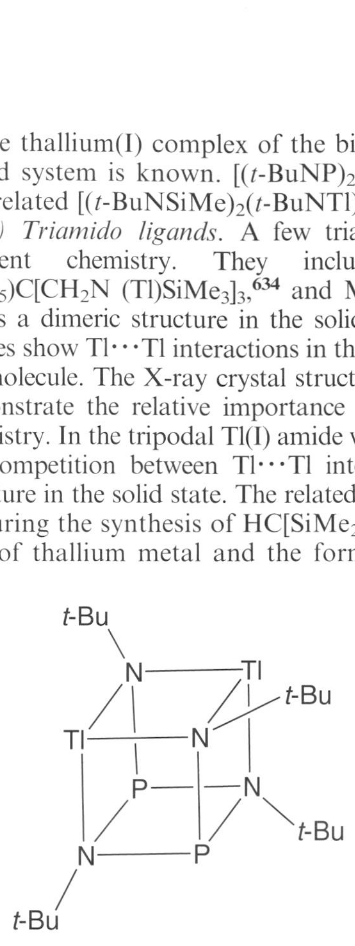

(120)

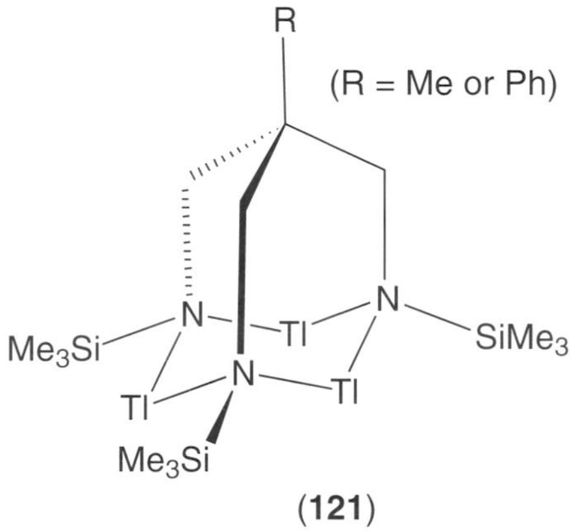

(121)

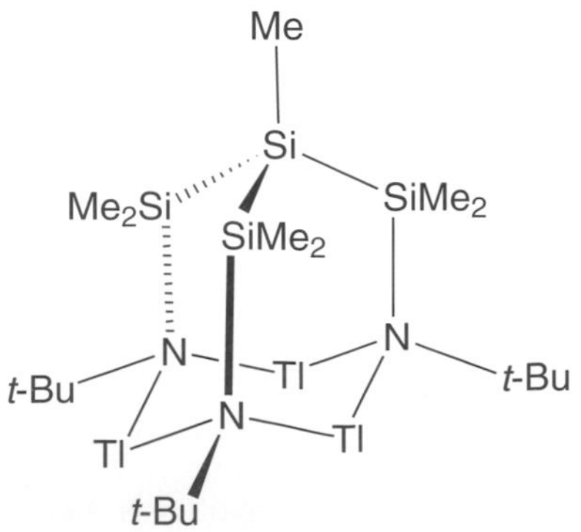

(122)

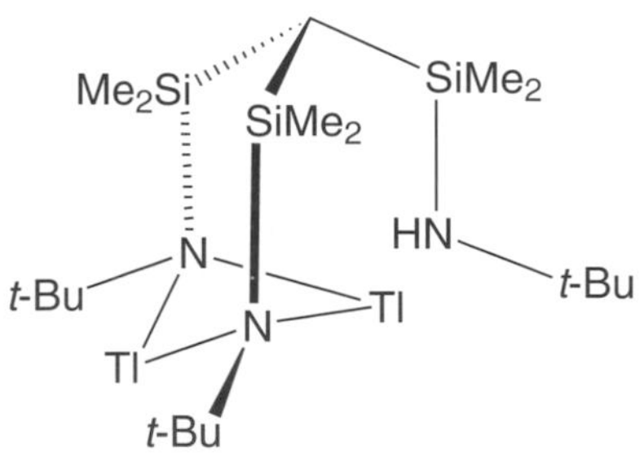

(123)

Interestingly, this does not happen during the synthesis of MeSi[SiMe$_2$N(Tl)t-Bu]$_3$. However, it is possible to obtain the partially demetalated thallium amide [MeSi{SiMe$_2$N(H)t-Bu}{SiMe$_2$N(Tl)t-Bu}$_2$] via the controlled thermolysis of MeSi[SiMe$_2$N(Tl)t-Bu]$_3$.[635]

The mixed-valent Tl(I)/Tl(II) species [MeC(CH$_2$NSiMe$_3$)$_3$Tl$_2$]$_2$, featuring rare Tl(II) sites, was described in a previous section.[417] The [HC{SiMe$_2$N(p-Tol)}$_3$]$^{3-}$ ligand system affords a mixed-valent Tl(I)/Tl(III) system [HC{SiMe$_2$N(p-Tol)}$_3$(TlBu)Tl)]. It also contains a donor-stabilized n-butylthallium(III) unit.[523]

(e) Pyrazolates and related ligands. Pyrazolate adducts of Tl(I) are of significant value for pyrazole-transfer reactions. Some of these reactions proceed with the reduction of Tl(I) to thallium metal. The reaction between pyrazoles 3,5-(Ph)$_2$PzH or 3-Me-5-PhPzH or 3-(2'-pyridyl)PzH with TlOEt proceeds with the elimination of ethanol to produce the corresponding Tl(I) pyrazolates.[638–640] In addition, compounds such as Tl(bin) (binH = 4,5-dihydro-2H-benz[g]indazole) and 4-Me-3,5-(Ph)$_2$PzTl, Tl(azin) (azin = 7-azaindazole) have been reported.[639,641] The synthesis of 3,5-(t-Bu)$_2$PzTl was not successful via the ethanol-elimination method.[641]

These compounds show a diverse range of thallium(I)-pyrazolate bonding modes. These include, μ-η^1:η^1, μ^3-η^1:η^1:η^1, μ^3-η^1:η^2:η^1, and η^5 bonding modes.[639] In addition, Tl$\cdots$Tl interactions are common. The compound 3-(2'-pyridyl)PzTl (**124**) displays a zigzag arrangement of pyrazalato-bridged thallium atoms.[640] Thallium(I) pyrazolates like 3,5-(Ph)$_2$PzTl served as important precursors for the synthesis of lanthanoid pyrazolate complexes via a redox transmetalation process.[638,641]

Tetrazole derivatives of Tl(I) may be prepared starting with Tl$_2$SO$_4$ or TlOEt.[642] Thallium complexes of nucleobases are reported, and they are of obvious interest for their biological relevance.[643,644] Early work involving imidazolate and benzotriazolate adducts is also known. Volatile thallium(I) pyrrole[645] derivatives have been reported as well.

(f) Other anionic nitrogen ligands with unsaturated backbones. Reaction of 1,3,5-triazine with (Me$_3$Si)$_2$NTl in toluene affords a novel product, 1,3,5,7-tetraazaheptatrienylthallium(I), involving both formally an α–ω Me$_3$Si shift and a ring opening.[646] The molecular structure consists of four units, each comprising of two thallium atoms and two [{Me$_3$SiNC(H)N}$_2$CH]$^-$ ligands (**125**).

A thallium(I) β-diketiminate, [HC{(Me)C(C$_6$H$_3$-2,6-Me$_2$)N}$_2$]Tl, and its use in the preparation of copper(I) complexes are reported.[647] No structural data are available on this compound. Thallium(I) complexes of 1,3-diphenyltriazenide[648] and 1,5-di-p-tolylpentaazadienes have been synthesized and structurally characterized.

Thallium derivatives of the tetracyanoethylene system are of interest as reagents for the introduction of [TCNE]$^{\bullet-}$ and [TCNE]$^{2-}$ (useful in the preparation of molecular-based magnets) via a halide-abstraction

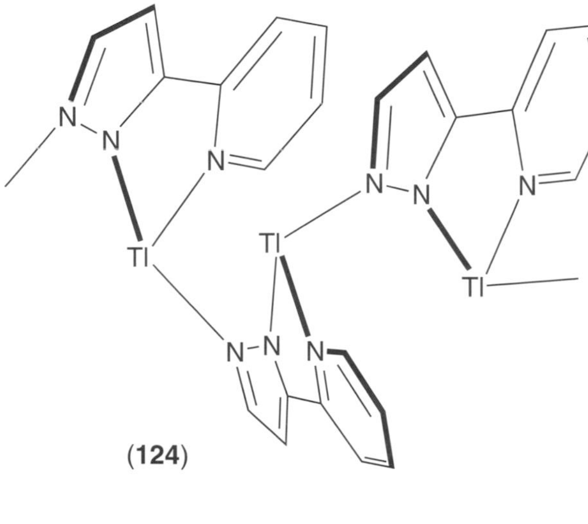

(124)

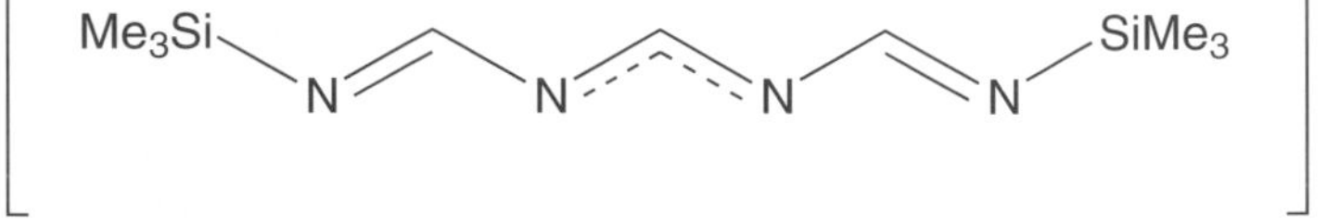

(125)

process. The synthesis and the reactivity of Tl[TCNE] and Tl$_2$[TCNE] have been investigated.[649] The structure of Tl[TCNE] consists of square-antiprismatic, eight-coordinate thallium sites.

(g) Poly(pyrazolyl)borate ligands. Thallium(I) complexes of bis-, tris-, and tetrakis(pyrazolyl)borates are clearly one of the largest, most well-characterized groups of thallium compounds containing Tl—N bonds.[439,440,650,651] The thallium(I) derivatives of poly(pyrazolyl)borates are used extensively as ligand-transfer reagents in the synthesis of various metal complexes.[650] They are usually milder, less reducing, and more stable than the corresponding alkali-metal salts. The thallium salts also facilitate the purification and characterization of the new poly(pyrazolyl)borate ligand systems. This choice is particularly valuable and commonly utilized in the synthesis of poly(pyrazolyl)borate ligands with bulky substituents.

Several thallium(I) complexes of the bis(pyrazolyl)borate ligand have been described. These include [H$_2$B(Pz)$_2$]Tl (**126**) (bis(pyrazolyl)hydroboratothallim(I)),[652] [H$_2$B(3-(9-triptycyl)Pz)$_2$]Tl,[653] [H$_2$B(3-(2-pyridyl)Pz)$_2$]Tl,[654] [H$_2$B(3-(2-pyrazinyl)Pz)$_2$]Tl,[655] [H$_2$B{3-[6-(2,2'-bipyridyl)]Pz}$_2$]Tl,[656] [H$_2$B(3-(*t*-Bu), 5-(Me)Pz)$_2$]Tl (**127**), and [H$_2$B(3-(*t*-Bu),5-(*i*-Pr)Pz)$_2$]Tl, [H$_2$B(3,5-(*t*-Bu)$_2$Pz)$_2$]Tl,[657] as well as thallium(I) adducts of asymmetric systems [H$_2$B(Pz)(3,5-(*t*-Bu)$_2$Pz)]Tl, [H$_2$B(3,5-(Me)$_2$Pz)(3,5-(*t*-Bu)$_2$Pz)]Tl, and [H$_2$B(3-(9-triptycyl)Pz)(3,5-(*t*-Bu)$_2$Pz)]Tl.[658] Typical synthetic procedures involve the metathesis reaction of a thallium(I) salt (e.g., thallium(I) formate, thallium(I) acetate, thallium(I) nitrate) with the appropriate alkali-metal bis(pyrazolyl)borate derivative.

(**126**)   (**127**)

Solid-state structures often show a monomeric structure with two-coordinate thallium(I) sites, with additional weak, secondary Tl···H–B interactions. The compounds [H$_2$B(3-(*t*-Bu),5-(Me)Pz)$_2$]Tl (**127**), [H$_2$B(3-(*t*-Bu),5-(*i*-Pr)Pz)$_2$]Tl, [H$_2$B(3-(9-triptycyl)Pz)$_2$]Tl, [H$_2$B(Pz)(3,5-(*t*-Bu)$_2$Pz)]Tl, and [H$_2$B(3,5-(*t*-Bu)$_2$Pz)$_2$]Tl adopt this type of structure. [H$_2$B(Pz)$_2$]Tl (**126**), in contrast, is dimeric in the solid state, with additional intermolecular Tl···Tl contacts of 3.70 Å length.[652] The close Tl···Tl contact observed in [H$_2$B{3-[6-(2,2'-bipyridyl)]Pz}$_2$]Tl has been attributed to π-staking.[656] The compounds [H$_2$B(3-(2-pyridyl)Pz)$_2$]Tl (**128**) and [H$_2$B(3-(2-pyrazinyl)Pz)$_2$]Tl contain additional nitrogen-donor sites on the ligand backbone.[654,655] The thallium(I) atoms in these compounds

(**128**)   (**129**)

prefer those nitrogen donors over Tl···H–B contacts. The ligand-transfer ability of some of these adducts has also been investigated.[659]

The tris(pyrazolyl)boratothallium(I) compounds are the most widely studied among thallium poly(pyrazolyl)borates. An excellent recent review article has appeared that covers the synthesis, structures, properties, and applications of tris(pyrazolyl)boratothallium(I).[650]

The Tl(I) adduct of the parent tris(pyrazolyl)borate [HB(Pz)$_3$]Tl (**129**)[660] is known, as well as many ligand varieties with different pyrazolyl groups and/or boron substituents. Substituents at the pyrazolyl ring 3-position are the closest to the thallium ion. They have the greatest influence both sterically and electronically on thallium (or any other metal ion coordinated to this ligand system). Compounds of the [HB(3-(R)Pz)$_3$]Tl type that have been reported include: R = cyclopropyl,[661] *i*-Pr,[661] *t*-BuCH$_2$,[662] cyclohexyl,[663] *t*-Bu,[664] Ph,[665] 2-pyridyl,[666] 2(pinene[4,5]-pyridyl,[667] 2-thienyl,[668] 4-MeC$_6$H$_4$,[668,669] 2-MeOC$_6$H$_4$,[670] 4-MeOC$_6$H$_4$,[668,669] 4-ClC$_6$H$_4$,[668] 2,4,6-Me$_3$C$_6$H$_2$,[671] 9-anthryl,[672] 1-napthyl,[673] 2-napthyl,[673] and 9-tryptycyl.[653]

In addition, thallium adducts featuring 3,4- and 3,5-disubstituted pyrazole-containing ligand systems are known. They include [HB(1,4-dihydroindeno[1,2-c]Pz)$_3$]Tl,[674] [HB(3-(i-Pr), 4-(Br)Pz)$_3$]Tl,[675] [HB(3,5-(Me)$_2$Pz)$_3$]Tl,[676] [HB(3-(CF$_3$),5-(Me)Pz)$_3$]Tl,[677] [HB(3-(CF$_3$),5-(2-thie-nyl)Pz)$_3$]Tl,[678] [HB(3,5-(CF$_3$)$_2$Pz)$_3$]Tl (**130**),[679] [HB(3,5-(i-Pr)$_2$Pz)$_3$]Tl,[680] [HB(3-(Ph), 5-(Me)Pz)$_3$]Tl,[674] [HB(3,5-(4-(t-Bu)C$_6$H$_4$)$_2$Pz)$_3$]Tl,[681] [HB(3-(t-Bu),5-(Me)Pz)$_3$]Tl,[664] [HB(3-(t-Bu), 5-(i-Pr)Pz)$_3$]Tl (**131**),[668] and [HB(3,5-(t-Bu)$_2$Pz)$_3$]Tl.[682] Although the substituents at the 4- or 5-position of the pyrazolyl ring are further away from the metal center, they also exert enough influence, and thus serve as valuable tools to control the chemistry of tris(pyrazolyl)borate metal adducts. The compounds [HB(Pz)$_2$(3,5-(t-Bu)$_2$Pz)]Tl (**132**),[683] and [HB(5-(Mes)Pz) (3-(Mes)Pz)$_2$]Tl[671] contain two different pyrazolyl ligands on the boron atom. Such ligand systems are rare. A few tetrakis(pyrazolyl)boratothallium(I) adducts are also known.[668]

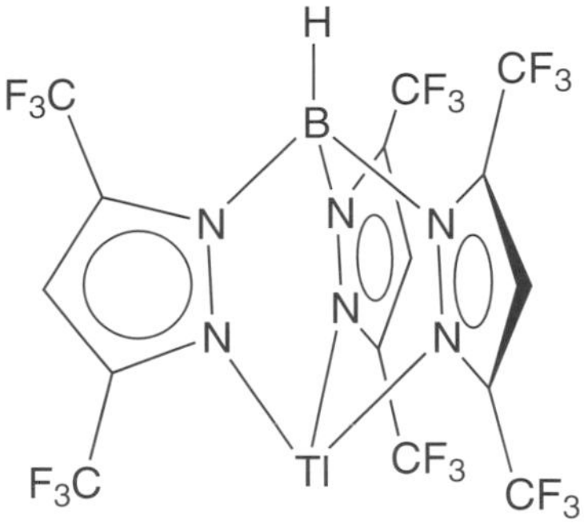

(**130**)

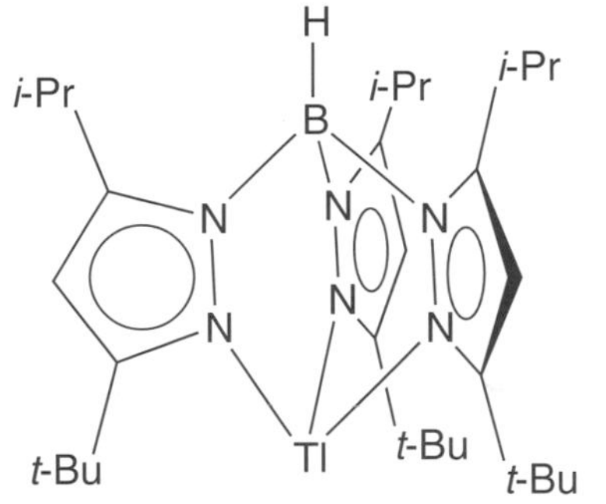

(**131**)

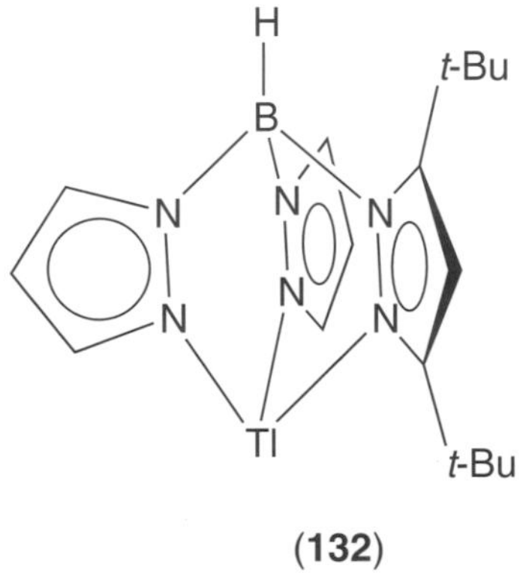

(**132**)

As with bis(pyrazolyl)borates, the synthesis of Tl(I) complexes involves the metathesis reaction between an alkali-metal tris(pyrazolyl)borate and a thallium salt. The triptycyl-substituted complex [HB(3-(9-triptycyl)Pz)$_3$]Tl has been synthesized from the reaction between [H$_2$B(3-(9-tripty-cyl)Pz)$_2$]Tl and 3-(9-triptycyl)PzH at 170 °C.[653] A promising, much milder route involving TlOEt is also available. The reaction of RBBr$_2$ (R = Me, cymentrenyl (Cym), methylcymentrenyl (Cym′), ferrocenyl (Fc)), pyrazole derivative, and NEt$_3$ in toluene at room temperature, followed by the addition of TlOEt, affords the thallium(I) tris(pyrazolyl)borate complex.[676,684,685] The compounds [MeB(Pz)$_3$]Tl, [MeB(3,5-(Me)$_2$Pz)$_3$]Tl, [MeB(3-(Me)Pz)$_3$]Tl, [MeB(3,5-(Me)$_2$Pz)$_3$]Tl, [CymB(Pz)$_3$]Tl, [CymB(4-C$_6$H$_{11}$CH$_2$)Pz)$_3$]Tl, [Cym'B(Pz)$_3$]Tl, and [FcB(Pz)$_3$]Tl have been prepared using this route.

The majority of the tris(pyrazolyl)boratothallium(I) adducts show monomeric structures with C_3 symmetry coordination of the tripodal ligand to the thallium(I) center. The Tl–N distances fall in the 3.50–2.73 Å range. The parent system [HB(Pz)$_3$]Tl[660] shows a structure in which [HB(Pz)$_3$]Tl units are arranged in a chain with long Tl···Tl separations. [HB(3-(4-MeC$_6$H$_4$)Pz)$_3$]Tl is dimeric with Tl···Tl distances of 3.86 Å.[686] However, the B–Tl–Tl–B sequence is collinear. The Tl(I) complex [HB(3-(cyclopropyl)Pz)$_3$]Tl (133) is a tetramer with a perfect tetrahedral Tl$_4$ core. The Tl···Tl distance is 3.6468(4) Å.[661] This distance is shorter than twice the van der Waals radius (3.92 Å), and only slightly longer than the Tl–Tl separation in elemental thallium (3.41 Å). It is not possible to predict the type of aggregation based on the ring substituents. The closely related [HB(3-(i-Pr)Pz)$_3$]Tl is a monomer.[661] The complex [PhB(3-(t-Bu)Pz)$_3$]Tl (134) shows an unusual structure in which one of the pyrazolyl groups is rotated by around 90°, and the Tl interacts with the pyrazolyl-ring nitrogen atom attached directly to the boron, via a p-orbital component of the aromatic π-system of the pyrazolyl ring.[687] In solution at room temperature, [PhB(3-(t-Bu)Pz)$_3$]Tl is stereochemically nonrigid on the NMR timescale. The repulsive methyl–methyl interaction forces [MeB(3,5-(Me)$_2$Pz)$_3$]Tl (135) to adopt the 2$_1$-helicoidal chain structure.[676] The [MeB (3,5-(Me)$_2$Pz)$_3$]$^-$ ligand shows a unique bridging coordination, rather than the expected trihapto, C_3-symmetrical thallium coordination.

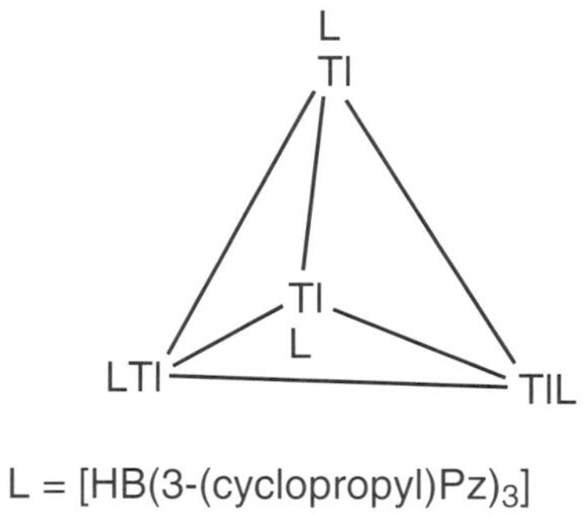

L = [HB(3-(cyclopropyl)Pz)$_3$]

(133)

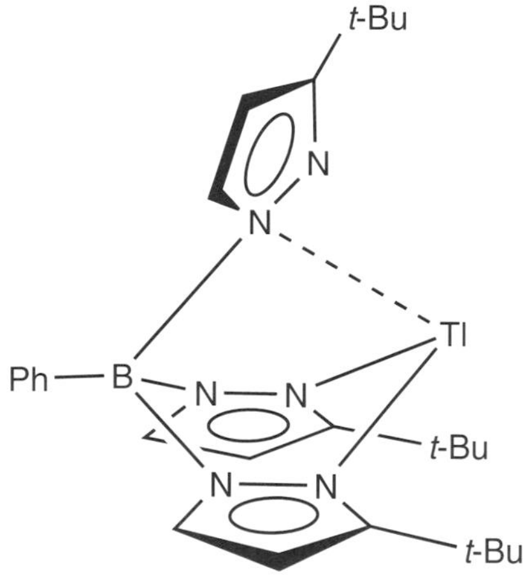

(134)

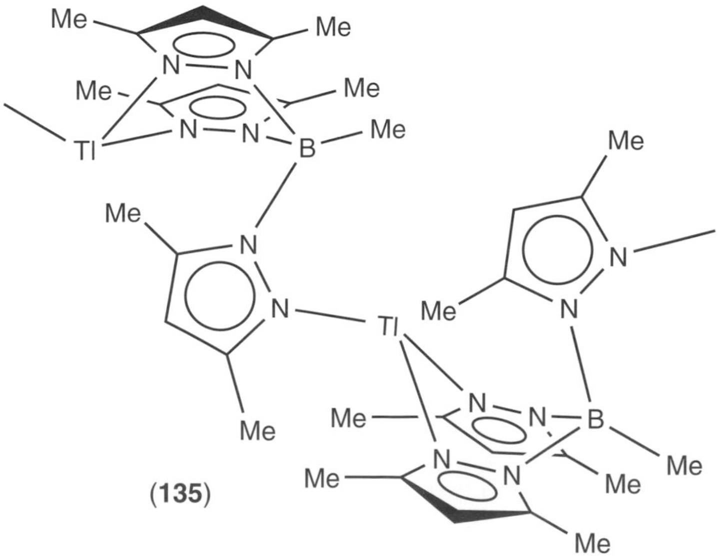

(135)

The compounds [CymB(Pz)₃]Tl (**136**)[685] and [FcB(Pz)₃]Tl[684] show polymeric structures, with bridging B(Pz)₃ fragments in the solid state. This is a result of unfavorable steric interaction between the substituent on boron and the hydrogen atoms on the pyrazolyl ring 5-position. The structure of [Cym′B(Pz)₃]Tl is somewhat related, but it adopts a macrocyclic tetrameric structure rather than a linear polymeric structure. Ligands with secondary donors on the backbone may form additional bonds to the thallium atom. For example, in [HB(3-(2-pyridyl)Pz)₃]Tl, weak Tl···N interactions between pyridyl nitrogens and the Tl atom have been observed.[666] [HB(3-(2-MeOC₆H₄)Pz)₃]Tl features close intramolecular Tl···O interactions.[670]

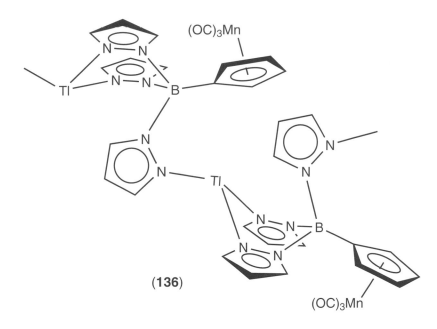

(**136**)

NMR spectroscopy also plays a large role in the characterization of poly(pyrazolyl)borato thallium adducts.[650,668,688] Coupling to spin-active ²⁰⁵Tl and ²⁰³Tl ($I = 1/2$) provides additional information about the solution structure. However, it is important to consider that nuclear relaxation due to chemical shift anisotropy has a significant effect on the apparent coupling constants to thallium.[677] It has been shown that higher applied magnetic field strengths and lower temperatures notably reduce the apparent J_{Tl-H} and J_{Tl-C} values. Some tris(pyrazolyl)borato thallium(I) adducts show photoluminescence, which originates from the metal-centered *sp* triplet of the Tl⁺¹ ion.[689] This technique provides information about the Tl···Tl interactions in the solid state.

The ligand-transfer chemistry of tris(pyrazolyl)boratothallium(I) compounds has been investigated extensively.[650] These adducts undergo metathesis reactions with a variety of metal halides or metal alkyl compounds, leading to the elimination of thallium halide (precipitates) or thallium alkyl products (which usually decompose to thallium metal) and the desired tris(pyrazolyl)borato metal adduct. The reaction of [HB(3-(*t*-Bu)Pz)₃]Tl with MeMgX offers the choice of an alkyl or a halide for the Tl(I).[690] The thallium(I) prefers the iodide when treated with MeMgI, but forms the thallium alkyl derivative when X = Cl or Br. Ligand-transfer reactions of [HB(3-(*t*-Bu)Pz)₃]Tl leading to a monovalent indium product have also been described.[442]

(h) Other anionic poly(azolyl) ligands. The closely related tris(pyrazolyl)methanesulfonate[691] and tris(indazolyl)borate (**137**) ligands also form Tl(I) complexes readily. The water-soluble, hydrolytically stable tris(pyrazolyl)methanesulfonate (Tpms) ligand adduct of Tl(I) has been prepared using [Tpms]Li and excess thallium(I) carbonate in water.[691,692] The Tpms ligand is a weakly coordinating ligand, and the donor properties are comparable to those of [HB(3,5-(CF₃)₂Pz)₃]⁻ or [HB(3-(CF₃),5-(CH₃)Pz)₃]⁻.[691] The tris(indazolyl)borates follow chemistry very similar to those of tris(pyrazolyl)borate relatives. The thallium adducts [HB(7-(*t*-Bu)indazolyl)₃]Tl,[693] [HB(7(*R*)-(*i*-Pr)-4(*R*)-(Me)-4,5,6,7-tetrahydro-2-indazolyl)₃]Tl,[694] [HB(7(*S*)-(*t*-Bu)-4(*R*)-(Me)-4,5,6,7-tetrahydro-2-indazolyl)₃]Tl,[694] [HB(2*H*-benz[g]indazol-2-yl)₃]Tl,[674] [HB(2*H*-benz[g]-4,5-dihydroindazol-2-yl)₃]Tl,[674] and [HB(3-Me-2*H*-benz[g]-4,5-dihydroindazol-2-yl)₃]Tl[674] have been synthesized, and some chemistry has been investigated. Most of these studies involving indazole derivatives are focused on the development of chiral ligand systems.

A tris(imidazolyl)borate (**138**) complex of thallium(I) has been synthesized.[695] The solid-state structure of hydrotris(imidazolyl)boratothallium(I) consists of one-dimensional, twisted, ladder-like strands, and three-coordinate thallium centers.[695] Due to the position of the nitrogen donors, the tris(imidazolyl)borate ligand is not capable of forming metal chelates as are observed in tris(pyrazolyl)borates. Poly(benzotriazolyl)borate ligands have some features of both tris(pyrazolyl)borate and tris(imidazolyl)borate systems. Thallium(I) complexes of bis-, tris-, and tetrakis (benzotriazolyl)borates are reported. These adducts have been synthesized by treating the corresponding potassium derivative with an equimolar quantity of thallium(I) formate.[696]

(137)   (138)

The reaction of 3-(2-pyridyl)PzH with POBr$_3$ in toluene–NEt$_3$ yields a hydrolysis product bis[3-(2-pyridyl)pyrazolyl]phosphinate, rather than the expected phosphines oxide OP(3-(2-pyridyl)Pz)$_3$.[697] The Tl(I) derivative of this ligand has been isolated and characterized.

(i) *Porphyrin and phthalocyanine ligands.* Thallium complexes of macrocyclic, nitrogen-based ligands such as porphyrin[97,698–700] and phthalocyanine[701–703] have been synthesized. The porphyrin adducts may be synthesized by treating the free ligand with TlOEt.[700] Thallium(I) complexes of 2,3,7,8,12,13,17,18-octaethylporphyrin (H$_2$OEP) and 5,10,15,20-tetraphenylporphyrin (H$_2$TPP) have two thallium(I) ions per porphyrin ligand. The crystal structure of [{Tl(THF)}$_2$(OEP)] shows that the thallium(I) atoms are four-coordinate, with bonds to three OEP nitrogens and one tetrahydrofuran molecule, and reside on opposite sides of the porphyrin.[698] Electrochemical studies on TPP complex of thallium(I) in DMF have been described.[699] The dithallium phthalocyanine Tl$_2$Pc (**139**) is a rare example of a group 13 dimetallophthalocyanine.[701] The most interesting feature of this material is its very high conductivity ($\sigma > 10^4$ Ω^{-1} cm^{-1}), which is comparable even to metals. The two-dimensional skeleton formed by the intramolecular Tl–Tl contacts and intermolecular Tl–N$_{aza}$ contacts is the key to this efficient charge transportation found in Tl$_2$Pc.[702].

(139)

(ii) *Phosphorus, arsenic, antimony, and bismuth ligands*

Coordination compounds containing thallium(I) and heavier group 15 elements are rare. Synthesis of a tris(phosphino)borate thallium adduct has been reported. The reaction between [PhB(CH$_2$PPh$_2$)$_3$]Li(TMEDA) and TlPF$_6$ affords [PhB(CH$_2$PPh$_2$)$_3$]Tl (**140**) as a yellow powder. The ^{31}P NMR spectrum shows two doublets with $^1J_{Tl–P}$ of 5,214 Hz and 5,168 Hz, as a result of coupling to ^{205}Tl and ^{203}Tl isotopes. Typical $^1J_{Tl-P}$ coupling constants for Tl(III) phosphine adducts are in the region of 1,500 Hz.[527,528] The [PhB(CH$_2$PPh$_2$)$_3$]$^-$ ligand binds to the thallium ion in a tridentate fashion. This thallium(I) adduct serves as a ligand-transfer agent to transition-metal ions. It is possible to synthesize the Co(II) adduct [PhB(CH$_2$PPh$_2$)$_3$]CoI using [PhB(CH$_2$PPh$_2$)$_3$]Tl and CoI$_2$. The use of lithium [PhB(CH$_2$PPh$_2$)$_3$]Li(TMEDA) with CoI$_2$ does not lead cleanly to the expected cobalt complex.

A heat- and air-stable diphosphastibilol complex of Tl(I) has been synthesized using [Li-(TMEDA)$_2$][1,4,2-P$_2$SbC$_2$(t-Bu)$_2$] and TlCl.[704] The solid-state structure of Tl[η^5-1,4,2-P$_2$SbC$_2$-(t-Bu)$_2$] (141) reveals a double-stranded, zigzag polymeric chain structure with intermolecular thallium–phosphorus interactions. The triphospholyl complex Tl[η^5-1,4,2-P$_3$C$_2$(t-Bu)$_2$] consists of weakly interacting monomeric half-sandwich units in the solid state.[705] The related Ga(I) and In(I) derivatives have also been synthesized.[448,449,705] Both Tl[η^5-1,4,2-P$_3$C$_2$(t-Bu)$_2$] and Tl[η^5-1,4,2-P$_2$SbC$_2$(t-Bu)$_2$] exit as monomeric species in the gas phase. The thallium–phosphorus coupling in the NMR spectra was not observed, perhaps indicating an ionic nature of bonding. Compounds Tl[η^5-1,4,2-P$_3$C$_2$(t-Bu)$_2$] and Tl[η^5-1,4,2-P$_2$SbC$_2$(t-Bu)$_2$] are useful oxidizing, ligand-transfer agents for lanthanide metals.[706]

(140)

(141)

3.5.2.4.3 *Group 16 ligands*

(i) Oxygen ligands

(a) Neutral oxygen ligands. Thallium(I) forms only weak interactions with most solvent molecules. Therefore, well-defined molecules with solvated Tl(I) ions are not common. The structure of solvated thallium(I) ion in aqueous, DMSO, and *N,N*′-dimethylpropyleneurea solutions has been investigated by large-angle X-ray scattering and EXAFS methods.[707] The Tl(I) coordinates to four water molecules with two short and two long distances. The solvation by DMSO and *N,N*′-dimethylpropyleneurea involves two short and four long solvent–Tl bonds. The different Tl—O bond lengths are believed to be due to the effects of stereochemically active lone pairs. More detailed information about the coordination geometry at Tl(I) in solution could not be obtained, because of weak Tl–solvent interactions. Thallium-ion binding by crown ethers and calixarenes has been investigated. One of the motivations for this work is to develop thallium(I) ion-selective analytical methods.[708–715] Owing to the toxic effects of thallium, the ability to quickly detect Tl$^+$ in biological fluids is important. Dibenzo-16-crown-4 has shown high selectivity for thallium(I) over sodium, potassium, and rubidium ions. Thallium-205 and carbon-13 NMR spectroscopy have been used to determine the stabilities of 18-crown-6 ligands with different structures and similar cavity sizes.[716] Several structurally characterized Tl(I) adducts of crown ethers are known, including those involving 12-crown-4 and even 30-crown-10.[357,551,713,717–722] In calixarenes, the π-coordination also contributes to the Tl(I) binding.

(b) Hydroxide ligands. Thallium complexes of anionic oxygen donors are relatively more common. Thallium(I) oxide is a hygroscopic solid and on contact with water forms TlOH. Solutions of TlOH are basic. The basic strength is about 10^5, 10 times greater than for NH$_3$ and calcium(II) hydroxide, respectively.[471] There is also evidence for the formation of [Tl(OH)$_2$]$^-$ species in solution.[723]

(c) Alkoxide ligands. Although thallium(I) alkoxides, TlOR, have been known since the 1800s, the detailed structural details became available only recently.[724,725] Their synthesis usually involves

a reaction between Tl and ROH, or TlOH and ROH, or Tl_2O and ROH.[1] They are useful as thallium(I) transfer agents, alkoxide donors, and for the preparation of mixed alkoxides.[451,725]

Based on Tl NMR spectroscopy and molecular-weight studies, cubane structures have been proposed for TlOR (R = Et, *i*-Pr, *t*-Bu).[726,727] Early work has revealed only the partial structure of $[Tl(OMe)]_4$ (**142**).[728] The crystal structure of $[Tl(OCH_2CMe_3)]_4$ has been reported as a $[Tl-O]_4$ cubane core.[724] Thallium(I) triphenylsilanolate also contains similar cubic units.[729] Reaction of poly(dimethylsiloxane) (silicone grease) with TlOEt led to the ladder polymer $[\{Tl_2(OSiMe_2)O\}_2]_n$ containing $[Tl-O]_4$ cuboids.[729] The compounds $[Tl\{\mu-O(C_6H_4)(C_6H_4OH)\}]_2$ and $[Tl\{OC_6H_2-2,4,6-(CF_3)_3\}]_2$ (**143**) are dimers,[730,731] whereas $[Tl\{OC_6H_3-2,6-(CH_3)_2\}]_n$ (**144**) and $[Tl\{OC_6H_3-2,6-(i\text{-}Pr)_2\}]_n$ feature polymeric chain structures.[724]

(**142**)   (**143**)

(**144**)

Mixed-metal alkoxide complexes of thallium are also known. For example, $Sn(\mu\text{-}t\text{-}BuO)_3Tl$ has both Sn(II) and Tl(I) ions.[451] The thallium site is unreactive as a donor for metal carbonyls. However, as indicated earlier, the indium(I) site of the indium analogue shows Lewis-base character. The Sn(IV)/Tl(I) mixed alkoxide $[Sn(EtO)_6Tl_2]$ exists as a one-dimensional polymer.[732] This adduct reacts quantitatively with $SnCl_2$ to form the homoleptic, mixed-valent $[Sn_2(OEt)_6]_n$. Thallium–titanium double alkoxides have been synthesized using thallium alkoxide as one of the starting materials.[725]

The study of reactions between TlOH and TlOEt with starch derived from different sources shows that potato starch binds thallium(I) chemically, whereas corn starch forms simple adducts with TlOH or TlOEt.[733] The iodination of thallium salts of phenols has been investigated.[734]

(d) β-diketonate ligands. Thallium(I) β-diketonates have been known for many years. These include the Tl(I) adducts of more common acetylacetonates, $[\{CH\{C(O)CH_3\}_2\}]^-$ and $[\{CH\{C(O)CF_3\}_2\}]^-$.[735,736] Recent studies reveal how simple thallium(I) β-diketonates can self-assemble to give discotic structures, via the formation of disk-like dimers, by means of Tl—Tl bonds reinforced with Tl—O bonds between the neighboring molecules.[737,738] This work is aimed at understanding the relationships between crystalline phases and liquid crystals. Ferroelectric liquid crystals containing palladium have also been prepared, using the thallium β-diketonato derivatives.[739]

Thallium β-diketonato complexes [Tl{CH{C(O)R}$_2$}] (R = Me, Ph) react with an excess of CS$_2$ to give 1,1-ethylenedithiolato complexes of thallium(I).[740] A volatile chelate (2,2,6,6-tetramethyl-3,5-heptanedionato)thallium(I) has been described.[741] This thallium adduct in CF$_2$Cl$_2$/O$_2$ and CF$_3$Br/O$_2$ gas mixtures has been used under CVD conditions to prepare TlX (X = Cl, Br). Solvent extraction of thallium(I) ions in aqueous solutions into chloroform was explored using several β-diketones.[742]

(e) Other anionic oxygen ligands. Attempts to prepare Tl(III) complexes of orthoquinone derivatives have been unsuccessful, leading only to Tl(I) adducts containing the semiquinonate anion.[452,731] These compounds are paramagnetic and colored. The paramagnetic property is due to the presence of the semiquinonate anion radical.

A large number of thallium(I) carboxylates have been synthesized and characterized by IR spectra.[743–745] X-ray diffraction methods have indicated that Tl(I) formate exists in solution as a tetramer.[746,747] Thallium saccharinate has been synthesized and characterized using crystallography.[748] It has a polymeric structure with eight- and five-coordinate thallium sites. Thallium(I) salts of the antibiotic lasalocid-A have been prepared.[749]

Many Tl(I) derivatives of polyoxy anions are known, and they are of interest for applications ranging from materials to chemical synthesis.[750–766]

Several molecular structures, including those of Tl(I) nitrate,[755,756] iodate,[757] borate,[758] germanate,[758] sulfate,[759,760] phosphates,[761–764] arsenate,[753] chromate,[765] and selenate,[766] have been reported. Heterobimetallic compounds with Tl–O interactions have been reported.[767,768] Their metal–thallium bonding and their photophysical properties are of particular interest.

(ii) Sulfur, selenium, tellurium ligands

(a) Neutral sulfur ligands. Thallium(I)-ion solvation by N,N-dimethylthioformamide has been investigated by large-angle X-ray acattering and EXAFS methods.[707] Although the solutions were prepared from the thallium(III) salt Tl(OTf)$_3$, the ^{205}Tl NMR measurements and the absence of typical Tl(III)—S bond distances suggest that the metal ion in the solution is in the monovalent form, i.e., thallium(III) has been completely reduced to Tl(I) by the solvent. Data show that there are two groups of Tl—S distances, two long and four short Tl—S bonds. This indicates that the lone pair of electrons on thallium plays a significant stereochemical role in the solvated Tl(I) ion. The stability and ligand-exchange properties of thiourea complexes of Tl(I) in aqueous solutions have been investigated.[769,770]

Just as do crown ethers, crown thioethers also form complexes with thallium(I) ion.[544,771–773] These include [Tl([9]aneS$_3$)][PF$_6$], [Tl([18]aneS$_6$)][PF$_6$], and [Tl([24]aneS$_8$)][PF$_6$] ([24]aneS$_8$ = 1,4,7,10,13,16,19,22-octathiacyclotetracosane).[771–773] These adducts can be synthesized by treating the crown thioether with TlPF$_6$, or by starting with TlNO$_3$ followed by the addition of NH$_4$PF$_6$. These studies provide useful information for the design of selective metal-complexing agents for the transport and uptake of toxic heavy metals like thallium.

Its crystal structure shows that [Tl([24]aneS$_8$)][PF$_6$] adopts a polymeric structure. The thallium atoms are eight-coordinate, and bridge two thioether crowns to give a sinusoidal chain.[772] The Tl(I) complex of the smaller ring adduct [Tl([18]aneS$_6$)][PF$_6$] shows that the thallium atom bonds strongly to the six sulfur donors of the macrocycle, with two additional weak interactions to the sulfurs of the neighboring ring.[773] The related thallium complex of the mixed-donor crown-[18]aneN$_2$S$_4$ (1,4,10,13-tetrathia-7,16-diazacyclooctadecane) has also been prepared.[773] The structure of [Tl([18]aneN$_2$S$_4$)][PF$_6$] (145) shows the Tl ion occupying the cradle formed by the macrocycle, leaving the top face of the metal center exposed. The adduct [Tl([9]aneS$_3$)][PF$_6$] (146)

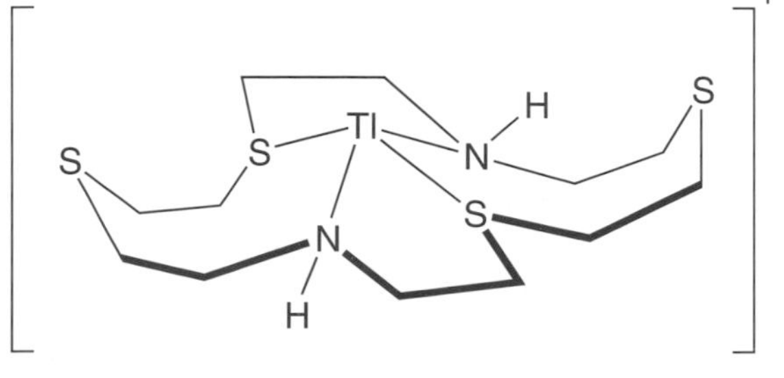

(145)

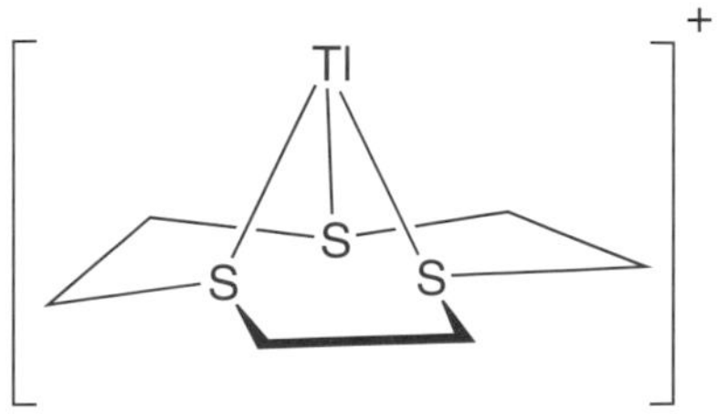

(146)

featuring the smallest crown, [9]aneS$_3$, shows that the thallium atoms are coordinated facially to the [9]aneS$_3$, with one additional secondary Tl—S bond formed between the Tl and a sulfur atom of the neighboring ring.[771] Further secondary interaction between the fluorine atoms of the anion results in overall eight-coordinate thallium sites.

(b) *Thiolate ligands.* A number of thallium(I) thiolates have been synthesized and characterized.[250,774–781] In addition to their fundamental interest, thallium thiolates are useful in analytical and materials chemistry fields, and for modeling the toxic effects of heavy metals.[776,782–786] The thallium thiolates TlSPh, TlSCH$_2$Ph, TlS(t-Bu), and 2,4,6-(CF$_3$)$_3$C$_6$H$_2$STl can be synthesized by treating Tl$_2$CO$_3$ with the corresponding sodium thiolate.[774,775,780] The reaction between (t-BuO)$_3$SiSH and TlNO$_3$, or 2,4,6-(CF$_3$)$_3$C$_6$H$_2$SH with EtOTl, affords (t-BuO)$_3$SiSTl,[781] or 2,4,6-(CF$_3$)$_3$C$_6$H$_2$STl,[780] respectively. Electrochemical methods have also been utilized to produce thallium thiolates such as o-CH$_3$C$_6$H$_4$STl, m-CH$_3$C$_6$H$_4$STl, 2-C$_{10}$H$_7$STl, and alkanedithiolates of thallium(I).[250] Alkanedithioles of the type Tl$_2$(S$_2$R) (R = alkane bridges –C$_n$H$_{2n}$–) have been synthesized via an electrochemical route as well.[456]

Thallium(I) thiolates display an interesting structural diversity. The thiophenol derivative is an ionic product, with [Tl$_7$(SPh)$_6$]$^+$ cations and anions [Tl$_5$(SPh)$_6$]$^-$.[775] The compound [(t-BuO)$_3$SiSTl]$_2$ (147) is a dimer.[781] The t-butyl derivative [TlS(t-Bu)]$_8$ is a covalent octamer.[775] The solid-state structure of TlS(i-Pr) contains [Tl$_4${S(i-Pr)}$_5$]$^-$ cages linked by additional thallium cations.[787] The compounds TlSCH$_2$Ph and TlSC$_6$H$_{11}$ contain Tl$_2$S$_2$ ring-coupled, two-dimensional polymers.[775,787] A similar, folded-ladder structure is adopted by the polymeric [2,4,6-(CF$_3$)$_3$C$_6$H$_2$STl]$_n$ (148).[780]

Under anaerobic conditions, the reaction of TlCl, TlNO$_3$, or Tl$_2$CO$_3$ with solutions of NaOMe and 1,2-(HS)$_2$C$_6$H$_4$ yields, after metathesis with [Et$_4$N]Br, yellow crystals of [Et$_4$N]$_2$[{Tl(1,2-(μ-S)$_2$C$_6$H$_4$)}$_2$] (149).[787] This compound contains rectangular-bipyramidal [TlS$_4$Tl] cages, with an S$_4$ rectangle sandwiched between two thallium atoms. Interestingly, only the Tl(III) product [Et$_4$N][Tl(1,2-S$_2$C$_6$H$_4$)$_2$] forms with the same 1:2 thallium salt to 1,2-(NaS)$_2$C$_6$H$_4$ stoichiometry, if the reaction is carried out under aerobic conditions. Thallium(I) derivatives of monocyclopentadienylbis(arene-1,2-dithiolato)-titanium(IV) have been synthesized and structurally characterized.[788]

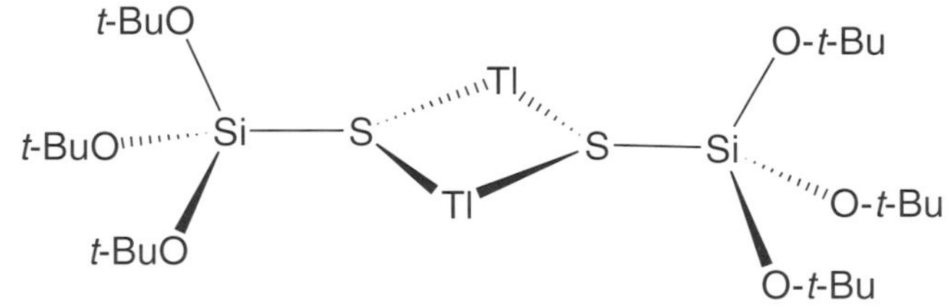

(147)

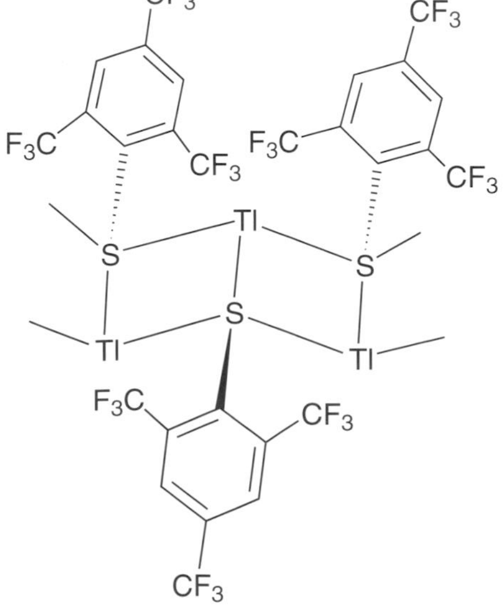

(148)

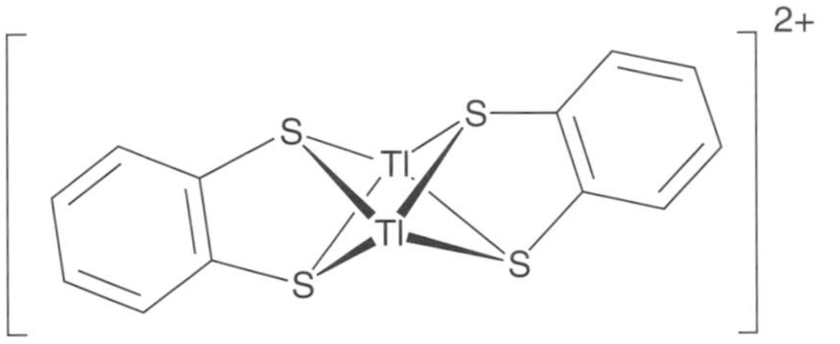

(149)

The thallium thiolate arising from the mixed donor tetrahydrofurfurylthiole (HSthff) has been prepared using NaSthff and TlPF₆.[778] It forms a polymeric thallium thiolate salt $[\{Tl_7(Sthff)_6\}_n][PF_6]_n$, and features an unusual, octahedrally coordinated Tl(I) linking novel Tl_6S_6 prismane units.

Acetylacetonato complexes of Tl(I) have been converted into 1,1-ethylenedithiolato complexes by treatment with carbon disulfide.[740] The compounds $[Tl_2\{S_2C=C\{C(O)R\}_2\}]_n$ (R = Me, Ph) have been obtained in quantitative yield.

(c) Other anionic sulfur ligands. Bis- and tris(mercaptoimidazolyl)borates are closely related to bis(pyrazolyl)borate and tris(pyrazolyl)borate ligands, but they contain softer sulfur-donor sites. Recently, a series of bis- and tris(mercaptoimidazolyl)borate ligands $[Bm^R]^-$ and $[Tm^R]^-$, as well as tris(mercaptothiazolyl)borate $[Tz]^-$, tris(mercaptobenzothiazolyl)borate $[Tbz]^-$, and bis(mercaptoimidazolyl)(pyrazolyl)hydroborato $[PzBm^R]^-$, ligands and their thallium(I) derivatives have been synthesized and characterized.[543,789–791] The bis(mercaptoimidazolyl)(pyrazolyl)hydroborato represents a hybrid [S₂N] system.

The reaction between $[Bm^{Me}]Li$ and CH_3CO_2Tl in MeOH provides the Tl(I) derivative $\{[Bm^{Me}]Tl\}_x$ with an oligonuclear solid-state structure.[789] It has four-membered Tl_2S_2 cores, and bridging thiazolyl groups. Related $\{[Tm^{Ph}]Tl\}_2$ is a dimer in the solid state.[543] The compound $\{[Bm^{Me}]Tl\}_x$ reacts with Me_2Zn or ZnI_2 to produce $[Bm^{Me}]ZnMe$ or $[Bm^{Me}]ZnI$. The Tl(I) adducts [Tz]Tl, [Tbz]Tl, and $[PzBm^{Me}]Tl$ have also been prepared, and $\{[Tbz]Tl\}_x$ has shown to be polymeric.[543,790]

Closely related to these systems is the tripodal sulfur-based donor, the phenyltris((*tert*-butylthio)methyl)borate ligand $[PhB(CH_2SBu^t)_3]^-$.[792] The thallium(I) adduct $[PhB(CH_2SBu^t)_3]^-$ **(150)** can be prepared by treating the product from the excess $LiCH_2SBu^t$ and $PhBCl_2$ reaction with aqueous $TlNO_3$. It forms a one-dimensional, extended structure with Tl–S and Tl–phenyl ring interactions. $[PhB(CH_2SBu^t)_3]^-$ serves as an excellent ligand-transfer agent.

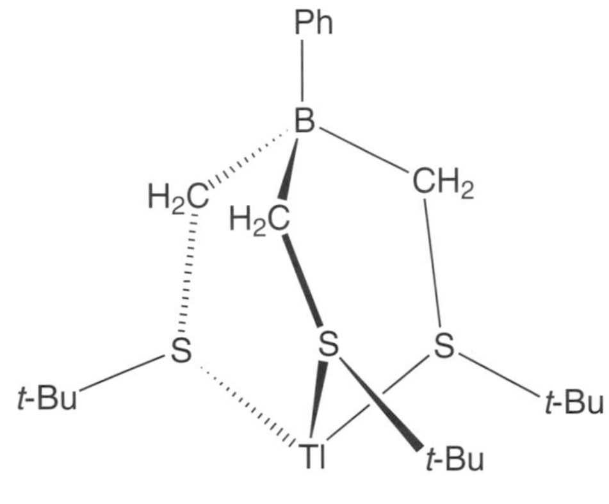

(150)

There are several early reports that describe the synthesis and characterization of monomeric $[Tl(1,1-S_2PEt_2)]$ and dimeric $[Tl(1,1-S_2CNR_2)]_2$ [R = Me, Et, *n*-Pr, *i*-Pr, *n*-Bu or *i*-Bu] thallium(I) adducts.[303,793–799] The synthesis of the first thallium(I) polysufides and chalcogenide cages has been described.[800–802] The two-coordinate Tl(I) compound $[Pt_2Tl(\mu_3-S)_2(PPh_3)_4]X$ (X = NO₃ or PF₆) is described as having a "Mexican hat-like" structure.[803] The heterobimetallic complex AuTl[CH$_2$P(S)Ph$_2$]$_2$, with short Au–Tl interactions and Tl—S bonds, is known.[804]

(d) Selenium and tellurium ligands. Eamples of thallium selenolates and tellurolates are less numerous. The electrochemical oxidation of thallium in nonaqueous solutions of PhSe–SePh leads to PhSeTl.[805] It is also possible to obtain the same product from an oxidative addition process, using a mixture of Tl and PhSe–SePh in refluxing toluene.[252] The related thiolate PhSTl could not be obtained using this latter method. With indium, both the thiolate and the selenolate could be obtained using the appropriate disulfide or diselenide, but the product is the In(III) derivative rather than the In(I) product. The thallium(I) selenolate TlSeSi(SiMe$_3$)$_3$ and tellurolate TlTeSi(SiMe$_3$)$_3$ have been prepared by chalcogenolysis of CpTl.[305] Attempts to prepare Tl(III) derivatives were unsuccessful, leading to the oxidation of the ligands to produce [MSi(SiMe$_3$)$_3$]$_2$ (M = Se or Te).

Selenium donor adducts of thallium(I), Tl(Et$_2$PSeS) and Tl(Et$_2$PSe$_2$) are known.[806,807] The 2,2-dicyanoethylene diselenolate-containing compound [AsPh$_4$]$_2$[Tl$_2$\{Se$_2$C=C(CN)$_2$\}$_2$] has been prepared.[808] This complex is a dimer with Se$_4$Tl$_2$ octahedral units. The synthesis of (*N,N*-diethyl-*N'*-benzoylselenoureato)thallium(I) has been achieved by treating thallium(I) acetate with *N,N*-diethyl-*N'*-benzoylselenourea.[809] It crystallizes as dimers forming Tl$_2$Se$_4$ rings.

3.5.2.4.4   *Group 17 ligands*

Fluoride, chloride, bromide, and iodide derivatives of thallium(I) are well known. Their solubilities and photosensitivity are similar to the corresponding silver(I) systems. TlF is water-soluble, whereas the chlorides, bromides, and iodides are water-insoluble solids. This property is exploited in ligand-transfer chemistry involving thallium precursors. Some solid-state structures of thallium(I) salts of weakly coordinated anions show Tl···halide interactions.[810–815] Selective abstraction of a fluoride from a C–F bond, leading to thallium fluoride, has been described.[816] The compound [\{P(CH$_2$CH$_2$PPh$_2$)$_3$\}RuH(η^1-ClTl)]PF$_6$ represents the first metal complex containing an η^1-Cl-bonded TlCl ligand.[817] This compound act as a thallium(I)-ion carrier.

3.5.2.4.5   *Hydride ligands*

Polymeric [TlH]$_n$ have been reported as one of the decomposition products of TlH$_3$. However, there is no convincing experimental evidence to support the existence of this species in the condensed phase.[365,366] It is possible to observe TlH in the gas phase. Reports containing theoretical studies on the bonding and stability of TlH and the related high-valent analogues are available.[467,469]

3.5.3   REFERENCES

1. Tuck, D. G. In *Comprehensive Coordination Chemistry*; Wilkinson, G., Gillard, R. D., McCleverty, J. A., Eds.; Pergamon: Oxford, UK, 1987; Vol. 3, pp 153–182.
2. Taylor, M. J.; Brothers, P. J. *Chem. Aluminum, Gallium, Indium, Thallium* **1993**, 111–247.
3. Tuck, D. G. *Chem. Aluminum, Gallium, Indium, Thallium* **1993**, 430–473.
4. Tuck, D. G. In *Encyclopedia of Inorganic Chemistry*; King, R. B., Ed.; Wiley: Chichester, UK, 1994; Vol. 3, pp 1513–1531.
5. Downs, A. J. *Chem. Aluminum, Gallium, Indium, Thallium* **1993**, 1–80.
6. Cotton, F. A.; Wilkinson, G.; Murillo, C. A.; Bochmann, M. *Advanced Inorganic Chemistry*, 6th ed.; Wiley: New York, 1999.
7. Greenwood, N. N.; Earnshaw, A. *Chemistry of the Elements*; 2nd ed.; Butterworth Heinemann: Oxford, 1997.
8. Davidson, G. *Coord. Chem. Rev.* **1983**, *49*, 117–192.
9. Carty, A. J.; Tuck, D. G. *Prog. Inorg. Chem.* **1975**, *19*, 243–337.
10. Tuck, D. G. *Coord. Chem. Rev.* **1966**, *1*, 286–291.
11. Tuck, D. G. In *Comprehensive Organometallic Chemistry*; Wilkinson, G., Stone, F. G. A., Abel, E. W., Eds.; Pergamon: Oxford, UK, 1982; Vol. 1, pp 683–723.
12. Starowieyski, K. B. *Chem. Aluminum, Gallium, Indium, Thallium* **1993**, 322–371.
13. Leman, J. T.; Barron, A. R.; King, R. B., Ed., *Encyclopedia of Inorganic Chemistry* **1994**, *3*, 1531–1542. Wiley: Chichester, UK.
14. Auner, N. *Synth. Methods Organomet. Inorg. Chem.* **1996**, *2*, 63–141.
15. Barron, A. R. *Comments Inorg. Chem.* **1993**, *14*, 123–153.
16. Schmidbaur, H. *Angew. Chem.* **1985**, *97*, 893–904.
17. Uhl, W. *Rev. Inorg. Chem.* **1998**, *18*, 239–282.

18. Neumuller, B. *Coord. Chem. Rev.* **1997**, *158*, 69–101.
19. Miller, J. A. *Chem. Aluminum, Gallium, Indium, Thallium* **1993**, 372–429.
20. Keh, C. C. K.; Li, C.-J. *Chemtracts* **1999**, *12*, 813–816.
21. Cintas, P. *Synlett* **1995**, 1087–1096.
22. Goggin, P. L.; McColm, I. J.; Shore, R. *J. Chem. Soc., A* **1966**, 1314–1317.
23. Williams, D.; Kouvetakis, J.; O'Keeffe, M. *Inorg. Chem.* **1998**, *37*, 4617–4620.
24. Arif, A. M.; Cowley, A. H.; Elkins, T. M.; Jones, R. A. *J. Chem. Soc., Chem. Commun.* **1986**, 1776–1777.
25. Wiberg, N.; Blank, T.; Lerner, H.-W.; Noeth, H.; Habereder, T.; Fenske, D. *Z. Naturforsch., B* **2001**, *56*, 652–658.
26. Kuhler, T.; Jutzi, P.; Stammler, A.; Stammler, H.-G. *Chem. Commun.* **2001**, 539–540.
27. Wiberg, N.; Amelunxen, K.; Lerner, H. W.; Noeth, H.; Knizek, J.; Krossing, I. *Z. Naturforsch., B* **1998**, *53*, 333–348.
28. Wiberg, N.; Amelunxen, K.; Blank, T.; Lerner, H.-W.; Polborn, K.; Noeth, H.; Littger, R.; Rackl, M.; Schmidt-Amelunxen, M.; Schwenk-Kircher, H.; Warchold, M. *Z. Naturforsch., B* **2001**, *56*, 634–651.
29. Wiberg, N.; Blank, T.; Amelunxen, K.; Noth, H.; Schnockel, H.; Baum, E.; Purath, A.; Fenske, D. *Eur. J. Inorg. Chem.* **2002**, 341–350.
30. Annan, T. A.; Tuck, D. G. *J. Organomet. Chem.* **1987**, *325*, 83–89.
31. Steevensz, R. S.; Tuck, D. G.; Meinema, H. A.; Noltes, J. G. *Can. J. Chem.* **1985**, *63*, 755–758.
32. Habeeb, J. J.; Said, F. F.; Tuck, D. G. *J. Chem. Soc., Dalton Trans.* **1981**, 118–120.
33. Ketchum, D. R.; Schimek, G. L.; Pennington, W. T.; Kolis, J. W. *Inorg. Chim. Acta* **1999**, *294*, 200–206.
34. Neumayer, D. A.; Ekerdt, J. G. *Chem. Mater.* **1996**, *8*, 9–25 and references therein.
35. Steffek, C.; McMurran, J.; Pleune, B.; Kouvetakis, J.; Concolino, T. E.; Rheingold, A. L. *Inorg. Chem.* **2000**, *39*, 1615–1617.
36. Sussek, H.; Stowasser, F.; Pritzkow, H.; Fischer, R. A. *Eur. J. Inorg. Chem.* **2000**, 455–461.
37. Fischer, R. A.; Miehr, A.; Ambacher, O.; Metzger, T.; Born, E. *J. Cryst. Growth* **1997**, *170*, 139–143.
38. Fischer, R. A.; Sussek, H.; Parala, H.; Pritzkow, H. *J. Organomet. Chem.* **1999**, *592*, 205–211.
39. Fischer, R. A.; Sussek, H.; Miehr, A.; Pritzkow, H.; Herdtweck, E. *J. Organomet. Chem.* **1997**, *548*, 73–82.
40. Golub, A. M.; Tsintsadze, G. V.; Makhatadze, T. L. *Soobshch. Akad. Nauk Gruz. SSR* **1971**, *61*, 57–60.
41. Patel, S. J.; Tuck, D. G. *J. Chem. Soc., A* **1968**, 1870–1873.
42. Petrosyants, S. P.; Molyarik, M. A.; Buslaev, Y. A. *Zh. Neorg. Khim.* **1990**, *35*, 1789–1792.
43. Malyarick, M. A.; Petrosyants, S. P. *Inorg. Chem.* **1993**, *32*, 2265–2268.
44. Mullica, D. F.; Kautz, J. A.; Sappenfield, E. *J. Chem. Crystallogr.* **1999**, *29*, 317–321.
45. Takahashi, R.; Suzuki, H.; Ishiguro, S. *J. Chem. Soc., Faraday Trans.* **1996**, *92*, 2715–2724.
46. Carmalt, C. J.; Norman, N. C.; Pember, R. F.; Farrugia, L. J. *Polyhedron* **1995**, *14*, 417–424.
47. Carmalt, C. J.; Clegg, W.; Elsegood, M. R. J.; Kneisel, B. O.; Norman, N. C. *Acta Crystallogr., C: Cryst. Struct. Commun.* **1995**, *C51*, 1254–1258.
48. Purdy, A. P. *Inorg. Chem.* **1994**, *33*, 282–286.
49. Himmel, H.-J.; Downs, A. J.; Green, J. C.; Greene, T. M. *J. Chem. Soc., Dalton Trans.* **2001**, 535–545.
50. Himmel, H.-J.; Downs, A. J.; Greene, T. M. *Chem. Commun.* **2000**, 871–872.
51. Carmalt, C. J. *Coord. Chem. Rev.* **2001**, *223*, 217–264.
52. Prust, J.; Muller, P.; Rennekamp, C.; Roesky, H. W.; Uson, I. *J. Chem. Soc., Dalton Trans.* **1999**, 2265–2266.
53. Roesky, H. W.; Seseke, U.; Noltemeyer, M.; Sheldrick, G. M. *Z. Naturforsch., B: Chem. Sci.* **1988**, *43*, 1130–1136.
54. Frey, R.; Gupta, V. D.; Linti, G. *Z. Anorg. Allg. Chem.* **1996**, *622*, 1060–1064.
55. Buerger, H.; Cichon, J.; Goetze, U.; Wannagat, U.; Wismar, H. J. *J. Organomet. Chem.* **1971**, *33*, 1–12.
56. Silverman, J. S.; Carmalt, C. J.; Cowley, A. H.; Culp, R. D.; Jones, R. A.; McBurnett, B. G. *Inorg. Chem.* **1999**, *38*, 296–300.
57. Rossetto, G.; Brianese, N.; Camporese, A.; Porchia, M.; Zanella, P.; Bertoncello, R. *Main Group Met. Chem.* **1991**, *14*, 113–122.
58. Pauls, J.; Chitsuz, S.; Neumuller, B. *Z. Anorg. Allg. Chem.* **2001**, *627*, 1723–1730.
59. Kim, J.; Bott, S. G.; Hoffman, D. M. *Inorg. Chem.* **1998**, *37*, 3835 3841.
60. Petrie, M. A.; Ruhlandt-Senge, K.; Hope, H.; Power, P. P. *Bull. Soc. Chim. Fr.* **1993**, *130*, 851–855.
61. Kopp, M. R.; Neumueller, B. *Z. Anorg. Allg. Chem.* **1998**, *624*, 361–363.
62. Grabowy, T.; Merzweiler, K. *Z. Anorg. Allg. Chem.* **2000**, *626*, 736–740.
63. Kim, J.; Bott, S. G.; Hoffman, D. M. *J. Chem. Soc., Dalton Trans.* **1999**, 141–146.
64. Veith, M.; Zimmer, M.; Müller-Becker, S. *Angew. Chem.* **1993**, *105*, 1771–1773 (See also *Angew. Chem., Int. Ed. Engl.* **1993**, *32(12)*, 1731–1733).
65. Leman, J. T.; Barron, A. R.; Ziller, J. W.; Kren, R. M. *Polyhedron* **1989**, *8*, 1909–1912.
66. Leman, J. T.; Roman, H. A.; Barron, A. R. *J. Chem. Soc., Dalton Trans.* **1992**, 2183–2191.
67. Zhou, Y.; Richeson, D. S. *Inorg. Chem.* **1996**, *35*, 1423–1424.
68. Zhou, Y.; Richeson, D. S. *Inorg. Chem.* **1996**, *35*, 2448–2451.
69. Ergezinger, C.; Weller, F.; Dehnicke, K. *Z. Naturforsch., B: Chem. Sci.* **1988**, *43*, 1621–1627.
70. Dehnicke, K.; Ergezinger, C.; Hartmann, E.; Zinn, A.; Hoesler, K. *J. Organomet. Chem.* **1988**, *352*, C1–C4.
71. Dias, H. V. R.; Wang, Z.; Jin, W. *Coord. Chem. Rev.* **1998**, *176*, 67–86.
72. Delpech, F.; Guzei, I. A.; Jordan, R. F. *Organometallics* **2002**, *21*, 1167–1176.
73. Dias, H. V. R.; Jin, W. *Inorg. Chem.* **1996**, *35*, 6546–6551.
74. Burgstein, M. R.; Euringer, N. P.; Roesky, P. W. *J. Chem. Soc., Dalton Trans.* **2000**, 1045–1048.
75. Stender, M.; Eichler, B. E.; Hardman, N. J.; Power, P. P.; Prust, J.; Noltemeyer, M.; Roesky, H. W. *Inorg. Chem.* **2001**, *40*, 2794–2799.
76. Zhou, Y.; Richeson, D. S. *Organometallics* **1995**, *14*, 3558–3561.
77. Ward, M. D.; Mann, K. L. V.; Jeffery, J. C.; McCleverty, J. A. *Acta Crystallogr., C: Cryst. Struct. Commun.* **1998**, *C54*, 601–603.
78. Phillips, P. R.; Wallbridge, M. G. H.; Barker, J. *J. Organomet. Chem.* **1998**, *550*, 301–308.
79. Atwood, D. A.; Atwood, V. O.; Cowley, A. H.; Atwood, J. L.; Roman, E. *Inorg. Chem.* **1992**, *31*, 3871–3872.
80. Reger, D. L. *Coord. Chem. Rev.* **1996**, *147*, 571–595 and references therein.
81. Nicholson, B. K.; Thomson, R. A.; Watts, F. D. *Inorg. Chim. Acta* **1988**, *148*, 101–104.

82. Reger, D. L.; Knox, S. J.; Rheingold, A. L.; Haggerty, B. S. *Organometallics* **1990**, *9*, 2581–2587.
83. Reger, D. L.; Mason, S. S.; Rheingold, A. L.; Ostrander, R. L. *Inorg. Chem.* **1994**, *33*, 1803–1810.
84. Cowley, A. H.; Carrano, C. J.; Geerts, R. L.; Jones, R. A.; Nunn, C. M. *Angew. Chem.* **1988**, *100*, 306–307.
85. Fraser, A.; Piggott, B. *J. Chem. Soc., Dalton Trans.* **1999**, 3483–3486.
86. Frazer, A.; Piggott, B.; Harman, M.; Mazid, M.; Hursthouse, M. B. *Polyhedron* **1992**, *11*, 3013–3017.
87. Reger, D. L.; Mason, S. S.; Reger, L. B.; Rheingold, A. L.; Ostrander, R. L. *Inorg. Chem.* **1994**, *33*, 1811–16.
88. Reger, D. L.; Coan, P. S. *Inorg. Chem.* **1995**, *34*, 6226–6227.
89. Kuchta, M. C.; Parkin, G. *Main Group Chem.* **1996**, *1*, 291–295.
90. Reger, D. L.; Mason, S. S.; Rheingold, A. L.; Haggerty, B. S.; Arnold, F. P. *Organometallics* **1994**, *13*, 5049–5053.
91. Klaui, W.; Liedtke, N.; Peters, W. *Magn. Reson. Chem.* **1999**, *37*, 867–870.
92. Klaui, W.; Peters, W.; Liedtke, N.; Trofimenko, S.; Rheingold, A. L.; Sommer, R. D. *Eur. J. Inorg. Chem.* **2001**, 693–699.
93. Wang, E.; Romero, C.; Santiago, D.; Syntilas, V. *Anal. Chim. Acta* **2001**, *433*, 89–95.
94. Nemykin, V. N.; Volkov, S. V. *Russ. J. Coord. Chem.* **2000**, *26*, 436–450.
95. Park, Y. C.; Lee, D. C.; Na, H. G.; Han, M. S. *J. Korean Chem. Soc.* **1998**, *42*, 454–457.
96. Coutsolelos, A. G.; Daphnomili, D.; Scheidt, W. R.; Ferraudi, G. *Inorg. Chem.* **1998**, *37*, 2077–2079.
97. Lemke, F. R.; Lorenz, C. R. *Recent Res. Dev. Electroanal. Chem.* **1999**, *1*, 73–89.
98. Park, Y. C.; Na, H. G. *Main Group Met. Chem.* **1997**, *20*, 269–276.
99. Steinle, E. D.; Schaller, U.; Meyerhoff, M. E. *Anal. Sci.* **1998**, *14*, 79–84.
100. Bedel-Cloutour, C. H.; Mauclaire, L.; Saux, A.; Pereyre, M. *Bioconjugate Chem.* **1996**, *7*, 617–627.
101. Hong, T.-N.; Sheu, Y.-H.; Jang, K.-W.; Chen, J.-H.; Wang, S.-S.; Wang, J.-C.; Wang, S.-L. *Polyhedron* **1996**, *15*, 2647–2654.
102. Mamardashvili, N. Z.; Semeikin, A. S.; Golubchikov, O. A. *Zh. Org. Khim.* **1994**, *30*, 770–773.
103. Lomova, T. N.; Berezin, B. D. *Koord. Khim.* **1993**, *19*, 171–184.
104. Park, S. B.; Matuszewski, W.; Meyerhoff, M. E.; Liu, Y. H.; Kadish, K. M. *Electroanalysis* **1991**, *3*, 909–916.
105. Guilard, R.; Jagerovic, N.; Tabard, A.; Richard, P.; Courthaudon, L.; Louati, A.; Lecomte, C.; Kadish, K. M. *Inorg. Chem.* **1991**, *30*, 16–27.
106. Yamazaki, K.; Hirata, S.; Nakajima, S.; Kubo, Y.; Samejima, N.; Sakata, I. *Jpn. J. Cancer Res. (GANN)* **1988**, *79*, 880–884.
107. Guilard, R.; Gerges, S. S.; Tabard, A.; Richard, P.; El Borai, M. A.; Lecomte, C. *J. Am. Chem. Soc.* **1987**, *109*, 7228–7230.
108. Cornillon, J. L.; Anderson, J. E.; Kadish, K. M. *Inorg. Chem.* **1986**, *25*, 991–995.
109. Hambright, P.; Adeyemo, A.; Shamim, A.; Lemelle, S. *Inorg. Synth.* **1985**, *23*, 55–59.
110. Kadish, K. M.; Cornillon, J. L.; Cocolios, P.; Tabard, A.; Guilard, R. *Inorg. Chem.* **1985**, *24*, 3645–3649.
111. Ebeid, E. Z. M.; El-Borai, M. A.; Morsi, S. E.; Guilard, R. *Inorg. Chim. Acta* **1984**, *86*, 71–74.
112. Hambright, P. *J. Coord. Chem.* **1983**, *12*, 297–301.
113. Cocolios, P.; Fournari, P.; Guilard, R.; Lecomte, C.; Protas, J.; Boubel, J. C. *J. Chem. Soc., Dalton Trans.* **1980**, 2081–2089.
114. Bhatti, M.; Bhatti, W.; Mast, E. *Inorg. Nucl. Chem. Lett.* **1972**, *8*, 133–137.
115. Ball, R. G.; Lee, K. M.; Marshall, A. G.; Trotter, J. *Inorg. Chem.* **1980**, *19*, 1463–1469.
116. Bedel-Cloutour, C. H.; Mauclaire, L.; Pereyre, M.; Adams, S.; Drager, M. *Polyhedron* **1990**, *9*, 1297–1303.
117. Guilard, R.; Tabard, A.; Zrineh, A.; Ferhat, M. *J. Organomet. Chem.* **1990**, *389*, 315–324.
118. Cornillon, J. L.; Anderson, J. E.; Kadish, K. M. *Inorg. Chem.* **1986**, *25*, 2611–2617.
119. Cocolios, P.; Guilard, R.; Bayeul, D.; Lecomte, C. *Inorg. Chem.* **1985**, *24*, 2058–2062.
120. Guilard, R.; Jagerovic, N.; Tabard, A.; Naillon, C.; Kadish, K. M. *J. Chem. Soc., Dalton Trans.* **1992**, 1957–1966.
121. Balch, A. L.; Noll, B. C.; Olmstead, M. M.; Reid, S. M. *J. Chem. Soc., Chem. Commun.* **1993**, 1088–1090.
122. Plater, M. J.; Jeremiah, A.; Bourhill, G. *J. Chem. Soc., Perkin Trans. 1* **2002**, 91–96.
123. Dini, D.; Barthel, M.; Hanack, M. *Eur. J. Org. Chem.* **2001**, 3759–3769.
124. Hanack, M.; Schneider, T.; Barthel, M.; Shirk, J. S.; Flom, S. R.; Pong, R. G. S. *Coord. Chem. Rev.* **2001**, *219–221*, 235–258.
125. Gorlach, B.; Dachtler, M.; Glaser, T.; Albert, K.; Hanack, M. *Chem. –Eur. J.* **2001**, *7*, 2459–2465.
126. Huckstadt, H.; Tutass, A.; Goldner, M.; Cornelissen, U.; Homborg, H. *Z. Anorg. Allg. Chem.* **2001**, *627*, 485–497.
127. Shirk, J. S.; Pong, R. G. S.; Flom, S. R.; Heckmann, H.; Hanack, M. *J. Phys. Chem. A* **2000**, *104*, 1438–1449.
128. Janczak, J.; Kubiak, R. *Inorg. Chim. Acta* **1999**, *288*, 174–180.
129. Janczak, J. *Pol. J. Chem.* **1998**, *72*, 1871–1878.
130. Hanack, M.; Heckmann, H. *Eur. J. Inorg. Chem.* **1998**, 367–373.
131. Schweiger, K.; Hueckstaedt, H.; Homborg, H. *Z. Anorg. Allg. Chem.* **1998**, *624*, 44–50.
132. Schweiger, K.; Kienast, A.; Latte, B.; Homborg, H. *Z. Anorg. Allg. Chem.* **1997**, *623*, 973–980.
133. Gavrilin, E. V.; Shishkina, O. V.; Shaposhnikov, G. P.; Maizlish, V. E.; Kulinich, V. P.; Smirnov, R. P. *Zh. Obshch. Khim.* **1996**, *66*, 1732–1735.
134. Perry, J. W.; Mansour, K.; Lee, I. Y. S.; Wu, X. L.; Bedworth, P. V.; Chen, C. T.; Ng, D.; Marder, S. R.; Miles, P.; Wada, P.; Tian, M.; Sasabe, H. *Science* **1996**, *273*, 1533–1536.
135. Ostendorp, G.; Homborg, H. *Z. Anorg. Allg. Chem.* **1996**, *622*, 1358–1364.
136. Assmann, B.; Franken, A.; Homborg, H. *Z. Naturforsch., B: Chem. Sci.* **1996**, *51*, 325–332.
137. Tomilova, L.; Podgaetsky, V.; Dyumaev, K.; Omel'chenko, A.; Sviridov, A.; Sobol, E. *Proc. SPIE-Int. Soc. Opt. Eng.* **1996**, *2623*, 62–65.
138. Assmann, B.; Ostendorp, G.; Homborg, H. *Z. Anorg. Allg. Chem.* **1995**, *621*, 1708–1714.
139. Janczak, J.; Kubiak, R.; Jezierski, A. *Inorg. Chem.* **1995**, *34*, 3505–3508.
140. Janczak, J.; Kubiak, R. *J. Chem. Soc., Dalton Trans.* **1994**, 2539–2543.
141. Janczak, J.; Kubiak, R. *J. Chem. Soc., Dalton Trans.* **1993**, 3809–3812.
142. Borovkov, N. Y.; Akopov, A. S. *Koord. Khim.* **1987**, *13*, 1358–1361.
143. Jennings, C.; Aroca, R.; Hor, A. M.; Loutfy, R. O. *Spectrochim. Acta, Part A* **1986**, *42A*, 991–995.
144. Linsky, J. P.; Paul, T. R.; Nohr, R. S.; Kenney, M. E. *Inorg. Chem.* **1980**, *19*, 3131–3135.

145. Schneider, T.; Heckmann, H.; Barthel, M.; Hanack, M. *Eur. J. Org. Chem.* **2001**, 3055–3065.
146. Janczak, J.; Idemori, Y. M. *Inorg. Chim. Acta* **2001**, *325*, 85–93.
147. Arif, A. M.; Benac, B. L.; Cowley, A. H.; Jones, R. A.; Kidd, K. B.; Nunn, C. M. *New J. Chem.* **1988**, *12*, 553–557.
148. Green, M.; O'Brien, P. *Chem. Commun.* **1998**, 2459–2460.
149. Guzelian, A. A.; Katari, J. E. B.; Kadavanich, A. V.; Banin, U.; Hamad, K.; Juban, E.; Alivisatos, A. P.; Wolters, R. H.; Arnold, C. C.; Heath, J. R. *J. Phys. Chem.* **1996**, *100*, 7212–7219.
150. Healy, M. D.; Laibinis, P. E.; Stupik, P. D.; Barron, A. R. *J. Chem. Soc., Chem. Commun.* **1989**, 359–360.
151. Yan, P.; Xie, Y.; Wang, W.; Liu, F.; Qian, Y. *J. Mater. Chem.* **1999**, *9*, 1831–1833.
152. Xie, Y.; Yan, P.; Lu, J.; Wang, W.; Qian, Y. *Chem. Mater.* **1999**, *11*, 2619–2622.
153. Lu, J.; Xie, Y.; Jiang, X.; He, W.; Yan, P.; Qian, Y. *Can. J. Chem.* **2001**, *79*, 127–130.
154. Merzweiler, K.; Spohn, J. *Z. Anorg. Allg. Chem.* **1993**, *619*, 318–320.
155. Douglas, T.; Theopold, K. H. *Angew. Chem.* **1989**, *101*, 1394–1395.
156. App, U.; Merzweiler, K. *Z. Anorg. Allg. Chem.* **1995**, *621*, 1731–1734.
157. Carrano, C. J.; Cowley, A. H.; Giolando, D. M.; Jones, R. A.; Nunn, C. M.; Power, J. M. *Inorg. Chem.* **1988**, *27*, 2709–2714.
158. Barron, A. R.; Cowley, A. H.; Jones, R. A.; Nunn, C. M.; Westmoreland, D. L. *Polyhedron* **1988**, *7*, 77–78.
159. Winkler, A.; Bauer, W.; Heinemann, F. W.; Garcia-Montalvo, V.; Moll, M.; Ellermann, J. *Eur. J. Inorg. Chem.* **1998**, 437–444.
160. Crea, J.; Lincoln, S. F. *Inorg. Chem.* **1972**, *11*, 1131–1132.
161. Fratiello, A.; Davis, D. D.; Peak, S.; Schuster, R. E. *Inorg. Chem.* **1971**, *10*, 1627–1632.
162. Fratiello, A.; Vidulich, G. A.; Cheng, C.; Kubo, V. *J. Solution Chem.* **1972**, *1*, 433–444.
163. Harrowfield, J. M.; Skelton, B. W.; White, A. H. *Aust. J. Chem.* **1990**, *43*, 759–763.
164. Beattie, J. K.; Best, S. P. *Coord. Chem. Rev.* **1997**, *166*, 391–415.
165. Beattie, J. K.; Best, S. P.; Skelton, B. W.; White, A. H. *J. Chem. Soc., Dalton Trans.* **1981**, 2105–2111.
166. Lindqvist-Reis, P.; Munoz-Paez, A.; Diaz-Moreno, S.; Pattanaik, S.; Persson, I.; Sandstroem, M. *Inorg. Chem.* **1998**, *37*, 6675–6683.
167. Armstrong, R. S.; Beattie, J. K.; Best, S. P.; Braithwaite, G. P.; Del Favero, P.; Skelton, B. W.; White, A. H. *Aust. J. Chem.* **1990**, *43*, 393–398.
168. Moeller, T. *J. Am. Chem. Soc.* **1941**, *63*, 1206–1207.
169. Moeller, T. *J. Am. Chem. Soc.* **1941**, *63*, 2625–2628.
170. Anderson, C. J.; Welch, M. J. *Chem. Rev.* **1999**, *99*, 2219–2234.
171. Avivi, S.; Mastai, Y.; Gedanken, A. *Chem. Mater.* **2000**, *12*, 1229–1233.
172. Wieghardt, K.; Kleine-Boymann, M.; Nuber, B.; Weiss, J. *Inorg. Chem.* **1986**, *25*, 1654–1659.
173. Mann, K. L. V.; Jeffery, J. C.; McCleverty, J. A.; Thornton, P.; Ward, M. D. *J. Chem. Soc., Dalton Trans.* **1998**, 89–98.
174. Abram, S.; Maichle-Mossmer, C.; Abram, U. *Polyhedron* **1997**, *17*, 131–143.
175. Hegetschweiler, K.; Ghisletta, M.; Faessler, T. F.; Nesper, R.; Schmalle, H. W.; Rihs, G. *Inorg. Chem.* **1993**, *32*, 2032–41.
176. Hegetschweiler, K.; Ghisletta, M.; Faessler, T. F.; Nesper, R. *Angew. Chem.* **1993**, *105*, 1514–1516 (See also *Angew. Chem., Int. Ed. Engl.,* **1993**, *32(10)*, 1426–1428).
177. Miinea, L. A.; Hoffman, D. M. *J. Mater. Chem.* **2000**, *10*, 2392–2395.
178. Suh, S.; Hoffman, D. M. *J. Am. Chem. Soc.* **2000**, *122*, 9396–9404.
179. Chatterjee, S.; Bindal, S. R.; Mehrotra, R. C. *J. Indian Chem. Soc.* **1976**, *53*, 867–869.
180. Bradley, D. C.; Chudzynska, H.; Frigo, D. M.; Hursthouse, M. B.; Mazid, M. A. *J. Chem. Soc., Chem. Commun.* **1988**, 1258–1259.
181. Bradley, D. C.; Chudzynska, H.; Frigo, D. M.; Hammond, M. E.; Hursthouse, M. B.; Mazid, M. A. *Polyhedron* **1990**, *9*, 719–726.
182. Miinea, L. A.; Suh, S.; Hoffman, D. M. *Inorg. Chem.* **1999**, *38*, 4447–4454.
183. Pauls, J.; Chitsaz, S.; Neumuller, B. *Z. Anorg. Allg. Chem.* **2000**, *626*, 2028–2034.
184. Soling, H. *Acta Chem. Scand., Ser. A* **1976**, *A30*, 163–170.
185. Haworth, D. T.; Beery, J. W.; Das, M. *Polyhedron* **1982**, *1*, 9–12.
186. Sreelatha, C.; Gupta, V. D.; Narula, C. K.; Noeth, H. *J. Chem. Soc., Dalton Trans.* **1985**, 2623–2628.
187. Saito, K.; Nagasawa, A. *Polyhedron* **1990**, *9*, 215–222.
188. Wakeshima, I.; Niikura, I.; Kijima, I. *Synth. React. Inorg. Met.-Org. Chem.* **1992**, *22*, 447–459.
189. Wakeshima, I.; Watanabe, S.; Kijima, I. *Bull. Chem. Soc. Jpn.* **1994**, *67*, 2583–2585.
190. Le, Q. T. H.; Umetani, S.; Matsui, M. *J. Chem. Soc., Dalton Trans.* **1997**, 3835–3840.
191. Utsunomiya, K. *Bull. Chem. Soc. Jap.* **1971**, *44*, 2688–2693.
192. Brain, P. T.; Buhl, M.; Robertson, H. E.; Jackson, A. D.; Lickiss, P. D.; MacKerracher, D.; Rankin, D. W. H.; Shah, D.; Thiel, W. *J. Chem. Soc., Dalton Trans.* **1998**, 545–551.
193. Mazurenko, E. A.; Novitskaya, G. N.; Bublik, Z. N.; Volkov, S. V. *Ukr. Khim. Zh. (Russ. Ed.)* **1984**, *50*, 227–229.
194. Maruyama, T.; Fukui, K. *J. Appl. Phys.* **1991**, *70*, 3848–3851.
195. Reich, S.; Suhr, H.; Waimer, B. *Thin Solid Films* **1990**, *189*, 293–302.
196. Maruyama, T.; Kitamura, T. *Jpn. J. Appl. Phys., Part 2* **1989**, *28*, L1096–L1097.
197. Wang, A.; Dai, J.; Cheng, J.; Chudzik, M. P.; Marks, T. J.; Chang, R. P. H.; Kannewurf, C. R. *Appl. Phys. Lett.* **1998**, *73*, 327–329.
198. Jablonski, Z.; Rychlowska-Himmel, I.; Dyrek, M. *Spectrochim. Acta, Part A* **1979**, *35A*, 1297–1301.
199. Sreelatha, C. H.; Gupta, V. D. *Curr. Sci.* **1984**, *53*, 858–860.
200. Singh, Y. P.; Rai, A. K. *Indian J. Chem., Sect. A* **1984**, *23A*, 350–351.
201. Haworth, D. T.; Das, M. *Synth. React. Inorg. Met.-Org. Chem.* **1982**, *12*, 721–730.
202. Nepveu, F.; Jasanada, F.; Walz, L. *Inorg. Chim. Acta* **1993**, *211*, 141–147.
203. Narbutt, J.; Czerwinski, M.; Krejzler, J. *Eur. J. Inorg. Chem.* **2001**, 3187–3197.
204. Brown, M. A.; McGarvey, B. R.; Tuck, D. G. *J. Chem. Soc., Dalton Trans.* **1998**, 3543–3548.
205. Brown, M. A.; McGarvey, B. R.; Ozarowski, A.; Tuck, D. G. *Inorg. Chem.* **1996**, *35*, 1560–1563.

206. Annan, T. A.; Brown, M. A.; El-Hadad, A.; McGarvey, B. R.; Ozarowski, A.; Tuck, D. G. *Inorg. Chim. Acta* **1994**, *225*, 207–213.
207. Tuck, D. G. *Coord. Chem. Rev.* **1992**, *112*, 215–225.
208. Faraglia, G.; Fregona, D.; Sitran, S. *Main Group Met. Chem.* **1994**, *17*, 649–657.
209. Matsuba, C. A.; Nelson, W. O.; Rettig, S. J.; Orvig, C. *Inorg. Chem.* **1988**, *27*, 3935–3939.
210. Ma, R.; Reibenspies, J. J.; Martell, A. E. *Inorg. Chim. Acta* **1994**, *223*, 21–29.
211. Beatty, E.; Burgess, J.; Patel, M. S. *Can. J. Chem.* **1994**, *72*, 1370–1375.
212. Li, Y. J.; Martell, A. E. *Inorg. Chim. Acta* **1993**, *214*, 103–111.
213. Clarke, E. T.; Martell, A. E. *Inorg. Chim. Acta* **1992**, *196*, 185–194.
214. Clarke, E. T.; Martell, A. E. *Inorg. Chim. Acta* **1992**, *191*, 56–63.
215. Simpson, L.; Rettig, S. J.; Trotter, J.; Orvig, C. *Can. J. Chem.* **1991**, *69*, 893–900.
216. Zhang, Z.; Rettig, S. J.; Orvig, C. *Inorg. Chem.* **1991**, *30*, 509–515.
217. Zhang, Z.; Hui, T. L. T.; Orvig, C. *Can. J. Chem.* **1989**, *67*, 1708–1710.
218. Tuck, D. G. *Chem. Soc. Rev.* **1993**, *22*, 269–276.
219. Annan, T. A.; Chadha, R. K.; Doan, P.; McConville, D. H.; McGarvey, B. R.; Ozarowski, A.; Tuck, D. G. *Inorg. Chem.* **1990**, *29*, 3936–3943.
220. Khan, G. M.; Imura, H.; Ohashi, K. *Solvent Extr. Res. Dev., Jpn.* **2000**, *7*, 106–117.
221. Korber, N.; Achour, B.; Nepveu, F. *J. Chem. Crystallogr.* **1994**, *24*, 685–688.
222. Addy, P.; Evans, D. F.; Sheppard, R. N. *Inorg. Chim. Acta* **1987**, *127*, L19–L20.
223. Green, M. A.; Huffman, J. C. *J. Nucl. Med.* **1988**, *29*, 417–420.
224. McAfee, J. G.; Thakur, M. L. *J. Nucl. Med.* **1976**, *17*, 480–487.
225. Thakur, M. L.; Welch, M. J.; Joist, J. H.; Coleman, R. E. *Thromb. Res.* **1976**, *9*, 345–357.
226. Lindel, W.; Huber, F. *Z. Anorg. Allg. Chem.* **1974**, *408*, 167–174.
227. Bulc, N.; Golic, L. *Acta Crystallogr., C* **1983**, *C39*, 174–176.
228. Kebede, T.; Ramana, K. V.; Rao, M. S. P. *Thermochim. Acta* **2001**, *371*, 163–168.
229. Kebede, T.; Ramana, K. V.; Rao, M. S. P. *J. Therm. Anal. Calorim.* **2001**, *66*, 439–447.
230. Kebede, T.; Ramana, K. V.; Prasada Rao, M. S. *Thermochim. Acta* **2002**, *381*, 31–36.
231. Yamamoto, M.; Seki, S.; Sawada, Y. *Trans. Mater. Res. Soc. Jpn* **2001**, *26*, 1223–1226.
232. Sengupta, A. K.; Sinha, K. *J. Fluorine Chem.* **1990**, *47*, 345–351.
233. Tang, X.; Jones, A.; Lachgar, A.; Gross, B. J.; Yarger, J. L. *Inorg. Chem.* **1999**, *38*, 6032–6038.
234. Lii, K.-H.; Huang, Y.-F. *Inorg. Chem.* **1999**, *38*, 1348–1350.
235. Yu, J.; Sung, H. H. Y.; Williams, I. D. *J. Solid State Chem.* **1999**, *142*, 241–246.
236. Tang, X.; Lachgar, A. *Inorg. Chem.* **1998**, *37*, 6181–6185.
237. Peltier, V.; Deniard, P.; Brec, R.; Marchand, R. *C. R. Acad. Sci., Ser. IIc: Chim.* **1998**, *1*, 57–62.
238. Chippindale, A. M.; Brech, S. J. *Chem. Commun.* **1996**, 2781–2782.
239. Chippindale, A. M.; Brech, S. J.; Cowley, A. R.; Simpson, W. M. *Chem. Mater.* **1996**, *8*, 2259–2264.
240. Dhingra, S. S.; Haushalter, R. C. *J. Chem. Soc., Chem. Commun.* **1993**, 1665–1667.
241. Huang, Y.-F.; Lii, K.-H. *J. Chem. Soc., Dalton Trans.* **1998**, 4085–4086.
242. Williams, I. D.; Yu, J.; Du, H.; Chen, J.; Pang, W. *Chem. Mater.* **1998**, *10*, 773–776.
243. Morizzi, J.; Hobday, M.; Rix, C. *J. Mater. Chem.* **2001**, *11*, 794–798.
244. Morizzi, J.; Hobday, M.; Rix, C. *J. Mater. Chem.* **2000**, *10*, 1693–1697.
245. Morizzi, J.; Hobday, M.; Rix, C. *J. Mater. Chem.* **1999**, *9*, 863–864.
246. Bollinger, J. E.; Roundhill, D. M. *Inorg. Chem.* **1993**, *32*, 2821–2826.
247. Hector, A. L.; Levason, W.; Webster, M. *J. Chem. Soc., Dalton Trans.* **1998**, 3463–3472.
248. Tuck, D. G.; Woodhouse, E. J. *Chem. Ind.* **1964**, 1363–1364.
249. Carty, A. J.; Tuck, D. G. *J. Chem. Soc., Suppl.* **1964**, 6012–6017.
250. Green, J. H.; Kumar, R.; Seudeal, N.; Tuck, D. G. *Inorg. Chem.* **1989**, *28*, 123–127.
251. Chadha, R. K.; Hayes, P. C.; Mabrouk, H. E.; Tuck, D. G. *Can. J. Chem.* **1987**, *65*, 804–809.
252. Kumar, R.; Mabrouk, H. E.; Tuck, D. G. *J. Chem. Soc., Dalton Trans.* **1988**, 1045–1047.
253. Ruhlandt-Senge, K.; Power, P. P. *Inorg. Chem.* **1993**, *32*, 3478–3481.
254. Suh, S.; Hoffman, D. M. *Inorg. Chem.* **1998**, *37*, 5823–5826.
255. Schluter, R. D.; Luten, H. A.; Rees, W. S., Jr. *Mater. Res. Soc. Symp. Proc.* **1996**, *410*, 97–101.
256. Perez-Lourido, P.; Romero, J.; Garcia-Vazquez, J. A.; Sousa, A.; Maresca, K.; Zubieta, J. *Inorg. Chem.* **1999**, *38*, 1293–1298.
257. Annan, T. A.; Kumar, R.; Mabrouk, H. E.; Tuck, D. G.; Chadha, R. K. *Polyhedron* **1989**, *8*, 865–871.
258. Bertel, N.; Noltemeyer, M.; Roesky, H. W. *Z. Anorg. Allg. Chem.* **1990**, *588*, 102–108.
259. Peppe, C.; Tuck, D. G. *Can. J. Chem.* **1984**, *62*, 2798–2802.
260. Hirpo, W.; Dhingra, S.; Kanatzidis, M. G. *J. Chem. Soc., Chem. Commun.* **1992**, 557–559.
261. Shang, G.; Kunze, K.; Hampden-Smith, M. J.; Duesler, E. N. *Chem. Vap. Deposition* **1996**, *2*, 242–244.
262. Avivi, S.; Palchik, O.; Palchik, V.; Slifkin, M. A.; Weiss, A. M.; Gedanken, A. *Chem. of Mater.* **2001**, *13*, 2195–2200.
263. Deivaraj, T. C.; Park, J.-H.; Afzaal, M.; O'Brien, P.; Vittal, J. J. *Chem. Commun.* **2001**, 2304–2305.
264. Dutta, D. P.; Jain, V. K.; Knoedler, A.; Kaim, W. *Polyhedron* **2002**, *21*, 239–246.
265. Dutta, D. P.; Jain, V. K.; Chaudhury, S.; Tiekink, E. R. T. *Main Group Met. Chem.* **2001**, *24*, 405–408.
266. Park, J.-H.; O'Brien, P.; White, A. J. P.; Williams, D. J. *Inorg. Chem.* **2001**, *40*, 3629–3631.
267. Clark, E. B.; Breen, M. L.; Fanwick, P. E.; Hepp, A. F.; Duraj, S. A. *J. Coord. Chem.* **2000**, *52*, 111–117.
268. Ng, S. W. *Main Group Met. Chem.* **1999**, *22*, 447–451.
269. Oliveira, M. M.; Pessoa, G. M.; Carvalho, L. C.; Peppe, C.; Souza, A. G.; Airoldi, C. *Thermochim. Acta* **1999**, *328*, 223–230.
270. Liu, X.-Z.; Xue, H.; Zhao, J.; Song, Y.-L.; Zang, S.-L. *Gaodeng Xuexiao Huaxue Xuebao* **1999**, *20*, 196–198.
271. Silaghi-Dumitrescu, L.; Silaghi-Dumitrescu, I.; Haiduc, I.; Toscano, R.-A.; Garcia-Montalvo, V.; Cea-Olivares, R. *Z. Anorg. Allg. Chem.* **1999**, *625*, 347–351.
272. Pahari, D.; Jain, V. K.; Patel, R. P. *Main Group Met. Chem.* **1998**, *21*, 261–270.

273. Bhattacharya, S.; Seth, N.; Srivastava, D. K.; Gupta, V. D.; Noeth, H.; Thomann-Albach, M. *J. Chem. Soc., Dalton Trans.* **1996**, 2815–2820.
274. Hepp, A. F.; Hehemann, D. G.; Duraj, S. A.; Clark, E. B.; Eckles, W. E.; Fanwick, P. E. *Mater. Res. Soc. Symp. Proc.* **1994**, *327*, 29–34.
275. Nomura, R.; Matsuda, H. *Trends Inorg. Chem.* **1991**, *2*, 79–89.
276. Casey, A. T.; Vecchio, A. M. *Inorg. Chim. Acta* **1987**, *131*, 191–194.
277. Geloso, C.; Kumar, R.; Lopez-Grado, J. R.; Tuck, D. G. *Can. J. Chem.* **1987**, *65*, 928–932.
278. Ahmad, R.; Srivastava, G.; Mehrotra, R. C.; Saraswat, B. S. *Indian J. Chem., A* **1985**, *24A*, 557–561.
279. Lindmark, A. F.; Fay, R. C. *Inorg. Chem.* **1983**, *22*, 2000–2006.
280. Dymock, K.; Palenik, G. J.; Slezak, J.; Raston, C. L.; White, A. H. *J. Chem. Soc., Dalton Trans.* **1976**, 28–32.
281. Zukerman-Schpector, J.; Haiduc, I.; Silvestru, C.; Cea-Olivares, R. *Polyhedron* **1995**, *14*, 3087–3094.
282. Landry, C. C.; Hynes, A.; Barron, A. R.; Haiduc, I.; Silvestru, C. *Polyhedron* **1996**, *15*, 391–402.
283. Coggon, P.; Lebedda, J. D.; McPhail, A. T.; Palmer, R. A. *J. Chem. Soc., D* **1970**, 78–79.
284. Ensslin, F.; Dreyer, H. *Z. Anorg. Allgem. Chem.* **1942**, *249*, 119–132.
285. O'Brien, P.; Otway, D. J.; Walsh, J. R. *Thin Solid Films* **1998**, *315*, 57–61.
286. Hoskins, B. F.; Tiekink, E. R. T.; Vecchiet, R.; Winter, G. *Inorg. Chim. Acta* **1984**, *90*, 197–200.
287. Bessdergenev, V. G.; Ivanova, E. N.; Kovalevskaya, Y. A.; Gromilov, S. A.; Kirichenko, V. N.; Larionov, S. V. *Inorg. Mater. (Transl. Neorg. Mater.)* **1996**, *32*, 592–596.
288. Horley, G. A.; Chunggaze, M.; O'Brien, P.; White, A. J. P.; Williams, D. J. *J. Chem. Soc., Dalton Trans.* **1998**, 4205–4210.
289. Chunggaze, M.; Horley, G. A.; O'Brien, P. *Top. Issues Glass* **1998**, *2*, 52–54.
290. Horley, G. A.; O'Brien, P.; Park, J.-H.; White, A. J. P.; Williams, D. J. *J. Mater. Chem.* **1999**, *9*, 1289–1292.
291. Svensson, G.; Albertsson, J. *Acta Chem. Scand.* **1989**, *43*, 511–517.
292. Cea-Olivares, R.; Toscano, R. A.; Carreon, G.; Valdes-Martinez, J. *Monatsh. Chem.* **1992**, *123*, 391–396.
293. Garcia-Montalvo, V.; Cea-Olivares, R.; Williams, D. J.; Espinosa-Perez, G. *Inorg. Chem.* **1996**, *35*, 3948–3953.
294. Cea-Olivares, R.; Garcia-Montalvo, V.; Novosad, J.; Woolins, J. D.; Toscano, R. A. *Chem. Ber.* **1996**, *129*, 919–923.
295. Cea-Olivares, R.; Toscano, R. A.; Hernandez-Ortega, S.; Novosad, J.; Garcia-Montalvo, V. *Eur. J. Inorg. Chem.* **1999**, 1613–1616.
296. Darwin, K.; Gilby, L. M.; Hodge, P. R.; Piggott, B. *Polyhedron* **1999**, *18*, 3729–3733.
297. Rose, D. J.; Chang, Y. D.; Chen, Q.; Kettler, P. B.; Zubieta, J. *Inorg. Chem.* **1995**, *34*, 3973–3979.
298. Garcia-Vazquez, J. A.; Romero, A.; Sousa, A. *Coord. Chem. Rev.* **1999**, *193–195*, 691–745.
299. Romero, J.; Duran, M. L.; Rodriguez, A.; Garcia-Vazquez, J. A.; Sousa, A.; Rose, D. J.; Zubieta, J. *Inorg. Chim. Acta* **1998**, *274*, 131–136.
300. Rodriguez, A.; Romero, J.; Garcia-Vazquez, J. A.; Sousa, A.; Zubieta, J.; Rose, D. J.; Maresca, K. *Inorg. Chim. Acta* **1998**, *281*, 70–76.
301. Jones, A. C.; O'Brien, P. *CVD of Compound Semiconductors* **1997**, VCH: Weinheim, Germany.
302. Arnold, J. *Prog. Inorg. Chem.* **1995**, *43*, 353–417.
303. Coucouvanis, D. *Prog. Inorg. Chem.* **1979**, *26*, 301–469.
304. Kuchta, M. C.; Rheingold, A. L.; Parkin, G. *New J. Chem.* **1999**, *23*, 957–959.
305. Wuller, S. P.; Seligson, A. L.; Mitchell, G. P.; Arnold, J. *Inorg. Chem.* **1995**, *34*, 4854–4861.
306. Cole, M. L.; Hibbs, D. E.; Jones, C.; Smithies, N. A. *J. Chem. Soc., Dalton Trans.* **2000**, 545–550.
307. Smith, D. M.; Ibers, J. A. *Polyhedron* **1998**, *17*, 2105–2108.
308. Hirpo, W.; Dhingra, S.; Sutorik, A. C.; Kanatzidis, M. G. *J. Am. Chem. Soc.* **1993**, *115*, 1597–1599.
309. Kuchta, M. C.; Parkin, G. *Coord. Chem. Rev.* **1998**, *176*, 323–372.
310. O'Brien, P.; Otway, D. J.; Walsh, J. R. *Chem. Vap. Deposition* **1997**, *3*, 227–229.
311. McAleese, J.; O'Brien, P.; Otway, D. J. *Chem. Vap. Deposition* **1998**, *4*, 94–96.
312. McAleese, J.; O'Brien, P.; Otway, D. J. *Mater. Res. Soc. Symp. Proc.* **1998**, *485*, 157–162.
313. Gysling, H. J.; Wernberg, A. A.; Blanton, T. N. *Chem. Mater.* **1992**, *4*, 900–905.
314. Cheng, Y.; Emge, T. J.; Brennan, J. G. *Inorg. Chem.* **1996**, *35*, 7339–7344.
315. Romero, J.; Duran, M. L.; Garcia-Vazquez, J. A.; Castineiras, A.; Sousa, A.; Christiaens, L.; Zubieta, J. *Inorg. Chim. Acta* **1997**, *255*, 307–311.
316. Kniep, R.; Blees, P.; Poll, W. *Angew. Chem.* **1982**, *94*, 370.
317. Kniep, R.; Blees, P. *Angew. Chem.* **1984**, *96*, 782–783.
318. Kopasz, J. P.; Hallock, R. B.; Beachley, O. T. *Inorg. Synth.* **1986**, *24*, 87–89.
319. Jungwirth, P.; Zahradnik, R. *Theochem* **1993**, *102*, 317–320.
320. Timoshkin, A. Y.; Suvorov, A. V.; Bettinger, H. F.; Schaefer, H. F., III. *J. Am. Chem. Soc.* **1999**, *121*, 5687–5699.
321. Black, S. J.; Hibbs, D. E.; Hursthouse, M. B.; Jones, C.; Abdul Malik, K. M.; Smithies, N. A. *J. Chem. Soc., Dalton Trans.* **1997**, 4313–4320.
322. Kuhn, N.; Henkel, G.; Kratz, T.; Kreutzberg, J.; Boese, R.; Maulitz, A. H. *Chem. Ber.* **1993**, *126*, 2041–2045.
323. Arduengo, A. J., III; Dias, H. V. R.; Calabrese, J. C.; Davidson, F. *J. Am. Chem. Soc.* **1992**, *114*, 9724–9725.
324. Li, X.-W.; Su, J.; Robinson, G. H. *Chem. Commun.* **1996**, 2683–2684.
325. Cole, M. L.; Davies, A. J.; Jones, C. *J. Chem. Soc., Dalton Trans.* **2001**, 2451–2452.
326. De Araujo Felix, L.; De Oliveira, C. A. F.; Kross, R. K.; Peppe, C.; Brown, M. A.; Tuck, D. G.; Hernandes, M. Z.; Longo, E.; Sensato, F. R. *J. Organomet. Chem.* **2000**, *603*, 203–212.
327. de Souza, A. C.; Peppe, C.; Tian, Z.; Tuck, D. G. *Organometallics* **1993**, *12*, 3354–3357.
328. Baker, L. J.; Kloo, L. A.; Rickard, C. E. F.; Taylor, M. J. *J. Organomet. Chem.* **1997**, *545–546*, 249–255.
329. Wells, R. L.; Aubuchon, S. R.; Kher, S. S.; Lube, M. S.; White, P. S. *Chem. Mater.* **1995**, *7*, 793–800.
330. Godfrey, S. M.; Kelly, K. J.; Kramkowski, P.; McAuliffe, A.; Pritchard, R. G. *Chem. Commun.* **1997**, 1001–1002.
331. Alcock, N. W.; Degnan, I. A.; Howarth, O. W.; Wallbridge, M. G. H. *J. Chem. Soc., Dalton Trans.* **1992**, 2775–2780.
332. Brown, M. A.; Tuck, D. G.; Wells, E. J. *Can. J. Chem.* **1996**, *74*, 1535–1549.
333. Self, M. F.; McPhail, A. T.; Wells, R. L. *Polyhedron* **1993**, *12*, 455–459.
334. Karia, R.; Willey, G. R.; Drew, M. G. B. *Acta Crystallogr., C: Cryst. Struct. Commun.* **1986**, *C42*, 558–560.

335. Beddoes, R. L.; Collison, D.; Mabbs, F. E.; Temperley, J. *Acta Crystallogr., C: Cryst. Struct. Commun.* **1991**, *C47*, 58–61.
336. Williams, D. J.; Bevilacqua, V. L. H.; Morson, P. A.; Dennison, K. J.; Pennington, W. T.; Schimek, G. L.; VanDerveer, D.; Kruger, J. S.; Kawai, N. T. *Inorg. Chim. Acta* **1999**, *285*, 217–222.
337. Degnan, I. A.; Alcock, N. W.; Roe, S. M.; Wallbridge, M. G. H. *Acta Crystallogr., C: Cryst. Struct. Commun.* **1992**, *C48*, 995–999.
338. Willey, G. R.; Aris, D. R.; Roe, S. M.; Haslop, J. V.; Errington, W. *Polyhedron* **2001**, *20*, 423–429.
339. Clegg, W.; Norman, N. C.; Pickett, N. L. *Acta Crystallogr., C: Cryst. Struct. Commun.* **1994**, *C50*, 36–38.
340. Malyarik, M. A.; Petrosyants, S. P.; Ilyukhin, A. B.; Buslaev, Y. A. *Zh. Neorg. Khim.* **1991**, *36*, 2816–2820.
341. Aris, D. R.; Errington, W.; Willey, G. R. *Acta Crystallogr., C: Cryst. Struct. Commun.* **1999**, *C55*, 1746–1748.
342. Wells, R. L.; Kher, S. S.; Baldwin, R. A.; White, P. S. *Polyhedron* **1994**, *13*, 2731–2735.
343. Robinson, W. T.; Wilkins, C. J.; Zhang, Z. *J. Chem. Soc., Dalton Trans.* **1990**, 219–227.
344. Brown, M. A.; Tuck, D. G. *Inorg. Chim. Acta* **1996**, *247*, 135–138.
345. Samsonenko, D. G.; Sokolov, M. N.; Virovets, A. V.; Pervukhina, N. V.; Fedin, V. P. *Eur. J. Inorg. Chem.* **2001**, 167–172.
346. Ziegler, M. L.; Schlimper, H. U.; Nuber, B.; Weiss, J.; Ertl, G. *Z. Anorg. Allg. Chem.* **1975**, *415*, 193–201.
347. Knop, O.; Cameron, T. S.; Adhikesavalu, D.; Vincent, B. R.; Jenkins, J. A. *Can. J. Chem.* **1987**, *65*, 1527–1556.
348. Clark, G. R.; Rickard, C. E. F.; Taylor, M. J. *Can. J. Chem.* **1986**, *64*, 1697–1701.
349. Taylor, M. J.; Tuck, D. G.; Victoriano, L. *J. Chem. Soc., Dalton Trans.* **1981**, 928–932.
350. Niu, W.; Wong, E. H.; Weisman, G. R.; Sommer, R. D.; Rheingold, A. L. *Inorg. Chem. Commun.* **2002**, *5*, 1–4.
351. Sigl, M.; Schier, A.; Schmidbaur, H. *Eur. J. Inorg. Chem.* **1998**, 203–210.
352. Sigl, M.; Schier, A.; Schmidbaur, H. *Z. Naturforsch., B: Chem. Sci.* **1999**, *54*, 1417–1419.
353. Abram, S.; Maichle-Mossmer, C.; Abram, U. *Polyhedron* **1997**, *16*, 2183–2191.
354. Abram, S.; Maichle-Mossmer, C.; Abram, U. *Polyhedron* **1997**, *16*, 2291–2298.
355. Willey, G. R.; Aris, D. R.; Errington, W. *Inorg. Chim. Acta* **2000**, *300–302*, 1004–1013.
356. Kloo, L. A.; Taylor, M. J. *J. Chem. Soc., Dalton Trans.* **1997**, 2693–2696.
357. von Arnim, H.; Dehnicke, K.; Maczek, K.; Fenske, D. *Naturforsch., B: Chem. Sci.* **1993**, *48*, 1331–1340.
358. Strel'tsova, N. R.; Ivanov, M. G.; Vashchenko, S. D.; Bel'skii, V. K.; Kalinichenko, I. I. *Koord. Khim.* **1991**, *17*, 646–651.
359. Ivanov, M. G.; Kalinichenko, I. I.; Vashchenko, S. D.; Gulyaeva, I. V.; Popov, A. N. *Koord. Khim.* **1993**, *19*, 499–504.
360. Shriver, D. F.; Wharf, I. *Inorg. Chem.* **1969**, *8*, 2167–2171.
361. Leone, S. R.; Swanson, B.; Shriver, D. F. *Inorg. Chem.* **1970**, *9*, 2189–2191.
362. Joy, G.; Gaughan, A. P., Jr.; Wharf, I.; Shriver, D. F.; Dougherty, J. P. *Inorg. Chem.* **1975**, *14*, 1795–1801.
363. Bubenheim, W.; Frenzen, G.; Mueller, U. *Acta Crystallogr., C: Cryst. Struct. Commun.* **1995**, *C51*, 1120–1124.
364. Ishihara, H.; Dou, S.-Q.; Gesing, T. M.; Paulus, H.; Fuess, H.; Weiss, A. *J. Mol. Struct.* **1998**, *471*, 175–182.
365. Aldridge, S.; Downs, A. J. *Chem. Rev.* **2001**, *101*, 3305–3365.
366. Downs, A. J.; Pulham, C. R. *Chem. Soc. Rev.* **1994**, *23*, 175–184.
367. Hunt, P.; Schwerdtfeger, P. *Inorg. Chem.* **1996**, *35*, 2085–2088.
368. Jones, C. *Chem. Commun.* **2001**, 2293–2298.
369. Hibbs, D. E.; Jones, C.; Smithies, N. A. *Chem. Commun.* **1999**, 185–186.
370. Francis, M. D.; Hibbs, D. E.; Hursthouse, M. B.; Jones, C.; Smithies, N. A. *J. Chem. Soc., Dalton Trans.* **1998**, 3249–3254.
371. Abernethy, C. D.; Cole, M. L.; Jones, C. *Organometallics* **2000**, *19*, 4852–4857.
372. Avent, A. G.; Eaborn, C.; Hitchcock, P. B.; Smith, J. D.; Sullivan, A. C. *J. Chem. Soc., Chem. Commun.* **1986**, 988–989.
373. Beachley, O. T., Jr.; Chao, S. H. L.; Churchill, M. R.; See, R. F. *Organometallics* **1992**, *11*, 1486–1491.
374. Churchill, M. R.; Lake, C. H.; Chao, S. H. L.; Beachley, O. T., Jr. *J. Chem. Soc., Chem. Commun.* **1993**, 1577–1578.
375. Kuemmel, C.; Meller, A.; Noltemeyer, M. *Z. Naturforsch., B: Chem. Sci.* **1996**, *51*, 209–219.
376. Aldridge, S.; Downs, A. J.; Parsons, S. *Chem. Commun.* **1996**, 2055–2056.
377. Hibbs, D. E.; Hursthouse, M. B.; Jones, C.; Smithies, N. A. *Organometallics* **1998**, *17*, 3108–3110.
378. Bakum, S. I.; Kuznetsova, S. F.; Tarasov, V. P. *Zh. Neorg. Khim.* **1999**, *44*, 346–347.
379. Yamada, M.; Tanaka, K.; Araki, S.; Butsugan, Y. *Tetrahedron Lett.* **1995**, *36*, 3169–3172.
380. Pullumbi, P.; Bouteiller, Y.; Manceron, L.; Mijoule, C. *Chem. Phys.* **1994**, *185*, 25–37.
381. Himmel, H.-J.; Downs, A. J.; Greene, T. M. *J. Am. Chem. Soc.* **2000**, *122*, 9793–9807.
382. Himmel, H.-J.; Downs, A. J.; Greene, T. M. *Inorg. Chem.* **2001**, *40*, 396–407.
383. Inoue, K.; Sawada, A.; Shibata, I.; Baba, A. *Tetrahedron Lett.* **2001**, *42*, 4661–4663.
384. Inoue, K.; Sawada, A.; Shibata, I.; Baba, A. *J. Am. Chem. Soc.* **2002**, *124*, 906–907.
385. Harris, W. R.; Chen, Y.; Wein, K. *Inorg. Chem.* **1994**, *33*, 4991–4998.
386. Weiner, R. E.; Thakur, M. L. *Radiochim. Acta* **1995**, *70*, 273–287.
387. Motekaitis, R. J.; Martell, A. E.; Koch, S. A.; Hwang, J.; Quarless, D. A., Jr.; Welch, M. J. *Inorg. Chem.* **1998**, *37*, 5902–5911.
388. Sun, Y.; Martell, A. E.; Welch, M. J. *Tetrahedron* **2000**, *56*, 5093–5103.
389. Chmura, A. J.; Orton, M. S.; Meares, C. F. *Proc. Natl. Acad. Sci. USA* **2001**, *98*, 8480–8484.
390. Li, M.; Meares, C. F.; Salako, Q.; Kukis, D. L.; Zhong, G.-R.; Miers, L.; DeNardo, S. J. *Cancer Res.* **1995**, *55*, 5726S–5728S.
391. Caravan, P.; Rettig, S. J.; Orvig, C. *Inorg. Chem.* **1997**, *36*, 1306–1315.
392. Caravan, P.; Orvig, C. *Inorg. Chem.* **1997**, *36*, 236–248.
393. Lowe, M. P.; Rettig, S. J.; Orvig, C. *J. Am. Chem. Soc.* **1996**, *118*, 10446–10456.
394. Wong, E.; Caravan, P.; Liu, S.; Rettig, S. J.; Orvig, C. *Inorg. Chem.* **1996**, *35*, 715–724.
395. Wong, E.; Liu, S.; Rettig, S.; Orvig, C. *Inorg. Chem.* **1995**, *34*, 3057–3064.
396. Figuet, M.; Averbuch-Pouchot, M. T.; du Moulinet d'Hardemare, A.; Jarjayes, O. *Eur. J. Inorg. Chem.* **2001**, 2089–2096.

397. Inoue, M. B.; Inoue, M.; Fernando, Q. *Inorg. Chim. Acta* **1998**, *271*, 207–209.
398. Bollinger, J. E.; Banks, W. A.; Roundhill, D. M. *Conf. Coord. Chem.* **1995**, 361–36615th.
399. Martell, A. E.; Hancock, R. D. *Metal Complexes in Aqueous Solutions* **1996**, Plenum: New York.
400. Sun, Y.; Anderson, C. J.; Pajeau, T. S.; Reichert, D. E.; Hancock, R. D.; Motekaitis, R. J.; Martell, A. E.; Welch, M. J. *J. Med. Chem.* **1996**, *39*, 458–470.
401. Davis, J.; Palenik, G. J. *Inorg. Chim. Acta* **1985**, *99*, L51–L52.
402. Craig, A. S.; Helps, I. M.; Parker, D.; Adams, H.; Bailey, N. A.; Williams, M. G.; Smith, J. M. A.; Ferguson, G. *Polyhedron* **1989**, *8*, 2481–2484.
403. Casas, J. S.; Garcia-Tasende, M. S.; Sordo, J. *Coord. Chem. Rev.* **2000**, *209*, 197–261.
404. Fischer, R. A.; Weiss, J. *Angew. Chem., Int. Ed.* **1999**, *38*, 2831–2850.
405. Wochele, R.; Schwarz, W.; Klinkhammer, K. W.; Locke, K.; Weidlein, J. *Z. Anorg. Allg. Chem.* **2000**, *626*, 1963–1973.
406. Wiberg, N.; Blank, T.; Purath, A.; Stosser, G.; Schnockel, H. *Angew. Chem., Int. Ed.* **1999**, *38*, 2563–2565.
407. Wiberg, N.; Amelunxen, K.; Noeth, H.; Schmidt, M.; Schwenk, H. *Angew. Chem., Int. Ed. Engl.* **1996**, *35*, 65–67.
408. Wiberg, N.; Blank, T.; Noth, H.; Ponikwar, W. *Angew. Chem., Int. Ed.* **1999**, *38*, 839–841.
409. Uhl, W. *Angew. Chem.* **1993**, *105*, 1449–1461 (See also Angew. Chem., Int. Ed. Engl A. 1993 32(10) 1386–1397).
410. Power, P. P. *J. Chem. Soc., Dalton Trans.* **1998**, 2939–2951.
411. Lomeli, V.; McBurnett, B. G.; Cowley, A. H. *J. Organomet. Chem.* **1998**, *562*, 123–125.
412. Klimek, K. S.; Cui, C.; Roesky, H. W.; Noltemeyer, M.; Schmidt, H.-G. *Organometallics* **2000**, *19*, 3085–3090.
413. Tyrra, W. E. *J. Fluor. Chem.* **2001**, *112*, 149–152.
414. Brothers, P. J.; Huebler, K.; Huebler, U.; Noll, B. C.; Olmstead, M. M.; Power, P. P. *Angew. Chem., Int. Ed. Engl.* **1996**, *35*, 2355–2357.
415. Veith, M.; Goffing, F.; Becker, S.; Huch, V. *J. Organomet. Chem.* **1991**, *406*, 105–18.
416. Grocholl, L.; Schranz, I.; Stahl, L.; Staples, R. J. *Inorg. Chem.* **1998**, *37*, 2496–2499.
417. Hellmann, K. W.; Gade, L. H.; Steiner, A.; Stalke, D.; Moeller, F. *Angew. Chem., Int. Ed. Engl.* **1997**, *36*, 160–163.
418. Hellmann, K.; Galka, C. H.; Rudenauer, I.; Gade, L. H.; Scowen, I. J.; McPartlin, M. *Angew. Chem., Int. Ed.* **1998**, *37*, 1948–1952.
419. Fryzuk, M. D.; Giesbrecht, G. R.; Rettig, S. J.; Yap, G. P. A. *J. Organomet. Chem.* **1999**, *591*, 63–70.
420. Von Hanisch, C.; Fenske, D.; Kattannek, M.; Ahlrichs, R. *Angew. Chem., Int. Ed.* **1999**, *38*, 2736–2738.
421. Gabbai, F. P.; Schier, A.; Riede, J.; Schmidbaur, H. *Inorg. Chem.* **1995**, *34*, 3855–3856.
422. Beck, H. P. *Z. Naturforsch., B: Chem. Sci.* **1987**, *42*, 251–252.
423. Beck, H. P. *Z. Naturforsch., B: Anorg. Chem., Org. Chem.* **1984**, *39B*, 310–313.
424. Ruck, M.; Barnighausen, H. *Z. Anorg. Allg. Chem.* **1999**, *625*, 577–585.
425. Tuck, D. G. *Polyhedron* **1990**, *9*, 377–386.
426. Okuda, T.; Shimoe, H.; Monta, M.; Nakata, A.; Terao, H.; Yamada, K. *J. Mol. Struct.* **1994**, *319*, 197–201.
427. Beck, H. P.; Wilhelm, D. *Angew. Chem.* **1991**, *103*, 897–898 (See also Angew. Chem., Int. Ed. Engl. 1991 30(7) 824–825).
428. Annan, T. A.; Gu, J.; Tian, Z.; Tuck, D. G. *J. Chem. Soc., Dalton Trans.* **1992**, 3061–3067.
429. Taylor, M. J.; Tuck, D. G.; Victoriano, L. *Can. J. Chem.* **1982**, *60*, 690–694.
430. Sinclair, I.; Worrall, I. J. *Can. J. Chem.* **1982**, *60*, 695–698.
431. Peppe, C.; Tuck, D. G. *Can. J. Chem.* **1984**, *62*, 2793–2797.
432. Khan, M. A.; Peppe, C.; Tuck, D. G. *Can. J. Chem.* **1984**, *62*, 601–605.
433. Scholten, M.; Dronskowski, R.; Staffel, T.; Meyer, G. *Z. Anorg. Allg. Chem.* **1998**, *624*, 1741–1745.
434. Haubrich, S. T.; Power, P. P. *J. Am. Chem. Soc.* **1998**, *120*, 2202–2203.
435. Schluter, R. D.; Cowley, A. H.; Atwood, D. A.; Jones, R. A.; Atwood, J. L. *J. Coord. Chem.* **1993**, *30*, 25–28.
436. Uhl, W.; Graupner, R.; Layh, M.; Schuetz, U. *J. Organomet. Chem.* **1995**, *493*, C1–5.
437. Uhl, W.; Melle, S. *Chem. – Eur. J.* **2001**, *7*, 4216–4221.
438. Himmel, II.-J.; Downs, A. J.; Greene, T. M.; Andrews, L. *Chem. Commun.* **1999**, 2243–2244.
439. Trofimenko, S. *Scorpionates: The Coordination Chemistry of Polypyrazolylborate Ligands* **1999**, Imperial College Press: London.
440. Trofimenko, S. *Chem. Rev.* **1993**, *93*, 943–980.
441. Frazer, A.; Piggott, B.; Hursthouse, M. B.; Mazid, M. *J. Am. Chem. Soc.* **1994**, *116*, 4127–4128.
442. Dias, H. V. R.; Huai, L.; Jin, W.; Bott, S. G. *Inorg. Chem.* **1995**, *34*, 1973–1974.
443. Kuchta, M. C.; Dias, H. V. R.; Bott, S. G.; Parkin, G. *Inorg. Chem.* **1996**, *35*, 943–948.
444. Dias, H. V. R.; Jin, W. *Inorg. Chem.* **1996**, *35*, 267–268.
445. Dias, H. V. R.; Jin, W. *Inorg. Chem.* **2000**, *39*, 815–819.
446. Frazer, A.; Hodge, P.; Piggott, B. *Chem. Commun.* **1996**, 1727–1728.
447. Kuchta, M. C.; Parkin, G. *J. Am. Chem. Soc.* **1995**, *117*, 12651–12652.
448. Callaghan, C.; Clentsmith, G. K. B.; Cloke, F. G. N.; Hitchcock, P. B.; Nixon, J. F.; Vickers, D. M. *Organometallics* **1999**, *18*, 793–795.
449. Clentsmith, G. K. B.; Cloke, F. G. N.; Francis, M. D.; Green, J. C.; Hitchcock, P. B.; Nixon, J. F.; Suter, J. L.; Vickers, D. M. *J. Chem. Soc., Dalton Trans.* **2000**, 1715–1721.
450. Scholz, M.; Noltemeyer, M.; Roesky, H. W. *Angew. Chem.* **1989**, *101*, 1419–1420.
451. Veith, M.; Kunze, K. *Angew. Chem., Int. Ed. Engl.* **1991**, *30*, 95–97.
452. Brown, M. A.; El-Hadad, A. A.; McGarvey, B. R.; Sung, R. C. W.; Trikha, A. K.; Tuck, D. G. *Inorg. Chim. Acta* **2000**, *300–302*, 613–621.
453. Red'kin, A. N.; Dubovitskaya, L. G.; Smirnov, V. A.; Dmitriev, V. S. *Zh. Neorg. Khim.* **1984**, *29*, 1955–1959.
454. Egorova, A. G.; Nefedov, A. N. *Izv. Akad. Nauk Kaz. SSR, Ser. Khim.* **1983**, 84–86.
455. Mabrouk, H. E.; Tuck, D. G. *Can. J. Chem.* **1989**, *67*, 746–50.
456. Geloso, C.; Mabrouk, H. E.; Tuck, D. G. *J. Chem. Soc., Dalton Trans.* **1989**, 1759–1763.
457. Tuck, D. G. *NATO ASI Ser., Ser. C* **1993**, *385*, 15–31.
458. Chandra, S. K.; Gould, E. S. *Chem. Commun.* **1996**, 809–810.
459. Swavey, S.; Gould, E. S. *Inorg. Chem.* **2000**, *39*, 1200–1203.

460. Fitz, H.; Muller, B. G. *Z. Anorg. Allg. Chem.* **1997**, *623*, 579–582.
461. Dronskowski, R. *Angew. Chem., Int. Ed. Engl.* **1995**, *34*, 1126–1128.
462. Dronskowski, R. *Z. Kristallogr.* **1995**, *210*, 920–923.
463. Staffel, T.; Meyer, G. *Z. Anorg. Allg. Chem.* **1988**, *563*, 27–37.
464. Dronskowski, R. *Inorg. Chem.* **1994**, *33*, 5960–5963.
465. Galvan-Arzate, S.; Santamaria, A. *Toxicol. Lett.* **1998**, *99*, 1–13.
466. Schomacker, K.; Schicha, H. *Eur. J. Nucl. Med.* **2000**, *27*, 1845–1863 and references therein.
467. Janiak, C.; Hoffmann, R. *J. Am. Chem. Soc.* **1990**, *112*, 5924–5946.
468. Janiak, C.; Hoffmann, R. *Angew. Chem.* **1989**, *101*, 1706–1708.
469. Schwerdtfeger, P. *Inorg. Chem.* **1991**, *30*, 1660–1663.
470. Budzelaar, P. H. M.; Boersma, J. *Recl. Trav. Chim. Pays-Bas* **1990**, *109*, 187–189.
471. Toth, I.; Gyori, B. *Encyclopedia of Inorganic Chemistry*; King, R. B., Ed., Wiley: Chichester, UK, 1994; Vol. 8, pp 4134–4142.
472. Lee, A. G. *Coord. Chem. Rev.* **1972**, *8*, 289–349.
473. Kurosawa, H. *Comprehensive Organometallic Chemistry*; Wilkinson, G., Stone, F. G. A., Abel, E. W., Eds.; Pergamon: Oxford, 1982; Vol. 1, pp 725–754.
474. Rees, W. S.; Krauter, G. *Encyclopedia of Inorganic Chemistry*; King, R. B., Ed., Wiley: Chichester, UK, 1994; Vol. 8, pp 4142–4151.
475. Casas, J. S.; Garcia-Tasende, M. S.; Sordo, J. *Coord. Chem. Rev.* **1999**, *193–195*, 283–359.
476. Siegal, M. P.; Venturini, E. L.; Morosin, B.; Aselage, T. L. *J. Mater. Res.* **1997**, *12*, 2825–2854 and references therein.
477. Blixt, J.; Glaser, J.; Mink, J.; Persson, I.; Persson, P.; Sandstroem, M. *J. Am. Chem. Soc.* **1995**, *117*, 5089–5104.
478. Banyai, I.; Glaser, J.; Toth, I. *Eur. J. Inorg. Chem.* **2001**, 1709–1717.
479. Blixt, J.; Glaser, J.; Solymosi, P.; Toth, I. *Inorg. Chem.* **1992**, *31*, 5288–5297.
480. Sheu, Y.-H.; Hong, T.-N.; Lin, C.-C.; Chen, J.-H.; Wang, S.-S. *Polyhedron* **1996**, *16*, 681–688.
481. Senge, M. O.; Ruhlandt-Senge, K.; Regli, K. J.; Smith, K. M. *J. Chem. Soc., Dalton Trans.* **1993**, 3519–3538.
482. Cheng, T. W.; Chen, Y. J.; Hong, F. E.; Chen, J. H.; Wang, S. L.; Hwang, L. P. *Polyhedron* **1994**, *13*, 403–408.
483. Jalilehvand, F.; Maliarik, M.; Sandstroem, M.; Mink, J.; Persson, I.; Persson, P.; Toth, I.; Glaser, J. *Inorg. Chem.* **2001**, *40*, 3889–3899.
484. Maliarik, M.; Berg, K.; Glaser, J.; Sandstroem, M.; Toth, I. *Inorg. Chem.* **1998**, *37*, 2910–2919.
485. Autschbach, J.; Ziegler, T. *J. Am. Chem. Soc.* **2001**, *123*, 5320–5324.
486. Russo, M. R.; Kaltsoyannis, N. *Inorg. Chim. Acta* **2001**, *312*, 221–225.
487. Lees, A. J. *Chem. Rev.* **1987**, *87*, 711–743.
488. Gade, L. H. *Angew. Chem., Int. Ed.* **2001**, *40*, 3573–3575.
489. Wiberg, N.; Blank, T.; Lerner, H.-W.; Fenske, D.; Linti, G. *Angew. Chem., Int. Ed. Engl.* **2001**, *40*, 1232–1235.
490. Glaser, J. *Advances in Thallium Aqueous Solution Chemistry*; Academic Press: New York, 1995; Vol. 43.
491. Lobov, B. I.; Kul'ba, F. Y.; Mironov, V. E. *Zh. Neorg. Khim.* **1967**, *12*, 341–346.
492. Lobov, B. I.; Kul'ba, F. Y.; Mironov, V. E. *Zh. Neorg. Khim.* **1967**, *12*, 334–340.
493. McWhinnie, W. R. *J. Chem. Soc., A, Inorg., Phys., Theoret.* **1966**, 889–892.
494. Sutton, G. J. *Aust. J. Chem.* **1958**, *11*, 120–124.
495. Ma, G.; Ilyukhin, A.; Glaser, J.; Toth, I.; Zekany, L. *Inorg. Chim. Acta* **2001**, *320*, 92–100.
496. Gramlich, V.; Lubal, P.; Musso, S.; Anderegg, G. *Helv. Chim. Acta* **2001**, *84*, 623–631.
497. Wieghardt, K.; Kleine-Boymann, M.; Nuber, B.; Weiss, J. *Inorg. Chem.* **1986**, *25*, 1309–1313.
498. Borgholte, H.; Dehnicke, K.; Goesmann, H.; Fenske, D. *Z. Anorg. Allg. Chem.* **1991**, *600*, 7–14.
499. Hoerner, M.; de Oliveira, A. B.; Beck, J. *Z. Anorg. Allg. Chem.* **1997**, *623*, 65–68.
500. Tung, J.-Y.; Chen, J.-H.; Liao, F.-L.; Wang, S.-L.; Hwang, L.-P. *Inorg. Chem.* **2000**, *39*, 2120–2124.
501. Tung, J.-Y.; Jang, J.-I.; Lin, C.-C.; Chen, J.-H.; Hwang, L.-P. *Inorg. Chem.* **2000**, *39*, 1106–1112.
502. Lu, Y.-Y.; Tung, J.-Y.; Chen, J.-H.; Liao, F.-L.; Wang, S.-L.; Wang, S.-S.; Hwang, L.-P. *Polyhedron* **1998**, *18*, 145–150.
503. Tung, J.-Y.; Chen, J.-H.; Liao, F.-L.; Wang, S.-L.; Hwang, L.-P. *Inorg. Chem.* **1998**, *37*, 6104–6108.
504. Lomova, T. N.; Mozhzhukhina, E. G. *Zh. Neorg. Khim.* **1997**, *42*, 1691–1696.
505. Coutsolelos, A. G.; Daphnomili, D. *Inorg. Chem.* **1997**, *36*, 4614–4615.
506. Tang, S.-S.; Liu, I. C.; Lin, C.-C.; Chen, J.-H. *Polyhedron* **1996**, *15*, 37–41.
507. Tang, S.-S.; Lin, Y.-H.; Sheu, M.-T.; Lin, C.-C.; Chen, J.-H.; Wang, S.-S. *Polyhedron* **1995**, *14*, 1241–1243.
508. Fuh, J.-J.; Tang, S.-S.; Lin, Y.-H.; Chen, J.-H.; Liu, T.-S.; Wang, S.-S.; Lin, J.-C. *Polyhedron* **1994**, *13*, 3031–3037.
509. Senge, M. O. *J. Chem. Soc., Dalton Trans.* **1993**, 3539–3549.
510. Coutsolelos, A. G.; Tsapara, A.; Daphnomili, D.; Ward, D. L. *J. Chem. Soc., Dalton Trans.* **1991**, 3413–3417.
511. Stanley, K. D.; Lopez de la Vega, R.; Quirke, J. M. E.; Beato, B. D.; Yost, R. A. *Chem. Geol.* **1991**, *91*, 169–183.
512. Coutsolelos, A. G.; Orfanopoulos, M.; Ward, D. L. *Polyhedron* **1991**, *10*, 885–892.
513. Guilard, R.; Zrineh, A.; Ferhat, M.; Tabard, A.; Mitaine, P.; Swistak, C.; Richard, P.; Lecomte, C.; Kadish, K. M. *Inorg. Chem.* **1988**, *27*, 697–705.
514. Kadish, K. M.; Tabard, A.; Zrineh, A.; Ferhat, M.; Guilard, R. *Inorg. Chem.* **1987**, *26*, 2459–2466.
515. Brady, F.; Henrick, K.; Matthews, R. W. *J. Organomet. Chem.* **1981**, *210*, 281–288.
516. Lee, W.-B.; Suen, S.-C.; Jong, T.-T.; Hong, F.-E.; Chen, J.-H.; Lin, H.-J.; Hwang, L.-P. *J. Organomet. Chem.* **1993**, *450*, 63–66.
517. Suen, S. C.; Lee, W. B.; Hong, F. E.; Jong, T. T.; Chen, J. H. *Polyhedron* **1992**, *11*, 3025–3030.
518. Yang, C.-H.; Tung, J.-Y.; Liau, B.-C.; Ko, B.-T.; Elango, S.; Chen, J.-H.; Hwang, L.-P. *Polyhedron* **2001**, *20*, 3257–3264.
519. Daphnomili, D.; Scheidt, W. R.; Zajicek, J.; Coutsolelos, A. G. *Inorg. Chem.* **1998**, *37*, 3675–3681.
520. Janczak, J.; Kubiak, R. *Acta Chem. Scand.* **1995**, *49*, 871–877.
521. Schweiger, K.; Hueckstaedt, H.; Homborg, H. *Z. Anorg. Allg. Chem.* **1998**, *624*, 167–168.
522. Schweiger, K.; Goldner, M.; Huckstadt, H.; Homborg, H. *Z. Anorg. Allg. Chem.* **1999**, *625*, 1693–1699.
523. Galka, C. H.; Gade, L. H. *Chem. Commun.* **2001**, 899–900.
524. Hellmann, K. W.; Bergner, A.; Gade, L. H.; Scowen, I. J.; McPartlin, M. *J. Organomet. Chem.* **1999**, *573*, 156–164.

525. Kritikos, M.; Ma, G.; Bodor, A.; Glaser, J. *Inorg. Chim. Acta* **2002**, *331*, 224–231.
526. Drew, M. G. B.; Howarth, O. W.; Martin, N.; Morgan, G. G.; Nelson, J. *J. Chem. Soc., Dalton Trans.* **2000**, 1275–1278.
527. Mueller, G.; Lachmann, J. *Z. Naturforsch., B: Chem. Sci.* **1993**, *48*, 1544–54.
528. Baldwin, R. A.; Wells, R. L.; White, P. S. *Main Group Chem.* **1997**, *2*, 67–71.
529. Glaser, J.; Johansson, G. *Acta Chem. Scand., Ser. A* **1981**, *A 35*, 639–644.
530. Banyai, I.; Glaser, J. *J. Am. Chem. Soc.* **1989**, *111*, 3186–3194.
531. Banyai, I.; Glaser, J. *J. Am. Chem. Soc.* **1990**, *112*, 4703–10.
532. Hinz, D. *Z. Anorg. Allg. Chem.* **2000**, *626*, 1012–1015.
533. Ma, G.; Molla-Abbassi, A.; Kritikos, M.; Ilyukhin, A.; Jalilehvand, F.; Kessler, V.; Skripkin, O. M.; Sandstroem, M.; Glaser, J.; Naeslund, J.; Persson, I. *Inorg. Chem.* **2001**, *40*, 6432–6438.
534. Faggiani, R.; Brown, I. D. *Acta Crystallogr., B* **1978**, *B34*, 2845–2486.
535. Brown, I. D.; Faggiani, R. *Acta Crystallogr., B* **1980**, *B36*, 1802–1806.
536. Ivanov-Emin, B. N.; Medvedev, Y. N.; Lin'ko, I. V.; Nevskii, N. N. *Zh. Neorg. Khim.* **1984**, *29*, 1417–1420.
537. Binsted, N.; Hector, A. L.; Levason, W. *Inorg. Chim. Acta* **2000**, *298*, 116–119.
538. Musso, S.; Anderegg, G.; Ruegger, H.; Schlaepfer, C. W.; Gramlich, V. *Inorg. Chem.* **1995**, *34*, 3329–3338.
539. Chen, B.; Lubal, P.; Musso, S.; Anderegg, G. *Anal. Chim. Acta* **2000**, *406*, 317–323.
540. Abrahamson, H.; Heiman, J. R.; Pignolet, L. H. *Inorg. Chem.* **1975**, *14*, 2070–2075.
541. Kepert, D. L.; Raston, C. L.; Roberts, N. K.; White, A. H. *Aust. J. Chem.* **1978**, *31*, 1927–1932.
542. Slavin, P. A.; Reglinski, J.; Spicer, M. D.; Kennedy, A. R. *J. Chem. Soc., Dalton Trans.* **2000**, 239–240.
543. Kimblin, C.; Bridgewater, B. M.; Hascall, T.; Parkin, G. *J. Chem. Soc., Dalton Trans.* **2000**, 1267–1274.
544. Levason, W.; Reid, G. *J. Chem. Soc., Dalton Trans.* **2001**, 2953–2960.
545. Levason, W.; Orchard, S. D.; Reid, G. *Coord. Chem. Rev.* **2002**, *225*, 159–199.
546. Hoyer, E.; Dietzsch, W.; Mueller, H.; Zschunke, A.; Schroth, W. *Inorg. Nucl. Chem. Lett.* **1967**, *3*, 457–461.
547. Olk, R.-M.; Dietzsch, W.; Kirmse, R.; Stach, J.; Hoyer, E. *Inorg. Chim. Acta* **1987**, *128*, 251–259.
548. Fields, R. O.; Waters, J. H.; Bergendahl, T. J. *Inorg. Chem.* **1971**, *10*, 2808–2810.
549. Cotton, F. A.; McCleverty, J. A. *Inorg. Chem.* **1967**, *6*, 229–232.
550. Casas, J. S.; Castellano, E. E.; Castineiras, A.; Sanchez, A.; Sordo, J.; Vazquez-Lopez, E. M.; Zukerman-Schpector, J. *J. Chem. Soc., Dalton Trans.* **1995**, 1403–1409.
551. Tebbe, K.-F.; El Essawi, M.; Abd El Khalik, S. *Z. Naturforsch., B: Chem. Sci.* **1995**, *50*, 1429–1439.
552. Domasevitch, K. V.; Rusanova, J. A.; Sieler, J.; Kokozay, V. N. *Inorg. Chim. Acta* **1999**, *293*, 234–238.
553. Geiser, U.; Schlueter, J. A.; Kini, A.; Achenbach, C. A.; Komosa, A. S.; Williams, J. M. *Acta Crystallogr., C: Cryst. Struct. Commun.* **1996**, *C52*, 159–162.
554. Riera, V.; Ruiz, M. A.; Villafane, F.; Bois, C.; Jeannin, Y. *J. Organomet. Chem.* **1989**, *375*, C23–C26.
555. Tebbe, K. F. *Acta Crystallogr., C: Cryst. Struct. Commun.* **1989**, *C45*, 180–2.
556. Geiser, U.; Wang, H. H.; Schlueter, J.; Chen, M. Y.; Kini, A. M.; Kao, I. H. C.; Williams, J. M.; Whangbo, M. H.; Evain, M. *Inorg. Chem.* **1988**, *27*, 4284–4289.
557. Beno, M. A.; Geiser, U.; Kostka, K. L.; Wang, H. H.; Webb, K. S.; Firestone, M. A.; Carlson, K. D.; Nunez, L.; Whangbo, M. H.; Williams, J. M. *Inorg. Chem.* **1987**, *26*, 1912–1920.
558. Thiele, G.; Rotter, H. W.; Zimmermann, K. *Z. Naturforsch., B: Anorg. Chem., Org. Chem.* **1986**, *41B*, 269–272.
559. Glaser, J.; Goggin, P. L.; Sandstroem, M.; Lutsko, V. *Acta Chem. Scand., Ser. A* **1983**, *A37*, 437–438.
560. Bermejo, M. R.; Castineiras, A.; Garcia-Vazquez, J. A.; Hiller, W.; Straehle, J. *J. Crystallogr. Spectrosc. Res.* **1991**, *21*, 93 96.
561. Bermejo, M. R.; Fernandez, B.; Fernandez, M. I.; Gomez, M. E. *An. Quim.* **1991**, *87*, 1052–1058.
562. Bermejo, M. R.; Fernandez, M. B.; Fernandez, M. I.; Gomez, M. E. *Synth. React. Inorg. Met.-Org. Chem.* **1991**, *21*, 915–929.
563. Castineiras, A.; Bermejo, M. R.; Garcia-Deibe, A.; Hiller, W. *Acta Crystallogr., Sect. C: Cryst. Struct. Commun.* **1991**, *C47*, 1738–1740.
564. Bermejo, M. R.; Castineiras, A.; Fernandez, M. I.; Gomez, M. E. *Acta Crystallogr., Sect. C: Cryst. Struct. Commun.* **1991**, *C47*, 1406–1408.
565. Bermejo, M. R.; Fernandez, B.; Fernandez, M. I.; Gomez, M. E.; Rey, M. *Synth. React. Inorg. Met.-Org. Chem.* **1995**, *25*, 639–652.
566. Bermejo, M. R.; Fernandez, B.; Fernandez, M. I.; Gomez, M. E.; Rey, M. *Synth. React. Inorg. Met.-Org. Chem.* **1994**, *24*, 1397–1410.
567. Bermejo, M. R.; Fernandez, M. I.; Fernandez, B.; Gomez, M. E. *Synth. React. Inorg. Met.-Org. Chem.* **1992**, *22*, 759–773.
568. Bermejo, M. R.; Gayoso, M.; Fernandez, M. I.; Hermida, A.; Gomez, E. *An. Quim., Ser. B* **1988**, *84*, 303–307.
569. Bermejo, M. R.; Fernandez, M. I.; Tajes, J.; Deibe, A. G. *An. Quim., Ser. B* **1988**, *84*, 298–302.
570. Bermejo, M. R.; Rodriguez, A.; Deibe, A. G.; Tajes, J. *An. Quim., Ser. B* **1988**, *84*, 293–297.
571. Bermejo, M. R.; Fernandez, A.; Gayoso, M.; Castineiras, A.; Hiller, W.; Straehle, J. *Polyhedron* **1988**, *7*, 2561–2567.
572. Bermejo, M. R.; Fernandez, M. I.; Fernandez, B.; Gomez, M. E.; Gayoso, Y. M. *An. Quim., Ser. B* **1988**, *84*, 52–56.
573. Bermejo, M. R.; Fernandez, M. I.; Varela, M. D.; Gomez, M. E.; Gayoso, M. *An. Quim., Ser. B* **1987**, *83*, 273–276.
574. Hiller, W.; Castineiras, A.; Garcia-Fernandez, M. E.; Bermejo, M. R.; Bravo, J.; Sanchez, A. *Z. Naturforsch., B: Chem. Sci.* **1988**, *43*, 132–133.
575. Bermejo, M. R.; Garcia Deibe, A.; Rodriguez, A.; Castineiras, A. *Synth. React. Inorg. Met.-Org. Chem.* **1987**, *17*, 693–707.
576. Fernandez, M. I.; Bermejo, M. R.; Fernandez, A.; Solleiro, E.; Gayoso, M. *An. Quim., Ser. B* **1987**, *83*, 26–30.
577. Fernandez, M. I.; Gomez, M. E.; Hermida, A.; Bermejo, M. R. *Acta Cient. Compostelana* **1985**, *22*, 749–763.
578. Castineiras, A.; Hiller, W.; Straehle, J.; Bermejo, M. R.; Gayoso, M. *An. Quim., Ser. B* **1986**, *82*, 282–286.
579. Bermejo, M. R.; Solleiro, E.; Rodriguez, A.; Castineiras, A. *Polyhedron* **1987**, *6*, 315–317.
580. Blanco, F.; Castano, M. V.; Bermejo, M. R.; Gayoso, M. *An. Quim., Ser. B* **1985**, *81*, 133–177.
581. Cole, M. L.; Haigh, R.; Jones, C. *Main Group Met. Chem.* **2001**, *24*, 819–820.
582. Walton, R. A. *Coord. Chem. Rev.* **1971**, *6*, 1–25.

583. Jeffs, S. E.; Small, R. W. H.; Worrall, I. J. *Acta Crystallogr., C: Cryst. Struct. Commun.* **1984**, *C40*, 1329–1331.
584. Jeffs, S. E.; Small, R. W. H.; Worrall, I. J. *Acta Crystallogr., C: Cryst. Struct. Commun.* **1984**, *C40*, 1827–1829.
585. Jeffs, S. E.; Small, R. W. H.; Worrall, I. J. *Acta Crystallogr., C: Cryst. Struct. Commun.* **1984**, *C40*, 381–383.
586. Jeffs, S. E.; Small, R. W. H.; Worrall, I. J. *Acta Crystallogr., C: Cryst. Struct. Commun.* **1984**, *C40*, 65–67.
587. Jeffs, S. E.; Small, R. W. H.; Worrall, I. J. *Acta Crystallogr., C: Cryst. Struct. Commun.* **1983**, *C39*, 1628–1630.
588. Lee, A. G. *The Chemistry of Thallium* **1971**, Elsevier: Amsterdam.
589. James, M. A.; Millikan, M. B.; James, B. D. *Main Group Met. Chem.* **1991**, *14*, 1–11.
590. James, M. A.; Clyburne, J. A. C.; Linden, A.; James, B. D.; Liesegang, J.; Zuzich, V. *Can. J. Chem.* **1996**, *74*, 1490–1502.
591. Linden, A.; James, M. A.; Millikan, M. B.; Kivlighon, L. M.; Petridis, A.; James, B. D. *Inorg. Chim. Acta* **1999**, *284*, 215–222.
592. Millikan, M. B.; James, B. D. *Inorg. Chim. Acta* **1980**, *44*, 93–L94.
593. James, B. D.; Millikan, M. B.; Skelton, B. W.; White, A. H. *Main Group Met. Chem.* **1993**, *16*, 335–343.
594. Millikan, M. B.; James, B. D. *Inorg. Chim. Acta* **1984**, *81*, 109–115.
595. Bastow, T. J.; James, B. D.; Millikan, M. B. *J. Solid State Chem.* **1983**, *49*, 388–390.
596. Boehme, R.; Rath, J.; Grunwald, B.; Thiele, G. *Z. Naturforsch., B: Anorg. Chem., Org. Chem.* **1980**, *35B*, 1366–1372.
597. Thiele, G.; Richter, R. *Z. Kristallogr.* **1993**, *205*, 129–130.
598. Thiele, G.; Richter, R. *Z. Kristallogr.* **1993**, *205*, 131–132.
599. Thiele, G.; Richter, R. *Z. Kristallogr.* **1993**, *207*, 142–144.
600. Hoard, J. L.; Goldstein, L. *J. Chem. Physics* **1935**, *3*, 199–202.
601. Colton, E.; Jones, M. M. *Z. Naturforsch.* **1956**, *11b*, 491–492.
602. Zimmermann, K.; Thiele, G. *Z. Anorg. Allg. Chem.* **1987**, *553*, 280–286.
603. Zimmermann, K.; Thiele, G. *Z. Naturforsch., B: Chem. Sci.* **1987**, *42*, 818–824.
604. Linden, A.; Nugent, K. W.; Petridis, A.; James, B. D. *Inorg. Chim. Acta* **1999**, *285*, 122–128.
605. Linden, A.; Petridis, A.; James, B. D. *Acta Crystallogr., C: Crystal Structure Communications* **2002**, *C58*, m53–m55.
606. Linden, A.; Petridis, A.; James, B. D. *Inorg. Chim. Acta* **2002**, *332*, 61–71.
607. Henkel, S.; Klinkhammer, K. W.; Schwarz, W. *Angew. Chem.* **1994**, *106*, 721–723 (See also *Angew. Chem., Int. Ed. Engl.*, **1994**, *33(6)*, 681–683).
608. Schwerdtfeger, P.; Boyd, P. D. W.; Bowmaker, G. A.; Mack, H. G.; Oberhammer, H. *J. Am. Chem. Soc.* **1989**, *111*, 15–23.
609. Uson, R.; Fornies, J.; Tomas, M.; Garde, R.; Alonso, P. J. *J. Am. Chem. Soc.* **1995**, *117*, 1837–1838.
610. Uhl, W.; Keimling, S. U.; Klinkhammer, K. W.; Schwarz, W. *Angew. Chem., Int. Ed. Engl.* **1997**, *36*, 64–65.
611. Niemeyer, M.; Power, P. P. *Angew. Chem., Int. Ed. Engl.* **1998**, *37*, 1277–1279.
612. Janiak, C. *Coord. Chem. Rev.* **1997**, *163*, 107–215.
613. Sawamura, M.; Iikura, H.; Nakamura, E. *J. Am. Chem. Soc.* **1996**, *118*, 12850–12851.
614. Stupko, T. V.; Mironov, V. E.; Pashkov, G. L.; Isaev, I. D. *Zh. Neorg. Khim.* **1996**, *41*, 275–277.
615. Stupko, T. V.; Isaev, I. D.; Mironov, V. E. *Zh. Neorg. Khim.* **1989**, *34*, 2441–2443.
616. Stupko, T. V.; Isaev, I. D.; Mironov, V. E. *Koord. Khim.* **1987**, *13*, 1467–1469.
617. Kogai, T. I.; Isaev, I. D.; Mironov, V. E. *Koord. Khim.* **1990**, *16*, 919–921.
618. Kogai, T. I. *Izv. Vyssh. Uchebn. Zaved., Khim. Khim. Tekhnol.* **1999**, *42*, 97–99.
619. Connelly, N. G.; Hicks, O. M.; Lewis, G. R.; Moreno, M. T.; Orpen, A. G. *J. Chem. Soc., Dalton Trans.* **1998**, 1913–1918.
620. Cassidy, J. M.; Whitmire, K. H. *Inorg. Chem.* **1989**, *28*, 1435–1439.
621. Casado, M. A.; Perez-Torrente, J. J.; Lopez, J. A.; Ciriano, M. A.; Lahoz, F. J.; Oro, L. A. *Inorg. Chem.* **1999**, *38*, 2482–2488.
622. Hazell, A.; McGinley, J.; Toftlund, H. *Inorg. Chim. Acta* **2001**, *323*, 113–118.
623. Reger, D. L.; Collins, J. E.; Layland, R.; Adams, R. D. *Inorg. Chem.* **1996**, *35*, 1372–1376.
624. Bylikin, S. Y.; Robson, D. A.; Male, N. A. H.; Rees, L. H.; Mountford, P.; Schroder, M. *J. Chem. Soc., Dalton Trans.* **2001**, 170–180.
625. Howarth, O. W.; Nelson, J.; McKee, V. *Chem. Commun.* **2000**, 21–22.
626. Klinkhammer, K. W.; Henkel, S. *J. Organomet. Chem.* **1994**, *480*, 167–171.
627. Haaland, A.; Shorokhov, D. J.; Volden, H. V.; Klinkhammer, K. W. *Inorg. Chem.* **1999**, *38*, 1118–1120.
628. Waezsada, S. D.; Belgardt, T.; Noltemeyer, M.; Roesky, H. W. *Angew. Chem.* **1994**, *106*, 1413–1414.
629. Peters, J. C.; Harkins, S. B.; Brown, S. D.; Day, M. W. *Inorg. Chem.* **2001**, *40*, 5083–5091.
630. Hellmann, K. W.; Gade, L. H.; Fleischer, R.; Stalke, D. *Chem. Commun.* **1997**, 527–528.
631. Galka, C. H.; Trosch, D. J. M.; Schubart, M.; Gade, L. H.; Radojevic, S.; Scowen, I. J.; McPartlin, M. *Eur. J. Inorg. Chem.* **2000**, 2577–2583.
632. Hellmann, K. W.; Galka, C.; Gade, L. H.; Steiner, A.; Wright, D. S.; Kottke, T.; Stalke, D. *Chem. Commun.* **1998**, 549–550.
633. Veith, M.; Goffing, F.; Huch, V. *Chem. Ber.* **1988**, *121*, 943–949.
634. Galka, C. H.; Gade, L. H. *Inorg. Chem.* **1999**, *38*, 1038–1039.
635. Galka, C. H.; Renner, P.; Gade, L. H. *Inorg. Chem. Commun.* **2001**, *4*, 332–335.
636. Veith, M.; Spaniol, A.; Poehlmann, J.; Gross, F.; Huch, V. *Chem. Ber.* **1993**, *126*, 2625–35.
637. Hellmann, K. W.; Gade, L. H.; Scowen, I. J.; McPartlin, M. *Chem. Commun.* **1996**, 2515–2516.
638. Deacon, G. B.; Delbridge, E. E.; Skelton, B. W.; White, A. H. *Eur. J. Inorg. Chem.* **1998**, 543–545.
639. Deacon, G. B.; Delbridge, E. E.; Forsyth, C. M.; Skelton, B. W.; White, A. H. *J. Chem. Soc., Dalton Trans.* **2000**, 745–751.
640. Singh, K.; Long, J. R.; Stavropoulos, P. *J. Am. Chem. Soc.* **1997**, *119*, 2942–2943.
641. Deacon, G. B.; Delbridge, E. E.; Skelton, B. W.; White, A. H. *Eur. J. Inorg. Chem.* **1999**, 751–761.
642. Bhandari, S.; Mahon, M. F.; Molloy, K. C.; Palmer, J. S.; Sayers, S. F. *J. Chem. Soc., Dalton Trans.* **2000**, 1053–1060.
643. Nafissi, S.; Aghabozorgh, H.; Sadjadi, S. A. S. *J. Inorg. Biochem.* **1997**, *66*, 253–258.
644. Renn, O.; Preut, H.; Lippert, B. *Inorg. Chim. Acta* **1991**, *188*, 133–137.

645. Ciliberto, E.; Di Bella, S.; Gulino, A.; Fragala, I. L. *Inorg. Chem.* **1992**, *31*, 1641–1644.
646. Boesveld, W. M.; Hitchcock, P. B.; Lappert, M. F.; Noth, H. *Angew. Chem., Int. Ed.* **2000**, *39*, 222–224.
647. Dai, X.; Warren, T. H. *Chem. Comm.* **2001**, 1998–1999.
648. Beck, J.; Straehle, J. *Z. Naturforsch., B: Anorg. Chem., Org. Chem.* **1986**, *41B*, 1381–1386.
649. Johnson, M. T.; Campana, C. F.; Foxman, B. M.; Desmarais, W.; Vela, M. J.; Miller, J. S. *Chem.-Eur. J.* **2000**, *6*, 1805–1810.
650. Janiak, C. *Main Group Met. Chem.* **1998**, *21*, 33–49.
651. Kitajima, N.; Tolman, W. B. *Prog. Inorg. Chem.* **1995**, *43*, 419–531.
652. Ghosh, P.; Rheingold, A. L.; Parkin, G. *Inorg. Chem.* **1999**, *38*, 5464–5467.
653. Fillebeen, T.; Hascall, T.; Parkin, G. *Inorg. Chem.* **1997**, *36*, 3787–3790.
654. Bardwell, D. A.; Jeffery, J. C.; McCleverty, J. A.; Ward, M. D. *Inorg. Chim. Acta* **1998**, *267*, 323–328.
655. Mann, K. L. V.; Jeffery, J. C.; McCleverty, J. A.; Ward, M. D. *Polyhedron* **1999**, *18*, 721–727.
656. Fleming, J. S.; Psillakis, E.; Couchman, S. M.; Jeffery, J. C.; McCleverty, J. A.; Ward, M. D. *J. Chem. Soc., Dalton Trans.* **1998**, 537–543.
657. Dowling, C.; Ghosh, P.; Parkin, G. *Polyhedron* **1997**, *16*, 3469–3473.
658. Ghosh, P.; Hascall, T.; Dowling, C.; Parkin, G. *J. Chem. Soc., Dalton Trans.* **1998**, 3355–3358.
659. Dowling, C. M.; Parkin, G. *Polyhedron* **2001**, *20*, 285–289.
660. Janiak, C.; Temizdemir, S.; Scharmann, T. G. *Z. Anorg. Allg. Chem.* **1998**, *624*, 755–756.
661. Rheingold, A. L.; Liable-Sands, L. M.; Trofimenko, S. *Chem. Commun.* **1997**, 1691–1692.
662. Calabrese, J. C.; Trofimenko, S. *Inorg. Chem.* **1992**, *31*, 4810–4814.
663. Rheingold, A. L.; Haggerty, B. S.; Trofimenko, S. *Angew. Chem.* **1994**, *106*, 2053–2056 (See also *Angew. Chem., Int. Ed. Engl.,* **1994**, *33(19)*, 1983–1985).
664. Yoon, K.; Parkin, G. *Polyhedron* **1995**, *14*, 811–821.
665. Trofimenko, S.; Calabrese, J. C.; Thompson, J. S. *Inorg. Chem.* **1987**, *26*, 1507–1514.
666. Amoroso, A. J.; Jeffrey, J. C.; Jones, P. L.; McCleverty, J. A.; Psillakis, E.; Ward, M. D. *J. Chem. Soc., Chem. Commun.* **1995**, 1175–76.
667. Motson, G. R.; Mamula, O.; Jeffery, J. C.; McCleverty, J. A.; Ward, M. D.; von Zelewsky, A. *J. Chem. Soc., Dalton Trans.* **2001**, 1802.
668. Lopez, C.; Sanz, D.; Claramunt, R. M.; Trofimenko, S.; Elguero, J. *J. Organomet. Chem.* **1995**, *503*, 265–276.
669. Trofimenko, S.; Calabrese, J. C.; Kochi, J. K.; Wolowiec, S.; Hulsbergen, F. B.; Reedijk, J. *Inorg. Chem.* **1992**, *31*, 3943–3950.
670. Jones, P. L.; Mann, K. L. V.; Jeffery, J. C.; McCleverty, J. A.; Ward, M. D. *Polyhedron* **1997**, *16*, 2435–2440.
671. Rheingold, A. L.; White, C. B.; Trofimenko, S. *Inorg. Chem.* **1993**, *32*, 3471–3477.
672. Han, R.; Parkin, G.; Trofimenko, S. *Polyhedron* **1995**, *14*, 387–391.
673. Rheingold, A. L.; Liable-Sands, L. M.; Trofimenko, S. *Inor. Chem.* **2001**, *40*, 6509–6513.
674. Rheingold, A. L.; Ostrander, R. L.; Haggerty, B. S.; Trofimenko, S. *Inorg. Chem.* **1994**, *33*, 3666–3676.
675. Trofimenko, S.; Calabrese, J. C.; Domaille, P. J.; Thompson, J. S. *Inorg. Chem.* **1989**, *28*, 1091–1101.
676. Janiak, C.; Braun, L.; Girgsdies, F. *J. Chem. Soc., Dalton Trans.* **1999**, 3133–3136.
677. Ghosh, P.; Desrosiers, P. J.; Parkin, G. *J. Am. Chem. Soc.* **1998**, *120*, 10416–10422.
678. Han, R.; Ghosh, P.; Desrosiers, P. J.; Trofimenko, S.; Parkin, G. *J. Chem. Soc., Dalton Trans.* **1997**, 3713–3717.
679. Renn, O.; Vananzi, L. M.; Marteletti, A.; Gramlich, V. *Helv. Chim. Acta* **1995**, *78*, 993–1000.
680. Akita, M.; Ohta, K.; Takahashi, Y.; Hikichi, S.; Moro-oka, Y. *Organometallics* **1997**, *16*, 4121–4128.
681. Libertini, E.; Yoon, K.; Parkin, G. *Polyhedron* **1993**, *12*, 2539–2542.
682. Dowling, C. M.; Leslie, D.; Chisholm, M. H.; Parkin, G. *Main Group Chem.* **1995**, *1*, 29–52.
683. Ghosh, P.; Churchill, D. G.; Rubinshtein, M.; Parkin, G. *New J. Chem.* **1999**, *23*, 961–963.
684. Jaekle, F.; Polborn, K.; Wagner, M. *Chem. Ber.* **1996**, *129*, 603–606.
685. Guo, S.; Bats, J. W.; Bolte, M.; Wagner, M. *J. Chem. Soc., Dalton Trans.* **2001**, 3572–3576.
686. Ferguson, G.; Jennings, M. C.; Lalor, F. J.; Shanahan, C. *Acta Crystallogr., C: Cryst. Struct. Commun.* **1991**, *C47*, 2079–2082.
687. Kisko, J. L.; Hascall, T.; Kimblin, C.; Parkin, G. *J. Chem. Soc., Dalton Trans.* **1999**, 1929–1936.
688. Sanz, D.; Claramunt Rosa, M.; Glaser, J.; Trofimenko, S.; Elguero, J. *Magn. Reson. Chem.* **1996**, *34*, 843–846.
689. Kunkely, H.; Vogler, A. *Chem. Phys. Lett.* **2000**, *327*, 162–164.
690. Han, R.; Parkin, G. *Organometallics* **1991**, *10*, 1010–1020.
691. Klaui, W.; Schramm, D.; Peters, W.; Rheinwald, G.; Lang, H. *Eur. J. Inorg. Chem.* **2001**, 1415–1424.
692. Klaui, W.; Berghahn, M.; Rheinwald, G.; Lang, H. *Angew. Chem., Int. Ed.* **2000**, *39*, 2464–2466.
693. Rheingold, A.; Liable-Sands, L. M.; Yap, G. P. A.; Trofimenko, S. *Chem. Commun.* **1996**, 1233–1234.
694. LeCloux, D. D.; Tokar, C. J.; Osawa, M.; Houser, R. P.; Keyes, M. C.; Tolman, W. B. *Organometallics* **1994**, *13*, 2855–2866.
695. Janiak, C.; Temizdemir, S.; Rohr, C. *Z. Anorg. Allg. Chem.* **2000**, *626*, 1265–1267.
696. Lalor, F. J.; Miller, S. M.; Garvey, N. *Polyhedron* **1990**, *9*, 63–68.
697. Psillakis, E.; Jeffery, J. C.; McCleverty, J. A.; Ward, M. D. *J. Chem. Soc., Dalton Trans.* **1997**, 1645–1651.
698. Lai, J.-J.; Khademi, S.; Meyer, E. F., Jr.; Cullen, D. L.; Smith, K. M. *J. Porphyrins and Phthalocyanines* **2001**, *5*, 621–627.
699. Filipek, S.; Wagner, E.; Darlewski, W.; Kalinowski, M. K. *Pol. J. Chem.* **1992**, *66*, 43–48.
700. Smith, K. M.; Lai, J. J. *Tetrahedron Lett.* **1980**, *21*, 433–436.
701. Janczak, J.; Kubiak, R. *J. Alloys Compd.* **1993**, *202*, 69–72.
702. Janczak, J.; Kubiak, R.; Zaleski, A.; Olejniczak, J. *Chem. Phys. Lett.* **1994**, *225*, 72–75.
703. Janczak, J. *Pol. J. Chem.* **1999**, *73*, 437–446.
704. Francis, M. D.; Jones, C.; Deacon, G. B.; Delbridge, E. E.; Junk, P. C. *Organometallics* **1998**, *17*, 3826–3828.
705. Francis, M. D.; Hitchcock, P. B.; Nixon, J. F.; Schnockel, H.; Steiner, J. *J. Organomet. Chem.* **2002**, *646*, 191–195.
706. Deacon, G. B.; Delbridge, E. E.; Fallon, G. D.; Jones, C.; Hibbs, D. E.; Hursthouse, M. B.; Skelton, B. W.; White, A. H. *Organometallics* **2000**, *19*, 1713–1721.
707. Persson, I.; Jalilehvand, F.; Sandstroem, M. *Inorganic Chemistry* **2002**, *41*, 192–197.

708. Ouchi, M.; Hakushi, T. *Coord. Chem. Rev.* **1996**, *148*, 171–181.
709. Couton, D.; Mocerino, M.; Rapley, C.; Kitamura, C.; Yoneda, A.; Ouchi, M. *Aust. J. Chem.* **1999**, *52*, 227–229.
710. Kimura, K.; Tatsumi, K.; Yokoyama, M.; Ouchi, M.; Mocerino, M. *Anal. Commun.* **1999**, *36*, 229–230.
711. Rounaghi, G.; Chamsaz, M.; Nezhadali, A. *Russ. J. Gen. Chem.* **2000**, *70*, 1358–1362.
712. Shamsipur, M.; Khayatian, G. *J. Inclusion Phenom. Macrocyclic Chem.* **2001**, *39*, 109–113.
713. Domasevitch, K. V.; Skopenko, V. V.; Sieler, J. *Inorg. Chim. Acta* **1996**, *249*, 151–155.
714. Fujiwara, M.; Matsushita, T.; Yamashoji, Y.; Tanaka, M.; Tuchi, M.; Hakushi, T. *Polyhedron* **1993**, *12*, 1239–44.
715. Buschmann, H. J. *Thermochim. Acta* **1986**, *107*, 219–226.
716. Lee, Y. C.; Allison, J.; Popov, A. I. *Polyhedron* **1985**, *4*, 441–445.
717. Jiang, Z.; Wang, G.; Wang, R.; Yao, X. *Jiegou Huaxue* **1989**, *8*, 163–167.
718. Domasevitch, K.; Mokhir, A.; Rusanov, E. *J. Coord. Chem.* **1995**, *36*, 15–22.
719. Domasevitch, K.; Ponomareva, V.; Rusanov, E. *J. Coord. Chem.* **1995**, *34*, 259–263.
720. Trush, V. A.; Domasevitch, K. V.; Amirkhanov, V. M.; Sieler, J. *Z. Naturforsch., B: Chem. Sci.* **1999**, *54*, 451–455.
721. Skopenko, V. V.; Domasevitch, K. V.; Mokhir, A.; Rusanov, E. B. *J. Coord. Chem.* **1997**, *41*, 13–18.
722. Gakh, A. A.; Sachleben, R. A.; Bryan, J. C.; Moyer, B. A. *Tetrahedron Lett.* **1995**, *36*, 8163–8166.
723. Sipos, P.; Capewell, S. G.; May, P. M.; Hefter, G. T.; Laurenczy, G.; Lukacs, F.; Roulet, R. *J. Solution Chem.* **1997**, *26*, 419–431.
724. Zechmann, C. A.; Boyle, T. J.; Pedrotty, D. M.; Alam, T. M.; Lang, D. P.; Scott, B. L. *Inorg. Chem.* **2001**, *40*, 2177–2184.
725. Boyle, T. J.; Zechmann, C. A.; Alam, T. M.; Rodriguez, M. A.; Hijar, C. A.; Scott, B. L. *Inorg. Chem.* **2002**, *41*, 946–957.
726. Burke, P. J.; Matthews, R. W.; Gillies, D. G. *J. Chem. Soc., Dalton Trans.* **1980**, 1439–1442.
727. Maroni, V. A.; Spiro, T. G. *Inorg. Chem.* **1968**, *7*, 193–197.
728. Dahl, L. F.; Davis, G. L.; Wampler, D. L.; West, R. *J. Inorg. Nucl. Chem.* **1962**, *24*, 357–363.
729. Harvey, S.; Lappert, M. F.; Raston, C. L.; Skelton, B. W.; Srivastava, G.; White, A. H. *J. Chem. Soc., Chem. Commun.* **1988**, 1216–1217.
730. Roesky, H. W.; Scholz, M.; Noltemeyer, M.; Edelmann, F. T. *Inorg. Chem.* **1989**, *28*, 3829–3830.
731. El-Hadad, A. A.; Kickham, J. E.; Loeb, S. J.; Taricani, L.; Tuck, D. G. *Inorg. Chem.* **1995**, *34*, 120–123.
732. Hampden-Smith, M. J.; Smith, D. E.; Duesler, E. N. *Inorg. Chem.* **1989**, *28*, 3399–3401.
733. Baran, W.; Sikora, M.; Tomasik, P.; Anderegg, J. W. *Carbohydr. Polym.* **1997**, *32*, 209–212.
734. Cambie, R. C.; Larsen, D. S.; Rutledge, P. S.; Woodgate, P. D. *Aust. J. Chem.* **1997**, *50*, 767–769.
735. Taylor, E. C.; Hawks, G. H. III.; McKillop, A. *J. Amer. Chem. Soc.* **1968**, *90*, 2421–2422.
736. Tachiyashiki, S.; Nakayama, H.; Kuroda, R.; Sato, S.; Saito, Y. *Acta Crystallogr., B* **1975**, *B31*, 1483–1485.
737. Atencio, R.; Barbera, J.; Cativiela, C.; Lahoz, F. J.; Serrano, J. L.; Zurbano, M. M. *J. Am. Chem. Soc.* **1994**, *116*, 11558–11559.
738. Barbera, J.; Cativiela, C.; Serrano, J. L.; Zurbano, M. M. *Adv. Mater.* **1991**, *3*, 602–605.
739. Baena, M. J.; Espinet, P.; Ros, M. B.; Serrano, J. L.; Ezcurra, A. *Angew. Chem.* **1993**, *105*, 1260–1262 (See also *Angew. Chem., Int. Ed. Engl.*, **1993**, *32(8)*, 1203–1205).
740. Vicente, J.; Chicote, M. T.; Gonzalez-Herrero, P.; Jones, P. G.; Humphrey, M. G.; Cifuentes, M. P.; Samoc, M.; Luther-Davies, B. *Inorg. Chem.* **1999**, *38*, 5018–5026.
741. Amano, R.; Shiokawa, Y. *Inorg. Chim. Acta* **1993**, *203*, 9–10.
742. Sekine, T.; Tsuda, J. *Bull. Chem. Soc. Jpn.* **1995**, *68*, 3429–3437.
743. Lysyak, T. V.; Rusakov, S. L.; Kolomnikov, I. S.; Kharitonov, Y. Y. *Zh. Neorg. Khim.* **1983**, *28*, 1339–1341.
744. Rusakov, S. L.; Lysyak, T. V.; Kharitonov, Y. Y.; Kolomnikov, I. S. *Koord. Khim.* **1984**, *10*, 566.
745. Lysyak, T. V.; Rusakov, S. L.; Kolomnikov, I. S.; Khitrova, A. V.; Kharitonov, Y. Y. *Zh. Neorg. Khim.* **1984**, *29*, 3035–3038.
746. Ozutsumi, K.; Ohtaki, H.; Kusumegi, A. *Bull. Chem. Soc. Jpn.* **1984**, *57*, 2612–2617.
747. Yamaguchi, T.; Tanaka, Y.; Ozutsumi, K.; Ohtaki, H.; Kusumegi, A. *Nippon Kagaku Kaishi* **1986**, 1484–1491.
748. Baran, E. J.; Wagner, C. C.; Rossi, M.; Caruso, F. *Z. Anorg. Allg. Chem.* **2001**, *627*, 85–89.
749. Aoki, K.; Suh, I. H.; Nagashima, H.; Uzawa, J.; Yamazaki, H. *J. Am. Chem. Soc.* **1992**, *114*, 5722–5729.
750. Zhuravlev, Y. N.; Poplavnoi, A. S. *Russian Physics Journal (Transl. Izvestiya Vysshikh Uchebnykh Zavedenii, Fizika)* **2001**, *44*, 391–397.
751. Marchand, R.; Piffard, Y.; Tournoux, M. *Can. J. Chem.* **1975**, *53*, 2454–2458.
752. Jeansannetas, B.; Thomas, P.; Champarnaud-Mesjard, J. C.; Frit, B. *Mater. Res. Bull.* **1998**, *33*, 1709–1716.
753. Effenberger, H. *Z. Kristallogr.* **1998**, *213*, 42–46.
754. Sali, S. K.; Iyer, V. S.; Jayanthi, K.; Sampath, S.; Venugopal, V. *J. Alloys Compd.* **1996**, *237*, 49–57.
755. Sastry, P. U. M.; Sequeira, A. *Philos. Mag. B* **1997**, *75*, 659–667.
756. Kulikov, V. A.; Ugarov, V. V.; Rambidi, N. G. *Zh. Strukt. Khim.* **1981**, *22*, 166–168.
757. Bergman, J. G.; Wood, J. S. *Acta Crystallogr., C: Cryst. Struct. Commun.* **1987**, *C43*, 1831–1832.
758. Touboul, M. *Phosphorus Sulfur* **1986**, *28*, 145–149.
759. Petrov, K. P.; Ugarov, V. V.; Rambidi, N. G. *Zh. Strukt. Khim.* **1980**, *21*, 159–161.
760. Diot, M.; Lachenal, G.; Vignalou, J. R. *Thermochim. Acta* **1981**, *44*, 203–211.
761. Zalkin, A.; Templeton, D. H.; Eimerl, D.; Velsko, S. P. *Acta Crystallogr., C: Cryst. Struct. Commun.* **1986**, *C42*, 1686–1687.
762. Rios, S.; Paulus, W.; Cousson, A.; Quilichini, M.; Heger, G. *Acta Crystallogr., B: Struct. Sci.* **1998**, *B54*, 790–797.
763. Rios, S.; Paulus, W.; Cousson, A.; Quilichini, M.; Heger, G.; Le Calve, N.; Pasquier, B. *J. Phys. I* **1995**, *5*, 763–769.
764. Narasaiah, T. V.; Choudhary, R. N. P.; Nigam, G. D.; Mattern, G. *Z. Kristallogr.* **1986**, *175*, 145–149.
765. Riou, A.; Gerault, Y.; Cudennec, Y. *Rev. Chim. Miner.* **1986**, *23*, 70–79.
766. Fabry, J.; Breczewski, T. *Acta Crystallogr., C: Cryst. Struct. Commun.* **1993**, *C49*, 1724–1727.
767. Crespo, O.; Laguna, A.; Fernandez, E. J.; Lopez-de-Luzuriaga, J. M.; Mendia, A.; Monge, M.; Olmos, E.; Jones, P. G. *Chem. Commun.* **1998**, 2233–2234.
768. Catalano, V. J.; Bennett, B. L.; Muratidis, S.; Noll, B. C. *J. Am. Chem. Soc.* **2001**, *123*, 173–174.
769. Golovnev, N. N.; Primakov, A. S.; Mulgaleev, R. F. *Zh. Neorg. Khim.* **1995**, *40*, 108–110.

770. Munivelu, T.; Seshaiah, K.; Rao, P. V. V. P.; Devi, P. R.; Naidu, G. R. K. *Proc. Indian Natl. Sci. Acad., Part A* **1986**, *52*, 685–688.
771. Blake, A. J.; Greig, J. A.; Schroder, M. *J. Chem. Soc., Dalton Trans.* **1991**, 529–532.
772. Blake, A. J.; Fenske, D.; Li, W.-S.; Lippolis, V.; Schroder, M. *J. Chem. Soc., Dalton Trans.* **1998**, 3961–3968.
773. Blake, A. J.; Reid, G.; Schroeder, M. *J. Chem. Soc., Dalton Trans.* **1992**, 2987–2992.
774. Krebs, B.; Broemmelhaus, A. *Angew. Chem.* **1989**, *101*, 1726–1728.
775. Krebs, B.; Broemmelhaus, A. *Z. Anorg. Allg. Chem.* **1991**, *595*, 167–182.
776. Krebs, B.; Broemmelhaus, A.; Kersting, B.; Nienhaus, M. *Eur. J. Solid State Inorg. Chem.* **1992**, *29*, 167–180.
777. Uemura, S.; Tanaka, S.; Okano, M. *Bull. Inst. Chem. Res., Kyoto Univ.* **1977**, *55*, 273–275.
778. Barclay, J. E.; Evans, D. J.; Davies, S. C.; Hughes, D. L.; Sobota, P. *J. Chem. Soc., Dalton Trans.* **1999**, 1533–1534.
779. Gilman, H.; Abbott, R. K. Jr. *J. Am. Chem. Soc.* **1949**, *71*, 659–660.
780. Labahn, D.; Pohl, E.; Herbst-Irmer, R.; Stalke, D.; Roesky, H. W.; Sheldrick, G. M. *Chem. Ber.* **1991**, *124*, 1127–1129.
781. Wojnowski, W.; Peters, K.; Peters, E. M.; Von Schnering, H. G. *Z. Anorg. Allg. Chem.* **1985**, *531*, 147–152.
782. Dhingra, S. S.; Kanatzidis, M. G. *Inorg. Chem.* **1993**, *32*, 2298–2307.
783. Clark, R. E. D. *Analyst* **1957**, *82*, 177–182.
784. Garcia-Tasende, M. S.; Suarez, M. I.; Sanchez, A.; Casas, J. S.; Sordo, J.; Castellano, E. E.; Mascarenhas, Y. P. *Inorg. Chem.* **1987**, *26*, 3818–3820.
785. Castano, M. V.; Macias, A.; Castineiras, A.; Sanchez Gonzalez, A.; Garcia Martinez, E.; Casas, J. S.; Sordo, J.; Hiller, W.; Castellano, E. E. *J. Chem. Soc., Dalton Trans.* **1990**, 1001–1005.
786. Garcia Bugarin, M.; Casas, J. S.; Sordo, J.; Filella, M. *J. Inorg. Biochem.* **1989**, *35*, 95–105.
787. Bosch, B. E.; Eisenhawer, M.; Kersting, B.; Kirschbaum, K.; Krebs, B.; Giolando, D. M. *Inorg. Chem.* **1996**, *35*, 6599–6605 and references therein.
788. Spence, M. A.; Rosair, G. M.; Lindsell, W. E. *J. Chem. Soc., Dalton Trans.* **1998**, 1581–1586.
789. Kimblin, C.; Bridgewater, B. M.; Hascall, T.; Parkin, G. *J. Chem. Soc., Dalton Trans.* **2000**, 891–897.
790. Ojo, J. F.; Slavin, P. A.; Reglinski, J.; Garner, M.; Spicer, M. D.; Kennedy, A. R.; Teat, S. J. *Inorg. Chim. Acta* **2001**, *313*, 15–20.
791. Reglinski, J.; Garner, M.; Cassidy, I. D.; Slavin, P. A.; Spicer, M. D.; Armstrong, D. R. *J. Chem. Soc., Dalton Trans.* **1999**, 2119–2126.
792. Schebler, P. J.; Riordan, C. G.; Guzei, I. A.; Rheingold, A. L. *Inorg. Chem.* **1998**, *37*, 4754–4755.
793. Esperas, S.; Husebye, S. *Acta Chem. Scand., Ser. A* **1974**, *A28*, 1015–1020.
794. Nilson, L.; Hesse, R. *Acta Chem. Scand.* **1951**, *23*, 1020–1965.
795. Elfwing, E.; Anacker-Eickhoff, H.; Jennische, P.; Hesse, R. *Acta Chem. Scand., Ser. A* **1976**, *A30*, 335–339.
796. Anacker-Eickhoff, H.; Jennische, P.; Hesse, R. *Acta Chem. Scand., Ser. A* **1975**, *A29*, 51–59.
797. Jennische, P.; Hesse, R. *Acta Chem. Scand.* **1973**, *27*, 3531–3544.
798. Jennische, P.; Olin, A.; Hesse, R. *Acta Chem. Scand.* **1972**, *26*, 2799–2812.
799. Hong, S.-H.; Jennische, P. *Acta Chem. Scand., Ser. A* **1978**, *A32*, 313–318.
800. Campbell, J.; Mercier, H. P. A.; Santry, D. P.; Suontamo, R. J.; Borrmann, H.; Schrobilgen, G. J. *Inorg. Chem.* **2001**, *40*, 233–254.
801. Borrmann, H.; Campbell, J.; Dixon, D. A.; Mercier, H. P. A.; Pirani, A. M.; Schrobilgen, G. J. *Inorg. Chem.* **1998**, *37*, 1929–1943.
802. Burns, R. C.; Corbett, J. D. *J. Am. Chem. Soc.* **1981**, *103*, 2627–2632.
803. Zhou, M.; Xu, Y.; Koh, L. L.; Mok, K. F.; Leung, P. H.; Hor, T. S. A. *Inorg. Chem.* **1993**, *32*, 1875–1876.
804. Wang, S.; Garzon, G.; King, C.; Wang, J. C.; Fackler, J. P. Jr. *Inorg. Chem.* **1989**, *28*, 4623–4629.
805. Kumar, R.; Tuck, D. G. *Can. J. Chem.* **1989**, *67*, 127–129.
806. Esperas, S.; Husebye, S. *Acta Chem. Scand.* **1973**, *27*, 1827–1828.
807. Esperas, S.; Husebye, S. *Acta Chem. Scand.* **1973**, *27*, 3355–3364.
808. Hummel, H. U.; Fischer, E.; Fischer, T.; Gruss, D.; Franke, A.; Dietzsch, W. *Chem. Ber.* **1992**, *125*, 1565–1570.
809. Bensch, W.; Schuster, M. *Z. Anorg. Allg. Chem.* **1993**, *619*, 1689–1692.
810. Hughes, R. P.; Lindner, D. C.; Rheingold, A. L.; Yap, G. P. A. *Inorg. Chem.* **1997**, *36*, 1726–1727.
811. Van Seggen, D. M.; Hurlburt, P. K.; Noirot, M. D.; Anderson, O. P.; Strauss, S. H. *Inorg. Chem.* **1992**, *31*, 1423–1430.
812. Barbarich, T. J.; Miller, S. M.; Anderson, O. P.; Strauss, S. H. *J. Mol. Catal. A: Chem.* **1998**, *128*, 289–331.
813. Hurlburt, P. K.; Anderson, O. P.; Strauss, S. H. *Can. J. Chem.* **1992**, *70*, 726–731.
814. Samuels, J. A.; Lobkovsky, E. B.; Streib, W. E.; Folting, K.; Huffman, J. C.; Zwanziger, J. W.; Caulton, K. G. *J. Am. Chem. Soc.* **1993**, *115*, 5093–5094.
815. Samuels, J. A.; Zwanziger, J. W.; Lobkovsky, E. B.; Caulton, K. G. *Inorg. Chem.* **1992**, *31*, 4046–4047.
816. Hughes, R. P.; Husebo, T. L.; Maddock, S. M.; Rheingold, A. L.; Guzei, I. A. *J. Am. Chem. Soc.* **1997**, *119*, 10231–10232.
817. Bianchini, C.; Masi, D.; Linn, K.; Mealli, C.; Peruzzini, M.; Zanobini, F. *Inorg. Chem.* **1992**, *31*, 4036–4037.

3.6
Arsenic, Antimony, and Bismuth

W. LEVASON and G. REID
University of Southampton, Southampton, UK

3.6.1 INTRODUCTION

This chapter deals with the coordination chemistry of the three heaviest elements of group 15, specifically arsenic, antimony, and bismuth. We have followed a working definition of coordination complexes as those containing group 15 compounds behaving as Lewis acids to either neutral or charged donor ligands and have not included simple halides, oxides, etc. which fall into the wider area of the inorganic chemistry of these elements. The distinction in some cases is not clear-cut and a pragmatic approach has been adopted, with borderline cases usually included.

In this section some general points about the area are made and previous major literature sources, including books and review articles dealing with all three elements are listed. Reviews dealing with only one element or particular ligand types are referred to in the appropriate sections below.

All three elements have long been known to chemists, despite their rarity in the Earth's crust (As ca. 1.8 ppm, Sb 0.02 ppm, and Bi 0.008 ppm), which places bismuth similar in abundance to Pt or Au.[1] The common oxidation states (III and V) are shared by all the group 15 elements, but apart from similar stoichiometries there is little resemblance between the properties of nitrogen compounds and those of the three heaviest elements.[2,3] The resemblance of the latter to phosphorus is closer, although trends down the group are irregular, resulting in a rich and diverse chemistry. Examples of the irregularities are the reduced stability of the V oxidation state for As and Bi compared with P and Sb (attributed respectively to the effects of insertion of the $3d$ elements and of the lanthanides), and in the electronegativities, which on the Allred-Rochow scale fall $N > As > P > Sb > Bi$, although the Pauling scale is more regular $N > P \geq As > Sb > Bi$.[1]

There are a number of compounds containing homoatomic bonds that fall outside the formal oxidation state III or V classification, e.g., Zintl anions and homoatomic cations of Sb or Bi, but these have little or no coordination chemistry.

A good general survey of the properties of As, Sb, and Bi is given by Carmalt and Norman.[4] Both common oxidation states exhibit Lewis acidity (and the III state also shows Lewis basicity) in both cases resulting in complexes in which the group 15 atom's outer electron count exceeds an octet. Although the traditional bonding model in such complexes invoked d-orbitals, the bonds are now more usually described in terms of delocalized 3-center-4-electron bonds, based upon s- and p-orbitals. For the Lewis acid complexes of the III oxidation state, the concept of primary E—X bonding in the parent Lewis acid (EX_3, $E = As$, Sb, or Bi; $X = $ halide) and secondary bonds to the Lewis base, utilizing E–X σ^* as acceptor orbitals, is a useful approach.[4] The primary/secondary bonding and the 3-center-4-electron bonds are not distinct models but may be shown to be variants of the same basic model.[5] The best description of this approach is in the review by Carmalt and Norman.[4]

There is an extensive literature on the chemistry of As, Sb, and Bi: in addition to recent editions of standard textbooks,[2,3] there is a book edited by Norman,[6] which includes a chapter[7] devoted to the coordination chemistry of these elements, as well as the article by McAuliffe in *Comprehensive Coordination Chemistry* (*CCC*, 1987).[8] The vast organic chemistry of these elements falls outside the scope of the present chapter. Sources providing recent coverage of the organic chemistry include chapters by Wardell,[9,10] in *Comprehensive Organometallic Chemistry I* and *II*, a volume in the Patai series (*The Chemistry of Organic Arsenic, Antimony and Bismuth Compounds*),[11] and chapters in Norman's book.[6] These texts also list many reviews on specific classes of organoderivatives.

Studies of the Lewis base complexes of ER_3 compounds remains an active area of modern coordination chemistry, and we have described recent developments in the synthesis of arsine, stibine, and bismuthine ligands elsewhere in the present work (see Chapter 1.16).

The classification adopted in this chapter is generally based upon the Periodic group of the donor atoms, describing sequentially complexes formed by the lightest through the heaviest donor ligands, with neutral donor ligands preceding charged anions, and with the E^{III} oxidation state complexes described before those of E^V. Mixed donor ligands create some problems, and the approach adopted has been pragmatic, including them with the nearest analogues. It is hoped that this will not prove problematic in practice. Finally, we have used the convention where the term "*pseudo*," prefixing the polyhedral geometry, refers to the overall geometry at the group 15 ion including a lone pair, i.e., *pseudo*-octahedral refers to a molecule with five ligands around the central ion and one lone pair, those six units being disposed in an octahedral array.

3.6.2 ARSENIC

In addition to the reviews listed in Section 3.6.1, there is an article detailing the structures, properties and bonding of penta-coordinate arsenic compounds.[12]

3.6.2.1 Group 14 Compounds

The conventional As^{III} and As^V compounds have As—C single bonds and are based upon AsR_3 and AsR_5 and their substituted variants with halogens, oxide, etc. In addition, there have been significant recent developments in the chemistry of compounds containing homo-element bonds

$(RAs)_n$ and multiple bonds to carbon such as $RC{\equiv}As$, $R^1_2C{=}AsR^2$, $RAs{=}AsR$, $R^1_3As{=}CR^2_2$, pyridine (C_5H_5As), and pyrrole (C_4H_4AsR) analogues and their transition metal complexes.[13–15]

Examples of bonds between arsenic and the heavier elements of group 14 are mostly of the type $AsR^1_{3-n}(YR^2_3)_n$ ($Y = Si$, Ge, Sn, or Pb) which are analogues of triorganoarsines and usually included in treatments of organoarsenic derivatives.[11]

3.6.2.2 Group 15 Compounds

The commonest examples of As—N bonds are found in aminoarsines, $As(NR_2)_3$, usually discussed together with organoarsines,[11] and arsenic complexes with nitrogen Lewis bases are few. The reaction of molten AsX_3 (X = Br or I) with NH_3, or heating As_2O_3 with the appropriate NH_4X gave $[AsX_3(NH_3)]$;[16] the $AsCl_3$ complex was not obtained although it has been described in older literature. The $[AsX_3(NMe_3)]$ (X = Cl or Br) formed by direct combination of AsX_3 and NMe_3 in the absence of solvent have a *pseudo*-trigonal bipyramidal geometry, with axial amine and with one equatorial position occupied by the lone pair.[17] The macrocycle 1,4,7-trimethyl-1,4,7-triazacyclononane ($Me_3[9]aneN_3$) reacts with $AsCl_3$ in MeCN to give white $[AsCl_3(Me_3[9]-aneN_3)]$ which probably has a half-sandwich structure.[18] Controlled hydrolysis in MeCN solution produces $[AsCl_2(Me_3[9]aneN_3)]^+$ isolated and structurally characterized as a salt with the unusual $[As_2OCl_5]^-$ anion. The cation has a *pseudo*-octahedral geometry with the lone pair occupying one vertex *trans* to N. In contrast, the simple adducts formed from $AsCl_3$ and $Me_4[14]aneN_4$ have not been characterized, although their hydrolysis products, including $[H_2Me_4[14]aneN_4][As_4O_2Cl_{10}]$, have been characterized structurally.[19]

The pincer anion $[2,6-(Me_2NCH_2)_2C_6H_3]^-$ (**1**) reacts with $AsCl_3$ to give colorless $[AsCl_2(\mathbf{1})]$, which probably has a square-pyramidal geometry (cf., the Sb analogue) with a N_2CCl_2 donor set.[20] Reduction of this complex with $LiAlH_4$ produces $[AsH_2(\mathbf{1})]$ a colorless, distillable liquid. In contrast, the 1:1 complex of (**2**) with $AsCl_3$ has a structure based upon a trigonal bipyramid with a vacant equatorial site and axial NMe_2 and Cl groups.[21] Using the ligands (**2**) and (**3**) (L) it is possible to isolate complexes $[AsL_3]$ which contain trigonal pyramidal AsC_3 skeletons, with As—C bond lengths typical of single bonds, and longer secondary As—N interactions completing a distorted octahedron (Figure 1).[22]

(1)   **(2)**   **(3)**

The instability of $AsCl_5$ would seem to preclude an extensive coordination chemistry, but AsF_5 is a very strong Lewis acid. Many of its reported reactions involve abstraction of fluoride from main group fluorides or oxide-fluorides to form cationic derivatives as $[AsF_6]^-$ salts (q.v.). Simple *N*-base adducts are more rare. The simplest, $[AsF_5(NH_3)]$, is formed from the constituents in solution in liquid SO_2, and is quantitatively converted into $[NH_4][AsF_6]$ by HF.[23] $[AsF_5(NHEt_2)]$ has been detected in solution in MeCN by ^{19}F NMR spectroscopy.[24] Klapötke and co-workers[23,25–27] have reported 1:1 adducts of AsF_5 with MeCN, pyridine, H_2NCN, C_2N_2, HCN, FCN, ClCN, BrCN, ICN, $CH_2(CN)_2$, and $CCl_2(CN)_2$; only $CH_2(CN)_2$ appears to form a $[L(AsF_5)_2]$ adduct. The complexes were characterized by analysis, vibrational, and NMR spectroscopy. In contrast, triazine ($C_3N_3H_3$) forms adducts $[(C_3N_3H_3)(AsF_5)_n]$ ($n = 1$, 2, or 3)[28] and the structure of the complex with $n = 1$ shows the expected six-coordinate As bonded to one nitrogen of the triazine ring. Other structurally characterized examples of *N*-coordinated ligand adducts are $[AsF_5L]$ (L = MeSCN,[29] NMe_2SOF_2,[30] benzo-2,1,3-thiadiazole, and benzo-1,2,3-thiadiazole[31]).

Highly explosive azides of As^{III} and As^V of types $[As(N_3)_4]^-$, $[As(N_3)_6]^-$, and $[As(N_3)_4]^+$ are obtained by reaction of the appropriate arsenic halide with NaN_3 or $TMSN_3$. Although $[As(N_3)_3]$ is known, attempts to isolate $[As(N_3)_5]$ have failed.[32,33]

The reactions of AsX_3 with PR_3 or AsR_3 were first examined many years ago and, depending upon the Lewis acid–Lewis base combination and the reaction conditions, were reported to

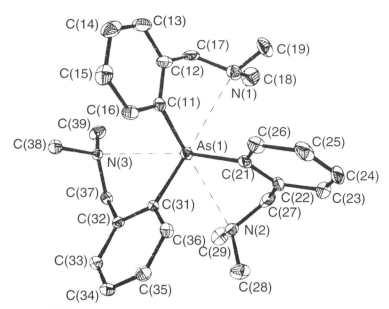

Figure 1   The structure of [As(2-Me$_2$NCH$_2$C$_6$H$_4$)$_3$] (reproduced by permission of the American Chemical Society from *Inorg. Chem.* **1996**, *35*, 6179–6183).

produce adducts [AsX$_3$L$_n$] (X = Cl, Br, or I, L = PR$_3$ or AsR$_3$, $n = 1$, (rarely) 2) or salts [R$_3$EAsX$_2$]X (E = P or As, R = alkyl or aryl). In contrast, SbR$_3$ usually caused reduction-forming SbR$_3$X$_2$.[34] In some cases RAsX$_2$ behaved similarly, but generally no reaction occurred with R$_2$AsX, consistent with reduced Lewis acidity as X was replaced by R. Reinvestigation of the reactions of AsX$_3$ (X = Cl, Br, or I) with PMe$_3$ (X ≠ Cl) or AsMe$_3$ in dry CH$_2$Cl$_2$,[35] and of AsCl$_3$ with AsEt$_3$,[34] found 1:1 adducts were formed irrespective of the ratio of the reactants. However, in the AsCl$_3$-PMe$_3$ system both 1:1 and 1:2 adducts could be isolated depending upon the conditions.[35] The X-ray structure of [{AsCl$_3$(AsEt$_3$)}$_2$] shows a dimer with asymmetric chlorine bridges and axial *anti*-AsEt$_3$ groups,[36] and the other 1:1 adducts are likely to be similar. The structure of [AsCl$_3$(PMe$_3$)] is based upon a similar dimer unit, but the lattice shows two crystallographically independent units, one with five-coordinate As, the second with [5 + 1]–coordination due to long-range interdimer Cl···As interactions (Figure 2).[35] Diphosphines and diarsines including *o*-C$_6$H$_4$(PMe$_2$)$_2$, *o*-C$_6$H$_4$(PPh$_2$)$_2$, and *o*-C$_6$H$_4$(AsMe$_2$)$_2$, and the triarsine MeC(CH$_2$AsMe$_2$)$_3$ also form 1:1 adducts with AsX$_3$.[35] The X-ray structures of [{AsX$_3$(*o*-C$_6$H$_4$(AsMe$_2$)$_2$}$_2$] (X = Br or I) show dimeric units with asymmetric dihalo-bridges (Figure 3).[35]

3.6.2.3   Group 16 Compounds

Arsenic(III) has a considerable affinity for charged O- or S-donor ligands, the latter including dithioacid chelates, but complexes with neutral O, S, or Se donor ligands are much rarer. Here complexes of neutral ligands are discussed first and then complexes with charged anions.

Tetrahydrofuran complexes [Ph$_4$P][AsX$_4$(THF)$_2$] (X = Cl or Br) were the unexpected major products of photolysis of W(CO)$_6$ + [Ph$_4$P]$_2$[As$_2$Cl$_8$] and Cr(CO)$_6$ + [Ph$_4$P][As$_2$SBr$_5$] in THF.[37] Both are regular octahedral anions with *trans*-THF ligands and hence a stereochemically inactive lone pair—although the As—O and As—X bonds are long, probably attributable to the effect of the lone pair. Crown ether adducts are also characterized by unusually long As—O bonds. The known examples are [AsCl$_3$(12-crown-4)][38] and [AsX$_3$(15-crown-5)] (X = Cl, Br, or I),[38,39] but 18-crown-6 failed to give a pure complex. The structures retain the pyramidal AsX$_3$ unit of the parent halide capped by the crown ether oxygens giving, respectively, seven- and eight-coordinate As, but with very long As—O bonds (Figure 4).

Arsenic(III) halides function as very weak Lewis acids towards thio- or seleno-ethers (no telluroether adducts are known). The products are hydrolytically unstable and extensively dissociated in solution. In all these complexes the As—S(Se) bonds are very long, indicative of weak,

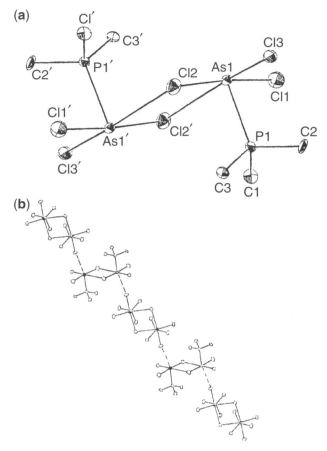

Figure 2 The structure of [AsCl$_3$(PMe$_3$)] and the packing showing the intermolecular interactions (reproduced by permission of the Royal Society of Chemistry from *J. Chem. Soc., Dalton Trans.* **2002**, 1188–1192).

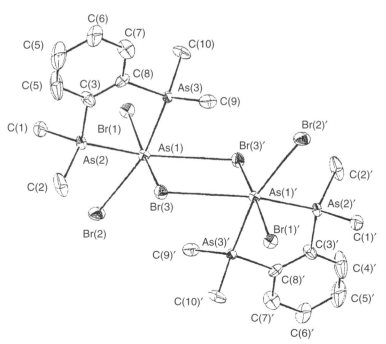

Figure 3 The structure of [AsBr$_3${*o*-C$_6$H$_4$(AsMe$_2$)$_2$}] (reproduced by permission of the Royal Society of Chemistry from *J. Chem. Soc., Dalton Trans.* **2002**, 1188–1192).

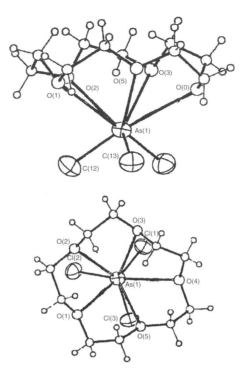

Figure 4 The structure of [AsCl₃(15-crown-5)] (reproduced by permission of the International Union of Crystallography from *Acta Crystallogr., Sect. B* **1993**, *49*, 507–514).

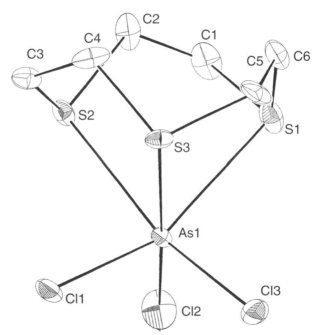

Figure 5 The structure of [AsCl₃{[9]aneS₃}] (reproduced by permission of the American Chemical Society from *Inorg. Chem.* **2002**, *41*, 2070–2076).

secondary interactions. Thus, AsX₃ (X = Br or I) react with MeSCH₂CH₂SMe in dry CH₂Cl₂ to form 1:1 adducts which are dihalo-bridged dimers with distorted octahedral arsenic coordinated to a chelating dithioether and two terminal and two asymmetrically bridging halides.[40] In contrast, the [AsX₃([9]aneS₃)] (X = Cl, Br, or I) are monomeric; the structure of the distorted octahedral chloride is shown in Figure 5.[40] The 1:1 complex with the tetrathioether macrocycle

[14]aneS$_4$, [AsCl$_3${[14]aneS$_4$}], is completely different, based upon six-coordinate As coordinated to two S atoms of different thioethers, two terminal and two bridging Cl (to different As), which produces an infinite sheet polymer.[40] Among the products of the reaction of AsI$_3$ with thioacetic acid were orange–red crystals shown by an X-ray structure determination to be the monomeric 1:1 AsI$_3$ adduct of 1,3,5,7-(tetramethyl)-2,4,6,8,9,10-(hexathia) adamantane, in which the pyramidal AsI$_3$ group is weakly bonded to three sulfurs in the adamantane (Figure 6).[41]

In contrast to SbX$_3$ or BiX$_3$ (q.v.), under similar reaction conditions, AsX$_3$ fail to give complexes with acyclic selenoethers such as MeSeCH$_2$CH$_2$SeMe or MeC(CH$_2$SeMe)$_3$. However, macrocyclic selenoethers are more effective ligands affording [AsX$_3$([8]aneSe$_2$)], [(AsX$_3$)$_2$([16]-aneSe$_4$)] (X = Cl, Br, or I), [(AsCl$_3$)$_4$([24]aneSe$_6$)], and [(AsBr$_3$)$_2$([24]aneSe$_6$)].[40,42] The [16]aneSe$_4$ complexes contain asymmetric dihalo-bridged As$_2$X$_6$ units linked into 3-D polymers by the tetraselenoether, with each Se bonded to a different As center. The unique structure of [(AsCl$_3$)$_4$([24]aneSe$_6$)] (Figure 7) shows a weakly associated As$_2$Cl$_6$ unit *endo* to the ring where it is coordinated to four seleniums (two per As), whilst the other two seleniums coordinate *exo* to pyramidal AsCl$_3$ groups which have a *pseudo*-trigonal bipyramidal geometry due to the stereo-chemically active lone pair.[42]

The only thioether complex with As^V is [AsF$_5$(Me$_2$S)], an involatile white solid made from Me$_2$S and AsF$_5$ at low temperature, although its properties were not described.[43]

Arsenic compounds with charged O-donor ligands include such diverse species as arsenite and arsenate esters, spiroarsoranes,[44] and arsenic carboxylates. Recent examples include the tetrahalocatecholate derivatives (**4**),[45] the triethanolamine derived (**5**),[46] and 2-Cl-4,4,6,6-tetra-methyl-1,3,2-dioxarsenane.[47] The reaction of As(NMe$_2$)$_3$ with *p*-Rcalix[4]arenes (R = Bu^t or H) gives mono- or di-arsenic derivatives (**6**) which have been structurally characterized.[48] In the presence of moisture the oxo-bridged (**7**) is formed. *N*-coordinated base adducts of cyclic arsenites are known with 8-hydroxyquinolate(1−) (**8**), formed from [ClAs(O-R-O)] (R = CH$_2$CMe$_2$CH$_2$, 2,2′-C$_6$H$_4$OC$_6$H$_4$, (Bu^t)$_2$C$_6$H$_2$CH$_2$C$_6$H$_2$(Bu^t)$_2$) and 8-hydroxyquinoline in the presence of base.[49–51] The reaction of [ClAs(OCH$_2$CMe$_2$CH$_2$O)] with Me$_2$C(CH$_2$OH)$_2$ and base gives the As^V compound (**9**), which also reacts with 8-hydroxyquinoline to give (**10**).[49] Organoarsenic(V) compounds of structure type (**9**), with R groups replacing the apical Cl,

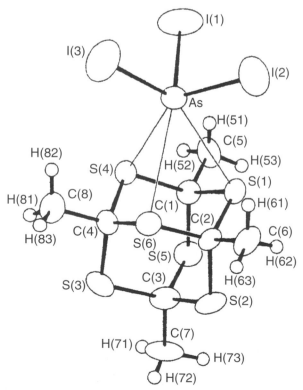

Figure 6 The structure of [AsI$_3$(1,3,5,7-Me$_4$-2,4,6,8,9,10-(hexathia)adamantane)] (reproduced by permission of Elsevier Science from *Inorg. Chim. Acta* **1982**, *64*, L83–L84).

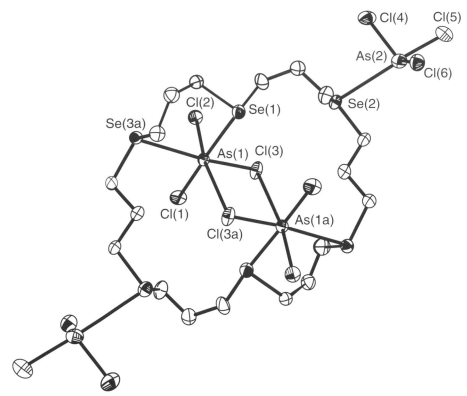

Figure 7 The structure of [(AsCl$_3$)$_4${[24]aneSe$_6$}] (reproduced by permission of the American Chemical Society from *J. Am. Chem. Soc.* **2001**, *123*, 11801–11802).

have been studied and the effects of substituents upon the distorted five-coordinate structures examined.[52] The structure of the tris(catecholato)arsenate(V) anion (as the [H$_7$O$_3$]$^+$ salt) has been determined.[53]

The reaction of acetic anhydride with As$_2$O$_3$ produces both [As(O$_2$CMe)$_3$] and [As$_2$O(O$_2$CMe)$_4$]. The arsenic environment in each is based upon a pyramidal AsO$_3$ core with longer As$\cdots$O contacts completing a distorted octahedron or distorted square pyramid, respectively.[54] Arsenic α-hydroxycarboxylates have more complicated structures. The arsenic(III)-tartaric acid complexes, related to the important tartrato-antimonates, have structures based upon the dimer unit (**11**) with a distorted *pseudo*-trigonal bipyramidal geometry at As with the lone pair equatorially disposed. In [Na$_8$As$_{10}$(C$_4$H$_2$O$_6$)$_8$(C$_4$H$_3$O$_6$)$_2$(H$_2$O)$_{19}$] the dimer units (**11**) are linked via Na$^+$ cations into a complex polymeric network.[55,56] In related silver salts [Ag$_9$As$_{10}$(C$_4$H$_2$O$_6$)$_9$(C$_4$H$_3$O$_6$)(H$_4$-As$_2$O$_5$)(H$_2$O)$_{10}$] and [Ag$_5$As$_4$(C$_4$H$_2$O$_6$)$_4$(H$_2$O)$_5$Y] (Y = NO$_3$ or ClO$_4$) the structures are based upon the dimer units but As—Ag bonds are also present.[55,56] 1,2-Dihydroxycyclohexane-1,2-dicarboxylic acid also forms an As^{III} complex based upon a similar dimer unit, and ^{13}C NMR spectroscopy shows the expected stereoisomers and some dissociation of the carboxylate from the arsenic in solution.[57]

Methanesulfonate complexes of As^{III} include [As(MeSO$_3$)$_3$], [As(MeSO$_3$)$_4$]$^-$, [AsO(MeSO$_3$)$_2$]$^-$, and [AsO(MeSO$_3$)].[58] The [As(MeSO$_3$)$_3$] forms adducts with pyridine or N,N-DMF. Mixed fluoride–fluorosulfonates of both As^{III} and As^V are known, including [AsF$_n$(SO$_3$F)$_{5-n}$] ($n = 2$–4).[59]

Arsenic(III) has great affinity for anionic sulfur chelates including xanthate (ROCS$_2$$^-$), dithiocarbamate (R$_2NCS_2$$^-$), and dithiophosphate ((RO)$_2$PS$_2$$^-$), and this area has been reviewed recently.[60] In contrast to antimony which forms compounds in oxidation state III and V, arsenic(V) is reduced by these ligands. The most popular synthetic route is reaction of AsCl$_3$ with the sodium, or sometimes the ammonium, salt of the dithioacid, although reactions of As(OR)$_3$ or As$_2$O$_3$ with the dithiophosphorus acid have also been used.[60] Replacement of AsCl$_3$ by RAsCl$_2$ or R$_2$AsCl results in the corresponding organoarsenic derivatives. The complexes synthesized are listed in Table 1 and here we discuss various points of interest.

In the [As(S$_2$COR)$_3$] (R = Me, Et, Pr^i, CH$_2$CH$_2$CMe$_3$) complexes distorted six-coordinate As is present and within each xanthate ligand there is one short and one longer As—S bond.[61,62,64,65] The distortion is greater in As compared to Sb or Bi analogues, consistent with greater stereochemical effect of the lone pair.

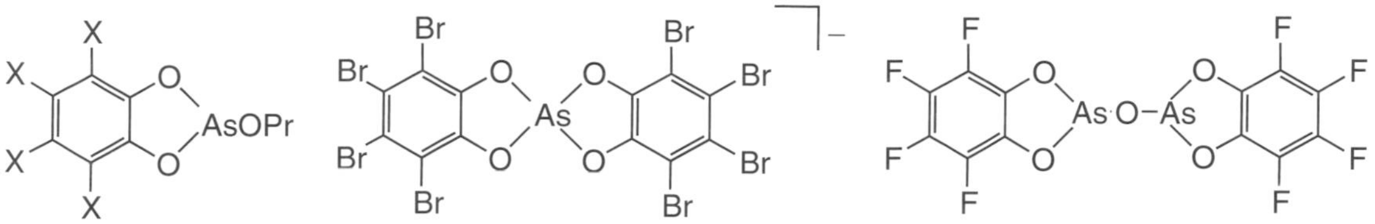

(4)

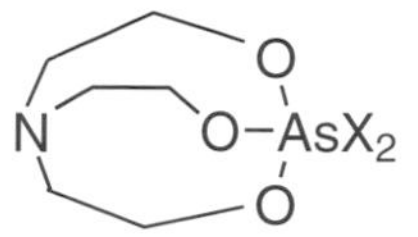

X = Cl, Br, 0.5 O

(5)   **(6)** 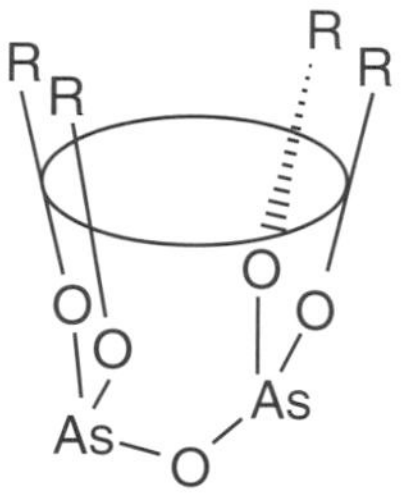   **(7)**

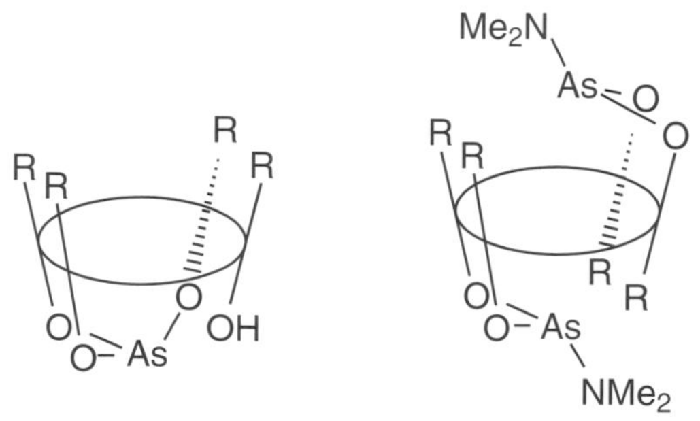

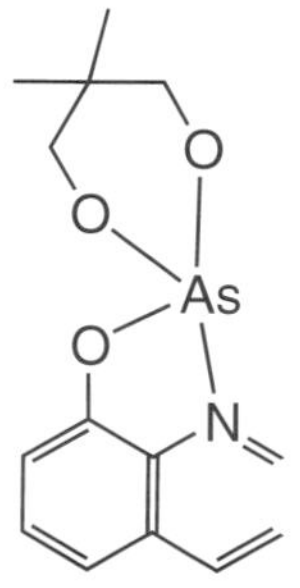

(8) 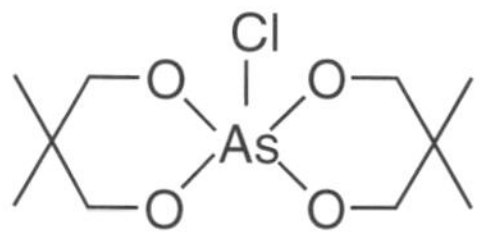   **(9)**

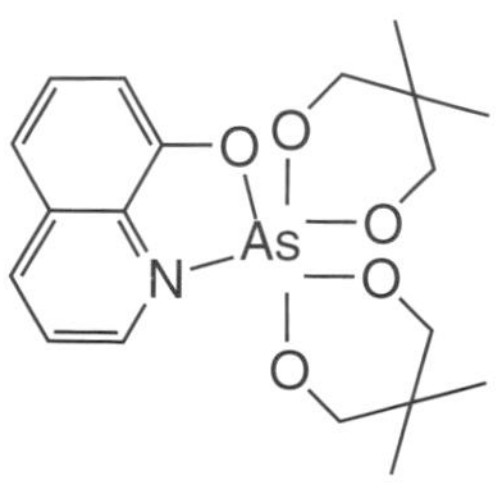

(10) 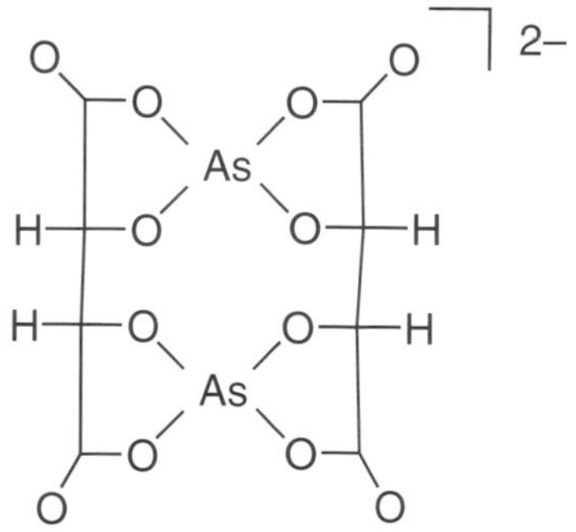   **(11)**

The arsenic dithiocarbamates also show distorted structures: in [As(S$_2$CN(Me)CH$_2$CH$_2$OH))$_3$] the structure is based on a distorted octahedron,[71] whereas in [As(S$_2$CN(CH$_2$CH$_2$OH)$_2$)$_3$] the geometry is a distorted trigonal prism (Figure 8) with three short As—S bonds (2.34(2) Å) and three much longer As—S interactions (2.84(2) Å).[74] The effect of the group 15 acceptor is also marked—the antimony analogue is best described as distorted pentagonal pyramidal (q.v.).[74] In

Table 1 Dithioacid compounds of arsenic(III).

Compound	Comments	References
$As(S_2COR)_3$	$R = Me, Et, Pr^i, CH_2CH_2CMe_3$	61,62,64–66
$AsPh(S_2COPr^i)_2$		67
$As(SCH_2CH_2S)(S_2COR)$	$R = Et, Pr, Pr^i, Bu, Bu^i$	68
$As(S_2CNR)_3$	$R = CHMeCH_2CH_2CH_2CH_2,$	69
	$CH_2CHMeCH_2CH_2CH_2,$	
	$CH_2CH_2CHMeCH_2CH_2$	
$As(S_2CNR)_3$	$R = 2\text{-alkylaminocyclopentene}$	70
$As(S_2CNR^1R^2)_3$	$R^1 = Me, R^2 = CH_2CH_2OH$	71
$AsBr(S_2CNEt_2)_2$		72
$As(SCH_2CH_2S)(S_2CNR_2)$	$R_2 = \text{pyrrolidyl, 4-morphoyl}$	73
$As(S_2CNR_2)_3$	$R = Et, N\text{-Methylaminoethanol},$	74
	$N,N'\text{-iminodiethanol}$	
$As(S_2CNR_2)_3$	$R = CH_2CH_2OH$	64
$AsR^2(S_2CNEt_2)_2$	$R^2 = Ph, Me$	75
$As(S_2CNPr^i_2)_3$		76
$As(\text{dithiolate})(S_2CNR_2)$	$R^1 = Et, R^2 = CH_2CH_2OCH_2CH_2$	77
$As(SCH_2CH_2S)(S_2CNR^1_2)$	$R^1 = Me_2, Et_2, CH_2CH_2CH_2CH_2$	68
$AsX(S_2CNMe_2)_2$	$X = Cl, Br, I$	78
$As[O(C_6H_4)_2](S_2CN(CH_2CH_2)_2)$		79
$As[S_2P(OR)_2]_3$	$R = Et, Pr^n, Pr^i, Bu^i, Ph$	80
$AsCl(S_2P(OR)_2)_2$	$R = Et, Pr^n, Pr^i, Bu^i$	80
$AsCl_2(S_2P(OR)_2)$	$R = Et, Pr^n, Pr^i, Bu^i$	80
$As(S_2P(O\text{-R-O}))_3$	$R = CHMeCHMe, CMe_2CMe_2,$	81
	$CMe_2CH_2CHMe, CH_2CMe_2CH_2,$	
	$CH_2CEt_2CH_2$	
$AsCl_{3-n}(S_2P(OCHMeCHMeO))_n$	$n = 1 \text{ or } 2$	81
$AsPh(S_2P(OR)_2)_2$	$R = Et, Pr^n, Pr^i, Ph$	82,83
$AsPh(S_2P(O\text{-R-O}))_2$	$R = CH_2CMe_2CH_2, CMe_2CMe_2,$	
	$CMe_2CH_2CMe_2,$	84
	$CMe_2CH_2CHMe, CHMeCHMe$	
$As\{O(C_6H_4)_2\}(S_2PR_2)$	$R = Me, Et, Ph$	85
$AsR^2(S_2PPh_2)$	$R^2 = Me, Ph$	86
$As(SCH_2CH_2S)(S_2P(O\text{-R-O}))$	$R = CH_2CMe_2CH_2, CH_2CEt_2CH_2$	87
$As\{Y(CH_2CH_2S)_2\}(S_2PR_2)$	$R = Me, Et, Ph, Y = O \text{ or } S$	88,89

mixed ligand complexes different motifs are found: in $[As(SCH_2CH_2S)(S_2CN\text{-morphyl-4})]$ the dithiocarbamate coordination is essentially monodentate,[73] whereas in $[AsPh(S_2CNEt_2)_2]$ or $[AsMe(S_2CNEt_2)_2]$ there are three short bonds (one As—C and two As—S) with two much longer As$\cdots$S interactions (Figure 9).[75]

Dithiophosphate and dithiophosphinate complexes of arsenic are listed in Table 1. Spectroscopic data have been reported for many examples but structural data are more rare. In $[AsPh\{S_2P(OPr^i)_2\}_2]$ the structure is an approximate square pyramid with an apical Ph group and very asymmetric chelation by the dithiophosphates, in which the As—S bonds differ by 0.082 Å (Figure 10).[82] The arsocane dithiophosphinates $[\{Y(CH_2CH_2S)_2\}As(S_2PPh_2)]$ (Y = O or S) show very distorted five-coordination with primary bonds to the thiolate sulfurs and to one sulfur of the dithiophosphinate, with weaker interactions to the second sulfur in the dithiophosphinate and an endocyclic *trans*-annular interaction to the O or S of the ring (Figure 11).[88,89] In addition to the mixed species noted above, simple dithiolate complexes are also known including $[AsCl(tdt)]$ (tdt = toluenedithiolate(2−)). The structure of the latter reveals an essentially three-coordinate pyramidal As center coordinated to a chelating dithiolate and a single chlorine, and the stacking appears to involve weak As$\cdots$Ph contacts.[90] In contrast to Sb or Bi, arsenic does not appear to form complexes with a higher ligand:metal ratio.

3.6.2.4 Group 17 Compounds

Haloarsenic anions are known in both III and V oxidation states, although the structural diversity is less than in the antimony and bismuth analogues. For the As^{III} species a variety of stoichiometries

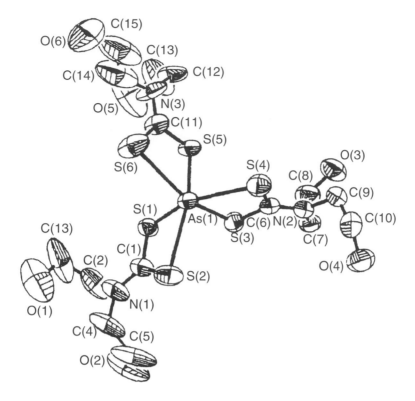

Figure 8 The structure of [As{S$_2$CN(CH$_2$CH$_2$OH)$_2$}$_3$] (reproduced by permission of Elsevier Science from *Polyhedron* **1997**, *16*, 1211–1221).

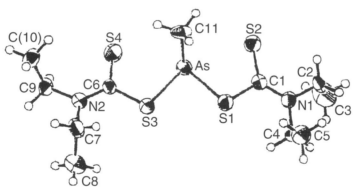

Figure 9 The structure of [AsMe(S$_2$CNEt$_2$)$_2$] (reproduced by permission of Elsevier Science from *J. Organomet. Chem.* **1997**, *538*, 129–134).

have been identified; it is also clear that the structural units present cannot be deduced simply from the stoichiometries.[91] In marked contrast to the As^V fluoroanions, those of As^{III} have been studied little. The structure of [AsF$_4$]$^-$ has been determined (as the hexamethylpiperidinium salt) and shows the expected bisphenoidal (SF$_4$-like) geometry, with longer axial (1.862(2), 1.878(8) Å) than equatorial (1.724(2), 1.727(2) Å) bonds.[92] The structure present in K$_2$As$_2$F$_7$ consists of [AsF$_4$]$^-$ anions weakly associated with AsF$_3$ molecules.[93]

The syntheses of the heavier haloanions are from AsX$_3$, X$^-$, and an appropriate cation, and the major feature of interest is the structural units present. The simplest stoichiometry is [AsX$_4$]$^-$ known for X = Cl, Br, or I, none of which contain monomeric anions. The chloro- and bromo-compounds are dimeric [As$_2$X$_8$]$^{2-}$ with edge-shared square pyramidal units with *anti*-apical halides and relatively symmetrical bridges. X-ray structures are available for [NPhMeH$_2$]$_2$ [As$_2$Cl$_8$],[94] [Ph$_4$P]$_2$[As$_2$Cl$_8$],[95] [Ph$_4$P]$_2$[As$_2$Br$_8$],[96] [Pr$_4$N]$_2$[As$_2$Br$_8$],[96] and [NPhMeH$_2$]$_2$[As$_2$Br$_8$].[97] The [Ph$_4$P]$_2$[As$_2$I$_8$] also belongs to this type.[98] However, with pyridinium cations the complexes

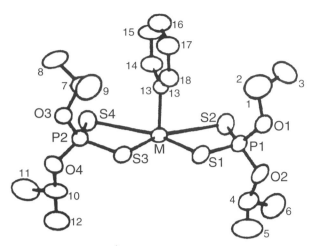

Figure 10 The structure of $[AsPh\{S_2P(OPr^i)_2\}_2]$ (reproduced by permission of the American Chemical Society from *Inorg. Chem.* **1985**, *24*, 3280–3284).

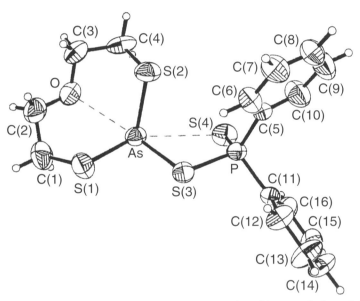

Figure 11 The structure of $[\{O(CH_2CH_2S)_2\}As(S_2PPh_2)]$ (reproduced by permission of the Royal Society of Chemistry from *J. Chem. Soc., Dalton Trans.* **1996**, 4135–4141).

of stoichiometry $[PyH][AsX_4]$ (X = Br or I) contain approximately octahedrally coordinated As, with an infinite polymer chain anion with *cis* dihalobridges (Figure 12).[97,98]

Discrete confacial bioctahedral anions are present in $[PyH]_3[As_2Cl_9]$,[94] $[PyH]_3[As_2Br_9]$[97] (Figure 13), and $[piperidineH]_4[As_2Br_9]Br$.[96] The environment about As is close to octahedral (although the bridging As–X are longer than terminal As–X as expected) and the bridges symmetric. In the $[Et_3NH]_3[As_3Br_{12}]$[99,100] and $[Me_3NH]_3[As_3I_{12}]$[100] there are discrete trimeric anions based upon face sharing octahedra with a common vertex (Figure 14). Two anions of formula $[As_8X_{28}]^{4-}$ are known, but with different structures. In the $[S_5N_5]_4[As_8Cl_{28}]\cdot2S_4N_4$, made serendipitously from $(NSCl)_3$ and As_2O_3 in CH_2Cl_2, a complex structure occurs which can be viewed as a cubane $[As_4Cl_{16}]^{4-}$ core (presently unknown as a discrete species) to which is attached four $AsCl_3$ units.[101] $[Et_3NH]_4[As_8I_{28}]$ has a different structure based upon AsI_6 edge-linked octahedra (Figure 15).[98]

The arsenic(V) fluoroanion $[AsF_6]^-$ which is a regular octahedron in the K^+ salt, is well known,[102] and often considered a "noncoordinating" anion. In fact, like other related species, it is better viewed as "weakly coordinating" and is known to bind to metals in the absence of other ligands.[103] In many cases where it is found as a product of fluoride abstraction from nonmetal

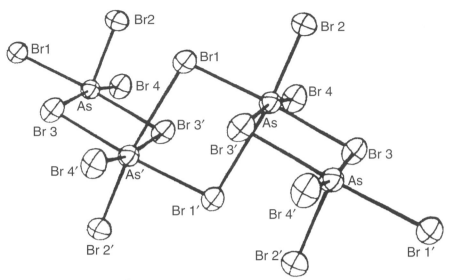

Figure 12 The structure of $[AsBr_4]_n^{n-}$ (reproduced by permission of the publishers from *Z. Naturforsch., B* **1984**, *39*, 1257–1261).

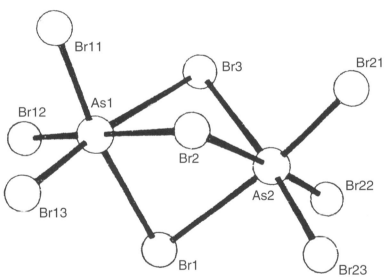

Figure 13 The structure of $[As_2Br_9]^{3-}$ (reproduced by permission of the publishers from *Z. Naturforsch., B* **1984**, *39*, 1257–1261).

fluorides or oxofluorides by AsF_5, the resulting $[AsF_6]^-$ "anion" is clearly associated with the cations through directional As–F–cation interactions. The anion $[F_5AsFAsF_5]^-$ is also well known.[104] Although the Sb and Bi analogues are known, the $[AsF_7]^{2-}$ anion has not been prepared.[105] The yellow $[AsCl_6]^-$ has been isolated as its PPh_4^+ salt by treatment of $[As_2Cl_8]^{2-}$ with Cl_2 or O_3 in CH_2Cl_2 at low temperatures.[106] A number of fluorochloroarsenates(V) $[AsF_{6-n}Cl_n]^-$ have been identified in MeCN solution by multinuclear NMR studies, as has $[AsF_5Br]^-$.[107]

Oxo-haloarsenates which have been obtained and characterized structurally include $[As_2OCl_5]^-$,[18] $[As_2OCl_6]^{2-}$,[94,108] and $[As_4O_2Cl_{10}]^{2-}$.[19,108,109] All contain both Cl and oxygen bridges, the last having the structure shown in Figure 16. Arsenic(V) anions include $[F_5AsOAsF_5]^{2-}$ and $[F_4As(O)_2AsF_4]^{2-}$ and the sulfur-bridged $[F_5AsSAsF_5]^{2-}$.[110]

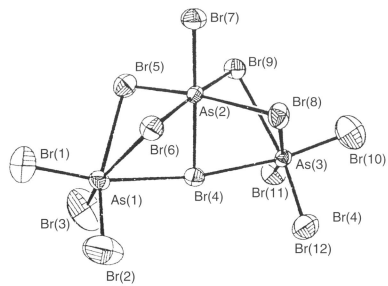

Figure 14   The structure of $[As_3Br_{12}]^{3-}$ (reproduced by permission of the publishers from *Z. Naturforsch., B* **1992**, *47*, 1079–1084).

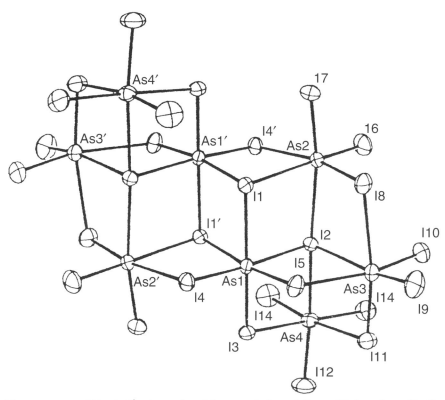

Figure 15   The structure of $[As_8I_{28}]^{4-}$ (reproduced by permission of the publishers from *Z. Naturforsch., B* **1988**, *43*, 789–794).

3.6.2.5   Arsenic in the Environment, Biology, and Medicine

Arsenic is widely distributed in nature and man-made distribution occurs through mining, smelt-ing, pesticides, and the use of fossil fuels. The vast majority of the forms identified in the environment are simple inorganic (oxide, oxo-anions) or organic (especially methylated forms)

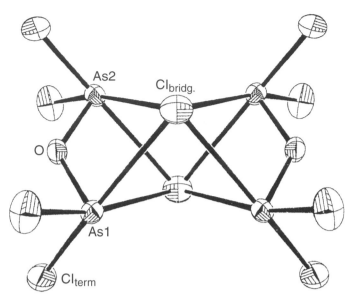

Figure 16 The structure of $[As_4O_2Cl_{10}]^{2-}$ (reproduced by permission of the publishers from *Z. Naturforsch., B* **2001**, *56*, 301–305).

and coordination chemistry plays only a small role. Several recent review articles should be consulted for details.[111–114] Medical uses of arsenic coordination complexes are similarly unimportant, in contrast to antimony and especially bismuth complexes (q.v.). Thiols such as 2,3-dithiopropanol (British anti-Lewisite) or 2,3-dithiosuccinic acid were developed many years ago for use in chelation therapy for heavy metal (including As) poisoning, and rely on the high affinity of As for sulfur ligands.[111]

3.6.3 ANTIMONY

The first major difference between the coordination chemistry of antimony compared to those of arsenic or bismuth is the significantly greater stability of the Sb^V state, which forms coordination complexes with a wide range of ligands. In contrast, for the other two elements the complexes of E^V are mostly haloanions or compounds with charged oxygen donor ligands. In general, the treatment follows the pattern established above with periodic group of the donor atom, and with Sb^{III} complexes described before Sb^V.

3.6.3.1 Group 14 Compounds

In addition to reviews of organoantimony chemistry noted in Section 3.6.1, there are articles dealing with low-coordination number species such as $(RSb)_n$, $RSb=SbR$, C_5H_5Sb, $RP=SbR$, and their transition metal derivatives.[13,14,115–118]

The silylstibines, $Sb(SiR_3)_3$, provide the only series of examples of Si—Sb bonds and are normally treated with other organostibines.[11,116] Such compounds have found use in the preparation of (III)–(V) materials via pyrolysis of their group 13 adducts.[119] There are also examples of $Sb(YR_3)_3$ (Y = Ge, Sn, or Pb) types.[116]

3.6.3.2 Group 15 Compounds

3.6.3.2.1 N-donor ligands

The reaction of molten SbX_3 (X = Br or I) with NH_3 produced $[SbX_3(NH_3)]$, whilst $SbCl_3$ and NH_3 in diethyl ether formed $[SbCl_3(NH_3)_2]$.[16] Trimethylamine forms both 1:1 and 1:2 adducts with $SbCl_3$ or $SbBr_3$, and 1:1 adducts have been described with NH_2Me, $NHMe_2$, and

PhNH$_2$.[16,17] Vibrational spectroscopy suggests the 1:1 complexes are *pseudo*-trigonal bipyramidal with axial amine and with an equatorial vertex occupied by the lone pair, whilst the 1:2 compounds probably have a structure based upon an octahedron where one vertex is occupied by the lone pair, as established by the X-ray crystal structure of [SbCl$_3$(PhNH$_2$)$_2$].[120] The structure of the yellow [SbCl$_3$(2,2'-bipyridyl)] is based upon a distorted five-coordinate geometry (N$_2$Cl$_3$ donor set) with Sb—Cl = 2.55 Å (av). This unit forms a long contact to a further Cl from a neighboring molecule (3.34 Å) completing a very distorted octahedron.[121] Distorted square pyramidal (N$_2$O$_2$X donor set) molecules are present in [SbX(1,10-phen)(cat)] (X = F, Cl, Br, or I; 1,10-phen = 1,10-phenanthroline, cat = phenylene-1,2-diolate(2−)),[122,123] whereas in [Sb(1,10-phen)$_2$(cat)]BPh$_4$ there is very distorted N$_4$O$_2$ coordination.[124]

The aza-macrocycle Me$_3$[9]aneN$_3$ produces a 1:1 complex with SbCl$_3$ of unknown structure,[18] but in the presence of SbCl$_5$ a similar reaction yields the complex [SbCl$_2$(Me$_3$[9] aneN$_3$)]SbCl$_6$, with a distorted square pyramidal cation.[125] Hydrolysis of [SbCl$_3$(Me$_3$[9] aneN$_3$)], or reaction of SbCl$_3$ and 1,4,8,11-tetramethyltetraazacyclotetradecane in wet MeCN, gave oxochloroantimonate anions (q.v.) with the protonated macrocycle as cations.[18,19] The phthalocyanine derivative [Sb(pc)$_2$]$^-$ has been isolated by heating together SbI$_3$, 1,2-C$_6$H$_4$(CN)$_2$, and KOMe. The structure as the Bu$_4$N$^+$ or PNP$^+$ salts show a distorted eight-coordinate antimony environment.[126]

The reaction of SbCl$_3$ with three equivalents of 2-(dimethylaminomethyl)phenyl lithium (Li$^+$(3)$^-$) produces [Sb{C$_6$H$_4$(CH$_2$NMe$_2$)}$_3$] which has a similar geometry to its arsenic analogue (Figure 1), but with rather stronger E—N coordination suggested by comparison of the bond lengths.[22] Using appropriate ratios of SbCl$_3$:(3) [SbCl$_2${C$_6$H$_4$(CH$_2$NMe$_2$)}] and [SbCl{C$_6$H$_4$(CH$_2$NMe$_2$)}$_2$] can be isolated. These again have structures based upon strong Sb—C bonds with weaker interactions with the amine functions completing distorted *pseudo*-trigonal bipyramidal geometry with an equatorially disposed lone pair.[21] Treatment of [SbCl{C$_6$H$_4$(CH$_2$NMe$_2$)}$_2$] with TlPF$_6$ in THF affords the related cation [Sb{C$_6$H$_4$(CH$_2$NMe$_2$)}$_2$]PF$_6$ which is also *pseudo*-trigonal bipyramidal with axial N and equatorial C atoms.[127] 8-(Dimethylamino)-1-naphthyl (2) also forms [SbCl$_2$(2)] and [SbCl(2)$_2$] complexes,[21] whereas with 2,6-bis[(dimethylamino)methyl]phenyl (1) the product is [SbCl$_2$(1)] which has a distorted square-pyramidal geometry with an apical C atom (Figure 17).[20] *Trans*-annular Sb···N coordination is present in the heterocyclic rings R^{1}Sb[(CH$_2$)$_3$]$_2$NR^2 (R^1 = Cl, I, Ph; R^2 = Me, Bz, Bu^i, etc.)[128,129] which have structures based upon a *pseudo*-trigonal bipyramid with axial N and R and an equatorial lone pair.

Schiff base ligands form complexes with both Sb^{III} and Sb^V.[130–134] Examples are known where the Schiff base coordinates as a neutral ligand bonded only via the azomethine nitrogen(s) to both *cis*- and *trans*-SbCl$_4$$^+$ units,[132] or as anions bonded both through the azomethine-N and deprotonated *o*-hydroxyphenyl groups.[134]

Antimony pentafluoride forms 1:1 adducts with HCN and C$_2$N$_2$ and the structure of the latter reveals a linear NCCN-SbF$_5$ linkage.[27] Related adducts of SbCl$_5$ including [SbCl$_5$(L)] (L = ICN, BrCN, ClCN, 1/2C$_2$N$_2$, NH$_2$CN, pyridine) have been prepared and the X-ray structures of

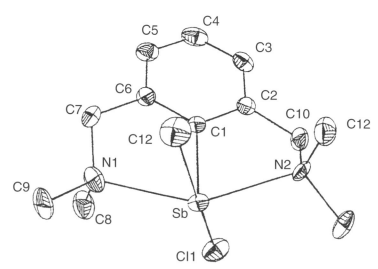

Figure 17   The structure of [SbCl$_2${2,6-(Me$_2$NCH$_2$)$_2$C$_6$H$_3$}] (reproduced by permission of Elsevier Science from *Inorg. Chim. Acta* **1992**, *198–200*, 271–274).

[SbCl$_5$(ClCN)] and [(SbCl$_5$)$_2$(C$_2$N$_2$)] determined.[135] A substantial range of organonitrile adducts of SbCl$_5$ have been described, including [SbCl$_5$(RCN)] (R = Me, Ph, various isomers of MeC$_6$H$_4$−, Me$_2$C$_6$H$_3$−, ClC$_6$H$_4$−, NH$_2$C$_6$H$_4$−).[136–139] The reaction of SbCl$_3$ with TMSNPR$_3$ (R = Ph or Me) formed the phosphine-iminato compounds [SbCl$_2$(NPR$_3$)], which react further with SbCl$_5$ in MeCN to form [SbCl(NPPh$_3$) (MeCN)$_2$]$_2$[SbCl$_6$]$_2$ or [Sb$_2$Cl$_5$(NPMe$_3$)$_2$-(MeCN)][SbCl$_6$].[140] The former contains dimeric cations with Sb^{III}Cl(MeCN)$_2$ units bridged by two NPPh$_3$, whereas the latter is a mixed-valence cation with Sb^{III}Cl(MeCN) and Sb^{V}Cl$_4$ units also bridged by two phosphine-iminato groups.

Unstable (often highly explosive) antimony azides have been synthesized recently.[32,141] The parent Sb(N$_3$)$_3$ is made from AgN$_3$ and SbI$_3$ in MeCN, whilst reaction of [SbCl$_4$]$^+$, [SbCl$_4$]$^-$, and [SbCl$_6$]$^-$ with TMSN$_3$ gave [Sb(N$_3$)$_4$]$^+$, [Sb(N$_3$)$_4$]$^-$, and [Sb(N$_3$)$_6$]$^-$, respectively. Attempts to isolate [Sb(N$_3$)$_5$] were unsuccessful, although some Lewis base adducts [Sb(N$_3$)$_5$(L)] (L = py, NH$_3$, quinoline, etc.) are known.[142]

3.6.3.2.2 *P-, As-, and Sb-donor ligands*

Early studies[32] reported that PR$_3$ or AsR$_3$ formed 1:1 or rarely 2:1 adducts with SbX$_3$. A reinvestigation[143] of the reaction of PMe$_3$ and SbI$_3$ in THF identified the yellow product as [Sb$_2$I$_6$(PMe$_3$)$_2$]·THF which has a structure based upon two edge-linked square pyramidal SbI$_4$P units with apical phosphines arranged *anti* to the plane. Weaker Sb···I contacts link the molecules into a polymer. In contrast, the reaction of SbBr$_3$ and PEt$_3$ in THF gave crystals of [PEt$_3$H][Sb$_2$Br$_7$(PEt$_3$)$_2$] which has the structure shown in Figure 18.[144] Bidentate diphosphines and diarsines (Me$_2$P(CH$_2$)$_2$PMe$_2$, o-C$_6$H$_4$(PMe$_2$)$_2$, o-C$_6$H$_4$(AsMe$_2$)$_2$, o-C$_6$H$_4$(PPh$_2$)$_2$, Ph$_2$As(CH$_2$)$_2$AsPh$_2$) form 1:1 complexes with SbX$_3$ (X = Cl, Br, or I), which are probably based upon edge-sharing dimers with *pseudo*-octahedral antimony centers.[144,145] An alternative description is in terms of primary SbX$_3$ units with weaker secondary bonding to the group 15 donor and bridging halides. The structure has been established for [Sb$_2$Br$_6${Me$_2$P(CH$_2$)$_2$PMe$_2$}$_2$] (Figure 19) and [Sb$_2$Br$_6${o-C$_6$H$_4$(PPh$_2$)$_2$}$_2$]. A polymorph of the former has been identified,[144] which contains a central Sb$_2$Br$_6${Me$_2$P(CH$_2$)$_2$PMe$_2$}$_2$ linked via single bromine bridges to two SbBr$_3$[Me$_2$P(CH$_2$)$_2$PMe$_2$] units. The 1:1 complexes of the triarsine MeC(CH$_2$AsMe$_2$)$_3$}, [SbX$_3${MeC(CH$_2$AsMe$_2$)$_3$}], may also be dimers.[145] Recrystallization of [Sb$_2$Cl$_6${o-C$_6$H$_4$(AsMe$_2$)$_2$}$_2$] from hot ethanol gave the 1:1 which has a polymeric structure composed of [SbCl$_2${o-C$_6$H$_4$(AsMe$_2$)$_2$}]$^+$ and (SbCl$_4$)$^-$ units linked into sheets through Cl-bridges (Figure 20).[145] A discrete distorted octahedral anion is present in [Py$_2$H][SbI$_4${Me$_2$P(CH$_2$)$_2$PMe$_2$}] formed by recrystallizing [SbI$_3${Me$_2$P(CH$_2$)$_2$PMe$_2$}] from pyridine.[146]

Adducts of Me$_3$Sb with SbI$_3$ and SbI$_2$Me have been characterized by X-ray crystallography. The former, isolated from THF solution as [Sb$_2$I$_6$(SbMe$_3$)$_2$(THF)$_2$], is a centrosymmetric dimer

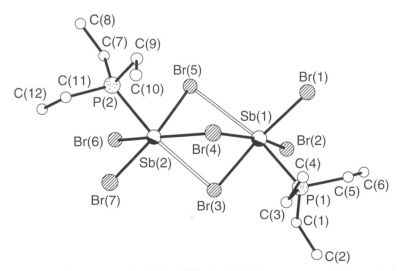

Figure 18 The structure of the anion in [PEt$_3$H][Sb$_2$Br$_7$(PEt$_3$)$_2$] (reproduced by permission of the Royal Society of Chemistry from *J. Chem. Soc., Dalton Trans.* **1994**, 1753–1757).

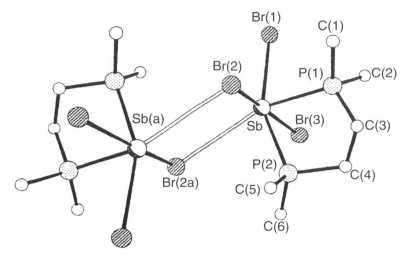

Figure 19   The structure of [Sb$_2$Br$_6$(Me$_2$PCH$_2$CH$_2$PMe$_2$)$_2$] (reproduced by permission of the Royal Society of Chemistry from *J. Chem. Soc., Dalton Trans.* **1994**, 1743–1751).

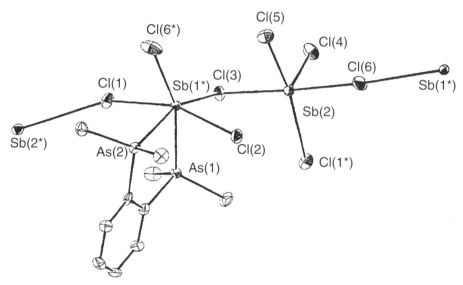

Figure 20   The structure of the asymmetric unit in [Sb$_2$Cl$_6${*o*-C$_6$H$_4$(AsMe$_2$)$_2$}] (reproduced by permission of the Royal Society of Chemistry from *J. Chem. Soc., Dalton Trans.* **2001**, 1007–1012).

based on a planar I$_2$Sb(μ^2-I)$_2$SbI$_2$ core with axial SbMe$_3$ and THF ligands arranged *anti* (Figure 21). The Sb—O(THF) bonds are weak, secondary interactions.[147] In contrast, in [SbI$_2$-Me(SbMe$_3$)], which is formed by the spontaneous rearrangement of SbMe$_2$I in the presence or absence of solvent, the structure is based upon a *pseudo*-trigonal bipyramidal antimony with the lone pair, the Me group, and SbMe$_3$ occupying equatorial positions (Figure 22).[148,149]

3.6.3.3   Group 16 Compounds

3.6.3.3.1   *O-donor ligands*

The crown ethers 12-crown-4, 15-crown-5, and 18-crown-6 form 1:1 adducts with SbCl$_3$, all of which have structures based upon a pyramidal SbCl$_3$ unit with much weaker interactions to 4, 5, or 6 crown ether oxygens respectively, completing a half sandwich structure (Figure 23).[38,150,151] The [SbCl$_3$(15-crown-5)] and [SbCl$_3$(12-crown-4)] have also been studied by EXAFS and these results are in good agreement with the single crystal X-ray data.[152] The complex [SbCl$_2$(18-crown-6)][SbCl$_6$]

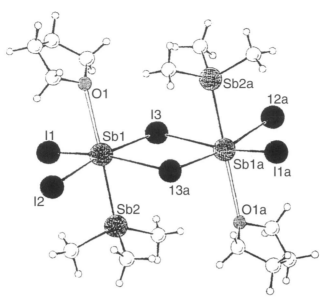

Figure 21 The structure of [Sb$_2$I$_6$(SbMe$_3$)$_2$(THF)$_2$] (reproduced by permission of Wiley-VCH from *Z. Anorg. Allg. Chem.* **1998**, *624*, 81–84).

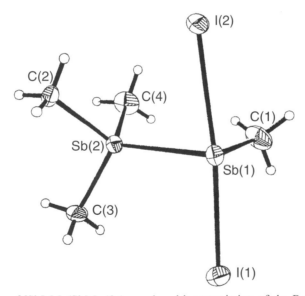

Figure 22 The structure of [SbI$_2$Me(SbMe$_3$)] (reproduced by permission of the Royal Society of Chemistry from *Chem. Commun.* **1994**, 875–876).

is formed by reaction of the crown ether with a mixture of SbCl$_3$ and SbCl$_5$ in MeCN.[153] The structure of the cation is shown in Figure 24 and is based upon primary coordination in a *pseudo*-trigonal bipyramid with axial oxygens, and notably these Sb-O$_{transO}$ are 0.2 Å shorter than the weak, secondary bonds to the other four oxygens. In marked contrast, [SbCl(15-crown-5)] [SbCl$_6$]$_2$ has a pentagonal pyramidal cation (Figure 25) with the lone pair occupying the vacant site *trans* to the chloride.[154] The complexes of dibenzo-24-crown-8 are neutral, of type [(SbX$_3$)$_2${dibenzo-24-crown-8}] (X = Cl or Br).[155] The structure of the chloride derivative shows the two antimony atoms bonded to opposite sides of the crown via three chlorines and five oxygens. In the bromide species the coordination is also via three bromines, but only four oxygens.[155]

The antimony(III) complexes of the maleonitriledithiolate derivatized crown ethers, mn-15S$_2$O$_3$ and mn-18S$_2$O$_4$ (**12**) have also been prepared.[156] In [SbCl$_3$(mn-15S$_2$O$_3$)] the structure is of the half-sandwich type (Figure 26) with rather longer Sb—O bonds than in the simple crown complexes

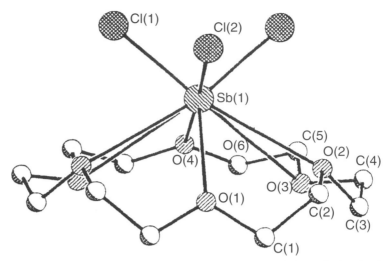

Figure 23 The structure of [SbCl₃(18-crown-6)] (reproduced by permission of Elsevier Science from *Inorg. Chim. Acta* **1990**, *167*, 115–118).

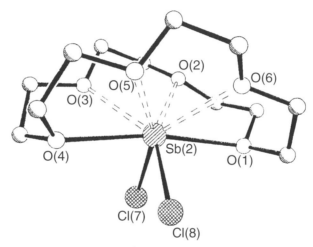

Figure 24 The structure of [SbCl₂(18-crown-6)]⁺ (reproduced by permission of Wiley-VCH from *Z. Anorg. Allg. Chem.* **1992**, *618*, 93–97).

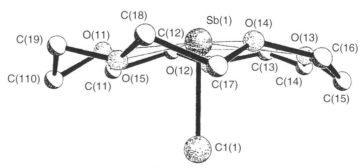

Figure 25 The structure of [SbCl(15-crown-5)]²⁺ (reproduced by permission of Wiley-VCH from *Angew. Chem., Int. Ed. Engl.* **1992**, *31*, 334–335).

and long Sb—S (ca. 3.4 Å). In [SbCl₃(mn-18S₂O₄)] the sulfurs are uncoordinated and the Sb is bonded only to the four oxygens.[156] In contrast, the reaction of Na[SbCl₆] with mn-18S₂O₄ produces [Na(mn-S₂O₄)₂][SbCl₆] in which the Na ion is sandwiched between the two crowns and coordinated to the eight oxygens.[156]

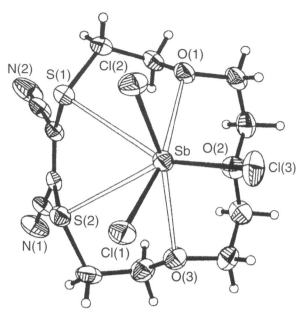

Figure 26 The structure of [SbCl$_3$(mn-15S$_2$O$_3$)] (reproduced by permission of the publishers from *Z. Naturforsch.*, B **1999**, *54*, 799–806).

mn–15S$_2$O$_3$

(12a)

mn–18S$_2$O$_4$

(12b)

Pnictogen or chalcogen oxides, ethers or amides form O-bonded adducts with both SbCl$_3$ and SbCl$_5$. Typical examples that have been characterized include [SbCl$_5$L] (L = PyNO,[157] Me$_3$NO,[157] (RO)$_3$PO,[138,139] Me$_2$O,[138,139] THF,[138,139] DMSO[138,139,158]) The structure of [SbCl$_5$(DMSO)] shows the sulfoxide is O-bonded to Sb^V.[158] Spectroscopic studies show that the stabilities of these complexes follow Gutmann's donor numbers, and NMR studies are consistent with a dissociative ligand exchange mechanism.[138,139] An organoantimony example is the cation in [Ph$_2$Sb{(Me$_2$N)$_3$PO}$_2$]PF$_6$, made from PhSbCl$_2$, (Me$_2$N)$_3$PO, and TlPF$_6$ (the reaction is accompanied by a phenyl migration).[127] The structure of the cation is *pseudo*-trigonal bipyramidal with the lone pair and Ph groups equatorially disposed.

A variety of antimony(III) and antimony(V) alkoxides and mixed halo-alkoxides are known, the majority of which are oligomeric via asymmetric Sb—O···Sb bridges. The Sb^{III} examples include [Sb(OMe)$_3$]$_n$ (six-coordinate Sb with a 3-D network—(**13**)),[160] [Sb(OPr^i)$_3$]$_2$ (four-coordinate Sb with a *pseudo*-trigonal bipyramidal geometry in a dimer—(**14**)),[161] and [Sb(2,6-Me$_2$C$_6$H$_3$O)$_3$] (trigonal pyramidal monomer).[162] The halo-alkoxides are also polymeric—[SbCl(OEt)$_2$]$_n$, [SbCl$_2$(OEt)]$_n$, and [SbCl$_2$(OEt)·NHMe$_2$]$_n$ contain six-coordinate antimony, whereas [SbCl(O-Pr^i)$_2$]$_2$ is five-coordinate.[160,161,163,164] The Sb^V alkoxides which have been structurally characterized are based upon six-coordinate antimony. The simplest, [Sb(OMe)$_5$]$_2$ is a dimer (Figure 27),[165] [Sb(OEt)$_5$(NH$_3$)] is monomeric (O$_5$N donor set),[162] and [SbBr$_2$Me(OMe)$_2$]$_2$[166] is dimeric with OMe bridges.

Antimony phenoxides are also readily prepared. Catechol (CatH$_2$,(1,2-C$_6$H$_4$(OH)$_2$)) and Sb(OPr^i)$_3$ form [Sb(cat)(OPr^i$_3$)] and [Sb(cat)(catH)], the latter being converted on reaction with M^{2}OMe (M^2 = Li, Na, or K) into M^2[Sb(cat)$_2$].[167] The structure of [NH$_4$][Sb(cat)$_2$] reveals a *pseudo*-trigonal bipyramidal geometry with an equatorial lone pair.[167] The [PyH][Sb-(o-C$_6$Cl$_4$O$_2$)Cl$_2$] is also pseudo-trigonal bipyramidal with asymmetric chlorine bridges giving an

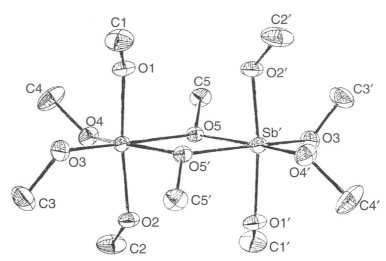

Figure 27 The structure of [Sb(OMe)$_5$] (reproduced by permission of Wiley-VCH from *Z. Anorg. Allg. Chem.* **1981**, *474*, 157–170).

(13)   (14)

extended structure.[168] The antimony(V) species [SbCl$_4$(ACAC)] (ACAC = acetylacetonate) is a discrete six-coordinate complex.[169] Adducts of SbCl$_5$ with methoxyethanol, 1,2-ethanediol, and 1,2-dimethoxyethane are readily prepared, but the alcohol complexes are prone to elimination of HCl to give stibocycles.[170] Organoantimony halides also react with O-, or mixed O/N-donor ligands, for example, R$_3$SbBr$_2$ (R = Ph or Me) react with LH (LH = acetylacetone, 8-hydroxyquinoline, salicylaldehyde, 2-hydroxyacetophenone) and NaOMe in benzene/methanol to give R$_3$Sb(OMe)L, which appear to be six-coordinate.[171]

Methanesulfonic anhydride (Me$_2$S$_2$O$_5$) dissolves Sb$_2$O$_3$ on prolonged heating to form [Sb(O$_3$SMe)$_3$], which reacts further with Cs[MeSO$_3$] in MeSO$_3$H to give Cs[Sb(O$_3$SMe)$_4$].[58] In contrast to the well-characterized [As(OTeF$_5$)$_5$] and [Bi(OTeF$_5$)$_5$], [Sb(OTeF$_5$)$_5$] is unstable and has not been isolated in a pure state.[172,173] However, the anion [Sb(OTeF$_5$)$_6$]$^-$ has been made from [NR$_4$][SbCl$_6$] and AgOTeF$_5$ in CH$_2$Cl$_2$, or by formation of [Sb^{III}(OTeF$_5$)$_4$]$^-$ from [Sb(OTeF$_5$)$_3$] and [NR$_4$][TeOF$_5$] followed by oxidation with [Xe(OTeF$_5$)$_2$].[173,174] The [Sb(OTeF$_5$)$_6$]$^-$ is a useful addition to the list of "weakly coordinating" anions.[103] Antimony(III) fluoride-fluorosulfates, [SbF$_2$(SO$_3$F)], [SbF(SO$_3$F)$_2$], and [Sb(SO$_3$F)$_3$], have been synthesized and characterized structurally.[175] The first is obtained from Sb and HSO$_3$F, the others from Sb and S$_2$O$_6$F$_2$ under appropriate conditions, and the structures of all three reveal triply bridging O-bound fluorosulfate groups.[175] Oxidation of elemental Sb with a large excess of S$_2$O$_6$F$_2$ in the presence of CsSO$_3$F gives Cs[Sb(SO$_3$F)$_6$], which has a discrete octahedral anion.[176,177] Oxidation of SbF$_3$ with S$_2$O$_6$F$_2$ yields Sb^V fluoride–fluorosulfates [SbF$_3$(SO$_3$F)$_2$], [SbF$_4$(SO$_3$F)], and [Sb$_2$F$_9$(SO$_3$F)].[178]

Antimony(V) forms complexes with organophosphorus acids.[179–186] These include [{SbCl$_4$(O$_2$PR$_2$)}$_2$] (R = Me, Cl, OPh, OMe) made from SbCl$_5$ and the acid in methanol,[179,180,182] which are oxo-bridged dimers (Figure 28). Other examples are [Cl$_3$Sb(O){R(MeO)PO$_2$}(OMe)SbCl$_3$], (R = 4-ClC$_6$H$_4$CH$_2$, Me, Et, PhCH$_2$4-O$_2$NC$_6$H$_4$CH$_2$) [Cl$_3$Sb(O){(PhO)$_2$PO$_2$}$_2$SbCl$_3$], and

$[Cl_3Sb(O)\{(PhO)_2PO_2\}(OMe)SbCl_3]$,[183–186] which also contain bridging organophosphorus anions. Organoantimony(V) phosphinates include $[\{Ph_2SbCl(O_2PR_2)\}_2O]$ (R = c-hexyl, c-octyl), made from Ph_2SbCl_3, $Ag(MeCO_2)$, and R_2PO_2H,[187,188] which have the structure shown in Figure 29. A tetramer of the dicyclohexylphosphinate complex has also been characterized structurally.[188] Triorganoantimony species $[R^1_3Sb(O_2PR^2_2)_2]$ are made from $R^1_3SbX_2$ and the silver salt of phosphinic acid,[189–191] and partial hydrolysis produces $[R^1_3Sb(OH)(O_2PR^2_2)]$, which may also be obtained directly from $R^1_3Sb(OH)_2$ and $R^2_2PO_2H$. There are also oxo-bridged complexes of type $[\{R^1_3Sb(O_2PMe_2)\}_2O]$ (R^1 = Ph, o-tolyl), $[\{Ph_3Sb(O_2AsR^2_2)\}_2O]$ (R^2 = Me or Ph) and $[(R_2Sb)_2(O)_2(O_2AsMe_2)_2]$ (Figure 30).[192]

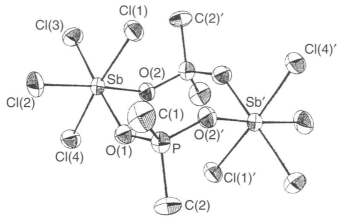

Figure 28 The structure of $[\{SbCl_4(O_2PMe_2)\}_2]$ (reproduced by permission of Wiley-VCH from *Z. Anorg. Allg. Chem.* **1981**, *472*, 102–108).

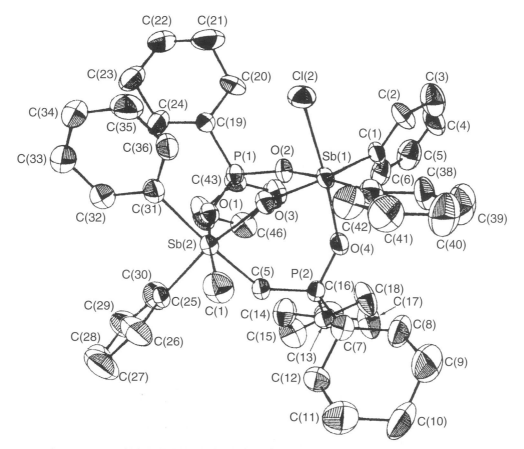

Figure 29 The structure of $[\{Ph_2SbCl(O_2P(C_6H_{11})_2)\}_2O]$ (reproduced by permission of the Royal Society of Chemistry from *J. Chem. Soc., Dalton Trans.* **1995**, 2151–2157).

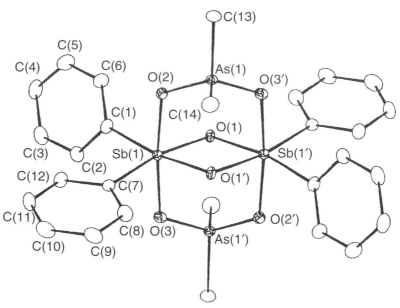

Figure 30   The structure of [(Ph$_2$Sb)$_2$(O)$_2$(O$_2$AsMe$_2$)$_2$] (reproduced by permission of the Royal Society of Chemistry from *J. Chem. Soc., Dalton Trans.* **1997**, 2785–2792).

There is a substantial literature dealing with antimony carboxylates and polyaminecarboxyl-ates. 1,2-Dihydroxycyclohexane-1,2-dicarboxylic acid (both ± and *meso* forms) form Na$_2$[Sb$_2$ (C$_8$H$_8$O$_6$)$_2$]·xH$_2$O which are dinuclear in solution.[56] In the medically important antimony tar-trates the usual building block is the dimeric anion **(15)** found in alkali and alkaline earth metal salts.[193,194] The silver(I) complex [Ag$_4$Sb$_4$(C$_4$H$_2$O$_6$)$_4$(H$_2$O)$_4$] contains this repeating tetramer unit linked into a polymeric network.[195] Antimony(III) citrates also exhibit a range of building blocks. In Li[Sb(C$_6$H$_6$O$_7$)$_2$(H$_2$O)]·2H$_2$O and Na[Sb(C$_6$H$_6$O$_7$)$_2$(H$_2$O)$_2$]·H$_2$O the antimony has a *pseudo*-trigonal bipyramidal geometry with the lone pair equatorial, and two citrate anions each coord-inating via one deprotonated carboxylate and one deprotonated hydroxy group.[196,197] The same basic antimony coordination is present in the isostructural M$_2$[Sb$_4$(C$_6$H$_4$O$_7$)$_2$(C$_6$H$_5$O$_7$)$_2$(C$_6$H$_6$O$_7$)$_4$(H$_2$O)$_2$] (M = K or Rb)[56,197] which are based upon tetrameric units with three differently charged citrate anions (Figure 31), in Ag$_2$[Sb$_2$(C$_6$H$_6$O$_7$)$_4$][196] and Cu[Sb(C$_6$H$_6$O$_7$)(C$_6$H$_5$O$_7$)(H$_2$O)$_2$]·2H$_2$O.[198]

Antimony(III) polyaminocarboxylates have also been studied in considerable detail.[199–205] In the EDTA^{4-} complexes the antimony is coordinated to two N- and four O-donors generating a *pseudo*-pentagonal bipyramid with the seventh vertex occupied by the lone pair. With hard, small cations (Li or Na) the lone pair usually occupies an equatorial position, whereas with large, soft cations (NR$_4$, Cs, aminoguanidinium) the lone pair is axially disposed. However, there is also evidence that H-bonding and packing interactions may affect the geometries adopted. Some of these metal complexes are useful precursors to metal-antimony oxides via pyrolysis in air. The propylenediaminetetra-acetate^{4-} (PDTA^{4-}) complexes, M[Sb(PDTA)]·H$_2$O (M = H, NH$_4$, or

(15)

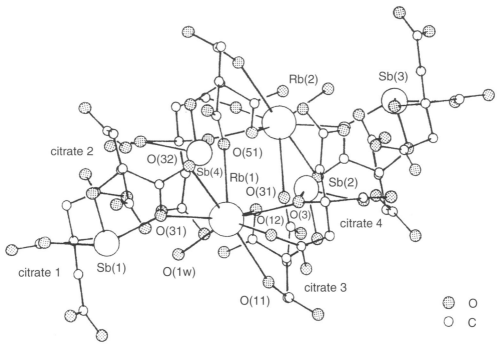

Figure 31 The structure of the citrate complex Rb$_2$[Sb$_4$(C$_6$H$_4$O$_7$)$_2$(C$_6$H$_5$O$_7$)$_2$(C$_6$H$_6$O$_7$)$_4$(H$_2$O)$_2$] (reproduced by permission of the Australian Chemical Society from *Aust. J. Chem.* **2000**, *53*, 917–924).

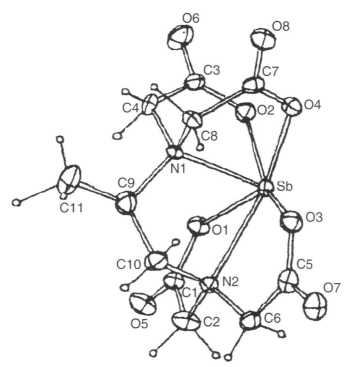

Figure 32 The structure of the propylenediaminetetracetate complex [Sb(PDTA)]$^-$ (reproduced by permission of Elsevier Science from *Inorg. Chim. Acta* **1995**, *232*, 161–165).

Na), also contain *pseudo*-pentagonal bipyramidal anions with an axial lone pair (Figure 32).[206,207] *Pseudo*-seven-coordinate antimony (six O/N-donors and a lone pair) is also present in complex anions derived from 1,2-cyclohexanediamine-*N,N,N'N'*-tetra-acetic acid,[208,209] diethylenetriamine-penta-acetic acid,[210–212] and triethylenetetraminehexa-acetic acid.[213] Several of these complexes show anti-tumor activity, which appears to vary with the fine detail of the geometry.

3.6.3.3.2   *S-, Se-, and Te-donor ligands*

Antimony(III) halides behave as weak Lewis acids towards neutral sulfur or selenium donor ligands. The products have a wide variety of structures but these are mostly built upon a pyramidal SbX_3 unit, which forms weak secondary bonds to the neutral donor and sometimes, weak asymmetric halide bridges. The antimony environments are often very asymmetric due to a combination of the constraints imposed by the ligands and varying degrees of stereochemical activity by the lone pair.[214] The reaction of SbX_3 (X = Cl, Br, or I) with $MeE(CH_2)_nEMe$ ($n = 2$ or 3), $MeC(CH_2EMe)_3$ (E = S or Se), [8]aneSe$_2$, [12]aneS$_4$, and [16]aneS$_4$ forms yellow, orange, or red complexes with a 1:1 stoichiometry.[214–216] $[SbCl_3\{MeS(CH_2)_2SMe\}]$ contains distorted octahedral antimony coordination based upon three terminal Cl and three S atoms from different dithioethers; of the three-coordinated S atoms, two bridge to neighboring antimony centers using both lone pairs available on S, and one is terminal, generating a 2-D network. The structure of $[SbBr_3\{MeS(CH_2)_3SMe\}]$ (Figure 33) is similar.[215] In contrast, the structure of $[SbCl_3\{MeSe(CH_2)_3SeMe\}]$ is based upon weakly associated Sb_2Cl_6 dimers linked by bridging diselenoethers (Figure 34). The tripodal tridentate $MeC(CH_2SMe)_3$ forms a 1:1 complex with $SbCl_3$ which is essentially five-coordinate with bridging bidentate trithioether ligands forming infinite chains.[216] In contrast, the $[SbI_3\{MeC(CH_2SMe)_3\}]$ is based upon six-coordinate antimony with Sb_2I_6 dimers linked into chains by bridging thioethers.[216] The selenoether $[SbBr_3\{MeC(CH_2SeMe)_3\}]$ is different again, with octahedral *fac* $SbBr_3Se_3$ units, with the selenoether ligands bidentate to one antimony and monodentate to a second.

   In $[SbCl_3([9]aneS_3)]$ there is seven-coordinate antimony, based upon three terminal chlorines, three sulfur donors from one macrocycle, and a bridging S atom from a neighboring molecule, producing a chain structure (Figure 35).[217] In contrast, $[SbI_3([9]aneS_3)]$, which involves the more sterically demanding iodo ligands, is a discrete octahedron with no significant evidence for a stereochemically active lone pair.[218] The complexes of [14]aneS$_4$ are of 2:1 Sb:ligand stoichiometry, and the structure of the bromide derivative shows (Figure 36) weakly associated Sb_2Br_6 units with distorted octahedral coordination completed by *cis* S_2-coordination at each antimony from different tetrathioethers.[216] The macrocyclic tetraselenoether complex $[(SbBr_3)_2([16]aneSe_4)]$ has a sheet structure[215] based upon each selenium atom bonded to a different $SbBr_3$ unit which are five-coordinate (Br_3Se_2). Finally, in $[(SbCl_3)_2([18]aneS_6)]$ there are two $SbCl_3$ units each coordinated to three sulfur atoms and disposed on opposite sides of the mean plane of the macrocycle.[217] Complexes with a $[SbX_3(L)]$ stoichiometry have been obtained for L = $MeTe(CH_2)_3TeMe$ or $MeC(CH_2TeMe)_3$, but their structures are not yet known.[214]

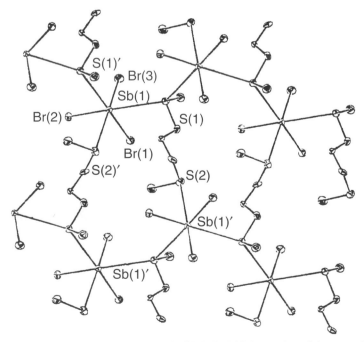

Figure 33   The polymeric structure of $[SbBr_3\{MeS(CH_2)_3SMe\}]$ (reproduced by permission of the Royal Society of Chemistry from *Chem. Commun.* **2001**, 95–96).

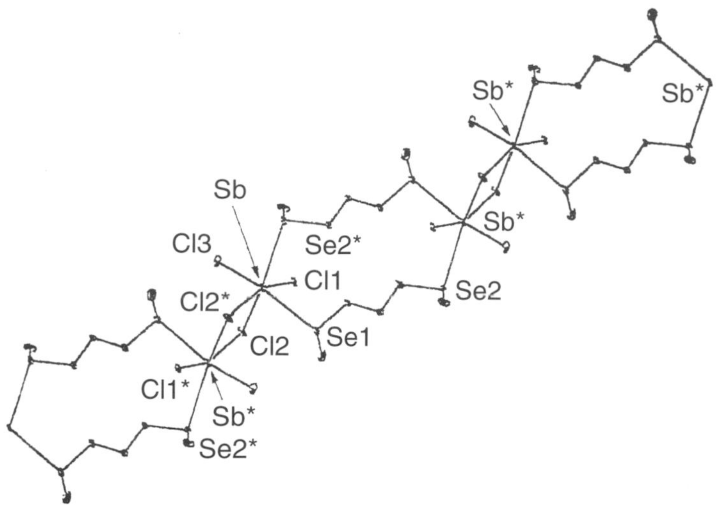

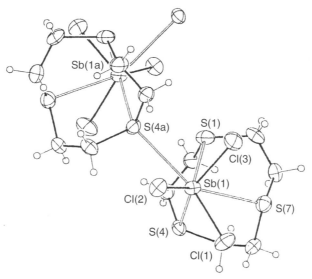

Figure 34 The chain structure of [SbCl₃{MeSe(CH₂)₃SeMe}] (reproduced by permission of the Royal Society of Chemistry from *Chem. Commun.* **2001**, 95–96).

Figure 35 The structure of [SbCl₃{[9]aneS₃}] (reproduced by permission of the Royal Society of Chemistry from *Chem. Commun.* **1991**, 271–272).

A variety of other neutral sulfur ligands form adducts with SbX₃ including thioureas, thiones, and thiophenes,[215–222] most of which appear to contain five-coordinate, *pseudo*-octahedral antimony with a stereochemically active lone pair. Dithio-oxamides (RHNC(S)C(S)NHR); R = Me, Et, Pr^i, Bu^n, c-C₆H₁₁) form [SbX₃(ligand)₁.₅] (X = Cl, Br) in which each ligand is bound to two antimony centers via bidentate (S₂) bridging and the lone pair on antimony is not stereochemically active,[224,225] whereas dithiomalonamides (RHNC(S)CH₂C(S)NHR) chelate, producing *pseudo*-octahedral complexes in which the lone pair clearly occupies one vertex.[226] In [Sb₂Br₆(SPPh₃)₂] and [Sb₂I₆(SePPh₃)₂], prepared from the constituents in CH₂Cl₂,[227] there are centrosymmetric halide-bridged dimers, and it appears that intramolecular Sb···Ph contacts complete the six-coordination about antimony. In contrast, [{SbBr₃(SPMe₂Ph)}₄] is a tetramer with both Br and S bridges (Figure 37).[227]

Antimony has a great affinity for charged sulfur ligands which include thiolates, xanthates (ROCS₂⁻), dithiocarbamates (R₂NCS₂⁻), and dithiophosphates ((RO)₂PS₂⁻).[56] In contrast to arsenic, where this chemistry is limited to oxidation state III, antimony forms compounds in oxidation states III and V. The xanthate, dithiocarbamate, and dithiophosphate complexes are mostly made by reaction of antimony(III) halides or organohalides with Na, NH₄, or Ag salts of the acids. Complexes

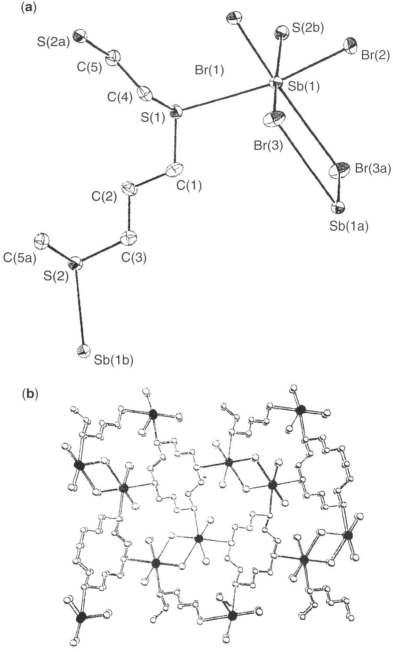

Figure 36 The asymmetric unit (a) and the 3-D network (b) of [(SbBr$_3$)$_2$\{[14]aneS$_4$\}] (reproduced by permission of the Royal Society of Chemistry from *J. Chem. Soc., Dalton Trans.* **2001**, 1621–1627).

reported in the last 20 years are listed in Table 2. The coordination of the dithioacid is often asymmetric and the geometry at the antimony center is distorted both by the constraints of the ligand structure and the effect of the lone pair. For example, in [SbBr(S$_2$COEt)$_2$][230] the structure is a zig-zag chain polymer with antimony in an S$_4$Br$_2$ environment. However, in [PhSb(S$_2$COEt)$_2$] there is one Sb—C and two Sb—S primary bonds, two weaker Sb—S secondary bonds from the asymmetrically chelated xanthate, and a weak intermolecular Sb···S contact.[232] In [Sb(oxine)$_2$(S$_2$COEt)], which is a *pseudo*-pentagonal bipyramid (N$_2$O$_2$S$_2$ plus the lone pair), the xanthate is close to symmetrically chelated. This contrasts with the very asymmetric coordination in [Sb(S$_2$COEt)$_3$] where the Sb—S bonds within each chelate differ by ca. 0.5 Å.[229]

Structures have also been determined for a variety of dithiocarbamate complexes, the dithiocarbamate groups usually coordinating as bidentate chelates containing markedly different Sb—S bond

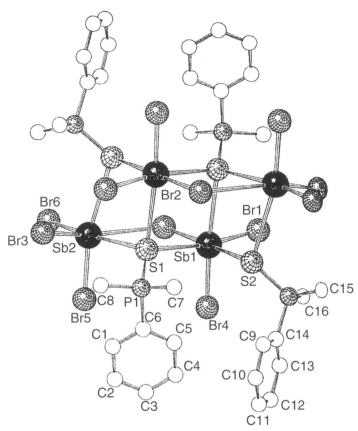

Figure 37 The tetrameric structure of [SbBr₃(SPMe₂Ph)] (reproduced by permission of the publishers from *Z. Naturforsch., B* **1990**, *45*, 1355–1362).

lengths.[74,76,240,243–245] The tris(N,N'-iminodiethanoldithiocarbamato)antimony is best described as a distorted pentagonal pyramid (Figure 38), which contrasts with the trigonal antiprismatic arsenic analogue.[74] The halo-dithiocarbamates [Sb(S₂CNEt₂)₂I],[246] and [Sb(S₂CNC₅H₈)₂I][247] have infinite chain structures with chelating dithiocarbamates and iodide bridges (Figure 39).

Similarly, the tris(dithiophosphate) complexes are based upon distorted octahedra with the differences in Sb—S within each ligand being up to 0.5 Å, and with the lone pair capping the open face associated with the long Sb—S bonds.[252,254] The [PhSb{S₂P(OPr^i)₂}₂] is square pyramidal and isostructural with its arsenic analogue.[82] The diphenyldithiophosphinate complex, [Ph₂Sb(S₂PPh₂)], is a dimer with square-pyramidal antimony centers (Figure 40),[255] but [Ph₂Sb(Y)] (Y = O₂PPh₂ or OSPPh₂) are *pseudo*-trigonal bipyramidal polymers (Figure 41).[257,259]

In contrast to other group 15 elements, antimony also forms dithioacid complexes in oxidation state V, although even these are readily reduced. Triorgano-dithiophosphate, -dithiocarbamate, and -xanthate complexes, [SbR₃(dithioanion)₂], are prepared from SbR₃Cl₂ and the sodium salt of the dithioacid[180,237,261,262] These have trigonal bipyramidal structures with equatorial R groups and monodentate, axially bound dithioanions (Figure 42). Notably, in the related [SbMe₃(OSPPh₂)₂] the anions are O-bonded to Sb^V.[261]

Antimony(III) also has a high affinity for thiolate ligands. The [Sb(SR)₃] complexes are formed by simple SR⁻ ligands including SPh⁻, S(4-MeC₆H₄)⁻, S(3,5-Me₂C₆H₃)⁻, S(2,4,6-Me₃C₆H₂)⁻, S(2,4,6-Pr^{i}₃C₆H₂)⁻, and there are more complex variants such as [Sb₂(SCH₂CH₂SCH₂CH₂S)₃].[265–268] Their syntheses are usually straightforward, from Li(Na)SR and SbCl₃, although other routes such as reaction of RSH with Sb(OR)₃ or Sb(NR₂)₃ are also used. The structures are based upon trigonal-pyramidal coordination at antimony (Figure 43) and with the smaller R-groups, weaker secondary interactions either intermolecular to other thiolate sulfur atoms or to aryl rings. The bulkier thiolate complexes are effectively three coordinate monomers.

Toluene-2,3-dithiol (tdtH₂) reacts with SbCl₃ to form [SbCl(tdt)] which is probably trigonal pyramidal like the As analogue (q.v.).[90] A 1:2 SbCl₃:tdtH₂ reaction ratio formed yellow [Sb(tdt)(tdtH)] from which base removes the final proton, but the product is the purple Sb^V

Table 2   Dithioacid compounds of antimony(III) and (V).

Compound	Comments	References
Antimony(III) compounds		
$Sb(S_2COR)_3$	$R = Me, Et, Pr^i,$	63–66,228,229
	$CH_2CH_2CMe_3$	
$SbBr(S_2COEt)_2$		230
$Sb(oxine)_{3-n}(S_2COEt)_n$	$n = 1, 2$	229
$RSb(S_2COEt)_2$	$R = Me, Ph$	231,232
$\{Sb(S_2COR)_2\}_2CH_2$	$R = Et, CHMe_2$	233
$Sb(S_2COEt)_2(S_2COMe)$		234
$SbClPh(S_2COR)$	$R = Me, Et, Pr^n, Pr^i,$	72
$Sb(SCH_2CH_2S)(S_2COR)$	$R = Et, Pr^n, Bu^n, Bu^i$	68,235
$Sb(SOCNR_2)_3$	$R_2 = Et_2,$ pyrolyl	236
$Sb(S_2CNR_2)_3$	$R = Et, Pr^i, Bz, CH_2CH_2OH$	
	$R_2 = (CH_2)_n \; n = 4, 5$	76,240
	$CH_2CH_2OCH_2CH_2$	
$Sb(S_2CNR_2)_3$	$R_2 = 2\text{-},3\text{-}, $ or 4- Me-piperidine	69
$Sb(S_2CNR_2)_3$	$R_2 = 2$-alkylaminocyclopentane	70
$Sb(S_2CNR_2)_3$	$R_2 = Et_2, N$-methylaminoethanol	
	N,N'-iminodiethanol	74,241
$Sb(S_2CNR^1R^2)_3$	$R^1 = Pr^i, R^2 = 2\text{-}HOC_2H_4$	244
$Sb(S_2CNR^1R^2)_3$	$R^1 = Me, R^2$ cyclohexyl	245
$Sb(S_2NMe_2)_2X$	$X = Cl, Br, I, SO_3CF_3$	78,237
$Sb(SCH_2CH_2S)(S_2CNR)$	$R =$ pyrrolidyl, 4-morphoyl	174
$Sb(SCH_2CH_2S)(S_2CNR_2)$	$R = Me, Et, Pr^i$	238
$\{Sb(S_2CNR_2)_2\}_2CH_2$	$R = Me, Et$	233
$MeSb(S_2NR_2)_2$	$R_2 = Me_2, Et_2,$ morphoyl	239
$Ph_2Sb(S_2CNEt_2)$		242
$Sb(S_2CNEt_2)_2I$		246
$Sb(S_2CNR_2)I$	$R =$ pyrrolidyl	247
$Sb[S_2P(OR)_2]_3$	$R = Me, Et, Pr^n, Pr^i$	80,252
$SbX\{S_2P(OR)_2\}_2$	$X = Cl, Br, I; R = Et, Pr^i, Pr^n$	80
$SbCl_2\{S_2P(OR)_2\}$	$R = Et, Pr^n, Pr^i$	80
$Sb(OPr^i)_{3-n}\{S_2P(OPr^i)_2\}_n$	$n = 1, 2$	80
$Sb[S_2P(OPr^i)_2]_2L$	$L = S_2CNR_2; R = Me, Et,$	248
$Sb(SCH_2CH_2S)(S_2PO_2R)$	$R = CH_2CR_2CH_2$	87
$\{Sb\{S_2P(OR)_2\}_2\}_2CH_2$	$R = Me, CHMe_2$	233
$Ph_nSb\{S_2P(OR)_2\}_{3-n}$	$R = Et, Pr^n, Pr^i, Ph$	82
$Sb(S_2PO_2Y)_3$	$Y = CHMeCHMe, CMe_2CMe_2,$	81
	$CH_2CMe_2CH_2, CH_2CEt_2CH_2,$ etc.	
$Sb(S_2PO_2Y)_{3-n}Cl_n$	Y as above, $n = 1, 2$	81
$Sb(S_2PO_2Y)_{3-n}Ph_n$	Y as above, $n = 1, 2$	84
$Sb(S_2PO_2Y)_{3-n}(OAc)_n$	Y as above, $n = 1, 2$	249
$Sb(S_2COR)[S_2P(OPr^i)_2]_2$	$R = Et, Pr, Pr^i, Bu^i, Bu$	250
$Sb(SCH_2CH_2S)\{S_2P(OR)_2\}$	$R = Et, Pr^i, Bu^i$	251
$SCH_2CH_2S\{Sb(S_2P(OR)_2)\}_2$		251
$Sb(S_2PR_2)_3$	$R = Me, Et, Ph, Pr^i$	252–254,260
$R^1_2Sb(S_2PR^2_2)$	$R^1 = Ph, Me; R^2 = Me,$	255,256
	Et, Pr, Ph	
$(p\text{-tol})_2Sb(S_2PEt_2)$		257
$R^1_2Sb(S_2AsR^2_2)$	$R^1 = Me, Ph$	256
$Sb(OSPR_2)_3$	$R = Ph,$ c-hexyl	258
$R_2Sb(OSPR_2)$	$R = Ph$	259
Antimony(V) Compounds		
$SbX_3(S_2CNMe_2)_2$	$X = Cl, Br$	237
$SbR^1_3(OSPR^2_2)_2$	$R^1 = Me, Et, Ph; R^2 = Et,$	181,261
	Ph, OEt, OPr^i	
$SbMe_3(S_2PR_2)_2$	$R = Me, Et, Ph$	262
$\{SbMe_3\{S_2P(OR)_2\}_2\}_2O$	$R = Me, Et$	263,264
$\{SbMe_3(S_2CNR_2)\}_2O$	$R = Me, Et$	263,264
$\{SbMe_3(S_2COR)\}_2O$	$R = Me, Et$	263,264

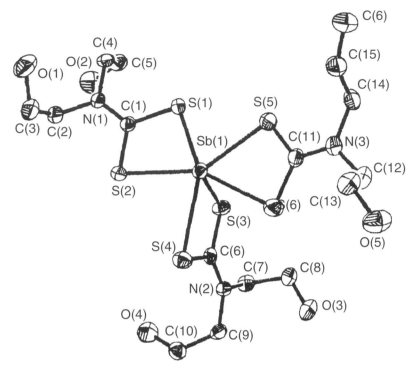

Figure 38 The structure of [Sb{S$_2$CN(CH$_2$CH$_2$OH)$_2$}$_3$] (reproduced by permission of Elsevier Science from *Polyhedron* **1997**, *16*, 1211–1221).

anion, [Sb(tdt)$_3$]$^-$, presumably formed by air oxidation.[90] The [Sb(tdt)$_3$]$^-$ has an octahedral geometry, although the Sb—S distances span 0.16 Å. Similar reactions using the Li salt of benzene-1,2-dithiol in the absence of air gave yellow Li[Sb(o-C$_6$H$_4$S$_2$)$_2$] which was shown to be a *pseudo*-trigonal bipyramid (Figure 44) and which reacted with further dithiol in dry oxygen to form the purple [Sb{o-C$_6$H$_4$S$_2$)$_3$]$^-$.[269] 2-Pyridinethiol forms [Sb(C$_5$H$_4$NS)$_3$], the structure of which shows a trigonal pyramidal SbS$_3$ unit (Sb—S = 2.472(2) Å), with weak association of the nitrogens (Sb$\cdots$N = 2.830(2) Å).[270] 2-Aminothiophenol (o-C$_6$H$_4$SH(NH$_2$)) forms [Sb{o-C$_6$H$_4$S(NH$_2$)}$_3$] which is probably similar.[271] The [SbI{S-(2,4,6-Me$_3$C$_6$H$_2$)}$_2$] has been isolated from the reaction of [SbI(OEt)$_2$] with the thiol in ethanol.[265]

There is a more limited number of selenolate and tellurolate analogues including [Sb{Se(2,4,6-R$_3$C$_6$H$_2$)}$_3$] (R − Me, Pr^i, Bu^i),[267] [Sb{SeSi(SiMe$_3$)$_3$}$_3$],[272] and [Sb{TeSi(SiMe$_3$)$_3$}$_3$],[272]

4,5-Dithio-1,3-dithiole-2-thione (dmitH$_2$) (**16**) complexes of Sb^{III} have been prepared from [Zn(dmit)$_2$]$^{2-}$, SbBr$_3$, and NaNCS in acetone, and have been isolated with a variety of cations.[273,274] The structure is based upon a *pseudo*-trigonal bipyramidal anion with an equatorial lone pair. The cations present control intermolecular Sb$\cdots$S interactions: in the [Et$_4$N]$^+$ salt two additional Sb$\cdots$S interactions lead to a *pseudo*-pentagonal bipyramidal arrangement in a 2-D network, whereas with [1,4-dimethylpyridinium]$^+$ dimers are present. The reaction of [Zn(dmit)$_2$]$^{2-}$, SbI$_3$, and I$_2$ in dry THF produced black [Sb(dmit)$_3$]$^-$ which contains octahedral anions, linked into a 3-D structure by S$\cdots$S contacts.[275] In [PhSb(dmit)]·THF, prepared from Na$_2$dmit and PhSbCl$_2$, there is a *pseudo*-pentagonal bipyramidal arrangement based upon a chelated dmit, O-coordinated THF, the lone pair, and secondary Sb$\cdots$S intermolecular contacts.[276] In the antimony(V) compounds, [R$_2$Sb(dmit)$_2$]$^-$ (R = Ph, p-tolyl) the R groups occupy *cis* positions on a slightly distorted octahedron.[277]

(**16**)

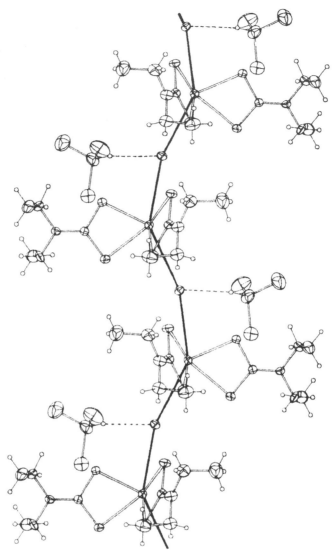

Figure 39   The polymeric chain in [Sb(S$_2$CNEt$_2$)$_2$I]·CHCl$_3$ (reproduced by permission of the Royal Society of Chemistry from *J. Chem. Soc., Dalton Trans.* **1981**, 1360–1365).

The only example of a diselenolate complex is [Sb(mns)$_2$]$^{3-}$ (mns = [Se$_2$C$_2$(CN)$_2$]$^{2-}$) made by combination of K, Sb$_2$Se$_3$, and Se in liquid ammonia in the presence of [2,2,2]crypt, followed by extraction with MeCN.[278] The anion is tetrahedral with formally Sb^I, although presumably the ligand is "noninnocent." Curiously, the corresponding reaction with arsenic gives an As^V selenide, [AsSe$_3$(CH$_2$CN)]$^{2-}$.[278]

There has been much recent interest in the synthesis and structures of antimony polychalcogenides. There are two comprehensive reviews[279,280] of these compounds which also place their structures in the context of those of related elements, and these should be consulted for details of work pre-1998. Examples reported since then include [Sb$_3$S$_5$]$^-$, [Sb$_4$S$_7$]$^-$, [Sb$_4$S$_8$]$^{4-}$, [Sb$_2$S$_5$]$^{4-}$, [Sb$_3$S$_{25}$]$^{3-}$, and [Sb$_2$S$_{15}$]$^{2-}$.[281–284] The stoichiometries provide little guide to the molecular units present or the connectivities. For example, in the material of stoichiometry [PPh$_4$]$_3$[Sb$_3$S$_{25}$], two quite different polythioantimonate ions are present, [Sb$_2$S$_{17}$]$^{2-}$ and [Sb$_2$S$_{16}$]$^{2-}$ (Figure 45).[281]

3.6.3.4   Group 17 Ligands

The haloanions of Sb^{III} have a surprisingly wide range of stoichiometries and structures, and several different structural motifs have been identified for particular Sb:X stoichiometries. In the

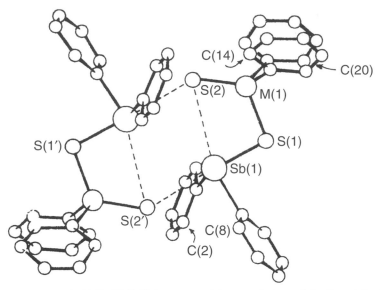

Figure 40 The structure of [Ph₂Sb(S₂SPPh₂)] (reproduced by permission of the Royal Society of Chemistry from *J. Chem. Soc., Dalton Trans.* **1986**, 1031–1034).

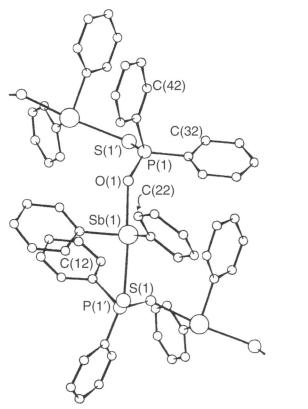

Figure 41 The chain structure of [Ph₂Sb(OSPPh₂)] (reproduced by permission of Elsevier Science from *J. Organomet. Chem.* **1986**, *316*, 281–289).

larger anions, there is often asymmetric halide bridging and the distinction between intra- and intermolecular Sb···X is not always clear. Most of the interest resides in the solid-state structures, and the aim here has been to summarize the major types known, giving representative references. Some of the anions with the heavier halides exhibit a variety of phases which show ferroelectric or ferroelastic behavior. The field has been reviewed twice.[91,285]

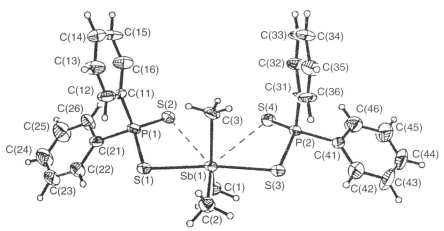

Figure 42 The structure of [SbMe₃(S₂PPh₂)₂] (reproduced by permission of Taylor & Francis Ltd. from *Main Group Met. Chem.* **1995**, *18*, 387–390).

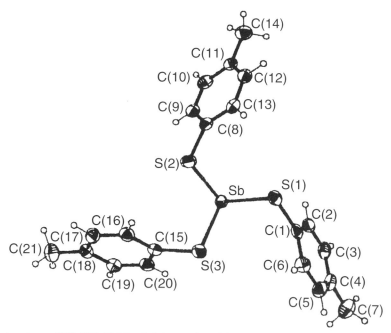

Figure 43 The structure of [Sb(SC₆H₄Me-4)₃] (reproduced by permission of the Royal Society of Chemistry from *J. Chem. Soc., Dalton Trans.* **1995**, 2129–2135).

Much of the systematics of the Sb^{III} fluoroanions were established pre-1980, including identification of [SbF₄]⁻, [Sb₂F₇]⁻, [Sb₃F₁₀]⁻, and [Sb₄F₁₃]⁻ and this work has been discussed by Sawyer and Gillespie.[285] There are also mixed-valence Sb^{III}–Sb^V fluoro-anions and cations, obtained from SbF₃ + SbF₅, by fluorination of Sb under controlled conditions, or during the syntheses of homoatomic cations of group 16 or 17 in superacid media containing SbF₅.[285] Further examples of the mixed-valence materials include Sb₇F₂₉ (=3SbF₃·4SbF₅, which is structurally described as [Sb^{III}F][Sb^{III}F₂]₂[Sb^{V}F₆]₄),[286] Sb₁₁F₄₃ (=6SbF₃·5SbF₅, [Sb^{III}₆F₁₃][Sb^{V}F₆]₅),[287] and two forms of Sb₈F₃₀ ([Sb^{III}₅F₁₂][Sb^{V}F₆]₃)[288] and [Sb^{III}₂F₅][Sb^{III}₃F₇][Sb^{V}F₆]₃).[289] (The formulations given are those of the authors, and those adopted for the cations involve a subjective judgment between intra- and intermolecular Sb^{III}—F bond lengths.) Typical of the complex products obtained with group 16 homoatomic cations are [[S₈][Sb₃F₁₄][SbF₆]] and [[S₄][Sb₂F₄][Sb₂F₅][SbF₆]₅].[290]

The simplest fluoroantimonate, [SbF₄]⁻ is found in [NMe₄]⁺,[291] guanidinium⁺,[292] and [H₃NCH₂CH₂NH₃]²⁺ salts;[293] it has the expected *pseudo*-trigonal bipyramidal geometry, with an equatorial lone pair and longer axial (av = 2.015 Å) than equatorial (av = 1.906 Å) Sb—F

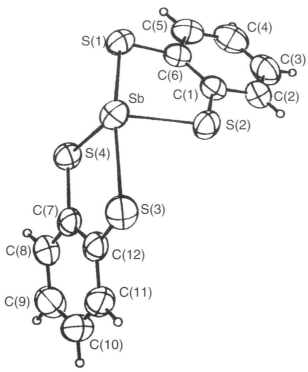

Figure 44 The structure of $[Sb(1,2-S_2C_6H_4)_2]^-$ (reproduced by permission of the Royal Society of Chemistry from *J. Chem. Soc., Dalton Trans.* **1994**, 1213–1218).

bonds.[291] Larger fluoroantimonate(III) anions include $[H_4([14]aneN_4)][Sb_4F_{16}]$, which contains tetrameric anions (Figure 46), $[H_2([14]aneN_4)][Sb_2F_{10}]\cdot2HF$ (Figure 47), $[H_4([14]aneN_4)]_2$-$[Sb_4F_{15}][HF_2]F_4$,[294] $[NH_4]_3[Sb_4F_{15}]$, and $Cs_3[Sb_4F_{15}]$ which have different structures,[295] and $NaCs_3[Sb_4F_{16}]\cdot H_2O$.[296]

The simplest stoichiometry for a chloroantimonate(III) is $[SbCl_4]^-$, but this is usually found in dimer or polymer units. The $[Sb_2Cl_8]^{2-}$ anion, composed of edge-sharing square pyramidal units with the apical chlorides *anti*, is found in the $[Bu^tH_3N]^+$,[297] $[Pr_4N]^+$,[298] and $[1,1'-Me_2-4,4'-$bipyridinium]$^{2+}$[299] salts, the last containing two crystallographically distinct anions which differ significantly in the asymmetry of the bridges. Curiously, in $[Bu^n_4N]_2[Sb_2Cl_8]$, which is also an edge-shared dimer, the apical chlorines are *syn* not *anti*.[298]

Polymeric zig-zag chains composed of distorted $SbCl_6$ units sharing adjacent edges are present in $[Mg(MeCN)_6][SbCl_4]_2$,[300] $[Fe([9]aneS_3)_2][SbCl_4]_2$,[301] $[Fe(Cp)_2]_2[SbCl_4]_2[SbCl_3]$,[302] and $[N,N',N'',N'''-Me_4$-guinidinium][SbCl_4]$.[303] Examples of $[Sb_4Cl_{16}]^{4-}$ units include those with $[Et_4N]^+$ (Figure 48)[298] and $[EtMe_2PhN]^+$,[304] whilst the $[H_2thiamine]^{2+}$ salt contains a chain of four antimony atoms $[Cl_3Sb(\mu-Cl)_2SbCl_2(\mu-Cl)_2SbCl_2(\mu-Cl)_2SbCl_3]^{4-}$, the outer two being five-coordinate.[305]

The mononuclear $[SbCl_5]^{2-}$ unit, with a square-pyramidal geometry, is present in $[NEt_4]^+$[306] and $[HMe_2NCH_2CH_2NH_3]^{2+}$[307] salts, whilst the edge-shared bi-octahedral $[Sb_2Cl_{10}]^{4-}$ was isolated with $[H_2thiamine]^{2+}$ cations.[308] Polymeric chains based upon vertex-linked $SbCl_6$ octahedra are present in $[NMe_2H_2]_2[SbCl_5]$,[309] $[C_5H_{12}N]_2[SbCl_5]$,[310] and $[4,4'$-bipyridinium][SbCl_5]$.[311] However, in $[2,2'$-bipyridinium][SbCl_5]$ there is a tetrameric unit (Figure 49).[311] The well-known hexachloroantimonate(III), $[SbCl_6]^{3-}$, is close to a regular octahedron with a stereochemically inactive lone pair.[298]

The $[Sb_2Cl_9]^{3-}$ ion is known both as a discrete confacial bioctahedron as in $[H_2([9]aneN_3)]_2[Sb_2Cl_9]Cl\cdot MeCN$,[18] and as a polymer in $[Y]_3[Sb_3Cl_9]$ (Y = Hpy or $C(NH_2)_3$).[312,313] Two examples of the $[Sb_2Cl_{11}]^{5-}$ anion have been structurally characterized, both of which contain octahedral $SbCl_6$ units sharing a vertex[161,314] The largest discrete chloro-antimonate(III) anion is in $[Me_4N]_4[Sb_8Cl_{28}]$, which consists of eight face-sharing octahedra.[315] Bromoantimonate(III) anions mostly resemble the chloro-analogues. Structurally characterized examples include the $[PPh_4]_2[Sb_2Br_8]$ (edge-shared dimer with *anti* disposed apical Br),[227,316] $[4-MeC_5H_4NH]_2[SbBr_5]$ (dimer),[317] and $[H_3N(CH_2)_6NH_3][SbBr_5]$ (*cis* edge-linked polymer).[318]

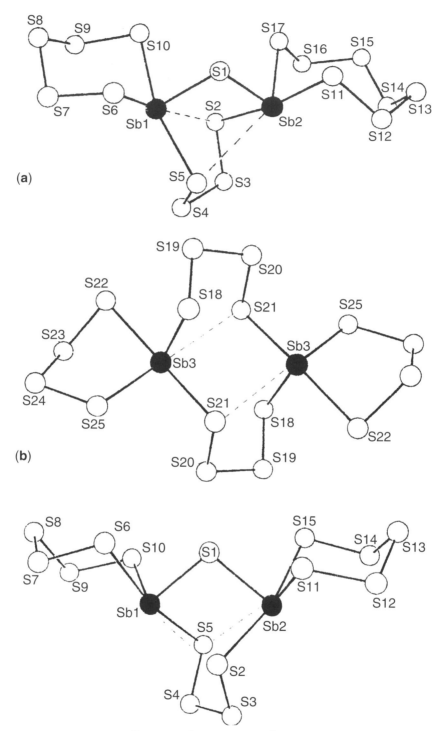

Figure 45 The structures of $[Sb_2S_{17}]^{2-}$, $[Sb_2S_{16}]^{2-}$, and $[Sb_2S_{15}]^{2-}$ (reproduced by permission of Wiley-VCH from *Z. Anorg. Allg. Chem.* **1998**, *624*, 310–314).

For the $[Sb_2Br_9]^{3-}$ type, those with $[NMe_4]^+$,[312] and $[EtMe_2PhN]^+$,[319] are confacial bioctahedra which contrast with the polymer structure present in $[Hpy]_3[Sb_2Cl_9]$ (above). Hall *et al.*[312] also prepared $[Hpy]_3[Sb_2Cl_{9-x}Br_x]$ ($x = 1–8$), $Cs_3[Sb_2Cl_6Br_3]$, and $Cs_3[Sb_2Cl_3Br_6]$, and a single crystal X-ray structure determination on $[NMe_4]_3[Sb_2Cl_6Br_3]$ showed a confacial bioctahedron with bromine bridges. X-ray powder patterns suggested the other bromochloroantimonates(III) were discrete dimers. In $[H_3NCH_2CH_2NH_3]_5[Sb_2Br_{11}]\cdot 4H_2O$, discrete dimeric anions are linked into chains by hydrogen-bonded water molecules.[320] Two larger anions are $[LH]_4[Sb_4Br_{16}]$

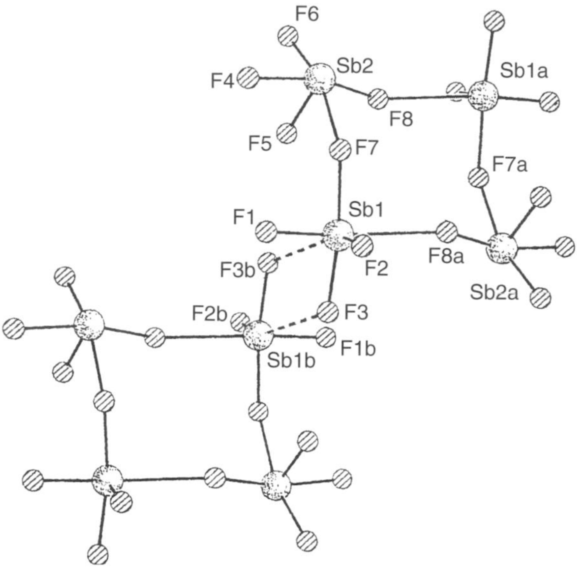

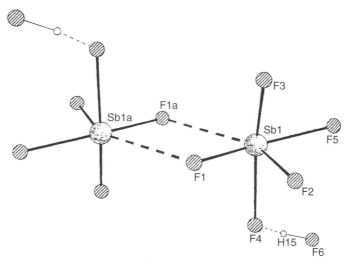

Figure 46 The structure of tetrameric $[Sb_4F_{16}]^{4-}$ (reproduced by permission of Wiley-VCH from *Z. Anorg. Allg. Chem.* **1996**, *622*, 105–111).

Figure 47 The structure of dimeric $[Sb_2F_{10}]^{2-} \cdot 2HF$ (reproduced by permission of Wiley-VCH from *Z. Anorg. Allg. Chem.* **1996**, *622*, 105–111).

(L = 2-amino-1,3,4-thiadiazolium),[321] and $[PPh_4]_4[Sb_8Br_{28}]$.[322] Both have antimony in distorted edge-shared octahedral environments. Other mixed halide species are $[Hpy][SbBr_2Cl_2]$, an infinite polymer,[312,323] and $[Hpy]_8[Sb_4Br_{12}Cl_8]$.[323]

Iodoantimonates(III) have shown an extraordinary structural diversity, particularly in the larger anions. The dinuclear examples are $[Me_3BzN][Sb_2I_7]$ (polymeric),[324] $[PPh_4]_2[Sb_2I_8]$ (dimer *anti*-axial iodides),[325] $Cs_3[Sb_2I_9]$,[326] and $[EtMe_2PhN]_3[Sb_2I_9]$ (both confacial bioctahedron),[327] whilst $[H_3N(CH_2)_6NH_3][SbI_5]$ is a polymer with SbI_6 octahedra sharing *cis* vertices.[318] Three types with an $[Sb_3I_{10}]^-$ stoichiometry are known: in $[PPh_4][Sb_3I_{10}]$ (Figure 50)[328] and $[Cu(MeCN)_3][Sb_3I_{10}]$ (Figure 51)[329] the distorted octahedral antimony centers are linked in different ways, whereas in $[Me_3\{2-(4-NO_2C_6H_4)CH_2CCH_2\}N][Sb_3I_{10}]$ there are close packed iodides with antimony disordered within the octahedral holes.[330] An $[Sb_3I_{11}]^{2-}$ anion is present in the $[Cu(MeCN)_4]^+$ salt.[331] In $[K(15\text{-crown-}5)_2][Sb_3I_{12}]$ there are three face-sharing SbI_6 units (alternatively described as an SbI_6 octahedron with two opposite faces capped by SbI_3 groups).[332] If four faces of SbI_6 are capped by SbI_3, the result is shown in Figure 52a and is found in

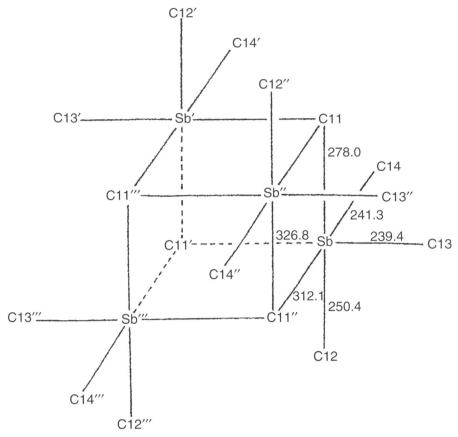

Figure 48 The structure of $[Sb_4Cl_{16}]^{4-}$ (reproduced by permission of the publishers from *Z. Naturforsch., B* **1982**, *37*, 1584–1589).

$[H\{OP(NMe_2)_3\}_2]_3[Sb_5I_{18}]$,[333] whereas an isomeric form in $[NMe_4]_3[Sb_5I_{18}]$ is an SbI_6 unit with four edges sharing a common vertex and bridged by SbI_3 units (Figure 52(b)).[331]

Three isomers of $[Sb_6I_{22}]^{4-}$ are known; in $[Fe(Cp)_2]_4[Sb_6I_{22}]$, $[EtMe_2PhN]_4[Sb_6I_{22}]$,[304] and [tetra-methylpyrazinium]$_4[Sb_6I_{22}]$[334] the unit is based upon Sb_4I_{16} with two opposite triangular faces capped by SbI_3,[331] whereas in $[Fe(1,10-phen)_3]_2[Sb_6I_{22}]\cdot2MeCN$,[335] and $[Et_3BzN]_4[Sb_6I_{22}]$[337] the two capping SbI_3 groups are differently placed. Three isomers of $[Sb_8I_{28}]^{4-}$ have also been identified.[324,333,335,337] In the $[PPh_4]^+$ or $[H(DMPU)_2]^+$ (DMPU $= N,N'$-dimethylpropylene urea)[333] a relatively regular unit is present (Figure 53), but the other two are based upon SbI_3 units edge- or face-bridging with smaller iodoantimonate anions.

A variety of organohaloantimonates(III) have been prepared and structurally characterized. Since the organo-groups normally occupy terminal positions, these anions are much less prone to polymerization than the homoleptic halo-analogues. The $[Ph_2SbX_2]^-$ (X = Cl, Br, or I)[338-341] are *pseudo*-trigonal bipyramidal with axial X groups and an equatorial lone pair. $[PhSbCl_4]^{2-}$ is monomeric,[338] but $[PhSbX_3]^-$ ($X_3 =$ Cl$_3$, ClBr$_2$, I$_3$) are $(\mu$-X$)_2$ dimers.[338,340,341] The $[Ph_2Sb_2X_7]^{3-}$ (X = Cl, Br, or I) contain square pyramidal $PhSbX_4$ units sharing a common vertex.[339,341]

Haloantimonate(V) chemistry is much simpler than that just discussed and is dominated by the $[SbF_6]^-$ and $[SbCl_6]^-$ anions. Both octahedral anions are popular choices as weakly coordinating anions in many areas of coordination chemistry,[103] but also commonly arise from abstraction of a halide ion from other reagents during the use of SbF_5 or $SbCl_5$ as powerful Lewis acids in organic, inorganic, and organometallic synthesis. Good examples are provided by the many main group and transition metal fluoro- or oxofluoro-cations, with $[SbF_6]^-$ produced in this way.[342] The singly F-bridged $[Sb_2F_{11}]^-$ and $[Sb_3F_{16}]^-$ ions are sometimes formed in these reactions. The pentagonal bipyramidal $[SbF_7]^{2-}$ ion has recently been prepared by heating two molar equivalents of MF and SbF_5 (M = K or Cs) or from NMe_4F and SbF_5 in MeCN.[105] Five of the eight possible isomers of mixed $[SbF_{6-n}Cl_n]^-$ ($n = 1–5$) have been identified in solution by ^{19}F and ^{121}Sb NMR spectroscopy, and some isolated (impure).[343] A range of $[SbBr_{6-n}Cl_n]^-$ anions has

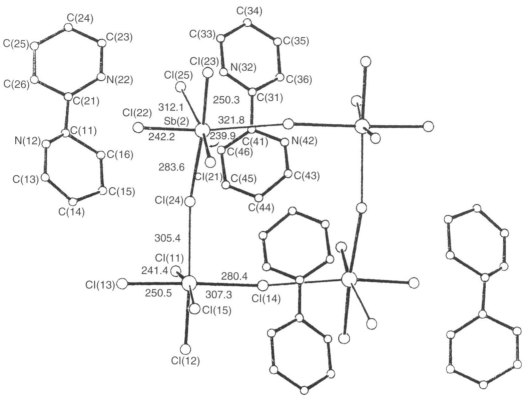

Figure 49 The tetrameric unit in [BipyH$_2$][SbCl$_5$] (reproduced by permission of the publishers from *Z. Naturforsch., B* **1983**, *38*, 1615–1619).

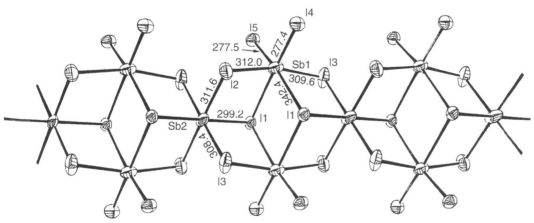

Figure 50 The polymer chain in one isomer of [Sb$_3$I$_{10}$]$^-$ (reproduced by permission of the publishers from *Z. Naturforsch., B* **1987**, *42*, 1493–1499).

also been detected in MeCN solution by NMR methods.[344,345] Organohaloantimonates(V) are similarly six-coordinate, including [SbPhX$_5$]$^-$ (X = Cl or Br), *trans*-[SbPh$_2$Cl$_4$]$^-$, and [SbRCl$_{5-n}$Br$_n$]$^-$ (R = Ph or Me).[346–349]

The [Sb$_2$OCl$_6$]$^{2-}$ and some related [Sb$_2$OCl$_{6-n}$Br$_n$]$^{2-}$ anions were prepared by cautious hydrolysis of [Sb$_2$Cl$_9$]$^{3-}$ or [SbCl$_{9-n}$Br$_n$]$^{3-}$.[350] The structure has been established in several salts,[18,19,351] and is shown in (17). The anion in [NH$_4$]$_3$[Sb$_2$OCl$_7$] has a third bridging Cl group forming a chain structure.[352] In the thiochloroanions [Sb$_2$SCl$_5$]$^-$ and [Sb$_2$SCl$_6$]$^{2-}$, obtained from Na$_2$S$_4$ and [Sb$_3$Cl$_{11}$]$^{2-}$ or [Sb$_2$Cl$_8$]$^{2-}$, respectively, the sulfur and one or two chlorines form the bridges.[353]

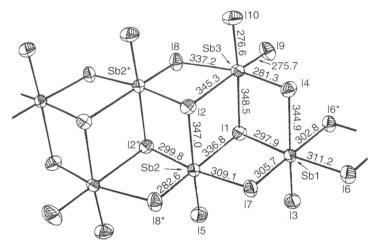

Figure 51 The structure of a second polymeric isomer of $[Sb_3I_{10}]^-$ (reproduced by permission of Wiley-VCH from *Angew. Chem., Int. Ed. Engl.* **1986**, *25*, 825).

(17)

3.6.3.5 Antimony in the Environment, Biology, and Medicine

The environmental distribution of antimony is wide, and it has been estimated that the atmospheric distribution resulting from man's activities (fossil fuels, mining, etc.) is greater than from natural sources. Most of the forms are simple inorganics in the form of oxides and oxoanions. Although some methylantimony species are found, their mode of production is not clear and attempts to demonstrate biological methylation have been inconclusive.[112]

However, in contrast to arsenic, antimony compounds remain of medicinal importance, particularly in the treatment of various parasitic infections including leishmaniasis, schistosomiasis, and trypanosomasis. Coordination compounds used include the sodium or potassium antimony tartrates and substituted catecholate complexes, which are O-donor complexes, whilst S-donors are utilized in 2,3-dithiosuccinate derivatives.[111,113] While the mode of action is unclear in many cases, the attraction of these compounds, in addition to their clinical effectiveness, is their simplicity of manufacture and low cost.

3.6.4 BISMUTH

In addition to the general reviews cited in Section 3.6.1, there is a detailed review by Briand and Burford[354] which describes bismuth complexes of groups 15 and 16 donor ligands. In contrast to antimony, there is very little coordination chemistry of bismuth in the V oxidation state, compounds being limited to fluoroanions and a few organobismuth species.

3.6.4.1 Group 14 Compounds

Despite the relative weakness and high reactivity of Bi—C bonds, there is an extensive chemistry of organobismuth compounds. In addition to the reviews noted in Section 3.6.1, other articles specifically focused on bismuth are available.[116,355,356] The review by Silvestri *et al.*[356] contains detailed descriptions of the structures of complexes of organobismuth compounds and contains much pertinent data on the stereochemistries adopted by bismuth. Organobismuth analogues of

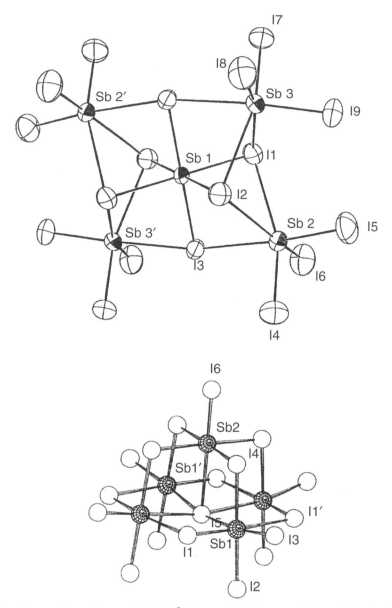

Figure 52 The structures of two forms of $[Sb_5I_{18}]^{3-}$ (reproduced by permission of (a) Elsevier Science from *Polyhedron* **1993**, *12*, 2081–2090, and (b) Wiley-VCH from *Angew. Chem. Int. Ed. Engl.* **1989**, *28*, 344–345).

pyridine, and examples of RBi = BiR and (RBi)$_n$ are the least stable of group 15 compounds of these types.[13,118] Like their antimony analogues, silylbismuthines have been prepared and explored as MOCVD precursors for (**III**)–(**V**) materials.[119]

3.6.4.2 Group 15 Compounds

3.6.4.2.1 *N-donor ligands*

There are many examples with nitrogen heterocycles, but aliphatic amine complexes are of low stability[357,358] and few are known. There has been recent interest in bismuth amides as precursors for both bismuth-containing semiconductors and superconductors. The simplest examples, $[Bi(NMe_2)_3]$ and $[Bi(NPh_2)_3]$, made from $LiNR_2$ and $BiCl_3$, are unstable to air and light. Both are trigonal pyramidal monomers.[359,360] In other systems, more complex products result: from lithiated 2,6-Pr$^i_2$C$_6$H$_3$NH$_2$ and BiCl$_3$ the cyclic dimer (**18**) is formed,[361] whilst 2,6-Me$_2$C$_6$H$_3$NH$_2$

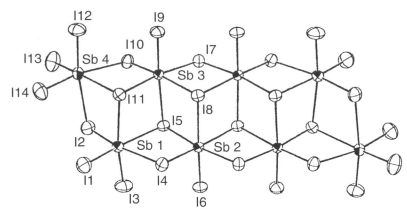

Figure 53 The structure of one form of $[Sb_8I_{28}]^{4-}$ (reproduced by permission of Elsevier Science from *Polyhedron* **1993**, *12*, 2081–2090).

produced the tetramer (**19**).[362] Using the bulkier "supermesityl" group in $NH_2(2,4,6\text{-Bu}^t{}_3C_6H_2)$ produces monomeric tris(amide) complexes with both antimony and bismuth, although neither arsenic nor phosphorus form analogous compounds.[363] The mononuclear compound with the chelating tripodal triamide, $[\{HC(SiMe_2NBu^t)_3\}Bi]$, is both air and light stable.[364]

R = $^iPr_2C_6H_3$

R = $Me_2C_6H_3$

(**18**)   (**19**)

The reaction of heated $BiBr_3$ or BiI_3 with gaseous NH_3 results in $[BiX_3(NH_3)]$ complexes, and $BiBr_3$ reacts with amines (NH_2Me, $NHMe_2$, NMe_3, or NH_2Ph) in benzene to give 1:1 complexes.[16] None have been structurally characterized. Bismuth complexes with aminoalcohols, aminothiols, and aminocarboxylic acids are discussed in Section 3.6.4.3. The carbanionic ligand (**1**) forms $[Bi(\mathbf{1})Cl_2]$ which almost certainly has the bismuth coordinated to a CN_2Cl_2 donor set.[20] From the reaction of $BiCl_3$ and three equivalents of (**3**), $(2\text{-}Me_2NCH_2C_6H_4)^-$, the product is colorless $[Bi(\mathbf{3})_3]$, which has the same structure as the antimony analogue with a trigonal pyramidal BiC_3 core and three weaker Bi–N interactions completing a distorted octahedron.[22] The same reagents in a 1:2 ratio produce $[Bi(\mathbf{3})_2Cl]$ which is best described as a *pseudo*-trigonal bipyramid with apical N and Cl atoms, two equatorial C and with the lone pair occupying the third equatorial position; the second dimethylamino-group is only very weakly associated.[21] Treatment of this complex with $TlPF_6$ removes the chlorine and forms $[Bi(\mathbf{3})_2]PF_6$ which is also *pseudo*-trigonal bipyramidal with axial amines and an equatorial lone pair.[127] From 8-(dimethylamino)-1-naphthyllithium, $[Li(\mathbf{2})]$, both $[Bi(\mathbf{2})Cl_2]$ and $[Bi(\mathbf{2})_2Cl]$ are formed depending upon the conditions,[21] but attempts to make $[Bi(\mathbf{3})Cl_2]$ directly failed. Comproportionation of $[Bi(\mathbf{3})_3]$ and $BiCl_3$ in Et_2O did give $[Bi(\mathbf{3})Cl_2]$, but this proved insoluble in common solvents. However, metathesis with NaI gave $[Bi(\mathbf{3})I_2]$ which has an iodide-bridged dimer structure with square-pyramidal geometry at bismuth (Figure 54).[21] There are closely related organobismuth species including $[Bi(4\text{-}MeC_6H_4)(2\text{-}Me_2NCH_2C_6H_4)Cl]$,[365] and $[Bi(Ph)\{2\text{-}(R)\text{-}1\text{-}Me_2N(Me)CHC_6H_4\}Cl]$.[366]

The triaza macrocycle $Me_3[9]aneN_3$ reacts with $BiCl_3$ in MeCN to form yellow $[Bi(Me_3[9]aneN_3)Cl_3]$ which has a discrete distorted octahedral structure.[18,367] A derivatized variant, 1-carboxymethyl-4,7-bis(1-methylimidazol-2-ylmethyl)-1,4,7-triazacyclononane, (LH) forms $[BiClL]BPh_4$, which is dimeric with each bismuth eight-coordinate, and bonded to the three macrocycle and two imidazole nitrogens, a chlorine, and two carboxylate oxygens, the carboxylates forming the bridges.[368] Bismuth compounds with tetra-azamacrocycles[357,358,369,370] also have

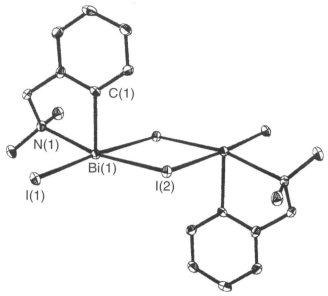

Figure 54 The structure of [BiI₂(2-Me₂NCH₂C₆H₄)] (reproduced by permission of the American Chemical Society from *Inorg. Chem.* **1997**, *36*, 2770–2776).

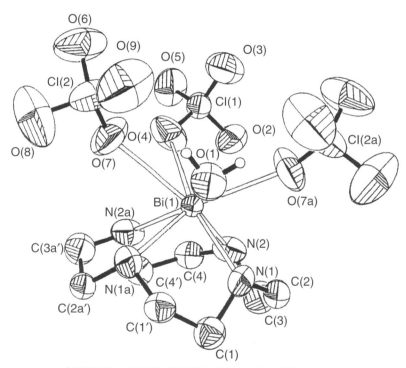

Figure 55 The structure of [Bi([12]aneN₄)(H₂O)(ClO₄)₃] (reproduced by permission of the Royal Society of Chemistry from *J. Chem. Soc., Dalton Trans.* **1997**, 901–908).

high coordination numbers. In [Bi([12]aneN₄)(H₂O)(ClO₄)₃], made from Bi₂O₃, HClO₄, and [12]aneN₄, the bismuth is eight-coordinate with a square antiprismatic geometry composed of the four macrocyclic nitrogens on one face and four oxygens (water and three η^1-ClO₄ groups) on the other (Figure 55).[369] The pendant arm analogue (**20**) forms [Bi(**20**)](ClO₄)₃·H₂O in which the bismuth is coordinated to an N₄O₄-donor set from (**20**), with a weakly coordinated perchlorate (3.34 Å) interacting through the O₄ face.[370] Bismuth(III) porphyrins have been little studied.[371–373] The dark green [Bi(TTP)][NO₃] (TTPH₂ = *meso*-tetra-*p*-tolylporphyrin) is made from TTPH₂ and bismuth nitrate in pyridine.[371] The [Bi(OEP)][CF₃SO₃] (OEPH₂ = octaethylporphyrin) has a dimeric structure with two Bi(OEP)$^+$ units linked by bridging $CF_3SO_3^-$ groups, producing seven-coordinate bismuth.[372] The complex of the pendant arm porphyrin (**21**) is also a dimer, with a distorted square antiprismatic bismuth coordination environment composed of the four nitrogens of the porphyrin, two oxygens from a nitrate group, a water molecule, and an ester oxygen.[373]

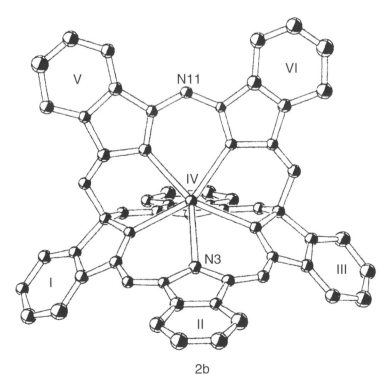

(20)   **(21)**

Bismuth phthalocyanines [Bi(pc)X] (X = Cl, Br, or NO_3, pcH_2 = phthalocyanine) are made by heating pcH_2 with the bismuth salt in Me_2CO or MeCN, or by heating BiX_3 with 1,2-dicyano-benzene.[374] However, when 1,2-dicyanobenzene and bismuth are heated in iodine vapor, the product is $[Bi(pc)]_4[Bi_4I_{16}]$.[375] Eight-coordinate bismuth is present in $[Bi(pc)_2]^-$ [126] which is anodically oxidized to green $[Bi(pc)_2]$, and by bromine to purple $[Bi(pc)_2]Br_x$ ($1.5 \leq x \leq 2.5$).[376] The triple-decker phthalocyanine $[Bi_2(pc)_3]$ consists of three pc rings with Bi atoms in a distorted square antiprismatic arrangement and interacting more strongly with the peripheral than the central pc ring.[377,378] The orange bicyclophthalocyanine shown in Figure 56 is made from $Bi(OAc)_3$ and 1,2-dicyanobenzene and contains bismuth in a very distorted trigonal prismatic arrangement.[379]

Crystallization of a 1:1 mixture of $BiCl_3$ and $SbCl_5$ from anhydrous MeCN produced $[BiCl_2(MeCN)_4][SbCl_6]$.[380] The reaction of $BiCl_3$ and pyridine afforded $[pyH]_2[BiCl_5(py)]$ which has a nearly octahedral anion.[381] The same anion has been obtained serendipitously in two other compounds,[382,383] and $[(4\text{-pic})H]_2[BiBrCl_5(4\text{-pic})]$.[381] The *cis*-$[BiI_4(py)_2]^-$,[384] and *trans*-$[BiBr_4(py)_2]^-$,[385] are also known. From excess pyridine and BiI_3 the product is *mer*-$[BiI_3(py)_3]$,[385] whilst $BiCl_3$ affords the pentagonal bipyramidal $[BiCl_3(py)_4]$ with axial Cls.[385]

2b

Figure 56 The structure of [Bi(1,24-bicyclophthalocyaninato)] (reproduced by permission of Wiley-VCH from *Eur. J. Inorg. Chem.* **2001**, 1343–1352).

Phenylbismuth halides form adducts with py, 4-MeC$_5$H$_4$N, and 4-Bu^{t}C$_5$H$_4$N(L) of type [BiPhX$_2$(L)$_2$] (X = Cl, Br, or I) which are square pyramidal with apical Ph and *trans*-X groups, which associate weakly into dimers (Figure 57) or chains via Bi$\cdots$X bridges.[385] Hydrolysis of [BiPhCl$_2$(Bu^{t}C$_5$H$_4$N)$_2$] produced [Bu^{t}C$_5$H$_4$NH][BiCl$_3$Ph(Bu^{t}C$_5$H$_4$N)], which is also square pyramidal with an apical Ph group. In none of these compounds is there much evidence of stereochemical activity by the bismuth lone pair. Diphenylbismuth halide adducts are rarer, but in [BiIPh$_2$(MeC$_5$H$_4$N)] the structure is a *pseudo*-trigonal bipyramid with equatorial Ph's and the lone pair, again with weak association into chains by long Bi$\cdots$I contacts.[385] Cationic [BiPh$_2$(py)$_2$]Y (Y = BF$_4$, PF$_6$) can be obtained from BiPh$_2$Br, py and, TlY.[384]

The reactions of 2,2'-bipyridyl or 1,10-phenanthroline with bismuth halides have been studied in detail.[386–391] Under some conditions the products are halobismuthate anions with protonated heterocycle cations (q.v. Section 3.6.4.4), but from reactions in acetonitrile or DMSO a range of structural types with coordinated diimines have been obtained. The 1:1 complexes [BiX$_3$L] (X = Br or I, L = 2,2'-bipy; X = Br, L = 1,10-phen) are dihalo-bridged dimers based upon distorted octahedral bismuth (Figure 58a).[389] The BiCl$_3$/2,2'-bipy system produces a [BiCl$_3$(2,2'-bipy)$_{1.5}$] which has an unusual structure containing (Figure 58b) a single chloride bridge between a seven-coordinate (N$_4$Cl$_3$) and a six-coordinate (N$_2$Cl$_4$) bismuth center.[389] The 2:1 complexes [BiX$_3$(L)$_2$] (X = Br or I, L = 2,2'-bipy, 1,10-phen; X = Cl, L = 1,10-phen) are all seven-coordinate monomers with distorted pentagonal bipyramidal geometries (the distortion appears to arise from the geometric constraints of the diimines).[390] The isolation of [BiCl$_3$(2,2'-bipy)$_2$] is problematic and a pure sample has not been obtained. Decomposition of the 2:1 bromo complex[387] produced [2,2'-bipyH][BiBr$_4$(2,2'-bipy)]. The chloro analogue has also been obtained. These are the expected (*cis*) octahedral monomers. Using DMSO as solvent produced [BiX$_3$(1,10-phen)(DMSO)$_2$]·DMSO (X = Cl or Br), [BiI$_3$(2,2'-bipy)(DMSO)], and [BiI$_3$(1,10-phen)(DMSO)$_{1.5}$].[388] The chloro- and bromo-compounds have seven-coordinate, pentagonal bipyramidal structures composed of one chelating diimine, three halides, and two O-bound DMSO ligands. The structure of [BiI$_3$(2,2'-bipy)(DMSO)] is a distorted octahedron with DMSO *trans* to (I), whilst [BiI$_3$(1,10-phen)(DMSO)$_{1.5}$] is ionic with the constitution [BiI$_2$(1,10-phen)(DMSO)$_3$][BiI$_4$(1,10-phen)].[388] Far IR and Raman spectra were reported for this extensive series of complexes.[387–390] Bismuth nitrate and 1,10-phenanthroline or 2,2'-bipyridyl produce [Bi(NO$_3$)$_3$(diimine)$_2$] which are

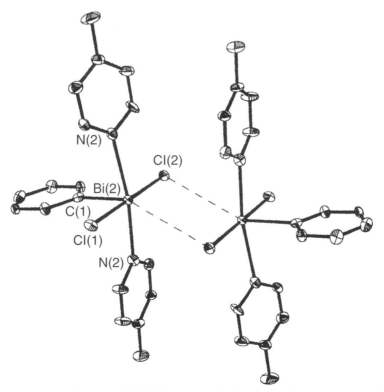

Figure 57 The structure of dimer of [BiPhCl$_2$(4-MeC$_5$H$_4$N)$_2$] (reproduced by permission of the Royal Society of Chemistry from *J. Chem. Soc., Dalton Trans.* **1999**, 2837–2843).

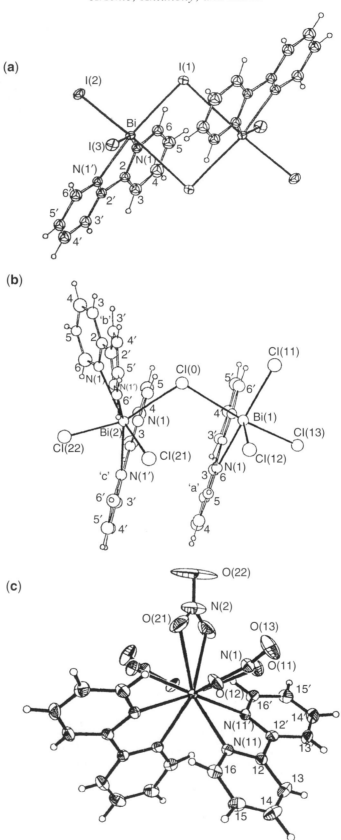

Figure 58 The structures of three 2,2-bipyridyl complexes: (a) the dimeric [BiI$_3$(2,2'-bipy)], (b) the dimeric [BiCl$_3$(2,2'-bipy)$_{1.5}$], and (c) mononuclear [Bi(NO$_3$)$_3$(2,2'-bipy)$_2$] (reproduced by permission of the Australian Chemical Society from *Aust. J. Chem.* **1998**, *51*, 325–330, and *Aust. J. Chem.* **1998**, *51*, 337–342).

10-coordinate with three bidentate nitrato-groups (Figure 58c).[391] From a lower Bi:1,10-phen ratio in DMSO, the product was [Bi₂(1,10-phen)₂(OH)₂(NO₃)₄] containing eight-coordinate bismuth (N₂O₆), based upon one diimine and two bidentate nitrates per bismuth, linked by two hydroxide bridges.[391] Other 2,2′-bipyridyl complexes are [Bi(2,2′-bipy)₂(NCS)₃], which is dimeric with eight-coordinate bismuth, linked by two bridging thiocyanates,[392] and [Bi(2,2′-bipy)(S₂C-NEt₂)I₂], a seven-coordinate dimer (N₂S₂I₃) with two bridging iodines.[393] The [Bi(terpy)(S₂C-NEt₂)I₂] is a pentagonal bipyramidal monomer with axial iodines,[393] but the structure of [Bi(terpy)(NCS)₃][394] is unknown.

There are 1:1 complexes [BiX₃(**22**)] (X = Cl, Br, I, NCS) of the Schiff base (**22**) of unknown structure.[386,394] More unusual examples of Bi–N coordination are found in the phosphine imine complexes [{BiF₂(NPEt₃)(HNPEt₃)}₂] and [Bi₂I(NPPh₃)₄]I₃.[395] The former, itself a very rare example of a coordination complex derived from BiF₃, has a structure based upon a Bi₂N₂ four-membered ring using the NPEt₃⁻ groups, with terminal F's and HNPEt₃ ligands.

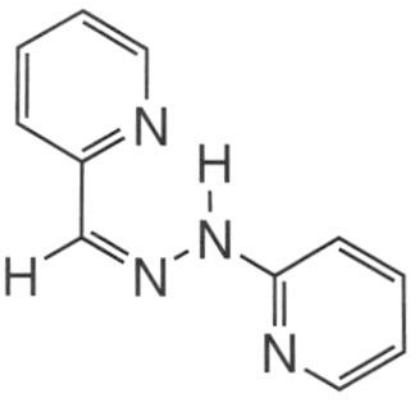

(22)

The azide chemistry of bismuth is limited to [Bi(N₃)₃].[396]

3.6.4.2.2 P- and As-donor ligands

Only a limited number of bismuth phosphines have been reported. The reaction of BiBr₃ with neat PMe₃ produced yellow [Bi₂Br₆(PMe₃)₄] which has a centrosymmetric structure (**23**), in which the bridges are very asymmetric.[143,397] The reaction of BiBr₃ and PMe₂Ph in THF produced [Bi₂Br₆(PMe₂Ph)₂(OPMe₂Ph)₂] (as a result of adventitious oxidation), which has the same basic structure.[397] However, from BiBr₃ and PEt₃, the product had a 1:1 stoichiometry and contained a tetrameric unit (Figure 59).[142] Two anionic species are known: the [PPh₄][BiI₄(PMe₂Ph)₂] is a distorted octahedron with *cis* phosphines,[144] whilst the anion in [PMe₃H][Bi₂Br₇(PMe₃)₂] is a chain polymer based upon a planar Br₂Bi(μ²-Br)₂BiBr₂ core with *anti*-axial PMe₃ groups, and with the axial bromines bridging the units.[398] Diphosphines including Me₂PCH₂CH₂PMe₂,[144] o-C₆H₄(PMe₂)₂,[145] o-C₆H₄(PPh₂)₂,[145] and Ph₂PCH₂CH₂PPh₂,[386,399,400] typically give 1:1 complexes all of which probably have the same structure or type as established for [Bi₂Br₆-(Me₂PCH₂CH₂PMe₂)₂],[144] and [Bi₂Cl₆(Ph₂PCH₂CH₂PPh₂)₂],[400] as halide-bridged dimers similar to those formed by antimony (Figure 19). However, in [Bi₂Cl₆(Ph₂PCH₂CH₂PPh₂)₂] (Figure 60) there are diphosphine bridges,[400] a motif common in transition metal complexes with this ligand attributed to the shorter interdonor linkage. A [Bi₂Cl₆(Ph₂PCH₂CH₂PPh₂)₃] complex has two BiCl₃(diphosphine) groups singly bridged by the third diphosphine giving six-coordinate bismuth.[400] In several of these systems adventitious oxygen produces diphosphine dioxide complexes[145,399] (see Section 3.6.4.3).

There seem to be no thoroughly characterized bismuth complexes with monodentate arsines, but diarsines include [BiX₃(diarsine)] (X = Cl, Br, or I; diarsine = Ph₂AsCH₂CH₂AsPh₂[145,399] and o-C₆H₄(AsMe₂)₂[145]). The structure of [Bi₂I₆{o-C₆H₄(AsMe₂)₂}₂] shows the same halide-bridged dimer type[145] (Figure 19) as found for the diphosphines and this is probably present in all. The

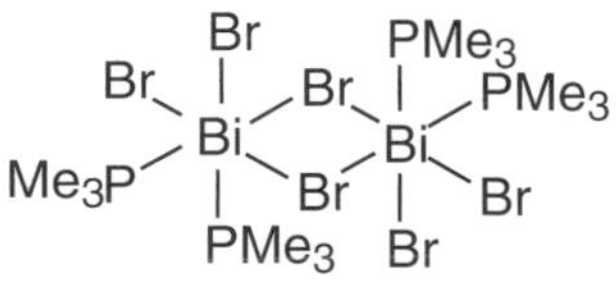

(23)

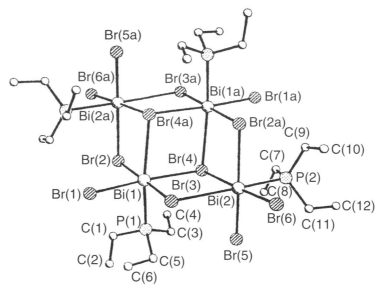

Figure 59 The tetrameric structure of [Bi$_4$Br$_{12}$(PEt$_3$)$_4$] (reproduced by permission of the Royal Society of Chemistry from *J. Chem. Soc., Dalton Trans.* **1994**, 1743–1751).

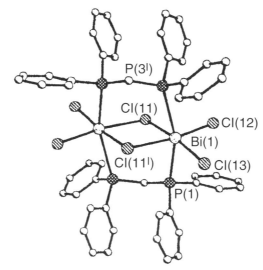

Figure 60 The structure of [Bi$_2$Cl$_6$(Ph$_2$PCH$_2$PPh$_2$)$_2$] (reproduced by permission of the Royal Society of Chemistry from *J. Chem. Soc., Dalton Trans.* **1996**, 1063–1067).

triarsine MeC(CH$_2$AsMe$_2$)$_3$ also forms 1:1 complexes with BiX$_3$.[145] The diarsine complexes show a tendency to oxidize to the corresponding diarsine dioxides.[399] NMR studies show that most of the phosphine and arsine complexes are labile in solution and extensively dissociated. There appear to be no examples of stibine complexes (contrast antimony Section 3.6.3.2.2), whilst the reaction of BiX$_3$ with BiR$_3$ typically yields the scrambled products BiX$_{3-x}$R$_x$.[355]

3.6.4.3 Group 16 Compounds

3.6.4.3.1 *O-donor ligands*

In aqueous solution and in the absence of coordinating ligands, arsenic and antimony are present either as oxides, oxoanions, or their protonated forms such as As(OH)$_3$.[2,3,6] However, for bismuth a wide range of basic salts are known and various polynuclear cations have been proposed or identified. The aquo-ion $[Bi(H_2O)_9]^{3+}$ has been isolated as the $CF_3SO_3^-$ salt by reaction of Bi_2O_3, CF_3SO_3H, and $(CF_3SO_2)_2O$.[401] The cation has a tricapped trigonal prismatic structure. The non-aquo ion appears to be specific to the triflate system, in $HClO_4$ or HNO_3 media the cation is the hexanuclear $[Bi_6O_4(OH)_4]^{6+}$, which has a bismuth octahedron with face-bridging oxide and hydroxide groups.[402] X-ray structures of $[Bi(H_2O)_9][CF_3SO_3]_3$ (O_9), $[Bi(DMSO)_8][ClO_4]_3$ (O_8), and $[Bi(N,N'$-dimethylpropyleneurea)_6][ClO_4]_3$ (O_6) provide examples of bismuth in homoleptic O-donor environments with different coordination numbers (Figure 61).[403] Bismuth L^{III} edge EXAFS and LAXS studies[403] of strongly acidic aqueous solutions of Bi^{3+} were consistent with $[Bi(H_2O)_8]^{3+}$; data were also reported for bismuth triflate solutions in various liquid organic

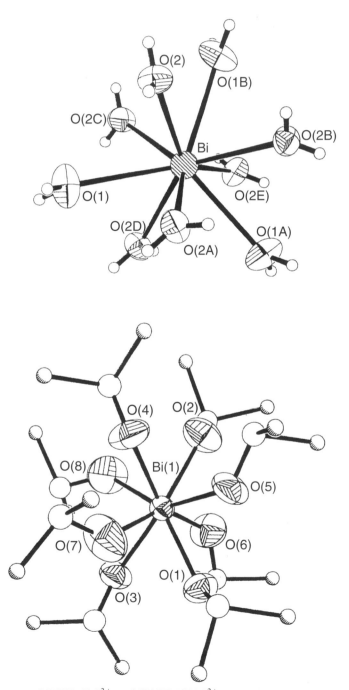

Figure 61 The structures of $[Bi(H_2O)_9]^{3+}$ and $[Bi(DMSO)_8]^{3+}$ (reproduced by permission of the American Chemical Society from *Inorg. Chem.* **2000**, *39*, 4012–4021).

ligands. The structures of various basic bismuth oxo-salts have been described, e.g., $[Bi_2(H_2O)_2-(SO_4)_2(OH)_2]$ and $[Bi_2O(OH)_2]SO_4$. The former has a planar $[Bi_2(OH)_2]^{2+}$ core with the bismuth coordination completed by a water molecule and one oxygen from each of three different sulfate groups.[404] In the second species, which is a further stage of hydrolysis, there are $[\{Bi(OH)^{2+}\}_n]$ chains bridged by oxides.[405]

Tetrahydrofuran adducts include *fac*-$[BiX_3(THF)_3]$ (X = Cl or Br),[406,407] $[BiCl_3(THF)_2]$ which is a polymer with pentagonal bipyramidal bismuth,[406] and $[Bi_2Cl_8(THF)_2]^{2-}$, an edge-shared bioctahedron with *anti*-axial THF groups.[147] The three $[BiPhX_2(THF)]$ (X = Cl, Br, or I) are isostructural with essentially square-pyramidal bismuth centers linked by single halide bridges, the sixth position being occupied by a weak π-Ph···Bi contact.[408] The polyethers $ROCH_2CH_2OCH_2CH_2OR$ (R = Me or Et) behave as tridentates in $[BiCl_3(ether)]$ which are dichloro-bridged dimers with pentagonal bipyramidal bismuth.[407] Longer chains in the polyethyleneglycols $HO(CH_2CH_2O)_nCH_2CH_2OH$ (n = 3, 4, 5, or 6) also produce $[BiX_3(glycol)]$ (X = Cl or Br), which are bicapped trigonal prisms (O_5X_3), although some ionic forms of type $[BiX_2(glycol)]^+$ are also known.[409] These polyethyleneglycols and bismuth nitrate form $[\{Bi(NO_3)_2(glycol-H)\}_2]$, in which one end of the ligand has been deprotonated, and the resulting alkoxides bridge the bismuth centers. Most of the ether oxygens coordinate to the Bi centers along with two bidentate nitrates (Figure 62).[410] An ionic form, $[Bi(NO_3)_2(glycol)][Bi(NO_3)_2(glycol-2H)]\cdot2H_2O$ (glycol = n = 4), contains a neutral glycol in the cation and a doubly deprotonated form in the anion.[410]

Structures have been determined for a variety of BiX_3-crown ether adducts; in general the structures are based upon a pyramidal BiX_3 group with the crown weakly capping the open face.[38,409,411,412] The bond lengths suggest that bismuth interacts more strongly with the crowns than either As or Sb with bismuth halides the smaller crowns 12-crown-4, 15-crown-5, and benzo-15-crown-5 generate mononuclear seven- and eight-coordinate bismuth respectively.[38,409,411,412] However, $[Bi(NO_3)_3(12\text{-crown-}4)]$ is 10-coordinate with three bidentate nitrate groups.[410] In the presence of $SbCl_5$, $BiCl_3$, and 12-crown-4 react in MeCN solution to form $[Bi(12\text{-crown-}4)_2(MeCN)][SbCl_6]_3$, which contains nine-coordinate bismuth,[413] whereas 15-crown-5 gives the eight-coordinate monocation $[BiCl_2(15\text{-crown-}5)(MeCN)][SbCl_6]$.[154] The larger ring in 18-crown-6 offers a number of bonding motifs. In $[Bi(NO_3)_3(H_2O)_3(18\text{-crown-}6)]$[410] the bismuth is coordinated to three waters and three bidentate nitrates, with the crown H-bonded to the water but not interacting directly with the bismuth. In the $BiCl_3$/18-crown-6 system, four different structures have been identified: $[BiCl_3(18\text{-crown-}6)]$ (nine-coordinate O_6Cl_3);[414] $[BiCl_3(18\text{-crown-}6)(H_2O)]$ (seven-coordinate $O_3Cl_3 + O(water))$;[415] $[BiCl_3(18\text{-crown-}6)(MeOH)]$ (seven-coordinate $O_3Cl_3 + O(methanol))$.[409] In $[BiCl_2(18\text{-crown-}6)]_2[Bi_2Cl_8]$ and $[BiBr_2(18\text{-crown-}6)][BiBr_4]$ the cation is a bicapped trigonal prism (O_6X_2) which can be regarded as the product of halide abstraction

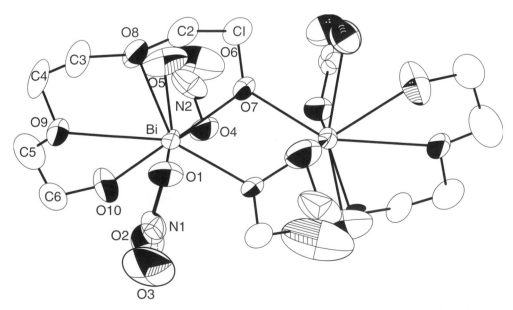

Figure 62 The structure of $[\{Bi(NO_3)_2(glycol-H)\}_2]$ (reproduced by permission of the American Chemical Society from *J. Am. Chem. Soc.* **1992**, *114*, 2967–2977).

by the weak Lewis acid BiX$_3$.[409,412] Using the stronger Lewis acid SbCl$_5$, results in the dication [BiCl(18-crown-6)(MeCN)$_2$][SbCl$_6$]$_2$ (O$_6$N$_2$Cl).[154] Finally, the large ring dibenzo-24-crown-8 coordinates two molecules of BiCl$_3$ on opposite faces of the crown (Figure 63), with each bismuth coordinated to five oxygens and three chlorines.[155] The maleonitrile-dithiacrown ether mn-15-S$_2$O$_3$ (**12a**) forms [BiCl$_3$(mn-15-S$_2$O$_3$)] which has bismuth in a Cl$_3$S$_2$O$_3$ environment, whilst the related mn-18-S$_2$O$_4$ (**12b**) binds only via the four oxygens in [BiCl$_3$(mn-18-S$_2$O$_4$)].[156]

Pnictogen- and chalcogen-oxides form a number of O-bonded bismuth complexes including *trans*-[BiI$_2${OP(NMe$_2$)$_3$}$_4$]$^+$, [Bi$_2$I$_6${OP(NMe$_2$)$_3$}$_2$],[415] [BiX$_3$(DMSO)$_3$] (X = Cl, Br), [Bi$_2$I$_4$ (μ-I)$_2$(DMSO)$_4$],[416] [BiX$_3$(diimine)(DMSO)$_2$] (see Section 3.6.4.2.1 above),[388] [Bi(DMSO)$_8$]$^{3+}$ (see Section 3.6.4.3.1 above),[403] [Bi(NO$_3$)$_3$(DMSO)$_3$] (nine-coordinate with three bidentate nitrates),[391] [Bi$_2$Ph$_2$Br$_4$(OPPh$_3$)$_2$],[417] [Bi(2,4,6-Me$_3$C$_6$H$_2$)$_2$BrL] (L = OSPh$_2$, OP(NMe$_2$)$_3$),[417] [BiR$_2$(L^2)$_2$]PF$_6$ (L^2 = OPPh$_3$, OP(NMe$_2$)$_3$; R = Ph, 4-MeC$_6$H$_4$, 2,4,6-Me$_3$C$_6$H$_2$),[384] and [BiPh{OP(N-Me$_2$)$_3$}$_4$][PF$_6$]$_2$.[384] Complexes with transition metal fragments which overall are isoelectronic with [BiX$_2$L$_2$]$^+$ are also known and have Bi-metal bonds, e.g., [Bi{OP(NMe$_2$)$_3$}$_2${Fe(CO)$_2$(Cp)$_2$}$_2$]$^+$.[418]

As described in Section 3.6.4.2.2, the adventitious oxidation of diphosphines or diarsines in the presence of BiX$_3$ result in complexes of the corresponding dioxide ligands.[145,400] Structurally characterized examples include [Bi$_2$Cl$_6${Ph$_2$P(O)CH$_2$P(O)Ph$_2$}$_2$], structure (**24**), which contrasts with that of the "parent" diphosphine complex (Figure 60),[400] and [BiCl$_3$(THF)-{o-C$_6$H$_4$(P(O)Ph$_2$)$_2$}],[145] but the product isolated from the reaction involving Ph$_2$AsCH$_2$CH$_2$AsPh$_2$ is [BiCl$_3${Ph$_2$As(O)CH$_2$CH$_2$As(O)Ph$_2$}{Ph$_2$MeAsO}], in which the Ph$_2$MeAsO apparently comes from cleavage of the diarsine.[400] The complexes can be made directly from the diphosphine dioxide and BiX$_3$, e.g., [BiX$_3$(THF)$_n${o-C$_6$H$_4$(P(O)Ph$_2$)$_2$}], X = Cl, n = 1; X = Br, n = 0.[145] Pyridine N-oxides also form complexes, e.g., ligands (**25**) function as tridentates in [Bi(NO$_3$)$_3$(**25**)] containing nine-coordinate bismuth.[419] In its deprotonated form imidobis(diphenylphosphine oxide) (**26**) coordinates to bismuth in the distorted octahedral [Bi{(Ph$_2$(O)P)$_2$N}$_3$].[420]

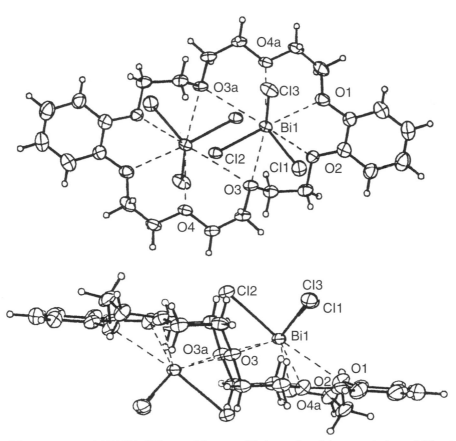

Figure 63 The structure of [(BiCl$_3$)$_2$(dibenzo-24-crown-8)] (reproduced by permission of Elsevier Science from *Inorg. Chem. Acta* **2000**, *300*, 1004–1013).

(24)   **(25)**   **(26)**

R = Ph, OEt

Bismuth alkoxides are usually prepared from Na(Li)OR and BiX_3 in benzene or THF, or by alcoholysis of $[Bi(NR_2)_3]$. Most examples are di- or polymeric and poorly soluble in organic solvents, although often sublimable in vacuum. The recent interest is due to the possible use of such complexes in CVD processes for bismuth oxide materials. Reaction of $[Bi(NMe_2)_3]$ with ROH (R = Pr^i, CH_2CH_2OMe, $CH_2CH_2NMe_2$, $CHMeCH_2NMe_2$, CMe_2Et) gave high yields of soluble and volatile $[Bi(OR)_3]$.[421,422] The structure of $[Bi(OCH_2CH_2OMe)_3]$ reveals a 1-D chain with square pyramidal bismuth composed of four bridging and one terminal alkoxides. In contrast, $[Bi(OCMe_2CH_2OMe)_3]$ is a six-coordinate monomer.[423] Other alkylalkoxides are $[Bi(O-Bu^t)_3]_n$,[422,423] and $[Bi(\mu\text{-}OCH_2CMe_3)(OCH_2CMe_3)_2(HOCH_2CMe_3)_2]$.[425] The phenoxide $[Bi\{O(2,4,6\text{-}Me_3C_6H_2)\}_3]$ is a trigonal pyramidal monomer, and $[Bi(OBu^t)_3]$ also appears to be mononuclear.[426] The latter reacts with $KOBu^t$ to give $K[Bu(OBi^t)_4]$ in which the bismuth has a *pseudo*-trigonal bipyramidal environment with $RO\cdots K$ interactions linking the units into a 1-D polymer.[427] In contrast, $NaOBu^t$ and $[Bi(OBu^t)_3]$ produce $Na_4[Bi_2O(OBu^t)_8]$.[427]

Fluorinated alkoxides have been examined in attempts to improve volatility.[428–431] The reaction of $BiCl_3$ with $NaOCH(CF_3)_2$ in THF produces dimeric $[Bi\{OCH(CF_3)_2\}_2\{\mu\text{-}OCH(CF_3)_2\}(THF)]_2$ which has a structure containing two square-pyramidal bismuth units bridged by alkoxides (**27**). Similar complexes are formed by $OC_6F_5^-$ and NMR studies show monomer–dimer equilibria occur in solution.[431] Under carefully controlled conditions, $NaOC_6F_5$ and $[Bi(OC_6F_5)_3]$ react to form $Na[Bi(OC_6F_5)_4\text{ (solvate)}]$ which is polymeric, based upon square-pyramidal bismuth. Under other conditions oxo-alkoxides form, such as $[Na_4Bi_2O(OC_6F_5)_8(THF)_4]$.[429] Oligomerization and oxide formation also occur when $[Bi(OC_6F_5)_3]$ is dissolved in various organic solvents, and the products have complex structures based upon $\mu^3\text{-O}$, $\mu^4\text{-O}$, $\mu^3\text{-OR}$, and $\mu^2\text{-OR}$ groups.[430] Mixed bismuth-transition metal or bismuth-alkaline earth metal alkoxides have also been synthesized, including $[\{BiCl_3OV(OC_2H_4OMe)_3\}_2]$,[432] $[Bi_4Ba_4O_2(OEt)_{12}(dpm)_4]$ (dpm = tetramethylheptane-3,5-dione),[433] and $[BiTi_2O(OPr^i)_9]$.[434]

(27)

Bismuth catecholates $M[Bi(cat)_2]\cdot nH_2O$ (M = Na, K, NH_4, etc.) have been known for many years, and a recent structure determination on the ammonium salt, $NH_4[Bi(O_2C_6H_4)_2]\cdot C_6H_4(OH)_2\cdot 2H_2O$ revealed a discrete dimer based upon *pseudo*-trigonal bipyramidal bismuth.[435] In contrast, the neutral $[Bi_2(OCH_2CH_2O)_3]$ derived from ethyleneglycol is polymeric with the basic dimer core linked into a 3-D polymer by alkoxide bridges.[435] Bismuth siloxides such as $[Bi(O-SiPh_3)_3]_n$ and $[Bi(OSiPh_3)_3(THF)_3]$ are known, the latter a discrete monomer.[422,423]

β-Diketonates include $[Bi(ACAC)_3]$,[437] and $[Bi(dpm)_3]$.[438] Tropolone derivatives have attracted interest as anti-*Helicobacter pylori* agents. Various types are known, including $[Bi_2(NO_3)_2(trop)_4]$ (tropH = tropolone) which is a dimer (Figure 64), $[Bi(trop)_2(H_2O)]NO_3$, $[BiPh(trop)_2]$, and $[Bi(trop)_4]^-$ all of which are essentially monomers.[389]

Methanesulfonates of bismuth include $[Bi(O_3SMe)_3]$, prepared from Bi_2O_3 and methanesulfonic anhydride. This is converted to $[Bi(O_3SMe)_4]^-$ by reaction with $M[O_3SMe]$ (M = Na, NH_4, etc.).[58]

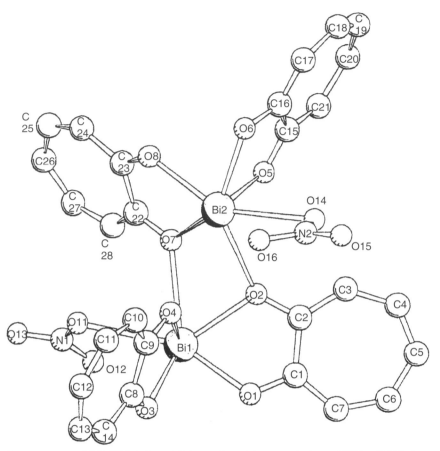

Figure 64 The structure of [Bi$_2$(NO$_3$)$_2$(trop)$_4$] (reproduced by permission of Wiley-VCH from *Chem. Ber.* **1995**, *128*, 335–342).

The bismuth(V) teflate, [Bi(OTeF$_5$)$_5$], is made from BiF$_5$ and B(OTeF$_5$)$_3$ in a freon and is stable at room temperature. It reacts with [NMe$_4$][OTeF$_5$] to form the octahedral [Bi(OTeF$_5$)$_6$]$^-$.[173]

Bismuth carboxylates show a wide variety of structural motifs. In K$_2$[Bi(HCO$_2$)$_5$] bidentate formate groups are present.[440] Bismuth acetate [Bi(O$_2$CMe)$_3$] has a layered structure,[441] but in the presence of thiourea (tu), two complexes can be isolated. [Bi(O$_2$CMe)$_3$(tu)$_3$] is a nine-coordinate monomer, but [Bi$_2$(O$_2$CMe)$_6$(tu)$_3$(H$_2$O)] is ionic with a [Bi$_2$(O$_2$CMe)$_4$(tu)$_6$]$^{2+}$ cation and a [Bi$_2$(O$_2$CMe)$_8$]$^{2-}$ anion, the latter with acetate bridges (28).[442] Bismuth pivalate, [Bi(O$_2$CCMe$_3$)$_3$], is a tetramer,[443] whereas [Bi(O$_2$CCF$_3$)$_3$]·CF$_3$CO$_2$H,[444] and [Bi(O$_2$CPh)$_3$],[445] are chain polymers in both cases with nine-coordinate bismuth centers. The subtle factors involved in bismuth carboxylate geometries are well illustrated by a series of complexes [diamineH$_2$][BiPh(O$_2$CCF$_3$)$_4$], where the anion geometry depends upon the cation present.[446] In K[Bi(C$_2$O$_4$)$_2$]·5H$_2$O, obtained by hydrolysis of squaric acid derivatives, a 3-D network polymer is present.[447]

—S = thiourea

(28)

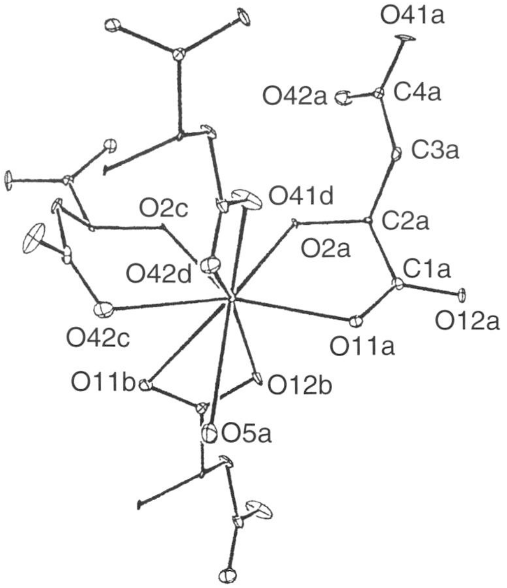

Figure 65 The structure of the bismuth malate complex (reproduced by permission of Wiley-VCH from *Chem. Ber.* **1993**, *126*, 51–56).

In contrast to antimony, few bismuth phosphonates have been described, and only two have been structurally characterized. The $[Bi(O_3PCH_2CH_2CO_2) \cdot H_2O]$ is polymeric with a layer structure.[448] However, Bu^tPO_3H reacts with $BiPh_3$ to form a $[Bi(O_3PBu^t)_3]$ phase as major product and as minor product, a 14-atom bismuth cluster $[Bi_{14}O_{10}(O_3PBu^t)_{10}(HO_3PBu^t)_2 \cdot 3C_6H_6 \cdot 4H_2O]$.[449]

Hydroxycarboxylic acid compounds of bismuth have been used in various medicines for many years, but the number and speciation of the complexes present is often far from clear. Recent work has produced structural characterizations of a range of examples, and solution studies are beginning to elucidate the complex equilibria present. In the tartrate complexes $[Bi(H_3tart)-(H_2tart) \cdot 3H_2O]$ and $NH_4[Bi(H_2tart)_2(H_2O)] \cdot H_2O$ $(H_4tart = HO_2CCH(OH)CH(OH)CO_2H)$, the bismuth is nine-coordinate, bonded to four bridging bidentate tartrate ligands (three bond via alkoxy/carboxy O-donors and one bonds via two carboxy oxygens) and a water molecule. The tartrates bridge neighboring bismuth centers to produce a polymeric network.[450,451] Bismuth malate $[Bi(mal) \cdot H_2O]$ $(H_3mal = HO_2CCH_2CH(OH)CO_2H)$ is also nine-coordinate with three different coordination modes exhibited by the chelating malate anions (Figure 65).[451] Rather similar structural features are present in bismuth lactate $[Bi(MeCH(OH)CO_2)_3]$.[452]

Bismuth citrate systems, often in the form of "colloidal bismuth subcitrate" are widely used medicinally, although the chemical speciation has been unclear. As a result of recent work,[453–458] some of the key structural features of crystalline bismuth citrates have been established, and the complex solution equilibria probed as a function of composition and pH. A common building block is the dimer unit shown in (**29**) (Figure 66) in which a citrate ligand functions as a tridentate chelate to one bismuth, and interacts more weakly through one carboxylate function to the second bismuth. In most of the structurally characterized bismuth citrates, this building block, in different degrees of protonation, is supplemented by H bonding to water molecules, and interaction with $K/Na/NH_4$ cations when present. A dodecanuclear cluster has been identified in $[NH_4]_{12}[Bi_{12}O_8(cit)_8] \cdot 10H_2O$ $(H_4cit = citric acid)$.[457]

3.6.4.3.2 *N/O-donor ligands*

Polydentate N,O-donor ligands form stable complexes with bismuth(III), usually with high coordination numbers. The aminocarboxylates have been examined in detail and a considerable amount of X-ray structural data are available (Table 3). The general synthetic route is reaction of Bi_2O_3, $Bi(OH)_3$, or basic bismuth carbonate with the ligand in water, followed by addition of the appropriate cations (where present) and adjustment of the pH. The bismuth coordination number varies between seven and 10 depending upon the complex, with water or other small ligands being incorporated to achieve this if necessary. For $EDTA^{4-}$ derivatives the ligand is always hexadentate with the structures varying from discrete anions with H_2O or thiourea co-ligands, through dimers with carboxylate bridges, to 1-D chains.

O^a = oxygen atom from citrate or water molecule

(29)

The description of the geometries at bismuth is not straightforward due to the distortions produced by the polydentate ligands (Figure 67). The highest coordination number observed is in the complex [Bi(H$_3$TTHA)]·3H$_2$O where the bismuth is 10-coordinate and approximates to a bi-capped square antiprism.[485] The nine-coordinate [Bi(HDTPA)(H$_2$O)]$^-$ and [Bi(DTPA)]$^{2-}$ complexes are mono-capped square antiprisms,[462,483] but for the eight-coordinate complexes the structures are more variable—[Bi(HEDTA)]·2H$_2$O (bicapped trigonal prism), [Bi(EDTA)]$^-$ (bicapped octahedron), and [Bi(HEDTA)] (square antiprism).[459,468]

Nitrilotriacetic acid, N(CH$_2$CO$_2$H)$_3$ (H$_3$NTA), iminodiacetic acid, HN(CH$_2$CO$_2$H)$_2$(H$_2$IDA), (2-hydroxyethyl)iminodiacetic acid, HOCH$_2$CH$_2$N(CH$_2$CO$_2$H)$_2$(H$_3$ONDA), and (N-hydroxy)-ethylethylenediaminetetra-acetic acid, (HO$_2$CCH$_2$)$_2$NCH$_2$CH$_2$N(CH$_2$CH$_2$OH)(CH$_2$CO$_2$H)-(H$_4$OEDTA), form similar complexes (Table 3).

Various other N,O-donor ligands complex with bismuth(III), including the heptadentate saltrenH$_3$, (LH$_3$ = N(CH$_2$CHN=CHC$_6$H$_4$OH)$_3$), which forms [BiL] containing seven-coordinate bismuth in an N$_4$O$_3$ environment.[493] The triaminetriol (taciH$_3$, (30)) forms [Bi$_3$(taci)$_2$](NO$_3$)$_3$ and [Bi$_3$(taci)$_2$]Cl$_3$·6H$_2$O the latter having the structure shown in Figure 68 in which each bismuth is coordinated N$_2$O$_4$Cl$_2$.[494] 2,6-Diacetylpyridinebis(2-thenoylhydrazone) (H$_2$DAPT, (31)) forms [Bi(HDAPT)X$_2$]·DMSO·H$_2$O (X = Cl, Br, I, NCS) and [Bi(DAPT)Y]·DMSO (Y = Cl, OH, N$_3$)[495,496] which are pentagonal bipyramidal and pentagonal pyramidal respectively with the ligand occupying the five equatorial positions. Other N,O-ligands are (32) which forms a [Bi(L-H)$_3$] complex,[497] and (33) which coordinates is a singly deprotonated form in the dinuclear [Bi$_2$((33)-H)$_2$(O$_2$CCF$_3$)$_4$(THF)$_2$].[498]

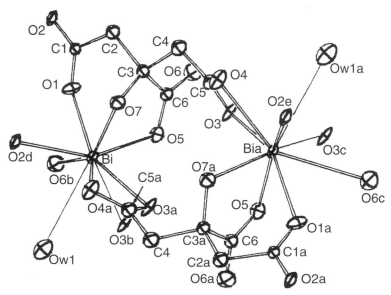

Figure 66 The dinuclear subunit in bismuth citrates (reproduced by permission of the American Chemical Society from *Inorg. Chem.* **1993**, *32*, 5322–5329).

Table 3 Bismuth aminocarboxylates.*

Complex	Comments	References
[Bi(HEDTA)]	1-D polymer	459
[Bi(HEDTA)(H$_2$O)$_2$]	Dodecahedral Bi coordination	460–462
[Bi(HETDA)(thiourea)$_2$]	Eight-coordinate Bi, N$_4$O$_2$S$_2$	463
Na[Bi(EDTA)]·3H$_2$O	Eight-coordinate Bi	464
[guanidinium][Bi(EDTA)(H$_2$O)]	Chain polymer	465,468
[aminoguanidinium][Bi(EDTA)]		468
[Hthiosemicarbazide][Bi(EDTA)(H$_2$O)]	Seven-coordinate Bi	466
[H(alanine)][Bi(EDTA)(H$_2$O)]	Dimer, eight-coordinate Bi	467
[H$_2$en][Bi(EDTA)(H$_2$O)$_2$]	Polymeric	469
[NH$_4$](Bi(EDTA)(H$_2$O)]		459
Li[Bi(EDTA)]·4H$_2$O	Seven-coordinate Bi	470
Na[Bi(EDTA)(H$_2$O)$_3$]		471
Cs[Bi(EDTA)]·H$_2$O		472
[Ca(H$_2$O)$_7$][Bi(EDTA)]$_2$·H$_2$O	Eight-coordinate Bi	473
[M(H$_2$O)$_6$][Bi(EDTA)]$_2$·3H$_2$O	(M = Co, Ni)	474
[aminoguanidinium]$_2$[Bi(EDTA)]Cl	Eight-coordinate Bi	475
Li[Bi(EDTA)(thiourea)$_2$]·5.5H$_2$O		476
K[Bi(EDTA)(thiourea)$_2$]	Eight-coordinate Bi	477
Cu[Bi(EDTA)]$_2$·9H$_2$O		478
[Co(C$_2$O$_4$)(NH$_3$)$_4$][Bi(EDTA)]·3H$_2$O		479,480
[Co(NH$_3$)$_5$(NCS)]$_2$[Bi$_2$(EDTA)$_2$ (μ-C$_2$O$_4$)]·12H$_2$O		481
[Bi$_5$(DTPA)$_3$]·10H$_2$O		482
Cu[Bi(DTPA)]·5H$_2$O		482
K[Bi(HDTPA)(H$_2$O)]·4H$_2$O	Eight-coordinate Bi	483
[guanidinium]$_2$[Bi(DTPA)]·4H$_2$O	Nine-coordinate Bi	462
[Bi(H$_2$DTPA)]·2H$_2$O	Chain polymer, eight-coordinate Bi	484
[Bi(H$_3$TTHA)]·H$_2$O	10-coordinate Bi	485
[guanidinium]$_2$[Bi(HTTHA)]·4H$_2$O		486
[guanidinium]$_2$[Bi(CYDTPA)]		483
[Bi(HCYDTA)·5H$_2$O	Eight-coordinate Bi	484
[Bi(NTA)(H$_2$O)$_2$]	Eight-coordinate Bi	462,486
[NH$_4$]$_3$[Bi(NTA)$_2$]	Eight-coordinate Bi, bicapped trigonal prism	487
K$_2$[Bi(NTA)(HNTA)]·H$_2$O		488
[Bi(HIDA)(IDA)]	Eight-coordinate Bi, N$_2$O$_6$	489
[Bi(ONDA)]·2H$_2$O	Eight-coordinate Bi	490,491
M[Bi(HONDA)$_2$]·nH$_2$O	M = K, Rb, Cs, NH$_4$, guanidinium	490,492
[guanidinium]$_2$[Bi(HONDA)(ONDA)]·3H$_2$O		490,492

*Abbreviations: H$_4$EDTA ethylenediaminetetraacetic acid; H$_5$DTPA diethylenetriaminepentaacetic acid; H$_6$TTHA triethylenetetra-aminehexaacetic acid; H$_5$CYDTA *N*-(2-aminoethyl)-*trans*-1,2-diaminocyclohexane-*N,N'N''*-pentaacetic acid; H$_4$CYDTA *trans*-cyclohex-ane-1, 2-tetraacetic acid; H$_3$NTA nitrilotriacetic acid; H$_2$IDA iminodiacetic acid; H$_3$ONDA (2-hydroxyethyl)iminodiacetic acid; H$_4$OEDTA (*N*-hydroxy)-ethylethylenediaminetetraacetic acid.

(30)   (31)

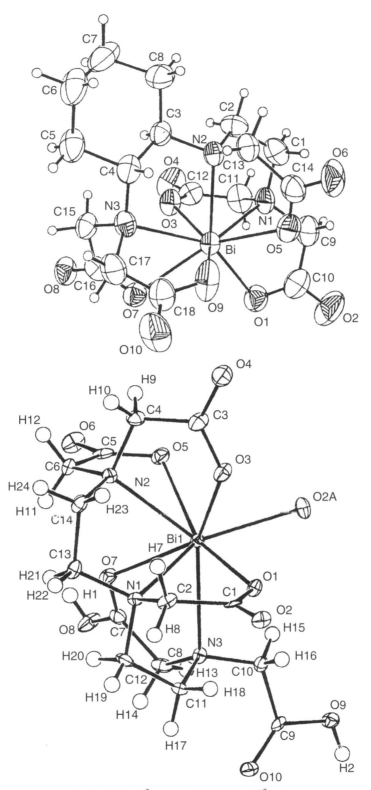

Figure 67 The structures of [Bi(CyDTPA)]²⁻ and [Bi(H₂DTPA)]²⁻ (reproduced by permission of the American Chemical Society from *Inorg. Chem.* **1996**, *35*, 6343–6348).

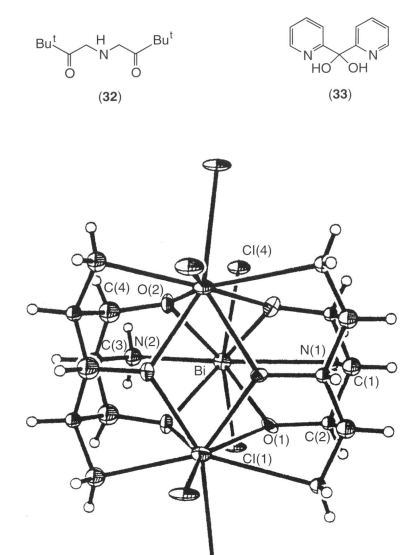

(32)   **(33)**

Figure 68   The structure of $[Bi_3(taci)_2]Cl_3$ (reproduced by permission of the American Chemical Society from *Inorg. Chem.* **1993**, *32*, 2699–2704).

3.6.4.3.3   S-, Se-, and Te-donor ligands

The first, and still the only structurally characterized, monodentate thioether complex of bismuth is $[SMe_3]_2[Bi_2I_8(SMe_2)_2]$, obtained in very poor yield from BiI_3 dissolved in a large excess of SMe_2. The structure of the anion is an edge-shared bioctahedron with *anti*-axial positioning of the SMe_2 ligands.[499] From $BiCl_3$ and $MeSCH_2CH_2CH_2SMe$ in CH_2Cl_2 the product is $[BiCl_3(MeSCH_2CH_2CH_2SMe)]$[214,500] which is a 3-D polymer based upon Bi_4Cl_4 "open-cradle" units linked by bridging dithioethers (Figure 69). From MeCN solutions of BiX_3 (X = Cl or Br) and $MeSCH_2CH_2CH_2SMe$ the products were $[BiX_3-(MeSCH_2CH_2CH_2SMe)]$, which have polymer sheet structures based upon planar Bi_2X_6 units cross-linked by dithioether bridges (Figure 70).[501] The shorter chain dithioether $MeSCH_2CH_2SMe$ forms $[BiX_3(MeSCH_2CH_2SMe)_2]$ (X = Cl, Br, or I) and the structure of the bromide derivative revealed a pentagonal bipyramidal monomer with axial bromines and two chelating dithioethers.[501] In contrast, the $[Bi_2Br_6(PhSCH_2CH_2SPh)]$ has infinite chains of Bi_2Br_6 groups linked via bromine-bridges with almost orthogonal thioether ligands cross-linking the

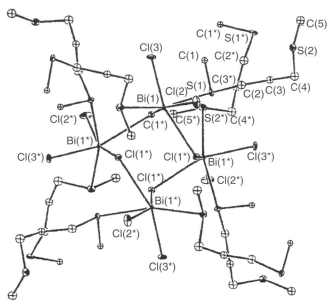

Figure 69 The tetramer unit in [BiCl$_3$\{MeS(CH$_2$)$_3$SMe\}] (reproduced by permission of the Royal Society of Chemistry from *Chem. Commun.* **1998**, 2159–2160).

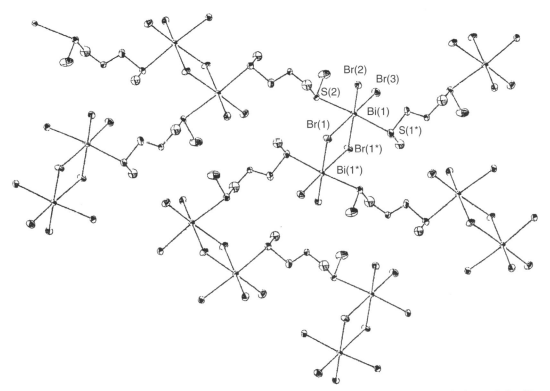

Figure 70 The polymeric structure of [BiBr$_3$\{MeS(CH$_2$)$_3$SMe\}] (reproduced by permission of the Royal Society of Chemistry from *J. Chem. Soc., Dalton Trans.* **2000**, 859–865).

chains. The structures of [BiX$_3$\{MeC(CH$_2$SMe)$_3$\}] are unknown,[501] but that of [BiCl$_3$\{Me-Si(CH$_2$SMe)$_3$\}][502] is polymeric with two different bismuth environments—one is BiS$_3$Cl$_3$ and the other BiS$_2$Cl$_4$ (**34**) and these units are linked by a single chloride bridge and by the trithioethers.

(34)

Bismuth trichloride complexes of the crown thioethers [9]aneS$_3$, [12]aneS$_4$, [15]aneS$_5$, and [18]aneS$_6$ have the common motif of a pyramidal BiCl$_3$ group capped by the weakly bound macrocycle using all the sulfur donors.[217,503,504] The larger [24]aneS$_8$ maintains the same basic motif in [(BiCl$_3$)$_2$([24]aneS$_8$)] with pyramidal BiCl$_3$ units coordinated to five sulfur donors (two of which are common to both bismuth) on opposite sides of the ring (Figure 71).[505]

There are no monodentate selenoether complexes, but the majority of those of the bidentate and polydentates differ in detail both from the antimony analogues and from the bismuth thioethers described above.[214] The 1:1 complexes [BiX$_3$(MeSeCH$_2$CH$_2$SeMe)] are of unknown structure, but the [BiX$_3$(MeSeCH$_2$CH$_2$CH$_2$SeMe)] complexes have analogous structures to the dithioether ligands with planar Bi$_2$X$_6$ units bridged by diselenoethers to give octahedral coordination at bismuth.[501]

The structures of the MeC(CH$_2$SeMe)$_3$ complexes differ with the halide present. In [BiCl$_3$-{MeC(CH$_2$SeMe)$_3$}] there are Bi$_2$Cl$_6$ units linked by tripodal selenoethers coordinated as bidentate chelates to one bismuth and monodentate to a second to produce seven-coordinate bismuth centers and a 2-D sheet polymer.[501] However, in [Bi$_2$I$_6${MeC(CH$_2$SeMe)$_3$}$_2$] there are discrete dimers with six-coordinate bismuth, the unit composed of a twisted Bi$_2$I$_6$ rhomboidal core further bound to two bidentate triselenoethers.[501]

The selenoether macrocycle complexes have completely different structures to those of the sulfur macrocycles. The [BiX$_3$L] (X = Cl or Br; L = [8]aneSe$_2$, [16]aneSe$_4$, and [24]aneSe$_6$) are deep orange–yellow solids.[506] The structures of [BiCl$_3${[8]aneSe$_2$}] and [BiBr$_3${[16]aneSe$_4$}] are ladder polymers with planar Bi$_2$X$_6$ units bridged (unusually) by *trans* selenoether ligands giving distorted octahedral coordination at bismuth (Figure 72).[506] Unfortunately, the structures of [BiX$_3$([16]aneS$_4$)] which would provide a direct comparison are unknown.[506]

Bismuth telluroether complexes are very rare; the structure of the first such example [BiBr$_3$(Ph-MeTe)] shows a planar Br$_2$Bi(μ-Br)$_2$BiBr$_2$ group with the PhMeTe ligands completing an *anti*-square pyramidal arrangement. These dimer units are then linked into chains via further long-range bromide bridges resulting in a distorted six-coordinate bismuth geometry.[507]

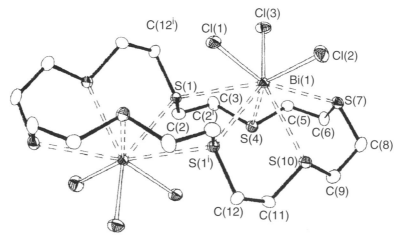

Figure 71 The structure of [(BiCl$_3$)$_2$([24]aneS$_8$)] (reproduced by permission of the Royal Society of Chemistry from *J. Chem. Soc., Dalton Trans.* **1998**, 3961–3968).

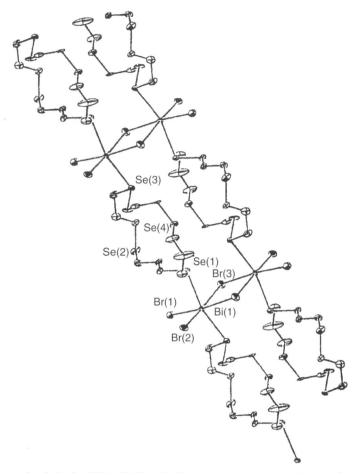

Figure 72 The polymeric chain in [BiBr$_3$([16]aneSe$_4$)] (reproduced by permission of the Royal Society of Chemistry from *J. Chem. Soc., Dalton Trans.* **2000**, 2163–2166).

Bismuth complexes of a variety of thiones are known and some have been discussed already in mixed ligand complexes.[442,463,476,477] For tu there are seven- and eight-coordinate bismuth nitrate complexes, [Bi(NO$_3$)(tu)$_5$][NO$_3$]$_2$ and [Bi(NO$_3$)$_3$(tu)$_2$], respectively.[508] Related ligands, which mostly form six-coordinate complexes with bismuth halides with the ligand S-bonded, include those of 1-allyl-3(2-pyridyl)thiourea,[509] 1-phenyl-3(2-pyridyl)thiourea,[510] 3,4,5,6-tetrahydropyrimidine-2(1H)-thione,[511] benzimadazole-2(3H)-thione,[511] and imidazolidine-2-thione.[512] The charged ligand (**35**) forms the S-bound zwitterion [BiCl$_5$(**35**)]$^-$,[513] and the anionic (**36**) forms a μ-Cl$_2$ dimer [Bi$_2$Cl$_4$(**36**)$_2$].[514] A related ligand (**37**) hydridotris(thioxotriazolyl)borate(1−) forms red crystals of [Bi(**37**)$_2$]Cl·H$_2$O in which the bismuth is coordinated to six thione S atoms.[515] Bismuth thiocyanate complexes, [Bi(SCN)(NCS)$_2$(L)$_3$] and [Bi(NCS)S(H$_2$O)(L)]·H$_2$O (L = 1,3-dimethyl-2(3H)-imidazolethione) have been prepared and characterized by X-ray crystallography; both have six-coordinate bismuth centers.[516] There are also semicarbazones,[517] various heterocyclic thiones[518,519] and dithio-oxamide,[224,520] which bond to bismuth as neutral S-donors. A few examples of phosphine sulfide complexes are known, for example, Pr^n$_2$P(S)P(S)Pr^n$_2$ forms [BiCl$_3${Pr^n$_2$P(S)P(S)Pr^n$_2$}] which is a μ-dibromide bridged dimer with chelating diphosphine disulfides.[521] The iminodiphosphinesulfide HN[P(S)Me$_2$]$_2$ forms a tris(chelate) in its monoanionic form in [Bi{N[P(S)Me$_2$]$_2$}$_3$],[522] and there is a related complex formed by the [Ph$_2$P(S)]HN[P(Se)PPh$_2$] which contains a very rare example of a bismuth-phosphine selenide linkage.[523]

Dithioacid complexes of bismuth are numerous (Table 4) and repeat many of the structural motifs of their arsenic and antimony analogues, most notably the very asymmetric coordination ("anisobidentate"), although, as a consequence of the larger bismuth atom, higher coordination numbers are more evident. In the xanthates, seven-coordinate bismuth appears to be favored, although this is achieved in different ways: in [Bi(S$_2$COR)$_3$] (R = PhCH$_2$, c-C$_6$H$_{11}$, Bu^n) there are centrosymmetric dimers,[526,531] [Bi(S$_2$COPr^i)$_3$] is a chain polymer,[528] whilst [Bi(S$_2$COEt)$_3$] is more highly polymerized.[526]

(35)   (36)   (37)

The compound [Bi{S$_2$CN(CH$_2$CH$_2$OH)$_2$}$_3$] is a dimer with square antiprismatic bismuth (Figure 73)[74] whereas the As and Sb analogues are mononuclear (q.v.). The synthesis and structures of a range of halo(dithiocarbamato)bismuth derivatives have been reported.[393,534–538] Generally these are polymeric. In [Bi(S$_2$CNEt$_2$)$_2$I] there is a zig-zag chain linking six-coordinate bismuth centers with *cis*-I$_2$S$_4$ donor sets, through single iodine bridges, whereas the bromide, [Bi(S$_2$CNEt$_2$)$_2$Br], is based upon a centrosymmetric tetramer.[533] The [Bi(S$_2$CNEt$_2$)I$_2$] is different

Table 4   Dithioacid compounds of bismuth(III).

Compound	Comments	References
Bi(SOCR)$_3$	R = Ph, 4-MeC$_6$H$_4$, 2-MeC$_6$H$_4$	524,526
Bi(S$_2$COR)$_3$	R = Me, Et, Pr^i, Bu^n,	63,64,66,526,528,530,531
	c-C$_6$H$_{11}$, PhCH$_2$	
Bi($_2$COCH$_2$CH$_2$CMe$_2$)$_3$		65
BiPh(S$_2$COR)$_2$	R = Me, Et, Pr^n, Pr^i, Bu^n, Bu^i	527,529
BiMe(S$_2$COR)$_2$	R = Me, Et, Pr^n, Pr^i, Bu^n, Bu^i	527
Bi(S$_2$COMe)(S$_2$COEt)$_2$		234
Bi(S$_2$COEt)$_2$X	X = Cl, Br	230
Bi(SOCNR$_2$)$_3$	R$_2$ = Et$_2$, pyrolyl	236
Bi(S$_2$CNR$_2$)$_3$	R$_2$ = N,N'-iminodiethanol	74,241
Bi(S$_2$CNR)$_3$	R = pyrolyl	532
Bi(S$_2$CNR)$_2$Cl(thiourea)		532
Bi(S$_2$CNPr^{i}R)$_3$	R = HOCH$_2$CH$_2$	244
Bi(S$_2$CNR)$_3$	R = CHMeCH$_2$CH$_2$CH$_2$CH$_2$,	69
	CH$_2$CHMeCH$_2$CH$_2$CH$_2$,	
	CH$_2$CH$_2$CHMeCH$_2$CH$_2$	
Bi(S$_2$CNEt$_2$)$_2$X	X = Br, I,	533
Bi(S$_2$CNEt$_2$)X$_2$	X = Cl, Br, I	383,534,535
Bi$_5$(S$_2$NEt$_2$)$_8$X$_7$(DMF)	X = Cl, Br, I	536
[Bi$_4$(S$_2$CNEt$_2$)$_4$Br$_{10}$]$^{2-}$		536
[Bi(S$_2$CNEt$_2$)X$_3$]$^-$	X = Cl, Br, I	537
Bi(S$_2$CNEt$_2$)X$_2$(py)$_3$	X = Cl, Br, I	538
Bi(S$_2$CNEt$_2$)I$_2$L	L = 2,2'-bipyridyl; 2,2',6',2''-terpyridyl	393
BiR(S$_2$CNEt$_2$)$_2$	R = Ph, 2-(2'-pyridyl)phenyl	539
Bi{S$_2$P(OR)$_2$}$_3$	R = Me, Et, Pr^i	539,541,543
Bi{S$_2$P(OR)$_2$}$_2$X	R = Et, Pr^i; X = Cl, Br, I	540
Bi{S$_2$P(OR)$_2$}Cl$_2$	R = Et, Pr^i, Pr^n, Bu^i	540
Bi{S$_2$PO$_2$R}$_3$	R = CHMeCHMe, CMe$_2$CH$_2$CHMe,	81,542
	CH$_2$CMe$_2$CH$_2$, CH$_2$CEt$_2$CH$_2$	
R^{1}Bi{S$_2$P(OR^2)$_2$}$_2$	R^1 = Me, Ph, p-tol; R^2 = Me, Et, Pr^i	541
R$^1_2$Bi{S$_2$P(OR^2)$_2$}	R^1 = Ph, p-tol, R^2 = Me, Et, Pr^i, Ph	541
Bi(S$_2$PR$_2$)$_3$	R = Me, Et, Ph	253,544–546
BiSMe(S$_2$PPh$_2$)$_2$		547
Bi(S$_2$AsR$_2$)$_3$	R = Me, Ph	548

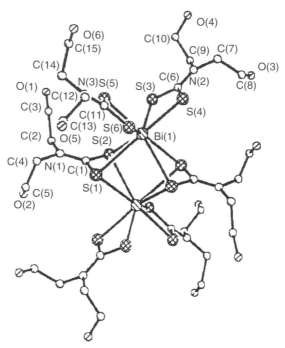

Figure 73 The structure of [Bi{S$_2$CN(CH$_2$CH$_2$OH)$_2$}$_3$] (reproduced by permission of Elsevier Science from *Polyhedron* **1997**, *16*, 1211–1221).

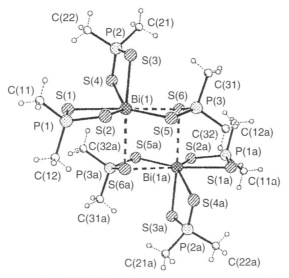

Figure 74 The dimer structure in [Bi(S$_2$PMe$_2$)$_3$] (reproduced by permission of Elsevier Science from *Polyhedron*, **1994**, *13*, 547–552).

again, an infinite polymer this time with I$_2$S bridges between each pair of bismuth centers giving facially bridged units.[535] The anion in [Et$_4$N][Bi(S$_2$NEt$_2$)I$_3$] is dimeric [I$_2$(S$_2$CNEt$_2$)Bi(μ-I)$_2$Bi(S$_2$-CNEt$_2$)I$_2$]$^{2-}$,[537] whereas in [Bi$_5$(S$_2$CNEt$_2$)$_8$X$_7$] (X = Cl, Br, or I) obtained by recrystallizing [Bi(S$_2$CNEt$_2$)X$_2$] from *N,N*-DMF, there is a central BiX$_6$ unit which links with four Bi(S$_2$CNEt$_2$)$_2$ groups, although in detail there are three different bismuth coordination environments.[439]

Of the three structurally characterized dithiophosphates with simple R groups, [Bi{S$_2$P(OEt)$_2$}$_3$] is six-coordinate (distorted octahedral), but [Bi{S$_2$P(OR)$_2$}$_3$] (R = Ph or Me) are dimeric with distorted pentagonal bipyramidal bismuth.[540–543] Similarly, in the dithiophosphinate series [Bi(S$_2$PR$_2$)$_3$] R = Et is distorted octahedral, whereas R = Me or Ph are dimeric (Figure 74).[544–547] One curious observation is that in [Bi(S$_2$PEt$_2$)$_3$] (Figure 75) the Bi—S bonds within the chelate rings are very similar (2.794(5), 2.782(5) Å) which contrasts with the anisobidentate coordination

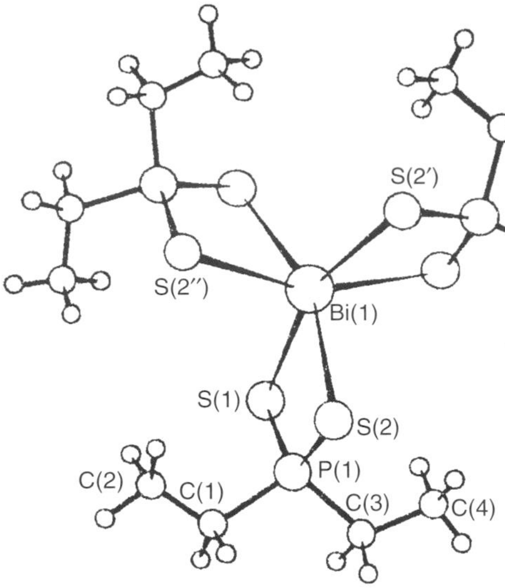

Figure 75   The monomeric [Bi(S$_2$PEt$_2$)$_3$] (reproduced by permission of the Royal Society of Chemistry from *J. Chem. Soc., Dalton Trans.* **1987**, 1257–1259).

(ca. 0.2 Å) found in many bismuth dithioacid complexes. The dimeric [Bi(S$_2$PR$_2$)$_3$] have Bi—S$_{terminal}$ bonds which range 3.641(3)–3.025(3) Å. The dithiophosphate, [Bi{S$_2$P[OCHMeCHMeO]}$_3$], also has very similar Bi—S bond lengths,[542] whereas the few other structurally characterized examples are rather more asymmetrically coordinated. The disparate bond lengths in the dithioacid chelates of As, Sb, and Bi have been variously attributed to a combination of the radius of the central atom, small chelate bite, and the degree of stereochemical activity of the lone pair on the central atom, but the fine details of the structures are not understood.

The high affinity of bismuth for charged sulfur ligands results in the ready generation of bismuth thiolates from a range of bismuth salts and an appropriate thiol, although reaction of Na(Li)SR with BiCl$_3$ or of BiPh$_3$ with RSH have also been used.[266,267,549–552] Examples are [Bi(SR)$_3$] (R = CH$_2$Ph,[550] 2,4,6-Bu$^t_3$C$_6$H$_2$,[267,549] 4-MeC$_6$H$_4$,[266] 2,6-Me$_2$C$_6$H$_3$,[266] 3,5-Me$_2$C$_6$H$_3$,[266] 2,4,6-Me$_3$C$_6$H$_2$,[267] 2,4,6-Pr$^i_3$C$_6$H$_2$,[266] C$_6$H$_4$F,[552] and C$_6$F$_5$[266,551]). The structure of [Bi{S(2,4,6-Bu$^t_3$C$_6$H$_2$)}$_3$] shows a pyramidal monomer,[549] but [Bi(SC$_6$F$_5$)$_3$] is a weakly associated dimer (**38**).[553] Unexpectedly, the reaction of NaSC$_6$F$_5$ and BiCl$_3$ in THF gave [Na$_2$(THF)$_4$][Bi(SC$_6$F$_5$)$_5$] with a square pyramidal anion.[553] This Lewis acidity of [Bi(SC$_6$F$_5$)$_3$] is demonstrated by the formation of a range of adducts with neutral and anionic ligands.[551] These include orange [Bi(SC$_6$F$_5$)$_3$(SPPh$_3$)] which is dimeric through asymmetric thiolate bridges (as in the parent), resulting in square pyramidal bismuth. In contrast, [Bi(SC$_6$F$_5$)$_3$(L)$_2$] (L = OPPh$_3$, *N,N'*-dimethyl-propyleneurea, OP(NMe$_2$)$_3$, OSPh$_2$) are square-pyramidal monomers with apical SC$_6$F$_5$ and *cis* basal SC$_6$F$_5$. The compound [Bi(SC$_6$F$_5$)$_3$\{SC(NHMe)$_2$\}$_3$] is close to a regular octahedron. Toluene-3,4-dithiol (tdtH$_2$) and BiCl$_3$ react in 1:1 ratio in CHCl$_3$ to form [BiCl(tdt)] whilst excess dithiol and NEt$_3$ give [Bi(tdt)$_2$]$^-$.[90] The structure of the dithiolato-anion [Bi{S$_2$C$_2$(CN)$_2$\}$_2$]$^-$ (as its AsPh$_4^+$ salt) shows polymeric chains with six-coordinate bismuth, with adjacent Bi atoms bridged by two dithiolate groups.[554] The dithione-dithiol dmitH$_2$ (**16**) forms [A][Bi(dmit)$_2$] (A = AsPh$_4$, NBu$_4$, NEt$_4$, etc.) by metathesis between BiBr$_3$ and [A][Zn(dmit)$_2$]. These have polymeric chain structures usually described as *pseudo*-pentagonal bipyramidal at bismuth (S$_6$ plus lone pair). The anion arrangements vary with the cation present.[555–557] Alkanedithiols HS(CH$_2$)$_n$SH (n = 2–4) react with slurries of BiCl$_3$ to form [BiCl{S(CH$_2$)$_n$S}] which have chelate structures (**39**).[558] If these complexes are reacted further with aqueous NaNO$_3$ and more dithiol, the final chloride is removed to give complexes of type (**40**).[558] The X-ray structures also reveal intermolecular secondary Bi···Cl and Bi···S interactions which link the bismuth heterocycles into chains with seven-coordinate bismuth centers. If the dithiol contains a further donor group in the chain, HSCH$_2$CH$_2$QCH$_2$CH$_2$SH (Q = O, S, or NR), *trans*-annular interactions produce bismocanes (**41**), which have structures based upon *pseudo*-trigonal bipyramidal geometry with axial Cl and Q.[558–560] There are organobismuth analogues where Ph or Me groups replace the Cl.[559,560] Dithiol-dithioether ligands have also been complexed with Bi^{III}.[561]

Selenolates and tellurolates are rarer and mostly have bulky R groups, e.g., [Bi(SeR^1)$_3$] (R^1 = 2,4,6-R$^2_3$C$_6$H$_2$, R^2 = Me, Pr^i or Bu^t),[267] [Bi{SeSi(SiMe$_3$)$_3$}$_3$],[272] and [Bi{TeSi (SiMe$_3$)$_3$}$_3$][272] which are probably pyramidal monomers, although no examples have been structurally characterized. The structure of [Bi(SeR)$_3$] (SeR = (42)) has been determined and contains primary Bi—Se bonds (3.691–2.745 Å) and secondary Bi···N interactions (2.827–2.952 Å) with the lone pair appearing to point into the N$_3$ face of the very distorted octahedron (Figure 76).[562] The small SePh$^-$ ligand, introduced via reaction of TMSSePh with BiBr$_3$, produced clusters [Bi$_4$(μ-SePh)$_5$-(SePh)$_8$] and [Bi$_6$(μ-SePh)$_6$(SePh)$_{10}$Br].[563] The SeCN$^-$ ligand forms [K$_3$(*N,N*'-dimethylpropyle-neurea)$_4$][Bi(SeCN)$_6$], which is Se-bonded (in [Bi(SCN)$_6$]$^{3-}$ the thiocyanate is S-bonded).[564,565]

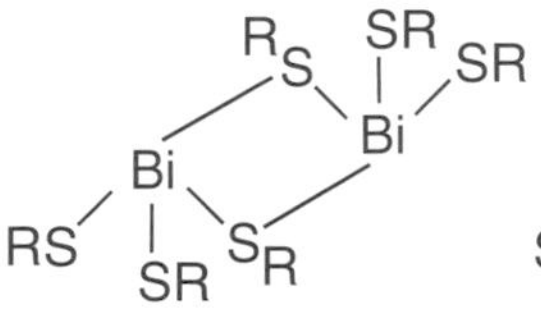

(38) 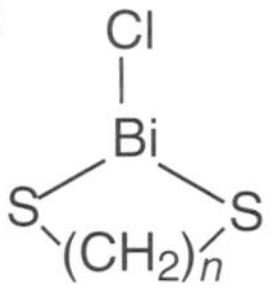   (39) 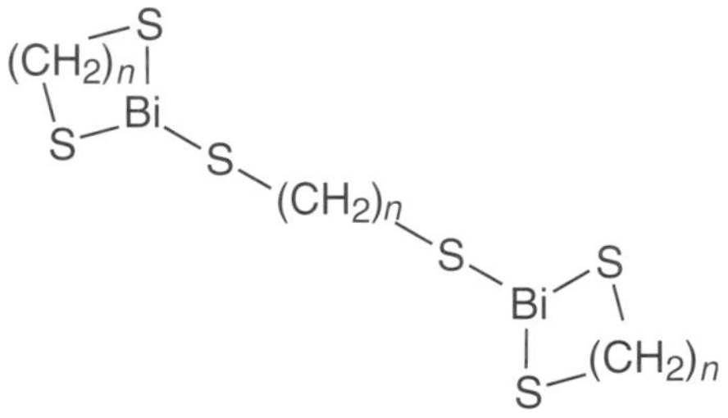   (40)

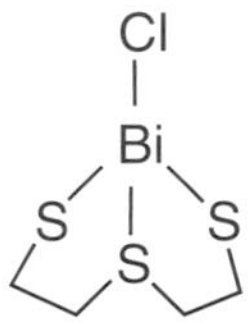

(41) 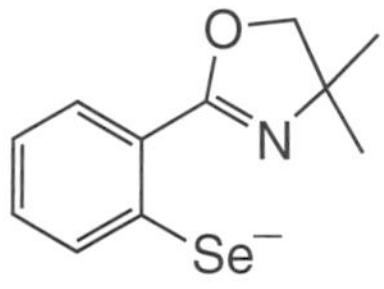   (42)

3.6.4.3.4 S/O- and S/N-donor ligands

Thioethanol HOCH$_2$CH$_2$SH, reacts with most bismuth salts to give [Bi(HOCH$_2$CH$_2$S)$_2$Y] (Y = Cl, Br, NO$_3$, ClO$_4$, MeCO$_2$, etc.).[566,567] The structures contain a BiS$_2$O$_2$ core which then link in various ways depending upon the anion. The nitrate ion weakly chelates and there are long intermolecular Bi···S linkages, whilst in the chloride the units weakly associate via Cl and S bridges (Figure 77).[566] Deprotonation of the alcohol function is rare but occurs in [Bi(S-CH$_2$CH$_2$O)(HOCH$_2$CH$_2$S)] which has a similar BiS$_2$O$_2$ core although with a much shorter Bi–O distance involving the alkoxide (2.195(9) Å) compared with the alcohol (2.577(9) Å). The structure is polymeric via alkoxide bridges.[567] The alkoxide is protonated by acetic acid or by excess thioethanol, the latter reaction producing [Bi(HOCH$_2$CH$_2$S)$_3$].[567] If the K$^+$ salt of (methyl-ester)methanethiolate reacts with BiCl$_3$ in ethanol the product is [Bi(SCH$_2$CO$_2$Me)$_2$Cl], which is polymeric with a seven-coordinate bismuth center (S$_4$O$_2$Cl) (Figure 78), whilst excess of the ligand formed [Bi(SCH$_2$CO$_2$Me)$_3$] which is a dimer with unsymmetrical thiolate bridges.[568] Ketothiolate complexes [Bi(L-H)$_3$] LH = benzoylthiobenzoylmethanes, PhC(S)CHC(O)R (R = C$_6$H$_4$H, C$_6$H$_4$OMe, C$_6$H$_4$Cl) are dimeric with *pseudo*-pentagonal bipyramidal bismuth with one thiolate donor and the lone pair occupying apical positions, and two asymmetric thiolate bridges.[569]

Aminoethanethiolates are, as might be expected, also good ligands for bismuth.[570–572] The reaction of BiCl$_3$ or Bi(NO$_3$)$_3$ with H$_2$NCH$_2$CH$_2$SH, depending upon the reaction conditions yields [Bi(H$_2$NCH$_2$CH$_2$S)$_3$] or [BiY(H$_2$NCH$_2$CH$_2$S)$_2$] (Y = Cl or NO$_3$), whilst from Me$_2$NCH$_2$CH$_2$SH it is possible to make analogues of these and also [Bi(Me$_2$NCH$_2$CH$_2$S)Cl$_2$] and [Bi(SCH$_2$CH$_2$NMe$_2$H)Cl$_3$]. In the latter, the ammonium function is (of necessity) uncoordin-ated and the structure is a zig-zag polymer with alternating (μ-SR)$_2$ and (μ-Cl)$_2$ bridges, with two terminal chlorines completing a six-coordinate bismuth environment.[570] In the other com-plexes, the aminoethanethiolates function as *N,S* chelates with longer intermolecular Bi···S coordination. The [Bi(Me$_2$NCH$_2$CH$_2$S)Cl$_2$] was isolated as an adduct with HCl and the structure (43) reveals four bismuth atoms arranged around the central chloride ion.[571]

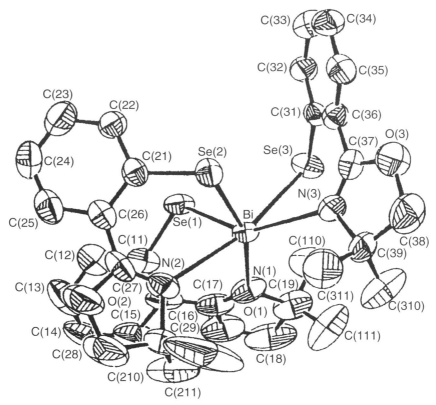

Figure 76 The structure of the bismuth selenolate [Bi(**42**)$_3$] (reproduced by permission of the Royal Society of Chemistry from *J. Chem. Res.* **1999**, 416–417).

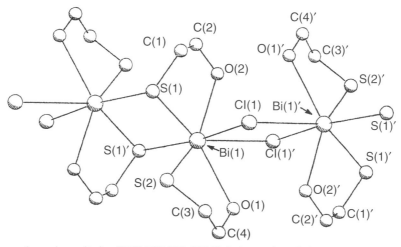

Figure 77 The polymeric unit in [BiCl(HOCH$_2$CH$_2$S)$_2$] (reproduced by permission of the American Chemical Society from *Inorg. Chem.* **1997**, *36*, 2855–2860).

Other N,S-donor ligands which have been complexed with bismuth include 2-aminobenzenethiol (L) as [Bi(L-H)$_3$] and [BiCl$_3$L$_3$],[271] 8-quinolinethiolate,[573] and 2-methyl-1-quinolinethiolate,[574] which form [BiL$_3$] (N$_3$S$_3$) with strongly bound pyramidal BiS$_3$ and weak Bi—N bonds. The hindered pyridinethiolate [Bi(2-SC$_5$H$_3$N-3-SiMe$_3$)$_3$] also shows strong Bi–S bonding, with a distorted pentagonal pyramidal S$_3$N$_3$ geometry with one apical S and a lone pair in the other apical site.[575] Dithizone LH (**44**) also binds as an *N,S* chelate in [Bi(L)$_3$].[576]

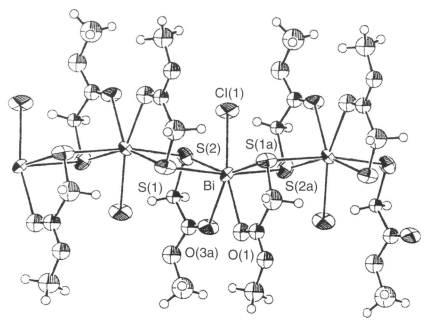

Figure 78 The polymeric unit in [Bi(SCH$_2$CO$_2$Me)$_2$Cl] (reproduced by permission of the Royal Society of Chemistry from *Chem. Commun.* **2000**, 13–14).

(43)   **(44)**

2,6-Diacetylpyridinebis(thiosemicarbazone) (dapsH$_2$), in its dianionic form, gives a 1:1 complex with bismuth azide, [BiN$_3$(daps)], in which the ligand coordinates to the equatorial plane of a *pseudo*-pentagonal bipyramid (N$_3$S$_2$) with the azide apical.[496]

3.6.4.4 Group 17 Compounds

Haloanions of bismuth(V) are limited to the fluorides. Halobismuthates(III) and organohalobismuthates(III) exhibit a range of structures and among the heavier halogens there is considerable similarity to their antimony(III) analogues. The halobismuthate(III) (not fluorides) structures have been reviewed by Fisher and Norman.[91]

Fluorobismuthates(III) with stoichiometries M[BiF$_4$], M$_2$[BiF$_5$], M$_3$[BiF$_6$], M[Bi$_3$F$_{10}$], and M$_3$[Bi$_2$F$_9$] (M = K, Rb, Cs, and sometimes Na, Li) have been prepared usually by melting together the components, although some have been isolated from aqueous HF solutions.[577–584] Much of the structural data are based on powder X-ray diffraction, but structures are typically polymeric with the bismuth in a distorted nine-coordinate environment similar to that in BiF$_3$ or the Tysonite structure. There has been considerable interest in these compounds as fast fluoride ion conductors.[585]

The M[BiF$_6$] (M = alkali metal) are made from MF and BiF$_5$ in anhydrous HF or by heating the constituents under fluorine pressure. There are also many examples with nonmetal cations [ClF$_2$]$^+$, [BrF$_2$]$^+$, etc. whilst [NR$_4$][BiF$_6$] are made from NR$_4$F and BiF$_5$ in anhydrous HF.[586,587]

The ^{209}Bi NMR spectrum of [BiF$_6$]$^-$ is a binomial septet confirming the regular O$_h$ geometry.[588] The hexafluorobismuthate(V) ion seems to be the only well-established six-coordinate form, and in contrast to antimony there seems to be no evidence for [Bi$_2$F$_{11}$]$^-$ anions.[586] The M$_2$[BiF$_7$] (M = Na, K, Rb, Cs) are made by combination of BiF$_5$ and excess MF under fluorine, whilst [NMe$_4$]$_2$[BiF$_7$] is made from NMe$_4$F and BiF$_5$ in MeCN.[105] The [BiF$_7$]$^{2-}$ ion decomposes in anhydrous HF to give [BiF$_6$]$^-$ and [HF$_2$]$^-$. The vibrational spectra are consistent with discrete pentagonal bipyramidal anions.[105] There is no evidence for [BiF$_8$]$^{3-}$.

The halobismuthates(III) are made by combination of BiX$_3$ and the appropriate cation in either organic solvents or aqueous acid. The major interest is in the structural units present in the anions and this section will illustrate the types known rather than list all the examples in the literature. There has also been some interest in the ferroelastic or ferroelectric properties of some of the halobismuthate phases.

The simplest chlorobismuthate(III) stoichiometry is [BiCl$_4$]$^-$ but this is never mononuclear. In [BiCl$_2$(18-crown-6)][Bi$_2$Cl$_8$] the anion is an edge-shared dimer with *anti*-apical Cl's and square-pyramidal bismuth coordination.[412] The reaction of MgCl$_2$ and BiCl$_3$ in MeCN produced [Mg(MeCN)$_6$]$_2$[Bi$_4$Cl$_{16}$] which has a discrete centrosymmetric anion (Figure 79).[380] Infinite polymers based upon BiCl$_6$ units sharing edges are present in [NEt$_2$H$_2$][BiCl$_4$][589] and [phenH][BiCl$_4$].[590] There appear to be no examples of a monomeric [BiCl$_5$]$^{2-}$ anion, the two motifs established are edge-linked bioctahedral [Cl$_4$Bi(μ-Cl)$_2$BiCl$_4$]$^{4-}$ found in K$^+$,[591] and (1H$^+$,5H$^+$-S-methylisothiocarbonohydrazidium),[592] salts and infinite polymers with BiCl$_6$ octahedra linked through two *cis* vertices into chains as in [phenH$_2$] [BiCl$_5$],[590] and [H$_3$N(CH$_2$)$_6$NH$_3$][BiCl$_5$].[318] A unique tetramer has recently been identified in [bipyH$_2$]$_4$[Bi$_4$Cl$_{20}$], also based upon BiCl$_6$ units linked via *cis* vertices but generating a square array.[590] The [BiCl$_6$]$^{3-}$ ion which is discrete and usually close to octahedral (consistent with minimal stereochemical activity of the lone pair[91]) is well established. Recent X-ray structures containing this anion include [(phenH)(phenH$_2$)(H$_2$O)$_2$][BiCl$_6$],[590] [NMe$_2$H$_2$]$_4$[BiCl$_6$]Cl,[593] and [2,6-Me$_2$C$_5$H$_3$NH]$_3$[BiCl$_6$].[590] The [Bi$_2$Cl$_9$]$^{3-}$ anion is a confacial bioctahedron in the [bipyH]$^+$,[590] [NMe$_4$]$^+$,[594] and [NPhMeEt$_2$]$^+$[595] salts, whereas 1-D double chains are present in the Cs$^+$[596] and [NMeH$_3$]$^+$ salts.[594]

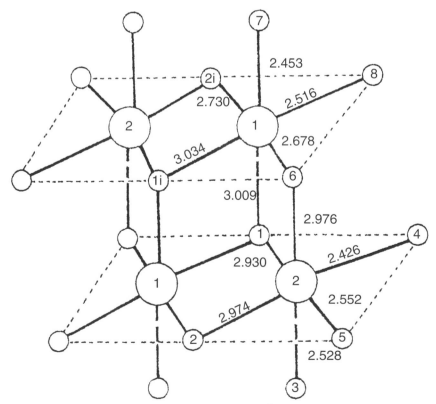

Figure 79 Representation of the tetranuclear anion [Bi$_4$Cl$_{16}$]$^{4-}$ (reproduced by permission of the Royal Society of Chemistry from *J. Chem. Soc., Dalton Trans.* **1991**, 961–965).

Two other chlorobismuthates, known in single examples are [NMeH$_3$]$_5$[Bi$_2$Cl$_{11}$], which has
BiCl$_6$ octahedra sharing a common vertex,[597] and the largest reported example is [NEt$_4$]$_6$[Bi$_8$Cl$_{30}$]
shown in Figure 80.[598]

Bromobismuthates(III) repeat many of the motifs of the chlorides. Thus [BiBr$_4$]$^-$ and [BiBr$_5$]$^{2-}$
appear never to be mononuclear[91] the former being either chain polymers or a tetramer
[Bi$_4$Br$_{16}$]$^{4-}$ with a structure of the type shown in Figure 79.[599] The bioctahedral [Bi$_2$Br$_{10}$]$^{4-}$ ion
is present in salts with [2,5-diamino-1,3,4-thiadiazolium]$^+$,[600] and [Sr(H$_2$O)$_8$]$^{2+}$.[601] What can be
seen either as a substituted variant of this or as a solvated [Bi$_2$Br$_8$]$^{2-}$ is found in [PPh$_4$]$_2$[Bi$_2$Br$_8$
(Me$_2$CO)$_2$], which has the structure shown in (45).[602] Octahedral [BiBr$_6$]$^{3-}$ is present in the
[NMe$_2$H$_2$]$^+$,[603] [PhCH$_2$CH$_2$NH$_3$]$^+$,[604] and [2,6-Me$_2$C$_5$H$_3$NH]$^+$ salts.[590] The [Bi$_2$Br$_9$]$^{3-}$ exists as
confacial bioctahedral and as chain polymeric forms.[594,596,605] Larger bromobismuthates are
[Bi$_2$Br$_{11}$]$^{5-}$ (isostructural with the chloride),[606] [Bi$_6$Br$_{22}$]$^{4-}$ (Figure 81), and [Bi$_8$Br$_{28}$]$^{4-}$.[605]

(45)

Neither [BiI$_4$]$^-$ nor [BiI$_5$]$^{2-}$ are known in monomeric forms. In [bipyH][BiI$_4$][590] and [2-amino-
1,3,4-thiadiazolium][BiI$_4$][607] there are chains of *cis* edged-shared BiI$_6$ octahedra, whilst
[(PhCH$_2$)$_4$P]$_2$[Bi$_2$I$_8$] is the first example of an iodobismuthate with square-pyramidal bismuth
coordination.[608] [BipyH]$_2$[BiI$_5$] has a discrete [I$_4$Bi(μ-I)$_2$BiI$_4$]$^{4-}$ anion,[590] and chains of [BiI$_6$]
octahedra are found in [H$_3$N-R-NH$_3$][BiI$_5$] (R various long chain organic groups).[318,609] Isolated
[BiI$_6$]$^{3-}$ octahedra are found in [PhCH$_2$CH$_2$NH$_3$]$_2$[BiI$_6$]I,[604] and [MeCOCH$_2$NC$_5$H$_5$]$_2$[C$_5$H$_4$NH]-
[BiI$_6$].[610] The [Bi$_2$I$_9$]$^{3-}$ ion is present as discrete confacial bioctahedra in the [NMe$_4$]$^+$ and
[NEt$_2$H$_2$]$^+$ salts, although polymeric in the Cs$^+$ salt.[611,612] Three face sharing BiI$_6$ octahedra
generate the [Bi$_3$I$_{12}$]$^{3-}$ anion found in salts with [NBu$^n_4$]$^+$,[614] and [*N,N'*-
dimethylpropyleneurea]$^+$[337] cations, whilst extension to a chain of five face sharing octahedra is
found in [Ph$_4$P]$_3$[Bi$_5$I$_{18}$].[336] If four BiI$_6$ octahedra share edges, one structure is [Bi$_4$I$_{16}$]$^{4-}$ which has
the geometry shown for the chlorine analogue in Figure79.[375,614] In [Q]$_4$[Bi$_6$I$_{22}$] (Q = PhCH$_2$Et$_3$N,
Ph$_4$P, Et$_4$P, EtMe$_2$PhN) the anion has the structure type shown in Figure 81;[336,399,595,615]
removal of a BiI$_3$ unit generates [Bi$_5$I$_{19}$]$^{4-}$,[615] whilst addition of two more BiI$_3$ groups generates
[Bi$_8$I$_{28}$]$^{4-}$, the largest discrete iodobismuthate presently known.[616] Infinite chains are present in

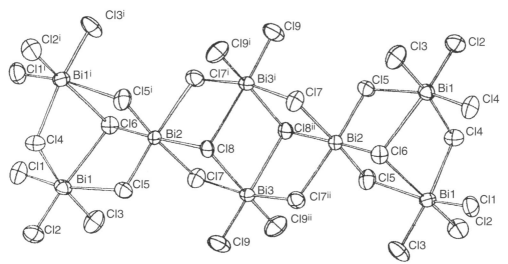

Figure 80 The structure of [Bi$_8$Cl$_{30}$]$^{6-}$ (reproduced by permission of Elsevier Science from *J. Phys. Chem.
Solids* **1989**, *50*, 1265–1269).

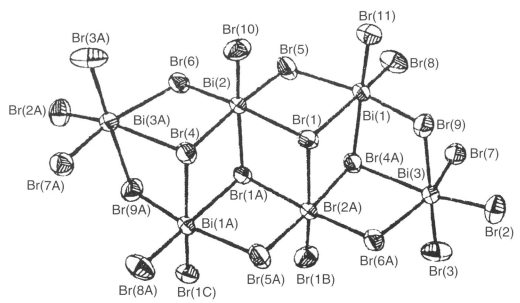

Figure 81 The structure of $[Bi_6Br_{22}]^{4-}$ (reproduced by permission of Wiley-VCH from *Z. Anorg. Allg. Chem.* **2001**, *627*, 2261–2268).

$[NBu^n_4][Bi_2I_7]$ which consists of Bi_4I_{16} units sharing common bridging iodines, and $[Ph_4P]_2[Bi_3I_{11}]$ is similarly related to Bi_6I_{24}.[616] Finally in $[Bi_4I_{14}(THF)_2]^{2-}$ there is the $[Bi_4X_{16}]^{4-}$ type (Figure 79) with two terminal halides replaced by tetrahydrofuran.[615]

Organohalobismuthine anions are also known. $[BiPh_2X_2]^-$ (X = Br or I) are *pseudo*-trigonal bipyramids with the lone pair occupying an equatorial position. $[BiPhI_2(THF)]$ reacts with $[NEt_4]I$ to form $[NEt_4]_2[Bi_2Ph_2I_6]$ which is a discrete dimer with edge-linked square pyramids.[408,617] Mixed halobismuthines, $[BiPhX_2Y]^-$ (X = Cl or Br; Y = Cl, Br, or I) are also known.[618]

3.6.4.5 Bismuth in the Environment, Biology, and Medicine

Bismuth is a rare element, mostly recovered as a by-product of lead and copper ore processing. Industrially, its major uses are as the element in various alloys, in simple inorganic chemicals and in medicine. Only the last of these relates to the coordination chemistry. Unusually for elements in this area of the periodic table, bismuth compounds are of low toxicity, in general markedly less so than arsenic and antimony analogues.[111–113] Indeed it seems that the majority of cases of bismuth poisoning have occurred during medical therapy rather than due to industrial exposure. Bismuth is used widely in treatment for intestinal disorders, anti-ulcer treatments, with much recent interest in the eradication of *Helicobacter pylori*.[619] There are three detailed reviews which describe bismuth compounds with biological or medicinal relevances,[620] bismuth anti-ulcer complexes,[621] and the biological and medicinal chemistry of bismuth,[622] which should be consulted for details. A further article discusses the coordination chemistry of metals in medicine-target sites for bismuth.[623] The medicinal preparations range from simple inorganic salts to bismuth complexes of carboxylic, hydroxo- and amino-carboxylates, of which the colloidal bismuth subcitrate is a widely used example. Many of the preparations are mixtures of complexes and the chemical speciation is ill-defined.

3.6.5 REFERENCES

1. Emsley, J. *The Elements* **1989**, Oxford University Press: Oxford.
2. Cotton, F. A.; Wilkinson, G.; Murillo, C. A.; Bochmann, M. *Advanced Inorganic Chemistry*, 6th ed.; Wiley, New York 1999.
3. Greenwood, N. N.; Earnshaw, A. *Chemistry of the Elements*, 2nd **1997**, Butterworth, Oxford.

4. Carmalt, C. J.; Norman, N. C. In *The Chemistry of Arsenic, Antimony and Bismuth*; Norman, N. C., Ed.; Blackie: London, 1998; Chapter 1, Arsenic, Antimony and bismuth: some general properties and aspects of periodicity, pp 1–38.
5. Landrum, G. A.; Hoffmann, R. *Angew. Chem., Int. Ed.* **1998**, *37*, 1887–1890.
6. Norman, N. C. *The Chemistry of Arsenic, Antimony and Bismuth* **1998**, Blackie: London.
7. Godfrey, S. M.; McAuliffe, C. A.; Mackie, A. G.; Pritchard, R. G. In *The Chemistry of Arsenic, Antimony and Bismuth*; Norman, N. C., Ed.; Blackie: London, 1998; Chapter 4, pp 159–205.
8. McAuliffe, C. A. Arsenic, antimony and bismuth. In *Comprehensive Coordination Chemistry*; Wilkinson, G., Ed.; Pergamon: Oxford, 1987; Chapter 28, pp 227–298.
9. Wardell, J. L. Arsenic, antimony and bismuth. In *Comprehensive Organometallic Chemistry*; Wilkinson, G.; Stone, F. G. A.; Abel, E. W., Eds.; Pergamon: Oxford, 1982; Vol. 2, pp 681–707.
10. Wardell, J. L. Arsenic, antimony and bismuth. In *Comprehensive Organometallic Chemistry II*; Abel, E. W.; Stone, F. G. A.; Wilkinson, G., Eds.; Pergamon: Oxford, 1995; Vol. 2, pp 321–347.
11. Patai, S., Ed. *The Chemistry of Organic Arsenic Antimony and Bismuth Compounds* **1994**, Wiley: New York.
12. Bohra, R.; Roesky, H. W. *Adv. Inorg. Chem. Radiochem.* **1984**, *28*, 203–254.
13. Jones, C. *Organometal. Chem.* **2000**, *28*, 138–152.
14. Breunig, H. J. In *Chemistry of Organic Arsenic, Antimony and Bismuth Compounds*; Patai, S. Ed.; Wiley: New York, 1994; Chapter 14, Organoarsenic and organoantimony heterocycles, pp 563–578.
15. Yamamoto, Y.; Akiba, K.-Y. In *Chemistry of Organic Arsenic, Antimony and Bismuth Compounds*, Patai, S. Ed.; Wiley: New York, 1994; Chapter 21, Synthesis of Organoarsenic Compounds, pp 813–882.
16. Biswas, A. K.; Hall, J. R.; Schweinsberg, D. P. *Inorg. Chim. Acta* **1983**, *75*, 57–64.
17. Wade, S. R.; Willey, G. R. *Inorg. Chim. Acta* **1979**, *35*, 61–63.
18. Willey, G. R.; Daly, L. T.; Meehan, P. R.; Drew, M. G. B. *J. Chem. Soc., Dalton Trans.* **1996**, 4045–4053.
19. Willey, G. R.; Asab, A.; Lakin, M. T.; Alcock, N. W. *J. Chem. Soc., Dalton Trans.* **1993**, 365–370.
20. Atwood, D. A.; Cowley, A. H.; Ruiz, J. *Inorg. Chim. Acta* **1992**, *198–200*, 271–274.
21. Carmalt, C. J.; Cowley, A. H.; Culp, R. D.; Jones, R. A.; Kamepalli, S.; Norman, N. C. *Inorg. Chem.* **1997**, *36*, 2770–2776.
22. Kamepalli, S.; Carmalt, C. J.; Culp, R. D.; Cowley, A. H.; Jones, R. A.; Norman, N. C. *Inorg. Chem.* **1996**, *35*, 6179–6183.
23. Buzek, P.; Schleyer, P. V. R.; Klapötke, T. M.; Tornieporth-Oetting, I. C. *J. Fluorine Chem.* **1993**, *65*, 127–132.
24. Il'in, E. G.; Buslaev, Yu. A.; Calov, U.; Kolditz, L. *Dokl. Akad. Nauk SSSR* **1983**, *270*, 1146–1148.
25. Broschag, M.; Klapötke, T. M. *Polyhedron* **1992**, *11*, 443–446.
26. Tornieporth-Oetting, I. C.; Klapötke, T. M. *Chem. Ber.* **1992**, *125*, 407–409.
27. Tornieporth-Oetting, I. C.; Klapötke, T. M.; Cameron, T. S.; Valkonen, J.; Rademacher, P.; Kowski, K. *J. Chem. Soc., Dalton Trans.* **1992**, 537–543.
28. Tornieporth-Oetting, I. C.; Klapötke, T. M.; Behrens, U.; White, P. S. *J. Chem. Soc., Dalton Trans.* **1992**, 2055–2058.
29. Minkwitz, R.; Koch, M.; Nowicki, J.; Borrmann, H. *Z. Anorg. Allg. Chem.* **1990**, *590*, 93–102.
30. Bellard, S.; Rivera, A. V.; Sheldrick, G. M. *Acta Crystallogr., Sect. B* **1978**, *34*, 1034–1035.
31. Apblett, A.; Chivers, T.; Richardson, J. F. *Can. J. Chem.* **1986**, *64*, 849–853.
32. Karaghiosoff, K.; Klapötke, T. M.; Krumm, B.; Noeth, H.; Schuett, T.; Suter, M. *Inorg. Chem.* **2002**, *41*, 170–179.
33. Klapötke, T. M.; Geisler, P. *J. Chem. Soc., Dalton Trans.* **1995**, 3365–3366.
34. Summers, J. C.; Sisler, H. H. *Inorg. Chem.* **1970**, *9*, 862–869.
35. Hill, N. J.; Levason, W.; Reid, G. *J. Chem. Soc., Dalton Trans.* **2002**, 1188–1192.
36. Baum, G.; Greiling, A.; Massa, W.; Hui, B. C.; Lorberth, J. *Z. Naturforsch., B* **1989**, *44*, 560–564.
37. Siewert, B.; Müller, U. *Z. Naturforsch., B* **1992**, *47*, 680–684.
38. Alcock, N. W.; Ravindran, M.; Willey, G. R. *Acta Crystallogr., Sect. B* **1993**, *49*, 507–514.
39. Borgsen, B.; Weller, F.; Dehnicke, K. *Chem.-Ztg.* **1990**, *114*, 111–112.
40. Hill, N. J.; Levason, W.; Reid, G. *Inorg. Chem.* **2002**, *41*, 2070–2076.
41. Kniep, R.; Reski, H. D. *Inorg. Chim. Acta* **1982**, *64*, L83–L84.
42. Barton, A. J.; Hill, N. J.; Levason, W.; Reid, G. *J. Am. Chem. Soc.* **2001**, *123*, 11801–11802.
43. Forster, A. M.; Downs, A. J. *J. Chem. Soc., Dalton Trans.* **1984**, 2827–2834.
44. Holmes, R. R. *Prog. Inorg. Chem.* **1984**, *32*, 119–235.
45. Gamayurova, V. S.; Niyazov, N. A.; Yusupov, R. L. *Zh. Obshch. Khim.* **1985**, *55*, 2497–2500.
46. Gamayurova, V. S.; Shabrukova, N. V.; Chechetkina, I. I.; Zyablikova, T. A.; Lipatova, I. P.; Chugunov, Yu. V. *Zh. Obshch. Khim.* **1994**, *64*, 1998–2002.
47. Van Nuffel, P.; Lenstra, A. T. H.; Geise, H. J.; Yuldasheva, L. K.; Chadaeva, N. A. *Acta Crystallogr., Sect. B* **1982**, *38*, 3089–3091.
48. Shang, S.; Khasnis, D. V.; Zhang, H.; Small, A. C.; Fan, M.; Lattman, M. *Inorg. Chem.* **1995**, *34*, 3610–3615.
49. Said, M. A.; Swamy, K. C. K.; Veith, M.; Huch, V. *Inorg. Chem.* **1996**, *35*, 6627–6630.
50. Swamy, K. C. K.; Musa, A.; Veith, M.; Huch, V. *Phosphorus, Sulfur Silicon Relat. Elem.* **1999**, *152*, 191–201.
51. Said, M. A.; Swamy, K. C. K.; Veith, M.; Huch, V. *J. Chem. Soc., Perkin Trans. 1*, **1995**, 2945–2951.
52. Poutasse, C. A.; Day, R. O.; Holmes, J. M.; Holmes, R. R. *Organometallics* **1985**, *4*, 708–713.
53. Borgias, B. A.; Hardin, G. G.; Raymond, K. N. *Inorg. Chem.* **1986**, *25*, 1057–1060.
54. Kamenar, B.; Bruvo, M.; Butumovic, J. *Z. Anorg. Allg. Chem.* **1993**, *619*, 943–946.
55. Bott, R. C.; Smith, G.; Sagatys, D. S.; Mak, T. C. W.; Lynch, D. E.; Kennard, C. H. L. *Aust. J. Chem.* **1993**, *46*, 1055–1065.
56. Bott, R. C.; Smith, G.; Sagatys, D. S.; Lynch, D. E.; Kennard, C. H. L. *Aust. J. Chem.* **2000**, *53*, 917–924.
57. Marcovich, D.; Duesler, E. N.; Tapscott, R. E.; Them, T. F. *Inorg. Chem.* **1982**, *21*, 3336–3341.
58. Kapoor, R.; Wadhawan, P.; Kapoor, P. *Can. J. Chem.* **1987**, *65*, 1195–1199.
59. Imoto, H.; Aubke, F. *J. Fluorine Chem.* **1980**, *15*, 59–66.
60. Chauhan, H. P. S. *Coord. Chem. Rev.* **1998**, *173*, 1–30.
61. Ito, T.; Hishino, H. *Acta Crystallogr., Sect. C* **1983**, *39*, 448–451.
62. Hoskins, B. F.; Piko, P. M.; Tiekink, E. R. T.; Winter, G. *Inorg. Chim. Acta* **1984**, *84*, L13–L14.
63. Hoskins, B. F.; Tiekink, E. R. T.; Winter, G. *Inorg. Chim. Acta* **1985**, *99*, 177–182.

536   *Arsenic, Antimony, and Bismuth*

64. Snow, M. R.; Tiekink, E. R. T. *Aust. J. Chem.* **1987**, *40*, 743–750.
65. Cox, M. J.; Tiekink, E. R. T. *Z. Kristallogr.* **1998**, *213*, 487–492.
66. Hounslow, A. M.; Lincoln, S. F.; Tiekink, E. R. T. *J. Chem. Soc., Dalton Trans.* **1989**, 233–236.
67. Karra, R.; Singh, Y. P.; Bohra, R.; Rai, A. K. *J. Crystallogr. Spectrosc. Res.* **1992**, *22*, 721–724.
68. Chauhan, H. P. S.; Chourasia, S.; Nahar, B.; Singh, R. K. *Phosphorus, Sulfur Silicon Relat. Elem.* **1998**, *134/135*, 345–353.
69. Fabretti, A. C.; Giusti, A.; Preti, C.; Tosi, G.; Zannini, P. *Polyhedron* **1986**, *5*, 871–875.
70. Singh, S. K.; Singh, Y. P.; Rai, A. K.; Mehrotra, R. C. *Indian J. Chem., Sect. A* **1989**, *28*, 585–587.
71. Venkatachalam, V.; Ramalingam, K.; Mak, T. C. W.; Luo, B. S. *J. Chem. Crystallogr.* **1996**, *26*, 467–470.
72. Gupta, R. K.; Rai, A. K.; Mehrotra, R. C. *Indian J. Chem., Sect. A* **1985**, *24*, 752–754.
73. Cea-Olivares, R.; Toscano, R. A.; Lopez, M.; Garcia, P. *Monatsh. Chem.* **1993**, *124*, 177–183.
74. Venkatachalam, V.; Ramalingam, K.; Casellato, U.; Graziani, R. *Polyhedron* **1997**, *16*, 1211–1221.
75. Garje, S. S.; Jain, V. K.; Tiekink, E. R. T. *J. Organomet. Chem.* **1997**, *538*, 129–134.
76. Kavounis, C. A.; Kokkou, S. C.; Rentzeperis, P. J.; Karagiannidis, P. *Acta Crystallogr., Sect. B* **1982**, *38*, 2686–2689.
77. Cea-Olivares, R.; Estrada, M. R.; Espinosa-Perez, G.; Haiduc, I.; Garcia, P. GarciaY.; Lopez-Cardoso, M.; Lopez-Vaca, M.; Cotero-Villegas, A. *Main Group Met. Chem.* **1995**, *18*, 159–164.
78. Engle, R.; Schmidt, A. *Z. Anorg. Allg. Chem.* **1994**, *620*, 539–544.
79. Cea-Olivares, R.; Toscano, R.; Silvestru, C.; Garcia-Garcia, P.; Lopez-Cardoso, M.; Blass-Amador, G.; Noeth, H. *J. Organomet. Chem.* **1995**, *493*, 61–67.
80. Chauhan, H. P. S.; Srivastava, G.; Mehrotra, R. C. *Polyhedron* **1983**, *2*, 359–364.
81. Chauhan, H. P. S.; Srivastava, G.; Mehrotra, R. C. *Polyhedron* **1984**, *3*, 1337–1345.
82. Gupta, R. K.; Rai, A. K.; Mehrotra, R. C.; Jain, V. K.; Hoskins, B. F.; Tiekink, E. R. T. *Inorg. Chem.* **1985**, *24*, 3280–3284.
83. Garje, S. S.; Jain, V. K. *Main Group Met. Chem.* **1997**, *20*, 217–222.
84. Gupta, R. K.; Rai, A. K.; Mehrotra, R. C.; Jain, V. K. *Inorg. Chim. Acta* **1984**, *88*, 201–207.
85. Cea-Olivares, R.; Alvarado, J.; Espinosa-Perez, G.; Silvestru, C.; Haiduc, I. *J. Chem. Soc., Dalton Trans.* **1994**, 2191–2195.
86. Silaghi-Dumitrescu, L.; Haiduc, I. *J. Organomet. Chem.* **1983**, *252*, 295–299.
87. Chauhan, H. P. S.; Porwal, B.; Singh, R. K. *Phosphorus, Sulfur Silicon Relat. Elem.* **2000**, *160*, 93–103.
88. Munoz-Hernandez, M.; Cea-Olivares, R.; Espinosa-Perez, G.; Hernandez-Ortega, S. *J. Chem. Soc., Dalton Trans.* **1996**, 4135–4141.
89. Munoz-Hernandez, M.; Cea-Olivares, R.; Hernandez-Ortega, S. *Inorg. Chim. Acta* **1996**, *253*, 31–37.
90. Kisenyi, J. M.; Willey, G. R.; Drew, M. G. B.; Wandiga, S. O. *J. Chem. Soc., Dalton Trans.* **1985**, 69–74.
91. Fisher, G. A.; Norman, N. C. *Adv. Inorg. Chem.* **1994**, *41*, 233–271.
92. Zhang, X.; Seppelt, K. *Z. Anorg. Allg. Chem.* **1994**, *623*, 491–500.
93. Edwards, A. J.; Patel, S. N. *J. Chem. Soc., Dalton Trans.* **1980**, 1630–1632.
94. Kaub, J.; Sheldrick, W. S. *Z. Naturforsch., B* **1984**, 39 1252–1256.
95. Mohammed, A. T.; Müller, U. *Acta Crystallogr., Sect. C* **1985**, *41*, 329–332.
96. Sheldrick, W. S.; Horn, C. *Z. Naturforsch., B* **1989**, *44*, 405–411.
97. Kaub, J.; Sheldrick, W. S. *Z. Naturforsch., B* **1984**, *39*, 1257–1261.
98. Sheldrick, W. S.; Haeusler, H. J.; Kaub, J. *Z. Naturforsch., B* **1988**, *43*, 789–794.
99. Sheldrick, W. S.; Haeusler, H. J. *Angew. Chem.* **1987**, *99*, 1184–1186.
100. Sheldrick, W. S.; Kiefer, J. *Z. Naturforsch., B* **1992**, *47*, 1079–1084.
101. Willing, W.; Müller, U.; Eicher, J.; Dehnicke, K. *Z. Anorg. Allg. Chem.* **1986**, *537*, 145–153.
102. Gafner, G.; Kruger, G. J. *Acta Crystallogr., Sect. B* **1974**, *30*, 250–251.
103. Strauss, S. H. *Chem. Rev.* **1993**, *93*, 927–942.
104. Minkwitz, R.; Hirsch, C.; Berends, T. *Eur. J. Inorg. Chem.* **1999**, 2249–2254.
105. Drake, G. W.; Dixon, D. A.; Sheehy, J. A.; Boatz, J. A.; Christie, K. O. *J. Am. Chem. Soc.* **1998**, *120*, 8392–8400.
106. Bebendorf, J.; Müller, U. *Z. Naturforsch., B* **1990**, *45*, 927–930.
107. Dove, M. F. A.; Sanders, J. C. P.; Lloyd Jones, E.; Parkin, M. *J. Chem. Commun.* **1984**, 1578–1581.
108. Czado, W.; Müller, U. *Z. Anorg. Allg. Chem.* **1998**, *624*, 103–106.
109. Klapötke, T. M.; Schütt, T. *Z. Naturforsch., B* **2001**, *56*, 301–305.
110. Il'in, E. G.; Buslaev, Yu. A.; Kalov, U.; Kolditz, L. *Dokl. Akad. Nauk SSSR* **1984**, *276*, 371–372.
111. Reglinski, J. In *The Chemistry of Arsenic, Antimony and Bismuth*; Norman, N. C., Ed.; Blackie: London, 1998; Chapter 8, Environmental and medicinal chemistry of arsenic, antimony and bismuth.
112. Wormser, U.; Nir, I. In *Chemistry of Organic Arsenic, Antimony and Bismuth Compounds*; Patai, S. Ed.; Wiley: New York, 1994; Chapter 18, Pharmacology and toxicology of organic bismuth, arsenic and antimony compounds, pp 715–723.
113. Maeda, S. In *Chemistry of Organic Arsenic, Antimony and Bismuth Compounds*; Patai, S. Ed.; Wiley: New York, 1994; Chapter 20, Synthesis of organoantimony and organobismuth compounds, pp 725–759.
114. Francesconi, K. A.; Edmonds, J. S. *Adv. Inorg. Chem.* **1997**, *44*, 147–189.
115. Jones, C. *Coord. Chem. Rev.* **2001**, *215*, 151–169.
116. Akiba, K.-Y.; Yamamoto, Y. In *Chemistry of Organic Arsenic, Antimony and Bismuth Compounds*; Patai, S. Ed.; Wiley: New York, 1994; Chapter 21, Snthesis of organoarsenic compounds, pp 761–812.
117. Breunig, H. J.; Roesler, R. *Coord. Chem. Rev.* **1997**, *163*, 35–53.
118. Breunig, H. J.; Roesler, R. *Chem. Soc. Rev.* **2000**, *29*, 403–410.
119. Schulz, S. *Coord. Chem. Rev.* **2001**, *215*, 1–37.
120. Cameron, J. U.; Killean, R. C. G. *Cryst. Struct. Comm.* **1972**, *1*, 31–33.
121. Lipka, A.; Wunderlich, H. *Z. Naturforsch., B* **1980**, *35*, 1548–1551.
122. Alonzo, G.; Bertazzi, N.; Maccotta, A. *Inorg. Chim. Acta* **1982**, *62*, 167–169.
123. Preut, H.; Huber, F.; Alonzo, G.; Bertazzi, N. *Acta Crystallogr., Sect. B* **1982**, *38*, 935–937.
124. Huber, F.; Preut, H.; Alonzo, G.; Bertazzi, N. *Inorg. Chim. Acta* **1985**, *102*, 181–186.
125. Willey, G. R.; Spry, M. P.; Drew, M. G. B. *Polyhedron* **1996**, *15*, 4497–4500.

126. Huckstadt, H.; Tutass, A.; Goldner, M.; Cornelissen, U.; Homberg, H. *Z. Anorg. Allg. Chem.* **2001**, *627*, 485–497.
127. Carmalt, C. J.; Walsh, D.; Cowley, A. H.; Norman, N. C. *Organometallics* **1997**, *16*, 3597–3600.
128. Brau, E.; Falke, R.; Ellner, A.; Beuter, M.; Kolb, U.; Dräger, M. *Polyhedron* **1994**, *13*, 365–374.
129. Brau, E.; Zickgraf, A.; Dräger, M.; Mocellin, E.; Maeda, M.; Takahashi, M.; Takeda, M.; Meali, C. *Polyhedron* **1998**, *17*, 2655–2668.
130. Sharma, P.; Cabrera, A.; Singh, S.; Jha, N. K. *Main Group Met. Chem.* **1997**, *20*, 551–565.
131. Di Bianca, F.; Bertazzi, N.; Alonzo, G.; Ruisi, G.; Gibb, T. C. *Inorg. Chim. Acta* **1981**, 50 235–237.
132. Aminabhavi, T. M.; Birandar, N. S.; Karajagi, G. V.; Banks, A. J. *Inorg. Chim. Acta* **1984**, *88*, 41–44.
133. Rastogi, R.; Parashar, G. K.; Kapoor, R. N. *Synth. React. Inorg. Met.-Org. Chem.* **1985**, *15*, 1061–1071.
134. Biradar, N. S.; Roddabasanagoudar, V. L.; Aminabhavi, T. M. *Indian J. Chem., Sect. A* **1985**, *24*, 701–702.
135. Klapötke, T. M.; Noth, H.; Schutt, T.; Suter, M.; Warchhold, M. *Z. Anorg. Allg. Chem.* **2001**, *627*, 1582–1588.
136. Alvarez-Valdes, A.; Gomez-Vaamonde, C.; Masaguer, J. R.; Garcia-Vazquez, J. A. *Z. Anorg. Allg. Chem.* **1985**, *523*, 227–233.
137. Alvarez-Valdes, A.; Masaguer, J. R.; Garcia-Vazquez, J. A. *Spectrochim. Acta, Part A* **1984**, *40*, 995–998.
138. Kessler, J. E.; Knight, C. T. G.; Merbach, A. E. *Inorg. Chim. Acta* **1986**, 115, 75–83.
139. Kessler, J. E.; Knight, C. T. G.; Merbach, A. E. *Inorg. Chim. Acta* **1986**, 115, 85–89.
140. Garbe, R.; Pebler, J.; Dehnicke, K.; Fenske, D.; Goesmann, H.; Baum, G. *Z. Anorg. Allg. Chem.* **1994**, *620*, 592–598.
141. Klapötke, T. M.; Schulz, A.; McNamara, J. *J. Chem. Soc., Dalton Trans.* **1996**, 2985–2987.
142. Klapötke, T. M.; Noth, H.; Schutt, T.; Warchhold, M. *Z. Anorg. Allg. Chem.* **2001**, *627*, 81–84.
143. Clegg, W.; Elsegood, M. R. J.; Graham, V.; Norman, N. C.; Tavakkoli, K. *J. Chem. Soc., Dalton Trans.* **1994**, 1743–1751.
144. Clegg, W.; Elsegood, M. R. J.; Norman, N. C.; Pickett, N. L. *J. Chem. Soc., Dalton Trans.* **1994**, 1753–1757.
145. Genge, A. R. J.; Hill, N. J.; Levason, W.; Reid, G. *J. Chem. Soc., Dalton Trans.* **2001**, 1007–1012.
146. Clegg, W.; Elsegood, M. R. J.; Graham, V.; Norman, N. C.; Pickett, N. L. *J. Chem. Soc., Dalton Trans.* **1993**, 997–998.
147. Breunig, H. J.; Denker, M.; Schulz, R. E.; Lork, E. *Z. Anorg. Allg. Chem.* **1998**, *624*, 81–84.
148. Breunig, H. J.; Ebert, K. H.; Gulec, S.; Drager, M.; Sowerby, D. B.; Begley, M. J.; Behrens, U. *J. Organomet. Chem.* **1992**, *427*, 39–46.
149. Breunig, H. J.; Denker, M.; Ebert, K. H. *Chem. Commun.* **1994**, 875–876.
150. Hough, E.; Nicholson, D. G.; Vasudevan, A. K. *J. Chem. Soc., Dalton Trans.* **1987**, 427–430.
151. Alcock, N. W.; Ravindran, M.; Roe, S. M.; Willey, G. R. *Inorg. Chim. Acta* **1990**, *167*, 115–118.
152. Beagley, B.; Endregard, M.; Nicholson, D. G. *Acta Chem. Scand.* **1991**, *45*, 349–353.
153. Neuhaus, A.; Frenzen, G.; Pebler, J.; Dehnicke, K. *Z. Anorg. Allg. Chem.* **1992**, *618*, 93–97.
154. Schaefer, M.; Frenzen, G.; Neumueller, B.; Dehnicke, K. *Angew. Chem. Int. Ed.* **1992**, *31*, 334–335.
155. Willey, G. R.; Aris, D. R.; Errington, W. *Inorg. Chim. Acta* **2000**, *300*, 1004–1013.
156. Drexler, H.-J.; Starke, I.; Grotjahn, M.; Reinke, H.; Kleinpeter, E.; Holdt, H.-J. *Z. Naturforsch., B* **1999**, *54*, 799–806.
157. Yamamoto, J.; Murakami, M.; Kameoka, N.; Otani, N.; Umezu, M.; Matsuura, T. *Bull. Chem. Soc. Jpn.* **1982**, *55*, 345–346.
158. Yamamoto, J.; Ito, S.; Tsuboi, T.; Tsuboi, T; Tsukihara, K. *Bull. Chem. Soc. Jpn.* **1985**, *58*, 470–472.
159. Neumuller, B.; Koeckler, R.; Meyer, R.; Dehnicke, K. *Z. Kristallogr.* **1994**, *209*, 90–91.
160. Ensinger, U.; Schwarz, W.; Schrutz, B.; Sommer, K.; Schmidt, A. *Z. Anorg. Allg. Chem.* **1987**, *544*, 181–191.
161. Fleischer, H.; Bayram, H.; Elzner, S.; Mitzel, N. W. *J. Chem. Soc., Dalton Trans.* **2001**, 373–377.
162. Horley, G. A.; Mahon, M. F.; Molloy, K. C.; Venter, M. M.; Haycock, P. W.; Myers, C. P. *Inorg. Chem.* **2002**, *41*, 1652–1657.
163. Binder, G. E.; Schwarz, W.; Rozdzinski, W.; Schmidt, A. *Z. Anorg. Allg. Chem.* **1980**, *471*, 121–130.
164. Edwards, A. J.; Leadbeater, N. E.; Paver, M. A.; Raithby, P. R.; Russell, C. A.; Wright, D. S. *J. Chem. Soc., Dalton Trans.* **1994**, 1479–1482.
165. Temple, N.; Schwarz, W.; Weidlein, J. *Z. Anorg. Allg. Chem.* **1981**, *474*, 157–170.
166. Wieber, M.; Walz, J.; Burschka, C. *Z. Anorg. Allg. Chem.* **1990**, *585*, 65–74.
167. Sen Gupta, A. K.; Bohra, R.; Mehrotra, R. C.; Das, K. *Main Group Met. Chem.* **1990**, *13*, 321–339.
168. Alamgir, M.; Allen, N.; Barnard, P. W. C.; Donaldson, J. D.; Silver, J. *Acta Crystallogr., Sect. B* **1981**, *37*, 1284–1286.
169. Korte, L.; Mootz, D.; Scherf, M.; Wiebocke, M. *Acta Crystallogr., Sect. C* **1988**, *44*, 1128–1130.
170. Binder, G. E.; Schmidt, A. *Z. Anorg. Allg. Chem.* **1981**, *462*, 73–80.
171. Jha, N. K.; Joshi, D. M. *Polyhedron* **1985**, *4*, 2083–2087.
172. Collins, M. J.; Schrobilgen, G. J. *Inorg. Chem.* **1985**, *24*, 2608–2614.
173. Mercier, H. P. A.; Sanders, J. C. P.; Schrobilgen, G. J. *J. Am. Chem. Soc.* **1994**, *116*, 2921–2937.
174. Van Seggen, D. M.; Hurlburt, P. K.; Anderson, O. P.; Strauss, S. H. *Inorg. Chem.* **1995**, *34*, 3453–3473.
175. Zhang, D.; Rettig, S. J.; Trotter, J.; Aubke, F. *Inorg. Chem.* **1995**, *34*, 3153–3164.
176. Zhang, D.; Rettig, S. J.; Trotter, J.; Aubke, F. *Inorg. Chem.* **1995**, *34*, 2269–2270.
177. Zhang, D.; Rettig, S. J.; Trotter, J.; Aubke, F. *Inorg. Chem.* **1996**, *35*, 6113–6130.
178. Wilson, W. W.; Aubke, F. *J. Fluorine Chem.* **1979**, *13*, 431–445.
179. Cooke, A. W.; Pebler, J.; Weller, F.; Dehnicke, K. *Z. Anorg. Allg. Chem.* **1985**, *524*, 68–74.
180. Knoedler, F.; Schwarz, W.; Schmidt, A. *Z. Naturforsch., B* **1987**, *42*, 1282–1290.
181. Sauvigny, A.; Faerber, J. E.; Rihm, A.; Thurn, A.; Schmidt, A. *Z. Anorg. Allg. Chem.* **1995**, *621*, 640–644.
182. Hornung, H. D.; Klinkhammer, K. W.; Faerber, J. E.; Schmidt, A.; Bensch, W. *Z. Anorg. Allg. Chem.* **1996**, *622*, 1038–1046.
183. Shihada, A. F.; Weller, F. *Z. Anorg. Allg. Chem.* **1981**, *472*, 102–108.
184. Hornung, H. D.; Klinkhammer, K. W.; Schmidt, A. *Z. Naturforsch., B* **1996**, *51*, 975–980.
185. Burchardt, A.; Klinkhammer, K. W.; Schmidt, A. *Z. Anorg. Allg. Chem.* **1998**, *624*, 35–43.
186. Lang, G.; Lauster, M.; Klinkhammer, K. W.; Schmidt, A. *Z. Anorg. Allg. Chem.* **1999**, *625*, 1799–1806.
187. Said, M. A.; Swamy, K. C. K.; Babu, K.; Aparna, K.; Nethaji, M. *J. Chem. Soc., Dalton Trans.* **1995**, 2151–2157.
188. Said, M. A.; Swamy, K. C. K.; Poojary, D. M.; Clearfield, A.; Veith, M.; Huch, V. *Inorg. Chem.* **1996**, *35*, 3235–3241.

189. Silvestru, C.; Haiduc, I.; Ebert, K. H.; Breunig, H. J.; Sowerby, D. B. *J. Organomet. Chem.* **1994**, *468*, 113–119.
190. Silvestru, C.; Silvestru, A.; Haiduc, I.; Sowerby, D. B.; Ebert, K. H.; Breunig, H. J. *Polyhedron* **1997**, *16*, 2643–2649.
191. Mahalakshmi, H.; Jain, V. K.; Teikink, E. T. R. *Main Group Met. Chem.* **2000**, *23*, 519–524.
192. Gibbons, M. N.; Sowerby, D. B. *J. Chem. Soc., Dalton Trans.* **1997**, 2785–2792.
193. Bohaty, L.; Frohlich, R.; Tebbe, K. F. *Acta Crystallogr., Sect. C* **1983**, *39*, 59–63.
194. Bohaty, L.; Frohlich, R. *Z. Kristallogr.* **1983**, *163*, 261–265.
195. Sagatys, D. S.; Smith, G.; Lynch, D. E.; Kennard, C. H. L. *J. Chem. Soc., Dalton Trans.* **1991**, 361–364.
196. Hartley, D. W.; Smith, G.; Sagatys, D. S.; Kennard, C. H. L. *J. Chem. Soc., Dalton Trans.* **1991**, 2735–2739.
197. Smith, G.; Sagatys, D. S.; Bott, R. C.; Lynch, D. E.; Kennard, C. H. L. *Polyhedron* **1993**, *12*, 1491–1497.
198. Smith, G.; Sagatys, D. S.; Bott, R. C.; Lynch, D. E.; Kennard, C. H. L. *Polyhedron* **1992**, *11*, 631–634.
199. Shimoi, M.; Orita, Y.; Uehiro, T.; Kita, I.; Iwamoto, T.; Ouchi, A.; Yoshino, Y. *Bull. Chem. Soc. Jpn.* **1980**, *53*, 3189–3194.
200. Marrot, B.; Brouca-Cabarrecq, C.; Mosset, A. *J. Mater. Chem.* **1996**, *6*, 789–793.
201. Fun, H.-K.; Raj, S. S. S.; Razak, I. A.; Ilyukhin, A. B.; Davidovich, R. L.; Huang, J.-W.; Hu, S.-Z.; Ng, S. W. *Acta Crystallogr., Sect. C* **1999**, *55*, 905–907.
202. Zhaoxiong, X.; Shengzhi, H. *Xiegou Huaxue* **1991**, *10*, 129–131.
203. Marrot, B.; Brouca-Cabarrecq, C.; Mosset, A. *J. Chem. Crystallogr.* **1998**, *28*, 447–452.
204. Davidovich, R. L.; Logvinova, V. B.; Kaidalova, T. A. *Russ. J. Coord. Chem.* **1998**, *24*, 399–404.
205. Ilyukhin, A. B.; Davidovich, R. L. *Kristallografiya* **1999**, *44*, 238–246.
206. Hu, S.-Z.; Lin, W. *Xiegou Huaxue* **1989**, *8*, 249–256.
207. Hu, S.-Z.; Tu, L.-D.; Huang, Y.-Q.; Li, Z.-X. *Inorg. Chim. Acta* **1995**, *232*, 161–165.
208. Fu, Y.-M.; Xie, Z.-X.; Hu, S.-Z.; Xu, B.; Tang, W.-D.; Yu, W.-J. *Xiegou Huaxue* **1997**, *16*, 91–96.
209. Hu, S.-Z.; Fu, Y.-M.; Toennessan, L. E.; Davidovich, R. L.; Ng, S. W. *Main Group Met. Chem.* **1998**, *21*, 501–505.
210. Hu, S.-Z.; Fu, Y.-M.; Xu, M.; Tang, W.-D.; Yu, W.-J. *Main Group Met. Chem.* **1997**, *20*, 169–180.
211. Shkol'nikova, L. M.; Fundamenski, V. S.; Davidovich, R. L.; Samsonova, I. N.; Dashevskaya, E. E. *Zh. Neorg. Khim.* **1991**, *36*, 2042–2047.
212. Gu, D.; Lu, B.; Lu, Y. *Xiegou Huaxue* **1989**, *8*, 311–315.
213. Hu, S.-Z.; Xie, Z.-X. *Xiegou Huaxue* **1991**, *10*, 81–83.
214. Levason, W.; Reid, G. *J. Chem. Soc., Dalton Trans.* **2001**, 2953–2960.
215. Barton, A. J.; Hill, N. J.; Levason, W.; Reid, G. *Chem. Commun.* **2001**, 95–96.
216. Barton, A. J.; Hill, N. J.; Levason, W.; Reid, G. *J. Chem. Soc., Dalton Trans.* **2001**, 1621–1627.
217. Willey, G. R.; Lakin, M. T.; Ravindran, M.; Alcock, N. W. *Chem. Commun.* **1991**, 271–272.
218. Pohl, S.; Haase, D.; Peters, M. *Z. Anorg. Allg. Chem.* **1993**, *619*, 727–730.
219. Berges, P.; Hinrichs, W.; Kopf, J.; Mandak; Klar, G. *J. Chem. Res.* **1985**, 218–219.
220. Mandak, D.; Klar, G. *J. Chem. Res.* **1984**, 76.
221. Williams, D. J.; Poor, P. H.; Ramirez, G.; Heyl, B. L. *Inorg. Chim. Acta* **1988**, *147*, 221–226.
222. Williams, D. J.; Vanderveer, D.; Jones, R. L.; Menaldino, D. S. *Inorg. Chim. Acta* **1989**, *165*, 173–178.
223. Korte, L.; Lipka, A.; Mootz, D. *Z. Anorg. Allg. Chem.* **1985**, *524*, 157–167.
224. Drew, M. G. B.; Kisenyi, J. M.; Wandiga, S. O.; Willey, G. R. *J. Chem. Soc., Dalton Trans.* **1984**, 1717–1721.
225. Drew, M. G. B.; Kisenyi, J. M.; Willey, G. R. *J. Chem. Soc., Dalton Trans.* **1982**, 1729–1721.
226. Kisenyi, J. M.; Willey, G. R.; Drew, M. G. B. *J. Chem. Soc., Dalton Trans.* **1985**, 1073–1075.
227. Pohl, S.; Saak, W.; Lotz, R.; Haase, D. *Z. Naturforsch., B* **1990**, *45*, 1355–1362.
228. Carrai, G.; Gottardi, G. *Z. Krystallogr.* **1960**, *113*, 373–384.
229. Hoskins, B. F.; Tiekink, E. R. T.; Winter, G. *Inorg. Chim. Acta* **1985**, *97*, 217–222.
230. Gable, R. W.; Hoskins, B. F.; Steen, R. J.; Tiekink, E. R. T.; Winter, G. *Inorg. Chim. Acta* **1983**, *74*, 15–20.
231. Wieber, M.; Wirth, D.; Burschka, C. *Z. Anorg. Allg. Chem.* **1983**, *505*, 141–146.
232. Blake, A. J.; Pearson, M.; Sowerby, D. B.; Woodhead, P. P. *Acta Crystallogr., Sect. C* **1997**, *53*, 583–585.
233. Kraft, S.; Wieber, M. *Z. Anorg. Allg. Chem.* **1992**, *607*, 164–168.
234. Hoskins, B. F.; Tiekink, E. R. T.; Winter, G. *Inorg. Chim. Acta* **1985**, *105*, 171–176.
235. Chauhan, H. P. S.; Chourasia, S. *Indian J. Chem., Sect. A* **1995**, *34*, 664–665.
236. Srivastava, D. K.; Singh, R. P.; Gupta, V. D. *Polyhedron* **1988**, *7*, 483–487.
237. Egle, R.; Kinkhammer, W.; Schmidt, A. *Z. Anorg. Allg. Chem.* **1992**, *617*, 72–78.
238. Cea-Olivares, R.; Wingartz, J.; Rios, E.; Valdes-Martinez, J. *Monatsh. Chem.* **1990**, *121*, 377–383.
239. Wieber, M.; Wirth, D.; Metter, J.; Burschka, C. *Z. Anorg. Allg. Chem.* **1985**, *520*, 65–70.
240. Nomura, R.; Takabe, A.; Matsuda, H. *Polyhedron* **1987**, *6*, 411–416.
241. Venkatachalam, V.; Ramalingham, K.; Bocelli, G.; Cantoni, A. *Inorg. Chim. Acta* **1997**, *261*, 23–28.
242. Meinema, H. A.; Noltes, J. G. *J. Organomet. Chem.* **1970**, *25*, 139–148.
243. Kavounis, C. A.; Kokkou, S. C.; Rentzeperis, P. J.; Karagiannidis, P. *Acta Crystallogr., Sect. B* **1980**, *36*, 2954–2958.
244. Low, K. Y.; Baba, I.; Farina, Y.; Othman, A. H.; Ibrahim, A. R.; Fun, H.-K.; Ng, S. W. *Main Group Met. Chem.* **2001**, *24*, 451–452.
245. Baba, I.; Ibrahim, S.; Farina, Y.; Othman, A. H.; Ibrahim, A. R.; Fun, H.-K.; Ng, S. W. *Acta Crystallogr., Sect. E* **2001**, *57*, m39–m40.
246. McKie, G.; Raston, C. L.; Rowbottom, G. L.; White, A. H. *J. Chem. Soc., Dalton Trans.* **1981**, 1360–1365.
247. Kello, E.; Kettmann, V.; Garaj, J. *Acta Crystallogr., Sect. C* **1985**, *41*, 520–522.
248. Chauhan, H. P. S.; Nahar, B.; Singh, R. K. *Synth. React. Inorg. Met.-Org. Chem.* **1998**, *28*, 1541–1549.
249. Pandey, S. K.; Srivastava, G.; Mehrotra, R. C. *Synth. React. Met.-Org. Chem.* **1989**, *19*, 795–807.
250. Nahar, B.; Chourasia, S.; Chauhan, H. P. S.; Rao, R. J.; Singh, M. S. *J. Ind. Chem. Soc.* **1997**, *74*, 711–712.
251. Chauhan, H. P. S.; Lunkad, S. *Main Group Met. Chem.* **1994**, *17*, 313–318.
252. Sowerby, D. B.; Haiduc, I.; Barbul-Rusu, A.; Salajan, M. *Inorg. Chim. Acta* **1983**, *68*, 87–96.
253. Begley, M. J.; Sowerby, D. B.; Haiduc, I. *J. Chem. Soc., Dalton Trans.* **1987**, 145–150.
254. Zuckerman-Schpector, J.; Haiduc, I.; Silvestru, C.; Cea-Olivares, R. *Polyhedron* **1995**, *14*, 3087–3094.
255. Silvestru, C.; Silaghi-Dumitrescu, L.; Haiduc, I.; Begley, M. J.; Nunn, M.; Sowerby, D. B. *J. Chem. Soc., Dalton Trans.* **1986**, 1031–1034.

256. Ebert, K. H.; Breunig, H. J.; Silvestru, C.; Haiduc, I. *Polyhedron* **1994**, *13*, 2531–2535.
257. Silvestru, C.; Haiduc, I.; Kaller, R.; Ebert, K. H.; Breunig, H. J. *Polyhedron* **1993**, *12*, 2611–2617.
258. Mattes, R.; Ruhl, D. *Z. Anorg. Allg. Chem.* **1984**, *508*, 19–25.
259. Begley, M. J.; Sowerby, D. B.; Wesolek, D. M.; Silvestru, C.; Haiduc, I. *J. Organomet. Chem.* **1986**, *316*, 281–289.
260. Gibbons, M. N.; Sowerby, D. B.; Silvestru, C.; Haiduc, I. *Polyhedron* **1996**, *15*, 4573–4578.
261. Garje, S. S.; Jain, V. K. *Main Group Met. Chem.* **1995**, *18*, 387–390.
262. Silvestru, C.; Sowerby, D. B.; Haiduc, I.; Ebert, K. H.; Breunig, H. J. *Main Group Met. Chem.* **1994**, *17*, 505–518.
263. Kraft, S.; Wieber, M. *Z. Anorg. Allg. Chem.* **1992**, *607*, 153–156.
264. Kraft, S.; Wieber, M. *Z. Anorg. Allg. Chem.* **1992**, *607*, 157–160.
265. Peters, M.; Saak, W.; Pohl, S. *Z. Anorg. Allg. Chem.* **1996**, *622*, 2119–2123.
266. Clegg, W.; Elsegood, M. R. J.; Farrugia, L. J.; Lawlor, F. J.; Norman, N. C.; Scott, A. J. *J. Chem. Soc., Dalton Trans.* **1995**, 2129–2135.
267. Bochmann, M.; Song, X.; Hursthouse, M. B.; Karaulov, A. *J. Chem. Soc., Dalton Trans.* **1995**, 1649–1652.
268. Hoffmann, H. M.; Dräger, M. *Z. Naturforsch., B* **1986**, *41*, 1455–1466.
269. Wegener, J.; Kirschenbaum, K.; Giolando, D. M. *J. Chem. Soc., Dalton Trans.* **1994**, 1213–1218.
270. Bozopoulos, A. P.; Kokkou, S. C.; Rentzeperis, P. J.; Karagiannidis, P. *Acta Crystallogr., Sect. C* **1984**, *40*, 944–946.
271. Alonzo, G. *Inorg. Chim. Acta* **1983**, *73*, 141–143.
272. Wuller, S. P.; Seligson, A. L.; Mitchell, G. P.; Arnold, J. *Inorg. Chem.* **1995**, *34*, 4854–4861.
273. Doidge-Harrison, S. M. S. V.; Irvine, J. T. S.; Spencer, G. M.; Wardell, J. L.; Wei, M.; Ganis, P.; Valle, G. *Inorg. Chem.* **1995**, *34*, 4581–4584.
274. Ganis, P.; Maston, D.; Spencer, G. M.; Wardell, J. L.; Wardell, S. M. S. V. *Inorg. Chim. Acta* **2000**, *308*, 139–142.
275. Spencer, G. M.; Wardell, J. L.; Aupers, J. H. *Polyhedron* **1996**, *15*, 2701–2706.
276. Avarvari, N.; Falques, E.; Fourmigue, M. *Inorg. Chem.* **2001**, *40*, 2570–2577.
277. Howie, R. A.; Low, J. N.; Spencer, G. M.; Wardell, J. L. *Polyhedron* **1997**, *16*, 2563–2571.
278. Smith, D. M.; Albrecht-Schmitt, T. E.; Ibers, J. A. *Angew. Chem. Int. Ed. Engl.* **1998**, *37*, 1089–1091.
279. Drake, G. W.; Kolis, J. W. *Coord. Chem. Rev.* **1994**, *137*, 131–178.
280. Sheldrick, W. S.; Wachhold, M. *Coord. Chem. Rev.* **1998**, *176*, 211–322.
281. Schur, M.; Bensch, W. *Z. Anorg. Allg. Chem.* **1998**, *624*, 310–314.
282. Stahler, R.; Bensch, W. *J. Chem. Soc., Dalton Trans.* **2001**, 2518–2522.
283. Stahler, R.; Nather, C.; Bensch, W. *Eur. J. Inorg. Chem.* **2001**, 1835–1840.
284. Bensch, W.; Nather, C.; Stahler, R. *Chem. Commun.* **2001**, 477–478.
285. Sawyer, J. F.; Gillespie, R. J. *Prog. Inorg. Chem.* **1986**, *34*, 65–113.
286. Fawcett, J.; Holloway, J. H.; Russell, D. R.; Edwards, A. J.; Khallow, K. I. *Can. J. Chem.* **1989**, *67*, 2041–2047.
287. Nandana, W. A.; Passmore, J.; White, P. S. *J. Chem. Soc., Dalton Trans.* **1985**, 1623–1632.
288. Nandana, W. A.; Passmore, J.; White, P. S.; Wong, C.-M. *J. Chem. Soc., Dalton Trans.* **1987**, 1989–1998.
289. Minkwitz, R.; Nowicki, J.; Borrmann, H. *Z. Anorg. Allg. Chem.* **1991**, *605*, 109–116.
290. Faggiani, R.; Gillespie, R. J.; Sawyer, J. F.; Verkis, J. E. *Acta Crystallogr., Sect. C* **1989**, *45*, 1847–1853.
291. Chitaz, S.; Dehnicke, K.; Frenzen, G.; Pilz, A.; Muller, U. *Z. Anorg. Allg. Chem.* **1996**, *622*, 2016–2022.
292. Udovenko, A. A.; Davidovitch, R. L.; Ivanov, S. B.; Antipin, M. Y.; Struchkov, Y. T. *Koord. Khim.* **1990**, *16*, 448–452.
293. Davodovitch, R. L.; Zemnukhova, L. A.; Semenova, T. L.; Kaidalova, T. A. *Koord. Khim.* **1986**, *12*, 924–928.
294. Becker, K.; Mattes, R. *Z. Anorg. Allg. Chem.* **1996**, *622*, 105–111.
295. Udovenko, A. A.; Gorbunova, Y. E.; Zemnukhova, L. A.; Mikhailov, Y. N.; Davidovitch, R. L. *Russ. J. Coord. Chem.* **2001**, *27*, 479–482.
296. Udovenko, A. A.; Zemnukhova, L. A.; Gorbunova, Y. E.; Mikhailov, Y. N.; Davidovitch, R. L. *Russ. J. Coord. Chem.* **1999**, *25*, 13–16.
297. Belz, J.; Weber, R.; Roloff, A.; Ross, B. *Z. Kristallogr.* **1992**, *202*, 281–282.
298. Ensinger, U.; Schwarz, W.; Schmidt, A. *Z. Naturforsch., B* **1982**, *37*, 1584–1589.
299. Jaschinski, B.; Blachnik, R.; Reuter, H. *Z. Naturforsch., B* **1998**, *53*, 565–568.
300. Drew, M. G. B.; Claire, P. P. K.; Willey, G. R. *J. Chem. Soc., Dalton Trans.* **1988**, 215–218.
301. Willey, G. R.; Palin, J.; Lakin, M. T.; Alcock, N. W. *Transition Met. Chem.* **1994**, *19*, 187–190.
302. Razak, I. A.; Raj, S. S.; Fun, H.-K.; Yamin, B. M.; Hashim, N. *Acta Crystallogr., Sect. C* **2000**, *56*, 664–665.
303. Bujak, M.; Osadczuk, P.; Zaleski, J. *Acta Crystallogr., Sect. C* **1999**, *55*, 1443–1447.
304. Jaschinski, B.; Blachnik, R.; Reuter, H. *Z. Anorg. Allg. Chem.* **1999**, *625*, 667–672.
305. Casa, J. S.; Castellano, E. E.; Couce, M. D.; Sanchez, A.; Sordo, J.; Taboada, C.; Vasquez-Lopez, E. M. *Main Group Met. Chem.* **1999**, *22*, 439–446.
306. Zaleski, J.; Pietraszko, A. *J. Phys. Chem. Solids* **1995**, *56*, 883–890.
307. Bujak, M.; Zaleski, J. *Z. Naturforsch., B* **2001**, *56*, 521–525.
308. Hursthouse, M. B.; Malik, K. M. A.; Bakshi, P. K.; Bhuiyan, A. A.; Ehsan, M. Q.; Haider, S. Z. *J. Chem. Crystallogr.* **1996**, *26*, 739–745.
309. Bujak, M.; Zaleski, J. *Acta Crystallogr., Sect. C* **1998**, *54*, 1773–1777.
310. Bednarska-Bolek, B.; Zaleski, J.; Bator, G. *J. Mol. Struct.* **2000**, *553*, 175–186.
311. Lipka, A. *Z. Naturforsch., B* **1983**, *38*, 1615–1619.
312. Hall, M.; Nunn, M.; Begley, M. J.; Sowerby, D. B. *J. Chem. Soc., Dalton Trans.* **1986**, 1231–1238.
313. Zaleski, J.; Pietraszko, A. *Z. Naturforsch., A* **1994**, *49*, 895–901.
314. Chaabouni, S.; Kamoun, S.; Daoud, A.; Jouini, T. *J. Chem. Crystallogr.* **1997**, *27*, 401–404.
315. Bujak, M.; Zaleski, J. *J. Mol. Struct.* **2000**, *555*, 179–185.
316. Mohammed, A. T.; Mueller, U. *Z. Naturforsch., B* **1985**, *40*, 562–564.
317. Ishihara, H.; Dou, S. Q.; Weiss, A. *Bull. Chem. Soc. Jpn.* **1994**, *67*, 637–640.
318. Mousdis, G. A.; Papavassiliou, G. C.; Terzis, A.; Raptopoulou, C. P. *Z. Naturforsch., B* **1998**, *53*, 927–931.
319. Ahmed, I. A.; Blachnik, R.; Reuter, H.; Eickmeier, H. *Z. Kristallogr.* **2001**, *216*, 207–208.
320. Chaabouni, S.; Kamoun, S.; Jaud, J. *Mater. Res. Bull.* **1998**, *33*, 377–388.
321. Antolini, L.; Benedetti, A.; Fabretti, A. C.; Giusti, A. *J. Chem. Soc., Dalton Trans.* **1988**, 2501–2503.

322. Czado, W.; Müller, U. *Z. Naturforsch., B* **1996**, *51*, 1245–1247.
323. Nunn, M.; Blake, A. J.; Begley, M. J.; Sowerby, D. B. *Polyhedron* **1998**, *17*, 4213–4217.
324. Pohl, S.; Lotz, R.; Haase, D.; Saak, W. *Z. Naturforsch., B* **1988**, *43*, 1144–1150.
325. Pohl, S.; Saak, W.; Haase, D. *Angew. Chem. Int. Ed.* **1987**, *26*, 467–468.
326. Novikova, M. S.; Makarova, I. P.; Blomberg, M. K.; Bagautdinov, B. S.; Aleksandrova, I. P. *Kristallografiya* **2001**, *46*, 33–36.
327. Ahmed, I. A.; Blachnik, R.; Reuter, H. *Z. Kristallogr.* **2000**, *215*, 253–254.
328. Pohl, S.; Saak, W.; Haase, D. *Z. Naturforsch., B* **1987**, *42*, 1493–1499.
329. Pohl, S.; Saak, W.; Mayer, P.; Schmidpeter, A. *Angew. Chem. Int. Ed. Engl.* **1986**, *25*, 825.
330. Carmalt, C. J.; Norman, N. C. *Polyhedron* **1994**, *13*, 1653–1658.
331. Pohl, S.; Lotz, R.; Saak, W.; Haase, D. *Angew. Chem. Int. Ed. Engl.* **1989**, *28*, 344–345.
332. Borgsen, B.; Weller, F.; Dehnicke, K. *Z. Anorg. Allg. Chem.* **1991**, *596*, 55–61.
333. Carmalt, C. J.; Farrugia, L. J.; Norman, N. C. *Polyhedron* **1993**, *12*, 2081–2090.
334. Smyth, M. V.; Bailey, R. D.; Pennington, W. T. *Acta Crystallogr., Sect. C* **1996**, *52*, 2170–2173.
335. Pohl, S.; Saak, W.; Haase, D. *Z. Naturforsch., B* **1988**, *43*, 1033–1037.
336. Pohl, S.; Peters, M.; Haase, D.; Saak, W. *Z. Naturforsch., B* **1994**, *49*, 741–746.
337. Carmalt, C. J.; Farrugia, L. J.; Norman, N. C. *Z. Anorg. Allg. Chem.* **1995**, *621*, 47–56.
338. Hall, M.; Sowerby, D. B. *J. Organomet. Chem.* **1988**, *347*, 59–70.
339. Sheldrick, W. S.; Martin, C. *Z. Naturforsch., B* **1992**, *47*, 919–924.
340. Sharma, P.; Rosas, N.; Toscano, A.; Hernandez, S.; Shankar, R.; Cabrera, A. *Main Group Met. Chem.* **1996**, *19*, 21–27.
341. Sheldrick, W. S.; Martin, C. *Z. Naturforsch., B* **1991**, *67*, 639–646.
342. Nakajima, T.; Zemva, B.; Tressaud, A, Eds. *Advanced Inorganic Fluorides* **2000**, Elsevier: Amsterdam; Chapters 2 and 4.
343. Dove, M. F. A.; Sanders, J. C. P. *J. Chem. Soc., Dalton Trans.* **1992**, 3311–3316.
344. Kidd, R. G.; Spinney, H. G. *Can. J. Chem.* **1981**, *59*, 2940–2944.
345. Goetz-Grandmont, G. J.; Leroy, M. J. F. *Z. Anorg. Allg. Chem.* **1983**, *496*, 40–46.
346. Zaitseva, E. G.; Medvedev, S. V.; Aslanov, L. A. *Zh. Strukt. Khim.* **1990**, *31*, 110–116.
347. Zaitseva, E. G.; Medvedev, S. V.; Aslanov, L. A. *Zh. Strukt. Khim.* **1990**, *31*, 104–109.
348. Wieber, M.; Walz, J. *Z. Anorg. Allg. Chem.* **1990**, *583*, 102–112.
349. Zaitseva, E. G.; Medvedev, S. V.; Aslanov, L. A. *Zh. Strukt. Khim.* **1990**, *31*, 133–138.
350. Hall, M.; Nunn, M.; Sowerby, D. B. *J. Chem. Soc., Dalton Trans.* **1986**, 1239–1242.
351. Jaschinski, B.; Blachnik, R.; Pawlak, R.; Reuter, H. *Z. Kristallogr.* **1998**, *213*, 543–545.
352. Rogers, R. D.; Jezl, M. L. *Acta Crystallogr., Sect. C* **1994**, *50*, 1527–1529.
353. Siewert, B.; Mueller, U. *Z. Anorg. Allg. Chem.* **1992**, *609*, 89–94.
354. Briand, G. G.; Burford, N. *Adv. Inorg. Chem.* **2000**, *50*, 285–357.
355. Suzuki, H.; Matano, Y. In *The Chemistry of Arsenic, Antimony and Bismuth*; Norman, N. C., Ed.; Blackie: London, 1998; Chapter 6, Organobisumth compounds, pp 283–343.
356. Silvestru, C.; Breunig, H. J.; Althaus, H. *Chem. Rev.* **1999**, *99*, 3277–3327.
357. Hancock, R. D.; Cukrowski, I.; Mashishi, J. *J. Chem. Soc., Dalton Trans.* **1993**, 2895–2899.
358. Hancock, R. D.; Cukrowski, I.; Antunes, I.; Cukrowska, E.; Mashishi, J.; Brown, K. *Polyhedron* **1995**, *14*, 1699–1707.
359. Clegg, W.; Compton, N. A.; Errington, R. J.; Fisher, G. A.; Green, M. E.; Hockless, D. C. R.; Norman, N. C. *Inorg. Chem.* **1991**, *30*, 4680–4682.
360. Clegg, W.; Compton, N. A.; Errington, R. J.; Norman, N. C.; Wishart, N. *Polyhedron* **1989**, *8*, 1579–1580.
361. Wirringa, U.; Roesky, H. W.; Noltemeyer, M.; Schmidt, H.-G. *Inorg. Chem.* **1994**, *33*, 4607–4608.
362. James, S. C.; Norman, N. C.; Orpen, A. G.; Quayle, M. J.; Weskenmann, U. *J. Chem. Soc., Dalton Trans.* **1996**, 4159–4161.
363. Burford, N.; Macdonald, C. L. B.; Robertson, K. N.; Cameron, T. S. *Inorg. Chem.* **1996**, *35*, 4013–4016.
364. Mason, M. R.; Phulpager, S. S.; Mshuta, M. S.; Richardson, J. F. *Inorg. Chem.* **2000**, *39*, 3931–3933.
365. Suzuki, H.; Murafuji, T.; Matano, Y.; Azuma, N. *J. Chem. Soc., Perkin Trans.* **1993**, *1*, 2969–2973.
366. Murafuji, T.; Azuma, N.; Suzuki, H. *Organometallics* **1995**, *14*, 1542–1544.
367. Willey, G. R.; Daly, L. T.; Rudd, M. D.; Drew, M. G. B. *Polyhedron* **1995**, *14*, 315–318.
368. Di Vaira, M.; Mani, F.; Stoppioni, P. *Eur. J. Inorg. Chem.* **1999**, 833–837.
369. Luckay, R.; Cukrovski, I.; Mashishi, J.; Reibenspies, J. H.; Bond, A. H.; Rogers, R. D.; Hancock, R. D. *J. Chem. Soc., Dalton Trans.* **1997**, 901–908.
370. Luckay, R.; Reibenspies, J. H.; Hancock, R. D. *Chem. Commun.* **1995**, 2365–2366.
371. Barbour, T.; Belcher, W. J.; Brothers, P. J.; Rickard, C. E. F.; Ware, D. C. *Inorg. Chem.* **1992**, *31*, 746–754.
372. Michaudet, L.; Fasseur, D.; Guilard, R.; Ou, Z.; Kadish, K. M.; Dahaoui, S.; Lecomte, C. *J. Porphyrins Phthalocyanines* **2000**, *4*, 261–270.
373. Michaudet, L.; Richard, P.; Boitrel, B. *Chem. Commun.* **2000**, 1589–1590.
374. Isago, H.; Kagaya, Y. *Bull. Chem. Soc. Jpn.* **1994**, *67*, 383–389.
375. Kubiaka, R.; Ejsmont, K. *J. Mol. Struct.* **1999**, *474*, 275–281.
376. Ostendorp, G.; Homberg, H. *Z. Anorg. Allg. Chem.* **1996**, *622*, 873–880.
377. Janczac, J.; Kubiak, R.; Richter, J.; Fuess, H. *Polyhedron* **1999**, *18*, 2775–2780.
378. Benihya, K.; Mossoyan-Deneux, M.; Hahn, F.; Boucharat, N.; Terzian, G. *Eur. J. Inorg. Chem.* **2000**, 1771–1779.
379. Benihya, K.; Mossoyan-Deneux, M.; Giorgi, M. *Eur. J. Inorg. Chem.* **2001**, 1343–1352.
380. Willey, G. R.; Collins, H.; Drew, M. G. B. *J. Chem. Soc., Dalton Trans.* **1991**, 961–965.
381. James, S. C.; Lawson, Y. G.; Norman, N. C.; Orpen, A. G.; Quayle, M. J. *Acta Crystallogr., Sect. C* **2000**, *56*, 427–429.
382. Raston, C. L.; Rowbottom, G. L.; White, A. H. *J. Chem. Soc., Dalton Trans.* **1981**, 1389–1391.
383. Bharadwaj, P. K.; Lee, A. M.; Skelton, B. W.; Srinivasan, B. R.; White, A. H. *Aust. J. Chem.* **1994**, *47*, 128–130.
384. Carmalt, C. J.; Farrugia, L. J.; Norman, N. C. *J. Chem. Soc., Dalton Trans.* **1996**, 443–454.

385. James, S. C.; Norman, N. C.; Orpen, A. G. *J. Chem. Soc., Dalton Trans.* **1999**, 2837–2843.
386. Alonzo, G.; Consiglio, M.; Bertazzai, N.; Preti, C. *Inorg. Chim. Acta* **1985**, *105*, 51–57.
387. Bowmaker, G. A.; Harrowfield, J. M.; Lee, A. M.; Skelton, B. W.; White, A. H. *Aust. J. Chem.* **1998**, *51*, 311–315.
388. Bowmaker, G. A.; Junk, P. C.; Lee, A. M.; Skelton, B. W.; White, A. H. *Aust. J. Chem.* **1998**, *51*, 317–324.
389. Bowmaker, G. A.; Hannaway, F. M. M.; Junk, P. C.; Lee, A. M.; Skelton, B. W.; White, A. H. *Aust. J. Chem.* **1998**, *51*, 325–330.
390. Bowmaker, G. A.; Hannaway, F. M. M.; Junk, P. C.; Lee, A. M.; Skelton, B. W.; White, A. H. *Aust. J. Chem.* **1998**, *51*, 331–336.
391. Barbour, L. J.; Belfield, S. J.; Junk, P. C.; Smith, M. K. *Aust. J. Chem.* **1998**, *51*, 337–342.
392. Bertazzi, N.; Alonzo, G.; Battaglia, L. P.; Corradi, A. B.; Pelosi, G. *J. Chem. Soc., Dalton Trans.* **1990**, 2403–2405.
393. Raston, C. L.; Rowbottom, G. L.; White, A. H. *J. Chem. Soc., Dalton Trans.* **1981**, 1383–1388.
394. Bertazzi, N.; Alonzo, G.; Consiglio, M. *Inorg. Chim. Acta* **1989**, *159*, 141–142.
395. Chitsaz, S.; Harms, K.; Neumuller, B.; Dehnicke, K. *Z. Anorg. Allg. Chem.* **1999**, *625*, 939–944.
396. Klapötke, T. M.; Schulz, A. *Main Group Met. Chem.* **1997**, *20*, 325–338.
397. Clegg, W.; Errington, R. J.; Flynn, R. J.; Green, M. E.; Hockless, D. C. R.; Norman, N. C.; Gibson, V. C.; Tavakkoli, K. *J. Chem. Soc., Dalton Trans.* **1992**, 1753–1754.
398. Clegg, W.; Errington, R. J.; Fisher, G. A.; Green, M. E.; Hockless, D. C. R.; Norman, N. C. *Chem. Ber.* **1991**, *124*, 2457–2459.
399. Willey, G. R.; Rudd, M. D.; Samuel, C. J.; Drew, M. G. B. *J. Chem. Soc., Dalton Trans.* **1995**, 759–764.
400. Willey, G. R.; Daly, L. T.; Drew, M. G. B. *J. Chem. Soc., Dalton Trans.* **1996**, 1063–1067.
401. Frank, W.; Reiss, G. J.; Schneider, J. *Angew. Chem. Int. Ed. Engl.* **1995**, *34*, 2416–2417.
402. Sundvall, B. *Inorg. Chem.* **1983**, *22*, 1906–1912.
403. Naeslund, J.; Persson, I.; Sanderstroem, M. *Inorg. Chem.* **2000**, *39*, 4012–4021.
404. Graunar, M.; Lazarini, F. *Acta Crystallogr., Sect. B* **1982**, *38*, 2879–2881.
405. Golic, L.; Graunar, M.; Lazarini, F. *Acta Crystallogr., Sect. B* **1982**, *38*, 2881–2883.
406. Carmalt, C. J.; Clegg, W.; Elsegood, M. R. J.; Errington, R. J.; Havelock, J.; Lightfoot, P.; Norman, N. C.; Scott, A. J. *Inorg. Chem.* **1996**, *35*, 3709–3712.
407. Eveland, J. R.; Whitmire, K. H. *Inorg. Chim. Acta* **1996**, *249*, 41–46.
408. Clegg, W.; Errington, R. J.; Fisher, G. A.; Hockless, D. C. R.; Norman, N. C.; Orpen, A. G.; Stratford, S. E. *J. Chem. Soc., Dalton Trans.* **1992**, 1967–1974.
409. Rogers, R. D.; Bond, A. H.; Aguinaga, S.; Reyes, A. *J. Am. Chem. Soc.* **1992**, *114*, 2967–2977.
410. Rogers, R. D.; Bond, A. H.; Aguinaga, S. *J. Am. Chem. Soc.* **1992**, *114*, 2960–2967.
411. Weber, R.; Koesters, H.; Bergerhoff, G. *Z. Krystallogr.* **1993**, *207*, 175–177.
412. Alcock, N. W.; Ravindran, M.; Willey, G. R. *Chem. Commun.* **1989**, 1063–1065.
413. Garbe, R.; Vollmer, B.; Neumueller, B.; Pebler, J.; Dehnicke, K. *Z. Anorg. Allg. Chem.* **1993**, *619*, 271–276.
414. Drew, M. G. B.; Nicholson, D. G.; Sylte, I.; Vasudevan, A. *Inorg. Chim. Acta* **1990**, *171*, 11–15.
415. Clegg, W.; Farrugia, L. J.; McCamley, A.; Norman, N. C.; Orpen, A. G.; Pickett, N. L.; Stratford, S. E. *J. Chem. Soc., Dalton Trans.* **1993**, 2579–2587.
416. Bowmaker, G. A.; Harrowfield, J. M.; Junk, P. C.; Skelton, B. W.; White, A. H. *Aust. J. Chem.* **1998**, *51*, 285–291.
417. Carmalt, C. J.; Cowley, A. H.; Decken, A.; Norman, N. C. *J. Organomet. Chem.* **1995**, *496*, 59–67.
418. Carmalt, C. J.; Farrugia, L. J.; Norman, N. C. *J. Chem. Soc., Dalton Trans.* **1996**, 455–459.
419. Engelhardt, U.; Rapko, B. M.; Duesler, E. N.; Frutos, D.; Paine, R. T.; Smith, P. H. *Polyhedron* **1995**, *14*, 2361–2369.
420. Garcia-Montalvo, V.; Cea-Olivares, R.; Williams, D. J.; Espinosa-Perez, G. *Inorg. Chem.* **1996**, *35*, 3948–3953.
421. Matchett, M. A.; Chiang, M. Y.; Buhro, W. E. *Inorg. Chem.* **1990**, *29*, 358–360.
422. Massiani, M. C.; Papiernik, R.; Hubert-Pfalzgraf, L. G.; Daran, J. C. *Chem. Commun.* **1990**, 301–302.
423. Massiani, M. C.; Papiernik, R.; Hubert-Pfalzgraf, L. G.; Daran, J. C. *Polyhedron* **1991**, *10*, 437–445.
424. Williams, P. A.; Jones, A. C.; Crosbie, M. J.; Wright, P. J.; Bickley, J. F.; Steiner, A.; Davies, H. O.; Leedham, T. J.; Critchlow, G. W. *Chem. Vap. Dep.* **2001**, *7*, 205–209.
425. Boyle, T. J.; Pedrotty, D. M.; Scott, B.; Ziller, J. W. *Polyhedron* **1998**, *17*, 1959–1974.
426. Evans, W. J.; Hain, J. H.; Ziller, J. W. *Chem. Commun.* **1989**, 1628–1629.
427. Veith, M.; Yu, E.-C.; Huch, V. *Chem. Eur. J.* **1995**, 26–32.
428. Jones, C. M.; Burkart, M. D.; Whitmire, K. H. *Angew. Chem. Int. Ed. Engl.* **1992**, *31*, 451–452.
429. Jolas, J. L.; Hoppe, S.; Whitmire, K. H. *Inorg. Chem.* **1997**, *36*, 3335–3340.
430. Whitmire, K. H.; Hoppe, S.; Sydora, O.; Jolas, J. L.; Jones, C. M. *Inorg. Chem.* **2000**, *39*, 85–97.
431. Jones, C. M.; Burkart, M. D.; Bachman, R. E.; Serra, D. L.; Hwu, S. J.; Whitmire, K. H. *Inorg. Chem.* **1993**, *32*, 5136–5144.
432. Pell, J. W.; Davies, W. C.; Loye, H. C. *Z. Inorg. Chem.* **1996**, *35*, 5754–5755.
433. Parola, S.; Papiernik, R.; Hubert-Pfalzgraf, L. G.; Bois, C. *J. Chem. Soc., Dalton Trans.* **1998**, 737–739.
434. Parola, S.; Papiernik, R.; Hubert-Pfalzgraf, L. G.; Jagner, S.; Hikansson, M. *J. Chem. Soc., Dalton Trans.* **1997**, 4631–4636.
435. Smith, G.; Reddy, A. N.; Byriel, K. A.; Kennard, C. H. L. *Aust. J. Chem.* **1994**, *47*, 1413–1418.
436. Cloutt, B. A.; Sagatys, D. S.; Smith, G.; Bott, R. C. *Aust. J. Chem.* **1997**, *50*, 947–950.
437. Fukin, G. K.; Pisarevskii, A. P.; Yanovskii, A. I.; Struchkov, Y. T. *Russ. J. Inorg. Chem.* **1993**, *38*, 1118–1123.
438. Armelao, L.; Bandoli, G.; Casarin, M.; Depaoli, G.; Tondello, E.; Vittadini, A. *Polyhedron* **1998**, *275–276*, 340–348.
439. Diemer, R.; Keppler, B. K.; Dittes, U.; Nuber, B.; Seifried, V.; Opferkuck, W. *Chem. Ber.* **1995**, *128*, 335–342.
440. Antsyshkina, A. S.; Porai-Koshits, M. A.; Ostrikova, V. N. *Koord. Khim.* **1983**, *9*, 1118–1120.
441. Troyanov, S. I.; Pisarevskii, A. P. *Russ. J. Coord. Chem.* **1991**, *17*, 489–492.
442. Bensch, W.; Blazso, E.; Dubler, E.; Oswald, H. R. *Acta Crystallogr., Sect. C* **1987**, *43*, 1699–1704.
443. Troyanov, S. I.; Pisarevsky, A. P. *Chem. Commun.* **1993**, 335–336.
444. Reiss, G. J.; Frank, W.; Schneider, J. *Main Group Met. Chem.* **1995**, *18*, 287–294.
445. Rae, A. D.; Gainsford, G. J.; Kemmitt, T. *Acta Crystallogr., Sect. B* **1998**, *54*, 438–442.
446. Breeze, S. R.; Chen, L.; Wang, S. *J. Chem. Soc., Dalton Trans.* **1994**, 2545–2557.
447. Ulrich, H.; Hinse, P.; Mattes, R. *Z. Anorg. Allg. Chem.* **2001**, *627*, 2173–2177.

448. Janvier, P.; Drumel, S.; Piffard, Y.; Bujoli, B. *C. R. Acad. Sci. Ser. II* **1995**, *320*, 29–35.
449. Mehring, M.; Schurmann, M. *Chem. Commun.* **2001**, 2354–2355.
450. Sagatys, D. S.; O'Reilly, E. J.; Patel, S.; Bott, R. C.; Lynch, D. E.; Smith, G.; Kennard, C. L. H. *Aust. J. Chem.* **1992**, *45*, 1027–1034.
451. Herrmann, W. A.; Herdtweck, E.; Scherer, W.; Kiprof, P.; Pajdla, L. *Chem. Ber.* **1993**, *126*, 51–56.
452. Kiprof, P.; Scherer, W.; Pajdia, L.; Herdtweck, E.; Herrmann, W. A. *Chem. Ber.* **1992**, *125*, 43–46.
453. Herrmann, W. A.; Herdtweck, E.; Padjla, L. *Z. Krystallogr.* **1992**, *198*, 257–264.
454. Herrmann, W. A.; Herdtweck, E.; Pajdla, L. *Inorg. Chem.* **1991**, *30*, 2579–2581.
455. Asato, E.; Driessen, W. L.; de Graaff, R. A. G.; Hulsbergen, F. B.; Reedijk, J. *Inorg. Chem.* **1991**, *30*, 4210–4218.
456. Asato, E.; Katsura, K.; Mikuriya, M.; Fujii, T.; Reedijk, J. *Inorg. Chem.* **1993**, *32*, 5322–5329.
457. Asato, E.; Katsura, K.; Mikuriya, M.; Turpeinen, U.; Mutikainen, I.; Reedijk, J. *Inorg. Chem.* **1995**, *34*, 2447–2454.
458. Barrie, P. J.; Djuran, M. J.; Mazid, M. A.; McPartlin, M.; Sadler, P. J. *J. Chem. Soc., Dalton Trans.* **1996**, 2417–2422.
459. Shkol'nikova, L. M.; Suyarov, K. D.; Davidovich, R. L.; Fundamenskii, V. S.; Dyatlova, N. M. *Koord. Khim.* **1991**, *17*, 253–261.
460. Shkol'nikova, L. M.; Porai-Koshits, M. A.; Davidovich, R. L.; Hu, C.-D.; Ksi, D.-K. *Koord. Khim.* **1994**, *20*, 593–596.
461. Davidovich, R. L.; Ilyukhin, A. B.; Hu, C. J. *Kristallografiya* **1998**, *98*, 653–655.
462. Summers, S. P.; Abboud, K. A.; Farrah, S. R.; Palenik, G. J. *Inorg. Chem.* **1994**, *33*, 88–92.
463. Shkol'nikova, L. M.; Porai-Koshits, M. A.; Davidovich, R. L.; Sadikov, G. G. *Koord. Khim.* **1993**, *19*, 633–636.
464. Starikova, Z. A.; Sysoeva, T. F.; Makarevich, S. S.; Ershova, S. D. *Koord. Khim.* **1991**, *17*, 317–321.
465. Shchelokov, R. N.; Mikhailov, Y. N.; Mistryukov, V. E.; Sergeev, A. V. *Dokl. Akad. Nauk. SSSR* **1987**, *293*, 642–644.
466. Davidovich, R. L.; Gerasimenko, A. V.; Logvinova, V. B. *Zh. Neorg. Khim.* **2001**, *46*, 1081–1086.
467. Davidovich, R. L.; Logvinova, V. B.; Ilyukhin, A. B. *Zh. Neorg. Khim.* **2000**, *45*, 1973–1977.
468. Ilyukhin, A. B.; Davidovich, R. L.; Logvinova, V. B.; Fun, H.-K.; Raj, S. S. S.; Razak, I. A.; Hu, S.-Z.; Ng, S. W. *Main Group Met. Chem.* **1999**, *22*, 275–281.
469. Davidovich, R. L.; Logvinova, V. B.; Ilyukhin, A. B. *Zh. Neorg. Khim.* **2001**, *46*, 73–76.
470. Davidovich, R. L.; Gerasimenko, A. V.; Logvinova, V. B. *Zh. Neorg. Khim.* **2001**, *46*, 1475–1480.
471. Jaud, J.; Marrot, B.; Brouca-Cabarrecq, C.; Mosset, A. *J. Chem. Crystallogr.* **1997**, *27*, 109–117.
472. Davidovich, R. L.; Gerasimenko, A. V.; Logvinova, V. B. *Zh. Neorg. Khim.* **2001**, *46*, 1673–1678.
473. Shkol'nikova, L. M.; Porai-Koshits, M. A.; Poznyak, A. L. *Koord. Khim.* **1993**, *19*, 683–690.
474. Porai-Koshits, M. A.; Antsyshkina, A. S.; Shkol'nikova, L. M.; Sadikov, G. G.; Davidovich, R. L. *Koord. Khim.* **1995**, *21*, 295–302.
475. Davidovich, R. L.; Gerasimenko, A. V.; Logvinova, V. B.; Zou, J.-X. *Zh. Neorg. Khim.* **2001**, *46*, 1305–1310.
476. Davidovich, R. L.; Gerasimenko, A. V.; Logvinova, V. B. *Zh. Neorg. Khim.* **2001**, *46*, 1297–1304.
477. Davidovich, R. L.; Gerasimenko, A. V.; Kovaleva, E. V. *Zh. Neorg. Khim.* **2001**, *46*, 623–628.
478. Sobanska, S.; Wignacourt, J. P.; Conflant, P.; Drache, M. BulimestruI.; Gulea, A. *Eur. J. Solid State Chem.* **1996**, *33*, 710–712.
479. Poznyak, A. L.; Ilyukhin, A. B. *Kristallografiya* **2000**, *45*, 50–51.
480. Antsyhkina, A. S.; Sadikov, G. G.; Poznyak, A. L.; Sergienko, V. S.; Mikhailov, Y. N. *Zh. Neorg. Khim.* **1999**, *44*, 727–742.
481. Stavila, V.; Gdanec, M.; Shova, S.; Simonov, Y. A.; Gulya, A.; Vignacourt, J.-P. *Koord. Khim.* **2001**, *26*, 741–747.
482. Martinenko, L. I.; Kupriyanova, G. N.; Kovaleva, I. B. *Zh. Neorg. Khim.* **1991**, *36*, 2449–2454.
483. Ilyukhin, A. B.; Shkol'nikova, L. M.; Davidovich, R. L.; Samsonova, I. N. *Koord. Khim.* **1991**, *17*, 903–908.
484. Brechbiel, M. W.; Gansow, O. A.; Pippin, C. G.; Rogers, R. D.; Planalp, R. P. *Inorg. Chem.* **1996**, *35*, 6343–6348.
485. Wullens, H.; Devilliers, M.; Tinant, B.; Declercq, J.-P. *J. Chem. Soc., Dalton Trans.* **1996**, 2023–2029.
486. Suyarov, K.; Shkol'nikova, L. M.; Porai-Koshits, M. A.; Fundamenskii, V. S. *Koord. Khim.* **1991**, *17*, 455–462.
487. Suyarov, K.; Shkol'nikova, L. M.; Porai-Koshits, M. A.; Fundamenskii, V. S.; Davidovich, R. L. *Dokl. Akad. Nauk. SSSR* **1990**, *311*, 1397–1400.
488. Ilyukhin, A. B.; Davidovich, R. L.; Logvinova, V. B. *Zh. Neorg. Khim.* **1999**, *44*, 1931–1934.
489. Davidovich, R. L.; Gerasimenko, A. V.; Logvinova, V. B. *Zh. Neorg. Khim.* **2001**, *46*, 1311–1316.
490. Davidovich, R. L.; Samsonova, I. N.; Logvinva, V. B.; Teplukhina, L. V. *Russ. J. Coord. Chem.* **1996**, *22*, 153–159.
491. Davidovich, R. L.; Shkol'nikova, L. M.; Huang, U.-Q.; Hu, S.-Z. *Russ. J. Coord. Chem.* **1996**, *22*, 858–862.
492. Huang, Y.-Q.; Hu, S.-Z.; Shkol'nikov, L. M.; Davidovich, R. L. *Russ. J. Coord. Chem.* **1995**, *21*, 853–857.
493. Bharadwaj, P. K.; Lee, A. M.; Mandal, S.; Skelton, B. W.; White, A. H. *Aust. J. Chem.* **1994**, *47*, 1799–1803.
494. Hegetschweiler, K.; Ghisletta, M.; Gramlich, V. *Inorg. Chem.* **1993**, *32*, 2699–2704.
495. Battaglia, L. P.; Corradi, A. B.; Pelosi, G.; Tarasconi, P.; Pelizzi, C. *J. Chem. Soc., Dalton Trans.* **1989**, 671–675.
496. Battaglia, L. P.; Corradi, A. B.; Pelizzi, C.; Pelosi, G.; Tarasconi, P. *J. Chem. Soc., Dalton Trans.* **1990**, 3857–3860.
497. Stewart, C. A.; Calabrese, J. C.; Arduengo, A. J. *J. Am. Chem. Soc.* **1985**, *107*, 3397–3398.
498. Breeze, S. R.; Wang, S.; Greedan, J. E.; Raju, N. P. *Inorg. Chem.* **1996**, *35*, 6944–6951.
499. Clegg, W.; Norman, N. C.; Pickett, N. L. *Polyhedron* **1993**, *12*, 1251–1252.
500. Genge, A. R. J.; Levason, W.; Reid, G. *Chem. Commun.* **1998**, 2159–2160.
501. Barton, A. J.; Genge, A. R. J.; Levason, W.; Reid, G. *J. Chem. Soc., Dalton Trans.* **2000**, 859–865.
502. Yim, H. W.; Lam, K.-C.; Rheingold, A. L.; Rabinovich, D. *Polyhedron* **2000**, *19*, 849–853.
503. Willey, G. R.; Lakin, M. T.; Alcock, N. W. *J. Chem. Soc., Dalton Trans.* **1992**, 591–596.
504. Willey, G. R.; Lakin, M. T.; Alcock, N. W. *J. Chem. Soc., Dalton Trans.* **1992**, 1339–1341.
505. Blake, A. J.; Fenske, D.; Li, W.-S.; Lippolis, V.; Schröder, M. *J. Chem. Soc., Dalton Trans.* **1998**, 3961–3968.
506. Barton, A. J.; Genge, A. R. J.; Levason, W.; Reid, G. *J. Chem. Soc., Dalton Trans.* **2000**, 2163–2166.
507. Hill, N. J.; Levason, W.; Reid, G. *J. Chem Soc., Dalton Trans.* **2002**, 4316–4317.
508. Jameson, G. B.; Blazso, E.; Oswald, H. R. *Acta Crystallogr., Sect. C* **1984**, *40*, 350–354.
509. Battaglia, L. P.; Corradi, A. B. *J. Chem. Soc., Dalton Trans.* **1981**, 23–26.
510. Battaglia, L. P.; Corradi, A. B. *J. Chem. Soc., Dalton Trans.* **1983**, 2425–2428.
511. Praekel, U.; Huber, F.; Preut, H. *Z. Anorg. Allg. Chem.* **1982**, *494*, 67–77.

512. Battaglia, L. P.; Corradi, A. B. *J. Cryst. Spectros. Res.* **1992**, *22*, 275–279.
513. Battaglia, L. P.; Corradi, A. B. *J. Chem. Soc., Dalton Trans.* **1984**, 2401–2407.
514. Reglinski, J.; Spicer, M. D.; Garner, M.; Kennedy, A. R. *J. Am. Chem. Soc.* **1999**, *121*, 2317–2318.
515. Bailey, P. J.; Lanfranchi, M.; Marchio, L.; Parsons, S. *Inorg. Chem.* **2001**, *40*, 5030–5035.
516. Williams, D. J.; Carter, T.; Fahn, K. L.; VanDerveer, D. *Inorg. Chim. Acta* **1995**, *228*, 69–72.
517. Singh, K.; Tandon, J. P. *Monatsch. Chem.* **1992**, *123*, 315–319.
518. Chauhan, H. P. S.; Srivastava, G.; Mehrotra, R. C. *Indian J. Chem., Sect. A* **1984**, 436–437.
519. Morsali, A.; Tadjarodi, A.; Mohammadi, R.; Mahjoub, A. *Z. Kristallogr.* **2001**, *216*, 379–380.
520. Drew, M. G. B.; Kisenyi, J. M.; Willey, G. R. *J. Chem. Soc., Dalton Trans.* **1984**, 1723–1726.
521. Willey, G. R.; Barras, J. R.; Rudd, M. D.; Drew, M. G. B. *J. Chem. Soc., Dalton Trans.* **1994**, 3025–3029.
522. Williams, D. J.; Travis, J. B.; Bergbauer, K. L. *J. Coord. Chem.* **1987**, *16*, 315–317.
523. Sekar, P.; Ibers, J. A. *Inorg. Chim. Acta* **2001**, *319*, 117–122.
524. Singh, P.; Singh, G.; Vishnu, D.; Noeth, H. *Z. Naturforsch., B* **1998**, *53*, 1475–1482.
525. Burnett, T. R.; Dean, P. A. W.; Vittal, J. J. *Can. J. Chem.* **1994**, *72*, 1127–1136.
526. Tiekink, E. R. T. *J. Crystallogr. Spectros. Res.* **1992**, *22*, 231–236.
527. Wieber, M.; Ruedling, H. G. *Z. Anorg. Allg. Chem.* **1983**, *505*, 150–152.
528. Hoskins, B. F.; Tiekink, E. R. T.; Winter, G. *Inorg. Chim. Acta* **1984**, *81*, L33–L34.
529. Burschka, C. *Z. Anorg. Allg. Chem.* **1982**, *485*, 217–224.
530. Tiekink, E. R. T. *Main Group Met. Chem.* **1994**, *17*, 727–736.
531. Cox, M. J.; Tiekink, E. R. T. *Z. Kristallogr.* **1998**, *213*, 533–534.
532. Battaglia, L. P.; Corradi, A. B. *J. Chem. Soc., Dalton Trans.* **1986**, 1513–1517.
533. Raston, C. L.; Rowbottam, G. L.; White, A. H. *J. Chem. Soc., Dalton Trans.* **1981**, 1352–1359.
534. Mandal, S.; Mandal, G. C.; Shukla, R.; Bharadwaj, B. R. *Indian J. Chem., Sect. A* **1992**, *31*, 128–130.
535. Raston, C. L.; Rowbottom, G. L.; White, A. H. *J. Chem. Soc., Dalton Trans.* **1981**, 1366–1368.
536. Raston, C. L.; Rowbottom, G. L.; White, A. H. *J. Chem. Soc., Dalton Trans.* **1981**, 1372–1378.
537. Raston, C. L.; Rowbottom, G. L.; White, A. H. *J. Chem. Soc., Dalton Trans.* **1981**, 1369–1371.
538. Raston, C. L.; Rowbottom, G. L.; White, A. H. *J. Chem. Soc., Dalton Trans.* **1981**, 1379–1382.
539. Ali, M.; McWhinnie, W. R.; West, A. A.; Hamor, T. A. *J. Chem. Soc., Dalton Trans.* **1990**, 899–905.
540. Chauhan, H. P. S.; Srivastava, G.; Mehrotra, R. C. *Phosphorus, Sulfur Silicon Relat. Elem.* **1983**, *17*, 161–167.
541. Wieber, M.; Schroepf, M. *Phosphorus, Sulfur Silicon Relat. Elem.* **1995**, *102*, 265–267.
542. Bohra, R.; Chauhan, H. P. S.; Srivastava, G.; Mehrotra, R. C. *Phosphorus, Sulfur Silicon Relat. Elem.* **1991**, *60*, 167–174.
543. Iglesias, M.; del Pino, C.; Martinez-Cabrera, S. *Polyhedron* **1989**, *8*, 483–489.
544. Sowerby, D. B.; Haiduc, I. *J. Chem. Soc., Dalton Trans.* **1987**, 1257–1259.
545. Svensson, G.; Johansson, J. *Acta Chem. Scand.* **1989**, *43*, 511–517.
546. Edelmann, F. T.; Noltemeyer, M.; Haiduc, I.; Silvestsru, R.; Cea-Olivares, R. *Polyhdron* **1994**, *13*, 547–552.
547. Ebert, K. H.; Schulz, R. E.; Breunig, H. J.; Silvestru, C.; Haiduc, I. *J. Organomet. Chem.* **1994**, *470*, 93–98.
548. Silaghi-Dumitrescu, L.; Avila-Diaz, L. A.; Haiduc, I. *Rev. Roum. Chim.* **1986**, *31*, 335–340.
549. Atwood, D. A.; Cowley, A. H.; Hernandez, R. D.; Jones, R. A.; Rand, L. L.; Bott, S. G.; Atwood, J. L. *Inorg. Chem.* **1993**, *32*, 2972–2974.
550. Boudjouk, P.; Remington, M. P.; Grier, D. G.; Jarabek, B. R.; McCarthy, G. J. *Inorg. Chem.* **1998**, *37*, 3538–3541.
551. Farrugia, L. J.; Lawlor, F. J.; Norman, N. C. *J. Chem. Soc., Dalton Trans.* **1995**, 1163–1171.
552. Hergett, S. C.; Peach, M. E. *J. Fluorine Chem.* **1988**, *38*, 367–374.
553. Farrugia, L. J.; Lawlor, F. J.; Norman, N. C. *Polyhedron* **1995**, *14*, 311–314.
554. Hunter, G.; Weakley, T. J. R. *J. Chem. Soc., Dalton Trans.* **1983**, 1067–1070.
555. Comerlato, N. M.; Costa, L. A. S.; Howie, R. A.; Pereira, R. P.; Rocco, A. M.; Silvino, A. C.; Wardell, J. L.; Wardell, S. M. S. V. *Polyhedron* **2001**, *20*, 415–421.
556. Comerlato, N. M.; Harrison, W. T. A.; Howie, R. A.; Silvino, A. C.; Wardell, J. L.; Wardell, S. M. S. V. *Inorg. Chem. Commun.* **2000**, *3*, 572–574.
557. Sheng, T.; Wu, X.; Ping, L.; Wenjian, Z.; Quanming, W.; Ling, C. *Polyhedron* **1999**, *18*, 1049–1054.
558. Agocs, L.; Burford, N.; Cameron, T. S.; Curtis, J. M.; Richardson, J. F.; Robertson, K. N.; Yhard, G. B. *J. Am. Chem. Soc.* **1996**, *118*, 3225–3232.
559. Brau, E.; Falke, R.; Ellner, A.; Beuter, M.; Kolb, U.; Dräger, M. *Polyhedron* **1994**, *13*, 365–374.
560. Dräger, M.; Schmidt, B. *J. Organomet. Chem.* **1985**, *290*, 133–145.
561. Sellman, D.; Fretberger, G.; Moll, M. *Z. Naturforsch., B* **1989**, *44*, 1015–1022.
562. Murgesh, G.; Singh, H. B.; Butcher, R. J. *J. Chem. Res.* **1999**, 416–417.
563. DeGrot, M. W.; Corrigan, J. F. *J. Chem. Soc., Dalton Trans.* **2000**, 1235–1236.
564. Farrugia, L. J.; Carmalt, C. J.; Norman, N. C. *Inorg. Chim. Acta* **1996**, *248*, 263–266.
565. Sieron, L.; Bukowska-Strrzyewska, M.; Cyganski, A.; Turek, A. *Polyhedron* **1996**, *15*, 3923–3931.
566. Agocs, L.; Briand, G. G.; Burford, N.; Cameron, T. S.; Kwiatkowski, W.; Robertson, K. N. *Inorg. Chem.* **1997**, *36*, 2855–2860.
567. Asato, E.; Kamamuta, K.; Akamine, Y.; Fukami, T.; Nukada, R.; Mikuriya, M.; Deguchi, S.; Yokota, Y. *Bull. Chem. Soc. Jpn.* **1997**, *70*, 639–648.
568. Briand, G. G.; Burford, N.; Cameron, T. S. *Chem. Commun.* **2000**, 13–14.
569. Mishra, A. K.; Gupta, V. D.; Linti, G.; Noth, H. *Polyhedron* **1992**, *11*, 1219–1223.
570. Briand, G. G.; Burford, N.; Cameron, T. S. *Chem. Commun.* **1997**, 2365–2366.
571. Briand, G. G.; Burford, N.; Cameron, T. S.; Kwiatkowski, W. *J. Am. Chem. Soc.* **1998**, *120*, 11374–11379.
572. Herrmann, W. A.; Kiprof, P.; Scherer, W.; Pajdla, L. *Chem. Ber.* **1992**, *125*, 2657–2660.
573. Silin, J.; Bankovskis, J.; Belskis, V.; Stash, A. I.; Peca, L.; Asaks, J. *Zh. Neorg. Khim.* **2000**, *45*, 1150–1155.
574. Silina, E. Y.; Bankovsky, Y. J.; Belsky, V. I.; Stass, A. I.; Asaks, J. V. *Latv. Khim. Zh.* **1996**, 57–62.
575. Block, E.; Ofori-Okai, G.; Kang, H.; Wu, J.; Zubieta, J. *Inorg. Chem.* **1991**, *30*, 4784–4788.
576. Niven, M. L.; Irving, H. M. N. H.; Nassimbeni, L. R.; Hutton, A. T. *Acta Crystallogr., Sect. B* **1982**, *38*, 2140–2145.
577. Matar, S.; Reau, J. M.; Grannec, J.; Rabardel, L. *J. Solid State Chem.* **1983**, *50*, 1–6.

578. Matar, S.; Reau, J. M.; Rabardel, L.; Grannec, J.; Hagenmuller, P. *Mater. Res. Bull.* **1983**, *18*, 1485–1492.
579. Schultheise, E.; Scharmann, A.; Schwabe, D. *J. Cryst. Growth* **1987**, *80*, 261–269.
580. Matar, S.; Reau, J. M.; Villeneuve, G.; Soubeyroux, J. L.; Hagenmuller, P. *Radiat. Eff.* **1983**, *75*, 55–60.
581. Zimina, G. V.; Zamanskaya, A. Y.; Sadokhina, L. A.; Spiridinov, F. M.; Fedorov, P. P.; Fedorov, P. I. *Zh. Neorg. Khim.* **1982**, *27*, 2800–2803.
582. Matar, S.; Reau, J. M.; Lucat, C.; Grannec, J.; Hagenmuller, P. *Mater. Res. Bull.* **1980**, *15*, 1295–1301.
583. Niznansky, D.; Rehspringer, J. L. *J. Mater. Res.* **1992**, *7*, 2511–2513.
584. Udovenko, A. A.; Gorbunova, Y. E.; Davidovich, R. L.; Mikhailov, Y. N.; Zemunukhova, L. A. *Russ. J. Coord. Chem.* **2000**, *26*, 97–100.
585. Reau, J. M.; Grannec, J. In *Inorganic Solid Fluorides*, Hagenmuller, P. Ed.; Academic Press: New York, 1985; Chapter 12, Fast fluoride ion conductors, pp 423–461.
586. Popov, A. I.; Scharabin, A. V.; Sukhoverkhov, V. F.; Tchumaevsky, N. A. *Z. Anorg. Allg. Chem.* **1989**, *576*, 242–254.
587. Popov, A. I.; Val'kovski, M. D.; Sukhoverkhov, V. F. *Zh. Neorg. Khim.* **1990**, *35*, 2831–2836.
588. Morgan, K.; Sayer, B. G.; Schrobilgen, G. J. *J. Magn. Res.* **1983**, *52*, 139–142.
589. Blazic, B.; Lazarini, F. *Acta Crystallog., Sect. C* **1985**, *41*, 1619–1621.
590. Bowmaker, G. A.; Junk, P. C.; Lee, A. M.; Skelton, B. W.; White, A. H. *Aust. J. Chem.* **1998**, *51*, 293–309.
591. Udovenko, A. A.; Davidovich, R. L.; Medkov, M. A.; Gerr, R. G.; Struchkov, Y. T. *Koord. Khim.* **1987**, *13*, 274–278.
592. Bigoli, F.; Lanfranchi, M.; Pellinghelli, M. A. *Inorg. Chim. Acta* **1984**, *90*, 215–220.
593. Herdtweck, E.; Kreusel, U. *Acta Crystallogr., Sect. C* **1993**, *49*, 318–320.
594. Ishihara, H.; Yamada, K.; Okuda, T.; Weiss, A. *Bull. Chem. Soc. Jpn.* **1993**, *66*, 380–383.
595. Eickmeier, H.; Jaschinski, B.; Hepp, A.; Juergen, N.; Reuter, H.; Blacknick, R. *Z. Naturforsch., B* **1999**, *54*, 305–313.
596. Meyer, G.; Schoenemund, A. *Z. Anorg. Allg. Chem.* **1980**, *468*, 185–192.
597. Lefebvre, J.; Carpenter, P.; Jakubas, R. *Acta Crystallogr., Sect. B.* **1991**, *47*, 228–234.
598. Zaleski, J.; Glowiak, T.; Jakubas, R.; Sobczyk, L. *J. Phys. Chem. Solids* **1989**, *50*, 1265–1269.
599. Rheingold, A. L.; Uhler, A. D.; Landers, A. G. *Inorg. Chem.* **1983**, *22*, 3255–3258.
600. Benedetti, A.; Fabretti, A. C.; Malavasi, W. *J. Crystallogr. Spectrosc. Res.* **1992**, *22*, 145–149.
601. Lazarini, F.; Leban, I. *Acta Crystallogr., Sect. B* **1980**, *36*, 2745–2747.
602. Ahmed, A. A.; Blachnik, R.; Reuter, H.; Eickmeier, H.; Schultze, D.; Brockner, W. *Z. Anorg. Allg. Chem.* **2001**, *627*, 1365–1370.
603. Lazarini, F. *Acta Crystallogr., Sect. C* **1985**, *41*, 1617–1619.
604. Papavassiliou, G. C.; Koutselas, I. B.; Terzis, A.; Ratapoulou, C. P. *Z. Naturforsch., B* **1995**, *50*, 1566–1569.
605. Ahmed, I. A.; Blachnik, R.; Kastner, G. *Z. Anorg. Allg. Chem.* **2001**, *627*, 2261–2268.
606. Matuszewski, J.; Jakubas, R.; Sobczyk, L.; Glowiak, T. *Acta Crystallogr., Sect. B* **1990**, *46*, 1385–1388.
607. Cornia, A.; Fabretti, C.; Grandi, R.; Malavasi, W. *J. Chem. Crystallogr.* **1994**, *24*, 277–280.
608. Krautscheid, H. *Z. Anorg. Allg. Chem.* **1999**, *625*, 192–194.
609. Mitzi, D. B.; Brock, P. *Inorg. Chem.* **2001**, *40*, 2096–2104.
610. Peng, Y.; Lu, S.; Wu, D. WuQ.; Huang, J. *Acta Crystallogr., Sect. C* **2000**, *56*, 183–184.
611. Lazarini, F. *Acta Crystallogr., Sect. C* **1987**, *43*, 875–877.
612. Feldmann, C. *Z. Krystallogr.* **2001**, *216*, 465–466.
613. Geiser, U.; Wade, E.; Wang, H. H.; Williams, J. M. *Acta Crystallogr., Sect. C* **1990**, *46*, 1547–1549.
614. Carmalt, C. J.; Farrugia, L. J.; Norman, N. C. *Z. Naturforsch., B* **1995**, *50*, 1591–1596.
615. Krautscheid, H. *Z. Anorg. Allg. Chem.* **1994**, *620*, 1559–1564.
616. Krautscheid, H. *Z. Anorg. Allg. Chem.* **1995**, *621*, 2049–2054.
617. Clegg, W.; Errington, R. J.; Fisher, G. A.; Flynn, R. J.; Norman, N. C. *J. Chem. Soc., Dalton Trans.* **1993**, 637–641.
618. Sharma, P.; Cabrera, A.; Rosas, N.; Arias, L.; Lemus, A.; Sharma, M.; Hernandez, S.; Garcia, J. L. *Z. Anorg. Allg. Chem.* **2000**, *626*, 921–924.
619. Scarpignato, C.; Pelosini, I. *Prog. Basic Clin. Pharmacol.* **1999**, *11*, 87–127.
620. Briand, G. G.; Burford, N. *Chem. Rev.* **1999**, *99*, 2601–2657.
621. Sun, H.; Sadler, P. J. *Top. Biol. Inorg. Chem.* **1999**, *2*, 159–185.
622. Sun, H.; Sadler, P. J. *Chem. Ber.-Recl.* **1997**, *130*, 669–681.
623. Sadler, P. J.; Li, K.; Sun, H. *Coord. Chem. Rev.* **1999**, *185–186*, 689–709.

Comprehensive Coordination Chemistry II
ISBN (set): 0-08-0437486

Volume 3, (ISBN 0-08-0443257); pp 465–544

3.7
Germanium, Tin, and Lead

J. PARR

Yale University, New Haven, CT, USA

3.7.1 INTRODUCTION

As has been cogently observed[1] the elements of Group 14 exhibit perhaps the most diverse chemical behavior seen for the members of any single group. This has the benefit of making Group 14 a fascinating area of study as well as a richly rewarding one—there is no such thing as "handle-turning" in the study of these elements. The variation in stability of oxidation states, the

Table 1   Significant properties of germanium, tin, and lead.[a]

Element	Electron configuration	Covalent radius (Å)	NMR nucleus (% abundance)
Germanium	$[Ar]3d^{10}4s^24p^2$	1.22	^{73}Ge (7.8)
Tin	$[Kr]4d^{10}5s^25p^2$	1.41	^{119}Sn (8.7) ^{117}Sn (7.7)
Lead	$[Xe]4f^{14}5d^{10}6s^26p^2$	1.54	^{207}Pb (22.6)

[a] Massey, A. G. *Main Group Chemistry*, 2nd. ed., Wiley: Chichester, 2000.

wide tolerance for coordination numbers, ligand types, and coordination geometries all work together to ensure that there is never a dull moment.

There are some overarching considerations that relate to these elements and offer some useful guidelines to their general behavior in terms of their coordination chemistry. The lightest of these elements, germanium, occurs predominantly in the M^{IV} oxidation state and where found as M^{II}, its reactions tend towards to its oxidation. The range of coordination numbers for germanium is narrower than those seen for tin or lead, four or five being the most commonly observed. Tin, equally content in either oxidation state, can form complexes of higher coordination number but may form perfectly stable low coordination number complexes, especially in the lower oxidation state. Lead, the heaviest of the triad, prefers the lower oxidation state and exhibits coordination numbers between two and 10, although some of these higher coordination numbers may be somewhat moot. There is no doubt, however, that lead spans the range of highest to lowest coordination number shown amongst these elements.

There is no marked preference for hard or soft donors, as all three elements are equally able to form complexes with both hard and soft donors in both oxidation states. Further, as post-transition elements, they adopt geometries in their complexes that do not follow regular patterns, such as transition metals do, but rather are governed by the number and nature of ligands present in their complexes.

These simple considerations in hand, the coordination chemistry of these elements becomes largely explicable. Fortunately, there are exceptions and surprises to keep the level of interest high and to drive the exploration of this area.

It is extremely fortunate that Group 14 also offers some potent spectroscopic tools for the investigation of its complexes. All three elements have at least one NMR active nucleus, and while ^{119}Sn has been widely studied for many years,[2,3] it is only recently that solution ^{207}Pb and ^{73}Ge spectra have become widely available to the synthetic chemist (Table 1). Additionally, Mössbauer spectroscopy is very useful in the assignment of coordination number, oxidation state, and geometry in tin complexes, and the expansion of crystallography has been of great utility in all areas.

While this is not a review of organometallic chemistry a great number of organo- substituted compounds are included where the remainder of the ligand set is of interest. This seems justified in that, for these compounds, the organic ligands are usually playing a spectator role and serve only to support and stabilize the metal.

3.7.2   COMPLEXES WITH CARBON DONOR LIGANDS

Excluding organic ligands, there are few examples of complexes of M(14) with carbon donor ligands. Most cyanide and cyanate complexes have been known for some time, although there are always new examples to be found. The simplest examples of complexes with M(14)—C bonds are the carbides, and a number of new routes to such compounds have been reported.

Thermolysis of $C(GeH_3)_4$, prepared from the four-fold insertion of $GeCl_2 \cdot diox$ into CBr_4[4] followed by reduction with $LiAlH_4$, gives the binary carbide Ge_4C, which exhibits a diamond structure.[5] Complex M(14) carbide-containing compounds $M_2M(14)C$ (M = Ti, Hf, Zr, Nb; M(14) = Sn: M = Zr, Hf; M(14) = Pb) have been prepared from heating the respective elements together at 1,200–1,325 °C for 4–48 h. The lead compounds, reported for the first time, are unstable under ambient conditions, and exhibit hardness and conductivity comparable with other such carbides.[6,7] Carbides M(3)M(14)C (M = Al, Sc, La–Nd, Sm, Gd–Lu; M(14) = Sn, Pb) have been prepared and all have been shown to exhibit Perovskite structures.[8] The direct combination of barium, germanium, and carbon in elemental form at 1,260 °C gives $Ba_3Ge_4C_2$, a moisture-sensitive carbide with semiconductor properties. The compound comprises $[Ge_4]^{4-}$ units with Ge—Ge bond lengths of 2.517 Å and $[C_2]^{2-}$ with C—C bond lengths of 1.20 Å. The carbide also reacts with NH_4Cl to give ethyne and a range of germanes.[9]

Bis(trimesitylgermylcarbodiimido)germylene is stable in the absence of oxygen or water but rapidly decomposes at 50 °C to give a mixture of bis(trimesitylgermyl)carbodiimide and poly (carbidodiimido)germylene, whereas hydrolysis leads to the monogermylated derivative of cyanamide as both trimesitylgermyl cyanamide and trimesitylgermyl carbodiimide.[10] The related $(mes)_3$-Ge(CN) has been characterized crystallographically and found to have a distorted tetrahedral geometry at the metal.[11]

Germanium(II) cyanide is a highly moisture-sensitive compound that is stable in solution as $Ge(CN)_2$ in the absence of air, moisture, or Lewis bases but forms an intractable liquid on isolation, possibly due to irreversible oligomerization. Prepared by the reaction of germanium(II) chloride and silver cyanide in refluxing THF, the complex was identified in solution by IR (ν CN 2,090 cm^{-1}) and by trapping reactions.[12]

Tin and tin cluster cyano complexes $SnCN^-$, $Sn(CN)_2^-$, Sn_2CN^-, Sn_3CN^-, and Sn_4CN^- have been studied by a combination of anion photoelectron spectroscopy and DFT calculations.[13] Further cyanide complexes can be prepared from the reaction of $SnCl_4$ with TMS(CN), forming Cl_3SnCN, or from the oxidative addition of X(CN) (X = Br, I) to $SnCl_2$, forming $SnCl_2X(CN)$. Both the IR and Mössbauer spectroscopic data indicate a polymeric structure for these compounds in the solid state with bridging ambidentate cyano groups. In the preparation of $SnCl_2Br(CN)$, a second species of composition $SnCl_2Br(CN)(THF)_{0.5}$ was isolated, which was identified spectroscopically as a mixture of $[SnCl_2Br(CN)]_n$ and monomeric $SnCl_2Br(CN)(THF)_2$, which has a near octahedral geometry with *trans* disposition of the solvent ligands.[14]

A rare example of a M(14) carbonyl complex has been formed in the gas-phase ion-molecule reaction of GeH_4 with CO, where ions of the type $[GeH_n(CO)]^+$ can be detected. Similarly, the carbonates $[GeH_n(CO_2)]^+$ can be prepared by the analogous reaction with CO_2.[15] These transient compounds are a tantalizing indication of what may be available to traditional synthetic chemists if the right approach can be found.

3.7.3 COMPLEXES WITH M(14)—M(14) BONDS

Catenation is a pronounced feature of Group 14 chemistry, especially for the lighter members and the ease with which stable M(14)—M(14) bonds may be formed has resulted in research involving the preparation and study of complexes that have such bonds. Examples of complexes that comprise bonding between atoms of the same element are numerous and range from dinuclear compounds to large polynuclear assemblies.

3.7.3.1 Complexes with M(14)—M(14) Homoelement Bonds

Larger molecules $(R_2Ge)_n$ are on the whole less stable than the equivalent silicon or carbon analogues, and are prone to thermal and photochemical reactions, including elimination of R_2Ge monomers and the concomitant formation of ring-contracted products. The photolysis of $(R_2Ge)_n$ (R = Me, $n = 6$[16]; R = Pr^i, $n = 4$[17]) gives the ring-contracted products, the R_2Ge monomer and $(R_2Ge)_2$ dimers, detected spectroscopically. The monomeric diorganogermanes are not stable but can be trapped, such as in reaction with carbon tetrachloride, forming R_2GeCl_2. Complexes $(R_2M)_2$, (M = Ge, Sn, Pb) where R is a very large group, are comparatively stable and the area has been reviewed.[18,19]

Bis(dimethylgermyl)methane reacts with $(Bu^t)_2Hg$—mercury to give the heterocyclic product 1,3,5,7,2,6-tetragermadimercurocane, which extrudes mercury to form 1,1,2,2,4,4,5,5-octamethyl-1,2,4,5-perhydrotetragermine, a six-membered ring with two germanium—germanium bonds.[20] The sulfur-containing cycloheptyne (**1**) reacts directly with $GeCl_2 \cdot diox$ to give (**2**), a digermacyclobutene with a Ge—Ge bond length of 2.380 Å (Scheme 1). There is no evidence either in the solid or in solution of any transannular Ge–S interaction.[21] The same alkyne reacts with tin(II) chloride to give a mononuclear complex.[22]

Trimetallacyclopropanes, germanium or tin triangles, have emerged as a fascinating subset of compounds with M(14)—M(14) bonds. The first fully characterized germanium example was $[(2,6-Me_2C_6H_3)_2Ge]_3$ prepared by the reaction of $(acac)_2GeCl_2$ (acac = acetylacetonate) with

Scheme 1

(2,6-Me$_2$C$_6$H$_3$)MgBr. The complex has a regular triangular array of the metals with intermetallic bond lengths 2.543–2.547 Å and in common with the larger arrays, the complex forms (R$_2$Ge)$_2$ on photolysis.[23] An improved general synthesis using R$_2$GeCl$_2$, magnesium and magnesium bromide subsequently broadened the field.[24]

Triangular [Cl(Bu$^t_3$Si)Ge]$_3$ has been prepared from the reaction of GeCl$_2$·diox with Na[SiBu$^t_3$] and its subsequent conversion to [(Bu$^t_3$Si)$_2$GeGe(SiBu$^t_3$)=Ge(SiBu$^t_3$)] has been examined by ^{29}Si NMR. This study implicates both Na[Ge(Cl)(SiBu$^t_3$)Ge(Cl)$_2$(SiBu$^t_3$)] and [(Bu$^t_3$Si)(Cl)Ge]$_2$ as intermediates in the reaction. These were trapped from the reaction mixture, in the former case by addition of further Bu$^t_3$SiCl and in the latter by addition of isoprene to give (3).[25] The presence of large organic groups seems to be essential to the formation of the ring structure, as the reaction of germanium(II) chloride with Li[2,4,6-Bu$^t_3$C$_6$H$_2$], Li[R] gives mononuclear RGeCl[26] or R$_2$Ge[27] complexes depending upon stoichiometry. Even though this is a fairly big group it seems that it is not large enough to promote the formation of the trimer.

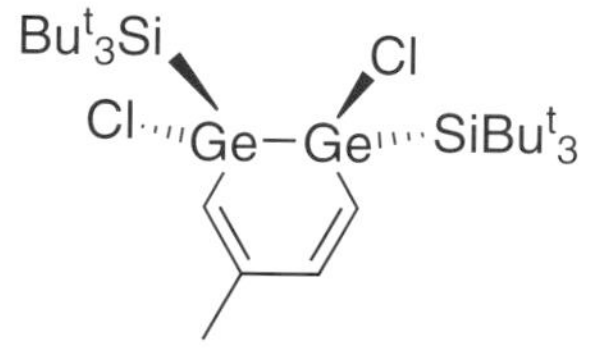

(3)

The reaction of GeCl$_2$·diox with Li[EBu$^t_3$] (E = Si, Ge) gives (4) (Scheme 2), that comprises an unsaturated Ge—Ge double bond. The geometry of (4) is an isosceles triangle, comprising one Ge—Ge double bond (2.239 Å) and two Ge—Ge single bonds (2.519 Å). The exocyclic Ge—Si bonds show a marked difference depending upon whether the germanium to which they are attached to is doubly (2.448 Å) or singly bonded (2.629 Å) to germanium.[28] The product (4) is liable to oxidation by [Ph$_3$C][BPh$_4$] yielding the monocation (5). Structural analysis of (5) shows that the intermetallic bonds within the triangle are all equivalent, 2.33 Å, a value intermediate between double (ca. 2.24 Å) and single (ca. 2.52 Å), indicating that the compound is aromatized.[29]

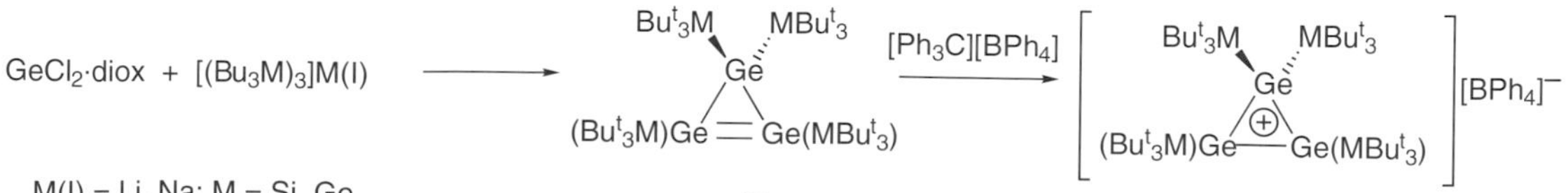

GeCl$_2$·diox + [(Bu$_3$M)$_3$]M(I) →

M(I) = Li, Na; M = Si, Ge

(4)   **(5)**

Scheme 2

The unsaturated triangulo germanium complexes are stable, but can undergo a number of reactions typical of double bonds. The complex (Bu$^t_3$Si)(mes)Ge{Ge(SiBu$^t_3$)}$_2$ (mes = 2,4,6-trimethylphenyl) undergoes (2+2) cycloaddition reactions with phenylalkyne to give 1,4,5-trigerma-5-mes-2-phenylbicyclo[2.1.0]pent-2-ene and (2+4) cycloaddition with isoprene to give 1,4,5-trigerma-7-mes-bicyclo[4.1.0]hept-3-ene.[30]

Triangulo (Bu$^t_2$Ge)$_3$ inserts PhNC to give trigermabutanimine (6), and in a similar fashion, the chalcogens sulfur and selenium insert to give the chalcogermetanes (7) and (8), the selenium compound being planar (Scheme 3).[31]

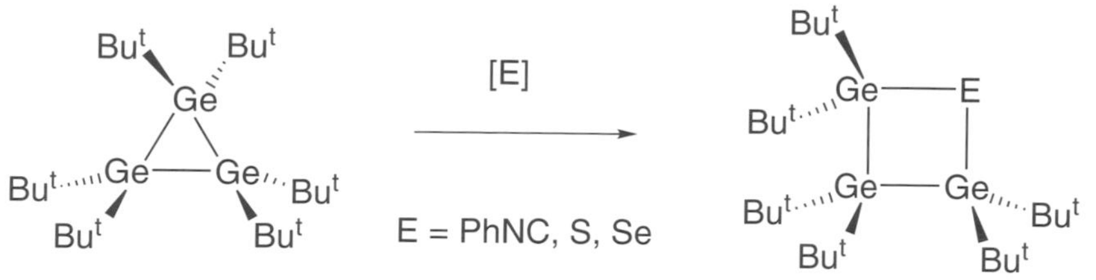

E = CNPh, (**6**); S, (**7**); Se, (**8**).

Scheme 3

Reaction of (2,6-mes$_2$C$_6$H$_3$)GeCl with KC$_8$ gives the cyclotrigermenyl radical (**9**), which again has all equivalent Ge—Ge bonds (2.35 Å). The blue crystalline product shows an EPR spectrum with low values of hyperfine coupling, indicating that the single electron is probably in a low-lying π antibonding orbital. Further reduction of (**9**) with an excess of KC$_8$ gives the ring-opened trigermenyl allyl anion (**10**), isolated as its deep green potassium salt (Scheme 4). The angle Ge–Ge–Ge is 159°, and the bond lengths Ge—Ge are 2.42 Å, slightly shorter than single bond length. These data all point to aromatization in this rather unusual product.[32]

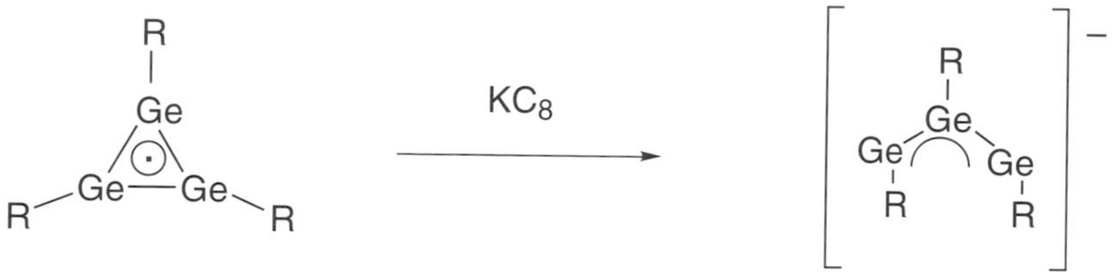

R = 2,6-mes$_2$-C$_6$H$_3$

(**9**)   (**10**)

Scheme 4

The same reducing agent has been used to convert (2,6-trip$_2$C$_6$H$_3$)SnCl (trip = 2,4,6-triisopropylphenyl) to the radical anionic dimer [(2,6-trip$_2$C$_6$H$_3$Sn)$_2$]$^{\cdot-}$ (**11**). Crystallographic analysis reveals

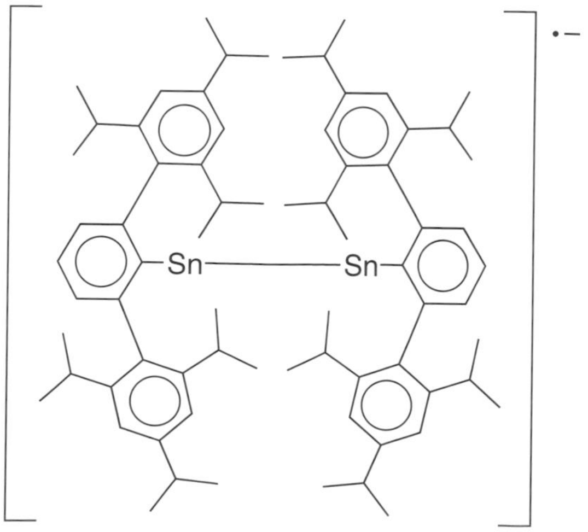

(**11**)

a moderately long Sn—Sn bond of 2.812 Å and this, taken with the angle C_{ipso}–Sn–Sn of 95.2° seem to preclude any Sn—Sn π bonding.[33] The deep red product $[(Ph_2Sn)_2]^-$ can be prepared by the reaction of Ph_2SnCl_2 with lithium in liquid ammonia. In this case, the compound is centrosymmetric and has a Sn—Sn bond length of 2.91 Å, significantly longer than the more sterically congested (11). The complex (12), comprising an asymmetric anionic chelating ligand, has a similar Sn—Sn bond (2.869 Å).[34] The reaction of LiMe with $[2,6-(trip)_2C_6H_3](Cl)Sn$ gives a stable heavy analogue of methyl methylene, as $Me_2(2,6-trip_2C_6H_3)SnSn(2,6-trip_2C_6H_3)$, with a tin(IV)—tin(II) bond.[35] Another example of a complex with a tin(IV)—tin(II) bond is available from the reaction of tin(II) chloride with the Grignard reagent prepared from 2-(diphenylphosphino)bromobenzene, (13).[36] Reduction of $(2,6-trip_2C_6H_2)PbBr$ with lithium aluminum hydride according to an unknown mechanism gives the dimer $[(2,6-trip_2C_6H_2)Pb]_2$ which has a trans bent geometry and a Pb—Pb bond of 3.118 Å, which seems closer to a single than a multiple bond.[37]

(12)   (13)

Tin complexes $(R_2Sn)_3$ have been less studied than their germanium counterparts but are no less readily prepared. The reaction of $(2,6-Et_2C_6H_3)_2SnCl_2$ with lithium naphthalenide gives $(R_2Sn)_3$ with intermetallic bond lengths of 2.854–2.870 Å.[38]

The first example of a molecular compound with a tetrahedral $[Ge_4]$ unit has been prepared in the reaction of $GeCl_2$·diox and $Na[SiBu^t_3]$, as $[GeSiBu^t_3]_4$. The bond lengths Ge—Ge are 2.44 Å, intermediate between double and single bond values.[39] Germanium(II) iodide reacts with $Na_2[Cr_2(CO)_{10}]$ in the presence of 2,2′-bipyridyl(bipy) to form $[Ge\{Cr(CO)_5\}]_6$ (14), an octahedral cluster of germanium substituted with organometallic ligands. The complex has as its core a Zintl ion unknown in the free state and has intermetallic bonds that are shorter than those seen in other Zintl ions such as $[Ge_9]$ (2.521–2.541 cf. 2.52–3.00 Å).[40] The corresponding tin complex is prepared from the reaction of $K_2[Cr(CO)_5)]$ and tin(II) chloride without the diimine ligand.[41]

(14)

An alternative geometry for a $[Ge_6]$ framework is trigonal prismatic, and an example of a complex with this structure is $(TMS_2CHGe)_6$ prepared from the corresponding Grignard reagent and germanium(IV) chloride. The complex has two distinct Ge—Ge bonds, on the triangular face (2.579 Å) and on the quadratic face (2.526 Å).[42]

The same framework is also seen in the tin compound $[Bu^t_3SiSn]_6$ (15), prepared from $(TMS_2N)_2Sn$ and 12 equivalents of $Na[SiBu^t_3]$.[43] Larger again, the octagermacubane $[(2,6-Et_2C_6H_3)Ge]_6$[44] and octastannacubane[45] have been prepared by dehalocoupling reactions of RMX_3 with Mg/MgBr. A complex of lower nuclearity with a different structural motif is

2,2,4,4,5,5,-hexakis(2,6-Et$_2$C$_6$H$_3$)pentastanna[1,1,1]propellane, prepared by the thermolysis of [(2,6-Et$_2$C$_6$H$_3$)$_2$Sn]$_3$ in refluxing xylene.[46] From the same thermolysis it is also possible to isolate the first heavy M(14) prismane (2,6-Et$_2$C$_6$H$_3$)$_{10}$pentacyclo[6.2.0.02,7.03,6.04,10.05,9]decastannadecane (16).[47]

R = Bu$^t_3$Si

(15)   (16)

Other homoelement assemblies can be found in the extensive family of Zintl anions. Of interest in this area is the recently reported synthesis of [Ge$_9$]$^{4-}$ from the direct combination of Cs and Ge at 900 °C. The structure of [Ge$_9$]^4 in Cs$_4$Ge$_9$ is a monocapped square antiprism. The same synthetic approach can be used to prepare the series of compounds M$_{12}$Ge$_{17}$ (M = Na, K, Rb, Cs) that comprises one [Ge$_9$]$^{4-}$ and two [Ge$_4$]$^{4-}$ clusters.[48,49] A polymeric assembly can be isolated from the en and 18-C-6 extraction of KGe$_4$ as a linear polymer of vertex linked ligand free [Ge$_9$]$^{2-}$ clusters.[50] The Zintl ions in the complexes M^{1}AuM(14)$_4$ (M^1 = K, Rb, Cs, M(14) = Sn, Pb) have been studied crystallographically and shown to comprise chains of tetrahedra of M(14) bridged by gold ions,[51] similar to the cadmium-bridged [Pb$_4$] tetrahedra in the structure of K$_6$Pb$_8$Cd.[52] A discussion of the cluster compounds of the heavier M(14) elements has been published.[53,54]

3.7.3.2 Complexes with M(14)—M(14) Heteroelement Bonds

The propensity for catenation may be exploited further to prepare M^1(14)–M^2(14) species that comprise direct bonds between different members of Group 14. This is a relatively young area of research but there are already many fascinating examples of such compounds in the literature.

Although there are many examples of compounds that have frameworks based upon Si$_n$ rings, there are very few examples of such rings incorporating heteroatoms. Substituted germatetrasilacyclopentanes [(R$^1_2$Si)$_4$GeR$^2_2$] (R^1 = Pr^i, R^2 = CH$_2$SiMe$_3$ or Ph; R^1 = CH$_2$Bu^t, R^2 = Ph) are stable compounds but can be photolyzed to give either R$^2_2$Ge and the ring contracted cyclosilanes or a range of silenes, disilenes, germens, and silagermenes, depending upon the organic substituents.[55]

The complexes TMSGePh$_3$, Me$_3$GeSiPh$_3$, FpMe$_2$SiGeMe$_3$, FpMe$_2$GeSiMe$_3$, IFpMe$_2$SiGeMe$_3$, IFpMe$_2$GeSiMe$_3$, IFpMe$_2$GeSiPh$_3$, IFpMe$_2$GeSiPh$_3$, and FcSiMe$_2$GeMe$_2$Fc (Fp = CpFe(CO)$_2$, IFp = (indenyl)Fe(CO)$_2$, Fc = ferrocene) were prepared and their decomposition under mass spectral conditions examined to probe the nature of the Si—Ge bond. For the bimetallic species, the main feature of the mass spectra is the presence of peaks due to products formed by R group scrambling. For the trimetallic species, the Fe–Si–Ge linkage cleaves predominantly at the Si—Ge bond, giving Fp silylene products, whereas the complexes with the Fe–Ge–Si linkage cleaves at the Fe—Ge bond, indicating that the Fe—Ge bond is less stabilized than the Fe—Si bond, which is in turn more stabilized than the Ge—Si bond.[56]

The disilane MeBu$^t_2$Si-SiBr$_3$ reacts with (MeBu$^t_2$Si)$_2$GeCl$_2$ in the presence of sodium metal to give the triangular complex (17). The Si=Si double bond is short at 2.146 Å giving a pronounced isosceles geometry to the ring. Thermal or photochemical rearrangement gives the cyclic germasilene (18) (Scheme 5).[57] Further germasilenes R$^1_2$Si=GeR$^2_2$ are available either from the reaction of Li$_2$[R$^1_2$Si] and R$^2_2$GeCl$_2$ (R^1 = Pr$^i_3$Si, Bu$^t_2$MeSi, R^2 = mes, 2,4,6-Pr$^i_3$C$_6$H$_2$), trapped by addition of methanol to give the silane hydride and germanium methoxide,[58] or from the photolysis of (mes$_2$Ge)$_2$Si(mes)$_2$. The product from this latter reaction, mes$_2$Ge=Si(mes)$_2$, rearranges to give the mixed valence complex (mes$_3$Si)Ge(mes). The M^{II} amides (19) react to form the tetranuclear (20) (Scheme 6) presumably by a similar mechanism, whereas the tin amide gives (21) (Scheme 7).[59]

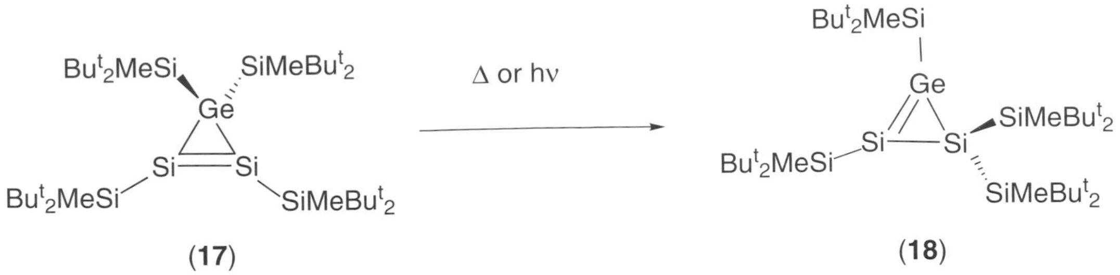

Scheme 5

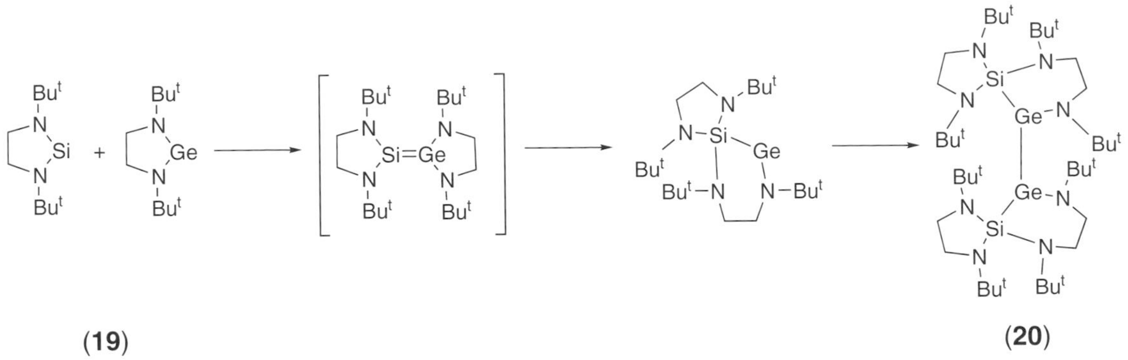

Scheme 6

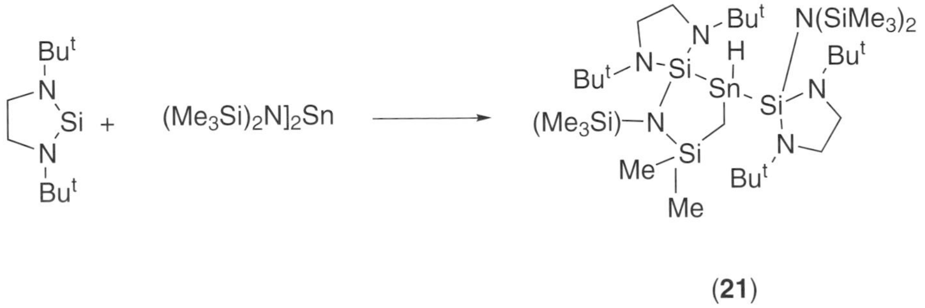

Scheme 7

The reaction of [mes$_2$GeCl]$_2$ with mes$_2$SiCl$_2$ in the presence of a reducing agent gives the heteroelement triangle (mes$_2$Ge)$_2$Si(mes)$_2$. Upon photolysis, the complex eliminates mes$_2$Ge in preference to mes$_2$Si and gives the germyl silylene mes$_2$GeSi(mes)$_2$.[60] The other member of the (M^1$_2$M^2) triangle series, R$_2$Ge(SiR$_2$)$_2$, can be prepared by the reaction of germanium(II) chloride with Li[Si(SiMe$_3$)$_3$], forming TMSGe[Si(SiMe$_3$)$_2$]$_2$. The mechanism by which this reaction proceeds is not clear but the preparation has a moderately high yield. The bond lengths Ge—Si are 2.35 and 2.391 Å, and Si—Si is 2.366 Å. In the same paper, the structure of (12-C-4-Li)(TMS$_3$Ge) is reported.[61]

The dihydride (22) can be lithiated and subsequently reacted with (trip)$_2$SnF$_2$ to give the heterobimetallic (23) (Scheme 8). The germanium can be lithiated again with the loss of the remaining hydride to form the heterotrimetallic (24). This reacts with alkyl halides, such as MeI, to give the corresponding germanium alkyl, or alternatively (24) can eliminate LiF to give the first stannagermene (25). The tin—germanium double bond is extremely reactive and so the complex is trapped, either by addition of alcohols to give the germanium hydride and tin alkoxide or by thermal rearrangement to give the triangular (R$_2$Sn)$_2$GeR$_2$.[62]

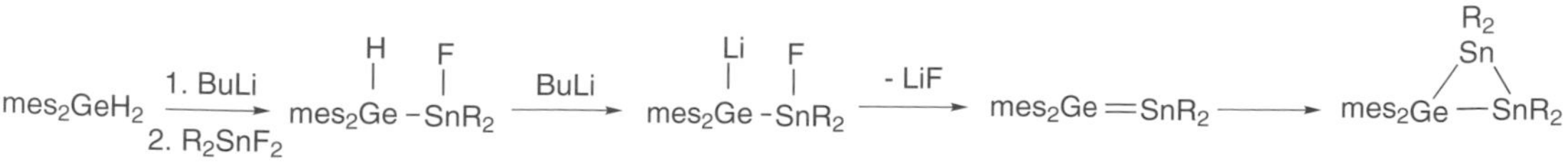

Scheme 8

Surprisingly, the first organosilyl plumbane Pb(SiMe₃)₄ was only reported as recently as 1983, the tetrahedral compound obtained from the reaction of lead(II) chloride with Mg(SiMe₃)₂.[63]

Amides (TMS₂N)₂M^{II} (M = Sn, Pb) react directly with K[Si(SiMe₃)₃] to give (TMS₃Si)₂M^{II}. The lead complex is monomeric with Pb—Si bond lengths of 3.70 Å, whereas the tin complex dimerizes to form a distannane with a *trans* configuration and a Sn—Sn bond length of 2.82 Å. This bond is long for a distannane and is close to the expected value for a single bond, which indicates the extent to which the steric and electronic effects exerted by the ligands influence the interaction of the metals.[64]

The tin(II) complex [1,8-{N(CH₂Bu^t)}₂C₁₀H₆]Sn reacts with [2,6-(NMe₂)₂C₆H₃]₂M (M = Ge, Sn) to form the dinuclear complexes (**26**) (Scheme 9). The bonding in these has been interpreted in terms of a donor–acceptor interaction between the more electron-rich metals with the aryl ligands to the more electron-deficient nitrogen-bound tin.[65]

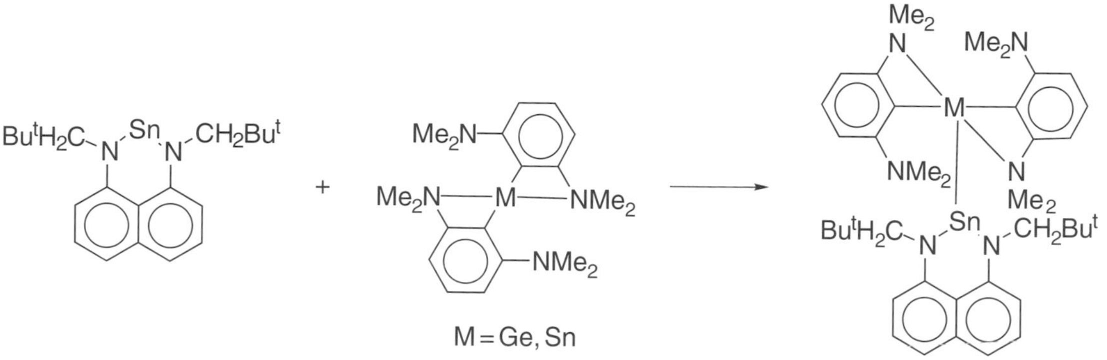

Scheme 9

3.7.4 COMPLEXES WITH GROUP 15 LIGANDS

The M(14) nitrides, phosphides, and arsenides are materials with industrially useful properties, and while these are fascinating areas and although there is a wealth of literature relating to these compounds they do not fall within the scope of this article. Some materials or methods of preparation that relate to this group of compounds that are new are covered because they are of more general interest.

3.7.4.1 Complexes with Neutral Monodentate Nitrogen Ligands

New complexes of monodentate nitrogen ligands are rare, as such ligands have been studied for many years. However, some seemingly simple ligands can be used to prepare complexes that have features that transcend expectations.

Neutral monodentate *N*-donor ligands have been used to prepare monomeric six-coordinate tin(IV) complexes all-trans R₂SnX₂(L)₂ (R = cyclohexyl (Cy), Ph, X = Br, L = pyrazole, imidazole,[66] R = Ph, vinyl, X = Cl, L = pyrazole[67]). However, the product of the reaction between Ph₂SnCl₂ and pyrazine has an overall stoichiometry of Ph₂SnCl₂(pyrazine)₀.₇₅, and consists in the solid state of alternating layers of polymeric [Ph₂SnCl₂(pyrazine)]$_n$, which has a six-coordinate tin center, and mononuclear Ph_2SnCl_2(pyrazine), which has a five-coordinate tin. If Me_2SnCl_2 is used, the product is exclusively $[(Me_2SnCl_2)_2$(pyrazine)$]_n$, an indication of the subtle variety of behavior exhibited by tin in its coordination compounds. Solid-state ^{119}Sn NMR can differentiate between the two discrete tin centers in the phenyl compounds. Crystallographic analysis reveals long Sn—N bonds of 2.961 and 3.783 Å in the polymeric compound and 2.683 Å in the dimer.[68]

Diphenyllead dichloride coordinates two equivalents of imidazole to give the six-coordinate complex with trans organic ligands and bond lengths Pb—N 2.45 Å.[69]

3.7.4.2 Complexes with Anionic Monodentate Nitrogen Ligands

The structure of γ-Ge_3N_4, prepared from either α-Ge_3N_4 or β-Ge_3N_4 at elevated temperatures and pressures (1,000 °C, 12 GPa) has a spinel structure and F_{d3m} symmetry. The structure comprises both octahedral and tetrahedral coordination of germanium by nitrogen (Ge–N 1.996 and 1.879 Å, respectively).[70] The binary tin nitride Sn_3N_4 has been prepared by the reaction of tin(IV) bromide with KNH_2 in liquid ammonia. The product itself is amorphous and decomposes directly to the elements at temperatures approaching 420 °C.[71] Ternary nitrides $Sr_3Ge_2N_2$ and Sr_2GeN_2 have been prepared from Na, NaN_3, Sr, and Ge at 750 °C. The former exhibits a structure with zig-zag chains of Ge^{2-} ions, and both have $[GeN_2]^{4-}$ ions, with angles N–Ge–N of 113.6° and bond lengths Ge–N ranging between 1.85–1.88 Å.[72] The complex lead nitride La_5Pb_3N is available from the reaction of La, La_4Pb_3, and LaN at 1,050–1,250 °C. The structure is an isopointal interstitial derivative of the Cr_5B_3 structure type. The Pb–Pb separation of 3.550 Å seems to preclude any significant Pb–Pb interaction.[73]

A number of azides of Group 14 metals have been prepared and reported following a growth in interest in such complexes, in part due to their potential application as precursors to metal nitrides. The germanium(IV) monoazide $(mes)_3Ge(N_3)$ has been prepared and shown to have a distorted tetrahedral geometry at the germanium and an almost perfectly linear azide.[11] Treating $(acac)_2GeCl_2$ with sodium azide in refluxing acetonitrile forms the *cis* diazide $(acac)_2Ge(N_3)_2$[74] and the first neutral octahedral triazide of germanium, $[HB(3,5-Me_2-pz)_3]Ge(N_3)_3$, has also been reported.[75] The impressive homoleptic hexaazide anion $[Ge(N_3)_6]^{2-}$ has been prepared for the first time and a number of its reactions explored. In the solid state, the anion has idealized S_2 symmetry as its $[\{(PPh_3)_2N\}^+]_2$ salt, with no close interactions between the ions, whereas the $[Na_2(THF)_3(Et_2O)]^{2+}$ salt has C_1 symmetry, with short Na···N contacts of 2.410–2.636 Å. Addition of the nitrogen donor ligands bipy or phen (L) (phen = 1,10-phenanthroline) gives the first neutral octahedral tetraazide complexes $LGe(N_3)_4$. A ^{14}N NMR study gave values of δ for N_α of −288.9 and for N_γ −208.[76]

Triphenyltin(IV) azide and a number of its 1:1 adducts (py, py-NO, HMPA, Ph_3PO) have been prepared and all of these complexes have been found to exhibit a five-coordinate geometry in the solid state. For $Ph_3Sn(N_3)$ this is achieved, in the absence of additional ligands, through 1,3 bridging azido groups that link two tin centers to form dimers through the formation of eight-membered $[Sn_2N_6]$ rings.[77] This difference in structure compared to $mes_3Ge(N_3)$ is more likely to be a result of the smaller radius of germanium rather than the larger co-ligands in that complex. In reaction with 1,4-$(SCN)_2$-C_6H_4 triorganotin azides $RSn_3(N)_3$ (R = Me, Et, Bu^n, Ph) form the dinuclear complexes (27), which can be converted to the lead analogue by reaction with triphenyllead chloride (Scheme 10).[78]

(27)

Scheme 10

Azides of the lower oxidation state are less stable but nonetheless several have been reported, including $[HB(3,5-Me_2-pz)_3]Ge(N_3)$ (pz = pyrazolyl),[75] $(K_L)Ge(N_3)$ (K_L = Klaui's ligand)[79] (aminotropinimate)$M(N)_3$ (M = Ge, Sn),[80] and (N,N'-mes_2-1,5-diazapentadienyl)$M(N_3)$ (M = Ge, Sn).[81] These complexes all exhibit geometries consistent with stereochemically active lone pairs and linear or near-linear azides. There is considerable ionic character in [HB(3,5-Me$_2$-pz)$_3$]Ge(N$_3$), dissociating in polar solvents to give well separated [HB(3,5-Me$_2$-pz)$_3$]Ge$^+$ and (N$_3$)$^-$ ions. The ^{14}NMR spectrum of the latter complex has δ N$_\alpha$ -291, N$_\beta$ -136, and N$_\gamma$ -215 for the germanium complex and -292, -136, and -223 for the tin complex. The recent advances in covalent azide coordination chemistry of main group elements have been reviewed.[82]

Tin isothiocyanato complexes (PhMe$_2$Si)Sn(Me)$_2$(NCS),[83] Ph$_3$Sn(pyridinium-2-carboylato)-(NCS),[84] and R^{1}R^{2}SnL(NCS)$_2$ (R^1,R^2 = Ph$_2$, (4-tolyl)$_2$, (3-ClC$_6$H$_4$)$_2$, (4-ClC$_6$H$_4$)$_2$, Me Et, Et Pr^n; L = neutral κ^1O-donor) have been reported.[85] The first of these is claimed to be the first example of a four-coordinate tin(IV) isothiocyanate. The second has an overall trigonal bipyramidal (tbp)geometry at the tin with all three phenyl groups equatorial and the isothiocyanate ligand axial, trans to the *O*-bound Zwitterionic pyridine carboxaldehyde. The series of complexes that are the third example have been examined by Mössbauer spectroscopy to explore the relationship between ligand type and the selectivity between *cis* and *trans* coordination of the organic ligands. This spectroscopic technique is useful in determining the coordination number, the geometry, and in favorable cases, the relative disposition of the ligands about the tin center.

The complex anions [SnX$_{6-n}$Y$_n$]$^{2-}$ (X = Cl, Br; Y = CN$^-$ or SCN$^-$) have been studied by ^{119}Sn NMR in solution and a correlation between the value of δ and the number and nature of the ligands on the anion established.[86]

The reaction between the seemingly simple reagents tin(IV) chloride and (TMSN)$_2$C gives the remarkable tetranuclear (**28**).[87] The compound is stable and isolated in good yield, an example of a bridging mode for a monodentate *N*-donor ligand a role that nitrogen ligands play in many complexes of the heavier M(14) congeners. A mixed oxidation state dimer (**29**) is formed from the reaction of [SnI(NPPh$_3$)]$_2$ with sodium metal in a reaction that seems to proceed with loss of NaSn$_x$. The complex has a planar [SnN]$_2$ ring with bond lengths Sn–N$_{terminal}$ 1.990 (ave.) Sn^{IV}–N$_{bridging}$ 1.957, and Sn^{II}–N 2.25 Å.[88]

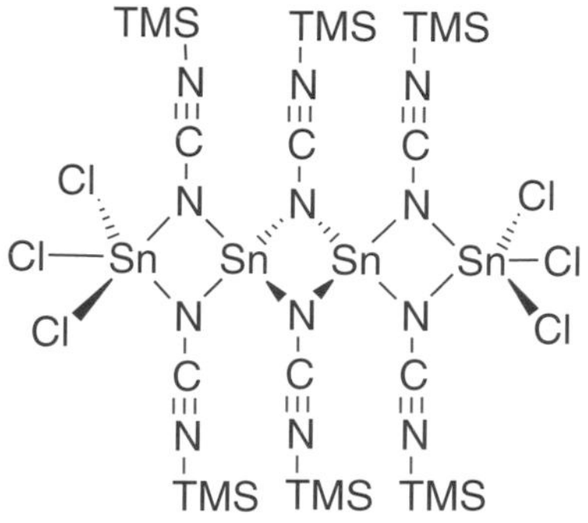

(28) 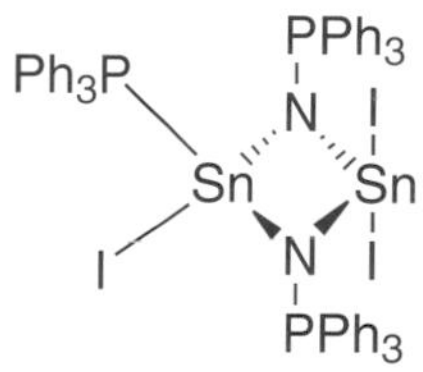   (29)

The triorganogermanium amine mes$_3$GeNH$_2$ is prone to Ge–N cleavage reactions by protic reagents, indicative of the relative polarity of the Ge—N bond, whereas reaction with Bu^{t}C(O)Cl gives an intact acylamino germane.[89] Complexes with bridging and terminal Ge–NH$_2$ functionality can be prepared from the reaction of H$_3$Ge[N(SiMe$_3$)(2-6-Pr$^i_2$-C$_6$H$_3$)] with ammonia, giving [(NH$_2$)$_2$Ge{N(SiMe$_3$)(2-6-Pr$^i_2$-C$_6$H$_3$)}]$_2$NH (**30**) (Scheme 11).[90] Tetrakis(trimethylhydrazido)-germanium(IV) (**31**) is available from the reaction of Li[N(Me)NMe$_2$] with germanium(IV) chloride.

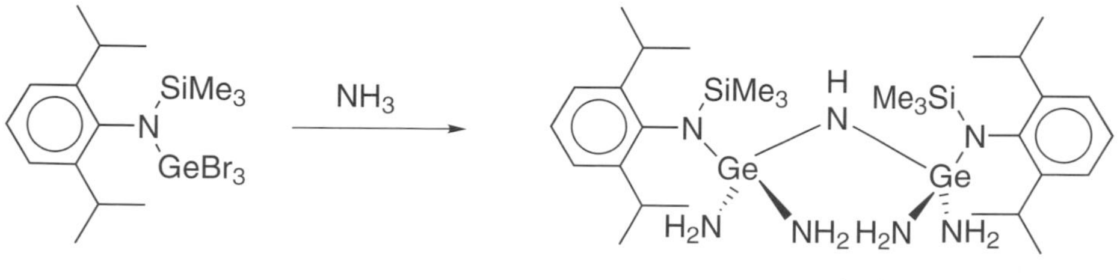

(30)

Scheme 11

Structural studies show that N_α is slightly pyramidal and that there is no N_β–Ge interaction,[91] a surprising result in light of the β interactions seen for closely related systems.[92]

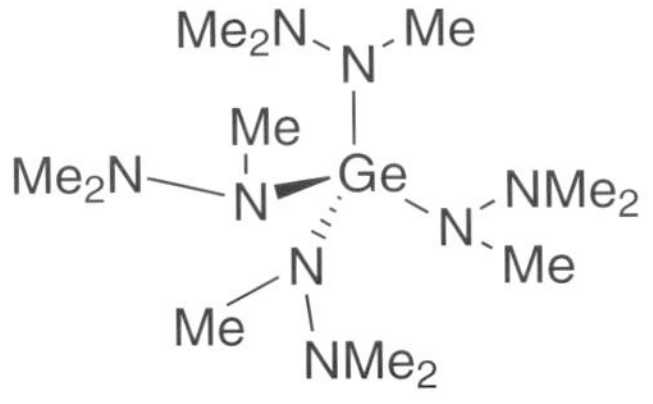

$$
\begin{array}{c}
Me_2N \diagdown _N \diagup ^{Me} \\
Me \mid \\
Me_2N{-}N{\blacktriangleright}Ge \diagdown _N \diagup ^{NMe_2} \\
\mid \diagup N \diagdown \quad Me \\
Me \mid \\
NMe_2
\end{array}
$$

(31)

The lead amide $Bu^t{}_3PbNH_2$ has been prepared by the reaction of $Bu^t{}_3PbCl$ with $LiNH_2$ and the amine protons shown to be liable to further exchange reactions. With $Li[SiMe_3]$, the asymmetric secondary amine $(Bu^t{}_3Pb)(SiMe_3)NH$ can be prepared.[93] Normal coordinate analysis of the series of organolead amines Me_3PbNH_2, $(Pr^i{}_3Pb)_2NH$, and $(Me_3Pb)_3N$, together with their isotopically labeled counterparts has been carried out. In each case, a force constant for Pb–N of 1.95×10^2 Nm^{-1} is found, a value that indicates the strongly ionic character of the Pb—N bond. This finding is in good agreement with the reaction behavior of organolead amides.[94]

Germanimine **(32)** has been prepared in the reaction of the sterically stabilized germanium(II) dialkyl $Ge[CH(SiMe_3)_2]_2$ with $Me_2Si(N_3)_2$. It rapidly rearranges to the more stable silanimine **(33)** which itself dimerizes to **(34)**, which comprises two terminal germanium azides (Scheme 12). Hydrolysis of **(33)** or **(34)** yields the silatetrazole **(35)**, comprising a single exocyclic germanium azide.[95] Taking $R_2Si(N_3)_2$ (R = mes, Bu^t) and two equivalents of the same germanium alkyl gives digermanimines **(36)**, and for the case where R is mes, the bond length $Ge–N_\alpha$ is 1.681 Å. In the case where R is Bu^t the corresponding rearrangement gives **(37)**, a cyclic silanimingermane (Scheme 13).[96] Similarly, germanimine **(38)** can be prepared from mes_2GeBr_2 and $Li[1-(NH)-2,4,6-(F)_3C_6H_2]$. In this example, the germanimine is sufficiently stable to be isolated, and the Ge=N double bond reactivity can be explored. The system is susceptible to a number of typical double bond reactions, such as the addition of chloroform to give the secondary amine $(mes)_2(CCl_3)GeN(H)C_6H_2F_3$, or of nitrone to give **(39)**, a $[GeN_2OC]$ germacycle (Scheme 14).[97] The reaction of $mesN_3$ with $Sn[2,4,6-(CF_3)_3C_6H_2]_2$ gives a product with a ring structure, with no Sn–N double bond **(40)**.[98] The sterically encumbered germanium(II) amine $Ge[N(SiMe_3)_2]_2$ reacts with primary arylamine $(2,6-Pr^i{}_2C_6H_3)NH_2$ to give the planar germazane **(41)**, where each germanium(II) ion has two (RN–) bridging groups, leading to a two-dimensional assembly.[99] The bond lengths Ge–N are 1.859 Å (ave.) and the angles N–Ge–N and Ge–N–Ge are 101.8 and 138.0°, respectively, showing a departure from the geometry of an ideal six-membered ring.

The simple homoleptic M^{II} amides $M^{II}(NMe_2)_2$ (M = Sn, Pb) preferentially dimerize with one terminal and two bridging amides on each metal. Both compounds are thermally unstable and decompose to a variety of intractable products.[100]

The tin amide $Sn^{II}(NMe_2)_2$ reacts with primary amines RNH_2 to form cubane complexes $[SnNR]_4$. These cubanes are three-dimensional arrays formed in contrast to the planar germazane **(41)** as a result of the ability of the larger tin(II) ion to accommodate three nitrogens in its coordination sphere. A range of amines can be use to prepare these cubes, including some relatively nonacidic examples.[101] The amines can also comprise some secondary functionality that can be further exploited, such as the complexes **(42)**, where the amines carry groups that promote further association of the cubes in a controlled fashion in the solid state.[102] Addition of sterically demanding amines RNH_2 to $Sn(NMe_2)_2$ (R = mes, $2,6-Pr^i{}_2C_6H_3$) gives complexes such

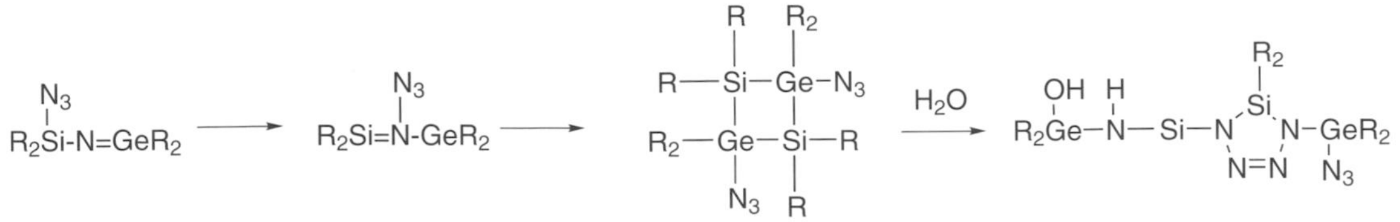

$$
\begin{array}{ccc}
N_3 & & N_3 \\
\mid & & \mid \\
R_2Si{-}N{=}GeR_2 & \longrightarrow & R_2Si{=}N{-}GeR_2
\end{array}
$$

(32)   **(33)**

$$
\begin{array}{c}
R \quad R_2 \\
\mid \quad \mid \\
R{-}Si{-}Ge{-}N_3 \\
\mid \quad \mid \\
R_2{-}Ge{-}Si{-}R \\
\mid \quad \mid \\
N_3 \quad R
\end{array}
\xrightarrow{H_2O}
\begin{array}{c}
\qquad\qquad R_2 \\
OH \; H \qquad \mid \\
\mid \quad \mid \quad Si \\
R_2Ge{-}N{-}Si{-}N \diagup \diagdown N{-}GeR_2 \\
\qquad\qquad N{=}N \quad N_3
\end{array}
$$

(34)   **(35)**

Scheme 12

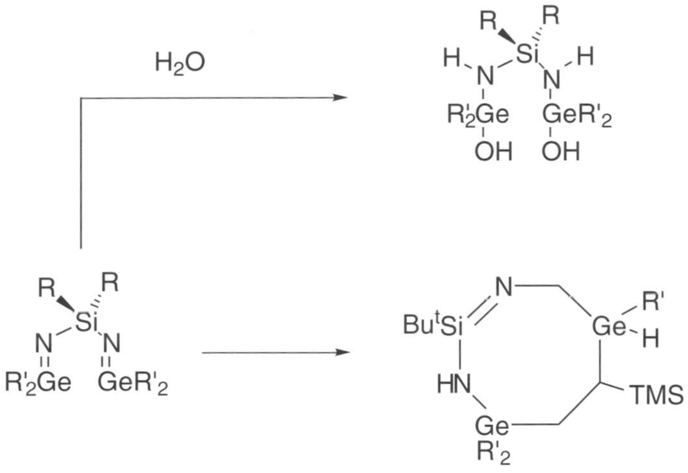

(36)

(37)

Scheme 13

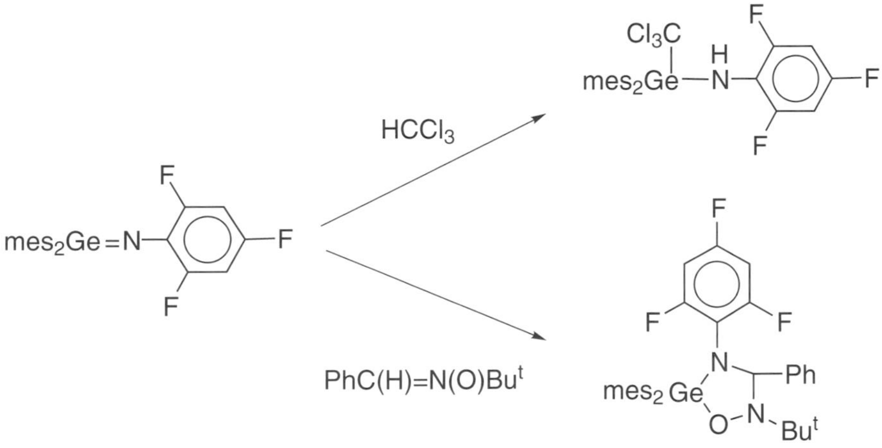

(38)

(39)

Scheme 14

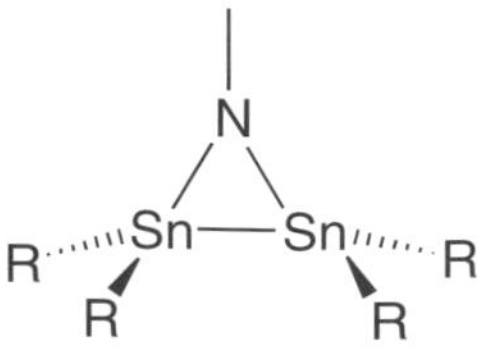

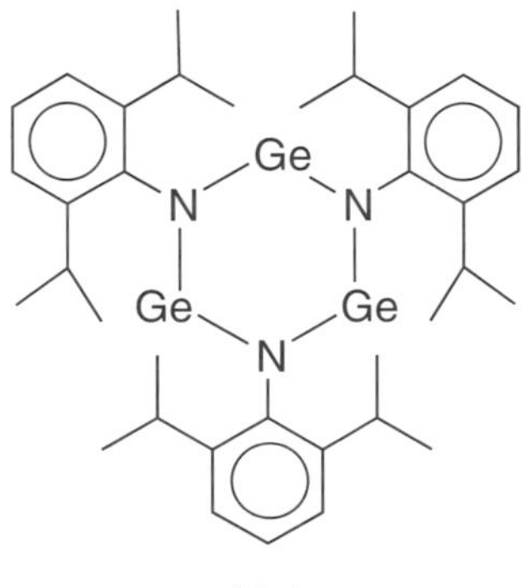

(40)

(41)

as (**43**), which may represent intermediates along the reaction pathway that leads to the formation of cubane complexes and as such may offer an insight into the reaction mechanism.[103]

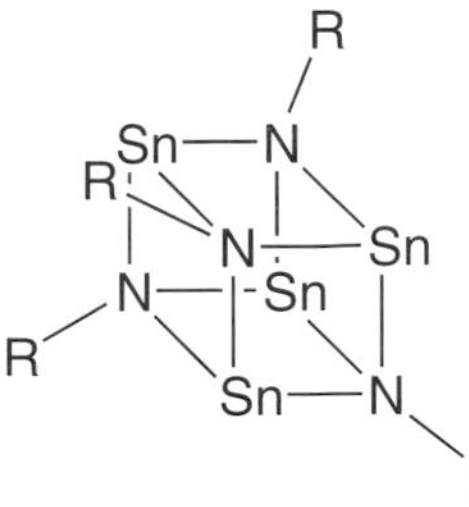

R

—⟨O⟩—OMe

MeO—⟨O⟩

(**42**)

—⟨O⟩—OMe

—⟨O⟩—OMe

MeO

—N⟨   ⟩O

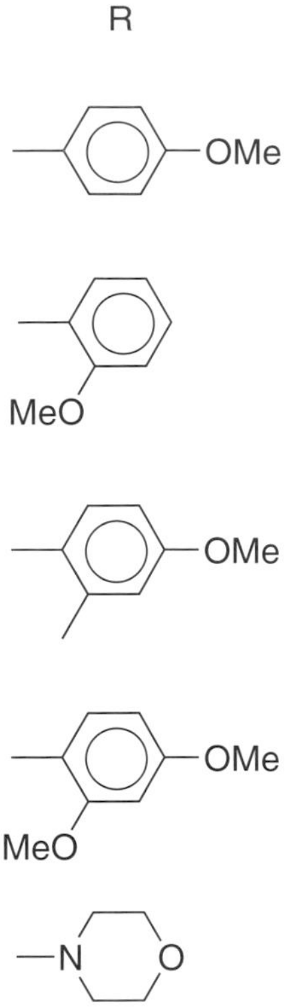

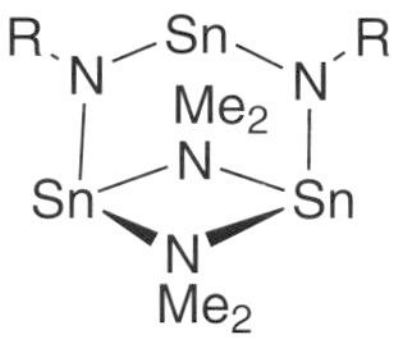

(**43**)

Such tin cubanes are prone to substitution reactions with oganolithium reagents, so the reaction of [(Bu^{t}N)Sn]$_4$ with six equivalents of Li[naphthyl] forms the vertex-substituted (**44**). The same cubane reacts with Li[(Cy)HP] to give [{(CyP)$_3$Sn$_2$}$_2$(Li(THF)$_4$]·THF$_2$, the first structurally characterized polyphosphinidine tin(II) anion.[104]

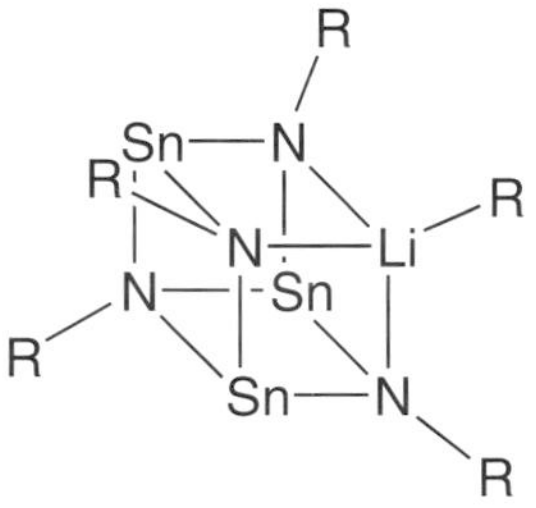

R = naphth

(**44**)

The reaction of benzonitrile, Bu^{t}Li and lead(II) chloride is the unexpected (THF)Li[{(Ph)-(Bu^t)C=N}$_3$Pb], a rare lead imino complex, in which the alkyl lithium reagent has added across the nitrile triple bond and generated a lithium imide that has reacted *in situ* with the lead(II) chloride. The anionic complex has a pyramidal geometry at the lead, and in the solid state, the lithium ion bridges the three nitrogens on the opposite face to the lead.[105]

3.7.4.3 Complexes with Neutral Bidentate Nitrogen Ligands

Ligands based upon bipyridyl have extensive application in coordination chemistry and have long been a popular subject for study. These ligands can form stable complexes with metals from

Table 2 NMR data for diimine complexes of germanium(IV).

Complex	$\delta^{73}Ge$	$\Delta\nu_{1/2}$(Hz)	$\delta^{14}N$
BipyGeCl$_4$	−313.7	35	
BipyGeCl$_3$(NCS)	−319.5	22	−266.1
BipyGeCl$_2$(NCS)$_2$	−327.1	35	−232.9
BipyGeCl(NCS)$_3$	−340.2	32	−237.5
BipyGe(NCS)$_4$	−351.8	48	−242.2
1,10-phenGeCl$_4$	−319.4	150	
[Ge(NCS)$_6$]$^{2-}$	−442.5		

throughout the periodic table, and Group 14 is certainly no exception, with a range of complexes available with such diimine ligands.

The enthalpies of formation of the complexes M(14)Cl$_4$(bipy) (M(14) = Ge, Sn) were determined by calorimetric methods.[106] The germanium complex reacts with KSCN to form the series of complexes (bipy)GeCl$_x$(NCS)$_{4-x}$[107] and the values of δ (^{73}Ge) for these complexes are given in (Table 2). Despite the great utility of ^{119}Sn NMR, there have been relatively few reports of research where ^{73}Ge NMR is used as an analytical technique, and the values given for these compounds are amongst the earliest such data reported. There is a linear change in δ with the change in electronic character of the donor set, a phenomenon well known in NMR but seen very rarely for this nucleus. In the same report, the value of δ for [Ge(NCS)$_6$]$^{2-}$ is also given, −442.5, the first value for a ^{73}Ge NMR chemical shift ever reported for a six-coordinate complex.[108]

The structure of SnCl$_4$(phen) closely approaches a regular octahedron[109] whereas the mixed organohalo complexes ROC(O)CH(Me)CH$_2$SnCl$_3$(L) (L = bipy, phen) exhibit a strongly distorted octahedral geometry at the tin despite having changed only one halide for a relatively sterically undemanding organic ligand.[110]

A range of complexes of general formula R$_2$SnCl$_2$(L) (L = bipy derivative) have been reported, research that is driven at least in part by the reported antitumor activity of complexes Me$_2$X$_2$Sn(L) (X = Cl, Br, I, L = bipy, phen).[111] For the majority of these a *trans* disposition of the two R groups is observed (L = bipy, R = Me, Bu^n, Pr^i, bn,[112] Pr^i[113], Bu^n, CH$_2$CH$_2$CN,[114] Ph;[115] L = 4,4′-Me$_2$-bipy, R = C$_5$H$_8$[116]). An exception to this seems to be (4-tolyl)$_2$SnCl$_2$(bipy), where crystallographic investigation reveals a *cis* disposition of the two organic ligands and *trans* halides.[117] For the complex (4-ClC$_6$H$_4$)$_2$SnCl$_2$(4,4′-Me$_2$-bipy), both isomers can be prepared by varying the reaction conditions. Addition of the ligand to an ethanolic solution of (4-ClC$_6$H$_4$)$_2$-SnCl$_2$ results in the *cis* form, which can be recrystallized from hot methanol to give the *trans* form exclusively.[118] The first example of *mer* coordination in a six-coordinate triorganotin compound is found in (3,4,7,8-Me$_4$-phen)Ph$_3$Sn(TfO).[119]

A less conventional, but nonetheless interesting, set of complexes of tin with bipy or phen can be prepared from the reaction of R$_3$SnCl with the diimines. The complexes [R$_3$SnCl(H$_2$O)(L)]$_2$ have two five-coordinate tin centers with equatorial disposition of the three phenyl groups and the chloro and the *O*-bound water in the *trans* positions. Each water is hydrogen bonded to the nitrogens of two diimine ligands, and these diimines are further hydrogen bonded to the protons on the aqua ligand of a second R$_3$SnCl(H$_2$O) to give a hydrogen-bond linked dimer (R = Ph,[120] 4-ClC$_6$H$_4$[121]).

The bimetallic species bis(dichloromethylstannyl)methane coordinates one equivalent of bipy to form the asymmetric (**45**) with both four- and six-coordinate tin centers (Scheme 15). Bis(chloro-dimethylstannyl)methane has been used to explore the bridging capability of pyrazine ligands (Scheme 16).[122]

Cationic complexes [R$_3$Sn(L)]ClO$_4$ (R = Ph, Bu^n; L = py$_2$, γ-picoline$_2$, bipy, phen) are available from the reaction of R$_3$SnCl with (L)AgClO$_4$ or directly from R$_3$SnClO$_4$ with (L). The complexes are 1,1 electrolytes in solution, and spectroscopic evidence indicates a tbp geometry with axial disposition of the *N*-donors for the monodentate ligands.[123]

Bisimidazoles also present a chelating diimine donor set and are similarly effective ligands. The six-coordinate complexes (**46**) (X = Cl or Br; R = Me,[124] Et, or Bu^n[125]) exhibit *trans* R groups in the same way as do complexes of bipy derivatives. For the complex where X is Br and R is Me, the average Sn—N bond length is 2.305 Å and the overall geometry is close to octahedral. In the same study, dimeric complexes (R$_2$SnX$_2$)$_2$(μ-*N*,*N*′-dimethyl-bisimidazole) comprising bridging bisimidazoles and five-coordinate tin centers are also reported, along

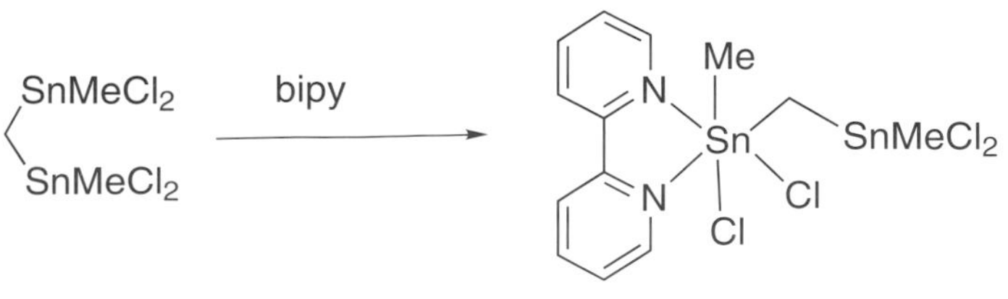

(45)

Scheme 15

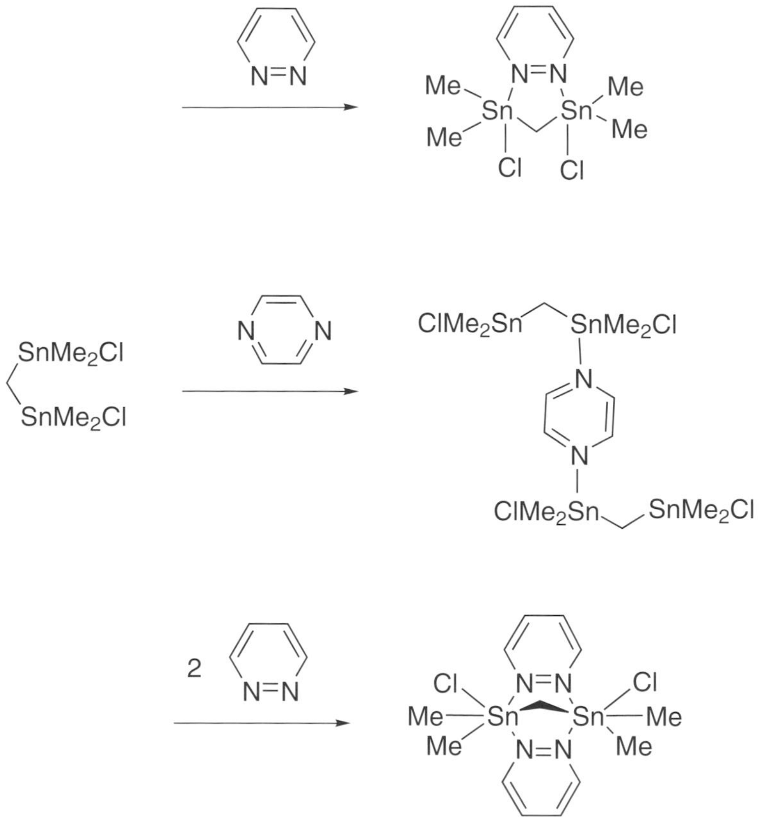

Scheme 16

with the complex ions $(NEt_4)_2[(R_2SnX_3)_2(\mu\text{-}N,N'\text{-dimethyl-bisimidazole})]$. The latter species also have bridging imidazoles but comprise six-coordinate tin centers in the solid state. The NMR data indicate that these dinuclear complex ions dissociate into monomers in solution. The ability to bridge metal centers in this way is a property of bis-imidazoles that is distinct from bipy derivatives.

Arylazo-2-pyridines coordinate R_2SnCl_2 (R = Me, Ph) to give the octahedral complexes (**47**). Spectroscopic evidence indicates that again a *trans* disposition of R groups is preferred. The angles C–Sn–C were calculated based on Parish's relationship as 148–155° (R = Me) and 148–150° (R = Ph).[126] The similar (**48**) shows trans methyl groups according to a crystallographic study, where the 2-MeO moiety has no bonding interaction with the tin center.[127]

Tin(II) chloride complexes bipy or phen to give mononuclear (L)SnCl$_2$ complexes with distorted tbp geometries.[128] Lead(II) perchlorate coordinates four phen ligands in the nine-coordinate $[(phen)_4Pb(\kappa^1\text{-}ClO_4)]ClO_4$. The geometry at the metal is best described as a monocapped square antiprism.[129] With lead(II) thiocyanate, phen gives a dimeric complex $[(phen)_2Pb(SCN)_2]_2$ where each lead coordinates two diimines, one monodentate, and two bridging thiocyanates, giving an overall seven-coordinate lead[130] and with the mixed thiocyanate nitrate system, a monomeric seven-coordinate complex $[(phen)_2Pb(SCN)(NO_3)]$ with a chelating nitrate.[131]

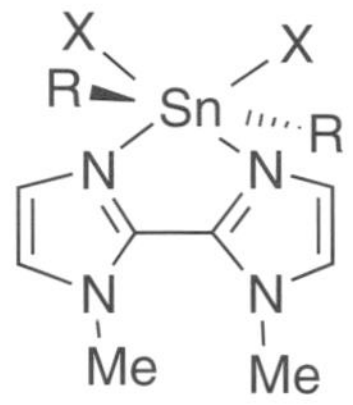

Cl
R////Sn
Cl   R
N=N
R'

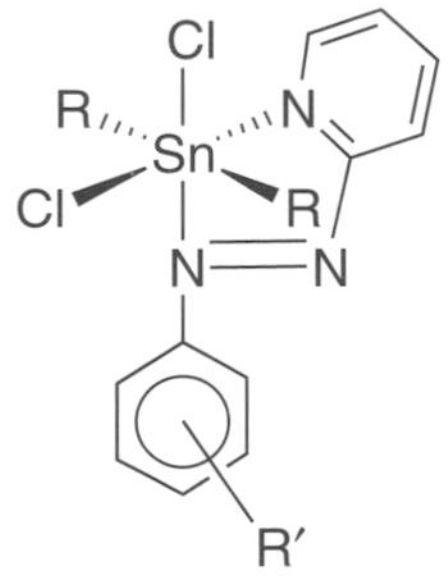

R' = H , 3-Me, 4-Me, 4-Cl

X   X
R   Sn   R
N   N
N   N
Me   Me

(46)

(47)

Me
Cl////Sn
Cl
Me
OMe

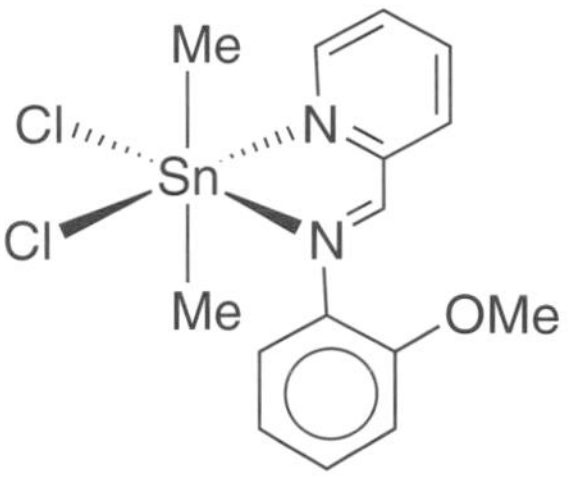

(48)

3.7.4.4 Complexes with Anionic Bidentate Nitrogen Ligands

The ligand (**49**) reacts with two equivalents of GeCl$_2$·diox to give the linear (**50**), with two-coordinate germanium centers (Ge–N 1.856 Å, N–Ge–N 80.93°) (Scheme 17). Tin, in the corresponding reaction, forms the complex (**51**), which has a geometry more closely related to a cubane structure (Sn–N 2.247 Å, N–Sn–N 110.8°).[132]

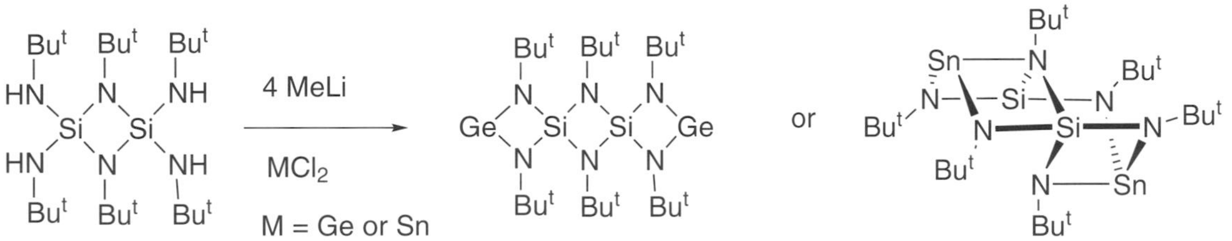

Bu^t Bu^t Bu^t
HN   N   NH
Si   Si
HN   N   NH
Bu^t Bu^t Bu^t

4 MeLi
——————→
MCl$_2$
M = Ge or Sn

Bu^t Bu^t Bu^t
N   N   N
Ge   Si   Si   Ge
N   N   N
Bu^t Bu^t Bu^t

or

Bu^t
Sn——N   Bu^t
N–Si–N
Bu^t N–Si—N–Bu^t
Bu^t N——Sn
Bu^t

(49)

(50)

(51)

Scheme 17

The dianionic ligand 1,3-Bu$^t_2$-2,2-Me$_2$-4,4-Cl$_2$-1,2,3,4-λ-4-diazasilide has been used to prepare stannetidine (**52**). Reaction with AgTfO yields the unexpected (**53**), where the product rearranges following halide-TfO metathesis to allow the formation of a thermodynamically favorable Si—O bond. In the solid state (**53**) is a coordination polymer and comprises tbp tin centers. The Sn—N bonds are short (2.005 Å) probably due to the geometric constraints of the [SiN$_2$Sn] ring.[133] Compound (**52**) also reacts with HX (X = Cl, Br) to give the three-coordinate addition product (**54**), formulated as a Zwitterion with no oxidation of the metal.[134] (RC(Ncy)$_2$)$_2$Sn (**55**)[135] is converted to the thione (**56**) by sulfide transfer from styrene sulfide (Sn–S 2.28 Å) (Scheme 18). The stannathione rapidly dimerizes with formation of a (SnS)$_2$ ring (Sn–S 2.42–2.47 Å).[136] Oxidative addition of diphenyl chalcogenide PhEEPh (E = S, Se) to [(CyN)$_2$CR]$_2$M^{II} (M = Ge, Sn; R = Me, Bu^t) proceeds with cleavage of the E—E bond and addition of both phenyl chalcogenide fragments to the metal, giving six-coordinate products [(CyN)$_2$CR]$_2$M^{IV}(EPh)$_2$. For the mixed amidinate amide complexes [(CyN)$_2$CR][N(SiMe$_3$)$_2$]M^{II} the reaction proceeds

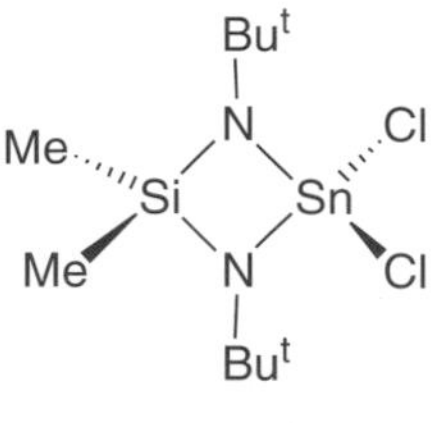

Bu^t
N   Cl
Me////Si   Sn
Me   N   Cl
Bu^t

Bu^t
N
CF$_3$SO$_3$////Si   Sn   O$_3$SCF$_3$
Me   N   Me
Bu^t

Bu^t + H
N   Cl
Me////Si   Sn—X
Me   N   Cl
Bu^t

(52)

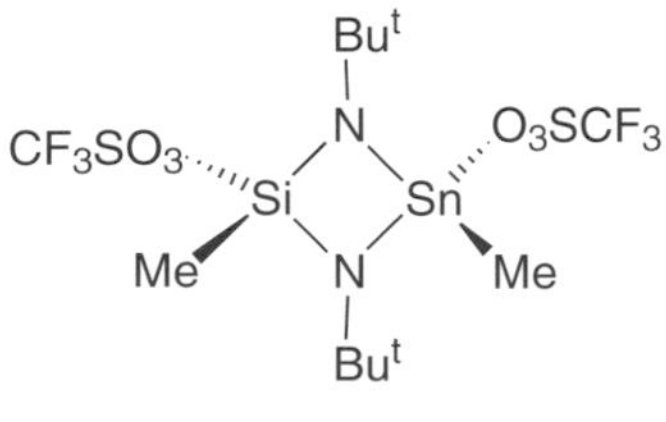

(53)

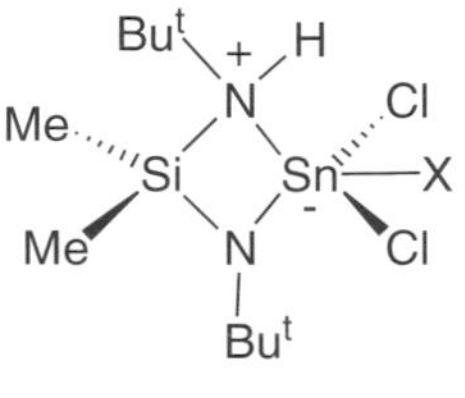

(54)

in a similar fashion, giving $[(CyN)_2CR][N(SiMe_3)_2]M^{IV}(EPh)_2$. For the germanium complex, spectroscopic and crystallographic data show that the complex is four-coordinate, in which the amidinate ligand is coordinated through one nitrogen only, whereas the tin complex is five-coordinate with the chelating amidinate intact.[137]

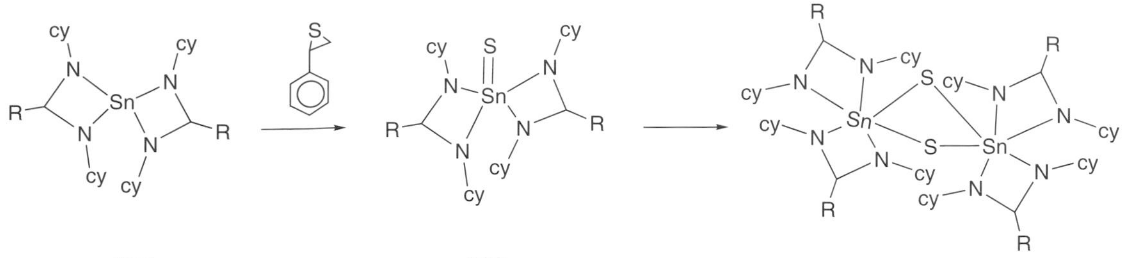

(55)   (56)

Scheme 18

Aminoiminophosphoranes $Bu^t_2P(NH)(NH_2)$ and $(H_2NPPh_2)(Ph_2PNH)N$ react with diamino-stannanes $R_2Sn(NEt_2)_2$ (R = Me, Bu^n) to give tricyclic stannaphosphazenes (57) and (58) comprising fused $[SnN_2]$, $[SnN_2P]$, and $[SnN_3P_2]$ rings.[138]

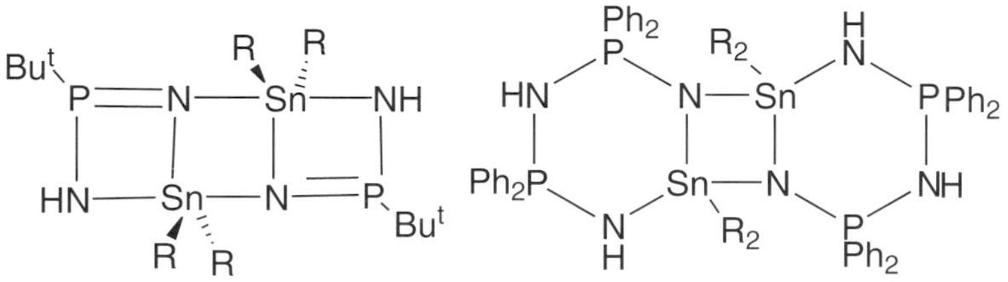

(57)   (58)

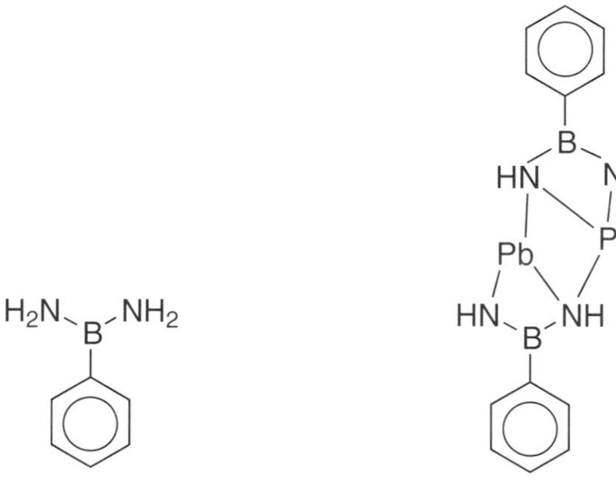

(59)   (60)

The dilithium salt of aminoborane (59) reacts with lead(II) chloride to give the dimeric (60) in which the lead is chelated by one ligand and has a bridging interaction with a nitrogen from a second ligand. The lead is three-coordinate and is stable in this configuration.[139] The geometry is similar to that found in $[Pb(NR_2)]_4$ complexes.[140]

3.7.4.5 Complexes with Polypyrazolyl Ligands

Polypyrazolylmethyl ligands and their anionic analogues polypyrazolylborohydrides exhibit a range of coordination behavior with M(14) ions. The complexes that they form with M(14) ions are usually mononuclear, where they coordinate through either two or three nitrogens in a chelate or facial tridentate fashion.

Bis(pyrazolyl)methanes are neutral chelating ligands that form six membered rings upon coordination. The complexes $(\kappa^2\text{-}H_2Cpz_2)R_{4-x}SnX_x$ (R = Me, Ph; X = Cl, Br; x = 4,3,2) show fluxional behavior in solution at ambient temperatures by NMR spectroscopy. Additionally, in acetone solution there is evidence from conductivity measurements that the pyrazolyl ligand fully dissociates.[141] Furthermore, in the range of complexes $(\kappa^2\text{-}L)R_xSnX_{4-m} \cdot yH_2O$ (L = H_2C (4-Me-pz)$_2$, $H_2C(3,4,5\text{-}Me_3\text{-}pz)_2$, $(H_2Cpz)_2$, $H_2C(3,5\text{-}Me_2\text{-}pz)_2$; R = Me, Et, Bu^n, Ph; X = Cl, Br, I; n = 0, 1, 2; y = 1, 1.5, 2) conductivity measurements taken on chloroform or acetone solutions suggest that when x is two, there is extensive dissociation of the chelating ligand, while in the cases where x is one or four, the six-coordinate structure is retained in solution. Structural analysis of the octahedral complex $[\kappa^2\text{-}H_2C(4\text{-}Me\text{-}pz)_2]Me_2SnCl_2$ shows again a trans disposition of the two methyl groups and Sn—N bond lengths of 2.436 Å (ave.).[142,143] The molecular structure of $[\kappa^2\text{-}H_2C(3,5\text{-}Me_2\text{-}pz)_2]Ph_2SnCl_2$ exhibits a distorted octahedral geometry with the phenyl groups disposed trans and the chelating ligand has typical Sn—N bond lengths (2.448 and 2.520 Å). The complex also shows moderate activity against L1210 mouse leukemia cells with LD$_{50}$ of 0.39 μM.[144]

Complexes $(L)R_2SnCl_2$ (L = H_2Cpz_2, $HCpz_3$, $HC(3,5\text{-}Me_2\text{-}pz)_3$, 1,2-py-3,5-Me$_2$-pz, $H_2C(2\text{-}py)_2$; R = Me, Et, Pr^n) are six-coordinate nonelectrolytes in acetonitrile solution with trans organic groups. In chloroform solution, all the bidentate ligands dissociate, whereas the tris(pyrazolyl)-methanes remain coordinated, but are chelated through two of the nitrogen donors.[145] Complexes $(L)R_nSnX_{4-n}$ (L = H_2Cpz_2, $H_2C(3,5\text{-}Me_2pz_2)$, Me_2Cpz_2, R = Me, Ph; X = Cl, Br; n = 0, 1, 2) have been examined by ^{119}Sn Mössbauer spectroscopy and the coordination number and geometry of the complexes assigned with success based on this technique. All of the complexes have six-coordinate geometry incorporating chelation of the pyrazolyl ligands with the exception of the complexes $(Me_2Cpz_2)R_2SnX_2$ where the decreased Lewis acidity of the metal and the increased size of the ligand combine to prevent the formation of this complex.[146]

The tris(pyrazolyl)methanes $HCpz_3$, $HC(4\text{-}Me\text{-}pz)_3$, $HC(3,5\text{-}Me_2\text{-}pz)_3$, $HC(3,4,5\text{-}Me_3\text{-}pz)_3$, and $HC(3\text{-}Me\text{-}pz)_2(5\text{-}Me\text{-}pz)$ react with RSnCl$_3$ (R = Me, Bu^n, Ph) to form complexes $[(L)SnRCl_2]^+[RSnCl_4]^-$ and $\{[(L)SnRCl_2]^+\}_2[RSnCl_5]^{2-}$ and with SnX$_4$ (X = Cl, Br, I) to form $[(L)SnX_3]^+[RSnCl_4]^-$ and $\{[(L)SnCl_3]^+\}_2[RSnCl_5]^{2-}$, respectively. The structures of $\{[HC(4\text{-}Me\text{-}pz)_3SnBuCl_2]^+\}_2[BuSnCl_5]^{2-}$, $[HC(3,5\text{-}Me_2\text{-}pz)_3SnMeCl_2]^+[MeSnCl_4]^-$, and $[HC(3,4,5\text{-}Me_3\text{-}pz)_3SnBr_3]^+[SnBr_5]^-$ show distorted octahedral environments for the tin centers in the cations, with bond lengths Sn—N of 2.22–2.32 Å, and tbp or distorted octahedral environments for the five- and six-coordinate anions, respectively.[147] The sterically demanding ligand $HC(3,5\text{-}Me_2\text{-}pz)_3$ forms complexes of general formula $[\kappa^3\text{-}HC(3,5\text{-}Me_2\text{-}pz)_3M^{II}]Y_2$ (M = Sn, Y = CF$_3$SO$_3^-$, X; M = Pb, Y = BF$_4^-$, X). In both cases, the metal is three coordinate with distances Pb—N of 2.379–2.434 Å from a crystallographic study. Treatment of PbX with excess $HC(3,5\text{-}Me_2\text{-}pz)_3$ in acetone solution gives $[\{\kappa^3\text{-}HC(3,5\text{-}Me_2\text{-}pz)_3\}_2Pb][BF_4^-]_2$, which exhibits a trigonally distorted octahedral geometry and Pb—N distances of 2.634 Å. This particular geometry indicates that the remaining lone pair is not stereochemically active in this complex, in contrast to the corresponding complex prepared using the unsubstituted ligand, which exhibits a distorted six-coordinate geometry with distance Pb—N of 2.609–3.789 Å. The closely related $[HB(3\text{-}Bu^t\text{-}5\text{-}Me\text{-}pz)_3]SnCl$ shows a tbp structure with an axial Sn—Cl of 2.601 Å and an equatorial stereochemically active lone pair.[148]

Reaction of PbX with $K[HB(pz)_3]$ or $K[HB(3,5\text{-}Me_2\text{-}pz)_3]$ leads to the mixed ligand complexes $[\{\kappa^2\text{-}HC(3,5\text{-}Me_2\text{-}pz)_3\}(\kappa^3\text{-}L)Pb]BF_4^-$. Where L = $[HB(3,5\text{-}Me_2\text{-}pz)_3]$ the complex has a five-coordinate geometry where the pyrazolylmethane is bidentate, and shows longer Pb—N bonds (3.745–2.827 Å) than the pyrazolylborohydride Pb—N (2.375–2.475 Å).[149]

Reaction of $Pb(ACAC)_2$ with two equivalents of HB(Arf)$_4$ in CH$_2$Cl$_2$ followed by HC(pz)$_3$ or $HC(3,5\text{-}Me_2\text{-}pz)_3$ gives the complexes $[L_2Pb]^{2+}[B(Arf)_4]_2^-$ (Arf = 3,5-(CF$_3$)$_2$-C$_6$H$_3$). For the HC(pz)$_3$ complex, the lead has a distorted octahedral geometry with a stereochemically active lone pair, while the complex with the more sterically encumbered $HC(3,5\text{-}Me_2\text{-}pz)_3$ has a trigonally distorted octahedral structure with a stereochemically inactive lone pair.[150]

The anionic polypyrazolylborohydride complexes might be expected to show some similarities to the neutral polypyrazolylmethanes. However, some differences are seen, e.g., the diethyl bis(pyrazolyl)borate $K[Et_2B(pz)_2]$ reacts with Et_2GeCl_2 with cleavage of the B—N bonds to yield $Et_2B(pz)BEt_2$ and $Et_2Ge(\kappa^1\text{-}pz)_2$.[151]

Di- and triorgano tin complexes of dihydro and diphenyl bis(pyrazolyl)borates are chelated through the two nitrogen donors to give five-coordinate complexes.[152] Bis- and tris(pyrazolyl)borohydride form the complexes [H$_m$B(pz)$_l$]Me$_n$SnCl$_{3-n}$ ($l = 2, 3$; $m = 1, 2$; $n = 0$–3) that are five-($m = 2$) or six-($m = 1$) coordinate. The solid-state structure of [H$_2$B(pz)$_2$]Me$_2$SnCl shows tbp geometry at the tin with axial Cl and nitrogen ligands.[153] [HB(3-Me-pz)$_3$]SnCl$_2$Ph is stereochemically rigid at ambient temperatures.[154]

The first tris(pyrazolyl)borohydride complex of germanium(II) was prepared directly from K[HB(3,5-Me$_2$-pz)$_3$] and GeCl$_2$·diox and isolated as its iodide [κ^3-HB(3,5-Me$_2$-pz)$_3$Ge]I[155] and cyanide salts.[156] The solid-state structure of the iodide reveals well separated germanium containing cations and iodide anions, and that the geometry at the germanium is distorted tetrahedral, with the lone pair occupying the fourth vertex.

The three-coordinate complex [κ^2-H$_2$B(pz)$_2$]SnCl has in the solid state a trigonal pyramidal geometry at the tin with the angles about the tin averaging 86°, indicating that the lone pair is again stereochemically active and occupies the fourth equatorial vertex. There are indications of a weak interaction between the tin and the chloro ligand of a neighboring molecule in the crystal. The structure of [κ^2-H$_2$B(pz)$_2$]$_2$Sn is approximately tbp, with the fifth vertex occupied by the lone pair. In solution, ^{1}H NMR experiments show that the molecule is stereochemically nonrigid, with axial and equatorial sites exchanging with a barrier of 10.2 Kcal mol^{-1} and boat-boat rearrangements taking place with similar energy barriers. Comparable fluxional processes to these are also seen for HB(3,5-Me$_2$-pz)$_3$]SnCl.[157] In the same study, [B(pz)$_4$]$_2$Sn was found to have two chelated ligands and a tbp structure in the solid state, with similar fluxional behavior. The low temperature limit NMR spectrum shows a 3:1 pattern indicative of a structure with three equilibrating pyrazolyl groups that interact with the tin and a fourth that does not.

A number of tin(II) complexes of general formula [HB(pz)$_3$](L)SnCl and (L)$_2$Sn have been reported (L = H$_2$B(pz)$_2$, H$_2$B(3-Me-pz)$_2$, Ph$_2$B(pz)$_2$, HB(pz)$_3$, HB(3,5-Me$_2$-pz)$_3$, B(pz)$_4$, B(3-Me-pz)$_4$), and their solution[158] and solid-state ^{119}Sn NMR spectra measured.[159] The values of δ are constant for [H$_2$B(pz)$_2$]$_2$Sn, and [HB(3,5-Me$_2$-pz)$_3$]$_2$Sn in both states indicating that there is a strong similarity between the structures of these complexes. The value of δ for [HB(pz)$_3$]$_2$Sn in the solid state is centered between these two complexes. Analysis of the spinning side band patterns for the MASNMR spectra of these complexes indicates a close similarity between [H$_2$B(pz)$_2$]$_2$Sn and [HB(pz)$_3$]$_2$Sn, whereas [HB(3,5-Me$_2$-pz)$_3$]$_2$Sn has a substantially different pattern. These results suggest that the two former complexes have the same geometry at the metal, the complex comprising a four-coordinate tin. The spectra for [H$_2$B(pz)$_2$]SnCl are distinctly different in the two states, presumably due to the presence of moderately strong bridging intermolecular Sn···Cl interactions in the solid state that are disrupted in solution, giving distinct coordination environments in the two different states. The bond lengths Sn—N in [HB{3,5-(CF$_3$)$_2$-pz}]SnCl are longer than the those in the protio complex, a reflection of the difference in the electronic character between the two ligands.[160]

The crystal structure of [HB(pz)$_3$]$_2$Sn has been determined and the tin found to exhibit an octahedral coordination geometry, with one tri- and one bidentate [HB(pz)$_3$], with the sixth coordination site occupied by the lone pair. The bond lengths Sn—N are in the range of 2.263–3.732 Å, and the nitrogens that lie *cis* to the lone pair deviate from the expected square plane, with three angles between 72.3–79.4° and one at 124.4°. This solid-state distortion correlates well with the solution ^{119}Sn NMR data which has δ between the values seen for four- and five-coordinate tin(II).[161]

The reaction of lead(II) chloride with K[H$_2$B(pz)$_2$], K[HB(pz)$_3$], K[HB(3,5-Me$_2$-pz)$_3$], and K[B(pz)$_4$] gives products (L)$_2$Pb in all cases. For [B(pz)$_4$]$_2$Pb, the structure is a distorted tbp, with two chelating ligands and an equatorial vertex occupied by the inert pair. As is seen with the tin analogue, the room temperature NMR indicates that the molecule is stereochemically nonrigid with four equilibrating pyrazolyl groups, but at 184 K, a 3:1 pattern is seen. For [HB(pz)$_3$]$_2$Pb the structure in the solid state is a monocapped octahedron, with the inert pair occupying the capping position.

A comparison of the structures of the neutral bispolypyrazolyl methane and -polypyrazolylborohydride complexes of lead(II) have similar structures, indicating that to some extent the structurally similar ligands do form similar complexes.[162]

The issue of the stereochemical activity of the inert pair is explored further in the complexes of tin(II) and lead(II) with [HB(pz)$_3$], [HB(3,5-Me$_2$-pz)$_3$], and [B(pz)$_4$]. For both metals, the complexes [B(pz)$_4$]$_2$M^{II} are four coordinate with two chelating ligands whereas the geometry of [HB(3,5-Me$_2$-pz)$_3$]$_2$Sn is close to octahedral with one tri- and one bidentate ligand, with the inert pair occupying the sixth position.[163] In comparison, [HB(3,5-Me$_2$-pz)$_3$]$_2$Pb is six coordinate with a trigonally distorted octahedral geometry, indicating the absence of a stereochemically active lone pair.[164,165] There is a complex interplay of factors which govern whether the inert pair is stereochemically active or not for complexes of $M(14)^{II}$ (M = Sn, Pb) and it seems clear from the experimental observations that lead(II) is more likely to be influenced by the steric effects of the ligand set than is tin(II).

An example of an eight-coordinate lead(II) complex is given by $[HB(1,2,4\text{-triazolyl})_3]_2$-$Pb(H_2O)_2$. From the same reaction mixture, a polymeric material $[\{HB(1,2,4\text{-triazolyl})_3\}$-$(\kappa^1\text{-}NO_3\text{-}O)(H_2O)Pb]_\infty$ was identified crystallographically and the geometry at the lead suggested that the metal had a stereochemically active lone pair.[166]

3.7.4.6 Complexes with Tridentate Nitrogen Ligands

Lead(II) thiocyanate forms a linear polymeric compound with $2,2':6',2''$-terpyridine(terpy), $[(\text{terpy})Pb(SCN)_2]_\infty$, comprising seven-coordinate lead centers coordinating three pyridyl nitrogens and two *N*-bound and two *S*-bound bridging thiocyanates.[167]

The tridenate 2,6-diacetylpyridine dihydrazone reacts directly with a solution of lead(II) nitrate to give $(L)Pb(NO_3)_2$ where all three nitrogens coordinate the lead (Pb–N_{py} 2.49 Pb–N 2.50, 2.59 Å) and each lead is further chelated by two nitrates in an asymmetric fashion (Pb–O 2.52, 2.86; 2.58, 2.93 Å) and has two monodentate interactions with the third nitrato oxygen of a neighboring molecule (Pb–O 3.08–3.19 Å). The overall coordination number of the lead in the solid is nine, of which five are short contacts, and four are comparatively long.[168] Triazacyclo-nonane forms 1:1 complexes with lead(II) of good stability, with irregular six-coordinate geometry at the lead in the case of the nitrate and perchlorate salts.[169]

3.7.4.7 Complexes with Tetradentate Nitrogen Ligands

Tetraaza macrocycles are tremendously powerful ligands and have wide application in coordination chemistry. The M(14) complexes of these ligands have a number of interesting applications, although the importance of these have only recently come to light. For this reason, the M(14) complexes of these ligands are not as widely studied as the corresponding complexes of metals from some other groups.

Porphyrins have a demonstrated propensity to accumulate in cancerous tissues, metal alkyls are powerful alkylating agents, and elemental germanium has been shown to have anticancer properties, so it is not to be wondered at that dimethyl-5,10,15,20-tetrakis($3',5'\text{-}Bu^t_2C_6H_3$)porphyrinato-germanium(IV) **(61)** has been prepared. The germanium is six coordinate, with *trans* methyl groups (Ge–C 1.99 Å, Ge–N 2.02 Å) and the complex has been shown to be active against neoplastic tissues both *in vitro* and *in vivo*.[170,171]

(61)

Germanium complexes $(\text{por})GeX_2$ (por = dianion of TPP, octaethylporphyrin (OEP); X = OH, ClO_4) have been shown to undergo single electron oxidation by electrochemical methods with a potential that varies according to the nature of the porphyrin. For the hydroxy complexes, the first electron is removed from the ligand, giving $(\text{por})Ge(OH)(Y)$, where Y is the anion of the supporting electrolyte, whereas for the perchlorate complexes, the first electron is removed from the π system of the porphyrin.[172,173]

Diorgano germanium porphyrins are photoactive, such that visible light irradiation of (TPP)GeR$_2$ (R = Me, Bu^n) in degassed chloroform solution gives (TPP)GeRCl.[173] EPR investigation of the mechanism of the photolysis of (OEP)GeR$_2$ (R = Ph, 4-HOC$_6$H$_4$, 4-ClC$_6$H$_4$) shows the existence of a Zwitterionic intermediate [(OEP)$^-$GeR]$^+$ R. The complex can be made photostable if R is a good quenching group, such as ferrocene, as in the photochemically robust complex (OEP)Ge(Fc)$_2$.[174]

In the absence of light, diorgano germanium porphyrin complexes are stable but have long been known to react with oxygen. The products of this oxidation process have been formulated speculatively as germanium-bound peroxyalkyl complexes[175] an assignment confirmed by the report of the crystal structure of (TPP)Ge(O$_2$R)$_2$ (R = Et, Bu^t).[176]

Polymeric [(tbp)GeO]$_n$ prepared from the thermolysis of (tbp)Ge(OH)$_2$, in turn obtained from the hydrolysis of (tbp)GeCl$_2$, is converted to a range of conducting polymers by the introduction of sub-stoichiometric quantities of iodine as [(tbp)Ge(O)$_x$(I)$_y$].[177]

Six-coordinate complexes (por)SnX$_2$ have been studied because of the ability of some of these to inhibit the enzyme, heme oxygenase, believed to be responsible for the disease hyperbilirubinemia in infants[178] and because of their potential application in photodynamic therapy.[179] Many complexes (por)SnX$_2$ are known where X is not an R group (X = F, NO$_3$,[180] OH, C$_6$H$_5$CO$_2$, 2-(OH)C$_6$HCO$_2$,[181] or N$_3$[182] but if X is a σ-bonded R group, the complexes are not generally stable. An example of a stable (por)Sn(R)(X) complex was found from the reaction of (por)Sn^{II} (por = OEP, TPP, TMP, TTP) (TTP = dianion of tetratolyporphyrin, TMP = dianion of tetramesitylporphyrin) with MeI forming (por)Sn(Me)(I).[183]

Typically complexes (por)SnR$_2$ exhibit a trans disposition of the two R groups, but *cis*-(TPP)SnPh$_2$ is available from the reaction of Ph$_2$SnCl$_2$ with the dilithium salt of the porphyrin. The complex is configurationally stable in the absence of light, but rapidly rearranges to the *trans* geometry on exposure to visible light.[184]

Bis-amido tin porphyrins *trans* (TTP)Sn(N(R)Ph)$_2$ (R = H, Ph) and *cis* (TTP)Sn(1,2-(NH)$_2$C$_6$H$_4$) have been prepared and shown to be more stable than the analogous Sn—C bonded alkyl or aryl complexes. The increased stability of these nitrogen bound ligands is probably a function of their increased basicity.[185]

The ^{119}Sn NMR spectra of a number of complexes (TPP)SnX$_2$ (X = CF$_3$SO$_3$, ClO$_4$, CF$_3$CO$_2$, NO$_3$, Cl$_2$CHCO$_2$, 2-(OH)C$_6$H$_4$CO$_2$, HCO$_2$, BnO, AcO, 4-NO$_2$C$_6$H$_4$O, 4-BrC$_6$H$_4$O, 4-MeC$_6$H$_4$O, OH, MeO, F, Cl, Br, I) have been examined, and a correlation between the change in δ with the change in axial ligand established. This is particularly interesting as it may ultimately help in establishing the nature of the axial ligands on tin porphyrin complexes *in vivo*.[186] Another spectroscopic property of tin porphyrins is exploited in the use of (por)Sn^{II}(H$_2$O)$_2$ as a shift reagent for carboxylates. Coordination of carboxylates to the tin leads to a large shift in δ and so is a sensitive and useful probe.[187]

The electrochemical activity of (por)Sn^{II} for a variety of porphyrins has been investigated and it has been shown that the first one electron reductions are all centered upon the ring system. Two-electron oxidation in the presence of supporting electrolyte comprising perchlorate leads to (por)Sn(sol)ClO$_4$ (sol = FHF, CH$_3$CN).[188]

Treating (TPP)SnE (E = S, Se) with (TTP)Sn in toluene results in the reversible transfer of the chalcogenide, forming (TPP)Sn^{II} and (TTP)SnE. The reactions proceed with second-order kinetics and seems to involve a bridging chalogenido intermediate.[186] The same result has also been found for amidinate complexes.[189]

The synthesis of lead(II) porphyrin complexes has been efficiently performed in a solid-state reaction. Grinding together equimolar amounts of (por)H$_2$ (porH$_2$ = meso-(4-HO-C$_6$H$_4$)$_4$-porphyrin, meso-(4-MeOC$_6$H$_4$)$_4$-porphyrin, or meso-(4-NO$_2$-C$_6$H$_4$)$_4$-porphyrin) with lead(II) acetate in a pestle and mortar with a trace of acetone leads to the (por)Pb^{II} complexes in excellent yield after chromatographic purification.[190]

Main group metal complexes of phthalocyanines (pc) are of some interest as one-dimensional conducting materials where the structures exhibit stacking. It is of interest that a recent reinvestigation of (pc)Ge^{II}, which shows quite different spectral characteristics from those expected, reveals that the literature preparative route does not yield this compound, as the structure of the ligand does not remain intact throughout the synthesis of the complex. The synthesis, which uses germanium(IV) chloride as a template to form (pc)GeCl$_2$ followed by borohydride reduction, yields not the anticipated (pc)Ge^{II} but the ring contracted α,β,γ-(triazatetrabenzcorrole)Ge^{IV} (**62**) a new tetrapyrrole macrocycle.[191]

Cofacially joined polymeric metallophthalocyanines with bridging oxo ligands [(pc)MO]$_n$ (M = Ge, Sn) have been prepared by a new route and are themselves precursors to electronically conductive polymers. The vibrational spectra of the polymers were investigated using isotopic substitution (^{18}O), and identification of the stretching modes has afforded a method for estimating the molecular weights of the polymers. For typical samples, the value of n for germanium is 70 and for tin is 100.[192]

Tetraaza macrocyclic ligands tetramethyl- and octamethyltetraazaanulene (TMTAA, OMTAA) (63) and (64) are analogous to porphyrins, in that they are dibasic, approximately planar N$_4$ donor ligands. Complexes (L)M^{II} (M = Ge, Sn; L = TMTAA)[193] are four coordinate and in the case of (OMTAA)Sn, crystallography shows the metal to be 1.12 Å out of the N$_4$ plane. These complexes are liable to oxidative addition reactions, such as with elemental chalcogens sulfur, selenium (Ge, Sn), and tellurium (Ge) that form the corresponding five-coordinate monochalcogenides.[194,195] In reaction with N$_2$O the tin(II) complex forms the oxo-bridged product [(OMTAA)Sn]$_2$O with no indication of a mononuclear product with a terminal oxo ligand.[196] Oxidative addition of I$_2$ to (OMTAA)Sn^{II} leads to the diiodide, which has a *trans* disposition of iodides, as shown by crystallography. This geometry is consistent with other known complexes (TMTAA)SnX$_2$, (X = Cl, ONO$_2$).[197] The bond lengths Sn-I are long (2.885 and 2.909 Å) and so it is perhaps unsurprising that one iodide is labile and, in the presence of excess I$_2$, in THF solution ionizes to form [(OMTAA)SnI(THF)]I$_3$. The tin(IV) is nearly coplanar with the four nitrogen donors, less than 0.2 Å out of plane[198] The only confirmed examples of *cis* coordination in such complexes is in the products of the reactions of (OMTAA)M(E) (M = Ge, Sn; E = S, Se) with C$_2$H$_4$S, (65).[199]

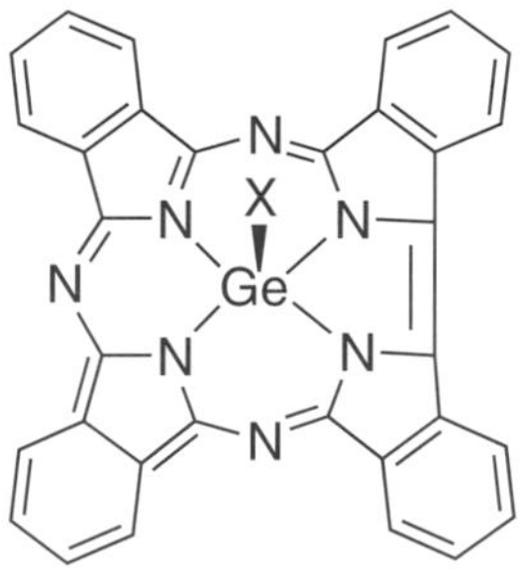

(62)

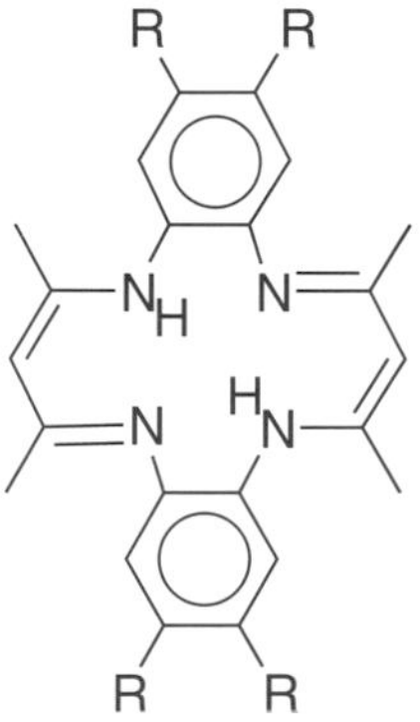

R = H, TMTAAH$_2$ (63); R = Me, OMTAAH$_2$ (64)

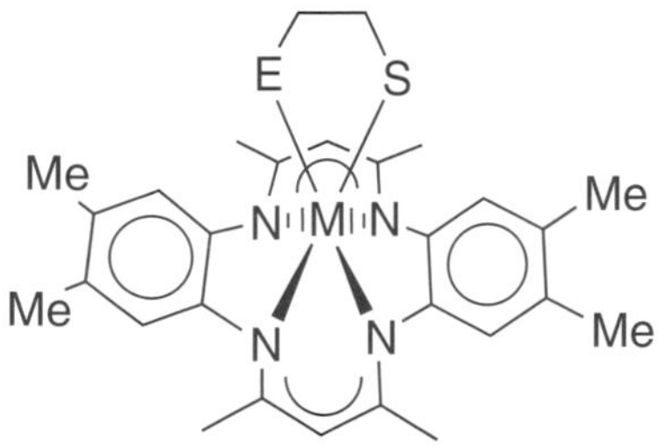

(65)

3.7.4.8 Complexes with Polydentate Nitrogen Ligands

Macrocyclic ligands with all-nitrogen donor sets are much studied and both tin and, in particular, lead are popular subjects in coordination studies of these ligands. Examples of such ligands used to complex tin include (66) and (67), prepared by Schiff-base condensations.[200] The complex (68) was isolated from an attempted template synthesis of a macrocycle[201] in which the condensation of the component parts of the ligand was incomplete.[202]

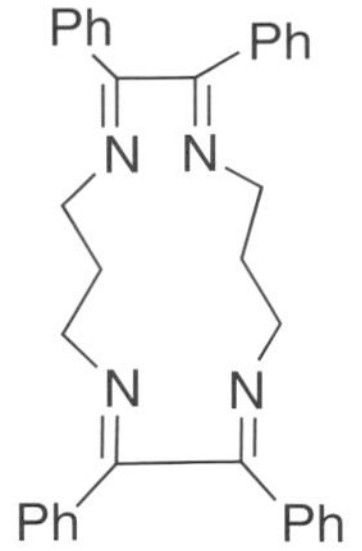

(66) 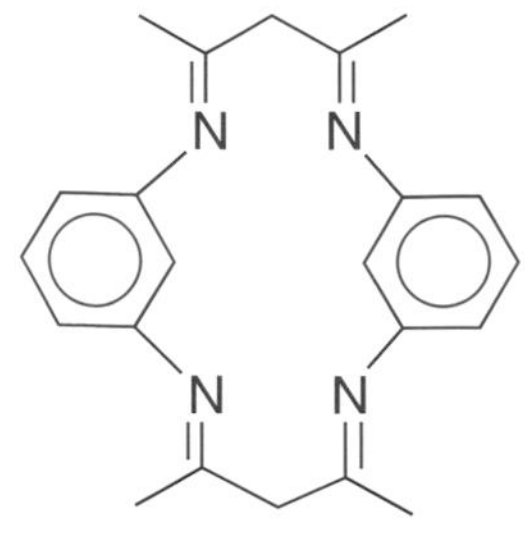   **(67)**

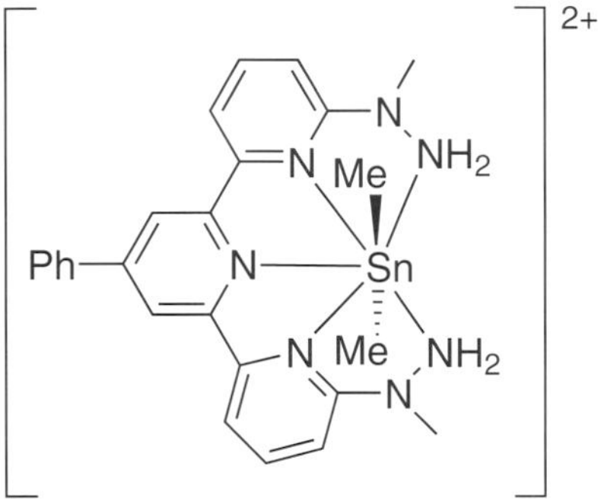

(68)

Lead(II) is a useful metal for such studies as it is relatively redox inert and has the ability to form complexes with a wide range of coordination numbers and with almost any donor atom type. Substituted triazacyclonanes (69)–(71) form 1:1 complexes with lead(II) and (72) forms a 1:2 complex in which two lead ions are coordinated, one in each of the distinct sites. In each case, the complexes were isolated as their (tetraphenyl)borate salts and in the cases where the complexes were characterized crystallographically, a close η^6 type interaction was seen with one of the phenyl groups of the counterion (Scheme 19).[203]

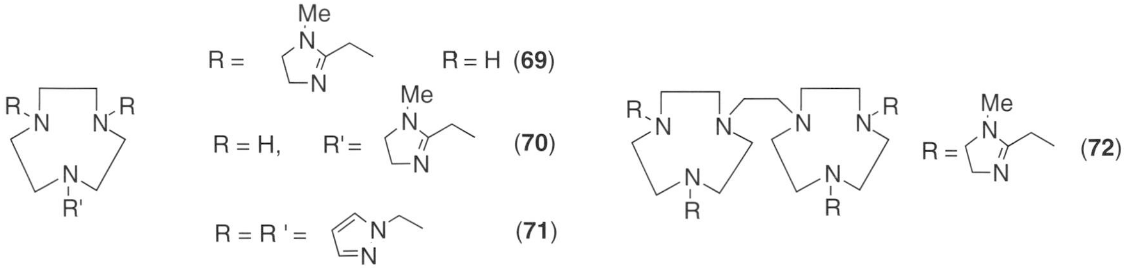

Scheme 19

Macrocyclic ligands will, under favorable conditions, form complexes of greater stability than an open-chain ligand with similar donor groups and geometry. A comparative study of the linear and cyclic polyamines (73)–(80) shows a maximum value for log K for the smallest cyclic polyamine under constant conditions.[204] A similar result is seen for tetraazacycles, where again the highest value for log K is seen a complex formed by a small ring ligand (81).[205] The related (82) coordinated lead through all six nitrogen donors in the solid, with an overall nine coordination completed by a chelating perchlorate and a molecule of water.[206]

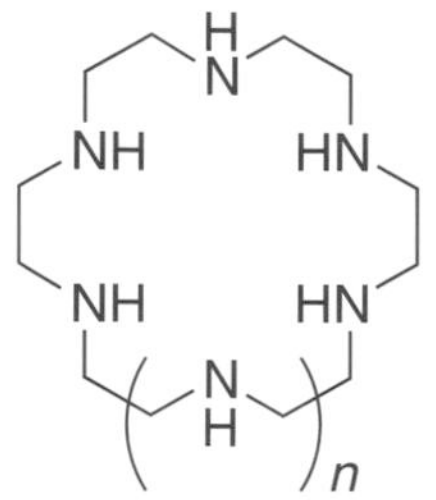

n = 1 (73)
n = 2 (74)
n = 3 (75)
n = 4 (76)

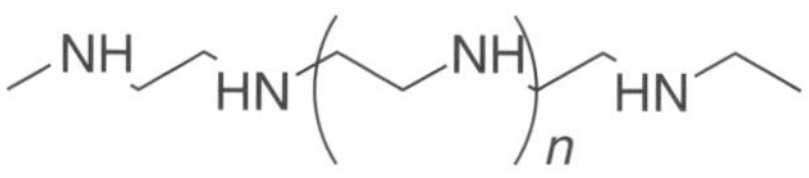

n = 3 (77)
n = 4 (78)
n = 5 (79)
n = 6 (80)

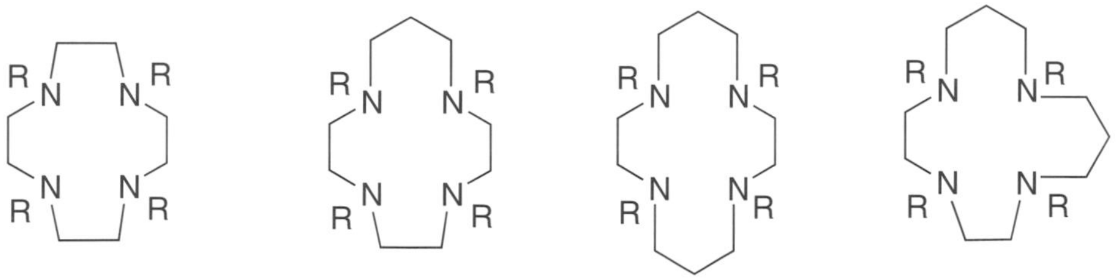

R = H, Me

(81)

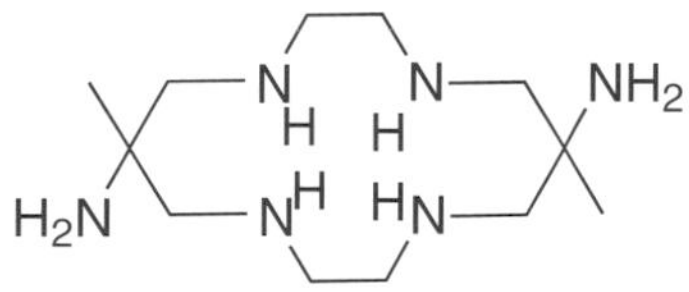

(82)

Larger polyazacycles with N_5–N_7 donor sets are found in the series of ligands (83) based upon phen. The ligand which forms the most stable complex is again the smallest example.[207] Similar design strategy produced the ligands (84) and (85), which coordinate lead within the ligand cavity through the pyridine nitrogens. The construction of the ligand is such that the aliphatic amines are not able to coordinate a metal ion bound to the bipy group because of steric constraints and so the ligand may be protonated at these nitrogens without disrupting the complex. Where the bipy moiety is oriented outward, the lead is bound within the ligand again this time by the aliphatic amines alone.[208] Very large polyazacycles, such as (86), can coordinate two lead ions.[209]

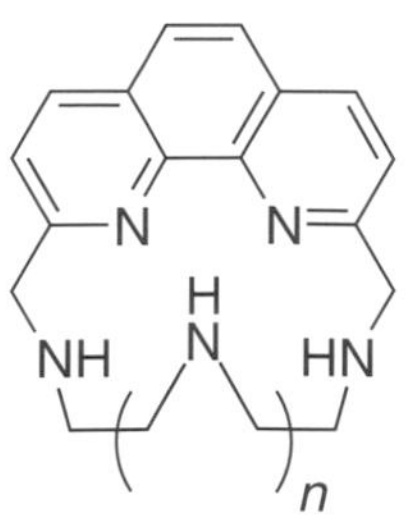

n = 1, 2, 3

(83)

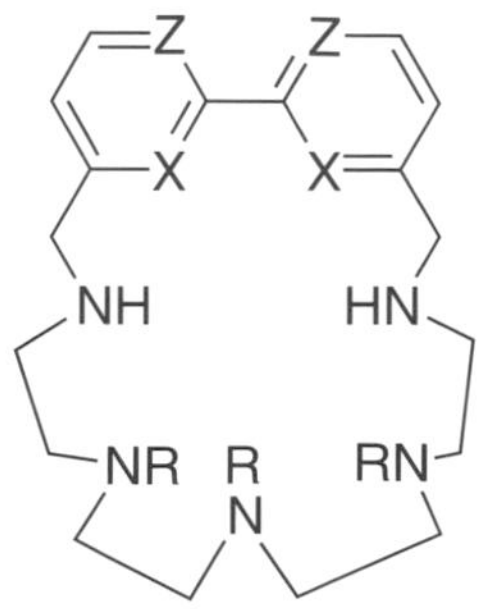

X = N, Z = CH (84)
X = CH, Z = N (85)

R = H, Me

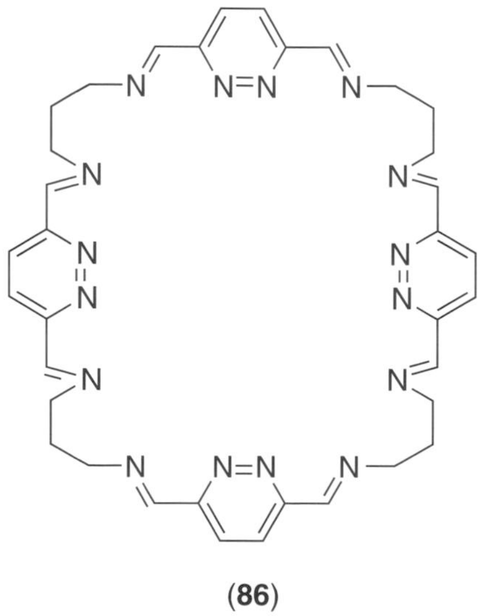

(86)

3.7.4.9   Complexes with Phosphorus or Arsenic Ligands

In contrast to nitrogen donor ligands, which have a rich and varied coordination chemistry with Group 14 metals, complexes of phosphorus and arsenic donors are encountered less commonly, especially for neutral donors. Descending Group 14 there is a pronounced decrease in Lewis acidity, most evident for tin(II) and lead(II), which may go some way to explaining this reduced affinity. There are, however, definite suggestions in the literature of further chemistry waiting to be uncovered in this area.

3.7.4.9.1   *Complexes of M^{IV} with phosphines or arsines*

The interaction of germanium(IV) chloride with a number of monodentate triorganophosphine ligands PR_3 (R = Me, Et, Pr^n, Bu^n, cy, 2,6-$(MeO)_2C_6H_3$, 2,4,6-$(MeO)_3C_6H_2$, Bn, Me_2N, Et_2N, Pr^i_2N) leads exclusively to the ionic germanium(II) complexes $[PR_3Cl][GeCl_3]$. This is in distinct contrast to the expected 1:1 or 1:2 adducts of germanium(IV), some of which have been previously claimed in the literature from this preparative route. The structure of $[Bu^n_3PCl][GeCl_3]$ shows no close interaction of the ion pair, and a trigonal pyramidal geometry at the germanium.[210] With triphenylphosphine, no reaction is seen, an observation of interest since the mixture of germanium(IV) chloride and triphenylphosphine has been used as a reagent for the reduction of α-bromo carboxylic acids.[211]

The first fully characterized germanium(IV) arsine complex, $(Me_3As)_2GeCl_4$, can be prepared from the direct reaction of the arsine with halide, and has *trans* structure with Ge–As of 2.472 Å.[210]

Tin(IV) chloride or bromide reacts with Bu^n_3P to give 1:2 complexes with octahedral geometry. These are particularly interesting subjects for ^{119}Sn NMR studies, as the ^{119}Sn–^{31}P spin interactions give information relating to the solution structure of the complexes that is not otherwise available. For $(Bu^n_3P)_2SnX_4$, δ ^{119}Sn is -575 and $^1J^{119}Sn–^{31}P$ is 2,395 (X = Cl) or -953 and 1,960 Hz (X = Br). Mixing equimolar amounts of these two in solution leads quickly to the mixed species $(Bu^n_3P)_2SnCl_xBr_{4-x}$ which show values of δ and J intermediate between the two single halide species. There is a clear additive change in the values of δ which is in turn related to the electronegativity of the halide ligands, and rapidity of the halide exchange is a common feature of complexes of tin(IV) with more than two halides.[212] These species can be characterized by NMR methods in solution to a degree that they cannot be in the solid state.

An attractive alternative preparation of halophosphine complexes of tin(IV) is the reaction of tin metal powder and triorganophosphorus(V) dihalides R_3PX_2 ($R_3 = Ph_3$, Ph_2Me, $PhMe_2$; X = Br, I). The products are both *cis* and *trans* $(R_3P)_2SnX_4$ suggesting that the formation takes place stepwise, initially forming $(R_3PX)(SnX_3)$ which would then react with a further phosphine to give either isomer with no preference.[213,214]

Analysis of coupling constants ^{119}Sn–^{31}P for complexes $(R_3P)_2SnX_4$ ($R_3 = Ph_2Me$, Bu^n_3) and $(Ph_2P)_2(CH_2)_nSnX_{4-m}Me_m$ ($n = 1$, 2, 3; X = Cl, Br; $m = 0$, 1) indicate that Sn—P bonds are strengthened when the bond is trans to an electron donating ligand.[215]

Triorganophosphines react readily with tin(IV) complexes providing the tin center is sufficiently Lewis acidic. For complexes R_nSnX_{4-n}, coordination of one phosphine usually proceeds readily for $n \leq 3$, but for the tetraorganotin complex, no coordination of phosphine is observed.

The complexes Ph_3SnCl, R_2SnCl_2, and (R = Et, Pr, Bu^n, Ph) coordinate Bu^n_3P to form 1:1 adducts, a complexation readily monitored by ^{119}Sn and ^{31}P NMR. Coordination is accompanied by a significant shift in the ^{119}Sn δ to lower field, and the change in multiplicity arising from ^{119}Sn–^{31}P coupling is an aid to determining stoichiometry and geometry in the complex. For $RSnCl_3$ (R = Bu^n, Ph) complexation is accompanied by a scrambling of the ligands between tin centers to give a number of complexes.[216] A similar scrambling of ligands is seen for the complex ions $[(Bu^n_3P)(Me)SnCl_4]^-$, which exhibits a single doublet in its ^{119}Sn NMR spectrum at intermediate temperatures, indicating either preferential formation of a single isomer in solution or a fluxional process that is rapid on the NMR timescale. Mixing equimolar amounts of $[(Bu^n_3P)(Me)SnCl_4]^-$ and $[(Bu^n_3P)(Me)SnBr_4]^-$ gives a solution that shows a ^{119}Sn NMR that indicates that all isomers $[(Bu^n_3P)(Me)SnCl_nBr_{4-n}]^-$ are formed. Since this must involve interionic transfer of halide, the solution behavior of these ions is clearly somewhat involved.[217]

The complexation of R_nSnX_{4-n} (R = Me, Et, Bu^n; X = Cl, Br, I) by Bu^n_3P ($n = 1–3$) shows increasing enthalpy of formation in the sequence Cl < Br < I for any given formulation, the reverse sequence of the acid strengthening effects arising from the increasing electronegativity of the halide. The equilibrium constants for formation increase in the order $I < Br < Cl$, indicating that the entropy term may be dominant for this complexation.[218]

Chelating bisphosphines bis(diphenylphosphino)methane (dppm) and 2-bis(diphenylphosphino) ethane (dppe) react with Ph_2SnX_2 ($X = Cl$, Br) to give five coordinate complexes with monodentate attachment of the phosphines. The more rigid $(+)-(R,R)$-1,2-bis(methylphenylphosphino) benzene chelates successfully to the same organotin halides, giving octahedral complexes. Phenyltrichloro tin(IV) reacts with each of the chelating phosphines to form six-coordinate complexes, again with scrambling of the ligands to form a number of products.[219]

Siladiarsine $(2,4,6-Pr^i_3-C_6H_2)(Bu^t)Si(AsH_2)_2$ can be lithiated and treated with $(mes)(Bu^t)GeF_2$ to give (87), a diarsenagermane which can be further converted to (88). The structure of (88) shows a degree of asymmetry in the As—M bonds, with distances As—Si of 2.39 Å and As—Ge of 2.45 Å.[220]

(87)   (88)

The reaction of $Bu^t_2SnCl_2$ with $Na[AsH_2]$ in liquid ammonia yields $[Bu^tSnAsH]_2$ with a central $(SnAs)_2$ ring. In the same solvent $Bu^t_2Sn(NHBu^t)_2$ reacts with $Na[AsH_2]$ to give $[Bu^tSnAsH]_3$, with a $(SnAs)_3$ ring.[221]

Further spectroscopic information can be gathered in the far IR, and the spectra of $LSnX_4$ ($L = (R_2P)_2(CH_2)_n$; $R = $ Me, Et, Ph; $n = 2$, $X = $ Cl, Br, I) have been measured between $400 \, cm^{-1}$ and $40 \, cm^{-1}$. In this region, the M–P stretching modes can be found between 116–110, and the M-X stretching vibrations at 310–295 (chloro), 210–208 (bromo), and 185–156 cm^{-1} (iodo) have been assigned.[222]

3.7.4.9.2 *Complexes of M^{II} with phosphines or arsines*

Triphenyl phosphine coordinates germanium(II) chloride[223] or iodide[224] to form 1:1 complexes with tbp geometry at the metal and bond lengths Ge—P 2.511 Å and 2.507 Å, respectively. These complexes are in some ways analogous to ylides ($R^1_3P = CR^2_2$) and have some properties in common with these lighter homologues.

The availability of solution ^{119}Sn and ^{207}Pb NMR for the study of complexes is a great boon to the coordination chemist. An elegant study of the interaction of a number of multidentate phosphine ligands with $M[SbF_6]_2$ ($M = Sn$, Pb) has been published and offers insight into the solution structure of complexes which is difficult to obtain in other ways. Solutions of $M[SbF_6]_2$ in nitromethane were treated with polydentate phosphines dppe, $PhP[(CH_2)_2PPh_2]$, $MeC(CH_2PPh_2)_3$, $P[(CH_2)_2PPh_2]_3$, and $[Ph_2P(CH_2)_2]_2P(CH_2)_2PPh_2$. The values of δ for ^{119}Sn cover the range -586 to -792 and for ^{207}Pb 60 to -269. The greatest change in shift for both nuclei is seen on coordination of any phosphorus donors (cf. δ for "free" M^{II}; $Sn = -1,540$, $Pb = -3,342$) where the subsequent changes in shift as the number of phosphorus ligands coordinated to the metal increases is small compared to the change associated with going from the effectively solvated $M[SbF_6]_2$ to the phosphine complex. The multiplicity of the peaks is of tremendous utility in assigning the number of phosphorus donors coordinated as the couplings between the phosphorus and the metal are well resolved.[225]

Alkynyl phosphine Bu^tCP coordinates $(TMS_2CH)_2Ge$ with a side-on κ^2 [P,C] interaction to give a *pseudo*-tetrahedral product (89), the first example of a phosphagermirane. The geometry is somewhat distorted because of the difference in steric demand of the phosphine in comparison to the organic ligands.[226] The corresponding tin compound reacts with the same alkynyl phosphine to form the phosphadistannacyclobutene (90).[227]

Reaction of the anion of dppm with MCl_2 ($M = Ge$, Sn, Pb) forms complexes $[CH(PPh_2)_2]_2M$ that exhibit three-coordinate geometry in the solid state, one ligand chelating through both

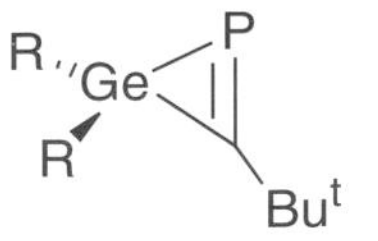

(89) 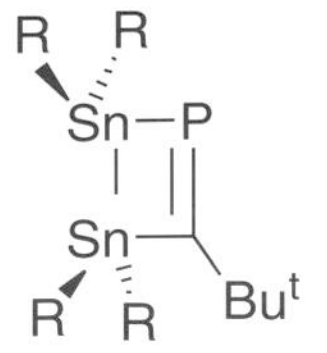   (90) 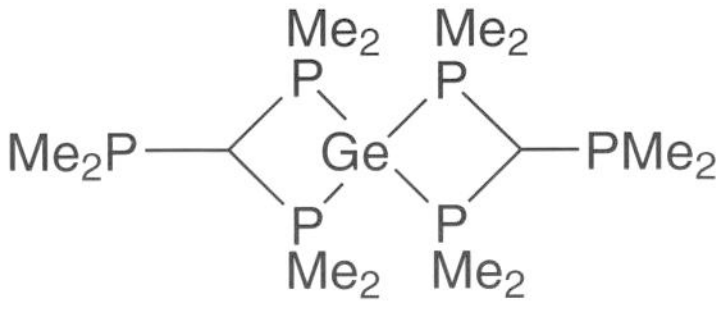   (91)

phosphorus and one coordinating in a monodentate fashion through the central carbon. In solution, the complexes are fluxional, where the ligands undergo a (κ^2-κ^1-κ^2) process. For the bulkier [C(SiMe₃)(PPh₂)₂]₂M, the bischelate structure is preferred.[228–231]

The potentially tridentate monobasic ligand [C(PMe₂)₃]⁻ forms the four coordinate (91) in reaction as its lithium salt with GeCl₂·diox. The geometry about the germanium(II) is tbp, with the lone pair occupying an axial position, and the complex has three short Ge—P bond lengths (2.359, 2.368, and 2.546 Å) and one longer interaction (2.926 Å). The corresponding tin complex has four similar bond lengths (2.602, 2.598, 3.790, and 2.839 Å), which suggests that the smaller germanium is less able to accommodate all four donor groups than the larger tin. In solution, the tin complex is stereochemically nonrigid and undergoes a *pseudo*-rotation that equilibrates the axial and equatorial sites according to NMR data. At elevated temperature all six phosphorus atoms equilibrate indicating that all phosphorus donors coordinate the tin during the fluxional process.[232, 233]

The reaction of GeCl₂·diox with [C(PMe₂)₂X]⁻ (X = PMe₂, SiMe₃) in the presence of magnesium gives the bisphosphide-supported Ge—Ge bonded dimer (92) which, on further reaction with additional GeCl₂·diox gives the remarkable pentagermane (93) (Scheme 20). With germanium(IV) chloride, the octahedral [C(PMe₂)₂X]₂GeCl₂ is formed, with a *trans* disposition of the halides.[234]

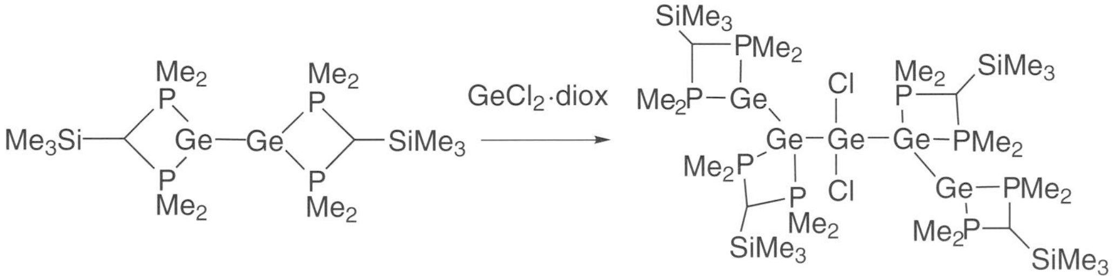

(92)   (93)

Scheme 20

3.7.4.9.3 *Complexes of M^IV with phosphides or arsinides*

The highly reactive phosphides Ge(PH₂)₄ and HGe(PH₂)₃ can be prepared in low yield from the reaction of germanium(IV) chloride with Li[Al(PH₂)₄]. The complexes were characterized by NMR spectroscopy as they are thermally unstable, decomposing to germanium(IV) phosphide and phosphine.[235] Germanium(IV) phosphides R₂PGeCl₃ are available from the oxidative addition of R₂PCl to GeCl₂. The reaction is reversible, the starting materials recoverable from the thermolysis of the product. The reaction proceeds through initial coordination of the GeCl₂ by the chlorodiorgano phosphine to form an intermediate complex (R₂ClP)GeCl₂. The corresponding reaction of GeCl₂ with RPCl₂ (R = Pr^i, Bu^t, Ad) gives the bis(trichlorogermyl)phosphines RP(GeCl₃)₂ and a number of cyclic products (94)–(96).[236]

The 1,2,3,4-diphosphadigermatane (97) is prepared by the reaction of (mes)(Bu^t)GeF₂ with [(dme)Li][PH₂]. The structure of the ring is trans with respect to the germanium. The bonds Ge—P are almost identical at 2.346 and 2.348 Å, and the internal angles Ge–P–Ge and P–Ge–P are 84.8° and 95.3°, respectively, a close approximation to a regular square.[237]

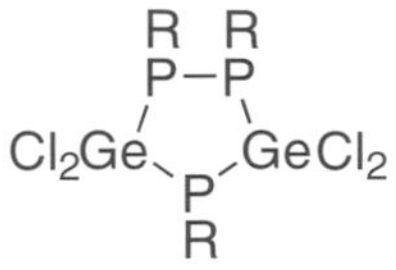

The near-tetrahedral diphosphagermiranes (**98**) have been prepared from the reaction of R_2GeCl_2 (R = Et, Ph) and $K_2[(Bu^tP)_2]$. The dianionic bisphosphide also reacts with germanium(IV) chloride to give the unexpected 1,2,4,5,6,7-hexa-Bu^t-1,2,4,5,6,7-hexaphospha-3-germaspiro[2,4]heptane (**99**).[238]

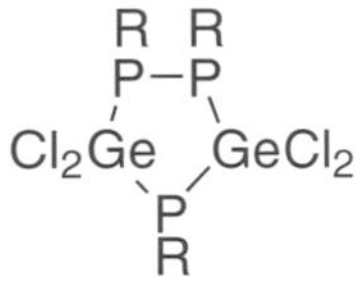

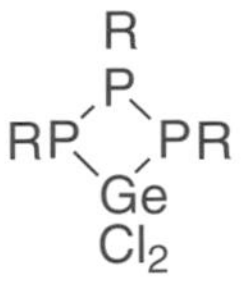

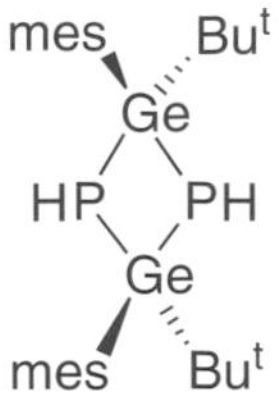

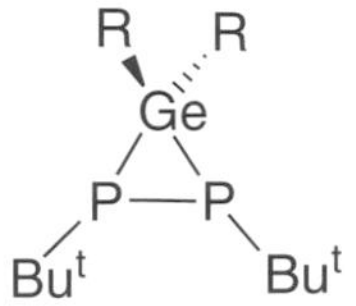

R = Et, Ph

(**98**)

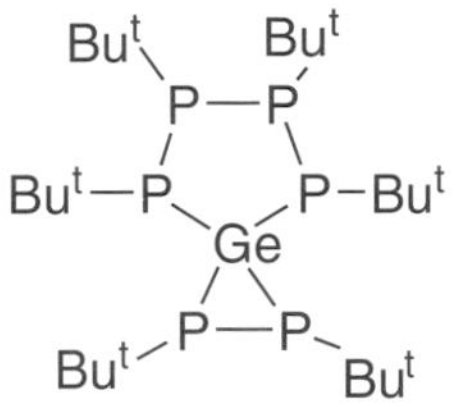

(**99**)

Diphosphinylmethanide $[C(PMe_2)_2SiMe_3]^-$ reacts with Me_2MX_2 (M = Ge, Sn; X = Cl Br) to form the six-coordinate complexes *cis*-$Me_2M[C(PMe_2)_2SiMe_3]_2$. Both complexes show inequivalent metal–phosphorus interactions, with two short and two long bonds. In the tin complex, which has a greater tendency to hypervalency, this difference is less than in the germanium complex, in which the structure has a greater degree of [4 + 2] character.[239]

The first examples of germanium–phosphorus double bonds have been reported for $(mes)_2$-Ge = P(2,4,6-$R_3C_6H_2$) (R = Pr^i,[240] Bu^t [241]) prepared by the reaction of $(mes)_2GeF_2$ with $Li[PH(2,4,6-Bu^t_3C_6H_2)]$ yielding the intermediate $(mes)_2Ge(F)P(H)(2,4,6-Pr^i_3C_6H_2)$. Subsequent lithiation of this product and elimination of LiF gives the germaphosphirane in good yield. The double bond is liable to addition reactions of RH (R = OH, MeO, Cl, Me_3P = CH)[241] or the chalcogens sulfur or selenium, giving the germathia- or germaselenaphosphines (**100**). Heating with excess chalcogen gives $[(mes)_2GeS]_2$ and $ArP(S)_2$.[242] The corresponding stannaphosphirane can be prepared by a similar route.[243]

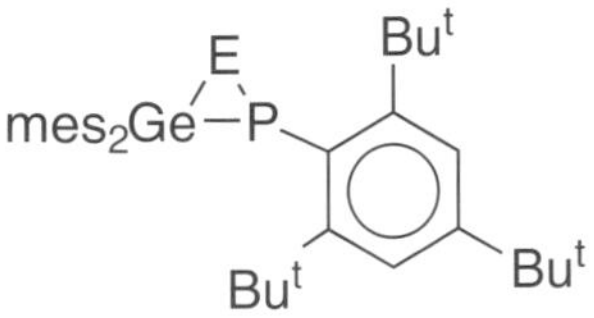

(**100**)

Bis(diorganochlorostanna)phosphine (**101**) can be converted to the Sn_2N_2SiP cyclohexane (**102**) by the sequence shown in (Scheme 21), forming a wholly inorganic six-membered ring.[244]

3.7.4.9.4 Complexes of M^{II} with phosphides or arsinides

Reaction of the primary silylalkyl phosphine $R^1_3SiPH_2$ with R^2_2Sn (R^2 = N(SiMe$_3$)$_2$, 2,4,6-$(CF_3)_3C_6H_2$) gives the hexanuclear complex (**103**) by a mechanism involving elimination of R^2H. The structure of (**103**) is a distorted hexagonal prism, with bond length Sn–P of 2.626 Å (ave.) and angles P–Sn–P ranging between 86.9° and 100.7°, values which are not greatly different

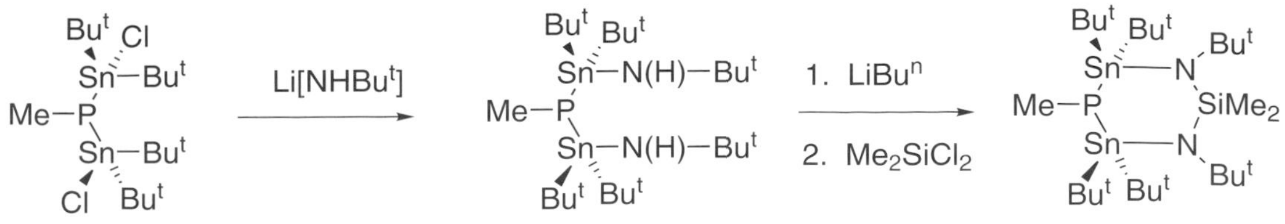

(101)   **(102)**

Scheme 21

for those seen for monomeric complexes $(R_2P)_2Sn$. The same reaction carried out in the presence of tin(II) chloride leads to the isolation of the $SnCl_2$ bridged dimer (**104**).[245] The tin–phosphorus bond lengths in the hexamer are longer than those found in $(R_2P)_2Sn$ compounds, such as those found in the series of complexes $[(Pr^i_3Si)(R)E]_2M$ ($R = trip_2SiF$, $E = P$, $M = Ge$, Sn; $R = trip_2SiF$, $E = As$, $M = Sn$; $R = (trip)Bu^tSiF$, $E = P$, $M = Pb$).[246] These complexes are monomeric, as is $[(Ph_3Si)_2P]_2Sn$,[247] where the complexes with a smaller ligand $(TMS_2P)_2M$ ($M = Sn$, Pb) are dimeric with bridging phosphorus groups.[248] Simple lead(II) phosphides $(Bu^t_2P)_2Pb$ can be prepared directly from the lithium phosphide and lead(II) chloride[249] where adjusting the stoichiometry gives $Li[(Bu^t_2P)_3Pb]$ with a three-coordinate lead and a central $[PbP_2Li]$ ring.[250]

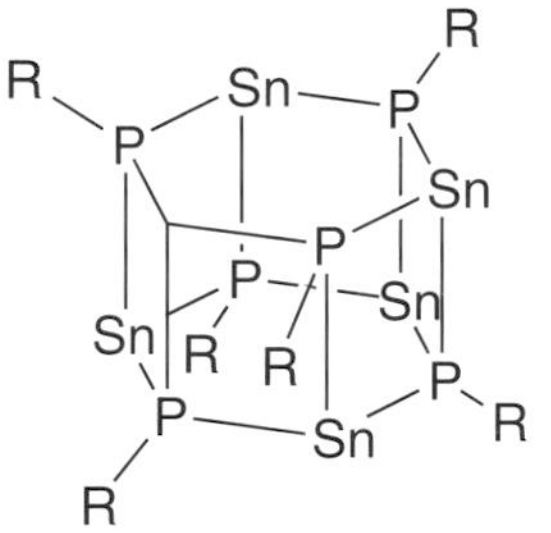

$R = Pr^i_3Si$

(103)   **(104)**

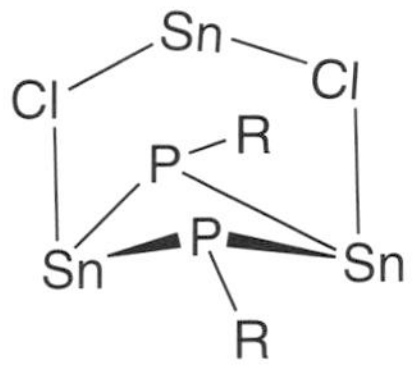

Pnictide complexes can also be formed by the elimination of TMSCl, as in the reaction of tin(II) chloride with $(Bu^t)_2TMSE$ ($E = P$, As), which forms $[Bu^t_2ESnCl]_2$ ($E = P$[251] or As[252]). The structure of the arsenide has been determined and exhibits bond length Sn—As 3.773 Å, a comparatively rare bond, and angles As–Sn–As of $77.8°$.[253]

The first example of a lithio arsinoorganogermane has been reported as the product of the reaction of Bu^tGeF_3 and six equivalents of $Li[(Pr^i_3Si)HAs]$ (**105**). The structure is a distorted rhombododecahedron and has As—Ge bonds of 2.442–2.447 Å.[254]

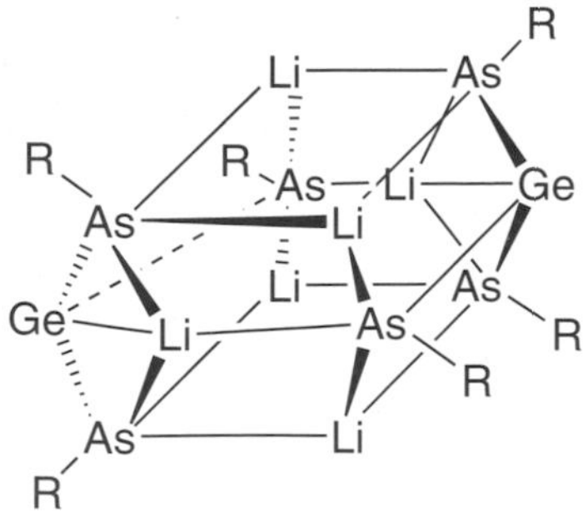

$R = Pr^i_3Si$

(105)

3.7.5 COMPLEXES WITH GROUP 16 LIGANDS

In much the same way that complexes of Group 14 metal ions with nitrogen ligands outnumber those with other Group 15 donor atoms, complexes of oxygen donors are the most numerous amongst Group 16 donor ligands. This is true for hard donors, such as hydroxy groups, as well as soft donors, such as crown ethers, for all three metals, though there are still enough complexes of the heavier chalcogens to make them a diverse and an interesting subject area.

3.7.5.1 Complexes with Neutral Oxygen Ligands

Sulfoxides are widely studied ambidentate ligands with donor properties that in some ways respond to the character of the metal to which they coordinate. In the majority of their complexes they coordinate through the oxygen, and the complex $(DMSO)_2GeCl_4$ is no exception. An IR study shows bands due to the Ge–O stretches at 506 and 495 cm^{-1}.[255] Structural studies of diphenylsulfoxide complexes $(Ph_2SO)_2SnI_4$ and $(Ph_2SO)_2Sn(Me)I_3$ show *cis* coordination of the sulfoxide ligands.[256] Dibenzylsulfoxide coordinates Me_2SnCl_2 to give a tbp complex with the sulfoxide oxygen and one chloride in the apical positions.[257] The nitrate ligands in $Me_2Sn(NO_3)_2$ readily dissociate, despite being potentially chelating, so that complexes $[Me_2Sn(L)_4](NO_3)_2$ are readily prepared (L = DMSO[258] or H_2O[259]).

Complexes of lead(II) with DMSO ligands $[(DMSO)_n(ClO_4)_2Pb]$ can be isolated from solutions of lead(II) perchlorate in DMSO where $n = 3$ or 5. For $[(DMSO)_3(ClO_4)_2Pb]_2$ two different isomers are formed with either perchlorate or DMSO oxygens acting as the bridging ligands (106). Interestingly, there is no indication of any sulfur bound sulfoxide on this apparently soft lead(II) center.[260] Diphenyllead dichloride coordinates two equivalents of DMSO or HMPA through the oxygen termini to give six-coordinate complexes with *trans* organic ligands and bonds Pb—O of 2.482 and 2.536 Å, respectively.[261]

(106)

Phosphine oxides are popular ligands for tin(IV), and many examples of monodentate R_3PO complexes are known.[262] Interaction of a range of organotin halides with $dppe(O)_2$ or $cis[Ph_2$-$P(O)CH]_2$ leads to monomeric six-coordinate complexes with chelation of the bisphosphine dioxide. The complexation of the related monodentate phosphine oxides Ph_2MePO and Ph_3PO follows a different pattern, with *trans* coordination of the two monodentate ligands preferred. The triaryl phosphine oxide is the least effective base, and there is NMR evidence that in solution the $dppe(O)_2$ is coordinated in a monodentate fashion. The structures show a typical *trans* arrangement of the two alkyl groups.[263,264] Cationic tin centers can be stabilized by $dppe(O)_2$ such as $[SnMe_2\{dppe(O)_2\}_2]^{2+}[(MeSO_2)_2N]_2^-$ prepared from the reaction of $Me_2Sn\{(MeSO_2)_2N\}_2$ with $dppe(O)_2$.[265]

The doubly oxidized form of dppm chelates $Ph_2Sn(NO_3)_2$ to give the seven-coordinate complex $(dppmO_2)Ph_2Sn(\kappa^2-NO_3)(\kappa^1-NO_3)$. Chelation of the bisphosphine dioxide gives rise to a six-membered ring with Sn—O bonds of 2.237 and 2.223 Å. The remainder of the ligand set is made up of a bidentate and a monodentate nitrate (Sn—O 2.472 and 2.350; 2.289 Å, respectively). The doubly oxidized arsenic analogue of dppe ($dpaeO_2$) coordinates in a rather different fashion, bridging two tin centers to give $[Ph_3Sn(\kappa^1-NO_3)]_2(\mu-\kappa^1,\kappa^1-dpaeO_2)$.[266] The potentially chelating ligand $Me_2NC(H)[P(O)(OEt)_2]_2$ reacts with Me_2SnCl_2 to form the robust dinuclear complex (107) comprising a 12-membered ring.[267]

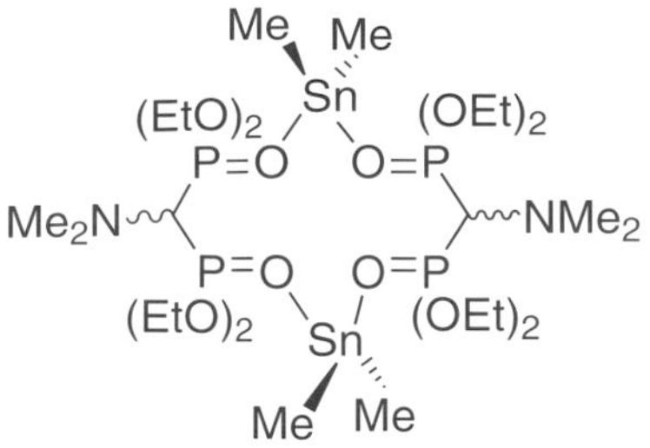

(107)

The complexation of lead(II) perchlorate by the Schiff-base ligands 3-MeO-salenH$_2$, 3-MeO-saltrenH$_3$, and saltrenH$_3$ (salen = N,N'-bis(salicylaldehydo)ethylenediamine, saltren = N,N',N''''-tris(salicylaldehydo)tris-(2-aminoethyl)amine) gives complexes in which they coordinate as innocent ligands through neutral phenolic oxygens. Structural analysis reveals *exo* coordination of the metal, with the ligand pocket occupied by the ionizable protons. In the case of 3-MeOsalenH$_2$, the lead is three-coordinate by the two oxygens of the Schiff-base and a molecule of methanol, and in the other cases is three coordinate through the ligand oxygens alone. Solution NMR studies show that the molecules are fluxional at ambient temperatures.[268] Nitriloacetamide is a neutral [O$_3$] donor ligand that coordinates tin to form a 10-coordinate complex cation [(NTA)$_2$(κ^2-NO$_3$)Pb]$^+$ in which the remaining lone pair is not stereochemically active.[269]

Macrocyclic *O*-donor ligands 1,3-xylyl 18-C-5 reacts with tin complexes Me$_{3-n}$SnCl$_{n+1}$ ($n = 0$, 1, 2) to form complexes that do not show inclusion of the tin inside the macrocycle, but rather coordination of the metal in an *exo* fashion. The structure of MeSnCl$_3$(H$_2$O)$_2$(1,3-xyly-18-C-6) shows a six-coordinate tin center with *cis* coordination of two *O*-bound water molecules. These water molecules are in turn hydrogen-bonded to four adjacent oxygens of the crown ether, in a manner similar to the coordination of diimine ligands seen previously.[270] The slightly larger 18-C-6 chelates tin(IV) chloride to give a six-coordinate tin complex (Sn—O 2.237 and 2.212 Å).[271] Lead(II) can be more successfully included in the cavity of a crown ether, as shown by the complexes [(15-C-5)(SCN)$_2$Pb], which has an eight coordinate lead bound by all five of the ether oxygens,[272] and both [(18-C-6)(SCN)$_2$Pb] and [(*cis-anti-cis*-cy$_2$-18-C-6) (SCN)Pb], which show a hexagonal bipyramidal geometry at the lead, with the median plane described by the macrocyclic ligand.[273] With lead(II) acetate, a different configuration results, whereby in the crystal the lead is 10-coordinate with two bidentate acetate ligands bound *cis* on one side of the lead and the hexadentate crown ether bound on the other face of the lead.[274]

3.7.5.2   Complexes with Anionic Monodentate Oxygen Ligands

Sterically stabilized germanium(II) alkyls (2,4,6-Bu$^t_3$C$_6$H$_2$)$_2$Ge and (2,4,6-Pr$^i_3$C$_6$H$_2$)[2,4,6-(CH-(SiMe$_3$)$_2$]$_3$C$_6$H$_2$)Ge undergo oxygen transfer reactions with various *N*-oxides to form the corresponding germanones that in turn react with isocyanates RNCO to give complexes (108) (Scheme 22). If the germanones are allowed to stand in solution (ca. 10 h) then they undergo intramolecular activations forming the diastereomeric (109).[275,276]

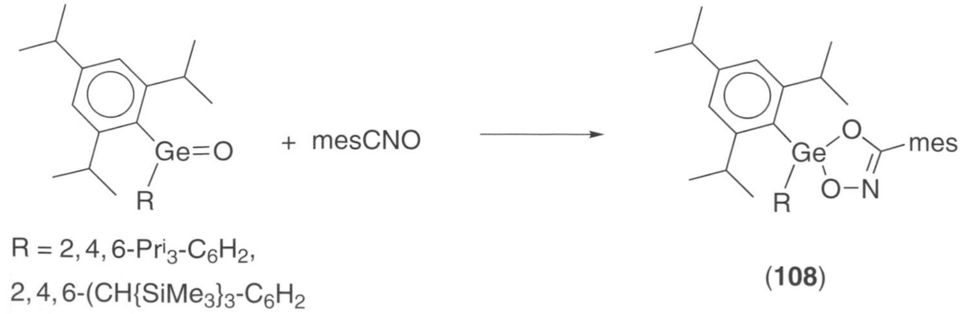

R = 2,4,6-Pr$^i_3$-C$_6$H$_2$,

2,4,6-(CH{SiMe$_3$}$_3$-C$_6$H$_2$

(108)

Scheme 22

(109)

Carbohydrates react with Bu^n_2SnO with elimination of water to give complexes that have geometries that depend upon the particular sugar. Of 19 different sugars investigated, Mössbauer spectroscopy identified products with octahedral, tbp, and tetrahedral geometries at the tin with a preponderance of tbp.[277]

Hydrolysis of monoorganogermanium chlorides $RGeCl_3$ ($R = Pr^i$, cy, mes) gives the germanium sesquioxanes $(RGe)_6O_9$, **(110)**. The core structure is a $(GeO)_3$ ring, linked through three bridging oxo ligands to a second such ring. The structural diversity of silsesquioxanes has attracted a great deal of attention and it is to be hoped that the germanium analogues will be as studied.[278] The new germanate $(C_2H_8N_2)(C_2H_{10}N_2)[Ge_9O_{18}(OH)_4]$ is synthesized by hydrothermal methods from germanium and TMEDA. The structure comprises $Ge_9O_{22}(OH)_4$ units with four-, five-, and six-coordinate germanium ions.[279] Solid argon matrix isolated germanium(II) oxides $(GeO)_n$ ($n = 1$–4) were studied by IR. The structure of $(GeO)_2$ is planar cyclic, $(GeO)_3$ has the highly symmetrical D_3h ring structure and contrary to previous ideas, $(GeO)_4$ is found to be a cubane.[280]

Alkoxides of the acidic triorganosilanol Ph_3SiOH reacts with $[M(OBu^t)_2]_n$ ($M = Ge$, Sn, Pb) to form discrete dimeric complexes $[(Ph_3SiO)MO]_2$ with a central $(MO)_2$ ring.[281] Hydrolysis of $RSnX_3$ gives cage complexes with structures related to the silsesquioxanes, such as $[(Pr^iSn)_{12}O_{14}(OH)_6]^{2+}$. The complex comprises a football-shaped framework of (SnO) units, where the tin atoms exhibit square pyramidal geometries and comprise half-chair $(SnO)_3$ rings. The related complex $[Sn(CH_2)_6Sn](ClCH_2CO_2)_4(OH)_2O_{10}$, prepared by the controlled hydrolysis of $[(ClCH_2CO_2)_3SnCH_2]_2CH_2$, has an almost planar array of all 12 tin atoms.[282] Prepared by an analogous route, $[(Pr^iSn)_9O_8(OH)_6]^{5+}$ has a pyramidal cage structure with both tbp and octahedral tin centers, linked by μ^3 oxo or μ^2 hydroxy ligands. The structure is further supported by intramolecular OH···Cl hydrogen bonds.[283,284]

A different class of cluster with an octahedral frame is exampled by $Sn_6(\mu^3\text{-}O)_4(\mu^3\text{-}OSiMe_3)_4$, which has all eight oxygen ligands bridging the faces of the Sn_6 octahedron.[285] The oxo cluster $Sn_6O_4(MeO)_4$ shows luminescence at 77 K, with λ_{max} at 565 nm, probably due to a metal-centered sp excited state.[286]

The reaction of Me_2SnO with $PhC(O)CH_2C(O)CF_3$ in the presence of CF_3CO_2H leads to the isolation of **(111)**, a tetrameric complex in which the central $(SnO)_2$ ring is planar and is itself almost coplanar with the chelate rings. The tin centers all have tbp geometry with equatorial methyl groups. The complex displays the structural motif of a μ^3 oxo ligand, seen in the structures of a number of tin oxo species.[287] The bridging $(SnO)_2$ structure is also present in the hydroxy bridged dimer $[(Bu^t)_2SnX(OH)]_2$.[288]

The nature of the species present in aqueous solutions of tin and lead salts has been the subject of much conjecture. Some information is now available following the crystallographic identification of two complexes from aqueous solutions of $M^{II}(NO_3)_2$. The open vertex cubane $[Sn_3(OH)_4](NO_3)_2$ (Sn—O_{OH} 2.149–2.345 Å) was crystallized from a solution of tin(II) nitrate[289] and the cubane $[\{Pb(OH)\}]_4(NO_3)_4$ from a solution of lead(II) nitrate (Pb—O_{OH} 2.387 Å ave.).[290] While these structures may not represent all the species present in solution under these conditions, it is an indication of the extent to which oligomerization can contribute in this speciation.

(110)   (111)

Lead also forms polynuclear assemblies supported by both monoanionic *O*-donor and oxo ligands. Lead alkoxides can be prepared by the alcoholysis of the labile amino complex $(TMS_2N)_2Pb$ with a range of alcohols to give corresponding complexes of overall formula $(RO)_2Pb$ ($R = Pr^i$, Bu^t, $C(Me)_2Et$, $C(Et)_3$, $CH(Me)CH_2NMe_2$). The larger alcohols form linear trinuclear complexes $[(RO)_2Pb]_3$ with each of the alcohol oxygens bridging two lead centers such as with Bu^tOH[291] where the smaller form linear polymers with four-coordinate lead centers bridged by alcohol oxygens in a distorted tbp geometry.[292] By modifying the reaction conditions, higher nuclearity complexes can be obtained such that the alcoholysis of $(TMS_2N)_2Pb$ by Bu^tOH also forms $(Bu^tO)_6OPb_4$ (112) and by ROH ($R = Et$, Pr^i) gives $(RO)_4O_4Pb_6$ (113). These complexes have structures that are related to adamantane.[293]

(112)   (113)

The reaction of $(TMS_2N)_2Pb$ with $Bu^t NCO$ gives both $[(TMS_2N)Pb(OSiMe_3)]_2$ and $Pb_7(\mu^3$-O)-$(\mu^4$-O)$(\mu$-OSiMe$_3)_{10}$.[294] The heptanuclear product seems at first sight to be a rather unusual product, but does in fact have a strong resemblance to the product of the hydrolysis of $(Pr^iO)_2Pb$.[295]

3.7.5.3 Complexes with Monoanionic Bidentate Oxygen Ligands

Monobasic bis-oxygen chelates offer the possibility of forming a variety of complexes with M(14) ions. The complex anions $[Ph_2(NO_3)_3Sn]^-$ and $[Ph_2(NO_3)_2ClSn]^-$ co-crystallize and the molecular structure of these show bidenate coordination of the nitrate groups in all cases, leading to pentagonal and hexagonal bipyramidal geometries with apical phenyl groups.[296] Cupferron, widely used as a chelating agent in analytical chemistry, forms complexes with tin that vary in nuclearity depending upon the starting material used. With tin(IV) chloride, an eight coordinate complex $(PhN(O)NO)_4Sn$ is formed, which has an irregular dodecahedral geometry in the crystal. With Me_3SnCl as the starting material, a tetrameric product $[\{PhN(O)NO\}Me_3Sn]_4$ is formed, which has a central 20-membered $[Sn_4O_8N_8]$ ring.[297]

Six-coordinate tin complexes of benzoylacetoacetonate are fluxional on the NMR timescale at ambient temperatures.[298] The mechanism of interconversion of isomers of $Ph(Cl)Sn(Bzacac)_2$ has been examined by one and two-dimensional NMR studies that show the isomerization can proceed by a Bailar twist, and that Ray-Dutt pathways and routes involving square planar intermediates can be excluded.[299] Triorganotin(IV) halides react with the sodium salts of 2,4,6-heptanetrione, 1-Ph-1,3,5-hexanetrione, and 1,5-Ph$_2$-1,3,5-pentanetrione to form mono- and dinuclear complexes (114) and (115) (R = Me, Et, Pr^n, Bu^n, Ph).[300] The acetylacetone analogue 4-acyl-2,4-dihydro-5-Me-2-Ph-3-H-pyrazol-3-one reacts with Me$_2$SnCl$_2$ in the presence of sodium methoxide to form (116) or (117) depending upon stoichiometry.[301,302] Homoleptic tropolonate complexes (trop)$_4$M (M = Ge, Sn) (trop = tropolonato) have been shown to comprise an ion pair in the case of germanium as [(trop)$_3$Ge](trop) and a genuine eight-coordinate tin in (trop)$_4$Sn.[303]

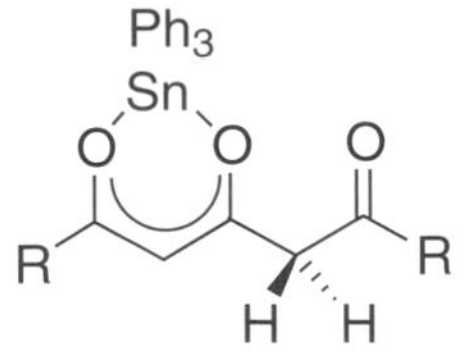

(114) 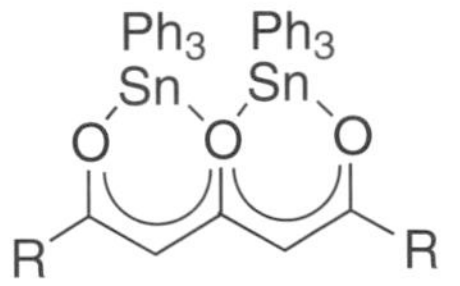   (115)

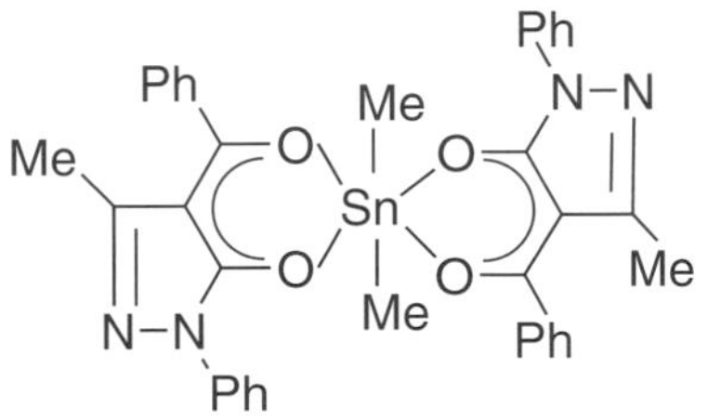

(116) 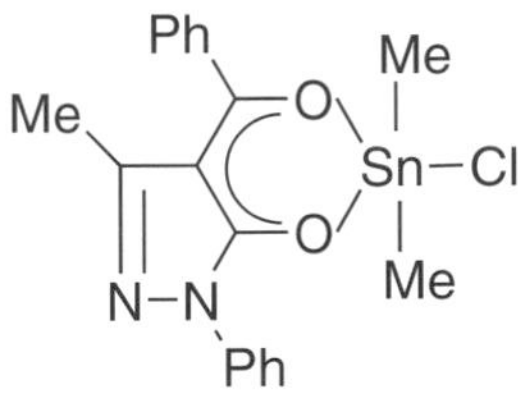   (117)

3.7.5.4 Complexes with Carboxylates or Phosphinates

Carboxylates can act as monodentate, chelating, or bridging ligands. Carboxylate complexes of Group 14 ions show all these coordination modes, depending upon the metal, the substitution on the metal, and the substitution on the carboxylate.

Triphenylgermanium chloride reacts with the sodium salts of carboxylates (2-furanyl, 2-furanyl vinyl, 2-(5-Bu^t)furanyl, 2-thiophenyl, 2-pyridinyl, 3-pyridinyl, 4-pyridinyl, 3-indinolyl, 3-indolylmethyl, and 3-indolylpropyl) to give in all cases four-coordinate germanium centers with monodentate carboxylate coordination. Interestingly the complexes all show high *in vitro* activity against human tumor cell lines MCF-7 and WiDr.[304]

Monodentate coordination is seen for the carboxylates in the complexes Ph$_3$Sn(O$_2$CC$_6$H$_4$X) (X = H, 2-Me, 2-NH$_2$, 2-NMe$_2$, 2-Cl, 4-Cl, 2-(OH), 4-(OH), 4-MeS, 2-MeO) in solution state by ^{119}Sn NMR and Mössbauer spectroscopy. This persists in the solid with the exception of the 2-Cl and 2-(OH) derivatives that both show Mössbauer spectra consistent with bridging structures.[305]

A different monodentate carboxylate is found in the complex Ph$_3$SnCl(quinolinium-2-carboxylate), where the five coordinate tin is bound in a monodentate fashion to the carboxylate, the proton having migrated to the heterocyclic nitrogen, forming a Zwitterionic ligand.[306] Picolinic acid and picolinic acid *N*-oxide also form complexes R$_2$Sn(pic)$_2$ and [R$_2$Sn(pic)]$_2$O (R = Me, Pr^i, *n*-octyl, bn) (pic = picolenate) with monodentate carboxylate coordination supported by pyridine-*N* or pyridine *N*-oxide *O*-coordination.[307] The structurally similar ligands nicotinic acid and nicotinic acid *N*-oxide form complexes R$_2$Sn(nic)$_2$ and [R$_2$Sn(nic)]$_2$O (nic = nicotinate) with chelated carboxylates seen in all cases[308] and this chelating mode is by far the most commonly encountered in carboxylate complexes of tin. Other examples of substituted carboxylates that form complexes of this type include 2-BrC$_6$H$_4$CO$_2$H[309] and 4-BrC$_6$H$_4$CO$_2$H.[310]

The carboxylate ligands in Me$_2$Sn(OAc)$_2$ are all chelating, giving a distorted octahedral geometry to this prototypical tin carboxylate. Reaction with [N(Me)$_4$][OAc] gives the triacetate complex [N(Me)$_4$][Me$_2$Sn(OAc)$_3$], which is seven-coordinate with one monodentate acetate.

At ambient temperatures, the anion is fluxional with rapid exchange between the mono- and bidentate acetates.[311]

The factors that influence the denticity of carboxylates are complex. As an illustration of the variation of coordination behavior, the reaction of amino acid derivatives N-benzyl glycinate (**1**) or N-benzoylglycylglycinate (**2**) with R_2SnO forms complexes $R_2Sn(L)_2$ and $[R_2Sn(L)]_2O$ ($R = Me$, Et, Pr^n, Bu^n, n-octyl). The mononuclear complexes have a distorted octahedral geometry when R is Me, Pr^n, L is 1, and R is n-octyl, L is 2. In all of the other $R_2Sn(L)_2$ complexes, the tin is four coordinate with monodentate carboxylate ligands. The dinuclear complexes have chelated carboxylates when R is Me, Pr^n, L is 1 or R is Me or n-octyl, L is 2, but have bidentate coordination of L through a monodentate carboxylate and monodentate amide carbonyl in all other cases. The ligands 1 and 2 react with Ph_3SnCl to form five coordinate complexes $Ph_3Sn(L)$ with bidentate coordination of L through monodentate carboxylate and the amide carbonyl.[312,313]

The influence of the size of the substituents of the tin and the carboxylic acid have been investigated using the series of complexes prepared from R_2SnO and R^1CO_2H ($R = Bu^n$, Bu^s, Bu^i, Bu^t; $R^1 = Me$, Et, Pr^i, Bu^t). In each case a reaction of stoichiometry tin:acid of 1:2 yielded the expected $R_2Sn(\kappa^2-O_2CR^1)_2$, and stoichiometry 1:1 gave $[R_2Sn(\kappa^2-O_2CR^1)]_2O$ except for the case were all R groups are Bu^t, where the product was $[Bu^t_2Sn(\kappa^2-O_2CBu^t)(OH)]_2$, with a central $[Sn(OH)]_2$ ring. These data seem to suggest that there is little steric influence over the course of these reactions.[314]

The dicarboxylic acids $HO_2C(R)CO_2H$ ($R = (CH_2)_{0-8}$, *trans*-CHCH, $1,4\text{-}C_6H_4$) react with Bu_2SnO to form complexes of general formula $Bu_2Sn(O_2CRCO_2)_2$ and have, by solution ^{119}Sn and ^{13}C NMR, oligo- and polymeric structures in which each tin is chelated by two carboxylates from two different molecules of the diacid.[315]

Complexes $[R_2Sn(L)]_2O$ and $R_2Sn(L)_2$ ($R = Me$, Et, Pr^n, Bu^n, $L =$ anion of 2-MeO-benzoic-acid;[316] $R = Et$, Bu^n, $L =$ anion of 2-MeS-nicotinic acid[317] or 2-NH_2-benzoic acid[318] prepared from R_2SnO and LH in 1:1 and 1:2 stoichiometry, respectively, show distinct structural features. The dinuclear complex (**118**) has a structure often seen for such carboxylates whereas the mononuclear complexes have a distorted octahedral geometry. Other examples of *ortho*-substituted benzoic acids that form complexes $R_2Sn(L)_2$ are $2\text{-}(OH)C_6H_4CO_2H$ and $2\text{-}ClC_6H_4CO_2H$, the first of which shows intermolecular hydrogen-bonding, forming dimers, and both of which show asymmetric chelation of the carboxylate. The structures are described in terms of bicapped tetrahedral geometry, with short Sn—C and two short Sn—O bonds and two long Sn—O bonds.[319] Thiophene-2-carboxylic acid also forms both 1:1 and 1:2 complexes ($R = Me$, Et, Pr^n, Bu^n, n-octyl) where the coordination of the ligand is exclusively through bidentate carboxylate groups, with no participation from the neighboring thiophene.[320]

Bridging ($\mu^2\text{-}\kappa^1\kappa^1$) carboxylates are a feature of polymeric $R_3Sn(O_2CR)$ species which tend to have extended linear oligo- or polymeric structures.[321] An example of a dinuclear complex with a bridging carboxylate that is not supported by other bridging groups is given by $[Ph_2Sn(O_2CCX_3)]_2$ (X = H, F, Cl) where the separation Sn—Sn is in the range 2.69–3.77 Å for these complexes.[322]

Higher nuclearity complexes of this type can be prepared from the reaction of organostannoic acids with carboxylic acids. The hexanuclear (**119**) is prepared from $PhSn(O)OH$ and $cyCO_2H$,

(**118**)

(**119**)

and was the first example of a drum-shaped tin(IV) molecule of this class, although this structural type is found for other main group metals.[323] The molecule comprises two $(SnO)_3$ rings linked by carboxylate bridges between six-coordinate tin centers.[324] This drum structure can be prepared using a wide range of carboxylic acids and the same structure is also seen in the product of the corresponding reactions with phosphinic and phosphoric acids. The structures of $(MeSn(O)O_2CMe)_6$, $[(MeSn(O)O_2CMe_3)(MeSn(O)O_2PBu^t_2)]_3$, and $[Bu^nSn(O)O_2P(OPh)_2]_6$ all exhibit the hexanuclear $(Sn_3O_3)_2$ core.[325]

The ladder structure is also a common structural motif for higher nuclearity tin carboxylate clusters and can be seen as an unrolled drum structure, and this structural type (120), seen in such complexes as $[(Bu^nSn(O)O_2CPh)_2(Bu^nSnClO_2CPh)_2]_2$[326] and $\{[Bu_2SnO(R)]_2O\}_2$ $(R = C(O)CH_2SC(O)N(CH_2CH_2)_2O)$ is commonly found. The latter compound was more active than *cis* platin against a number of cancer cell lines.[327]

Other complexes with different geometries can be prepared by varying the acid used, and though there are no known examples of the drum structure for polynuclear tin complexes of phosphinic acids, other structural types such as cubes $[\{Bu^nSn(O)O_2PBu^t_2\}_4]$ and butterflies $[\{Bu^nSn(OH)O_2P(OPh)_2O\}\{(PhO)PO_2\}]$ have been observed.[328,329] Mixed complexes of phosphinic and thiophosphinic acids show yet further structural types, as shown by the linear tetramer (121), prepared from $Bu^nSn(O)OH$, Ph_2PO_2H and $Ph_2P(S)OH$[330] and the double cube (122) prepared from the reaction of $Bu^nSn(O)(OH)$ and Ph_2PO_2H in the presence of elemental sulfur.[331] Triphenylmetal monothiophosphinato complexes $Ph_3M(OSPR_2)$ $(M = Ge, Sn; R = Me, Et, Ph)$ show monodentate coordination of the monothiophosphinate through oxygen to form the four-coordinate germanium complexes and five-coordinate tbp tin complexes $[\{R_2P(S)O\}Ph_2SnOH]_2$ with bridging hydroxy groups.[332]

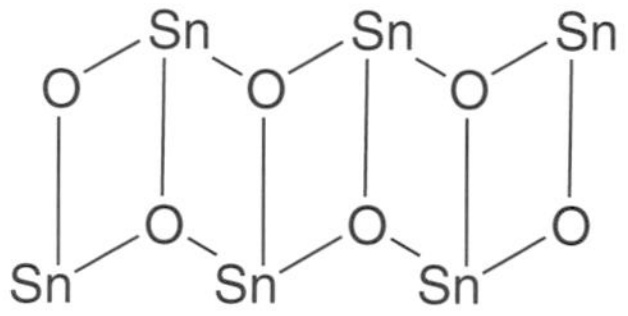

(120)

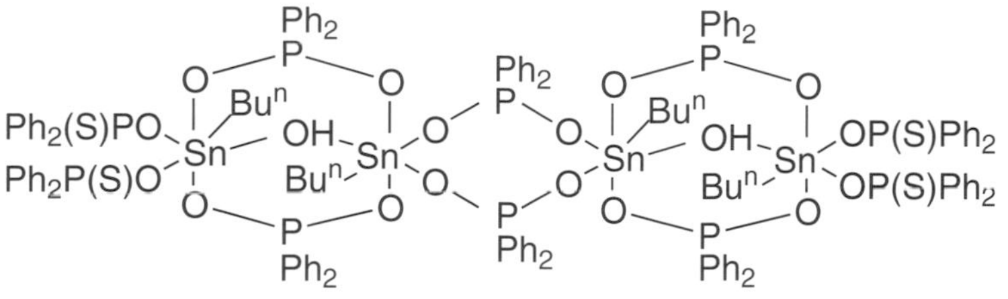

(121)

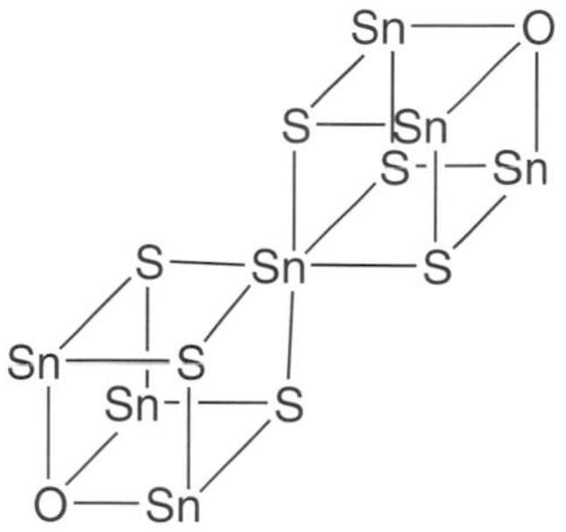

(122)

Carbon dioxide reversibly inserts into the Sn—O bond of compounds Bu^n_3SnOR $(R = Me, Pr^i, Bu^t, SnBu^n_3)$ to give Ph_3SnOCO_2R. Treating these insertion products with caesium fluoride and methyl iodide yields Bu^n_3SnF and dimethyl carbonate.[333]

Lead(IV) carboxylates exhibit a less diverse coordination behavior, so that complexes of N-protected amino acids $Ph_2Pb(L)_2$ (L = R-LLeu-OH), $ClCH_2CO$-X-OH (X = Gly, DLAla, LLeu), $Cl_3CC(O)$-DLAla-OH, $F_3CC(O)$-X-OH (X = DLAla, LPhe) are polymeric six-coordinate and $Ph_3Pb(L)$ five coordinate with tbp chain structures, showing bidentate carboxylates in both cases.[334,335] The pentanuclear $\{2,4,6-(CF_3)_3C_6H_2S\}_8(O)Pb_5$, (123) was isolated after adventitious oxidation of the thiol complex $[2,4,6-(CF_3)_3C_6H_2S]_2Pb$ during isolation.[336]

Tin(II) carboxylates $M_2Sn(C_2O_4)_2 \cdot nH_2O$ (M = NH_4, Na, K, Rb, Cs; n = 0, 1) all exhibit distorted square planar geometry as determined by Mössbauer spectroscopy. The molecular structure of $K_2[Sn(O_2C)_2CH_2)_3] \cdot H_2O$ is polymeric with malonates that bridge tin centers (124).[337] Partial oxidation of $(CF_3CO_2)_2Sn$ allows isolation of the mixed oxidation state penta-nuclear assembly $Sn^{IV}_4Sn^{II}(O)_3(CF_3CO_2)_8$.[338]

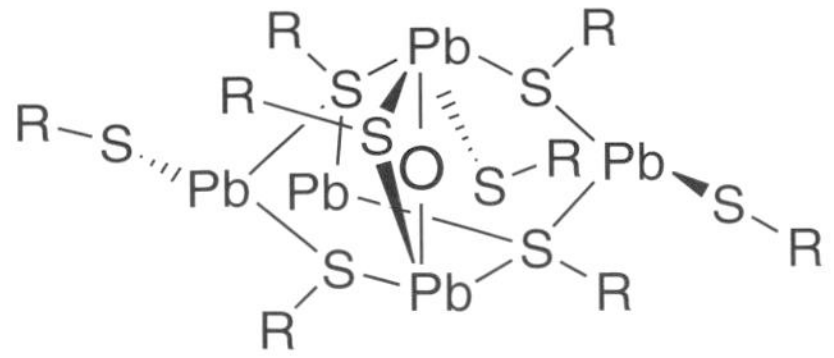

(123)

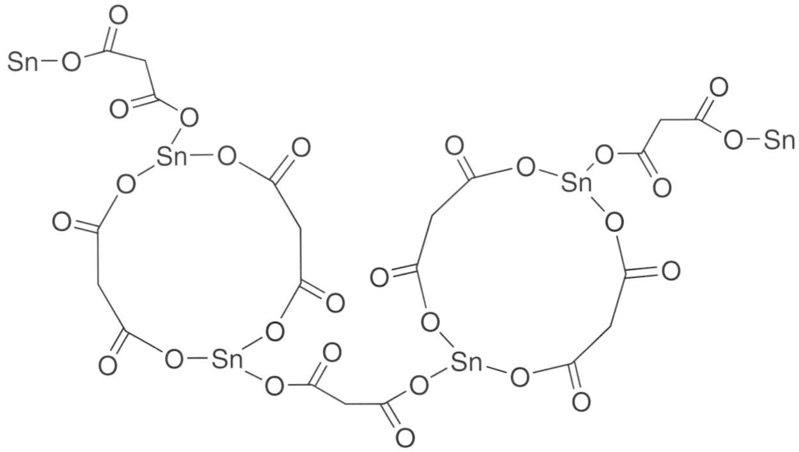

(124)

3.7.5.5 Complexes with Dianionic Bidentate Oxygen Ligands

Germanium powder reacts with 3,5-Bu$^t_2$ benzoquinone in refluxing toluene to give products that comprise corresponding 3,5-Bu$^t_2$catecholato complexes of germanium(IV). With an initial ratio Ge:benzoquinone of 1:2, the product isolated is a neutral oligomeric species [Ge(3,5-Bu$^t_2$cat)]$_n$ (cat = dianion of catechol). Addition of the chelating ligand bipy allows isolation of the mononuclear (bipy)Ge(3,5-Bu$^t_2$cat)$_2$. If the ratio of metal: ligand is 3:1, the product is the six-coordinate diradical species Ge(3,5-Bu$^t_2$cat)$_2$(3,5-Bu$^t_2$cat).[339] Other complexes (3,5-Bu$^t_2$cat)GeX$_2$ can be prepared from the reaction of the benzoquinone with GeX$_2$ (X$_2$ = F$_2$, Cl$_2$, OMe$_2$, Cl(OMe), F(OMe), Et$_2$, Ph$_2$) or from either R$_2$GeX$_2$ or Ge(OMe)$_4$ with the catechol.[340] Dimethyl germanium(II), generated by the thermolysis of 7-germanabornadienes[341] reacts with linear, acyclic, or orthoquinone diketones to form the corresponding complexes Me$_2$Ge(diol) (125)–(129).[342]

Catechol, tetrachlorocatechol, or 3,4-Me$_2$-thiocatechol reacts with RGeCl$_3$ (R = Me, Ph) to form the five coordinate anions [R(κ^2-L-E_2)$_2$Ge]$^-$ with geometries close to tbp.[343] The related ethane-1,2-dithioate (edt) containing anion [Ph(edt)$_2$Ge]$^-$ has also been prepared[344] as have the mixed catechol-thiocatecholate complexes [NR$_4$][X(C$_6$H$_4$SO)$_2$Ge] (X = F, Cl, Br) prepared from (C$_6$H$_4$SO)$_2$Ge and [NR$_4$]X. The geometries of these complexes are distorted from tbp toward square pyramidal and the extent of distortion is dependent upon the nature of the chelate ring. Square pyramidal geometry seems to be stabilized in the cases where there are two unsaturated five-membered chelate rings comprising like atoms within each ring.[345] An X-ray crystallographic study of K$_2$[Ge(cat)$_3$] has shown the geometry at the germanium to be close to octahedral.[346]

Six-coordinate (NH$_4$)$_2$[Sn(cat)$_3$] complexes with a range of substituted catechols have been prepared and characterized. The complexes are six-coordinate as determined by the value of δ in the ^{119}Sn NMR. In the ^{1}H NMR it is possible to observe well resolved long range $^4J^{119}$Sn–^{1}H and $^5J^{119}$Sn–^{1}H couplings.[347]

Catecholato complexes of tin may also be prepared by the reaction of SnX$_2$ (X = Cl, Br, I) with tetrachlorobenzoquinone in the presence of phen, yielding (phen)SnX$_2$(C$_6$Cl$_4$O$_2$). If TMEDA is used in place of phen, a variety of products are formed, including the corresponding complexes and (C$_6$H$_{18}$N$_2$)[Sn(C$_6$Cl$_4$O$_2$)$_3$].[348]

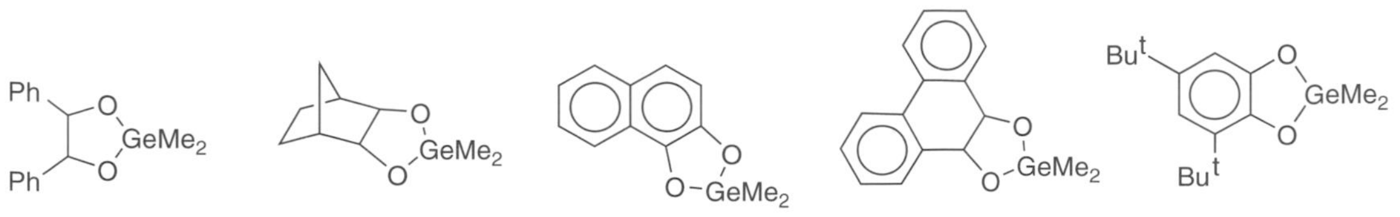

(125)   **(126)**   **(127)**   **(128)**   **(129)**

Anodic oxidation of tin in the presence of catechol and derivatives (catH$_2$, Br$_4$catH$_2$, 2,3-(HO)$_2$-naphth, 2,2'-(OH)$_2$biphenyl) gives complexes [Sn^{II}(L)]$_n$ which can be further converted to a range of tin(II) and tin(IV) diolate complexes.[349] The structure of [(4-NO$_2$-cat)Sn·THF]$_n$ has been determined and the geometry at the tin shown to be distorted square pyramidal with chelation by one catechol (Sn—O 2.112, 2.208 Å), two intermolecular tin–oxygen bonds with two different [(4-NO$_2$cat)Sn·THF] units (Sn—O 2.430 Å) and coordination of a molecule of THF through the ether oxygen (Sn—O 2.535 Å).[350]

The structure of (Me$_3$Sn)$_2$CO$_3$ is polymeric in the solid state arising from the tridentate coordination of the carbonate ion as (130). The C—O bonds show distinct differences in bond lengths (1.267, 1.264, and 1.315 Å) which suggests that there is some localization of the charge despite all three oxygens coordinating tin.[351]

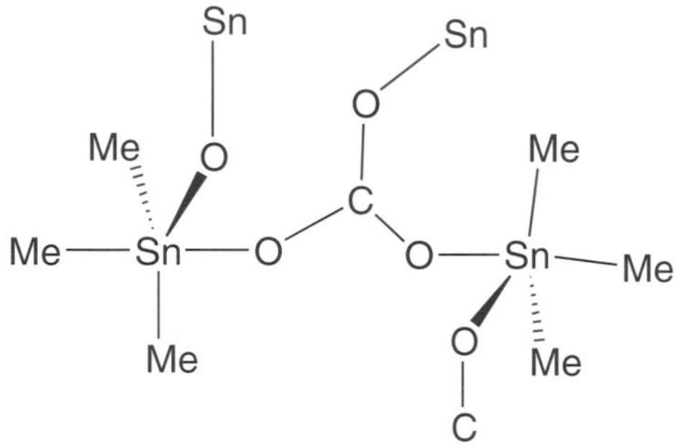

(130)

A number of examples of Zwitterionic five-coordinate germanium(IV) complexes with two dianionic chelating ligands and an alkyl group with a remote basic nitrogen have been isolated reported including mononuclear complexes with 2,3-(HO)$_2$naphthalene[352] or 2-(HO)-carboxylates[353] as the chelate.

Piperazine reacts with two equivalents of (chloromethyl)trimethoxygermanium(IV) to give [1,4-bis(trimethoxygermyl)methyl]piperazine, which is further reacted with 2-Me-2-HO-propionic acid to give the first dispirocyclic Zwitterionic germanium(IV) complex λ^5Ge,λ^5Ge'-digermanate meso[1,4-piperaziniumdiylbis(methylene)-{bis[bis-2-Me-2-OH-proprionate$O$,O]germanate}(131). The complex comprises two pentacoordinate germanium(IV) centers with formal negative charges and distorted tbp geometries with carboxylate oxygens in the axial positions (bond lengths Ge—O 1.769–1.919 Å).[354]

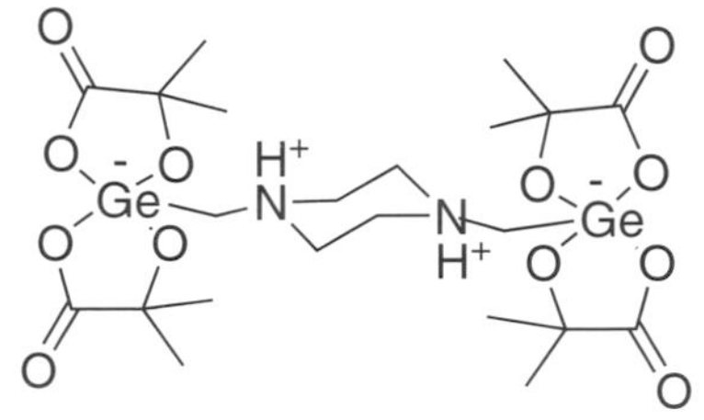

(131)

The complexes (132) (Ge) and (133) (Sn) of the sterically demanding aryloxide react with oxidizing agents to form the oxo-bridged dimer (134) and with 3,5-Bu$^t_2$ benzoquinone to form the four coordinate (135) in which the chelate is acting as a catecholate (Scheme 23).[355]

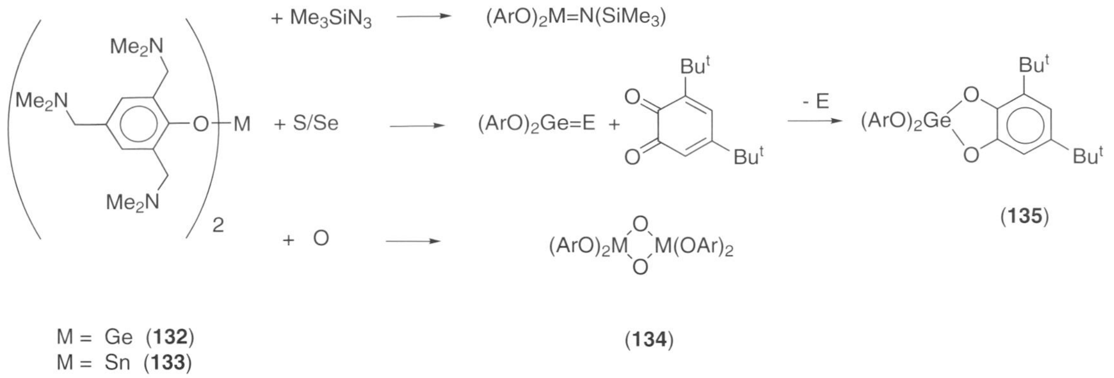

Scheme 23

Germanium and tin complexes of $1,3\text{-}(SiMe_3)_2\text{-}4\text{-}Bu^t$ calix(4) arene can be prepared in both *exo* (Ge, Sn) and *endo* (Ge) forms. In the *endo* form, the metal is two coordinate, but the extent to which the ether oxygens are involved in bonding even in the *exo* forms is not clear. The distances M–O(SiMe$_3$) are not over long (Ge 2.421, 2.486; Sn 2.521, 2.532 Å) but their donating ability is in doubt. The *endo* isomer is thermodynamically preferred for germanium, and is converted to the *exo* form only on prolonged heating at $>80\,^\circ C$. The corresponding lead complex is unstable to light and has not been characterized thoroughly.[356,357]

The general area of germanium and tin coordination by bidentate oxygen donor ligands has been reviewed.[358]

3.7.5.6 Complexes with Neutral Sulfur Ligands

Diorganotin nitrate readily forms cationic complexes with a range of neutral ligands upon dissociation of the nitrates. The thione Hmimt coordinates to tin giving the complexes $[R_2Sn(Hmimt)_4](NO_3)_2(R = Me,$[359] $Et,$[360] or Ph[361] where all complexes have *trans* organic groups.

The sulfur ligands 1,4-dithiane and 6aneS$_3$ react with tin(IV) chloride or bromide to give complexes $(\kappa^2\text{-}1,4\text{-dithiane})SnX_4$, $(\kappa^n\text{-}6aneS_3)_2SnX_4$,[362] and ligands 9aneS$_3$ and 18aneS$_6$ react with tin(IV) chloride to give $[\kappa^3\text{-}(9aneS_3)SnCl_3][SnCl_5]$ and $(\kappa^2,\kappa^2\text{-}\mu\text{-}18aneS_6)(SnCl_4)_2$. In the ionic complex, the bond lengths Sn—Cl are 2.369 Å in the cation and 2.448 Å in the anion, and in the latter complex, the two tin centers are symmetrically bound to two thioether sulfurs giving overall six-coordinate tin centers.[363] The larger thioether macrocycle 28aneS$_8$ forms a dinuclear complex $[(28aneS_8)Pb_2](ClO_4)_4$ with inclusion of the two lead(II) within the ligand leading to an $[S_4O_4]$ coordination of each lead by the ligand and two chelating perchlorates.[364]

3.7.5.7 Complexes of Anionic Monodentate Sulfur, Selenium, or Tellurium Ligands

The sterically stabilized aryloxides (132) and (133) have been reported and for the germanium complex, shown to react with sulfur or gray selenium to afford the corresponding germathione or selenone. The tin complex did not react with either chalcogen. The germathione and selenone each react with 3,5-Bu^t-1,2-benzoquinone with extrusion of the sulfur or selenium to give the four-coordinate catecholato bis aryloxy germanium compound.[365]

Direct reaction of sulfur with $[2,4,6\text{-}\{CH(SiMe_3)_2\}_3\text{-}C_6H_2](mes)MH_2$ (M = Ge, Sn) gives the tetrasulfur ring compounds which are converted to the germa- or stannathione on heating.[365–367] The germanium complex (R)(mes)GeS$_4$ reacts further with Ph$_2$CN$_2$ to give (R)(mes)Ge(S$_4$CPh$_2$) and two isomers of (R)(mes)Ge(S$_4$CPh$_2$).[368] When $[2,4,6\text{-}\{CH(SiMe_3)_2\}_3C_6H_2](mes)GeBr_2$ is treated with lithium naphthide and gray selenium (R)(mes) GeSe$_4$ is formed, comprising a five membered [GeSe$_4$] ring. The ring can be contracted by reaction with three equivalents of triphenylphosphine, forming triphenylphosphine selenide and the germaselenone R(mes)GeSe (136) (Ge—Se 2.180 Å). The germaselenone reacts with mesCNO or PhNCS to give products with the corresponding (GeSeCNO) (**137**) and (GeSeCS) heterocycles (**138**) (Scheme 24).[369–371] The tin selenones $[2,4,6-\{CH(SiMe_3)_2\}_3C_6H_2](R)SnSe$ (R = $2,4,6-Pr^i_3C_6H_2$, mes) (**139**) are available from the corresponding diaryl tin(II) and gray selenium[372] and these selenones each react with other chalcogen bearing molecules to yield products with tin-bound ring structures. Phenyl isothiocyante gives a mixture of the phenyldiselena- and dithiastannanes (**140**) and (**141**) rather than the expected mixed thiaselenastannane (Scheme 25). This is an intriguing reaction, which seems to require a bimolecular intermediate to give rise to the observed products. The same diaryl tin(II) reacts with carbon disulfide giving a product that is a symmetrical tetrathiaethylene-bridged dimer.[373] The corresponding lead(II) aryls react with sulfur to form $R^1R^2PbS_4$[374] whereas $(2,4,6-Pr^i_3C_6H_2)_2Pb$ gives $R^1_2PbS_4$ and both $[R^1_2PbS]_2$ and $R^1_4Pb_2S_3$ that have a central $[Pb_2S_3]$ ring. Interestingly, $[2,4,6-\{CH(SiMe_3)_2\}_3C_6H_2]_2Pb$ reacts with sulfur to give no products containing lead and sulfur but principally R^1S_nR ($n = 6$, 8).[375] The lead thione R^1R^2PbS is stable at temperatures below $-20\,^\circ C$, above which it dimerizes, forming a $(PbS)_2$ ring. Reaction with phenyl isothiocyanate below $-20\,^\circ C$ gives the phenyldithiaplumbane (**142**).[376]

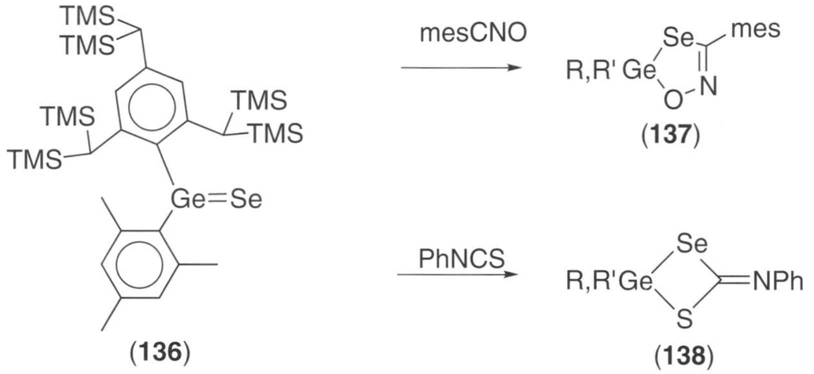

Scheme 24

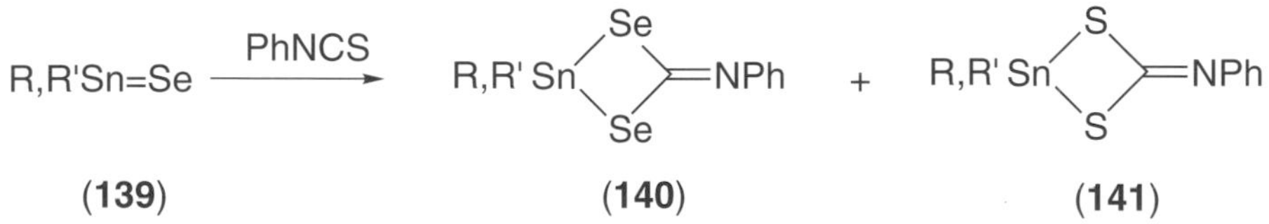

Scheme 25

The tetraselenium ring in $[2,4,6-\{CH(SiMe_3)_2\}_3C_6H_2](R)SnSe_4$ (R = $2,4,6-(Cy_3)C_6H_2$, $2,4,6-(CHEt_2)C_6H_2$, $2,6-(2-Pr^iC_6H_4)_2C_6H_3$) can be contracted to form the selenone again by reaction with three equivalents of triphenylphosphine. Unusual among such monochalcogenides, these complexes are monomeric under ambient conditions, presumably by virtue of the enormous ligands, where the other known examples all dimerize by forming $[SnE]_2$ bridges. By using only two equivalents of triphenylphosphine in the ring contracting deselenation, the remarkable perselenide $[2,4,6-\{CH(SiMe_3)_2\}_3C_6H_2](R)SnSe_2$ can be isolated (Se—Se $2.524\,\text{Å}$, Sn—Se 2.530 (ave.) Å).[377]

The area of germanium sulfide cluster anions has been enriched by the publication of a number of new compounds. The structure of the quaternary germanium sulfide $AgLa_3GeS_9$ has been elucidated and shown to comprise La_3GeS_4 cubes linked through Ge—S bonds to form a three dimensional array.[378] In contrast, the anion $[CuGe_2S_5]^-$ comprises $[Ge_4S_{10}]^{4-}$ units.[379]

The combination of germanium sulfide, silver acetate, and DABCO leads to the formation of the complex sulfide $[(DABCO)_2(H_5O_2)]AgGe_4S_{10}$, which has a three dimensional array of Ge_4S_{10} clusters linked by triply bridging silver ions.[380] Germanium, selenium, and silver acetate react in the presence of M_2CO_3 (M = Rb, Cs) to give $M_3AgGe_4S_{10}$. This structure is somewhat different, with a four-fold Ag-$[Ge_4S_{10}]$ interaction in the solid.[381] The nonadamantane [5.1.1.1] tetragerma-hexachalcogenanes (**143**) are prepared from the reaction of $RGeCl_3$ with the appropriate lithium chalcogenide[382] or $(NH_4)_2S_5$.[383] These rearrange to the more stable [3.3.1.1] adamantane structure on heating. In reaction with hydrogen sulfide, Bu^tGeCl_3 forms the cyclic tetramer (**144**) which again rearranges thermally to the adamantane structure.[383] Adamantane $(RGe)_4E_6$ structures can also be prepared from the reactions of $RGeCl_3$ with $(H_3Si)_2E$ (R = CF$_3$, Et, E = S; R = CF$_3$, E = Se).[384] Corresponding tin complexes $(RSn)_4S_6$ can be prepared from $RSnCl_3$ and either Na_2S or $(Me_3Si)_2S$, but using R_2SnCl_2 leads to the cyclotrimeric $(R_2SnS)_3$.[385]

(142)　　　　　　　　　　　(143)　　　　　　　　　　　(144)

The selenide or telluride Rb_2GeE_4, prepared from Rb_2CO_3, germanium, and the chalcogen, has been reported. When E = Se, the structure comprises tetrahedral [GeSe$_4$] units with terminal (2.27–2.30 Å) and bridging (2.42–2.44 Å) Ge—Se bonds. Also isolated from the same reaction, $Rb_4Ge_4Se_{10}$ has an adamantane Ge_4S_{10} with terminal (2.25 Å) and cage (2.40–2.39 Å) Ge—Se bonds.[386]

Sodium sulfide reacts with $[PPh_4][SnCl_3]$ to give $[PPh_4]_2[Sn(S_4)_3]$ with discrete six-coordinate tin centers, whereas $\{NHMe_3][Sn_3S_7]$ comprises sheets of 24-membered rings having six [Sn$_3$S$_4$] units connected through sulfide bridges at each tin.[387,388] Hydrothermal reaction of caesium carbonate with tin(IV) sulfide at 130 °C gives $Cs_4Sn_5S_{12}\cdot2H_2O$ which comprises polythiostannate(IV) sheet anions $[Sn_5S_{12}]^{4-}$ with octahedral [SnS$_6$] and pyramidal [SnS$_5$] units.[389]

Tin selenide $Bi_2Sn_3Se_6$ can be reduced by potassium in the presence of $[PPh_4]^+$ to give the anionic $[Sn_2Se_4Ph_2]^-$ which has a planar (SnSe)$_2$ ring, analogous to (SnO)$_2$, substituted with *trans* phenyl groups and selenides.[390] The tin chalcogenides $[Sn_2E_6]^{4-}$ (**145**) and $[Sn_2E_7]^{4-}$ (**146**) (E = Se, Te) can be isolated by extracting the alloys $K_3Sn_2Se_6$ or $K_3Sn_2Te_5$ with alkaline solutions containing [2,2,2].[391] From the same reaction, the first mixed hydroxychalcogeno anion of tin was isolated $[(HO)Te_3Sn]^{3-}$. The telluride version of (**145**) can also be prepared from the reaction of the Zintl anion $[Sn_9]^{4-}$ with elemental tellurium.[392] The reaction of potassium or rubidium carbonate with tin and selenium in aqueous methanol gives $M^I_6Sn_4Se_{11}\cdot8H_2O$, which comprises $[Sn_4Se_{11}]^{6-}$ ions where the corresponding reaction with caesium carbonate gives a product containing the $[Sn_2Se_5]^{2-}$ ion, which has a chair configuration. Chains of $[Sn_3Se_7]^{2-}$ are formed in the reaction using tetraethyl ammonium in place of an alkali metal ion.[393]

(145)　　　　　　　　　　　(146)

Cyclic trimeric $[Bu^n_2SnTe]_3$ is formed in the reaction of $[NH_4]_2Te$ and $Bu^n_2SnCl_2$ and can be used as a single source precursor to cubic tin selenide.[394]

3.7.5.8 Complexes of Anionic Bidentate Sulfur, Selenium, or Tellurium Ligands

Carbon disulfide inserts into the tin–carbon bonds of $(2,4,6-Bu^t_3C_6H_2)_2Sn$ to form both $(2,4,6-Bu^t_3C_6H_2)(\kappa^2-2,4,6-Bu^t_3C_6H_2CS_2S,S)Sn$ and $(\kappa^2-2,4,6-Bu^t_3C_6H_2CS_2S,S)_2Sn$.[395] A similar insertion is also seen for $(RS)_2Pb$ complexes (R = 2,6-CH(SiMe$_3$)$_2$-4-C(SiMe$_3$)$_3$C$_6$H$_2$, 2,4,6-CH-(SiMe$_3$)$_3$C$_6$H$_2$, the first examples of thiocarbonate complexes of lead(II).[396]

Organotin complexes $R_3Sn(L)$ (R = Me, Ph; L = S$_2$CNEt$_2$, S$_2$COEt, S$_2$P(OEt)$_2$), $R_2Sn(L)_2$ (R = Me, Bu^n, Bu^t, Ph), and $R_2SnX(L)$ (R = Me, Bu^n, Bu^t, X = Cl; R = Ph, X = Cl, Br) were studied by NMR spectroscopy. For triorganotin derivatives, only dithiocarbamate shows spectra consistent with a chelation by the sulfur ligand at ambient temperatures, the other ligands being involved in rapid interconversion between monodentate and chelate attachment. In solution, the dithiocarbamate ligands are chelating in $Me_2Sn(S_2CNEt_2)_2$ but are monodentate in $(Bu^t)_2Sn$ $(S_2CNEt_2)_2$. The diorganotin derivatives are more effectively chelated, and the extent to which the molecule is nonrigid in solution is dependent upon the nature of the organic ligand, such that at $-100\,°C$ $Ph_2SnCl(S_2CNEt_2)$ is stereochemically rigid in solution and $(Bu^t)_2SnCl(S_2CNEt_2)$ is not.[397]

Diorganotin bis(xanthates) $R^1_2Sn(S_2COR^2)_2$ ($R^1 = Me$, Et, Bu^n, Ph; $R^2 = Et$, $CHMe_2$, cy) also exhibit the asymmetric bonding of the two sulfur atoms to the extent that the six-coordinate complexes are skewed trapezoidal rather than octahedral in geometry.[398] Dimethyl bis(ethoxy-xanthato)tin(IV) exhibits a solution ^{119}Sn NMR indicating that the complex is four-coordinate whereas the crystal structure shows a six-coordinate geometry albeit with markedly asymmetric Sn–S interactions.[399] In a similar vein, the complex $BuPhSn(S_2CNMe_2)$ also shows an asymmetric coordination of the chelate with bond lengths Sn–S of 2.466 and 3.079 Å. The diethyldithiocar-bamate complex has Sn—S bonds that are closer in length (2.454 and 3.764 Å) but still show an asymmetry.[400] A useful qualitative discussion of the observed asymmetry in Sn–S bonds for tin(IV) dithiolates has appeared.[401] Thiocarboxylic acids form five-coordinate complexes with M(14) such as $(4\text{-}MeC_6H_4CS_2)MPh_3$ ($M = Ge$, Sn, Pb) that also show anisobidentate coordination of the $[S_2]$ donor set.[402]

With dimethyl dithiophosphinic acid, germanium(IV) chloride forms a tetrahedral complex (Ge—S 2.218–2.236 Å) with monodentate coordination of the ligand[403] as do the organotin dithiophosphates $Me_3Sn(S_2P(OEt)_2)$ and $MeSn(S_2P(OEt)_2)_3$.[404]

Tin(II) complexes of dithiophosphates $[(RO)_2PS_2]_2Sn$ ($R = Me$, Et, Pr^i, Ph) have a dimeric structure comprising five-coordinate tin centers, each coordinated by an approximately symmetrical chelating ligand (Sn—S 2.830, 2.623 Å) and one short and two long intermolecular bridging interactions (Sn—S 2.651, 3.042, 3.391 Å).[405]

Bis(diorganophosphorylchalcogeno)amides $[\{R_2P(E)\}_2N]^-$ are compounds with excellent ligand properties and there are a correspondingly large number of complexes known. Complexes of the Group 14 metals are known mostly for the M^{II} state although $[\{Ph_2P(S)\}_2N]_2SnMe_2$ has been reported.[406] For M^{II}, the complexes $[\{R_2P(E)\}_2N]_2M$ ($R = Ph$, $E = O$, $M = Sn$;[407] $R = Ph$, $E = S$, $M = Pb$;[408] $R = Ph$, $E = Se$, $M = Sn$, Pb^{409}) a distorted tbp geometry is observed. The complex $[\{Ph_2P(Se)\}_2N]Sn$ also crystallizes in a second form giving the first example of a square planar spiro tin(II) complex.[410] Unsymmetrical examples can also be prepared and can be used to form complexes such as $[\{Ph_2P(S)\}\{Ph_2P(O)\}N]_2M$ ($M = Sn$, Pb).[411]

Alkyl and aryltin(IV) diphenyldithioarsenates $R_nSn(S_2AsPh_2)_{4-n}$ ($n = 2$, $R = Me$, Bu^n, Ph; $n = 3$, $R = Me$, cy, Ph) are available from the organotin halides and the sodium salt of diphenyldithio-arsenates. The dialkyl and trialkyltin species are four-coordinate by spectroscopy whereas the phenyl derivatives are six-coordinate. A structural study of $Me_2Sn(S_2AsMe_2)_2$ reveals a four-coordinate tin center with monodentate coordination of the dithioarsenate.[412]

3.7.5.9 Complexes of Dianionic Bidentate Sulfur, Selenium, or Tellurium Ligands

Lawesson's reagent reacts with germanium amines to give products that vary according to the nature of the starting material. Germanium(II) amine $[HCN(Pr^i)]_2Ge$ gives the oxidized bis chelated product (147) whereas $(TMS_2N)_2Ge$ gives (148), with the monodentate amines intact.[413]

Germanium complexes $(C_2H_4E_2)Ge$ ($E_2 = S_2$, SO) react with 3,5-Bu^t_2benzoquinone to give the mixed $(3,5\text{-}Bu^t_2cat)Ge(C_2H_4E_2)$ complexes. These rearrange rapidly to give the homoleptic complexes $(3,5\text{-}Bu^t_2cat)_2Ge$ and $(C_2H_4E_2)_2Ge$.[414] The reaction of $K_2[edt]$ with R_2SnCl_2 or tin(IV) chloride gives the complexes $R_2Sn(edt)$ or $(edt)_2Sn$, respectively. The complexes $Bu^n_2Sn(edt)$ have a six-coordinate geometry in the solid state with two intermolecular Sn···S interactions complet-ing the coordination sphere and forming a linear polymer. If the R group is smaller, the geometry at the tin is five-coordinate tbp, with only one strong intermolecular Sn–S interaction.[415] The dianion of toluene 3,4-dithiolate (tdt) forms analogous complexes $(tdt)_2Sn$, which has a similar solid-state structure with intermolecular Sn···S interactions making a six-coordinate geometry at the tin. Addition of bases DMSO or triphenylphosphine oxide gives the mononuclear $(tdt)_2Sn(base)_2$ complexes with *trans* disposition of the monodentate ligands.[416] Anionic edt complexes $[Et_4N][(edt)_2SnR]$ ($R = Bu^n$, Ph) and $[Et_4N]_2[(edt)_2R(Cl)SnSCH_2]_2$ show square pyram-idal geometries for the mononuclear complexes and distorted tbp for the dimeric complexes.[417]

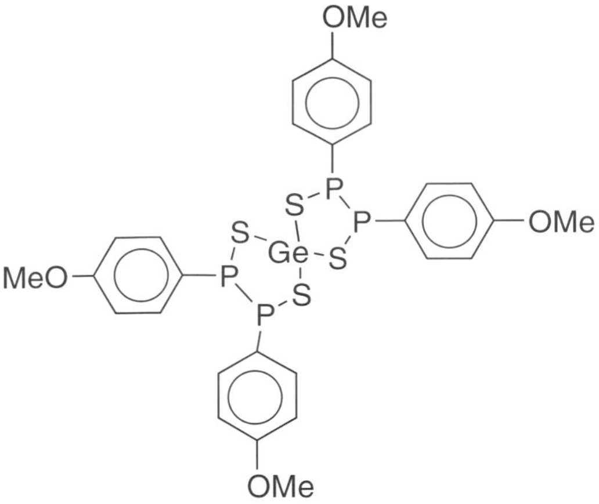

(147)

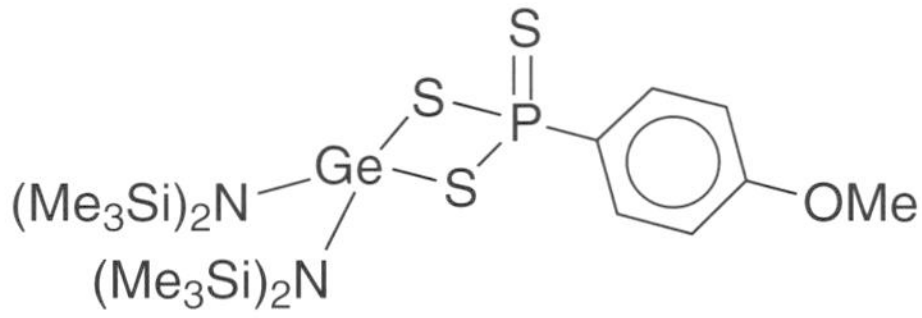

(148)

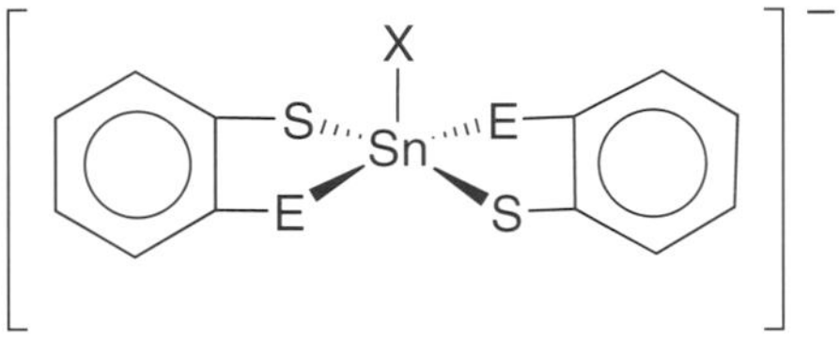

E = S, R = Me, X = Cl (149)
E = O, R = H, X = F   (150)
E = O, R = H, X = Cl   (151)

The five-coordinate complexes [A][(L)$_2$SnX] (A = Ph$_3$MeP, H$_2$L = 3,4-tdt, X = Cl (149); A = Et$_4$N, H$_2$L = 2-O-SC$_6$H$_4$, X = F (150), Cl (151)) can be prepared by addition of [Ph$_3$MeP]Cl to Sn(3,4-toluenedithiolate)$_2$ or by direct reaction of tin(IV) acetate, 2-OH-thiophenol, and [Et$_4$N]X. The sulfur-coordinated (149) has a square pyramidal geometry at the tin, where the mixed oxygen–sulfur complexes (150) and (151) have tbp geometries. These complexes can be hydrolyzed to form six-coordinate tin species, such as [Et$_4$N][H]$_3$[(L)$_3$Sn] from (150), which have distorted octahedral geometry.[418]

Dithiolates (R$_3$M)$_2$(L) (L = 3,4-tdt, M = Sn, Pb, R = Ph; L = 1,2-Me$_2$-bdt, M = Sn, R = Me, Ph; M = Pb, R = Ph), R$_2$M(L) (L = 1,2-bdt, M = Pb, R = Me, Et, Pb; L = 3,4-tdt, M = Sn, R = Me, Ph, M = Pb, R = Me, Et, Ph; L = 2,3-dithioquinoxaline, M = Pb, R = Ph), and Pb(L)$_n$ (L = 1,2-bdt, n = 2; L = 3,4-tdt, n = 2; L = 1,2-Me$_2$-bdt, n = 1 or 2) all exhibit spectral properties consistent with mononuclear complexes having four-coordinate geometries.[419]

In an attempt to prepare new dithiolate complexes Ph$_2$PbCl$_2$ was allowed to react with (NR$_4$)$_2$[Zn(MNT)$_2$], giving Ph$_2$Pb(MNT)$_2$, which further reacts with (NR$_4$)I to give (NR$_4$)[Ph$_2$Pb(MNT)$_2$I]. Triphenyllead chloride reacts with (NR$_4$)$_2$[Zn(MNT)$_2$] to give (Ph$_3$Pb)$_2$(MNT) which has symmetrical monodentate coordination of [MNT]$^{2-}$ (Pb—S 2.523, 2.580 Å).[420]

Lead(II) ethane-1,2-dithiolate is polymeric in the solid state with each lead is chelated by a dithiolate and has a further four close interactions with other neighboring sulfur atoms, giving an overall six-coordinate geometry.[421] The chelating ligands K$_2$[E$_2$C$_2$(CN)$_2$] (E = S, Se) react with lead to give the complexes M^I$_2$[{(CN)$_2$C$_2$E$_2$}$_2$Pb] (M^I = K, E = S; M^I = Ph$_4$As, E = Se).[422,423]

3.7.6 COMPLEXES WITH GROUP 17 LIGANDS

Halide complexes of Group 14 metals continue to offer surprises in their structural chemistry. A new fluoro complex of germanium Ge$_7$F$_{16}$ has been isolated from the decomposition of germanium(IV) fluoride and shown by crystallography to comprise sheets of [Ge$_6$F$_{10}$]$^{2+}$ clusters interspersed by $[GeF_6]^{2-}$ anions.[424] A different structural motif is seen in $[Ge_5Cl_{12} \cdot GeCl_4]$, a product isolated from the thermal decomposition of germanium(IV) chloride. The pentanuclear cluster has a neopentyl arrangement of germanium atoms and has Ge—Cl bond lengths that are longer than those in the $GeCl_4$ unit (2.119 cf. 2.081 Å).[425]

Addition of excess fluoride ions to aqueous or acetonitrile solutions of $(CF_3)_3GeX$ (X = F, Cl, Br) or $(CF_3)_4Ge$ gives the tbp complex $[(CF_3)_3GeF_2]^-$, and octahedral *fac* $[(CF_3)_3GeF_3]^{2-}$ or *cis* $[(CF_3)_4GeF_2]^{2-}$, respectively. The structures of the anions have been elucidated by ^{19}F NMR and a crystallographic study of $[N(Me)_4][(CF_3)_3GeF_2]$ shows the anion to have axial fluorides and equatorial CF_3 groups.[426]

Addition of $[Et_4N]F$ to a solution of Me_2SnF_2 leads to the formation of the organofluorostannate $[Me_4Sn_2F_5]^-$. The dimeric structure is derived from $[Me_2SnF_3]^-$, and even at low temperatures there is no evidence of coupling between the fluorine and tin nucleii, indicating a rapid fluxional process.[427]

Structural studies on anionic heptafluoro complexes $[X]^{3+}[F_7M]^{3-}$ (X = (NH$_4$)$_3$, M = Sn;[428,429] X = Ln, Tl, M = Sn, Pb[430] show that the complexes comprise octahedral $[F_6M]^{2-}$ ions and isolated fluoride ions rather than any seven-coordinate metals. Other tin fluorides have more complex structures, such as the dimeric $[Sn_2F_4]^{2-}$ ion present in $[NH_4]_4[Sn_2F_4](NO_3)_2$[431] and in mixed halide complexes, such as $Cs_2Sn_6Br_3F_{11}$ which has in the crystal three distinct tin sites, each of which has close contacts with the fluoride ions only.[432] The electronic effects of adduct formation of halides of germanium(IV) and tin(IV) have been reviewed with particular reference to the geometry of the complexes formed.[433]

Complexes of simple *N*-donor ligands with germanium(IV) fluoride have been studied at low temperatures by matrix isolation techniques. Complexes $RCN \cdot GeF_4$ (R = H, Me) and $py \cdot GeF_4$ give IR spectra consistent with simple complexation by coordination through the nitrogen, even in the case of HCN. For the pyridine complex, the shift in the bands associated with the pyridine were comparable with those seen in pyridine complexes of transition metal ions and are greater than those seen for the corresponding silicon complex, giving an indication of the acidity of the germanium in germanium(IV) fluoride.[434,435]

Six-coordinate mixed halide complexes of tin(IV) supported by pyridine can be prepared from the addition of X_2 (X = Br, I) and $XI \cdot py$ (X = Cl, Br) to $SnCl_2$ in the presence of excess pyridine as $SnCl_2X_2py_2$ (X = Br, I), $SnCl_3Ipy_2$, and $SnCl_2BrIpy_2$. From the IR data of the complexes it is possible to extract a linear regression from the change in the values of the frequency of some of the bands arising from the pyridine ligands in the IR spectra and the electronegativity of the halides.[436]

Tin(II) difluoride oxidatively adds X_2 (X = Cl, Br) in acetonitrile solution to form the mixed monomeric halide $(MeCN)_2SnF_2Cl_2$ and oligomeric $[(MeCN)_2SnF_2Br_2]_n$. For the reaction with I_2, the product obtained is $(MeCN)_2SnF_4$, and as such represents a new and convenient route to the tetrafluoride, by the elimination of the solvent molecules. These may also be exchanged for a range of other ligands, and complexes with DMSO, DMF, THF, and pyridine were reported. In DMSO solution the same reactions lead to disproportionation products in preference to the mixed halides prepared in acetonitrile.[437]

Correlation of the ^{35}Cl and ^{79}Br NQR spectra of four- and five-coordinate organohalides of germanium and tin with structural studies on the same compounds has shown that NQR can be a rapid and effective method for determining the structure of such compounds.[438,439]

The valuable report of a route to the important starting material $GeCl_2 \cdot diox$ and a range of other germanium(II) chloride adducts has appeared. Easy access to these useful compounds is very likely to increase their application in a range of reactions.[440] Variable temperature solid-state NMR studies of PbF_2 have been used to probe the mechanism of fluoride mobility in the lattice. The pathway of the motion of the fluoride has been made on the basis of the lowest resistance to mobility associated with the largest lattice holes.[441] The reaction of $[PPh_4]Cl$ with $PbCl_2$ forms the trichloroplumbate(II) ion, which has a similar structure to the triiodoplumbate(II).[442]

The structural features of iodo complexes are somewhat more complex than those of the lighter halogens, and a range of new iodo complexes of M(14) have been reported. The tin complex $[(NH_3)(CH_2)_3][SnI_4]$ has six six-coordinate tin ions in the asymmetric unit, four edge-sharing and two face-sharing,[443] whereas the complex iodides $[(Me_2NCH_2)_2][SnI_4]$ and $[PPh_4][Sn_2I_6]$ comprise chains of weakly associated $[SnI_4]^{2-}$ and $[SnI_3]^-$ ions in the solid state.[444]

The complex $[(Bu_3NCH_2)_2][Pb_5I_{16}] \cdot 4DMF$ has in its crystal structure an iodoplumbate ion with D_{5h} symmetry. Five nearly octahedral $[PbI_6]$ units are disposed in a planar ring, each sharing a single iodide at the center of the ring and each having two sets of two bridging iodides that make up the central square plane. The coordination is completed by a single terminal iodide *trans* to the central shared iodide.[445] In [PPh$_4$][Pb$_2$I$_6$] and [{N(Bu$^n_3$)}$_2$(CH$_2$)$_3$][PbI$_4$] the lead ions are all four coordinate.[446] The product of the reaction of lead(II) iodide, sodium iodide, and 1,1'-Me$_2$-4,4'-bipyridinium dichloride dihydrate in acetone comprises a linear polymer of face-sharing [PbI$_6$] octahedra.[447] The reaction of lead(II) iodide, sodium iodide, and [(Pr$^n_3$N)$_2$(CH$_2$)$_3$]$^{2+}$ in DMF leads to [(Pr$^n_3$N)$_2$(CH$_2$)$_3$][Pb$_6$I$_{14}$]·4(DMF) and [(Pr$^n_3$N)$_2$(CH$_2$)$_3$][{Pb(DMF)$_6$}Pb$_5$I$_{14}$]·DMF. The former comprises [PbI$_6$]$^{4-}$ and [PbI$_5$(DMF)]$^{3-}$ octahedra sharing edges forming a one-dimensional polymeric structure, the latter comprises lead surrounded by either six bridging iodides, in [Pb$_5$I$_{16}$]$^{4-}$, or six DMF molecules. In the structure of [(Me$_3$N)$_2$(CH$_2$)$_3$][Pb$_5$I$_7$] there are layers of six-coordinate iodoplumbate ions interspersed with noncoordinating iodide ions.[448] However, the giant of this family of compounds is the truly extraordinary (Bu$_4$N)$_8$[Pb$_{18}$I$_{44}$], characterized crystallographically and shown to comprise lead ions coordinated in six-coordinate environments by iodides in a structure reminiscent of an octahedral section of the NaCl lattice.[449] The strategy of including hydrogen-bonding counterions or the inclusion of such solvents in order to partially influence the structure of polyhalo complexes has been discussed.[450]

Lead tetrafluorostannate has a range of useful conducting properties and has been much studied as a result. A crystallographic investigation reveals a structure in which there are two distinct sites for fluoride coordination to tin and a large number of partially occupied sites. These results suggest that the high fluoride mobility may be due to the existence of near-equivalent sites that serve to lower the energy barrier to ion mobility.[451]

3.7.7 COMPLEXES OF HYDRIDE LIGANDS

There has been a marked increase in interest in the hydrides of Group 14 metals arising from the potential use of these metals, principally germanium, in the electronics industry, and the need to find routes to volatile pure compounds for vapor deposition processes. Accordingly, a number of new methods of preparation of both low molecular weight reactive germanes and stable primary germanes have appeared.

The copper-catalyzed reaction of germanium metal with dibromo- or dichloromethane gives mixtures of products depending upon the organohalide used (Scheme 26). All of these organo-germanium halides can be converted to germanes (Scheme 26) making this a very productive approach to germane synthesis.[452,453] Similar reactions with suitable organohalosilanes gives access to mixed volatile germasilanes which are precursors to GeSi materials.[454]

$$Ge \xrightarrow[CH_2Cl_2]{Cu} MeGeCl_3 + CH_2(GeCl_3)_2 + (Cl_2GeCH_2)_3 \xrightarrow{H^-} MeGeH_3 + CH_2(GeH_3)_2 + (H_2GeCH_2)_3 + H_2Ge(CH_2GeH_2)_2$$

$$Ge \xrightarrow[CH_2Br_2]{Cu} MeGeBr_3 + CH_2(GeBr_3)_2 + Br_3GeCH_2Br + (Br_2GeCH_2)_3 \xrightarrow{H^-} MeGeH_3 + CH_2(GeH_3)_2 + H_3GeCH_2Br + (H_2GeCH_2)$$

Scheme 26

General routes to primary and secondary germanes have been developed, either by the oxidative addition of RX to GeCl$_2$·diox followed by hydride reduction of the RGeCl$_2$X formed or by the reaction of Grignard reagents R^{1}MgX with Ge(OR^2)$_4$ compounds, and hydride reduction of the R^{1}Ge(OR^2)$_3$ (Scheme 27). For some of these germanes the ^{73}Ge NMR spectra show well-resolved spectra and $^1J^1$H-^{73}Ge of ca. 100 Hz.[455] Insertion of GeH$_2$, prepared from the flash photolysis of 3,4-Me$_2$-germacyclopentane, into GeH$_4$ gives Ge$_2$H$_6$ in high yield.[456]

$$(EtO)_4Ge + RMgX \longrightarrow (EtO)_3GeR \xrightarrow{H^-} H_3GeR$$

$$GeCl_2 + RX \longrightarrow RGeCl_2 \xrightarrow{H^-} H_3GeR$$

Scheme 27

The first stable crystalline primary germane (152) has been prepared. The complex is monomeric in the solid state.[457] Primary germanes are also liable to dehydrocoupling in the presence of Cp$_2$ZrCl$_2$/Bu^{n}Li forming poly(organogermanes) with moderate (3×10^4–7×10^4) molecular weight.[458]

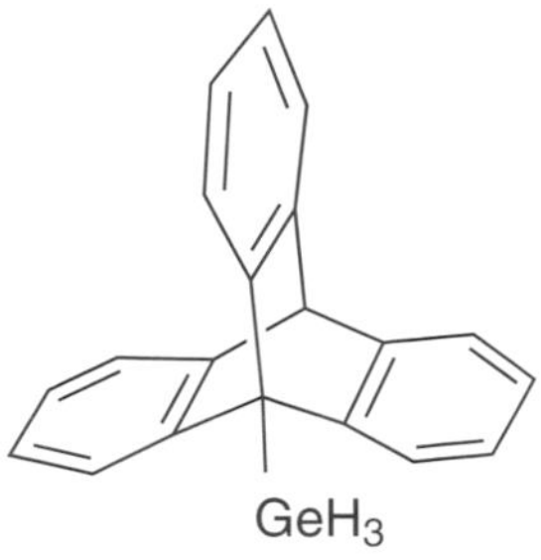

GeH₃

(152)

The first well-characterized tin(II) hydride has been prepared from the DIBAH reduction of [2,6-(trip)$_2$C$_6$H$_3$]SnCl. The hydride is isolated as orange crystals from a blue solution, and has a dimeric structure with a central (SnH)$_2$ ring. The geometry at the tin centers is distinctly pyramidal, indicating that the lone pair is stereochemically active.[459]

3.7.8 COMPLEXES OF LIGANDS WITH MIXED DONOR SETS

3.7.8.1 Complexes of Heterobidentate Ligands

An important group of compounds in this class are those with tethered [C,X] ligands, where an organic group bound to the metal comprises a functional group at an appropriate distance from the *ipso* carbon to allow the coordination of this group to the same metal center, forming a ring structure. In some cases, the tethered group supports the M(14)—C bond, and in some, the M(14)—C bond supports the coordination of an indifferent ligand.

Germanium—aryl bonds can be stabilized by intramolecular coordination of the germanium by secondary donor groups on the organic ligand, such as the methoxy substituent on the naphthalide (**153**). The supporting role played by this oxygen donor (Ge—O ca. 2.357 Å) seems to facilitate the formation of a range of stable complexes.[460]

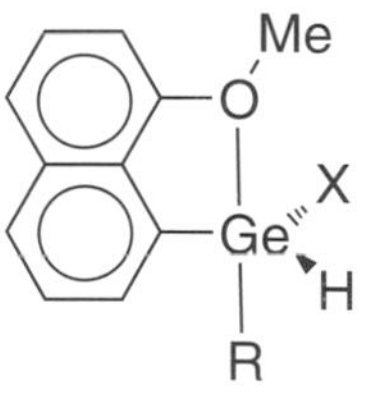

R = H, X = TfO
R = Ph, X = TfO
R = Ph, X = I
R = 8-MeO-Naphth, X = TfO
R = 8-MeO-Naphth, X = I

(153)

A series of complexes of general formula [Me$_2$N(CH$_2$)$_3$]M^{IV}Ph$_y$X$_z$ (M = Ge, Sn, Pb; $y = 0$–3; X = Cl, Br, I, OPh; $z = 0$–3) have been prepared and shown to exhibit intramolecular coordination of the dimethyamino group to the metal, forming a five-membered ring structure centered on a distorted tbp metal. The complexes were studied by a range of spectroscopic techniques to establish the correlation between the electronegativity of the complementary ligand set and the strength of the metal to nitrogen interaction. The ^{13}C NMR spectra are particularly useful in assessing the strength of this interaction because of the strong dependence of the value of δ for the α-methylene carbon upon the geometry of the ring formed by the intramolecular chelation.[461] A further example is given by the bicyclic complex (Me$_2$SnCH$_2$CH$_2$)$_2$P(O)Ph wherein the oxygen coordinates to both tin centers in chloroform solution but is displaced from one tin on addition of a coordinating solvent molecule such as pyridine.[462] In the same way, the complex ClMe$_2$Sn{CH$_2$-SiH$_2$CH$_2$P(O)(OEt)$_2$} shows intramolecular coordination of the phosphine oxide (Sn—O 2.371 Å), giving rise to a six membered ring with a chair conformation in the solid state[463] and similarly the dimethyldithiocarbamate complex (MeCO$_2$CH$_2$CH$_2$)SnCl$_2$(S$_2$CNMe$_2$) has intramolecular coordination of the ester carbonyl.[464,465] The rings remain intact in solution according to NMR experiments.[463] The γ-alkoxytin trichlorides $(OH)(CH_2)_nSnCl_3$ ($n = 3$–5) have five-coordinate tin centers. When n is 5 there is intermolecular Sn–(OH) coordination, and the crystal structure shows a bond length Sn—O of 2.365 Å. For the cases where n is three or four, the coordination is intramolecular, forming five- and six-membered rings.[466]

Dimeric intramolecularly coordinated organotin sulfides $[(Me_2NCH_2CH_2CH_2)_2SnS]_2$ have an octahedral *trans-cis-cis* $[C_2N_2S_2]$ donor set in which the alkyl derivatives are *trans*, the bridging sulfur ligands are obliged to be *cis* and the dimethylamino nitrogens are necessarily *cis* to complete the octahedral geometry. The $(SnS)_2$ ring is planar as is seen for all cases, and although the intramolecular coordination of the nitrogen is temperature dependent, the $(SnS)_2$ ring remains intact in solution.[467]

Tin(IV) coordinates 2-thienyl pyridine not through a [N,S] chelate but rather a [C,N] donor set, with activation of the proton on the 2-thienyl position, such as bis[3-(2-pyridyl)-2-thienyl]Ph$_2$Sn, (154).[468]

Bis(lactamoylmethyl)germanium dichlorides react with TMSX (X = Br, I, TfO), LiZ (Z = Br, I, ClO$_4$), or AgA (A = F, BF$_4$) to yield products where the extent to which the halides are exchanged is dependent upon the nature of the anion rather than stoichiometry. For the noncoordinating anions TfO, ClO$_4$, and BF$_4$ only one chloride is exchanged, giving products (155) with an all *trans* disposition of ligands. In reactions with the more coordinating anions, both chlorides may be exchanged giving products with the oxygens and monodentate ligands both *cis* and the carbons *trans*.[469]

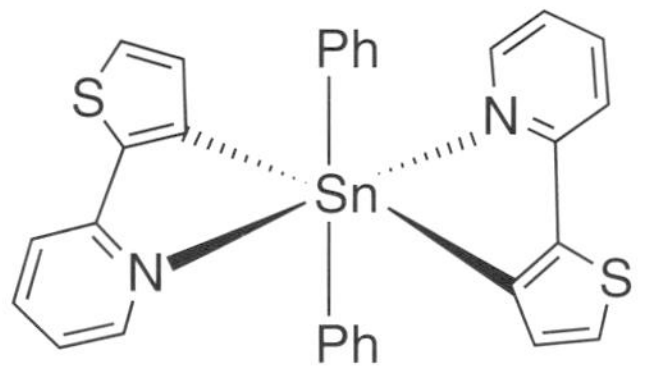

(154)

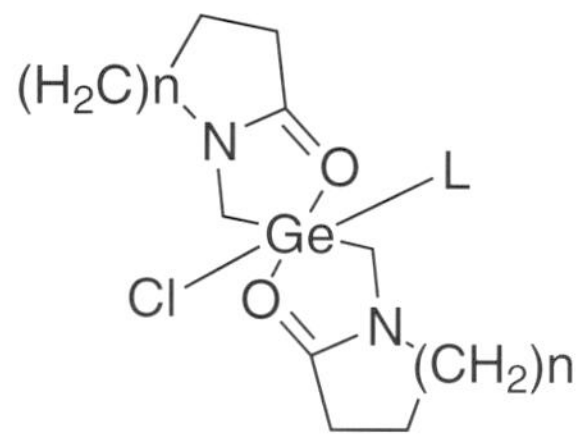

(155)

Stable five-coordinate anionic complexes bis[α,α-bis(CF$_3$)benzenemethanolato]stannates (156) have been reported. Reaction of (156) (R = Ph, 4-MeC$_6$H$_4$) with SO$_2$Cl$_2$ gives the corresponding chlorostannates which were metathesized to the fluoro complexes with [Bu$_4$N]F.[470]

Lead(II) chloride reacts with lithiated dimethylamino(ferrocenyl)methane to give (157). In the solid state, the complex exists in the *meso* form, but in solution it rapidly converts to a mixture of both the *meso* and *rac* forms.[471] The general area of intramolecular coordination chemistry of tethered [C,X] donor ligands has been reviewed.[472]

Thioacetic acid reacts with Bu^{t}GeCl$_3$ to form the five-coordinate tbp (158), with one doubly and one singly deprotonated thioacetate ligand.[473] The disodium salt of 2-thioethanol reacts with Bu^{n}SnCl$_3$ to give the trinuclear (159) in which the bridges are again of the (SnO)$_2$ type. Attempts to replace the remaining chloro group by reaction with a Grignard reagent lead instead to the isolation of (160), which dimerizes in the solid state through Sn$\cdots$O interactions.[474] The structure of Me$_2$Sn(2-pyridinethiolato-*N*-oxide)$_2$ exhibits very asymmetric chelation of the [S,O] donor and shows a skewed trapezoidal bipyramidal geometry with the methyl groups in axial positions.[475]

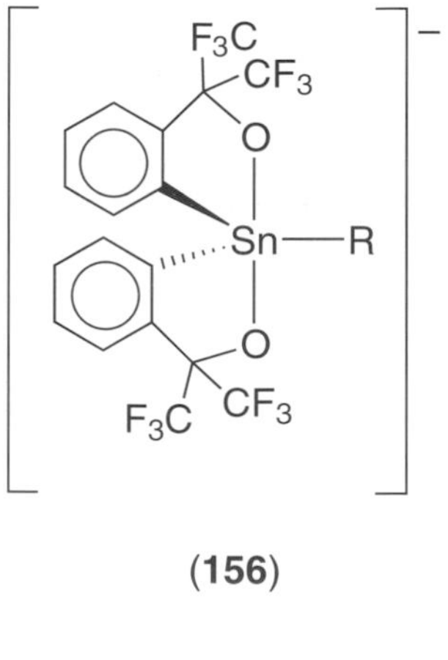

(156)

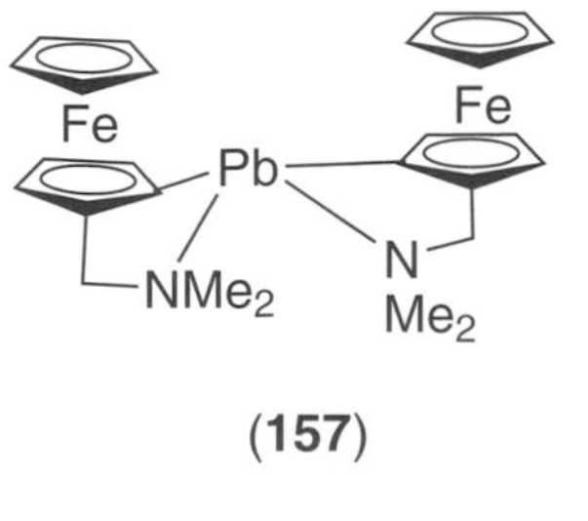

(157)

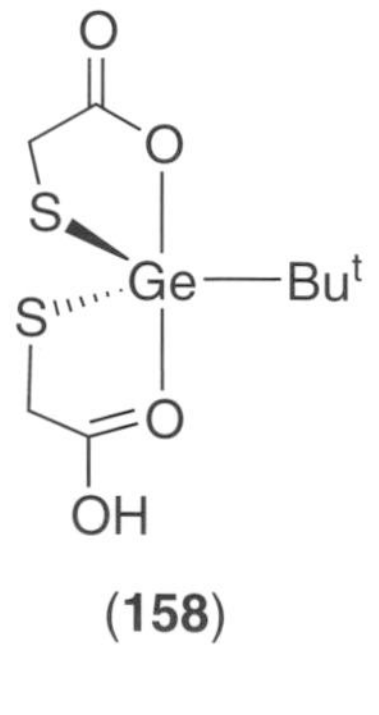

(158)

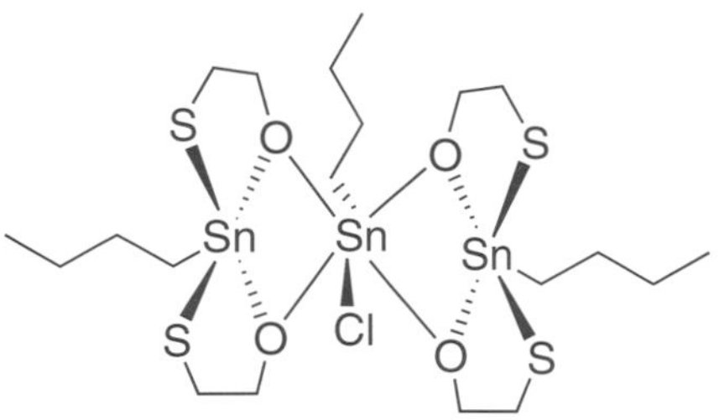

(159)

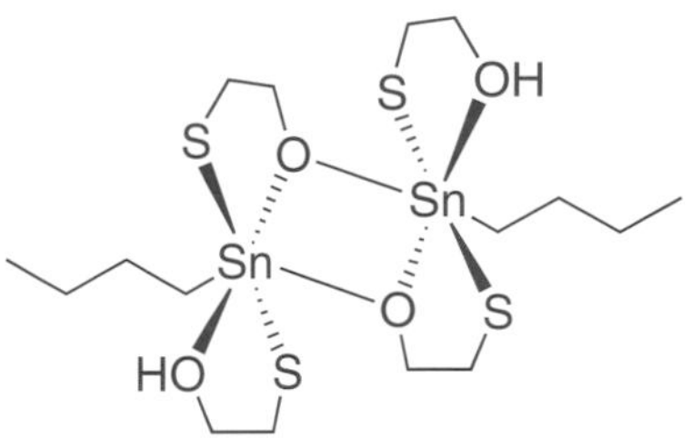

(160)

In the majority its complexes, 8-HO-quinoline (LH) is a chelating [N,O] donor, but in complexes $R_3Sn(L)$ ($R = Me$, Et, Pr^i, Bu^n, Ph) the ligand is coordinated in a monodentate fashion through the oxygen alone in solution according to ^{119}Sn NMR data[476] while for tricyclohexyl stannyl complexes of substituted 8-hydroxyquinolines chelation is observed.[477] The substituted ligand in complexes (161) acts as a chelate.[478]

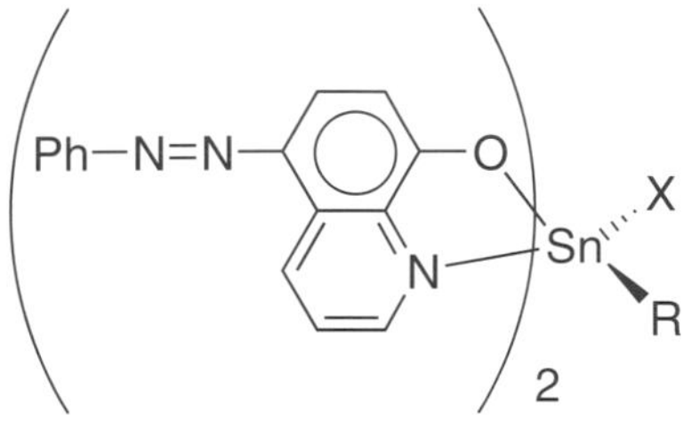

(161)

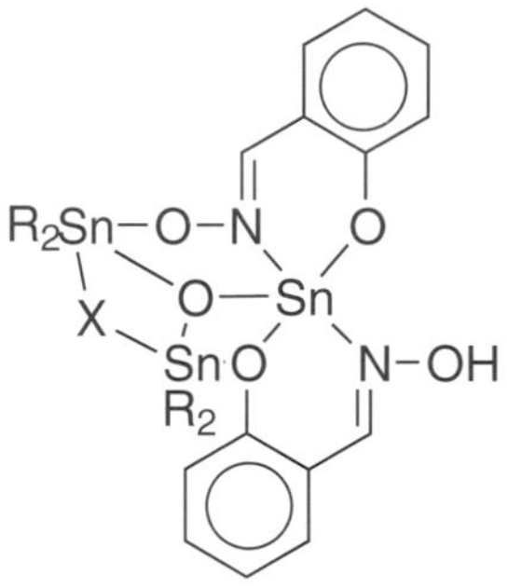

(162)

The Schiff-base salicylaldoximate forms two trinuclear complexes with tin upon refluxing with dimethyltin oxide (162). In both cases, the complex has a $[L_2Me_6Sn_3]$ unit with a bridging group X, which is either a fluoride or the oxygen of a second salicylaldoxime.[479] Other bidentate [N,O] Schiff base ligands form stable complexes with tin that have antifungal activity.[480]

3.7.8.2 Complexes of Heterotridentate Ligands

The tolerance for diverse ligand type and coordination number makes the heavier members of Group 14 especially liable to coordinate ligands of higher denticity and with more varied donor atom type.

The dianions of *N*-substituted diethanolamines coordinate germanium through an [NO$_2$] donor set in the five coordinate complex $\{RN(CH_2CH_2O)_2\}Ge(OH)_2$ ($R = H$, Me). The two hydroxy ligands are labile and readily displaced by bidentate ligands LH_2 (diols, α-hydroxy carboxylic acids, oxalic acid, 2-NH$_2$phenol) to give the neutral five-coordinate complexes $HN(CH_2CH_2O)_2$-$Ge(L)$. The coordination of nitrogen is confirmed by NMR data and in the case where LH_2 is $Ph_2C(OH)CO_2H$, by crystallography (Ge—N 2.08 Å).[481]

Five coordinate germanium(IV) complexes 1,1,5-trimethyl-2,8-dioxa-5-aza-1-germa-bicyclo [3.3.01,5]octane diones (163) can be prepared from the reaction of R_2GeX_2 ($R = Me$, Ph; $X = Cl$, OR) with $MeN(CH_2CO_2H)_2$.[482] Tridentate ligand 5-aza-2,2,8,8-tetramethylnonane-3,7-dione reacts with tin(IV) chloride to form (164), comprising a tin(II) center which is liable to oxidation by SO_2Cl_2 to form (165).[483]

The pyridine-based stannatrane (166) can be prepared from Bu^n_2SnO and 2,6-(CH$_2$OH)$_2$ pyridine.[484] Aldimino alcohols react with $Bu^n_2Sn(NMe_2)_2$ to form initially *N,O*-chelated complexes that quickly dimerize to give five-coordinate geometries such as (167).[485]

The fluxional behavior of five-coordinate tin complexes of *N*-methyl diethanolamine and *N*-methyl diethylthiolate has been investigated. In the solid state, the structure of Bu^t_2Sn-$(OCH_2CH_2)_2NMe$ has oxygen in the axial positions and equatorial alkyl and nitrogen ligands

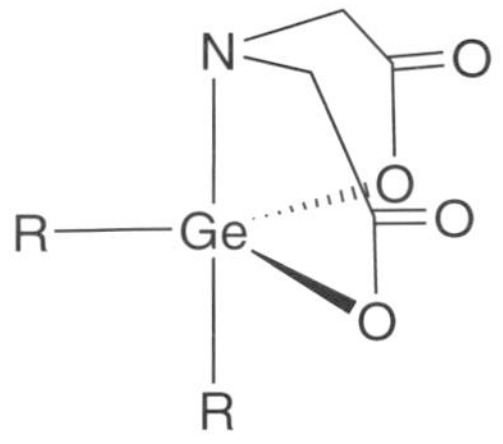

(163)

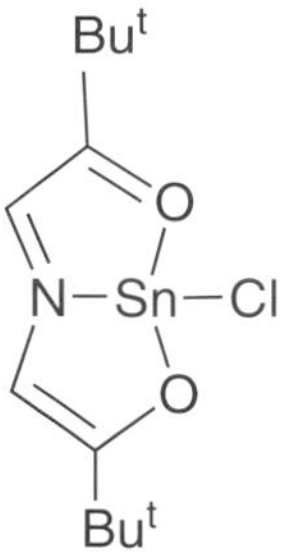

(164)

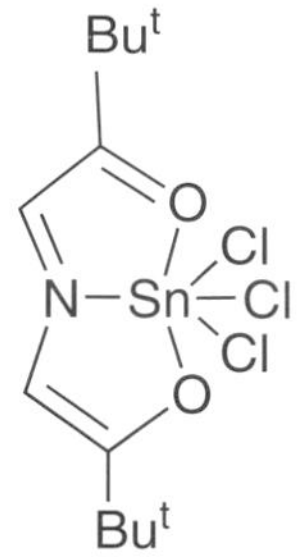

(165)

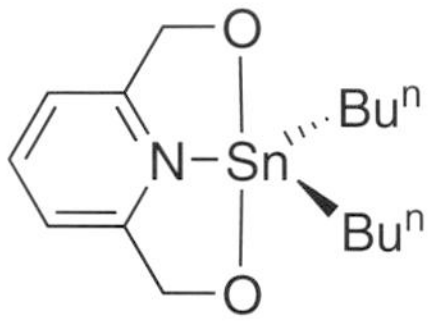

(166)

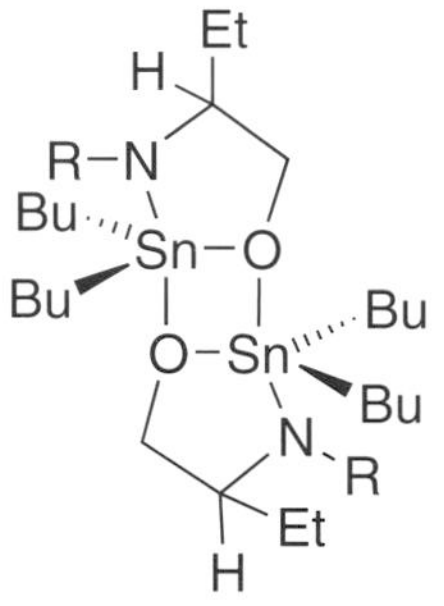

(167)

(Sn—O 2.58 Å) where the sulfur analogue $Me_2Sn(SCH_2CH_2)_2NMe$ has equatorial sulfur coordination and an axial alkyl and nitrogen ligand (Sn—S 2.431 Å, ave.). This difference in disposition of ligands may arise from the change in electronegativity on going from oxygen to sulfur donors, or from the change in size of the alkyl ligand. The exchange of ligand positions in the fluxional processes in solution proceeds through a Berry pseudo-rotation process at low temperatures and by a dissociation–inversion pathway at higher temperatures.[486]

The complex $[PhSn(SCH_2CH_2)_2NMe]_2CH_2$ has two five-coordinate tin centers, each coordinated by a tridentate [N,S,S] and phenyl and the bridging methylene. In the solid state, the phenyl groups occupy axial positions, but in solution the molecule is fluxional, and NMR data indicate that the phenyl groups can symmetrically occupy axial or equatorial positions or asymmetric axial and equatorial isomers.[487] Complexes $[XSn(CH_2CH_2CH_2)NMe]_2$ (X = Cl, Me) both have directly linked five-coordinate tin centers and are fluxional in solution. The fluxional processes that depend upon a rotation about the Sn—Sn bond are lower in energy for the methyl complex than for the chloro, a difference in behavior arising from the variation in the selectivity for apical positioning for the two ligands.[488]

The tridentate monobasic ligand 1-(2-pyridylazo)-2-naphtholate (pan) has been used to prepare seven-coordinate complexes R_2(pan)M(L) (L = ACAC derivative, M = Sn, R = Bu, Me; M = Pb, R = Me). For both metals the difference in δ is distinct from lower coordination numbers, both showing an upfield shift in δ of ca. 200[489] typical for such a change in coordination number.[490] The dianionic tetracyclic ligand (168) is prepared from the lithiation of 2-Me-benzoxazole and reacts with tin(II) chloride to form (169).[491]

Acetylacetonato complexes $Sn(ACAC)_2Cl_2$ react with $2\text{-}NH_2C_6H_4OH$, $2\text{-}NH_2C_6H_4SH$, benzoylhydrazine, and thiobenzoylhydrazine gives the bis-tridentate complexes SnL_2 (L = acetylacetone-*o*-iminophenol, -*o*-iminothiophenol, -benzoylhydrazone, and -thiobenzoylhydrazone) that exhibit distorted octahedral geometries.[492]

Tridentate Schiff-base ligands formed from the reaction of salicylaldehydes with either 2-(NH_2)-phenol[493–495] anthranilic acid,[496] or amino acids[497,498] coordinate M^{IV} (M = Ge, Sn, Pb) as tridentate $[NO_2]$ donor ligands to form stable complexes that are either tbp or octahedral depending upon stoichiometry. The lead complex (170) is associated into a dimeric unit through Pb⋯O bridging interactions, forming a $[PbO]_2$ ring.[494] Using salicylaldehyde-5-sulfonic acid and 2-(NH_2)-phenol-5-sulfonate, the water soluble version of the ligand can be prepared and used to form complexes of germanium in aqueous solution.[499]

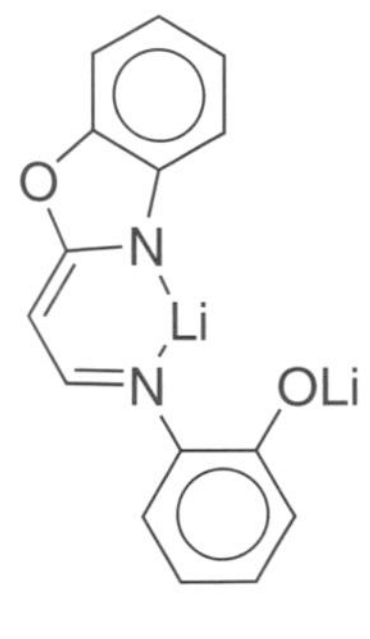

(168) 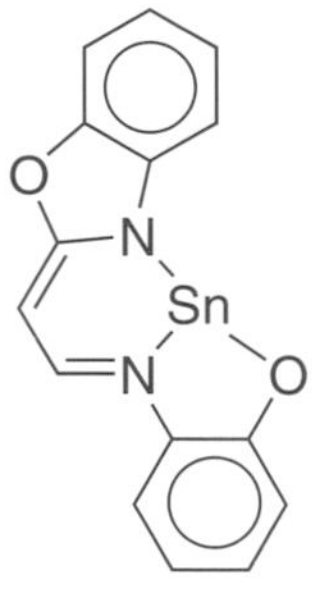　　　　(169) 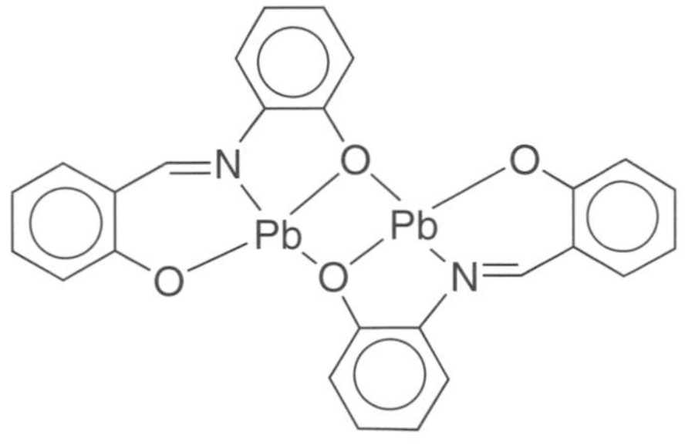　　　　(170)

The reaction of 3,5-Bu$^t_2$catH$_2$ with ammonia under oxidizing conditions gives 3,5-Bu$^t_2$-1, 2-quinone-1-(2-hydroxy-3,5-Bu$^t_2$phenyl)imine anion which acts as a tridentate [NO$_2$] ligand in reaction with MCl$_2$ (M = Sn, Pb). In these complexes, the ligand responds to the nature of the metal, such that the tin complex comprises tin(IV) and the lead complex comprises lead(II).[500]

Electrochemical oxidation of a tin anode in the presence of Schiff-base ligands derived from substituted salicylaldehydes and bis(2-aminophenyl)disulfide (L$_2$H$_2$) gives complexes SnL$_2$. The tin shows a distorted octahedral geometry for these complexes, and the structure of bis[2-(2-thiophenyl)imino-4,6-(MeO)$_2$C$_6$H$_3$O]tin(IV) has averaged bond lengths of Sn—N 2.17, Sn—O 2.07, and Sn—S 2.47 Å.[501]

Nickel porphyrazineoctathiolate has four [S$_2$N] sites that can be used to coordinate further metal ions, and the crystal structure of (R$_2$Sn)$_4$S$_8$(porphyrazine)Ni^{II} (171) shows symmetrical coordination of the four tin centers onto the periphery of the ring.[502]

The flexibility in coordination number and geometry exhibited by M(14) in comparison to transition metal ions sometimes leads to the formation of complexes in which the coordination of a particular ligand takes an unexpected form. The ligand (172) H$_2$ would seem to present an [O$_2$] donor set, but in its complexes with germanium acts as a tridentate [O$_2$Se] ligand, coordinating in a *fac* configuration in both the tbp (172)GeMe$_2$ and octahedral (172)$_2$Ge. The complexes are configurationally stable in solution by NMR and they seem to be the first examples of selenoether coordination to neutral germanium(IV) centers.[503]

Tridentate ligands with pincer arrays have become popular subjects for study. The reaction of Li[2,6-(Me$_2$NCH$_2$)C$_6$H$_3$] with tin(II) chloride leads to the pincer complex [κ^3-2,6-(Me$_2$NCH$_2$)C$_6$H$_3$-*C,N,N*]SnCl. The complex has a tbp geometry with a stereochemically active lone pair in an equatorial position and both axial positions taken by nitrogens. The complex is stereochemically nonrigid at temperatures above −70 °C but the nature of the fluxional process was not unambiguously determined. The chloro ligand is liable to substitution in reaction with aryllithium reagents.[504] Another pincer stannylene (173) has been reported and again shown to be readily substituted with a range of ligands to form further stannylenes.[505]

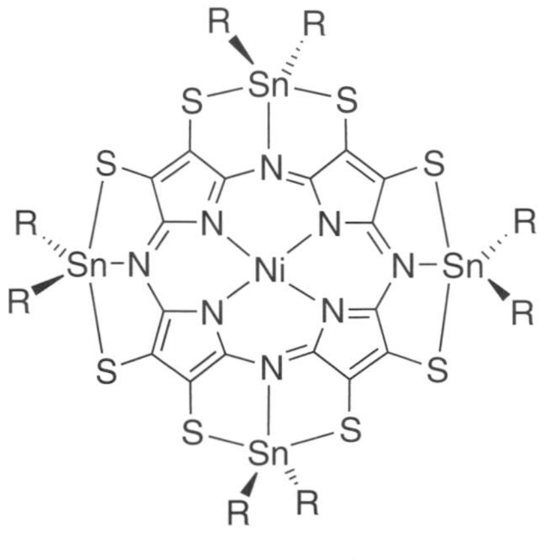

(171) 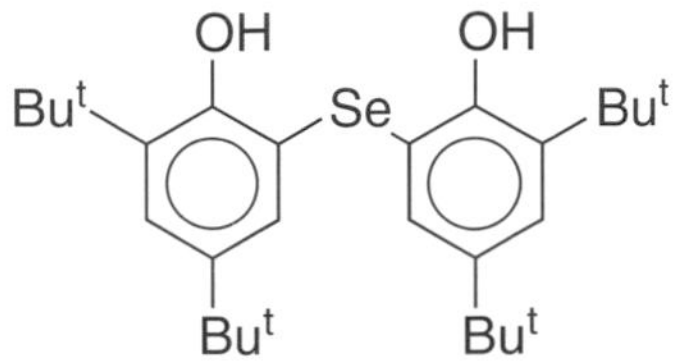　　　　(172) 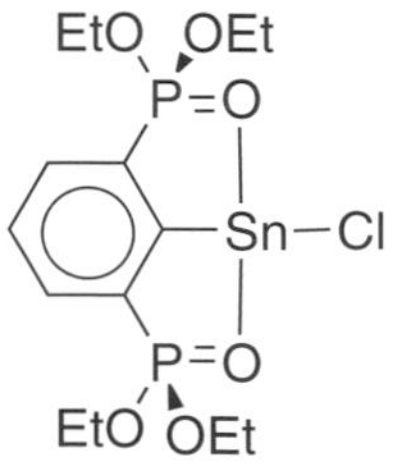　　　　(173)

3.7.8.3 Complexes of Heterotetradentate Ligands

Schiff-base ligands prepared form the condensation of salicylaldehyde with diamines offer [N$_2$O$_2$] donor sets that are capable of coordinating a wide range of metals. Tin(IV) acetate reacts with salenH$_2$[506] and R$_2$SnCl$_2$ reacts with either (3-MeO)salphen (R = Ph, Bu^n, Me)[507] or the ligand prepared by the condensation of 1,2-(NH$_2$)$_2$C$_6$H$_4$ and 2-(HO)-1-naphthaldeyde (R = Ph)[508] to give in all cases the corresponding six-coordinate products (L)SnR$_2$. Divalent M(14) Schiff-base complexes can be prepared by the reaction of the amines (TMS$_2$N)$_2$M (M = Ge, Sn, Pb) with the Schiff bases directly. These M^{II} complexes are prone to oxidation, by iodine to give the diiodide or by 3,5-Bu$^t_2$-benzoquinone to give the catecholate.[509,510] In these M^{II} complexes, the metals do not sit in the plane defined by the [N$_2$O$_4$] donors but rather are displaced to one face of the ligand. With lead(II) perchlorate, salenH$_2$ forms a trinuclear complex [(salen)$_3$Pb$_3$](ClO$_4$)$_2$, with the same out of plane coordination of the two lead(II) centers within the ligand pockets and with the third lead linking the two [(salen)Pb] monomers by coordinating in an *exo* fashion to all four phenolic oxygens.[511] In the same paper, a four-coordinate complex of lead(II) with the potentially heptadentate ligand saltrenH$_3$ is also reported, where the lead(II) coordinates to the ligand through only four donor atoms.

Amine phenol (**174**) related to reduced salen reacts with R$_2$SnO (R = Me, Bu^n, Bu^t, Ph) to give six-coordinate complexes with [N$_2$O$_2$] coordination of the ligand[512] whereas reduced salenH$_2$ reacts with lead(II) acetate to give a dimeric complex [(κ^2-HL-*N,O*)(κ^1-OAc)Pb]$_2$ linked through bridging phenolic oxygens and reduced saltrenH$_3$ gives a dimeric complex [{(κ^4-HL-*N,N,O*)Pb}$_2\mu$-OAc](OAc).[513]

(**174**)

Tripodal ligands (CH$_2$CH$_2$OH)(CH$_2$CO$_2$H)$_2$N and (CH$_2$CONH$_2$)(CH$_2$CO$_2$H)$_2$N react with R$_2$SnO (R = Bu^n, *n*-octyl) to give complexes (L)SnR$_2$. The crystal structure of [(CH$_2$CH$_2$OH)(CH$_2$CO$_2$)$_2$N]SnBu$^n_2$ has a distorted octahedral geometry in which the hydroxy group is coordinated as an innocent ligand.[514] The unsubstituted iminodiacetate complexes HN(CH$_2$CO$_2$)$_2$SnR$_2$ (R = Me, Bu^n) prepared in the same way crystallizes as a dimer with a seven-coordinate tin center, in which the alkyl groups are in the *trans* positions.[515]

Despite the proven depressant neurotropic influence of furan- or thiophene-substituted germatranes,[516] germatranes and stannatranes have been studied to develop routes for their synthesis, from triethanolamine[517,518] or tristannyl ethers[519–521] their substitution reactions,[522,523] their structures,[524,525] and iododestannation.[526]

The series of germatranes RC$_6$H$_4$Ge(OCH$_2$CH$_2$)$_3$N (R = H, 4-Me, 3-Me, 2-Me) have been prepared by the insertion of germanium(II) bromide into a carbon-halide bond on the aryl group to give RC$_6$H$_4$GeBr$_3$, which can be converted to the alkoxy derivative RC$_6$H$_4$Ge(OR)$_3$. Reaction with triethanolamine gives a good yield of the phenyl germatranes, some of which were characterized by crystallography. Inclusion of a group in the ortho position decreases the angle N–Ge–C$_{ipso}$ from 177.5° in the unsubstituted complex to 144.2°, and for these complexes the transannular Ge—N bond is found to be in the range 2.212–2.230 Å.[517]

Unsymmetrical stannatranes R$_2$Sn(XCH$_2$CH$_2$)$_2$Y, RSn(XCH$_2$CH$_2$)Y, and R$_2$Sn(OC(O)CH$_2$)$_2$Y (X = O, NMe, S, Y = O, S, NR) have been studied by Mössbauer spectroscopy to establish the coordination geometries. The preferred geometry is tbp for all cases where the apical atom is nitrogen and distorted four-coordinate where the apical atom is a chalcogen, indicating that in these cases the apical group is not coordinated.[527]

3.7.8.4 Complexes of Heterodonor Ligands of Higher Denticity

Seven-coordinate complexes of tin are not unusual, and the structure of Et$_2$Sn{2,6-diacetylpyridine bis(2-thienyoyl)}hydrazone is an example having a pentagonal bipyramidal geometry with axial alkyl groups where the pentagonal plane is defined by the pentadentate ligand.[528]

Diethylenetriaminepentaacetic acid (H_5dtpa) forms complexes (Hdtpa)Sn·$3H_2O$, (Hdtpa)Sn, and Na[(dtpa)Sn]. The structure of (Hdtpa)Sn·$3H_2O$ shows an eight-coordinate tin with an [N_5O_3] donor set, one of the highest coordination numbers seen for tin(IV).[529]

The tripodal Schiff-base ligand 3-MeO-saltrenH$_3$ reacts with lead(II) chloride to give a dinuclear complex [(3-MeO-satren)Pb$_2$]Cl, crystallized as its perchlorate salt. The complex comprises two distinct lead(II) centers, one coordinated within the ligand cavity having an [N_4O_3] donor set, the other coordinated in an *exo* fashion to the three phenolic oxygens, an example of the breadth of tolerance for coordination number and donor atom type even within a single complex.[530]

Lead complexes of a range of mixed [N,O] donor macrocycles and substituted macrocycles have been prepared. Schiff-base condensation of a range of amines with pyridine-2,6-dicarboxaldehyde gives the ligands (175)–(177). The mononuclear complex of lead with (175) comprises a 10-coordinate lead, bound to all six of the macrocycle donors and two chelating nitrates, whereas the complex of the reduced version (176) shows a different conformation associated with the greater degree of flexibility in the ring of the macrocycle.[531] Ligand (177) complexes lead to give a linear polymer[532] where the octadentate (178) forms mononuclear complexes.[533] The structurally related (179) forms both mono- and dinuclear complexes depending upon the reaction stoichiometry.[534]

(175)   (176)   (177)   (178)

(179)

The substituted cyclam (180) forms a mononuclear lead complex that has a six-coordinate geometry in the solid state and is fluxional in solution. A ^{13}C NMR study shows that the four ring donors stay coordinated throughout and that all four pendant groups are involved in coordinating the lead.[535,536]

In order to investigate preferences for ligand configuration and donor atom type, families of related complexes have been prepared and their complexes compared. For the family of complexants (**181**) the most successful ligand for lead(II) was found to be the smallest example[537] and for the complexes (**182**), the most successful was that with an all-nitrogen donor set.[538,539]

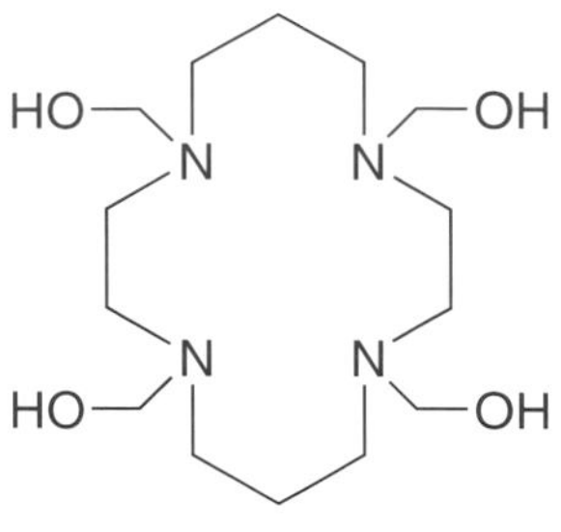

(**180**)

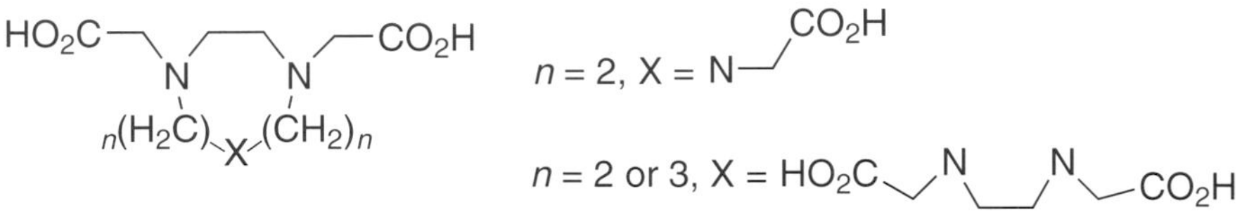

$n = 2$, X = N

$n = 2$ or 3, X = HO$_2$C

(**181**)

$n = 1, 2, 3, 4$
X = (NH), O, S
Y = (NH), O, S

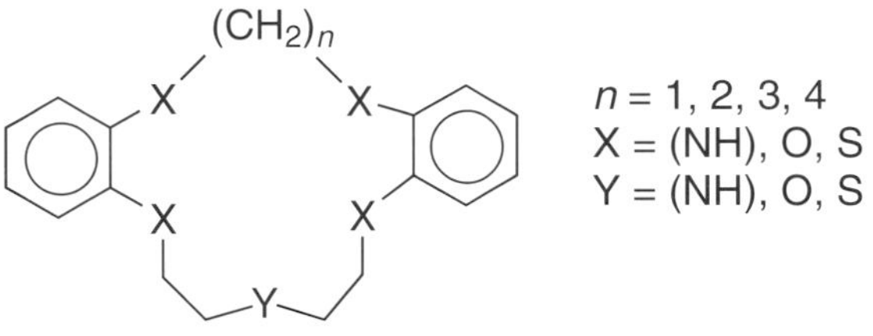

(**182**)

Introduction of even a single nitrogen donor enhances the stability of complexes of lead(II) with macrocycles such as aza crown ethers (**183**) and (**184**), which both form lead(II) complexes that show markedly higher stability than those of the corresponding oxygen donors 15-C-5 and 18-C-6. Structurally, the complexes are distinct, with the former showing a nine-coordinate lead with *cis* coordination of two chelating nitrates and the coordination of the macrocycle on the opposite face of the lead, and the latter 10-coordinate with an equatorial macrocycle and *trans* disposed chelating nitrates.[540–542] The mixed donor substituted macrocycles (**185**) and the related macrobicycles (**186**) all complex lead(II) by coordination within the ligand ring. The structure of the complexes with (**185**) $a = b = 1$ comprises a six-coordinate lead, with one primary amine nitrogen, two tertiary amine nitrogens and three ether oxygens coordinating, whereas the complex with (**186**) has donor set made up of a pyridine nitrogen, one imine nitrogen, two tertiary amine nitrogens, two ether oxygens, and a monodentate perchlorate.[543]

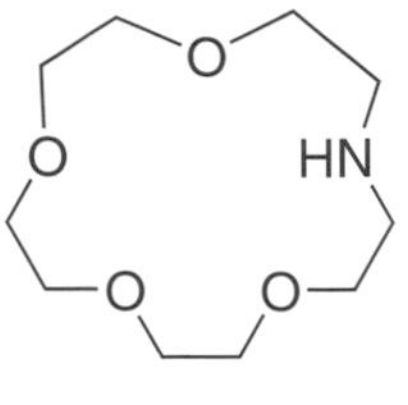

(**183**)

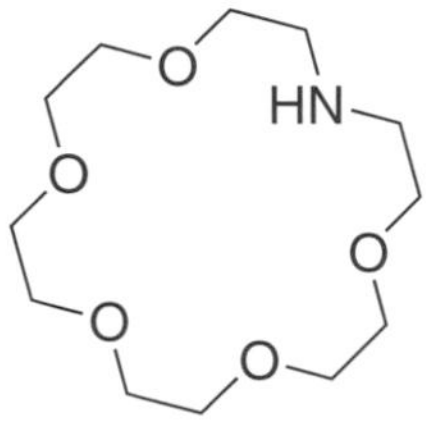

(**184**)

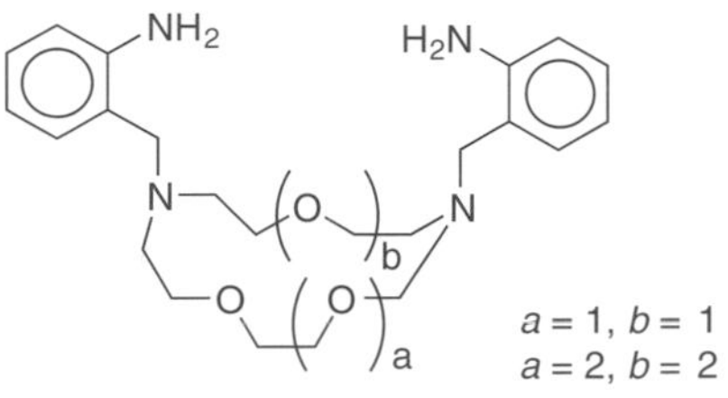

(185)

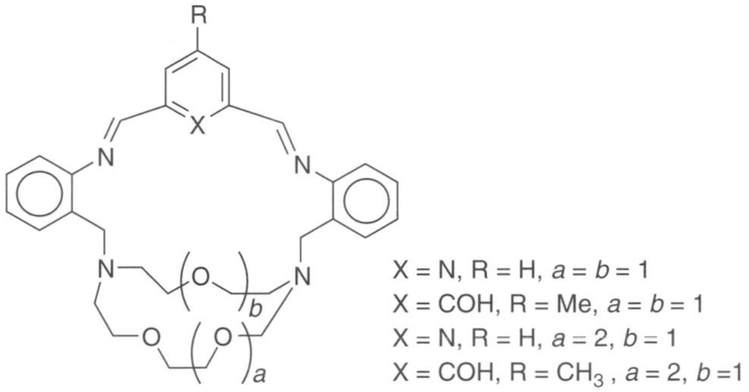

a = 1, b = 1
a = 2, b = 2

X = N, R = H, a = b = 1
X = COH, R = Me, a = b = 1
X = N, R = H, a = 2, b = 1
X = COH, R = CH$_3$, a = 2, b =1

(186)

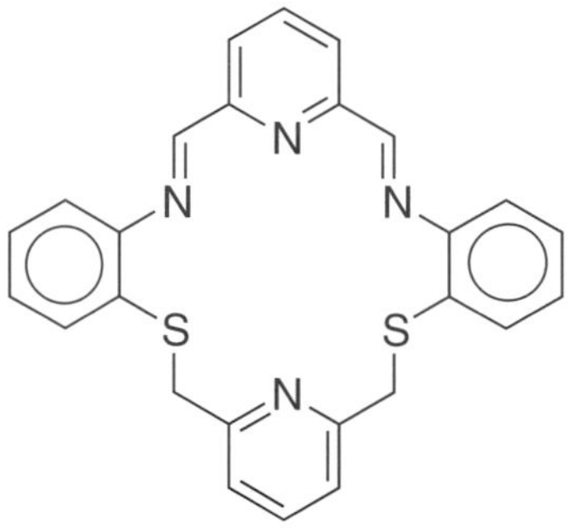

(187)

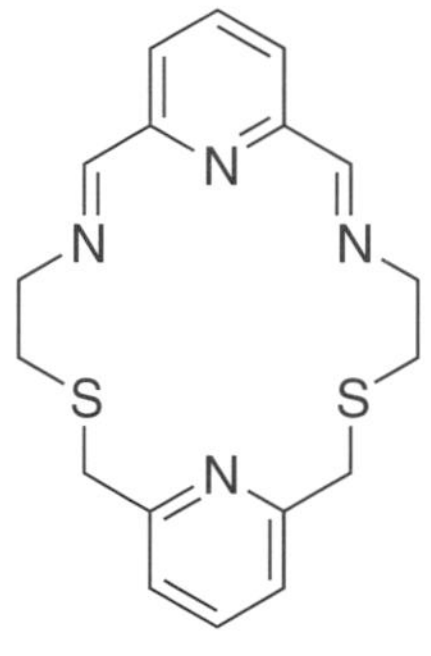

(188)

Mixed nitrogen–sulfur donor macrocycles show relatively low affinities for lead(II) despite assumptions based upon hard–soft arguments. The ligands (**187**) and (**188**) complex lead with moderate efficiency.[544]

3.7.9 REFERENCES

1. Harrison, P. G. Silicon, germanium, tin, and lead. In *Comprehensive Coordination Chemistry*; Wilkinson, G.; Gillard, R. D.; McCleverty, J. A, eds.; Pergamon: Oxford, 1987; Vol. 3, Chapter 26.
2. Martins, J. C.; Biesemans, M.; Willem, R. *Prog. Nucl. Mag. Reson. Spectrosc.* **2000**, *36*, 271–322.
3. Wrackmeyer, B. *Annu. Rep. NMR Spectrosc.* **1999**, *38*, 203–264.
4. Haaland, A.; Shorokhov, D. J.; Strand, T. G.; Kouvetakis, J.; O'Keeffe, M. *Inorg. Chem.* **1997**, *36*, 5198–5201.
5. Kouvetakis, J.; Haaland, A.; Shorokhov, D. J.; Volden, H. V.; Girichev, G. V.; Sokolov, V. I.; Matsunaga, P. *J. Am. Chem. Soc.* **1998**, *120*, 6738–6744.
6. El-Raghy, T.; Chakraborty, S.; Barsoum, M. W. *J. Eur. Ceram. Soc.* **2000**, *20*, 2619–2625.
7. Barsoum, M. W.; Yaroschuk, G.; Tyagi, S. *Scr. Mater.* **1997**, *37*, 1583–1591.
8. Gesing, T. M.; Wachtmann, K. H.; Jeitschko, W. *Z. Naturforsch. B: Anorg. Chem. Org. Chem.* **1997**, *52*, 176–182.
9. Von Schnering, H. G.; Baitinger, M.; Bolle, U.; Carrillo-Cabrera, W.; Curda, J.; Grin, Y.; Heinemann, F.; Llanos, J.; Peters, K.; Schmeding, A.; Somer, M. *Z. Anorg. Allg. Chem.* **1997**, *623*, 1037–1039.
10. Riviere-Baudet, M.; Dahrouch, M.; Gornitzka, H. *J. Organomet. Chem.* **2000**, *595*, 153–157.
11. Hihara, G.; Hynes, R. C.; Lebuis, A.-M.; Riviere-Baudet, M.; Wharf, I.; Onyszchuk, M. *J. Organomet. Chem.* **2000**, *598*, 276–285.
12. Onyzschuk, M.; Castel, A.; Riviere, P.; Satge, J. *J. Organomet. Chem.* **1986**, *317*, C35–C37.
13. Moravec, V. D.; Jarrold, C. C. *J. Chem. Phys.* **2000**, *113*, 1035–1045.
14. Tudela, D.; Fernandez, V.; Tornero, J. D. *Inorg. Chem.* **1985**, *24*, 3892–3895.
15. Benzi, P.; Operti, L.; Vaglio, G. A.; Volpe, P.; Speranza, M.; Gabrielli, R. *Int. J. Mass Spectrom.* **1990**, *100*, 647–663.
16. Mochida, K.; Kanno, N.; Kato, R.; Kotani, M.; Yamauchi, S.; Wakasa, M.; Hayashi, H. *J. Organomet. Chem.* **1991**, *415*, 191–201.
17. Mochida, K.; Tokura, S. *Organometallics* **1992**, *11*, 2752–2754.
18. West, R. *Pure Appl. Chem.* **1984**, *56*, 163–173.
19. Raabe, G.; Michl, J. *Chem. Rev.* **1985**, *85*, 419–509.
20. Barrau, J.; Ben Hamida, N.; Agrebi, A.; Satge, J. *Organometallics* **1987**, *6*, 659–662.
21. Espenbetov, A. A.; Struchkov, Yu.T.; Kolesnikov, S. P.; Nefedov, O. M. *J. Organomet. Chem.* **1984**, *275*, 33–37.
22. Sita, L. R.; Bickerstaff, R. D. *J. Am. Chem. Soc.* **1988**, *110*, 5208–5209.

23. Masamune, S.; Hanzawa, Y.; Williams, D. J. *J. Am. Chem. Soc.* **1982**, *104*, 6136–6137.
24. Ando, W.; Tsumuraya, T. *J. Chem. Soc., Chem. Commun.* **1987**, 1514–1515.
25. Ichinohe, M.; Sekiyama, H.; Fukaya, N.; Sekiguchi, A. *J. Am. Chem. Soc.* **2000**, *122*, 6781–6782.
26. Jutzi, P.; Leue, C. *Organometallics* **1994**, *13*, 2898–2899.
27. Lange, L.; Meyer, B.; Du Mont, W. W. *J. Organomet. Chem.* **1987**, *329*, C17–C20.
28. Sekiguchi, A.; Yamazaki, H.; Kabuto, C.; Sakurai, H.; Nagase, S. *J. Am. Chem. Soc.* **1995**, *117*, 8025–8026.
29. Sekiguchi, A.; Tsukamoto, M.; Ichinohe, M. *Science* **1997**, *275*, 60–61.
30. Fukaya, N.; Ichinohe, M.; Sekiguchi, A. *Angew. Chem., Int. Ed. Engl.* **2000**, *39*, 3881–3884.
31. Weidenbruch, M.; Ritschl, A.; Peters, K.; Von Schnering, H. G. *J. Organomet. Chem.* **1992**, *438*, 39–44.
32. Olmstead, M. M.; Pu, L.; Simons, R. S.; Power, P. P. *Chem. Commun.* **1997**, 1595–1596.
33. Olmstead, M. M.; Simons, R. S.; Power, P. P. *J. Am. Chem. Soc.* **1997**, *119*, 11705–11706.
34. Benet, S.; Cardin, C. J.; Cardin, D. J.; Constantine, S. P.; Heath, P.; Rashid, H.; Teixeira, S.; Thorpe, J. H.; Todd, A. K. *Organometallics* **1999**, *18*, 389–398.
35. Eichler, B. E.; Power, P. P. *Inorg. Chem.* **2000**, *39*, 5444–5449.
36. Jurkschat, K.; Abicht, H. P.; Tzschach, A.; Mahieu, B. *J. Organomet. Chem.* **1986**, *309*, C47–C50.
37. Pu, L.; Twamley, B.; Power, P. P. *J. Am. Chem. Soc.* **2000**, *122*, 3524–3525.
38. Masamune, S.; Sita, L. R.; Williams, D. J. *J. Am. Chem. Soc.* **1983**, *105*, 630–631.
39. Wiberg, N.; Hochmuth, W.; Noth, H; Appel, A.; Schmidt-Amelunxen, M. *Angew. Chem., Int. Ed. Engl.* **1996**, *35*, 1333–1334.
40. Kircher, P.; Huttner, G.; Heinze, K.; Renner, G. *Angew. Chem., Int. Ed. Engl.* **1998**, *37*, 1664–1666.
41. Schiemenz, B.; Huttner, G. *Angew. Chem., Int. Ed. Engl.* **1993**, *32*, 297–298.
42. Sekiguchi, A.; Kabuto, C.; Sakurai, H. *Angew. Chem., Int. Ed. Engl.* **1989**, *28*, 55.
43. Wiberg, N.; Lerner, H.-W.; Noth, H.; Ponikwar, W. *Angew. Chem., Int. Ed. Engl.* **1999**, *38*, 1103–1105.
44. Sekiguchi, A.; Yatabe, T.; Kamatani, H.; Kabuto, C.; Sakurai, H. *J. Am. Chem. Soc.* **1992**, *114*, 6260–6262.
45. Sita, L. R.; Kinoshita, I. *Organometallics* **1990**, *9*, 2865–2867.
46. Sita, L. R.; Kinoshita, I. *J. Am. Chem. Soc.* **1992**, *114*, 7024–7029.
47. Sita, L. R.; Kinoshita, I. *J. Am. Chem. Soc.* **1991**, *113*, 1856–1857.
48. Queneau, V.; Sevov, S. C. *J. Am. Chem. Soc.* **1997**, *119*, 8109–8110.
49. Von Schnering, H. G.; Baitinger, M.; Bolle, U.; Carrillo-Cabrera, W.; Curda, J.; Grin, Y.; Heinemann, F.; Llanos, J.; Peters, K.; Schmeding, A.; Somer, M. *Z. Anorg. Allg. Chem.* **1997**, *623*, 1037–1039.
50. Downie, C.; Tang, Z.; Guloy, A. M. *Angew. Chem. Int. Ed. Engl.* **2000**, *39*, 338–340.
51. Zachwieja, U.; Mueller, J.; Wlodarski, J. *Z. Anorg. Allg. Chem.* **1998**, *624*, 853–858.
52. Todorov, E.; Sevov, S. C. *Angew. Chem., Int. Ed. Engl.* **1999**, *38*, 1775–1777.
53. Nagase, S. *Angew. Chem.* **1989**, *101*, 340–341.
54. Sita, L. R. *Acc. Chem. Res.* **1994**, *27*, 191–197.
55. Suzuki, H.; Tanaka, K.; Yoshizoe, B.; Yamamoto, T.; Kenmotsu, N.; Matuura, S.; Akabane, T.; Watanabe, H.; Goto, M. *Organometallics* **1998**, *17*, 5091–5101.
56. Guerrero, A.; Cervantes, J.; Velasco, L.; Gomez-Lara, J.; Sharma, S.; Delgado, E.; Pannell, K. *J. Organomet. Chem.* **1992**, *430*, 273–86.
57. Lee, V. Ya.; ; Ichinohe, M.; Sekiguchi, A.; Takagi, N.; Nagase, S. *J. Am. Chem. Soc.* **2000**, *122*, 9034–9035.
58. Ichinohe, M.; Arai, Y.; Sekiguchi, A.; Takagi, N.; Nagase, S. *Organometallics* **2000**, *20*, 4141–4143.
59. Schaefer, A.; Saak, W.; Weidenbruch, M.; Marsmann, H.; Henkel, G. *Chem. Ber. Recl.* **1997**, *130*, 1733–1737.
60. Baines, K. M.; Cooke, J. A. *Organometallics* **1991**, *10*, 3419–3421.
61. Heine, A.; Stalke, D. *Angew. Chem., Int. Ed. Engl.* **1994**, *33*, 113–115.
62. Chaubon, M.-A.; Escudie, J.; Ranaivonjatovo, H.; Satge, J. *Chem. Commun.* **1996**, 2621–2622.
63. Rosch, L.; Storke, U. *Angew. Chem., Int. Ed. Engl.* **1983**, *22*, 557–558.
64. Klinkhammer, K. W.; Schwarz, W. *Angew. Chem., Int. Ed. Engl.* **1995**, *34*, 1334–1336.
65. Drost, C.; Hitchcock, P. B.; Lappert, M. F. *Angew. Chem., Int. Ed. Engl.* **1999**, *38*, 1113–1116.
66. Bandoli, G.; Dolmella, A.; Peruzzo, V.; Plazzogna, G. *J. Organomet. Chem.* **1993**, *452*, 47–53.
67. Peruzzo, V.; Plazzogna, G.; Valle, G. *J. Organomet. Chem.* **1989**, *375*, 167–171.
68. Cunningham, D.; McCardle, P.; McManus, J.; Higgins, T.; Molloy, K. C. *J. Chem. Soc., Dalton Trans.* **1988**, 2621–2627.
69. Yatsenko, A. V.; Schenk, H.; Aslanov, L. A. *J. Organomet. Chem.* **1994**, *474*, 107–111.
70. Leinenweber, K.; O'Keeffe, M.; Somayazulu, M.; Hubert, H.; McMillan, P. F.; Wolf, G. H. *Chem. Eur. J.* **1999**, *5*, 3076–3078.
71. Maya, L. *Inorg. Chem.* **1992**, *31*, 1958–1960.
72. Clarke, S. J.; Kowach, G. R.; DiSalvo, F. J. *Inorg. Chem.* **1996**, *35*, 7009–7012.
73. Guloy, A. M.; Corbett, J. D. *Z. Anorg. Allg. Chem.* **1992**, *616*, 61–66.
74. Koroteev, P. S.; Egorov, M. P.; Nefedov, O. M.; Alexandrov, G. G.; Nefedov, S. E.; Eremenko, I. L. *Russ. Chem. Bull.* **2000**, *49*, 1800–1801.
75. Filippou, A. C.; Portius, P.; Kociok-Kohn, G. *Chem. Commun.* **1998**, 2327–2328.
76. Filippou, A. C.; Portius, P.; Neumann, D. U.; Wehrstedt, K.-D. *Angew. Chem., Int. Ed. Engl.* **2000**, *39*, 4333–4336.
77. Wharf, I.; Wojtowski, R.; Bowes, C.; Lebuis, A.-M.; Onyszchuk, M. *Can. J. Chem.* **1998**, *76*, 1827–1835.
78. Bhandari, S.; Mahon, M. F.; McGinley, J. G.; Molloy, K. C.; Roper, C. E. E. *J. Chem. Soc., Dalton Trans.* **1998**, 3425–3430.
79. Filippou, A. C.; Portius, P.; Kociok-Kohn, G.; Albrecht, V. *J. Chem. Soc., Dalton Trans.* **2000**, 1759–1768.
80. Ayers, A. E.; Marynick, D. S.; Dias, H. V. R. *Inorg. Chem.* **2000**, *39*, 4147–4151.
81. Ayers, A. E.; Klapötke, T. M.; Dias, H. V. R. *Inorg. Chem.* **2001**, *40*, 1000–1005.
82. Tornieporth-Oetting, I. C.; Klapotke, T. M. *Angew. Chem., Int. Ed. Engl.* **1995**, *34*, 511–520.
83. Al-Juaid, S. S.; Al-Rawi, M.; Eaborn, C.; Hitchcock, P. B.; Smith, J. D. *J. Organomet. Chem.* **1993**, *446*, 161–166.
84. Gabe, E. J.; Lee, F. L.; Khoo, L. E.; Smith, F. E. *Inorg. Chim. Acta* **1986**, *112*, 41–46.
85. Das, V. G. K.; Yap, C. K.; Smith, P. J. *J. Organomet. Chem.* **1987**, *327*, 311–326.
86. Dillon, K. B.; Marshall, A. *J. Chem. Soc., Dalton Trans.* **1987**, 315–317.

87. Reischmann, R.; Hausen, H. D.; Weidlein, J. *Z. Anorg. Allg. Chem.* **1988**, *557*, 123–133.
88. Chitsaz, S.; Neumuller, B.; Dehnicke, K. *Z. Anorg. Allg. Chem.* **2000**, *626*, 813–815.
89. Riviere-Baudet, M.; Morere, A.; Britten, J. F.; Onyszchuk, M. *J. Organomet. Chem.* **1992**, *423*, C5–C8.
90. Wraage, K.; Lameyer, L.; Stalke, D.; Roesky, H. W. *Angew. Chem., Int. Ed. Engl.* **1999**, *38*, 522–523.
91. Mitzel, N. W.; Smart, B. A.; Blake, A. J.; Parsons, S.; Rankin, D. W. H. *J. Chem. Soc., Dalton Trans.* **1996**, 2095–2100.
92. Losehand, U.; Mitzel, N. W. *Inorg. Chem.* **1998**, *37*, 3175–3182.
93. Herberhold, M.; Trobs, V.; Zhou, H.; Wrackmeyer, B. *Z. Naturforsch. B: Anorg. Chem. Org. Chem.* **1997**, *52*, 1181–1184.
94. Goetze, H. J.; Garbe, W. *Spectrochim. Acta A* **1982**, *38*, 665–669.
95. Ohtaki, T.; Ando, W. *Chem. Lett.* **1994**, 1061–1064.
96. Ando, W.; Ohtaki, T.; Kabe, Y. *Organometallics* **1994**, *13*, 434–435.
97. Riviere-Baudet, M.; Satge, J.; El Baz, F. *J. Chem. Soc., Chem. Commun.* **1995**, 1687–1688.
98. Gruetzmacher, H.; Pritzkow, H. *Angew. Chem., Int. Ed. Engl.* **1991**, *30*, 1017–1018.
99. Bartlett, R. A.; Power, P. P. *J. Am. Chem. Soc.* **1990**, *112*, 3660–3662.
100. Olmstead, M. M.; Power, P. P. *Inorg. Chem.* **1984**, *23*, 413–415.
101. Allan, R. E.; Beswick, M. A.; Edwards, A. J.; Paver, M. A.; Rennie, M.-A; Raithby, P. R.; Wright, D. S. *J. Chem. Soc., Dalton Trans.* **1995**, 1991–1994.
102. Bashall, A.; Feeder, N.; Harron, E. A.; McPartlin, M.; Mosquera, M. E. G.; Saez, D.; Wright, D. S. *J. Chem. Soc., Dalton Trans.* **2000**, 4104–4111.
103. Allan, R. E.; Beswick, M. A.; Coggan, G. R.; Raithby, P. R.; Wheatley, A. E. H.; Wright, D. S. *Inorg. Chem.* **1997**, *36*, 5202–5205.
104. Allan, R. E.; Beswick, M. A.; Cromhout, N. L.; Paver, M. A.; Raithby, P. R.; Steiner, A.; Trevithick, M.; Wright, D. S. *Chem. Commun.* **1996**, 1501–1502.
105. Edwards, A. J.; Paver, M. A.; Raithby, P. R.; Russell, C. A.; Wright, D. S. *J. Chem. Soc., Chem. Commun.* **1993**, 1086–1088.
106. Anatsko, O. E.; Sevast'yanova, T. N.; Suvorov, A. V.; Kondrat'ev, Yu. V. *Russ. J. Gen. Chem.* **1999**, *69*, 1262–1265.
107. Kupce, E.; Upena, E.; Trusule, M.; Lukevics, E. *Latv. PSR Zinat. Akad. Vestis Kim. Ser.* **1988**, 359–360.
108. Kupce, E.; Ignatovich, L. M.; Lukevics, E. *J. Organomet. Chem.* **1989**, *372*, 189–191.
109. Hall, V. J.; Tiekink, E. R. T. *Z. Kristallogr.* **1996**, *211*, 247–250.
110. Tian, L.; Zhao, B.; Fu, F. *Synth. React. Inorg. Met.-Org. Chem.* **1998**, *28*, 175–190.
111. Crowe, A. J.; Smith, P. J.; Atassi, G. *Inorg. Chim. Acta* **1984**, *93*, 179–184.
112. Tiekink, E. R. T.; Hall, V. J.; Buntine, M. A.; Hook, J. *Z. Kristallogr.* **2000**, *215*, 23–33.
113. Bhushan, V.; Gupta, K. L.; Saxena, G. C. *Synth. React. Inorg. Met.-Org. Chem.* **1990**, *20*, 363–375.
114. Hall, V. J.; Tiekink, E. R. T. *Z. Kristallogr.* **1998**, *213*, 403–404.
115. Cox, M. J.; Tiekink, E. R. T. *Z. Kristallogr.* **1994**, *209*, 291–292.
116. Das, V. G. K.; Yap, C. K.; Smith, P. J. *J. Organomet. Chem.* **1987**, *327*, 311–326.
117. Das, V. G. K.; Wei, C.; Keong, Y. C.; Mak, T. C. W. *J. Organomet. Chem.* **1987**, *299*, 41.
118. Das, V. G. K.; Keong, Y. C.; Wei, C.; Smith, P. J.; Mak, T. C. W. *J. Chem. Soc., Dalton Trans.* **1987**, 129–137.
119. Ng, S. W. *Z. Kristallog.* **1999**, *214*, 424–426.
120. Gabe, E. J.; Lee, F. L.; Smith, F. E. *Inorg. Chim. Acta* **1984**, *90*, L11–L13.
121. Ng, S. W.; Das, V. G. K. *J. Organomet. Chem.* **1996**, *513*, 105–108.
122. Austin, M.; Gebreyes, K.; Kuivila, H. G.; Swami, K.; Zubieta, J. A. *Organometallics* **1987**, *6*, 834–842.
123. Basu Baul, T. S.; Dcy, D.; Mishra, D. D.; Basaiawmoit, W. L.; Rivarola, E. *J. Organomet. Chem.* **1993**, *447*, 9–13.
124. Lopez, C.; Sanchez Gonzalez, A.; Garcia, M. E.; Casas, J. S.; Sordo, J.; Graziani, R.; Casellato, U. *J. Organomet. Chem.* **1992**, *434*, 261–268.
125. Sanchez Gonzalez, A.; Casas, J. S.; Sordo, J.; Russo, U.; Lareo, M. I.; Regueiro, B. J. *J. Inorg. Biochem.* **1990**, *39*, 227–235.
126. Baul, T. S. B.; Dey, D.; Mishra, D. D. *Synth. React. Inorg. Met. -Org. Chem.* **1993**, *23*, 53–65.
127. Chattopadhyay, T. K.; Kumar, A. K.; Roy, A.; Batsanov, A. S.; Shamuratov, E. B.; Struchkov, Y. T. *J. Organomet. Chem.* **1991**, *419*, 277–282.
128. Archer, S. J.; Koch, K. R.; Schmidt, S. *Inorg. Chim. Acta* **1987**, *126*, 209–218.
129. Engelhardt, L. M.; Kepert, D. L.; Patrick, J. M.; White, A. H. *Aust. J. Chem.* **1989**, *42*, 329–334.
130. Engelhardt, L. M.; Furphy, B. M.; Harrowfield, J. M.; Patrick, J. M.; Skelton, B. W.; White, A. H. *J. Chem. Soc., Dalton Trans.* **1989**, 595–599.
131. Engelhardt, L. M.; Patrick, J. M.; White, A. H. *Aust. J. Chem.* **1989**, *42*, 335–338.
132. Veith, M.; Lisowsky, R. *Angew. Chem., Int. Ed. Engl.* **1988**, *27*, 1087.
133. Veith, M.; Royan, B. W.; Huch, V. *Phosphorus, Sulfur Silicon Relat. Elem.* **1993**, *79*, 25–31.
134. Veith, M.; Jarczyk, M.; Huch, V. *Chem. Ber.* **1988**, *121*, 347–355.
135. Zhou, Y.; Richeson, D. S. *Inorg. Chem.* **1997**, *36*, 501–504.
136. Zhou, Y.; Richeson, D. S. *J. Am. Chem. Soc.* **1996**, *118*, 10850–10852.
137. Foley, S. R.; Yap, G. P. A.; Richeson, D. S. *J. Chem. Soc., Dalton Trans.* **2000**, 1663–1668.
138. Doering, U.; Haenssgen, D.; Jansen, M.; Nieger, M.; Tellenbach, A. *Z. Anorg. Allg. Chem.* **1998**, *624*, 965–969.
139. Heine, A.; Fest, D.; Stalke, D.; Habben, C. D.; Meller, A.; Sheldrick, G. M. *J. Chem. Soc., Chem. Commun.* **1990**, 742–743.
140. Chen, H.; Bartlett, R. A.; Dias, H. V. R.; Olmstead, M. M.; Power, P. P. *Inorg. Chem.* **1991**, *30*, 3390–3394.
141. Lobbia, G. G.; Cingolani, A.; Leonesi, D.; Lorenzotti, A.; Bonati, F. *Inorg. Chim. Acta* **1987**, *130*, 203–207.
142. Pettinari, C.; Lorenzotti, A.; Sclavi, G.; Cingolani, A.; Rivarola, E.; Colapietro, M.; Cassetta, A. *J. Organomet. Chem.* **1995**, *496*, 69–85.
143. Lobbia, G.; Zamponi, S.; Marassi, R.; Berrettoni, M.; Stizza, S.; Cecchi, P. *Gazz. Chim. Ital.* **1993**, *123*, 589–592.
144. Cox, M. J.; Rainone, S.; Siasios, G.; Tiekink, E. R. T.; Webster, L. K. *Main Group Met. Chem.* **1995**, *18*, 93–99.
145. Visalakshi, R.; Jain, V. K.; Kulshreshtha, S. K.; Rao, G. S. *Inorg. Chim. Acta* **1986**, *118*, 119–124.
146. Lobbia, G. G.; Cingolani, A.; Cecchi, P.; Calogero, S.; Wagner, F. E. *J. Organomet. Chem.* **1992**, *436*, 35–42.

147. Pettinari, C.; Pellei, M.; Cingolani, A.; Martini, D.; Drozdov, A.; Troyanov, S.; Panzeri, W.; Mele, A. *Inorg. Chem.* **1999**, *38*, 5777–5787.
148. Reger, D. L.; Mason, S. S.; Takats, J.; Zhang, X. W.; Rheingold, A. L.; Haggerty, B. S. *Inorg. Chem.* **1993**, *32*, 4345–4348.
149. Reger, D. L.; Collins, J. E.; Rheingold, A. L.; Liable-Sands, L. M.; Yap, G. P. A. *Inorg. Chem.* **1997**, *36*, 345–351.
150. Reger, D. L.; Wright, T. D.; Little, C. A.; Lamba, J. J. S.; Smith, M. D. *Inorg. Chem.* **2001**, *40*, 3810–3814.
151. Dungan, C. H.; Maringgele, W.; Meller, A.; Niedenzu, K.; Noeth, H.; Serwatowska, J.; Serwatowski, J. *Inorg. Chem.* **1991**, *30*, 4799–4806.
152. Dey, D. K.; Das, M. K.; Bansal, R. K. *J. Organomet. Chem.* **1997**, *535*, 7–15.
153. Lee, S. K.; Nicholson, B. K. *J. Organomet. Chem.* **1986**, *309*, 257–265.
154. Gioia Lobbia, G.; Calogero, S.; Bovio, B.; Cecchi, P. *J. Organomet. Chem.* **1992**, *440*, 27–40.
155. Reger, D. L.; Coan, P. S. *Inorg. Chem.* **1996**, *35*, 258–260.
156. Filippou, A. C.; Portius, P.; Kociok-Kohn, G. *Chem. Commun.* **1998**, 2327–2328.
157. Reger, D. L.; Knox, S. J.; Huff, M. F.; Rheingold, A. L.; Haggerty, B. S. *Inorg. Chem.* **1991**, *30*, 1754–1759.
158. Hansen, M. N.; Niedenzu, Kurt; Serwatowska, J.; Serwatowski, J.; Woodrum, K. R. *Inorg. Chem.* **1991**, *30*, 866–868.
159. Reger, D. L.; Huff, M. F.; Knox, S. J.; Adams, R. J.; Apperley, D. C.; Harris, R. K. *Inorg. Chem.* **1993**, *32*, 4472–4473.
160. Dias, H. V. R.; Jin, W. *Inorg. Chem.* **2000**, *39*, 815–819.
161. Reger, D. L.; Ding, Y. *Polyhedron* **1994**, *13*, 869–871.
162. Reger, D. L. *Comm. Inorg Chem.* **1999**, *21*, 1–28.
163. Cowley, A. H.; Geerts, R. L.; Nunn, C. M.; Carrano, C. J. *J. Organomet. Chem.* **1988**, *341*, C27–C30.
164. Reger, D. L.; Huff, M. F.; Rheingold, A. L.; Haggerty, B. S. *J. Am. Chem. Soc.* **1992**, *114*, 579–584.
165. Reger, D. L. *Synlett* **1992**, 469–475.
166. Janiak, C.; Temizdemir, S.; Scharmann, T. G.; Schmalstieg, A.; Demtschuk, J. *Z. Anorg. Allg. Chem.* **2000**, *626*, 2053–2062.
167. Engelhardt, L. M.; Furphy, B. M.; Harrowfield, J. M.; Patrick, J. M.; White, A. H. *Inorg. Chem.* **1989**, *28*, 1410–1413.
168. Radecka-Paryzek, W.; Gdaniec, M. *Polyhedron* **1997**, *16*, 3681–3686.
169. Wieghardt, K.; Kleine-Boymann, M.; Nuber, B.; Weiss, J.; Zsolnai, L.; Huttner, G. *Inorg. Chem.* **1986**, *25*, 1647–1650.
170. Miyamoto, T. K. *Main Group Met. Chem.* **1994**, *17*, 145–150.
171. Miyamoto, T. K.; Sugita, N.; Matsumoto, Y.; Sasaki, Y.; Konno, M. *Chem. Lett.* **1983**, 1695–1698.
172. Kadish, K. M.; Xu, Q. Y.; Barbe, J. M.; Anderson, J. E.; Wang, E.; Guilard, R. *Inorg. Chem.* **1988**, *27*, 691–696.
173. Kadish, K. M.; Xu, Q. Y.; Barbe, J. M.; Anderson, J. E.; Wang, E.; Guilard, R. *J. Am. Chem. Soc.* **1987**, *109*, 7705–7714.
174. Maiya, G. B.; Barbe, J. M.; Kadish, K. M. *Inorg. Chem.* **1989**, *28*, 2524–2527.
175. Cloutour, C.; Lafargue, D.; Pommier, J. C. *J. Organomet. Chem.* **1980**, *190*, 35–42.
176. Balch, A. L.; Cornman, C. R.; Olmstead, M. M. *J. Am. Chem. Soc.* **1990**, *112*, 2963–2969.
177. Hanack, M.; Zipplies, T. *J. Am. Chem. Soc.* **1985**, *107*, 6127–6129.
178. Cannon, J. B. *J. Pharm. Sci.* **1993**, *82*, 435–446.
179. Kessel, D.; Morgan, A.; Garbo, G. M. *Photochem. Photobiol.* **1991**, *54*, 193–196.
180. Arnold, D. P.; Tiekink, E. R. T. *Polyhedron* **1995**, *14*, 1785–1789.
181. Smith, G.; Arnold, D. P.; Kennard, C. H. L.; Mak, T. C. W. *Polyhedron* **1991**, *10*, 509–516.
182. Guilard, R.; Barbe, J. M.; Boukhris, M.; Lecomte, C. *J. Chem. Soc., Dalton Trans.* **1988**, 1921–1925.
183. Kadish, K. M.; Dubois, D.; Koeller, S.; Barbe, J. M.; Guilard, R. *Inorg. Chem.* **1992**, *31*, 3292–3294.
184. Dawson, D. Y.; Sangalang, J. C.; Arnold, J. *J. Am. Chem. Soc.* **1996**, *118*, 6082–6083.
185. Chen, J.; Woo, K. *Inorg. Chem.* **1998**, *37*, 3269–3275.
186. Arnold, D. P.; Bartley, J. P. *Inorg. Chem.* **1994**, *33*, 1486–1490.
187. Hawley, J. C.; Bampos, N.; Sanders, J. K. M.; Abraham, R. J. *Chem. Commun.* **1998**, 661–662.
188. Kadish, K. M.; Dubois, D.; Barbe, J. M.; Guilard, R. *Inorg. Chem.* **1991**, *30*, 4498–4501.
189. Foley, S. R.; Richeson, D. S. *Chem. Commun.* **2000**, 1391–1392.
190. Zhang, Y.-H.; Liu, Y.-P.; Fan, S.-H. *Synth. React. Inorg. Met. -Org. Chem.* **1999**, *29*, 279–288.
191. Fujiki, M.; Tabei, H.; Isa, K. *J. Am. Chem. Soc.* **1986**, *108*, 1532–1536.
192. Dirk, C. W.; Inabe, T.; Schoch, K. F., Jr.; Marks, T. J. *J. Am. Chem. Soc.* **1983**, *105*, 1539–1550.
193. Atwood, D. A.; Atwood, V. O.; Cowley, A. H.; Atwood, J. L.; Roman, E. *Inorg. Chem.* **1992**, *31*, 3871–3872.
194. Kuchta, M. C.; Parkin, G. *J. Chem. Soc., Chem. Commun.* **1994**, 1351–1352.
195. Kuchta, M. C.; Parkin, G. *J. Am. Chem. Soc.* **1994**, *116*, 8372–8373.
196. Kuchta, M. C.; Hascall, T.; Parkin, G. *Chem. Commun.* **1998**, 751–752.
197. Belcher, W. J.; Brothers, P. J.; Meredith, A. P.; Rickard, C. E. F.; Ware, D. C. *J. Chem. Soc., Dalton Trans.* **1999**, 2833–2836.
198. Kuchta, M. C.; Parkin, G. *Polyhedron* **1996**, *15*, 4599–4602.
199. Kuchta, M. C.; Parkin, G. *Chem. Commun.* **1996**, 1669–1670.
200. Varshny, A. K.; Varshny, S.; Singh, H. L. *Synth. React. Inorg. Met. -Org. Chem.* **1999**, *29*, 245–254.
201. Constable, E. C.; Khan, F. K.; Lewis, J.; Liptrot, M. C.; Raithby, P. R. *J. Chem. Soc., Dalton Trans.* **1985**, *2*, 333–335.
202. Constable, E. C.; Holmes, J. M. *Polyhedron* **1988**, *7*, 2531–2536.
203. Di Vaira, M.; Mani, F.; Stoppioni, P. *J. Chem. Soc., Dalton Trans.* **1998**, 3209–3214.
204. Andres, A.; Bencini, A.; Carachalios, A.; Bianchi, A.; Dapporto, P.; Garcia-Espana, E.; Paoletti, P.; Paoli, P. *J. Chem. Soc., Dalton Trans.* **1993**, 3507–3513.
205. Amorim, M. T. S.; Chaves, S.; Delgado, R.; Frausto da Silva, J. J. R. *J. Chem. Soc., Dalton Trans.* **1991**, 3065–3072.
206. White, A. H. *J. Chem. Soc., Dalton Trans.* **1994**, 793–798.
207. Bazzicalupi, C.; Bencini, A.; Fusi, V.; Giorgi, C.; Paoletti, P.; Valtancoli, B. *J. Chem. Soc., Dalton Trans.* **1999**, 393–400.

208. Arranz, P.; Bazzicalupi, C.; Bencini, A.; Bianchi, A.; Ciattini, S.; Fornasari, P.; Giorgi, C.; Valtancoli, B. *Inorg. Chem.* **2001**, *40*, 6383–6389.
209. Brooker, S.; Kelly, R. J. *J. Chem. Soc., Dalton Trans.* **1996**, 2117–2122.
210. Godfrey, S. M.; Mushtaq, I.; Pritchard, R. G. *J. Chem. Soc., Dalton Trans.* **1999**, 1319–1324.
211. Kagoshima, H.; Hashimoto, Y.; Oguro, D.; Kutsuna, T.; Saigo, K. *Tetrahedron Lett.* **1998**, *39*, 1203–1206.
212. Colton, R.; Dakternieks, D.; Harvey, C. *Inorg. Chim. Acta* **1982**, *61*, 1–7.
213. Bricklebank, N.; Godfrey, S. M.; McAuliffe, C. A.; Pritchard, R. G. *J. Chem. Soc., Chem. Commun.* **1994**, 695–696.
214. Bricklebank, N.; Godfrey, S. M.; McAuliffe, C. A.; Molloy, K. C. *J. Chem. Soc., Dalton Trans.* **1995**, 1593–1596.
215. Reutov, O. A.; Petrosyan, V. S.; Yashina, N. S.; Gefel, E. I. *J. Organomet. Chem.* **1988**, *341*, C31–C34.
216. Yoder, C. H.; Margolis, L. A.; Horne, J. M. *J. Organomet. Chem.* **2001**, *633*, 33–38.
217. Colton, R.; Dakternieks, D. *Inorg. Chim. Acta* **1988**, *143*, 151–159.
218. Spencer, J. N.; Ganunis, T.; Zafar, A.; Eppley, H.; Otter, J. C.; Coley, S. M.; Yoder, C. H. *J. Organomet. Chem.* **1990**, *389*, 295–300.
219. Dakternieks, D.; Zhu, H.; Tiekink, E. R. T. *Main Group Met. Chem.* **1994**, *17*, 519–535.
220. Driess, M.; Pritzkow, H. *Chem. Ber.* **1993**, *126*, 1131–1133.
221. Haenssgen, D.; Jeske, R.; Korber, N.; Mohr, C.; Nieger, M. *Anorg. Allg. Chem.* **1998**, *624*, 1202–1206.
222. Sarikahya, F. *Synth. React. Inorg. Met.-Org. Chem.* **1989**, *19*, 641–650.
223. Bokii, N. G.; Struchkov, Yu. T.; Kolesnikov, S. P.; Rogozhin, I. S.; Nefedov, O. M. *Izv. Akad. Nauk SSSR, Ser. Khim.* **1975**, 812–815.
224. Inoguchi, Y.; Okui, S.; Mochida, K.; Itai, A. *Bull. Chem. Soc. Jpn.* **1985**, *58*, 974–977.
225. Dean, P. A. W. *Can. J. Chem.* **1983**, *61*, 1795–1799.
226. Cowley, A. H.; Hall, S. W.; Nunn, C. M.; Power, J. M. *J. Chem. Soc., Chem. Commun.* **1988**, 753–754.
227. Cowley, A. H.; Hall, S. W.; Nunn, C. M.; Power, J. M. *Angew. Chem., Int. Ed. Engl.* **1988**, *100*, 874–875.
228. Balch, A. L.; Oram, D. E. *Inorg. Chem.* **1987**, *26*, 1906–1912.
229. Karsch, H. H.; Deubelly, B.; Riede, J.; Mueller, G. *J. Organomet. Chem.* **1987**, *336*, C37–C40.
230. Balch, A. L.; Oram, D. E. *Organometallics* **1986**, *5*, 2159–2161.
231. Karsch, H. H.; Appelt, A.; Hanika, G. *J. Organomet. Chem.* **1986**, *312*, C1–C5.
232. Karsch, H. H.; Deubelly, B.; Hanika, G.; Riede, J.; Mueller, G. *J. Organomet. Chem.* **1988**, *344*, 153–161.
233. Karsch, H. H.; Appelt, A.; Mueller, G. *Organometallics* **1986**, *5*, 1664–1670.
234. Karsch, H. H.; Deubelly, B.; Riede, J.; Mueller, G. *Angew. Chem., Int. Ed. Engl.* **1987**, *26*, 673.
235. Driess, M.; Monse, C.; Boese, R.; Blaser, D. *Angew. Chem., Int. Ed. Engl.* **1998**, *37*, 2257–2259.
236. Karnop, M.; Du Mont, W. W.; Jones, P. G.; Jeske, J. *Chem. Ber. Recl.* **1997**, *130*, 1611–1618.
237. Driess, M.; Pritzkow, H.; Winkler, U. *Chem. Ber.* **1992**, *125*, 1541–1546.
238. Baudler, M.; De Riese-Meyer, L.; Schings, U. *Z. Anorg. Allg. Chem.* **1984**, *519*, 24–30.
239. Karsch, H. H.; Deubelly, B.; Keller, U.; Bienlein, F.; Richter, R.; Bissinger, P.; Heckel, M.; Mueller, G. *Chem. Ber.* **1996**, *129*, 759–764.
240. Draeger, M.; Escudie, J.; Couret, C.; Ranaivonjatovo, H.; Satge, J. *Organometallics* **1988**, *7*, 1010–1013.
241. Escudie, J.; Couret, C.; Satge, J.; Andrianarison, M.; Andriamizaka, J. D. *J. Am. Chem. Soc.* **1985**, *107*, 3378–3379.
242. Andrianarison, M.; Couret, C.; Declercq, J. P.; Dubourg, A.; Escudie, J.; Ranaivonjatovo, H.; Satge, J. *Organometallics* **1988**, *7*, 1545–1548.
243. Couret, C.; Escudie, J.; Satge, J.; Raharinirina, A.; Andriamizaka, J. D. *J. Am. Chem. Soc.* **1985**, *107*, 8280–8281.
244. Haenssgen, D.; Stahlhut, E.; Aldenhoven, H.; Doerr, A. *J. Organomet. Chem.* **1992**, *425*, 19–25.
245. Driess, M.; Martin, S.; Merz, K.; Pintchouk, V.; Pritzkow, H.; Grutzmacher, H.; Kaupp, M. *Angew. Chem., Int. Ed. Engl.* **1997**, *36*, 1894–1896.
246. Driess, M.; Janoschek, R.; Pritzkow, H.; Rell, S.; Winkler, U. *Angew. Chem., Int. Ed. Engl.* **1995**, *34*, 1614–1616.
247. Matchett, M. A.; Chiang, M. Y.; Buhro, W. E. *Inorg. Chem.* **1994**, *33*, 1109–1114.
248. Goel, S. C.; Chiang, M. Y.; Rauscher, D. J.; Buhro, W. E. *J. Am. Chem. Soc.* **1993**, *115*, 160–169.
249. Cowley, A. H.; Giolando, D. M.; Jones, R. A.; Nunn, C. M.; Power, J. M. *Polyhedron* **1988**, *7*, 1909–1910.
250. Arif, A. M.; Cowley, A. H.; Jones, R. A.; Power, J. M. *J. Chem. Soc., Chem. Commun.* **1986**, 1446–1447.
251. Du Mont, W. W.; Grenz, M. *Chem. Ber.* **1985**, *118*, 1045–1049.
252. Du Mont, W. W.; Rudolph, G. *Z. Naturforsch. B: Anorg. Chem. Org. Chem.* **1981**, *36*, 1215–1218.
253. Cowley, A. H.; Giolando, D. M.; Jones, R. A.; Nunn, C. M.; Power, J. M.; Du Mont, W. W. *Polyhedron* **1988**, *7*, 1317–1319.
254. Zsolnai, L.; Huttner, G.; Driess, M. *Angew. Chem., Int. Ed. Engl.* **1993**, *32*, 1439–1440.
255. Voronkov, M. G.; Gavrilova, G. A.; Basenko, S. V. *Russ. J. Gen. Chem.* **2001**, *71*, 210–212.
256. Yatsenko, A. V.; Medvedev, S. V.; Paseshnichenko, K. A.; Aslanov, L. A. *J. Organomet. Chem.* **1985**, *284*, 181–188.
257. Ng, S. W.; Rheingold, A. L. *J. Organomet. Chem.* **1989**, *378*, 339–345.
258. Blaschette, A.; Hippel, I.; Krahl, J.; Wieland, E.; Jones, P. G.; Sebald, A. *J. Organomet. Chem.* **1992**, *437*, 279–297.
259. Hippel, I.; Jones, P. G.; Blaschette, A. *J. Organomet. Chem.* **1993**, *448*, 63–67.
260. Harrowfield, J. M.; Skelton, B. W.; White, A. H. *J. Chem. Soc., Dalton Trans.* **1993**, 2011–2016.
261. Yatsenko, A. V.; Aslanov, L. A.; Schenk, H. *Polyhedron* **1995**, *14*, 2371–2377.
262. Cunningham, D.; Landers, E. M.; McArdle, P.; Ni Chonchubhair, N. *J. Organomet. Chem.* **2000**, *612*, 53–60.
263. Yoder, C. H.; Coley, S. M.; Kneizys, S. P.; Spencer, J. N. *J. Organomet. Chem.* **1989**, *362*, 59–62.
264. Pelizzi, G.; Tarasconi, P.; Pelizzi, C.; Molloy, K.; Waterfield, P. *Main Group Met. Chem.* **1987**, *10*, 353–362.
265. Wirth, A.; Moers, O.; Blaschette, A.; Jones, P. G. *Z. Anorg. Allg. Chem.* **1999**, *625*, 982–988.
266. Dondi, S.; Nardelli, M.; Pelizzi, C.; Pelizzi, G.; Predieri, G. *J. Organomet. Chem.* **1986**, *308*, 195–206.
267. Lorberth, J.; Shin, S. H.; Otto, M.; Wocadlo, S.; Massa, W.; Yashina, N. S. *J. Organomet. Chem.* **1991**, *407*, 313–318.
268. Parr, J.; Ross, A. T.; Slawin, A. M. Z. *Polyhedron* **1997**, *16*, 2765–2770.
269. Smith, D. A.; Sucheck, S.; Pinkerton, A. *J. Chem. Soc., Chem. Commun.* **1992**, 367–368.
270. Johnson, S. E.; Knobler, C. B. *Organometallics* **1994**, *13*, 4928–4938.
271. Bott, S. G.; Prinz, H.; Alvanipour, A.; Atwood, J. L. *J. Coord. Chem.* **1987**, *16*, 303–309.
272. Bruegge, H. J.; Foelsing, R.; Knoechel, A.; Dreissig, W. *Polyhedron* **1985**, *4*, 1493–1498.
273. Nazarenko, A. Y.; Rusanov, E. B. *Polyhedron* **1994**, *13*, 2549–2553.

274. Shin, Y. G.; Hampden-Smith, M. J.; Kodas, T. T.; Duesler, E. N. *Polyhedron* **1993**, *12*, 1453–1458.
275. Jutzi, P.; Schmidt, H.; Neumann, B.; Stammler, H.-G. *Organometallics* **1996**, *15*, 741–746.
276. Tokitoh, N.; Matsumoto, T.; Okazaki, R. *Chem. Lett.* **1995**, 1087–1088.
277. Burger, K.; Nagy, L.; Buzas, N.; Vertes, A.; Mehner, H. *J. Chem. Soc., Dalton Trans.* **1993**, 2499–2504.
278. Puff, H.; Braun, K.; Franken, S.; Koek, T. R.; Schuh, W. *J. Organomet. Chem.* **1988**, *349*, 293–303.
279. Sun, K.; Dadachov, M. S.; Conradsson, T.; Zou, X. *Acta Cryst.* **2000**, *56*, C1092–C1094.
280. Zumbusch, A.; Schnockel, H. *J. Chem. Phys.* **1998**, *108*, 8092–8100.
281. Veith, M.; Mathur, C.; Huch, V. *J. Chem. Soc., Dalton Trans.* **1997**, 995–999.
282. Zobel, B.; Costin, J.; Vincent, B. R.; Tiekink, E. R. T.; Dakternieks, D. *J. Chem. Soc., Dalton Trans.* **2000**, 4021–4022.
283. Puff, H.; Reuter, H. *J. Organomet. Chem.* **1989**, *373*, 173–184.
284. Puff, H.; Reuter, H. *J. Organomet. Chem.* **1989**, *368*, 173–183.
285. Sita, L. R.; Xi, R.; Yap, G. P. A.; Liable-Sands, L. M.; Rheingold, A. L. *J. Am. Chem. Soc.* **1997**, *119*, 756–760.
286. Kunkely, H.; Vogler, A. *Chem. Phys. Lett.* **1991**, *187*, 609–612.
287. Agarwal, B. K.; Singh, Y. P.; Bohra, R.; Srivastava, G.; Rai, A. K. *J. Organomet. Chem.* **1993**, *444*, 47–51.
288. Puff, H.; Hevendehl, H.; Hoefer, K.; Reuter, H.; Schuh, W. *J. Organomet. Chem.* **1985**, *287*, 163–178.
289. Donaldson, J. D.; Grimes, S. M.; Johnston, S. R.; Abrahams, I. *J. Chem. Soc., Dalton Trans.* **1995**, 2273–2276.
290. Grimes, S. M.; Johnston, S. R.; Abrahams, I. *J. Chem. Soc., Dalton Trans.* **1995**, 2081–2086.
291. Veith, M. *Chem. Rev.* **1990**, *90*, 3–16.
292. Goel, S. C.; Chiang, M. Y.; Buhro, W. E. *Inorg. Chem.* **1990**, *29*, 4640–4646.
293. Papiernik, R.; Hubert-Pfalzgraf, L. G.; Massiani, M. C. *Polyhedron* **1991**, *10*, 1657–1662.
294. Weinert, C. S.; Guzei, I. A.; Rheingold, A. L.; Sita, L. R. *Organometallics* **1998**, *17*, 498–500.
295. Teff, D. J.; Caulton, K. G. *Inorg. Chem.* **1999**, *38*, 2240.
296. Pelizzi, C.; Pelizzi, G.; Tarasconi, P. *J. Organomet. Chem.* **1984**, *277*, 29–35.
297. Parkanyi, L.; Kalman, A.; Deak, A.; Venter, M.; Haiduc, I. *Inorg. Chem. Commun.* **1999**, *2*, 265–268.
298. Searle, D.; Smith, P. J.; Bell, N. A.; March, L. A.; Nowell, I. W.; Donaldson, J. D. *Inorg. Chim. Acta* **1989**, *162*, 143–149.
299. Willem, R.; Gielen, M.; Pepermans, H.; Hallenga, K.; Recca, A.; Finocchiaro, P. *J. Am. Chem. Soc.* **1985**, *107*, 1153–1160.
300. Kumar, A.; Bachlas, B. P.; Maire, J. C. *Polyhedron* **1983**, *2*, 907–916.
301. Jain, A.; Saxena, S.; Rai, A. K. *Ind. J. Chem., Sect. A* **1991**, *30*, 881–885.
302. Marchetti, F.; Pettinari, C.; Cingolani, A.; Leonesi, D. *Synth. React. Inorg. Met.-Org. Chem.* **1993**, *23*, 1485–1505.
303. Kira, M.; Zhang, L. C.; Kabuto, C.; Sakurai, H. *Organometallics* **1998**, *17*, 887–892.
304. Yin, H.-D.; Zhang, R.-F.; Wang, C.-H.; Ma, C.-L.; Wang, Y.; Tao, X.-Q. *Chin. J. Chem.* **2001**, *19*, 783–787.
305. Molloy, K. C.; Quill, K.; Blunden, S. J.; Hill, R. *Polyhedron* **1986**, *5*, 959–965.
306. Gabe, E. J.; Lee, F. L.; Khoo, L. E.; Smith, F. E. *Inorg. Chim. Acta* **1985**, *105*, 103–106.
307. Sandhu, G. K.; Boparov, N. S. *J. Organomet. Chem.* **1991**, *411*, 89–98.
308. Sandhu, G. K.; Boparov, N. S. *J. Organomet. Chem.* **1990**, *420*, 23–34.
309. Ng, S. W.; Das, V. G. K.; Yip, W. H.; Wang, R. J.; Mak, T. C. W. *J. Organomet. Chem.* **1990**, *393*, 201–204.
310. Ng, S. W.; Das, V. G. K.; Skelton, B. W.; White, A. H. *J. Organomet. Chem.* **1989**, *377*, 221–225.
311. Lockhart, T. P.; Calabrese, J. C.; Davidson, F. *Organometallics* **1987**, *6*, 2479–2483.
312. Sandhu, G. K.; Kaur, G. *Main Group Met. Chem.* **1990**, *13*, 149–165.
313. Sandhu, G. K.; Kaur, G. *J. Organomet. Chem.* **1990**, *388*, 63–70.
314. Mokal, V. B.; Jain, V. K. *J. Organomet. Chem.* **1992**, *441*, 215–226.
315. Holecek, J.; Lycka, A.; Nadvornik, M.; Handlir, K. *Collect. Czech. Chem. Commun.* **1991**, *56*, 1908–1915.
316. Parulekar, C. S.; Jain, V. K.; Kesavadas, T.; Tiekink, E. R. T. *J. Organomet. Chem.* **1990**, *387*, 163–173.
317. Gielen, M.; El Khloufi, A.; Biesemans, M.; Willem, R.; Meunier-Piret, J. *Polyhedron* **1992**, *11*, 1861–1868.
318. Narula, S. P.; Bharadwaj, S. K.; Sharma, H. K.; Mairesse, G.; Barbier, P.; Nowogrocki, G. *J. Chem. Soc., Dalton Trans.* **1988**, 1719–1723.
319. Narula, S. P.; Bharadwaj, S. K.; Sharda, Y.; Day, R. O.; Howe, L.; Holmes, R. R. *Organometallics* **1992**, *11*, 2206–2211.
320. Sandhu, G. K.; Boparoy, N. S. *Synth. React. Inorg. Met.-Org. Chem.* **1990**, *20*, 975–988.
321. Molloy, K. C.; Quill, K.; Nowell, I. W. *J. Chem. Soc., Dalton Trans.* **1987**, 101–106.
322. Adams, S.; Draeger, M.; Mathiasch, B. *J. Organomet. Chem.* **1987**, *326*, 173–186.
323. Holmes, R. R.; Day, R. O.; Chandrasekhar, V.; Shafeizad, S.; Harland, J. J.; Rau, D. N.; Holmes, J. M. *Phosphorus, Sulfur Silicon Relat. Elem.* **1986**, *28*, 91–98.
324. Chandrasekhar, V.; Day, R. O.; Holmes, R. R. *Inorg. Chem.* **1985**, *24*, 1970–1971.
325. Day, R. O.; Chandrasekhar, V.; Swamy, K. C. K.; Holmes, J. M.; Burton, S. D.; Holmes, R. R. *Inorg. Chem.* **1988**, *27*, 2887–2893.
326. Chandrasekhar, V.; Schmid, C. G.; Burton, S. D.; Holmes, J. M.; Day, R. O.; Holmes, R. R. *Inorg. Chem.* **1987**, *26*, 1050–1056.
327. Ng, S. W.; Hook, J. M.; Gielen, M. *Appl. Organomet. Chem.* **2000**, *14*, 1–7.
328. Day, R. O.; Holmes, J. M.; Chandrasekhar, V.; Holmes, R. R. *J. Am. Chem. Soc.* **1987**, *109*, 940–941.
329. Holmes, R. R.; Swamy, K. C. K.; Schmid, C. G.; Day, R. O. *J. Am. Chem. Soc.* **1988**, *110*, 7060–7066.
330. Swamy, K. C. K.; Schmid, C. G.; Day, R. O.; Holmes, R. R. *J. Am. Chem. Soc.* **1988**, *110*, 7067–7076.
331. Swamy, K. C. K.; Day, R. O.; Holmes, R. R. *J. Am. Chem. Soc.* **1988**, *110*, 7543–7544.
332. Silvestru, A.; Silvestru, C.; Haiduc, I.; Drake, J. E.; Yang, J.; Caruso, F. *Polyhedron* **1996**, *16*, 949–961.
333. Ballivet-Tkatchenko, D.; Douteau, O.; Stutzmann, S. *Organometallics* **2000**, *19*, 4563–4567.
334. Sandhu, G. K.; Kaur, H. *Main Group Met. Chem.* **1990**, *13*, 29–50.
335. Sandhu, G. K.; Kaur, H. *Appl. Organomet. Chem.* **1990**, *4*, 345–352.
336. Edelmann, F. T.; Buijink, J.-K. F.; Brooker, S. A.; Herbst-Irmer, R.; Kilimann, U.; Bohnen, F. M. *Inorg. Chem.* **2000**, *39*, 6134–6135.
337. Arifin, Z.; Filmore, E. J.; Donaldson, J. D.; Grimes, S. M. *J. Chem. Soc., Dalton Trans.* **1984**, 1965–1968.

338. Birchall, T.; Faggiani, R.; Lock, C. J. L.; Manivannan, V. *J. Chem. Soc., Dalton Trans.* **1987**, 1675–1682.
339. El-Hadad, A. A.; McGarvey, B. R.; Merzougui, B.; Sung, R. G. W.; Trikha, A. K.; Tuck, D. G. *J. Chem. Soc., Dalton Trans.* **2001**, 1046–1052.
340. Riviere, P.; Castel, A.; Satge, J.; Guyot, D. *J. Organomet. Chem.* **1986**, *315*, 157–164.
341. Schriewer, M.; Neumann, W. P. *Angew. Chem., Int. Ed. Engl.* **1981**, *20*, 1019.
342. Michels, E.; Neumann, W. P. *Tetrahedron Lett.* **1986**, *27*, 2455–2458.
343. Holmes, R. R.; Day, R. O.; Sau, A. C.; Holmes, J. M. *Inorg. Chem.* **1986**, *25*, 600–606.
344. Holmes, R. R.; Day, R. O.; Sau, A. C.; Poutasse, C. A.; Holmes, J. M. *Inorg. Chem.* **1986**, *25*, 607–611.
345. Holmes, R. R.; Day, R. O.; Sau, A. C.; Poutasse, C. A.; Holmes, J. M. *Inorg. Chem.* **1985**, *24*, 193–199.
346. Parr, J.; Slawin, A. M. Z.; Woollins, J. D.; Williams, D. J. *Polyhedron* **1994**, *13*, 3261–3263.
347. Denekamp, C. I. F.; Evans, D. F.; Parr, J.; Woollins, J. D. *J. Chem. Soc., Dalton Trans.* **1993**, 1489–1492.
348. Annan, T. A.; Chadha, R. K.; Tuck, D. G.; Watson, K. D. *Can. J. Chem.* **1987**, *65*, 2670–2676.
349. Mabrouk, H. E.; Tuck, D. G. *J. Chem. Soc., Dalton Trans.* **1988**, 2539–2543.
350. Machell, J. C.; Mingos, D. M. P.; Stolberg, T. L. *Polyhedron* **1989**, *8*, 2933–2935.
351. Tiekink, E. R. T. *J. Organomet. Chem.* **1986**, *302*, C1–C3.
352. Tacke, R.; Sperlich, J.; Becker, B. *Chem. Ber.* **1994**, *127*, 643–646.
353. Tacke, R.; Heermann, J.; Puelm, M. *Organometallics* **1997**, *16*, 5648–5652.
354. Tacke, R.; Heermann, J.; Pfrommer, B. *Inorg. Chem.* **1998**, *37*, 2070–2072.
355. Barrau, J.; Rima, G.; El Amraoui, T. *J. Organomet. Chem.* **1998**, *570*, 163–174.
356. Hascall, T.; Rheingold, A. L.; Guzei, I.; Parkin, G. *Chem. Commun.* **1998**, 101–102.
357. McBurnett, B. G.; Cowley, A. H. *Chem. Commun.* **1999**, 17–18.
358. Wong, C. Y.; Woollins, J. D. *Coord. Chem. Rev.* **1994**, *130*, 175–241.
359. Garcia Martinez, E.; Sanchez Gonzalez, A.; Casas, J. S.; Sordo, J.; Casellato, U.; Graziani, R. *Inorg. Chim. Acta* **1992**, *191*, 75–79.
360. Casas, J. S.; Martinez, E. G.; Gonzalez, A. S.; Sordo, J.; Casellato, U.; Graziani, R.; Russo, U. *J. Organomet. Chem.* **1995**, *493*, 107–111.
361. Casa, J. S.; Castineiras, A.; Garcia Martinez, E. G.; Gonzalez, A. S.; Sordo, J.; Vazquez Lopez, E. M.; Russo, U. *Polyhedron* **1996**, *15*, 891–902.
362. Wade, S. R.; Willey, G. R. *Inorg. Chim. Acta* **1983**, *72*, 201–204.
363. Willey, G. R.; Jarvis, A.; Palin, J.; Errington, W. *J. Chem. Soc., Dalton Trans.* **1994**, 255–258.
364. Blake, A. J.; Fenske, D.; Li, W.-S.; Lippolis, V.; Schroder, M. *J. Chem. Soc., Dalton Trans.* **1998**, 3961–3968.
365. Tokitoh, N.; Matsumoto, T.; Ichida, H.; Okazaki, R. *Tetrahedron Lett.* **1991**, *32*, 6877–6878.
366. Tokitoh, N.; Matsumoto, T.; Okazaki, R. *Tetrahedron Lett.* **1991**, *32*, 6143–6146.
367. Tokitoh, N.; Matsuhashi, Y.; Okazaki, R. *Tetrahedron Lett.* **1991**, *32*, 6151–6154.
368. Matsumoto, T.; Tokitoh, N.; Okazaki, R.; Goto, M. *Organometallics* **1995**, *14*, 1008–1015.
369. Matsumoto, T.; Tokitoh, N.; Okazaki, R. *Angew. Chem., Int. Ed. Engl.* **1994**, *33*, 2316–2317.
370. Tokitoh, N.; Matsumoto, T.; Okazaki, R. *Tetrahedron Lett.* **1992**, *33*, 2531–2534.
371. Matsumoto, T.; Kishikawa, K.; Tokitoh, N.; Okazaki, R. *Phosphorus, Sulfur Silicon Relat. Elem.* **1994**, *93–94*, 177–180.
372. Saito, M.; Tokitoh, N.; Okazaki, R. *J. Organomet. Chem.* **1995**, *499*, 43–48.
373. Saito, M.; Tokitoh, N.; Okazaki, R. *Organometallics* **1995**, *14*, 3620–3622.
374. Tokitoh, N.; Kano, N.; Shibata, K.; Okazaki, R. *Organometallics* **1995**, *14*, 3121–3123.
375. Kano, N.; Shibata, K.; Tokitoh, N.; Okazaki, R. *Organometallics* **1999**, *18*, 2999–3007.
376. Kano, N.; Tokitoh, N.; Okazaki, R. *Chem. Lett.* **1997**, 277–278.
377. Saito, M.; Tokitoh, N.; Okazaki, R. *J. Am. Chem. Soc.* **1997**, *119*, 11124–11125.
378. Hwu, S.-J.; Bucher, C. K.; Carpenter, J. D.; Taylor, S. P. *Inorg. Chem.* **1995**, *34*, 1979–1980.
379. Tan, K.; Darovsky, A.; Parise, J. B. *J. Am. Chem. Soc.* **1995**, *117*, 7039–7040.
380. Parise, J. B.; Tan, K. *Chem. Commun.* **1996**, 1687–1688.
381. Loose, A.; Sheldrick, W. S. *Z. Naturforsch. B: Anorg. Chem. Org. Chem.* **1997**, *52*, 687–692.
382. Unno, M.; Kawai, Y.; Shioyama, H.; Matsumoto, H. *Organometallics* **1997**, *16*, 4428–4434.
383. Ando, W.; Kadowaki, T.; Kabe, Y.; Ishii, M. *Angew. Chem., Int. Ed. Engl.* **1992**, *31*, 59–61.
384. Haas, A.; Kutsch, H. J.; Krueger, C. *Chem. Ber.* **1987**, *120*, 1045–1048.
385. Berwe, H.; Haas, A. *Chem. Ber.* **1987**, *120*, 1175–1182.
386. Sheldrick, W. S.; Schaaf, B. *Z. Naturforsch. B: Anorg. Chem. Org. Chem.* **1995**, *50*, 1469–1475.
387. Bubenheim, W.; Muller, U. *Z. Naturforsch. B: Anorg. Chem. Org. Chem.* **1995**, *50*, 1135–1136.
388. Tan, K.; Ko, Y.; Parise, J. B. *Acta Crystallogr.* **1995**, *51*, C398–C401.
389. Sheldrick, W. S. *Z. Anorg. Allg. Chem.* **1988**, *562*, 23–30.
390. Sportouch, S.; Tillard-Charbonnel, M.; Belin, C. *J. Chem. Soc., Dalton Trans.* **1995**, 3113–3116.
391. Campbell, J.; Devereux, L. A.; Gerken, M.; Mercier, H. P. A.; Pirani, A. M.; Schrobilgen, G. J. *Inorg. Chem.* **1996**, *35*, 2945–2962.
392. Fassler, T. F.; Schutz, U. *J. Organomet. Chem.* **1997**, *541*, 269–276.
393. Loose, A.; Sheldrick, W. S. *Z. Anorg. Allg. Chem.* **1999**, *625*, 233–240.
394. Boudjouk, P.; Remington, M. P., Jr.; Grier, D. G.; Triebold, W.; Jarabek, B. R. *Organometallics* **1999**, *18*, 4534–4537.
395. Weidenbruch, M.; Grobecker, U.; Saak, W.; Peters, E.-M.; Peters, K. *Organometallics* **1998**, *17*, 5206–5208.
396. Kano, N.; Tokitoh, N.; Okazaki, R. *Organometallics* **1998**, *17*, 1241–1244.
397. Dakternieks, D.; Zhu, H.; Masi, D.; Mealli, C. *Inorg. Chem.* **1992**, *31*, 3601–3606.
398. Donoghue, N.; Tiekink, E. R. T.; Webster, L. *Appl. Organomet. Chem.* **1993**, *7*, 109–117.
399. Pelizzi, C.; Pelizzi, G.; Tarasconi, P. *J. Organomet. Chem.* **1984**, *277*, 29–35.
400. Das, V. G. K.; Wei, C.; Sinn, E. *J. Organomet. Chem.* **1985**, *290*, 291–299.
401. Harcourt, R. D.; Tiekink, E. R. T. *Aust. J. Chem.* **1987**, *40*, 611–618.
402. Kato, S.; Tani, K.; Kitaoka, N.; Yamada, K.; Mifune, H. *J. Organomet. Chem.* **2000**, *611*, 190–199.
403. Chadha, R. K.; Drake, J. E.; Sarkar, A. B. *Inorg. Chem.* **1987**, *26*, 2885–2888.

404. Clark, H. C.; Jain, V. K.; Mehrotra, R. C.; Singh, B. P.; Srivastava, G.; Birchall, T. *J. Organomet. Chem.* **1985**, *279*, 385–394.
405. Lefferts, J. L.; Molloy, K. C.; Hossain, M. B.; Van der Helm, D.; Zuckerman, J. J. *Inorg. Chem.* **1982**, *21*, 1410–1416.
406. Haiduc, I.; Silvestru, C.; Roesky, H. W.; Schmidt, H. G.; Noltemeyer, M. *Polyhedron* **1993**, *12*, 69–75.
407. Day, R. O.; Holmes, R. R.; Schmidpeter, A.; Stoll, K.; Howe, L. *Chem. Ber.* **1991**, *124*, 2443–2448.
408. Casas, J. S.; Castineiras, A.; Haiduc, I.; Sanchez, A.; Sordo, J.; Vazquez-Lopez, E. M. *Polyhedron* **1994**, *13*, 2873–2879.
409. Garcia-Montalvo, V.; Novosad, J.; Kilian, P.; Woollins, J.; Slawin, A. M. Z.; Garcia, P. G. Y.; Lopez-Cardoso, M.; Espinosa-Perez, G.; Cea-Olivares, R. *J. Chem. Soc., Dalton Trans.* **1997**, 1025–1029.
410. Cea-Olivares, R.; Novosad, J.; Woollins, J.; Slawin, A. M. Z.; Garcia-Montalvo, V.; Espinosa-Perez, G.; Garcia, P. G. Y. *Chem. Commun.* **1996**, 519–520.
411. Garcia-Montalvo, V.; Cea-Olivares, R.; Espinosa-Perez, G. *Polyhedron* **1996**, *15*, 829–834.
412. Dumitrescu, L. S.; Haiduc, I.; Weiss, J. *J. Organomet. Chem.* **1984**, *263*, 159–165.
413. Carmalt, C. J.; Clyburne, J. A. C.; Cowley, A. H.; Lomeli, V.; McBurnett, B. G. *Chem. Commun.* **1998**, 243–244.
414. Laurent, C.; Mazieres, S.; Lavayssiere, H.; Mazerolles, P.; Dousse, G. *J. Organomet. Chem.* **1993**, *452*, 41–45.
415. Davies, A. G.; Slater, S. D.; Povey, D. C.; Smith, G. W. *J. Organomet. Chem.* **1988**, *352*, 283–294.
416. Sau, A. C.; Holmes, R. R.; Molloy, K. C.; Zuckerman, J. J. *Inorg. Chem.* **1982**, *21*, 1421–1427.
417. Holmes, R. R.; Shafieezad, S.; Holmes, J. M.; Day, R. O. *Inorg. Chem.* **1988**, *27*, 1232–1237.
418. Holmes, R. R.; Shafieezad, S.; Chandrasekhar, V.; Sau, A. C.; Holmes, J. M.; Day, R. O. *J. Am. Chem. Soc.* **1988**, *110*, 1168–1174.
419. Graetz, K.; Huber, F.; Silvestri, A.; Alonzo, G.; Barbieri, R. *J. Organomet. Chem.* **1985**, *290*, 41–51.
420. Doidge-Harrison, S. M. S. V.; Irvine, J. T. S.; Spencer, G. M.; Wardell, J. L.; Ganis, P.; Valle, G.; Tagliavini, G. *Polyhedron* **1996**, *15*, 1807–1815.
421. Dean, P. A. W.; Vittal, J. J.; Payne, N. C. *Inorg. Chem.* **1985**, *24*, 3594–3597.
422. Hummel, H. U.; Meske, H. *J. Chem. Soc., Dalton Trans.* **1989**, 627–630.
423. Hummel, H. U.; Fischer, E.; Fischer, T.; Gruss, D.; Franke, A.; Dietzsch, W. *Chem. Ber.* **1992**, *125*, 1565–1570.
424. Hummel, H.-U.; Fischer, E.; Fischer, T.; Gruß, D.; Franke, A.; Dietche, W. *Chem. Ber.* **1992**, *125*, 1565–1570.
425. Beattie, I. R.; Jones, P. J.; Reid, G.; Webster, M. *Inorg. Chem.* **1998**, *37*, 6032–6034.
426. Brauer, D. J.; Wilke, J.; Eujen, R. *J. Organomet. Chem.* **1986**, *316*, 261–269.
427. Lambertsen, T. H.; Jones, P. G.; Schmutzler, R. *Polyhedron* **1992**, *11*, 331–334.
428. Plitzko, C.; Meyer, G. *Z. Anorg. Allg. Chem.* **1997**, *623*, 1347–1348.
429. Plitzko, C.; Meyer, G. *Z. Kristallogr.* **1998**, *213*, 475.
430. Graudejus, O.; Mueller, B. G. *Z. Anorg. Allg. Chem.* **1996**, *622*, 1601–1608.
431. Kokunov, Yu. V.; Detkov, D. G.; Gorbunova, Yu. E.; Ershova, M. M.; Mikhailov, Yu. N.; Buslaev, Yu. A. *Doklady Akademii Nauk* **2001**, *378*, 347–350.
432. Abrahams, I.; Donaldson, J. D.; Grimes, S. M. *J. Chem. Soc., Dalton Trans.* **1992**, 669–673.
433. Kravchenko, E. A.; Buslaev, Y. A. *Russ. Chem. Rev* **1999**, *68*, 709–726.
434. Ault, B. S. *J. Mol. Struct.* **1985**, *130*, 215–226.
435. Ault, B. S. *J. Mol. Struct.* **1985**, *129*, 287–298.
436. Tornero, J. D.; Tudela, D.; Fernandez, V. *An. Quim., Ser. B* **1986**, *82*, 145–149.
437. Tudela, D.; Rey, F. *Z. Anorg. Allg. Chem.* **1989**, *575*, 202–208.
438. Feshin, V. P. *Z. Naturforsch., A* **1992**, *47*, 141–146.
439. Feshin, V. P.; Dolgushin, G. V.; Lazarev, I. M.; Voronkov, M. G. *Z. Naturforsch., A* **1990**, *45*, 219–223.
440. Shcherbinin, V. V.; Shvedov, I. P.; Pavlov, K. V.; Komalenkova, N. G.; Chernyshev, E. A. *Russ. J. Gen. Chem.* **1998**, *68*, 1013–1016.
441. Wang, F.; Grey, C. P. *J. Am. Chem. Soc.* **1998**, *120*, 970–980.
442. Czado, W.; Mueller, U. *Z. Anorg. Allg. Chem.* **1998**, *624*, 925–926.
443. Guan, J.; Tang, Z.; Guloy, A. M. *Chem. Commun.* **1999**, 1833–1834.
444. Lode, C.; Krautscheid, H. *Z. Anorg. Allg. Chem.* **2000**, *626*, 326–331.
445. Krautscheid, H.; Vielsack, F. *Z. Anorg. Allg. Chem.* **2000**, *626*, 3–5.
446. Krautscheid, H.; Vielsack, F. *Z. Anorg. Allg. Chem.* **1999**, *625*, 562–566.
447. Tang, Z.; Guloy, A. M. *J. Am. Chem. Soc.* **1999**, *121*, 452–453.
448. Krautscheid, H.; Vielsack, F.; Klaassen, N. *Z. Anorg. Allg. Chem.* **1998**, *624*, 807–812.
449. Krautscheid, H.; Vielsack, F. *Angew. Chem., Int. Ed. Engl.* **1995**, *34*, 2035–2037.
450. Corradi, A. B.; Ferrari, A. M.; Pellacani, G. C.; Saccani, A.; Sandrolini, F.; Sgarabotto, P. *Inorg. Chem.* **1999**, *38*, 716–721.
451. Chernov, S. V.; Moskvin, A. L.; Murin, I. V. *Solid State Ionics* **1991**, *47*, 71–73.
452. Schmidbaur, H.; Rott, J.; Reber, G.; Mueller, G. *Z. Naturforsch. B: Anorg. Chem. Org. Chem.* **1988**, *43*, 727–732.
453. Schmidbaur, H.; Rott, J. *Z. Naturforsch. B: Anorg. Chem. Org. Chem.* **1989**, *44*, 285–287.
454. Schmidbaur, H.; Rott, J. *Z. Naturforsch. B: Anorg. Chem. Org. Chem.* **1990**, *45*, 961–966.
455. Riedmiller, F.; Wegner, G. L.; Jockisch, A.; Schmidbaur, H. *Organometallics* **1999**, *18*, 4317–4324.
456. Becerra, R.; Boganov, S. E.; Egorov, M. P.; Faustov, V. I.; Nefedov, O. M.; Walsh, R. *J. Am. Chem. Soc.* **1998**, *120*, 12657–12665.
457. Brynda, M.; Geoffroy, M.; Bernardinelli, G. *Chem. Commun.* **1999**, 961–962.
458. Brynda, M.; Dutan, C.; Berccaz, T.; Geoffroy, M.; Bernardinelli, G. *J. Phys. Chem. Solids* **2003**, *64*, 939–946.
459. Eichler, B. E.; Power, P. P. *J. Am. Chem. Soc.* **2000**, *122*, 8785–8786.
460. Cosledan, F.; Castel, A.; Riviere, P.; Satge, J.; Veith, M.; Huch, V. *Organometallics* **1998**, *17*, 2222–2227.
461. Zickgraf, A.; Beuter, M.; Kolb, U.; Drager, M.; Tozer, R.; Dakternieks, D.; Jurkschat, K. *Inorg. Chim. Acta* **1998**, *275–276*, 203–214.
462. Dargatz, M.; Hartung, H.; Kleinpeter, E.; Rensch, B.; Schollmeyer, D.; Weichmann, H. *J. Organomet. Chem.* **1989**, *361*, 43–51.
463. Kolb, U.; Draeger, M.; Fischer, E.; Jurkschat, K. *J. Organomet. Chem.* **1992**, *423*, 339–350.
464. Jung, O. S.; Jeong, J. H.; Sohn, Y. Soo. *Polyhedron* **1989**, *8*, 1413–1417.

465. Ng, S. W.; Wei, C.; Das, V. G. K.; Jameson, G. B.; Butcher, R. J. *J. Organomet. Chem.* **1989**, *365*, 75–82.
466. Biesemans, M.; Willem, R.; Damoun, S.; Geerlings, P.; Tiekink, E. R. T.; Jaumier, P.; Lahcini, M.; Jousseaume, B. *Organometallics* **1998**, *17*, 90–97.
467. Jurkschat, K.; Van Dreumel, S.; Dyson, G.; Dakternieks, D.; Bastow, T. J.; Smith, M. E.; Draeger, M. *Polyhedron* **1992**, *11*, 2747–2755.
468. Das, V. G. K.; Mun, L. K.; Wei, C.; Mak, T. C. W. *Organometallics* **1987**, *6*, 10–14.
469. Bylinkin, S. Yu.; Shipov, A. G.; Kramarova, E. P.; Negrebetskii, Vad. V.; Smirnova, L. S.; Pogozhikh, S. A.; Ovchinnikov, Yu.E.; Baukov, Yu. I. *Russ. J. Gen. Chem.* **1997**, *67*, 1742–1756.
470. Akiba, K.; Ito, Y.; Kondo, F.; Ohashi, N.; Sakaguchi, A.; Kojima, S.; Yamamoto, Y. *Chem. Lett.* **1992**, 1563–1566.
471. Seidel, N.; Jacob, K.; van der Zeijden, A. A. H.; Menge, H.; Merzweiler, K.; Wagner, C. *Organometallics* **2000**, *19*, 1438–1441.
472. Jastrzebski, J. T. B. H.; Van Koten, G. *Adv. Organomet. Chem.* **1993**, *35*, 241–294.
473. Takeuchi, Y.; Tanaka, K.; Tanaka, K.; Ohnishi-Kameyama, M.; Kalman, A.; Parkanyi, L. *Chem. Commun.* **1998**, 2289–2290.
474. Cea-Olivares, R.; Gomez-Ortiz, L. A.; Garcia-Montalvo, V.; Gavino-Ramirez, R. L.; Hernandez-Ortega, S. *Inorg. Chem.* **2000**, *39*, 2284–2288.
475. Ng, S. W.; Wei, C.; Das, V. G. K.; Mak, T. C. W. *J. Organomet. Chem.* **1987**, *334*, 283–293.
476. Clark, H. C.; Jain, V. K.; McMahon, I. J.; Mehrotra, R. C. *J. Organomet. Chem.* **1983**, *243*, 299–303.
477. Blunden, S. J.; Patel, B. N.; Smith, P. J.; Sugavanam, B. *Appl. Organomet. Chem.* **1987**, *1*, 241–244.
478. Deb, B. K.; Ghosh, A. K. *Can. J. Chem.* **1987**, *65*, 1241–1246.
479. Mercier, F. A. G.; Meddour, A.; Gielen, M.; Biesemans, M.; Willem, R.; Tiekink, E. R. T. *Organometallics* **1998**, *17*, 5933–5936.
480. Dwivedi, B. K.; Bhatnagar, K.; Srivastava, A. K. *Synth. React. Inorg. Met.-Org. Chem.* **1986**, *16*, 841–855.
481. Chen, D. H.; Chiang, H. C. *J. Chin. Chem. Soc.* **1993**, *40*, 373–377.
482. Tandura, S. N.; Khromova, N. Y.; Gar, T. K.; Alekseev, N. V.; Mironov, V. F. *Zh. Obshch. Khim.* **1983**, *53*, 1199–2000.
483. Bettermann, G.; Arduengo, A. J., III. *J. Am. Chem. Soc.* **1988**, *110*, 877–879.
484. Picard, C.; Tisnes, P.; Cazaux, L. *J. Organomet. Chem.* **1986**, *315*, 277–285.
485. Nebout, B.; De Jeso, B.; Marchand, A. *J. Organomet. Chem.* **1986**, *299*, 319–330.
486. Swisher, R. G.; Holmes, R. R. *Organometallics* **1984**, *3*, 365–369.
487. Willem, R.; Gielen, M.; Meunier-Piret, J.; Van Meerssche, M.; Jurkschat, K.; Tzschach, A. *J. Organomet. Chem.* **1984**, *277*, 335–350.
488. Jurkschat, K.; Tzschach, A.; Muegge, C.; Piret-Meunier, J.; Van Meerssche, M.; Van Binst, G.; Wynants, C.; Gielen, M.; Willem, R. *Organometallics* **1988**, *7*, 593–603.
489. Otera, J.; Kusaba, A.; Hinoishi, T.; Kawasaki, Y. *J. Organomet. Chem.* **1982**, *228*, 223–228.
490. Holecek, J.; Nadvornik, M.; Handlir, K.; Lycka, A. *J. Organomet. Chem.* **1986**, *315*, 299–308.
491. Kerschl, S.; Wrackmeyer, B. *J. Organomet. Chem.* **1987**, *332*, 25–33.
492. Bansse, W.; Ludwig, E.; Uhlemann, E.; Mehner, H.; Weller, F.; Dehnicke, K. *Z. Anorg. Allg. Chem.* **1992**, *607*, 177–182.
493. Pettinari, C.; Marchetti, F.; Pettinari, R.; Martini, D.; Drozdov, A.; Troyanov, S. *Inorg. Chim. Acta* **2001**, *325*, 103–114.
494. Diamantis, A. A.; Gulbis, J. M.; Manikas, M.; Tiekink, E. R. T. *Phosphorus, Sulfur Silicon Rel. Elem.* **1999**, *150–151*, 251–259.
495. Tastekin, M.; Kenar, A.; Atakol, O.; Tahir, M. N.; Ulku, D. *Synth. React. Inorg. Met.-Org. Chem.* **1998**, *28*, 1727–1741.
496. Dey, D. K.; Saha, M. K.; Gielen, M.; Kemmer, M.; Biesemans, M.; Willem, R.; Gramlich, V.; Mitra, S. I. *J. Organomet. Chem.* **1990**, *590*, 88.
497. Cai, D.; Li, J.; Yang, L.; Lou, Q.; Shi, Z.; Lin, K. *Chin. Chem. Lett.* **1994**, *5*, 155–156.
498. Nath, M.; Sharma, C. L.; Sharma, N. *Synth. React. Inorg. Met.-Org. Chem.* **1991**, *21*, 807–824.
499. Evans, D. F.; Jakubovic, D. A. *Polyhedron* **1988**, *7*, 2723–2726.
500. McGarvey, B. R.; Ozarowski, A.; Tian, Z.; Tuck, D. G. *Can. J. Chem.* **1995**, *73*, 1213–1222.
501. Labisbal, E.; De Blas, A.; Garcia-Vazquez, J. A.; Romero, J.; Duran, M. L.; Sousa, A.; Bailey, N. A.; Fenton, D. E.; Leeson, P. B.; Parish, R. V. *Polyhedron* **1992**, *11*, 227–233.
502. Velazquez, C. S.; Broderick, W. E.; Sabat, M.; Barrett, A. G. M.; Hoffman, B. M. *J. Am. Chem. Soc.* **1990**, *112*, 7408–7410.
503. Thompson, T.; Pastor, S. D.; Rihs, G. *Inorg. Chem.* **1999**, *38*, 4163–4167.
504. Jastrzebski, J. T. B. H.; Van der Schaaf, P. A.; Boersma, J.; Van Koten, G.; Zoutberg, M. C.; Heijdenrijk, D. *Organometallics* **1989**, *8*, 1373–1375.
505. Mehring, M.; Loew, C.; Schuermann, M.; Uhlig, F.; Jurkschat, K.; Mahieu, B. *Organometallics* **2000**, *19*, 4613–4623.
506. Sakuntala, E. N.; Vasanta, E. N. *Z. Naturforsch. B: Anorg. Chem. Org. Chem.* **1985**, *40*, 1173–1176.
507. Dey, D. K.; Das, M. K.; Noth, H. *Z. Naturforsch. B: Anorg. Chem. Org. Chem.* **1999**, *54*, 145–154.
508. Teoh, S.-G.; Yeap, G.-Y.; Loh, C.-C.; Foong, L.-W.; Teo, S.-B.; Fun, H.-K. *Polyhedron* **1997**, *16*, 2213–2221.
509. Kuchta, M. C.; Hahn, J. M.; Parkin, G. *J. Chem. Soc., Dalton Trans.* **1999**, 3559–3563.
510. Agustin, D.; Rima, G.; Gornitzka, H.; Barrau, J. *J. Organomet. Chem.* **1999**, *592*, 1–10.
511. Parr, J.; Ross, A. T.; Slawin, A. M. Z. *J. Chem. Soc., Dalton Trans.* **1996**, 1509–1512.
512. Mancilla, T.; Farfan, N.; Castillo, D.; Molinero, L.; Meriem, A.; Willem, R.; Mahieu, B.; Gielen, M. *Main Group Met. Chem.* **1989**, *12*, 213–223.
513. Bhattacharyya, P.; Parr, J.; Slawin, A. M. Z. *Inorg. Chem. Commun.* **1999**, *2*, 113–115.
514. Meriem, A.; Willem, R.; Meunier-Piret, J.; Gielen, M. *Main Group Met. Chem.* **1989**, *12*, 187–198.
515. Lee, F. L.; Gabe, E. J.; Khoo, L. E.; Leong, W. H.; Eng, G.; Smith, F. E. *Inorg. Chim. Acta* **1989**, *166*, 257–261.
516. Lukevics, E.; Ignatovich, L.; Porsyurova, N.; Germane, S. *Appl. Organomet. Chem.* **1988**, *2*, 115–120.
517. Lukevics, E.; Ignatovich, L.; Belyakov, S. *J. Organomet. Chem.* **1999**, *588*, 222–230.
518. Gevorgyan, V.; Borisova, L.; Vyater, A.; Ryabova, V.; Lukevics, E. *J. Organomet. Chem.* **1997**, *548*, 149–155.

519. Zaitseva, G. S.; Siggelkow, B. A.; Karlov, S. S.; Pen'kovkov, G. V.; Lorberth, V. *Z. Naturforsch. B: Anorg. Chem. Org. Chem.* **1998**, *53*, 1255.
520. Zaitseva, G. S.; Siggelkow, B. A.; Karlov, S. S.; Pen'kovkov, G. V.; Lorberth, V. *Z. Naturforsch. B: Anorg. Chem. Org. Chem.* **1998**, *53*, 1255–1258.
521. Zaitseva, G. S.; Karlov, S. S.; Siggelkow, B. A.; Avtomonov, E. V.; Churakov, A. V.; Howard, J. A. K.; Lorberth, J. *Z. Naturforsch. B: Anorg. Chem. Org. Chem.* **1998**, *53*, 1247–1254.
522. Narula, S. P.; Soni, S.; Shankar, R.; Chadha, R. K. *J. Chem. Soc., Dalton Trans.* **1992**, 3055–3056.
523. Nasim, M.; Livantsova, L. I.; Zaitseva, G. S.; Lorberth, J. *J. Organomet. Chem.* **1991**, *403*, 85–91.
524. Lukevics, E.; Ignatovich, L.; Belyakov, S. *J. Organomet. Chem.* **1999**, *588*, 222–230.
525. Korecz, L.; Saghier, A. A.; Burger, K.; Tzschach, A.; Jurkschat, K. *Inorg. Chim. Acta* **1982**, *58*, 243–249.
526. Ravenscroft, M. D.; Roberts, R. M. G. *J. Organomet. Chem.* **1986**, *312*, 33–43, Ravenscroft, M. D.; Roberts, R. M. G. *J. Organomet. Chem.* **1986**, *312*, 45–52.
527. Jurkschat, K.; Tzschach, A.; Weichmann, H.; Rajczy, P.; Mostafa, M. A.; Korecz, L.; Burger, K. *Inorg. Chim. Acta* **1991**, *179*, 83–88.
528. Carini, C.; Pelizzi, G.; Tarasconi, P.; Pelizzi, C.; Molloy, K. C.; Waterfield, P. C. *J. Chem. Soc., Dalton Trans.* **1989**, 289–293.
529. Iyer, R.; Krishna; Deshpande, S. G.; Amirthalingam, V. *Polyhedron* **1984**, *3*, 1099–1104.
530. Bhattacharyya, P.; Parr, J.; Ross, A. T.; Slawin, A. M. Z. *J. Chem. Soc., Dalton Trans.* **1998**, 3149–3150.
531. Bashall, A.; McPartlin, M.; Murphy, B. P.; Fenton, D. E.; Kitchen, S. J.; Tasker, P. A. *J. Chem. Soc., Dalton Trans.* **1990**, 505–509.
532. Brooker, S.; Croucher, P. D. *J. Chem. Soc., Chem. Commun.* **1993**, 1278–1280.
533. Adams, H.; Bailey, N. A.; Fenton, D. E.; Good, R. J.; Moody, R.; Rodriguez de Barbarin, C. O. *J. Chem. Soc., Dalton Trans.* **1987**, 207–218.
534. Tandon, S. S.; McKee, V. *J. Chem. Soc., Dalton Trans.* **1989**, 19–24.
535. Clarke, P.; Lincoln, S. F.; Wainwright, K. P. *Inorg. Chem.* **1991**, *30*, 134–139.
536. Pittet, P. A.; Laurence, G. S.; Lincoln, S. F.; Turonek, M. L.; Wainwright, K. P. *J. Chem. Soc., Chem. Commun.* **1991**, 1205–1206.
537. Kumar, K.; Magerstaedt, M.; Gansow, O. A. *J. Chem. Soc., Chem. Commun.* **1989**, 145–146.
538. Adam, K. R.; Baldwin, D. S.; Duckworth, P. A.; Lindoy, L. F.; McPartlin, M.; Bashall, A.; Powell, H. R.; Tasker, P. A. *J. Chem. Soc., Dalton Trans.* **1995**, 1127–1131.
539. Adam, K. R.; Baldwin, D. S.; Bashall, A.; Lindoy, L. F.; McPartlin, M.; Powell, H. R. *J. Chem. Soc., Dalton Trans.* **1994**, 237–238.
540. Buschmann, H.-J. Germanium, Tin and Lead. In *Stereochemistry of Organometallic and Inorganic Compounds*; I. Bernal, ed., Elsevier: Amsterdam, 1987, Vol. 2, p 103.
541. Buschmann, H.-J. *Thermochim. Acta* **1986**, *107*, 219–226.
542. Byriel, K.; Dunster, K. R.; Gahan, L. R.; Kennard, C. H. L.; Latten, J. L.; Swann, I. L. *Polyhedron* **1992**, *11*, 1205–1212.
543. Esteban, D.; Banobre, D.; De Blas, A.; Rodriguez-Blas, T.; Bastida, R.; Macias, A.; Rodriguez, A.; Fenton, D. E.; Adams, H.; Mahia, J. E. *J. Inorg. Chem.* **2000**, 1445–1456.
544. Bashall, A.; McPartlin, M.; Murphy, B. P.; Powell, H. R.; Waikar, S. *J. Chem. Soc., Dalton Trans.* **1994**, 1383–1390.

Comprehensive Coordination Chemistry II
ISBN (set): 0-08-0437486

Volume 3, (ISBN 0-08-0443257); pp 545–608

Volume Subject Index

The index is in letter-by-letter order, whereby hyphens and spaces within index headings are ignored in the alphabetization. An entry with a prefix/locant is filed after the same entry without any attachments, and in alphanumerical sequence. Bold page numbers indicate major discussion of a subject.

Antimony(V) complexes coordination
  compounds
  haloantimonate, 502–503
  phosphonate, 486–487
ANTRIN
  properties and applications, 157
Aqua ligands
  bismuth
    O-donor, 512–514
  indium(III) complexes, 410
  lanthanides
    group 16 oxygen, 127
  scandium complexes
    anhydrous/hydrated salts, 98–99
    structural properties, 101
Arsenic complexes/coordination
  compounds, **465–534**
  biological distribution, 478–479
  environmental distribution, 478–479
  medical applications, 478–479
  s-block elements, 47
Arsenic(III) complexes/coordination
  compounds
  dithioacid, 472
  group 16 metals, 468
Arsenic ligands
  aluminum complexes, 355
  antimony coordination chemistry,
    481
  bismuth complexes, 511
  gallium complexes, 372
  group 14 metals, 466–467
  group 15 metals, 467
  group 17 metals, 474–475
  indium(III) complexes, 395
  thallium(I) complexes, 444–445
  thallium(III) complexes, 429
Arsenic(V) complexes/coordination
  compounds
  fluoroanions, 476–477
Arsine ligands
  antimony complexes, 481
  bismuth complexes, 511
    arsine oxides, 515
  lanthanides, 131–132
    nitrate complexes, 132
  oxides
    bismuth complexes, 515
    scandium complexes, 102
Arylamides
  lanthanides, 123–124
Aryl complexes/coordination
  compounds
  aluminum, 348
  lanthanum²⁺ ion
    group 14 ligands, 163–164
  scandium, 94
Aryloxides
  lanthanides, 109, 138–139
  oxo-centered clusters, 141
  s-block elements
    oxygen ligands, 52–53
"Ate" complexes
  indium(III) nitrogen ligands,
    386–387
Atomic layer epitaxy (ALE). *See also*
  Chemical vapor deposition
  (CVD); Metal oxide chemical
  vapor deposition (MOCVD)
Azaindazoles
  thallium(I) ligands, 439
Azides
  antimony complexes, 481

arsenic complexes
  group 14 metals, 467
  indium(III) complexes, 385
Azidoindium(III)
  porphyrin ligands, 391–392
Azolate ligands
  indium(III) complexes, 391–392

B

Backbone structures
  thallium(1) complexes
    carbon 15 ligands, 439
Barium complexes/coordination
  compounds
  s-block elements
    alkoxides/aryl oxides, 62
    chemical vapor deposition, 65–66
    nitrogen donor ligands, 38, 40
Benzamide ligands
  scandium complexes, 96
Benzamidinate ligands
  structure and function, 124
Benzotriazolylborates
  thallium(I) complexes, 444
Benzoylacetoacetonates
  tin complexes, isomer
    interconversion, 532
Bidentate ligands
  indium(III) complexes, 392
Bimetallic complexes/coordination
  compounds
  s-block elements
    nitrogen donor ligands, 38–39
Biologically active compounds
  antimony distribution, 504
  arsenic, 478–479
  bismuth complexes, 534
  s-block elements
    cation-π interaction, 4
Bipyramidal structures
  pseudo-trigonal
    of antimony complexes with
      nitrogen donor ligands, 480
Bipyridyl ligands
  antimony complexes, 479–480
  bismuth complexes, 509–511
  lanthanides, 112
2,6-Bis(benzimidazol-2'-yl)pyridine
  lanthanides, 118–119
2,6-Bis(1'-ethyl-5'-methyl-
  benzimidazol-2'-yl)pyridine
  lanthanides, 118–119
2,6-Bis(5-methyl-1,2,4-triazol-3-yl)-
  pyridine (DMTZP)
  lanthanides, 120
Bismuth complexes/coordination
  compounds, **465–534**
  coordination chemistry, 504
  environmental, biological, and
    medical distribution, 534
  medical applications, 534
  P- and As-donor ligands, 511
Bismuth ligands
  group 14 metal complexes, 504–505
  group 15 metal complexes
    N-donor ligands, 505–506
  group 16 metal complexes, 512–514
    N/O-donor ligands, 518–519
    O-donor ligands, 512–514
    S/O and S/N donor ligands, 529
  group 17 metal complexes, 531
  indium(III) complexes, 395

thallium(I) complexes, 444–445
thallium(III) complexes, 429
Bis(1-phenyl-3-methyl-4-acylpyrazole-
  5-one) derivatives
  in uranium(VI) extraction, 298
Bispyrazolylborates
  thallium(I) complexes, 440
Bis(*tert*-butylaamido)-
  cyclodiphosphazane
  indium(II) group 15 ligands, 417
  thallium(I) ligands, 438
Bistriaznido ions
  thallium(III) ligands, 428
Bis(trimethylsilyl)amides
  organoimido complex formation
    uranium(IV), 205
  s-block elements
    arsenic ligands, 47
    nitrogen ligands, 27
  thorium(IV) complexes, 205
  uranium(IV) complexes, 205
Blue-emitting conjugated polymers
  lanthanide diketonates, 137
Bonding structures
  indium(II) group 17 ligands,
    420–421
Borohydrides
  thorium(III)–plutonium(III), 203
  thorium(IV)–plutonium(IV), 252
"Breastplate" structures
  s-block elements
    alkoxides/aryl oxides, 62–64
Bridging ligands
  lanthanides, 109–110
Bromide complexes/coordination
  compounds
  indium(II) group 17 ligands, 420
  lanthanides
    group 17 halide ligands, 163
    hydrated salts, 128–129
    terpyridyl complexes, 115
  scandium
    group 17 ligands, 107
Bromoantimonate compounds
  antimony(III), 499–501
Bromo-arsenate(III) compounds
  haloanions, 475–476
Bromobismuthate(III) compounds,
  533
Bromomethallate(III) complexes/
  coordination compounds
  thiallium(III) ligands, 432–433
Brønsted acids
  aqueous media
    thorium(IV)–plutonium(IV), 204
Butylthallium(III) unit
  thallium(I) ligands, 439

C

Calcium complexes/coordination
  compounds
  s-block elements
    alkoxides/aryl oxides, 62
    macrocyclic porphyrin, 9
Calixarenes
  arsenic(III), 471–472
  germanium complexes, 584
  lanthanides, 161
  s-block elements
    macrocyclic complexes, 15–16
    oxygen ligands, 63–64
  scandium crown ethers, 106